OPERATION AND MAINTENANCE OF WASTEWATER COLLECTION SYSTEMS

VOLUME II

Sixth Edition

A Field Study Training Program

prepared by

Office of Water Programs
College of Engineering and Computer Science
California State University, Sacramento

in cooperation with the
California Water Environment Association

★★★★★★★★★★★★★★★★★★★★★★★★★★★★★★★

Kenneth D. Kerri, Project Director
John Brady, Co-Director

★★★★★★★★★★★★★★★★★★★★★★★★★★★★★★★

for the

U.S. Environmental Protection Agency
Office of Water Program Operations
Municipal Permits and Operations Division
First Edition, Grant No. T900494 (1976)
Spanish Edition, NETA Grant No. X-999658-01-1 (1999)

2004

NOTICE

This manual is revised and updated before each printing based on comments from persons using the manual.

FIRST EDITION, First Printing, 1976	5,000
SECOND EDITION, First Printing, 1983	5,000
THIRD EDITION, First Printing, 1987	5,000
Second Printing, 1990	5,000
FOURTH EDITION, First Printing, 1993	5,000
FIFTH EDITION, First Printing, 1995	6,000
Second Printing, 1998	10,000
Third Printing, 2002	5,000
SIXTH EDITION, First Printing, 2004	10,000

ISBN 1-884701-44-2 (Volume II)
ISBN 1-884701-45-0 (Volumes I and II)

OPERATOR TRAINING MANUALS

OPERATOR TRAINING MANUALS AND VIDEOS IN THIS SERIES are available from the Office of Water Programs, California State University, Sacramento, 6000 J Street, Sacramento, CA 95819-6025, phone: (916) 278-6142, e-mail: water-office@csus.edu, FAX: (916) 278-5959, or website: www.owp.csus.edu.

1. *OPERATION AND MAINTENANCE OF WASTEWATER COLLECTION SYSTEMS*, 2 Volumes,*
2. *OPERATION OF WASTEWATER TREATMENT PLANTS*, 2 Volumes,
3. *SMALL WASTEWATER SYSTEM OPERATION AND MAINTENANCE*, 2 Volumes,
4. *ADVANCED WASTE TREATMENT*,
5. *INDUSTRIAL WASTE TREATMENT*, 2 Volumes,
6. *TREATMENT OF METAL WASTESTREAMS*,
7. *PRETREATMENT FACILITY INSPECTION*,**
8. *WATER TREATMENT PLANT OPERATION*, 2 Volumes,
9. *SMALL WATER SYSTEM OPERATION AND MAINTENANCE*,
10. *WATER DISTRIBUTION SYSTEM OPERATION AND MAINTENANCE*, and
11. *UTILITY MANAGEMENT*.

* Other training materials and training aids developed by the Office of Water Programs to assist operators in improving collection system operation and maintenance and overall performance of their systems include:

1. *COLLECTION SYSTEMS: METHODS FOR EVALUATING AND IMPROVING PERFORMANCE*. This handbook presents detailed benchmarking procedures and worksheets for using performance indicators to evaluate the adequacy and effectiveness of existing O & M programs. It also describes how to identify problems and suggests many methods for improving the performance of a collection system.
2. *OPERATION AND MAINTENANCE TRAINING VIDEOS*. This series of six 30-minute videos demonstrates the equipment and procedures collection system crews use to safely and effectively operate and maintain their collection systems. These videos complement and reinforce the information presented in Volumes I and II of *OPERATION AND MAINTENANCE OF WASTEWATER COLLECTION SYSTEMS*.

** *PRETREATMENT FACILITY INSPECTION TRAINING VIDEOS*. This series of five 30-minute videos demonstrates the procedures to effectively inspect an industry, measure flows, and collect samples. These videos complement and reinforce the information presented in *PRETREATMENT FACILITY INSPECTION*.

The Office of Water Programs at California State University, Sacramento, has been designated by the U.S. Environmental Protection Agency as a *SMALL PUBLIC WATER SYSTEMS TECHNOLOGY ASSISTANCE CENTER*. This recognition will provide funding for the development of training videos for the operators and managers of small public water systems. Additional training materials will be produced to assist the operators and managers of small systems.

PREFACE TO THE THIRD EDITION
VOLUME II

Volume II is a continuation of Volume I. Volume I emphasizes the knowledge and skills needed by new collection system operators and the operators of smaller collection systems. Topics covered in Volume I stress the important role of being a collection system operator and the operator's responsibilities with respect to protecting the community's public health, investment in collection system and wastewater treatment facilities and ultimately the environment in which we live. Volume I discusses the objectives of collection systems, describes collection systems and stresses the importance of following safe procedures at all times. A section of Volume I provides details on the installation and inspection of sewers so operators will know how they are built and what is underground. Major subjects covered include how to inspect old and new collection systems and how to conduct TV inspections and interpret the results. Both hydraulic and mechanical means of cleaning and maintaining sewers, as well as underground repair procedures are covered extensively.

Volume II shifts emphasis from tasks performed by line maintenance crews to the duties of operating and maintaining lift stations, maintenance of equipment and sewer rehabilitation. Topics related to the duties of supervisors are also covered with emphasis on administration and organization of the collection system operation and maintenance activities.

Safety is stressed throughout both Volumes I and II. In Volume II an entire chapter is devoted to safety programs for collection system operators. The safety record to date in the collection system field is dismal. All of us must take responsibility for ourselves and all of our co-workers and do everything possible to eliminate safety hazards and follow safe procedures. None of us want the burden of knowing that we could have prevented an injury to ourselves or to a personal friend.

Regardless of your experience in the collection system profession or the size, age or type of collection system that you operate and maintain, all of the chapters in Volumes I and II contain information you need to know to do your job. Every collection system needs operators who know how to operate and maintain the facilities as well as competent administrators to manage the program.

Please refer to Volume I for information on:

1. Objectives of these manuals,
2. Scope of these manuals,
3. Instruction to participants in the home-study course,
4. Uses of these manuals, and
5. Summary of procedures.

1987

KENNETH D. KERRI
JOHN BRADY

OPERATION AND MAINTENANCE OF WASTEWATER COLLECTION SYSTEMS

VOLUME I — COURSE OUTLINE

VOLUME II — COURSE OUTLINE

TECHNICAL CONSULTANTS

First and Second Editions	John Carvoretto, Ira Cotton, George Gardner, William Garber, James Kenmir, and Warren Prentice
Third Edition	John Brady and Russ Armstrong
Fourth and Fifth Editions	Rick Arbour
Sixth Edition	Gary Batis

CHAPTER 8

LIFT STATIONS

by

Manuel Muñoz

and

John Brady

Revised by

Rick Arbour

and

Gary Batis

TABLE OF CONTENTS

Chapter 8. LIFT STATIONS

LESSON 4

LESSON 5

OBJECTIVES

Chapter 8. LIFT STATIONS

Following completion of Chapter 8, you should be able to:

1. Determine the locations of lift stations,
2. Describe the requirements of a lift station,
3. Discuss the components of a lift station,
4. Indicate the advantages and disadvantages of the different types of controllers,
5. Review lift station prints and specifications,
6. Inspect a new lift station,
7. Keep a lift station operating as intended,
8. Determine the frequency of visits to a lift station,
9. Perform necessary lift station maintenance tasks, and
10. Prepare record forms for a lift station, complete and file them.

WORDS

Chapter 8. LIFT STATIONS

CAVITATION (CAV-uh-TAY-shun) CAVITATION

The formation and collapse of a gas pocket or bubble on the blade of an impeller or the gate of a valve. The collapse of this gas pocket or bubble drives water into the impeller or gate with a terrific force that can cause pitting on the impeller or gate surface. Cavitation is accompanied by loud noises that sound like someone is pounding on the impeller or gate with a hammer.

COMMINUTOR (com-mih-NEW-ter) COMMINUTOR

A device used to reduce the size of the solid chunks in wastewater by shredding (comminuting). The shredding action is like many scissors cutting to shreds all the large solids in the wastewater.

DISCHARGE HEAD DISCHARGE HEAD

The pressure (in pounds per square inch or psi) measured at the centerline of a pump discharge and very close to the discharge flange, converted into feet. The pressure is measured from the centerline of the pump to the hydraulic grade line of the water in the discharge pipe.

Discharge Head, ft = (Discharge Pressure, psi)(2.31 ft/psi)

DRY PIT DRY PIT

(See DRY WELL)

DRY WELL DRY WELL

A dry room or compartment in a lift station, near or below the water level, where the pumps are located.

DYNAMIC HEAD DYNAMIC HEAD

When a pump is operating, the vertical distance (in feet) from a point to the energy grade line. Also see TOTAL DYNAMIC HEAD, STATIC HEAD, and ENERGY GRADE LINE.

ELECTROLYTE (ee-LECK-tro-LITE) SOLUTION ELECTROLYTE SOLUTION

A special solution that is capable of conducting electricity.

ENERGY GRADE LINE (EGL) ENERGY GRADE LINE (EGL)

A line that represents the elevation of energy head (in feet) of water flowing in a pipe, conduit or channel. The line is drawn above the hydraulic grade line (gradient) a distance equal to the velocity head ($V^2/2g$) of the water flowing at each section or point along the pipe or channel. Also see HYDRAULIC GRADE LINE.

ENTRAIN ENTRAIN

To trap bubbles in water either mechanically through turbulence or chemically through a reaction.

FLAT FLAT

A flat is the length of one side of a nut.

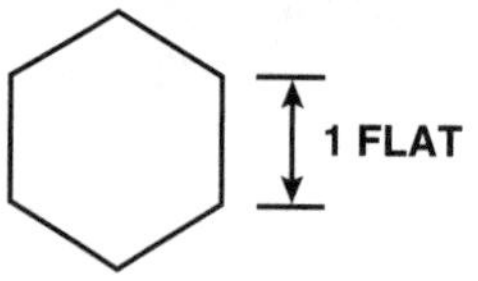

FORCE MAIN FORCE MAIN

A pipe that carries wastewater under pressure from the discharge side of a pump to a point of gravity flow downstream.

HEAD HEAD

The vertical distance, height or energy of water above a point. A head of water may be measured in either height (feet) or pressure (pounds per square inch (psi)). Also see DISCHARGE HEAD, DYNAMIC HEAD, STATIC HEAD, SUCTION HEAD, SUCTION LIFT and VELOCITY HEAD.

HYDRAULIC GRADE LINE (HGL) HYDRAULIC GRADE LINE (HGL)

The surface or profile of water flowing in an open channel or a pipe flowing partially full. If a pipe is under pressure, the hydraulic grade line is at the level water would rise to in a small tube connected to the pipe. To reduce the release of odors from sewers, the water surface or hydraulic grade line should be kept as smooth as possible. Also see ENERGY GRADE LINE.

IMPELLER IMPELLER

A rotating set of vanes in a pump or compressor designed to pump or move water or air.

LIFT STATION LIFT STATION

A wastewater pumping station that lifts the wastewater to a higher elevation when continuing the sewer at reasonable slopes would involve excessive depths of trench. Also, an installation of pumps that raise wastewater from areas too low to drain into available sewers. These stations may be equipped with air-operated ejectors or centrifugal pumps. Sometimes called a PUMP STATION, but this term is usually reserved for a similar type of facility that is discharging into a long FORCE MAIN, while a lift station has a discharge line or force main only up to the downstream gravity sewer. Throughout this manual when we refer to lift stations, we intend to include pump stations.

LUBRIFLUSHING (LOOB-rah-FLUSH-ing) LUBRIFLUSHING

A method of lubricating bearings with grease. Remove the relief plug and apply the proper lubricant to the bearing at the lubrication fitting. Run the pump to expel excess lubricant.

NAMEPLATE NAMEPLATE

A durable metal plate found on equipment which lists critical operating conditions for the equipment.

PNEUMATIC EJECTOR (new-MAT-tik ee-JECK-tor) PNEUMATIC EJECTOR

A device for raising wastewater, sludge or other liquid by compressed air. The liquid is alternately admitted through an inward-swinging check valve into the bottom of an airtight pot. When the pot is filled compressed air is applied to the top of the liquid. The compressed air forces the inlet valve closed and forces the liquid in the pot through an outward-swinging check valve, thus emptying the pot.

PUMP PIT PUMP PIT

A dry well, chamber or room below ground level in which a pump is located.

STATIC HEAD STATIC HEAD

When water is not moving, the vertical distance (in feet) from a specific point to the water surface is the static head. Also see DYNAMIC HEAD.

STILLING WELL STILLING WELL

A well or chamber which is connected to the main flow channel by a small inlet. Waves and surges in the main flow stream will not appear in the well due to the small-diameter inlet. The liquid surface in the well will be quiet, but will follow all of the steady fluctuations of the open channel. The liquid level in the well is measured to determine the flow in the main channel.

SUCTION HEAD SUCTION HEAD

The *POSITIVE* pressure (in feet or pounds per square inch (psi)) on the suction side of a pump. The pressure can be measured from the centerline of the pump *UP TO* the elevation of the hydraulic grade line on the suction side of the pump.

SUCTION LIFT SUCTION LIFT

The *NEGATIVE* pressure (in feet or inches of mercury vacuum) on the suction side of a pump. The pressure can be measured from the centerline of the pump *DOWN TO* (lift) the elevation of the hydraulic grade line on the suction side of the pump.

TOTAL DYNAMIC HEAD (TDH) TOTAL DYNAMIC HEAD (TDH)

When a pump is lifting or pumping water, the vertical distance (in feet) from the elevation of the energy grade line on the suction side of the pump to the elevation of the energy grade line on the discharge side of the pump.

VELOCITY HEAD VELOCITY HEAD

The energy in flowing water as determined by a vertical height (in feet or meters) equal to the square of the velocity of flowing water divided by twice the acceleration due to gravity ($V^2/2g$).

VOLUTE (vol-LOOT) VOLUTE

The spiral-shaped casing which surrounds a pump, blower, or turbine impeller and collects the liquid or gas discharged by the impeller.

WATER HAMMER
WATER HAMMER

The sound like someone hammering on a pipe that occurs when a valve is opened or closed very rapidly. When a valve position is changed quickly, the water pressure in a pipe will increase and decrease back and forth very quickly. This rise and fall in pressures can cause serious damage to the system.

WET PIT
WET PIT

(See WET WELL)

WET WELL
WET WELL

A compartment or tank in which wastewater is collected. The suction pipe of a pump may be connected to the wet well or a submersible pump may be located in the wet well.

CHAPTER 8. LIFT STATIONS

(Lesson 1 of 5 Lessons)

8.0 PURPOSE OF LIFT STATIONS

Lift stations are used to lift or raise wastewater or storm water from a lower elevation to a higher elevation. Lifting of the wastewater is accomplished by centrifugal pumps or air-operated *PNEUMATIC EJECTORS*.[1] The term "lift station" usually refers to a wastewater facility with a relatively short discharge line up to the downstream gravity sewer. A "pump station" commonly is a similar type of facility that is discharging into a long *FORCE MAIN*.[2] Throughout this manual when we refer to lift stations, we intend to include pump stations.

In many areas regional agencies are being created to serve more than one community. As a result, lift stations and force mains tend to be larger. The pumps can range in size up to several hundred horsepower and the force mains can be several miles long.

Lift stations represent a major capital expenditure for an agency or community and they require an adequate budget to operate and maintain them properly. Failures of lift stations and force mains can have a significant impact on the environment when raw wastewater is discharged over land or into lakes, streams, or rivers. Backups into private residences caused by a lift station failure can easily cost thousands of dollars to clean up, replace, and repair damaged homes or businesses. Significant lift station failures make headlines in newspapers and tarnish the image of your collection system agency.

Collection system operators therefore play a vital role in preventing catastrophic failures affecting the community. Because pump stations contain complex electrical, mechanical, and hydraulic systems, collection system operators must have a wide range of training to properly operate and maintain them. This obviously requires a high level of knowledge, skills, and professionalism by all operators. If operators do not have the level of knowledge and skills needed to understand the complex systems in pump stations, they will be unable to develop adequate maintenance programs for the equipment and systems installed in the lift stations or to troubleshoot and diagnose lift station operation and maintenance problems. As a result, the lift stations will not operate reliably. Unfortunately, this is the case in many agencies today. When lift station operation and maintenance are not understood, lift stations are allowed to deteriorate resulting in system failures, bypasses and overflows.

In many agencies today, collection system operators participate at the design stage of new lift stations because of their knowledge and experience with various types of lift stations. Operators can help engineers produce designs which result in improved maintenance and operation of lift stations and lower operating costs while minimizing bypasses and backups.

8.00 Location

Location and design of lift stations depend on economic factors that are analyzed by the design engineer. Lift stations are installed at low points in the collection system at the end of gravity sewers where the following conditions exist:

---

[1] *Pneumatic Ejector (new-MAT-tik ee-JECK-tor). A device for raising wastewater, sludge or other liquid by compressed air. The liquid is alternately admitted through an inward-swinging check valve into the bottom of an airtight pot. When the pot is filled compressed air is applied to the top of the liquid. The compressed air forces the inlet valve closed and forces the liquid in the pot through an outward-swinging check valve, thus emptying the pot.*

[2] *Force Main. A pipe that carries wastewater under pressure from the discharge side of a pump to a point of gravity flow downstream.*

1. Excavation costs to maintain gravity flow and sufficient velocity become excessive,
2. Soil stability is unsuitable for trenching,
3. Groundwater table is too high for installing deep sewers, and
4. Present wastewater flows are not sufficient to justify extension of large trunk sewers and a lift station offers an economical short-term solution.

Other factors that influence the location of lift stations include the location of other utilities and also the location of surface and overhead structures (buildings and transit systems).

Lift station pumps are designed to move the wastewater with a minimum of energy consumption. The pumps are selected to provide a flow as continuous or constant as possible to minimize surges of wastewater in the downstream sewer and the wastewater treatment plant. Consideration also must be given to installing pumps that require a minimum of maintenance.

The general public today is much more aware of facilities that are installed in their neighborhoods, particularly those that are associated with wastewater collection systems. Existing lift stations must be operated and maintained in a manner that they are acceptable to the neighborhood and are not creating a public safety or health hazard or a nuisance to people living nearby. When constructing new lift stations, particularly in residential areas, it is critical that the concerns of those who are living in the neighborhood are taken into consideration during the design stage as well as in the future operation and maintenance of the station. The appearance of lift station buildings and grounds should blend in with the surrounding environment. Odors must be controlled and the noise must be kept to a minimum to prevent the lift station from being a nuisance to nearby neighbors.

Safety of the operators involved in operating and maintaining the lift station must be considered, as well as the safety of the public. If chemicals are added to the wastewater at the lift station, care must be exercised when storing, handling, and applying these chemicals.

Collection system operators should have easy access to the lift station during all types of weather conditions so the station can be properly operated and maintained at all times. Utilities required to operate the station must be readily available.

Reliability is the most important requirement of a lift station. Pumps or ejectors, controls, and the maintenance program must be designed to minimize the chances of failures to prevent the flooding of homes and streets. What this means is that lift stations must be designed to be able to provide continuous operation. Provisions must be made for lift station operation during power failures and equipment failures in the lift station. Standby portable generators, portable lifts, and multiple pumps that provide backup for other equipment in the station are typical examples of considerations to ensure continuous operation of the lift station.

QUESTIONS

Write your answers in a notebook and then compare your answers with those on page 118.

8.0A How is wastewater raised or lifted in a lift or pump station?

8.0B What factors influence the location of lift stations?

8.0C What are the two major safety items that must be considered around a lift station?

8.01 Types of Lift Stations

Wastewater lift stations may be constructed in various sizes and shapes depending upon the volume of wastewater or storm water to be handled, the elevation water must be lifted, and the distance water must be pumped before the water returns to flowing in a gravity system. The lift station may range from a standard manhole equipped with a submersible or other wet well type pump, to a factory prefabricated package station, or to an elaborately designed and constructed station capable of pumping large volumes of wastewater. When larger stations are constructed and require the continuous presence of an operating staff, they are often referred to as pumping plants.

Stations may be classified as wet well or dry well installations, depending on the location of the pumping units.

8.010 Manhole Used as Wet Well Station (*Figure 8.1*)

Precast concrete manhole sections with submersible pumps have been used as lift stations to serve smaller communities. The lift station shown in Figure 8.1 has guide bars made of two-inch pipe which allow the pump to be removed from the top of the manhole by pulling the pump up the guide bars. When the pump is lowered, a flanged, gasketed connection seals the pump to the discharge. The weight of the pump keeps it in place and there are no bolts to remove. Therefore, an operator does not need to enter the wet well to remove or install the pump. Figure 8.2 contains the details of a submersible pump and the connection to the discharge pipe or force main.

In this type of lift station configuration, there is a minimum of equipment located in the wet well. Depending on the size of the system, the electrical controls are either mounted on a utility pole outside the station or on a pedestal mounted at ground level on a concrete slab with the controls contained in a weatherproof enclosure. Typically the pumps are controlled by one of several different types of level controlling devices such as multiple floats, enclosed-electrode controllers, pneumatic bubbler systems, level pressure transducers, and ultrasonic transducers. These devices are located in the wet well for sensing wet well levels and low- or high-level alarms. Because this station is not designed for maintenance activities to take place in the wet well, check valves and isolation valves should be installed in a valve vault outside the station. Also, any necessary flowmeters may be installed in the valve pit.

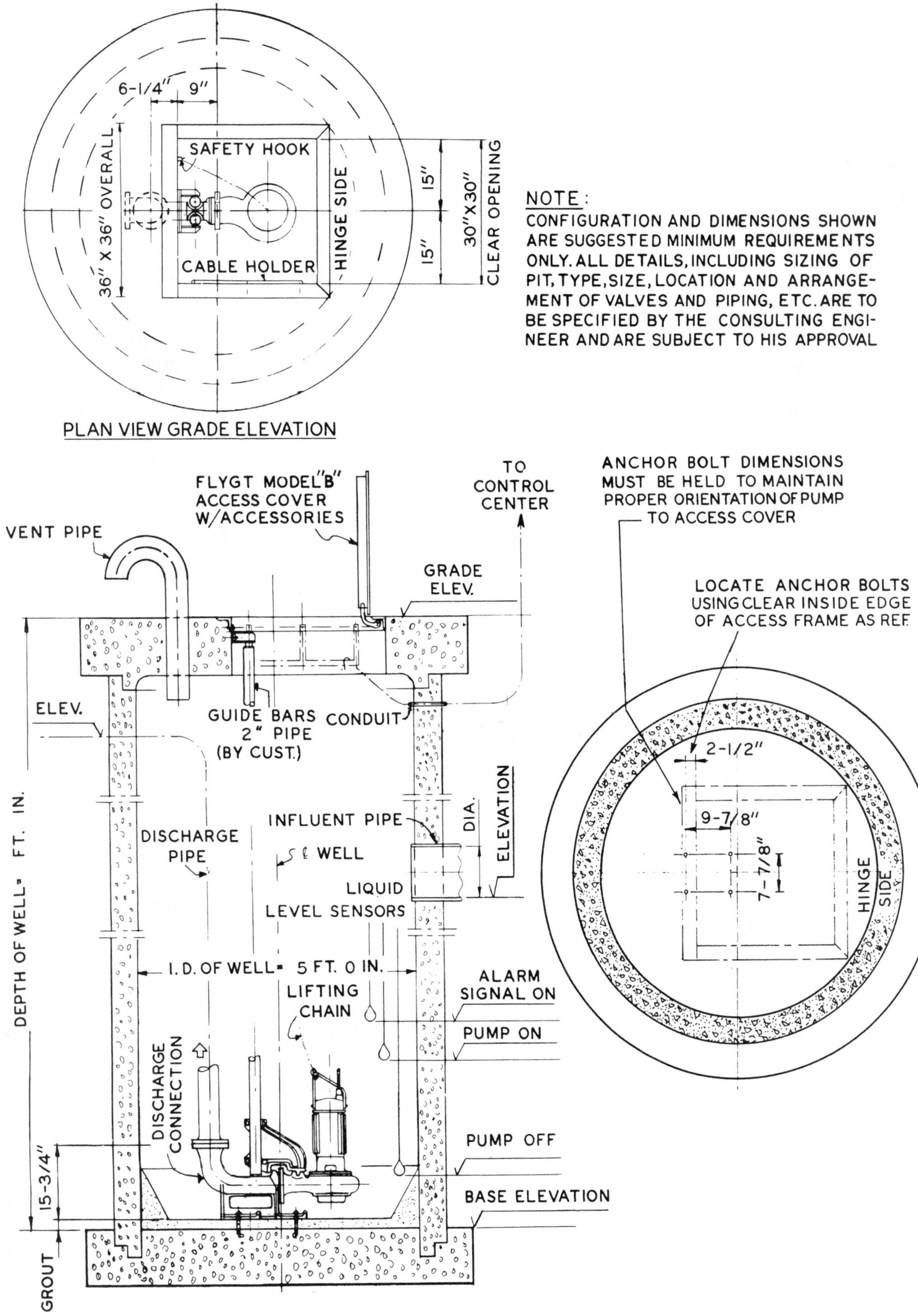

Fig. 8.1 Submersible pump in wet well

(Courtesy of Flygt)

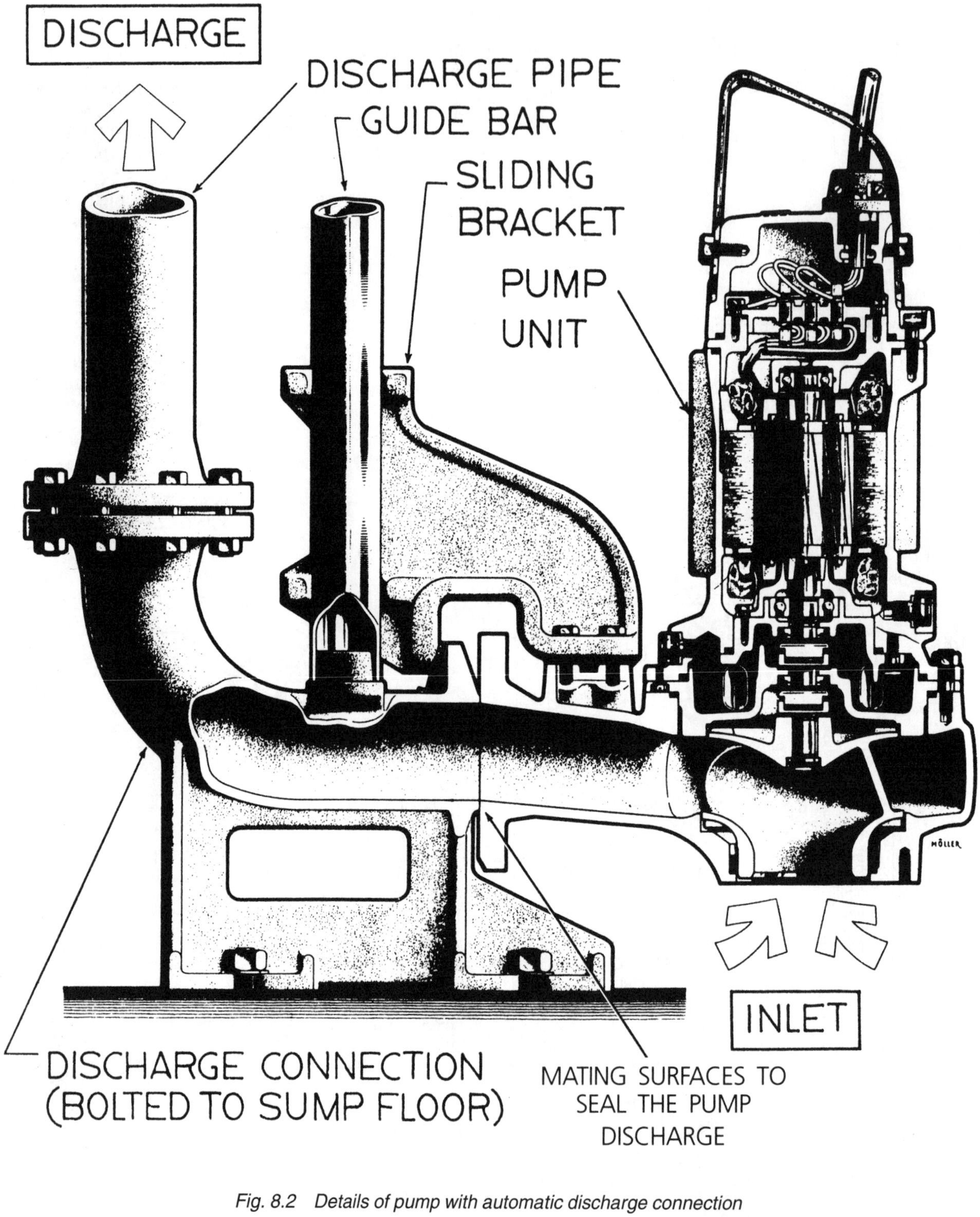
DISCHARGE
DISCHARGE PIPE
GUIDE BAR
SLIDING BRACKET
PUMP UNIT
INLET
DISCHARGE CONNECTION (BOLTED TO SUMP FLOOR)
MATING SURFACES TO SEAL THE PUMP DISCHARGE

Fig. 8.2 Details of pump with automatic discharge connection

(Courtesy of Flygt)

The design of the station should take into consideration the humid and corrosive atmosphere which may exist in the station and materials such as lifting chains, guide rails, and fasteners should be selected for their resistance to the corrosive atmosphere in the wet well. Electrical systems and equipment that are installed in the wet well must conform to the National Electrical Code section on hazardous locations which defines various classes, divisions and groups for electrical equipment.

Note: This chapter does not discuss any of the alternative types of sewer systems that use small grinder pumps or systems such as septic tank effluent pump (STEP) systems. Additional information on these systems can be found in the Office of Water Programs' *SMALL WASTEWATER SYSTEM OPERATION AND MAINTENANCE*, Volume I, price $33.00.

8.011 Wet Well Station

Large wet well stations today can provide capacities up to 25,000 GPM at discharge heads of 80 to 100 feet or more. These large stations are in wide use and two different types of configurations are frequently constructed:

1. Those that use larger submersible pumps, and
2. Those that use an above-ground pump station. In this type, a fiberglass enclosure usually sits on top of the wet well with the suction pipe extending down into the wet well (this is known as a pump which pulls a suction lift).

Because the above-ground station must "lift" the water up to the pump suction, the above-ground type stations use self-priming centrifugal pumps. The maximum pump size is usually six or eight inches, whereas the submersible pump can be as large as twenty-four inches. Depending on the pump size and number of pumps, the wet well could be constructed of precast concrete manhole sections, rectangular box culvert sections, or poured-in-place concrete structures for the larger pumps. Wet well installations typically have limited access to the pumping equipment. If anything goes wrong, they can be difficult to repair.

Figure 8.3 illustrates a typical duplex (two pumps) submersible pumping station and its major components while Figure 8.4 illustrates an above-ground pump station with a fiberglass-reinforced plastic enclosure.

Equipment located in the wet well should be minimized, including suction and discharge valves, check valves, or other equipment which requires routine, periodic maintenance. This equipment can be located in small equipment manholes located adjacent to the wet well to make accessibility and maintenance much easier for the operator.

8.012 Dry Well Station

Dry well stations are another common type of lift station found in the collection system. This type of station isolates the pumps, motors, electrical control, and auxiliary equipment from the wet well in a separate dry well. This feature provides a cleaner and safer environment for operation and maintenance personnel. These stations will range in size from smaller prefabricated pump stations constructed of steel or fiberglass that come already assembled from the factory (Figure 8.5) to much larger concrete cast-in-place stations which may or may not include a building to house the equipment, depending on environmental conditions (Figures 8.6, 8.7, and 8.8).

The third type of wet well/dry well lift station that has been used in smaller areas is the pneumatic ejector (Figure 8.9), which uses air pressure to eject the wastewater; however, this type of station is limited in capacity and size.

QUESTIONS

Write your answers in a notebook and then compare your answers with those on page 118.

8.0D What type of pump is often installed in a wet well?

8.0E What is a limitation of installing a pump inside a wet well?

8.02 Lift Station Requirements

The most desirable operation of a lift station would be the situation where all the flow and solids that discharge into the wet well from the gravity sewer are lifted to the higher elevation and continue to the wastewater treatment plant without delay. This would occur with the highest use of equipment and energy efficiency possible. Also operational and maintenance problems would be minimized.

Some lift stations almost meet these ideal requirements, but there are many more that do not. Usually a lift station is designed to handle expected peak flows. Often this means that long detention or holding times occur in wet wells during low flow periods. Provisions for wet well aeration during low flow periods could help keep the detained wastewater fresh.

One of the most frequently overlooked considerations during the pump station design is the addition of air and vacuum release valves on the force main. Force mains, unlike gravity sewers, are normally installed to follow the contour of

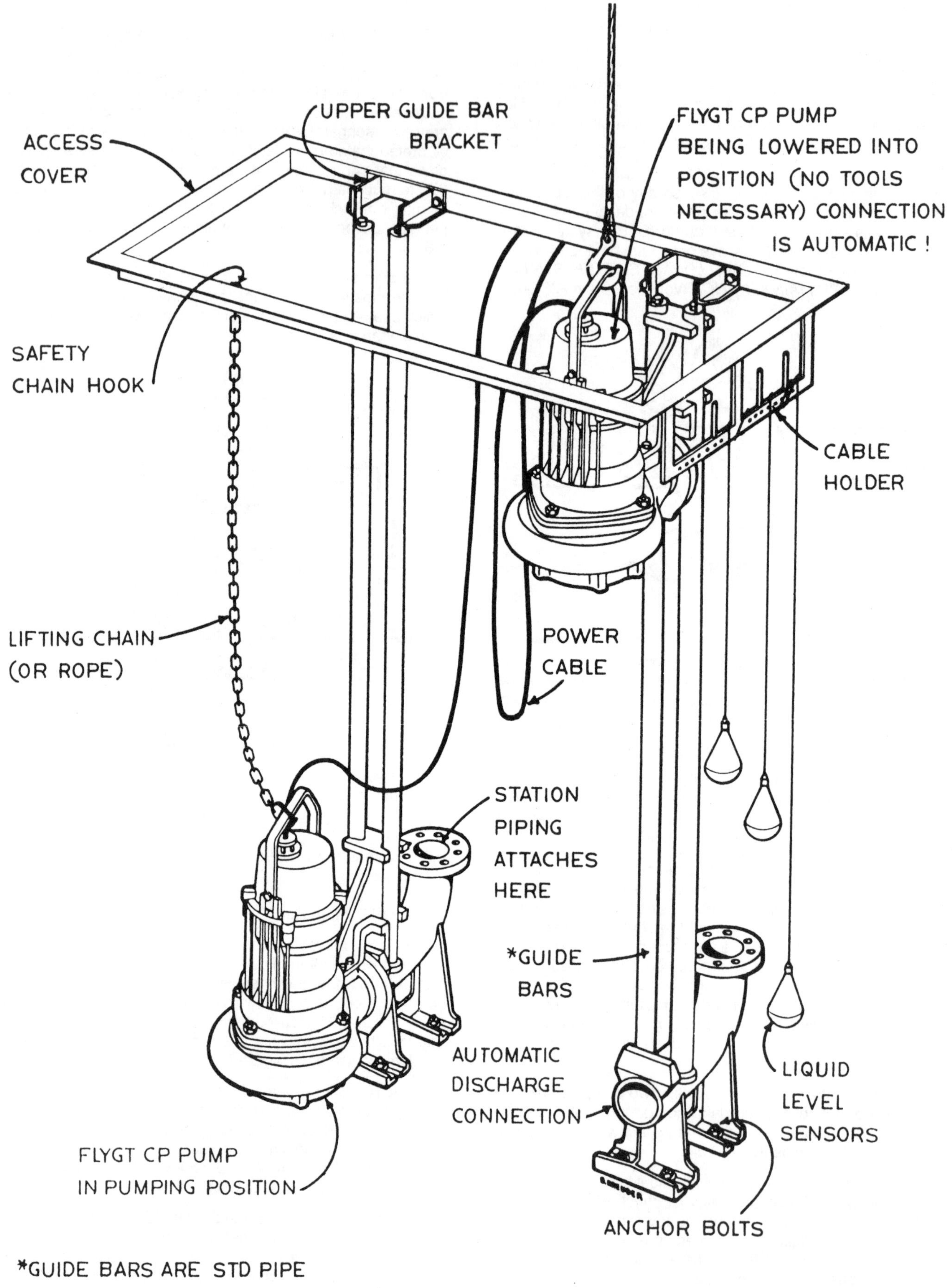
UPPER GUIDE BAR BRACKET
FLYGT CP PUMP BEING LOWERED INTO POSITION (NO TOOLS NECESSARY) CONNECTION IS AUTOMATIC !
ACCESS COVER
SAFETY CHAIN HOOK
CABLE HOLDER
LIFTING CHAIN (OR ROPE)
POWER CABLE
STATION PIPING ATTACHES HERE
*GUIDE BARS
AUTOMATIC DISCHARGE CONNECTION
LIQUID LEVEL SENSORS
FLYGT CP PUMP IN PUMPING POSITION
ANCHOR BOLTS
*GUIDE BARS ARE STD PIPE

Fig. 8.3 Duplex submersible pumping station

(Courtesy of Flygt)

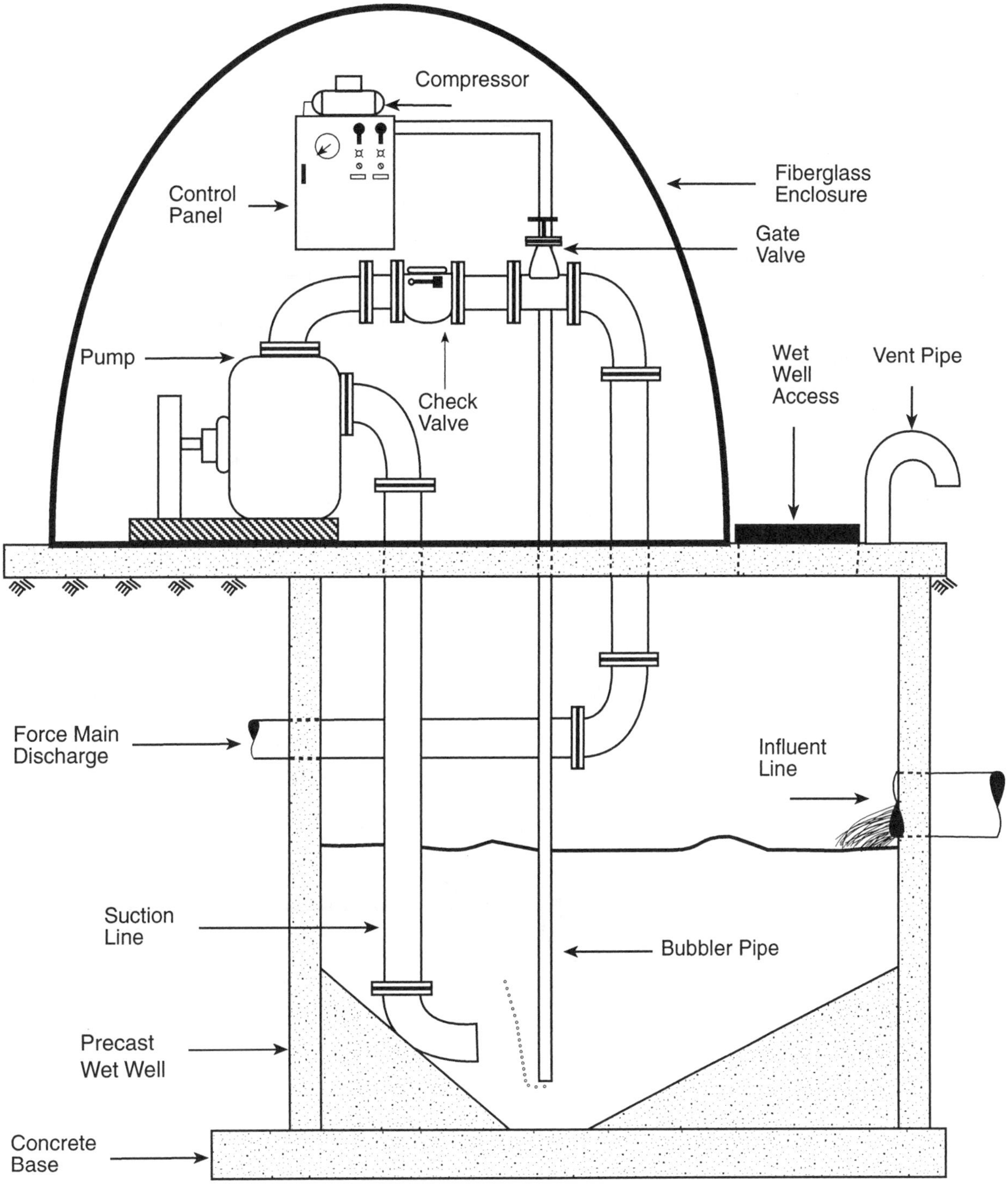

Fig. 8.4 Above-ground pump station with a fiberglass-reinforced plastic enclosure

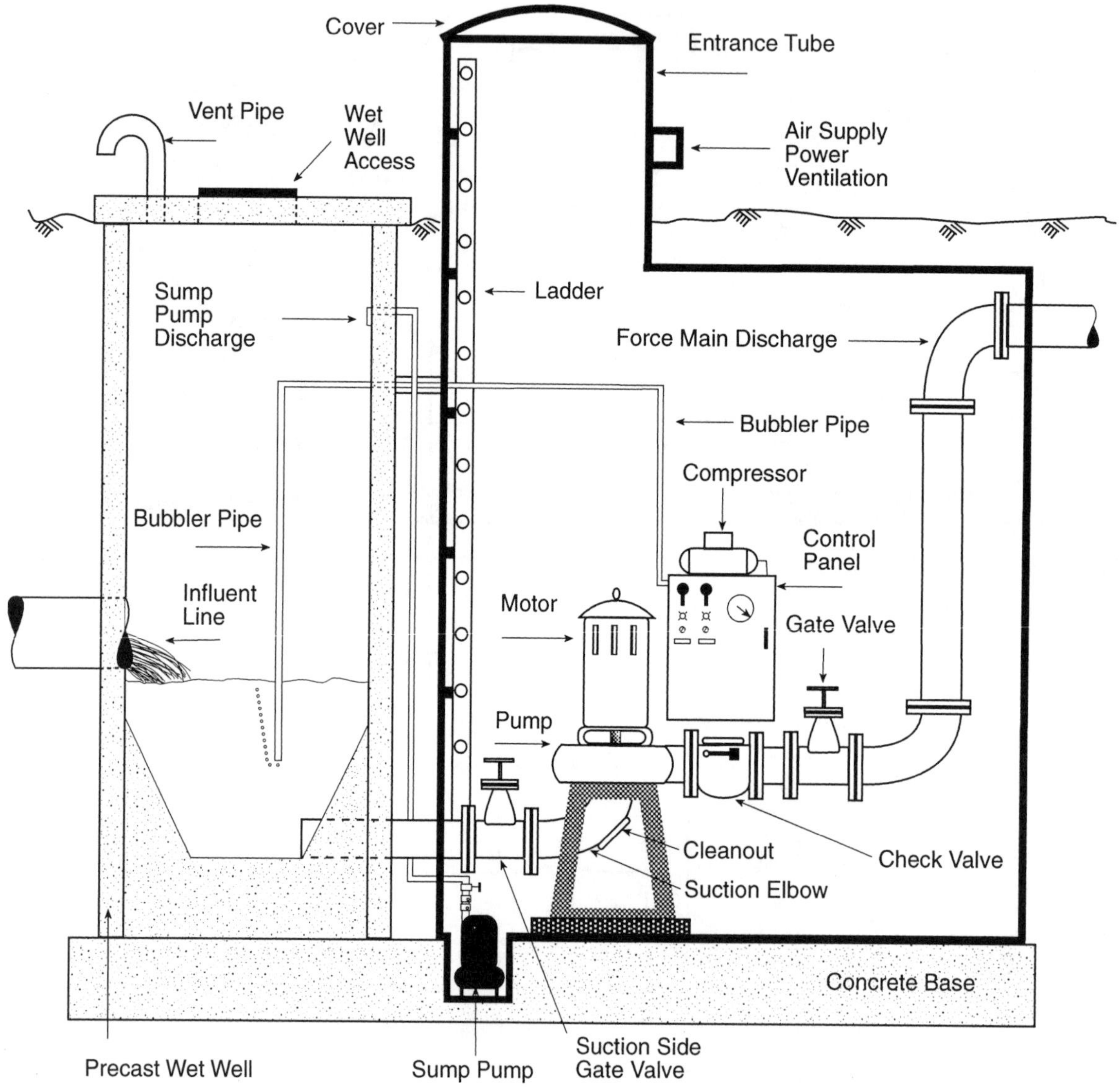

Fig. 8.5 Prefabricated factory lift station

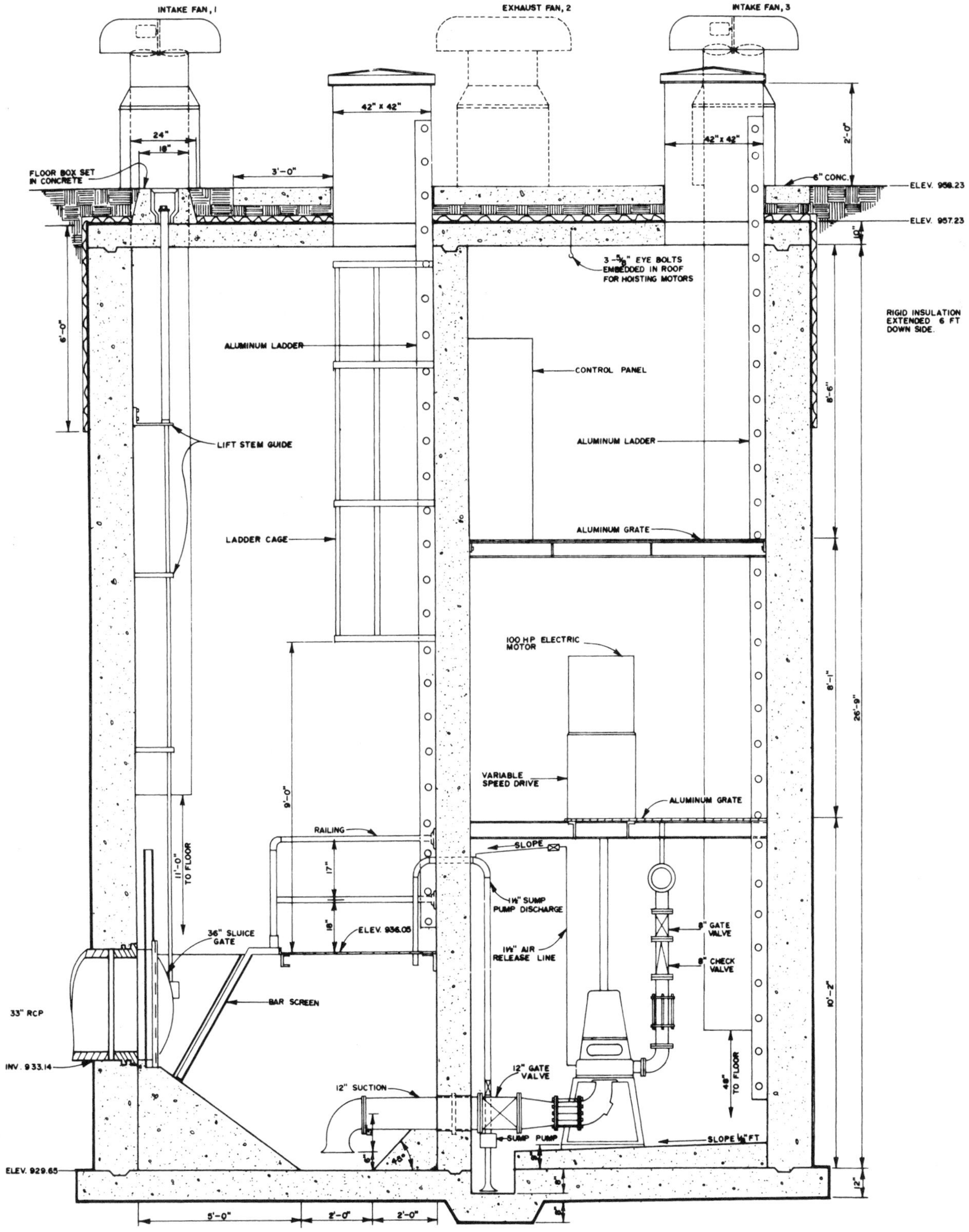

SECTION A A LIFT STATION

Fig. 8.6 Section of cast-in-place lift station

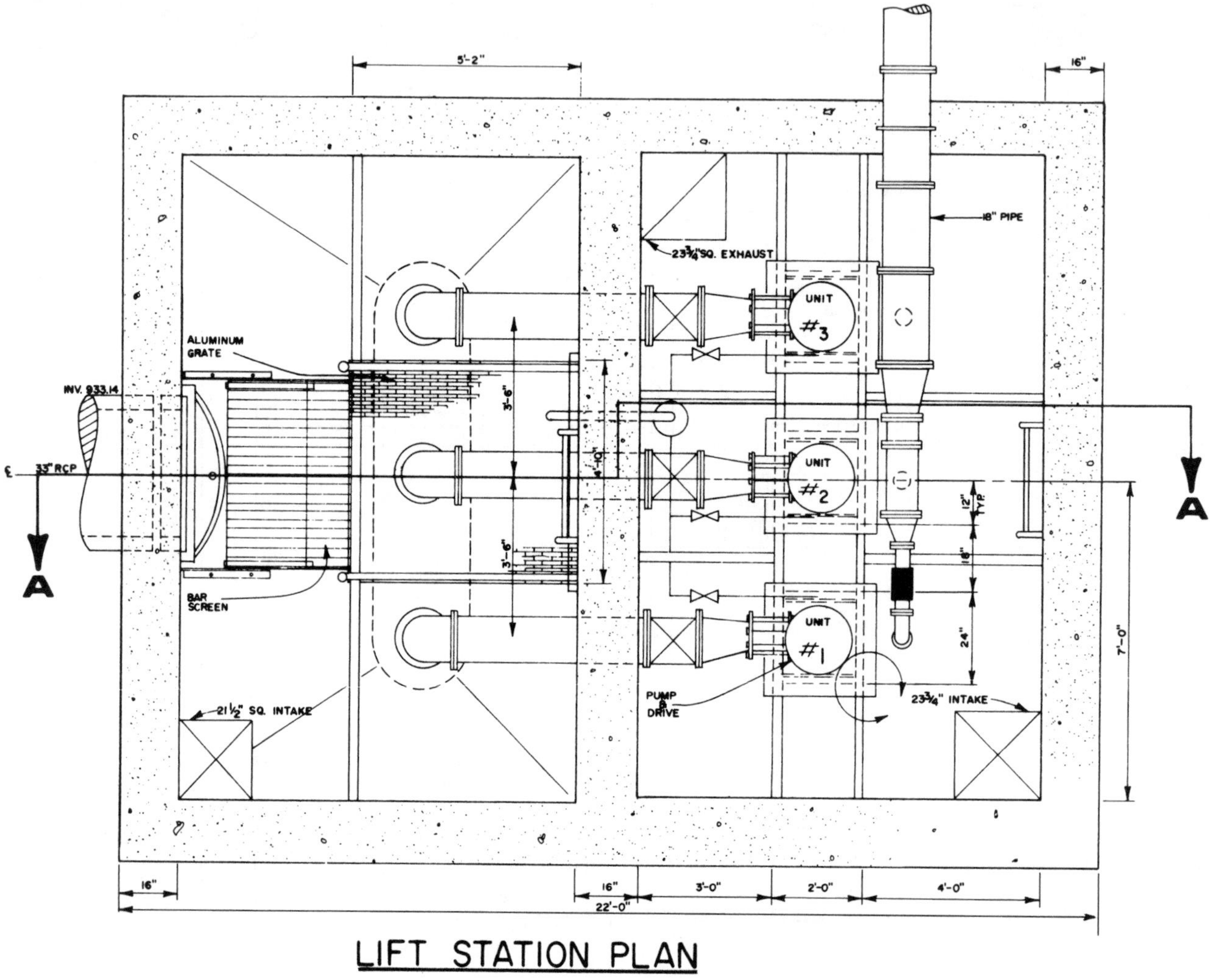

Fig. 8.7 Plan view of Figure 8.6, cast-in-place lift station

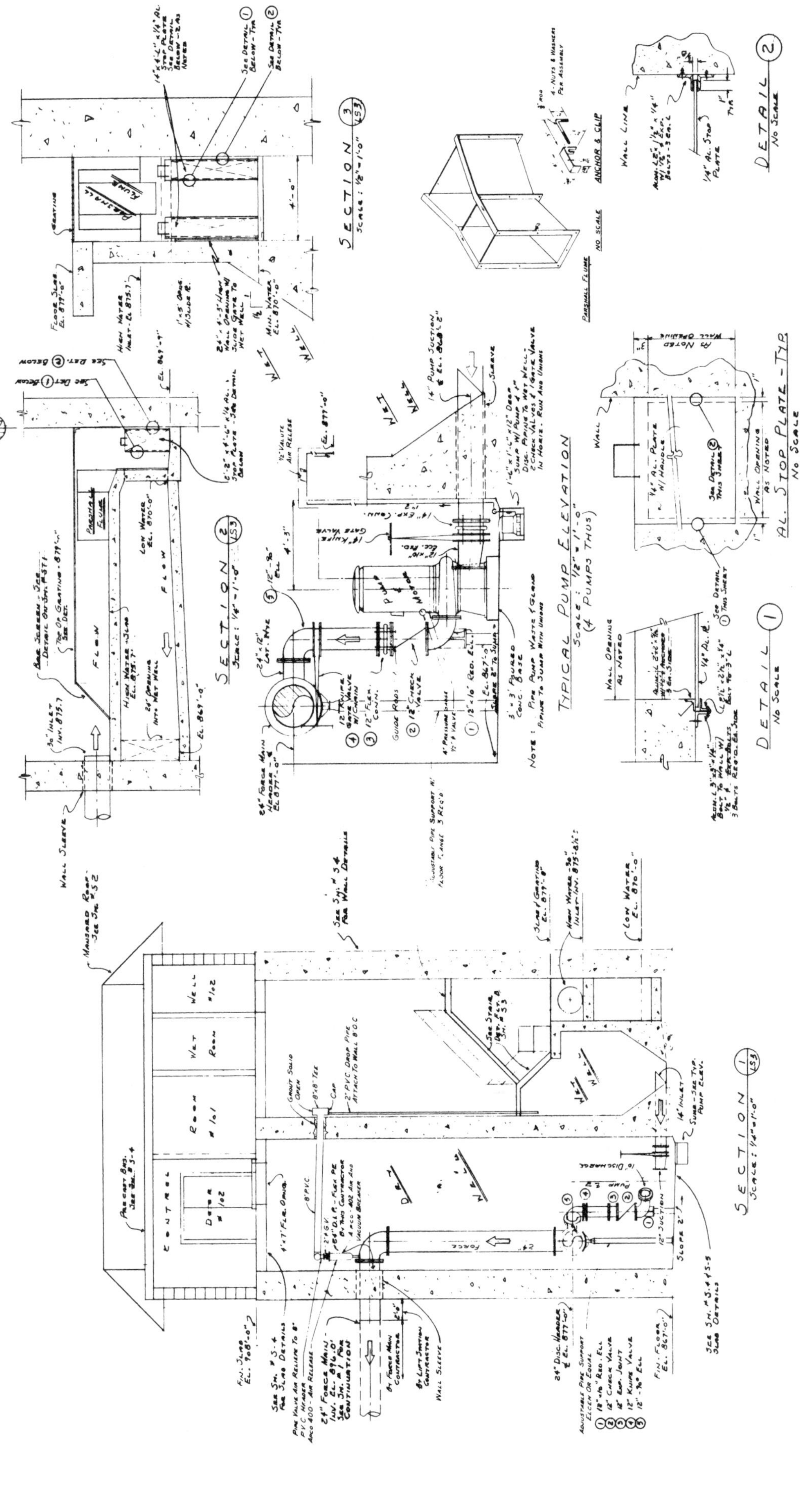

Fig. 8.8 Cast-in-place lift station with an above-ground structure

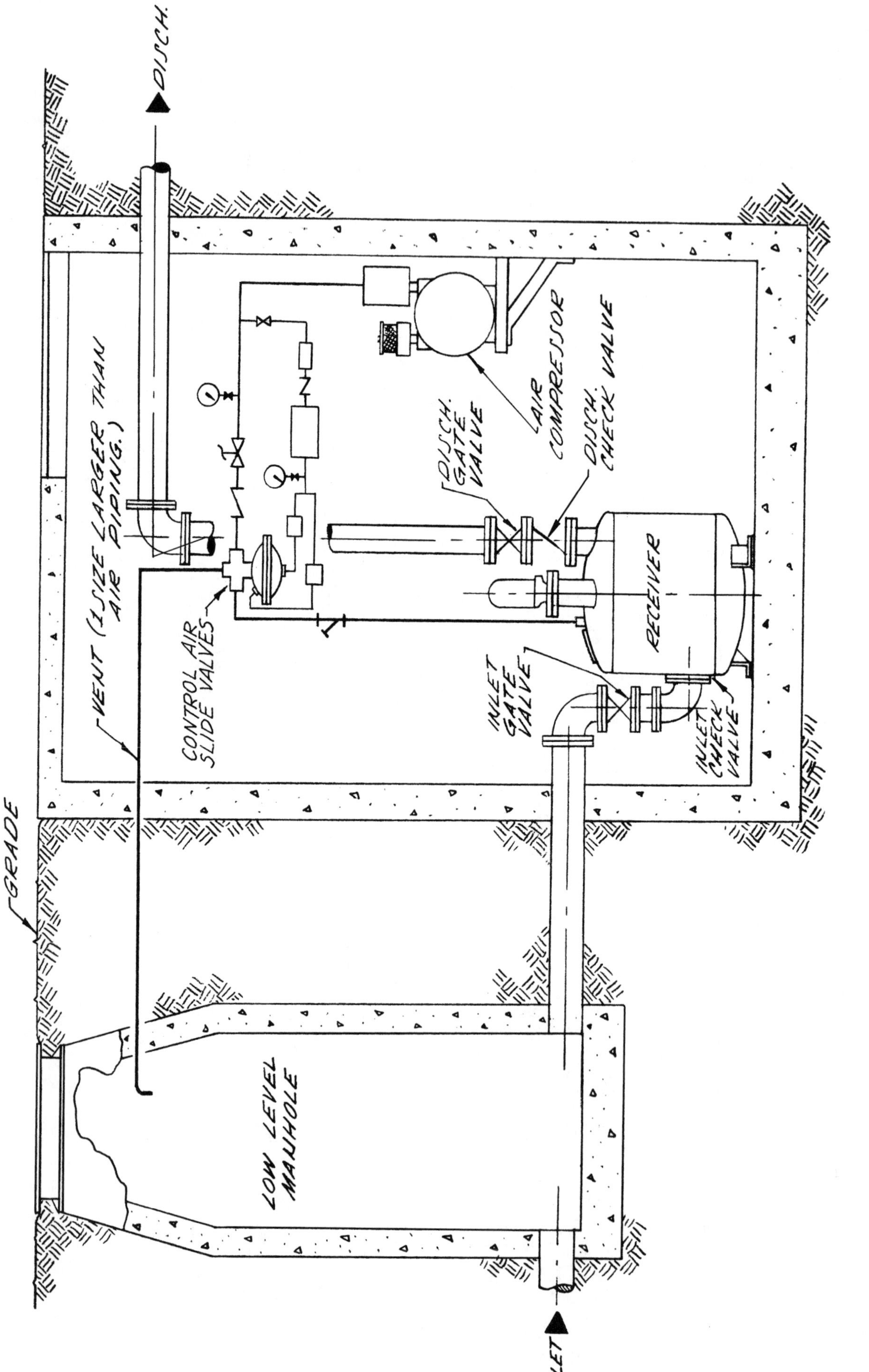

Fig. 8.9 Manhole wet well and pneumatic ejector dry well station
(Courtesy of James Equipment and Manufacturing Company)

the ground surface. (Remember that raw wastewater is moved through the force main by the pressure or head developed by the pump as opposed to gravity flow in a gravity sewer system.) When force mains follow the ground surface slopes, they will often have high points where air can accumulate and reduce the effective inner diameter of the pipe. This will reduce pipe capacity and increase pumping costs.

Raw wastewater contains varying amounts of *ENTRAINED*[3] air and may also contain gases such as methane and hydrogen sulfide from the anaerobic digestion of wastewater in the collection system. Entrained air and gases will separate from the raw wastewater and, because they are lighter than water, will accumulate at the high points in the force main. Air and gas bubbles create two problems: (1) they reduce the effective diameter of the pipe and (2) they can cause a condition known as *WATER HAMMER*[4] in which extremely high pressure shock waves travel up and down the force main. Water hammer associated with check valve "slamming" can produce a force strong enough to damage check valves and piping in the station. In severe cases, water hammer can break force mains and thereby cause a system failure and flooding. In some cases, depending on the layout of the force main, trapped pockets of air cause negative pressures to occur when the pumps are shut off. A negative pressure can also occur at the high point in a line as the column of water downstream from the high point continues to move. Negative pressure sufficient to collapse the pipe can develop and can also contribute to the destructive effects of water hammer. The solution to these problems is to install air relief valves or, where required, combination air relief and vacuum release valves at the high points in the force main. A typical air relief valve is shown in Figure 8.10.

An air release valve installed on a high point on the force main will fill with wastewater, shut off and then be subjected to the same pressure as that inside the force main. When wastewater flows through the force main, the entrained air separates from the wastewater and enters the body of the valve. Air entering the valve displaces an equal amount of water causing the float to drop as the wastewater level drops. As the float drops, the valve at the top of the unit opens and air is allowed to discharge to the atmosphere. When the air is released, the wastewater level within the valve chamber once again rises, which lifts the float and closes the valve orifice at the top. A vacuum relief valve (Figure 8.11) is also float operated but works in the reverse. As the wastewater drains from the high point in the force main, the valve also drains. However, when the float drops, instead of closing the valve at the top it opens the valve and allows air to enter.

Figure 8.12 illustrates a combination air release, vacuum and cleanout manhole on a force main. This configuration shows a combination air release and vacuum valve that is installed on the blind flange of a T-section of pipe. Under normal operation, the valve operates as previously described with the upstream and downsteam 14-inch plug valves in the open position. In the event access to the force main is required, it can be isolated by operating the plug valves and removing the blind flange.

Air release and vacuum relief valves need to be maintained on a periodic basis which is best determined by experience. Both of these valves may fail to operate reliably if grease is allowed to accumulate in the valve body or on the operating mechanism. As shown in Figure 8.10, fittings can be attached at the top and the bottom of the valve to permit backflushing. To clean the valve, attach the backflushing hose to a pressurized water source using a quick disconnect coupling, open the ½-inch shutoff valve and backflush the valve through the 1-inch blowoff valve at the bottom. If using a potable (drinking) water source, it is important that the system be provided with an anti-siphon device to prevent contamination of the potable water. During the backflushing operation, the valve is isolated from the force main by closing the 2-inch shutoff valve shown in Figure 8.10. Provisions should be made during the maintenance of the valve to collect the backflush water from the 1-inch blowoff valve since this is contaminated wastewater that should not be discharged onto the street or into the valve pit.

The most critical component of a lift station is the pumping unit, but pipes, valves, controls and power supply also are essential. All of these items must be operational or the lift station will fail.

1. *FLOW*

Most pump capacities are rated in GPM (gallons per minute). The design engineer must carefully estimate flows under operating conditions to select the proper pump or pumps for the lift station. Important operating conditions include:

Flow, Q_1 = Average daily dry weather flow,

Flow, Q_2 = Seasonal dry weather flow, (to handle industrial flows that produce high flows for 1 to 3 month periods such as canneries), and

Flow, Q_3 = Peak wet weather flow, (based upon a one-in-ten year occurrence (also known as a 10-year storm) or some other severe storm frequency that produces conditions where inflow/infiltration may enter the collection system, especially important in combined wastewater and storm water collection systems).

Under certain circumstances, Q_2 and Q_3 may range from 1.5 to 10.0 times or more above Q_1. These high flows present greater problems in smaller lift stations than in the larger stations with multiple pumping units. All lift stations should have sufficient pump capacity to handle peak flows. During usual or average flow conditions, the pumps that handle peak flows can serve as backup units during periods when the other pumps are shut down for maintenance and repairs.

[3] *Entrain. To trap bubbles in water either mechanically through turbulence or chemically through a reaction.*

[4] *Water Hammer. The sound like someone hammering on a pipe that occurs when a valve is opened or closed very rapidly. When a valve position is changed quickly, the water pressure in a pipe will increase and decrease back and forth very quickly. This rise and fall in pressures can cause serious damage to the system.*

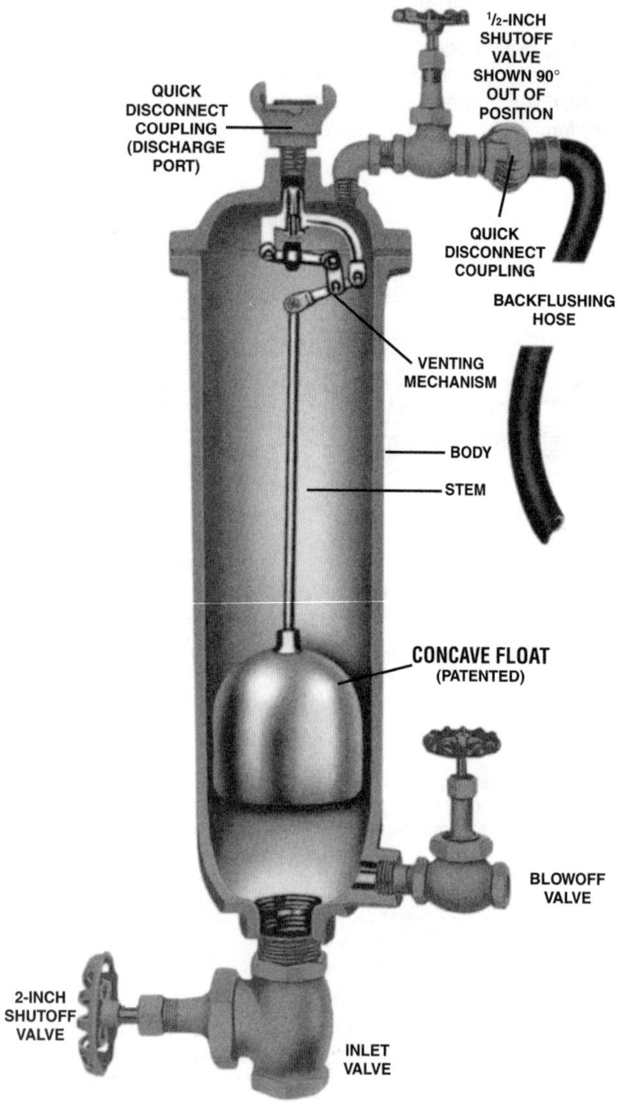

Fig. 8.10 Air release valve
(Permission of APCO)

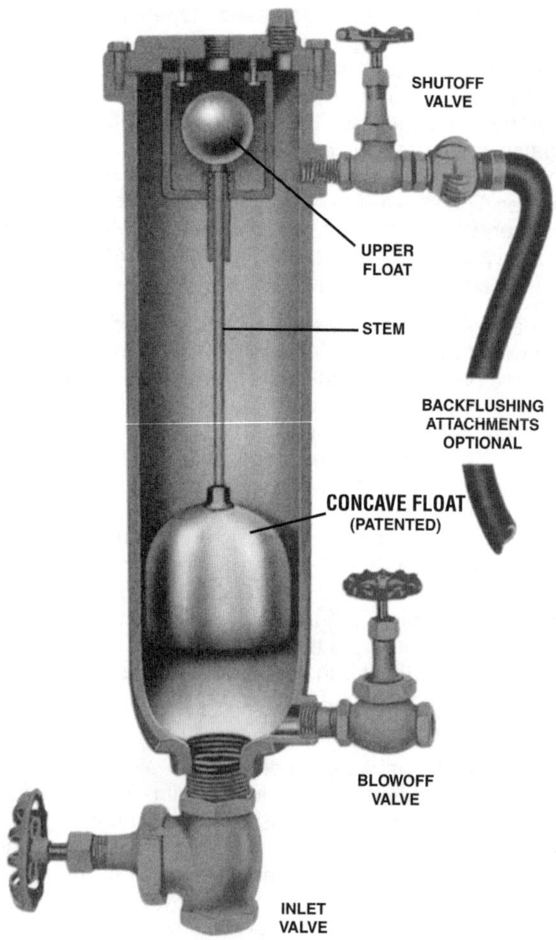

Fig. 8.11 Combination air release and vacuum relief valve
(Permission of APCO)

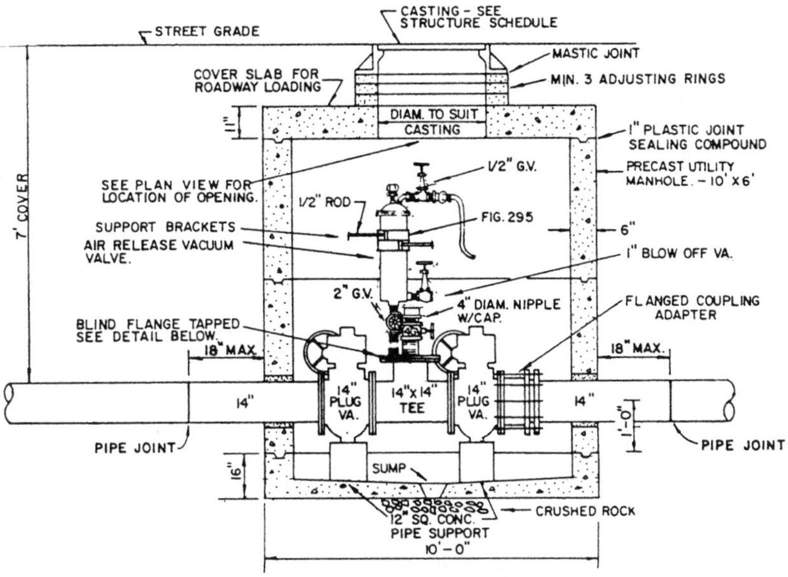

COMBINATION AIR RELEASE, VACUUM 8 CLEANOUT
MANHOLE TYPE S-6
NO SCALE

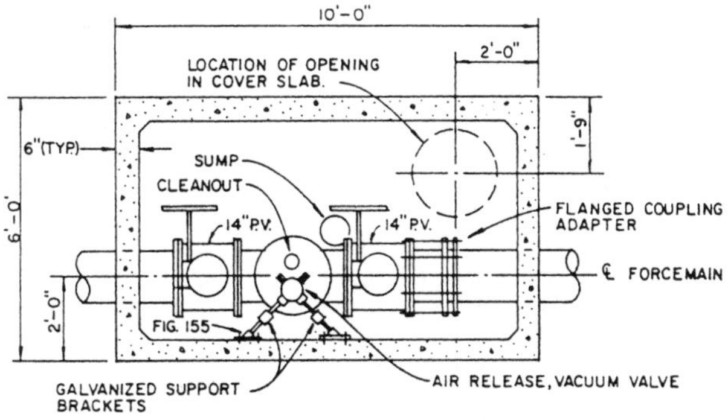

PLAN VIEW MANHOLE TYPE S-6
NO SCALE

Fig. 8.12 Installation of a combination air release/vacuum relief valve and cleanout in a force main

2. *HEAD*

The term *HEAD* refers to pressure head (pounds per square inch or psi) or elevation head (feet). Head is a vertical distance and is measured as the difference in elevation between two points. Head may be expressed in feet or psi.[5] Important head terms that should be known when pumps are selected include *SUCTION HEAD, DISCHARGE HEAD, TOTAL STATIC HEAD,* and *TOTAL DYNAMIC HEAD.*

Suction heads may be positive (suction head) or negative (suction lift), depending on whether the water level in the wet well is above (positive) or below (negative) the center line of the pump impeller. Discharge heads are measured on the discharge side of a pump from the center line of the pump to an elevation on the discharge side. Static heads are shown on Figure 8.13 and dynamic heads on Figure 8.14. Static heads are measured when the pump is not operating and are the differences in elevation (in feet) between the surface of the water on the suction side of the pump and the surface of the water on the discharge side of the pump. Dynamic heads are measured when the pump is operating and depend on the velocity of water in the pump suction and discharge pipes.

Dynamic heads are greater than static heads because they include the static heads plus the friction losses in the suction and discharge pipes. The energy required to start a pump is greater than the total dynamic head (TDH) during normal operating conditions because additional energy is required to start the motor and the pump and to start the water flowing through the pipes, the check valves, and the pump.

Friction losses result from pipe friction and the friction losses due to the water flowing through pipe fittings such as valves, reducers and elbows. The greater the velocity of flow in pipes, the greater the friction losses. Suction pipes are often larger in diameter than discharge pipes to reduce friction losses in the suction pipe and thus also to reduce *CAVITATION*[6] problems. However, cavitation problems are more likely to be caused by excessive tip speed of the impeller, air leaks on the suction side of the pump, and restrictions in the suction line. Friction losses may be calculated from tables in pump and pipe handbooks by knowing the flow (in GPM) and the diameter, type, and length of pipe and also the size and type of valves and fittings. See Volume I, "Applications of Arithmetic to Collection Systems," Section A.88, "Friction or Energy Losses," for an explanation and examples of how to calculate friction losses.

Design engineers try to minimize friction losses by careful layout of pipes, selection of pipe size (diameter) and length, and the number of valves and fittings. Consideration must be given to friction losses at expected flows and suction conditions when determining desired pump characteristics and selecting a pump. If the lift station piping and pumps are not properly designed, problems can develop from vibrations, cavitation, and insufficient pumping capacity, thus requiring excessive maintenance.

QUESTIONS

Write your answers in a notebook and then compare your answers with those on page 118.

8.0F What is the most desirable operational situation for a wet well?

8.0G What are the effects of water hammer?

8.0H What is the difference between static and dynamic head?

8.0I Friction losses are caused by what factors in a lift station?

Please answer the discussion and review questions next.

[5] *A column of water 2.31 feet high creates a pressure of 1 pound per square inch (psi) at the bottom of the column. Therefore 1 psi = 2.31 ft and 1 ft = 0.433 psi.*

[6] *Cavitation (CAV-uh-TAY-shun). The formation and collapse of a gas pocket or bubble on the blade of an impeller or the gate of a valve. The collapse of this gas pocket or bubble drives water into the impeller or gate with a terrific force that can cause pitting on the impeller or gate surface. Cavitation is accompanied by loud noises that sound like someone is pounding on the impeller or gate with a hammer.*

NOTE: This figure illustrates a pump with a suction *LIFT.* Pumps should have a suction *HEAD* which means the wet well water level should be higher than the pump impeller. This pump will have difficulty starting unless it is a self-priming pump because the water level in the wet well is below the pump. Also, if air gets into the suction line, the only way it can get out is through the pump. Controls may be modified to allow the pump to operate only when a suction head exists if flooding of the service area will not result.

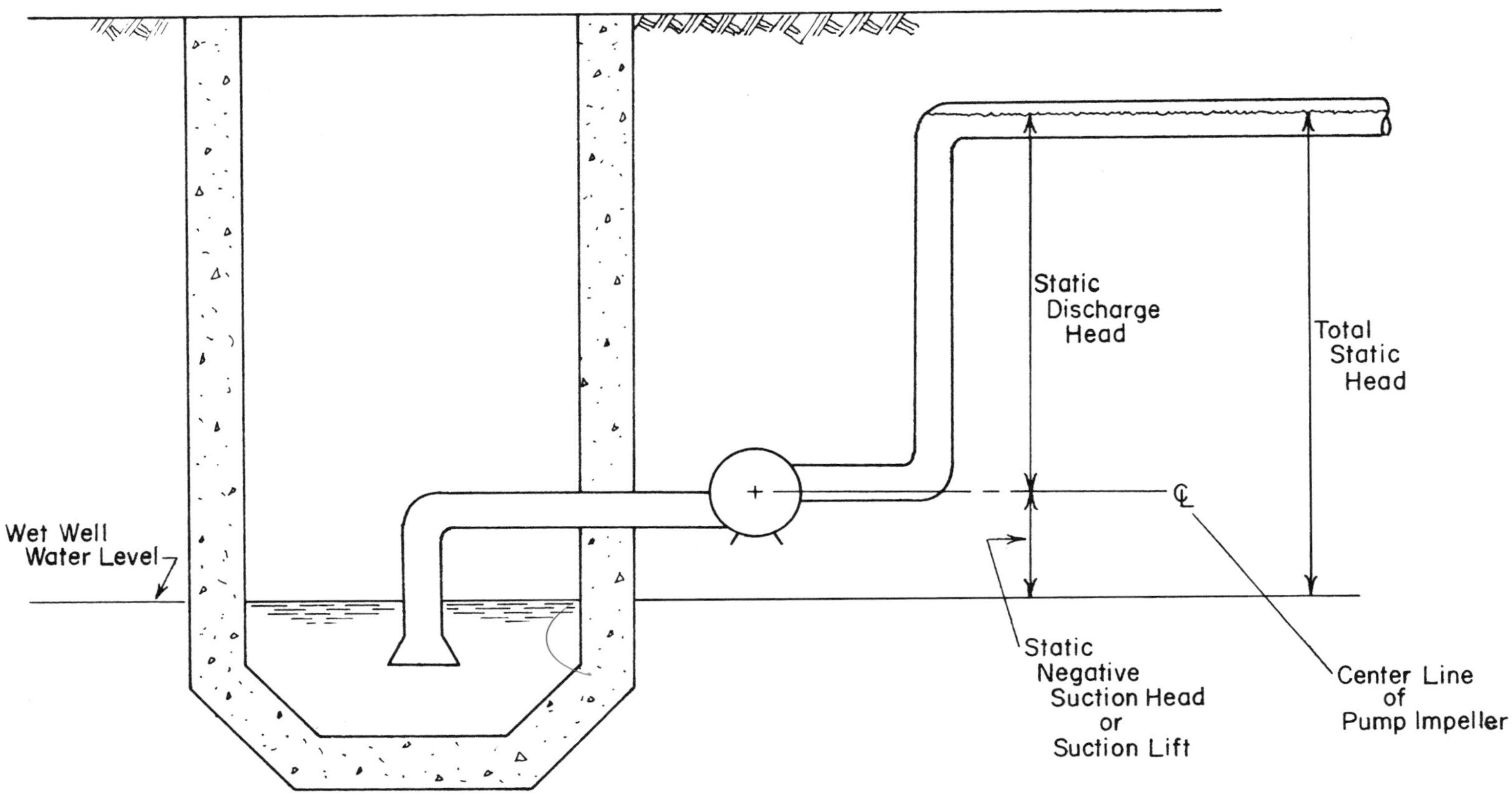

Fig. 8.13 Static heads (pump is not operating)

NOTE: This figure illustrates a pump with a suction *LIFT*. Pumps should have a suction *HEAD* which means the wet well water level should be higher than the pump impeller. This pump will have difficulty starting unless it is a self-priming pump because the water level in the wet well is below the pump. Also, if air gets into the suction line, the only way it can get out is through the pump. Controls may be modified to allow the pump to operate only when a suction head exists if flooding of the service area will not result.

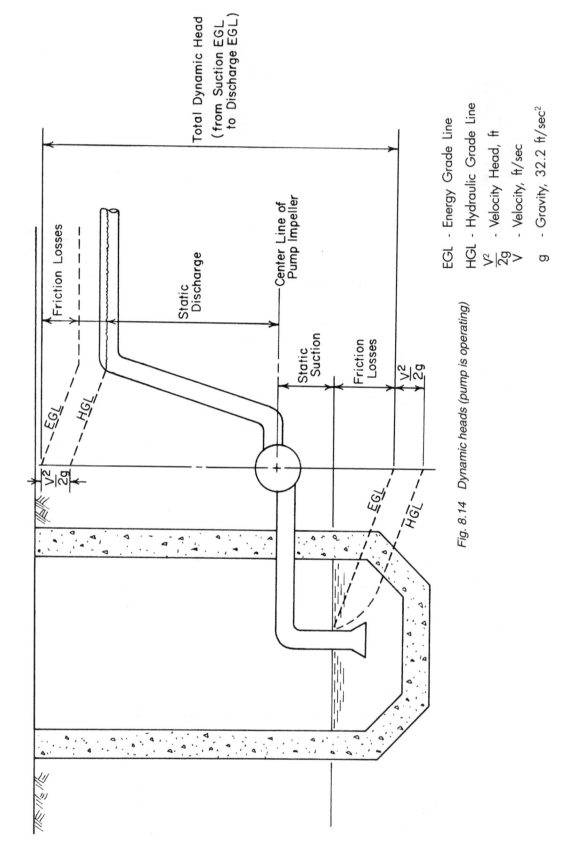

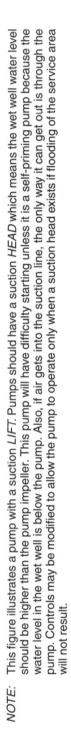

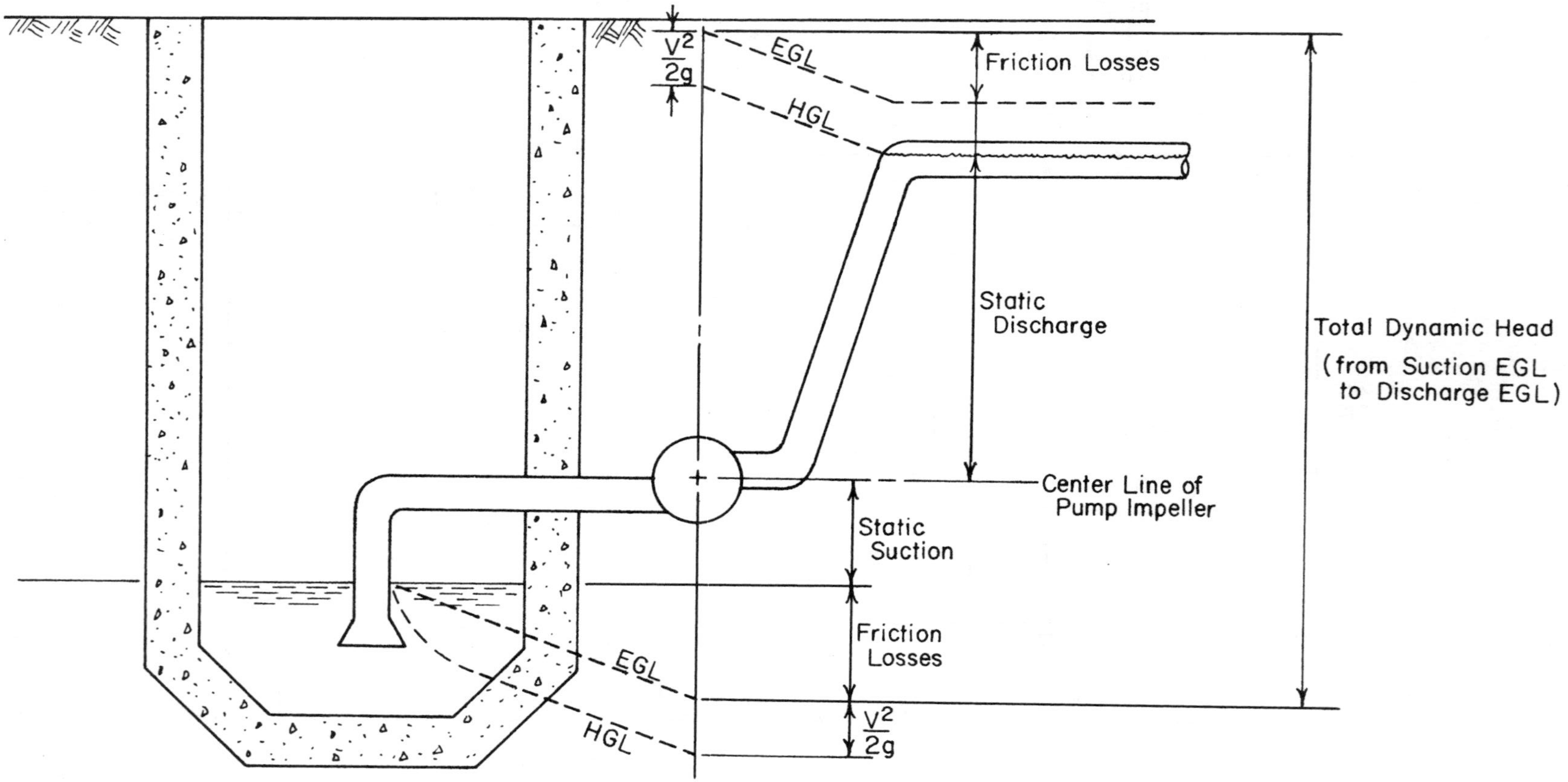

Fig. 8.14 Dynamic heads (pump is operating)

EGL - Energy Grade Line
HGL - Hydraulic Grade Line
$\frac{V^2}{2g}$ - Velocity Head, ft
V - Velocity, ft/sec
g - Gravity, 32.2 ft/sec^2

DISCUSSION AND REVIEW QUESTIONS

Chapter 8. LIFT STATIONS

(Lesson 1 of 5 Lessons)

At the end of each lesson in this chapter you will find some discussion and review questions. The purpose of these questions is to indicate to you how well you understand the material in the lesson. Write the answers to these questions in your notebook before continuing.

1. What is the purpose of a lift station?
2. Where are lift stations located?
3. Describe the ideal result of operating a lift station.
4. Why should air release valves be installed at high points in force mains?
5. Why does a pump require more energy (draw more power) to start than it does during normal operating conditions?

CHAPTER 8. LIFT STATIONS

(Lesson 2 of 5 Lessons)

8.1 COMPONENTS OF A LIFT STATION

8.10 Wet Wells

The wet well size and shape and the equipment located in it must be carefully designed in order to allow proper operation of the pumping equipment and to allow the operator to maintain the wet well area.

8.100 Wet Well Dimensions

The wet well size affects many operating conditions of the lift station; however, many of these are fixed at the time of design which limits the improvements an operator can make if problems develop or exist.

The length, width and usable depth of a wet well will affect the cycle ON/OFF time of the pumps since the wet well acts as a storage basin. If the capacity of the wet well is too small, excessive starting and stopping of the pump motors will occur which will ultimately cause a premature failure of the motor winding insulation.

Cycle times should be calculated for minimum, maximum, and peak flow periods. The ability of a motor to withstand frequent starting is generally limited by the motor size and, in the larger horsepowers, may be limited to as few as five starts per hour (see Chapter 9, Section 9.1, "Electrical Equipment Maintenance").

The operating cycle time is a function of the pump capacity, the storage that is available between the high level and low level, and the flow rate of raw wastewater into the wet well. This can be expressed as the following equation:

$$t = \frac{v}{d - q}$$

Where:

d = the actual pump capacity in gallons per minute

v = the storage volume between high and low level in gallons

q = inflow into the wet well in gallons per minute

t = total time between successive pump starts in minutes — this is also called the cycle time.

EXAMPLE 1

The storage volume in a wet well between the high and low levels for pump starting and stopping is 1,000 gallons. The actual pump capacity is 50 GPM and the inflow to the wet well is 25 GPM. What is the pump operating cycle time?

Known		**Unknown**
Storage Volume, gal	= 1,000 gal	Cycle Time, min
Pump Capacity, GPM	= 50 GPM	
Wet Well Inflow, GPM	= 25 GPM	

Calculate the pump operating cycle time.

$$\text{Cycle Time, min} = \frac{\text{Storage Volume, gal}}{\text{Pump Capacity, GPM} - \text{Wet Well Inflow, GPM}}$$

$$= \frac{1{,}000 \text{ gal}}{50 \text{ GPM} - 25 \text{ GPM}}$$

$$= 40 \text{ minutes}$$

If the wet well is too large, solids will tend to accumulate on the bottom since the wet well will be acting as a settling tank, similar to what occurs in a treatment plant sedimentation tank or clarifier. Accumulated solids could partially or totally block the pump suction pipes and, over a period of time, microorganisms decomposing the organic material will generate hydrogen sulfide (toxic) and methane (explosive) gas. Therefore, the design engineer may have to compromise between motor cycle time, minimum and maximum flows, and the detention time period. This may require annual, semi-annual, or even more frequent removal of the solids from the wet well using a vacuum-type cleaning machine.

Another method of cleaning the wet well is the use of high pressure water from the hydraulic cleaning (jet) machine. **WARNING:** As the solids are disturbed they may release hydrogen sulfide and methane gas. Also, the introduction of large quantities of solids in a short period of time into the pump suction could cause pumps to become partially or fully plugged and actually damage the pumps.

8.101 Flow Distribution

Flow distribution from the influent pipe in the wet well can create operating problems if the wastewater flow discharges directly opposite one or more of the pump suction pipes. Normally, this flow is aerated and contains large amounts of air. As the entrained air rises to the surface, it can accumulate in the suction pipe causing the pump to become air-bound. Then, when the pump is turned on, it will not pump the wastewater. Entrained air frequently contributes to pump cavitation which can rapidly destroy the impeller or wear rings and other pump components in the liquid end of the pump. This situation is usually caused by improper design and the operator has limited options available for correction; however, construction of a baffle to divert the influent flow away from the pump suction pipes may be effective.

8.102 Operating Levels

The operating levels in a lift station are determined by the high and low water levels in the wet well. The high water level is usually limited to the invert (bottom) of the incoming gravity pipe, since it is not desirable to allow "backup" into the incoming gravity pipe. The lowest water level is usually defined by the center line of the pump impeller. When using pumps that are not self-priming, the wet well water level should not be drawn down below the center line of the pump impeller.

Too small a distance between the gravity pipe invert and the center line of the pump contributes to the short cycling of the pump motors. Automatic lead/lag alternation of the pumps, which selects a different pump each time a complete pump-on/pump-off cycle is completed, may be an effective way to minimize the effects of short cycling.

The minimum required wet well level (submergence), in some cases, may be higher than the center line of the pump impeller depending on the suction line size and pump capacity. Even though the minimum water level is above the center line of the impeller, a condition called "vortexing" may occur. This happens when the water is being drawn into the pump so rapidly that a vortex or whirlpool forms on the surface of the water. This turbulence draws excessive amounts of air into the pump suction along with the wastewater. Although the vortex is usually visible to the naked eye and can be observed while the pump is pumping, it can be difficult to detect. The only solution is to raise the minimum operating level for the pumps; however, this will then affect the number of times per hour that the pump motors start. Submergence is illustrated in Figure 8.15 and examples of how to calculate submergence are provided.

Aeration systems are sometimes necessary to reduce odors in the wet well if detention times are excessive or if the wastewater reaching the wet well is septic due to warm temperatures and long travel times in the sewers. These systems may be another source of entrained air creating pump operating problems and/or air accumulation in the force main. Therefore, do not introduce excessive air into wet wells of pumping stations with long force mains or numerous high points.

Many lift stations built since the 1960s have variable-speed pumps that adjust pump speed and pump capacity in an attempt to match the pumping rate with the flow rate into the wet well. In many cases, however, the wet well in these types of installations may not be of adequate size, since there are a number of factors in the pumping system that determine whether or not the pump's speed will actually be able to match the incoming flow. If the variable-speed pumping system works as intended, it does provide an almost constant flow to the treatment plant which results in an easier operation of the plant. On the other hand, if pumps located near a treatment plant come on and off according to levels in a wet well, the treatment plant receives surges of wastewater that are difficult to handle.

8.103 Wet Well Access

Larger lift stations which require frequent access by the operator to perform routine maintenance should have stairs constructed in them. Stairs should be built of a corrosion-resistant material such as fiberglass or aluminum, since steel will deteriorate in the corrosive atmosphere of the wet well. Similarly, doors or access hatches should be made of a noncorrosive material.

Environmental conditions can also create problems for wet well access, particularly below-freezing weather in the north-

SUBMERGENCE
DESCRIPTION:

Submergence is the height of water or wastewater above a component or specific point, such as the center line of a pump. Air may be entrained in the pumped liquid if the pump suction is located too close to the water surface in the suction source. Pumping wastewater with entrained air can cause a reduction of capacity, rough and noisy operation, vibration, loss of efficiency and wasted power. Excessive wear of close-running parts, bearing stresses and shaft damage are also subsequent effects.

EXAMPLES (see drawing below):

1. When the pipe size is known, the minimum submergence required for 2,000 GPM through an 8-inch diameter pipe is 9.6 feet.
2. When the inlet area is known, the minimum submergence required for 2,000 GPM, 8-inch pipe through a 50 square inch outlet is 9.6 feet. Minimum submergence requirements may exceed the available space requirements. When this occurs a larger pipe size or inlet will reduce the required submergence.
3. The minimum submergence required for 2,000 GPM through a 10-inch pipe is 4.6 feet (5.0 feet less than required for an 8-inch pipe).

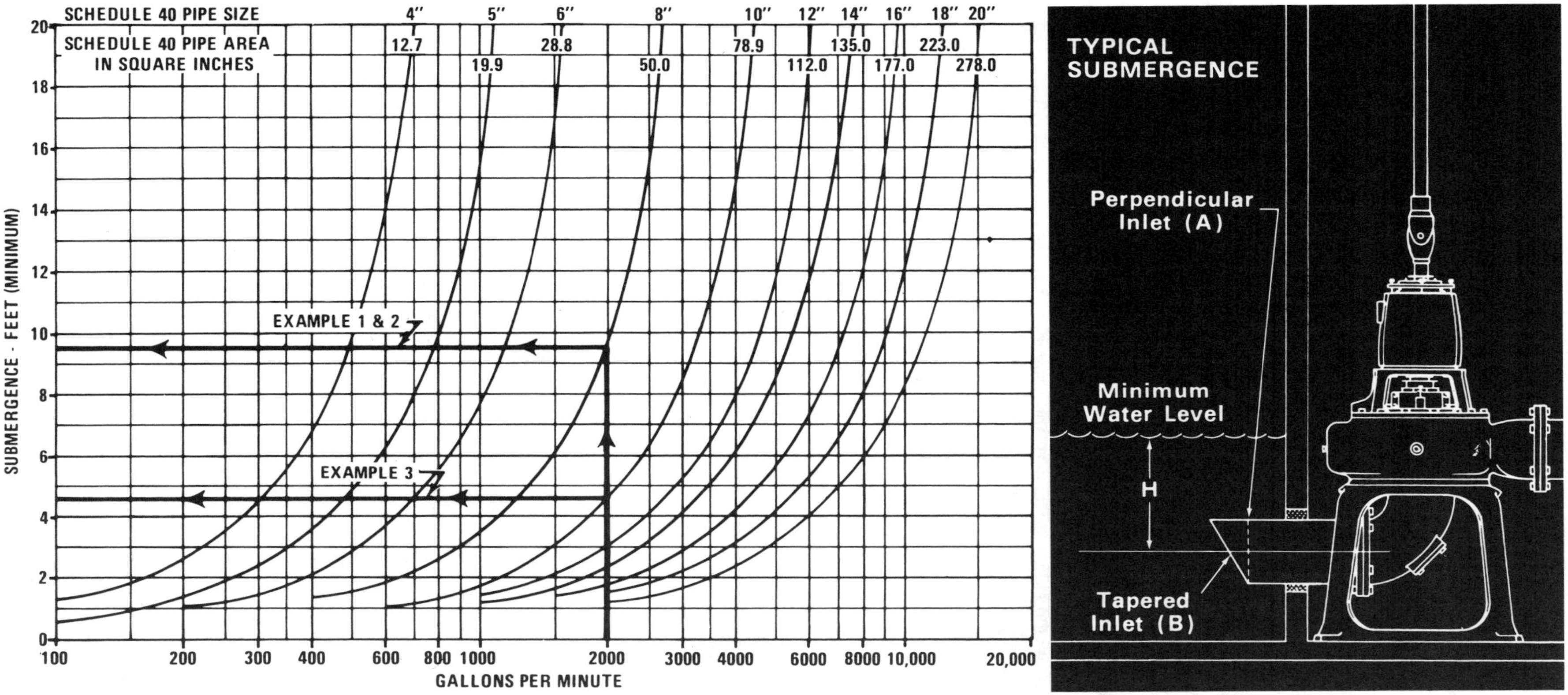

Fig. 8.15 Determination of minimum submergence depth

ern climates. Frost and ice buildups tend to accelerate rusting of metal doors. Ice buildup may actually freeze the door shut.

If the wet well contains a bar screen, a method of material handling must be provided so that screenings can be removed from the wet well. Floats, bubblers, fixtures, metering flumes or other equipment which require maintenance in the wet well should be easily accessible.

8.104 Wet Well Inlet Channel

The inlet channel of larger stations must be designed to allow proper flow distribution throughout the wet well and to minimize problems of solids settling and submergence described in previous sections.

A split wet well which allows isolation of one-half of the wet well volume and pump suction pipes is a desirable feature, particularly in larger stations. Frequently, a flume-type flow metering device is also located in the inlet channel to measure the flows being pumped.

8.105 Wet Well Safety

Wet wells are defined by most regulatory agencies as confined spaces. All the safety procedures and equipment recommended in Chapters 4 and 11 must be reviewed carefully and implemented when working in the wet well area.

Hydrogen sulfide and other toxic gases, oxygen deficiency, and explosive atmospheres are all conditions which can exist in the wet well. In addition, slippery ladders or stairs caused by the buildup of slime and the continual presence of moisture can create slipping and falling hazards.

If the wet well has standard manhole rungs for access, they should not be depended on to be structurally sound. After a period of time, corrosion will cause the rungs to fail and this could be fatal to the operator.

8.106 Wet Well Hardware

All hardware in the wet well used for fastening piping, platforms, or stairs should use stainless steel anchor bolts or other noncorroding material. Aluminum ladders and platforms are popular because of their corrosion resistance features. However, aluminum must not come in direct physical contact with concrete since this will result in corrosion of the aluminum. For example, handrail supports set into concrete without any protective coating will eventually corrode at the point where the aluminum enters the concrete, resulting in a failure of the handrail when an operator or a visitor leans against it.

8.107 Wet Well Electrical Systems

The wet well atmosphere can contain explosive or flammable gases and is classified as a Class 1, Division 1 area (National Electrical Code). Any electrical systems must either be explosion-proof or be rated as intrinsically (by its very nature) safe. Equipment installed in the wet well should be kept to a minimum since maintenance can be very difficult. Changing a light bulb in an explosion-proof fixture may be impossible, for example, because of corrosion of the machined surfaces and threads in the fixture. Replacement of the entire fixture would then be necessary.

Standard galvanized conduit and fittings will corrode rapidly in the wet well atmosphere; therefore, PVC-coated conduit should be used. Fixtures and alarm floats should be easily accessible and not require the operator to increase the possibility of slipping or falling. If possible, do not install fans, blowers, or motors in the wet well because of the corrosive conditions.

QUESTIONS

Write your answers in a notebook and then compare your answers with those on page 119.

8.1A Why should the pumps in a lift station not start and stop frequently?

8.1B What problems develop in a wet well that is too large?

8.1C What is the main advantage of variable-speed pumping equipment?

8.11 Bar Racks

Installation of bar racks or screening devices in the wet well used to be a common practice, especially on combined wastewater and storm water collection systems due to the debris entering the system from the surface runoff drains containing leaves, sticks, cans, and trash. In collection systems for only domestic and industrial wastewaters, bar racks or screens at the lift station have been eliminated by some agencies because most pumping units are equipped with open *IMPELLERS*[7] or closed two-port impellers sized to pass solids up to the size of a 2.5-inch diameter sphere. This is the largest size sphere which may pass through most home toilets and disposal systems.

Bar racks are installed to prevent any large debris from entering and plugging or damaging a pump. In areas where vandals remove manhole covers and throw debris into sewers, bar racks may be helpful. When bar racks are used, they should be cleaned frequently so there is no substantial restriction of wastewater flow to the pumps. The cleaning of bar racks is accomplished with forks, screen baskets, or any type of tool capable of picking up the debris and removing it from the bar racks. A limitation of bar racks is the fact that screenings cause odors and attract flies.

Screenings are usually removed from the station in buckets and transported to a disposal site. Agencies that have removed bar screens from lift stations which pump sanitary wastewater have found the pump impellers occasionally become plugged with rags (two to three times a year). However, this problem is preferred to the necessity of visiting a station one or two times a day to clean a screen and struggle with hauling the screenings to a disposal site. Many pumps are designed to "chew" the rags into a size small enough to be pumped.

[7] *Impeller.* *A rotating set of vanes in a pump or compressor designed to pump or move water or air.*

In larger lift stations, some agencies are installing *COMMINUTORS*[8] and barminutors to shred rags and debris ahead of wet wells and pumps. A bypass channel with a rack is needed for use when either device is being repaired.

8.12 Dry Wells

A dry well is that portion of the lift station which is isolated from the wastewater and is used to house the necessary equipment for the lift station to function. The structure commonly has two or more floor levels. The lower floor usually contains the pumping units, isolation valves, discharge manifold (pipes), and a sump pump to remove excess water such as seal water leakage and cleanup water.

Electrical control panels, motors, ventilation equipment, necessary station controllers and auxiliary equipment are housed on the upper floor. This provides protection to station equipment in case of a broken valve or fitting or a leak that would permit wastewater to flow from the wet well or force main into the dry well side of the station. Flooding of the lower level could occur, but the electrical systems, motors and expensive control equipment would be protected from water damage and would allow the station to be put back into service very quickly.

8.13 Electrical Systems

8.130 Typical Systems

With the exception of very small lift stations such as grinder pump stations that serve one or two residential services, electrical power is three-phase utility power. Single-phase systems are limited to about five horsepower since beyond that, wiring, controller equipment and other components increase dramatically in size to carry the amount of current required for a single-phase system. Three-phase electrical systems are able to operate higher horsepower loads because the current is distributed over the three phases and three individual sets of conductors.

Figure 8.16 illustrates a block diagram of a typical lift station electrical system. Figure 8.17 is a one line riser diagram of a power and control center for a lift station which contains four pumps, an alarm system, and a standby emergency generator.

8.131 Power Transformers

Power is supplied by the utility company through one or more pole-mounted transformers or a pad-mounted transformer. The power company usually supplies a high voltage; for example, 13,800 volts (13.8 KV). The transformer(s) steps this voltage down to 3-phase 208, 220, 240, 440, 460, or 480 volt systems at a frequency of 60 cycles per second (60 Hz). In larger pump stations the operating voltage for the motors may be higher than 480 volts; however, this is not common except in extremely large pump stations. For purposes of discussion, let us assume that in this particular pump station the utility company transformer steps the voltage down to 480 volts, 3-phase.

8.132 Metering

Power is fed into the power company metering which is usually a watt/hour meter. A demand meter which measures the peak electrical demand required by the pump station, usually on a monthly basis, may also be installed. Peak demand in a multiple-pump station usually occurs when all pumps are operating at peak flow conditions. In some cases, the power company may impose a penalty on the user for high demand conditions. Coordinating pump operation to reduce maximum electrical demand is one area an agency should examine to reduce station electrical operating costs.

8.133 Main Disconnect

The main disconnect switch may be fused, a circuit breaker, or simply a disconnect switch to allow isolation of the entire electrical system within the pump station. In newer stations, this section also may contain a ground-fault circuit interrupter (GFCI).

A ground-fault circuit interrupter is *NOT* an over-current device. A GFCI is used to open a circuit if the current flowing to the load does not return by the prescribed route. In a simple 120-volt circuit, we usually think of the current flowing through the black (ungrounded) wire to the load and returning to the source through the white (grounded) wire. If it does not return through the grounded wire, then it must have gone somewhere else, usually to ground. The GFCI is designed to limit electric shock to a current level and time duration value below that which can produce serious injury. Several types of GFCIs are available, with some variations between types. Although all types will provide ground-fault protection, the specific application may dictate one type over another.

Ground-fault circuit interrupters are divided into two classes, Class A and Class B. The Class A device is designed to trip when current flow in other than the normal path is 6 milliamperes or greater. The Class B device will trip when current flow in other than the normal path is 20 milliamperes or greater. Class B devices are approved for use on underwater swimming pool lighting installed prior to the adoption of the 1996 National Electrical Code. The most commonly used types of GFCIs are circuit breaker, receptacle, permanently mounted, portable, and cord-connected.

8.134 Transfer Switch

If the pump station is supplied with either a permanently installed emergency generator or provisions for a portable unit, a transfer switch is used which may either be manually operat-

[8] *Comminutor (com-mih-NEW-ter). A device used to reduce the size of the solid chunks in wastewater by shredding (comminuting). The shredding action is like many scissors cutting to shreds all of the large solids in the wastewater. Barminutor (bar-mih-NEW-ter). A device similar to a comminutor.*

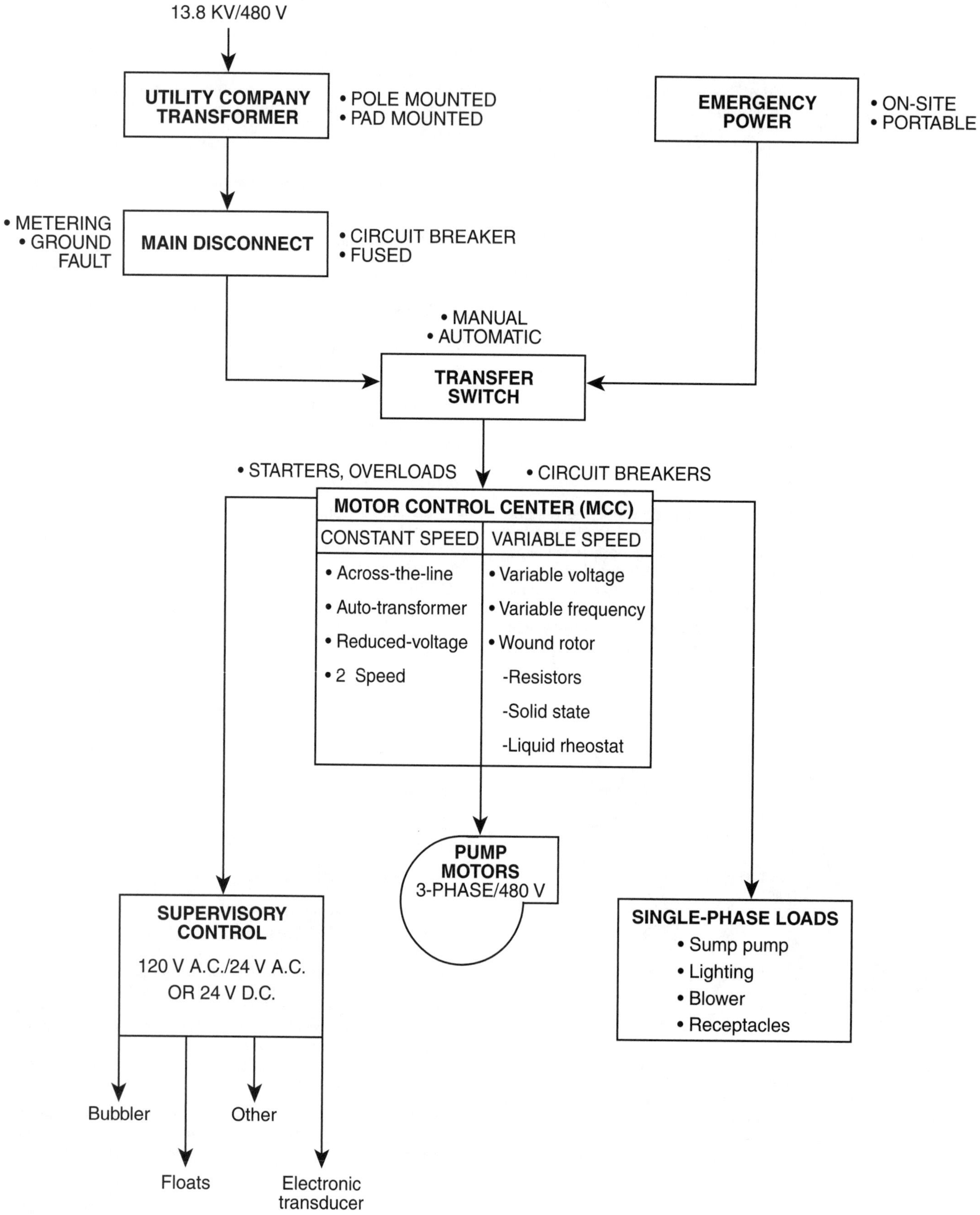

Fig. 8.16 Block diagram, typical pump station electrical system

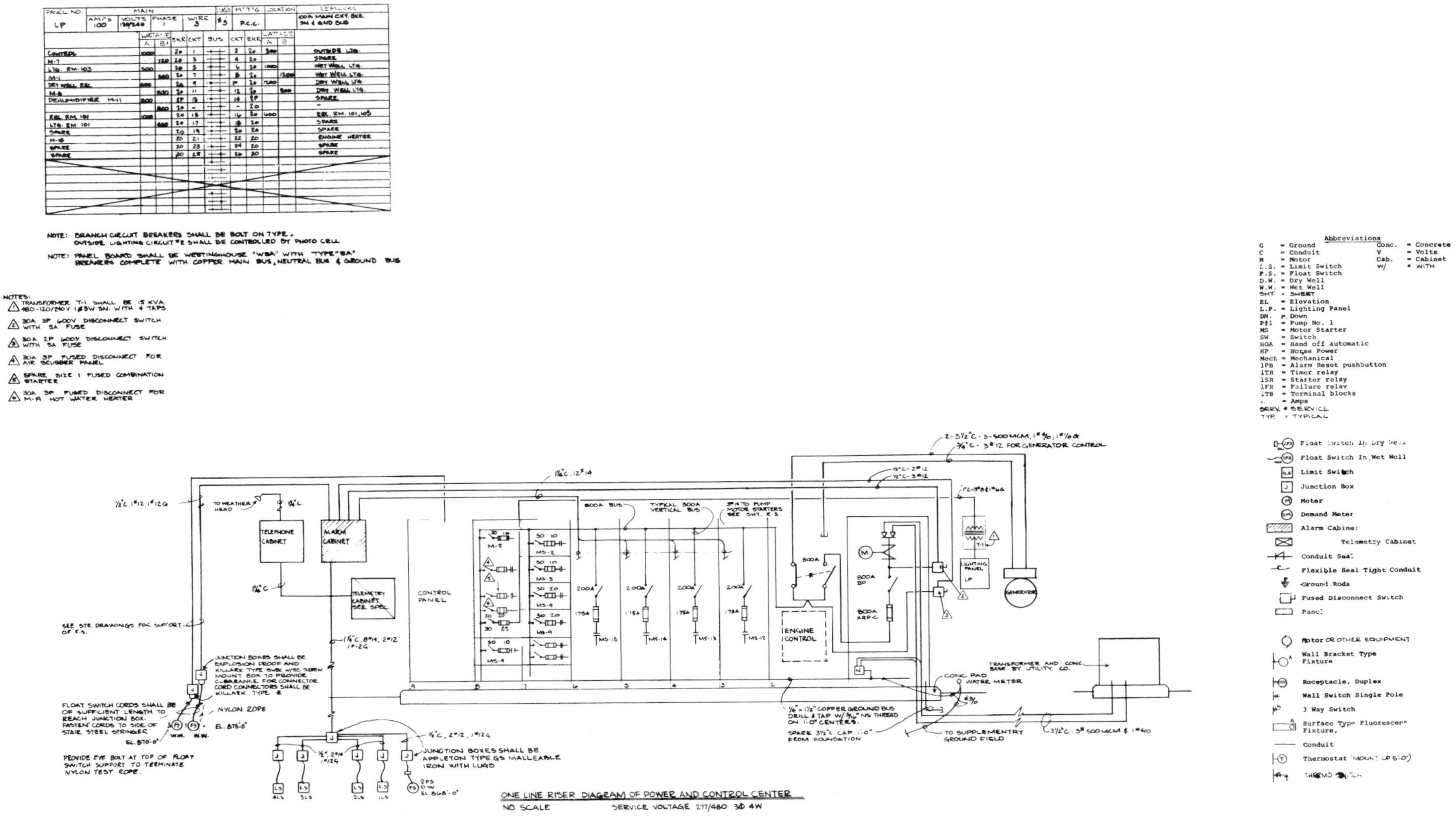

Fig. 8.17 One line riser diagram of a power and control center

ed or automatically operated when the utility power source is lost. Pump stations with an automatic transfer switch will also have an automatic control system to start up and operate the emergency engine-generator set.

Pump stations may also have provisions for connecting a portable standby power system which, again, is an engine-driven generator that can be towed to the site and plugged in.

8.135 Motor Control Center (MCC)

The motor control center, or main panel as it is sometimes called, houses the controls for all motors and electrical equipment operated within the station. These controls include a starter, fuses, heater strips and all the coils and relays necessary for any particular pump or motor to operate.

The motor control center is fed 480 V, three-phase power and contains branch circuit breakers, starter devices, three-phase overload relays, and a low voltage control circuit for starter coil operation and indicating lights. Each three-phase motor will have these control components in the circuit. (See Chapter 9 for component details.)

Variable-speed motors and drives are often used in an attempt to match the pumping rate to the incoming wet well flow. Types of variable-speed systems typically used are:

1. Variable voltage,
2. Variable frequency (VFDs), and
3. Wound-rotor motor using variable resistances in the secondary winding of the motor to change the operating speed of the motor. This is accomplished by:
   a. Metal resistors where different values of metal resistors are switched by a controller,
   b. A liquid rheostat which varies the resistance in the secondary by varying the level of a conducting liquid on electrodes, and
   c. Solid state devices which vary the resistance in the secondary using silicon-controlled rectifiers (SCRs).

Variable frequency drives have become more popular in recent years because of advances in technology that have improved reliability while offering advantages in operating efficiencies over the other types of drives. However, variable frequency drives are more complex and require specialized electrical/electronic skills to maintain them.

8.136 110-Volt/220-Volt Loads

In addition to the power levels discussed in Section 8.131, there are also single-phase, 110-volt/220-volt loads in most stations. An additional transformer is required to convert the 480 volt, 3-phase to 110/220 volt, 1-phase. Typically, 120-volt loads would be used for lighting systems, sump pumps, ventilation blowers, and convenience receptacles.

8.137 Supervisory Controls

The supervisory control panel (Figure 8.18), which controls the operation of the pump motors, is usually powered with single-phase, 120-volt current. An additional step-down transformer may be used to reduce control voltage to 24-volt A.C. or 24-volt D.C. (24 V A.C., 24 V D.C.).

Since pumps are controlled by the level of water in the wet well, it is important to consider how to measure the level. There are several different methods of measuring the liquid level in the wet well. Older legacy systems may use several float switches or a single float connected to a pulley and mechanism with several mercury switches. These signals are used to sequence relays that operate the pump starters. Another method used in older systems is a bubbler system. Compressed air is forced down a tube or pipe (usually ½") inserted in the wet well. Several pressure switches are connected to this circuit and the switches are connected to relays that sequence the pumps on and off. A variation of this method substitutes an electronic pressure transducer connected to the

Fig. 8.18 Supervisory control center

(Permission of the EDCCO Group www.edcco.com)

bubbler system. The electronic signal may be connected to electronic trip relays, an electronic pump controller, or a PLC (programmable logic controller). Three other methods of obtaining an electronic signal are: (1) a submersible pressure transducer with a milliamp output, (2) a capacitance probe, or (3) an ultrasonic level transducer, all of which produce an electrical signal that can be connected to trip relays, a dedicated controller, or to a PLC. In each method, a supervisory control system is used to turn the pump motors on and off. A popular method of controlling and sequencing pumps is a combination ultrasonic transducer (Figure 8.19) and controller with a digital readout. A keypad is used to program the starting and stopping levels for each of the pumps.

Programmable logic controllers (PLCs) (Figure 8.20) can be found in most supervisory control systems in new pump sta-

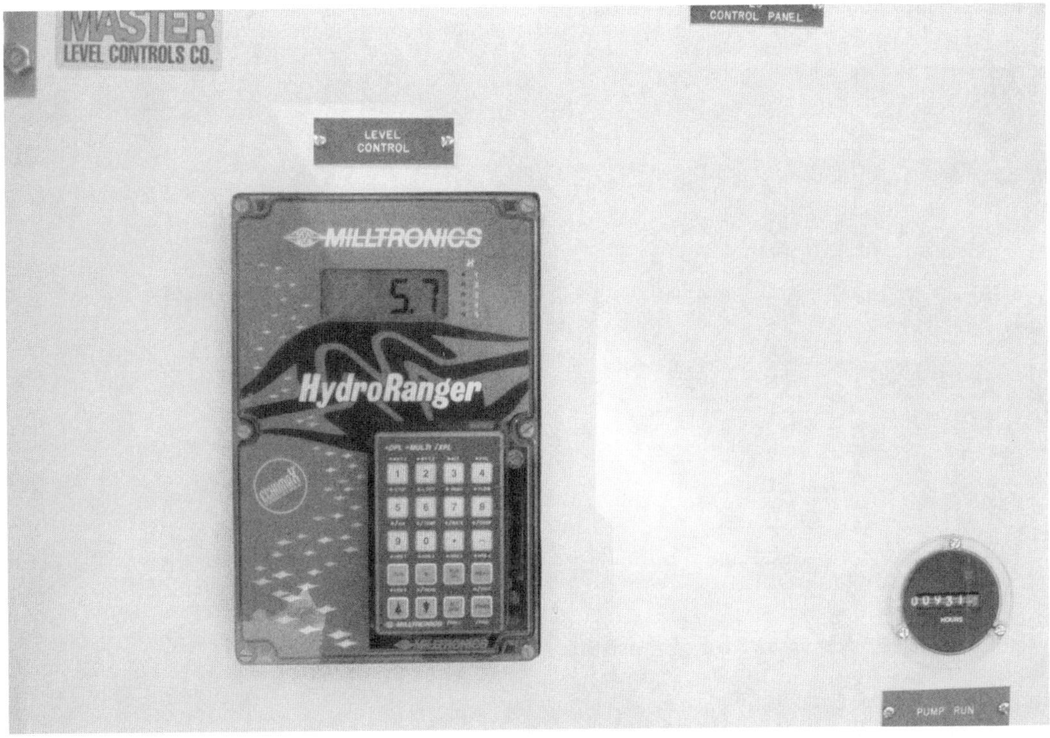

Mounted level controller with keypad stored in place at pump station
(wet well level displayed on screen)

Fig. 8.19 Combination keypad, ultrasonic transducer, and level controller

tions or existing stations that have been upgraded or retrofitted. Programmable logic controllers use microprocessor-based control unit input modules and output modules to automatically control the sequencing of pump motors and other equipment in pump stations. PLCs can accept either digital or analog signals from remote devices such as float switches and pressure switches or analog inputs from level transducers, flowmeters, or other process instruments. The PLC is usually programmed using a laptop computer (Figure 8.21), which allows the user to customize control functions such as pump sequencing, pump ON and OFF points, alternating sequence, and alarm functions.

The PLC replaces multiple electrical-mechanical relays previously used in supervisory control systems in pump stations. They offer several advantages in that changes to the control scheme can be made very easily using the programming device without having to rewire the circuit. Because electromechanical devices are eliminated, maintenance is reduced and reliability is increased. Programmable logic controllers

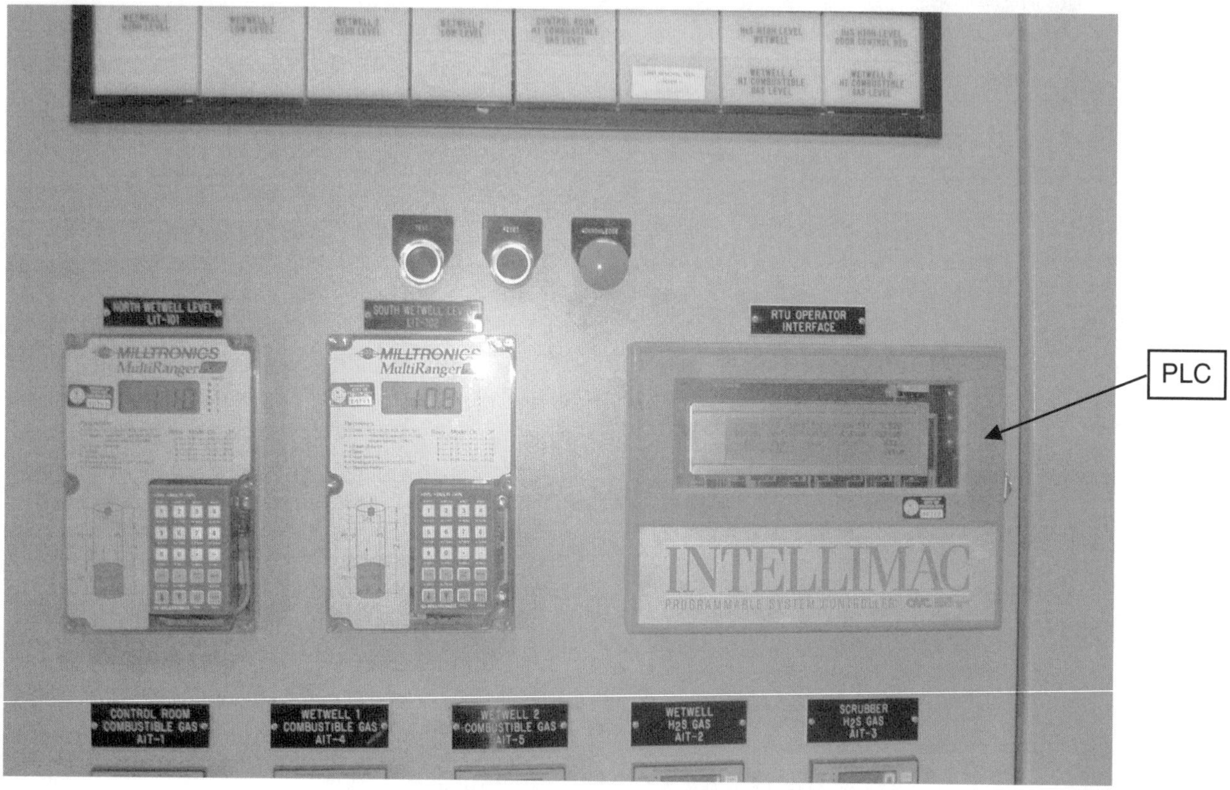

Fig. 8.20 Programmable logic controller (PLC) mounted on control panel

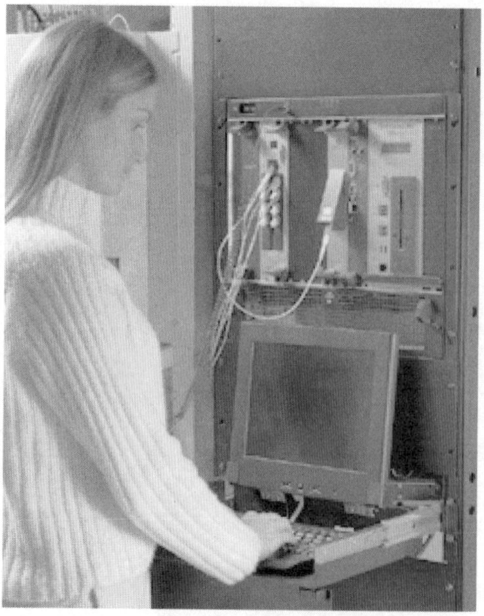

Fig. 8.21 PLC being programmed using laptop computer

(Permission of the EDCCO Group www.edcco.com)

range from very small, simple modules with a fixed number of programmable inputs and outputs to much larger, more complex systems designed to control variable frequency drives, alarm systems, remote process control systems, and graphic displays of pump station components, flow, and other systems in the pump station. The operator can supervise or control operations from the field, at the master control center, or from a mobile device with a wireless modem. The PLC will display information and give the operator access in controlling and maintaining facilities. The operator could have access to current data and historical trends (Figure 8.22) at the station or remotely by radio communications. If a communications interface is made with PLCs at other stations or to a central station, the PLC becomes the front end of a simple or complex SCADA (Supervisory Control and Data Acquisition) system.

PLCs, variable frequency drives, and the other computer-based control devices are examples of ongoing advances in technology that are being used in pump stations. The use of "smart technology" is becoming the norm rather than the ex-

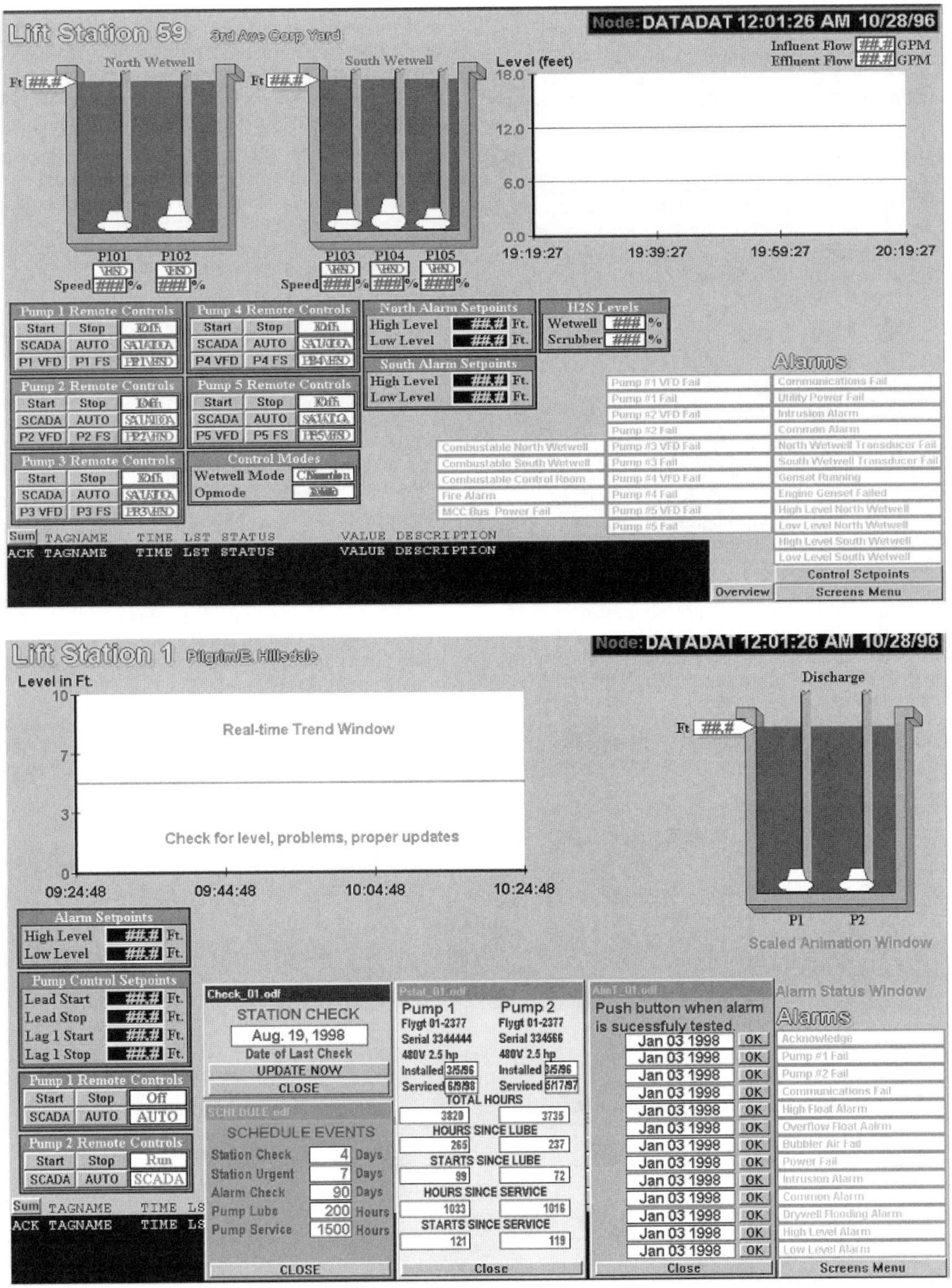

Fig. 8.22 SCADA monitor screen displays historical and maintenance trends

(Permission of the EDCCO Group www.edcco.com)

ception in most stations due to the lower installed costs and increased reliability over older systems. The expanding use of this technology underscores the need for increased skills and knowledge by the persons responsible for the pump station operation and maintenance.

8.14 Motors (See Chapter 9, Section 9.1, "Electrical Equipment Maintenance")

8.140 Types of Motors

Pump station motors are one of the key equipment components in the pump station. Electric motors are generally the most common driver used for pumps used in the pump station. Unfortunately, even though electric motors are very common, they are one of the least understood components in the pump station.

There are two types of motor classifications: the first is direct current (D.C.), the second is alternating current (A.C.) (see Figure 8.23). D.C. motors are seldom, if ever, found in pump stations as the prime mover for the raw wastewater pumps. If they are found at all, they are normally small (less than one horsepower) motors used for operating metering pumps or other similar applications.

The alternating current motor, therefore, is the most common type of motor found in the pump station. A.C. motors are further classified as either induction or synchronous. Synchronous motors are very seldom found in raw wastewater pump stations. The other classification, the induction motor, also has two subclassifications: squirrel cage and phase wound. A phase wound motor is more commonly known as a wound rotor motor. It has a rotor with windings and connections are brought out through slip rings and brushes so that external controls can be used in the rotor circuit. The other type of induction motor, the squirrel cage, is the type of motor most commonly used to drive raw wastewater pumps. The majority of squirrel cage motors are polyphase rather than single-phase motors since single-phase motors are horsepower limited. Since operators are only dealing with a single type of motor in most cases, understanding how they operate and how to maintain them is not too difficult.

Motors can have different voltages, horsepowers and ampere readings. Also, they will be rated at 60 cycles and single-phase or three-phase. Voltage ratings depend on the local power source and may be 208, 220, 440, 460, 480 volts or higher for very large stations, 60 cycles, and three-phase. Occasionally in a very small lift station a single-phase motor is used.

Submersible pump motors are an integral part of the pump and are encased in the pumping unit. When a seal fails on a submersible pump, the wastewater penetrates the motor compartment and burns or shorts out the motor. If a maintenance operator resets the unit and it operates properly for a short period of time with no overload, the operator can assume that a seal failure was not the reason for the shutoff, but that the pump had overloaded the motor. Determine the load on the motor and check for proper voltage. Also look for plugged lines and be sure the wet well and pump are clear of rags, mud and debris that could cause a temporary overload.

As noted earlier, the squirrel cage induction motor (SCIM) is by far the most common motor used in lift stations. Figure 8.24 illustrates a typical horizontal squirrel cage induction motor and the various motor components. The term squirrel cage originated from earlier induction motors in which the rotor (rotating part) looked like a squirrel cage.

Three-phase power from the Motor Control Center (MCC) is applied to the stator winding (item 17 on Figure 8.24) which generates a rotating magnetic field. The rotor assembly (item 13), because of the way it is constructed, behaves like a transformer. Voltage is induced into the rotor which causes current flow and generates a second magnetic field in the rotor circuit. There are *NO* electrical connections between the stator windings and the rotor. It is the interaction of the two magnetic fields that causes the rotor to turn and to develop torque.

Some variable-speed systems in lift stations use a variation of the squirrel cage induction motor called the wound-rotor induction motor (WRIM). In this case, the rotor circuit is connected externally through the use of slip rings and brushes which allows the external connection of various values of resistance. This causes the rotor to change speed ranging from zero RPM when the rotor circuit is "open circuited" to full RPM when the rotor circuit is "short circuited."

Depending on space requirements in the station, motors and pumps may be mounted horizontally or vertically. The mounting configuration is unique for each design.

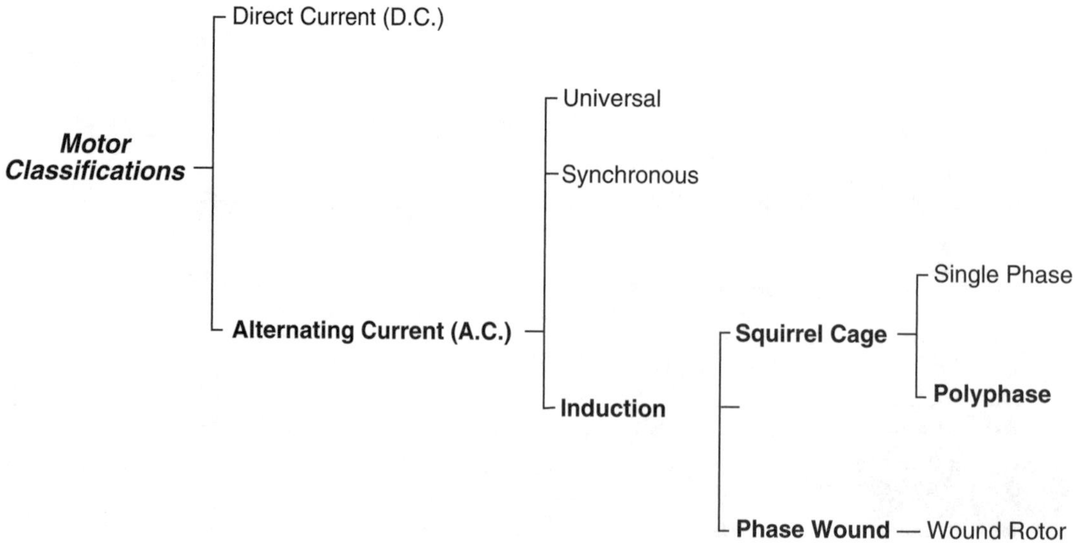

Fig. 8.23 Classification of motor types

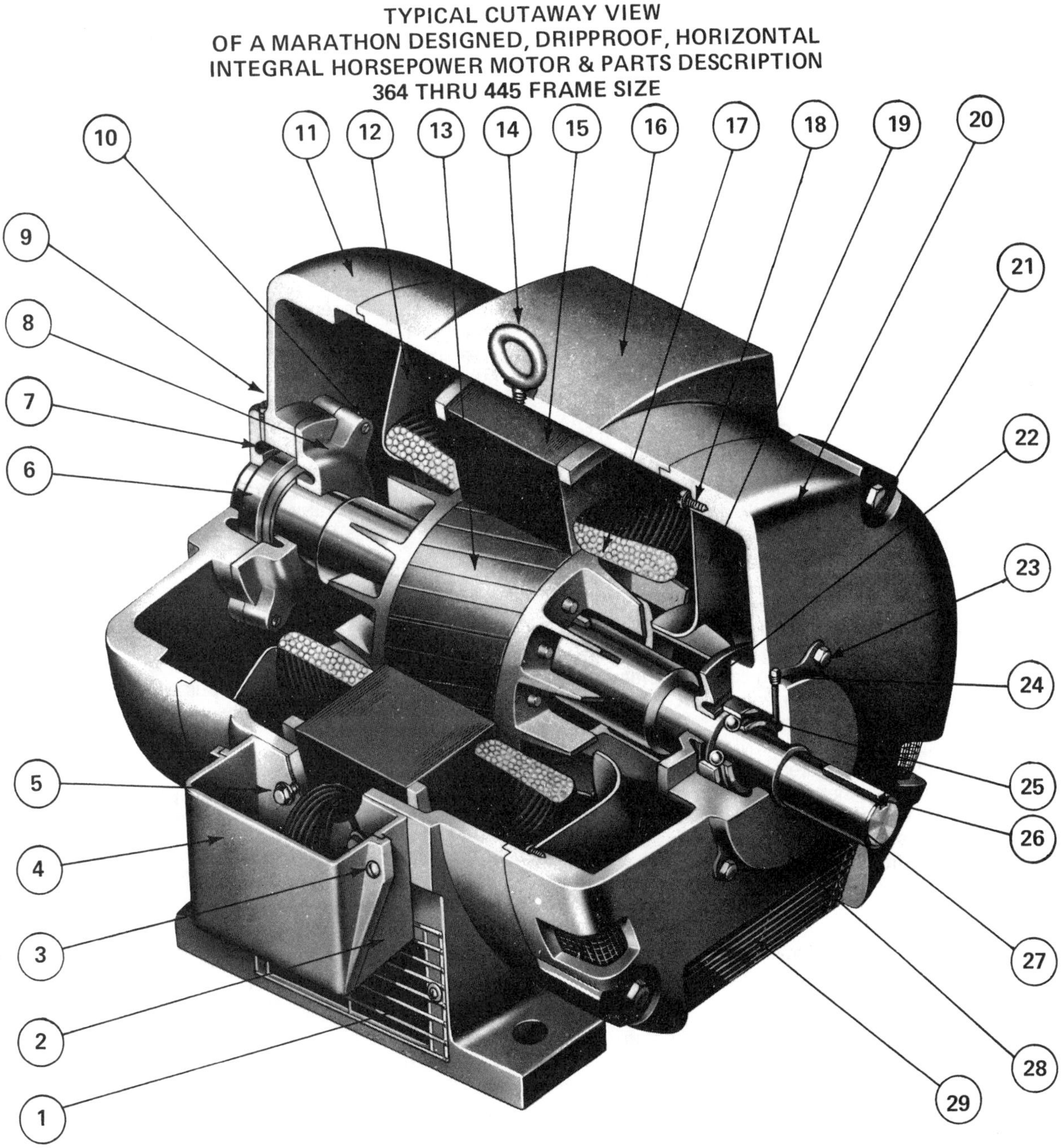

ITEM	DESCRIPTION	ITEM	DESCRIPTION	ITEM	DESCRIPTION
1.	*Frame Vent Screen	11.	Bracket O.P.E.	21.	Bracket Holding Bolt
2.	Conduit Box Bottom	12.	Baffle Plate O.P.E.	22.	Inner Bearing Cap P.E.
3.	Conduit Box Top-Holding Screw	13.	Rotor Core	23.	Inner Bearing Cap Bolt
4.	Conduit Box Top	14.	Lifting Eye Bolt	24.	Grease Plug
5.	Conduit Box Bottom Holding Bolt	15.	Stator Core	25.	**Ball Bearing P.E.
6.	**Ball Bearing O.P.E.	16.	Frame	26.	Shaft Extension Key
7.	Pre-Loading Spring	17.	Stator Winding	27.	Shaft
8.	Inner Bearing Cap O.P.E.	18.	Baffle Plate Holding Screw	28.	Drain Plug (grease)
9.	Grease Plug	19.	Baffle Plate P.E.	29.	*Bracket Screen
10.	Inner Bearing Cap Bolt	20.	Bracket P.E.		

P.E. = Pulley End
O.P.E. = Opposite Pulley End
* = Bracket and frame screens are optional.
** = Bearing Numbers are shown on motor nameplate. When requesting information or parts, always give complete motor description, model and serial numbers.

Fig. 8.24 Horizontal squirrel cage induction motor

(Courtesy of Marathon Electric)

Motors are selected based on a number of factors as listed below:

DESIGN — NEMA A, B, C, OR D

HORSEPOWER

SYNCHRONOUS SPEED — 3,600, 1,800, 1,200 RPM

FRAME SIZE

VOLTAGE

FREQUENCY

PHASES

ENCLOSURE

SERVICE FACTOR

AMBIENT TEMPERATURE

ENVIRONMENTAL CONDITIONS

SPECIAL ELECTRICAL FEATURES

SPECIAL MECHANICAL FEATURES

MOUNTING

TYPE BEARINGS REQUIRED

DUTY: CONTINUOUS, 1 HOUR, ETC.

These features are discussed in more detail below. *NOTE:* Because the use of induction motors dates back to the early 1900s, and because they are in such widespread use, they are very well standardized. Electrical equipment standards including motors are specified by organizations such as the National Electrical Manufacturers Association (NEMA) and the Edison Electric Institute. Therefore, many of the things which apply to a motor of one manufacturer, such as dimensional information, will be identical to that of any other manufacturer. If, for example, you were to replace a motor that was constructed with a 445 T frame, this machine would be dimensionally the same as a 445 T frame from a different manufacturer.

Power Supply

Voltage. Motor *NAMEPLATE*[9] voltages will be less than the utility power system voltage because of the following established recommended standard for 60 Hz power systems:

Nominal Power System Volts	Motor Nameplate Volts
208	200
240	230
480	460
600	575

Motors rated 200 volts and below will operate at rated load with a voltage variation of ±10 percent of nameplate value. If frequency varies as well, the sum of the voltage and frequency variation should not exceed ±10 percent. In any case, frequency should not vary more than ±5 percent.

Figure 8.25 is a typical motor nameplate for a dual-speed A.C. induction motor. The nameplate indicates the motor has 6 leads. In the low speed connection, 3-phase power is connected to lines 1, 2, and 3; lines 4, 5, and 6 are not connected. The high speed connection requires the power to be connected to lines 4, 5, and 6, and leads 1, 2, and 3 are connected together. In addition, the nameplate indicates serial number, type and model, the horsepower rating, full load amps, operating RPM, duty shaft and bearing number, motor frame size, operating voltage and frequency, ambient operating temperature, opposite end bearing number, KVA code, service factor, insulation rating, efficiency in percent, name and design, and operating instruction numbers. The last piece of information this nameplate provides is the phase sequence which determines rotation in conjunction with the phase sequence of the utility supplied power.

Motor Type

Although there are four basic motor designs (A, B, C, D), the NEMA design B is most common in wastewater lift stations and is characterized by normal starting torque, low starting current, and low slip. In the case of variable-voltage and variable-speed drives, high-slip NEMA design D motors are generally used and their full load speed is 8 to 15 percent lower than the synchronous speed.

Enclosure

The motor enclosure used depends on the ambient or environmental conditions where the motor is installed; there are two general classifications:

1. Open, and
2. Totally enclosed.

An open machine has ventilation openings to permit passage of external air over and around the windings of the motor. A totally enclosed machine is constructed to prevent the free exchange of air between the inside and outside of the motor frame, but is not airtight.

The two general categories of enclosures are broken down further as described in the following paragraphs.

Open Drip Proof (ODP) The ventilation openings are constructed so that drops of liquid or solid material falling on the motor at an angle of not greater than 15 degrees from the vertical cannot enter the motor directly or by striking and running along the motor frame surfaces.

Open Drip Proof, Fully Guarded An open motor having all air openings that allow direct access to electrically live or rotating parts, has either limited size openings or has screens to prevent the accidental contact of parts.

Totally Enclosed Ventilated An enclosed motor which is not equipped for cooling by external means.

Totally Enclosed Fan Cooled (TEFC) An enclosed motor equipped with an integral fan.

Explosion Proof An enclosed motor whose enclosure is designed and constructed to withstand an explosion of a gas or vapor if it occurs within the motor and prevent the ignition of the gas or vapor surrounding the motor (hazardous atmospheres are classified by the National Electrical Code).

Motor Windings/Insulation

In certain environments such as very damp or warm weather climates, additional insulation treatment is desirable to improve the ability of the motor insulation to resist moisture. Standard motor insulation will have one or more coats of varnish applied during construction. Additional coats will improve the resistance to moisture. In certain cases, a special insulation system called nonhygroscopic insulation is available; it is made of materials that will not absorb or retain moisture.

---

[9] *Nameplate. A durable metal plate found on equipment which lists critical operating conditions for the equipment.*

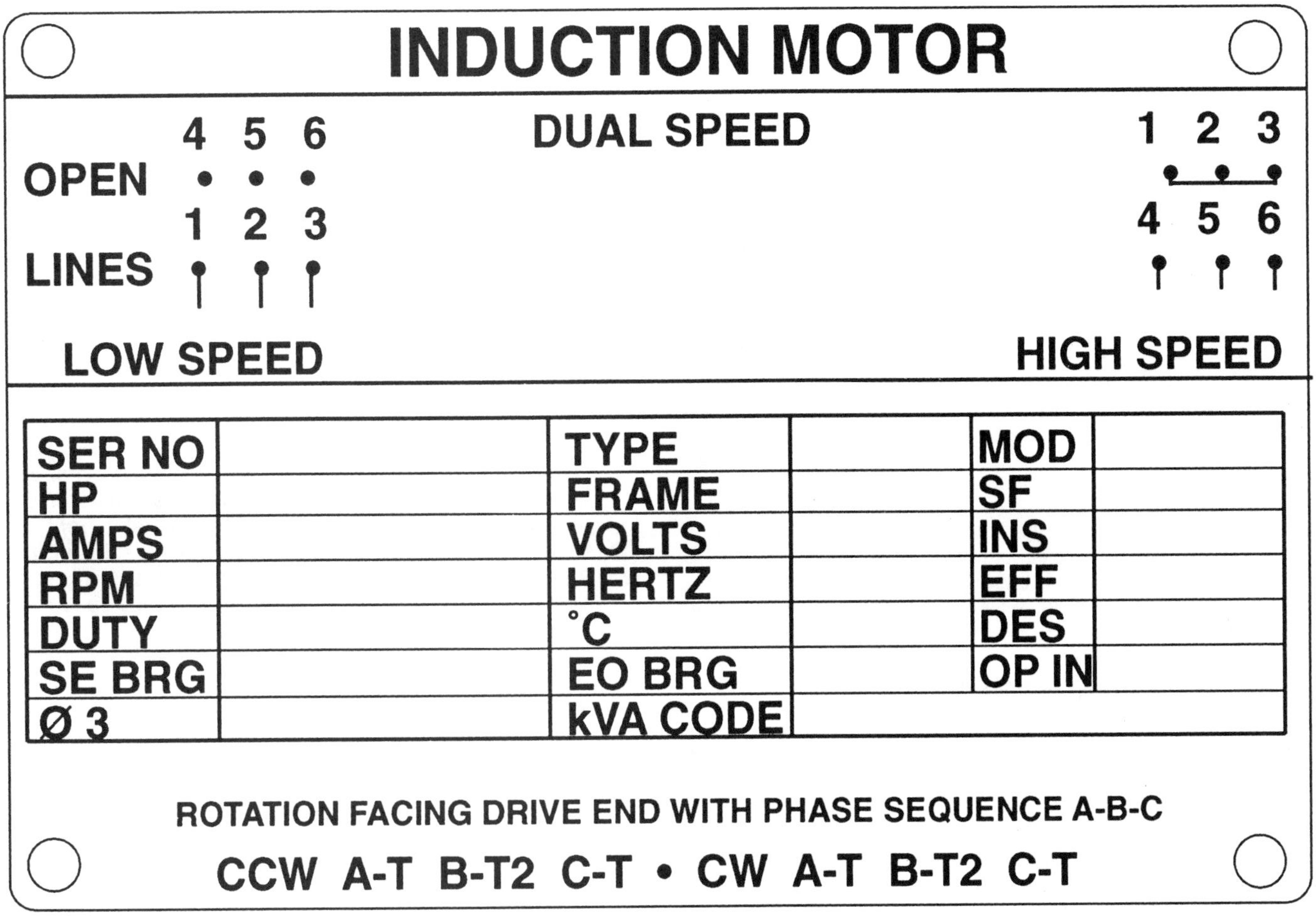

Fig. 8.25 Typical motor nameplate

Motors can also be selected for hot, humid environments by specifying tropical insulation treatment, including:

1. Use of a higher class of insulation for higher ambient operating temperatures,
2. Special insulating materials such as varnish to prevent fungus growth, and
3. Screens over all air openings on open, drip-proof motors to prevent the entrance of rodents and snakes.

Altitude

Motors depend on the ambient air to remove heat and this heat transfer is a function of air density. Altitude can decrease the effectiveness of this process. Accepted guidelines state that air density has little effect on cooling between sea level and 3,300 feet. As a rule of thumb, motor operating temperature will increase one percent for each 330-foot increase in altitude over 3,300 feet.

Motor Mounting

There are three types of standardized machine mounting configurations for motors. These allow you to attach pumps or other equipment to the motor.

1. Type C face end shield. Provides a male rabbet and tapped holes for mounting bolts (sump pumps).
2. Type D flange. Has a male rabbet, but with holes for through bolts in the flange.
3. The Type P base. Has a female rabbet and through bolts for mounting in the flange and is used to mount motors in a vertical position.

NEMA Code Letters for Locked Rotor Kilovolt-Ampere

Motor Codes are particularly important when considering the sizing of pump station standby power since the generator's code size directly relates to its ability to start the motor.

The motor code represents the locked rotor kilovolt-ampere (KVA) per horsepower of the motor and is calculated by the following formula:

$$\frac{KVA}{HP} = \frac{1.73 \times 1 \times E}{1{,}000 \times HP}$$

The table below lists the code letter designations and the KVA per horsepower required for each.

Letter Designation	KVA Per Horsepower
A	0.00 - 3.15
B	3.15 - 3.55
C	3.55 - 4.00
D	4.00 - 4.50
E	4.50 - 5.00
F	5.00 - 5.60
G	5.60 - 6.30
H	6.30 - 7.10
J	7.10 - 8.00
K	8.00 - 9.00

Typically, lift station motors are Code G or Code H motors.

Internally Thermally Protected Motors

Many agencies require motors to be supplied with internal thermal protection which constantly monitors the winding temperature of the motor. If the temperature is exceeded, a switch, which is connected to the control circuit in the motor control center, opens and stops the motor. (These standard connections are labeled P1 and P2 in the motor conduit box.) The circuit can be wired so that when the motor winding cools down and the thermostatic switch resets, the motor restarts; however, it is preferable to lock the motor out through the control circuit to prevent the motor from restarting until the problem is investigated.

Unless the motor is specially designed, the motor ambient operating temperature should not exceed 40°C (104°F). See Chapter 9, Section 9.23 for more information on motor operating temperatures and insulation characteristics.

Service Factor

Service factors greater than 1.0 allow an added margin of safety to account for higher voltage or frequency variations. A 1.0 or 1.15 service factor is usually found on lift station motors. The service factor should not be used for continuous motor overloading which exceeds the nameplate horsepower.

8.141 Motor Starting Devices and Methods

For small, constant-speed pump motors, power is supplied to the motor by means of the starter contacts. This is referred to as across-the-line starting (ACL). In larger motors it is common practice to use an additional control section to reduce starting currents, since across-the-line starting will require motor starting currents of six to eight times the normal running currents. Frequent starts for a motor are not good because high starting currents generate heat in the windings which cannot be dissipated. If the winding temperature is exceeded, the insulation will break down and cause a short within the motor windings.

Typically, full voltage starting of squirrel cage induction motors requires a starting current up to 600 percent of full load current and from 200 percent to over 400 percent of full load torque. The electrical distribution system must be capable of supplying this high starting current without excessive voltage drop at the motor. To reduce the mechanical and electrical stress resulting from this high in-rush current and high starting torque, several reduced voltage starting methods have been developed.

Reduced voltage starting is frequently used when the line current available from the distribution system is limited. Reduced voltage starting provides a cushioned start and has the effect of stabilizing the line voltage. Almost all applications benefit from the reduced mechanical and electrical stress of a reduced voltage start. Reduced voltage starting can also be used where hydraulic system signal problems occur with pumping systems.

Reduced voltage starting methods and devices include

- Wye Delta,
- Part Winding,
- Auto Transformers,
- Primary Resistor, and
- Solid State Starters.

Table 8.1 compares the voltage, current and torque of across-the-line starting with the voltage, current and torque produced by reduced voltage starting methods.

TABLE 8.1 COMPARISON OF REDUCED VOLTAGE STARTING METHODS

	Across-the-Line	Part Winding	Wye Delta	Auto-Transformer	Primary Resistor	Soft Start
Voltage	100%	100%	100%	50, 65, 80%	70%	Varies to maintain linear current
Current	600% or more	200%	200%	300, 390, 480%	420%	Adjustable 150-450%
Torque	150%	50%	50%	25, 42, 64%	74%	10-150%

1. Part winding motors and starters

Part winding starting uses a specially wound motor with two motor windings. Power is applied to only one winding to achieve reduced voltage starting. The normal current and torque are approximately 50 percent of locked rotor full voltage condition. After a time delay, power is applied to the second winding which brings the motor to full speed and torque.

Limitations of part winding starting include:

- Maintenance of electro-mechanical contacts,
- Initial starting torque is not adjustable to load conditions,
- A motor suitable for part winding starts and a special starter are required, and
- The transition from initial start to full voltage is stepped instantaneously which causes additional mechanical and electrical stress.

2. Wye-Delta motor and starter

The Wye-Delta motor is specially wound and connected for use with a Wye-Delta starter. The motor is first connected in Wye configuration to achieve reduced voltage starting with current and torque approximately $^2/_3$ of locked rotor values. After a time delay, the windings are switched to a Delta configuration which brings the motor to full speed and torque.

Limitations of Wye-Delta starting include:

- Maintenance of electro-mechanical contacts,
- Initial starting torque is not adjustable to load conditions,
- A special Wye-Delta motor and starter are required, and
- The transition from Wye to Delta connection (full voltage) is stepped instantaneously, which causes additional mechanical and electrical stress.

3. Auto-transformer starting

This method uses a standard induction motor. The starter may have up to three taps (usually 50 percent, 65 percent and 85 percent of full voltage). Reduced voltage is accomplished by selecting the desired starting voltage tap. After a time delay, the starter will switch to full voltage which brings the motor to full speed and torque.

Limitations of auto-transformer starting include:

- Maintenance of electro-mechanical contacts,

- Thermal limitations of the auto-transformer,
- Limited number of starts per minute, and
- The transition to full voltage is stepped instantaneously which causes additional mechanical and electrical stress.

4. Primary resistor starter

Primary resistance starting is done by inserting power resistors that reduce the input voltage to the motor. After the starting interval, the resistors are removed from the circuit bringing the motor to full voltage.

Limitations of Primary Resistance starting include:

- Maintenance of electro-mechanical contacts,
- Reduced efficiency due to power consumed by the resistors,
- Maintenance cost is relatively high compared to other methods of reduced voltage starting, and
- The transition to full voltage is instantaneous which causes additional mechanical and electrical stress.

5. Solid state soft start controls

Solid state starters use silicon-controlled rectifiers (SCRs) that are microprocessor controlled to control motor starting. They are frequently furnished with a wide variety of functions that boost performance and increase reliability. These features include:

- Precise overcurrent protection,
- Current limiting,
- Single phasing protection,
- Transient (brief power fluctuation) protection,
- Shunt trip,
- Precise starting and stopping times, and
- Power factor control.

One feature that is available is a "power factor corrector" that permits significant energy savings when motors are operated at very low loadings, up to 50 to 69 percent of full load. Solid state switching devices give a smooth transition from initial voltage to full voltage and eliminate switching transients, thus reducing both electrical and mechanical stress on the motor windings.

Solid state starters allow the user to adapt the motor starting characteristics to the load by controlling starting current and torque. Mechanical stress can be minimized and starting current can be reduced to the minimum required to start the load. The control logic and protection section is a microprocessor-based device that switches SCRs based on adjustments. In addition, solid state starters can be used where hydraulic problems such as water hammer or pressure surges exist in pumping systems.

Just as computers have changed the way control systems operate and the way maintenance functions are performed in pump stations, computer technology has also improved the protection of motors. The conventional motor starter is a combination electro-mechanical device which incorporates a means of connecting the 3-phase power to the motor through mechanical relay contacts. The motor starter also provides thermal overload protection which disconnects the motor if excessive current is applied. Since this is an electro-mechanical device, it is subject to certain limitations.

With the development of the microprocessor, motor starters now incorporate several microprocessor-based functions to improve pump motor control and overload protection (Figure 8.26). Typically the solid state motor starters incorporate current sensing devices in each phase of the power being sup-

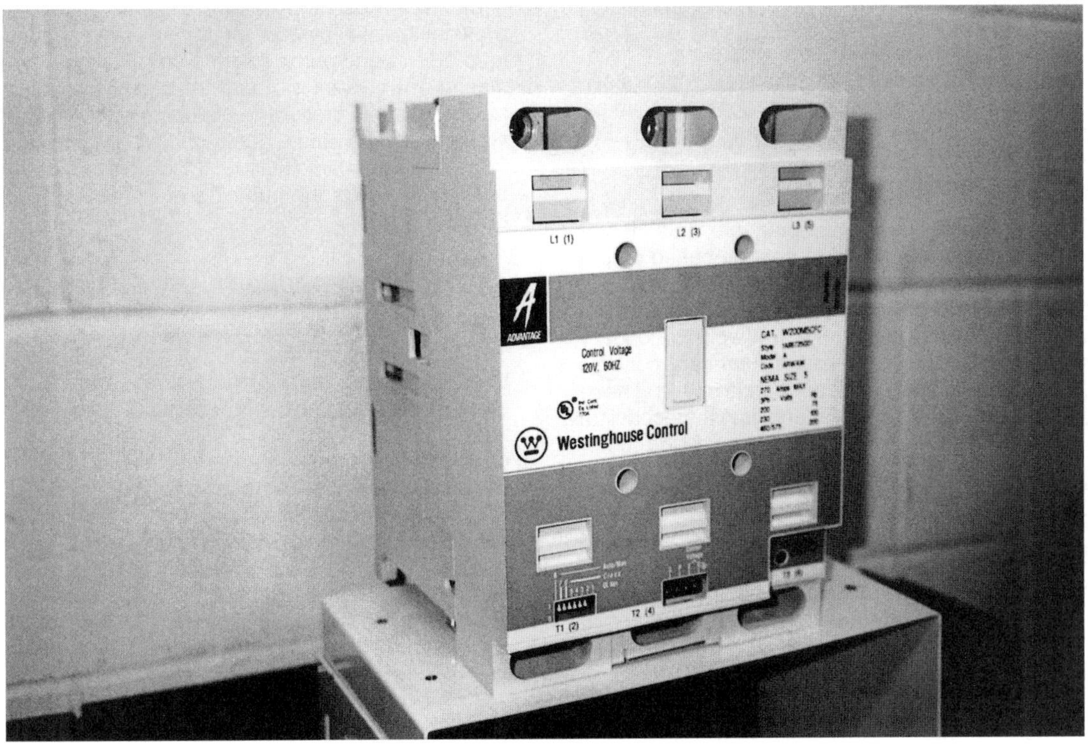

Fig. 8.26 Motor starter with microprocessor control

plied to the motor. This allows much more precise overload protection for the motor. The system can also detect nuisance type control inputs from floats, pressure switches or other input devices. Mechanical wear on the contacts is also reduced which prolongs contact life and reduces maintenance. Other features of solid state starters include phase unbalance and phase loss protection (see Chapter 9, Section 9.231) as well as ground-fault protection. Because the starter uses current transformers to sense current, the motor protection provided is much more accurate and can be adjusted to maximum motor protection while eliminating nuisance tripping. Solid state starters can also be used with programmable logic controllers in more complex control schemes.

QUESTIONS

Write your answers in a notebook and then compare your answers with those on page 119.

8.1D Why have bar racks been eliminated or not installed in lift stations pumping only domestic and industrial wastewater?

8.1E What is the purpose of a sump pump in a dry well?

8.1F What happens when a seal fails on a submersible pump?

8.1G What are some advantages of solid state motor starters compared to conventional electro-mechanical starters?

8.15 Supervisory Controls

Controls to start, stop or change pumping rates in a lift station are used to tell the pumps when to operate based on the level of the wastewater to be pumped from the wet well. Primary controls such as floats, bubblers or pressure-sensitive devices measure the level of water in a wet well. Secondary controls convert the measurement from the primary controls into a signal for a pump to start, stop, or change speed. These secondary controls convert a sensing signal into a mechanical or electrical signal which, in turn, actuates low voltage motor relays to start or stop motors or milliamp signals to change ranges on variable-speed equipment.

8.150 Float Controllers

Float controllers are one of the oldest methods used to start and to stop pumps or to indicate the level of water surface in the wet well. The float may be a four- to eight-inch diameter ball manufactured from copper or stainless steel, or it may be a flat plate that is six to twelve inches in diameter and several inches thick manufactured from copper, stainless steel, or a ceramic material, or it may be a cylinder six to ten inches in diameter and six to ten inches in depth and made of glass, stainless steel, or a ceramic material. In some instances old brown jugs were used for floats. The jugs were partially filled with sand or lead shot to provide stability and to counteract a portion of the float buoyancy.

All of these devices float on the water surface. When the wet well fills and the water surface rises, the float rises; and as the wet well is pumped down, the float drops with the water surface. The float must be physically attached by steel rods, steel tapes, cables, or ropes to transmit the rise and fall of the float to a recording or signaling device.

Float controllers are used for many applications due to their economical cost and ease of maintenance. The floats also provide flexibility by allowing the lift station operator to change lead pumps and pump start-up sequences in a multiple-pump station every week or month or on some other prescheduled frequency. This procedure distributes wear fairly equally on all equipment by a single pump control circuit plug being moved from cam to cam output. Also this procedure could be adjusted so all of the pumping equipment will not wear out at once.[10]

8.1500 ROD-ATTACHED FLOATS (Figure 8.27)

Ball floats are usually attached to $^1/_4$- or $^3/_8$-inch steel rods. The rod is attached to the top of the ball float and extends up to or through the cover of the wet well and passes through the eye of an actuator arm that is connected to a microswitch. The float rod is equipped with brass stops on both sides of the microswitch arm. When the wet well fills, the bottom stop on the float rod pushes the microswitch arm up and starts the pump. When the wet well is pumped down, the float drops and the rod travels downward. When the top stop on the float rod pushes the microswitch arm down, it turns the pump off. The wet well rise and fall levels are selected or changed by moving the rod's stops to new positions on the float rod.

[10] *"Lift Station Design and Maintenance: Room for Improvement," by Glenn Folk, Deeds & Data, Water Pollution Control Federation, Washington, DC, June 1975, p. D-1 to 4.*

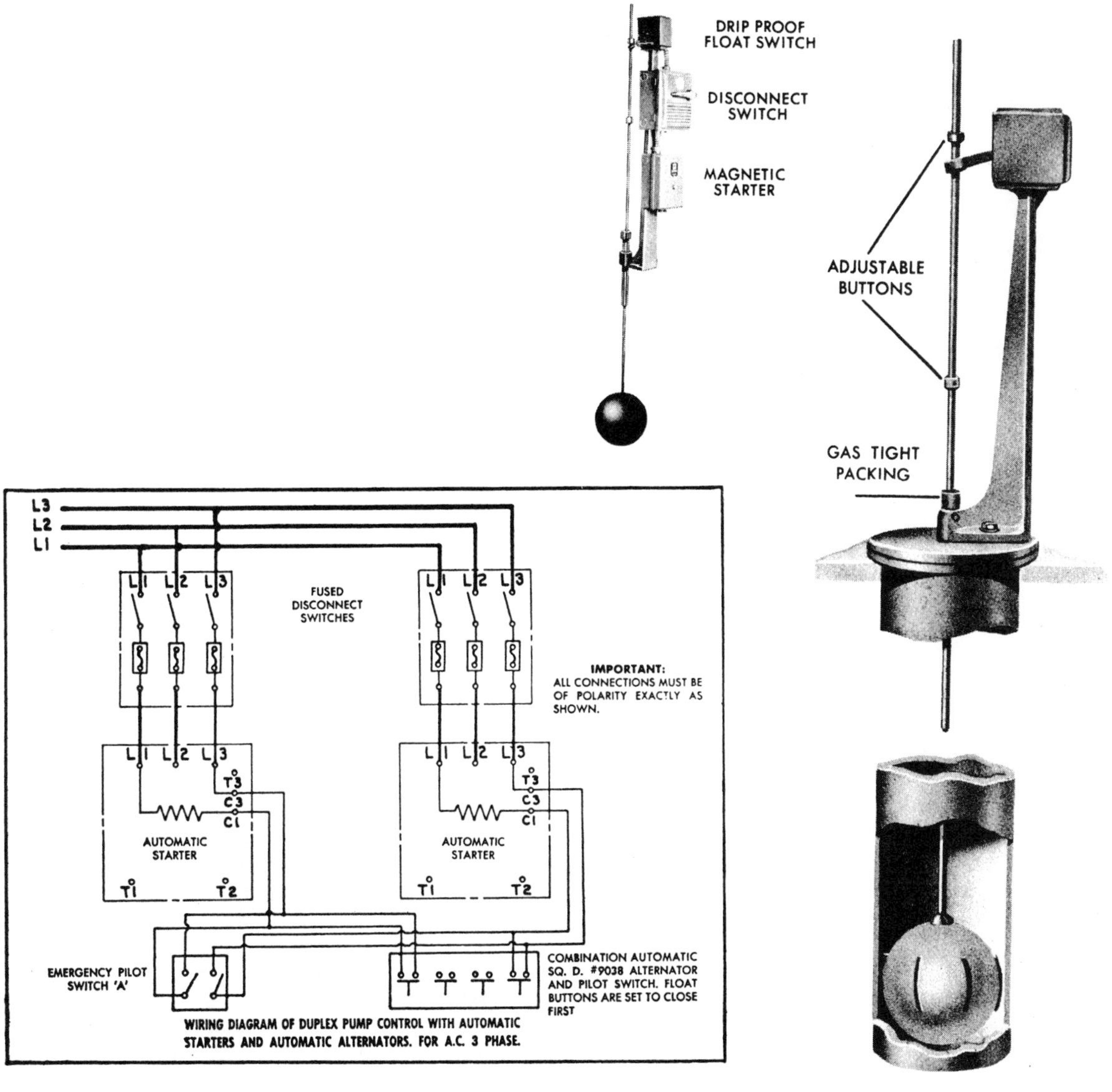

Fig. 8.27 Rod-attached floats and wiring diagram

(Courtesy of Well Pump Company)

Lowering the bottom stop on the float rod permits a higher wet well water level by starting the pump later. Raising the bottom stop will lower the wet well level at which pumping is started. Once the top stop on the float rod is positioned, it should not be changed because it is set to turn the pump off before it loses suction. If a pump loses its suction and is allowed to operate, the pump could be seriously damaged.

The most common application of ball-type float controllers is on sump pumps and on small wet well lift stations having only one pump.

8.1501 STEEL TAPE-, CABLE-, OR ROPE-ATTACHED FLOATS (Figure 8.28)

This type of float is a little more complicated than the rod-attached float. Generally this type of float requires a stilling well for the float to prevent excessive movement of the float. Too much movement of the float will produce incorrect readings. A stilling well is usually a section of pipe ten to twelve inches in diameter which extends several feet above and below the normal wet well operating water level. Usually the stilling well stands vertically in a corner or against a wall of the wet well. The stilling well is open at the bottom or is provided with a smaller inlet pipe of 2.5 to 4 inches in diameter near the bottom to permit water in the stilling well to follow the rise or fall of the water in the wet well.

The float is housed inside the stilling well and is provided one to two inches of free clearance to permit the float to travel up and down the stilling well. At the top of the float is attached the steel tape, cable, or rope that goes up the stilling well, through the cover of the wet well, and is wound onto a small drum which is counterbalanced to maintain tension on the float. When the wet well water level rises, the float rises and the float line is wound onto the drum or sheave. This permits the counterbalance to drop and rotate the drum holding the cams and mercury switch (Figure 8.28). When the drum rotates sufficiently to drop the mercury switch from contact with the cam, this allows the mercury to run down to the contact end. When this occurs, the switch closes, the pump motor starter is energized, and the pump starts. When the wet well water level drops, the float drops, reverses drum rotation and unwinds the line attached to the float. If the float drops far enough, the cam on the drum will raise the mercury switch, tilt the switch so the mercury leaves the contact end, opens the circuit and shuts off the pump.

In a multiple-pump lift station, if the first pump cannot handle the wet well inflow, the float will continue rising and at a predetermined level established by the cam on the drum shaft, the second pump starts (Figure 8.28). If the inflow is greater than the capacity of the two pumps and a third pump is available, the third pump will start next as called for by another cam on the drum shaft. After all pumps available are running and the wet well level continues to rise, the last position on the cam often is used to transmit an alarm signal that the wet well is flooded or will be flooded soon.

Limitations of rope- or cable-attached float control systems include the following:

1. Grease and debris enter the stilling well and hinder or stop the up and down movement of the float. Debris can be removed from the top of the stilling well. Grease can be controlled by running clear water (hot water is better than cold water) into the stilling well at a rate of 1 to 2 gallons per minute to keep the float and stilling well clear of grease and solids. Be sure that this flow of water does not produce false readings of the wet well level unless the openings to the wet well become plugged. Another method of controlling grease is to pour a gallon of detergent down the float tube and use a half-moon scraper.

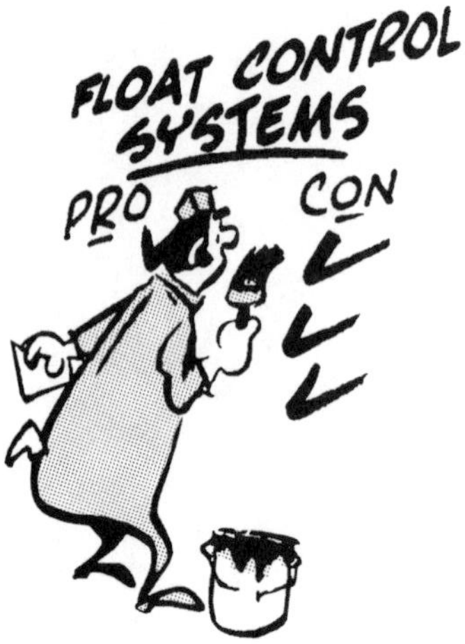

2. The float attachment line or counterbalance line may break and unwind from the drum sheave, thus allowing the counterbalance to fall and all or none of the pumps could be called on to operate.
3. Floats develop leaks which change or stop pump operation.
4. Cables or ribbons attached to the float can become twisted.

8.1502 MERCURY SWITCH FLOAT CONTROLS

The types of float controllers described in Sections 8.1500 and 8.1501 are no longer installed in new pump stations, largely because of the maintenance required for the electromechanical components. The more current system of float control uses stainless steel or plastic floats that are sealed with a mercury switch inside (Figure 8.29).

Mercury is a conducting metal that is liquid in its normal state. It has very good conducting characteristics and is used to close contacts in the float switch. The bulb containing mercury is permanently mounted at a slightly offset angle so that when the float is in the vertical position, mercury makes contact between two of the switch contacts. As the float tips in response to changing water levels, the mercury flows to the other side of the bulb and makes electrical contact between the other switch contacts. Thus, the position of the float determines which contacts the mercury touches. In Figure 8.29, for example, the float is in the normal vertical position and the mercury is making contact between the switches connected to T1 and T2. This closes the electrical circuit between T1 and T2. As the water level rises, the float will tip causing the mercury to flow to the other side of the switch and make contact between T2 and T3. Each of the three contacts is connected to a switch lead from a three-wire electrical cable leading up to the pump station control system.

Generally, each pump in the pump station will have a start float and stop float. A third float is usually used to signal a high wet well condition. In some cases, an additional float is used to indicate the low level cutoff or a low level condition in the wet well. When used for alarms, it is convenient for operators to attach an eye bolt to the bottom of the float, tie a rope to the

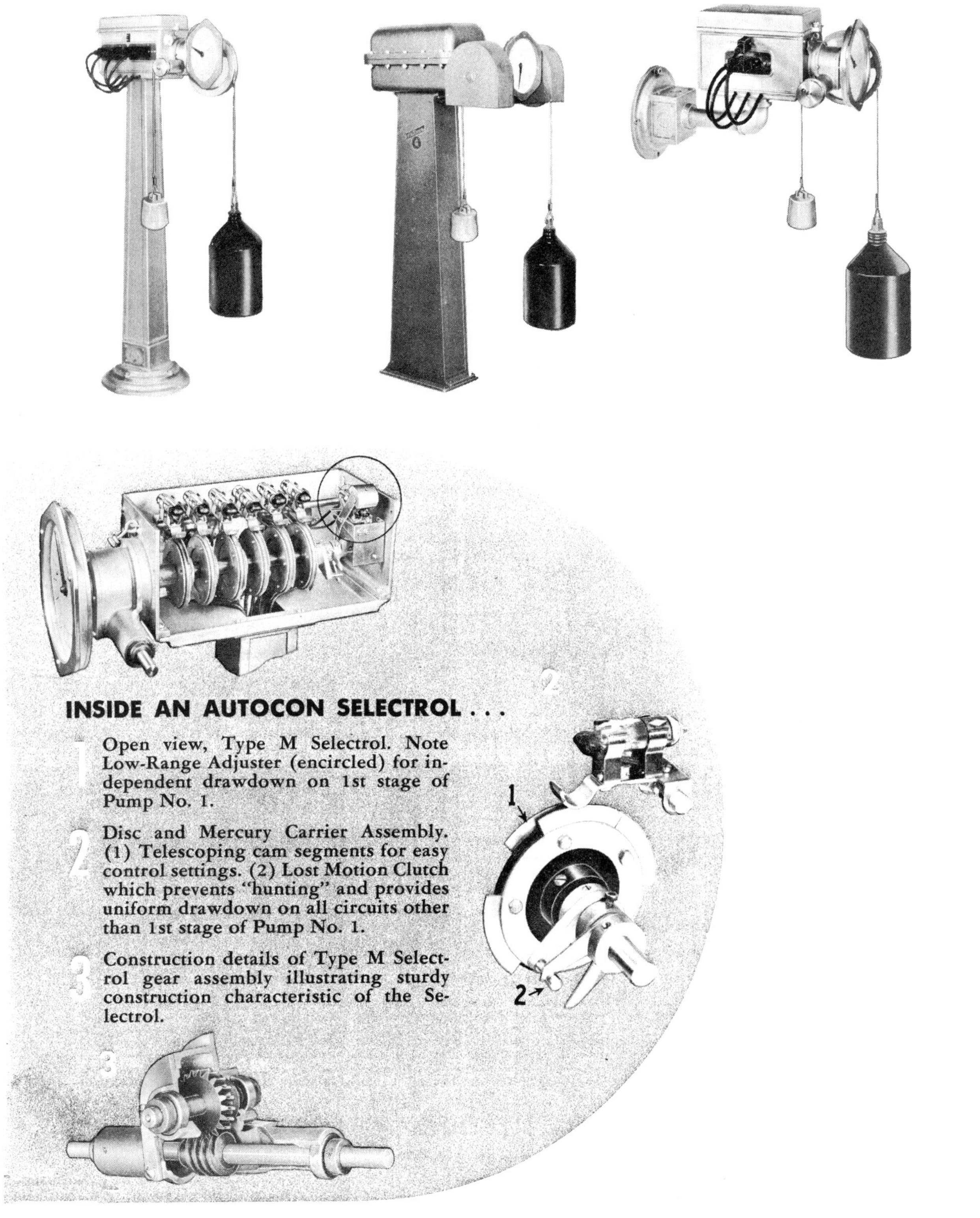
INSIDE AN AUTOCON SELECTROL . . .
1 Open view, Type M Selectrol. Note Low-Range Adjuster (encircled) for independent drawdown on 1st stage of Pump No. 1.
2 Disc and Mercury Carrier Assembly. (1) Telescoping cam segments for easy control settings. (2) Lost Motion Clutch which prevents "hunting" and provides uniform drawdown on all circuits other than 1st stage of Pump No. 1.
3 Construction details of Type M Selectrol gear assembly illustrating sturdy construction characteristic of the Selectrol.
1
2
2
3

Fig. 8.28 Cable-attached floats
(Courtesy of Autocon)

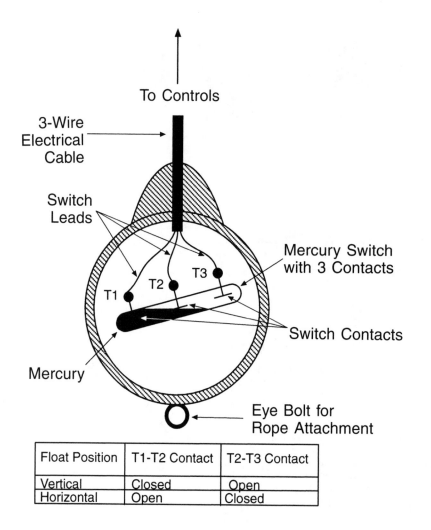

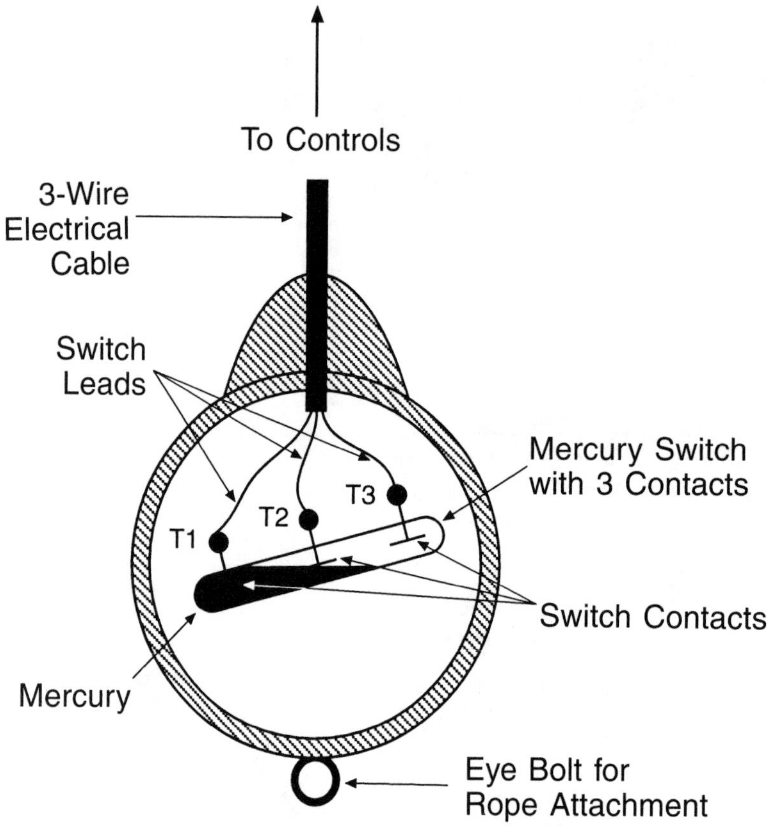
To Controls
3-Wire Electrical Cable
Switch Leads
Mercury Switch with 3 Contacts
T1
T2
T3
Switch Contacts
Mercury
Eye Bolt for Rope Attachment

Float Position	T1-T2 Contact	T2-T3 Contact
Vertical	Closed	Open
Horizontal	Open	Closed

Fig. 8.29 Typical mercury float switch for pump control

eye bolt and bring the end of the rope out the top of the wet well. This allows you to test the float from the top of the wet well without physically removing it from the wet well. Normally these floats are weighted internally to provide some degree of stabilization. Floats can be easily adjusted for different elevations in the wet well; simply raise or lower the float by means of the cable and tie off the cable at the top. Like other types of floats, mercury switch floats are also affected by grease and rags and may become entangled with other floats and cables. They can be purchased with pipe-mounted attachments to reduce the possibility of cable fouling, but this type of mounting makes adjustment of the float elevation more difficult.

Figure 8.30 shows a typical float arrangement for a duplex pump station with a high level alarm float and a low level cutoff float. As the wet well rises, float switch FS-1, the low level cutoff float, tips and activates the control circuit. As the level continues to rise, float switch FS-2, the lead pump OFF float, also tips; however, no control action takes place. As the level continues to rise, float switch FS-3, lead pump ON float, turns on the lead pump. Under normal circumstances, the wet well level is then pumped down until float switch FS-2, the lead pump OFF float, returns to the vertical position. At this point, the lead pump turns off. The purpose of the low level cutoff float is to ensure that the wastewater level in the wet well remains above the center line of the pump volute. If, for example, the system was being operated in manual control and the operator forgot to turn the pump off, the low level cutoff float switch would turn off the pump before pumping the wet well down below the center line on the pump volute. The low level cutoff float also provides additional protection in the event that the lead or lag pump OFF float switches fail.

If the lead pump is unable to pump down the wet well and the level continues to rise, float switch FS-4 moves to the horizontal position; this float switch is the lag pump OFF float switch and it does not cause any control action. However, as the wet well continues to rise, float switch FS-5, the lag pump ON float switch, will move to the horizontal position and turn on the lag pump. Now as the wet well is pumped down and the water level falls, float switch FS-4 will turn the lag pump off when the falling water level returns float FS-4 to the vertical position. In the event that two pumps are unable to pump down the wet well, the high level alarm float switch (FS-6), will activate when the level in the wet well moves it to a horizontal position.

8.151 Level Transducer

Figure 8.31 illustrates a level sensor which is mounted with the sensor below the water surface in the wet well. The transducer section, which is continually submerged, is actually a pressure transducer. Pressure created by the *STATIC HEAD*[11] of water level in the wet well is sensed by a flexible membrane at the bottom of the transducer. This pressure is converted into an electrical signal which is transmitted through the electrical cable to the control system. The electrical signal is therefore proportional to the level in the wet well. The transducer can be permanently mounted with brackets or it can be suspended from above the wet well. The controls for this type of transducer offer wide flexibility. Operating levels in the wet well can be changed at the control panel, and this type of level control console can be operated by programmable logic controls or other forms of microprocessor-based supervisory control systems.

8.152 Ultrasonic Level Detector

Figure 8.32 illustrates an ultrasonic level detector. A transducer is permanently mounted above the highest water level in the wet well. The transducer generates ultrasonic pulses which hit the surface of the water in the wet well and are reflected back to the transducer. The system measures the time required for a pulse of energy to leave the transducer and return. (This is the same principle on which radar and sonar operate.) The time information is then converted into an electrical signal in the control system. As with the level transducer, the controls provide a wide range of flexibility in pump station operation, and the device can be wired to operate in coordination with other pump station controls. The ultrasonic transducer converts the electrical energy of the transmit pulse from the transceiver into acoustical energy. It then converts the acoustical energy of the echo back into electrical energy for the transceiver receive period. When mounting the transducer, it should be placed above the maximum material level by at least the blanking distance.

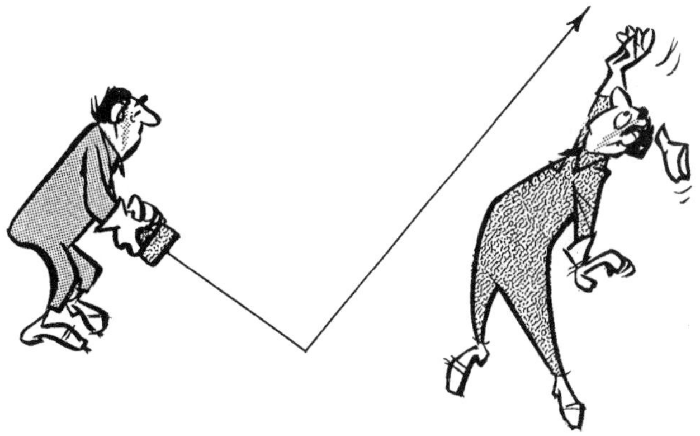

QUESTIONS

Write your answers in a notebook and then compare your answers with those on page 119.

8.1H What are the three types of float controllers?

8.1I If a float is attached to a rod, what will happen if the top float rod stop is raised?

8.1J What maintenance does a stilling well require?

8.1K How does an ultrasonic level detector measure water levels?

8.153 Electrode Controllers

Use of this method of control requires that electrodes be installed and maintained in compliance with safety regulations. Be sure that power to the electrodes is turned off and properly tagged before performing any maintenance on the equipment. Operators should be cautioned not to touch electrodes with bar screen rakes and other tools.

The water level sensing or detecting device is simply a series of weighted electric leads or electrodes hanging in the wet well at staggered elevations, usually six to eight inches apart. As the wet well water level rises and submerges an electrode, the water surrounding the electrode completes the

[11] *Static Head. When water is not moving, the vertical distance (in feet) from a specific point to the water surface is the static head.*

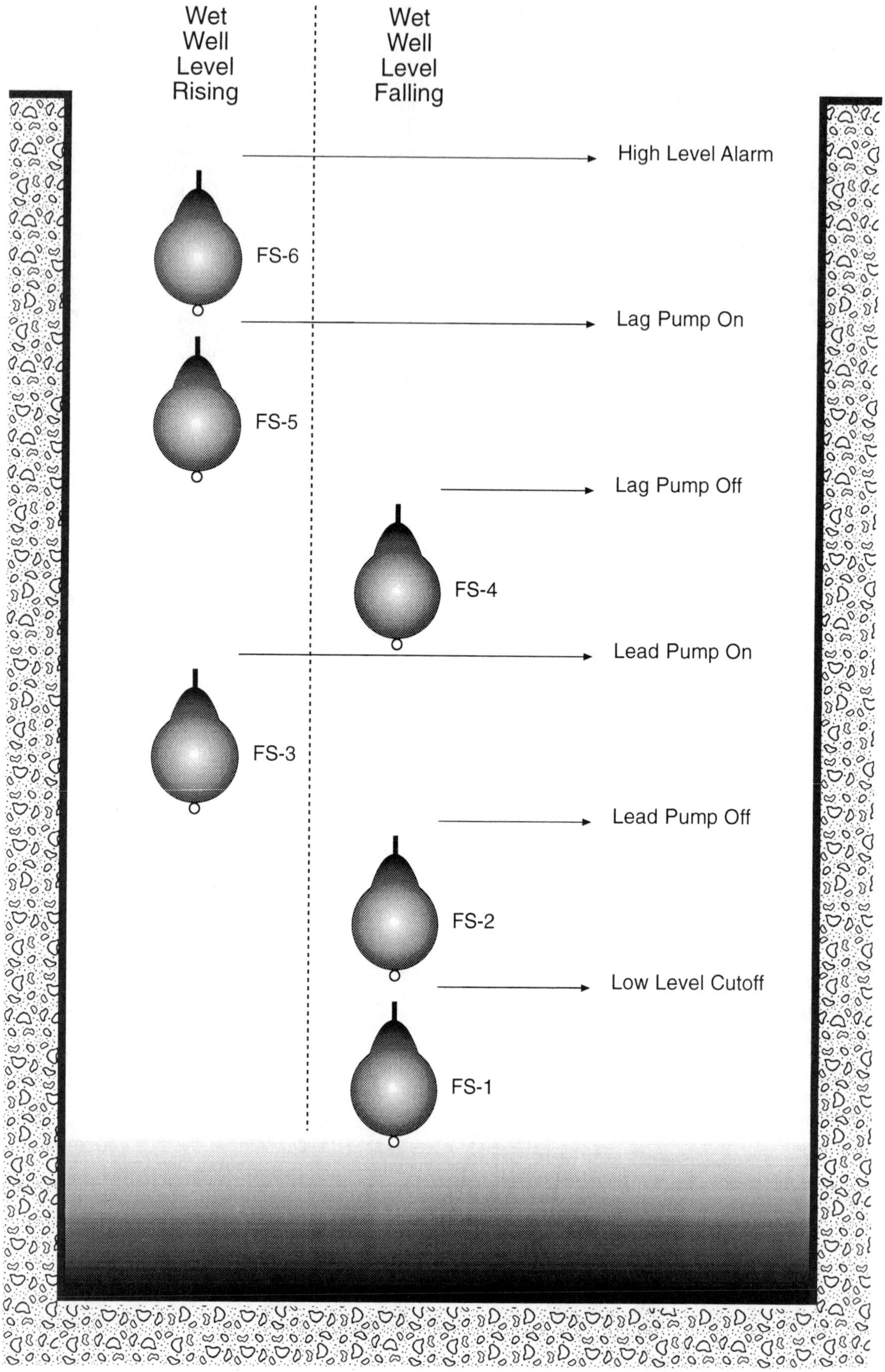

Fig. 8.30 Typical float arrangement for duplex pump station with high level alarm and low level cutoff

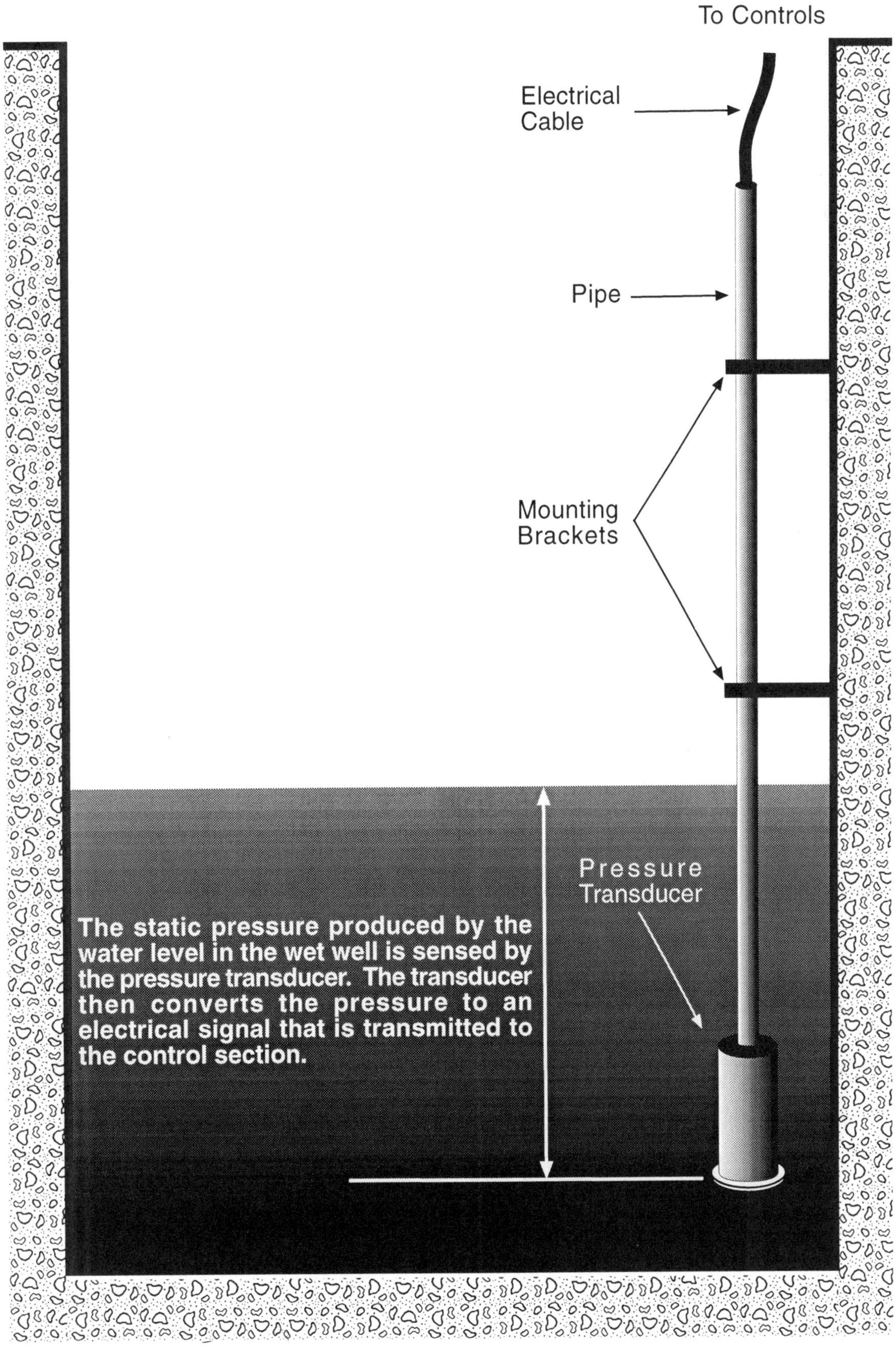
To Controls
Electrical Cable
Pipe
Mounting Brackets
Pressure Transducer
The static pressure produced by the water level in the wet well is sensed by the pressure transducer. The transducer then converts the pressure to an electrical signal that is transmitted to the control section.

Fig. 8.31 Level transducer for sensing wet well level

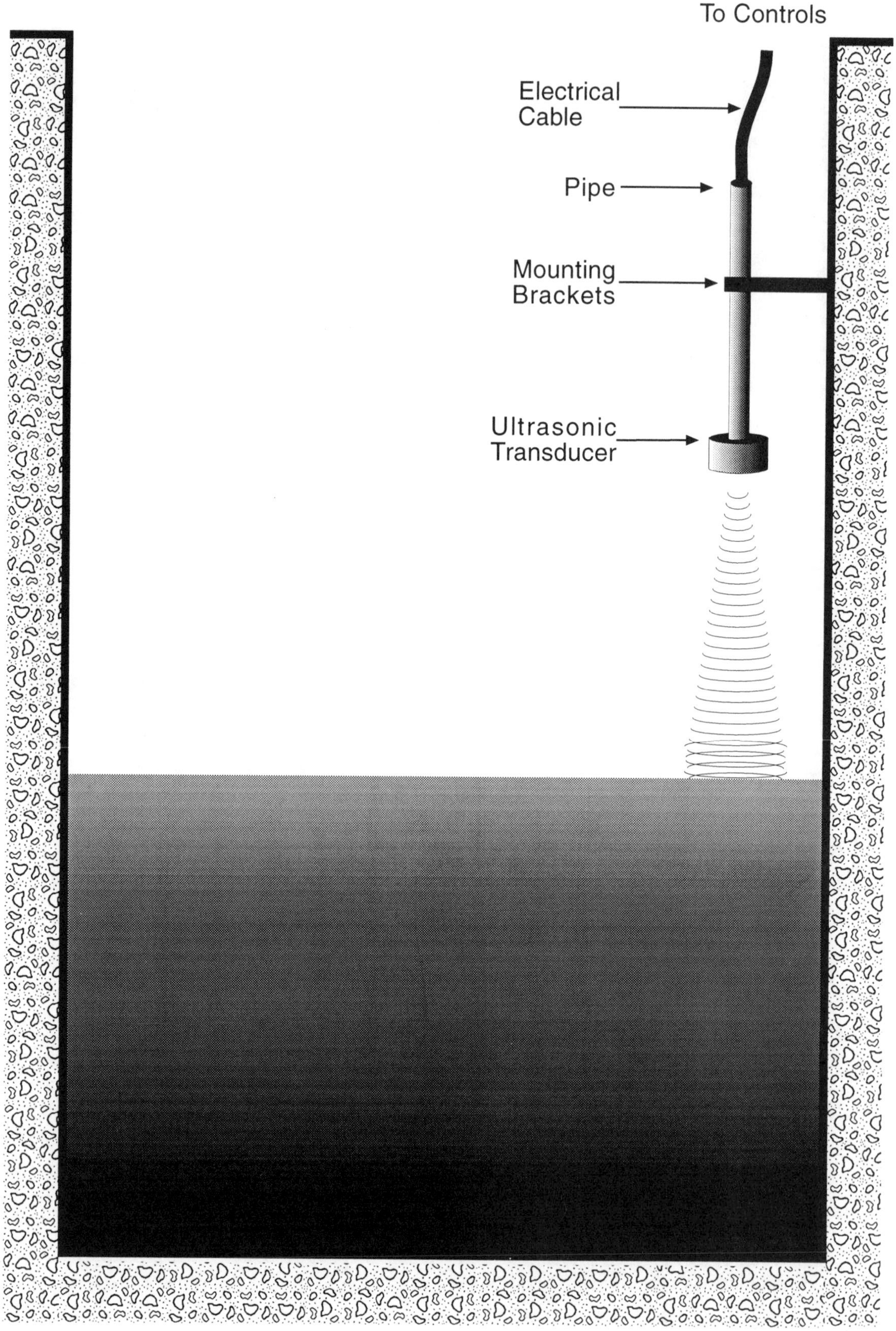

Fig. 8.32 Ultrasonic transducer for sensing wet well level

control circuit and starts the pump. When the wet well level is pumped down and the electrode exposed to air, it opens the electrical circuit and shuts off the pump. One electrode controls a single pump, and each pump usually has its own electrode. A high wet well elevation electrode is used to indicate flooding of the wet well; it activates an alarm to signal a high water level or flooding.

Limitations of electrode controllers include the following:

1. Rags and debris can wrap around electrodes and alter desired start and stop water levels and sequences of pumps, and
2. Grease and/or slime can cover an electrode, thus preventing good conductivity and causing intermittent or unreliable operation.

8.154 Enclosed-Electrode Controllers (Seal Trode Units)

These controllers are a refinement of the old free-floating electrodes. They work in the same manner except they are enclosed in a two- to three-inch diameter pipe with a bulb-type container on the bottom end containing an electrolyte solution (Figure 8.33). The unit is attached to the floor of the wet well.

Electrodes (one for each pump and a high level alarm) are set at different elevations in the pipe housing. The bulb at the end of the pipe holds from three to eight quarts of electrolyte solution. This volume is sufficient so that water pressure on the outside of the bulb will force the electrolyte solution up the tube to an electrode that completes the circuit and turns on the pump. As the water level in the wet well rises, it compresses the bulb and forces the electrolyte up the pipe to the electrode. When the wet well level is high enough to develop the pressure required to force the electrolyte solution up the tube to the electrode, the circuit is completed and the pump is started. When the water level in the wet well drops, the pressure on the bulb is reduced and the electrolyte solution falls and opens the circuit. The electrodes in this system are spaced much closer in elevation (1.5 to 2.0 inches) because the change in water pressure on the bulb is small compared to changes in water depth in the wet well.

Limitations of enclosed-electrode controllers include the following:

1. When bulb breaks, electrolyte solution is lost,
2. Shorts occur in electrode wiring, and
3. Bulb life is three to five years.

8.155 Pressure-Sensing Controllers (Air Bubblers)

Air sensing (pneumatic) controllers are being used in many new lift stations being constructed today. They are adaptable to several different control systems and may be used in simple diaphragm and mercury switches for control inputs, in complex fluidics (air) systems using orifices and chambers to control outputs, or in analog equipment with differential transmitter square root extraction (a method of measuring pressure) and averaging relays that convert signals from pneumatic to electronic (4 to 20 milliamp) outputs for use by controllers and integrators.

Bubblers provide constant, low-volume, low-pressure air through a vertically mounted pipe in the wet well. The bubbler system is aptly named since, when working properly, bubbles can be observed at the surface of the wet well.

The principle of the bubbler system is actually fairly simple. If you have ever blown air through a straw into a glass of water, you know that it requires a certain amount of air pressure to displace the water in the straw and to force air out the bottom of the straw. The amount of force required is determined by the depth of the liquid over the end of the straw. For example, if the depth of the water is 2.3 feet, it will require a pressure of 1 psi (pounds per square inch) to push the air through the straw and out the bottom. (We know this since 1 psi is equal to a column of water 2.3 feet high.) Since the air pressure in the straw is directly related to the level of water over the end of the straw, it stands to reason that by measuring the air pressure we can determine the actual water level or depth.

Figure 8.34 illustrates a pneumatic bubbler system which because of its redundancy is extremely reliable. Described below are the individual components of this system.

1. Two $^1/_6$ or $^1/_{12}$ horsepower air compressors are mounted on a small ten-gallon air receiver tank which acts as an air reservoir. The tank is desirable since it allows the compressors to shut off instead of running continuously. Additionally, in the event of a failure of both compressors, the system would continue to operate because of the reservoir for some period of time (depending on the pressure in the tank).

Note: Tank condensate must be bled off from the tank drain periodically.

2. Pressure in the discharge side of the tank is sensed by two pressure switches which turn the compressor on and off in a lead/lag sequence. If the lead compressor fails, the lag compressor comes on when pressure in the system drops to a lower preset value at the lag pressure switch.
3. A two-inch, 0-to-100 psi gage is mounted in the line so that high pressure air can be monitored.
4. A desiccant-type air dryer (one that uses silica gel as the dehumidifying agent and that can be regenerated) is located in the line to remove moisture and prevent condensation farther downstream in the instrumentation or in the bubbler line.
5. A shutoff valve permits isolation of the high-pressure air side from the low-pressure air side for maintenance purposes and is in the normally open position during operation.

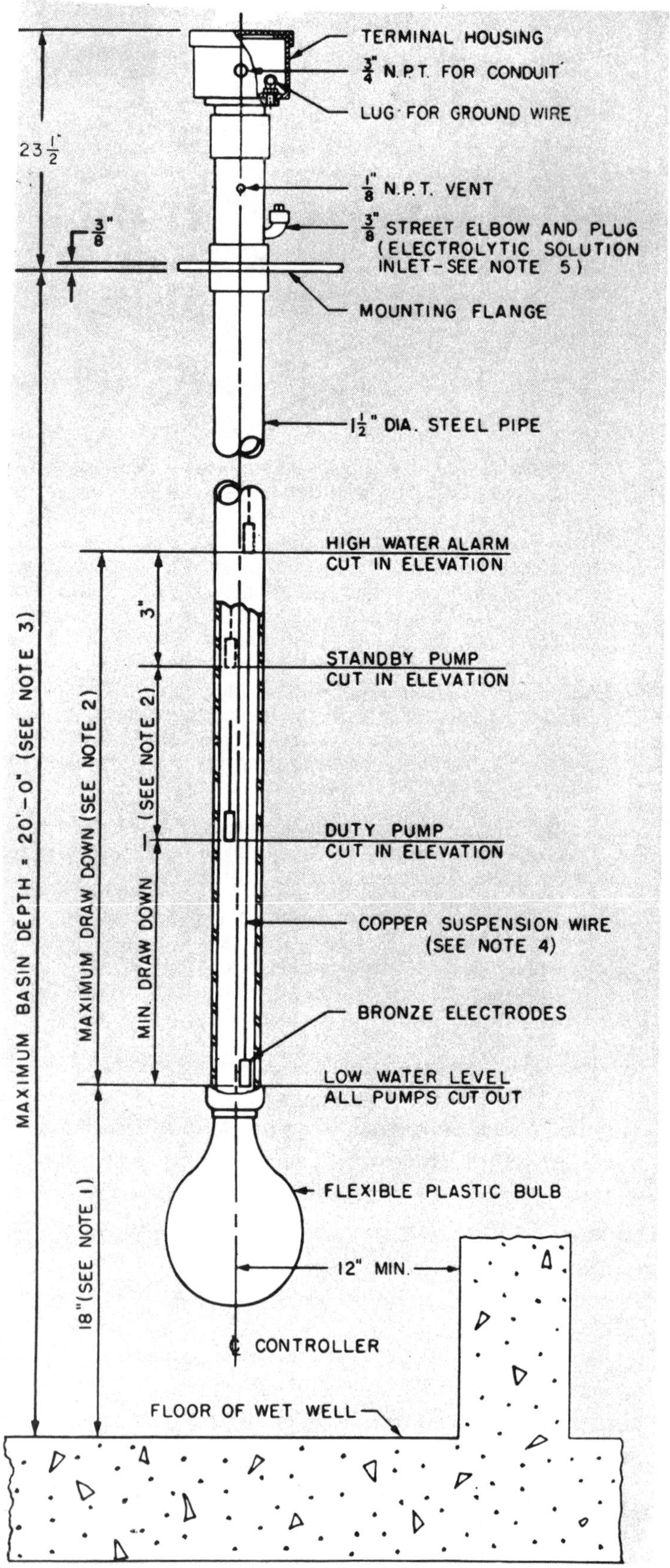
TERMINAL HOUSING
3/4" N.P.T. FOR CONDUIT
LUG FOR GROUND WIRE
23 1/2"
1/8" N.P.T. VENT
3/8" STREET ELBOW AND PLUG (ELECTROLYTIC SOLUTION INLET-SEE NOTE 5)
3/8"
MOUNTING FLANGE
1 1/2" DIA. STEEL PIPE
HIGH WATER ALARM CUT IN ELEVATION
3"
STANDBY PUMP CUT IN ELEVATION
DUTY PUMP CUT IN ELEVATION
COPPER SUSPENSION WIRE (SEE NOTE 4)
BRONZE ELECTRODES
LOW WATER LEVEL ALL PUMPS CUT OUT
FLEXIBLE PLASTIC BULB
12" MIN.
℄ CONTROLLER
FLOOR OF WET WELL
MAXIMUM BASIN DEPTH = 20'-0" (SEE NOTE 3)
MAXIMUM DRAW DOWN (SEE NOTE 2)
MIN. DRAW DOWN (SEE NOTE 2)
18" (SEE NOTE 1)

ENGINEERING INFORMATION

NOTE 1—18″ is the minimum low water level setting for the controller. However, the actual low water setting must be set at a height to keep the pump casing flooded.

NOTE 2—The maximum overall draw down is 7′. The minimum recommended draw down setting is 6″ per pump.

NOTE 3—When the basin depth exceeds 20′, a 2″ steel pipe must be supported by wall brackets.

NOTE 4—Suspension wires in the terminal housing have identifying labels, "A," "B," "C," etc. Wire "A" is attached to the lowest, or pump cut out electrode. Wire "B" is the next lowest, or pump cut in electrode. Wire "C", etc.

NOTE 5—Electrolytic solution consists of a mixture of 3½ quarts of clear water and one package of sodium bicarbonate.

NOTE 6—Organic acids or alcohol should not be used with the Flexible Plastic Bulb. Temperature limitations are 35°F. minimum . . . 140°F. maximum.

NOTE 7—Wires to terminals 3 and 5 must be same phase. Wires to terminals 4 and 6 must be same phase.

Fig. 8.33 Enclosed-electrode controller

(Courtesy of Chicago Pump)

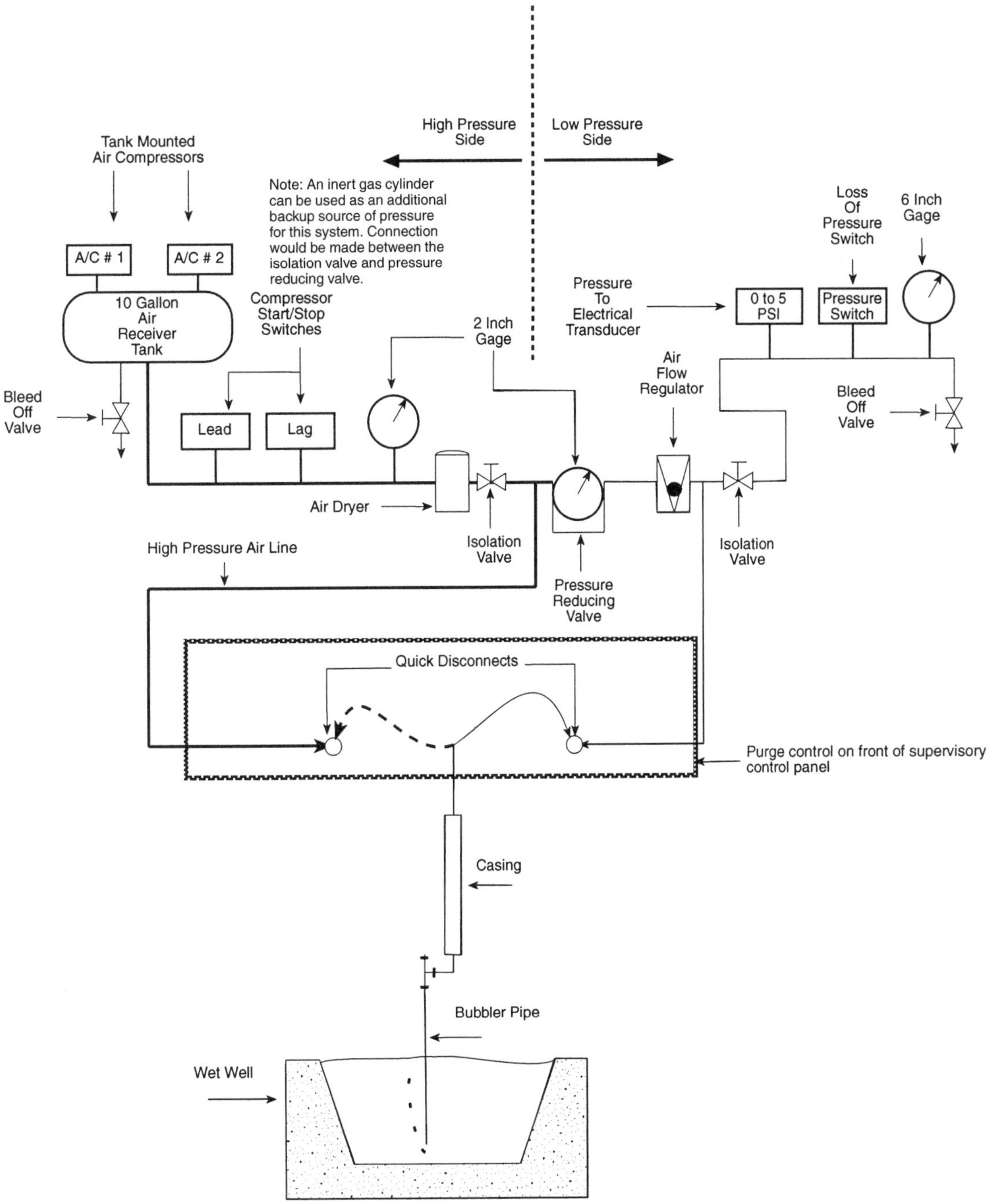

Fig. 8.34 Pneumatic bubbler system

6. A pressure-reducing valve and a two-inch gage reduce the high pressure from 40 to 80 pounds (psi) to generally less than 5 psi (remember that the pressure will be determined by the depth of water in the wet well). Usually this is established by the highest level that can be reached under high wet well conditions. The two-inch gage monitors the downstream or low pressure side of the pressure-reducing valve.
7. An air flow regulator maintains a constant low volume flow of air into the system; typically, one-half to one cubic foot per hour.
8. A zero-to-five psi pressure transducer senses the pressure in the line and converts the pressure to an electrical signal which is then sent to the motor control center and sequences pumps, alarms, and other supervisory control functions.
9. A low pressure switch is also installed at this point so that a loss of pressure can be sensed and alarmed.
10. A six-inch to 160-inch gage, which is mounted on the supervisory control panel door, reflects actual level in inches in the wet well.
11. The last bleed-off valve shown in the system is in the normally closed position; this valve is used for bleeding off condensation.
12. This particular system incorporates a manual purging panel which allows high-pressure blowdown of the bubbler pipe in the wet well to clear grease or other accumulated debris from the system.

The advantage of a manual system versus an automatic purging system (which uses solenoid valves) is positive isolation of the low pressure instruments, specifically the pressure transducer, low pressure switch, and gages. Permanent damage to these instruments will occur if they become overpressurized as might happen with automatic solenoid valves if high-pressure air leaks past the seat.

The manual purging panel has quick disconnect pneumatic fittings on the front of the supervisory control. The operator physically removes the bubbler system low pressure side and connects it to the high pressure side.

Bubbler System Maintenance

System maintenance consists of the following tasks:

1. Draining condensation from air receiver weekly or monthly,
2. Changing dryer desiccant,
3. Weekly purging,
4. Weekly checks of pump/alarm sequencing, and
5. Semiannual checks of bubbler pipe in wet well for corrosion or looseness.

Many pump stations have been flooded when high wet well levels caused flow through the bubbler pipe wall hole and back into the dry well. The hole should be well sealed. If wastewater enters the system, it will damage the instruments. A loop in the bubbler pipe in the dry well *ABOVE* the highest wet well level possible will eliminate backsiphoning/draining into the system under high wet well conditions. In northern climates during cold weather, if moisture accumulates in the loop it can freeze and cause a plugged bubbler. Do not allow condensate to accumulate.

Symptoms of Plugged or Broken Bubbler

	Wet Well Level	Air Pressure	Pumps
Plugged	minimum	maximum (will indicate a high wet well)	full on—all pumps called for (pumps may be airbound)
Broken or severe air leak	maximum	minimum (will indicate a low wet well)	off—no pumps called for

Many manufacturers and instrumentation companies use pneumatic systems for control. Consult their O & M manuals for your type of equipment and application.

Limitations of pressure differential controllers include the following:

1. Air compressor failure,
2. Bubbler blockage (line must be purged regularly), tube breakage, or outlet elevation shifts, and

3. Complex equipment provides more opportunity for failures and requires a higher level of training for operators to troubleshoot, repair and keep instruments calibrated. Often the more reliable an alarm system is, the more complex it becomes and thus the more difficult to maintain because of backup systems.

8.156 Microprocessor-Based Supervisory Controllers

The development of microelectronics has also impacted lift station control systems. Many manufacturers today produce microprocessor-based supervisory controllers, which typically offer the following features:

1. Easily adjustable pump on and off levels,
2. Alternating sequence,
3. Auto/manual functions,
4. Alarms at both high level and low level set points, and
5. Digital readouts of wet well level.

The microprocessor-based systems incorporate small computers into the supervisory control system. They usually receive input from a bubbler-type wet well sensing system that translates air pressure into an electrical signal or from an electronic transducer mounted in the wet well.

These systems are very flexible, easy to adjust and are consistent. Care must be taken, however, in their selection, installation and maintenance since microprocessors are susceptible to voltage surges on the line from lightning strikes, utility company switching and power failures. Figure 8.35 illustrates a microprocessor-based liquid level controller.

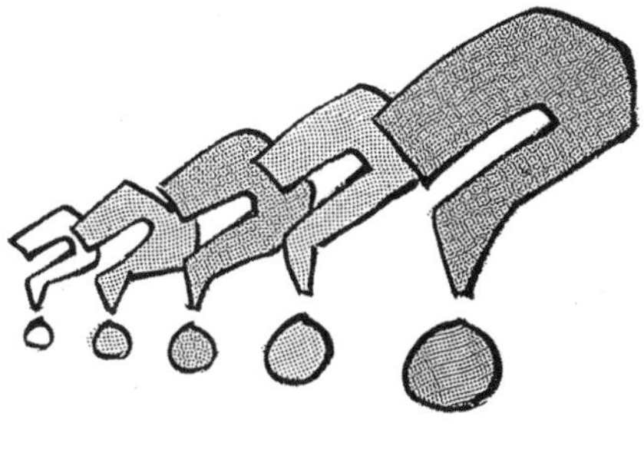

QUESTIONS

Write your answers in a notebook and then compare your answers with those on page 119.

8.1L What types of controls are used to start, stop or change pumping rates?

8.1M What safety considerations are necessary when using electrode controllers?

8.1N What are the limitations of electrode controllers?

8.1O How are pressure changes detected in a pressure-sensing controller?

8.1P What kinds of problems can develop using a seal trode sensing controller?

8.1Q How is the depth of water in a wet well measured using a pressure-sensing controller?

8.1R What kinds of problems can occur with pressure-sensing controllers?

8.16 Pumps

8.160 Centrifugal Pumps

Pumps constitute major pieces of equipment in the lift station and, obviously, one of the most important components. Yet they remain one of the least understood pieces of equipment in the station. Reliable and efficient pump operation depends heavily on the collection system operator's knowledge and understanding of the principles of operation.

Chapter 9 (Section 9.30, "Let's Build A Pump") can be used as a reference for this section; however, it should be pointed out that the pump in the example is a multi-stage centrifugal pump normally used for clean water applications and is not the type of pump found in a station pumping raw wastewater. The operating principles, however, are generally the same.

Pumps are one of either two types: positive displacement or kinetic pumps. Each of these two types has several subclassifications as shown on Figure 8.36. Reciprocating pumps are used as metering pumps; for example, to add chlorine for disinfection and for pumping liquids with high concentrations of solids such as sludge from a wastewater plant. By far the most common type of pump in the water and wastewater industry is the group of pumps classified as kinetic, also called dynamic. Of the three major subclassifications, centrifugal, peripheral and special, the most common are those classified as centrifugal pumps.

Centrifugal pumps are further divided into three subclassifications based on impeller design and the method used to develop energy in the pump. In a radial flow design, the pressure is developed primarily by the action of centrifugal force. Fluid enters the eye of the impeller and is directed radially outward from the shaft. Energy is imparted to the fluid in the process. Radial flow pumps are used in a wide variety of applications.

Axial flow design, also known as a propeller pump, develops most of the pressure by the lifting or propelling action of the impeller vanes on the liquid. Fluid flow in a propeller pump is parallel to the shaft. Axial flow pumps are usually found in low-head, high-capacity applications, and the impellers may be open, semi-open or closed, depending on the application and liquid being pumped.

Mixed flow design uses both centrifugal force and the lifting action of the impeller vanes to develop pressure. Mixed flow designs are used in low-head, high-capacity applications, and the impellers may be open, semi-open or closed, depending on the application and liquid being pumped.

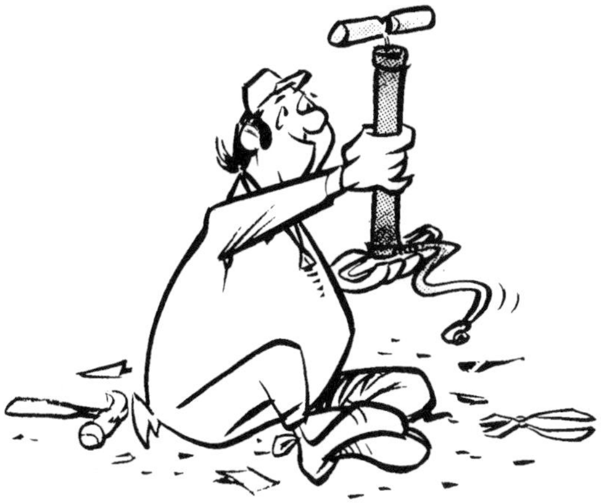

By far the most common type of pump found in raw wastewater lift stations is centrifugal; therefore, this section will focus on centrifugal pumps. It is preferable to design pump stations with multiple pumps so that at any given time, one pump can be taken out of operation for routine, periodic maintenance. Stations must be capable of operating even during peak flow periods with the possibility of one pump being out of service for maintenance; therefore, a station designed for capacity at peak flows will normally have one additional pump sized to handle peak flow pumping requirements. If two pumps are required under peak flow circumstances, the station should have three pumps and so on.

Pumps are installed with a piping layout that allows isolation of any pump from the rest of the hydraulic system through suction and discharge isolation valves (discussed later in this chapter). Check valves in the discharge piping for each pump prevent backflow through the pump back into the wet well.

8.161 Pump Layouts

In dry well/wet well types of lift stations, raw wastewater pumps are found in one of the following types of layouts:

1. Flexibly Coupled. In this layout the pump and the motor shafts are connected by a flexible-type coupling, as illustrated in Figure 8.37, which shows a horizontal type of centrifugal pump, and Figure 8.38, which shows a vertically mounted centrifugal pump. Note the adapter between the motor and pump required on the vertical installation.

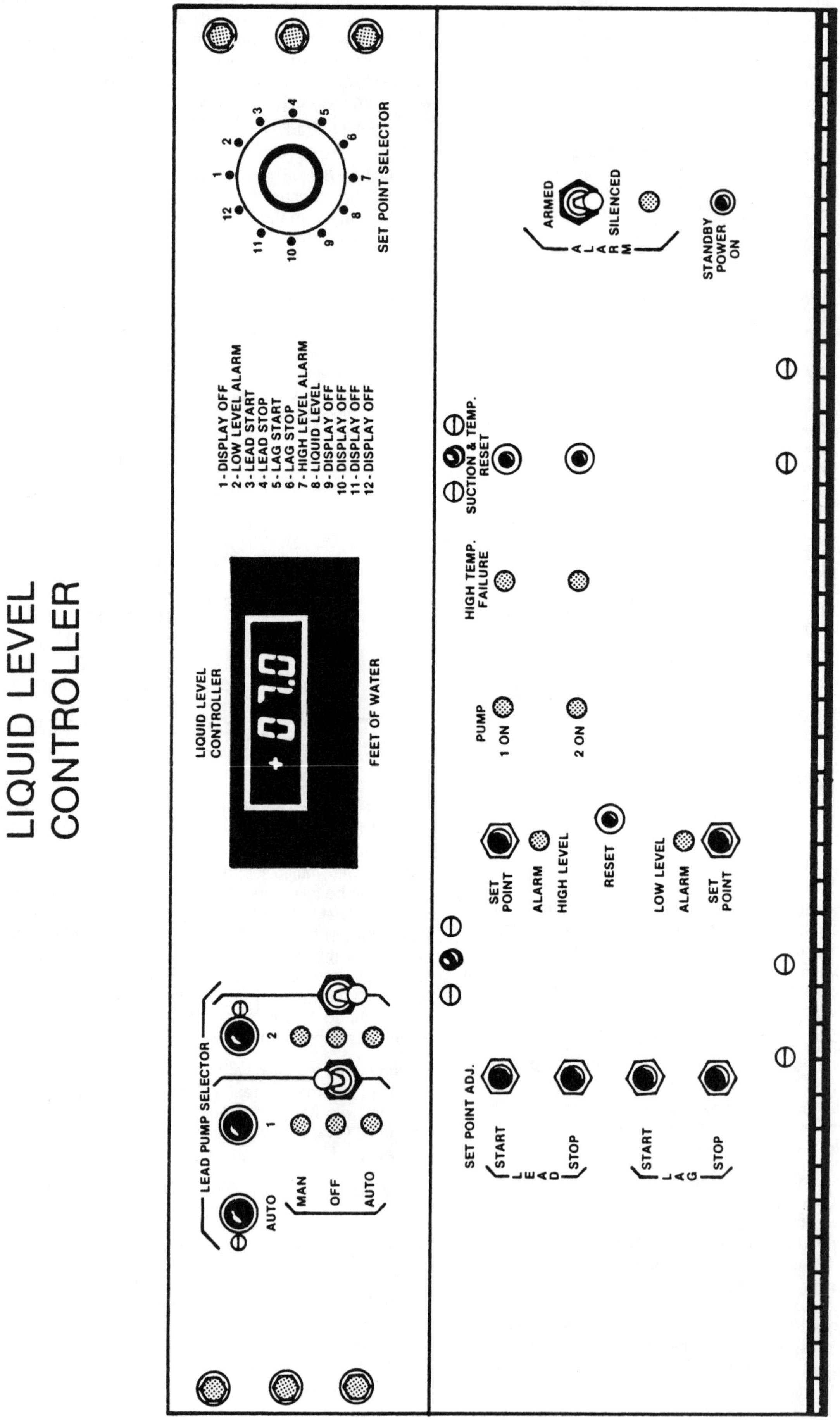

Fig. 8.35 Microprocessor-based liquid level controller
(Courtesy of Goman-Rupp)

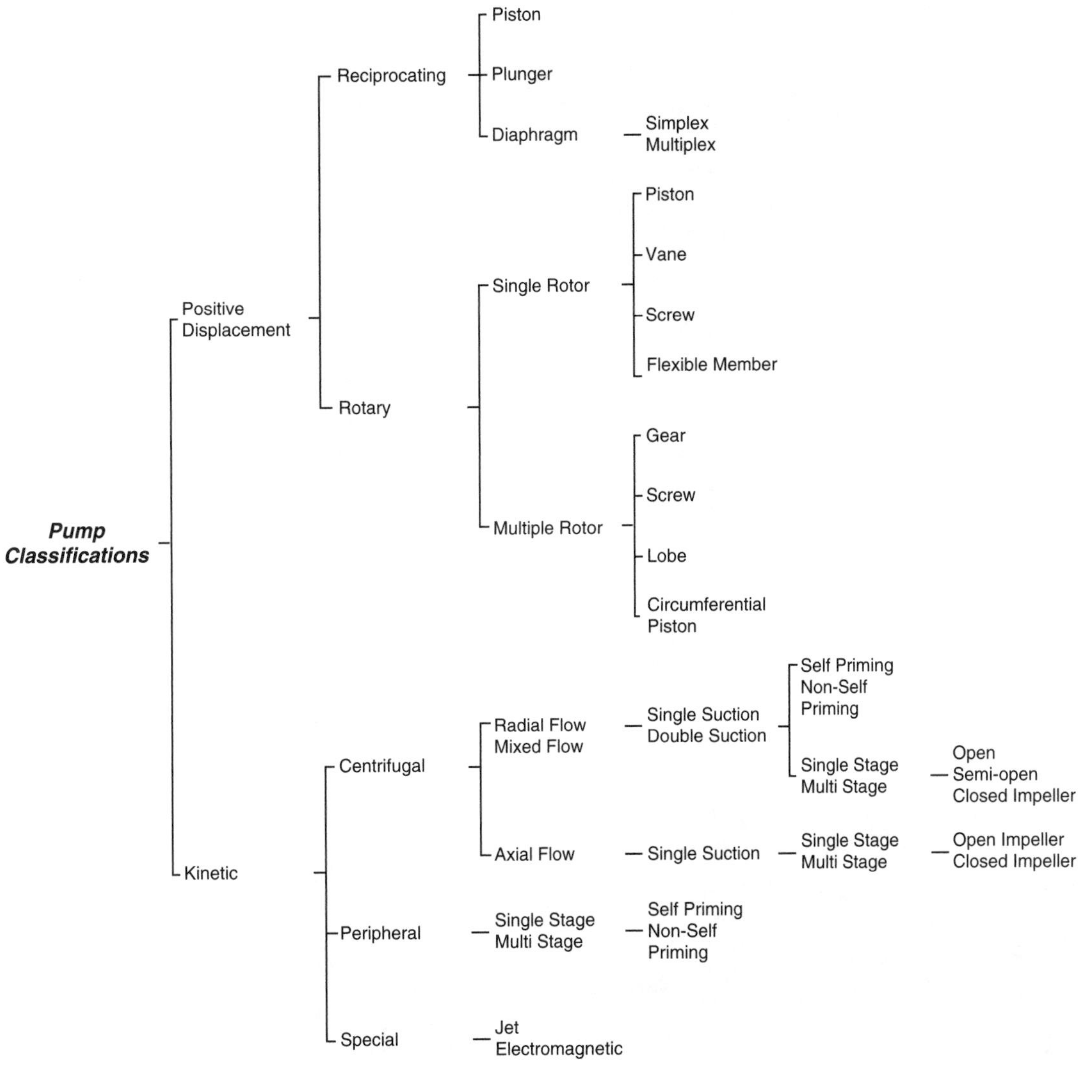

Fig. 8.36 Classification of pump types

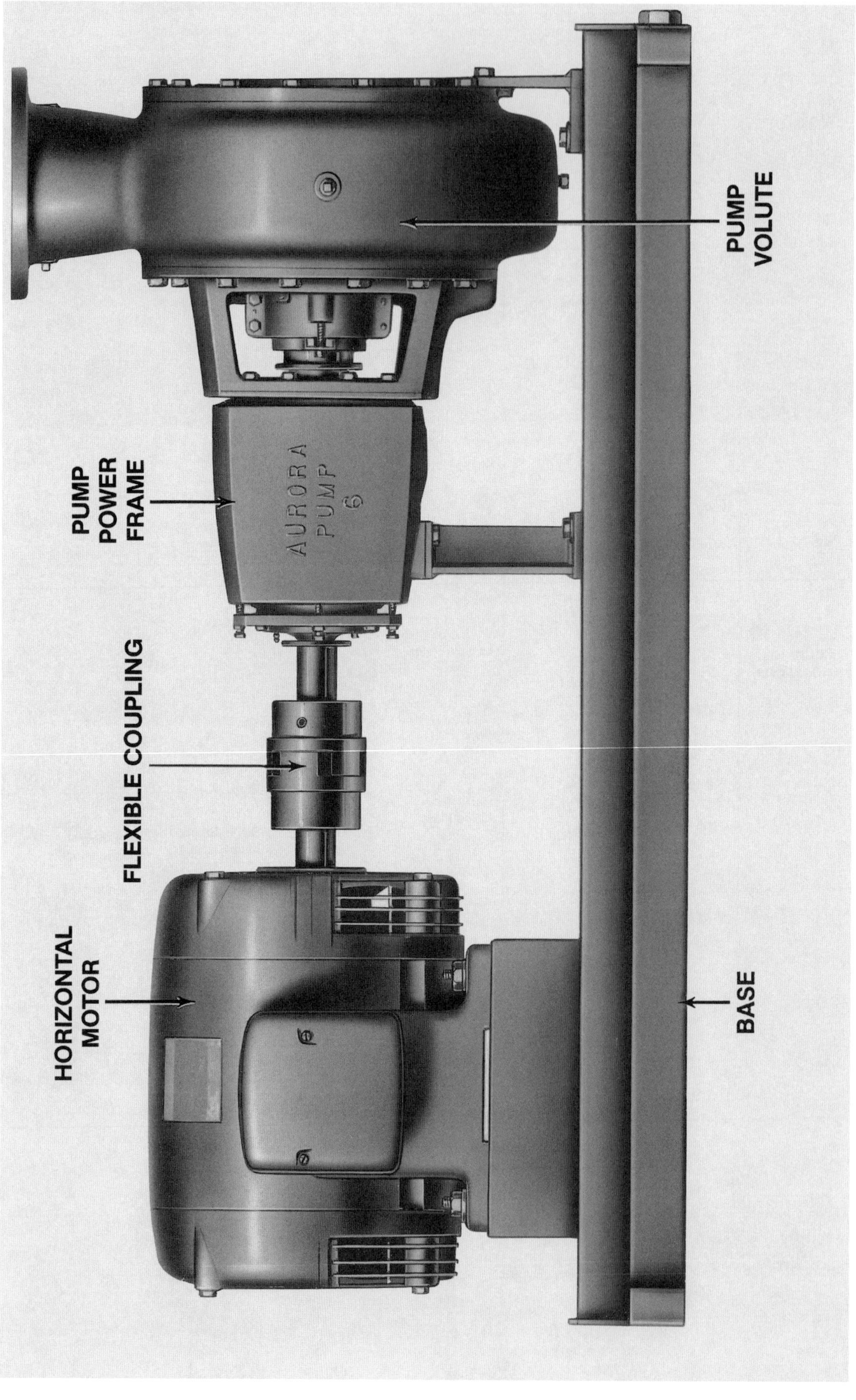

Fig. 8.37 Flexibly coupled, horizontally mounted centrifugal pump
(Courtesy of Aurora Pump)

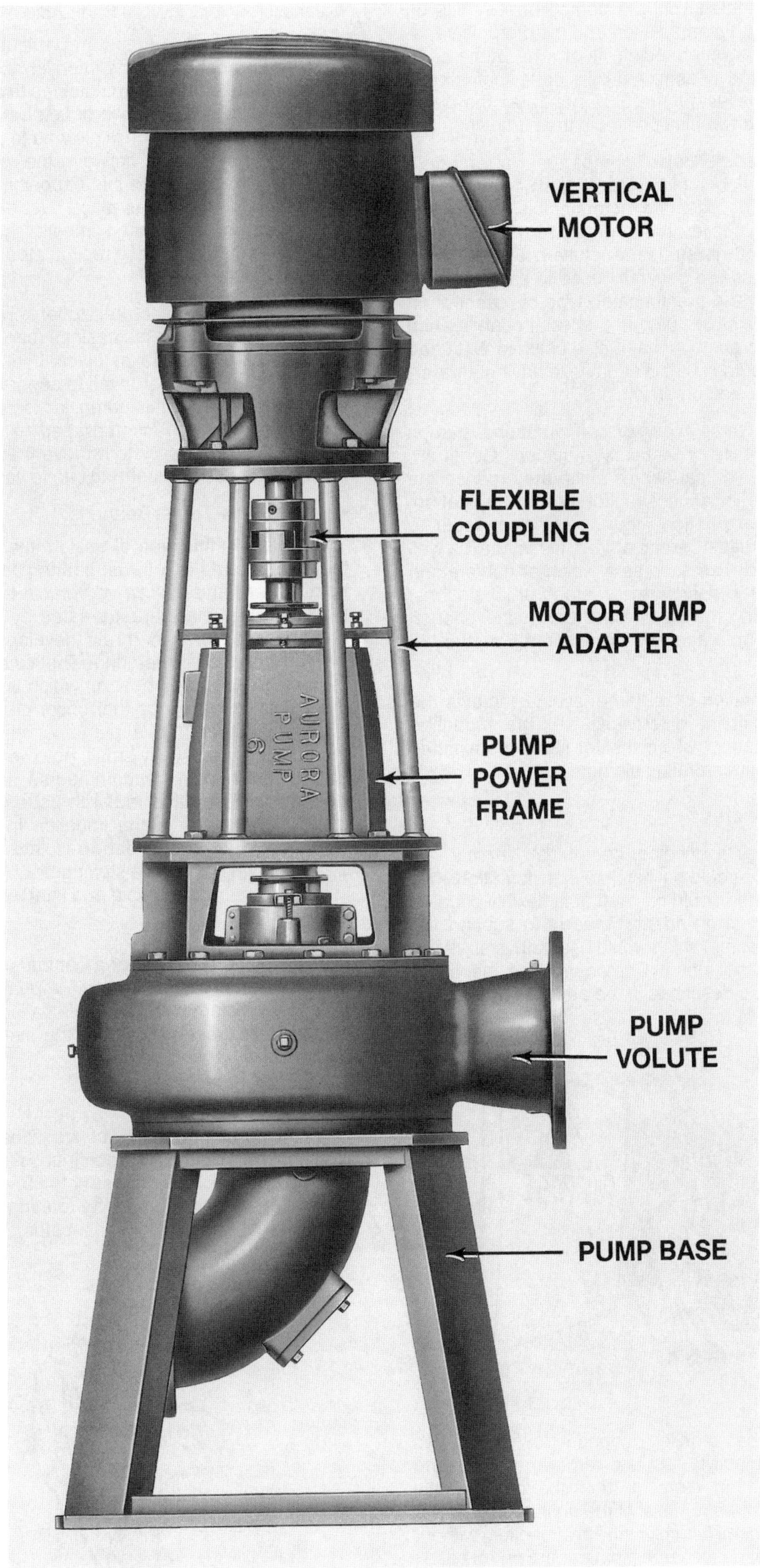

Fig. 8.38 Flexibly coupled, vertically mounted centrifugal pump

(Courtesy of Aurora Pump)

2. Flexibly Shafted. In this layout the pump normally sits on the bottom floor of the pump station and the motor is separately mounted on an intermediate floor. The pump shaft and the motor shaft are connected by a drive shaft similar to that found in vehicles. Figure 8.39 illustrates the pump layout. The drive shaft would connect to the pump shaft.
3. Close-Coupled. Close-coupled pumps use a standard squirrel cage induction motor with a shaft that extends beyond the motor. The extended motor shaft is mounted directly to the pump impeller through the use of an adapter bracket. Figure 8.40 illustrates a close-coupled pump. Close-coupled pumps are generally used in smaller pump stations, particularly the prefabricated type of "canned" lift stations. The submersible pump is a special configuration of the close-coupled pump in that it also uses an extended motor shaft for the pump shaft in addition to having a waterproof motor capable of submersion.

In addition to these three types of configurations, pumps may either be horizontally or vertically mounted. Generally, pump stations use the vertical pumps since they reduce the size and, therefore, the initial construction cost of the station. The vertical pump configuration requires sufficient room to remove the pump and rotating element from the volute, as well as room for the lifting device such as a winch or come-along on the ceiling. During the design review stage of lift stations, be sure there is a lifting device for the pumps and enough space to remove the pump and rotating element from the volute.

The principle of operation of all three layouts remains the same. The only significant differences are the mounting method and the method in which the torque is transmitted from the motor to the pump rotating element.

8.162 Pump Components

Figure 8.41 illustrates a nonclog, centrifugal, vertical raw wastewater pump which could be installed in a lift station. The particular pump illustrated could be used as a flexibly coupled pump with the addition of an adapter bracket to support the motor, or as a shafted pump which would use a drive shaft between the pump and the motor. Components of the pump shown in Figure 8.41 are described in the paragraphs below.

- Keyway

The top of the shaft has a drilled, tapped hole which allows the insertion of an eye bolt to facilitate removal of the pump power frame and rotating element from the permanently mounted volute. This allows removal of the entire pump power frame and rotating assembly, including the impeller, without disturbing the suction and discharge piping or the pump base.

- Jackscrew for External Shaft Adjustment

Raw wastewater pumps, in particular, are subject to wear between the bottom of the impeller and the bottom of the volute. Clearance must be maintained to a few thousandths of an inch (generally 0.0015 inch or less), depending on the manufacturer. If this clearance is allowed to open up, rags and other debris may be caught between the rotating impeller and the stationary volute. When this happens, pump efficiency is lost since recirculation takes place. Not all of the flow being discharged from the rotating impeller goes into the discharge pipe, but rather, some of it recirculates back around to the suction and the impeller eye.

Most pump manufacturers offer replaceable wear rings for both the impeller and the casings (see Figure 8.42) which can be replaced as wear takes place. In addition, some method of adjustment of the clearances is generally provided on the outside of the pump. In this pump, jackscrews are provided so the adjustment can be accomplished externally. In some pump designs it is necessary to remove the upper bearing cap and shim underneath the bearing cap to vary the clearances.

- Double Row Thrust Bearings

Normally, in this type of design, the upper pump bearing in the power frame is a thrust bearing which can accommodate both upward and downward vertical thrusts along the axis of the shaft, including thrust created by the weight of the shaft and the impeller and thrust developed by hydraulic forces when the pump is operating. The lower bearing in the power frame is a radial-type bearing which is designed to accommodate radial thrust loads that occur perpendicular to the shaft (side-ways thrust). If a pump is operating normally, radial thrust loads are usually small; however, the bearings are generally designed to accommodate unusual circumstances which would increase radial thrust such as occasional plugging or unbalance in the impeller. In some cases, a shield called a slinger is provided to isolate the lower bearing from water spraying out of the stuffing box. The majority of bearings in pumps are regreaseable and fitted with zerk fittings.

- Packing Box

The stuffing box area of a centrifugal pump provides isolation between the liquid end of the pump and the atmosphere. One of two different methods of sealing may be used: packing or mechanical seals. Both sealing methods have advantages and disadvantages.

Packing

A stuffing box that uses conventional packing is characterized primarily by the slow, continuous leakage at the top of the stuffing box which is necessary to provide lubrication between the stationary packing and the rotating shaft element. Packing material can be graphite asbestos, Teflon, or some other braided material.

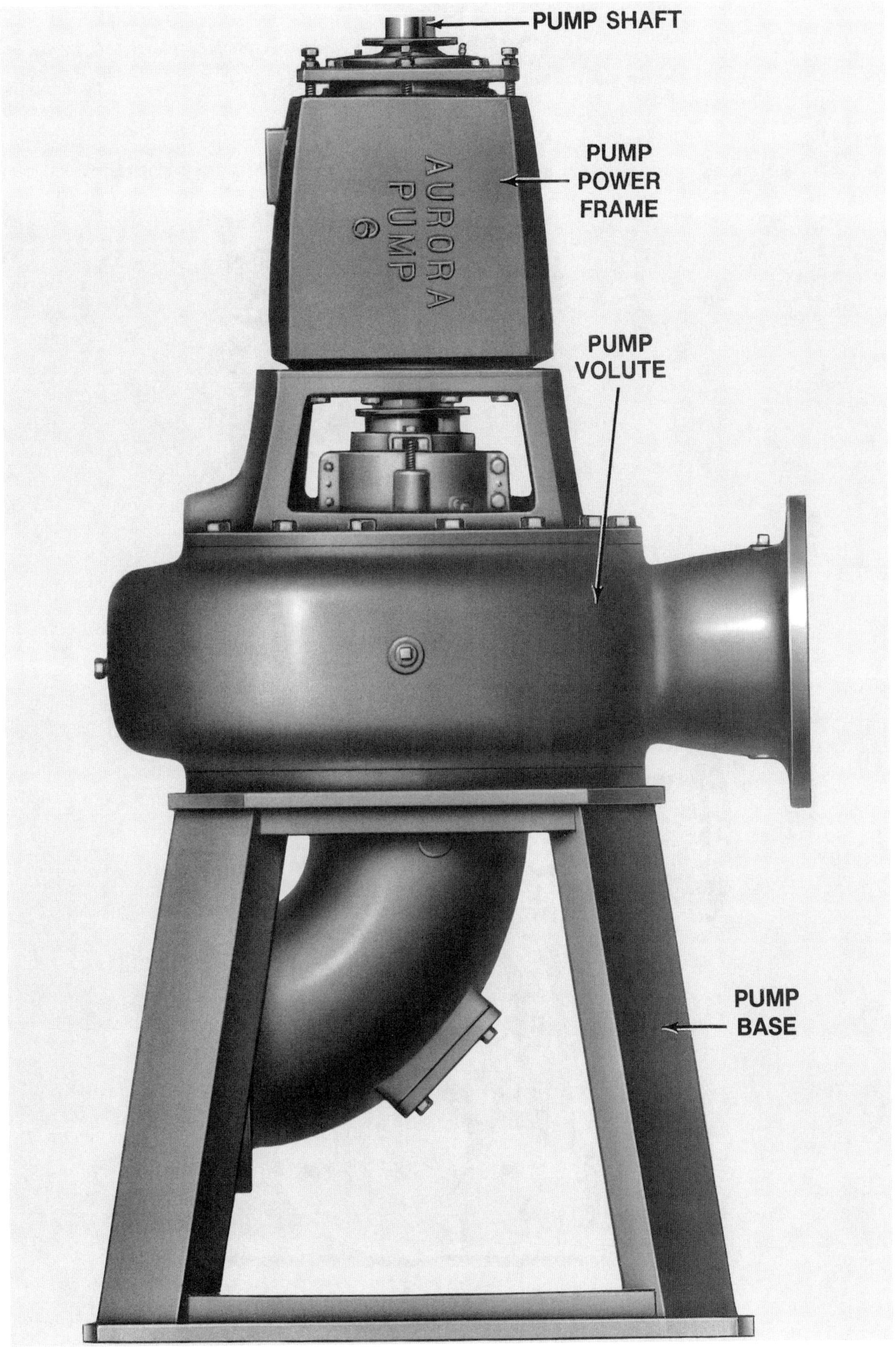

Fig. 8.39 Flexibly shafted vertical pump
(Courtesy of Aurora Pump)

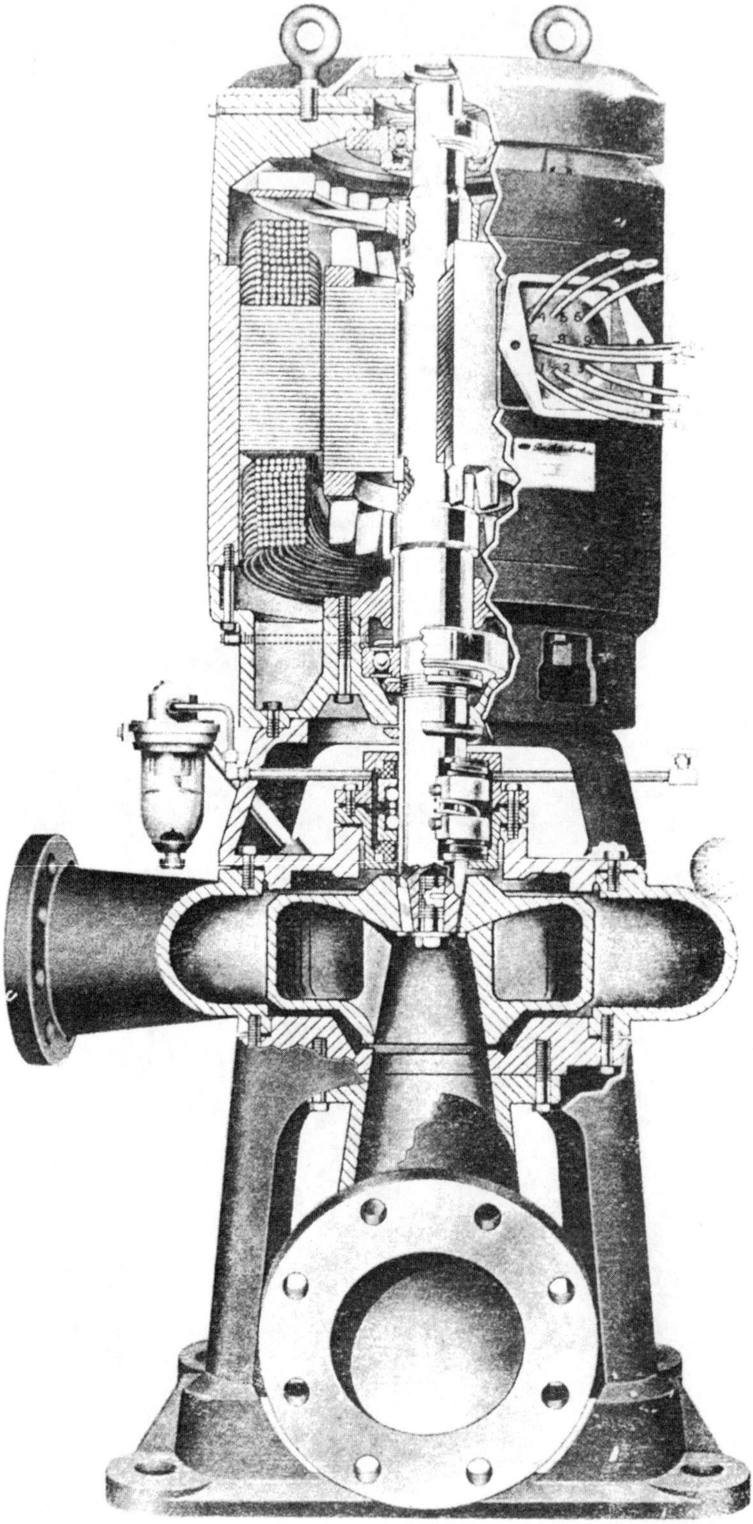

Fig. 8.40 Close-coupled motor pump

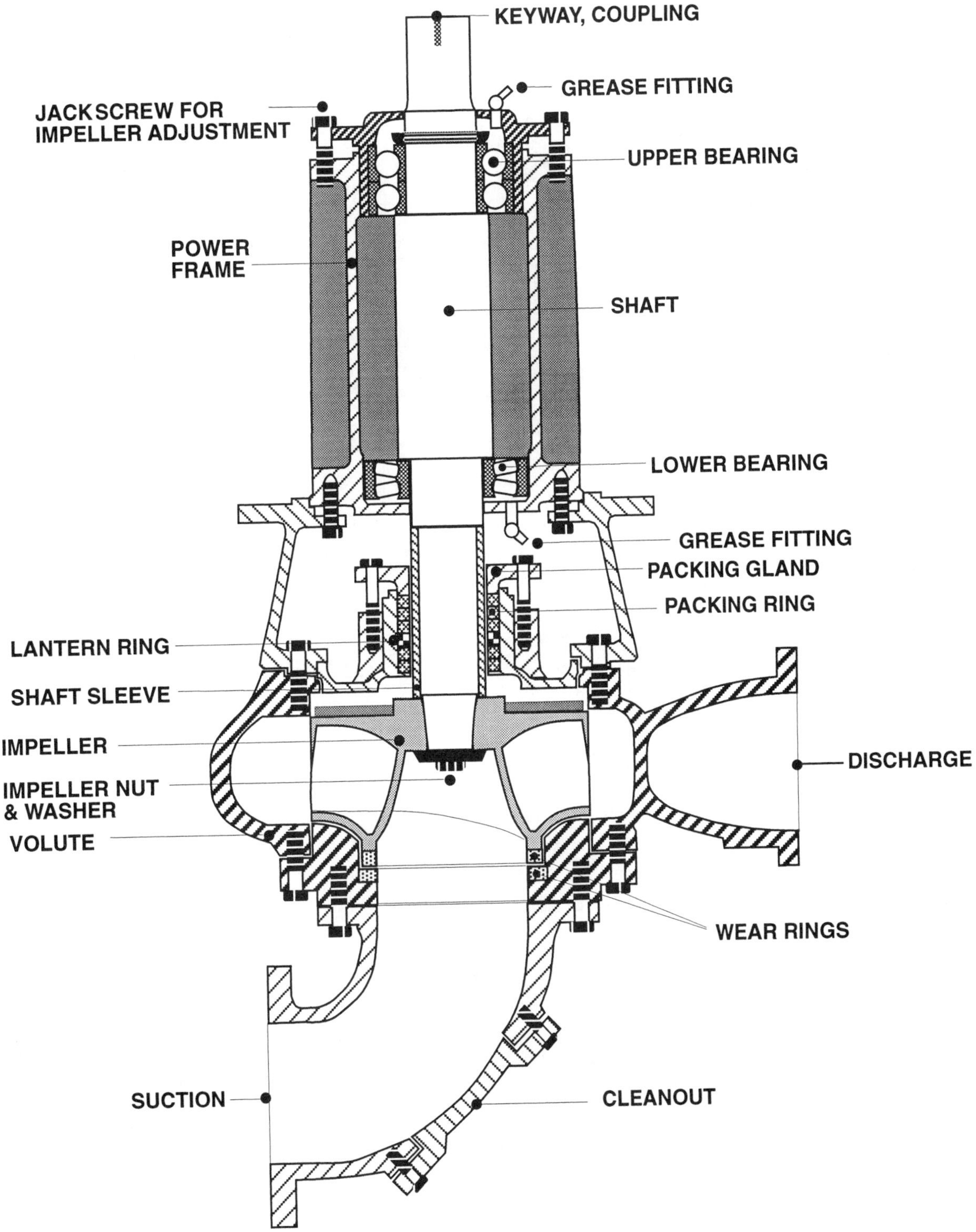

Fig. 8.41 Typical nonclog, vertical, raw wastewater centrifugal pump

Fig. 8.42 Wear rings

Some manufacturers have stopped making graphite asbestos packing because of current health concerns over asbestos; however, the asbestos in that type of packing is "bound" into the material as opposed to friable asbestos which breaks loose easily. There is little danger to the health of operators from the use of graphite asbestos packing.

A lantern ring is usually installed in the stuffing box. This is a metal or plastic perforated ring, completely surrounding the shaft, that distributes water into the stuffing box when an external water source is used as a method of lubrication. The alignment of the lantern ring must be maintained so that water continues to be distributed evenly throughout the stuffing box and the shaft. In the case where pumps are pulling a suction lift (where the normal water level of the wet well is below the center line of the impeller), the water distributed throughout the lantern ring serves the important function of providing a seal between the atmosphere and the liquid end of the pump. If this were not the case, air would be drawn into the volute and the pump would be unable to maintain a prime.

Stuffing boxes that use packing require frequent adjustments to minimize the amount of leakage through the top of the stuffing box while still providing adequate lubrication. Packing can also harden during long use and require replacement. When it loses its flexibility, it loses its ability to seal the stuffing box area. When packing replacement is required, it is necessary to remove the old packing and install new rings of packing.

The top of the stuffing box is called a gland and is normally held in place by adjustable cap screws. These screws allow the operator to increase the amount of pressure the gland exerts on the packing as well as remove the gland to install new packing.

A desirable feature of stuffing boxes is that they be manufactured in two parts or split (see Figure 8.43). This simplifies the job of packing replacement and permits shaft inspection at the point where the packing makes contact with the shaft or shaft sleeve.

Collection system operators must understand that the use of conventional braided packing requires that the stuffing box area clearances be maintained in terms of wear, surface roughness, or other factors which will affect the life of the packing and its ability to seal. For example, if the lower pump bearing allows radial movement, or if the shaft deflection occurs at the stuffing box, the packing will be "pounded" each time the shaft rotates and will lose its ability to seal effectively. Surface roughness on the stuffing box also affects the ability to seal. If the clearance between the liquid end and the bottom of the stuffing box is excessive, then as pressure is applied to the gland, packing can be extruded into the pump area. This results in the need for more frequent replacement of the packing and a reduced ability of the packing to seal the stuffing box area.

PACKING ADVANTAGES

1. Over the short term, packing may be less expensive than a mechanical seal; however, over the long run, when labor and material costs are considered, this may not be the case.
2. Pumps can generally accommodate some mechanical looseness from bad bearings, tolerances, or clearances in the stuffing box area. Under these circumstances, however, more frequent replacement/adjustment of the packing will be required.

PACKING LIMITATIONS

1. Increased wear on the shaft or the shaft sleeve.
2. Continuous leakage of raw wastewater used as the lubricating method or clean water to the atmosphere which requires increased housekeeping. If the seal water comes from a domestic water supply, there must be an "air gap" between the domestic supply and the seal water piping to prevent contamination of the public water supply from backflows.
3. Increased labor required for adjustment and replacement of packing.

Mechanical Seals

A second method of sealing the stuffing box area is a mechanical seal, the most common being the double mechanical seal with two sealing areas (see Figure 8.44). Double mechanical seals consist of the following components:

1. A stationary ring inserted into the lower part of the stuffing box and held in place by o-rings. Typically, this is a carbon graphite material.
2. A rotating element installed on the shaft using o-rings to secure the element to the shaft which then rotates with the shaft. This element is normally ceramic and may have some type of hardened metal material on the face. Both the stationary element and the rotating element have very highly polished faces, which contact each other through a very thin film of lubricating liquid, usually clean water. As the shaft rotates, it is the interface between the rotating and the stationary elements which provides the method of sealing. In a double mechanical seal, the rotating and stationary elements are duplicated at the top of the stuffing box, but reversed.
3. The next element is a compressed stainless steel spring which exerts a continuous pressure against the upper and lower rotating elements and maintains a positive pressure of the rotating faces against the stationary faces.

Most mechanical seals are not designed to run dry; they need water for lubrication and to dissipate heat generated at the faces of the seal. The fluid must be clean since any abrasive material introduced into the stuffing box would cause rapid wear and premature seal failure. Water may be supplied from an external clean water source such as a utility's drinking water system or a well. An appropriate backflow-prevention device or air gap must be installed to prevent contamination of the clean water source with raw wastewater in the event of a seal failure. Filtered raw wastewater piped from the pump discharge through a filtering mechanism into the stuffing box area is common. Most seals must be lubricated at all times when the pump is operating.

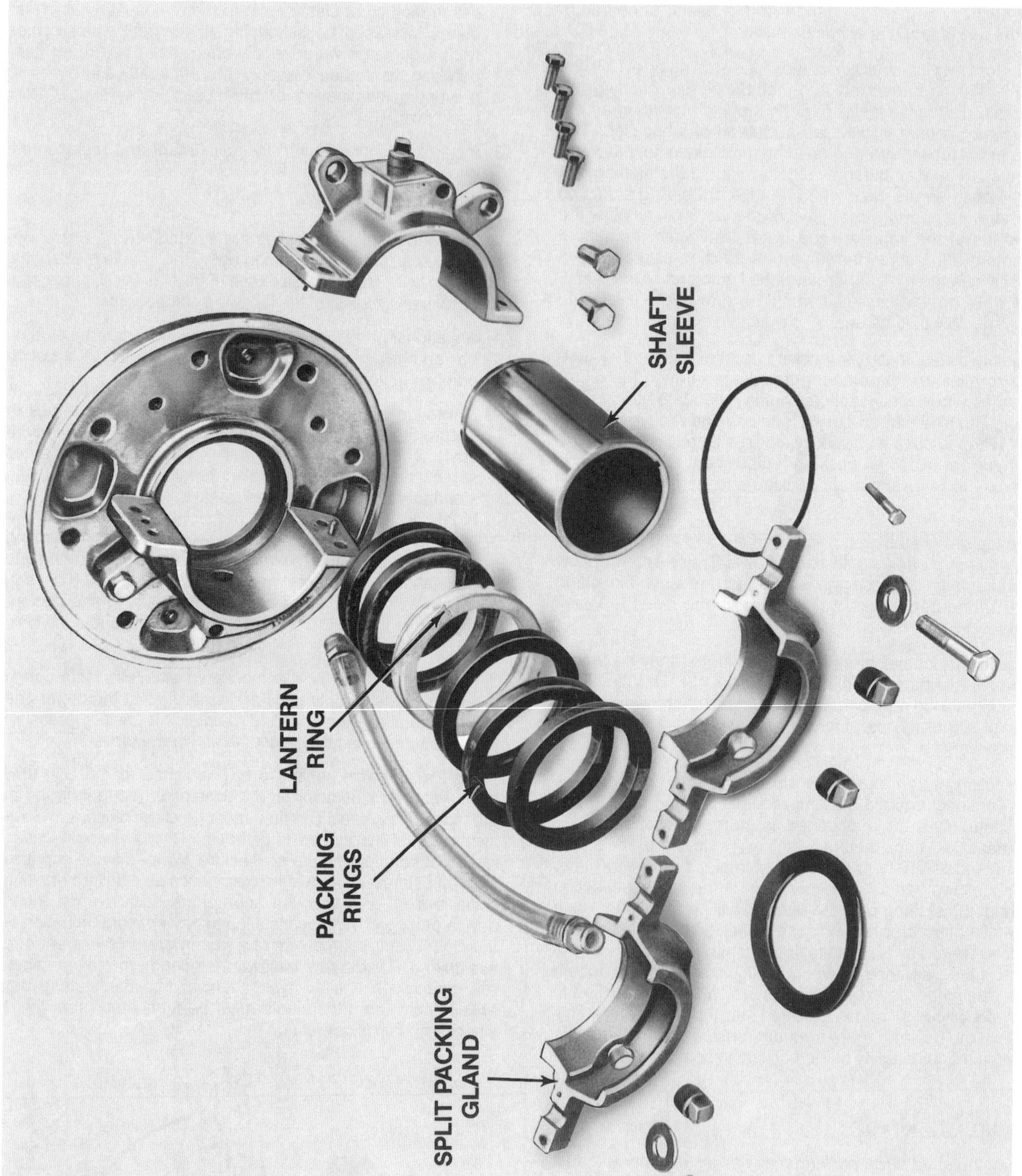

Fig. 8.43 Split packing box

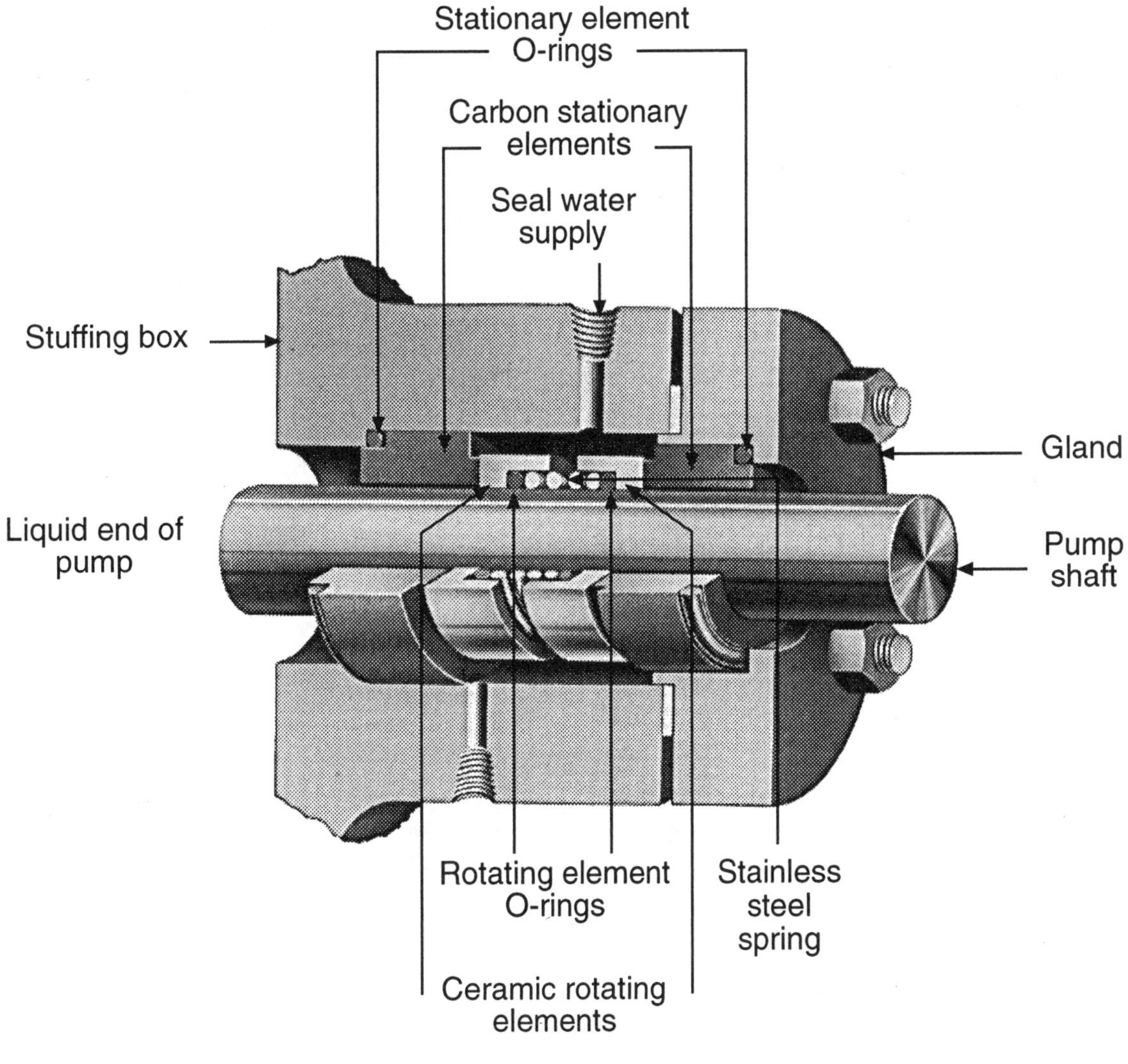

Fig. 8.44 Double mechanical seal

ADVANTAGES OF MECHANICAL SEALS

1. Mechanical seals are virtually maintenance free with no adjustment required or normally available,
2. Leakage of raw wastewater into the pump station out of the top of the stuffing box is eliminated,
3. Less wear on the pump shaft or shaft sleeve,
4. Less power consumption, and
5. When installed and operated properly, mechanical seals will last for several years before replacement is required.

DISADVANTAGES OF MECHANICAL SEALS

1. Initial cost is significantly higher than conventional packing and can range from several hundred to several thousand dollars, depending on the size, and
2. More skill and care by the operator are required when replacing mechanical seals.

The following is a checklist for cost-effective installation and use of mechanical seals:

1. Use cartridge-mounted seals. They are easy to install and do not damage either the seal or the pump.
2. Purchase standard off-the-shelf seals to ensure seals will be available from suppliers, even in an emergency.
3. Return seals to the original manufacturer for repair.
4. Obtain bearing isolators (labyrinths) or magnetic seals to protect the bearing from intrusion of moisture or debris.
5. Use synthetic lubricants in all pumps, motors, bearings, and compressors. Synthetic lubricants are very cost-effective because they lower the operating temperature, tolerate water intrusion better, lower high-frequency vibration, and last two or three times longer than the low-priced mineral oils.
6. Every time the motor or pump is worked on, be sure to perform precision coupling alignment by using a dial indicator or laser to ensure long-term service life of bearings and mechanical seals.

- Shaft Sleeve (Figure 8.41)

A desirable feature of a raw wastewater pump is a replaceable shaft sleeve which is keyed to the shaft in the stuffing box area. This allows for replacement of the sleeve rather than the entire pump shaft when wear occurs in the stuffing box area because of packing. Under certain circumstances wear will occur when using mechanical seals, particularly if the lower seal face fails, thus allowing raw wastewater to fill the stuffing box.

Different types of corrosion and pitting of the shaft sleeve can also occur even when the seal is operating normally, so that when seal replacement is required, it may be necessary to replace the shaft sleeve or resurface it in order to provide a positive sealing surface for the rotating element o-rings. Shaft sleeves may be manufactured from brass, hardened stainless steel (440 series), or conventional stainless steel (316 series).

It is also possible in some cases to have local machine shops manufacture shaft sleeves with a customized coating such as ceramic or hardened metal materials through flame spray technology. This method of manufacture uses a base material such as brass for the sleeve. The shaft sleeve is slightly undercut and then either molten ceramic or hardened alloy is sprayed onto the surface. The sleeve is then machined and polished. Ceramic material is frequently used in pumps with conventional packing since it significantly increases the life of the shaft sleeve. Polished hardened alloys are used with mechanical seals since the polished surface provides a positive sealing method between the o-rings used for the installation of the rotating element portion of the mechanical seal.

- Shaft

Two key elements of the pump shaft are adequate size to prevent breaking, and a tapered end (which most manufacturers use) where the impeller attaches to the shaft. Shaft dimensional tolerances must be maintained because of critical clearances at the upper and lower bearing, the stuffing box area and the tapered portion at the impeller end. The shaft must be carefully handled during pump assembly and disassembly to prevent damage which would cause the shaft to deflect; for example, 0.002 inch of deflection is the maximum limit at the stuffing box for long packing and mechanical seal life.

- Impeller

Raw wastewater pump impellers are designed with very large openings at the eye of the impeller passages and discharge areas in order to pump solids in the wastewater. Each pump should be equipped with a bottom *VOLUTE*[12] drain valve and line to the sump and a top or middle volute access inspection port for cleaning the pump. Large horizontal pumps should have a vent valve on top of the volute for priming purposes.

8.163 Basic Pump Hydraulics

In order to better understand hydraulic concepts that relate to pumps, operators need to be familiar with the terms normally used in working with hydraulic concepts.

- Head

Head is a measurement of the energy possessed by the water at any point in a hydraulic system. This is an indication of the pressure or force exerted by the water. Head can be expressed in feet or meters to represent the height of water above some reference point or it can be expressed as pressure in pounds per square inch gage (psig). Feet of head divided by 2.31 equals pounds per square inch of head. Three types of head are important in pumping: pressure head, velocity head, and elevation head.

- Pressure Head

Pressure head is the amount of energy in water due to water pressure. The reading on the pressure gage in psi can be converted to feet by multiplying it by 2.31 ft/psi.

- Velocity Head

Velocity head is the amount of energy in water due to its velocity or motion. The greater the velocity, the greater the energy and the greater the velocity head. The velocity head may be determined by the following formula:

$$\text{Velocity Head, ft} = V^2/2g$$

Where:

V = Velocity, ft/sec

g = acceleration due to gravity = 32.2 ft/second^2

Or:

$$V^2/2g = V^2/64.4 \text{ ft/second}^2$$

[12] *Volute (vol-LOOT). The spiral-shaped casing which surrounds a pump, blower, or turbine impeller and collects the liquid or gas discharged by the impeller.*

For example, the water velocity in a pipeline on the discharge side of a pump is 6 feet/second. The velocity head is:

$$\text{Velocity Head, ft} = \frac{V^2}{2g}$$

$$= \frac{(6 \text{ ft/sec})^2}{2(32.2 \text{ ft/sec}^2)}$$

$$= 0.56 \text{ ft of head}$$

A centrifugal pump's volute or diffuser vanes change the velocity head produced by the rotating impeller in the fluid to pressure head.

- Elevation Head

Elevation head is a measurement of the energy that water possesses because of elevation.

- Friction Head Loss

Friction head loss is the energy that water loses from friction while moving from one point in the system to another through the pipelines and valves. It can be significant if the pipe surface is roughened by corrosion, slime growths, or sediment. Other head losses are caused by the water suddenly changing direction or velocity as a result of valves, bends, and reducers. Friction head loss is one of the simpler head losses to calculate or it simply can be determined using tables.

Friction head loss in the piping system is a function of the velocity head multiplied by a loss coefficient for each pipe fitting, valve, and length of pipe. Tables are available that give the loss coefficients for various fittings. Charts are also available which list pipe friction losses for varying flow rates in different pipe sizes and materials or in terms of equivalent pipe lengths. (See Figures A.1 and A.2 in the Appendix of *OPERATION AND MAINTENANCE OF WASTEWATER COLLECTION SYSTEMS*, Volume I, for examples of a friction loss chart and a conversion table.)

- Total Dynamic Head

Total dynamic head (TDH) is a measurement of the amount of energy that a pump must develop to move a liquid. TDH varies for each pumping system and for flow conditions within a specific pumping system. The term "dynamic" indicates that the water is in motion rather than static (not moving). This head is a measure of the total energy that the pump must impart to the water to move it from one point to another.

- Brake Horsepower

The input or brake horsepower (BHP) required by a centrifugal pump to discharge a given amount of water against a known total dynamic head can be calculated as follows:

$$\text{BHP} = \frac{(\text{Q x TDH})}{(3{,}960 \text{ x Eff})}$$

Where:

BHP = brake horsepower required

Q = pump discharge, in GPM

TDH = Total Dynamic Head, in feet

Eff = pump efficiency, decimal

3,960 = English units constant: 33,000 ft-lb/min/HP/8.34 lb/gal

- Cavitation

Cavitation occurs in pumps when the absolute pressure at the pump inlet drops below the vapor pressure of the water being pumped. Cavitation is likely to occur if modifications over time change the basic balance of the hydraulic system, if the system was not properly designed to begin with, or if the wrong pump is used for a particular application. Cavitation causes severe vibration in the pump and sounds like the pump is pumping gravel or marbles. Vibration can be severe enough to damage mechanical seals and bearings. Damage occurs at the eye of the impeller, the back shroud of the impeller and the casing.

Two types of cavitation can occur: suction or eye cavitation and impeller vane tip cavitation. Suction or eye cavitation occurs when a partial vacuum is created at the eye of the impeller and vapor bubbles form at the pump inlet. When the pressure at the eye drops below the vapor pressure of the liquid, entrained (trapped) air in the liquid expands and forms bubbles since it is no longer under sufficient pressure. These vapor bubbles are carried into the high pressure zone inside the impeller where they collapse or implode under the higher pressure. Water surrounding the vapor bubble rushes to fill the cavity left by its collapse with such force that a hammering action occurs. The implosion can create pressures in excess of 10,000 psi which cause pitting on the case and impeller surfaces. Suction or eye cavitation is caused by (1) excessive suction lift, (2) partially blocked suction line, or (3) insufficient Net Positive Suction Head (NPSH).

Impeller vane tip or discharge cavitation occurs when the discharge head required is too high. This causes the pump to operate on the left side of the performance curve outside its operating range. As the impeller tip is passing through the water, a large pressure differential develops on the two sides of the impeller tip. Vacuum bubbles develop on the tip of the impeller. High water pressures cause the bubbles to collapse with great force on the front side of the tip of the impeller causing pitting at the tip of the impeller vane.

- Net Positive Suction Head

Net positive suction head (NPSH) can be defined as the total absolute suction head at the inlet of the pump minus the vapor pressure of the water. Cavitation occurs when the NPSH available ($NPSH_A$) at the pump inlet is less than the minimum required NPSH ($NPSH_R$). The minimum required NPSH is determined by pump manufacturers in tests of each pump model. $NPSH_R$ and $NPSH_A$ are critical factors in the selection of pumps. In many existing pump installations, inadequate NPSH contributes to high maintenance and poor performance of pump systems.

QUESTIONS

Write your answers in a notebook and then compare your answers with those on page 119.

8.1S What are the two major types of pumps?

8.1T Which type of pump is most commonly used in raw wastewater lift stations?

8.1U Under what conditions is cavitation likely to occur?

8.1V Why must pump seal water from a domestic water supply have an air gap?

8.1W Why should lift stations be equipped with at least two pumps?

8.164 How to Read and Understand a Centrifugal Pump Selection Curve

Operating and maintenance problems in pump stations are frequently the result of the operator's inability to understand centrifugal pumps. They are basically simple machines that do not require extensive technical knowledge to understand and interpret their operation.

The first step in understanding centrifugal pumps is to understand a pump selection curve. Although pump curves look complicated and confusing (see Figure 8.45), they contain valuable information about how a pump will operate under specific conditions. The ability to read, understand and interpret centrifugal pump curves is necessary to correctly diagnose problems and ensure that the hydraulic systems are operating at maximum efficiency. Operators must be able to read and interpret pump curves in order to verify pump operating conditions, troubleshoot existing pump systems, evaluate pump repairs, and evaluate pump replacement needs.

Pump curves are normally available from two sources:

1. Manufacturers' catalogs contain standard pump curves. These curves may vary slightly from pump to pump for a given pump size and therefore actual pump performance is not guaranteed to match the standard curve; and
2. Factory test pump curves are supplied with new pumps when a certified factory test is requested. This curve is based on an actual test at the factory under laboratory conditions using the actual pump supplied.

Centrifugal pumps will discharge a certain quantity of water against a given total dynamic head (TDH) or head. If the head changes, the quantity of water that the pump discharges will also change. A pump curve graphically shows the quantity of water a particular centrifugal pump will discharge at several different heads. The pump curve also shows a plot of pump discharge versus required horsepower and pump discharge versus pump efficiency. As with total dynamic head (TDH), there is only one point on the horsepower curve for a given discharge. A greater or lesser discharge will require a corresponding greater or lesser horsepower input to the pump, called brake horsepower.

Every centrifugal pump operates at less than 100 percent efficiency, generally between 50 percent and 85 percent. Some input power is lost in operating the pump and does not reach the water. The pump's efficiency varies with discharge and can decrease dramatically if the pump is operated at other than its optimum discharge.

To better understand how the performance characteristics of a specific pump influence each other, let's look at the pump curve shown in Figure 8.45.

This particular curve is for a nonclog, end suction solids pump that would be used in a wastewater pump station where solids are normally encountered. The bold numbers **1 – 10** each represent a characteristic of this particular pump.

1 12 x 14 x 22 is the size of the pump:

- 12 is the discharge size in inches,
- 14 is the suction size in inches, and
- 22 is the volute size in inches.

In some pumps, the suction and discharge size may be the same. The suction size must be at least as big as the discharge size and is always the larger of the first two numbers when they are not the same. The reason for this is that the suction side of the pump must be able to draw as much water into the pump as is discharged. Also a larger suction diameter has lower flow velocities, thus lower suction friction losses which reduce the negative suction lift on the pump.

2 Factory Information. Max Spheres - 6: This indicates that the impeller is capable of handling a maximum solid sphere size of 6 inches. Imp Patt No: This is the impeller casting pattern number for this model pump. Case Patt No: This is the casting pattern number for the pump casing.

3 700 RPM. This is the operating RPM (revolutions per minute) of the pump that the curve data is based on. The 700 RPM is based on a typical Design B squirrel cage induction motor as the driver for the pump. The RPM of the motor is determined by the slip or the difference between synchronous speed and actual full load operating RPM for the motor. In this case, the synchronous speed is 720 RPM for a 10 pole motor and the slip is approximately 3 percent, which results in a full load operating RPM of 700 for the motor and therefore the pump. If the motor is operating at a different RPM than 700, then the curves will have to be adjusted accordingly. A NEMA Design D motor can have as much as a 15 percent slip, which would result in an operating RPM of 612 instead of 700. Always verify the motor operating RPM being used with the pump when using curves.

4 Total Dynamic Head (TDH). This is the total energy the pump is capable of delivering to the water as it passes through the pump. TDH is normally expressed in feet of water as opposed to pounds per square inch (psi). The conversion from psi to feet of water is:

$$\text{Head, ft} = (\text{Pressure, psi})(2.31\,\text{ft/psi})(\text{Specific Gravity of Liquid Pumped})$$

Normally for water and wastewater, a specific gravity of 1 is used.

5 Capacity in Gallons per Minute is the flow the pump will produce.

6 This is the Brake Horsepower curve for the pump. In this case, depending on the head and capacity of the pump, a 30, 40, 50, 60, 75, or 100 horsepower motor could be used.

7 This is the efficiency curve for the pump. In this case, the pump efficiency shown can range from 70 percent to 84 percent, a significant difference depending on where the pump is operating. Normally a pump is selected based on the best efficiency of the pump. This is called the Best

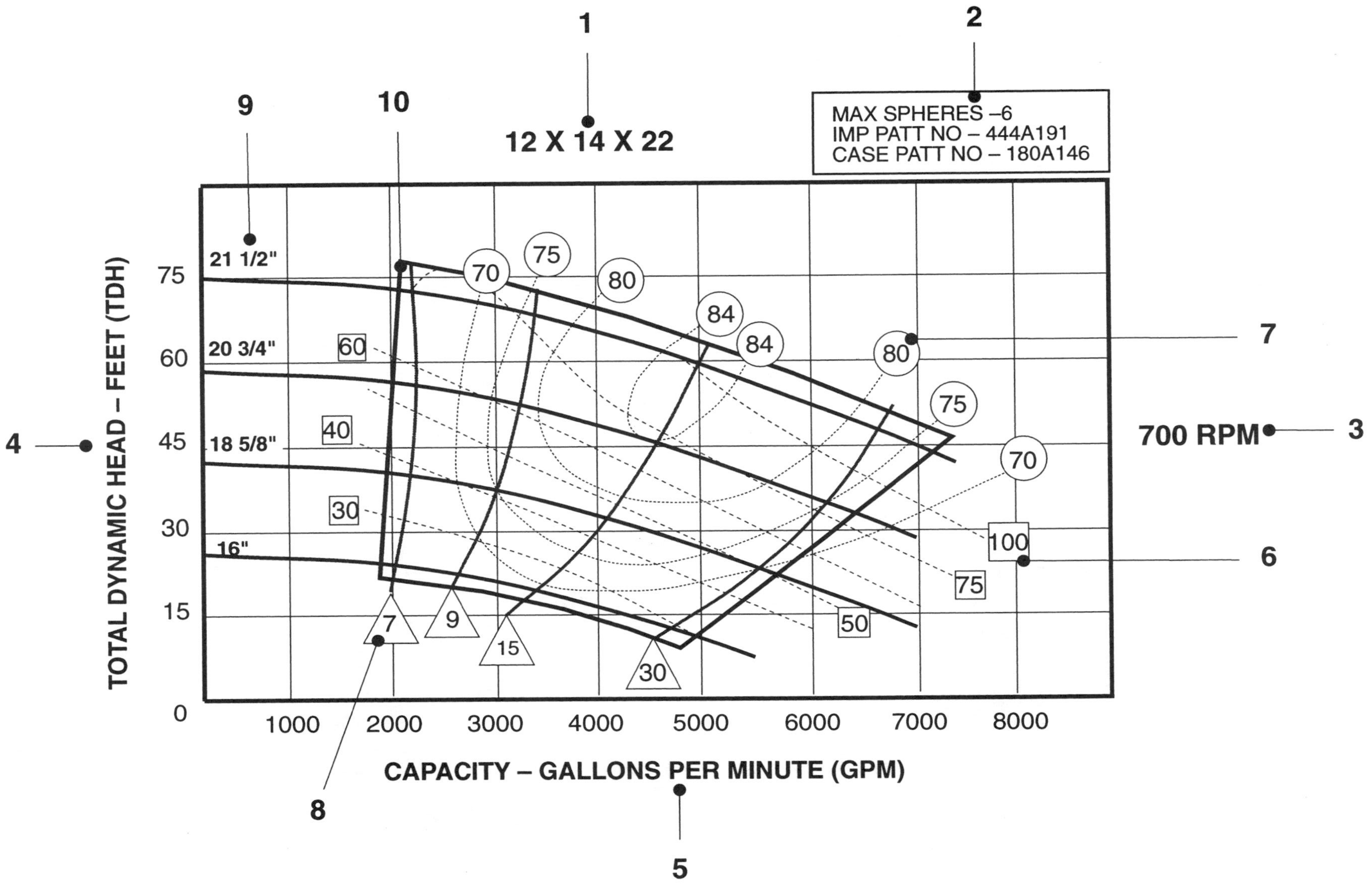

Fig. 8.45 Typical manufacturer's pump performance curve

Efficiency Point (BEP) of the pump, in this case 84 percent. You may not always be able to match the required head and capacity of your system with a pump operating at the BEP. The options are to see if a different manufacturer has a higher efficiency pump for those conditions; modify the system hydraulics to the BEP of the pump which is difficult to do in existing installations; see if the same pump is more efficient at a different RPM; or see if a different size pump is more efficient.

8 This is the Net Positive Suction Head required by the pump ($NPSH_R$), expressed in feet. $NPSH_R$ for this pump ranges from 7 feet to 30 feet.

9 $21\frac{1}{2}$ inches is the impeller diameter. Each pump has the ability to operate with different impeller diameters in order to match the head and capacity required to be delivered by the pump. This pump can have impellers ranging in diameter from 16 inches to $21\frac{1}{2}$ inches. Although this curve only shows 4 different diameters for the impeller, they can be trimmed to any diameter between 16 inches and $21\frac{1}{2}$ inches. If we trimmed an impeller to a diameter of $21\frac{1}{8}$ inches, for example, this would produce a curve exactly halfway between the curves shown for the $20\frac{3}{4}$-inch diameter and the $21\frac{1}{2}$-inch diameter impellers.

10 This outline represents the operating area of the pump recommended by the manufacturer. Operation to the left (toward shutoff) will result in higher bearing loads, possibly resulting in the early failure of the pump bearings, or tip cavitation may occur. Operation to the right of the outline will most likely result in suction cavitation and pump damage. Operation outside the envelope also results in operating in a much less efficient area of the pump.

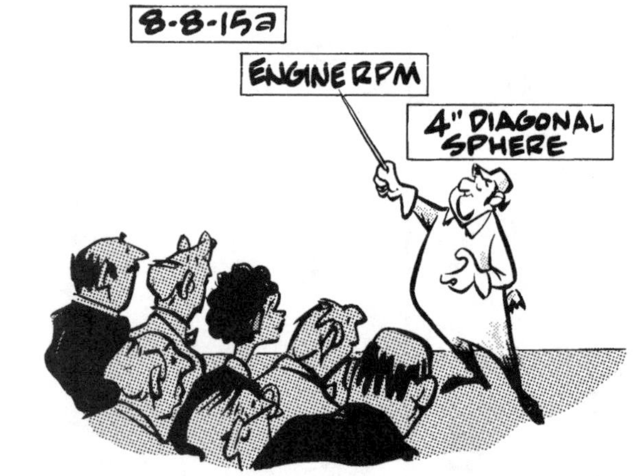

SIMPLIFIED PUMP CURVE

Figure 8.46 is the same pump as shown in Figure 8.45 but showing the performance curve for a $20\frac{3}{4}$-inch diameter impeller only. It is the exact same pump curve except that it has been simplified in the following ways:

- Only one impeller diameter is shown, $20\frac{3}{4}$ inches,
- $NPSH_R$ curves for 7 and 9 feet have been removed,
- All horsepower curves except 75 have been removed, and
- The 70 percent efficiency curve has been removed.

This is what the curve for a 12 x 14 x 22 pump, operating at 700 RPM, with a $20\frac{3}{4}$-inch diameter impeller would look like. This type of curve would be used for a pump actually installed in the field.

If we now take the same pump curve and develop a "system" curve, we can analyze pump performance under different operating conditions. Using Figure 8.47 as an example, we will develop the operating conditions of the pump based on the static head, dynamic head, and capacity, and then calculate the BHP. Next we will change the suction head conditions. This change will affect the static head and therefore the operating point of the pump, and we will be able to see how BHP, efficiency and $NPSH_R$ are affected by the change in static head.

1 This is the beginning point in developing what is called a "system curve." A system curve is a representation of the combination of static and dynamic heads and determines where the pump will operate on the curve. The initial point is approximately 33 feet, which represents the static head. Static head can be thought of as the vertical distance from the water surface elevation on the suction of the pump to the highest elevation that the pump must pump to in the system, as shown in Figure 8.48. In this case, the vertical distance is 33 feet.

2 This part of the system curve is developed by calculating the dynamic losses through the piping system and pump. Tables which have been developed on the basis of friction losses through pipe are normally used to determine dynamic losses. Notice that the system curve is not in a straight line. Points on the curve are developed by determining the friction head loss at different capacities, for example, at 1,000 GPM, 3,000 GPM, and 6,500 GPM. This loss is added to the static head and the resulting curve is superimposed on the pump curve.

3 The point where the system curve intersects the impeller curve is where the pump will operate, in this case 4,600 GPM at 45 feet of head. One of the principal concepts in understanding pump curves is the fact that the pump will operate along the impeller curve depending on the resistance in the system. For example, imagine a valve on the discharge side of the pump that is completely closed. Obviously, there will be no flow and a discharge gage on the pump would read approximately 59 feet, which corresponds to the impeller curve for no flow. (This is referred to as shutoff head.) As we open the valve, the pump begins to pump and the discharge pressure drops as the pump "slides" down the impeller curve, again depending on the amount of resistance offered by the valve. As we continue to open the valve, the flow increases and pressure continues to decrease. What we can conclude is that the operating point of a pump is not determined by the head and capacity specified on the nameplate of the pump. In the field, the pump will operate along the impeller curve depending on the static and dynamic conditions, that is, how high the pump has to deliver the water and the resistance due to friction losses of the system.

Let's go back now to **3**, the point where the system curve intersects the impeller curve. Under the conditions we have established, this pump in this system

- Will deliver 4,600 GPM, **4**,
- At a TDH of 45 feet, **5**.

This is known as the duty point of the pump. The pump will operate at 84 percent efficiency. It will require a net positive suction head (NPSH) of 15 feet, **6**, and require a 75 HP motor, **11**.

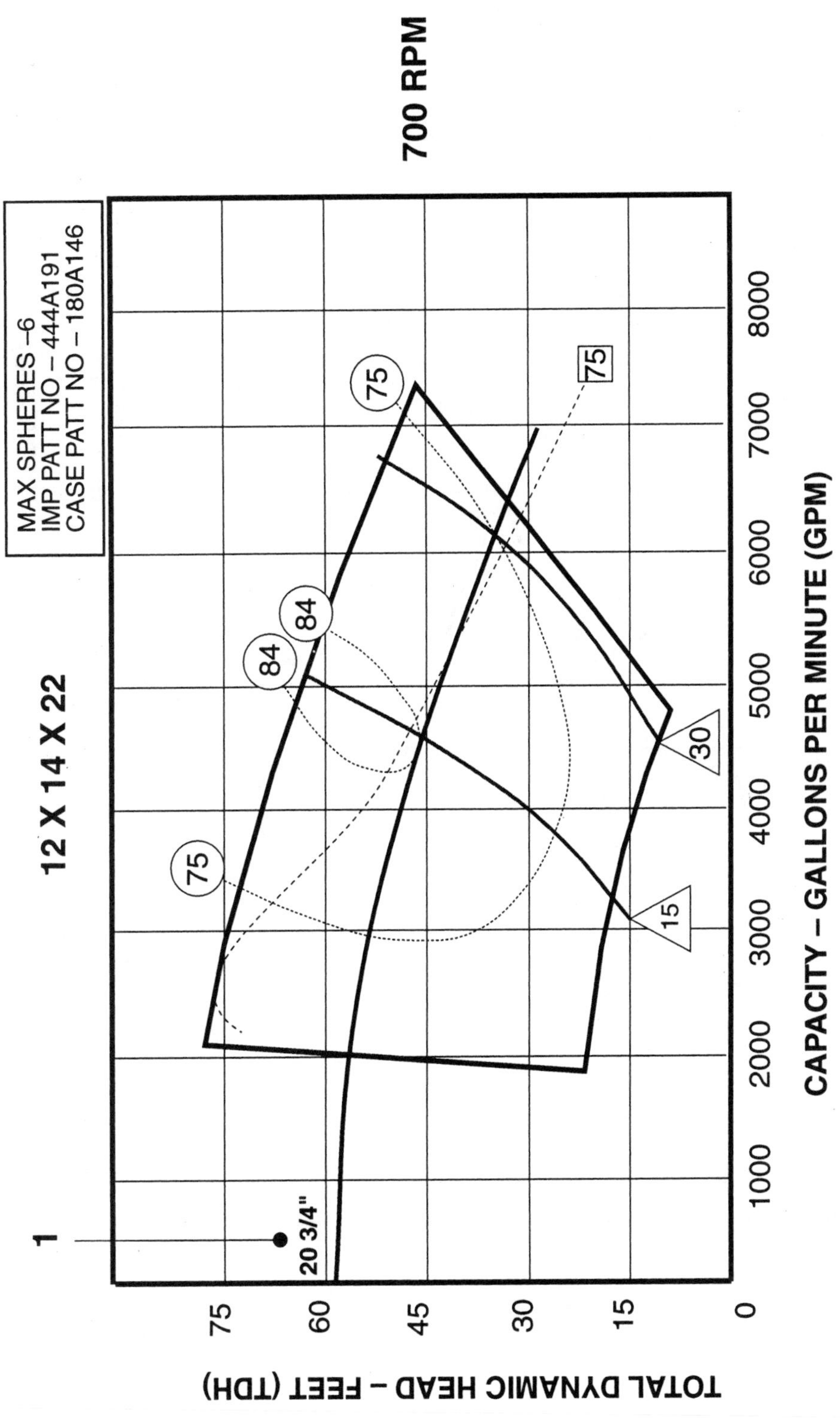

Fig. 8.46 Catalog pump curve for a 20¾-inch diameter impeller

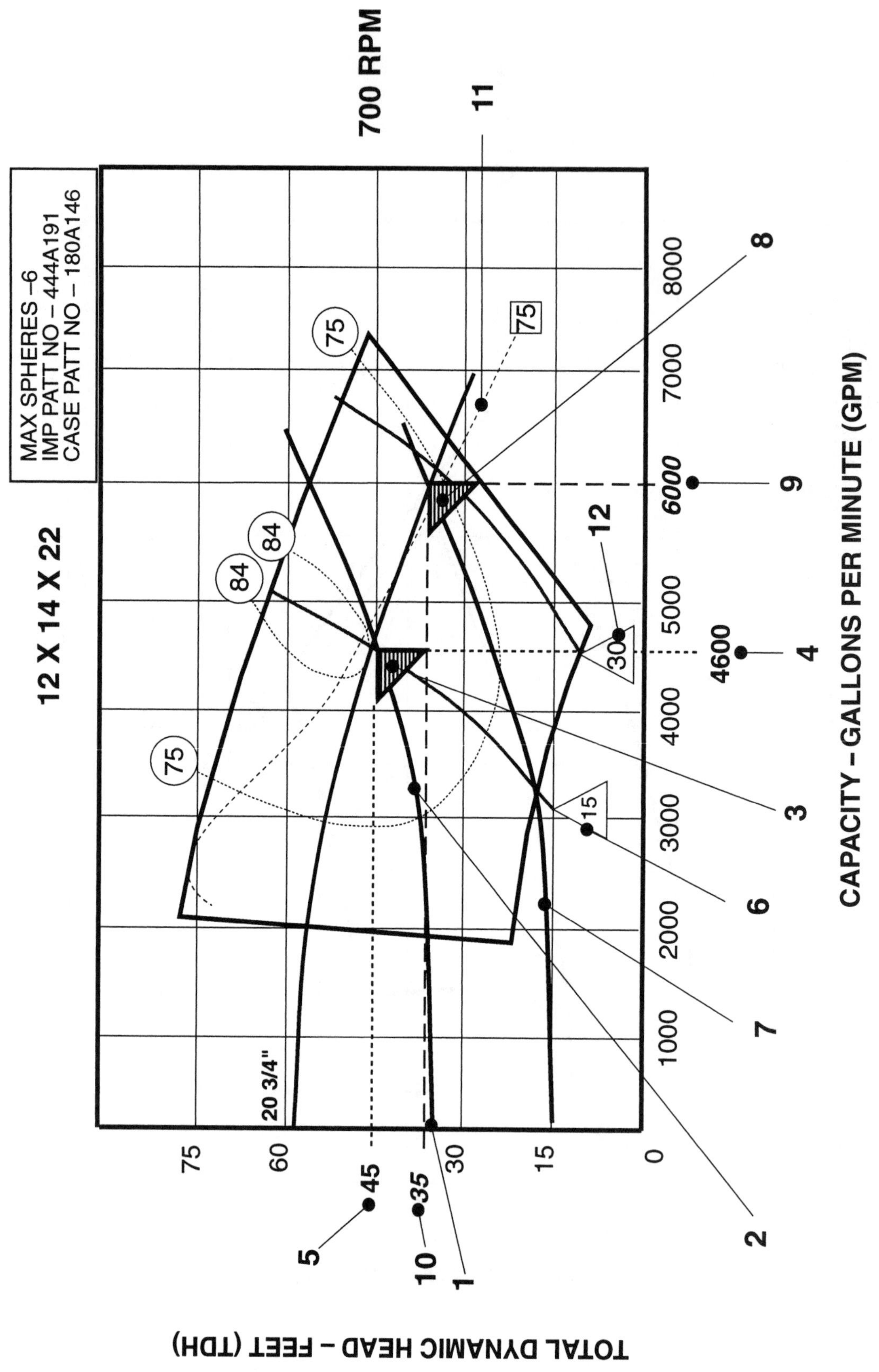

Fig. 8.47 Catalog curve with system curves

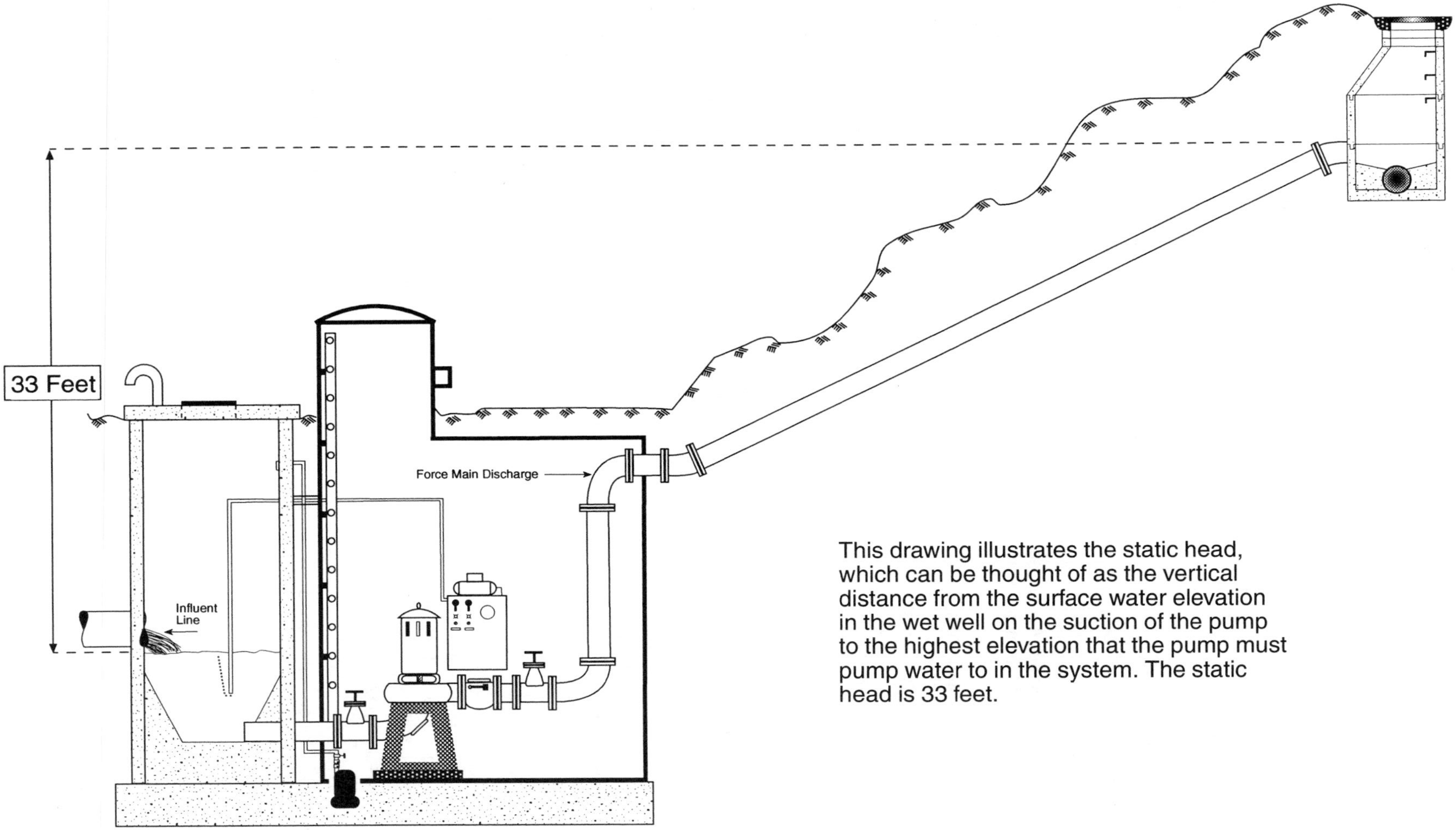

This drawing illustrates the static head, which can be thought of as the vertical distance from the surface water elevation in the wet well on the suction of the pump to the highest elevation that the pump must pump water to in the system. The static head is 33 feet.

Fig. 8.48 Static head for a pump system

The BHP of the motor can be calculated to determine actual operating HP required at this duty point as follows:

$$BHP = \frac{TDH \times Q}{3{,}960 \times E}$$

Where:

TDH = TDH, feet

Q = Flow, GPM

3,960 = Conversion Factor Constant

E = Pump Efficiency, decimal percent

$$BHP = \frac{(TDH, ft)(Q, GPM)}{(3{,}960)(E)}$$

$$= \frac{(45\ ft)(4{,}600\ GPM)}{(3{,}960)(0.84)}$$

$$= 62.2\ BHP\ required$$

Therefore, a 75 HP motor is adequate. Always select the next larger available motor. A 62.2 BHP motor is not manufactured.

7 Now we are going to change the static head conditions in the system. This is a situation which occurs in the field when the suction head is changed by a change in the water elevation on the suction side, or it can occur if field construction changes the highest elevation on the discharge side. In this case, we will raise the level in the wet well 18 feet. This is not an unusual condition during peak wet weather flows in a system that has excessive I/I. Raising the level in the wet well reduces the static head by 18 feet, which means the head or pressure delivered by the pump is 18 feet less. The system curve now shifts down to an initial point which is 18 feet less than the original 33 feet, or 15 feet, as shown on Figure 8.49.

The dynamic losses are again calculated and as shown, the new system curve becomes slightly steeper beyond 4,600 GPM. The steeper curve reflects a short piping system and/or pipe of large enough diameter that friction losses do not increase significantly with an increase in flow through the pipe.

8 The new system curve now intersects the impeller curve much farther down and establishes the new duty point of 6,000 GPM, at a TDH of 38 feet, **9**, pump efficiency is 75.5 percent, **10**, and NPSH required is 30 feet, **12**. If we calculate the BHP required by the pump at this point:

$$BHP = \frac{(TDH, ft)(Q, GPM)}{(3{,}960)(E)} = \frac{(38\ ft)(6{,}000\ GPM)}{(3{,}960)(0.755)} = 76.2$$

Therefore, a 75 HP motor is now overloaded and inadequate.

This example illustrates how the performance of a pump is defined by the suction and discharge conditions. This may or may not be the same operating point that is listed on the pump nameplate. The nameplate indicates only what the pump is designed for and capable of pumping under specified conditions.

MOTOR AND CONTROL TERMS

- Ambient Temperature

  Temperature of the surrounding environment, such as the normal air temperature within a pump station room.

- Ampere (amp)

  An ampere is the unit used to measure the flow of electric current. If no current is flowing, only electrical pressure (voltage) is present. Once flow starts, amperes can be measured.

- Bearings

  Usually stainless steel balls set in grooved rings over one or both ends of the pump shaft. Pump bearings are usually regreasable; smaller motor bearings may be sealed bearings.

- Control Relay

  A device that allows low-power electrical signals to operate the ON/OFF switch for high-power equipment.

- Coupling

  A mechanical device, usually fabricated of synthetic material, used to connect the ends of the motor shaft and the pump shaft.

- Insulation

  Material used to separate electric components, such as the resin material surrounding the windings of the stator in an electric motor.

- Locked-Rotor Current

  The current drawn by the motor the instant the power is turned on while the rotor is still at rest. Also called the motor-starting current or inrush current.

- Ohm

  A measure of the ability of a path to resist or impede the flow of electric current.

- Overload Relay

  A switch device used to sense an overloaded electric motor and disconnect it from the power source before the motor is damaged.

- Relay

  An electrical device in which an input signal, usually of low power, is used to operate a switch that controls another circuit, often of higher power.

- Sensor

  The part of an instrument that responds directly to changes in whatever is being measured, then sends a signal to the receiver and indicator or recorder.

- Starter

  A motor control device that uses a push button switch to activate a control relay, which sends electrical current to the motor.

- Thermistor

  A semiconductor type of sensor that measures temperature.

- Thermocouple

  A sensor, made of two wires of dissimilar metals, that measures temperature.

- Volt

  A volt is a measurement of electrical pressure. This can be compared to the measurement of water pressure in pounds per square inch (psi). Common voltages are 110/

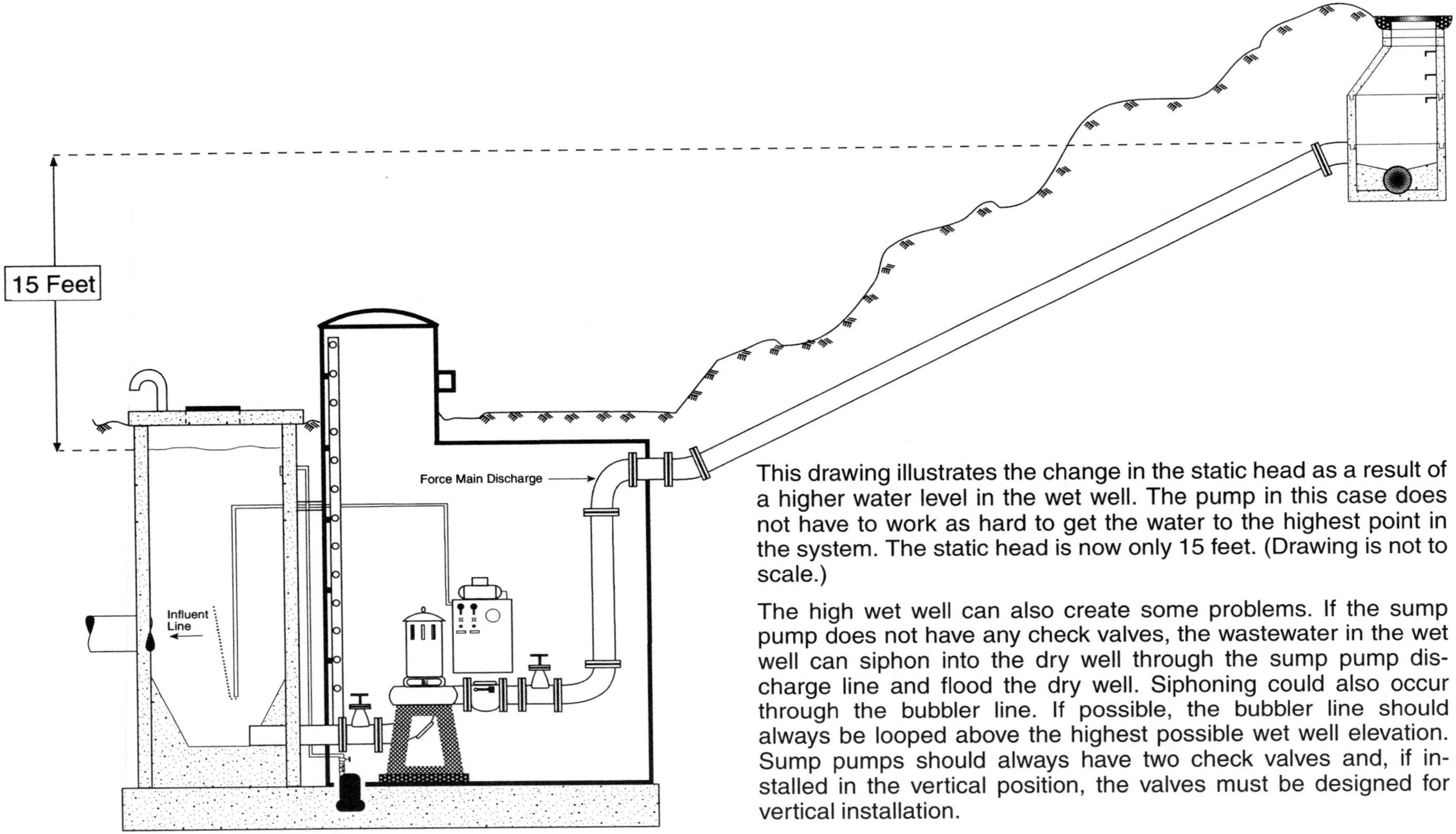

This drawing illustrates the change in the static head as a result of a higher water level in the wet well. The pump in this case does not have to work as hard to get the water to the highest point in the system. The static head is now only 15 feet. (Drawing is not to scale.)

The high wet well can also create some problems. If the sump pump does not have any check valves, the wastewater in the wet well can siphon into the dry well through the sump pump discharge line and flood the dry well. Siphoning could also occur through the bubbler line. If possible, the bubbler line should always be looped above the highest possible wet well elevation. Sump pumps should always have two check valves and, if installed in the vertical position, the valves must be designed for vertical installation.

Fig. 8.49 Static head resulting from changed wet well water level

120 and 220/240 V, although large motors may be supplied with 440/480 V and higher.

- Windings

  The wiring in an electric motor laid around the stator and embedded in an insulating resin material.

8.165 Sump Pump

The lower floor of the dry well should be provided with a sump and submersible pump to collect seal water and water used to hose down and clean up the area. This drainage water should be pumped out of the dry well sump and discharged at a high elevation into the wet well.

The discharge pipe of the sump pump must include at least one check valve or, for added safety, two check valves. In addition, if the check valve is located in the vertical piping run, it should be a valve designed specifically for a vertical installation. Generally it is desirable to have the sump pump discharge into the wet well above the highest possible elevation that water could reach under flooded conditions in the wet well. This minimizes siphoning back through the piping into the dry well. Normally this is not possible but the double check valve installation provides added protection. The dry wells of many pump stations have been flooded because of check valve failure which did not prevent backflow from the wet well.

It is a good idea to install a wire mesh screen around the sump area to screen out large pieces of material which could and will clog the sump pump. In addition, installing a union on the discharge piping will facilitate easy removal of the pump for maintenance and cleaning of the sump.

8.17 Lift Station Valves

Valves are of critical importance to lift station operation and maintenance but are frequently neglected during preventive maintenance. The major valves found in wastewater lift stations are:

1. Pump suction and discharge isolation valves (gate, plug, knife),
2. Discharge check valves (swing or ball check, wafer), and
3. Cross connect or pump discharge pipe manifold isolation valves (gate, plug).

Different types of valves are used for each of the above applications and several of them are discussed individually in detail below.

8.170 Gate Valves

Pump suction and discharge isolation gate valves are found immediately before and after the pump to allow isolation of the pump from the wet well and the force main during pump or check valve maintenance. The most common type of valve used for this application is a gate valve in which discs or gates are used to provide a shutoff. Figures 8.50 and 8.51 illustrate a non-rising stem gate valve (NRS) and an outside screw and yoke valve (OSY).

Normally, the non-rising stem valve is used in lift stations because, as the name implies, additional headroom is not required to accommodate a non-rising stem as the valve operates. The principle of operation for non-rising stem valves requires that sliding wedges, which are attached to a threaded shaft, are also attached to the discs or gates. The stationary portion of the valve has machined faces upon which the discs seat. As the wedge and disc assembly is lowered on the shaft into the flow area, the wedges slide against each other and provide wedging pressure on the back of the discs, which then seat against machined faces in the valve body.

In the open position, the wedge and disc assembly is raised into the upper bonnet portion of the valve where it is out of the flow and does not interfere with the flow through the valve. The threaded shaft normally is sealed, either through o-ring type seals or with conventional packing.

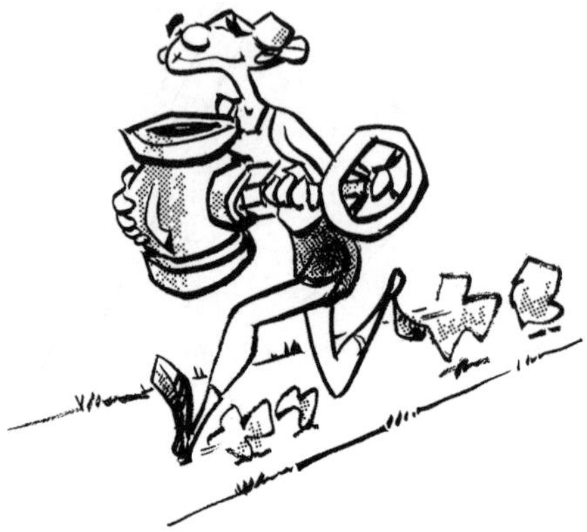

The valve should be frequently exercised, at least semiannually, in order to keep the threads on the shaft stem and the collar operational.

As can be seen from Figures 8.50 and 8.51, the threaded shaft, gate, and wedge and disc assembly, although it is out of the flow, can still accumulate solids in the bonnet area since the bonnet cavity is not isolated from the flow.

Solids and grease often accumulate on the shaft, thus making it difficult to operate the valve. In the event this occurs to the suction valve, disassemble the valve bonnet after isolating the valve from the wet well and the force main. This normally requires placing a pneumatic plug or bag in the suction line in the wet well. Drain the suction pipe and the pump volute of water, and then disassemble the valve. In some cases it may be feasible to attach a high-pressure water source or an air source to the bonnet using a threaded hole with a pipe plug installed in the bonnet; however, this is not always successful.

CAUTION! Before disassembling the bonnet, you must ensure that the pump discharge valve is shut and holding (don't depend on the check valve alone) and that the pneumatic plug or bag in the suction pipe is secure. Use an auxiliary pump to install the pneumatic bag or plug in the suction line. It may be necessary to schedule this work during the low flow periods and, depending on the upstream storage and other conditions, it may be necessary to provide a tanker to pump into when using the auxiliary pump. Normally, a thorough cleaning of the valve's threaded stem and the gate assembly is all that is required unless you observe severe deterioration of the faces on the discs and on the seat of the valve body. To disassemble the discharge valve, it will be necessary to drain the force main in addition to isolating it from the wet wall.

The outside screw and yoke valve is similar in operation with the exception that the threaded stem actually rises and is not exposed to the accumulation of solids from the wastewater. Since the threaded portion is located outside of the bonnet, however, additional room is required to accommodate the rising stem when the valve is in the open position.

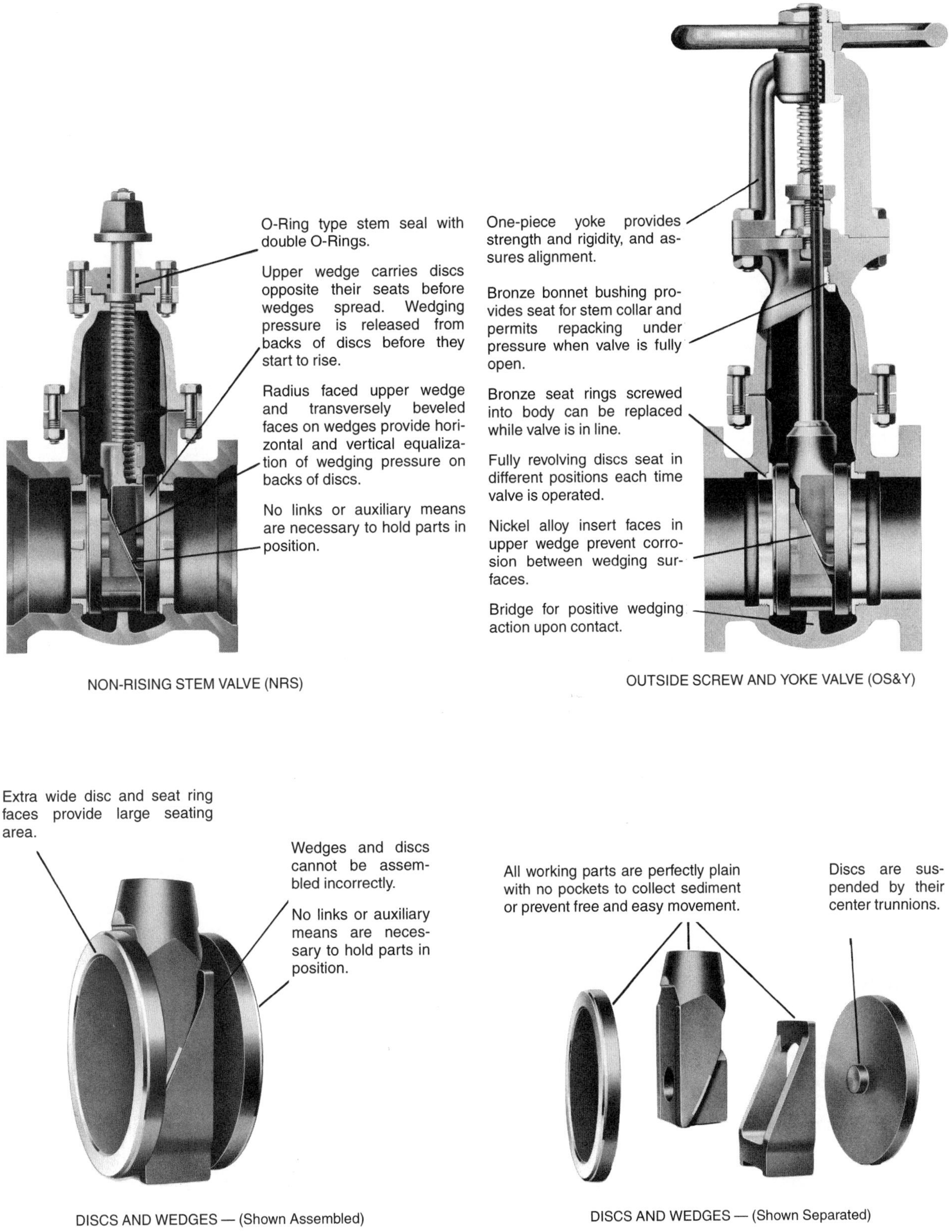
O-Ring type stem seal with double O-Rings.
Upper wedge carries discs opposite their seats before wedges spread. Wedging pressure is released from backs of discs before they start to rise.
Radius faced upper wedge and transversely beveled faces on wedges provide horizontal and vertical equalization of wedging pressure on backs of discs.
No links or auxiliary means are necessary to hold parts in position.
One-piece yoke provides strength and rigidity, and assures alignment.
Bronze bonnet bushing provides seat for stem collar and permits repacking under pressure when valve is fully open.
Bronze seat rings screwed into body can be replaced while valve is in line.
Fully revolving discs seat in different positions each time valve is operated.
Nickel alloy insert faces in upper wedge prevent corrosion between wedging surfaces.
Bridge for positive wedging action upon contact.
NON-RISING STEM VALVE (NRS)
OUTSIDE SCREW AND YOKE VALVE (OS&Y)
Extra wide disc and seat ring faces provide large seating area.
Wedges and discs cannot be assembled incorrectly.
No links or auxiliary means are necessary to hold parts in position.
All working parts are perfectly plain with no pockets to collect sediment or prevent free and easy movement.
Discs are suspended by their center trunnions.
DISCS AND WEDGES — (Shown Assembled)
DISCS AND WEDGES — (Shown Separated)

Fig. 8.50 Sectional views of gate valves

(Permission of American-Darling)

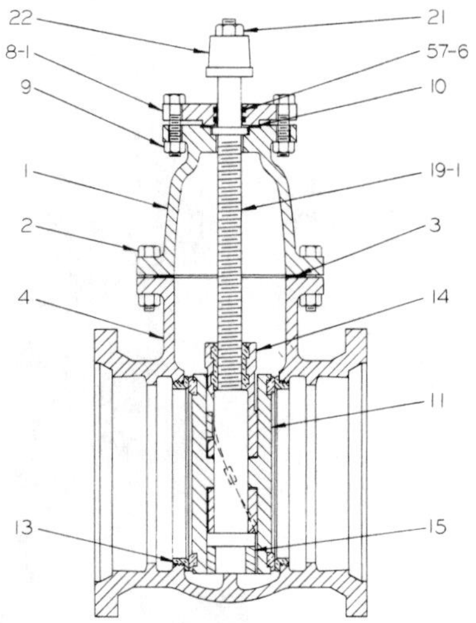

NON-RISING STEM (NRS)

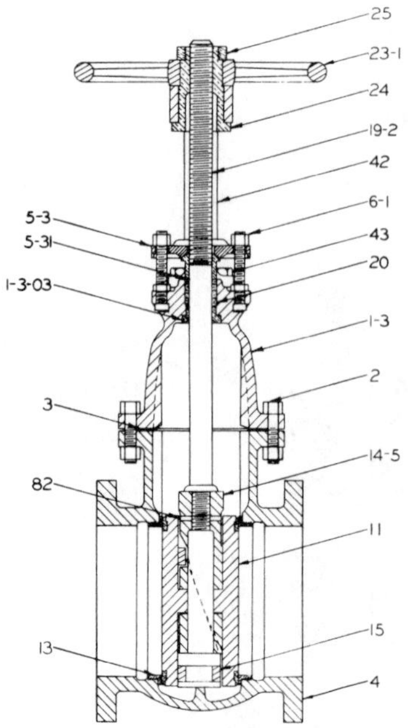

OUTSIDE SCREW AND YOKE (OS&Y)

Part No.	Name	Material	Quantity Per Valve
1	Bonnet	Cast iron (See Note 4)	1
2	Bonnet bolts and nuts	Plated steel	*
3	Bonnet gasket	Asbestos	1
4	Body	Cast iron	1
8-1	Bonnet cover	Cast iron (See Note 4)	1
9	Cover bolts and nuts	Plated steel	4
10	Cover gasket	Asbestos	1
11	Disc with ring	(See Note 3)	2
13	Seat ring	Bronze	2
14	Upper wedge	(See Note 1)	1
15	Lower wedge	(See Note 2)	1
19-1	Stem	Manganese bronze	1
21	Clamp nut	Plated steel	1
22	Wrench nut	Cast iron	1
57-6	O-Rings	BUNA-N rubber	2
23	Handwheel (Optional)	Cast iron	1

*Quantity required per valve depends on valve size.

NOTES:

1. Upper wedges on valves 2½" and smaller are ASTM B-62 bronze; 3"-12" sizes, iron with stainless steel shoes; 14" and larger sizes, iron with monel shoes. 3" and larger non-rising stem upper wedge has integrally cast ASTM B-62 bronze stem bushing.

2. Lower wedges on valves 2½" and smaller are ASTM B-62 bronze. 3" through 14", iron. 16" and larger, iron with monel shoes.

Part No.	Name	Material	Quantity Per Valve
1-3	Yoke bonnet	Cast iron	1
1-3-03	Bonnet bushing	Bronze	1
2	Bonnet bolts and nuts	Plated steel	*
3	Bonnet gasket	Asbestos	1
4	Body	Cast iron	1
5-31	Follower packing gland	Bronze	1
5-3	Follower flange	Cast iron	1
6-1	Follower bolts and nuts	Plated steel bolts bronze nuts	2
11	Discs with ring	(See Note 3)	2
13	Seat rings	Bronze	2
14-5	Yoke upper wedge	(See Note 1)	1
15	Lower wedge	(See Note 2)	1
19-2	Yoke stem	Manganese bronze	1
20	Packing	Asbestos	1
23-1	Yoke handwheel	Cast iron	1
24	Yoke stem nut	Bronze	1
25	Yoke clamp nut	Bronze	1
42	Yoke	Cast iron	1
43	Yoke bolts and nuts	Plated steel	4
82	Wedge pin	Stainless steel	1

*Quantity required per valve depends on valve size.

3. Discs on valves 3" and smaller are ASTM B-148 Alloy 954 aluminum bronze. 4" and larger, iron with ASTM B-62 bronze face.

4. Bonnet and bonnet cover of NRS valves have a bronze ASTM B-62 stem bushing on valves 16" and larger.

Fig. 8.51   Basic parts of gate valves

(Permission of American-Darling)

Under no circumstances should gate valves be buried, as is common practice in drinking water distribution systems. At some point in time a gate valve will require disassembly and maintenance. Submersible station valves should also be located out of the wet well area, since corrosion occurs and access is difficult when maintenance is required. A valve manhole next to the station is a better design.

Gate valves are also commonly used in the discharge manifold and cross-connect valves in the pump station.

8.171 Plug Valves

Plug valves are another type of valve that are successfully used as suction and discharge isolation valves in lift stations and are frequently used in wastewater plants where wastestreams with a high solids content are encountered, such as in sludge pumping systems.

As Figure 8.52 illustrates, a plug valve consists of the valve body and a rotating plug that operates through 90 degrees. Figure 8.53 illustrates the open, closing and closed positions of a plug valve. In the open position, the plug normally resides in the side of the valve out of the flow; in the closed position the plug seats up against the valve body providing a positive shutoff.

Many agencies specify plug valves as opposed to gate valves since they are less susceptible to plugging.

8.172 Check Valves

Normally a check valve is installed in the discharge of each pump to provide a positive shutoff from force main pressure when the pump is shut off and to prevent the force main from draining back into the wet well.

The most common type of check valve is the swing check valve which is shown in Figure 8.54. This valve consists of a valve body with a clapper arm attached to a hinge that opens when the pump comes on and closes to seat when the pump is shut off.

Check valves must close before the water column in the pipe reverses flow; otherwise, severe water hammer can occur when the clapper arm slams against the valve body seat. If this occurs, an adjustment of the outside weight or spring is usually required. A traditional clapper type of check valve has a lever on the extended shaft which allows adjustment of the weight on the arm and/or spring to vary the closing time. Wear occurs within the valve primarily on the clapper hinge and shaft assemblies and should be checked annually for looseness.

Other types of valves can also be used as check valves in larger lift stations. These valves may be air-operated or electrically operated. Plug valves are frequently used. Ball check valves are also used and rubber flapper-type check valves are used in some facilities.

8.18 Ventilation and Auxiliary Equipment

The dry well must be equipped with ventilation equipment to maintain the dry well portion of the station atmosphere in a dry and safe condition. Ventilation equipment should operate continuously to assure a safe working access for operation and maintenance operators. A lift station dry well is considered a confined space by state industrial safety agencies and requires frequent air changes. Follow all confined space entry procedures. An exception is those dry wells with fixed stairways and railings leading to lower floors. Monitor for oxygen deficiency and explosive and toxic gases during visits to the station. Wet wells also require provisions for ventilation.

Precautions that can be taken to protect operator health and safety when working in dry wells include:

A. Ventilating fans which are activated as soon as the light switch is turned on.

B. To minimize any engulfment hazard, regular preventive maintenance activities are done, for example, changing pump seals.

C. Have air monitoring capability. This could be done using portable atmospheric meters, worn by the employees. These meters would run continuously, sampling the air in the immediate work area of the employee. Alarms would be preset to detect hazardous conditions, for example, high hydrogen sulfide level or an oxygen deficiency.

D. Monitoring fan exhaust for hazardous atmospheric conditions.

In regions with high humidity, dehumidifiers may be necessary to keep condensation under control. This problem is most serious in the smaller, package-type lift stations which are made of prefabricated steel. Usually the dehumidifier runs constantly and outside air is used for ventilation only when the access cover is open. A ventilation fan can be turned on or off by a switch at the access cover. This procedure keeps moist, outside air from entering a lift station when it is unoccupied and provides the driest possible conditions.

Depending upon agency requirements or preference, lift stations also may include telemetering systems for monitoring or controlling the station operation. A flow measuring device may consist of either a Parshall flume, a weir, a magnetic meter, or a Venturi meter.

The station may have emergency generator units for continued operation during electrical power failures. Large lift stations may use gas or diesel engines to operate pumping equipment due to the ease and reliability of regulating pump speed.

Some stations are equipped with individual water wells to supply seal water, water for the application of chemicals such as chlorine, or cooling systems for heat exchangers on equipment. A lift station may vary from a simple and efficient telemetered design requiring a visit once a month to a highly complex installation demanding continuous staffing.

QUESTIONS

Write your answers in a notebook and then compare your answers with those on page 120.

8.1X List the major valves found in wastewater lift stations.

8.1Y Where are pump suction and discharge isolation valves located and why?

8.1Z Where are check valves installed and why?

Please answer the discussion and review questions next.

1. Corrosion resistant plug Plug materials resist corrosion and prolong seat life. Available materials are as follows:

Bronze (½"-2" only)

Electroless nickel plated cast iron (2½"-4" valves only).

2. Double-seal for tight shutoff and safety A resilient seal molded into a groove in the plug face assures dead-tight shutoff on liquids and gases without the use of sealing lubricants. When the plug is closed, the resilient seal is compressed against the seat. The metal on the plug face also makes contact with the metal seat to provide a second seal for safety. Gas industry fire tests showed that this second metal seal provides nearly bubble-tight shutoff with the resilient seal burned away.

3. O-Ring stem seal A variety of stem seal materials provides maintenance-free sealing that matches valve performance and assures long life and reliability.

4. Corrosion resistant bearings Permanently lubricated bearings in the upper and lower plug journals resist corrosion and assure easy operation without lubrication. And, operation is just as easy whether the valve is operated once a day or once a year.

5. Corrosion resistant seal A plastic seat coating provides extra corrosion resistance to minimize plug wear.

6. Meets ANSI 125 lb. standard Body walls and end connections conform to all applicable ANSI 125 pound standards. Valves meet MSS standards in ½" and ¾" sizes where ANSI standards are not available.

7. Choice of end connections A complete choice of end connections includes screwed, flanged, Dresser and Victaulic. A complete listing of availability for each valve size and FIG number is shown in the ordering section.

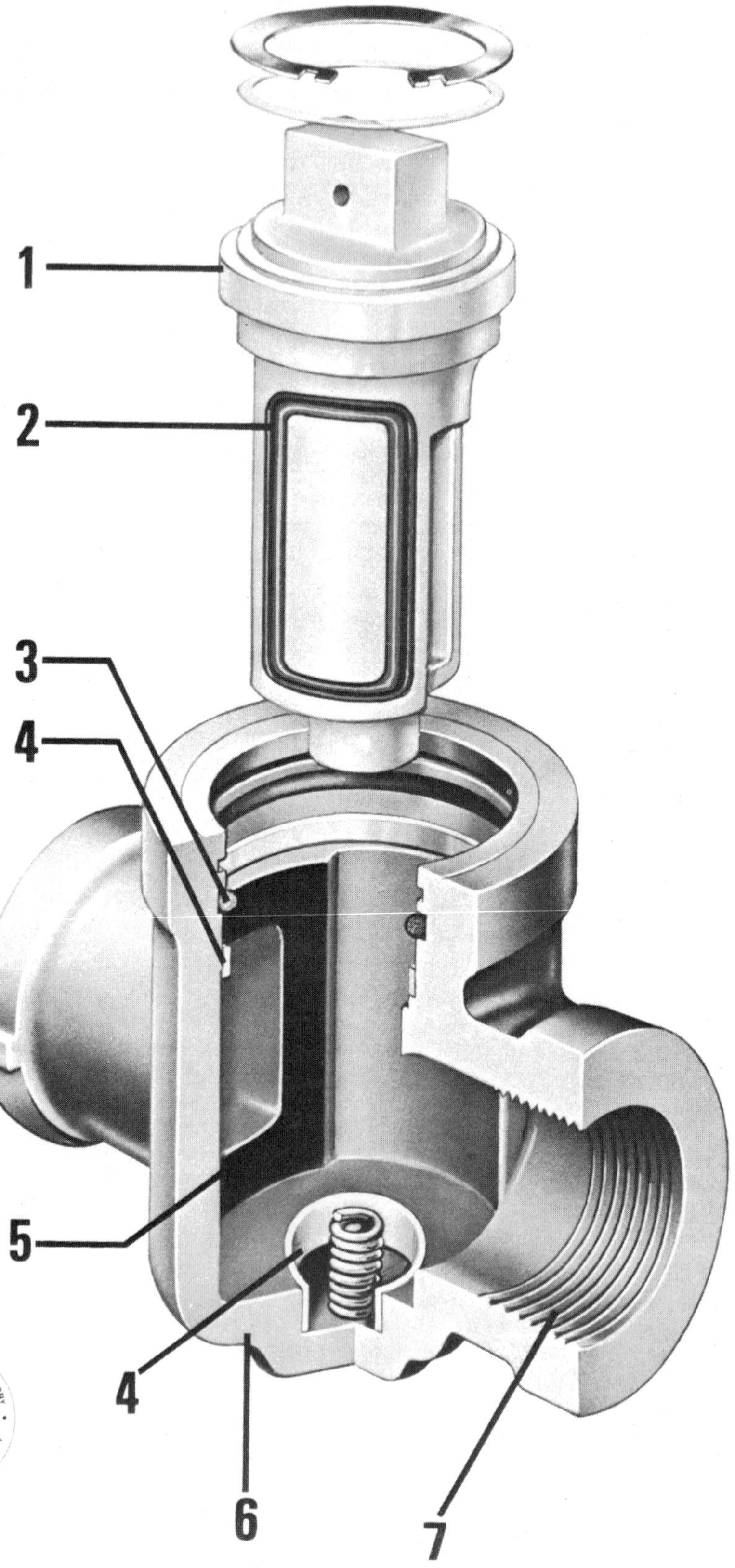

Fig. 8.52 Eccentric plug valve

(Permission of DeZurik)

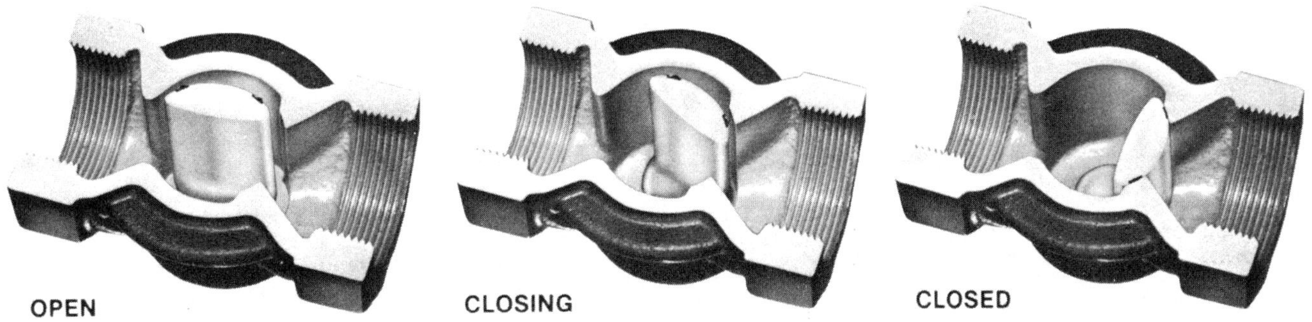

DeZURIK eccentric action and resilient seating assure lasting dead-tight shutoff. As the eccentric plug rotates 90 degrees from open to closed, it moves into a raised eccentric seat.

In the open position, flow is straight through and flow capacity is high.

As the plug closes, it moves toward the seat without scraping the seat or body walls so there is no plug binding or wear. Flow is still straight through making the throttling characteristic of this valve ideal for manual throttling of gases and liquids.

In the closed position, the plug makes contact with the seat. The resilient plug seal is pressed firmly into the seat for dead-tight shutoff. Eccentric plug and seat design assure lasting shutoff because the plug continues to move into the seat until firm contact and seal is made.

Fig. 8.53 Open, closing and closed eccentric plug valve
(Permission of DeZurik)

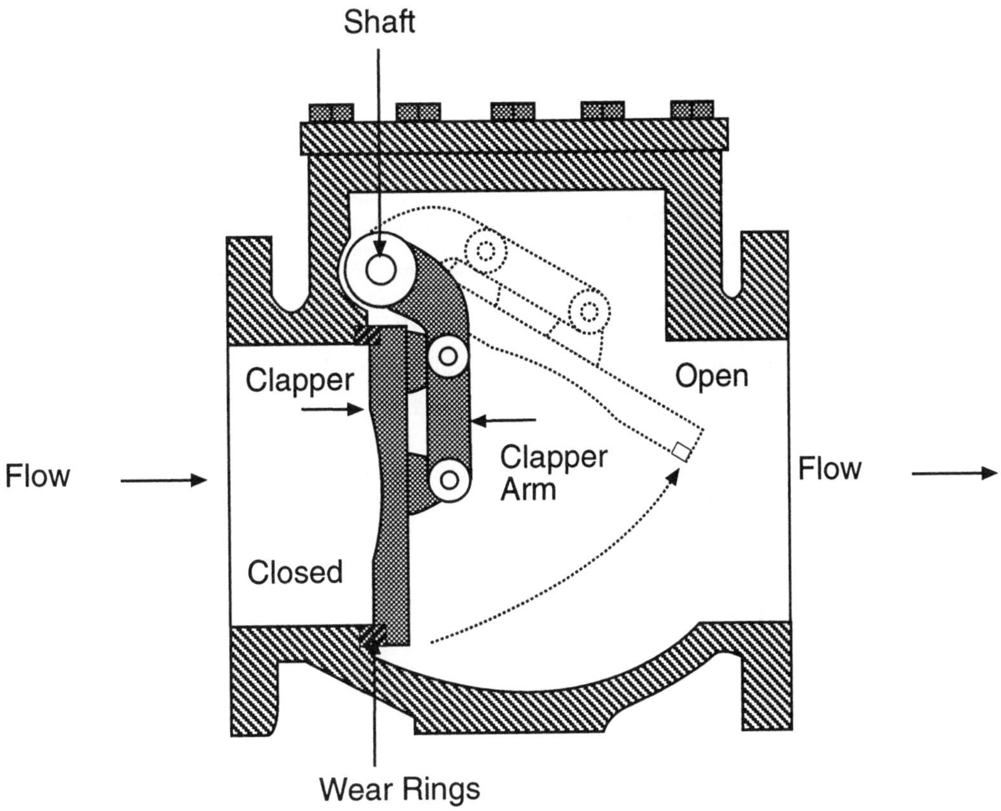

Fig. 8.54 Typical horizontal swing check valve

DISCUSSION AND REVIEW QUESTIONS

Chapter 8. LIFT STATIONS

(Lesson 2 of 5 Lessons)

Write the answers to these questions in your notebook before continuing. The question numbering continues from Lesson 1.

6. What are the advantages and limitations of a wet well being too large or too small?
7. What are the advantages and limitations of having bar racks in wet wells?
8. What is the purpose of a stilling well in a wet well?
9. How is the depth of water in a wet well determined?
10. Why should isolation valves in lift stations be gate valves instead of plug valves?
11. Why must lift stations have adequate ventilation?
12. Why is it important to use a clean water source to lubricate both stuffing boxes and mechanical seals?
13. What information is shown on a pump curve?
14. What types of auxiliary equipment may be installed in lift stations?

CHAPTER 8. LIFT STATIONS

(Lesson 3 of 5 Lessons)

8.2 THE NEWLY CONSTRUCTED LIFT STATION

Many large municipal agencies do their own design work for lift stations. Smaller agencies usually rely upon a consulting engineering firm for station design. In either case, operation and maintenance operators should be given the opportunity to review the prints and specifications of a new lift station before the award of a contract for construction. This review is very important to be sure adequate provisions have been made for the station to be easily and properly operated and maintained.

When it comes to operation and maintenance of a lift station, the knowledgeable, experienced operator is an expert and can contribute to new lift station construction projects during the planning and design phase. The operator's input can have a long-term effect on the operating requirements and, therefore, the operation and maintenance budget of the station. Participation by the operator during planning and design will result in the following positive contributions:

1. Identify problems in existing stations to ensure that those problems are eliminated in new stations,
2. Make the station more reliable, reducing the number of emergency call outs,
3. Increase operator safety,
4. Provide for sufficient space to safely operate and maintain the facility, and
5. Reduce maintenance required, including emergency maintenance (EM), corrective maintenance (CM), and preventive maintenance (PM).

Keep in mind that it is not the operator's job to tell the engineer how to design a facility, but rather to provide positive input on improving the operation and maintenance of the station. If possible, be very specific and make recommendations. A carefully kept log of existing station operation and maintenance requirements will assist the engineer. The operator

should have an opportunity to participate in the project design in the following stages:

1. Planning stage,
2. 20% design review,
3. 40% design review, and
4. 90% design review.

8.20 Examination of Prints

When examining the prints, operation and maintenance operators should look for accessibility not only for equipment, but for operators to get to the station. Is there sufficient space for vehicles to park and not restrict vehicles passing on streets or pedestrians on sidewalks? Is there room to use hydrolifts, cranes and high velocity cleaners as needed at the lift station? Are overhead clearances of power lines, trees and roofs adequate for a crane to remove the largest piece of equipment? Are lifting hooks or overhead rails available where needed in the structure? Has sufficient overhead and work room been provided around equipment and control panels to work safely? Is lighting inside and outside the station provided and is it adequate? Does the alarm system signal high water levels in the wet well and water on the floor of the dry well? Is there sufficient fresh water at a high enough pressure to adequately wash down the wet well? Are there any operator traps or head knockers such as low-hanging projections or pipes, unprotected holes, or unsafe stairs or platforms? What happens when the power goes off and backup or standby systems fail? Is there access to the wet well (if you have to clean out incoming lines, it may be necessary to put a temporary pump in the wet well)? All of these questions must be answered satisfactorily if the lift station is to be easily operated and maintained.

Equipment should be laid out in an orderly sequence with sufficient work room and access to valves and other station equipment, controls, wiring and pipes. If there is any possibility of future growth and the station may be enlarged, be sure provisions are made to allow for pumping units to be changed or for the installation of additional pumping units. If additional pumping units will be necessary, be sure spools and valves are built now for ease of expansion. Be sure there is sufficient room to add electrical switchgear for future units. If stationary standby power units are not provided, make certain there are external connections and transfer switches for a portable generator.

8.21 Reading Specifications

Review the specifications for the acceptability of the equipment, piping, electrical system, instrumentation, and auxiliary equipment. Determine if the equipment is familiar to your agency and if its reliability has been proven. Find out what warranties, guarantees, and operation and maintenance aids will be provided with the equipment and the lift station. Require a list of names, addresses and phone numbers of persons to contact in case help is needed regarding supplies or equipment during start-up and shake-down runs. Services of factory representatives should be available during the initial start-up of all major pieces of equipment. Require factory representatives to demonstrate the operation of equipment under all possible operating conditions. Be sure that the equipment brochures and other information apply to the equipment supplied. Sometimes new models are installed and you are provided with old brochures.

Be sure the painting and color coding on pipes and electrical circuits meet with your agency's practice. Try to standardize electrical equipment and components as much as possible so one manufacturer can't blame the other when problems develop. Standardization also can help to reduce the inventory of spare parts necessary for replacement. A few hours spent reviewing plans and specifications will save many days of hard and discouraging labor in the field in the future when it is a major job to make a change. Very often changes on paper are relatively simple.

QUESTIONS

Write your answers in a notebook and then compare your answers with those on page 120.

8.2A Why should operation and maintenance operators be given an opportunity to review the prints and specifications of a new lift station before the award of a construction contract?

8.2B Why is accessibility to a lift station important?

8.2C What information should be determined for equipment to be installed in a lift station?

8.2D Why should prints and specifications be carefully reviewed?

8.22 Preliminary Inspection of a New Lift Station

After a lift station has been constructed, it is inspected by the contractor, the engineer, and the operating agency before final acceptance. This inspection should be planned and conducted by the people responsible for the operation and maintenance of the lift station in cooperation with the contractor and the engineer.

One way to accomplish this is to include in the specifications a specific schedule that identifies the section of specification, the item to be checked and who is responsible; for example, electrical contractor, mechanical contractor, or other specific contractor. Then check off each item by noting who accepted each item and the date. All equipment checkout and adjustments should be performed *PRIOR* to any equipment training as this will maximize the amount of time spent on training as opposed to adjustment and troubleshooting.

Documentation is particularly important for proper operation and maintenance of the lift station equipment. The equipment manufacturers' O & M manuals should include the following:

1. A suitable binder that allows for quick and easy reference to all mechanical/electrical and hydraulic equipment provided in the pump station. Generally, photocopies should not be accepted, since they may be illegible and therefore of limited value to the operator.
2. Information about each piece of electrical and mechanical equipment should include nameplate data, operation and maintenance instructions, spare parts list, recommended spare parts stock list, part numbers, troubleshooting information, assembly/disassembly instructions, tolerances, tools required, and safety precautions. In other words, a sufficient amount of information to allow adequate maintenance of the equipment.
3. The manual should be available when you accept the station from the contractor, not three months later.

Now, during the preliminary inspection, is the appropriate time to develop and record both routine and emergency information and procedures. Use the following outline as a guide in assembling your documentation.

1. As-built specification of the facility including:
   a. Size, length and details of force main (discharge point)
   b. Available power and fuse size
   c. Pump and motor sizes and rated capacities
   d. Actual capacities of pumps and rates of flow throughout the day
   e. Normal pump levels (start/stop/high and low levels)
2. Elevation of each manhole invert and depth of the manhole
3. Lowest homes on the system (most probable backup points)
4. Alternate route of temporary pumping:
   a. Elevation and length of temporary pumping
   b. Type of equipment needed
5. Number of services, plus commercial hookups
6. Available equipment and methods of operating station when a power outage occurs
7. Equipment needed for pump and motor removal
8. List of private companies providing emergency pumping equipment and personnel
9. Station data:
   a. Wet well size
   b. Storage time, average flow
   c. Average flow, GPM
   d. Peak flow, GPM
   e. Average flow, GPD
   f. Wet well depth
   g. Distance of bubbler tube off wet well bottom
   h. Type of wet well suction

8.23 Pump Station Calibration

As part of the start-up and acceptance procedure, a pump station calibration should be performed to:

1. Verify operating conditions of the pumps in accordance with the nameplate, and
2. Establish a base line that can be used for comparison when future pump station calibrations are performed.

   The calibration of a pump station is a fairly quick and easy method to verify the operating efficiency of the pumps, including capacities and discharge heads.

Operation under test conditions will reveal only the immediate equipment and construction problems. These problems should be recorded on a punch list during the preliminary inspections and a copy given to the contractor. This punch list of problems should be quickly corrected so the station can be ready for service. Do not accept any portion of a new lift station, because official acceptance of completed work is the engineer's job.

To obtain maximum life and use from a new lift station, begin the station preventive maintenance program now when it is new. Start the program by filing data in a station record book. Record station identification code number, location by map and street numbers, date of construction, size of station by flow capacity and other important information required by your agency. Also file a set of plans and specifications for reference purposes.

Start the preventive maintenance program by making a complete list of all equipment at the station. See Chapter 9, Section 9.28, "Records," for typical data sheets. Information recorded should include equipment nameplate data. Mark each piece of equipment with an identification code number and record the number. There should be a minimum of two copies of this information. Keep one copy at the lift station and the other in the station record book at the agency office. Once this has been done, permanent identification of all equipment involved in the station will be available when needed. Prepare a spare parts list and order any extra parts now.

Recordkeeping must begin when the station is first started so that any peculiarities of the equipment are known from the beginning. These records should include any of the contractor's data collected during construction and also data required according to the specifications. These specifications include:

1. Equipment manufacturers' O & M manuals,
2. Engineers' O & M instructions, including detailed emergency procedures,
3. Pump and motor operational characteristics,
4. Pump and motor coupling alignment readings,
5. Wet well control levels, and
6. Auxiliary equipment data.

QUESTIONS

Write your answers in a notebook and then compare your answers with those on page 120.

8.2E Who should inspect a lift station after it is completed, but before it is accepted?

8.2F Where should information regarding lift station equipment and performance be filed?

8.2G Why should data from initial tests on equipment be recorded and filed?

8.24 Inspection of a Lift Station

Hopefully you had the opportunity to review the plans and specifications for the new lift station. Now is the time to determine if the station was built as you wished. Look at those items you previously reviewed. Ask yourself what problems will be faced when maintenance and repairs are to be carried out under adverse conditions. Think of the worst situations so that your preparations will be adequate when needed.

Inspect the building for access so equipment can be removed. To ensure that equipment can be worked on, the following items must be considered. Is there an emergency lighting system and is it adequate? Is there enough head room to pull pumps, motors, gear heads or other pieces of equipment? If head room is not needed, are the passageways wide enough to allow for the removal of gear boxes, engines, control panels, and standby generators? If pumps have to be removed through the roof of a building, is there enough clearance for a crane or other lifting device to get within reaching distance? Where are the electrical lines located that supply the building? Will they restrict use of the lifting equipment? The time to consider all of the above is now, not when conditions are adverse. If there are potential problems related to equipment removal, the time to correct these problems is now. Plan to conduct corrective work before a crisis occurs. Don't wait for an emergency to arise to discover that a critical piece of equipment can't be moved when a unit needs repair. Dismantling equipment in place under restrictive conditions is slow and difficult work.

The next unit to inspect is the wet well. What kind of conditions will be encountered when you attempt to clean the sump or possibly to enter it and remove grit? In a sanitary wet well, are there enough openings for washing the sump walls and floor? Can it be ventilated easily? What type and size of ladder will you need for access? If there is a bar rack, or comminuting device, consider the conditions you will encounter to clean the rack and how you will dispose of the debris. Also consider the tools needed and whether or not the necessary tools will be stored at the site. If tools must be transported, make notes of what will be needed and how transportation will be accomplished. Inspect the equipment that measures the water level in the wet well.

Before entering any lift station or other wastewater facility, safety procedures must be followed to ensure that the hazards encountered will be eliminated or brought to a minimum. The hazards are as follows:

1. Insufficient oxygen,
2. Explosive and toxic gases,
3. Poor footing caused by grease or slimes,
4. Unsafe stairs and walkways,
5. Dangerous electrical gear, and
6. Inadequate drainage.

If the wet well is to be entered, begin by monitoring the wet well for toxic or poisonous gases, explosive gases, and lack of oxygen. *IF ALL CONDITIONS ARE SATISFACTORY, THEN AND ONLY THEN IS IT PERMISSIBLE TO ENTER THE WET WELL.* All of the safety equipment and clothing listed under safety for confined spaces should be available for use at this time (see Volume I, Chapter 4, Sections 4.4 and 4.5, "Classification and Description of Manhole Hazards" and "Safety Equipment and Procedures for Confined Space Entry"). Next isolate the influent flow and set up a blower that will give a minimum of two air changes a minute within the wet well. The wet well must be washed thoroughly using a hose with a nozzle. Monitor the wet well gases and oxygen continuously until the job is completed and everyone is out of the sump.

QUESTIONS

Write your answers in a notebook and then compare your answers with those on page 120.

8.2H What problems may be encountered when cleaning a wet well?

8.2I What are the hazards found in confined spaces that handle wastewater?

8.2J What can you do to eliminate or reduce to a minimum the hazards encountered in sumps?

The next item to inspect is the electrical equipment. Begin by recording all nameplate data. Put code numbers on each panel beginning with the main breaker. Branch panels that supply power to major circuits carry the code number of the piece of equipment that it serves. A lighting branch panel has its own listing or numbers and the lights or plugs it serves also carry this number. With a system like this, the problem of locating controls, plugs, and lights is quite simple.

Individual inspection of the major circuits is necessary. Record the parts that most likely will need attention during an emergency. All fuses must be listed and spares kept at the station to minimize equipment downtime. Data on switches, relays and heater strips must be recorded. Be sure that all overload heaters are the proper size. Control circuit equipment must be listed. Identify all related equipment with the proper code number.

Make this inspection with the main breaker to the equipment being inspected locked in the *OFF* position. No attempt should be made to inspect or repair electrical equipment unless a basic knowledge of electricity and the safety precautions needed are fully understood and applied. Request an electrician to help if you are not qualified or not authorized to inspect electrical equipment, circuits and controls.

After listing all necessary data and putting code numbers in place, read the operating instructions and examine the controls for their proper operation. Move all switches with the power off to see how they operate. Check reset buttons and know what they are supposed to do. Before the station becomes operational, it is essential to learn how things are supposed to work.

The approach to solving any problem must be made with logic and caution. Be deliberate and think each step out before making any changes. Mark all settings, if possible, so that you can return to the starting point. Wait a short period when changing settings to obtain and observe the results. Do not make a change if you think it is not safe.

Before leaving the electrical equipment, tag any dangerous area with "not to be tampered with or touched by inexperienced hands" and lock out.

QUESTIONS

Write your answers in a notebook and then compare your answers with those on page 120.

8.2K Where should the inspection of the electrical equipment begin?

8.2L When inspecting an electrical circuit or equipment, in what position should the main breaker be positioned?

8.2M Should extra fuses be kept at the station? Why?

8.2N When should you learn how the electrical circuits work?

The next point of inspection is the pumps. All nameplate data must be recorded. Lock out pump(s) and determine direction of rotation. If the local electrical utility company has worked on their lines near the station, be sure the power leads (legs) are hooked up correctly and the rotation of the pumps is in the proper direction. Inspect the backspin preventer if the pumps are so equipped by trying to turn the pump shaft in the direction against the indicated rotation. Backspin equipment operation also may be detected by listening to the pump motor when the unit is shut down after a run. When the unit comes to a stop, the backspin unit begins to engage and it can be heard as a ratcheting sound. If a right-angle gear head is used, the backspin equipment may be located on this unit. Equipment specifications will have to be checked to determine if there is anti-rotation equipment provided in the gear head. Examine all gages for location and determine at this time what they are to indicate. Any valves located on the suction or discharge should be opened and closed to examine for correct and easy operation so that you are familiar with them. Place equipment numbers on these valves so that a preventive maintenance card can be made to service them. Maintenance record cards should be prepared for each piece of equipment inspected. Inspect the drive and pump for proper alignment. If belt-driven, record the sizes of the belts and numbers so that spares will be available when needed. Inspect all rotating parts for missing guards and secureness. The guards must be built for easy access to the equipment being shielded, but still meet all safety requirements.

Inspect the lubrication equipment used and note all pertinent information needed to keep it operating. Solenoids used to operate the oilers should be standardized when reviewing the specifications in order to reduce the inventory of spares necessary for replacement.

Inspect suction and discharge lines. The suction line must be kept clear of debris. If a check valve is supplied on the pump discharge pipe, this valve can become a problem when stringy material gets caught on the flapper and it stays in the open position.

If the lift station uses gasoline or diesel engines, they may be in the form of stand-by generators or as a pump drive. When used as a pump drive, the operation may go through a right-angle gear unit (Figure 8.55). These units are used to keep a large engine from being located on a lower deck where the pump must be located for efficient operation. The engine is kept on the top for easy access, cooling, ventilation, and exhaust, and to keep it above the level of flooding.

There will be very few problems with the right-angle gear if a unit of the correct design is used for a given job. The most important point is to have the proper lubrication for the gears. Preventive maintenance will keep the unit free of leaks. Cooling is supplied to some units, but only when the gear head is to run continuously for a long period of time. You must inspect the drive shaft and its universal joints. Lubrication is necessary, but should be kept to a minimum because too much will rupture the seal on the universal and cause failure by allowing the entrance of foreign material.

The engine must be inspected for sufficient oil, water and fuel to operate properly. Examine the battery for water, tight cables and clean terminals. If the engine is large and requires compressed air for starting, inspect the auxiliary air system. Inspect and adjust the alarm system for proper operation at the predetermined points of alarm or shutdown. Failure of the alarm system to shut down devices could cause an engine to burn up. Make sure that all hoses are secure and clamped tight. Loose nuts and bolts must be tightened. Be aware of the ones that continually become loose. A different type of fastener may solve this problem.

Leakage of any sort should not be tolerated. Keep cleaning equipment at the station to wipe up leaks so that you know when a new leak starts. Vibration will continually cause problems of leakage and loose parts, so you must never let up on your preventive maintenance program.

If there is a clutch to disengage the engine from the drive shaft, test its operation. It should be a little hard to engage and disengage because the unit must snap over center when operated.

Lubrication of the clutch throw-out bearing should not be excessive because the friction plates will fail if contaminated with grease. The engine should be maintained as you maintain your automobile engine. Change the oil on a regular schedule

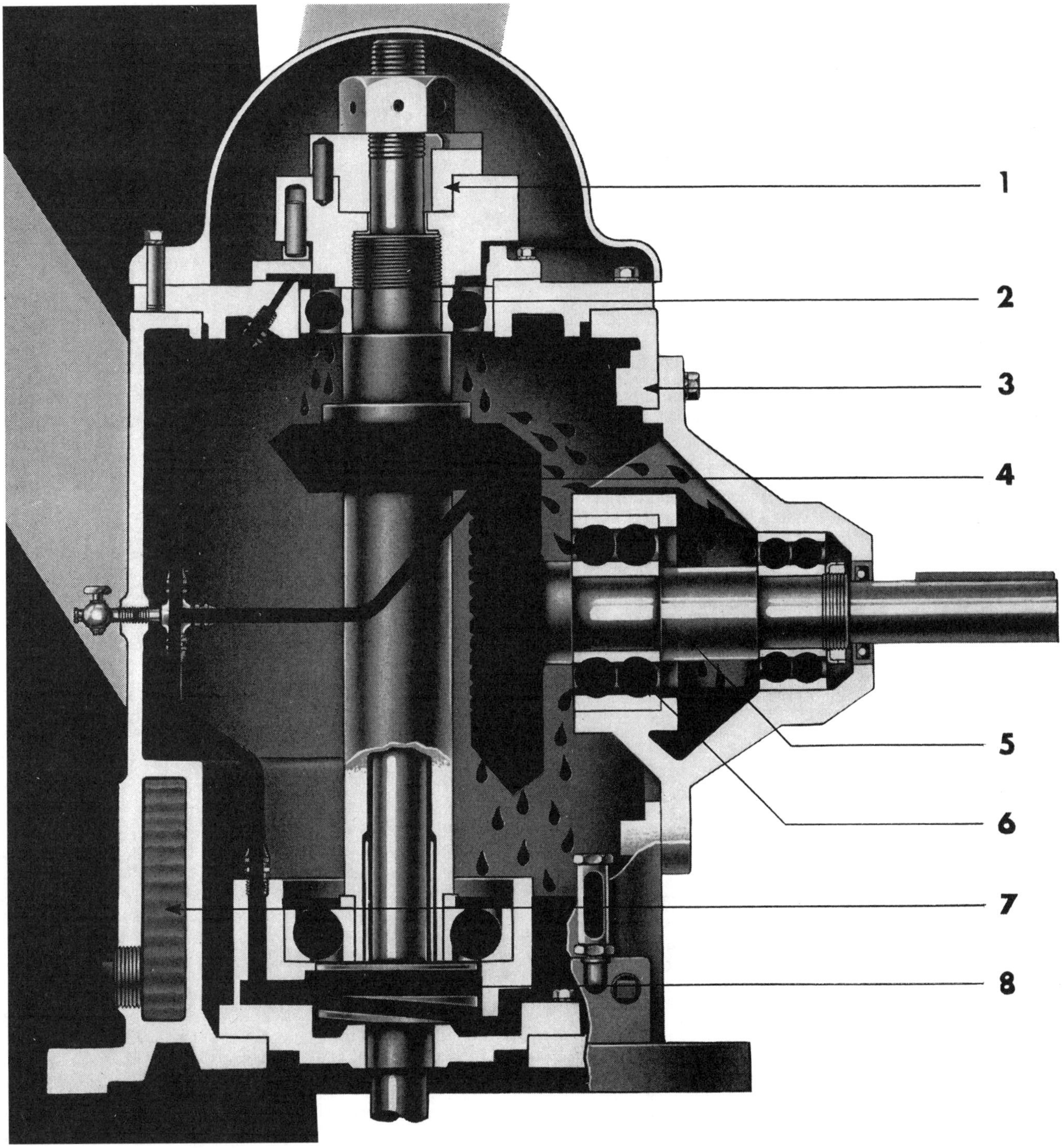

1. Clutch
2. Vertical output shaft assembly
3. Housing
4. Gear and pinion
5. Horizontal input shaft assembly
6. Anti-friction bearings
7. Cooling
8. Lubrication

Fig. 8.55 Right-angle gear drive
(Courtesy of Fairbanks-Morse)

to remove contamination from the engine's inner parts. Inspect the oil for metal filings and sludge when changing the oil to stay aware of the engine condition. If filings are noted, a failure of bearings may be assumed. Look for water or an oil-water emulsion that indicates water in the wrong place. If a qualified lab is available, oil may be tested to determine when a change is necessary.

Sludge indicates that excessive dirt or carbon has found its way into the crankcase. This may be a failure of the engine's breathing system or just plain blow-by at the rings. Engines should be run from two to four hours during extensive test runs to permit complete heating of all parts and to observe performance during continuous operation. Inspect the engines' heating and cooling systems. Short runs of 5 to 15 minutes should be avoided (especially with large engines) because this is when the most damage can be done to a cold or inadequately heated engine. If the engines are not being used, they should be test run carrying a typical load for one hour every two weeks.

Determine what type of fuel is to be used and the storage capacity, the engine's consumption rate, and when the fuel will be delivered. The fuel tank should be kept full so condensate won't form on top of the tank.

Inspect all controls for free and easy operation. Make a record of their settings before and after operation. This becomes the standard and a starting point for troubleshooting when problems occur in the future. When operating, set marks can be made with fingernail polish at the set point. The question may arise at this point as to why bother to check everything on the engine so thoroughly? Inspecting and obtaining a standard for the engine will help tremendously when problems do occur. The jobs of troubleshooting and repairing become easy when you have good data for reference. This data can be obtained only when the engine is new or rebuilt and is being operated for the first time.

The final point of inspection is the housekeeping needed to maintain an acceptable station which will satisfy the public or surrounding area. Inspect all fences to see that they are complete and have a neat appearance. If made of wood, be sure the paint has good color and is protecting the wood. Chain link fences usually require little maintenance, but should still be inspected for damage or poor installation. Metal chain link fences must be grounded in case a power line drops across them and an unsuspecting operator goes to open the gate. Examine all paved areas for proper drainage and good workmanship when an area is paved. Inspect unpaved areas for weed control.

Examine all exterior parts of buildings. Look for poor installation or possible future problem areas. At this time, get a ladder and inspect the building's roof. Be sure that all openings are adequately covered or louvered.

QUESTIONS

Write your answers in a notebook and then compare your answers with those on page 120.

8.2O What is the purpose of anti-rotation equipment on a pump?

8.2P What can happen if a universal joint is overlubricated?

8.2Q What three items must we have for engine operation?

8.2R What can happen to an engine if its internal alarm system fails?

8.2S What must one look for when changing engine oil?

8.2T Why is the engine examined so thoroughly?

8.2U What precautions should you take when lubricating the clutch throw-out bearing?

8.2V Why should an engine's fuel tank be kept full?

8.25 Operational Inspection

We are now ready to operate the lift station. Open any valves on the influent line to fill the wet well. Make certain that the pump volute drain lines and vent lines are closed. If there is no wet well, open the valve to the suction side of the pump. Inspect the sump and bar rack at this point for debris and free flow into the station. Remove the floating debris left by the contractor in the sump or off the bar rack that you should have removed *BEFORE* the station was placed in service. Problem materials include small pieces of wood, like grade stakes, or pieces of plastic of any size, cans, and bottles. Some objects can plug the eye of the impeller.

With the sump filling, inspect your sump level indicator to see that it is working properly. If there is no indicator, visually check the sump level or watch the rise in the pump control stilling well. If a bubbler system is used, watch the pressure gage indicator and verify wet well level readings with actual readings. Once the station is in operation, the determination of the sump level will become a part of the normal inspection for system maintenance.

Inspect the discharge side of the pump for clear passage of the liquid to be pumped. All valves must be open. Bleed air from the volute of the pump. Sumps usually fill slowly unless a means of causing rapid filling can be found, such as using water from a fire hydrant.

When the sump is filling, recheck for power to the pump and proper rotation of the pump. If an engine is the power source, it must be checked out before operating. Put the pump on manual operation and momentarily start the pump. Watch rotation of the pump shaft to see that rotation is correct. There is usually an arrow attached to the side of the pump to indicate the proper direction. If there is not, you will have to refer to the manufacturer's manual to find out how to determine correct rotation. If the pump is a centrifugal pump, the rotation is in the direction of the volute. Another means of determining rotation is to take a load check of the amperage drawn by the pump motor and compare this value with motor nameplate data. If the rotation is in the wrong direction, the amp reading will be lower than the motor rating due to less work being done by the pump by rotating backwards. If the discharge flow is low or the wet well is being drawn down slowly, the rotation could be in the wrong direction. Higher amp readings can result from head conditions lower than design head conditions. Low head conditions can be corrected temporarily by throttling a discharge valve or permanently by installing an orifice plate in the discharge piping.

Now that the rotation is in the proper direction, the next job is to see that all operating controls are put into the automatic position for the test operation. Circuit breakers must be checked to see that they are in the ON position. Oilers or solenoids operating the oilers should have been inspected during previous inspections. The electrical alarm systems must be examined to see that they are in the activated position. The electrical components of an engine must be inspected to be sure that there are no drains on the battery. The mechanical inspections should include inspecting all valves to see that they are open or closed according to the operation desired. You must be sure that all oilers have been filled and are not dripping if solenoid operated. If a bubbler system is used, an inspection should be made to determine that it is functioning properly and that there is enough air pressure to make the control system operate.

Other mechanical checks should be made on the engine. Determine that the linkage is ready to operate. Inspect the clutch to be sure that it is engaged. Examine all cooling water valves to see that they are in the proper positions.

Check to see that preheaters (if provided) on the engine are operating. The heaters inside the building should be examined to see that they are functioning to maintain the proper humidity and temperature within the station. Up to this point we have been recording all the data that have been observed. This information becomes your base for future operation reference.

Notify operators of downstream lift stations and wastewater treatment plants that the lift station is being tested and tell them the volumes of water that can be expected. Allow the sump to fill and activate the station. Watch the level of water in the sump to be sure that the pump starts at the desired level. If not, adjust this operating point by whatever means are available.

The pump must be stopped after being started so as not to pull the sump level down too far. When the starting point has been set by a visual or simulated means, we can continue to the next point of operation and that is to allow the sump to continue to rise. If more than one pump is involved, each start position for each pump must be determined and recorded as the level rises until all start points have been activated. Allow the level to rise and activate the high level alarm.

At this point you may now start the pumps in their reverse order. Each pump when started must be inspected for its operation and all data recorded. The first item to inspect is the electrical load on motors. Record any electrical malfunctions and have them corrected. The pump should be inspected for proper operation and any unusual noises or vibrations noted along with excessive heating.

Inspect the packing gland or mechanical seal. If the unit has a mechanical seal, there should be no leakage. If it is a packing gland, get the data on the size and type of packing needed. Set packing so that there will be leakage of approximately one drop per second. If the packing is Teflon, you must adjust the gland tight with your fingers and run the pump 15 to 30 minutes while you watch the packing gland. Teflon will expand. Do not tighten the gland during this period of operation. You are watching to see that it does not get hot. If Teflon gets hot it will glaze and stop working properly. After this period of operation, adjust the packing gland nut one *FLAT*[13] tight by turning the bolt one flat on the nut. Maintain one to three drops of water per second on the packed glands. No more adjustments are necessary. If leakage continues excessively, something wrong was done when the unit was packed and it must be repacked.

Most mechanical seals have flushing water supplied to the seal. The pressure needed here is 3 to 5 psi above the pump discharge pressure. There should be a filter in the flushing line if the water being pumped is used for this flushing water. Obtain all information needed to service the unit and to have spare parts available.

Do not operate any unit if you think it is not functioning properly. Shut it down and have the contractor examine the problem. When the problem is corrected to your satisfaction, you may proceed with the test operation. All gages should be inspected and their set points recorded.

Continue with the individual operation of all pumps until they have been checked out to your satisfaction. They must comply with the specifications and operate on the performance curves as intended by the design engineer. If a flowmeter has not been installed, use the following procedures to calibrate the lift station pumps.

[13] *Flat. A flat is the length of one side of a nut.*

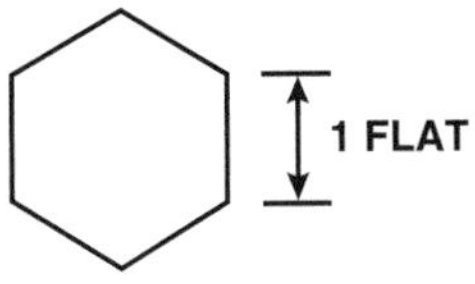

PUMP HEAD-CAPACITY CALIBRATION, TIME/VOLUME METHOD

A. PURPOSE

This procedure is a method for field checking the capacity of lift station pumps when a flowmeter is not available. When suction and discharge pressure readings are obtained at the same time, the pump head and capacity can be checked against the manufacturer's catalog curve. (Since field testing does not use rigid lab test procedures, field results that are within plus or minus 10 percent of certified data are acceptable.) This procedure should be performed before initial start-up of a lift station and again annually to check the performance of the pumps.

B. METHOD (Refer to Figure 8.56.)

1. Determine the rate of flow into the station from the gravity line or lines.
2. Determine the wet well capacity in gallons over a fixed operating level. (*NOTE:* It is important that the operating level be less than the invert of the gravity pipe to prevent storing wastewater in the pipe, but above the level where the taper begins in the wet well. Either condition will affect the time required to drain or fill the wet well, resulting in an error.)
3. Measure the time required to pump the wet well down over the fixed operating level.
4. Calculate the pump capacities.
5. Record test data (see Figure 8.57 for a sample record sheet).
   a. Pump discharge pressure and pump suction pressure when the pump is running;
   b. Pump suction pressure with the pump off (will measure static suction head which is the wet well level above the gage); and
   c. Pump discharge pressure with the pump off, the suction valve closed, the discharge gate open and the check valve open (will measure static discharge head of the system).

CAUTION: When checking the force main static pressure with the suction gate valve closed and the discharge gate valve open, the check valve may be difficult to open or you may not be able to open it at all if there is any leakage back through the suction gate valve. Depending on the static head on the clapper of the valve, it may not open under any circumstances. An alternate to this method is to review the as-built drawings of the force main to determine the highest point in elevation from the drawings and then calculate the static head from the drawing.

C. PROCEDURE

1. Pump wet well down to a level lower than the turn off point which you have selected. (*NOTE:* A piece of flexible plastic tubing on the discharge of the pump not being used will give an accurate indication of actual wet well levels. The bubble pipe will not do this, since it is located off the wet well floor.) Turn pump off and allow the wet well level to start increasing again.
2. As the level passes through the low level point you have selected, start timing with a stopwatch.
3. When the level reaches your selected high level point, record the fill time, which can be called T1. Start pump and restart timing immediately. When the water level reaches your selected low level point, record the pump down time as T2. Record suction and discharge pressure gage readings while pump is operating.

This sequence constitutes one cycle and for accuracy should be repeated several times.

D. FORMULAS

1. Wet Well Volume
   a. For a round wet well, the volume (V) is equal to the cross-sectional area (A) times the depth (d), where the cross-sectional area is equal to 0.785 x D^2 and D^2 = diameter squared. The depth is equal to the operating level which you select, that is, the difference between the turn on and turn off points for the pump.
   b. To calculate the volume in cubic feet, use the formula 0.785 x D^2 x 1 foot. This will give you the volume in cubic feet for each foot of operating depth. To convert from cubic feet to gallons, multiply the volume in cubic feet by 7.5 gallons per cubic foot. If the operating depth is more than one foot, simply multiply by the number of feet to get the total volume in gallons over the operating level which you have selected.

   NOTE: The greater the operating depth, the more accurate the results should be because changes in the rate of flow into the wet well caused by such things as pump stations turning on and off upstream of the station will be averaged out.

2. Pump Capacity

$$\text{Q, in GPM} = \frac{\text{Wet Well Volume}}{\text{T1}} + \frac{\text{Wet Well Volume}}{\text{T2}}$$

Where:

Wet well volume is the volume in gallons of the section of wet well the test is run over.

T1 = Fill time in minutes

T2 = Pump down time in minutes

3. Total Dynamic Head (TDH)

$$\text{TDH, ft} = \text{(Suction Gage, psi)(2.31 ft/psi)} + \text{Elevation Difference, ft} + \text{(Discharge Gage, psi)(2.31 ft/psi)}$$

Where:

Elevation difference is the vertical distance (difference in elevation) between the suction gage and discharge gage pressure taps.

EXAMPLE 2

The pump station shown in Figure 8.56 consists of a wet well 42 inches in diameter. The depth or difference in elevation between level 1 and level 2 is 2 feet. Water flowing into the wet well requires 5 minutes and 24 seconds to rise from level 1 to level 2. The first pump is turned on immediately when the water reaches level 2 and it takes 18 minutes and 45 seconds to pump the water level down to level 1. Calculate the pump capacity in gallons per minute (GPM).

Known		Unknown
Wet Well Diameter, in	= 42 in	1. Pump Capacity, GPM
Depth, ft	= 2 ft	2. TDH, ft
Fill Time, min	= 5 min 24 sec	
Pump Down Time, min	= 18 min 45 sec	

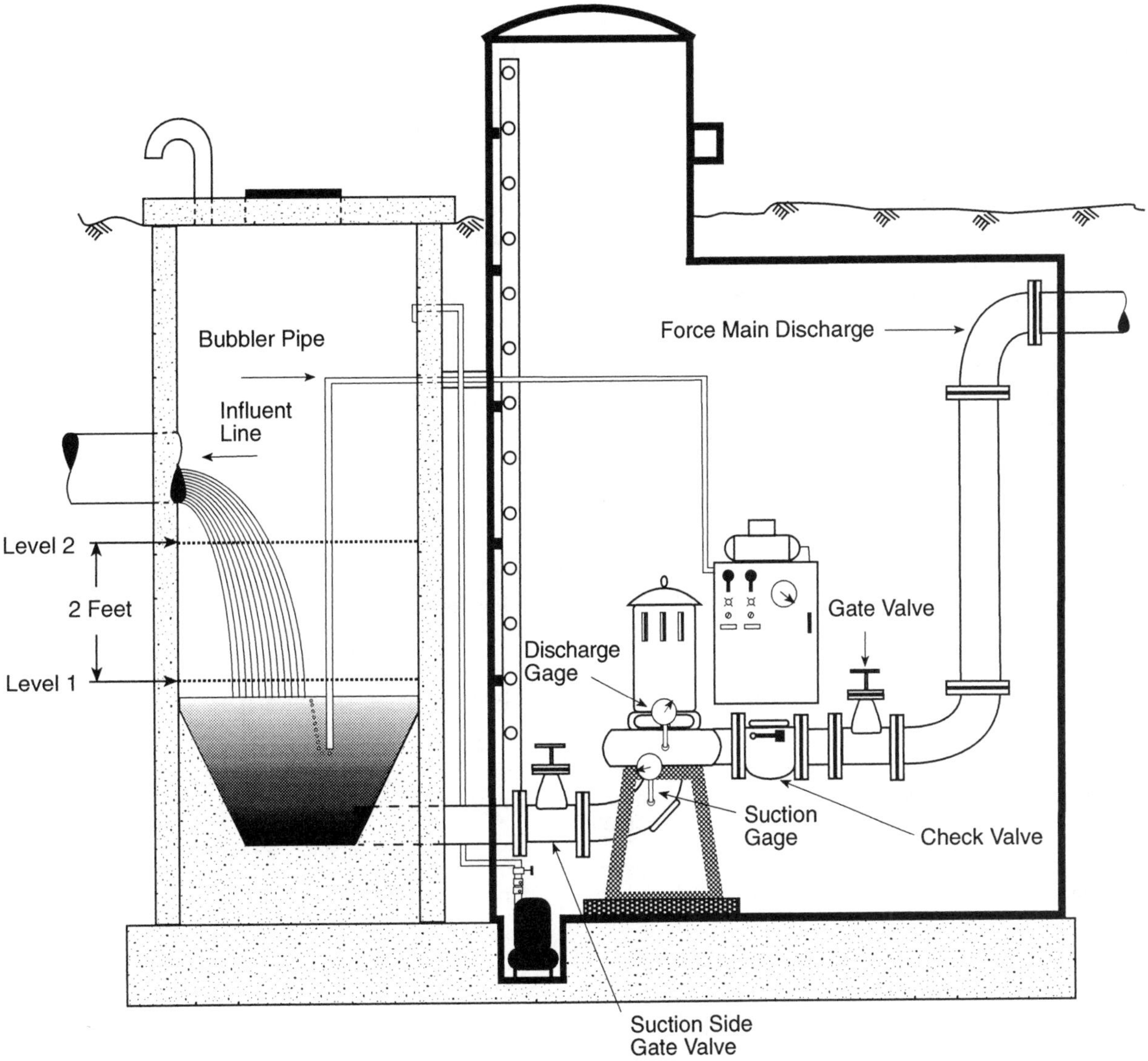

Fig. 8.56 Illustration of time/volume method of pump head-capacity calibration

EQUIPMENT STATUS AND DATA SHEET

Pressure Measurement	Suction Side Gate Valve	Discharge Side Gate Valve	Check Valve	Pump Condition	Gage Location	Gage Readings	Wet Well Operating Range, inches
Static, wet well	open	closed	closed	ALL pumps off	volute or plastic tubing		
Static, force main	closed	open	open	ALL pumps off	pump volute		
Dynamic, "shutoff"	open	closed	closed	1 pump on	pump volute discharge		
Dynamic discharge	open	open	normal operation	1 pump on	pump volute discharge		
Dynamic suction	open	open	normal operation	1 pump on	pump volute suction		
Dynamic discharge	open	open	normal operation	2 pumps on	pump volute discharge		
Dynamic suction	open	open	normal operation	2 pumps on	pump volute suction		
Dynamic discharge	open	open	normal operation	3 pumps on	pump volute discharge		
Dynamic suction	open	open	normal operation	3 pumps on	pump volute suction		

Comments:

Fig. 8.57 Typical data sheet for field calibration of lift station pumps

1. Convert wet well diameter to feet.

$$\text{Diameter, ft} = \frac{\text{Diameter, in}}{\text{12 in/ft}}$$

$$= \frac{\text{42 in}}{\text{12 in/ft}}$$

$$= \text{3.5 ft}$$

2. Convert fill and pump down times to minutes.

$$\text{Fill Time, min} = \text{5 min} + \frac{\text{24 sec}}{\text{60 sec/min}}$$

$$= \text{5.4 min}$$

$$\text{Pump Down Time, min} = \text{18 min} + \frac{\text{45 sec}}{\text{60 sec/min}}$$

$$= \text{18.75 min}$$

3. Calculate the wet well volume between levels 1 and 2 in gallons.

$$\text{Wet Well Vol, gal} = 0.785(\text{Dia, ft})^2(\text{Depth, ft})(\text{7.48 gal/cu ft})$$

$$= (0.785)(\text{3.5 ft})^2(\text{2 ft})(\text{7.48 gal/cu ft})$$

$$= \text{144 gallons}$$

4. Calculate the pump capacity in gallons per minute (GPM).

$$\text{Pump Capacity, GPM} = \frac{\text{Wet Well Vol, gal}}{\text{Fill Time, min}} + \frac{\text{Wet Well Vol, gal}}{\text{Pump Down Time, min}}$$

$$= \frac{\text{144 gal}}{\text{5.4 min}} + \frac{\text{144 gal}}{\text{18.75 min}}$$

$$= \text{26.7 GPM} + \text{7.7 GPM}$$

$$= \text{34.4 GPM}$$

To determine the total dynamic head (TDH) for the pump, use the pump discharge and suction pressure readings and the difference in elevation of the pressure taps (assume pressure gages read actual pressure at tap). In the example problem the dynamic (pump running) suction gage reading is 1.5 psi (a negative pressure) and the dynamic discharge gage reading is 5 psi. The difference in elevation between the pressure gage taps is one foot.

Known		Unknown
Suction Gage, psi	= 1.5 psi	Total Dynamic Head, ft
Discharge Gage, psi	= 5 psi	
Gage Tap Elevation Difference, ft	= 1 ft	

5. Calculate the total dynamic head (TDH) in feet.

$$\text{TDH, ft} = (\text{Suction Gage, psi})(\text{2.31 ft/psi}) + \text{Elevation Difference, ft} + (\text{Discharge Gage, psi})(\text{2.31 ft/psi})$$

$$= (\text{1.5 psi})(\text{2.31 ft/psi}) + \text{1 ft} + (\text{5 psi})(\text{2.31 ft/psi})$$

$$= \text{3.5 ft} + \text{1 ft} + \text{11.6 ft}$$

$$= \text{16.1 ft}$$

Compare the calculated capacity of 35 GPM against a head of 16 feet with the manufacturer's catalog head-capacity curve to determine if the pump is performing at the expected head and capacity.

NOTE: If the suction gage reads a positive pressure, the TDH, ft, would be

$$\text{TDH, ft} = -\text{3.5 ft} + 1.0 + \text{11.6 ft}$$

$$= \text{9.1 ft}$$

EXAMPLE 3

Another way to measure the capacity of the pump shown in Figure 8.56 is to plug the influent line into the wet well. This procedure is acceptable *PROVIDED* influent flows are low and there is sufficient wastewater storage capacity in the influent line. Do not attempt this method if the influent line will fill with wastewater, back up into homes or overflow manholes into streets and yards.

Plug the influent line and fill the wet well as full as possible with water. For example, fill the wet well with water to a level or depth 6 feet above level 1. Turn the pump on and record a time of 12 minutes and 18 seconds to drop the water 6 feet down to level 1. Calculate the pump capacity in gallons per minute (GPM). The wet well diameter is 42 inches or 3.5 feet.

Known		Unknown
Wet Well Diameter, ft	= 3.5 ft	Pump Capacity, GPM
Depth, ft	= 6 ft	
Pump Down Time, min	= 12 min 18 sec	

1. Convert the pump down time to minutes.

$$\text{Pump Down Time, min} = \text{12 min} + \frac{\text{18 sec}}{\text{60 sec/min}}$$

$$= \text{12.3 min}$$

2. Calculate the wet well volume pumped down during the time interval.

$$\text{Wet Well Vol, gal} = (0.785)(\text{Dia, ft})^2(\text{Depth, ft})(\text{7.48 gal/cu ft})$$

$$= (0.785)(\text{3.5 ft})^2(\text{6 ft})(\text{7.48 gal/cu ft})$$

$$= \text{432 gallons}$$

3. Calculate the pump capacity in gallons per minute (GPM).

$$\text{Pump Capacity, GPM} = \frac{\text{Wet Well Vol, gal}}{\text{Pump Down Time, min}}$$

$$= \frac{\text{432 gal}}{\text{12.3 min}}$$

$$= \text{35.1 GPM}$$

NOTE: Pump capacity is approximately 35 GPM by both methods. Results in the field may be within ± 10% or ± 3 or 4 GPM.

EXAMPLE 4

The wire to water efficiency of a pumping system (pump and electric motor) is important information to record. The information needed is listed under "Known."

Known		**Unknown**
Flow, GPM	= 470 GPM	Efficiency, %
TDH, ft	= 58 ft	
Voltage, volts	= 220 volts	
Current, amps	= 36 amps	

1. Calculate the wire to water efficiency.

$$\text{Efficiency, \%} = \frac{(\text{Flow, GPM})(\text{TDH, ft})(100\%)}{(\text{Voltage, volts})(\text{Current, amps})(5.308)}$$

$$= \frac{(470 \text{ GPM})(58 \text{ ft})(100\%)}{(220 \text{ volts})(36 \text{ amps})(5.308)}$$

$$= 65\%$$

When all data have been recorded, shut off the pumps and allow the wet well to refill. When full again, operate the pumps to adjust the shutdown points. You may have to operate more than one pump to pull the sump level down. While pumping the level down, observe the pumps in operation and record any unusual conditions. With all set points adjusted, operate the station on automatic a number of times through the full range, observing what happens and recording the results. Too much information is better than not enough. When you are satisfied that the automatic system is operating in a consistent manner, notify the engineer that the lift station is acceptable to you and to put it on line.

During the first few weeks of operation, frequently inspect the equipment. Bearing failures and other problems may develop after a few days of operation.

QUESTIONS

Write your answers in a notebook and then compare your answers with those on pages 120 and 121.

8.2W How much leakage should there be from a packing gland?

8.2X Are there any special precautions with Teflon packing?

8.2Y What minimum pressure is needed for the flushing water on a mechanical seal?

8.2Z What would be the first thing that you should do before operating a lift station?

8.2AA Why should you bother to remove the debris left in the sump?

8.2BB What are the ways or means used to determine whether the sump level is rising or not?

8.2CC How do you determine the direction of pump rotation?

8.2DD Is it necessary to inspect all points of operation if there is more than one pump located within a station?

8.2EE Should you continue operating a unit if you feel it is not functioning to your satisfaction?

8.2FF Why is it necessary to record all data when test operating the station?

END OF LESSON 3 OF 5 LESSONS

on

LIFT STATIONS

Please answer the discussion and review questions next.

DISCUSSION AND REVIEW QUESTIONS

Chapter 8. LIFT STATIONS

(Lesson 3 of 5 Lessons)

Write the answers to these questions in your notebook before continuing. The question numbering continues from Lesson 2.

15. How would you examine or review a set of prints and specifications for a new lift station? or What items would you be sure were satisfactory from the viewpoint of operation and maintenance operators?
16. Who should inspect a newly completed lift station? and What items should be inspected?
17. When should a preventive maintenance program be started for a new lift station?
18. What items would you inspect in a new engine in a lift station?
19. How would you make an operational inspection of a new lift station?
20. What would you do if an operating unit is not functioning properly during the check-out inspection?

CHAPTER 8. LIFT STATIONS

(Lesson 4 of 5 Lessons)

8.3 OPERATION OF WASTEWATER LIFT STATIONS

After a lift station has been constructed and put into operation, it is the responsibility of the operating agency to ensure the continuous and efficient operation and maintenance of the lift station, including the structures and the grounds. This responsibility includes preventing failures in operation that would result in flooding upstream homes, businesses or streets. Responsible design includes no facilities for bypassing wastewater to rivers, streams, lakes or drainage courses. When emergencies occur, portable emergency equipment must be used to pump the wastewater to a functioning section of the downstream collection system and not to the environment. When untreated wastewater is discharged to the environment, public health hazards and pollution of adjacent receiving waters result.

Lift stations may be located throughout a community and must be neat in appearance, blend with the architecture and landscaping of the neighborhood, and not create a nuisance to neighbors through odors or noise. Complaints from the public will be few if the operators responsible for the lift station maintain the facility in top operating condition and respond to questions or complaints from the public in a positive and concerned manner. When responding to a complaint, be sure to tell the public what has been done or will be done to correct the complaint.

8.30 Lift Station Visits by Operators

ONE RULE THAT SHOULD APPLY TO ALL LIFT STATION VISITS IS THAT FOR SAFETY PRECAUTIONS, THERE MUST ALWAYS BE TWO OPERATORS MAKING THE STATION VISIT. Many state OSHA agencies consider a wastewater lift station a confined space. Always follow confined space entry procedures. Safety precautions regarding the potential presence of hazardous gases apply not only in the wet well, but in the dry well area of the station too. This rule must be obeyed during off-duty hours such as nights, weekends and holidays when operators are responding to lift station telemetry alarms. Always take the few extra minutes to pick up the required additional crew member. The additional effort is critical and worthwhile when compared to the sorrow and costs that result from an injury or a lost life.

The two-operator crew usually consists of:

1 Maintenance Mechanic I, and

1 Maintenance Mechanic Helper.

QUESTIONS

Write your answers in a notebook and then compare your answers with those on page 121.

8.3A How can nearby streams and rivers be protected from wastewater when a lift station fails?

8.3B How can complaints from the public about a lift station be minimized?

8.3C What is the most important rule that always must be considered when visiting a lift station?

8.31 Frequency of Visits to Lift Stations

A rule cannot be developed for determining the frequency of visits to lift stations for operational inspections. The frequency of lift station visits varies by community and may range from continuously staffed pumping plants, twice daily visits, once a day, two to four times a week, once a week, to once a month visits.

Frequency of lift station visits depends on the following:

1. Number of stations in the community,
2. Type of wastewater being conveyed,
3. Potential damage resulting from storms flooding station,
4. Condition of equipment, such as equipment temporarily repaired and waiting for replacement parts,
5. Design of facility and equipment installed in the lift station,
6. Adequacy of preventive maintenance and overhaul program,
7. Type, adequacy, and reliability of telemetry system,
8. Attitude of operating agency toward operation and maintenance, and
9. Number of employees available to visit lift station.

The attitude of the operating agency is probably the most significant factor today in determining the frequency of lift station visits. Many operating agencies make daily visits to their stations because they have always visited lift stations daily and also because they know that a great number of other communities make daily visits. However, have the people responsible for station operation and maintenance analyzed their collection system and their operation and maintenance programs and proved that daily station visits are required? To conduct this type of analysis, records must be kept and studied in order to answer the following questions:

1. How many station failures are found each day?

   a. Electrical power failures.

   b. Failure of level sensing or other control equipment.

   c. Flooded wet wells.

d. Plugged pump impellers or lines.
e. Overheated and tripped-out motor thermals.
f. Sump pump off and sumps or dry wells flooded.
g. Air- or gas-bound pump.
h. Stuck or blocked check valve.
i. Pump control systems not functioning.
j. Failure of lift station telemetry system.
k. Ventilation fans burned out.

2. What functions were performed at each lift station that kept the station operating, and what would have occurred if the function had not been performed during the daily station visit?

As an example, assume a pump impeller was plugged with rags and not pumping. The station crew during their visit and inspection discover the problem, derag the pump and restore the pump to service. However, if the pump had plugged immediately following their visit, the telemetry system would have indicated a failure. This crew or another crew would be dispatched to return to the station. If the pump was not telemetered, the backup or alternate pump would have pumped the wastewater until a high wet well level was reached. At this point a high water level alarm would have been activated or the other pump would have handled the flow until the next day when the crew again routinely visited the station. Therefore, with an adequate telemetry system, a routine visit is not essential under these conditions.

3. How many actual hours of work are performed at each station during the visit? Do not include crew travel time to the station.

4. What is the critical period of time if the station does fail?

This time period is the length of time from when the failure occurs at high water level in the wet well to when back-up flooding or overflow will occur. When this critical time is determined (it should be reevaluated annually because flow conditions change), it provides the time available for a station crew to respond to an alarm condition. Any critical time under one hour creates a serious problem for the operating agency and corrections should be made by installation of larger pumps. If power failures are a problem, an emergency generator should be installed to operate the station during power failures. Another possibility is to staff the station on a continuous basis and to operate with manual controls or overrides during power failures.

Lift stations operate automatically and, if properly designed and maintained, do not require daily visits under most circumstances. Typically the daily visit is usually a quick inspection of the power panel for tripped breakers, indicating lights of operating equipment, flow data and recorded elapsed time meter readings. Motors in the station are usually examined to see if they are noisy or running hot. Items typically checked during a visit include the pump packing or seal system, check valves, and suction and discharge pressures, routine lubrication and cleanup. The sump pump switch is flipped to see if it is operating. The visiting crew may bleed condensate from the air bubbler system for wet well and pump control and glance at the wet well indicator to see if it is reading properly. The crew may look into the wet well for sticks and even observe that there is water in it.

Unfortunately this description of a lift station inspection is not uncommon even when performed by competent and well qualified operators. Daily visits tend to cause operators to give the station a quick glance in a noncritical manner because the crew was here yesterday and will be back tomorrow. For this reason, important maintenance items are often delayed until tomorrow which sometimes doesn't come for a month. Stations that are visited daily may be dirtier and more in need of maintenance than stations visited less frequently. If a lift station requires a daily inspection, the station book should have a daily task check-off sheet indicating the tasks to be performed and a space for the initials of the operator who performed the task to indicate who did the job.

A very important aspect regardless of the frequency of visits is the fact that operating and maintenance operators must be provided the time and training to adequately perform each task. Also supervisors must occasionally inspect stations after a crew visit to see that the tasks were performed and not merely signed off. Once procedures and objectives have established that specified tasks must be done on a routine frequency (whether daily, weekly, monthly, quarterly or annually), the operators responsible for the lift station must accept the responsibility and be sure the task is completed when scheduled so the operation of the lift station will be as reliable as possible. Analysis of records can indicate if adjustments in the frequency of tasks are necessary.

Small communities with one or two lift stations may find it more economical to visit stations twice daily rather than to install and to maintain a telemetry system. When the operating agency has 15 to 50 lift stations to operate and maintain, then the time consumed for the single daily visit becomes very large and quite expensive. If it is possible to reduce station visits from once a day to once a week, the savings are substantial. Monthly station visits are practical, but visits at greater time intervals are not cost effective.

This chapter is not attempting to convince every operating agency to extend lift station visit frequencies, because this is not applicable to every station. It is attempting to indicate that many people have little confidence or no faith in their lift station's equipment and programs and do not believe that a station can operate for a week, let alone 30 days, by itself. Sacramento County, California, has been successfully operating lift stations with monthly visits for over 35 years with relatively few problems. In fact there are fewer problems than when the stations were visited on a daily basis.

The following is a list of requirements that must be met to permit limited visits to lift stations of once a week to once a month.

1. Equipment lubrication reservoirs must be large enough to hold lubricant supplies to meet needs between visits to the station.
2. Telemetry of the station is required. Most lift stations constructed today are telemetered at least for the high water alarm. Usually other items may be sensed and alarmed on the same high water alarm signal, for example:
   a. Water level in sump pump pit of dry well,
   b. Power failure,
   c. Air compressor failure,
   d. Lift pump check valves (when pump motor starts and check valve does not open, an alarm sounds indicating a pump failure),
   e. Emergency generator run or fail alarm,
   f. Station entry or intrusion alarm, and
   g. In stations using other auxiliary equipment necessary for station operation, the following additional items also may be sensed and telemetered:
      (1) Water supply pressure,
      (2) Chlorine leaks,
      (3) Low chlorine pressure,
      (4) High temperature of motors and engines, and
      (5) Worker inside stuck elevator.

With the recent advances in solid state electronic equipment, telemetered monitoring and supervisory control systems have become much more commonplace in collection systems. They range in complexity from very basic monitoring systems to more sophisticated computer-based alarm, telemetry, supervisory control, and flow-monitoring systems.

These systems consist of an electronic interface device located in the lift station which transmits alarm or monitored information by radio signals or over leased telephone lines. In some cases, an agency may run its own transmission lines or microwave systems; however, these methods currently are not common practice.

Information is received at a central location. This could be a 24-hour, seven-days-a-week dispatch center in the community such as the public safety office, or in the case of a utility that has its own 24-hour, seven-day-a-week dispatcher, to that office.

A simple system may be nothing more than an annunciator board that indicates the lift station and type of alarm. However, microcomputers are now frequently used in this application and perform logging functions to record various information such as date, time and nature of alarm.

In some systems the computer automatically polls various lift stations at a predetermined sequence and inquires the status of the station. If an alarm occurs, the computer can be programmed to automatically dial a preprogrammed number to a collection system operator and relay a message over the phone.

These monitoring and telemetry systems are frequently referred to as SCADA, which stands for Supervisory Control and Data Acquisition system. Figures 8.58 through 8.64 illustrate recent state-of-the-art systems that are used for monitoring lift station alarms, flow monitoring, rainfall information, and river quality monitoring information. Figures 8.58 and 8.59 illustrate the components of the system; Figure 8.60 shows a SCADA communication monitor screen; and Figure 8.61 shows an aerial view of the entire collection system as it appears on the SCADA monitoring screen.

Analog or digital inputs (AI, DI) are fed into a remote terminal unit (RTU) which is in fact a microprocessor located in the lift station. Information is transmitted over a leased telephone line, into a communication interface, and ultimately into a minicomputer. Other auxiliary equipment includes printers, XY plotters, video copier, fixed disc for memory, and tape backup unit.

Although these are fairly sophisticated systems, the complexity is dictated by the numbers of field inputs and the needs of this particular agency. However, the basic concept would apply to small systems without the optional auxiliary equipment.

Figures 8.62 and 8.63 illustrate graphic information available to the dispatcher when a lift station alarm occurs and includes lift station number, flowmeter number, time, date, alarm acknowledge indicator, power failure, transmission line failure, station power failure, standby generator running, high wet well, high dry well, pump 1 fail or pump 2 fail. The screen shows that flow is coming into the station through a bar screen and gives the actual flow reading occurring at the Parshall flume at the time of the alarm. The flume discharges into the wet well and the station has two pumps located in the dry well, as well as a standby generator permanently located at the site. The screen also indicates current station status.

Figure 8.64 is additional information that can be pulled up on the individual lift station and includes critical utility telephone numbers, who is responsible for the station, daily comments, what operator response is required and what dispatcher action was taken.

Virtually any type of alarm telemetry system improves the options available to the collection system operator when delegating personnel and equipment resources, particularly during emergency situations. Events such as tornadoes, extremely high intensity rainfall, thunderstorms, loss of power, flooding or other emergency conditions place severe strains on any collection system. If the collection system operator has a thorough knowledge of the system, a telemetered supervisory system which indicates station status will be invaluable in preventing backups and/or bypasses, station flooding or other severe problems during an emergency.

Chapter 12, Section 12.712, "SCADA Systems," includes additional information on the application of SCADA systems and hand-held computers by field crews.

QUESTIONS

Write your answers in a notebook and then compare your answers with those on page 121.

8.3D What items or factors influence the frequency of lift station visits?

8.3E What are the different types of lift station failures?

8.3F What is the critical time for a lift station after failure has occurred?

8.3G When is a daily visit to a lift station required?

8.3H What requirements must be met for a lift station to perform effectively and only be visited from once a week to once a month?

8.3I How is an alarm situation transmitted from a lift station to a dispatcher at a control center?

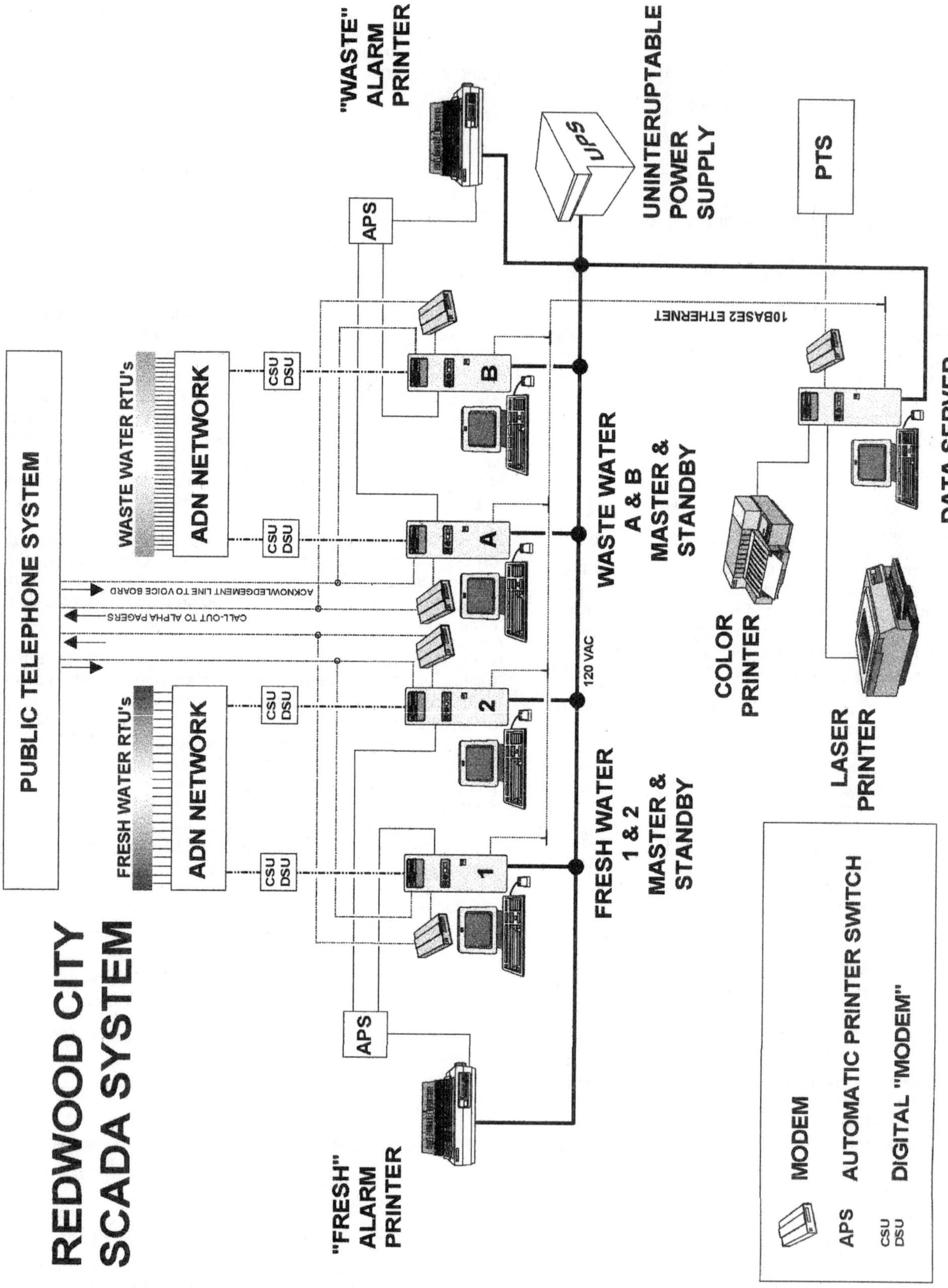

Fig. 8.58 Components of a SCADA system
(Permission of the EDCCO Group www.edcco.com)

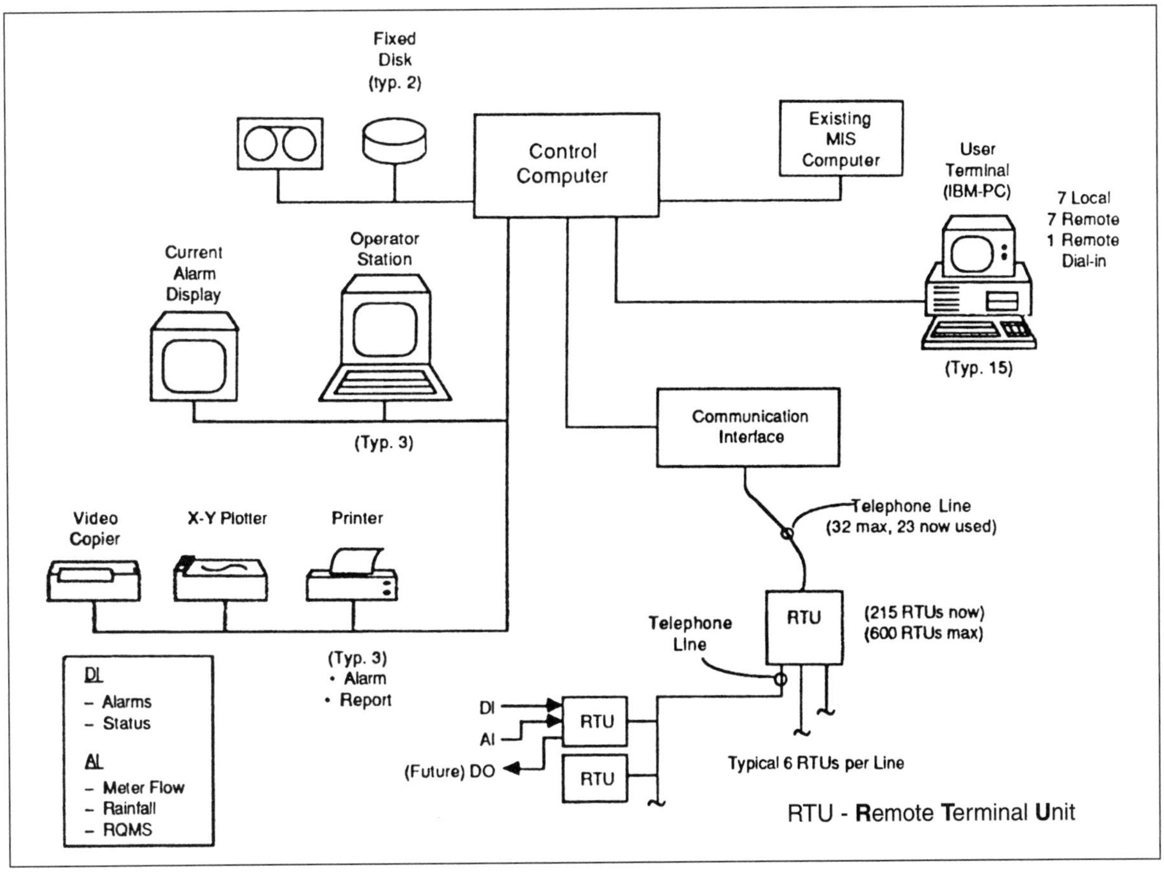

Fig. 8.59 Components of a monitoring and telemetry system

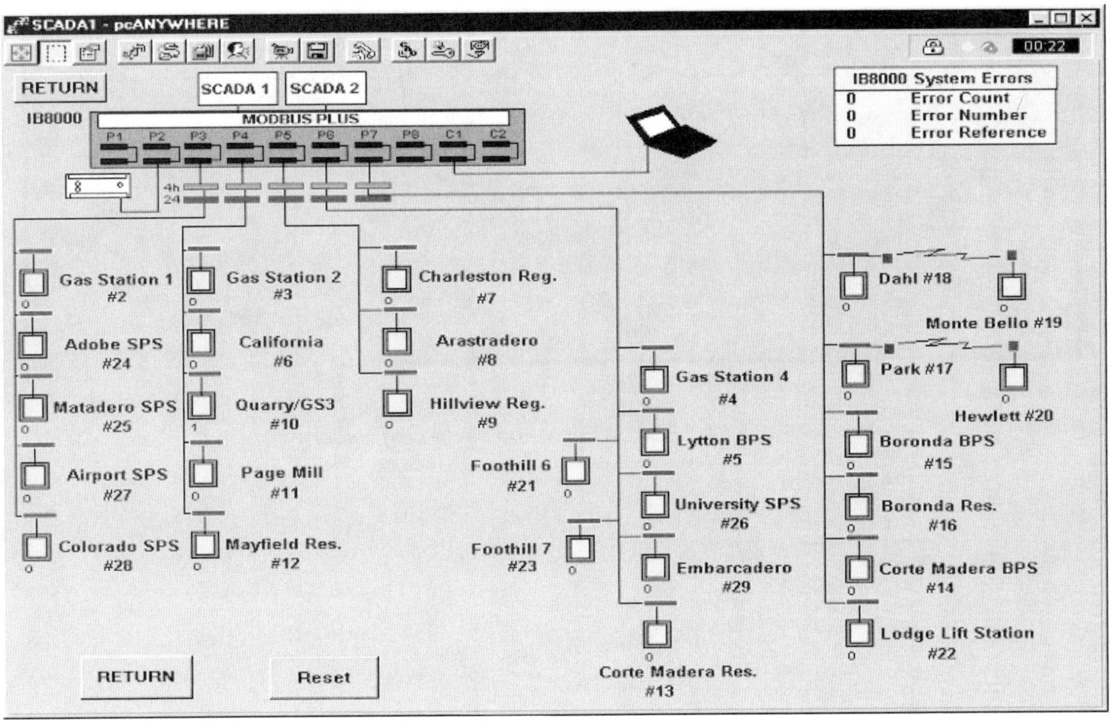

Fig. 8.60 SCADA communication monitor screen

(Permission of the EDCCO Group www.edcco.com)

Fig. 8.61 SCADA aerial and monitoring screen
(Permission of the EDCCO Group www.edcco.com)

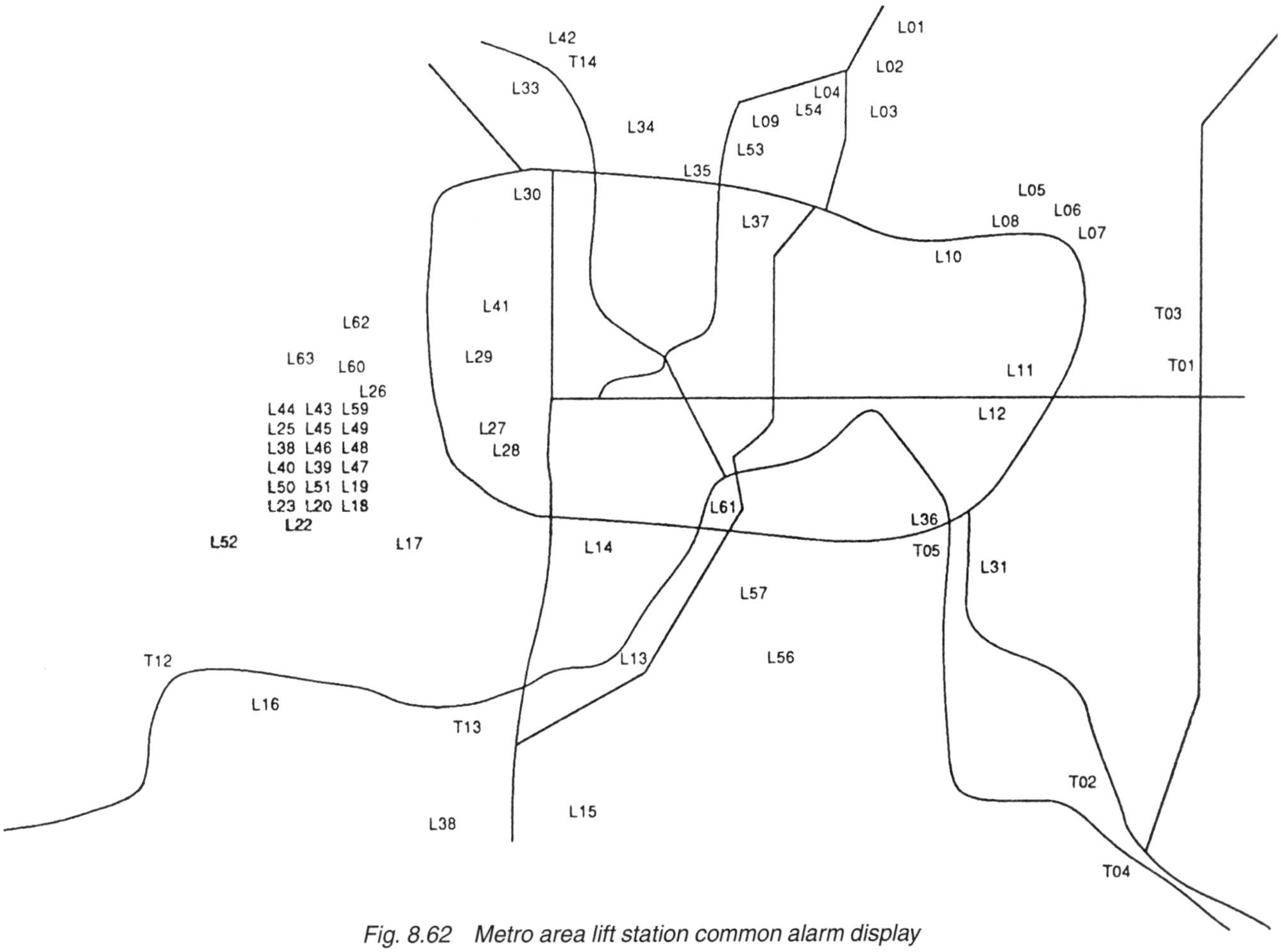

Fig. 8.62 Metro area lift station common alarm display

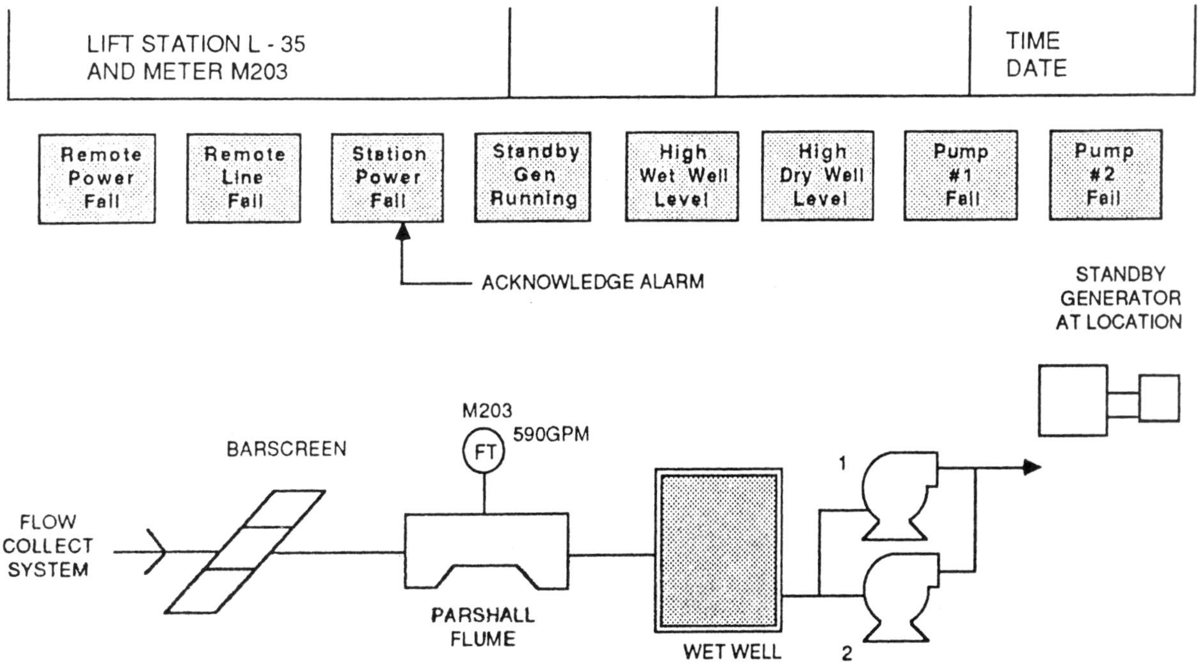

DAILY NOTES: PUMP NO. 1 IS TURNED OFF UNTIL 4/17.
IF PUMP NO. 2 FAILS, IMMEDIATELY CALL
SERVICE PERSONNEL.

Fig. 8.63 Typical individual lift station information

LIFT STATION L-35 AND METER M203 MOUNDS VIEW MM/DD/YY HH:MM:SS

ADDRESS: 2345 W. COUNTY ROAD H, MOUNDS VIEW SERVICE AREA: 3

TELEPHONE COMPANY: NWB
TELEPHONE NUMBER AT STATION: 784-0910
TELEPHONE LINE NUMBER: 3TC XX 0044
TELEPHONE REPAIR NUMBER: 611

POWER COMPANY: NSP
OFFICE NUMBER: 221-4411
EMERGENCY NUMBER: 221-4411

NOTES: DURING NORMAL WORK DAY HOURS RESPOND TO ALL ALARMS IMMEDIATELY.
BOB McCABE AND BERNIE MURPHY SERVICE THIS STATION DAILY
UNIT 104 IS FAMILIAR WITH STATION
DETENTION TIME FOR L-35 IS ABOUT 1 HOUR
STATION IS EQUIPPED WITH AN EMERGENCY GENERATOR
THIS STATION HAS 2 PUMPS

DAILY
NOTES: PUMP NO. 1 WILL BE TURNED OFF FOR TWO DAYS 4/16 AND 4/17
IF PUMP NO. 2 FAILS, IMMEDIATELY CALL
SERVICE PERSONNEL.

WEEKDAY SERVICE: UNIT 104 OR 109
7:00 AM - 3:30 PM

OFF HOURS SERVICE
ON CALL: BOB McCABE 789-2964
STANDBY:

ALARM MESSAGE: STATION POWER FAIL

RESPONSE REQUIRED: NONE, WAIT FOR 10 MINUTE TIME-OUT ALARM MM/DD/YY HH:MM:SS

OPERATOR RESPONSE: CALLED BOB McCABE AT 789-2964

Fig. 8.64 Telemetered alarm report

8.32 Essential Tasks During Lift Station Visits

The only time a lift station should be visited is when it is toured by visiting dignitaries from out of town or by management inspecting the facilities. Neither of these two groups nor any other tour group tend to significantly add to the functional capability of a well maintained lift station. So as not to be misleading, let's define the term *VISIT.* There are two other types of station visits that are significant:

1. To perform scheduled routine preventive maintenance upon the equipment and the station, and
2. To respond to a lift station alarm or emergency condition to restore the station to operation.

If there is a *GOOD PREVENTIVE MAINTENANCE PROGRAM* and the *FREQUENCY SCHEDULE* is at the proper intervals with clear duty assignments in writing and completion verified by management, then there will be only a few occasions when a crew must respond to an alarm.

In this section we have attempted to convey a very important fact. Wastewater lift stations can be capable of operating automatically. Proper maintenance will allow these stations to operate automatically for long time periods.

Why have many lift stations been visited daily? Probably because they were run by operators of wastewater treatment plants. Wastewater treatment plants are complex facilities requiring daily observation and regulation of treatment processes. Fortunately most small lift stations are not as complex and can be designed and maintained in a manner that allows the station to perform its functions automatically and without daily visits. The frequency of visits to a station depends on the equipment and station conditions. If a lift station overflows after five minutes of down time, this station may require an operator on duty at all times. The remainder of this section outlines what should be done to maintain the operational capabilities of a lift station.

8.320 Station Sign-In Log

Every lift station should have a sign-in log (Figure 8.65). The larger the operating agency, the more important a sign-in log becomes. People who enter the station must sign in and log their time of arrival and departure. This rule includes managers, supervisors and operation and maintenance operators. This practice reduces the number of mysterious changes in station operation such as pumps turned off, control system changes, station doors or gates left unlocked, and provides a

PUMP STATION SIGN-IN LOG

DATE	IN	OUT	NAME	DATE	IN	OUT	NAME

Fig. 8.65 Station sign-in log

method for reaching the appropriate person when corrective instructions need to be issued. The sign-in log produces a higher reliability factor in operator performance because operators do not wish to sign their name to a log at a given time indicating that they had made a station visit and a few minutes after their departure have a station fail because they had neglected a minor or major item. The station's sign-in log is kept inside the station on a small shelf or cabinet with the station book. When the last space is used on the sign-in sheet, that person or crew turns the sheet in to the agency office to be filed in the station book maintained at the office. New sign-in sheets are kept in the station as part of the preventive maintenance schedule. Some agencies use log books where agency operators sign in and record their activities at the station. The log books are generally maintained at the station for an entire year. Under these conditions, the agency's management may not see these logs and may not be kept informed of station problems and conditions.

8.321 The Station Book

The station book is a loose leaf binder so that changes in the contents can easily be made, thus keeping the station's book up to date. The lift station book should contain the following information.

1. Station identification number

   Most lift stations are given a name, such as "River Side" or "Park Road." In larger agencies and with the use of computer systems for cost accounting, stations are usually given a code consisting of letters and numbers (RS-25) for identification. This identification code must be used by operating and maintenance operators (see Section 8.5, "Recordkeeping").

2. Map showing important facilities

   This map should show the location of the lift station and force main, including valves, manholes, and discharge point of the station force main. Other important features such as access route and other utilities are helpful.

3. Station description

   This is a brief general description of the lift station and includes type, pump layout, control system and auxiliary systems.

4. Safety instructions

   Station safety instructions for working in the station and on its equipment are included in the book. These instructions must spell out specific hazards encountered in this specific lift station.

5. Equipment data sheets

   Detailed equipment data sheets that were completed during the lift station inspection after completion of construction are important. These sheets should contain nameplate data and equipment operating characteristics.

6. Emergency conditions

   Action to be taken when an emergency occurs must be clearly outlined.

   Emergency conditions include:

   a. Power failure,

   b. Stoppage of inlet line,

   c. Simultaneous failure of pumps, and

   d. Flooding of dry pit.

   Critical time until homes or streets are flooded must be included. If auxiliary power generators are not part of the lift station, directions for obtaining one of adequate size and the minimum size should be included in the station book. Also included are procedures to electrically isolate the lift station and to connect a portable generator. If the station must be bypassed, the location of manholes for pump suction and discharge pipes, length and size of pipe required, and size of pump needed to pump around the station until repairs can be made must be noted in the station book.

7. Station preventive maintenance schedule

   The lift station preventive maintenance schedule should contain every piece of station equipment and include the following information:

   a. Task description,

   b. Frequency each task is to be performed,

   c. Whether isolation of equipment is required, and

   d. Maintenance and/or overhaul instructions or number of paragraph containing instructions that apply to that station equipment or maintenance functions which are scheduled for that visit.

Development of the station book is not any easy job and the book is not prepared overnight. The book is never completed, because it must be updated and revised due to changing flow conditions, equipment changes and modifications, and continuous attempts to improve operating and preventive maintenance programs.

When your first station book has been prepared, the chore becomes much easier for the next station, because most of the preventive maintenance and overhaul descriptions have been prepared for common equipment and only new equipment that is not in the program requires the development of maintenance procedures. The procedure described in this section has been developed over the past ten years and is continuously being expanded and improved.

8.322 Operation and Maintenance Duties

See Section 8.4, "Lift Station Maintenance."

Experience with many lift stations for many years indicates that most lift stations have very few problems. The following factors greatly affect the dependability of a lift station:

1. At a minimum, inspect the station on a weekly basis and during each inspection: (a) clean the control floats or bubbler tube, (b) inspect the wet well and note the condition, (c) activate the alarm float to test the alarm system and the response of the monitoring company, and (d) clean the bar screen, if present.
2. Once a month, simulate a power failure at the stations equipped with a genset by throwing the incoming main. This fully tests the genset and transfer switch.
3. At least twice a year (more frequently if conditions dictate), have a jet or high-velocity sewer collection system cleaning crew degrease the station by using the high-pressure jet spray to break up any grease accumulation and wash down the interior of the station.
4. The majority of the stations many agencies maintain have submersible pumps. Pull each pump once a year and change the oil, checking to see if there is any indication of seal failure. At this time, determine if there is any accumulation of grit/debris in the bottom of the wet well and, if so, have a sewer collection system vacuum unit respond to remove the grit.
5. If a station is equipped with equipment such as a macerator or genset, service these at the manufacturer recommended intervals. For gensets, many agencies use a third party technician recommended by the manufacturer to perform a "minor" and "major" service (six-month intervals) each year.
6. Once a year, have an electrician inspect the station controls and cabling. Clean and tighten the contacts.

If you have very large, submersible pump stations (100 HP) that have hydraulic macerators at the influent inlet point, these are usually quite reliable and contribute greatly to the dependability of the station while lowering labor costs significantly.

Monitoring motor draw amperages is very important in order to detect anomalous loading "ragging" of pumps and signs of other problems such as seal failure. Seal failure probes/indicators on submersible pumps can be lifesavers and save or greatly lower repair costs, if acted upon quickly.

Most lift stations failures occur due to lack of attention/preventive maintenance. Additionally, an effective sanitary sewer line cleaning program can greatly contribute to lowering the loading on a lift station from grit and other debris.

QUESTIONS

Write your answers in a notebook and then compare your answers with those on page 121.

8.3J What are the types of significant lift station visits?

8.3K How can the number of responses to lift station alarms or emergency conditions be minimized?

8.3L Why should lift stations have a sign-in log?

8.3M What information is contained in the station book?

8.323 Responding to Station Alarms

Only a few agencies can afford to use telemetry systems having a dial or a meter with numbers that indicate the piece of equipment and nature of the problem in a remote lift station. However, many agencies use telemetry alarm systems that transmit one signal indicating that the lift station is either operational or it needs to be visited. The signal is received at a facility that is staffed 24 hours a day. In many cities this is the local police department or emergency dispatch center. When the station has a failure, the alarm is transmitted (usually through leased phone lines) to a control panel at the center. An indicator light for the particular station goes on when a problem develops, and a horn or some other type of noisy alarm may also be activated. When trouble is indicated, the dispatcher notifies the operator on duty by phone or radio of the station identification number. The dispatcher must record on an appropriate log sheet the time and identification number of the station that sent the alarm signal and the time and the name of the operator notified (see Figure 8.66).

Usually during off-duty hours, the operator notified by the dispatcher picks up the other operator on the crew with an agency repair vehicle enroute to the problem lift station. When this emergency crew arrives at the lift station, the dispatcher should be notified and given an estimate of when the crew will call back in again. This procedure notifies the dispatcher that the crew has arrived at the lift station and protects the crew in case a serious injury occurs and they are unable to contact the dispatcher. If the crew does not call in again and the dispatcher cannot contact the crew, rescue operations can be started. During stormy and adverse weather conditions and especially at night, crews should be very cautious. Overhead power lines must be inspected *BEFORE* attempting to open the gates of a lift station. Fallen power lines across a chain link fence could electrocute a person trying to open a gate. When an emergency crew attempts to locate the cause of an alarm, they must inspect the station in the following sequence (also see Section 8.324, "Typical Lift Station Problems"):

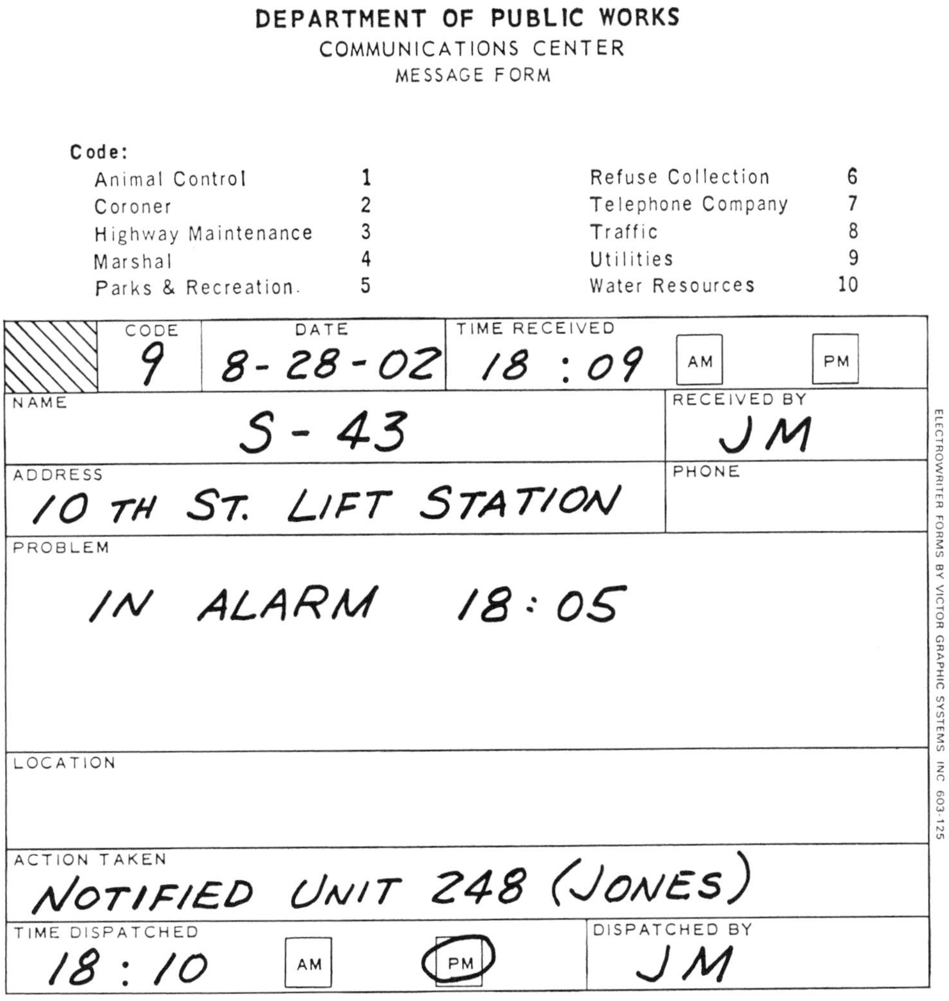

DEPARTMENT OF PUBLIC WORKS
COMMUNICATIONS CENTER
MESSAGE FORM

Code:

Animal Control	1	Refuse Collection	6
Coroner	2	Telephone Company	7
Highway Maintenance	3	Traffic	8
Marshal	4	Utilities	9
Parks & Recreation	5	Water Resources	10

CODE: 9 DATE: 8-28-02 TIME RECEIVED: 18:09 AM PM

NAME: S-43 RECEIVED BY: JM

ADDRESS: 10 TH ST. LIFT STATION PHONE:

PROBLEM: IN ALARM 18:05

LOCATION:

ACTION TAKEN: NOTIFIED UNIT 248 (JONES)

TIME DISPATCHED: 18:10 AM PM DISPATCHED BY: JM

ELECTROWRITER FORMS BY VICTOR GRAPHIC SYSTEMS INC 603-125

Fig. 8.66 Typical dispatcher's record

1. Power supply to station

   If lack of power supply to the station is the problem, the control center dispatcher should be notified by radio to request power company assistance in restoring power to the station. Dispatcher should determine estimated time of arrival of power company crew and then notify the field crew. Since lift stations are essential to the community's welfare, power companies place a high priority on restoring service. Some power companies provide special phone numbers for critical utility agencies to use in emergencies to contact their dispatchers in order to restore power as soon as possible to the area where the station is located.

2. Open access to dry well side of station and inspect:

   a. Interior lighting, and

   b. Operation of ventilation equipment.

   If the interior lighting in the station is out, use flashlights from the repair truck.

   If the ventilation equipment is off, measure the atmosphere in the lift station and work areas for sufficient oxygen and the presence of explosive and toxic gases. Continue to test the atmosphere in the station until the ventilation equipment has been returned to operation for at least 30 minutes.

   If during the test of the station atmosphere the gas detection meter indicates an adverse condition, the crew must leave the station dry well and ventilate the dry well area with a portable blower. If the station must be entered to correct a serious problem, additional help must be requested. Also needed will be a safety harness and self-contained breathing equipment for reentry to the station under hazardous conditions. All confined space entry procedures must be followed in these situations.

3. Station well pit areas

   Examine this area carefully to be sure that all levels of the dry well area are not flooded *BEFORE* restoring high voltage power to operating equipment if power is out.

   If electric motors are flooded and/or will not start, open main power switches to lower area electrical circuits. Request help in setting up emergency pumping equipment. Locate and repair leak to dry pit. Pump the water out of the dry pit with portable pumps. Conduct electrical megger checks (see Chapter 9, Section 9.12, "Tools, Meters, and Testers") on electrical circuits and motors that were flooded to make sure they can be operated, or to determine whether they must be removed and repaired.

4. Motor control center and power panel

   If the dry well area is not flooded and equipment looks normal (no broken pump shafts, pumps not rotating backwards, check valves closed and seated), then return to the motor control center and power panel.

   NOTE. In some lift stations, full power may be restored to all operating equipment at one time. In other stations, especially where large pumps are used, the equipment should be put back on line one unit at a time.

   The station book should have the proper starting sequence for all of the station equipment. A good practice is to open equipment breakers and bring the station back on line by starting one unit at a time. A suggested starting sequence is as follows:

   a. Station lights,

   b. Station ventilation equipment,

   c. Station sump pumps,

   d. Station control system (air compressors),

   e. Essential station auxiliary system (water wells and hydraulic system pumps),

   f. Station last pump to start,

   g. Station second pump,

   h. Station lead pump, and

   i. All other station systems.

   As each piece of equipment is restored to service it should be inspected for proper operation, lubrication, and if appropriate, operating pressures and flow rates. The station book should indicate the operating ranges of the equipment.

   The reason for bringing the last pump on first, the second pump next, and the lead pump last is that this procedure provides each unit with the opportunity to operate and to be inspected before the wet well level returns to normal levels. If the lead pump is started first, the wet well level may fall quickly, and by the time the crew is ready to test run the last pump, the wet well level is too low for automatic operation of the pump. The pump may be inspected in the manual mode of operation, but this does not guarantee it will function in the automatic sequence.

5. Wet well

   If the wet well is flooded and has inlet gates, the gates should be closed or throttled in order to store wastewater in the gravity portion of the collection system until the wastewater level in the wet well can be restored to normal elevations. The station book should list the procedure for closing the inlet gates and specify the critical time the gates can remain closed before damage will result from flooding.

6. After alarm condition has been corrected

   Before the crew departs from the station, they must make sure that all equipment is functioning properly or that damaged equipment is isolated and the station will perform until repairs are made. The following tasks should be completed before leaving the station:

   a. Station restored to service,

   b. All equipment inspected,

   c. Observe wet well and inflow,

   d. Station problem and correction recorded in station book along with identification of crew doing repairs, and

   e. Dispatcher contacted to verify that alarm condition is cleared from telemetry alarm panel.

7. Report

   Complete a report form listing crew members, vehicle and equipment numbers, time dispatched, time of arrival at station, work performed, and time of departure from station.

8. Secure lift station

   Lock lift station and yard gates. Contact dispatcher for next assignment.

When a crew responds to a lift station alarm, the worst possible conditions should be expected. Extreme caution must be exercised by the crew for their own safety, as well as for protection of the equipment and the station. Any number of serious problems may be encountered from downed power lines and/or flooding to broken pipes, valves, or pump shafts. Chlorine leaks can develop if chlorine is applied at the lift station. A coiled rattlesnake on the third step down to the lower pump room is a possible hazard in some areas.

8.324 Typical Lift Station Problems

A properly designed pump used in a wastewater lift station, although properly maintained, can fail to operate when least expected and usually when needed most. There are at least three reasons why a pump will fail to function. The pump may fail due to a malfunction in either the electrical system or in the mechanical system, or by debris plugging an impeller or jamming a check valve.

Electrical failure may be the result of a power failure or an outside power source being switched off due to a circuit overload on the main power source. A circuit breaker tripped due to a circuit overload in the station itself could cause an electrical failure. A quick check of the reset button or breaker switch could identify the electrical failure caused by an in-station overload trip. When manual reset solves the problem temporarily but trips the reset or breaker switches (thermal overloads) again after a few moments of lift station operation, do not attempt to restart the pumps until a thorough inspection has been made of the pumping equipment in the lift station to find the problem. Correct the situation before attempting to restart.

When the problem has been located and corrected, reset the electrical breaker and put the lift station back on line. If the reset breakers kick out again, shut down associated equipment.

An overload may be caused by an electrical short in the motor or wiring control system, a mechanical part failure putting a load on the motor, or a plug in the impeller causing a load on the motor.

A mechanical failure could result from the breakdown of a bearing, poor shaft alignment, vibrations, mechanical seal failure, or malfunction of a mechanical float switch unit. Bearing failure can be noticed by a noise, vibration or heat. Mechanical breakdown could also be caused by malfunctions in the check valves, air bubbler compressor, and switching units associated with the centrifugal pump.

In a pneumatic ejector lift station, mechanical problems are generally caused by a check valve malfunction on the intake or discharge pipe valves. The air compressor and the controls for air bypass valves are other items which may fail. The shorting of the control electrode by grease or corrosion is another electrical problem found at times in pneumatic ejector stations, and the cycle timing device unit also can cause problems.

In the centrifugal pump, most failures are caused by debris piled up in the impeller area. Common material found in stopped impellers includes lodged sticks, wooden blocks, bricks, rocks, rags, mops, plastic bags, sacks, stringy material, and roots.

If a lift station cannot be put back on line before flooding occurs, emergency pumps should be set up and piping arranged to pump the station's influent from the wet well into the force main until the problem can be corrected.

Lift station problems may be grouped into the following four categories:

1. Power
   a. Power failures
   b. Electrical circuit failures
      (1) Thermal overloads tripped
      (2) Fuses blown
      (3) Relays burned out
   c. Motors burned out
2. Control systems
   a. Telemetry system failure
   b. Wet well controller
      (1) Float type
         (a) Float stuck in stilling well
         (b) Float line broken
         (c) Power off and cam override stops
         (d) Mercury switches failed
      (2) Bubbler type
         (a) Air compressor failed
         (b) Bubbler line plugged or restricted
         (c) Leak in air piping system
         (d) Diaphragm ruptured
         (e) Mercury switches failed
         (f) Condensed water in air lines
         (g) Bubbler line broken off in wet well
      (3) Electrode systems
         (a) Short in electrode leads
         (b) Electrodes coated with grease or rags
         (c) Electrodes tangled when wet well changes levels
      (4) Seal trode units
         (a) Bulb ruptured
         (b) Electrolyte leaked out
         (c) Short in electrode leads
3. Pumping systems
   a. Pump impeller plugged with rags or sticks
   b. Pump suction blocked
   c. Check valve stuck open or flap broken from shaft
   d. Pump drive shaft from motor broken
   e. Failure of packing gland or seal water supply
   f. Pump air- or gas-bound
4. Structures, including wet wells and force mains
   a. Corrosion of concrete and metal materials
   b. Force main restricted or plugged
   c. Grit deposits in wet well
   d. Grease and floating debris in wet well
   e. Unusual and hazardous materials floating and deposited in wet wells (remove before they damage or plug pumps and pipes)

Most of these problems can be prevented through a good preventive maintenance program.

8.33 Troubleshooting Force Main Odors

The pumping of wastewater through force mains or pressurized sewers creates a condition where hydrogen sulfide can be released when the wastewater is no longer under pressure. The release of hydrogen sulfide can cause obnoxious odors, toxic atmospheres, and corrosive conditions.

The amount of hydrogen sulfide released or the concentration in the atmosphere will depend on the amount of turbulence in the wastewater when it is no longer under pressure, the pH of the wastewater (lower pH, more hydrogen sulfide gas released), the strength (BOD) of the wastewater, the wastewater temperature (high temperature, more hydrogen sulfide), the time the wastewater is in the sewers before the force main, the time the wastewater is in the force main, and the amount of slimes in the force main.

The methods available for controlling hydrogen sulfide include:

1. Dosing air into the force main immediately downstream from the lift or pumping station,
2. Dosing gaseous pure oxygen into the force main immediately downstream from the pumping station,
3. Dosing a nitrate solution (source of oxygen) into the sump at the pumping station,
4. Dosing a caustic solution (increase pH) into the sump at the pumping station or farther upstream in the sewer,
5. Dosing a solution of an iron salt (react with hydrogen sulfide to produce a sulfide precipitate) into the sump at the pumping station or near the force main outlet, and
6. Installing ventilation equipment at the force main outlet.

Some methods may be combined. For example, iron dosing and ventilation are commonly combined. Oxygen dosing combined with nitrate dosing has also been used with success. Where sufficient oxygen cannot be dosed at a single point, and subsequently dissolved in the wastewater, the option of dosing at several points along the force main is a possibility.

Also see Section 6.54, "Chemical Control of Hydrogen Sulfide," in Volume I.

QUESTIONS

Write your answers in a notebook and then compare your answers with those on page 121.

8.3N Why should an emergency crew at a lift station notify the dispatcher when they will call in again?

8.3O Why should overhead power lines be inspected by emergency crews upon arrival at a lift station?

8.3P What are the major categories of lift station problems?

Please answer the discussion and review questions next.

DISCUSSION AND REVIEW QUESTIONS

Chapter 8. LIFT STATIONS

(Lesson 4 of 5 Lessons)

Write the answers to these questions in your notebook before continuing. The question numbering continues from Lesson 3.

21. What action would you take if a lift station failed and the wastewater in the collection system was rising rapidly?
22. How would you determine the frequency of visits to a lift station?
23. Discuss the advantages and limitations of daily versus weekly to monthly visits to lift stations.
24. How would you respond to a lift station alarm failure and how would you attempt to locate or identify the cause of the failure?
25. What would you do during a lift station visit (what items should be inspected)?
26. What types of books or records should be kept in a lift station? Why?
27. What safety precautions would you take *BEFORE* attempting to determine the cause of a lift station failure?
28. What kinds of problems or failures can occur in lift stations?

CHAPTER 8. LIFT STATIONS

(Lesson 5 of 5 Lessons)

8.4 LIFT STATION MAINTENANCE

Chapter 9 of this manual describes *EQUIPMENT MAINTENANCE PROCEDURES.* Chapter 13 explains how to *ORGANIZE* personnel and equipment for efficient and effective maintenance. The intent of this section is to describe how to establish a maintenance *PROGRAM* and a *FREQUENCY SCHEDULE* for lift stations. The maintenance program consists of two major parts:

1. Scheduling the work. Necessary work must be scheduled in advance and must include items that must be performed during a given time period.
2. Performing the work. Maintenance work must be performed in a specific manner. Every task must be done the same way by every operator.

Both scheduling and performing are important, but the development of a schedule is the most difficult job. Ensuring consistent procedures for performing the work is accomplished by providing the operators with written instructions, proper training, necessary tools and evaluation of completed work by supervisors.

8.40 Scheduling Maintenance

Preparation of a lift station maintenance program requires consideration of three important factors:

1. Recommendations of equipment manufacturers,
2. Requirements of lift station, and
3. Knowledge previously gained from experience by the operating agency.

8.400 Recommendations of Equipment Manufacturers

Equipment manufacturers provide maintenance and overhaul recommendations for each piece of equipment they install in a lift station. Usually this information is contained in a convenient binder or manual. Information includes frequency of oil changes and lubrication of bearings, types of lubricants, operating temperature ranges, pressures, flow rates, and disassembly procedures for specific equipment maintenance or parts replacement.

During the first year of operation of new equipment, the manufacturers' recommendations must be closely followed to maintain equipment warranties. The equipment maintenance schedule is developed by listing all of the manufacturers' recommendations in sequence according to time periods, such as daily (D), weekly (W), monthly (M), quarterly (Q), semiannually (SA), and annually (A) according to when these functions or tasks are to be performed.

When all of the equipment in the station and the frequency of each maintenance task has been tabulated, then the procedures for performing the tasks should be indexed. If you do not have a standard procedure, refer to the manufacturers' recommendations. Many tasks follow the same procedure regardless of the manufacturer of the equipment. For example, adjustment of packing glands, *LUBRIFLUSHING*[14] motor and pump bearings, or alignment of drive couplings usually follow the same procedures, even though the tolerance specifications may be different.

8.401 Lift Station Requirements

Each individual lift station may have different requirements. These differences result from the design and location of the station. The items listed here are the tasks that are required to be performed at given frequencies to keep the station operating. These items should be developed primarily from station operating experience and should be reevaluated annually by field operators, supervisors and management. Reevaluation includes changes in frequency of tasks, operational methods, or lift station revisions or redesign in order to increase station reliability and reduce station failures, alarms and crew work loads.

For example, let's examine the situation where a lift station uses a bar screen, a pump operation control system using a stilling well and float, and the lift station is serving a community with several restaurants which produce an excessive grease load on the station. These conditions cause the station's maintenance demands to be higher than necessary because of the following conditions:

1. The bar screen should be cleaned daily or at least two or three times a week.

---

[14] *Lubriflushing (LOOB-rah-FLUSH-ing). A method of lubricating bearings with grease. Remove the relief plug and apply the proper lubricant to the bearing at the lubrication fitting. Run the pump to expel excess lubricant.*

2. Grease will create operational problems with the float travel in the stilling well. Hosing and flushing of the stilling well will be required once or twice a week.
3. The wet well will have to be cleaned quarterly instead of annually due to the excessive grease. Clean the wet well walls manually, or use a high-velocity cleaner to wash down the wet well walls to prevent grease buildup and the generation of hydrogen sulfide. If the grease is not thoroughly removed regularly, the wet well concrete walls will deteriorate and could require patching and protective coatings within five years.

What kinds of alternatives or choices are available? You may not have any and may be forced to schedule the frequent visits to this particular lift station. Or you may have some of the following options:

1. Remove the bar screen and allow the station to operate without it. If the piping system or pump starts plugging, the bar screen can be replaced. Hopefully the station will operate without problems after the bar screen is removed. However, you may have to remove floating debris on the water surface of the wet well. Remove this debris before it clogs the pumps or suction piping intakes.
2. Enforce the local sewer-use ordinance and have the discharger remove the grease at the source and thus prevent the grease from entering the collection system.
3. Record each station alarm and the cause of the failure. Estimate the cost of labor, equipment and materials to correct each failure. Determine the best solution by comparing costs of repairs with the cost of a more frequent and extensive maintenance program and possibly more alarm calls. Consideration also must be given to your objectives, such as reducing stoppages and odor problems.

Remember that flows usually increase in the future and wastewater characteristics also change. In many cases maintenance requirements of wet wells and force mains are greater on new lift stations handling low flows because detention times in wet wells and force mains can become very long. When flows through a new lift station increase, conditions may improve and the frequency of performing certain maintenance tasks can and should be reduced. Field crews should be able to make important contributions to decisions regarding the frequency and types of maintenance tasks required by a lift station. These crews are familiar with the station, notice changing conditions, and are capable of comparing conditions and operational characteristics with other lift stations.

8.402 Knowledge Gained From Experience

A very important factor in maintenance scheduling is the knowledge and experience gained by an agency regarding how to deal with local conditions, the ability of operators to perform tasks and the reliability of existing equipment. The major problem with applying knowledge and experience is that occasionally procedures other than those recommended by the manufacturer appear better. Remember that the manufacturer prepared the manual for use throughout the world and not for your specific conditions and problems. Perhaps the recommended lubricant will not perform satisfactorily due to high or low temperatures in your area. Dust conditions or humidity may require equipment protection filters to be changed at twice the recommended frequency or possibly the filter life can be extended three months longer than recommended. If lift station wet wells can be safely cleaned only in July and August, don't schedule cleaning during February when there is three feet of snow on the ground and the temperatures are below freezing.

8.41 The Actual Maintenance Schedule

The supplement to this chapter consists of pages from a typical lift station (identification code S-14). The purpose of this supplement is to illustrate a preventive maintenance program in a station book. This lift station is visited once a month for scheduled maintenance. See page 123.

8.410 Crew Notification of Required Maintenance

Once the preventive maintenance program has been defined, there must be a procedure that notifies a crew to perform the required maintenance at the station. There also must be a method of recording when the work is completed by field crews. This information should be checked off against the scheduled work and filed in the lift station book in the main office. If maintenance items are overlooked or not performed, they must be rescheduled and completed.

One method of ensuring that work gets completed is by the use of a form such as the one shown in Figure 8.67. This form has a hard copy and two soft copies. The forms are produced by the agency office staff using an addressograph plate that is coded to a specific time schedule or interval, such as weekly, monthly, quarterly, semiannually, annually, two years, and five years. At the beginning of each month, the forms are produced by the addressograph machine with the maintenance functions that must be performed during that month at each facility, as shown in Figure 8.67. This activity can also be performed by a computer.

The top soft copy of the form is retained in the office with a soft and hard copy going to the supervisors of the field crews. The supervisors assign the maintenance jobs required at each lift station to the field crews who do the work. The form contains the tasks to be performed and the sections in the maintenance and overhaul manual where the task directions and instructions are located. The maintenance and overhaul manual is developed from the manufacturers' instructions and the operating agency's actual experience in maintaining and overhauling the various types of equipment.

When new stations are placed in service, the equipment maintenance procedures are obtained from the appropriate paragraphs in the existing maintenance and overhaul manuals. If new equipment requires different maintenance items or procedures than those already in the existing manual, new instructions are prepared and inserted in the station book and in

SCHEDULED MAINTENANCE N° 1

ASSIGNED TO:

North Area

FEB 03

CREW CHIEF
GILBERT, J

DATE ASSIGNED
2-5-03

CLASS	HRS
MECH. REPAIRMAN	5
ASS'T REPAIRMAN	5

TOTAL MAN HOURS 10

VEH.	HRS
153-462	5

TOTAL EQUIP. HOURS 5

DATE COMPLETED 2-5-03
COMPLETED BY: James Gilbert

PARK ROAD MONTHLY 1 OF 9
NO. 1 SEWAGE PUMP YH-S14 010137
CHECK MOTOR 14 A-B
CHECK PUMP 22 E-1
CHECK PACKING 22 A
CK. DRAIN LINE 22 O-2

PARK ROAD MONTHLY 2 OF 9
NO. 2 SEWAGE PUMP YH-S14 010137
CHECK MOTOR 14 A-B
CHECK PUMP 22 E-1
CHECK PACKING 22 A
CK. DRAIN LINE 22 O-2

PARK ROAD MONTHLY 3 OF 9
YH-S14 010137
ALTERNATE PUMPS 22 G-1
CHECK PUMPING RANGE 22 P
RECORD HOURS-GALLONS 15 E
RECORD SMUD METER 15 F

PARK ROAD MONTHLY 4 OF 9
AIR COMPRESSOR YH-S14 010137
CHECK MOTOR 14 A-B
CHECK COMPRESSOR 74 A
CHANGE OIL 74 B-2
BLEED TANK 74 F-3

PARK ROAD MONTHLY 5 OF 9
WATER WELL YH-S14 010137
CHECK TANK 69 B-1-2
CK. PUMPING RANGE 22 P
SOUND WELL 69 C

PARK ROAD MONTHLY 6 OF 9
YH-S14 010137
SUMP PUMP 22 L-1
VENT. BLOWER 2 113 A
DEHUMIDIFIER 2 53 A
SM. AIR COMPRESSOR 74 F-3

PARK ROAD MONTHLY 7 OF 9
YH-S14 010137
FILTER REGULATOR 2 74 H-3
FILTER STA. DRY 2 74 H-4
ROTOMETER 16 D
LIGHTS 57 C

PARK ROAD MONTHLY 8 OF 9
YH-S14 010137
DOORS 57 D-1
LOCKS 57 A
CH. ORINE 101 A
CHECK CONTROLER CLOCK 15-G

PARK ROAD MONTHLY 9 OF 9
CLEAN YH-S14 010137
MOTORS-PUMP UNITS
BUILDING INTERIOR
YARD-DRIVEWAY

REMARKS: DRIVE WAY NEED TO BE SCHEDULED FOR PATCHING.

PW 74-10208

Fig. 8.67 Maintenance task forms

the field crews' maintenance and overhaul manuals. When a specific maintenance item or job is scheduled, the form (Figure 8.67) will show the section of the maintenance and overhaul manual where the directions are located that specify how to perform the scheduled work. These procedures can also be performed by a computer.

8.411 Recording Completed Work

When each task is completed, the crew leader signs it off. When the work has been completed and the form filled out, it is returned to the office where it is matched with the cover copy. A match-up of work required with work completed indicates that the scheduled work was done. On the bottom of the form is an area for comments so that if a field crew finds a discrepancy or a pending problem, it is noted and brought to the attention of a supervisor for corrective action.

Not shown on the station book are preventive maintenance jobs scheduled for two- to five-year intervals. These jobs include equipment overhauls and lift station painting.

QUESTIONS

Write your answers in a notebook and then compare your answers with those on page 122.

8.4A What are the two major parts of a maintenance program?

8.4B Why should equipment manufacturers' recommendations be followed, especially during the first year?

8.4C Why are maintenance requirements sometimes greater for new lift stations with low flows than for older stations with greater flows?

8.4D Why should field crews be allowed to contribute to decisions regarding the frequency and types of maintenance tasks required by a lift station?

8.5E Why should a form showing work completed be compared with the work assignment?

8.5 RECORDKEEPING

8.50 Filing of Records

Records are an important part of a lift station operation and maintenance program. They should be filed at the agency's main office. Records left in the lift station are of little value to supervisors or managers. Frequently records left in lift stations get lost or misplaced. Active records (current year) should be kept in the office copy of the station book. This procedure keeps all information at one source. At the end of each operating year or the end of a designated period selected by the agency, the past year's operational and maintenance records are removed from the station book and placed in a station file. This method keeps the station book at the office from becoming too bulky and keeps only pertinent data in the book for the station's operation and maintenance programs.

8.51 Cost Records

Records maintained on each lift station should show at least the costs of operating and maintaining the facility. The costs include the following:

1. Electrical power cost,
2. Fuel costs such as gasoline, diesel, natural or bottled gas,
3. Operational and maintenance costs on basis of labor, vehicles, equipment and supplies, including water and chlorine used,
4. Scheduled preventive maintenance and repair costs on the basis of labor, parts, shop expenses, vehicles, tool and equipment rental, paint, lubricants and other supplies,
5. Unscheduled repair costs, including responding to the station telemetry alarms, power failures and other problems that require a visit to the station that was not regularly scheduled, and
6. Repair costs to station caused by vandalism or accidents. Record whether or not the costs of the repairs were recovered from individuals who caused the damage.

Record all goods received. A simple and convenient log sheet is as follows:

Date Goods Received	Order Number	Description of Goods	Cost	Date Bill Received

When a bill is received, be sure that the goods were received and that the order is complete before paying the bill. Sometimes orders are delivered that are not complete and when a bill arrives six months after the goods were received, it can be difficult to remember if the order was complete.

By keeping accurate cost records, these costs for operating and maintaining the station provide the information needed to prepare next year's annual budget. Cost data on a lift station may be kept in the office of a supervisor where the supervisor is expected to maintain up-to-date records or the information may be recorded on forms supplied by the agency's accounting division.

8.52 Station Identification Number

Auditors and computer people seem to delight in using long strings of numbers and letters for identification purposes. They cannot accept obvious descriptive names such as "Park Road" or "Riverside" for lift stations. By using letters and numbers, they describe the "Park Road" lift station as "L14." L indicates lift and 14 is the number assigned to the Park Road facility. "Riverside" lift station may be identified as "L68." Money from the budget to operate the station is assigned a work authorization number so that L14 becomes L14-010691 and Riverside is L68-011274. This code or designation only indicates at which facility and when the money is available. By adding an activity code, you can identify how the money is spent. For example, a typical list of activity codes is as follows:

GX = Operations

YH = Scheduled preventive maintenance

YZ = Unscheduled repairs

AA = Responding to telemetry alarms or emergency conditions

PE = Electrical power charges

PD = Purchase of diesel fuel

PN = Purchase of natural gas

PG = Purchase of gasoline

PC = Purchase of chlorine

XC = Delivery of chlorine

Depending upon the system and agency, the activity code list may consist of a hundred or more items. A list this long can

become a bother, if not confusing to field operators. If the proper identification code is not used by field operators, then the accounting system is of little use because one station may be charged with costs from another station. Therefore, any agency wishing to use a computer accounting system effectively must use a simple system that field operators understand and use.

If a crew is making a visit to the Park Road lift station for an operational inspection, they would record their time as:

1.80 hours GX — L14-01069

If another crew was performing scheduled preventive maintenance at the Riverside station, they would record their time as:

6.50 hours YH — L68-011274

Computer systems can become very complex, depending on the size of agency, information desired from the computer, and number of agencies using the computer. The computer can provide printouts at regular frequencies that show data on costs for labor, energy, equipment, materials, parts, lubricants and supplies. This information goes to a supervisor who must determine if the operation and maintenance programs are effective and within the budget.

8.53 Other Records

Records other than cost data that are important include the items listed in this section.

8.530 Preventive Maintenance Schedule for Lift Stations

The equipment and preventive maintenance schedule for tasks to be performed on a monthly, quarterly, semiannual and annual basis must be properly identified. When this work has been completed and signed off by the appropriate crews, the forms should be filed in the station book at the office.

8.531 Unscheduled Work Order Requests

A report form for unscheduled work shows the type of unscheduled work, why it was performed, and the costs, including labor, equipment and materials. Also the form should indicate where materials were bought and where used to perform the job or repairs.

8.532 Modifications Made to the Station or Force Main

All drawings and plans must be kept up to date to facilitate future work and to evaluate station performance.

8.533 Written Reports

Especially important are written reports providing details of unusual conditions or repairs made to the lift station. These reports should indicate how repairs were made, time required, unusual conditions encountered, special equipment or materials needed, or additional maintenance scheduling required in case the job needs special attention or must be repeated in the future. Unfortunately we sometimes have to do jobs in the future that are similar and no one is available who remembers exactly how a previous job was done or the problems encountered.

8.534 Operational Data

Important operational data includes flow records, equipment lapsed time meter readings, chlorination rates, and other operating data that may be required by your agency.

8.6 SUMMARY

We hope this chapter has stimulated some thought regarding the operation and maintenance of wastewater lift stations. Our experience has shown that proper design and maintenance are the key to lift station reliability. Lift stations can effectively operate themselves if properly designed and given the care they need.

QUESTIONS

Write your answers in a notebook and then compare your answers with those on page 122.

8.5A Why should records of a lift station not be filed in the lift station?

8.5B What items should be included in lift station costs?

8.5C Why are activity codes used in accounting or to keep track of lift station costs?

8.5D Why should reports be written detailing how repairs are made?

Please answer the discussion and review questions next.

DISCUSSION AND REVIEW QUESTIONS

Chapter 8. LIFT STATIONS

(Lesson 5 of 5 Lessons)

Write the answers to these questions in your notebook. The question numbering continues from Lesson 4.

29. How would you develop a maintenance schedule for a lift station?

30. Why should a maintenance schedule be reviewed from time to time?

31. Review the advantages and limitations of the different possible solutions to a grease problem in a lift station.

32. What records would you keep of a lift station and how and where would you file them?

33. Why are lift station costs important?

34. What types of written reports should be prepared for lift stations? Why?

SUGGESTED ANSWERS

Chapter 8. LIFT STATIONS

ANSWERS TO QUESTIONS IN LESSON 1

Answers to questions on page 10.

8.0A Wastewater is raised or lifted in a lift station by centrifugal pumps or pneumatic ejectors.

8.0B The location of lift stations depends on economic factors. They are installed at low points in the collection system at the end of gravity sewers where the following conditions exist:

1. Excavation costs to maintain gravity flow and sufficient velocity become excessive,
2. Soil stability is unsuitable for trenching,
3. Groundwater table is too high for installing deep sewers, or
4. Present wastewater flows do not justify extension of large trunk sewers.

Lift station locations are also influenced by the location of other utilities, buildings and transit systems.

8.0C The two major safety items that must be considered around lift stations are:

1. Safety of operation and maintenance operators, and
2. Safety of the public.

Answers to questions on page 13.

8.0D A submersible pump is often installed in a wet well.

8.0E A limitation of installing a pump inside a wet well is limited access to the pumping equipment, which creates a problem when the pump has to be repaired.

Answers to questions on page 24.

8.0F The most desirable operational situation for a wet well is when all the flow and solids that discharge into the wet well from the gravity sewer are lifted to the higher elevation and continue to the wastewater treatment plant without delay.

8.0G Water hammer causes extremely high pressure shock waves to travel up and down a force main. The pressure may damage check valves, collapse pipelines, and, in severe cases, break a force main.

8.0H Static head is the difference between the water surface elevation in a wet well on the suction side of a pump and the water surface elevation on the discharge side of a pump. Dynamic head is the difference in elevation between the energy grade line on the suction side and the discharge side of the pump. The dynamic head is the static head plus the friction or energy losses that result from water flowing through the pipes, valves and fittings in the lift station.

8.0I Friction losses are caused by water flowing through pipes, valves and fittings in a lift station. The greater the velocity of the flowing water, the greater the friction losses.

ANSWERS TO QUESTIONS IN LESSON 2

Answers to questions on page 30.

8.1A Pumps should not start and stop frequently in lift stations because frequent motor starts cause excessive wear on equipment and may cause surges in the wastewater flow. Also frequent starting will increase power costs.

8.1B If a wet well is too large, wastewater is stored too long and solids will settle out. This creates the opportunity for the wastewater to turn anaerobic and produce hydrogen sulfide, methane, and nuisance odors.

8.1C The main advantage of variable-speed pumping equipment is that the pumping rate can be adjusted to match the inflow rate.

Answers to questions on page 44.

8.1D Bar racks usually are not used in lift stations pumping only domestic and industrial wastewaters because:

1. Screenings are a problem to remove and to transport to a disposal site, and
2. Pumping units are equipped with open impellers or closed two-port impellers sized to pass most solids found in the collection system.

8.1E Sump pumps in dry wells are used to remove excess water, such as seal water leakage and cleanup water.

8.1F When a seal fails on a submersible pump, wastewater can leak into the motor compartment and the unit will short or burn out the motor.

8.1G Solid state motor starters provide more precise overload protection and greater pump motor control than conventional electro-mechanical motor starters. Mechanical wear on the contacts is reduced with solid state starters and they provide phase unbalance protection, phase loss protection and ground-fault protection.

Answers to questions on page 49.

8.1H The three types of float controllers are:

1. Rod-attached floats,
2. Steel tape-, cable-, or rope-attached floats, and
3. Mercury switch float controls.

8.1I If the upper stop on the float rod is raised, the wet well will be pumped farther down, because the float will have to be dropped lower before the pump will shut off.

8.1J Maintenance on a stilling well includes removing grease and debris that hinder or stop the up and down movement of the float. Other maintenance or repairs include float attachment line breaking or unwinding from drum sheave and float developing a leak.

8.1K In an ultrasonic level detector, a transducer is mounted above the highest water level in the wet well. The transducer generates ultrasonic pulses which hit the water surface and bounce back to the transducer. The system measures the pulse's travel time and converts this information into an electrical signal in the control system.

Answers to questions on page 57.

8.1L Controls used to start, stop or change pumping rates include float controllers, electrode controllers and pressure-sensing controllers.

8.1M Electrical safety regulations must be complied with when using electrode controllers. Be sure that power to the electrodes is turned off and properly tagged before performing any maintenance on the equipment. Operators should be cautioned not to touch electrodes with bar screen rakes or other tools.

8.1N Limitations of electrode controllers include:

1. Rags and debris can wrap around electrodes and alter desired pump start and stop level in a wet well, and
2. Grease and/or slime can cover an electrode, thus preventing good conductivity and causing intermittent or unreliable operation.

8.1O Pressure-sensing controllers detect pressure changes by measuring the back pressure required to overcome the pressure created by the level of the water above the outlet of the bubbler.

8.1P Problems that can develop using seal trode sensing controllers include:

1. Bulb can break,
2. Shorts occur in electrode wiring, and
3. Bulb life is 3 to 5 years.

8.1Q Pressure-sensing controllers measure the depth of water in a wet well on the basis of air pressure or back pressure in the air bubbler line.

8.1R Problems that can develop with pressure-sensing controllers include:

1. Air compressor failure,
2. Bubbler blockage, tube breakage, or outlet elevation shifts,
3. Many opportunities for equipment failure, and
4. Need for a higher level of training for operators to troubleshoot, repair and keep instruments calibrated.

Answers to questions on page 71.

8.1S The two major types of pumps are positive displacement pumps and kinetic (or dynamic) pumps.

8.1T Kinetic (or dynamic) pumps, usually of the centrifugal type, are most commonly used in raw wastewater lift stations.

8.1U Cavitation is likely to occur if modifications over time change the basic balance of the hydraulic system, if the system was improperly designed, or if the wrong pump is being used for a particular application.

8.1V Pump seal water from a domestic water supply must have an air gap to prevent contamination of the public water supply from backflows.

8.1W Lift stations should be equipped with at least two pumps to provide continuous operation if one pump fails or requires repairs.

Answers to questions on page 83.

8.1X The major valves found in wastewater lift stations are:

1. Pump suction and discharge isolation valves (gate, plug, knife),
2. Discharge check valves (swing or ball check, wafer), and
3. Cross-connect or pump discharge pipe manifold isolation valves (gate, plug).

8.1Y Pump suction and discharge isolation valves are found immediately before and after the pump to allow isolation of the pump from the wet well and the force main during pump or check valve maintenance.

8.1Z Check valves are installed in the discharge of each pump to provide a positive shutoff from force main pressure when the pump is shut off and to prevent the force main from draining back into the wet well.

ANSWERS TO QUESTIONS IN LESSON 3

Answers to questions on page 87.

8.2A Operation and maintenance operators should review prints and specifications of a new lift station before construction to be sure adequate provisions have been made for the station to be easily and properly operated and maintained.

8.2B Accessibility to a lift station is important because operators and equipment need to be able to easily reach the lift station and have room to safely park vehicles and work.

8.2C Before equipment is installed in a lift station, determine its familiarity to your agency, its reliability, warranties, guarantees and the operation and maintenance aids that will be provided.

8.2D Prints and specifications should be carefully reviewed before construction and installation because changes are easily made on paper and are much more difficult in the field.

Answers to questions on page 88.

8.2E After a lift station is completed but before it is accepted, it should be inspected by the engineer, the contractor, and the operators responsible for its operation and maintenance.

8.2F Information regarding lift station equipment and performance should be filed with a copy in the lift station and a duplicate copy at the agency office.

8.2G Data from initial tests on equipment should be recorded and filed to document that the equipment meets specifications and to show how it performed when installed.

Answers to questions on page 89.

8.2H Problems that may be encountered when cleaning a wet well include slippery floors, inadequate ventilation, cleaning bar rack, disposal of debris, lack of oxygen, explosive gases, and toxic gases.

8.2I Hazards found in confined spaces that handle wastewater include lack of oxygen, explosive and toxic gases.

8.2J Hazards encountered in sumps can be reduced by adequate hosing of sludges and slimes on sump floor and ventilation to provide a satisfactory environment.

Answers to questions on page 90.

8.2K When inspecting electrical equipment, start by recording all nameplate data.

8.2L When inspecting an electrical circuit or equipment, the main circuit breaker must be locked in the *OFF* position.

8.2M Extra fuses should be kept at the station so blown fuses can be replaced immediately and equipment downtime is minimized.

8.2N The working of the electrical circuits should be learned before the station becomes operational.

Answers to questions on page 92.

8.2O Anti-rotation equipment is used to prevent the pump from rotating in the wrong direction.

8.2P If a universal joint is overlubricated, the seal on the universal will blow and cause failure in a very short time.

8.2Q We must have oil, water and fuel for engine operation.

8.2R If an engine's internal alarm system fails, the engine could burn up.

8.2S When changing engine oil, look for metal filings and sludge as indicators of engine condition. Look for water or an oil-water emulsion that indicates water in the wrong place.

8.2T Engines must be inspected thoroughly in an attempt to prevent problems and to learn what to do if problems develop. The jobs of troubleshooting and repairing become easy when there are good data for reference.

8.2U When lubricating the clutch throw-out bearing, lubrication should not be excessive.

8.2V An engine's fuel tank should be kept full to prevent condensate from forming on top of the tank.

Answers to questions on page 98.

8.2W Leakage from a packing gland should be approximately one drop per second.

8.2X Teflon packing should not get hot because it will glaze and stop working properly.

8.2Y The minimum pressure for the flushing water on a mechanical seal is 3 to 5 psi above the pump discharge pressure.

8.2Z The first thing to do before operating a lift station is to open any valves on the influent line to fill the sump. The pump suction and discharge valves must be open before the pump is started.

8.2AA The debris left in the sump should be removed to prevent damage to the pump and clogging of pipelines.

8.2BB Sump level can be determined by:

1. Sump level indicator,
2. Visual inspection, and
3. Observation of the level in the pump control stilling well.

8.2CC Pump rotation can be determined by noting:

1. If rotation is in direction of arrow on side of pump,
2. If rotation is in direction of volute in centrifugal pumps,
3. Size of flow and discharge pressure, and
4. Amperage drawn by pump motor.

8.2DD Yes, all points of operation for all pumps must be inspected.

8.2EE Do not operate any unit if you think it is not functioning properly.

8.2FF All data must be recorded when test operating the station so you have a record of conditions when the station started.

ANSWERS TO QUESTIONS IN LESSON 4

Answers to questions on page 99.

8.3A Nearby streams and rivers can be protected from wastewater when a lift station fails by the use of portable emergency equipment to pump wastewater to a functioning section of the downstream collection system.

8.3B Complaints from the public about a lift station can be minimized if the operators responsible for the lift station maintain it in top operating condition and respond to complaints and questions in a positive and concerned manner. When responding to a complaint, be sure to tell the public what has been done or what will be done to correct the complaint.

8.3C The most important rule that always must be considered when visiting a lift station is *SAFETY*.

Answers to questions on page 101.

8.3D The frequency of lift station visits depends on the following:

1. Number of stations in community,
2. Type of wastewater being conveyed,
3. Potential damage from storms flooding station,
4. Condition of equipment,
5. Design of facility and equipment installed,
6. Adequacy of preventive maintenance and overhaul program,
7. Type, adequacy, and reliability of telemetry system,
8. Attitude of operating agency toward operation and maintenance, and
9. Number of employees available to visit lift station.

8.3E Lift station failures include:

1. Electrical power failures,
2. Failure of level sensing or other control equipment,
3. Flooded wet wells,
4. Plugged pump impellers or lines,
5. Overheated and tripped out motor thermals,
6. Sump pump off and sumps or dry wells flooded,
7. Air- or gas-bound pump,
8. Stuck or blocked check valve,
9. Pump control systems not functioning,
10. Failure of lift station telemetry system, and
11. Ventilation fans burned out.

8.3F The critical time for a lift station after failure is the time interval before flooding or damage occurs.

8.3G Daily visits to a lift station are required when essential operational and maintenance tasks must be performed on a daily basis to keep the station operable and reliable.

8.3H For a lift station to be visited only from once a week to once a month, the following requirements must be met:

1. Equipment lubrication reservoirs must be large enough,
2. Telemetry of the station must include wet wells, dry well sump pump pit level, power, compressor, check valves and auxiliary equipment, and
3. Crews must be able to respond to alarms before flooding or damage occurs.

8.3I Alarm situations may be transmitted over leased telephone lines or by a radio or microwave signal.

Answers to questions on page 109.

8.3J There are two types of significant lift station visits:

1. To schedule routine preventive maintenance upon the equipment and the station, and
2. To respond to a lift station alarm or emergency condition to restore the station to operation.

8.3K The number of responses to lift station alarms or emergency conditions can be minimized by a good preventive maintenance program with the frequency schedule at the proper intervals, clear duty assignments in writing, and completion of maintenance tasks verified by management.

8.3L Lift stations should have a sign-in log to reduce the number of mysterious changes in station operation such as pumps turned off, control system changes and station doors or gates left unlocked. Also this practice provides a method for reaching the appropriate person when the need arises to issue corrective instructions.

8.3M The station book should contain the station identification number, a map showing important facilities, station description, equipment data sheets, safety instructions, responses to emergency conditions and the station's preventive maintenance schedule.

Answers to questions on page 112.

8.3N An emergency crew at a lift station should notify the dispatcher when they will call in again so in case a serious injury occurs and they are unable to contact the dispatcher, rescue operations can be started.

8.3O Overhead power lines should be inspected by emergency crews upon arrival at a lift station because if they are broken, they could come in contact with a chain link fence or other conductor and electrocute a crew member opening a gate.

8.3P The major categories of lift station problems are:

1. Power,
2. Control systems,
3. Pumping systems, and
4. Structures.

ANSWERS TO QUESTIONS IN LESSON 5

Answers to questions on page 116.

8.4A The two major parts of a maintenance program are:

1. Scheduling the work, and
2. Performing the work.

8.4B Equipment manufacturers' recommendations must be closely followed to maintain equipment warranties.

8.4C Maintenance requirements may be greater on new lift stations handling low flows because detention times in wet wells and force mains can be very long and allow the production of hydrogen sulfide.

8.4D Field crews can make valuable contributions because they are familiar with the station, notice changing conditions, and are capable of comparing conditions and operational characteristics.

8.4E A form showing work completed must be compared with the work assignment to be sure all scheduled work is completed.

Answers to questions on page 117.

8.5A Records should not be filed in a lift station because they can get lost or misplaced. Also they are of little value to supervisors or managers.

8.5B Lift station costs include power, fuel, operation and maintenance (labor, vehicles, equipment and supplies), scheduled preventive maintenance, unscheduled repair costs, responses to alarms and vandalism.

8.5C Activity codes are used to help identify exactly how money is spent on a lift station.

8.5D Reports should be written detailing how repairs are made so if a similar job must be done in the future, this report can serve as a helpful reference.

SUPPLEMENT TO CHAPTER 8

LIFT STATIONS

A PREVENTIVE MAINTENANCE PROGRAM

FOR

PARK ROAD LIFT STATION

NOTE: For another type of operation and maintenance (O & M) manual for a typical lift station, see "Standard Operating Job Procedures (SOJP) No. 3, Pump Stations" from "A Guide for the Development of Standard Operating Job Procedures for Wastewater Treatment Plant Unit. Operations," developed by Charles County Community College, La Plata, Maryland, for the Manpower Development Staff, U.S. Environmental Protection Agency, Washington, D.C. 20460, 1973.

EXPLANATION OF SUPPLEMENT

This supplement is intended to serve only as an example of a portion of a typical lift station book. The actual station book for a typical lift station (S-14) consists of two, three-inch thick binders which contain equipment and agency data that are described in this chapter. Material contained in this supplement (starting on page 125) is from Sacramento County's maintenance and overhaul manual. The manual was developed and written by the County and serves 128 separate facilities.

The actual maintenance and overhaul manual contains several hundred pages and prescribes maintenance procedures for equipment in 128 facilities including wastewater lift stations, storm water lift stations, wastewater treatment plants, and domestic water wells and distribution systems. The entire maintenance and overhaul manual is not presented. Only portions that pertain to some of the equipment listed in the typical station (S-14) are presented to illustrate a method used by one agency to accomplish its maintenance program.

The mechanical maintenance field crews that use the maintenance and overhaul manual are skilled operators who perform all work with the exceptions of electrical power services and instrumentation that consists of analog or digital components. This work is performed by other operators qualified with the necessary skills in the electrical and instrumentation fields.

TABLE OF CONTENTS

EQUIPMENT	MANUAL REFERENCE	DRAWING	PAGE NO.
PIT — WET	7.10		142
Inspection	7.10		142
Definition Of Wet Pit	(footnote)		142
PUMPS — CENTRIFUGAL	7.20	7.20-A	142
General	7.20		142
Packing Glands	7.20-A-B		142
Condition Check	7.20-E-1		142
Operate Pumps Alternately	7.20-G		144
Lubrication	7.20-H-1		144
Water Lube Line	7.20-N		144
Pumping Range	7.20-P		144
PUMPS SUMP	7.20	7.20-C	144
Operation Check	7.20-L-1		144
Inspection	7.20-L-2		144
STRAINERS	7.94		146
Inspection And Cleaning	7.94-A		146
VALVES — AIR RELIEF	8.10		146
VALVES — GATE	8.30		146
Operation	8.30-A		146
Packing Adjustment	8.30-B		146
Packing Lube	8.30-C		146
VALVES — BALL	8.30		146
Operation Check	8.30-D-1		146
Inspect/Clean	8.30-D-2		146
VALVES — GLOBE	8.30-E		146
VALVES — BURIED	8.30		146
Inspect/Lubricate	8.30-F		146
Hydraulic Gate Operator	8.40-C		146
WATER WELLS	9.01		146
Clean/Inspect Tank	9.01-A		146
Tank Maintenance	9.01-B		147
Standing Water Level	9.01-C-1		147
Sounding Well	9.01-C-2		147

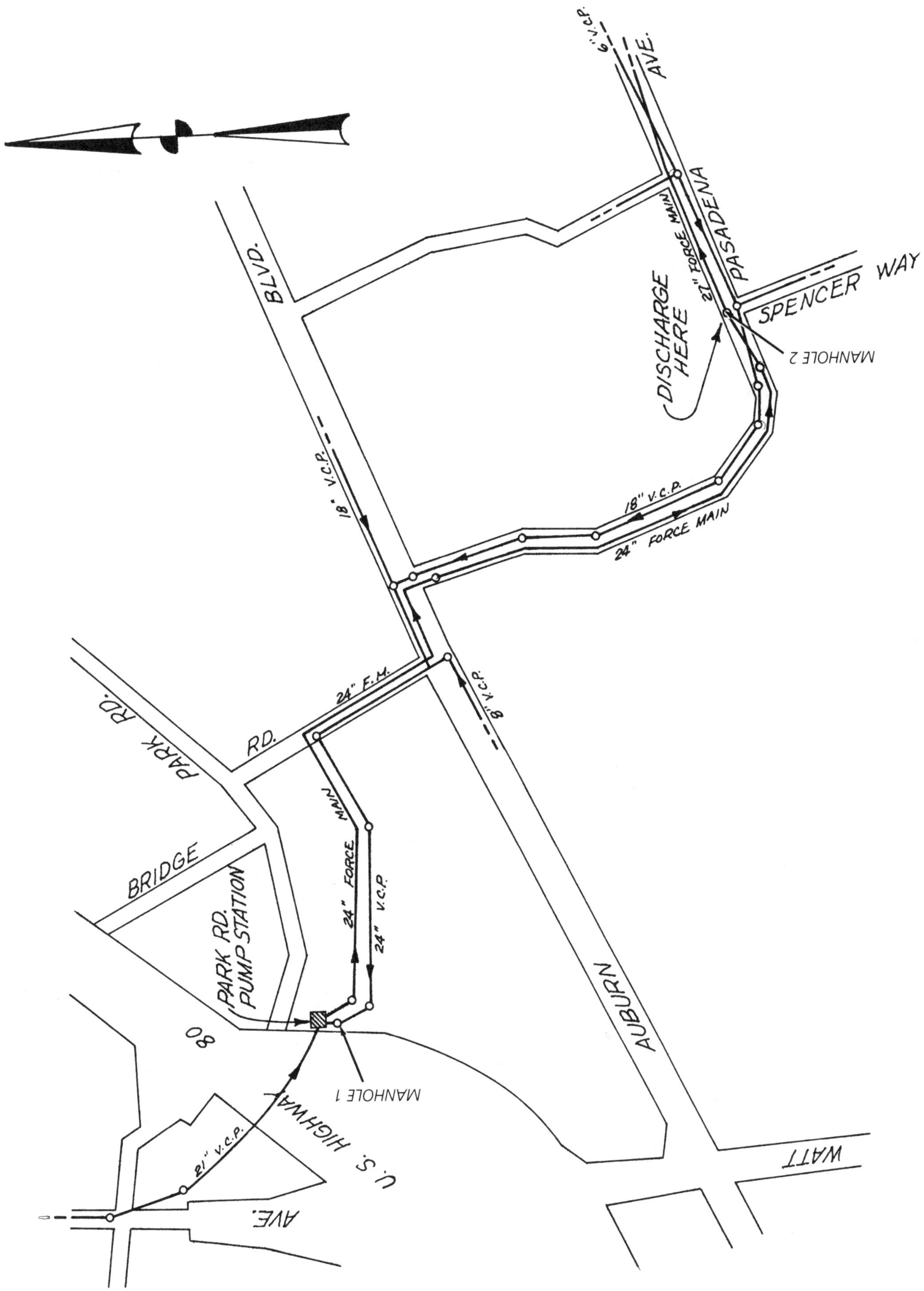
6" V.C.P.
AVE.
PASADENA
27" FORCE MAIN
SPENCER WAY
DISCHARGE HERE
MANHOLE 2
BLVD.
18" V.C.P.
18" V.C.P.
24" FORCE MAIN
8" V.C.P.
24" F.M.
RD.
PARK RD.
BRIDGE
PARK RD. PUMP STATION
24" FORCE MAIN
24" V.C.P.
AUBURN
80
MANHOLE 1
U.S. HIGHWAY
21" V.C.P.
AVE.
WATT

GENERAL DESCRIPTION

1. *STATION DESCRIPTION*

The design of the Park Road Wastewater Lift Station S-14 includes the following features:

a. A dual centrifugal pump lift system discharging into a common force main.

b. An automatic sample-speed flow match control system.

c. A chlorine injection system into the discharge flow.

d. A telemeter monitoring system which telemeters to the central office.

2. *SAFETY INSTRUCTIONS*

a. The station uses 115/440 volt electricity in various applications. Use caution when near wires or connection.

b. Use adopted safety measures when in areas or working on equipment containing chlorine.

c. Air pressure lines within the station contain pressures up to 60 PSI. Care should be taken to isolate any air-operated component and bleed residual pressure before removing it from the system.

d. Working in an enclosed area presents safety hazards to personnel (wet pit). Unless there is a way of constantly testing the station atmosphere, a ventilation blower must operate at all times while the wet pit is occupied. If the main ventilation blower is out of service and work must be performed in an enclosed area, a blower of equivalent capacity must be placed in operation during occupancy of the station. An *ENCLOSED AREA* (confined space) is defined by the State of California, Division of Industrial Safety as: The interior of storm drains, sewers, vaults, utility pipelines, manholes, and any other structure which is similarly surrounded by confining surfaces so as to permit the accumulation of dangerous gases or vapors.

e. Return all equipment to normal operating condition before leaving the station. Verify that the station lights have been turned off and the door is locked.

f. In the event of station flooding, three operators must be present during maintenance, one in the pit and two above flooded area.

A safety harness and safety line, and at least one explosion-proof light must be available during the performance of confined space maintenance. Confined space entry procedures must be followed at all times.

3. *EMERGENCY CONDITIONS*

The Park Road Lift Station S-14 will be put out of operation under the following conditions:

a. Power failure,

b. Inlet line stoppage,

c. Simultaneous pump system failure, and

d. Station flooding dry pit area.

In the event of an emergency condition whereby the station becomes inoperative due to power failure, an auxiliary generator can be substituted for normal power. This generator must have a minimum capacity output of 400 volts–135 K.W. to operate one wastewater lift pump and small power demands for station operation.

The isolation of the station from normal power source to operation on emergency generator conditions shall be performed by an electrician.

In the improbable situation the station would have to be bypassed, the following procedure would be used, using an auxiliary pump. The suction line must be placed in manhole 1 and the pump discharge line placed in manhole 2 at the intersection of Pasadena Ave. and Spencer Way, approximately 3,550 ft., that is, a little under 3/4 of a mile. For these locations, refer to the drawing on page 127.

In order to satisfy the flow demand, it will be necessary to use a 10" discharge line. (Pump (10" Jaeger) and hose available at maintenance yard.)

MISCELLANEOUS EQUIPMENT — S14

Wastewater Pumps

PACKING 3/4" WATER LUBE

LUBE LINE

3/8" line, solenoid, press gage, 1/4" Foxboro purge meter model NA 1501A scale 1-5 GPM. Check and gate valve. Strainer screen in main line.

DRAIN LINE: 3/4"

COUPLING: WEEDS SUR. FLEX #14F

SUCTION: 12" LINE

Gate valve, 12" Crane, non rising stem, wheel control.

DISCHARGE: 10" LINE

Ball valve automatic stop and check control. Bottom coupling 1026DV2815-10". Top coupling 101131. Willamette emergency shut off control size 10. Emergency close switch A540031B. Micro adj. pulse switch #1PD1-2-64 serial 2. 2 mercoid control switches on 3/4" pipes. Leslie strainer.

CONTROL BUBBLE LINE, AIR COMP.

Alternation-lead pump manual.

PANEL

Wet well level
Hour counters (2)
Fold Chart GPM & PSI
Pump motor control switches (2)
Pump motor tachometers (2)

FORCE MAIN

14" line in the station. Foxboro pressure. Transmitter Model 144B.P. Foxboro filter Regulator type 65-R. Magnetic flowmeter on 24" discharge line.

GATE VALVE #1

Location across the creek in the rear of the station. Raised manhole cement box outside of manhole with buried valve stem, wrench in the station.

GATE VALVE #2

Location on the southeast side of Pasadena Ave. at Auburn Blvd., intersection large manhole cover, small valve stem cover nearby. Wrench in station.

SLIDE GATE

Influent, power failure automatic, manual hydraulic water power norpack compression cylinder, model XDCSS 3/4" water lines, valves, solenoids and Leslie strainer.

TELEMETER

Pit high water.

Lower floor sump pump failure.
Low water pressure.
Influent gate.
Panel door.
Low chlorine pressure.

Motor bypass pin out to relieve alarm signal during intentional motor power shutoff for mechanical repair.

PREVENTIVE MAINTENANCE SCHEDULE — S-14

EQUIPMENT	TASK DESCRIPTION	FREQUENCY*	ISOLATION REQUIRED	MAINT./OVERHAUL INST. NO.	PAGE NO.
AIR COMPRESSOR	LUBRIFLUSH MOTOR	A	YES	6.01-C	139
AIR COMPRESSOR	SCOPE MOTOR BEARINGS	A	NO	6.01-F	139
AIR COMPRESSOR	SCOPE COMP. BEARINGS	A	NO	6.01-F	139
AIR COMPRESSOR	CHECK BELTS	S/A	YES	2.02-B	133
AIR COMPRESSOR	CHECK GAGES	S/A	NO	1.10-J	133
AIR COMPRESSOR	CHECK RELIEF VALVE	S/A	NO	1.10-I	133
AIR COMPRESSOR	CLEAN AIR FILTER	S/A	YES	1.10-H-2	133
AIR COMPRESSOR	CHECK MOTOR	M	NO	6.01-A-B	139
AIR COMPRESSOR	CHECK COMPRESSOR	M	NO	1.10	134
AIR COMPRESSOR	CHANGE LUBE OIL	M	YES	1.10-B-2	133
AIR COMPRESSOR	BLEED TANK	M	YES	1.10-F-3	133
AIR COMPRESSOR SMALL	LUBE COMPRESSOR	S/A	YES	1.10-F-1	133
AIR COMPRESSOR SMALL	DRAIN CONDENSATE	M	YES	1.10-F-3	133
VENT BLOWERS (2)	CHECK MOTOR LUBE	A	YES	2.20-A-4-5	134
VENT BLOWERS (2)	CHECK OPERATION	M	NO	2.20-A	133
BUILDING MAINTENANCE	CHECK LOCKS	M	NO	2.40-A	134
BUILDING MAINTENANCE	CHECK LIGHTS	M	NO	2.40-B	134
BUILDING MAINTENANCE	CHECK DOORS	M	NO	2.40-C	134
CHLORINATOR	CLEAN VALVE SEAT & PLUG	M	YES	3.20-A-16	134
CHLORINATOR	CLEAN PRIMARY AND SECONDARY FILTER	M	YES	3.20-A-17	135
CHLORINATOR	SHUT DOWN CHLORINATOR	M	YES	3.20-A-1	134

*A — Annually; S/A — Semi-Annually; M — Monthly.

EQUIPMENT	TASK DESCRIPTION	FREQUENCY*	ISOLATION REQUIRED	MAINT./OVERHAUL INST. NO.	PAGE NO.
CHLORINATOR	CLEAN METER TUBE	M	YES	3.20-A-18	136
CHLORINATOR	CHECK C.P.R. VALVE	M	YES	3.20-A-7	134
CHLORINATOR	CHECK FOR LEAKS	M	NO	3.20-A-9	134
CHLORINATOR	BODY CLEANING	M	NO	3.20-A-15	134
CHLORINATOR	CHECK VALVES	M	NO	3.20-A-11	134
CONTROL ROTAMETER	CHECK BUBBLE LINE	M	NO	3.31-D	137
DEHUMIDIFIERS (2)	INSPECT LUBE UNIT	A	YES	3.50-B	139
DEHUMIDIFIERS (2)	CHECK OPERATION	M	NO	3.50-A	139
WET PIT	CLEAN PIT	A	NO	7.10-A	142
WASTEWATER NO 1 PUMP	LUBRIFLUSH MOTOR	A	YES	6.01-C	139
WASTEWATER NO 1 PUMP	SCOPE MOTOR BEARINGS	A	NO	6.01-F	139
WASTEWATER NO 1 PUMP	SCOPE PUMP BEARINGS	A	NO	6.01-F	139
WASTEWATER NO 1 PUMP	LUBE PUMP BEARINGS	S/A	NO	7.20-H	144
WASTEWATER NO 1 PUMP	CHECK WATER LUBE LINE	S/A	NO	7.20-N-1	144
WASTEWATER NO 1 PUMP	CHECK COUPLING	S/A	YES	3.40-B	138
WASTEWATER NO 1 PUMP	CHECK BALL VALVE	S/A	YES	8.30-D	146
WASTEWATER NO 1 PUMP	CHECK STRAINER SCREEN	S/A	YES	7.94-A	146
WASTEWATER NO 1 PUMP	CHECK MOTOR	M	NO	6.01-A-B	139
WASTEWATER NO 1 PUMP	CHECK PUMP	M	NO	7.20-E	142
WASTEWATER NO 1 PUMP	CHECK PACKING	M	NO	7.20-A	142
WASTEWATER NO 1 PUMP	CHECK DRAIN LINE	M	NO	7.20-E-1	142
WASTEWATER NO 2 PUMP	LUBRIFLUSH MOTOR	A	YES	6.01-C	139
WASTEWATER NO 2 PUMP	SCOPE MOTOR BEARINGS	A	NO	6.01-F	139
WASTEWATER NO 2 PUMP	SCOPE PUMP BEARINGS	A	NO	6.01-F	139
WASTEWATER NO 2 PUMP	LUBE PUMP BEARINGS	S/A	NO	7.20-H	144
WASTEWATER NO 2 PUMP	CHECK WATER LUBE LINE	S/A	NO	7.20-N	144
WASTEWATER NO 2 PUMP	CHECK COUPLING	S/A	YES	3.40-B	138
WASTEWATER NO 2 PUMP	CHECK BALL VALVE	S/A	YES	8.30-D	146

EQUIPMENT	TASK DESCRIPTION	FREQUENCY*	ISOLATION REQUIRED	MAINT./OVERHAUL INST. NO.	PAGE NO.
WASTEWATER NO 2 PUMP	CHECK STRAINER SCREEN	S/A	YES	7.94-A	146
WASTEWATER NO 2 PUMP	CHECK MOTOR	M	NO	6.01-A-B	139
WASTEWATER NO 2 PUMP	CHECK PUMP	M	NO	7.20-E	142
WASTEWATER NO 2 PUMP	CHECK PACKING	M	NO	7.20-A	142
WASTEWATER NO 2 PUMP	CHECK DRAIN LINE	M	NO	7.20-E-1	142
SUMP PUMP	INSPECT SUMP PUMP	S/A	YES	7.20-L-2	144
SUMP PUMP	CHECK OPERATION	M	NO	7.20-L-1	144
MAIN WATER STRAINER	CLEAN STRAINER	A	YES	7.94-A	146
GATE VALVES (2)	REPAIR GATE VALVES (2)	A	YES	8.30	146
FORCE MAIN BURIED VALVES (2)	OPERATE LUBE-VALVES	A	NO	8.30-F	146
SLIDE GATE	CHECK SLIDE GATE	A	NO	8.40-C	146
WATER WELL	CHECK TANK	S/A	YES	9.01-B	147
WATER WELL	CHECK AIR RELIEF VALVES	S/A	NO	8.10-C-4	146
WATER WELL	CHECK TANK	M	NO	9.01-B	147
WATER WELL	CHECK PUMPING RANGE	M	NO	7.20-P	144
WATER WELL	SOUND WELL	M	YES	9.01-C-1-2	147
REGULATOR FILTERS (2)	CLEAN FILTERS	S/A	YES	1.10-H-2	133
REGULATOR FILTERS (2)	CLEAN FILTERS	M	YES	1.10-H-2	133
STA-DRY FILTERS (2)	BLEED FILTERS	S/A	YES	1.10-H-4	133
STA-DRY FILTERS	BLEED FILTERS	M	NO	1.10-H-4	133
WASTEWATER PUMP CONTROLS	ALTERNATE PUMPS	M	NO	7.20-G-1	144
WASTEWATER PUMPS	CHECK PUMPING RANGE	M	NO	7.20-P	144
WASTEWATER PUMPS	RECORD HOURS/GALLONS	M	NO	6.10-J	139
STATION POWER SUPPLY	RECORD ELECTRIC METER	M	NO	6.10-K	139
HOUSEKEEPING	CLEAN MOTORS	M	NO	2.40	134
HOUSEKEEPING	CLEAN PUMP UNITS	M	NO	2.40	134
HOUSEKEEPING	CLEAN STA. INTERIOR	M	NO	2.40	134
HOUSEKEEPING	CLEAN YARD/DRIVEWAY	M	NO	2.40	134

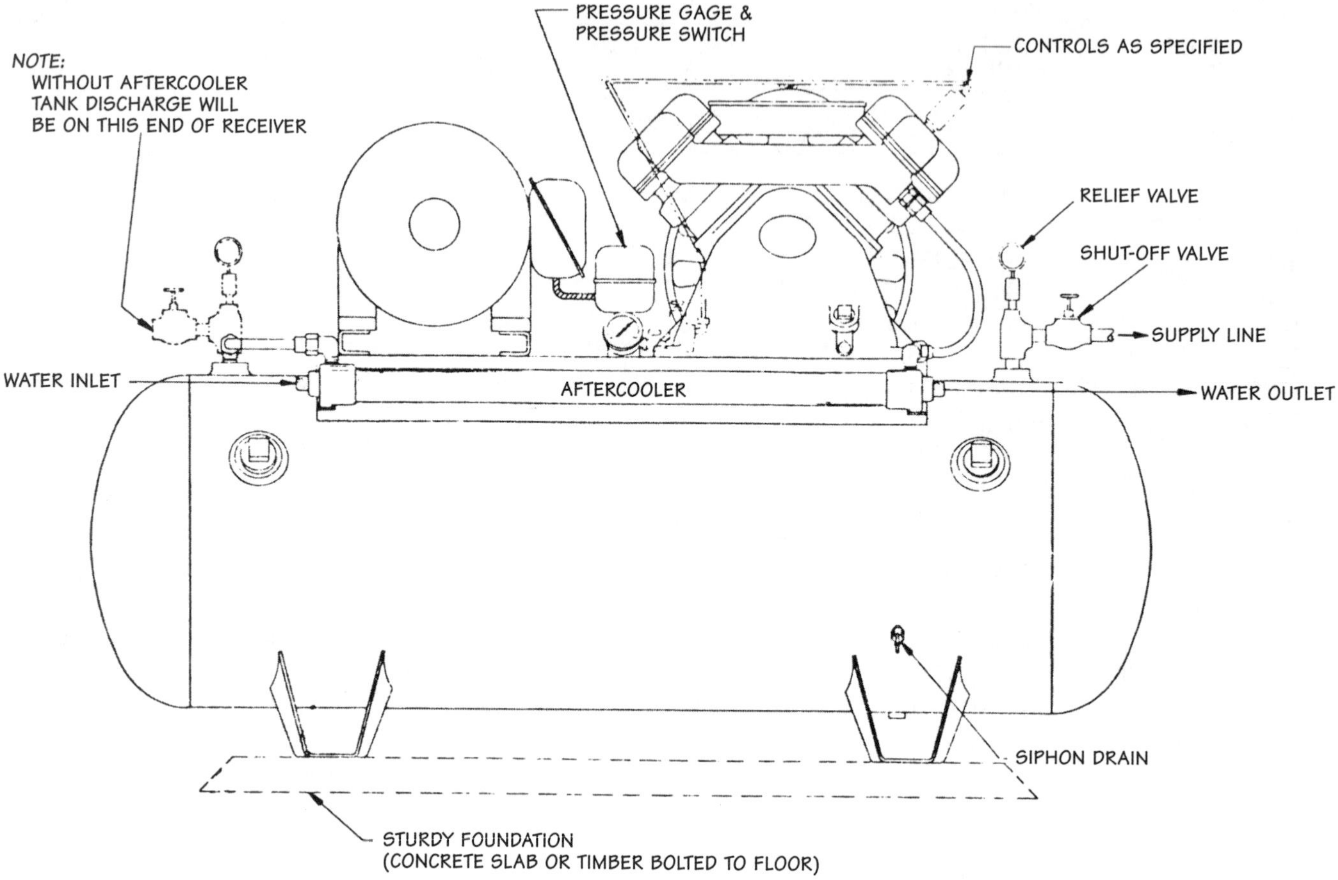
PRESSURE GAGE &
PRESSURE SWITCH
CONTROLS AS SPECIFIED
NOTE:
WITHOUT AFTERCOOLER
TANK DISCHARGE WILL
BE ON THIS END OF RECEIVER
RELIEF VALVE
SHUT-OFF VALVE
SUPPLY LINE
WATER INLET
AFTERCOOLER
WATER OUTLET
SIPHON DRAIN
STURDY FOUNDATION
(CONCRETE SLAB OR TIMBER BOLTED TO FLOOR)

Pressure Tank Compressor
Drawing No. 1.10-G

PREVENTIVE MAINTENANCE ON AIR COMPRESSOR INCLUDES THE FOLLOWING:

A. Keep unit clean

B. Check oil level in crankcase, add if needed

C. Drain condensate from receiver

D. Test relief valves

E. Clean suction air filter

F. Scheduled crankcase oil change

G. Check condition and tension of belts

H. Check all screws and nuts for tightness

AIR COMPRESSOR — 1.10 ISOLATION INSTRUCTION

A. PREREQUISITES:

Prior to the electrical isolation of an air compressor, ensure that the air compressor is operating normally.

B. ELECTRICAL ISOLATION:

Place the compressor motor power supply switch in the off position, lock in off position, and tag with a "Do Not Turn On" card to prevent unit operation.

C. MECHANICAL ISOLATION:

Close the air compressor discharge valve.

D. RESTORATION:

Mechanical Restoration: Open the air compressor discharge valve.

E. ELECTRICAL RESTORATION:

Unlock or untag the power switch and place in the ON position.

F. OPERABILITY DEMONSTRATION:

Demonstrate the air compressor is operable by starting the air compressor and observing its operation through one complete cycle. If any portion of this system was opened, inspect that portion for leakage, that is, oil, air, or water.

NOTE: Air receivers (unfired pressure vessels) require inspection and certification every five years in many areas. Contact your local safety regulatory agency for regulations.

AIR COMPRESSOR — 1.10 MAINTENANCE INSTRUCTIONS

1.10A *CONDITION CHECK*

If compressor shows evidence of overheating or excessive noise, stop immediately until repairs are made. A compressor will run hot if running continuously. Check for vibrations. Keep compressor free from dirt and dust. Check for leaks of oil or grease.

1.10B 2 *OIL CHANGE*

1. Check crankcase oil by turning compressor motor switch to off position and locking out if possible. Remove filler plug, oil should be at the level of the filler opening or never below the low mark on the dipstick if so equipped.
2. To change crankcase oil in a compressor, turn motor switch to off position, lock out if possible. Remove both drain and filler plugs. Drain out old oil thoroughly, replace drain plug and fill crankcase with new oil to the proper line. See paragraph 1.10B-1. Replace fill plug. Run the compressor 3 or 4 minutes. Shut it off and allow the oil to settle. Recheck, adding more oil if necessary.

1.10F 1 *SMALL COMPRESSOR LUBRICATION*

Small compressor bearing lube can be either oil cups or grease fittings. If equipped with oil cups, five or six drops of oil are enough. If equipped with grease fittings, lube very sparingly.

1.10F 3 *DRAIN CONDENSATE*

Drain condensate from the tank through the drain valve at the bottom.

1.10H 2 *AIR FILTER*

Air cleaners with felt pads and screens: remove cover, wipe out interior, wipe the oil-impregnated pad with a solvent-dampened rag. Add three or four drops of oil to the felt. Clean the wire screen with solvent and dry thoroughly.

1.10H 4 *STA-DRY FILTERS*

Bleed sta-dry filters periodically by slowly opening, in turn, all drain cocks on filters or the air line.

1.10I *SAFETY VALVE*

When a compressor is controlled by a pressure switch for a high and low limit, a safety valve is installed. To test the safety valve, lift the lever allowing air to escape; this will ensure the valve is operating.

1.10J *GAGE CHECK*

Check pressure gage for correct pressure of this particular line or unit. If you have reason to believe the pressure gage is not accurate or the demand of the pressure switch does not coincide with pressure gage and you want to check which is at fault, replace with a pressure gage known to be accurate. If the pressure gage is combined with a dampener, remove, disassemble and clean. On the pulsation-type dampener for air, set the pin in hole 5.

BELTS — 2.02

2.02B *BELT CHECK*

1. Check tension and wear. Rubber wearings near the drive are signs of improper tension, incorrect alignment or damaged sheaves. If tightening a belt does not correct slipping, check for overload, oil on the belt or other possible causes. Never use belt dressing to stop a V-belt slipping.
2. When belts need replacing, never replace one V-belt on a multiple-belt drive. Replace the complete set with a set of matching belts.

VENTILATION BLOWER — 2.20 MAINTENANCE INSTRUCTIONS

PREREQUISITE

Removal of the ventilation blower from service requires the substitution of an equivalent ventilation blower to maintain adequate ventilation inside Station S-14. Observe all pertinent safety practices when performing maintenance under this condition.

2.20A *INSPECTION*

1. Open the ventilation blower circuit at the wall panel.
2. Disassemble the unit and check the frame, cover, and damper for rust or corrosion. Clean with solvent. Paint if needed.
3. Check that the vibration damper and springs are operating freely.

4. If the motor is equipped with sealed bearings, no lubrication is required.
5. If the motor bearing housing is equipped with plugs, remove and check if lubricant is grease or oil. Lubricate accordingly. (If grease refer to Par. 6.01C 2-7.)
6. Check for worn or frayed electrical wires. When reassembled, close the blower circuit at the wall panel and test for operation and vibration.

ISOLATION

Turn the ventilation blower power switch in control panel to the off position. If possible, lock in the OFF position to prevent accidental power supply to unit.

BUILDING MAINTENANCE AND INSPECTION — 2.40

A. *LOCKS*

Oil locks with graphite. Make sure all electrical panel locks, padlocks, and door locks work freely. Replace all defective locks.

B. *LIGHTS*

Check for broken or burned out light fixtures. Replace burned out or broken bulbs. Report broken or damaged switches.

C. *DOORS*

Check hinges, catches, and seals on all doors. Check that they swing freely and close tightly.

HOUSEKEEPING

A. Equipment and wall surface cleaning can best be done with a rag moistened with a cleaning solution. Avoid contact with electric terminals. If this is not possible, isolate the system according to the applicable isolation instructions before cleaning. Floors must be mopped with a cleaning solution.
B. Gather the accumulated debris from the exterior yard area and place it in a waste container carried in the service vehicle.

ELECTRICAL PREVENTIVE MAINTENANCE

Isolate all circuits prior to performing maintenance according to the applicable isolation instruction. Inspect electric connectors for looseness and insulation cracking. Tighten and repair when necessary. Do not attempt to maintain or work on live circuits.

CHLORINATOR — 3.20-A MAINTENANCE INSTRUCTIONS

3.20A 1 *ISOLATION*

Turn off chlorine supply; close tank valve, close containment valve. Turn off injector water supply. (Be sure all chlorine is exhausted from unit before turning off water.)

3.20A 7 *PRESSURE-REDUCING VALVE*

Remove protective cap nut, remove cap, remove silver lock nuts (2), remove needle assembly, clean seat assembly. Reassemble new gasket as needed.

3.20A 9 *CHECK FOR CHLORINE LEAKS*

Use ammonia and check all chlorine connections for leaks. (If white smoke appears, a leak is present.)

3.20A 11 *CHECK VALVES*

Check valves for leaks around stems and for damaged threads. Check flex lines for kinks and/or signs of corrosion. Replace if needed.

3.20A 15 *CLEANING CHLORINATOR BODY*

Use a damp cloth, then wipe dry. Apply a small amount of polishing wax and wipe with dry cloth.

3.20A 16 *CLEANING VALVE SEAT AND VALVE PLUG*

See Drawings Nos. 3.20-P and 3.20-Q.

To disassemble capsule, remove large hexagonal nut (T-229) on back of capsule and remove the secondary inlet filter assembly (A-187). Insert small screwdriver into the inlet to hold the inlet valve plug (V-140) against valve seat, unscrew the vent screw (W-182); the valve plug, spring retainer, spring and vent plug may now be removed. Unscrew the adaptor seal plug (U-545) and remove the valve seat (U-543).

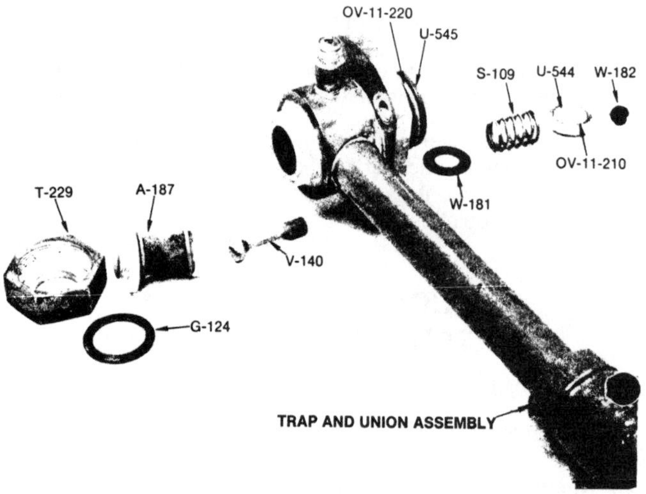

Drawing No. 3.20-P

Drawing No. 3.20-Q

CHLORINATOR — 3.20A
MAINTENANCE INSTRUCTIONS (cont'd)

3.20A 17 *CLEANING PRIMARY FILTER*

Remove the four (4) bolts holding the filter cover on and remove the cover and screen from body. Clean the screen by submerging in hot water with a detergent and remove dirt with a coarse bristle brush. Dry thoroughly.

Secondary filter (A-187). Clean in the same manner.

Drawing No. 3.20-R

CHLORINATOR — 3.20A
MAINTENANCE INSTRUCTIONS (cont'd)

3.20A 18 *METER TUBE CLEANING*

To remove metering tube loosen the 4 screws at the bottom meter fitting. Press downward on the tube and tilt out at the top until tube is clear of top meter fitting. Remove float carefully and clean tube by pushing through a clean lint-free swab dampened with a cleanser. Use clean dry swab to dry tube.

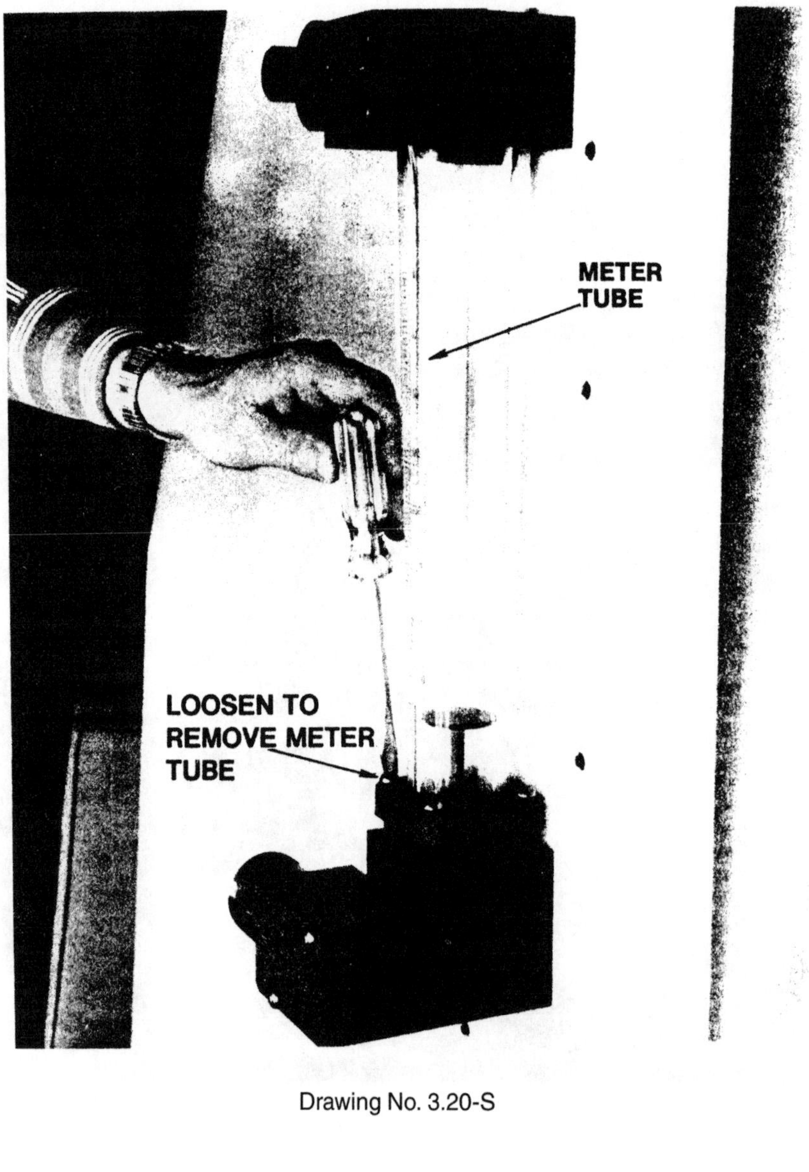

Drawing No. 3.20-S

CONTROLS — 3.31

3.31D *ROTAMETER*

Check bubble line. If rotameter has a marking on the panel, check that the setting corresponds to the mark.

MANUAL PURGE

1. Depressing a button or moving a switch or valve on or in the panel marked "purge" will automatically shut off the air to the pressure mechanism and provide a direct flow of air to the compression bell, thereby purging the line. When released or switch thrown back to operate, controls go back to normal operation.

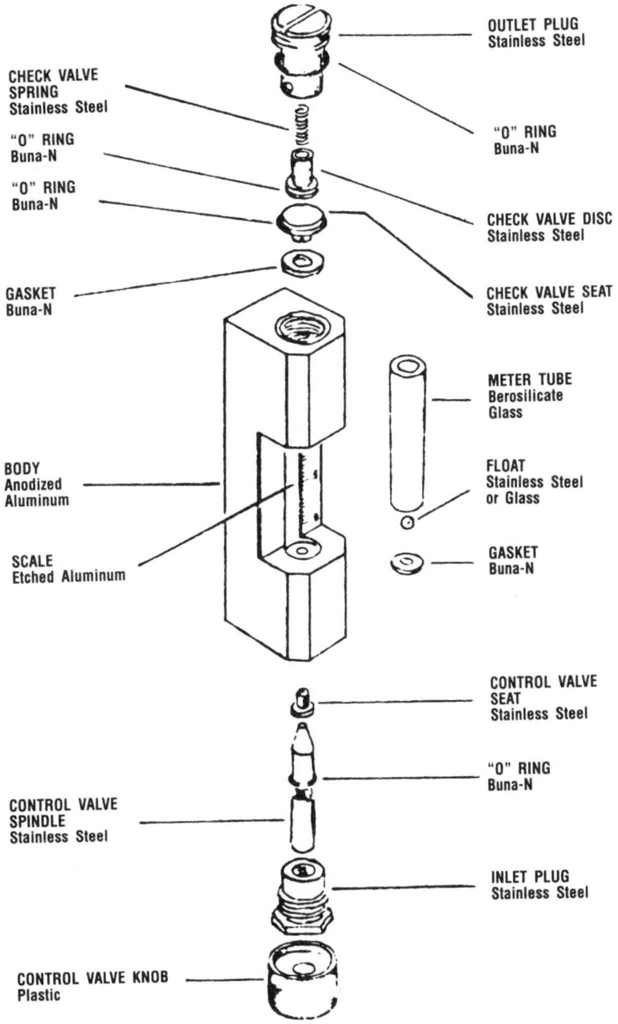

Rotameter
Drawing No. 3.31A

COUPLING — 3.40 MAINTENANCE INSTRUCTIONS

3.40A *GENERAL*

Unless couplings between the ends of the driving shaft and driven shaft of a pump or any other piece of equipment are kept in proper alignment, breaking or excessive wear result in either or both the driven machinery and the driver. Burned out bearings and sprung or broken shafts are some of the damages caused by misalignment. To prevent these damages, periodic checking and maintenance must be performed. Any maintenance performed on couplings, either checking or realignment, must be done with a motor switch in OFF position and a special lock on the panel switch.

1. Flexible couplings permit easy assembly of equipment, but must be checked for both horizontal and vertical alignment. Distance between coupling halves must be exactly the same at all points. A straightedge must have continuous contact when laid across both couplings. Set screws must be tight and if found continually loose must be replaced. Realign when necessary.
2. If there is a shield or guard over the coupling, check that bolts are tight and guard will not in any way touch the coupling or shaft.

3.40B *CHECK FLEXIBLE COUPLING*

Realign when necessary, using a straightedge and thickness gage or a dial indicator.

1. Remove coupling pins, rigidly tighten driven equipment, and slightly tighten bolts holding the drive. To correct horizontal and vertical alignment, shift and shim the drive to bring the coupling halves into position so no light can be seen under a straightedge laid across them. Place straightedge in four positions. Hold a light in back of the straightedge to help ensure accuracy. Check angular misalignment with a thickness or feeler gage inserted at four places to make certain space between the coupling halves is exactly the same at all points. When proper alignment has been secured, coupling pins can be put in place easily using only finger pressure. Never hammer pins into place.
2. In couplings where pins are solid on the coupling halves and fit into rubber eyelets, or couplings which mesh together with rubber or nylon inserts, if properly aligned and loose on the shafts, coupling halves should slide together with finger pressure.
3. To check coupling alignment with a dial indicator, disconnect the coupling halves. Fix a dial indicator to one of the shafts or coupling hubs. Span the indicator arm across the mating shaft or coupling hub. Scribe index lines on the coupling halves or mark where the indicator point rests. Set indicator dial to zero. Slowly turn both coupling halves so that the index lines match, or indicator point is always on the mark. Observe dial reading to determine whether driven or driver needs adjustment. Acceptable parallel alignment and acceptable angular alignment occur when one-half of the total indicator reading (complete turn) does not exceed manufacturer's limits.

These limits depend upon size and speed of the coupling. Refer to manufacturer's specification limits. Correct parallel and angular misalignment by slightly shifting the leveling shims under the base plate of the driver. Retest after each shifting of shims.

COUPLING ALIGNMENT

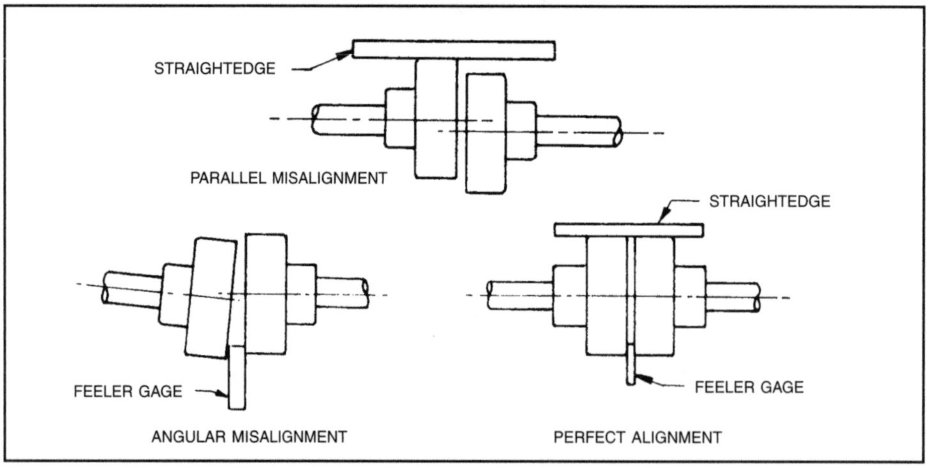

Testing alignment, feeler gage and straightedge

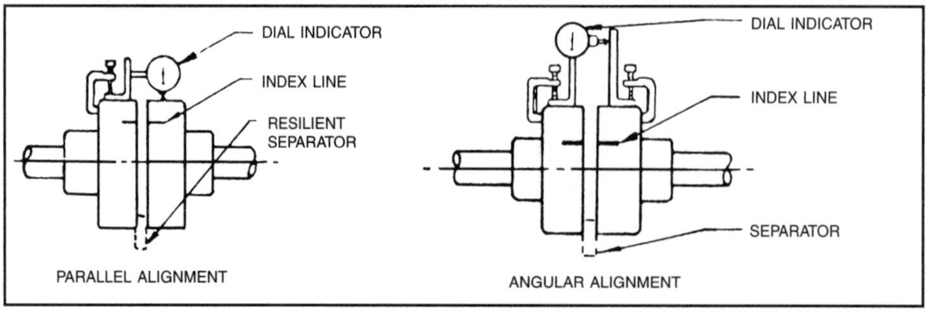

Testing alignment, dial indicator
Drawing No. 3.40C

DEHUMIDIFIER — 3.50 MAINTENANCE INSTRUCTIONS

3.50A *OPERATIONAL CHECK*

Set dehumidifier dial on dry for normal operation. Check for condensate on the walls of the station. If condensate is noted, reset dehumidifier to very dry until moisture is cleared up. If the coils of the dehumidifier are frozen or frosted, turn the unit off until they defrost.

3.50B *LUBE AND INSPECT UNIT*

Disconnect the cord from the wall outlet. Disassemble the dehumidifier. Clean throughout, including drip pan and drain hose. Lube fan motor with about three drops of light oil. Reassemble. Replace the cord plug in the wall outlet and test.

ELECTRIC MOTORS — 6.01 MAINTENANCE INSTRUCTIONS

6.01A *CONDITION CHECK*

1. Keep motors free from dirt and dust.
2. Keep operating space free from articles which may obstruct air circulation.
3. Check for excessive grease or oil leakage from bearing fill lines and drain lines.

6.01B *UNUSUAL CONDITION CHECK*

Note and record in the station log any of the following conditions:

1. Unusual operating noises,
2. Motor failing to start or come to speed normally,
3. Motor or bearings which feel or smell hot,
4. Continuous or excessive sparking of the commutator or brushes,
5. Intermittent sparking at brushes,
6. Hot or blackened commutator,
7. Brush chatter,
8. Smoke, charred insulation, or solder whiskers extending from the armature,
9. Excessive humming,
10. Regular clicking,
11. Rapid knocking,
12. Vibration, and
13. Fine dust under couplings having rubber buffers or pins.

6.01C *LUBE FLUSH*

1. Turn motor to OFF position. If there is a lock out switch, lock out the panel switch of the motor you are preparing for lubrication.
2. Wipe pressure gun fitting, bearing housing and relief plug with a clean rag.
3. Before using grease gun always remove relief plug on the opposite side of the bearing from the grease fitting. This will prevent excessive pressure in housing which might rupture the bearing seals.
4. Use a clean screwdriver or similar tool to remove hardened grease from the relief hole and permit excess grease to run freely from the bearing.
5. Remove all fittings, tools, and rags from the motor.
6. Place the unit back in service. Start the motor. While the motor is running, add grease with a hand-operated pressure gun unit until it flows from the relief hole.

6.01F *SCOPE BEARINGS*

Check all bearings using a stethoscope. Listen while the motor is being started and shut off. Listen also during motor operation for whines, gratings or uneven noises in several locations around the bearing housing as close as possible to the bearing. If unusual noises are heard, pinpoint the location, record it in the station log, and report the condition to the responsible maintenance supervisor.

ELECTRIC MOTOR ISOLATION

Place the electrical control switch which is in the control panel in the OFF position and lock it in the OFF position with the special lock. If there is a secondary switch at the motor, lock this in the OFF position also. Both switches in the OFF position will ensure against accidental power supply.

METERS — 6.10

6.10J *COUNTER — GALLONS/HOUR*

1. Where a unit is equipped with a counter or a register to record gallons pumped or a register to record hours the unit has run, take the reading from the counter or register and record it on a special log sheet provided for that purpose by following directions on the sheet. When reading meters with a revolving register, note if there are zeros (00) printed on face of the meter at the right of the revolving figures. If there are zeros, add them to the register number. Also check to see what the meter reading measure is such as cu. ft. or gallons. Log the units of measurement. For example, if the meter reading is in gallons and there are two zeros, a register number of 7,439 would be logged on the form as 743,900 gal. Also, for any meters or gages registering amps, volts, or air, take the reading from the panel gage or meter if so requested.

FLOW RECORDER

2. The flow recorder is regulated by a magnetic flowmeter on discharge line. A fold chart receives the resultant signal and is recorded by pen and ink system. There is a recording of psi on the same fold chart.

6.10K *ELECTRIC POWER METERS*

1. There are various types of electric meters. The most common and more frequently used meter has four or five registers or dials with pointers. Two of the pointers move in one direction, the others in the opposite direction. To read and log a meter of this type, write down from left to right the numbers the pointers *HAVE PASSED.* Before writing down any number the pointer has passed, always check the direction of the pointer to get the correct number. Suppose one of the pointers points directly to the "4." Before writing down 4, look at the pointer on the dial to the right, if it has not quite reached the "0" then the pointer which seemed to point directly to 4 has not actually reached the 4, so write down 3 instead. If the pointer to the right has passed the "0" then the correct number would be 4.

ELECTRIC MOTOR — 6.01

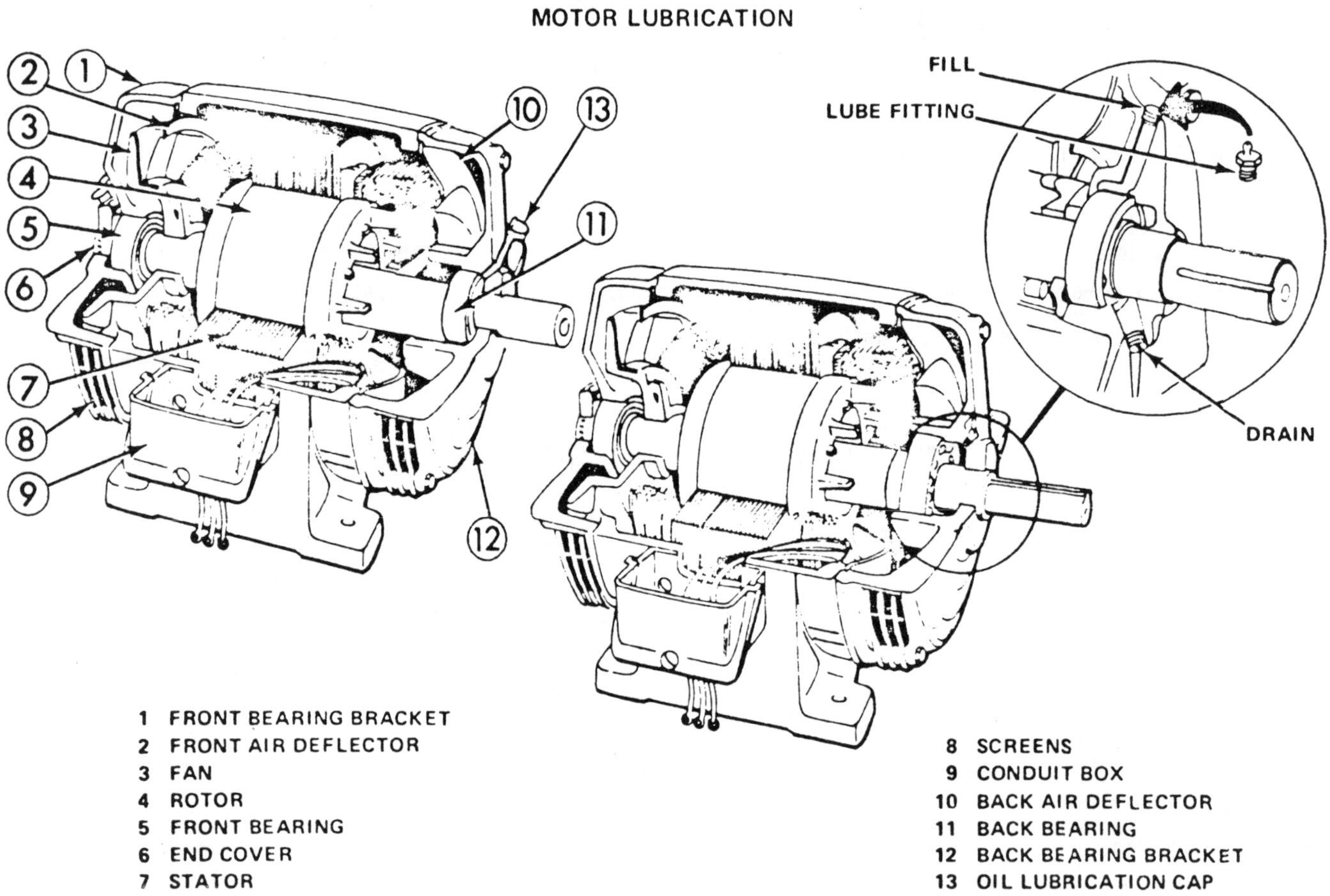
MOTOR LUBRICATION
FILL
LUBE FITTING
DRAIN
1 FRONT BEARING BRACKET
2 FRONT AIR DEFLECTOR
3 FAN
4 ROTOR
5 FRONT BEARING
6 END COVER
7 STATOR
8 SCREENS
9 CONDUIT BOX
10 BACK AIR DEFLECTOR
11 BACK BEARING
12 BACK BEARING BRACKET
13 OIL LUBRICATION CAP

Electric Motor
Lubrication

Drawing No. 6.01B

ELECTRIC MOTOR — 6.01

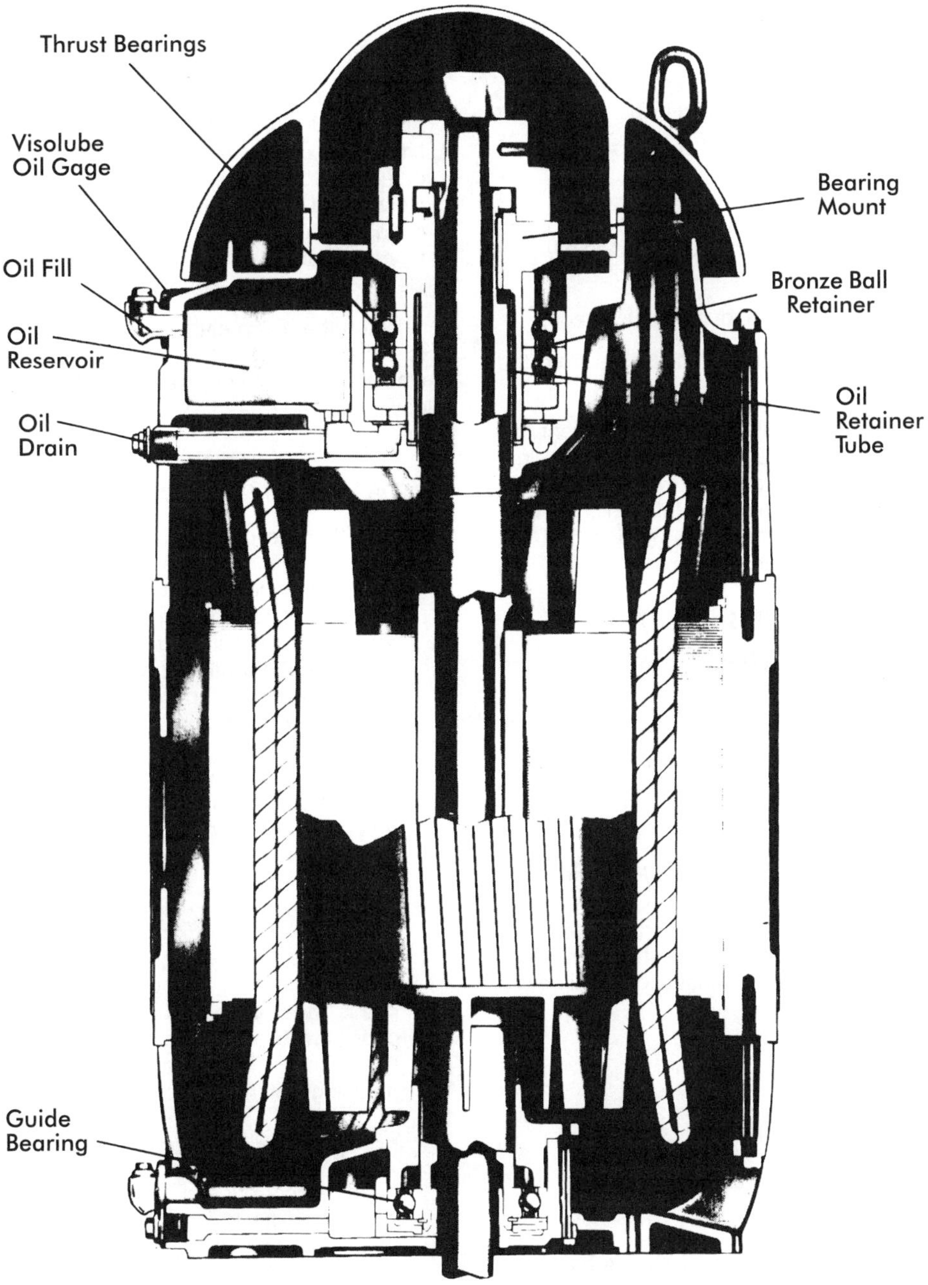
Thrust Bearings
Visolube Oil Gage
Oil Fill
Oil Reservoir
Oil Drain
Bearing Mount
Bronze Ball Retainer
Oil Retainer Tube
Guide Bearing

Motor Bearing
Lubrication

Drawing No. 6.01C

2. A special log sheet can be provided to log the meter reading. On some meters there may be a notation to multiply the reading by 40 or 1,600 or some other figure. Put this multiplication factor in the space provided. If previous and present readings are required, use direct meter readings and subtract before multiplying by the multiplication factor.

WET PITS — 7.10 MAINTENANCE INSTRUCTION

7.10A *INSPECTION AND CLEANING*

1. When cleaning a wet pit, manhole or sump (where water pressure is available either at the station or portable water tank with pressure), check the pit for any floating debris which might plug or damage a pump.

   If there is debris, remove it.

   Use a hose and nozzle to wash the solids off the pit walls while the pit is being pumped down.

   If solids are heavier than normal, pump out by stages.

   If water pressure does not remove the solids from the walls, the pit walls must be scraped.

DEFINITION: A WET PIT OR ENCLOSED AREA (confined space) is defined by the State of California, Division of Industrial Safety as: The interior of storm drains, sewers, vaults, utility pipelines, manholes, and any other structure which is similarly surrounded by confining surfaces so as to permit the accumulation of dangerous gases or vapors.

WASTEWATER PUMPS — 7.20 ISOLATION INSTRUCTIONS

PREREQUISITES:

Prior to the electrical isolation of a wastewater pump, ensure that the motor and pump are operating.

ELECTRICAL ISOLATION:

Place the electrical control switch in the OFF position and lock in the OFF position with the special lock. There is a secondary lockout switch at the motor for the purpose of eliminating power to unit without using the switch panel. Both switches in OFF position will ensure against accidental power supply.

MECHANICAL ISOLATION:

Close all valves.

RESTORATION:

Open all valves.

ELECTRICAL:

Unlock power supply switches and place them in the ON position.

OPERABILITY DEMONSTRATION:

Demonstrate the wastewater pump is operable by starting the pump and observing if it is operating normally.

WASTEWATER PUMPS — 7.20 MAINTENANCE INSTRUCTIONS

After the unit has run for a short period under normal operating conditions, it should be shut down and checked for alignment. All alignment checks must be made with the coupling halves disconnected and again after they are connected. Remember that the correction of alignment in one direction may alter the alignment in the other direction; therefore, it is necessary to recheck in all directions after making any adjustment. As in any piece of precision machinery, it is economical to keep your pumping unit in the best operating condition by checking the alignment periodically.

The stuffing box packing is subject to wear and should be given regular attention. If normal inspection indicates that the stuffing box is operating satisfactorily with no excessive heat or leakage of the sealing liquid, it should not be tampered with. However, if leakage of the sealing liquid has increased and cannot be reduced by a slight tightening of the glands, it is time to shut down the unit and repack the box or check the sleeve surface for roughness.

When the pump starts to show normal wear, it is advisable to order replacement parts, thus avoiding excessive lost time which may occur if parts are not ordered until the pump breaks down. Scoping bearings will determine pending bearing failure.

7.20A *WATER SEAL PACK GLANDS*

1. See that the packing box is protected with a clean water supply at all times. Make sure that pressure at water seal valve entrance is not less than pump shutoff head. Check water line pressure gage. Check packing gland for leakage. The packing is subject to wear and should be given regular attention. If normal inspection indicates that the packing is operating satisfactorily with no excessive heat or leakage of the sealing liquid, it should not be tampered with. However, if excessive leakage is found, tighten gland nuts evenly just enough to stop excessive leakage. After adjusting gland, be sure that the shaft turns freely by hand. A strap wrench may be needed on large pumps. If serious leakage continues, renew packing.

7.20B *GREASE SEAL PACK GLANDS*

1. When grease is used as a packing seal, maintain constant pressure on the packing during operation. When a grease cup is used, keep it loaded with proper lubricant. Turn the cap down two or three turns until resistance is felt. If this uses part of the grease in the cup, remove and refill. When greasing a fitting, two or three strokes of the gun is enough.

7.20E *CONDITION CHECK*

1. Keep exterior of pump clean. Keep drains open. Run pump manually. Check if pumping at full capacity. This can be done numerous ways, depending on the design of the station. Check the discharge if convenient. Watch the check valve. Watch the lowering of the pit. Check for excessive grease or oil leakage from the bearings. Look for black powdery substances near bearings or shaft. Note if pump feels or smells hot. Note any unusual noises or vibrations. Check guards and shields.

PUMP CENTRIFUGAL — 7.20

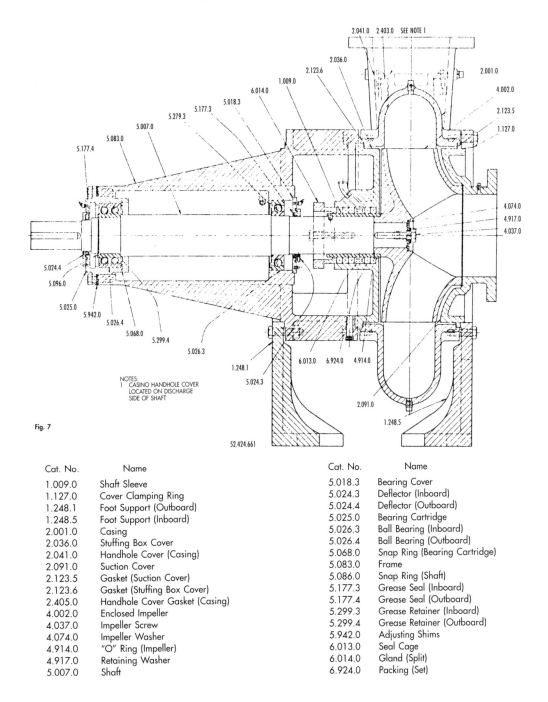

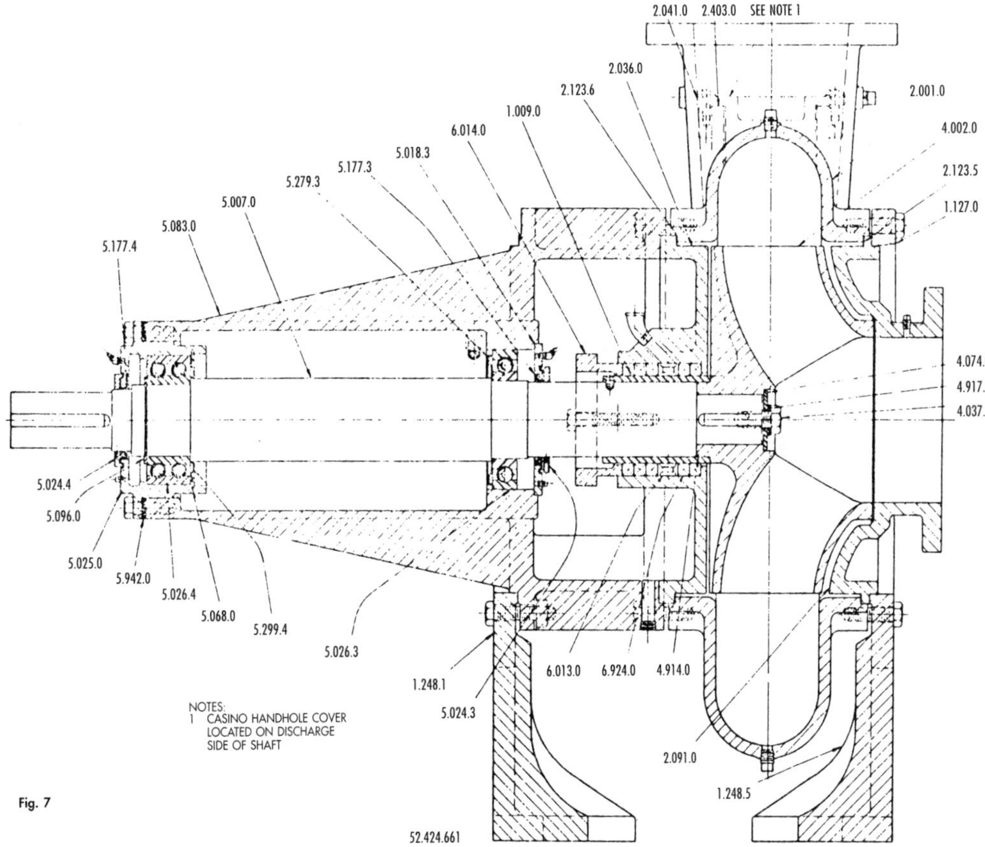

Fig. 7

Cat. No.	Name	Cat. No.	Name
1.009.0	Shaft Sleeve	5.018.3	Bearing Cover
1.127.0	Cover Clamping Ring	5.024.3	Deflector (Inboard)
1.248.1	Foot Support (Outboard)	5.024.4	Deflector (Outboard)
1.248.5	Foot Support (Inboard)	5.025.0	Bearing Cartridge
2.001.0	Casing	5.026.3	Ball Bearing (Inboard)
2.036.0	Stuffing Box Cover	5.026.4	Ball Bearing (Outboard)
2.041.0	Handhole Cover (Casing)	5.068.0	Snap Ring (Bearing Cartridge)
2.091.0	Suction Cover	5.083.0	Frame
2.123.5	Gasket (Suction Cover)	5.086.0	Snap Ring (Shaft)
2.123.6	Gasket (Stuffing Box Cover)	5.177.3	Grease Seal (Inboard)
2.405.0	Handhole Cover Gasket (Casing)	5.177.4	Grease Seal (Outboard)
4.002.0	Enclosed Impeller	5.299.3	Grease Retainer (Inboard)
4.037.0	Impeller Screw	5.299.4	Grease Retainer (Outboard)
4.074.0	Impeller Washer	5.942.0	Adjusting Shims
4.914.0	"O" Ring (Impeller)	6.013.0	Seal Cage
4.917.0	Retaining Washer	6.014.0	Gland (Split)
5.007.0	Shaft	6.924.0	Packing (Set)

Drawing No. 7.20-A

7.20G *OPERATE PUMPS ALTERNATELY*

1. If alternator is not installed and two or more pumps of the same size are installed, alternate their use to equalize their wear.
2. Switches are marked either alternator or selector. In some electrical installations a pump will not run on manual unless the selector switch is turned to that particular pump. Thus, if you wish to run #1 pump on manual, turn the selector switch to #1.

7.20H *LUBRICATION*

1. The importance of proper lubrication cannot be overemphasized. It is difficult to say how often a bearing should be greased since that depends on the conditions of operation. One ounce of grease should be added at regular intervals, but it is of equal importance to avoid adding too much grease. Excess grease is the most common cause of overheating. If possible remove the bearing cover and visually inspect the grease in the housing. If you are unable to remove the cover, remove the relief plug and add grease cautiously (five or six strokes of the gun are enough).

7.20N *WATER LUBE LINE CHECK*

1. Close the gate valve or shutoff cock in the water supply line to the pump packing box. Disconnect the water lube line at the union and remove the section of the line from the union to the packing box. In case of a flexible hose, remove it at the packing box inlet. Check the inlet to the lantern ring for grease or particles of packing. Start the pump, slowly open the supply line valve. The solenoid valve in the lube line should now be open and water flowing through the line. Shut off the pump; this should stop the flow of water as the solenoid valve should automatically close. Reinstall the lube line. If it consists of pipe, use suitable thread compound. Make sure the gate valve or shutoff cock is fully open. Put the unit back in service. Run the pump. Check the packing leakage; adjust if necessary.

7.20P *PUMPING RANGE (ON and OFF levels)*

1. Wastewater lift, storm drain and other types of pumps may have an automatic set pumping range. The control for the range can be a bubble line, float, electrode or in case of a water well pump, a pressure switch. The test is to check the on and off levels which are set for all pumps unless manually operated. There are several ways of checking pumping levels. One way is with a moderate or good flow of water, allow the pit to fill until the lead pump comes on automatically. Check that the pump starts at the correct level. Shut this pump off and let the pit keep filling up until the second pump starts. Check the starting level of this pump and continue with the same procedure until all pumps have been tested for starting level. Next turn all pumps on automatic and check the stop level as each one stops. Variations of this procedure will have to be made due to installation of the controls in a particular station. For example one pump may be set not to run while two others are running, thus this pump would have to be checked independently. Make sure all pump switches are on automatic before leaving the station. With water well pumps and pressure switches, test the pressure on and off range. As there is a delay on water well pumps before starting, it would be more accurate to watch the mercoid switch to see that it trips at the correct pressure. Should it be necessary, the well can be isolated by closing the tank outlet valve. Then the water can be drained by opening the tank drain valve until the pressure is down and the pump comes on automatically. Close the drain valve. Check both on and off settings. After checking the range, open the outlet valve. Anytime pumping levels do not correspond with the settings for the station pumps and the reason cannot be readily found and corrected, report it to your supervisor.

SUMP PUMPS — 7.20L MAINTENANCE INSTRUCTIONS

7.20L 1 *OPERATIONAL CHECK*

Fill the sump basin and verify that the pump starts automatically and stops at the proper level.

7.20L 2 *INSPECTION*

A. Pump all the water possible out of the basin. Isolate the sump pump according to isolation instruction.

B. Disconnect the pump at the union and remove from the basin.

C. Remove the bottom cover or strainer and check the impeller. Clean the entire unit of rust, grease or corrosion.

D. Check the float for pin holes or wear spots. Replace if necessary.

E. Put a thin film of grease or oil on the float rod and rod guides.

F. Before returning the sump pump to the basin, test the check valve by slowly opening the discharge gate valve. If water flows from the line, the check valve is leaking. In this situation, close the gate valve and repair or replace the check valve.

G. Clean the basin thoroughly and reinstall the pump.

H. Restore the pump to service according to isolation instruction.

Fill the sump basin until the pump automatically starts. Verify that the pump operates smoothly and then stops at the proper level.

ISOLATION INSTRUCTION

A. *PREREQUISITES*

Prior to the isolation of the sump pump, pump the basin as dry as possible.

B. *ELECTRICAL ISOLATION*

Remove the sump pump fuse or remove the electrical plug from the wall outlet, whichever is applicable.

C. *MECHANICAL ISOLATION*

Close the discharge gate valve.

D. *RESTORATION*

1. Open the discharge gate valve.
2. Replace the sump pump fuse or reconnect the electrical plug to the wall outlet.

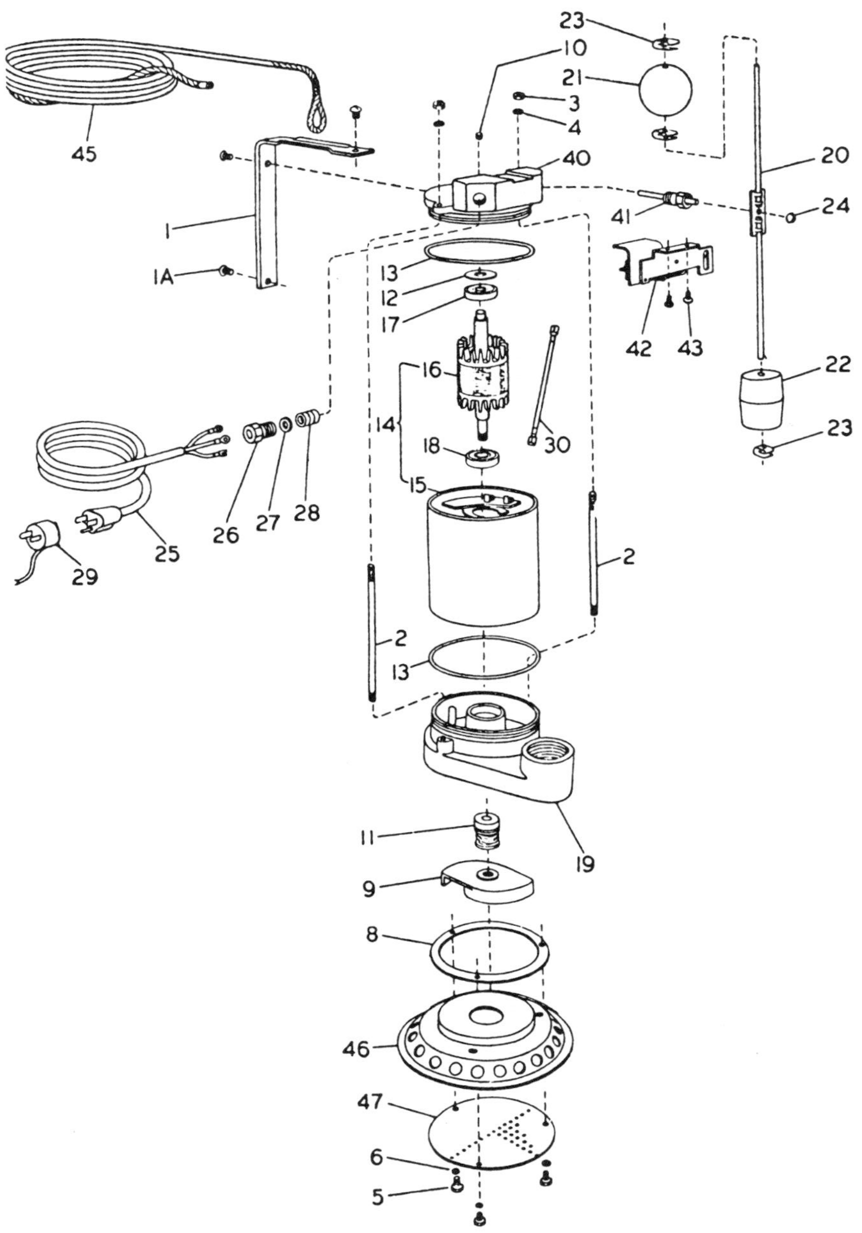

Drawing No. 7.20-C

Model PCD-6 ⅓ HP, 115V.

REPAIR PARTS LIST

Key No.	Part Description	No. Used	Part Symbol
1	Handle	1	PS54-5
1A	Machine Screw #10-24 x 5/16" Lg. Truss Head	3	U30-583C
2	Stud — ¼-20 x 7⅜" Lg.	2	U30-619C
3	Nut — ¼-20 Hex	2	U36-36C
4	Lockwasher — ¼"	2	U43-10C
5	Capscrew — ¼-20 x ½" Lg. Hex Hd.	3	U30-50C
6	Washer — Plain — ¼"	3	U43-60C
8	Gasket (as Req'd.)	1	PS20-3
9	Impeller	1	PS5-2P1C
10	Pipe Plug — ¼" NPT Soc. Hd.	1	U78-95C
11	Shaft Seal — John Crane Type 6	1	U9-103
12	Bearing Loading Spring 202	1	PS18-7
13	"O" Ring 5⅛ x 5⅜ x ⅛	2	U9-114
14	Motor — ⅓ HP, 115 V	1	PS118-15
15	Stator	1	PS18-35
16	Rotor	1	PS18-34
17	Bearing — MRC #7109	1	U18-682
18	Bearing — MRC #203SFZ	1	U18-628
19	Volute	1	PS1-7BB
20	Float Rod Assembly	1	PS128-3
21	Float — 2½" Dia. x ⅛" Hole	1	PS28-7
22	Weight (Control #502)	1	PS28-6
23	Speed Clip #C26046SS-014-27	3	PS28-8
24	Float Rod Ring	1	PS28-3
25	Cord & Plug (25' Lg.)	1	U17-337
26	Packing Gland	1	PS16-1
27	Washer	1	PS43-1
28	Packing Ring	3	PS21-1
29	Adapter Plug	1	U17-99
30	Lead — Switch to Motor	1	PS18-30
40	Upper End Bell	1	PS3-8B
41	Lever Arm Assembly	1	PS178-2
42	Micro Switch Assembly	1	U117-310
43	Machine Screw #8-32 x 1/4" Lg. Rd. Hd.	2	U30-111C
45	Plastic Rope (24' Lg.)	1	U97-21
46	Base	1	S4-7D
47	Strainer	1	S8-7

STRAINERS — 7.94 MAINTENANCE INSTRUCTIONS

7.94A *INSPECTION AND CLEANING*

1. Single basket strainers are only used where flow can be interrupted as equipment must be shut off and valves closed to remove and clean the basket.
2. Twin strainers are cleaned without interrupting the flow. There are several types, one having a flow switch valve which will divert the flow from one basket to the other, thus permitting removal of the out-of-service basket for cleaning. The access covers of the chambers may have an air release cock. If so, open the pet cock before removing the cover. Another type of strainer has two gate type sliding valves. In this type both valves are always turned the same way, thus opening one passage through the strainer and closing the other.
3. If there is a valve at the base of the strainer or sand trap, open this valve and flush until water runs clear.

VALVES — AIR RELIEF — 8.10

8.10C 4 *CHECK AIR RELIEF VALVE*

To inspect and clean water tank air relief valves, close the gate valve in the air line and the gate valve in the water line to the air relief valve. Open the drain valve to relieve pressure. Remove the air valve from the line. Disassemble. Check float ball, spring, needle valve and orifice. Clean out all rust and corrosion. Reassemble. Check that the needle valve and orifice are aligned and that the spring is operating correctly. Flush the water line before replacing the air relief valve.

VALVES — 8.30 MAINTENANCE INSTRUCTIONS

8.30 *VALVES*

A. Operate inactive gate valves to prevent sticking.

B. Adjust the stem stuffing box packing tight enough that it will not leak, but not so tight that the stem turns with difficulty.

C. Lubricate the packing with a few drops of oil to eliminate excessive friction between the valve stem and the packing.

Modern gate valves can be repacked without removing them from service. Before repacking, verify the gate valve is in the fully opened position. This will prevent excessive leaking when the packing is removed.

REPACKING

REPACK GATE VALVE AS FOLLOWS:

A. Remove all packing from the stuffing box,

B. Clean the valve stem with fine emery cloth,

C. Insert new split ring packing in the stuffing box, staggering the joints and tamping down while inserting, and

D. Place a few drops of oil on the stem and tighten down the packing just enough to keep it from leaking.

8.30D *BALL VALVE (AUTO. STOP AND CHECK)*

1. Check operation. Lightly oil the stem. Exercise the hand control and lightly oil the stem threads.
2. To clean and inspect the strainer screen in the water line to the control valve of the Ball Valve, close the incoming line gate valve, remove the screen, clean and inspect. Replace if necessary. Flush the line by momentarily opening the gate valve. Replace the screen and open the gate valve.

8.30E *GLOBE VALVES*

Operate globe valves to prevent sticking. Check valve for leakage. Replace discs if necessary. Lube stem and packing with one or two drops of oil.

8.30F *BURIED VALVES*

Operate inactive valves. Lubricate buried valves, if possible, by pouring oil down a pipe.

SLIDE GATE — 8.40

8.40C *HYDRAULIC GATE OPERATOR*

1. Periodically check tank fluid level. Level should be at the middle of the gage. Check the system pressure. Inspect piping for leaks.
2. Replace hydraulic fluid. Drain the reservoir and fill to the middle of the sight gage. Oil should be filtered and all containers and funnels clean. If equipped with pump and hydraulic motor cases with external drain connections, remove the plug (or tubed drain connection where there is no plug) and fill the case with hydraulic fluid. Then replace the plug. Clean suction strainers and replace filter elements.

WATER WELLS — 9.01

A. *CLEAN/INSPECT WATER TANK INTERIOR*

Place motor switch in OFF position, and lock out the controls. Close the tank discharge gate valve. Open the tank drain gate valve and drain the tank. Remove the inspection cover and 8-inch blind flange on the opposite end of the tank. In large tanks that can be entered, install an electric fan or blower in the hole where the blind flange was removed. Purge the tank ten minutes before entering. Remove sand, clean and inspect the tank. Record amount of sand removed and condition of the tank. Remove the fan and reinstall the blind flange. Close the drain gate valve, pour one gallon of chlorine solution into the tank. Replace the inspection cover and put the unit back in service with the selector switch on automatic. Let the tank fill to its normal capacity, then let water and chlorine solution stand in the tank for one-half hour. Drain and flush the tank until chlorine residual is below one-half mg/*L*. After this is accomplished, fill the tank with water and open the tank discharge gate valve.

B. *WATER TANK MAINTENANCE*

1. Check for leaks, water and air, all lines and valves.

2. Check tank contents by the glass tube sight glass. Water level would be gaged by the length of the air relief valve line. Clean the sight glass tube if necessary. If the water tank is waterlogged (too much water and not enough air), isolate the tank by closing the tank outlet valve. Turn the pump switch to OFF position. Open the tank drain valve and drain enough water out to leave the tank about $^1/_3$ full of water. Close the drain valve and turn the pump switch to MANUAL. When the pump starts, watch the check valve lever. When it shows the flow is full and continuous, shut off the pump and let it back spin. This draws air in through the inlet valve. As soon as the pump stops back spinning, start it again. This will force air into the tank that was drawn in by the backspin. Continue this procedure until the proportion of air and water is correct. Open the tank outlet valve and place the switch on automatic. If there is a clock on the well pump, reset the clock. If the proportion of air is too great, see Par. 8.10 C-4.

3. When leaks occur around the glass tube, renew the packing. Renew the paint stripe on the back of the glass tube if worn or faded. If sight gage draining cock leaks, repair or replace.

C. *STANDING WATER LEVEL*

1. If there is an air line at the base of the water well pump motor frame with a water level gage attached, remove the cap from the valve stem and attach a hand air pump. Pump air into the line until water level indicator hand stops rising and remains constant. Record the reading on the maintenance sheet.

2. To sound a well with battery operated equipment, use the following procedure. This equipment will have either one long and one short wire or two long wires taped together and wound on a reel or drum, connected in series with a battery and ammeter. Remove the cap from the pipe into the well casing. If the equipment has a short wire, clamp this wire to metal (metal base plate or discharge pipe) for a ground. Lower the long wire slowly through the pipe into the well casing. When the wire touches the water, this will complete the circuit through the ground and the ammeter hand will move. The same holds true with two wires. When they touch the water, the circuit will be complete through the two wires and the ammeter hand will move. Mark the wire with a piece of tape or string at the top, at the center of the discharge. Remove it from the casing and measure from wire ends to the place you marked on the wire. If the wire has been previously marked as to footage, measure from the footage mark below your mark to your mark; the total is the standing water level or distance from the center of the discharge to the water. Always shut off and lock out the pump before sounding the well.

CHAPTER 9

EQUIPMENT MAINTENANCE

by

Lee Doty

Revised by

Rick Arbour

TABLE OF CONTENTS

Chapter 9. EQUIPMENT MAINTENANCE

LESSON 4

OBJECTIVES

Chapter 9. EQUIPMENT MAINTENANCE

Following completion of Chapter 9, you should be able to:

1. Explain the serious consequences that could occur when inexperienced, unqualified or unauthorized persons attempt to troubleshoot or repair electrical panels, controls, circuits, wiring or equipment,
2. Communicate with electricians by indicating possible causes of problems in electrical panels, controls, circuits, wiring and motors,
3. Properly select and use the following pieces of equipment (if qualified and authorized):
   a. Multimeter,
   b. Ammeter,
   c. Megger, and
   d. Ohmmeter,
4. Describe how a pump is put together,
5. Discuss the application or use of different types of pumps,
6. Maintain the various types of pumps,
7. Operate and maintain a compressor, and
8. Develop and conduct an equipment lubrication program.

ATTENTION

I. DO NOT ATTEMPT TO INSTALL, TROUBLESHOOT, MAINTAIN, REPAIR OR REPLACE ELECTRICAL EQUIPMENT, PANELS, CONTROLS, WIRING OR CIRCUITS UNLESS YOU

A. KNOW WHAT YOU ARE DOING,

B. ARE QUALIFIED, AND

C. ARE AUTHORIZED.

Section 9.1, "Electrical Equipment Maintenance," is presented to provide you with an understanding and awareness of electricity. *THE PURPOSE OF THE SECTION IS TO HELP YOU PROVIDE ELECTRICIANS WITH THE INFORMATION THEY WILL NEED WHEN YOU CONTACT THEM AND REQUEST THEIR ASSISTANCE.*

II. Due to the wide variety of equipment and manufacturers in the wastewater collection field, detailed procedures for the maintenance of some types of equipment were very difficult to include in this chapter. Also manufacturers are continually improving their products and some details would soon be out of date. *FOR DETAILS CONCERNING THE OPERATION, MAINTENANCE AND REPAIR OF A PARTICULAR PIECE OF EQUIPMENT, REFER TO THE ENGINEER'S O & M MANUAL OR CONTACT THE MANUFACTURER.*

III. Effective equipment and sewer maintenance is the key to successful system performance. The better your maintenance, the better your facilities will perform. Abuse your equipment and facilities and they will abuse you. Everyone must realize that if the equipment can't work, no one can work.

WORDS

Chapter 9. EQUIPMENT MAINTENANCE

ALTERNATING CURRENT (A.C.) | ALTERNATING CURRENT (A.C.)

An electric current that reverses its direction (positive/negative values) at regular intervals.

AMPERAGE (AM-purr-age) | AMPERAGE

The strength of an electric current measured in amperes. The amount of electric current flow, similar to the flow of water in gallons per minute.

AMPERE (AM-peer) | AMPERE

The unit used to measure current strength. The current produced by an electromotive force of one volt acting through a resistance of one ohm.

ANALOG READOUT | ANALOG READOUT

The readout of an instrument by a pointer (or other indicating means) against a dial or scale.

BRINELLING (bruh-NEL-ing) | BRINELLING

Tiny indentations (dents) high on the shoulder of the bearing race or bearing. A type of bearing failure.

CAUTION | CAUTION

This word warns against potential hazards or cautions against unsafe practices. Also see DANGER, NOTICE, and WARNING.

CAVITATION (CAV-uh-TAY-shun) | CAVITATION

The formation and collapse of a gas pocket or bubble on the blade of an impeller or the gate of a valve. The collapse of this gas pocket or bubble drives water into the impeller or gate with a terrific force that can cause pitting on the impeller or gate surface. Cavitation is accompanied by loud noises that sound like someone is pounding on the impeller or gate with a hammer.

CIRCUIT | CIRCUIT

The complete path of an electric current, including the generating apparatus or other source; or, a specific segment or section of the complete path.

CIRCUIT BREAKER | CIRCUIT BREAKER

A safety device in an electric circuit that automatically shuts off the circuit when it becomes overloaded. The device can be manually reset.

CONDUCTOR | CONDUCTOR

(1) A pipe which carries a liquid load from one point to another point. In a wastewater collection system, a conductor is often a large pipe with no service connections. Also called a conduit, interceptor or interconnector.

(2) In plumbing, a line conducting water from the roof to the storm drain or other means of disposal. Also called a downspout.

(3) In electricity, a substance, body, device or wire that readily conducts or carries electric current.

COULOMB (COO-lahm) | COULOMB

A measurement of the amount of electrical charge carried by an electric current of one ampere in one second. One coulomb equals about 6.25×10^{18} electrons (6,250,000,000,000,000,000 electrons).

CURRENT

A movement or flow of electricity. Water flowing in a pipe is measured in gallons per second past a certain point, not by the number of water molecules going past a point. Electric current is measured by the number of coulombs per second flowing past a certain point in a conductor. A coulomb is equal to about 6.25×10^{18} electrons (6,250,000,000,000,000,000 electrons). A flow of one coulomb per second is called one ampere, the unit of the rate of flow of current.

DANGER

The word *DANGER* is used where an immediate hazard presents a threat of death or serious injury to employees. Also see CAUTION, NOTICE, and WARNING.

DIGITAL READOUT

The use of numbers to indicate the value or measurement of a variable. The readout of an instrument by a direct, numerical reading of the measured value. The signal sent to such readouts is usually an analog signal.

DIRECT CURRENT (D.C.)

Electric current flowing in one direction only and essentially free from pulsation.

ELECTROMOTIVE FORCE (E.M.F.)

The electrical pressure available to cause a flow of current (amperage) when an electric circuit is closed. Also called VOLTAGE.

ELECTRON

(1) A very small, negatively charged particle which is practically weightless. According to the electron theory, all electrical and electronic effects are caused either by the movement of electrons from place to place or because there is an excess or lack of electrons at a particular place.

(2) The part of an atom that determines its chemical properties.

FUSE

A protective device having a strip or wire of fusible metal which, when placed in a circuit, will melt and break the electric circuit if heated too much. High temperatures will develop in the fuse when a current flows through the fuse in excess of that which the circuit will carry safely.

HERTZ

The number of complete electromagnetic cycles or waves in one second of an electric or electronic circuit. Also called the frequency of the current. Abbreviated Hz.

JOGGING

The frequent starting and stopping of an electric motor.

LEAD (LEE-d)

A wire or conductor that can carry electric current.

MANDREL (MAN-drill)

(1) A special tool used to push bearings in or to pull sleeves out.

(2) A testing device used to measure for excessive deflection in a flexible conduit.

MEGGER (from megohm)

An instrument used for checking the insulation resistance on motors, feeders, bus bar systems, grounds, and branch circuit wiring.

MEGOHM

Meg means one million, so 5 megohms means 5 million ohms. A megger reads in millions of ohms.

NAMEPLATE

A durable metal plate found on equipment which lists critical operating conditions for the equipment.

NOTICE

This word calls attention to information that is especially significant in understanding and operating equipment or processes safely. Also see CAUTION, DANGER, and WARNING.

OSHA (O-shuh) OSHA

The Williams-Steiger **O**ccupational **S**afety and **H**ealth **A**ct of 1970 (OSHA) is a federal law designed to protect the health and safety of industrial workers and collection system operators. The Act regulates the design, construction, operation and maintenance of industrial plants and wastewater collection and treatment facilities. The Act does not apply directly to municipalities, *EXCEPT* in those states that have approved plans and have asserted jurisdiction under Section 18 of the OSHA Act. *HOWEVER, CONTRACT OPERATORS AND PRIVATE FACILITIES DO HAVE TO COMPLY WITH OSHA REQUIREMENTS.* Wastewater collection systems have come under stricter regulation in all phases of activity as a result of OSHA standards. OSHA also refers to the federal and state agencies which administer the OSHA regulations.

OHM OHM

The unit of electrical resistance. The resistance of a conductor in which one volt produces a current of one ampere.

POLE SHADER POLE SHADER

A copper bar circling the laminated iron core inside the coil of a magnetic starter.

RESISTANCE RESISTANCE

That property of a conductor or wire that opposes the passage of a current, thus causing electric energy to be transformed into heat.

SHEAVE (SHE-v) SHEAVE

V-belt drive pulley which is commonly made of cast iron or steel.

SHIM SHIM

Thin metal sheets which are inserted between two surfaces to align or space the surfaces correctly. Shims can be used anywhere a spacer is needed. Usually shims are 0.001 to 0.020 inch thick.

VOLTAGE VOLTAGE

The electrical pressure available to cause a flow of current (amperage) when an electric circuit is closed. Also called ELECTROMOTIVE FORCE (E.M.F.).

WARNING WARNING

The word *WARNING* is used to indicate a hazard level between *CAUTION* and *DANGER*. Also see CAUTION, DANGER, and NOTICE.

CHAPTER 9. EQUIPMENT MAINTENANCE

(Lesson 1 of 4 Lessons)

9.0 BEWARE OF ELECTRICITY

RECOGNIZE YOUR LIMITATIONS

In the wastewater collection and treatment system maintenance departments of all cities, there is a need for maintenance operators to know something about electricity. Duties could range from repairing a taillight on a trailer or vehicle to repairing complex pump controls and motors. *VERY FEW MAINTENANCE OPERATORS DO THE ACTUAL ELECTRICAL REPAIRS OR TROUBLESHOOTING BECAUSE THIS IS A HIGHLY SPECIALIZED FIELD AND UNQUALIFIED PEOPLE CAN SERIOUSLY INJURE THEMSELVES AND DAMAGE COSTLY EQUIPMENT.* For these reasons you must be familiar with electricity, *KNOW THE HAZARDS*, and *RECOGNIZE YOUR OWN LIMITATIONS* when you must work with electrical equipment.

Most municipalities employ electricians or contract with a "commercial electrical company" that they call when major problems occur. However, the maintenance operator should be able to *EXPLAIN HOW THE EQUIPMENT IS SUPPOSED TO WORK AND WHAT IT IS DOING OR IS NOT DOING WHEN IT FAILS.*

After this lesson you should be able to tell an electrician what appears to be the problem with electrical panels, controls, circuits, and equipment. Even though operators may only be assigned maintenance tasks on electrical systems, the more operators know about electricity the better equipped they are to diagnose electrical problems, especially during emergency situations.

When an outside electrical contractor performs maintenance on your system, ask the contractor to assign the same person each time. This allows one person to become familiar with the pump station controls and electrical systems, and therefore, troubleshoot and repair much more rapidly during failures. This person should be accompanied by an assistant so backup people will be available with some knowledge of your system when the regular person is not available.

The need for safety should be apparent. If proper safe procedures are not followed in operating and maintaining the various electrical equipment used in wastewater collection and treatment facilities, accidents can happen that cause injuries, permanent disability, or loss of life. Some of the serious accidents that have happened, and could have been avoided, occurred when machinery was not shut off, locked out, and tagged properly (Figure 9.1) as required by *OSHA*[1] requirements. Possible accidents include:

1. Maintenance operator could be cleaning pump and have it start, thus losing an arm, hand, or finger,
2. Electric motors or controls not properly grounded could lead to possible severe shock, paralysis, or death, and
3. Improper circuits created by mistakes such as wrong connections, bypassed safety devices, wrong fuses or improper wire can cause fires or injuries due to incorrect operation of machinery.

Another reason for having a basic working knowledge of electricity is to prevent financial losses resulting from motors burning out and from damage to equipment, machinery and control circuits. Additional costs result when damages have to be repaired, including payments for outside labor.

WARNING

Never touch electrical panels, controls, circuits, wiring or equipment unless you are *QUALIFIED AND AUTHORIZED.* By the time you find out what you don't know about electricity, you could find yourself too *DEAD* to use the knowledge.

---

[1] *OSHA (O-shuh). The Williams-Steiger **O**ccupational **S**afety and **H**ealth **A**ct of 1970 (OSHA) is a federal law designed to protect the health and safety of industrial workers and collection system operators. The Act regulates the design, construction, operation and maintenance of industrial plants and wastewater collection and treatment facilities. The Act does not apply directly to municipalities, EXCEPT in those states that have approved plans and have asserted jurisdiction under Section 18 of the OSHA Act. HOWEVER, CONTRACT OPERATORS AND PRIVATE FACILITIES DO HAVE TO COMPLY WITH OSHA REQUIREMENTS. Wastewater collection systems have come under stricter regulation in all phases of activity as a result of OSHA standards. OSHA also refers to the federal and state agencies which administer the OSHA regulations.*

DANGER

OPERATOR
WORKING
ON LINE

DO NOT CLOSE THIS
SWITCH WHILE THIS
TAG IS DISPLAYED

TIME OF: ______________________

DATE: ______________________

SIGNATURE: ______________________

This is the ONLY person authorized to remove this tag.

INDUSTRIAL INDEMNITY/INDUSTRIAL UNDERWRITERS/
INSURANCE COMPANIES

4E210–R66

Fig. 9.1 Typical warning tag

(Source: Industrial Indemnity/Industrial Underwriters/Insurance Companies)

QUESTIONS

Write your answers in a notebook and then compare your answers with those on page 264.

9.0A Why must unqualified or inexperienced people be extremely careful when attempting to troubleshoot or repair electrical equipment?

9.0B What could happen when machinery is not shut off, locked out, and tagged properly?

9.1 ELECTRICAL EQUIPMENT MAINTENANCE

9.10 Introduction

This section contains a basic introduction to electrical terms and information plus directions on how to troubleshoot problems with electrical equipment.

Most electrical equipment used in wastewater collection systems and treatment plants is labeled with *NAMEPLATE*[2] information indicating the proper voltage and allowable current in amps.

9.11 Volts, Amps, Watts, and Power Requirements

VOLTS

Voltage (E) is also known as Electromotive Force (E.M.F.), and is the electrical pressure available to cause a flow of current (amperage) when an electric circuit is closed.[3] This force can be compared with the pressure or force that causes water to flow in a pipe. Some pressure in a water pipe is required to make the water move. The same is true of electricity. A force is necessary to push electricity or electric current through a wire. This force is called voltage.

There are two types of voltage: Direct Current (D.C.) and Alternating Current (A.C.).

DIRECT CURRENT

Direct current (D.C.) is flowing in one direction only and is essentially free from pulsation. Direct current is seldom used in lift stations and wastewater treatment plants except in motor-generator sets, some control components of pump drives and standby lighting. Direct current is used exclusively in automotive equipment, certain types of welding equipment, and a variety of portable equipment. Direct current is found in various voltages, such as 6 volts, 12 volts, 24 volts, 48 volts, and 110 volts. All batteries are direct current. D.C. is tested by holding the multimeter leads on the positive and negative poles on a battery. These poles are usually marked Positive (+) and Negative (–). Direct current usually is not found in higher voltages (over 24 volts) around plants and lift stations unless in motor-generator sets. Care must be taken when installing battery cables and wiring that Positive (+) and Negative (–) poles are connected properly to wires marked (+) and (–). If not properly connected, you could get an arc across the unit that could cause an explosion.

ALTERNATING CURRENT

Alternating current (A.C.) is periodic current that has alternating positive and negative values. In other words, it goes from zero to maximum strength, back to zero and to the same strength in the opposite direction, which comprises a cycle. Our A.C. voltage is 60-cycle frequency, or "hertz," which means that this happens 60 times per second. Alternating current is classified as:

- Single Phase,
- Two Phase,
- Three Phase or Polyphase.

The most common of these are single phase and three phase. The various voltages you probably will find on your job are 110 volts, 120 volts, 208 volts, 220 volts, 240 volts, 277 volts, 440 volts, 460 volts, 480 volts, and 550 volts.

Single-phase power is found in lighting systems, small pump motors, various portable tools, and throughout our homes. It is usually 120 volts or 240 volts. Single phase means that only one phase of power is supplied to the main electrical panel at 240 volts and the power supply has three wires or leads. Two of these leads have 120 volts each, the other lead is neutral and usually is coded white. The neutral lead is grounded. Many appliances and power tools have an extra ground (commonly a green wire) on the case for additional protection.

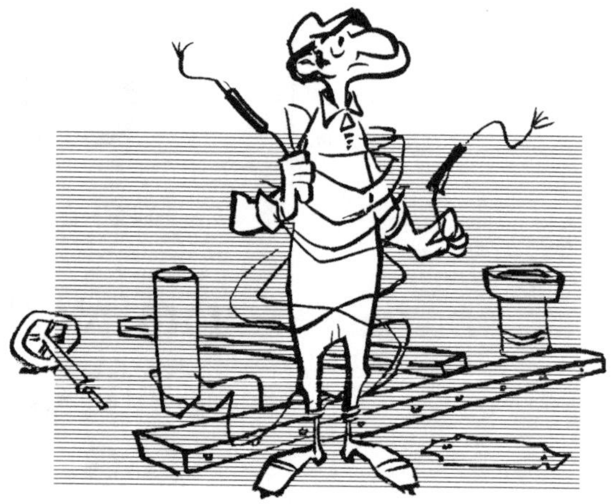

Three-phase power is generally used with motors and transformers found in lift stations and wastewater treatment plants. This power generally is 208, 220, 240 volts, or 440, 460, 480 and 550 volts. Higher voltages are used in some lift stations. Three-phase power is used when higher power requirements or larger motors are used because efficiency is usually higher and motors require less maintenance. Generally, all motors above two horsepower are three-phase unless there is a problem with the power company getting three-phase to the installations. Quite a few residential lift stations are on single-phase power due to their remote locations. Three-phase power usually is brought in to the point of use with three leads and there is power in all three leads.

When taking a voltage check between any two of the three leads, you measure 208, 220, 240 volts, or 440, 460, 480 volts depending on the supply voltage. There are some instances where you might measure a difference between the leads due to the use of different transformers. If there is power in three leads and if a fourth lead is brought in, it is a neutral lead.

---

[2] *Nameplate. A durable metal plate found on equipment which lists critical operating conditions for the equipment.*

[3] *Electricians often talk about closing an electrical circuit. This means they are closing a switch that actually connects circuits together so electricity can flow through the circuit. Closing an electrical circuit is like opening a valve on a water pipe.*

Two-phase and polyphase systems will not be discussed because they generally are not found in wastewater collection and treatment facilities.

AMPS

An ampere (I) is the practical unit of electric current. This is the current produced by a pressure of one volt in a circuit having a resistance of one ohm. Amperage is the measurement of current or electron flow and is an indication of work being done or "how hard the electricity is working."

In order to understand amperage, one more term must be explained. The *OHM* is the practical unit of electrical resistance (R). "Ohm's Law" states that in a given electrical circuit the amount of current in amperes (I) is equal to the pressure in volts (E) divided by the resistance (R) in ohms. The following three formulas are given to provide you with an indication of the relationships among current, resistance and E.M.F. (electromotive force).

$$\text{Current, amps} = \frac{\text{E.M.F., volts}}{\text{Resistance, ohms}} \qquad I = \frac{E}{R}$$

$$\text{E.M.F., volts} = \text{(Current, amps)(Resistance, ohms)} \qquad E = IR$$

$$\text{Resistance, ohms} = \frac{\text{E.M.F., volts}}{\text{Current, amps}} \qquad R = \frac{E}{I}$$

These equations are used by electrical engineers for calculating circuit characteristics. If you memorize the following relationship, you can always figure out the correct formula.

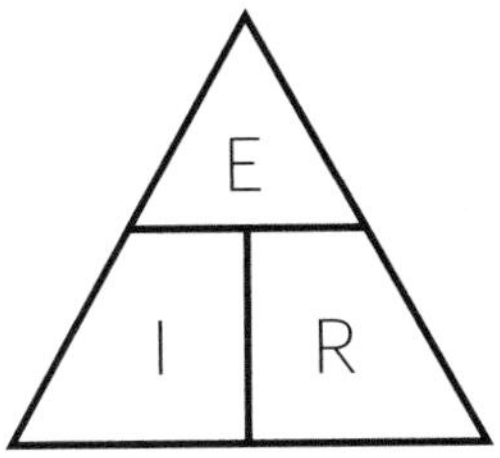

To use the above triangle you cover up with your finger the term you don't know or are trying to find out. The relationship between the other two known terms will indicate how to calculate the unknown. For example, if you are trying to calculate the current, cover up I. The two knowns (E and R) are shown in the triangle as E/R. Therefore, I = E/R. The same procedure can be used to find E when I and R are known or to find R when E and I are known.

WATTS

The watt and kilowatt (one thousand watts) are measures of power units used to rate electrical machines or motors or to indicate power in an electric circuit. The watts or power (P) required by a machine is obtained by multiplying the amps (I) required (or pulled) times the potential drop in volts (E) or the electromotive force (E.M.F.). The amount of watts in a circuit is equal to the voltage (E) times the amperage (I). The total power in a circuit, measured in watts, at any given moment is similar to the power produced by a motor, measured in horsepower, at any given moment.

$$\text{Power, watts} = \text{(Electromotive Force, volts)(Current, amps)}$$

$$\text{or P, watts} = \text{(E, volts)(I, amps)}$$

For the sake of comparisons

1 horsepower = 746 watts = 0.746 kilowatt = 550 ft-lbs/sec

The actual horsepower output of a motor also depends on the power factor (a number less than one) and the efficiency of the motor.

$$\text{Output, HP} = \frac{\text{(P, watts)(Power Factor)(Efficiency, \%)}}{\text{(746 watts/horsepower)(100\%)}}$$

POWER REQUIREMENTS

Power requirements (PR) are expressed in kilowatt hours: 500 watts for two hours or one watt for 1,000 hours equals one kilowatt hour (kW-hr or kWh). The power company charges so many cents per kilowatt hour.

$$\text{Power Req, kW-hr} = \text{(Power, kilowatts)(Time, hours)}$$

$$\text{PR, kW-hr} = \text{(P, kW)(T, hr)}$$

CONDUCTORS AND INSULATORS

A material, like copper, which permits the flow of electric current is called a conductor. Material that will not permit the flow of electricity, like rubber, is called an insulator. When such a material is wrapped or cast around a wire, it is called insulation. Insulation is commonly used to prevent the loss of electrical flow by two conductors coming into contact with each other.

QUESTIONS

Write your answers in a notebook and then compare your answers with those on page 264.

9.1A How can you determine the proper voltage and allowable current in amps for a piece of equipment?

9.1B What are two types of voltage?

9.1C Amperage is a measurement of what?

9.12 Tools, Meters, and Testers

> **WARNING**
>
> Never enter an electrical panel or attempt to troubleshoot or repair any piece of electrical equipment or any electric circuit unless you are qualified and authorized.

A wide variety of instruments are used to maintain lift station electrical systems. These instruments measure current, voltage and resistance. They are used not only for troubleshooting, but for preventive maintenance as well. These instruments may have either an *ANALOG READOUT*,[4] which uses a

[4] *Analog Readout. The readout of an instrument by a pointer (or other indicating means) against a dial or scale.*

pointer and scale, or a *DIGITAL READOUT*,[5] which gives a numerical reading of the measured value.

To check for "VOLTAGE," a *MULTIMETER* is needed. There are several types on the market and all of them work. They are designed to be used on energized circuits and care must be exercised when testing. By holding one lead on ground and the other on a power lead, you can determine if power is available. You also can tell if it is A.C. or D.C. and the intensity or voltage (110, 220, 480, or whatever) by testing the different leads.

A multimeter can also be used to measure voltage, current and resistance. A digital multimeter is shown in Figure 9.2 and an analog clamp-on multimeter is shown in Figure 9.3.

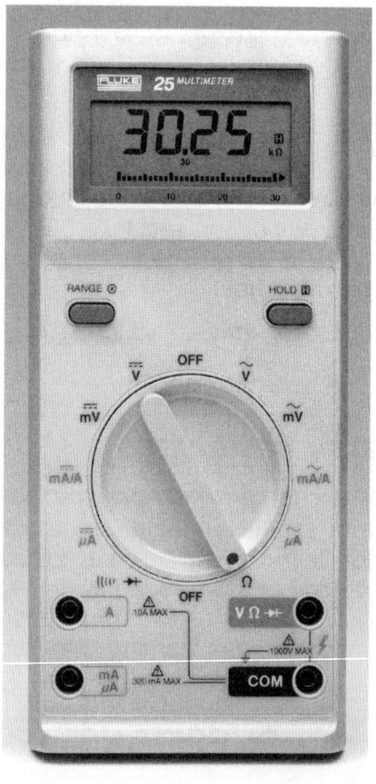

Fig. 9.2 Digital multimeter

(Reproduced with permission of Fluke Corporation)

Do not work on any electric circuits unless you are qualified and authorized. (Refer also to Chapters 4 and 11 for more information on electrical safety and OSHA requirements.) Use a multimeter and other circuit testers to determine if the circuit is energized, or if all voltage is off. This should be done after the main switch is turned off to make sure it is safe to work inside the electrical panel. Always be aware of the possibility that even if the disconnect to the unit you are working on is off, the control circuit may still be energized if the circuit originates at a different distribution panel. Check with a multimeter before and during the time the main switch is turned off to have a double-check. This procedure ensures that the multimeter is working and that you have good continuity to your tester. Use circuit testers to measure voltage or current characteristics to

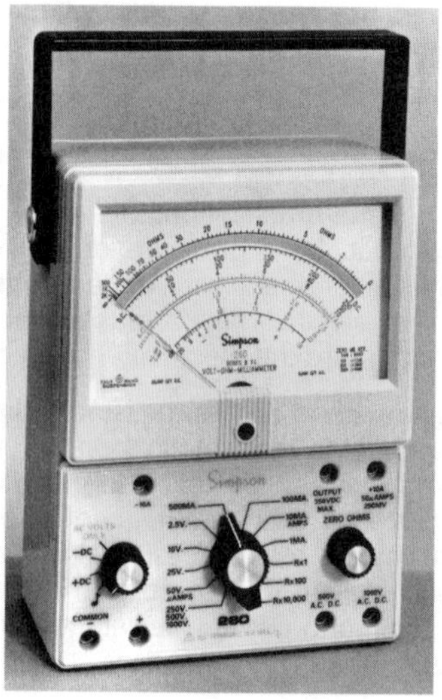

Fig. 9.3 Analog clamp-on multimeter

(Permission of Simpson Electric)

a given piece of equipment for making sure that you have or do not have a "live" circuit. *WARNING:* Switches can fail and the only way to ensure that a circuit is dead is to test the circuit.

Besides using the multimeter for checking power, it can be used to test for open circuits, blown fuses, single phasing of motors, grounds, and many other uses. Some examples are illustrated in the following paragraphs.

In the circuit shown in Figure 9.4, test for power by holding one lead of the multimeter on point "A" and the other at point "B." If no power is indicated, the switch is open or faulty. The sketch shows the switch in the "open" position.

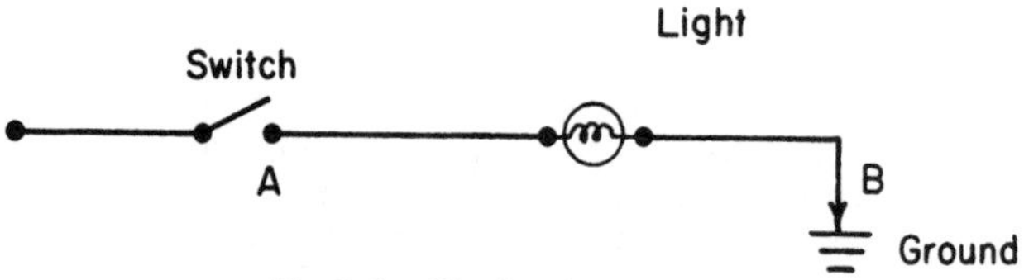

Fig 9.4 Single-phase circuit

To test for power at Point "A" and Point "B" in Figure 9.5, open the switch as shown. Using a multimeter with clamp-on leads, clamp a lead on L1 and a lead on L2 between the fuses and the load. Bring the multimeter and leads out of the panel and close the panel door as far as possible without cutting or damaging the meter leads. Some switches cannot be closed if the panel door is open. The panel door is closed when testing because hot copper sparks could seriously injure you when the circuit is energized and the voltage is high. Close the switch.

1. Multimeter should register at 220 volts. If there is no reading at points "A" and "B," the fuse or fuses could be "blown."

[5] *Digital Readout.* *The use of numbers to indicate the value or measurement of a variable. The readout of an instrument by a direct, numerical reading of the measured value. The signal sent to such readouts is usually an analog signal.*

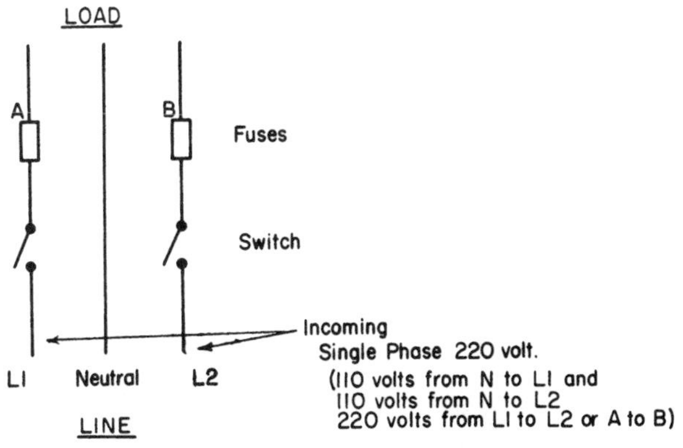
LOAD
A
B
Fuses
Switch
Incoming
Single Phase 220 volt.
(110 volts from N to L1 and
110 volts from N to L2
220 volts from L1 to L2 or A to B)
L1
Neutral
L2
LINE

Fig. 9.5 Single-phase, three-lead circuit

2. Move multimeter down below fuses to line 1 and line 2 (L1, L2). If there is still no reading on the multimeter, check for an open switch in another location, or call the power company to find out if power is out.
3. If a 220 volt reading is registered at line 1 and line 2, move the test leads to point "A" and the neutral lead. If a reading of 110 volts is observed, the fuse on line "A" is OK. If there isn't a voltage reading, the fuse on line "A" is blown. Move the lead from line "A" to line "B." Observe the reading. If 110 volt power is recorded, the fuse on line "B" is OK. If there isn't a voltage reading, the fuse on line "B" is blown. Another possibility to consider is that the neutral line could be broken.

> **WARNING**—*TURN OFF POWER AND BE SURE THAT THERE IS NO VOLTAGE IN EITHER POWER LINE BEFORE CHANGING FUSES.* Use a *FUSE PULLER.* Test circuit again in the same manner to make sure fuses or circuit breakers are OK. 220 volts power or voltage should be present between points "A" and "B." If fuse or circuit breaker trips again, shut off and determine the source of the problem.

Test for power at points "A," "B," and "C" in a three-phase circuit (Figure 9.6). Place multimeter leads on lines "A" and "B." Close all switches. 220 volts should register on multimeter. Check between lines "A" and "C," and between lines "B" and "C." 220 volts should be recorded between all of these points. If voltage is not present, one or all of the fuses are blown or the circuit breaker has been tripped. First check for voltage above the fuses at all of these points, "A" to "B," "A" to "C," and "B" to "C," to make sure power is available (see 220 readings in Figure 9.6). If voltage is recorded, move leads back down to bottom of fuses. If voltage is present from "A" to "B," but not at "A" to "C" and "B" to "C," the fuse on line "C" is blown. If there weren't any voltage readings at any of the test points, all the fuses could be blown.

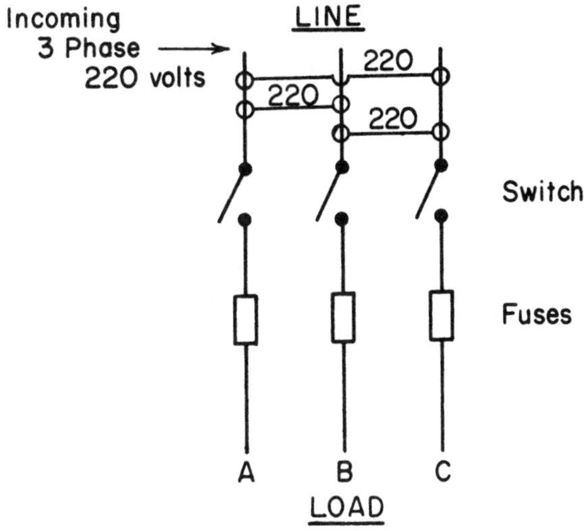
Incoming
3 Phase
220 volts
LINE
220
220
220
Switch
Fuses
A
B
C
LOAD

Fig. 9.6 Three-phase circuit, 220 volts

Another way of checking the fuses on this three-phase circuit would be to take your multimeter and place one lead on the bottom and one lead on the top of each fuse. You should *NOT* get a voltage reading on the multimeter. This is because electricity takes the path of least resistance. If you get a reading across any of the fuses (top to bottom), that fuse is bad.

ALWAYS MAKE SURE THAT WHEN YOU USE A MULTIMETER IT IS SET FOR THE PROPER VOLTAGE. IF VOLTAGE IS UNKNOWN AND THE METER HAS DIFFERENT SCALES THAT ARE MANUALLY SET, ALWAYS START WITH THE HIGHEST VOLTAGE RANGE AND WORK DOWN. Otherwise the multimeter could be damaged. Look at the equipment instruction manual or nameplate for the expected voltage. Actual voltage should not be much higher than given unless someone goofed when the equipment was wired and inspected.

Voltage readings are important because they determine how you connect motor leads, relays, and transformers. Low or high voltages can drastically affect motors. Operators of small wastewater collection systems should also be aware of the common but little understood problem of unbalanced current. Operating a pumping unit with unbalanced current can seriously damage three-phase motors and cause early motor failure. Unbalanced current reduces the starting torque of the motor and can cause overload tripping, excessive heat, vibrations, and overall poor performance. (Section 9.231, Problem 7, describes how to test circuits for unbalanced current.)

Another meter used in electrical maintenance and testing is the *AMMETER.* The ammeter records the current or "amps" in the circuit. There are several types of ammeters, but only two will be discussed in this chapter. The ammeter generally used for testing is called a "clamp-on" type (Figures 9.7 and 9.8). The term "clamp-on" means that it can be clamped around a lead or each lead supplying a motor lead, and no direct electrical connection needs to be made. These are used by clamping the meter over only one of the power leads to the motor or other apparatus and taking a direct reading. Each "leg" or lead on a three-phase motor must be checked by itself.

The first step should be to read the motor nameplate data and find what the amperage reading should be for the particular motor or device you are testing. After you have this information, set the ammeter to the proper scale. Set it on a higher scale than necessary if the expected reading is close to the top of the meter scale. Place the clamp around one lead at a time. Record each reading and compare with the nameplate rating. If the readings don't compare with the nameplate rating, find the cause, such as low voltage, bad bearings, poor connections, plugging, or excessive load. If the ammeter readings are higher than expected, the high current could produce overheating and damage the equipment. Try to find the problem and correct it.

When using a clamp-on ammeter, be sure to set the meter on a high enough range or scale for the starting current if you are testing during start-up. Starting currents range from 110 to 150 percent higher than running currents and using too low a

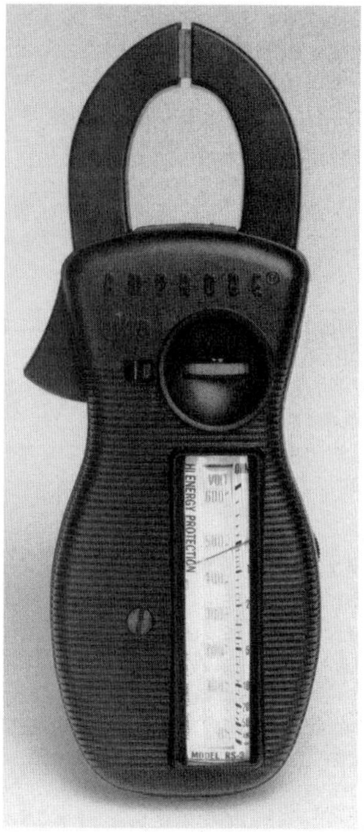

Fig. 9.7 Analog clamp-on type ammeter
(Permission of Amprobe)

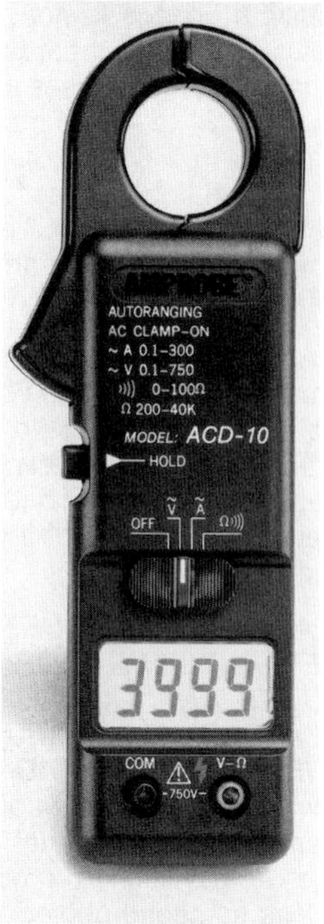

Fig. 9.8 Digital clamp-on type ammeter
(Permission of Amprobe)

range can ruin an expensive and delicate instrument. Newer clamp-on ammeters automatically adjust to the proper range and can measure both starting or peak current and normal running current.

Another type ammeter is one that is connected in line with the power lead or leads. Generally, they are not portable and are usually installed in a panel or piece of equipment. They require physical connections to put them in series with the motor or apparatus being tested. These ammeters are usually more accurate than the clamp-on type and are used in motor control centers and pump panels.

A *MEGGER* or *MEGOHMMETER* is used for checking the insulation resistance on motors, generators, feeders, bus bar systems, grounds, and branch circuit wiring. This device actually applies a D.C. test voltage which can be as high as 5,000 volts D.C., depending on the megohmmeter selected. The one shown in Figure 9.9 is a hand-held, hand-cranked system that applies 500 volts D.C. and is particularly useful for testing motor insulation. Battery-operated and instrument-style meggers are also available in both analog and digital models.

WARNING

Turn off circuit breaker when using a megger.

To use a megger there are two leads to connect. One lead is clamped to a ground lead and the other to the lead you are

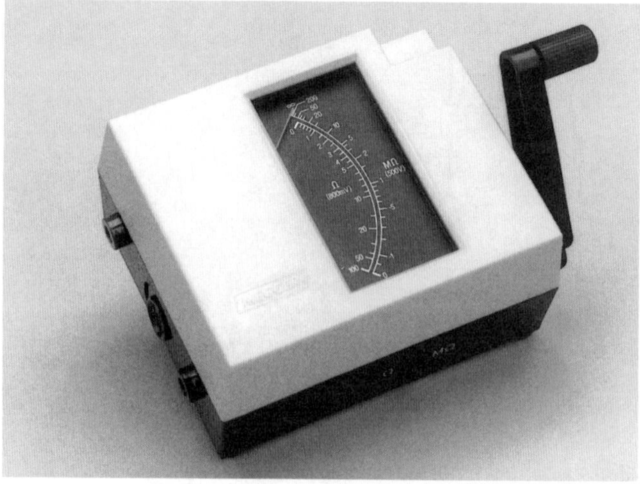

Fig. 9.9 Hand-cranked megohmmeter
(Permission of AVO International)

testing. The readings on the megger will range from "0" (ground) to infinity (perfect), depending on the condition of your circuit.

The megger is usually connected on the motor terminals, one at a time, at the starter, and the other lead to the ground lead. Results of this test indicate if the insulation is deteriorating or cut.

If a low reading is obtained, disconnect motor leads from power or line leads. Meg motor and if low reading is observed, the motor winding insulation is breaking down. If a good reading is obtained, meg the circuit or branch wiring. If this reading is low, the wiring to the motor is bad. A rule of thumb is not to run a motor if it is less than one *MEGOHM*[6] per horsepower. This means a 5 horsepower motor should have a 5 megohm reading; 10 horsepower, a 10 megohm reading; 20 horsepower, 20 megohm reading; and so forth.

Motors and wirings should be megged at least once a year, and twice a year, if possible. The readings taken should be recorded and plotted to determine when insulation is breaking down. Meg motors and wirings after a pump station has been flooded. If insulation is wet, excessive current could be drawn and cause pump motors to "kick out" (shut off).

OHMMETERS, sometimes called circuit testers, are valuable tools used for checking electric circuits. An ohmmeter is used only when the electric circuit is *OFF*, or de-energized. The ohmmeter supplies its own power by using batteries. An ohmmeter is used to measure the resistance (ohms) in a circuit. These are most often used in testing the control circuit components such as coils, fuses, relays, resistors, and switches. They are used also to check for continuity. An ohmmeter has several scales that can be used. Typical scales are: R x 1, R x 10, R x 1,000, and R x 10,000. Each scale has a level of sensitivity for measuring different resistances. To use an ohmmeter, set the scale, start at the low point (R x 1), and put the two leads across the part of the circuit to be tested such as a coil or resistor and read the resistance in ohms. A reading of

[6] *Megohm. Meg means one million, so 5 megohms means 5 million ohms. A megger reads in millions of ohms. For additional information, see A STITCH IN TIME: THE COMPLETE GUIDE TO ELECTRICAL INSULATION TESTING. To order, write to AVO Training Bookstore, 4573 South Westmoreland Road, Dallas, Texas 75237-3526. Order No. AVOB001. Price, $5.00.*

infinity would indicate an open circuit, and a "0" would indicate no resistance. Ohmmeters usually would be used only by skilled technicians because they are very delicate instruments.

The motor rotation indicator illustrated in Figure 9.10 is another specialized instrument used in electrical maintenance. This device is useful for determining the phase rotation of utility power when connecting 3-phase motors to ensure that the motor is connected properly for correct rotation.

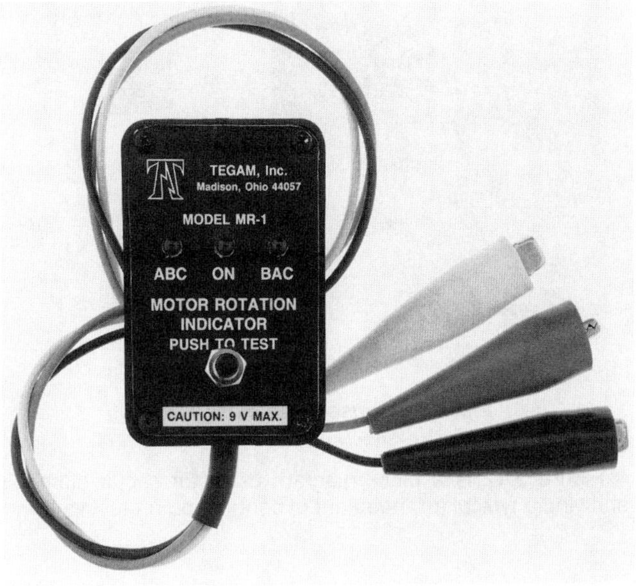

Fig. 9.10 Motor rotation indicator
(Permission of Tegam Inc.)

All meters should be kept in good working order and calibrated periodically. They are very delicate, susceptible to damage, and should be well protected during transportation. When readings are taken, they should always be recorded on a machinery history card for future reference. Meters are a good way to determine pump and equipment performance. *NEVER USE A METER UNLESS YOU ARE QUALIFIED AND AUTHORIZED.* The risk of electric shock and/or electrocution is very high if meters are used by unqualified personnel.

QUESTIONS

Write your answers in a notebook and then compare your answers with those on page 264.

9.1D How can you determine if there is voltage in a circuit?

9.1E What are some of the uses of a multimeter?

9.1F What precautions should be taken before attempting to change fuses?

9.1G How do you test for voltage with a multimeter when the voltage is unknown?

9.1H What could be the cause of amp readings different from the nameplate rating?

9.1I How often should motors and wirings be megged?

9.1J An ohmmeter is used to check the ohms of resistance in what control circuit components?

9.13 Electrical System Equipment

9.130 Need for Maintenance

Electrical system equipment is frequently the least understood and, therefore, most neglected equipment in a lift station. Usually we do not think of it as a system that requires frequent inspection or maintenance; however, the reverse of this is actually true. Electrical equipment can be damaged more readily by operating conditions than almost any other kind of equipment. Water, dust, heat, cold, humidity, corrosive atmospheres, and vibration are all common lift station conditions that can affect the performance and the life of electrical equipment.

This section will discuss various elements in the electrical system of a lift station including motor control center components such as circuit breakers, contactors, protective devices, transformers and control relays. The section will also explain in detail important maintenance and operating aspects of the A.C. induction motor, the most common pump driver used in lift stations.

Electrical equipment should be inspected and maintained on at least an annual basis or more frequently depending on the equipment and the application. Inspection should include a thorough examination, replacement of worn and expendable parts, and operational checks and tests.

Listed below are examples of electrical equipment maintenance tasks. Check and inspect each of these items annually.

1. All switch gear and distribution equipment. Look for worn parts and note general condition.
2. Wiring integrity.
3. Terminal connections. Tighten, if necessary (often needed with aluminum wire conductors).
4. Interlocking devices to prevent unauthorized entry.
5. Control circuits. Check operation and verify sequencing, including actual sequencing of pump motors as a function of wet well level.
6. All panel instruments. Clean and check for accuracy. Permanently installed, panel-mounted current and multimeters can be added to existing electrical systems and should be specified on new projects. This eliminates the need for routine access to the inside of the electrical system to obtain voltage current readings and exposure to hazardous voltages. Select a multimeter that measures both phase-to-phase and phase-to-neutral voltage. Current meters should also allow switching to measure current in each of the three phases.
7. Mechanical disconnect switches. Service and lubricate all switches, fuses, disconnects, and transfer switches.
8. Fuses. Verify proper application, size, and general condition. Check inventory of spare fuses.
9. Circuit breakers. Cycle each breaker and check for proper response and performance.
10. Contacts. Check for response (especially those in motor starters that carry high switching current); replace, if necessary.
11. Enclosures. Clean and vacuum.

If these basic maintenance procedures are done at least annually, it will help minimize lift station electrical system failures including:

1. Current imbalances or unbalances which ultimately result in motor failure,
2. Loose contacts or terminals in control and power circuits causing high-resistance contacts,
3. Overheating resulting in arcing, fire, and electrical system damage,
4. Dirty enclosures and components which allow a conductive path to build up between incoming phases causing phase-to-phase shorts, and
5. Corrosion that causes high-resistance contacts and heating.

9.131 Equipment Protective Devices

Electricity, by its very nature, is extremely hazardous and safety devices are needed to protect operators and equipment. Water systems have pressure valves, pop-offs, and different safety equipment to protect the pipes and equipment. So must electricity have safety devices to contain the voltage and amperage that comes in contact with the wiring and equipment.

Lift station electrical equipment protective devices may consist of fuses, circuit breakers, and motor and circuit overload devices, and grounds. These devices usually are found in combination so that all three are present. For example, a fused disconnect switch (see Figure 9.11) may protect the entire electrical distribution system within the lift station; circuit breakers may be used to protect branch motor circuits from short circuits; and overload elements would be installed to protect the motor from overloads.

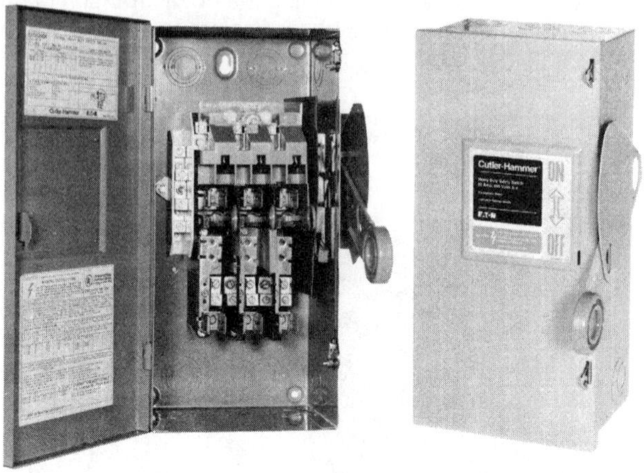

Fig. 9.11 Fused disconnect switch
(Permission of Eaton)

Figure 9.12 is a block diagram of motor circuit elements showing a typical arrangement of control and protective devic-

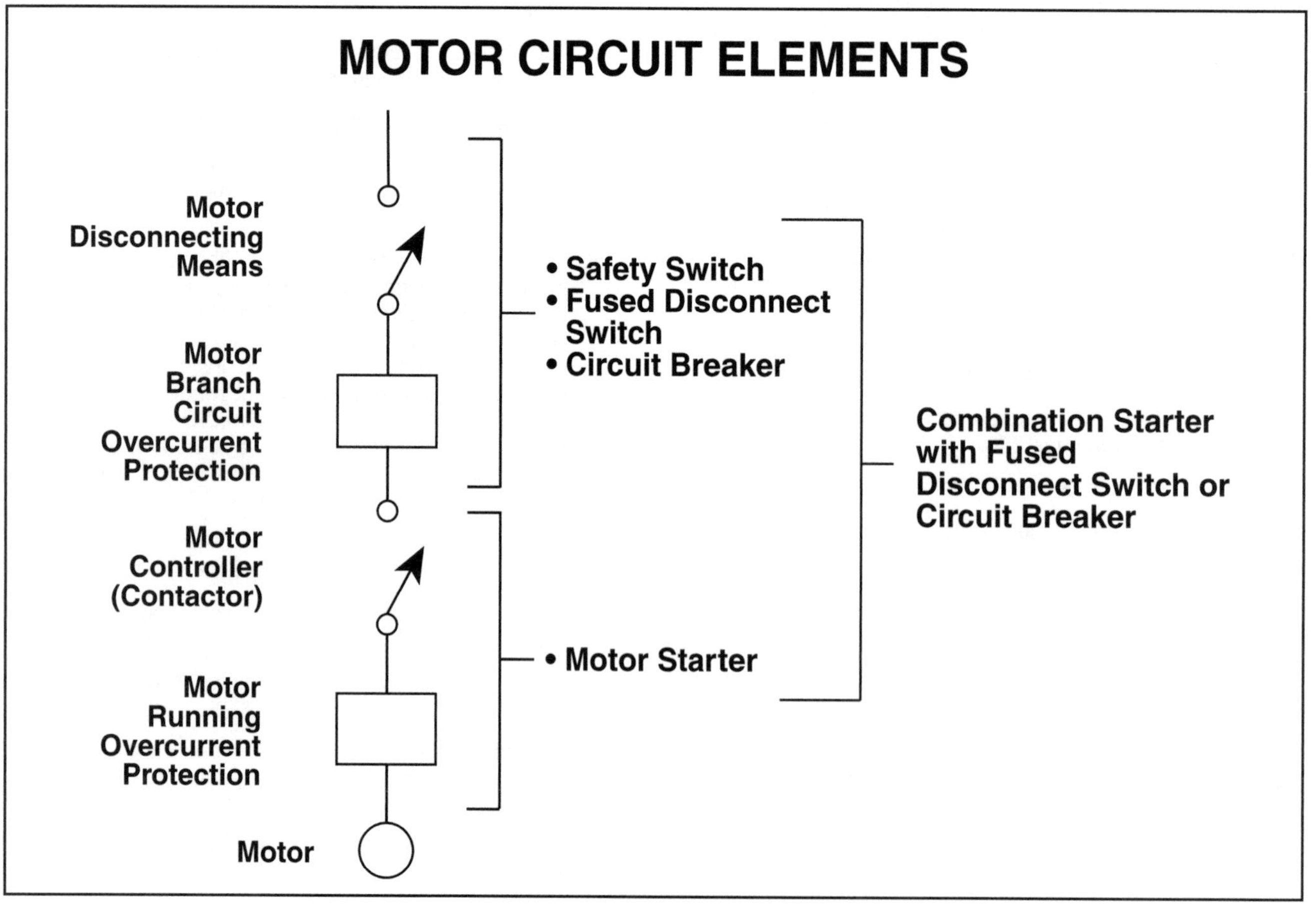

Fig. 9.12 Block diagram of motor circuit elements

es used to apply 3-phase line voltage to the motor. The first component in the circuit is some means of disconnecting the motor. This can be either a safety switch, a fused disconnect switch, or a circuit breaker. When using a fused disconnect switch or circuit breaker (the second element in the block diagram), the motor branch circuit overcurrent protection is incorporated. The next element in the block diagram is the motor controller, also referred to as a contactor. This is a 3-phase switch that connects 3-phase line voltage from the supply to the motor terminals. The final element in the diagram is the motor running overcurrent protection device. This device can be one that uses thermal elements that sense motor running current and trip under overload conditions or one of the newer solid state motor control devices that also sense motor running current and provide overload protection. When the motor controller, or contactor, and the motor overload device are combined in a single unit, they are referred to as a motor starter.

The motor branch circuit breaker may provide only short-circuit fault current protection if it is a magnetic only breaker or, in the case of a thermal magnetic circuit breaker, it may provide overload protection as well. Typically the motor branch circuit breaker is a molded-case circuit breaker (described later in Section 9.133).

Regardless of which type of protective device you are working with, if it blows or trips, the potential source of the problem should be investigated, identified, and corrected. All too frequently a fuse is simply replaced or a circuit breaker simply reset without any investigation into the problem. If the protective device has operated reliably for long periods of time (for example, no nuisance tripping), then tripping is almost a sure indication that a problem exists somewhere in the circuit and the device is trying to protect the circuit. Simply resetting the circuit breaker or replacing a fuse and then restarting the system may result in damage or more severe damage to the equipment that is being protected.

Once again, if you are not qualified to perform electrical troubleshooting, you should not attempt to proceed any further than providing diagnostic analysis. Maintenance must always be performed by trained personnel. It does not take high voltage or extreme currents to seriously injure an operator or cause a fatality.

9.132 Fuses (*Figure 9.13*)

The power company installs fuses on their power poles to protect their equipment from damage. We also must install something to protect the main control panel and wiring from damage due to excessive voltage or amperage.

A *FUSE* is an electrical device that opens a circuit when the current flowing through it exceeds the rating of the fuse. Fuses are designed to protect operators, main circuits, branch circuits, and equipment such as heaters and motors from excessive current.

The "heart" of a fuse is a special metal strip (or wire) designed to melt and blow out when its rated amperage is exceeded. Over-current devices (fuses, circuit breakers) are always placed in the "hot" side of a circuit (usually a black wire) and in series with the load so that all the current in the circuit must flow through them. If the current flowing in the circuit exceeds the rating of the fuse, the metal strip will melt and open the circuit so that no current can flow. A fuse cannot be re-used and must be replaced after eliminating the cause of the over-current. It is important to always replace fuses with the proper type and current rating. Too low a rating will result in unnecessary blowouts, while too high a rating may allow dangerously high currents to pass and damage equipment or injure operators.

There are several types of fuses, each being used for a certain type of protection. Some of these are:

1. *CURRENT-LIMITING FUSES:* Used to protect against current in circuits.
2. *DUAL-ELEMENT FUSES:* Used for motor protection.
3. *TIME-DELAY FUSES:* Used in electronic and motor starting circuits.
4. *SAND-FILLED FUSES:* Used on high voltage.
5. *PHASE FUSES:* Used to protect phase sequence.
6. *VOLTAGE-SENSITIVE FUSES:* Used where close voltage control is needed.

Since fuses are one-time-use devices, little can be done to check them during maintenance procedures; however, the following tasks should be performed at least annually:

1. Inspect bolted connections at the fuse clip or fuse holder for signs of looseness,
2. Check connections for any evidence of corrosion from moisture or atmosphere (air pollution),
3. Tighten connections,
4. Check fuse for obvious overheating,
5. Inspect insulation on the conductors coming into the fuses on the line side and out of the fuses on the load side for evidence of discoloration or bubbling which would indicate overheating of the conductors.

Definition Line Side/Load Side

Line side/load side are terms frequently used to describe the incoming and outgoing conductors of circuit breakers, motor starters, and other devices. The line side of the device is where incoming power is fed into the device. The load side is the terminal where power is fed to the load, for example, a motor.

Usually when a fuse blows it is not visibly apparent and it is necessary to check the fuse with a meter. The following procedure should be followed to determine whether or not a fuse has blown:

1. Ensure the main disconnect or circuit breaker is in the OFF position and locked out;
2. Check for live voltage in the panel (power may feed into a control system from other sources);
3. Remove fuse with a fuse-pulling device;
4. Test resistance of fuse using an ohmmeter. An open circuit (infinite resistance) indicates a blown fuse, whereas, a short circuit indicates a good fuse); and
5. Only in the event it is not possible to disconnect incoming power should the fuse be checked with power applied. In this case the procedure as outlined in Section 9.12, "Tools, Meters, and Testers," may be used.

9.133 Circuit Breakers (*Figure 9.14*)

Circuit breakers provide protection from excessive current for equipment and conductors without the inconvenience of changing fuses. Circuit breakers trip (open the circuit) when the current flow is excessive. There are two primary types of

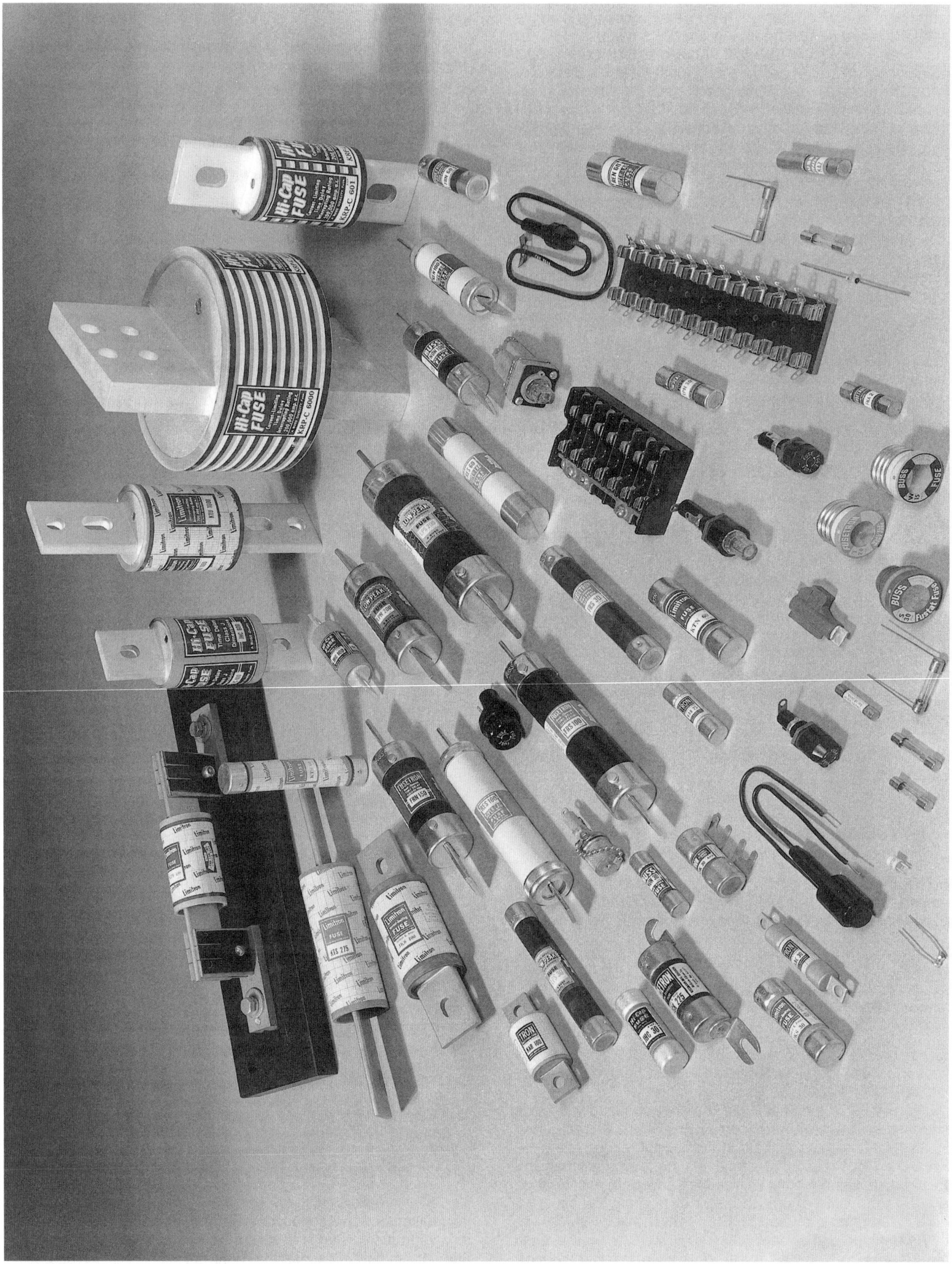

Fig. 9.13 Fuses
(Courtesy of Bussman Manufacturing, McGraw-Edison Company Division)

Fig. 9.14 Circuit breaker panelboard

(Permission of Square D Company)

circuit breakers; the difference between them is the type of current sensing mechanism. In the magnetic circuit breaker, the current is sensed by a coil that forms an electromagnet. When the current is excessive, the electromagnet activates a small armature that pulls the trip mechanism, thus opening the circuit breaker. In the thermal type of circuit breaker, the current heats a bi-metallic strip; when heated sufficiently, the strip bends enough to allow the trip mechanism to operate.

Circuit breakers are used as disconnecting devices and to protect electrical systems primarily from short-circuit conditions that may occur on the load side of the circuit breaker. In some cases, circuit breakers can also be used in conjunction with motor starters to provide motor overload protection. In most cases, however, they are sized to protect the entire system as opposed to specific components in the circuit from short-circuit fault currents.

MOLDED-CASE CIRCUIT BREAKERS (Figure 9.15)

Lift station motor control center circuit breakers are typically housed in a molded case unless the motor and/or other electrical equipment is high horsepower. Molded-case circuit breakers can be thermal-magnetic trip or magnetic only.

Thermal-magnetic trip circuit breakers provide both overload protection and short-circuit protection. If the circuit is overloaded, a thermal sensing element will detect this and cause the circuit breaker to open or trip. In addition, the magnetic sensing element of the circuit breaker rapidly senses extremely high currents, which flow under a short-circuit condition, and will also trip the circuit breaker.

Fig. 9.15 Molded case circuit breaker
(Permission of Eaton Corporation)

The amount of current that can flow during a short-circuit condition is a function of many things. For this reason, circuit breaker replacement should be done very carefully and by personnel who are qualified to evaluate circuit characteristics.

Molded-case circuit breakers require little maintenance other than:

1. Manually trip and operate the mechanism,
2. Check connections for tightness,
3. Inspect for evidence of overheating on the line and load side conductors, and
4. Inspect the circuit breaker case for evidence of overheating.

The following additional tests can be performed on molded-case circuit breakers, but require specialized equipment.

1. Insulation resistance test. This test uses a high-voltage D.C. megger device (described previously in this chapter) to check the internal resistance of the circuit breaker.
   a. Phase-to-phase check. Apply high voltage between the three phases of the circuit breaker, which ultimately could lead to a phase-to-phase short. This check is performed by applying the high voltage to each combination of two phases of the three phases (1 and 2, 1 and 3, and 2 and 3) one at a time. At least one megohm of resistance in the insulation is adequate. Inadequate insulation resistance could lead to a phase-to-phase short.
   b. Phase-to-ground. To check the insulation resistance between each individual phase to ground, connect one phase on the circuit breaker at a time to ground, apply a high voltage, and measure the resistance. At least one megohm of resistance in the insulation is desirable.
2. Contact resistances. With a D.C. power supply, measure the voltage drop from the line side to the load side of the circuit breaker with the circuit breaker closed, but power disconnected. An excessive voltage drop indicates a high-resistance condition between the contacts, and the circuit breaker must be replaced.
3. Overload trip test. This test requires a specialized circuit breaker tester that is capable of generating currents in the range of the trip devices of the circuit breaker. Connect the test device between the line and the load and run a current equal to the circuit breaker capacity through the circuit breaker. A current in the range of the trip devices that causes them to trip will indicate that the circuit breaker is functioning properly.

Because the interior components of molded-case circuit breakers are not accessible, repair or replacement of internal parts is not possible. The entire circuit breaker must be replaced when it fails.

MOTOR PROTECTION DEVICES

Since circuit breakers are normally used as a disconnecting device and as a protective device for the entire circuit, including the conductors that are connected to the breaker, the circuit breaker must be sized for the current carrying capacity of all components or equipment connected to it. For example, a thermal-magnetic circuit breaker must have a continuous current rating of at least 115 percent of motor full-load amps and the rating may be as great as 250 percent. This means that the thermal overload sensing device will not trip the breaker unless there is 115 percent overload present. In most cases, by the time this occurs, the motor will have been seriously damaged.

To overcome this problem and to provide further protection for motors, a device called a motor starter is usually installed in lift station motor control centers. The device operates through the supervisory control system responding automatically to the level in the wet well. A typical configuration is referred to as a magnetic motor starter; it can be operated manually as well as automatically. Section 9.24 of this chapter contains a discussion of motor starters.

NEVER INCREASE THE RATING OF THE OVERLOAD HEATERS BECAUSE OF TRIPPING. YOU SHOULD FIND THE PROBLEM AND REPAIR IT.

Internal thermal protection devices are another form of motor and insulation protection. Such devices are especially desirable on motors that are likely to experience occasional overload conditions or voltage imbalances. Thermal protection devices may include the following features:

1. Built-in thermal switches that automatically open when the motor overheats. This type of device will not reset until the motor cools, at which time, the switch resets automatically. (In some cases there may be a reset button on the motor enclosure.) This type of protection is normally limited to small, single-phase motors.

2. Three-phase motors with internal thermal protection have thermostats bedded in each phase of the three-phase windings. The thermostats are connected in series and brought out into the motor conduit box and labeled P1 and P2. The switches are connected into the motor control circuit which will open the magnetic starter contacts when an over-temperature condition occurs in the windings.

   Two circuit configurations are available:

   a. One in which the control circuit automatically restarts when the thermal switches and the motor reset, and

   b. A circuit that is locked out and must be manually reset in order to restart.

   Figure 9.16 illustrates a typical control circuit in which normally closed thermostats are connected in series with the motor starter coil in the control circuit. *NOTE:* This is the type of control circuit that will automatically reset when the motor thermostat's switches are reset.

There are many other protective devices for electricity such as motor winding thermostats, phase protectors, low voltage protectors, and ground-fault circuit interrupters. Each has its own special applications and should never be tampered with or jumped (bypassed).

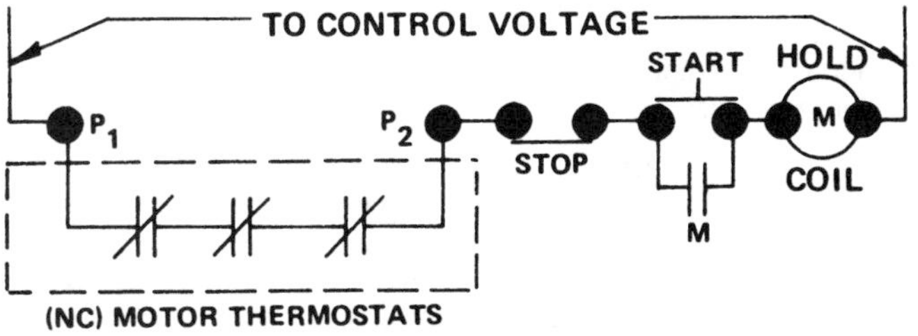

Fig. 9.16 Closed thermostats connected in series with the motor starter coil

9.134 Ground

GROUND is an expression representing an electrical connection to earth or a large conductor that is at the earth's potential or neutral voltage. Motor frames and all electrical tools and equipment enclosures should be connected to ground. This is generally referred to simply as grounding, or equipment ground.

Connecting motor frames, tools and electrical equipment to ground is a safety precaution that protects *YOU* and the motors, tools, and equipment. If one of the conductors opens and is not connected to ground, then *YOU* become the ground, the current could flow through *YOU*, and *YOU* could receive a severe or fatal electric shock. If the current flows through motors, tools, or equipment, severe damage could occur.

The third prong on cords from electric hand tools is the equipment ground and must never be removed. When an adapter is used with a two-prong receptacle, the green wire on the adapter should be connected under the center screw on the receptacle cover plate. Many times equipment grounding, especially at home, is achieved by connecting onto a metal water pipe or drain rather than a rod driven into the ground. This practice is not recommended when plastic pipes and other nonconducting pipe materials are used. Also corrosion can be accelerated if pipes of different metals are used. A rod driven into dry ground isn't very effective as a ground.

Grounding

Grounding must be taken into account wherever electric current flows. It can never be stressed too strongly that proper grounding and bonding must be correctly applied if the system, the equipment, and the people who come in contact with them are to be protected. Effective grounding means that the path to ground (1) is permanent and continuous, (2) has ample current-carrying capacity to conduct safely any currents liable to be imposed on it, and (3) has impedance (opposition to current flow) sufficiently low to limit the potential above ground and to facilitate the operation of the over-current devices in the circuit. Effective bonding means that the electrical continuity of the grounding circuit is ensured by proper connections between service raceways, service cable armor, all service equipment enclosures containing service entrance conductors, and any conduit or armor that forms part of the grounding conductor to the service raceway. The requirements for effective grounding are among the most frequently cited violations of OSHA's electrical standards. Effective grounding has no function unless and until there is electrical leakage from a current-carrying conductor to its enclosure. When such a ground fault occurs, the equipment grounding conductor goes into action to provide the following safeguards:

- It prevents voltages between the electrical enclosure and other enclosures or surroundings, and
- It provides a path for large amounts of fault or overload current to flow back to the service entrance, thus blowing the fuse or tripping the circuit breaker.

How does grounding do its job?

Proper grounding requires connecting all of the enclosures (equipment housings, boxes, conduit) together, back to and including the service entrance enclosure. This is accomplished by means of the green wire in the cord on portable equipment and the conduit system or a bare wire in the fixed wiring of the building.

When a ground fault occurs, as in a defective tool, the grounding conductor must carry enough current to immediately trip the circuit breaker or blow the fuse. This means that the ground-fault path must have low impedance. The only low-impedance path is the green wire (in portable cords) and the metallic conduit system (or an additional bare wire if conduit is not used). However, when the insulation on the black (ungrounded) conductor fails and the copper conductor touches the case of the tool, the ground-fault current flows through the green (grounding) conductor and the conduit system back to the service entrance. If the equipment-grounding conductors are properly installed, this current will be perhaps ten times or more greater than normal current, so the circuit breaker will trip out immediately.

But what happens if the grounding does not do the job?

If the ground-fault path is not properly installed, it may have such high impedance that it does not allow a sufficiently large amount of current to flow. Or, if the grounding conductor continuity has been lost (such as when the U-shaped grounding prong has been broken off the plug), no fault current will flow. In these cases, the circuit breaker will not trip out, the case of the tool will be energized, and persons touching the tool may be shocked. The hazard created is that persons touching the tool may provide a path through their body and eventually back to the source of voltage. This path may be through other surfaces in the vicinity, through building steel, or through earth. The dangerous ground-fault current flowing through this high-impedance path will not rise to a high enough value to immediately trip the circuit breaker. Only the metallic equipment-grounding conductor, which is carried along with the supply conductors, will have impedance low enough that the required large amount of fault current will flow.

The only way to ensure that the equipment-grounding conductor does its job is to be certain that the grounding wire, the grounding prong, the grounding receptacle, and the conduit system are intact and have electrical continuity from each electrical tool back to the service entrance. Effective grounding, coupled with the use of over-current devices such as fuses and circuit breakers, not only protects equipment and facilities, but also protects operators from electric shock in most situations. However, the only protective device whose sole purpose is to protect people is the ground-fault circuit interrupter (GFCI).

QUESTIONS

Write your answers in a notebook and then compare your answers with those on page 264.

9.1K What is the most common pump driver used in lift stations?

9.1L Basic lift station maintenance procedures performed at least annually will help minimize what types of lift station failures?

9.1M What should be done when a fuse or circuit breaker blows or trips?

9.1N What are two types of safety devices found in main electrical panels or control units?

9.1O What are fuses used to protect?

9.1P Why must a fuse never be bypassed or jumped?

9.1Q What types of annual maintenance should be performed with regard to fuses?

9.1R How does a circuit breaker work?

9.14 Motor Control/Supervisory Control and Electrical System

The motor and supervisory control systems are composed of the auxiliary electrical equipment such as relays, transformers, lighting panels, pump control logic, alarms, and other electrical equipment typically found in a lift station electrical system over and above the protective devices and the motor starters.

In general, annual maintenance should be performed on all these systems as follows:

1. Control Transformers
   a. Check primary, secondary, and ground connections
   b. Check for loose windings/coils
   c. Inspect insulation for signs of overheating as a result of overloading
   d. Check mounting for tightness
   e. Check primary/secondary fusing and fuse clips for tightness
2. Motor Control Centers
   a. Check panel lights for operation
   b. Check control knobs/switches for freedom of movement and contact condition
   c. Check horizontal and vertical bus and supports for evidence of heating or arcing and tighten. (Bus refers to the copper or aluminum bars that run horizontally and vertically in the motor control center; they feed the three-phase power to the branch circuits.)
3. Control Relays
   a. Check mounting for looseness
   b. Tighten all screw terminal connections
   c. Check for evidence of overheating or arcing indicated by carbon buildup or discoloration of plastic housing
4. Clean and vacuum enclosure.

The use of aluminum wire as a conductor has become very common because of its economic advantage over copper. Copper traditionally had been used for virtually all wiring applications in wastewater lift stations including the conductors feeding the station from the utility transformers, bus bar, motor control centers, control wiring, as well as power wiring to motors.

The greatest advantage of copper is that it oxidizes very slowly. Even though this reaction with air or moisture results in a surface layer of impurities, the surface is soft and is easily penetrated by the connecting device. With the exception of the annual tightening of the terminals, copper requires very little maintenance.

Aluminum is much softer and reacts much more rapidly with the air so aluminum begins oxidizing almost immediately when exposed to air. This oxide, as opposed to that found on copper, tends to form an insulating layer over the aluminum wire. Aluminum also expands and contracts 36 percent more than copper, which will result in loose connections, high-resistance contacts, heat formation, and failure of the connection. When using aluminum conductors, the following rules must be observed:

1. The connecting terminal must be specifically designed to accommodate aluminum conductors (connectors that are rated for this use are stamped with the letters CU/AL indicating that they are suitable for use with copper (CU) or aluminum (AL) conductors), and
2. The termination point of the aluminum conductor must be coated with a compound to prevent the formation of the insulating oxides and should be wire-pressure scraped before the compound is applied.

Because of these limitations, aluminum conductors and bus bars should be inspected and maintained more frequently than the traditional copper conductors. Conduct a visual inspection of the terminal, conductor, and insulation for any evidence of discoloration and overheating.

Three basic factors contribute to the reliable operation of electrical systems found in lift stations. They are:

1. An adequate preventive maintenance program must be implemented,
2. A knowledge of the system by the collection system operator, even though the operator does not perform the actual maintenance, and
3. KEEP IT CLEAN!

   KEEP IT DRY!

   KEEP IT TIGHT!

QUESTIONS

Write your answers in a notebook and then compare your answers with those on page 265.

9.1S Motor and supervisory control systems are composed of what types of auxiliary electrical equipment?

9.1T What are the three basic factors that contribute to the reliable operation of electrical systems found in lift stations?

9.2 MOTORS

9.20 Types

Electric motors are the machines most commonly used to convert electrical energy into mechanical energy. Although a multitude of different types of motors are produced today, the most common pump motor is the A.C. (Alternating Current) induction motor.

Two types of induction motor construction are typically encountered when dealing with A.C. induction motors (refer to Figure 8.23 on page 38):

1. Squirrel Cage Induction Motor (SCIM, by far the most numerous), and
2. Wound Rotor Induction Motor (WRIM, used for variable-speed applications).

Figure 9.17 illustrates a horizontal squirrel cage induction motor with a drip-proof enclosure. Major components of the motor are:

1. Stator winding with connection to the three-phase power supply,
2. Shaft end and opposite end ball bearings,
3. Rotor assembly including rotor, shaft, and fans on the rotor cage, and
4. Enclosure.

Stator construction is generally the same in both types of motors. The primary difference is that the squirrel cage induction motor has no electrical connections to the rotor circuit. In contrast, a wound rotor motor has a slip ring assembly and brushes that connect the rotor circuit to an external electric circuit. This arrangement varies the resistance, thus causing the speed characteristics of the wound rotor motor to change. Therefore, a wound rotor is usually found in a variable-speed pump station application. This section will deal specifically with the squirrel cage induction motor since it is the type you will most often encounter.

9.21 Nameplate Data (Figure 9.18)

As part of your preventive maintenance program, record the motor nameplate information in the file for each motor or piece of equipment as recommended in NEMA (National Electrical Manufacturers Association) publication MG1, Section 10.3A. Motor nameplate data must be recorded and filed so the information is available when needed to repair the motor or to obtain replacement parts.

A brief description of the information you will find on the nameplate and how it relates to motor performance is provided in the following paragraphs.

Serial Number. This is a unique number assigned by the manufacturer based on the manufacturer's numbering system to identify that specific motor. (This number should be available and used whenever it is necessary to communicate with the manufacturer.)

Type. This may be a combination of letters and numbers, established by the manufacturer to identify the type of enclosure and any modifications to it.

Model Number. The manufacturer's model number.

Horsepower. Rated horsepower is the horsepower that the motor is designed to produce at the shaft when power is applied at the rated frequency and voltage and the motor is operating at a service factor of 1.0.

Frame. This identifies the frame size in accordance with established NEMA Standards and does not vary from manufacturer to manufacturer. Therefore, the motor from one manufacturer with a NEMA frame is dimensionally identical to that same frame size from another manufacturer.

Service Factor. Service factors of 1.0 and 1.15 are commonly found on pump station motors. A service factor of 1.0 indicates that the motor may be run continuously at its rated horsepower without causing damage to the insulation system. A 1.15 service factor indicates that the motor may occasionally be run at a horsepower equal to the rated horsepower times the service factor without serious injury to the insulation system. This allows for intermittent variations in voltage which will cause internal heating on the motor windings. A service factor above 1.0 should never be relied upon to accommodate continuous loads because such use will quickly degrade the insulation system.

Amps. The current drawn by the motor at rated voltage and frequency at rated horsepower.

Volts. This is the voltage that would be measured at the terminals of the motor as opposed to the voltage of the supply.

Class of Insulation. Various classes of insulation material (described in Section 9.23) are available and the listed class will determine the operating temperatures at which the motor can safely operate.

RPM. Speed of the motor shaft, in revolutions per minute, at rated horsepower with rated voltage and frequency being supplied.

TYPICAL CUTAWAY VIEW
OF A MARATHON DESIGNED, DRIPPROOF, HORIZONTAL
INTEGRAL HORSEPOWER MOTOR & PARTS DESCRIPTION
364 THRU 445 FRAME SIZE

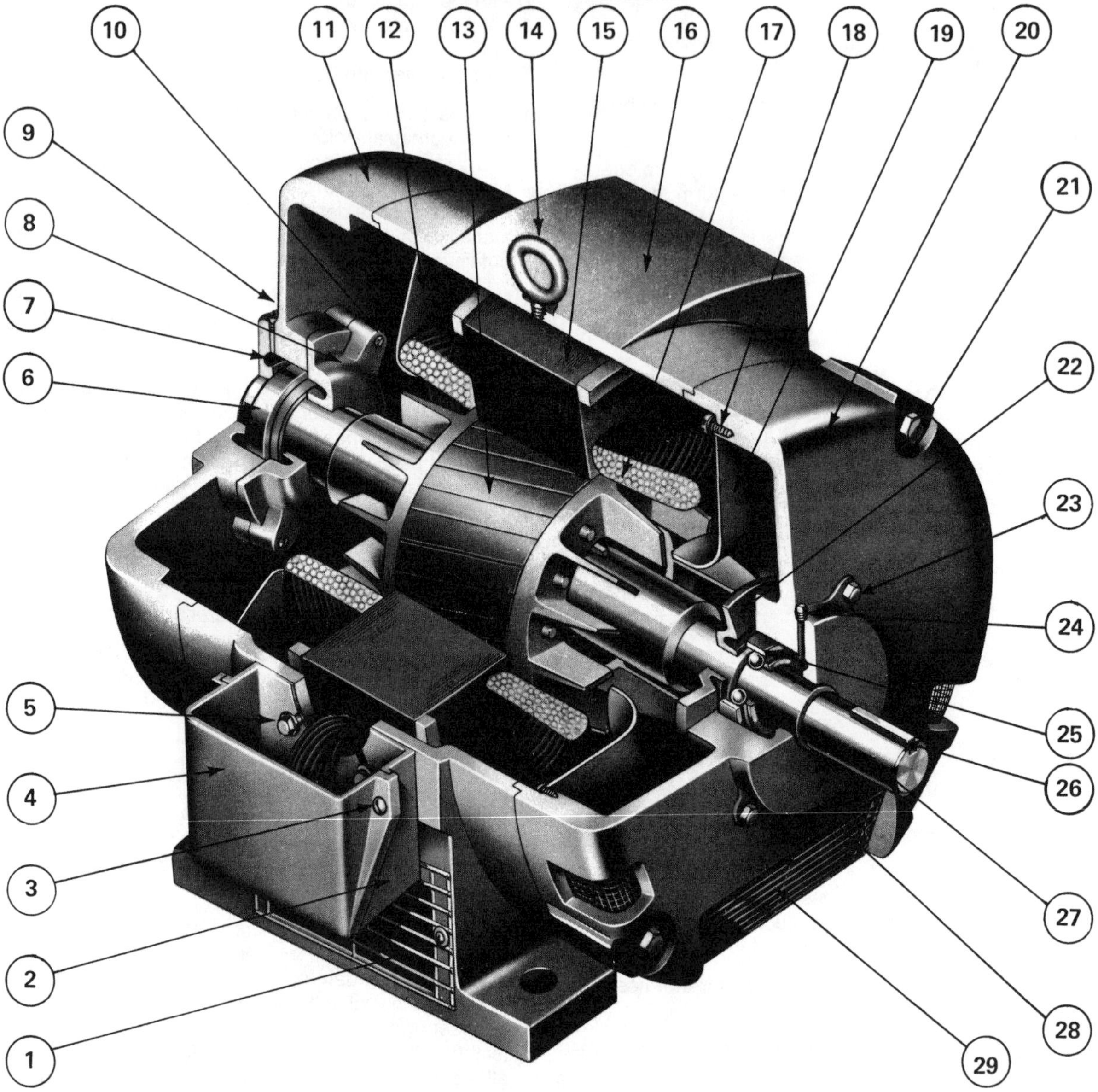

ITEM	DESCRIPTION	ITEM	DESCRIPTION	ITEM	DESCRIPTION
1.	*Frame Vent Screen	11.	Bracket O.P.E.	21.	Bracket Holding Bolt
2.	Conduit Box Bottom	12.	Baffle Plate O.P.E.	22.	Inner Bearing Cap P.E.
3.	Conduit Box Top-Holding Screw	13.	Rotor Core	23.	Inner Bearing Cap Bolt
4.	Conduit Box Top	14.	Lifting Eye Bolt	24.	Grease Plug
5.	Conduit Box Bottom Holding Bolt	15.	Stator Core	25.	**Ball Bearing P.E.
6.	**Ball Bearing O.P.E.	16.	Frame	26.	Shaft Extension Key
7.	Pre-Loading Spring	17.	Stator Winding	27.	Shaft
8.	Inner Bearing Cap O.P.E.	18.	Baffle Plate Holding Screw	28.	Drain Plug (grease)
9.	Grease Plug	19.	Baffle Plate P.E.	29.	*Bracket Screen
10.	Inner Bearing Cap Bolt	20.	Bracket P.E.		

P.E. = Pulley End
O.P.E. = Opposite Pulley End
* = Bracket and frame screens are optional.
** = Bearing Numbers are shown on motor nameplate. When requesting information or parts, always give complete motor description, model and serial numbers.

Fig. 9.17 Horizontal squirrel cage induction motor
(Courtesy of Marathon Electric)

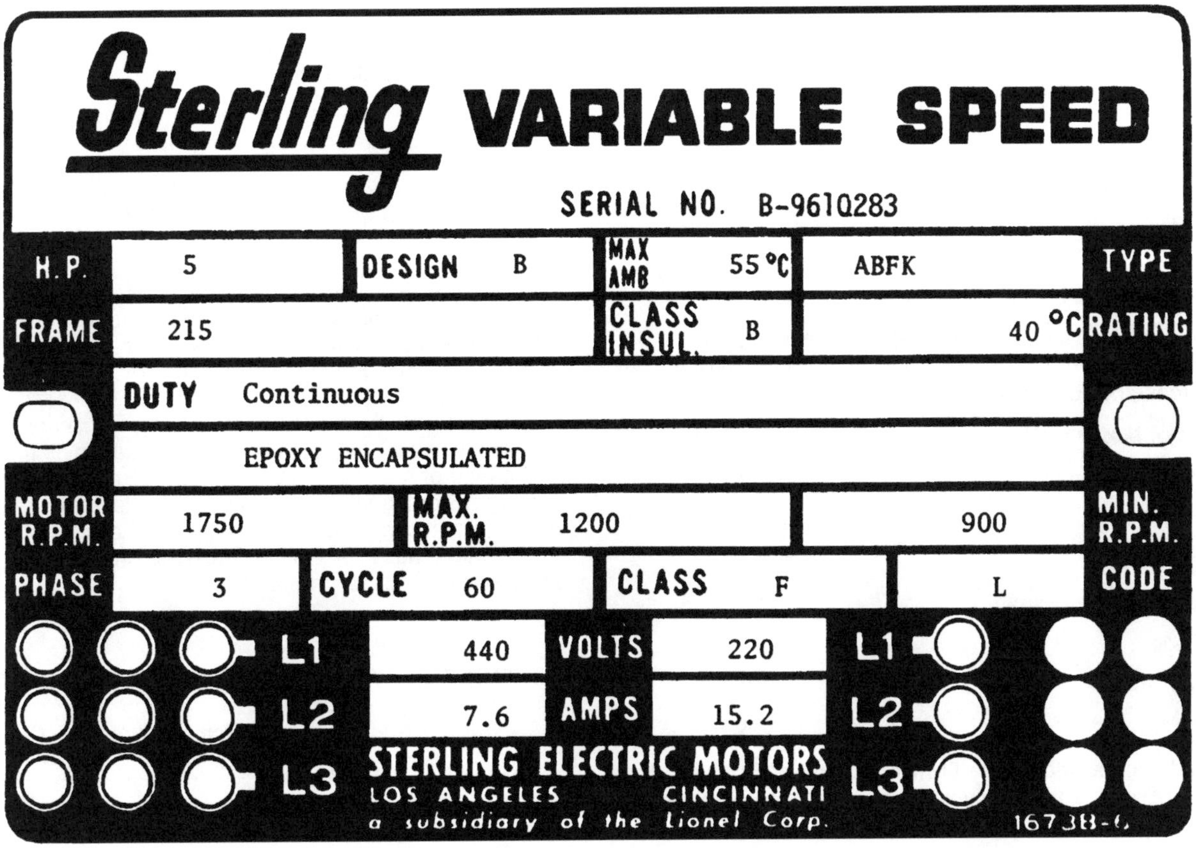
Sterling VARIABLE SPEED
SERIAL NO. B-9610283
H.P. 5
DESIGN B
MAX AMB 55 °C
ABFK
TYPE
FRAME 215
CLASS INSUL. B
40 °C RATING
DUTY Continuous
EPOXY ENCAPSULATED
MOTOR R.P.M. 1750
MAX. R.P.M. 1200
900
MIN. R.P.M.
PHASE 3
CYCLE 60
CLASS F
L
CODE
L1
440
VOLTS
220
L1
L2
7.6
AMPS
15.2
L2
L3
STERLING ELECTRIC MOTORS
LOS ANGELES
CINCINNATI
a subsidiary of the Lionel Corp.
L3

NOTE:

1. The motor for this unit is rated at 1,750 RPM and the maximum speed for the variable drive unit is 1,200 RPM.
2. The 40°C rating is the allowable operating temperature above ambient temperature.

Fig. 9.18 Example nameplate

(Courtesy of Sterling Power Systems, Inc.)

Hertz. Frequency, in cycles per second, of the utility company power.

Duty. Normally stamped "continuous" which means the motor can operate 24 hours a day, 365 days a year. In some cases, the motor nameplate will indicate "intermittent duty" for a specified time interval. This means the motor can operate at full load for the time interval specified, and then must be shut down and allowed to cool before restarting.

Ambient Temperature. Specifies the maximum ambient (surrounding) temperature at which the motor can safely operate. If the ambient temperature is exceeded, a corresponding increase of operating temperatures in the winding will degrade the insulation system and cause premature failure of the motor.

Phase. This indicates the number of phases; for example, 3 or three-phase which indicates the phase for which the motor is designed.

KVA Code. Starting inrush current which relates KiloVolt Amps to HorsePower (KVA/HP).

Design. This letter indicates the electrical design characteristics, and therefore, torque, speeds, inrush current, and slip values (these values and codes are also specified by NEMA).

Bearings. This number and letter sequence is designated by the Anti-Friction Bearing Manufacturers' Association Standards (AFBMA). It specifies looseness of the bearing fit, type of retainer, degree of protection, and dimension of the bearings. In most cases, unless the bearing is an extremely specialized type, the standard allows the use of bearings of different manufacturers.

Efficiency. This indicates the nominal operating efficiency of the motor at full load (Power Out/Power In)(100%).

9.22 Causes of Failure

Over the years, a considerable amount of research has been (and continues to be) conducted by various associations and manufacturers of electrical equipment to identify types of motor failures and the reasons for failures in A.C. induction motors. This section will deal with some of the common causes of motor failure and will suggest what you can do to minimize motor failures in lift stations.

In a survey of 9,000 motor failures by the Electrical Research Association, Letterhead, England, the following causes and percentages of motor failures were identified:

Causes of Motor Malfunction	Frequency, %
1. Overload (thermal)	30
2. Contaminants	19
3. Single phasing	14
4. Bearing failures	13
5. Old age	10
6. Miscellaneous	9
7. Rotor failures	5

A study conducted by the Dymac Division of Scientific Atlanta identifies the frequency with which specific components failed.

Failed Component	Frequency, %
1. Stator windings	70
2. Rotor windings	10
3. Bearings	10
4. Other	10

This information suggests that a large percentage of motor failures could be prevented. Sixty-three percent of the failures examined were primarily due to overload, single phasing, and contaminants, and all of these factors can be controlled by the operator. In addition, since 70 percent of the failures occur in the stator, we can assume that many of the failures are related to the insulation system used to protect the motor stator windings. For this reason, the next section will focus on the motor insulation system and show how a variety of conditions can lead to excessive heat buildup that will eventually result in motor failure.

9.23 Insulation

9.230 Types and Specifications

Four classes or levels of insulation systems are available:

1. Class A,
2. Class B,
3. Class F, and
4. Class H.

Each class of insulation is defined by the temperature limitation of the insulation itself. If the temperature limitation is exceeded, the insulation will deteriorate and, ultimately, cause a premature motor failure. The temperature rating consists of three components:

1. Ambient operating temperature—the air temperature where the motor is operating; for example, air temperature inside the lift station;
2. Temperature rise—the maximum inside temperature of the windings during normal operation; and
3. Hot-spot allowance—because of inconsistencies in the manufacturing process and the winding materials, a 10°C hot-spot allowance is included.

Older motors, those over 20 to 25 years old, are likely to have Class A insulation systems. A standard motor today has a Class B system or, in some cases, Class F. In extreme environments, a Class H insulation system may be used.

Table 9.1, A.C. Motor Temperature Limits, lists the temperature limitations for the four classes of insulation as they relate to the type of motor construction. For example, for all motors with a 1.15 or higher service factor, the total temperature that the insulation can withstand is 140°C for Class F insulation, or 165°C with Class H insulation. Table 9.2 lists typical materials used for motor insulation.

TABLE 9.1 A.C. MOTOR TEMPERATURE LIMITS[a]

	Temperature (Deg. C)				
	Class A		Class B	Class F	Class H
1.0 Service Factor					
Drip-proof					
Ambient Temperature	40	40	40	40	40
Rise by Thermometer	40	..	..	..	..
Rise by Resistance	..	50	80	105	125
Service-Factor Margin	10	10	..	..	..
Hot-Spot Allowance	15	5	10	10	15
Total Temperature	105	105	130	155	180
TEFC					
Ambient Temperature	40	40	40	40	40
Rise by Thermometer	55	..	..	..	..
Rise by Resistance	..	60	80	105	125
Hot-Spot Allowance	10	5	10	10	15
Total Temperature	105	105	130	155	180
TENV[b]					
Ambient Temperature	40	40	40	40	40
Rise by Thermometer	55	..	85	..	..
Rise by Resistance	..	65	..	110	135
Hot-Spot Allowance	10	0	5	5	5
Total Temperature	105	105	130	155	180
Encapsulated[c]					
Ambient Temperature				40	40
Rise by Thermometer				85	110
Hot-Spot Allowance				5	5
Total Temperature				130	155
1.15 or Higher Service Factor					
All Motors					
Ambient Temperature				40	40
Rise by Thermometer[d]				90	115
Hot-Spot Allowance				10	10
Total Temperature				140	165

[a] Adapted from National Electrical Manufacturers Association (NEMA) publication MG 1-12.39 and 12.40.
[b] Including all fractional-horsepower totally enclosed motors and fractional-horsepower motors smaller than frame 42.
[c] Enclosed.
[d] At service-factor load.

TABLE 9.2 INSULATION MATERIALS

	Class A Systems-105°C	Class B Systems-130°C	Class F Systems-155°C	Class H Systems-180°C
Varnish	Modified phenolic Modified asphalt Alkyd polyester	Unmodified polyester Epoxy Modified phenolic	Modified polyester Epoxy	Silicone Polyimide
Wire Insulation	Vinyl acetal enameled	Modified polyester enameled Epoxy enameled Enamel plug glass yarn	Modified polyester enameled Epoxy enameled	Glass yarn, silicone varnish covered Polyimide
Other	Rag paper Kraft paper Polyester film Acetate film Varnished cambric Wood Fiber Cotton cord	Polyester film Polyester mat Varnished glass Mica flake or mica paper Polyester glass Laminated glass Asbestos Melamine Glass cord	Varnished glass Laminated glass Mica flake or mica paper Glass cord Polyester glass Tetrafluoro-ethylene resin	Glass cord Mica flake or mica paper Silicone varnished glass Laminated glass Polyimide film Polyimide varnished glass Polyimide filament

9.231 Causes of Failure

Obviously, induction motors are a critical and integral part of the operating ability, efficiency, and reliability of the lift station. When failures occur they can be extremely serious because the loss of pumping capacity creates critical operating problems. In addition, repair or replacement of a motor is an expensive, labor-intensive job.

If a motor is operated in a clean, dry environment and within its specified nameplate load and operating characteristics, there is no reason why the motor should not operate for years and years without major maintenance. Unfortunately, a number of different factors can cause a motor to operate outside its specified ranges. When this occurs, the life of the motor is shortened significantly.

In the paragraphs that follow, these conditions will be discussed in terms of problems/solutions. You will see that, in one way or another, failure to adhere to motor nameplate guidelines frequently results in excessive heat buildup, damaged motor insulation, and eventual motor failure.

1. *PROBLEM:* Contaminants. Insulation failure can occur when deposits of dust, grease, or other foreign material accumulate on the windings and prevent the dissipation of the heat generated in the motor winding during normal operation. This causes local hot spots in the winding and, when the insulation breaks down completely, it will cause a phase-to-phase short characterized by arcing and melting between the phases' windings.

Vertical motors are susceptible to improper greasing methods and materials. Frequently, grease escapes from the bearing and the bearing housing and then contaminates the upper end turns on the stator winding, resulting in winding failure.

1. *SOLUTION:* Keep motors clean and free of dirt or grease accumulations. Follow the manufacturer's recommended methods and materials for greasing the equipment.

2. *PROBLEM:* Short cycling or excessive starts. Short cycling occurs when the automatic control system triggers frequent starts and stops of the pump and motor in response to fluctuating wet well elevations or because of a failure in the control system.

When induction motors start, the current (called locked rotor amps) required to magnetize the windings and start the rotor can be five to eight times the normal running current. For example, with a motor rated at 100 amps for full-load current, the starting sequence will require anywhere from 500 to 800 amps to start the motor rotating.

When frequent starts occur, the heat generated by the locked rotor currents never has a chance to dissipate and the internal winding temperature increases with each successive start of the motor. Typically, motors up to 100 horsepower should not exceed more than four or five starts per hour and as the horsepower increases, motor starting limitations may reduce the number to as few as one start per hour.

2. *SOLUTION:* To overcome excessive heat buildup from frequent starts and stops of a motor, use a reduced-voltage method of starting the motor. Table 9.3 compares different reduced-voltage starting methods. Also see Chapter 8, Section 8.141, "Motor Starting Devices and Methods."

TABLE 9.3 REDUCED-VOLTAGE STARTING[a]

Type Of Starting	Relative Starting Current	Relative Starting Torque
1. Across-the-Line	100%	100%
2. Resistors/Reactor (at 65% voltage)	65%	65%
3. Auto Transformer (at 65% voltage)	42%	42%
4. Wye-Delta Winding	33%	33%
5. Two-Part Winding	50%	50%

[a] Compared with full-voltage, across-the-line starting which typically draws 6.5 times the full-load current.

3. *PROBLEM:* High ambient operating temperature. If the ambient operating temperature exceeds that specified on the nameplate, it will contribute to a higher internal operating temperature. For all classes of insulation, the maximum ambient temperature is specified as 40°C (approximately 104°F).

3. *SOLUTION:* Provide adequate ventilation in lift stations and outdoor motor installations, particularly in southern climates where temperatures can be expected to exceed 104°F.

4. *PROBLEM:* Obstructed enclosure vents. All motor enclosures are designed for maximum dissipation of internally generated heat.

4. *SOLUTION:* In open, drip-proof motors, do not allow obstructions of the ventilation openings. Similarly, in totally enclosed, fan-cooled motors, a buildup of dirt, grease, or dust on the ribs of the enclosure will decrease the ability of the enclosure to dissipate heat. Keep the enclosure clean.

In outside installations, construct the enclosure to prevent entry by rodents. Frequently check rodent screens, if present, to ensure they are not clogged with foreign material. In areas of high humidity, treat the enclosure with a fungicide to prevent formation of fungus on the insulation surface.

5. *PROBLEM:* Single phasing. Single phasing refers to the condition that occurs when one phase of the power source to the motor is lost, either from the utility company or from a fuse blowing in one phase in the motor control center. Under a single-phase condition, an induction motor which is already rotating will continue to rotate; however, it will be characterized by increased noise and vibration. If the motor is not rotating, a single-phase condition will not start the motor, will cause excessive noise and vibration, and will continue to do so until the protective device senses the condition and trips the overloads. Single-phase condition causes unbalanced currents to circulate in the rotor causing increases in internal motor heating.

5. *SOLUTION:* To correct a single-phasing problem, determine why one phase is missing. Did the utility company lose a phase (a common problem) or is a fuse blown or a circuit breaker tripped in the motor control center? Once the source of the problem is identified, then it can be corrected by notifying the utility company, replacing the blown fuse, or resetting the tripped circuit breaker.

6. *PROBLEM:* Motor overloading. This is operation of the motor in a way that causes it to draw current in excess of the motor nameplate current. Overloading can happen inadvertently through improper operation on the pump curve by changing impeller diameters or through a change in the dynamic operating conditions of the pump which changes the total dynamic head (TDH). Other conditions that can cause motor overloading include bearing problems and jamming of material between the rotating impeller and the stationary pump housing. More energy is required to operate the pump when rags, rocks or timbers interfere with free rotation of the impeller.

The heat generated by continuous operation of a motor above its design rating is extremely damaging to the insulation. For example, a relatively minor overload of 6 percent will cause a 10°C increase in temperature; continuous operation at this elevated temperature will reduce the insulation life by 50 percent. A continuous 12 percent overload cuts insulation life to one-quarter of its design life.

6. *SOLUTION:* A thorough understanding of hydraulics and the existing operating conditions is required before changes in the pump can be made since improper changes can have disastrous effects on the motor. Know what you are doing before making any changes. Also, avoid continuous operation of a motor above its design rating. For additional protection, consider installing an overload protective device (described in Section 9.131, "Equipment Protective Devices").

7. *PROBLEM:* Voltage imbalance. A common problem found in pump stations with a high rate of motor failure is voltage imbalance or unbalance. Unlike a single-phase condition, all three phases are present but the phase-to-phase voltage is not equal in each phase.

Voltage imbalance can occur in either the utility side or the pump station electrical system. For example, the utility company may have large single-phase loads (such as residential services) which reduce the voltage on a single phase. This same condition can occur in the pump station if a large number of 120/220 volt loads are present. Slight differences in voltage can cause disproportional current imbalance; this may be six to ten times as large as the voltage imbalance. For example, a two percent voltage imbalance can result in a 20 percent current imbalance. A 4.5 percent voltage imbalance will reduce the insulation life to 50 percent of the normal life. This is the reason a dependable voltage supply at the motor terminals is critical. Even relatively slight variations can greatly increase the motor operating temperatures and burn out the insulation.

It is common practice for electrical utility companies to furnish power to three-phase customers in open delta or wye configurations. An open delta or wye system is a two-transformer bank that is a suitable configuration where *LIGHTING LOADS ARE LARGE AND THREE-PHASE LOADS ARE LIGHT.* This is the exact opposite of the configuration needed by most pumping facilities where *THREE-PHASE LOADS ARE LARGE.* (Examples of three-transformer banks include Y-delta, delta-Y, and Y-Y.) In most cases three-phase motors should be fed from three-transformer banks for proper balance. The capacity of a two-transformer bank is only 57 percent of the capacity of a three-transformer bank. The two-transformer configuration can cause one leg of the three-phase current to furnish higher amperage to one leg of the motor, which will greatly shorten its life.

Operators should acquaint themselves with the configuration of their electric power supply. When an open delta or wye configuration is used, operators should calculate the degree of current imbalance existing between legs of their polyphase motors. If you are unsure about how to determine the configuration of your system or how to calculate the percentage of current imbalance, *ALWAYS* consult a qualified electrician. *CURRENT IMBALANCE BETWEEN LEGS SHOULD NEVER EXCEED 5 PERCENT UNDER NORMAL OPERATING CONDITIONS* (NEMA Standards MGI-14.35).

Loose connections will also cause voltage imbalance as will high-resistance contacts, circuit breakers, or motor starters.

Another serious consideration for operators is voltage fluctuation caused by neighborhood demands. A pump motor in near perfect balance (for example, 3 percent unbalance) at 9:00 AM could be as much as 17 percent unbalanced by 4:00 PM on a hot day due to the use of air conditioners by customers on the same grid. Also, the hookup of a small market or a new home to the power grid can cause a significant change in the degree of current unbalance in other parts of the power grid. Because energy demands are constantly changing, wastewater system operators should have a qualified electrician check the current balances between legs of their three-phase motors at least once a year.

7. *SOLUTION:* Motor connections at the circuit box should be checked frequently (semiannually or annually) to ensure that the connections are tight and that vibration has not rubbed through the insulation on the conductors. Measure the voltage at the motor terminals and calculate the percentage imbalance (if any) using the procedures below.

Do not rely entirely on the power company to detect unbalanced current. Complaints of suspected power problems are frequently met with the explanation that all voltages are within the percentages allowed by law and no mention is made of the percentage of current unbalance which can be a major source of problems with three-phase motors. A little research of your own can pay large benefits. For example, a small water company in Central California configured with an Open Delta system (and running three-phase unbalances as high as 17 percent as a result) was routinely spending $14,000 a year for energy and burning out a 10 HP motor on the average of every 1.5 years (six 10 HP motors in 9 years). After consultation, the local power utility agreed to add a third transformer to each power board to bring the system into better balance. Pump drop leads were then rotated, bringing overall current unbalances down to an average of 3 percent, heavy duty three-phase capacitors were added to absorb the frequent voltage surges in the area, and computerized controls were added to the pumps to shut them off when pumping volumes got too low. These modifications resulted in a saving in energy costs the first year alone of $5,500.00.

FORMULAS

Percentage of current unbalance can be calculated by using the following formulas and procedures:

$$\text{Average Current} = \frac{\text{Total of Current Value Measured on Each Leg}}{3}$$

$$\text{\% Current Unbalance} = \frac{\text{Greatest Amp Difference from the Average}}{\text{Average Current}} \times 100\%$$

PROCEDURES

A. Measure and record current readings in amps for each leg. (Hookup 1.) Disconnect power.

B. Shift or roll the motor leads from left to right so the drop cable lead that was on terminal 1 is now on 2, lead on 2 is now on 3, and lead on 3 is now on 1. (Hookup 2.) Rolling the motor leads in this manner will not reverse the motor rotation. Start the motor, measure and record current reading on each leg. Disconnect power.

C. Again shift drop cable leads from left to right so the lead on terminal 1 goes to 2, 2 goes to 3, and 3 to 1. (Hookup 3.) Start pump, measure and record current reading on each leg. Disconnect power.

D. Add the values for each hookup.

E. Divide the total by 3 to obtain the average.

F. Compare each single leg reading to the average current amount to obtain the greatest amp difference from the average.

G. Divide this difference by the average to obtain the percentage of unbalance.

H. Use the wiring hookup which provides the lowest percentage of unbalance.

CORRECTING THE THREE-PHASE POWER UNBALANCE

Example: Check for current unbalance for a 230 volt, 3-phase 60 Hz submersible pump motor, 18.6 full load amps.

Solution: Steps 1 to 3 measure and record amps on each motor drop lead for Hookups 1, 2 and 3 (Figure 9.19).

	Step 1 (Hookup 1)	Step 2 (Hookup 2)	Step 3 (Hookup 3)
(T_1)	DL_1 = 25.5 amps	DL_3 = 25 amps	DL_2 = 25.0 amps
(T_2)	DL_2 = 23.0 amps	DL_1 = 24 amps	DL_3 = 24.5 amps
(T_3)	DL_3 = 26.5 amps	DL_2 = 26 amps	DL_1 = 25.5 amps
Step 4	Total = 75 amps	Total = 75 amps	Total = 75 amps

Step 5 Average Current = $\frac{\text{Total Current}}{\text{3 readings}} = \frac{75}{3}$ = 25 amps

Step 6	Greatest amp difference from the average:	(Hookup 1) = 25 – 23	= 2
		(Hookup 2) = 26 – 25	= 1
		(Hookup 3) = 25.5 – 25	= .5
Step 7	% Unbalance	(Hookup 1) = 2/25 x 100	= 8
		(Hookup 2) = 1/25 x 100	= 4
		(Hookup 3) = 0.5/25 x 100	= 2

As can be seen, Hookup 3 should be used since it shows the least amount of current unbalance. Therefore, the motor will operate at maximum efficiency and reliability on Hookup 3.

By comparing the current values recorded on each leg, you will note the highest value was always on the same leg, L_3. This indicates the unbalance is in the power source. If the high current values were on a different leg each time the leads were changed, the unbalance would be caused by the motor or a poor connection.

If the current unbalance is greater than 5 percent, contact your power company for help.

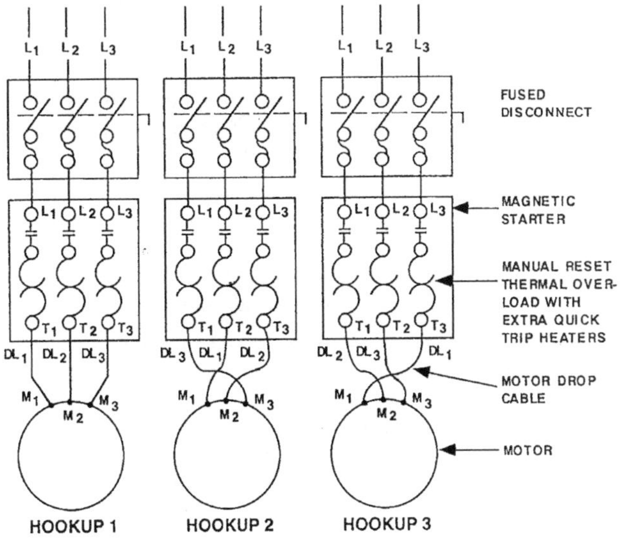

Fig. 9.19 Three hookups used to check for current unbalance

Acknowledgment

Material on unbalanced current was provided by James W. Cannell, president, Canyon Meadows Mutual Water Company, Inc., Bodfish, CA. His contribution is greatly appreciated.

9.232 Increasing Resistance Value

The minimum insulation value in megohms is calculated by the following procedures. Divide the rated voltage by 1,000 and add 1 to the result. For example, for a motor operating at 460 volts, the minimum resistance value is 1.46 megohms.

$$\frac{460}{1{,}000} + 1 = 1.46$$

Measure the actual insulation resistance using a megger and compare the measured megger value with the calculated value; then any one of the three following procedures can be used to increase the actual (measured) insulation resistance value. The time required to increase the insulation resistance value depends on the wetness of the insulation and the size of the motor.

1. Remove the motor and bake in an oven at a temperature of not more than 194°F (90°C) until the resistance reaches an acceptable level.

2. Cover motor with a canvas or tarp and insert heating units or lamps. Heat until the resistance reaches an acceptable level. The heating time depends on the number and size of the heating units.
3. Provide a low-voltage current at the motor terminals which will generate heat within the windings. Heat until the resistance reaches an acceptable level.

9.24 Starters

A motor starter is a device or group of devices which are used to connect the electrical power to a motor. Motor starters can be either manually or automatically controlled.

Manual and magnetic starters range in complexity from a single ON/OFF switch to a sophisticated automatic device using timers and coils. The simplest motor starter is used on single-phase motors where a circuit breaker is turned on and the motor starts. This type of starter also is used on three-phase motors of smaller horsepower as well as on fan motors, machinery motors, and other motors where it isn't necessary to have automatic control.

Magnetic starters (Figure 9.20) are usually used to start pumps, compressors, blowers, and anything where automatic or remote control is desired. They permit low-voltage circuits to energize the starter of equipment at a remote location or to start larger starters (Figure 9.21). A magnetic starter is operated by electromagnetic action. It has contactors and these operate by energizing a coil which closes the contact, thus starting the motor. The circuit that energizes the starter is called the control circuit; it is usually operated on a lower voltage (115 volts) than the motor. Whenever a starter is used as a part of an integrated circuit (such as for flow, pressure or temperature control), a magnetic starter or controller is necessary.

Magnetic starters are sized for their voltage and horsepower ratings. Additional information can be found in electrical catalogs, manuals and manufacturers' brochures.

A magnetic starter actually consists of two distinct sections:

1. The contacts which connect and disconnect the power to the motor, and
2. Overload protection.

Figure 9.20 shows a three-phase magnetic starter. The replaceable contacts (in the upper portion of the illustration) close when the motor is required to start, thus closing the electrical circuit to the motor. Similarly, when the wet well level gets to a point where the pump is no longer required, a signal is sent to the motor starter and the contacts open, thus breaking the electrical circuit to the motor. Each time this occurs, an arc takes place between the movable contact and the stationary contact and pitting occurs. This is why contacts must be replaced as a regular part of your preventive maintenance program.

The control coil usually uses a lower voltage than the line voltage to the motors. The magnetic coil is the device that actually causes the contacts to energize and de-energize.

In addition, each phase has an overload protection device that operates as a function of the length of the overload and the amount of the overload. Two types of devices are used, both based on the heating principle:

1. A bimetallic strip which is precisely calibrated to open under higher temperature conditions to de-energize the coil, or
2. A small solder pot within the coil that melts because of the heat and will de-energize the system.

A more common term for the overload protection devices is "heater elements." They are replaceable and can be selected and changed to correspond to the desired trip setting. Typically, the overload heater is selected for a trip setting that will de-energize the contactor when a 10 percent overload occurs.

Figure 9.22 shows a typical schematic diagram (referred to as a ladder diagram) for the control of one pump. This diagram is intended only for illustrative purposes. It does not include required grounding. Electrical systems must conform with the National Electrical Code in all cases.

Three-phase 480 volt power is fed into terminals L1, L2 and L3 on the line side of the main circuit breaker. The main circuit breaker is normally located in the motor control center and provides circuit protection for all electrical equipment on the load side of the circuit breaker. On the load side of the circuit breaker, connections are made for a branch circuit breaker for the motor circuit. The load side of the branch circuit breaker feeds into motor starter 1M which is a combination contactor

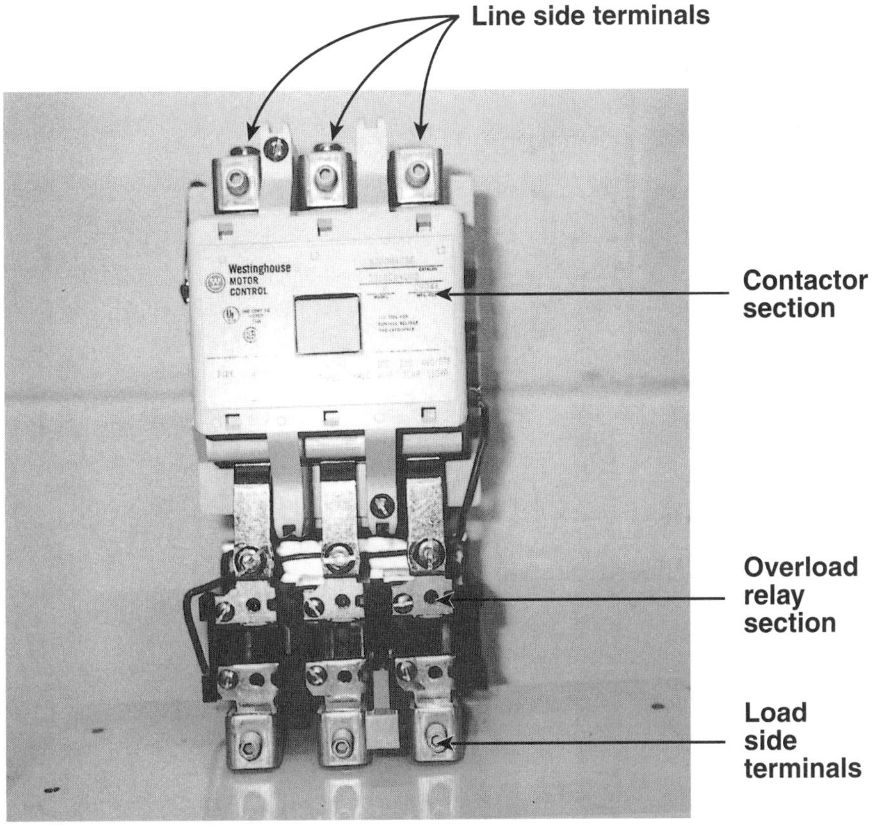

Fig. 9.20 Magnetic starter

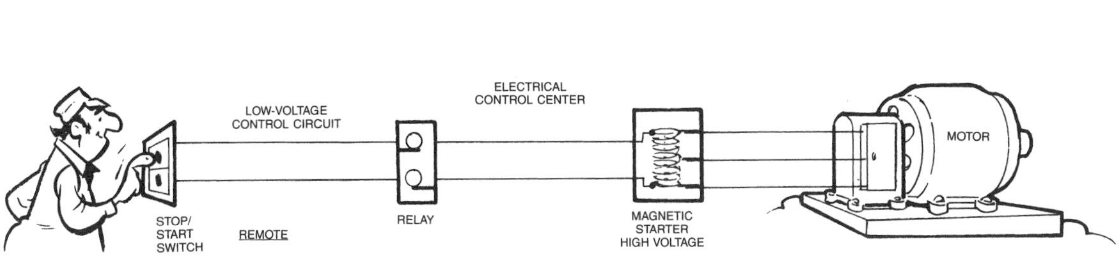

Fig. 9.21 Application of magnetic starter

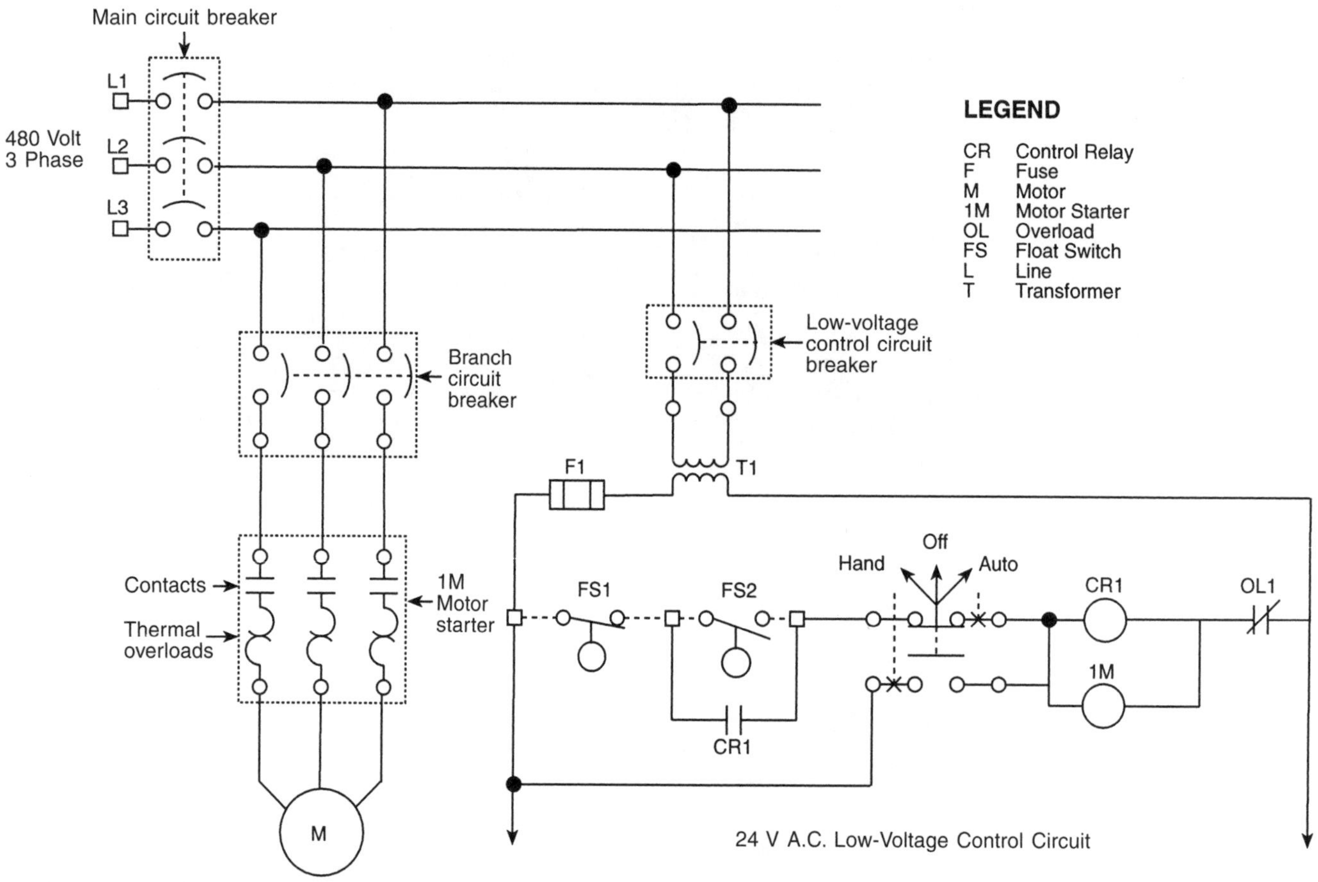

Fig. 9.22 Schematic diagram for control of one pump

and thermal overload. The load side of the motor starter 1M is then connected to the motor terminals.

Figure 9.22 shows the circuit components needed to get the 3-phase 480 volt power to the motor terminals. However, some means must be provided for turning the motor on and off in response to wet well levels. The 24 volt A.C. low-voltage control circuit illustrates how this is accomplished. Two phases of the 480 volt power are fed to a low-voltage control circuit breaker. The load side of the circuit breaker is connected to transformer T1 which reduces the line voltage to 24 volts A.C. Power is then supplied to the low-voltage control circuit through fuse F1. Under a rising wet well condition, float switch FS1 is energized (as shown in Figure 9.22). This means that the float has tipped and the contacts are made in the float switch.

FS1 is the low level or pump shutoff switch. As the level rises in the wet well, the pump start float FS2 tips. The contacts close and the circuit is completed through the hand-off-auto (HOA) switch; this energizes control relay CR1 and the motor starter contactor coil 1M. In a non-overload condition, thermal overload relay contacts OL1 are in the closed position allowing the circuit to be completed. When CR1 energizes, the contacts in parallel with FS2 close. This is necessary because as the wet well is pumping down, float switch FS2 will de-energize; however, we want the pump to continue operating until float switch FS1 de-energizes. When 1M energizes, this closes the contacts in the motor starter connecting the 3-phase power to the motor terminals. In the event of an overload condition, which would be sensed by the thermal overloads in motor starter 1M, the overload relay contact OL1 in the low-voltage control circuit will open, thereby shutting off the motor. As the wet well continues to drop, float switch FS1 will return to the normal position when the water level drops below the float. This opens the control circuit and de-energizes relay coil 1M which in turn opens the contacts in motor starter 1M.

If more than one motor is installed in the pump station, which is usually the case, additional branch circuit breakers, motor starters, float switches and control devices are required. Additional control functions are easily added such as indicating lights, alarms and other control elements. As previously discussed, a programmable logic controller (PLC) could be installed to control pump functions. The PLC would essentially replace the control circuit elements thus reducing the number of relays and the wiring required in pump motor control circuits.

Figure 9.23 illustrates how a motor contactor operates. The coil, when energized from the low-voltage control circuit, pulls the armature up in a vertical movement. The armature is attached to a bell crank lever that translates the vertical motion into horizontal motion, moving the contacts to the right where they make contact with the stationary contacts. Three-phase voltage is brought in on the line side connection and, in the closed position, the contacts allow that voltage to be applied through the load side connection to the motor. Both the stationary contacts and the moving contacts are replaceable. *NOTE:* This diagram does not illustrate the thermal overload protection.

Starters are also available without the overload function. They provide only the motor disconnecting means and are referred to as contactors rather than motor starters. Manual contactors are available as well as magnetic contactors which are capable of automatic control.

Magnetic starter maintenance consists of at least an annual inspection of the equipment, including the following tasks:

1. Inspect line and load conductors for evidence of high temperatures as indicated by bubbling or discoloring of the wire insulation;
2. Tighten all line and load connection terminals, including all low-voltage terminals (control, auxiliary switches); and
3. Inspect and replace, if necessary, the stationary and movable contacts. See Figure 9.24 which illustrates the appearance of new contacts, contacts that are used but still suitable for use, and contacts that are used and should be replaced. (A troubleshooting guide for magnetic starters is presented in Section 9.272.)

CAUTION: No attempt should be made to file down contacts to restore the surface to a new condition, since this will result in an uneven surface and uneven distribution of electrical energy across the face of the contacts.

9.25 Safety

1. The eye bolt used for lifting and moving motors is designed for the weight of the motor alone without other equipment attached.
2. Whenever physically working on rotating equipment, including the pump and the motor, open, lock out and tag the electrical disconnect switch. This will prevent accidental energization of the motor, either by a careless operator or as a result of motor thermal switches automatically resetting and restarting the motor.
3. Ground all motors in accordance with requirements of National Electrical Code, Article 340.
4. Always keep hands and clothing away from moving parts.
5. Discharge all capacitors including power factor direction capacitors before servicing motor or motor controller.
6. Be sure that required safety guards are always in place.

9.26 Other Motor Considerations

9.260 Alignment

Horizontal motors should be mounted so that all four mounting feet are aligned to within 0.010 inch of each other for NEMA Frame 56 to NEMA Frame 210 and within 0.015 inch for NEMA Frame 250 through NEMA Frame 680. This alignment ensures a good, rigid foundation and also will make alignment of the motor and pump easier.

Whenever two pieces of rotating equipment such as a pump and a motor are used, there must be some means of transmit-

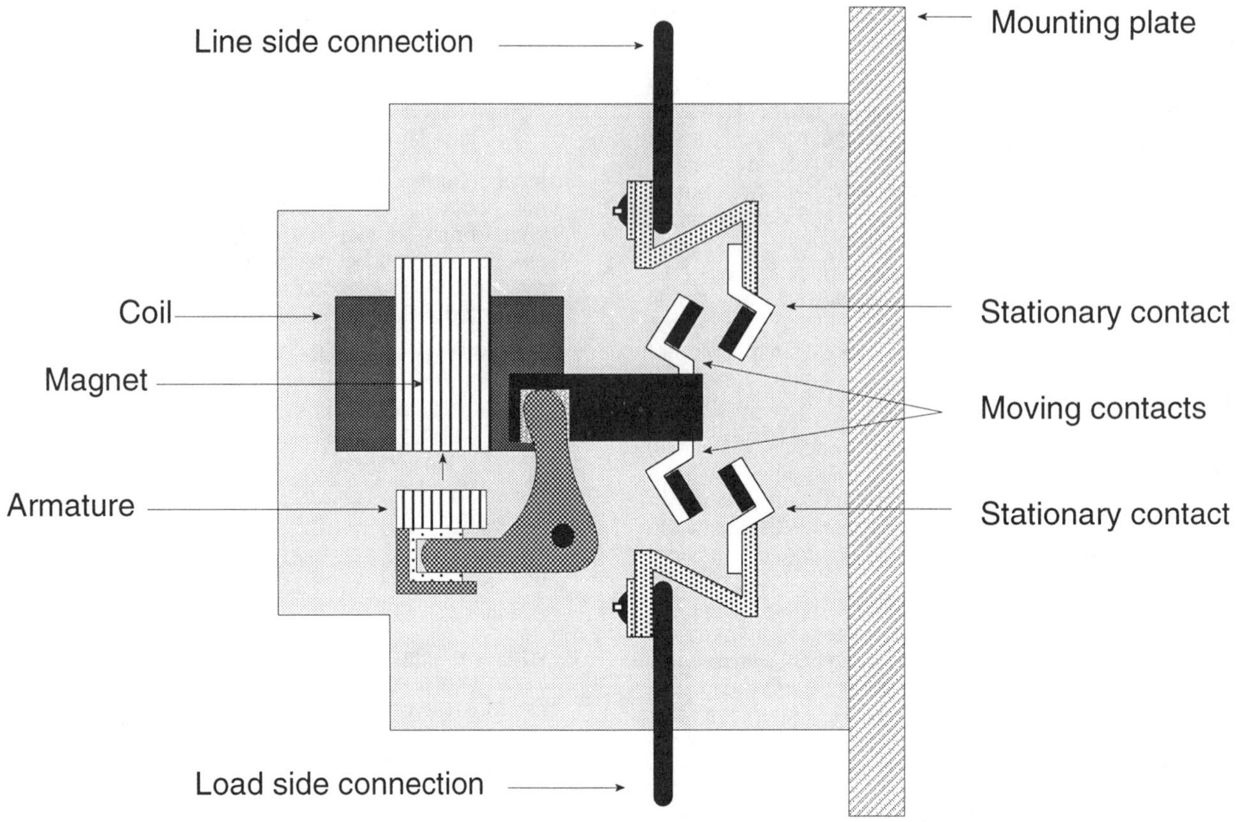
Line side connection
Mounting plate
Coil
Magnet
Armature
Stationary contact
Moving contacts
Stationary contact
Load side connection

Fig. 9.23 How a motor contactor operates

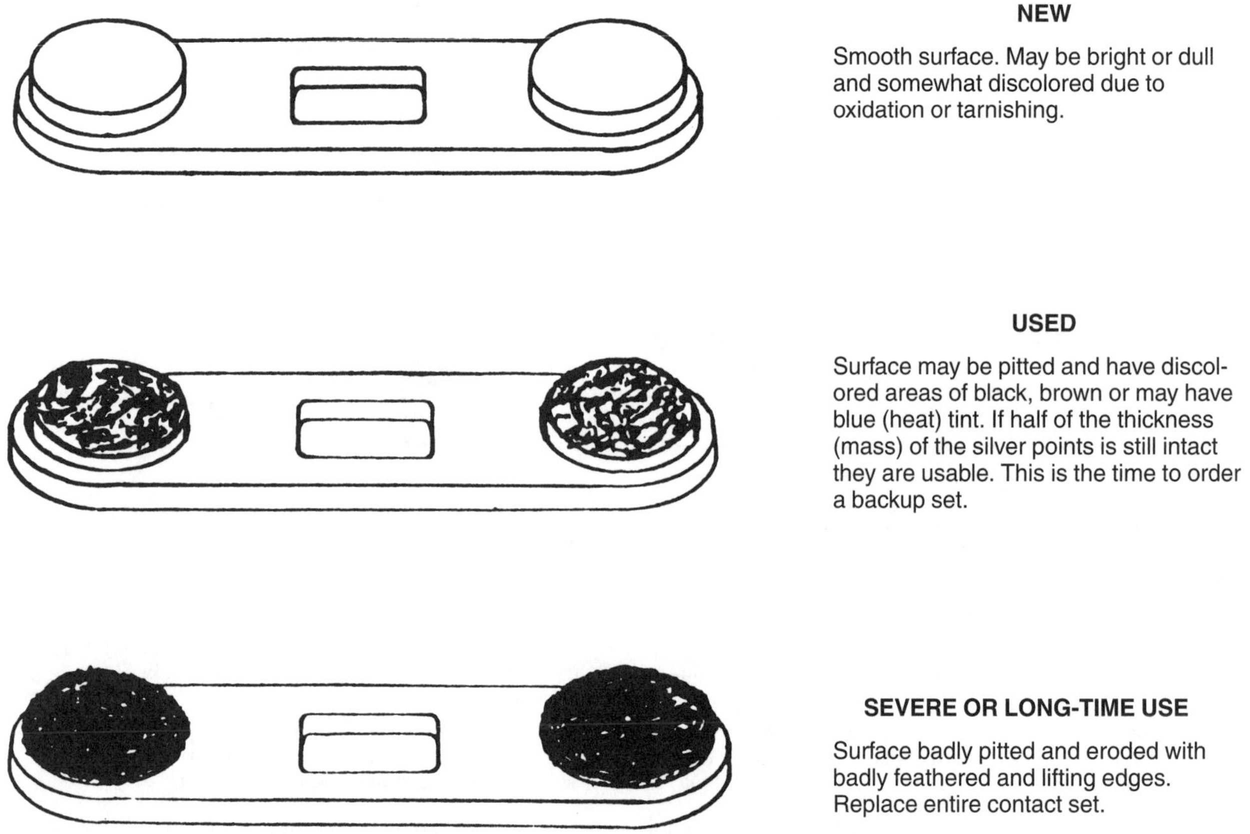
NEW
Smooth surface. May be bright or dull and somewhat discolored due to oxidation or tarnishing.
USED
Surface may be pitted and have discolored areas of black, brown or may have blue (heat) tint. If half of the thickness (mass) of the silver points is still intact they are usable. This is the time to order a backup set.
SEVERE OR LONG-TIME USE
Surface badly pitted and eroded with badly feathered and lifting edges. Replace entire contact set.

Fig. 9.24 Visual inspection of contact points

ting the torque from the motor to the pump. Couplings are designed to do this. To function as intended, the equipment must be properly aligned at the couplings. Misalignment of the pump and the motor, or any two pieces of rotating equipment, can seriously damage the equipment and shorten the life of both the pump and the motor. Misalignment can cause excessive bearing loading as well as shaft bending which will cause premature bearing failure, excessive vibration or permanent damage to the shaft. Remember that the purpose of the coupling is to transmit power and, unless the coupling is of special design, it is not to be used to compensate for misalignment between the motor and the pump.

When connecting a pump and a motor, there are two important types of misalignment: (1) parallel, and (2) angular. Parallel misalignment occurs when the center lines of the pump shaft and the motor shaft are offset. Figure 9.25 illustrates parallel misalignment. The pump and the motor shafts remain parallel to each other but are offset by some amount. Parallel misalignment can be detected very easily by holding a straightedge on one hub of the coupling and measuring the gap between the straightedge and the other hub of the coupling. Feeler gages can be used to measure the amount of offset misalignment.

The second type of alignment problem is angular misalignment, also shown in Figure 9.25 (page 186). In this case the shaft center lines are not parallel but instead form an angle, which represents the amount of angular misalignment. This type of misalignment can also be detected with a feeler gage, calipers or more sophisticated laser alignment equipment by measuring the distance between the coupling hubs at the point of maximum and minimum openings in the hubs, which would be 180° from each other. When using a dial indicator to measure angular misalignment, a general rule of thumb is that angular misalignment of the shafts must not exceed a total indicator reading of 0.002 inch for each inch of diameter of the coupling hub. To check for angular misalignment, mount a dial indicator on one coupling hub as shown in Diagram 1, Figure 9.26, with the finger or the button of the indicator against the finished face of the other hub and the dial set at zero. While rotating the shaft, note the reading on the indicator dial at each revolution.

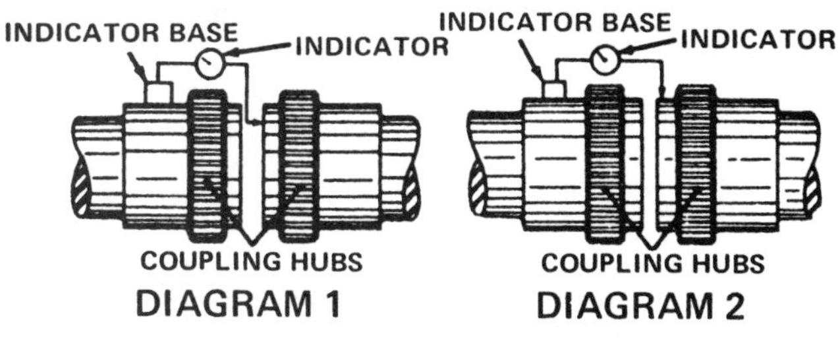

Fig. 9.26 Use of dial indicator to check for shaft angular alignment and trueness

In reality, misalignment usually includes both parallel and angular misalignment. The goal when aligning machines is to reduce the angular and parallel misalignment to a minimum. The purpose of this is two-fold: (1) couplings are not designed to accommodate large differences in parallel or angular misalignment, and (2) most alignment takes place when the machines being aligned are cold. However, all metal has a coefficient of expansion, which means that the metal expands as it heats up during operation. Since various types of metals expand at different rates and to different degrees, some additional misalignment may occur during operation. Couplings are designed to accommodate this type of misalignment.

In addition to misalignment, we are also concerned with end float in the shafts on the pump and the motor and with runout. End float is an in and out movement of the shaft along the axis of the shaft (see Figure 9.25).

Runout should also be checked. This checks the trueness or straightness of the shaft. Diagram 2, Figure 9.26, illustrates how the trueness is checked with the dial indicator mounted, again, on one coupling hub and the finger or button mounted on the outside surface of the second hub. As the shaft is rotated, note the indicator reading. The reading should not exceed 0.002 inch. A reading in excess of 0.002 inch indicates a bent shaft or a shaft that is eccentrically machined.

Misalignment is one of the most frequent causes of vibration problems in lift station motors and pumps. It frequently causes premature failure of bearings, mechanical seals and packing. While these failures in and of themselves are expensive and affect equipment availability and reliability, the failures can also cause more catastrophic related failures such as broken shafts, destruction of the rotor element in the motor and/or damage to motor windings.

9.261 Changing Rotation Direction

Changing the rotation direction of a three-phase motor is accomplished by simply changing any two of the power leads. Generally, this is done on the load side of a magnetic starter. The direction of motor rotation must be changed if a pump is rotating in the wrong direction.

9.262 Allowable Voltage and Frequency Deviations

1. Voltage 10 percent above or below the value stamped on the nameplate.
2. Frequency 5 percent above or below the value stamped on the nameplate.
3. Voltage and frequency together within 10 percent providing frequency is less than 5 percent above or below the value stamped on the nameplate.

As mentioned previously, both voltage and frequency have a direct effect on the performance and life of the motor as illustrated in Tables 9.4 and 9.5. A 10 percent increase in the rated voltage, for example, results in a 0 to 17 percent increase in temperature, which will have a significant effect on the insulation life of the motor (see Table 9.5, Voltage DP, 110% of Rated Voltage, 1-200 HP, Temperature Rise, Full-Load, 0 to 17%).

9.263 Maximum Vibration Levels Allowed by NEMA

SPEED, RPM	MAXIMUM VIBRATION AMPLITUDE, INCHES
3,000 - 4,000	0.001
1,500 - 2,999	0.002
1,000 - 1,499	0.0025
0 - 999	0.003

9.264 Lubrication

The correct procedure for lubricating motors is as follows:

1. Stop motor, lock out and tag;
2. Wipe all grease fittings;
3. Remove filler and drain plugs (CAUTION, Zerk fittings should *NOT* be installed in both the filler and the drain holes);
4. Free drain hole of any hard grease using a piece of wire, if necessary;

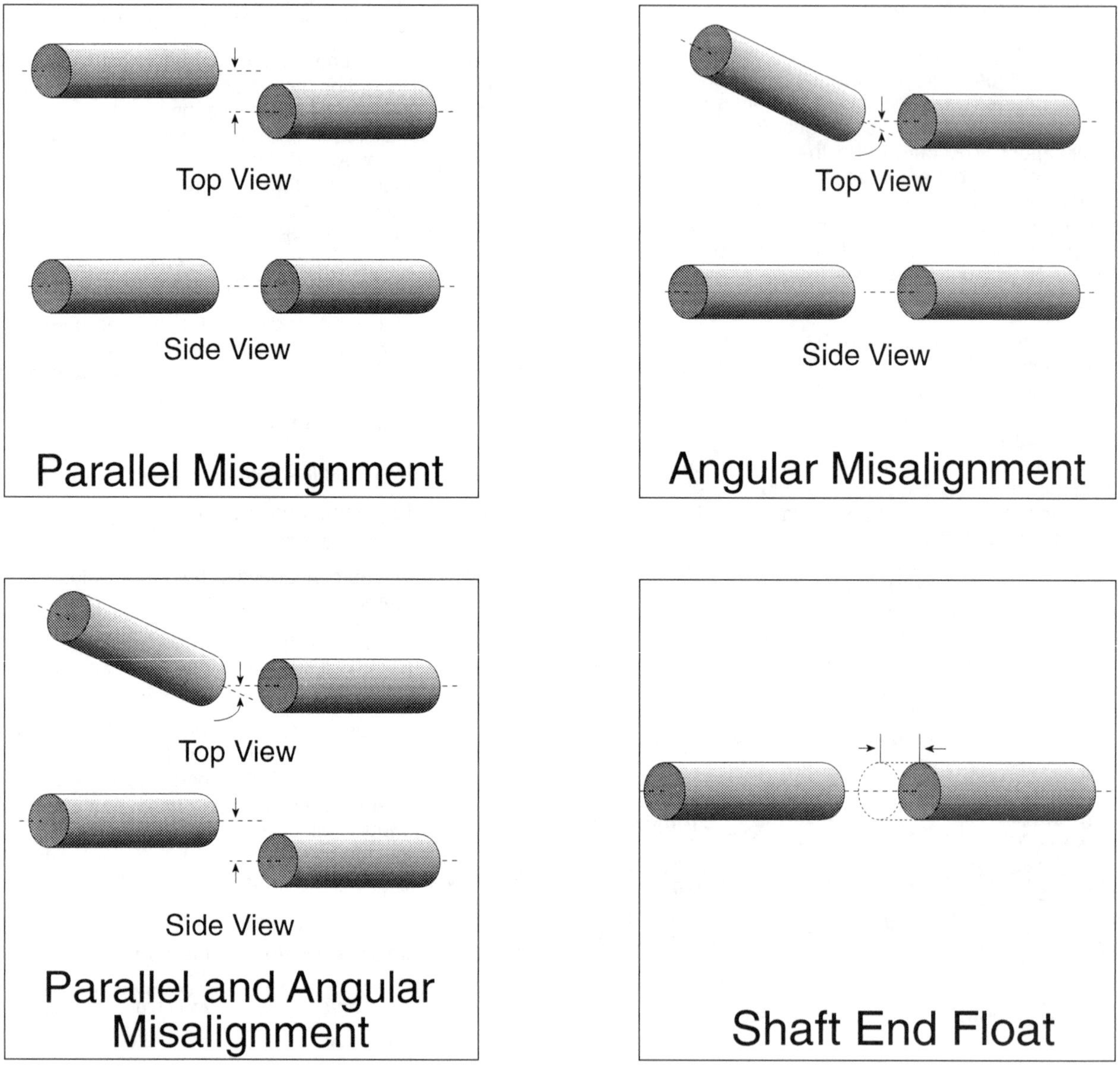

Fig. 9.25 Types of shaft misalignment and end float

TABLE 9.4 TYPICAL MOTOR PERFORMANCE VARIATIONS DUE TO POWER SUPPLY VARIATIONS

POLYPHASE · INTEGRAL HORSEPOWER

				TORQUE				Speed	Power Factor
				Starting (Locked Rotor)	DIP	Break-Down	Full-Load	Full-Load	Full-Load
		1% Unbalance		Slight –	Slight –	Slight –	Slight +	Slight –	–5.5%
		2% Unbalance		Slight –	Slight –	Slight –	Slight +	Slight –	–7.1%
VOLTAGE	DP	110% of Rated Volt.	1-10 HP 15-30 HP 40-75 HP 100-200 HP	+21 to 23%	+21 to 23%	+21 to 23%	–1.0% –0.6% –0.5% –0.3%	+1.0% +0.6% +0.5% +0.3%	–13 to 10% –9 to 8% –8 to 6% –6 to 4%
VOLTAGE	DP	90% of Rated Volt.	1-10 HP 15-30 HP 40-75 HP 100-200 HP	–17 to 19%	–17 to 19%	–17 to 19%	+1.5% +1.0% +0.6% +0.3%	–1.5% –1.0% –0.6% –0.3%	+11 to 7% +6 to 3% +3 to 2% +2 to 3%
VOLTAGE	TEFC	110% of Rated Volt.	1-10 HP 15-30 HP 40-75 HP 100-200 HP	+21 to 23%	+21 to 23%	+21 to 23%	–1.0% –0.6% –0.5% –0.3%	+1.0% +0.6% +0.5% +0.3%	–13 to 6% –5 to 3% –3 to 2% +2 to 0%
VOLTAGE	TEFC	90% of Rated Volt.	1-10 HP 15-30 HP 40-75 HP 100-200 HP	–17 to 19%	–17 to 19%	–17 to 19%	+1.5% +1.0% +0.6% +0.3%	–1.5% –1.0% –0.6% –0.3%	+11 to 4% +2 to 0% +1 to 0% +1 to 0%
FREQ	DP & TEFC	105% of Rated Freq.	1 to 200 HP	–10%	–10%	–10%	–5%	+5%	Slight +
FREQ	DP & TEFC	95% of Rated Freq.	1 to 200 HP	+11%	+11%	+11%	+5%	–5%	Slight –

+ Increase
– Decrease

Reprinted with permission of Marathon Electric, Wausau, WI

TABLE 9.5 TYPICAL MOTOR PERFORMANCE VARIATIONS DUE TO POWER SUPPLY VARIATIONS

				AMPS				
				Starting (Locked Rotor)	Full-Load	No-Load	Effic. Full-Load	Temp. Rise Full-Load
		1% Unbalance		+1.5%	+8%	+13%	−2%	+2%
		2% Unbalance		+3%	+17%	+27%	−8%	+8%
VOLTAGE	DP	110% of Rated Volt.	1-10 HP	+10 to 12%	+8 to 4%	+25 to 37%	−6 to 1%	+17 to 9%
			15-30 HP		+4 to 1%	+32 to 37%	−1 to 0%	+7 to 3%
			40-75 HP		−0 to 2%	+37%	No Change	+2 to 0%
			100-200 HP		−3 to 5%	+36 to 30%	+0 to 0.3%	−0 to 4%
		90% of Rated Volt.	1-10 HP	−10 to 12%	−3 to +5%	−20 to 19%	+1 to 0%	−6 to +12%
			15-30 HP		+6 to 9%	−19 to 18%	−0 to 0.4%	+15 to 19%
			40-75 HP		+9 to 10%	−18 to 17%	−0.4 to 0.1%	+19 to 20%
			100-200 HP		+9 to 10%	−17 to 16%	−0.1%	+19 to 20%
	TEFC	110% of Rated Volt.	1-10 HP	+10 to 12%	+8 to −4%	+37 to 27%	−6 to 1%	+17 to −5%
			15-30 HP		−5 to 6%	+26 to 25%	−1 to 0%	−7 to 9%
			40-75 HP		−6 to 7%	+24 to 21%	No Change	−9 to 10%
			100-200 HP		−7 to 9%	+20 to 13%	+0 to 0.3%	−10 to 11%
		90% of Rated Volt.	1-10 HP	−10 to 12%	−3 to +5%	−20 to 16%	+1 to 0%	−6 to +15%
			15-30 HP		+6 to 11%	−14 to 12%	−0 to 0.4%	+19 to 23%
			40-75 HP		+12%	−12 to 10%	−0.4 to 0.1%	+24 to 23%
			100-200 HP		+11 to 9%	−10%	−0.1%	+24 to 23%
FREQ	DP & TEFC	105% of Rated Freq.	1 to 200 HP	−5 to 6%	Slight −	−5 to 6%	Slight +	Slight −
		95% of Rated Freq.	1 to 200 HP	+5 to 6%	Slight +	+5 to 6%	Slight −	Slight +

\+ Increase
− Decrease

Reprinted with permission of Marathon Electric, Wausau, WI

5. Add appropriate amount and type of grease using low-pressure grease gun;
6. Start motor and let run for approximately 30 minutes (this allows excess grease to drain out. If this is not done, pressure buildup in the bearing can blow out the bottom seal, allowing grease to run down the shaft onto the top of the motor windings.); and
7. Stop motor, wipe off any drained grease, replace filler and drain plugs.

Figure 9.27 shows a cross section of the bearing cap and filler/drain plugs.

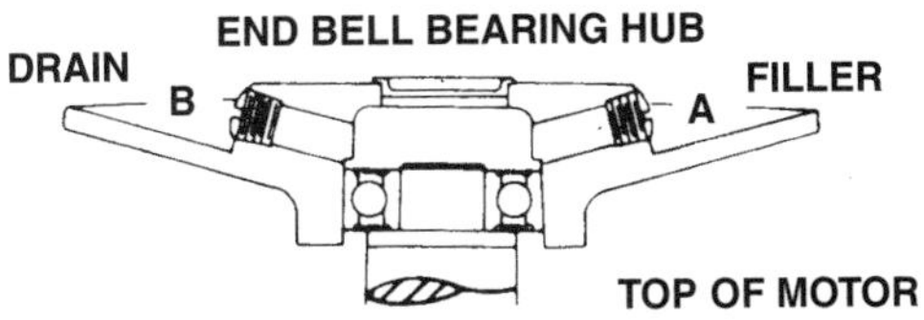

Fig. 9.27 Cross section of bearing cap and filler/drain plugs

Excessive greasing can cause as much damage to motor bearings as undergreasing. Therefore, manufacturer's recommendations should be followed without exception. The amount and frequency of greasing depends on a number of factors including RPM, operating temperature, duty cycle, and environmental conditions.

QUESTIONS

Write your answers in a notebook and then compare your answers with those on page 265.

9.2A What are the two types of induction motor construction typically encountered when dealing with A.C. induction motors?

9.2B List the five most common causes of electric motor failure.

9.2C List the three components of an insulation temperature rating.

9.2D How can the direction of rotation of a three-phase motor be changed?

9.2E How are motor starters controlled?

9.2F When are magnetic starters used?

9.2G What two types of overload protection devices are used on magnetic starters?

9.27 Troubleshooting

9.270 Step-by-Step Procedures

The key to effective troubleshooting is the use of practical step-by-step procedures combined with a common sense approach.

"NEVER TAKE ANYTHING FOR GRANTED"

1. Gather preliminary information. The first step in troubleshooting any motor control which has developed trouble is to understand the circuit operation and other related functions. In other words, what is supposed to happen, operate, and so forth when it's working right? Also what is it doing now? The qualified maintenance operator should be able to do the following:
   a. *KNOW WHAT SHOULD HAPPEN WHEN A SWITCH IS PUSHED:* When switches are pushed or tripped, what coils go in, contacts close, relays operate, and motors run?
   b. *EXAMINE ALL OTHER FACTORS:* What other unusual things are happening in the lift station (facility) now that this circuit doesn't work properly? Lights dimmed, other pumps ran faster, lights went out when it broke, everything was flooded, operators were hosing down area, and many other possible factors.
   c. *ANALYZE WHAT YOU KNOW:* What part of it is working correctly? Is switch arm tripped? Everything but this is all right, except pump gets plugged with rags frequently. Is it a mechanical failure or an electrical problem caused by a mechanical failure?
   d. *SELECT SIMPLE PROCEDURES:* To localize the problem, select logical ways that can be simply and quickly accomplished.
   e. *MAKE A VISUAL INSPECTION:* Look for burned wires, loose wires, area full of water, coil burned, contacts loose, or strange smells.
   f. *CONVERGE ON SOURCE OF TROUBLE:* Mechanical or electrical. Motor or control, whatever it might be. Electrical problems result from some type of mechanical failure.
   g. *PINPOINT THE PROBLEM:* Exactly where is the problem and what do you need for repair?
   h. *FIND THE CAUSE:* What caused the problem? Moisture, wear, poor design, voltage, or overloading?
   i. *REPAIR THE PROBLEM AND ELIMINATE THE CAUSE IF POSSIBLE:* If the problem is inside the switch gear or motors, call an electrician. Give the electrician the information you have regarding the equipment. Do not attempt electrical repairs unless qualified and authorized, otherwise you could cause additional damage to the equipment and possibly injure yourself.
2. Some of the things to look for when troubleshooting are given in the remainder of this section.

9.271 Troubleshooting Guide for Electric Motors

SYMPTOM	CAUSE	RESULT*	REMEDY
1. Motor does not start (switch is on and not defective)	a. Incorrectly connected	a. Burnout	a. Connect correctly per diagram on motor.
	b. Incorrect power supply	b. Burnout	b. Use only with correctly rated power supply.
	c. Fuse out, loose or open connection	c. Burnout	c. Correct open circuit condition.
	d. Rotating parts of motor may be jammed mechanically	d. Burnout	d. Check and correct: 1. Bent shaft 2. Broken housing 3. Damaged bearing 4. Foreign material in motor.
	e. Driven machine may be jammed	e. Burnout	e. Correct jammed condition.
	f. No power supply	f. None	f. Check for voltage at motor and work back to power supply.
	g. Internal circuitry open	g. Burnout	g. Correct open circuit condition.
2. Motor starts but does not come up to speed	a. Same as 1-a, b, c above	a. Burnout	a. Same as 1-a, b, c above.
	b. Overload	b. Burnout	b. Reduce load to bring current to rated limit. Use proper fuses and overload protection.
	c. One or more phases out on a 3-phase motor	c. Burnout	c. Look for open circuits.
3. Motor noisy electrically	a. Same as 1-a, b, c above	a. Burnout	a. Same as 1-a, b, c above.
4. Motor runs hot (exceeds rating)	a. Same as 1-a, b, c above	a. Burnout	a. Same as 1-a, b, c above.
	b. Overload	b. Burnout	b. Reduce load.
	c. Impaired ventilation	c. Burnout	c. Remove obstruction.
	d. Frequent starts or stops	d. Burnout	d. 1. Reduce number of starts or reversals. 2. Secure proper motor for this duty.
	e. Misalignment between rotor and stator laminations	e. Burnout	e. Realign.
5. Noisy (mechanically)	a. Misalignment of coupling or sprocket	a. Bearing failure, broken shaft, stator burnout due to motor drag	a. Correct misalignment.
	b. Mechanical unbalance of rotating parts	b. Same as 5-a	b. Find unbalanced part, then balance.
	c. Lack of or improper lubricant	c. Bearing failure	c. Use correct lubricant, replace parts as necessary.
	d. Foreign material in lubricant	d. Same as 5-c	d. Clean out and replace bearings.

9.271 Troubleshooting Guide for Electric Motors (continued)

SYMPTOM	CAUSE	RESULT*	REMEDY
5. Noisy (mechanically) (continued)	e. Overload	e. Same as 5-c	e. Remove overload condition. Replace damaged parts.
	f. Shock loading	f. Same as 5-c	f. Correct causes and replace damaged parts.
	g. Mounting acts as amplifier of normal noise	g. Annoying	g. Isolate motor from base.
	h. Rotor dragging due to worn bearings, shaft or bracket	h. Burnout	h. Replace bearings, shaft or bracket as needed.
6. Bearing failure	a. Same as 5-a, b, c, d, e	a. Burnout, damaged shaft, damaged housing	a. Replace bearings and follow 5-a, b, c, d, e.
	b. Entry of water or foreign material into bearing housing	b. Same as 6-a	b. Replace bearings and seals and shield against entry of foreign material (water, dust, etc.). Use proper motor.

SYMPTOM	CAUSED BY	APPEARANCE
1. Shorted motor winding	a. Moisture, chemicals, foreign material in motor, damaged winding	a. Black or burned coil with remainder of winding good.
2. All windings completely burned	a. Overload	a. Burned equally all around winding.
	b. Stalled	b. Burned equally all around winding.
	c. Impaired ventilation	c. Burned equally all around winding.
	d. Frequent reversal or starting	d. Burned equally all around winding.
	e. Incorrect power	e. Burned equally all around winding.
3. Single-phase condition	a. Open circuit in one line. The most common causes are loose connection, one fuse out, loose contact in switch.	a. If 1,800 RPM motor — four equally burned groups at 90° intervals. b. If 1,200 RPM motor — six equally burned groups at 60° intervals. c. If 3,600 RPM motor — two equally burned groups at 180° intervals. *NOTE:* If Y connected, each burned group will consist of two adjacent phase groups. If delta connected, each burned group will consist of one-phase group.
4. Other	a. Improper connection	a. Irregularly burned groups or spot burns.
	b. Ground	

* Many of these conditions should trip protective devices rather than burn out motors. Also, many burnouts occur within a short period of time after motor is started up. This does not necessarily indicate that the motor was defective, but usually is due to one or more of the above-mentioned causes. The most common causes of failure shortly after start-up are improper connections, open circuits in one line, incorrect power supply or overload.

9.272 Troubleshooting Guide for Magnetic Starters

TROUBLE	POSSIBLE CAUSE	REMEDY
CONTACTS		
Contact chatter	1. Broken *POLE SHADER*[7]	1. Replace.
	2. Poor contact in control circuit	2. Improve contact or use holding circuit interlock.
	3. Low voltage	3. Correct voltage condition. Check momentary voltage drop.
Welding or freezing	1. Abnormal surge of current	1. Use larger contactor and check for grounds, shorts, or excessive motor load current.
	2. Frequent *JOGGING*[8]	2. Install larger device rated for jogging service or caution operator.
	3. Insufficient contact pressure	3. Replace contact spring; check contact carrier for damage.
	4. Contacts not positioning properly	4. Check for voltage drop during start-up.
	5. Foreign matter preventing magnet from seating	5. Clean contacts.
	6. Short circuit	6. Remove short fault and check that fuse and breaker are right.
Short contact life or overheating of tips	1. Contacts poorly aligned, spaced or damaged	1. Do not file silver-faced contacts. Rough spots or discoloration will not harm contacts. Replace.
	2. Excessively high currents	2. Install larger device. Check for grounds, shorts, or excessive motor currents.
	3. Excessive starting and stopping of motor	3. Caution operators. Check operating controls.
	4. Weak contact pressure	4. Adjust or replace contact springs.
	5. Dirty contacts	5. Clean with approved solvent.
	6. Loose connections	6. Check terminals and tighten.
Coil, overheated	1. Starting coil may not kick out	1. Repair coil.
	2. Overload won't let motor reach minimum speed	2. Remove overload.
	3. Overvoltage or high ambient temperature	3. Check application and circuit.
	4. Incorrect coil	4. Check rating; if incorrect, replace with proper coil.
	5. Shorted turns caused by mechanical damage or corrosion	5. Replace coil.
	6. Undervoltage, failure of magnet to seal it	6. Correct system voltage.
	7. Dirt or rust on pole faces increasing air gap	7. Clean pole faces.

[7] *Pole Shader. A copper bar circling the laminated iron core inside the coil of a magnetic starter.*
[8] *Jogging. The frequent starting and stopping of an electric motor.*

9.272 ***Troubleshooting Guide for Magnetic Starters*** *(continued)*

TROUBLE	POSSIBLE CAUSE	REMEDY
Overload relays tripping	1. Sustained overload	1. Check for grounds, shorts or excessive motor currents. Mechanical overload.
	2. Loose connection on all or any load wires	2. Check, clean, and tighten.
	3. Incorrect heater	3. Replace with correct size heater unit.
	4. Fatigued heater blocks	4. Inspect and replace.
Failure to trip	1. Mechanical binding, dirt or corrosion	1. Clean or replace.
	2. Wrong heater, or heaters omitted and jumper wires used	2. Check ratings. Apply heaters of proper rating.
	3. Motor and relay in different temperatures	3. Adjust relay rating accordingly, or install temperature compensating relays.
MAGNETIC AND MECHANICAL PARTS		
Noisy magnet (humming)	1. Broken shading coil	1. Replace shading coil.
	2. Magnet faces not mating	2. Replace magnet assembly or realign.
	3. Dirt or rust on magnet faces	3. Clean and realign.
	4. Low voltage	4. Inspect system voltage and voltage dips or drops during start-up.
Failure to pick up and seal	1. Low voltage	1. Inspect system voltage and correct.
	2. Coil open or shorted	2. Replace.
	3. Wrong coil	3. Check coil number and voltage rating.
	4. Mechanical obstruction	4. With power off, check for free movement of contact and armature assembly. Repair.
Failure to drop out	1. Gummy substance on pole	1. Clean with solvent.
	2. Voltage not removed from coil	2. Check coil circuit.
	3. Worn or rusted parts causing binding	3. Replace or clean parts as necessary.
	4. Residual magnetism due to lack of air gap in magnet path	4. Replace worn magnet parts or align if possible.
	5. Welded contacts	5. Shorted circuit, grounded, overloaded.

9.273 *Trouble/Remedy Procedures for Induction Motors*

1. Motor will not start.

   Overload control tripped. Wait for overload to cool, then try to start again. If motor still does not start, check for the causes outlined below.

   a. Open fuses: test fuses.

   b. Low voltage: check nameplate values against power supply characteristics. Also check voltage at motor terminals when starting motor under load to check for allowable voltage drop.

   c. Wrong control connections: check connections with control wiring diagram.

   d. Loose terminal-lead connection: turn power off and tighten connections.

   e. Drive machine locked: disconnect motor from load. If motor starts satisfactorily, check driven machine.

   f. Open circuit in stator or rotor winding: check for open circuits.

   g. Short circuit in stator winding: check for short.

   h. Winding grounded: test for grounded wiring.

   i. Bearing stiff: free bearing or replace.

   j. Overload: reduce load.

2. Motor noisy.

   a. Three-phase motor running on single phase: stop motor, then try to start. It will not start on single phase. Check for open circuit in one of the lines.

   b. Electrical load unbalanced: check current balance.

   c. Shaft bumping (sleeve-bearing motor): check alignment and conditions of belt. On pedestal-mounted bearing check for play and axial centering of rotor.

   d. Vibration: driven machine may be unbalanced. Remove motor from load. If motor is still noisy, rebalance.

   e. Air gap not uniform: center the rotor and if necessary replace bearings.

   f. Noisy ball bearing: check lubrication. Replace bearings if noise is excessive and persistent.

   g. Rotor rubbing on stator: center the rotor and replace bearings if necessary.

   h. Motor loose on foundation: tighten hold-down bolts. Motor may possibly have to be realigned.

   i. Coupling loose: insert feelers at four places in coupling joint before pulling up bolts to check alignment. Tighten coupling bolts securely.

3. Motor at higher than normal temperature or smoking. (Measure temperature with thermometer or thermister and compare with nameplate value.)

   a. Overload: measure motor loading with ammeter. Reduce load.

   b. Electrical load imbalance: check for voltage imbalance or single-phasing.

   c. Restricted ventilation: clean air passage and windings.

   d. Incorrect voltage and frequency: check nameplate values with power supply. Also check voltage at motor terminals with motor under full load.

   e. Motor stalled by driven tight bearings: remove power from motor. Check machine for cause of stalling.

   f. Stator winding shorted or grounded: test windings by standard method.

   g. Rotor winding with loose connection: tighten, if possible, or replace with another rotor.

   h. Belt too tight: remove excessive pressure on bearings.

   i. Motor used for rapid reversing service: replace with motor designed for this service.

4. Bearings hot.

   a. End shields loose or not replaced properly: make sure end shields fit squarely and are properly tightened.

   b. Excessive belt tension or excessive gear side thrust: reduce belt tension or gear pressure and realign shafts. See that thrust is not being transferred to motor bearing.

   c. Bent shaft: straighten shaft or send to motor repair shop.

5. Sleeve bearings.

   a. Insufficient oil: add oil—if supply is very low, drain, flush, and refill.

   b. Foreign material in oil or poor grade of oil: drain oil, flush, and relubricate using industrial lubricant recommended by a reliable oil manufacturer.

   c. Oil rings rotating slowly or not rotating at all: oil too heavy; drain and replace. If oil ring has worn spot, replace with new ring.

   d. Motor tilted too far: level motor or reduce tilt and realign if necessary.

   e. Rings bent or otherwise damaged in reassembling: replace rings.

   f. Rings out of slot (oil ring retaining clip out of place): adjust or replace retaining clip.

   g. Defective bearings or rough shaft: replace bearings. Resurface shaft.

6. Ball bearings.

   a. Too much grease: remove relief plug and let motor run. If excess grease does not come out, flush and relubricate.

b. Wrong grade of grease: flush bearing and relubricate with correct amount of proper grease.

c. Insufficient grease: remove relief plug and grease bearing.

d. Foreign material in grease: flush bearing, relubricate; make sure grease supply is clean (keep can covered when not in use).

9.28 Records

Records are a very important part of electrical maintenance. They must be accurate and complete. Pages 196 and 197 are examples of typical record sheets. Most of the information you will need to complete these forms can be found on the manufacturer's data sheet or in the instruction manual.

Whenever a piece of equipment is changed, repaired, or tested, the work performed should be recorded on an equipment history card of some type. Complete, up-to-date equipment records will enable you to evaluate the reliability of your equipment and will provide the basis for a realistic preventive maintenance program.

QUESTIONS

Write your answers in a notebook and then compare your answers with those on page 265.

9.2H What is the key to effective troubleshooting?

9.2I What are some of the steps that should be taken when troubleshooting electrical equipment?

9.2J What kind of information should be recorded regarding electrical equipment?

9.29 Additional Reading

1. *BASIC ELECTRICITY* by Van Valkenburgh, Nooger & Neville, Inc. Obtain from Sams Technical Publishing Company, 5436 West 78th Street, Indianapolis, IN 46268, or call 800-428-7267. ISBN 0-7906-1041-8. Price, $32.95.

2. "Instrumentation" by Leonard Ainsworth, Chapter 9 in *ADVANCED WASTE TREATMENT*. Obtain from the Office of Water Programs, California State University, Sacramento, 6000 J Street, Sacramento, CA 95819-6025. Price, $45.00.

3. "Maintenance" by Parker Robinson, Chapter 18 in *WATER TREATMENT PLANT OPERATION*, Volume II. Obtain from the Office of Water Programs, California State University, Sacramento, 6000 J Street, Sacramento, CA 95819-6025. Price, $45.00.

4. *MAINTENANCE ENGINEERING HANDBOOK* by Higgins. Obtain from the McGraw-Hill Companies, Order Services, PO Box 182604, Columbus, OH 43272-3031. ISBN 0-07-028819-4. Price, $150.00, plus nine percent of order total for shipping and handling.

END OF
LESSON 1 OF 4
LESSONS
ON EQUIPMENT MAINTENANCE

Please answer the discussion and review questions next.

PUMP RECORD CARD

NAME______ MAKE______ MODEL______

TYPE______ SIZE______ SERIAL #______

ORDER NUMBER______ SUPPLIER______ DATE PURCHASED______

DATE INSTALLED______ APPLICATION______ PLANT #______

Name Plate Data and Pump Info	Stuffing Box Data	Motor Data
GPM ______	Diameter______ Depth______	Name______ Serial #______
TDH ______	Pack. Size___ Type______	H.P.______ Speed______
RPM ______	Length______ No. Rings______	Ambient° ______
Gage Press Disc___ ______	Lantern Ring______ Flushed______	RPM______ Frame______
Gage Press Suc ___ ______	Mech. Seal Name______ Size___	Volts______ Amps______
Shut off Press ___ ______	Type______	Phase______ Cycle______
Suction Head ___ ______	**Pump Materials**	Shaft Size______ Key______
Rotation ___ ______	Casing ______	Bearing Front______
Impeller Type______	Shaft______	Rear______
Impeller Dia.______	Wearing Rings Casing ______	Code______ Type______
Impeller Clear______	Wearing Rings Impeller______	Amps @ Max. Speed______
Coupl Type & Size______	Shaft Sleeve______	Amps @ Shut Off______
Front Brg #______	Slinger______	Control Data Info
Rear Brg #______	Shims______	Starter______
Lub Interval______	Gaskets______	NEMA Size______
Lubricant______	"O" Rings______	Cat. #______
Wearing Rings______	Brg. Seals Front______	Heater Size______
Shaft Sleeve Size______	Rear______	Rated @______
Pump Shaft Size______	Casing Wear Ring Size ID______	Control Voltage______
Pump Keyway______	OD______	Variable Speed Type______
______	Width______	Speed Max______
Other Related Information:	Impeller Wear Ring ID______	Speed Min______
	OD______	
	Width______	

MOTOR STARTERS Number________________

Title:____

Mfg.:____________________Address____________________

Style:______________Class______________Size______________

Type:______________ ______________ ______________

O.L. HEATERS

Style______________Code__________

Amps______________ __________

______________ __________

O.L. TRIP UNITS

Mfg:______________Style:__________

Type:______________ __________

Amps Range:__________ __________

CIRCUIT BREAKER

Mfg:______________Address____________________

Style:__________Frame:__________Volts__________Amps Setting__________

Cat. No.__________ __________ __________ __________

MOTOR Number________________

TITLE____

Mfg:____________________Address____________________

HP:__________Volts:__________Ser. No.__________Duty:__________

Phase:__________Amps:__________Frame:__________Temp:__________

Cycles:__________RPM:__________Type__________Class:__________

Code:__________S.F.:__________Model__________Spec.:__________

SO#__________S#__________Style:__________CSA App:__________

Form__________Spec.__________Shft. Brg.__________Rear Brg.__________

50 Cycle Data____

Suitable for 208V Network:__________

Additional data__________________

____

____

____

Connection Diagram

(6)	(5)	(4)		(6)	(5)	(4)
(7)	(8)	(9)		(7)	(8)	(9)
(1)	(2)	(3)		(1)	(2)	(3)

DISCUSSION AND REVIEW QUESTIONS

Chapter 9. EQUIPMENT MAINTENANCE

(Lesson 1 of 4 Lessons)

At the end of each lesson in this chapter you will find some discussion and review questions. The purpose of these questions is to indicate to you how well you understand the material in the lesson. Write the answers to these questions in your notebook before continuing.

1. Why should inexperienced, unqualified or unauthorized persons and even qualified and authorized persons be extremely careful around electrical panels, circuits, wiring and equipment?
2. What is the difference between direct current (D.C.) and alternating current (A.C.)?
3. What meters and testers are used to maintain, repair and troubleshoot electrical circuits and equipment? Discuss the use of each meter and tester.
4. What protective or safety devices are used to protect operators and equipment from being harmed by electricity?
5. Why must motor nameplate data be recorded and filed?
6. How would you attempt to find the cause of a pump motor that won't start?

CHAPTER 9. EQUIPMENT MAINTENANCE

(Lesson 2 of 4 Lessons)

9.3 PUMPS

Pumps are used to move wastewater in collection systems when a gravity system is not feasible or too costly. In the collection system they are used in lift stations where it is necessary to lift the water from a low elevation to a higher elevation. Pumps also are used in high-velocity cleaners for cleaning collection systems. They are used to dewater trenches or excavations while installing pipelines or repairing them. Automobiles have oil pumps and water pumps, while backhoes have hydraulic pumps.

These pumps vary in type, construction, and use. There are centrifugal pumps, reciprocating pumps, turbine pumps, diaphragm pumps, gear pumps, screw pumps, and many others. Each type of pump is designed for a particular job or application.

Pump selection is based on several factors. Important considerations should include the three factors listed below.

1. Type of material to be pumped:
   a. Wastewater,
   b. Storm water,
   c. Sludge, grit,
   d. Mud, rocks, sand,
   e. Oil, and
   f. Clear water.
2. Type of service required:
   a. High pressure—low volume,
   b. High pressure—high volume,
   c. Low pressure—low volume, and
   d. Low pressure—high volume.

3. Other factors should include:
   a. Reliability,
   b. Required maintenance,
   c. Parts availability,
   d. Efficiencies, and
   e. Operating conditions.

Generally, centrifugal pumps are used for pumping wastewater; piston or diaphragm pumps for heavy solids; gear pumps and piston pumps for high pressures; and turbine or propeller pumps for mixing air or chemicals.

Pumps are so varied in their construction, operation, and maintenance that it is necessary to consult each manufacturer's instruction book for information regarding your particular pump. Each and every type and manufacturer of pumps cannot be discussed in this section, but several of the common types used in wastewater collection will be studied.

To better understand pumps and parts, the following section "Let's Build a Pump" is presented. The pump described is a clear water centrifugal pump, not a wastewater pump, but there are many similarities. The major difference is that wastewater pumps have open impellers (big openings) instead of closed impellers (small openings).

9.30 Let's Build a Pump

This section is reproduced with the permission of Allis-Chalmers Corporation. Originally the material was printed in Allis-Chalmers Bulletin No. OBX62568.

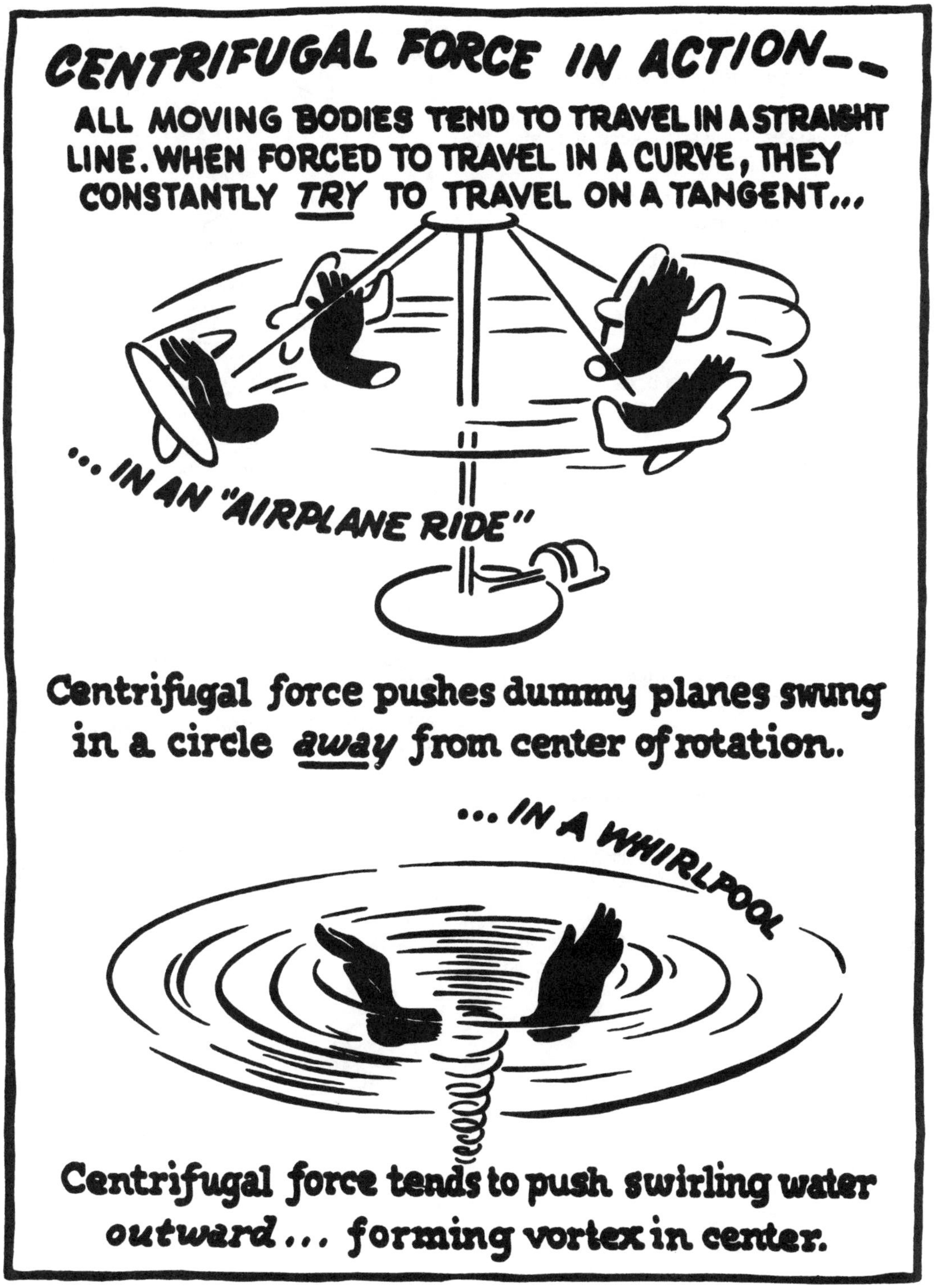

Let's Build a Pump!

A student of medicine spends long years learning exactly how the human body is built before attempting to prescribe for its care. Knowledge of PUMP anatomy is equally basic in caring for centrifugal pumps!

But whereas the medical student must take a body apart to learn its secrets, it will be far more instructive to us if we put a pump *TOGETHER* (on paper, of course). Then we can start at the beginning—adding each new part as we need it in logical sequence.

As we see *WHAT* each part does, *HOW* it does it . . . we'll see how it must be *CARED FOR!*

Another analogy between medicine and maintenance: there are various types of human bodies, but if you know basic anatomy, you understand them all. The same is true of centrifugal pumps. In building one basic type, we'll learn about *ALL* types.

Part of this will be elementary to some maintenance people . . . but they will find it a valuable "refresher" course, and, after all, maintenance just can't be too good.

So with a glance at the centrifugal principle on the previous page, let's get on with building our pump . . .

FIRST WE REQUIRE A DEVICE TO SPIN LIQUID AT HIGH SPEED . . .

That paddle wheel device is called the "impeller" . . . and it's the heart of our pump.

Note that the blades curve out from its hub. As the impeller spins, liquid between the blades is impelled outward by centrifugal force.

Note, too, that our impeller is open at the center—the "eye." As liquid in the impeller moves outward, it will suck more liquid in behind it through this eye . . . *PROVIDED IT'S NOT CLOGGED!*

That brings up Maintenance Rule No. 1: if there's any danger that foreign matter (sticks, refuse, etc.) may be sucked into the pump—clogging or wearing the impeller unduly—*PROVIDE THE INTAKE END OF THE SUCTION PIPING WITH A SUITABLE SCREEN.*

1ST. LINE OF PUMP DEFENSE IN
WATER CARRYING STICKS, ETC!

NOW WE NEED A SHAFT TO SUPPORT AND TURN THE IMPELLER . . .

Our shaft looks heavy—and it *IS.* It must maintain the impeller in precisely the right place.

But that ruggedness does *NOT* protect the shaft from the corrosive or abrasive effects of the liquid pumped . . . so we must protect it with sleeves slid on from either end.

THESE
SLEEVES
WILL
PROTECT THE SHAFT

What these sleeves—and the impeller, too—are made of depends on the nature of the liquid we're to pump. Generally they're bronze, but various other alloys, ceramics, glass, or even rubber-coating are sometimes required.

Maintenance Rule No. 2: *NEVER PUMP A LIQUID FOR WHICH THE PUMP WAS NOT DESIGNED.*

Whenever a change in pump application is contemplated and there's any doubt as to the pump's ability to resist the different liquid, *CHECK WITH YOUR PUMP MANUFACTURER!*

WE MOUNT THE SHAFT ON SLEEVE, BALL OR ROLLER BEARINGS . . .

As we'll see later, clearances between moving parts of our pump are *QUITE SMALL.*

If bearings supporting the turning shaft and impeller are allowed to wear excessively and lower the turning units within a pump's closely fitted mechanism, the life and efficiency of that pump will be seriously threatened.

Maintenance Rule No. 3: *KEEP THE RIGHT AMOUNT OF THE RIGHT LUBRICANT IN BEARINGS AT ALL TIMES. FOLLOW YOUR PUMP MANUFACTURER'S LUBRICATION INSTRUCTIONS TO THE LETTER.*

Main points to keep in mind are . . .

1. Although too much oil won't harm sleeve bearings, too much grease in antifriction type bearings (ball or roller) will *PROMOTE* friction and heat. *THE MAIN JOB OF GREASE IN ANTIFRICTION BEARINGS IS TO PROTECT STEEL ELEMENTS AGAINST CORROSION, NOT FRICTION.*
2. Operating conditions vary so widely that no one rule as to frequency of changing lubricant will fit all pumps. So play safe: if anything, change lubricant *BEFORE* it's too worn or too dirty.

TO CONNECT WITH THE MOTOR, WE ADD A COUPLING FLANGE . . .

Some pumps are built with pump and motor on one shaft and, of course, offer no alignment problem.

But our pump is to be driven by a separate motor . . . and we attach a flange to one end of the shaft through which bolts will connect with the motor flange.

Use a dial indicator to ensure shaft alignment (see pages 186 and 185, Figures 9.25 and 9.26).

Maintenance Rule No. 4: *SEE THAT PUMP AND MOTOR FLANGES ARE PARALLEL VERTICALLY AND AXIALLY . . . AND THAT THEY'RE KEPT THAT WAY!*

If shafts are eccentric or meet at an angle, every revolution throws tremendous extra load on bearings of both pump and motor. Flexible couplings will *NOT* correct this condition if excessive.

Checking alignment should be a regular procedure in pump maintenance. Foundations can settle unevenly, piping can change pump position, bolts can loosen. Misalignment is a *MAJOR* cause of pump and coupling wear.

NOW WE NEED A "STRAW" THROUGH WHICH LIQUID CAN BE SUCKED . . .

Notice two things about the suction piping: (1) the horizontal piping slopes UPWARD toward the pump; (2) any reducer which connects between the pipe and pump intake nozzle should be horizontal at the top (*ECCENTRIC,* not concentric).

This up-sloping prevents air pocketing in the top of the pipe . . . where trapped air might be drawn into the pump and cause loss of suction.

Maintenance Rule No. 5: *ANY DOWN-SLOPING TOWARD THE PUMP IN SUCTION PIPING (AS EXAGGERATED IN THE DIAGRAMS ABOVE) SHOULD BE CORRECTED.*

This rule is *VERY* important. Loss of suction greatly endangers a pump . . . as we'll see shortly.

WE CONTAIN AND DIRECT THE SPINNING LIQUID WITH A CASING . . .

We got a little ahead of our story in the previous paragraphs . . . because we didn't yet have the casing to which the suction piping bolts. And the manner in which it is attached is of great importance.

Maintenance Rule No. 6: *SEE THAT PIPING PUTS ABSOLUTELY NO STRAIN ON THE PUMP CASING.*

THE WEIGHT OF PIPING CAN EASILY RUIN A PUMP!

When the original installation is made, all piping should be in place and self-supporting before connection. Openings should meet with no force. Otherwise the casing is apt to be cracked . . . or sprung enough to allow closely fitted pump parts to rub.

It's good practice to check the piping supports regularly to see that loosening, or settling of the building, hasn't put strains on the casing.

NOW OUR PUMP IS ALMOST COMPLETE, BUT IT WOULD LEAK LIKE A SIEVE . . .

We're far enough along now to trace the flow of water through our pump. It's not easy to show suction piping in the cross-section view above, so imagine it stretching from your eye to the lower center of the pump.

Our pump happens to be a "double suction" pump, which means that water flow is divided inside the pump casing . . . reaching the eye of the impeller from either side.

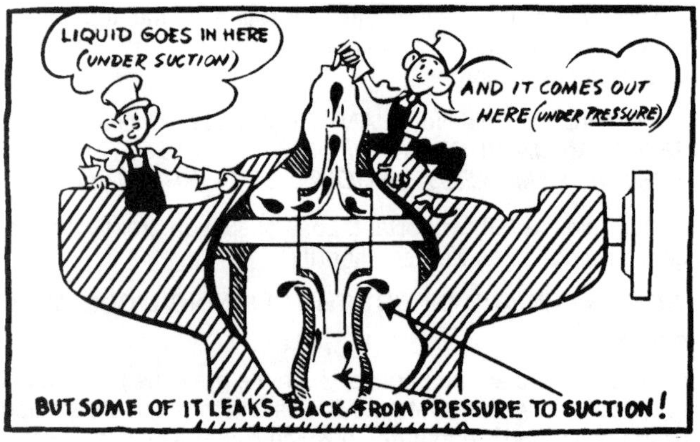

As water is drawn into the spinning impeller, centrifugal force causes it to flow outward . . . building up high pressure at the outside of the pump (which will force water *OUT*) and creating low pressure at the center of the pump (which will draw water *IN*). This situation is diagrammed in the upper half of the pump, above.

So far so good . . . except that water tends to be drawn back from pressure to suction through the space between impeller and casing—as diagrammed in the lower half of the pump, above—and our next step must be to plug this leak, if our pump is to be very efficient!

SO WE ADD WEARING RINGS TO PLUG INTERNAL LIQUID LEAKAGE . . .

You might ask why we didn't build our parts closer fitting in the first place—instead of narrowing the gap between them by inserting wearing rings (see Figure 9.28, page 206).

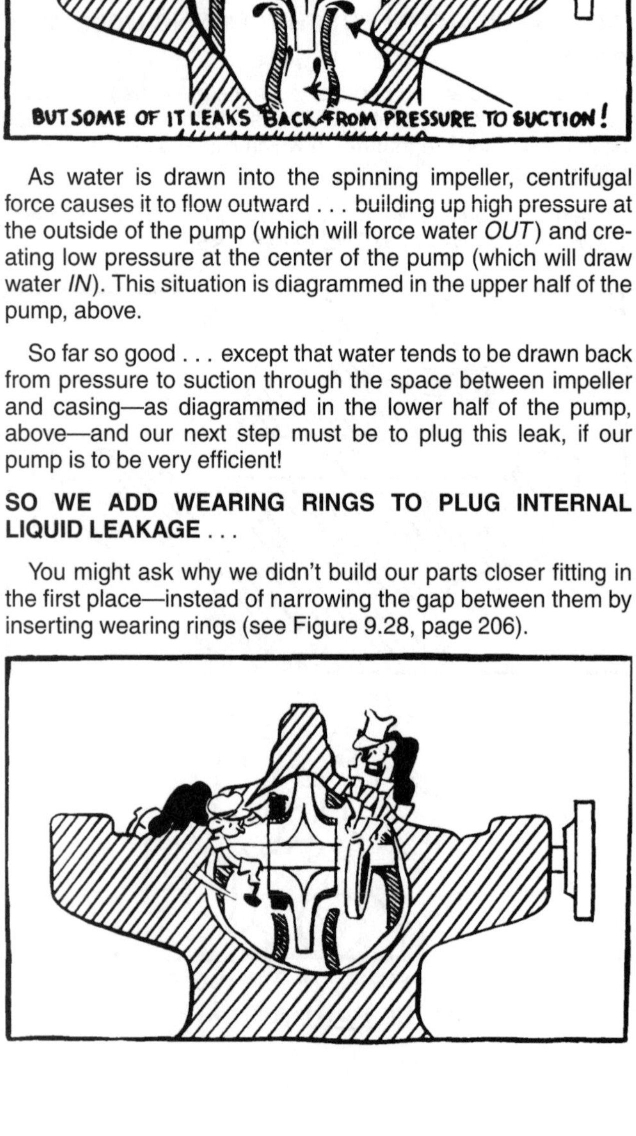

The answer is that those rings are removable and *REPLACEABLE* . . . when wear enlarges the tiny gap between them and the impeller. (Sometimes rings are attached to impeller rather than casing—or rings are attached to *BOTH* so they face each other.)

Maintenance Rule No. 7: *NEVER ALLOW A PUMP TO RUN DRY* (either through lack of proper priming when starting or through loss of suction when operating). Water is a *LUBRICANT* between rings and impeller.

Maintenance Rule No. 8: *EXAMINE WEARING RINGS AT REGULAR INTERVALS.* When seriously worn, their replacement will greatly improve pump efficiency.

TO KEEP AIR FROM BEING DRAWN IN, WE USE STUFFING BOXES . . .

We have two good reasons for wanting to keep air out of our pump: (1) we want to pump water, not air; and (2) air leakage is apt to cause our pump to lose suction.

Each stuffing box we use consists of a casing, rings of packing and a gland at the outside end. A mechanical seal may be used instead.

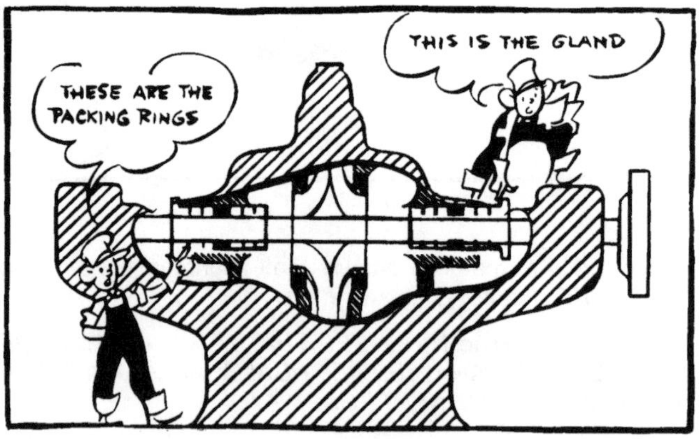

Maintenance Rule No. 9: *PACKING SHOULD BE REPLACED PERIODICALLY—DEPENDING ON CONDITIONS —USING THE PACKING RECOMMENDED BY YOUR PUMP MANUFACTURER.* Forcing in a ring or two of new packing instead of replacing worn packing is *BAD PRACTICE.* It's apt to displace the seal cage (see next page).

Put each ring of packing in separately, seating it firmly before adding the next. Stagger adjacent rings so the points where their ends meet do not coincide.

Maintenance Rule No. 10: *NEVER TIGHTEN A GLAND MORE THAN NECESSARY* . . . as excessive pressure will wear shaft sleeves unduly.

Maintenance Rule No. 11: *IF SHAFT SLEEVES ARE BADLY SCORED, REPLACE THEM IMMEDIATELY* . . . or packing life will be entirely too short.

TO MAKE PACKING MORE AIRTIGHT, WE ADD WATER SEAL PIPING . . .

In the center of each stuffing box is a "seal cage." By connecting it with piping to a point near the impeller rim, we bring liquid *UNDER PRESSURE* to the stuffing box.

This liquid acts both to block out air intake and to lubricate the packing. It makes both packing and shaft sleeves wear longer . . . *PROVIDING IT'S CLEAN LIQUID!*

WATER IS A LUBRICANT!

Maintenance Rule No. 12: *IF THE LIQUID BEING PUMPED CONTAINS GRIT, A SEPARATE SOURCE OF SEALING LIQUID SHOULD BE OBTAINED* (for example, it may be possible to direct some of the pumped liquid into a container and settle the grit out).

To control liquid flow, draw up the gland just tight enough so a *THIN* stream (approximately one drop per second) flows from the stuffing box during pump operation.

DISCHARGE PIPING COMPLETES THE PUMP INSTALLATION—AND NOW WE CAN ANALYZE THE VARIOUS FORCES WE'RE DEALING WITH . . .

SUCTION. At least 75% of centrifugal pump troubles trace to the suction side. To minimize them . . .

1. Total suction lift (distance between center line of pump and liquid level when pumping, plus friction losses) generally should not exceed 15 feet.
2. Piping should be at least a size larger than pump suction nozzle.
3. Friction in piping should be minimized . . . use as few and as easy bends as possible . . . avoid scaled or corroded pipe.

DISCHARGE lift, plus suction lift, plus friction in the piping from the point where liquid enters the suction piping to the end of the discharge piping equals total head.

PUMPS SHOULD BE OPERATED NEAR THEIR RATED HEADS.

Otherwise, the pump is apt to operate under unsatisfactory and unstable conditions which reduce efficiency and operating life of the unit.

Note the description of "cavitation" on the next page. Cavitation can seriously damage your pump.

PUMP CAPACITY generally is measured in gallons per minute. A new pump is guaranteed to deliver its rating in capacity and head.

But whether a pump *RETAINS* its actual capacity depends to a great extent on its maintenance.

Wearing rings must be replaced when necessary—to keep internal leakage losses down.

Friction must be minimized in bearings and stuffing boxes by proper lubrication . . . and misalignment must not be allowed to force scraping between closely fitted pump parts.

POWER of the driving motor, like capacity of the pump, will not remain at a constant level without proper maintenance.

Starting load on motors can be reduced by throttling or closing the pump discharge valve (*NEVER* the suction valve!) . . . but the pump must not be operated for long with the discharge valve closed. Power then is converted into friction—overheating the water with serious consequences.

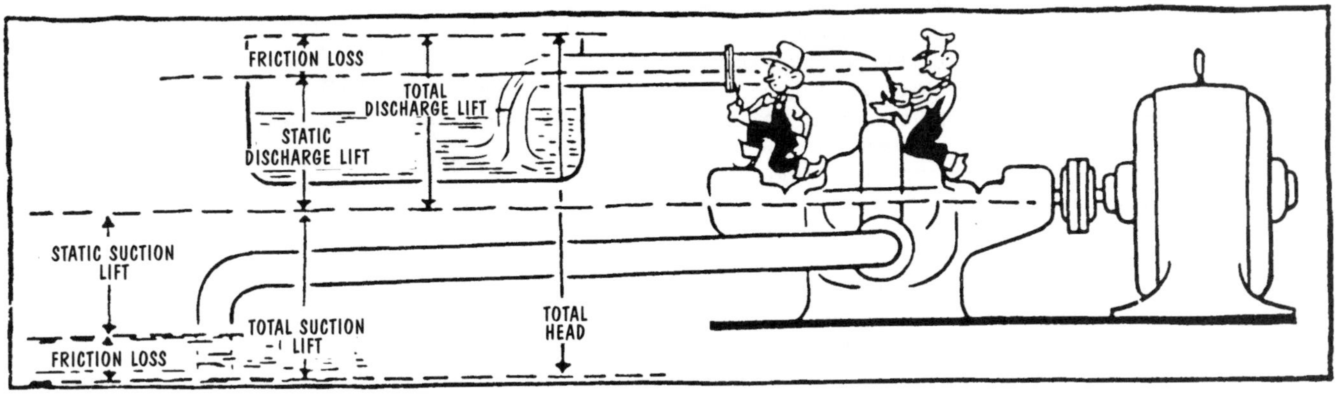

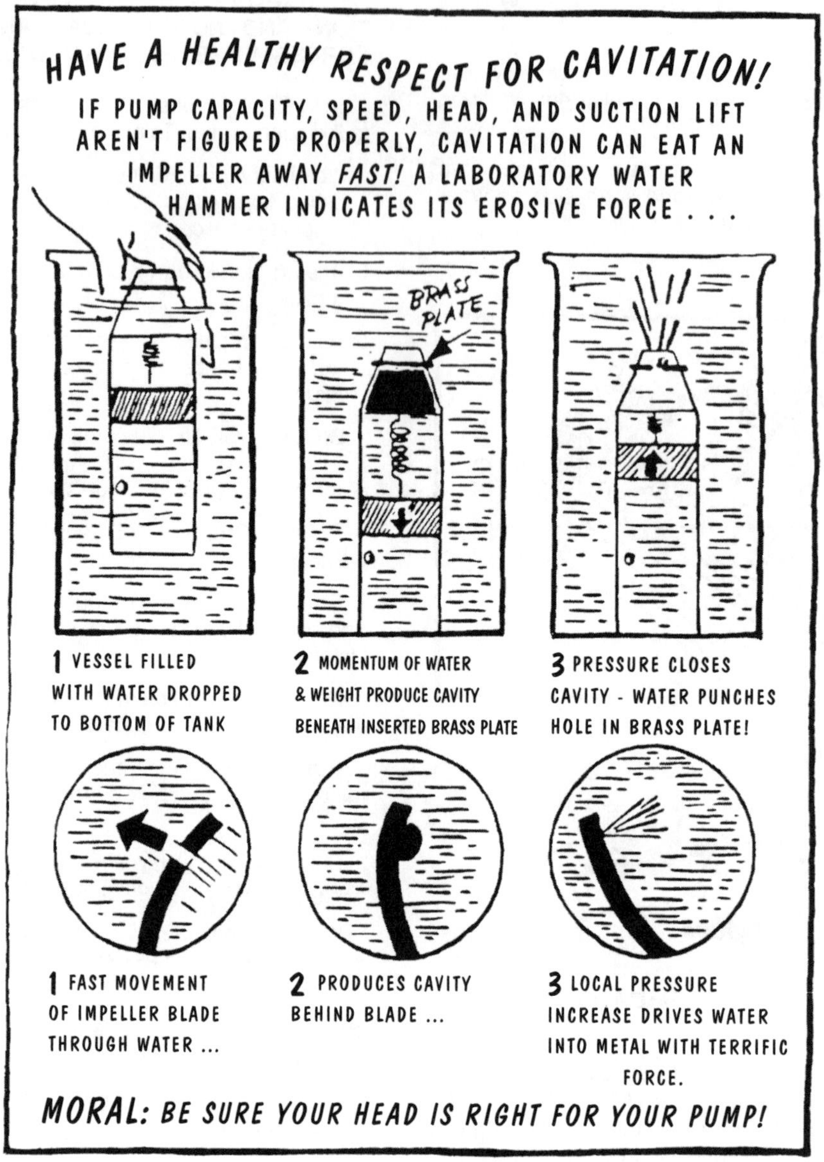
HAVE A HEALTHY RESPECT FOR CAVITATION!
IF PUMP CAPACITY, SPEED, HEAD, AND SUCTION LIFT AREN'T FIGURED PROPERLY, CAVITATION CAN EAT AN IMPELLER AWAY FAST! A LABORATORY WATER HAMMER INDICATES ITS EROSIVE FORCE . . .
BRASS PLATE
1 VESSEL FILLED WITH WATER DROPPED TO BOTTOM OF TANK
2 MOMENTUM OF WATER & WEIGHT PRODUCE CAVITY BENEATH INSERTED BRASS PLATE
3 PRESSURE CLOSES CAVITY - WATER PUNCHES HOLE IN BRASS PLATE!
1 FAST MOVEMENT OF IMPELLER BLADE THROUGH WATER ...
2 PRODUCES CAVITY BEHIND BLADE ...
3 LOCAL PRESSURE INCREASE DRIVES WATER INTO METAL WITH TERRIFIC FORCE.
MORAL: BE SURE YOUR HEAD IS RIGHT FOR YOUR PUMP!

CAVITATION

Cavitation is a condition that can cause a drop in pump efficiency, vibration, noise and rapid damage to the impeller of a pump. Cavitation occurs due to unusually low pressures within a pump. These low pressures can develop when pump inlet pressures drop below the design inlet pressures or when the pump is operated at flow rates considerably higher than design flows. When the pressure within the flowing water drops very low, the water starts to boil and vapor bubbles form. These bubbles then collapse with great force which knocks metal particles off the pump impeller. This same action can and does occur on pressure reducing valves and partially closed gate and butterfly valves. See Chapter 8, Section 8.163, "Basic Pump Hydraulics," for more information about the causes and effects of cavitation.

QUESTIONS

Write your answers in a notebook and then compare your answers with those on page 265.

9.3A What are some uses of pumps?

9.3B Why is it important to know how a pump is built?

9.3C When should pump lubricants be changed?

9.3D How can you tell if a pump is properly aligned?

9.3E Why must horizontal suction lift piping always slope upward toward the pump?

9.3F How can piping be prevented from putting a strain on the pump casing?

9.3G Why should a pump never be allowed to run dry?

9.31 Pump Types and Parts

After understanding the pump construction, you know that there are several components and parts. Most of these are similar in construction, material, and location in all manufacturers' pumps. To help you find the location of the different parts and troubleshoot pumps that don't perform properly, the following figures show cut-away views of different types of pumps, and the location and name of the pump parts.

1. Figure 9.28 Horizontal Nonclog Wastewater Pump with Open Impeller
2. Figure 9.29 Vertical Ball-Bearing Type Wastewater Pump
3. Figure 9.30 Propeller Pump
4. Figure 9.31 Incline Screw Pump
5. Figure 9.32 Pneumatic Ejector
6. Figure 9.33 Piston Pump on High-Velocity Cleaner
7. Figure 9.34 Submersible Wastewater Pump
8. Figure 9.38 Diaphragm (Pneumatic) Pump
9. Figure 9.39 Diaphragm (Pneumatic) Pump

Every manufacturer has an operation, maintenance, and repair manual for its particular pump. These manuals contain installation instructions, start-up procedures, operation information, maintenance program, lubrication schedules, repair procedures, parts list, and other valuable information. These should be in the file for each pump in your system. If you need a brochure for a particular pump, ask your sales representative or supplier, or write to the manufacturer. Very seldom is there a charge for these manuals, but, if so, it is minimal and well worth the investment.

9.310 Centrifugal Pumps

Centrifugal pumps designed for pumping wastewater usually have smooth channels and impellers with large-sized openings to prevent clogging. Figures 9.28 through 9.30 illustrate different types of centrifugal pumps.

9.311 Screw Pumps (*Figure 9.31*)

Screw pumps are another type of pump used to lift wastewater to a higher elevation. This pump consists of a screw operating at a constant speed within a housing or trough. The screw has a pitch and is set at a specific angle. When revolving, it carries the wastewater up the trough to a discharge point. The discharge is usually a free discharge. The screw is constructed of a steel tube with flights welded around it and looks very similar to a screw conveyor. The trough can be either steel or concrete depending upon size of screw and preference.

The screw is supported by two bearings, one at the top and one on the bottom. The upper bearing is generally a ball or roller bearing and the lower bearing is a sleeve bearing designed for underwater service. The bearings must be lubricated according to the manufacturer's recommendations. If either bearing fails, it must be replaced immediately before the screw rubs on the trough and causes damage.

Screw pumps are usually driven by helical (spiral) gear speed reducers powered by electric motors. The speed reducers must be filled with a proper oil recommended by the equipment manufacturer. The oil level should be checked according to the manufacturer's recommendations, such as a weekly measurement, and changed at least every six months.

Screw pumps should be inspected daily for bent flights, unusual noises, and general operating condition.

Screw pumps have fairly high maximum operating efficiencies, ranging as high as 75 to 80 percent. A unique characteristic is that they generally do not lose more than 10 to 15 percent of the maximum efficiency throughout the entire 85 to 90 percent of their upper capacity pumping range. They are a constant-speed pump which has the capability to pump from no flow to maximum flow directly dependent upon the rate of incoming flow to the lower end of the screw. Greater pumping rates are the result of increased submergence which will occur when higher flows are present, and lesser pumping rates are achieved by just the opposite conditions. Screw pumps are capable of handling large solids and rags without plugging. Wherever these pumps are installed, proper handrails, guards, and barriers should be installed to protect operators and equipment from entering the pumping area. Special precautions must be taken to avoid problems caused by splashing on trough sidewalls and steps. Good cleanup procedures are needed to avoid a slippery and odorous mess.

An anti-reverse rotational back stop is usually incorporated in the gear reducer to prevent the liquid trapped in the screw from driving it in the reverse direction when it runs back down the trough at the time of power cutoff.

WARNING

Before working on any part of this pump, be sure the electrical panel is tagged off and the motor main breaker is locked out.

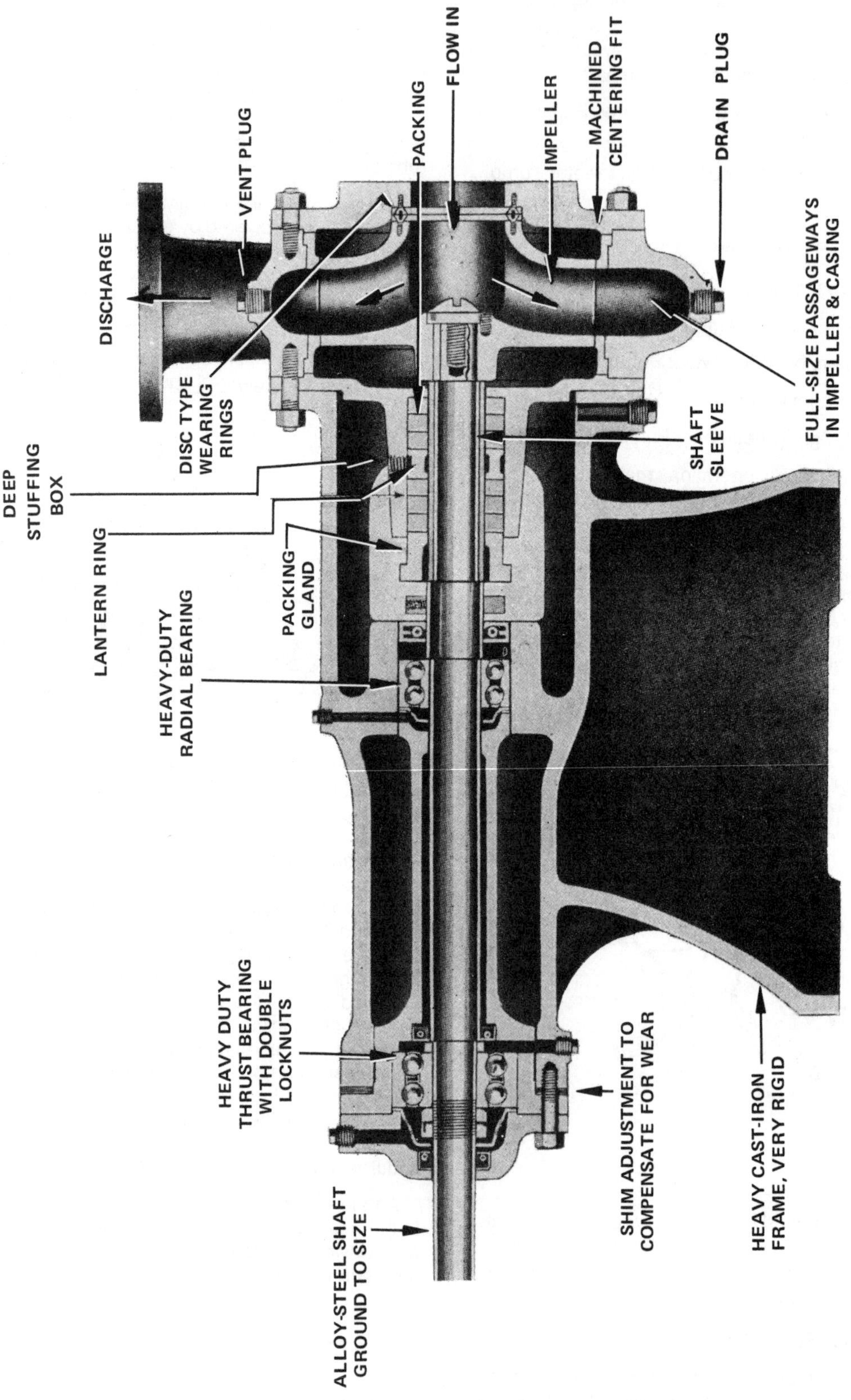
DISCHARGE
VENT PLUG
PACKING
FLOW IN
IMPELLER
MACHINED CENTERING FIT
DRAIN PLUG
FULL-SIZE PASSAGEWAYS IN IMPELLER & CASING
SHAFT SLEEVE
DISC TYPE WEARING RINGS
DEEP STUFFING BOX
LANTERN RING
PACKING GLAND
HEAVY-DUTY RADIAL BEARING
HEAVY DUTY THRUST BEARING WITH DOUBLE LOCKNUTS
ALLOY-STEEL SHAFT GROUND TO SIZE
SHIM ADJUSTMENT TO COMPENSATE FOR WEAR
HEAVY CAST-IRON FRAME, VERY RIGID

Fig. 9.28 Horizontal nonclog wastewater pump with open impeller
(Source: War Department Technical Manual TM5-666)

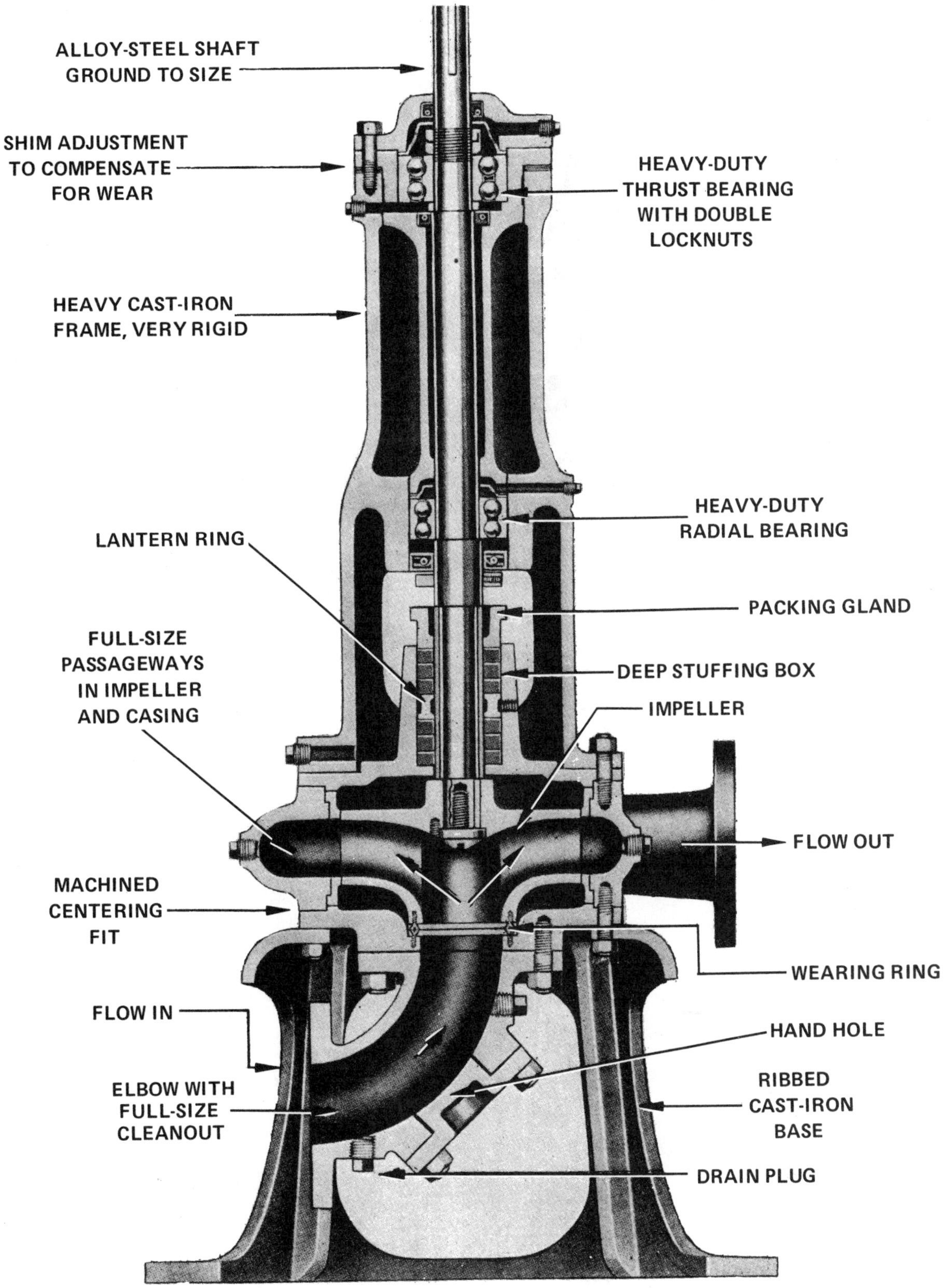
ALLOY-STEEL SHAFT
GROUND TO SIZE
SHIM ADJUSTMENT
TO COMPENSATE
FOR WEAR
HEAVY-DUTY
THRUST BEARING
WITH DOUBLE
LOCKNUTS
HEAVY CAST-IRON
FRAME, VERY RIGID
HEAVY-DUTY
RADIAL BEARING
LANTERN RING
PACKING GLAND
FULL-SIZE
PASSAGEWAYS
IN IMPELLER
AND CASING
DEEP STUFFING BOX
IMPELLER
FLOW OUT
MACHINED
CENTERING
FIT
WEARING RING
FLOW IN
HAND HOLE
ELBOW WITH
FULL-SIZE
CLEANOUT
RIBBED
CAST-IRON
BASE
DRAIN PLUG

Fig. 9.29 Vertical ball bearing-type wastewater pump

(Source: War Department Technical Manual TM5-666)

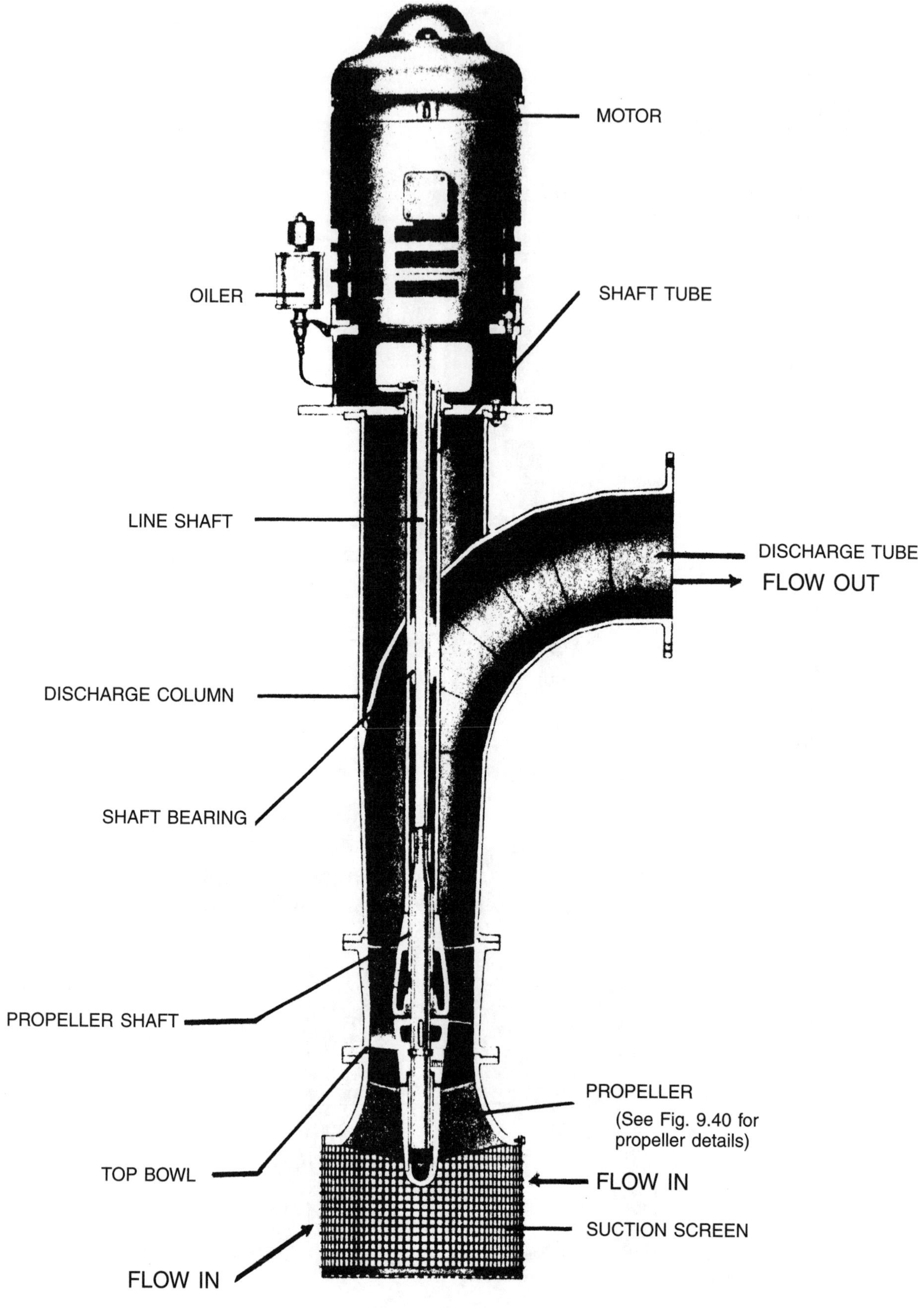

Fig. 9.30 Propeller pump
(Source: Unknown)

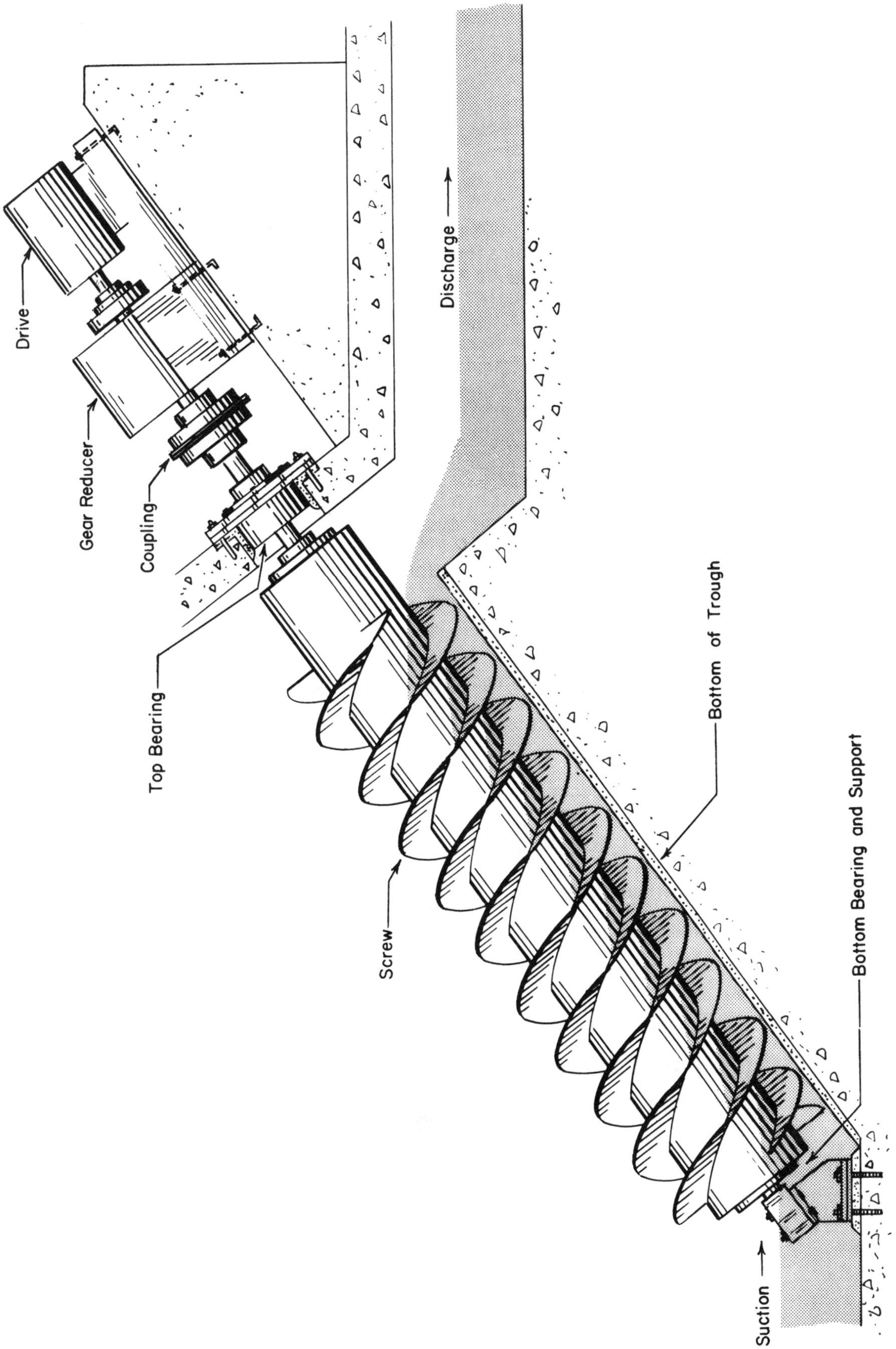

Fig. 9.31 Incline screw pump
(Courtesy of FMC Corporation, Environmental Equipment Division)

9.312 Pneumatic Ejectors (*Figure 9.32*)

Pneumatic ejectors are used when it is necessary to handle limited or low flows with relatively large solids. Centrifugal pumps are highly efficient for pumping large flows; however, when scaled down for lower flows, they tend to plug easily. There is a restriction of flow in a centrifugal pump because the impeller opening is too small to pass unstrained solids. Pneumatic ejectors, on the other hand, are capable of passing solids up to the size of the inlet and discharge valves, and there is nothing on the inside of the ejector-receiver to restrict the flow.

The pneumatic ejector is a clean, reliable device for pumping wastewater. The ejector consists simply of two check valves, an air valve, and a receiver or tank. Some designs may contain a float device or ball. Wastewater flows into the receiver by gravity. When the ejector is filled, a control admits compressed air to force out the wastewater in the receiver.

Pneumatic ejector systems fall into four categories:

1. Tankless System. A compressor with no intermediate air tank is supplied with each ejector, and starts each time the unit is filled.
2. Package System. This is an ejector with the compressor mounted above the receiver and without an air storage tank. The controls and compressor are prewired and mounted at the factory.
3. Stored Air System. In this system, an air tank is used between the compressor and the ejector. This arrangement has the advantage that more than one compressor can be used for standby.
4. Plant Air System. If an ample supply of air is available from a wastewater treatment plant, this air can be used rather than installing another air compressor.

The pneumatic ejector has a built-in level control device that produces an intermittent discharge. This control is either electrically or pneumatically operated, depending on the job requirements.

Maintenance of pneumatic ejectors is not as complicated as maintenance of most pumps; however, maintenance must be performed when scheduled. Clean the interior of the pots to remove grease. Inspect valves, floats and/or electrical controls for proper operation. If a failure occurs and wastewater enters the air lines or air control valves, dismantle and clean. The motor and air compressor require regular preventive maintenance and overhaul.

OPERATION PROBLEM CHECKLIST

Ejector Will Not Discharge

Cause:	Procedure and Cure:
1. Ejector not filling	Inlet gate valve and check valve closed or plugged. Open and/or clean.
2. Insufficient air pressure or volume	One psi of pressure for each 2 feet of discharge head is the minimum air pressure needed for discharge. The amount of air determines the discharge time. Remember that the inlet check valve is closed during discharge.
3. Pilot valve not functioning	a. Counterweight must be in the down position to discharge ejector. The counterweight is a weight on the end of a lever on the check valve that helps the check valve operate properly. b. Accumulation of grease and foreign matter on upper bell preventing proper travel to shift the pilot valve. Check top of bell and clean as necessary through access opening in wastewater receiver. c. Check air supply.
4. Strainer and air line to receiver plugged	Remove and clean.
5. Discharge gate or check valve closed or plugged	Open and clean.
6. Discharge line plugged	Hand rod discharge line.
7. Counterweight stuck	Packing too tight. Loosen and finger-tighten only.
8. Grease on probes	Clean probes.
9. Faulty switching circuit	Find cause of problem and repair.

Short Cycling of Pneumatic Ejector

Cause:	Procedure and Cure:
1. Inlet check valve leaks and allows liquid to be pumped back into low level manhole causing excess head in manhole	Inspect check valve for foreign matter under valve seat and/or worn seat. Remove and repair or replace.
2. Discharge check valve leaks permitting liquid discharged to flow back to the ejector receiver	Same as above except for discharge check valve.
3. Counterweight too close to guide	Adjust counterweight out farther on the lever arm.
4. Packing around float linkage too loose	Finger-tighten packing and check adjustment of counterweight.

Ejector Will Not Fill

Cause:	Procedure and Cure:
1. Inlet check valve or gate valve closed or clogged	Open and clean.
2. Clogged strainer between main valve and receiver preventing venting of receiver	Clean or replace screen.
3. Air exhaust line clogged	Clean.
4. Sewer plugged upstream from ejector	Clear sewer.
5. No flow in sewer	Discharge water from water truck or fire hydrant into upstream manhole when you wish to operate ejector.

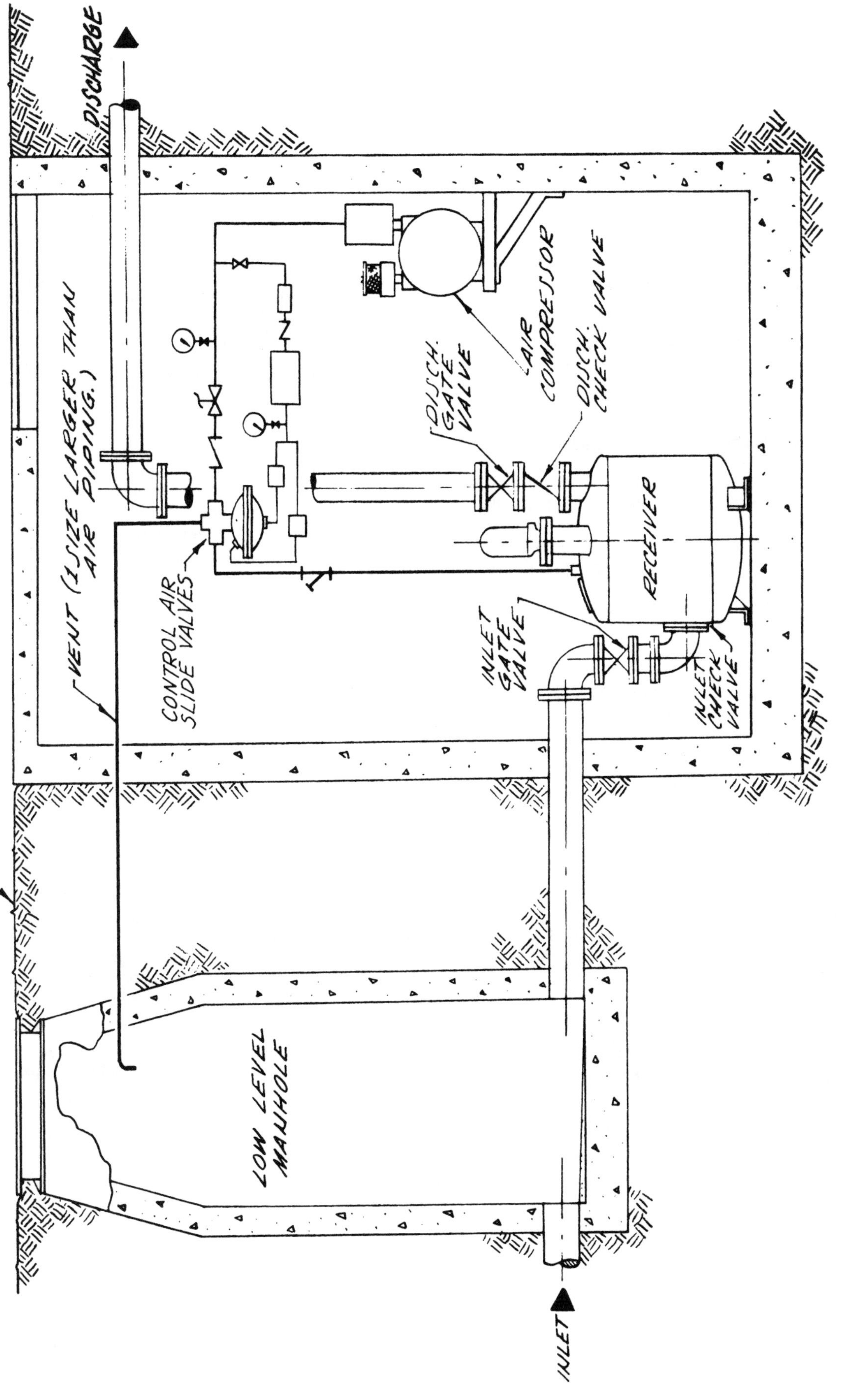

Fig. 9.32 Pneumatic ejector system
(Courtesy of James Equipment and Manufacturing Company)

RECOMMENDED PREVENTIVE MAINTENANCE SCHEDULE

DAILY: Observe several cycles assuring that counterweight action is proper.

WEEKLY: Hold down counterweight approximately 2 to 3 minutes and check general operation.

QUARTERLY: Inspect operation of all valves.

SEMIANNUALLY: Clean air strainer between ejector receiver and main air valve.

ANNUALLY: General inspection of all equipment, cleaning and painting all items as necessary.

Every five years, open the inspection port on the ejector receiver. Inspect and clean as necessary. Examine inlet and discharge check valves, main air valve and pilot valve.

9.313 Piston Pumps (Figure 9.33)

Piston pumps are commonly used in high-velocity cleaners. The maintenance operator will be required to operate, maintain, and service these pumps. Piston pumps, no matter what manufacturer, operate on the same principle. There is a fluid cylinder, piston, intake valve, and discharge valve. The piston is connected by a rod to a crankshaft. As the crankshaft is turned, the piston travels to the bottom of the fluid cylinder. When this happens, it draws water into the fluid cylinder due to the negative pressure. As the piston travels to the top, it pushes or pumps the water out the discharge valve. A pump may have two fluid cylinders (duplex), three fluid cylinders (triplex), four fluid cylinders (quadruplex), or five fluid cylinders (quintuplex) and the cylinder may be mounted either horizontally or vertically. The number of fluid cylinders varies with pressure and gallons per minute needed.

Three other factors also determine the pump capacity—the number of strokes per minute, length of stroke, and diameter of the piston or plunger. The larger piston or stroke allows more water to enter, and the "Speed" gives it more strokes per minute, therefore pumping more water. Unlike a centrifugal pump, the pressure is not determined by the speed. A piston pump is a positive displacement pump. This means that what enters the pump on the suction stroke must be discharged on the discharge stroke, if the valves open and close properly. Because of this, *THE DISCHARGE VALVE ON AN OPERATING PISTON PUMP MUST NEVER BE CLOSED.* If it is, the pressure will continue to build until the relief valve is activated, the engine is stalled or something breaks. The pressure developed by a piston pump depends on the size of the piston, the force applied to the piston, and the lack of leakage around the valves. Piston pumps should have suction lines at least as large as the pump suction.

On the discharge side of the pump there is a relief valve. This is installed to protect pumps from failure due to excessive pressure. Flows are usually bypassed back to the suction side of the pump when excessive pressures develop. This relief valve must be set according to factory specifications for each particular pump. Failure to have a properly set relief valve can result in loss of life, property, or irreparable damage. Also located on the pump is a pulsation dampener. This is a precharged unit in which an air pocket absorbs the shock of pulsations that occur when the pump is operating. This unit provides longer life to valves, valve springs, and plungers by reducing surges.

Piston pumps also have packing, although it is not usually a problem. They use a hard core packing that must be kept tight at all times and lubricated. The packing is usually greased daily with a lithium-base grease, or with water pump grease. Use a manually operated grease gun. If a pressure grease gun is used, the pressure can become too great and you could lose the packing seals. Some pumps of this type are equipped with force feed oilers, and these must be kept full. A universal-type packing is sometimes used, and this type of packing must be installed so the lips of each sealing ring face the fluid pressure.

Disc or flap valves and valve springs should be inspected periodically for wear. These valves usually consist of a valve seat, valve disc, and spring. The valves can be replaced in most models of pumps, although special tools are required. By removing the valve cover plug, the valves can be inspected. Look for broken springs; worn, bent, or broken discs; or seat faces which are corroded, cut, or grooved. Sometimes the discs can be flipped over when worn excessively on one side. Seat faces may show uniform wear and still be satisfactory. The springs and discs are usually easily replaced, but seats require special pullers, *MANDRELS*[9] and tools. If you replace seats, take special care not to damage the seating face. This can be done by holding an old disc over the new seat when tapping the new disc into the pump body. Valve replacement or repair will be necessary when loss of pressure, volume, or an unusual knocking is noticed.

The crankshaft, pistons, rods, and other parts of the piston pump usually don't cause problems when properly used and maintained. To completely overhaul the pump, each manufacturer's literature must be followed. Overhauling a pump requires a high degree of mechanical aptitude and experience.

---

[9] *Mandrel (MAN-drill) (1) A special tool used to push bearings in or to pull sleeves out. (2) A testing device used to measure for excessive deflection in a flexible conduit.*

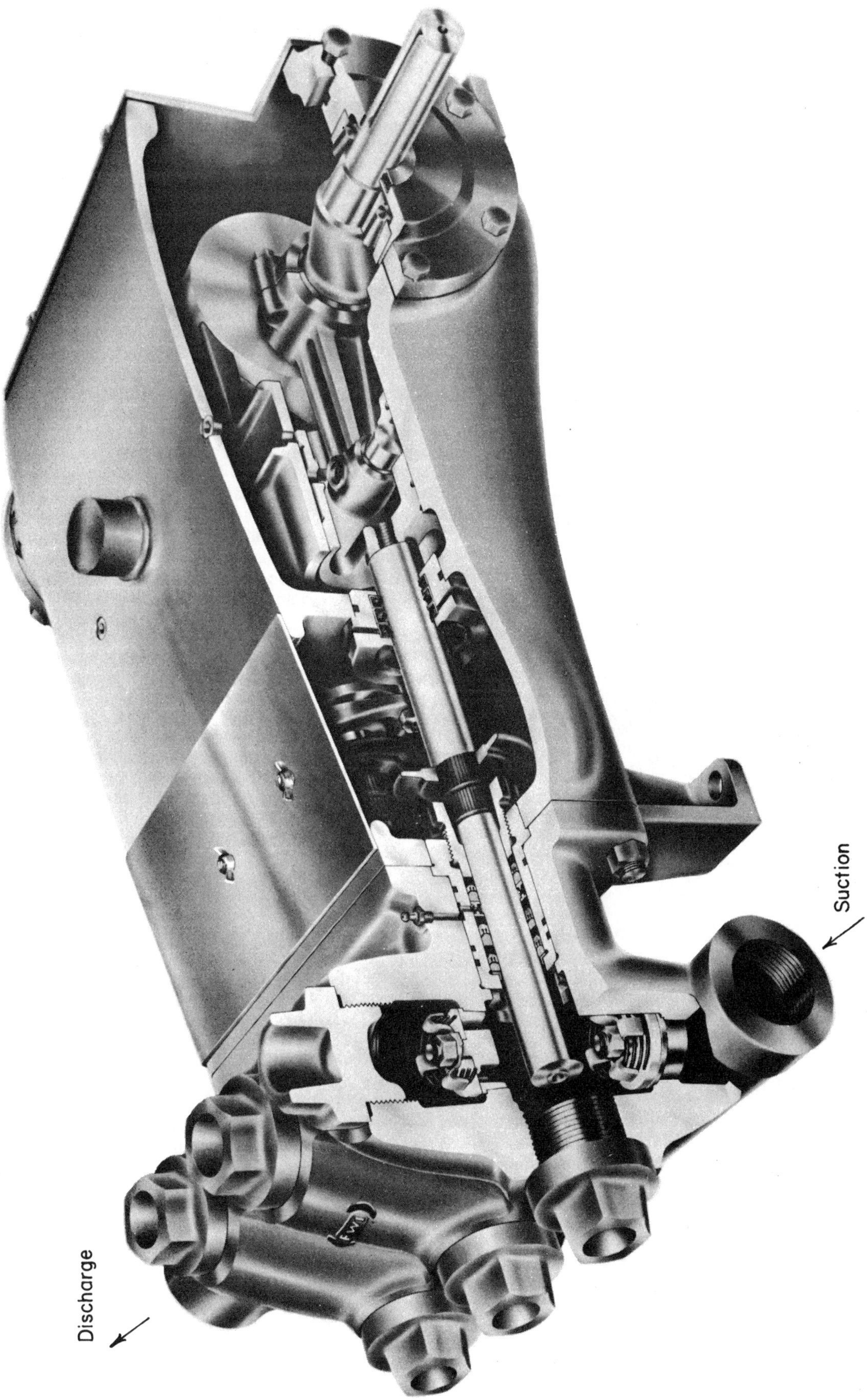

Fig. 9.33 Piston pump on high-velocity cleaner
(Courtesy Wheatley Company, Tulsa, OK)

Maintenance required on piston pumps:

DAILY
1. Determine oil level in crankcase,
2. Lubricate packing,
3. Inspect packing for leakage, and
4. Keep pump clean.

MONTHLY
1. Wash crankcase filter,
2. Inspect and tighten all nuts and bolts, if needed,
3. Inspect gaskets for oil leaks, and
4. Thoroughly observe operation of all equipment.

SEMIANNUALLY
1. Change oil in crankcase,
2. Inspect valves,
3. Inspect for excessive play in crankshaft bearings and connecting rod bearings, and
4. Clean and touch up paint.

Each manufacturer has lubrication and maintenance brochures and these must be followed for your particular pump.

9.314 Close-Coupled Pumps

Close-coupled pumps differ from the other pumps because the motor and pump are one unit. The motor has an extended shaft and has provisions for mounting the volute surrounding the impeller on the end of the motor. The extended shaft has the impeller mounted on it. On this type of pump there is only one shaft which is the motor shaft; therefore, there isn't any coupling. The pump runs at the same speed as the motor.

The advantages of this pump include the fact that the size permits installation in tight spaces and the combined unit is less expensive than a separate motor and pump. Some common uses of close-coupled pumps are in freshwater systems, recirculating water systems and hot water systems.

Close-coupled pumps may have packed or mechanical seals. Mechanical seals probably are more common. These types of pumps are supported by the motor bearings so the only lubrication needed is on the motor. Some of these pumps have three points of lubrication: one point at each end of the motor and one point on the packing. When working on close-coupled pumps, special care must be taken to keep the mating flanges clean when assembling so the shaft is properly installed in the housing.

Some of the limitations of close-coupled pumps are:

1. If the shaft is scored on the pump and can't be repaired, a rotor for the motor must be purchased since it is all one piece (some smaller motor shafts can be forced out and a new shaft installed), and
2. Costs for disassembly and reassembly of a damaged motor may be more than the costs of a new pump and motor.

9.315 Submersible Pumps (*Figure 9.34*)

The submersible pump is similar to the close-coupled pump in many ways. The motor used is waterproof and mechanical seals are used for keeping out the water being pumped. This pump is completely submersed in the liquid being pumped, and also uses this liquid for cooling the motor. Some older installations require water over the motor at all times. New submersible pumps have thermal switches embedded in the windings to shut off the pump if it gets too hot. These thermal switches must be connected into the control circuit for this purpose.

Submersible pumps need very little daily maintenance. Since they are submerged in water, you cannot easily get to them for cleaning. They all have mechanical seals so there isn't any packing adjustment. If they become plugged with rags or debris, usually the pump has to be removed from the pit. Some of the newer submersible installations have lifting mechanisms and slide fittings on the discharge so they can be removed without pumping the wet well empty.

One of the limitations with submersible pumps is burnout of the motors. When the mechanical seal starts leaking, water can enter the motor and cause a motor to burn out. Repair is very expensive. Proper preventive maintenance can prevent motor burnout and includes the items listed below.

1. Check the motor monthly with a megger to see if moisture is entering and damaging the insulation.
2. If the submersible pump ever trips the main circuit breaker, do not reset it until cause is found.
3. Remove the submersible pump at regular intervals and replace the mechanical seal before it fails.
4. Keep grit in wet well to a minimum to reduce wear. Bucket out the grit on a weekly basis if necessary.
5. Take amperage tests every three months to determine if pump is starting to overload. Higher than normal amperage readings can indicate an overload.
6. If the motor operates in an oil bath and has an inspection plug, be sure that the bath is full of oil and not an oil-water emulsion (droplets of water in and on the oil). If the oil is emulsified, the seals are leaking. When this happens, the seals and bearings need attention.

QUESTIONS

Write your answers in a notebook and then compare your answers with those on page 265.

9.3H List three types of centrifugal pumps.

9.3I How does a screw pump work?

9.3J Under what conditions would a pneumatic ejector be used?

9.3K What type of pump is used in a high-velocity cleaner?

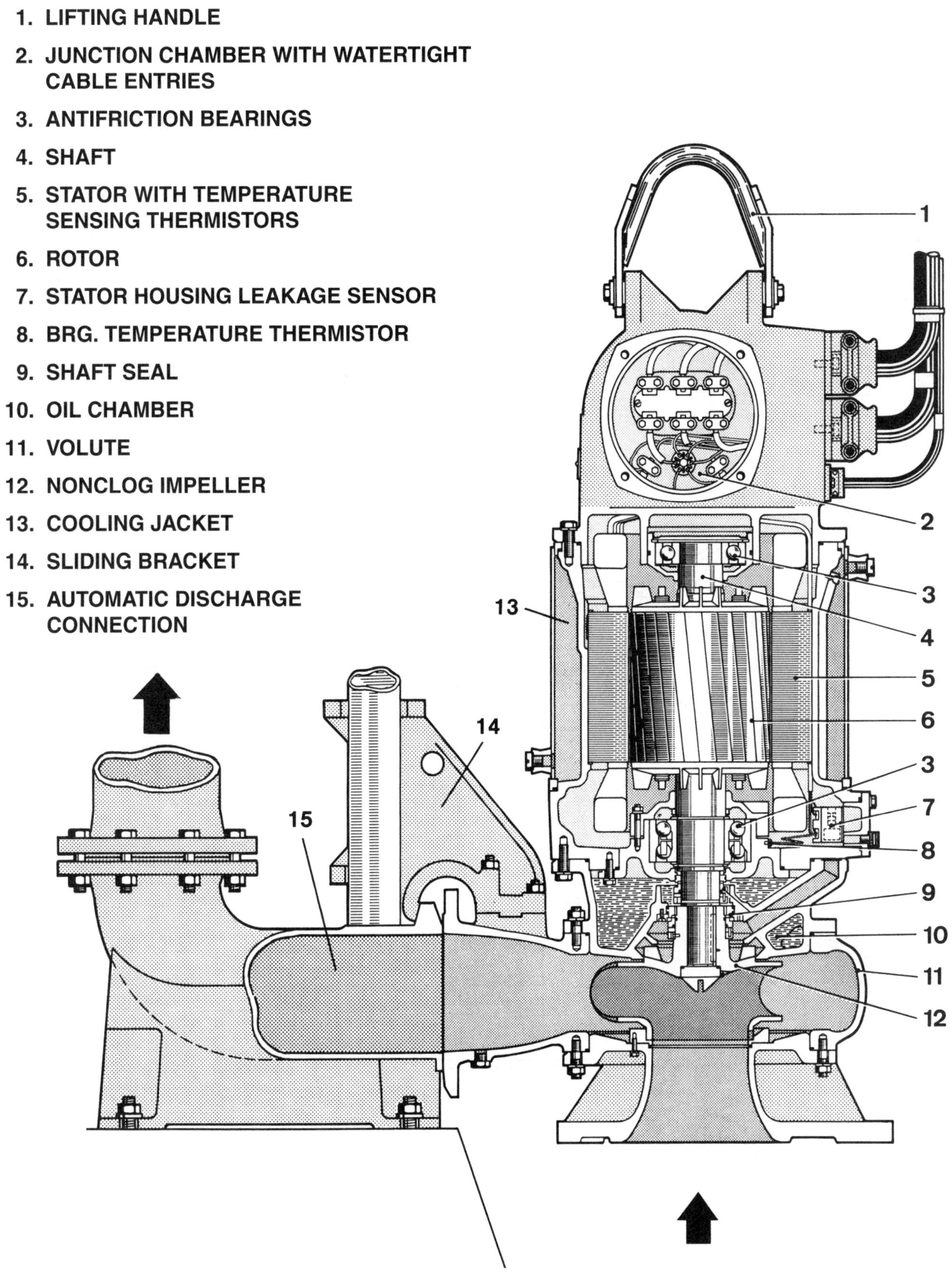
1. LIFTING HANDLE
2. JUNCTION CHAMBER WITH WATERTIGHT CABLE ENTRIES
3. ANTIFRICTION BEARINGS
4. SHAFT
5. STATOR WITH TEMPERATURE SENSING THERMISTORS
6. ROTOR
7. STATOR HOUSING LEAKAGE SENSOR
8. BRG. TEMPERATURE THERMISTOR
9. SHAFT SEAL
10. OIL CHAMBER
11. VOLUTE
12. NONCLOG IMPELLER
13. COOLING JACKET
14. SLIDING BRACKET
15. AUTOMATIC DISCHARGE CONNECTION
1
2
3
4
5
6
3
7
8
9
10
11
12
13
14
15

Fig. 9.34 Submersible wastewater pump

(Courtesy of Flygt Corporation)

9.316 Centrifugal Trash Pumps

Centrifugal pumps designed for use as portable pumps are often referred to as trash pumps because the water being pumped is not clean and may contain suspended solids of various sizes. When setting up your pump, always prime the pump with clean water. If the priming water contains soap or detergents, problems can develop if a high suction lift exists.

Always locate the pump as near as possible to the surface of the water being pumped (Figure 9.35). A high suction lift will dramatically reduce pump discharge volume (Figure 9.36). A centrifugal pump in good condition can perform satisfactorily with a suction lift or vacuum up to 18 inches of mercury. This corresponds to a possible suction lift of 20 feet. Considerable effort is required to start portable centrifugal pumps with excessive suction lift; however, these pumps can operate with considerable suction lift at decreasing flow rates as the water level drops.

When setting up a portable pump, lay out the suction and discharge hoses as straight as possible to reduce friction losses through kinks and bends in the hoses. The suction hose has a strainer (Figure 9.37) attached to the entrance to prevent pulling rocks and debris into the pump to avoid damaging the pump or plugging the hoses or pipes. Placing the end of the suction hose in a coarse rock sump or bucket will prevent material from building up on the strainer.

The most frequent pumping problem is an air leak in the suction hose or a connection on the suction side of the pump. To test that the problem is not in the pump, prime the pump, start the engine or motor, and place your hand over the suction inlet. The vacuum created by the pump should draw on your palm with a strong force if there are no leaks. You can also cap the suction inlet and install a vacuum gage on the suction side of the pump to indicate pump operation.

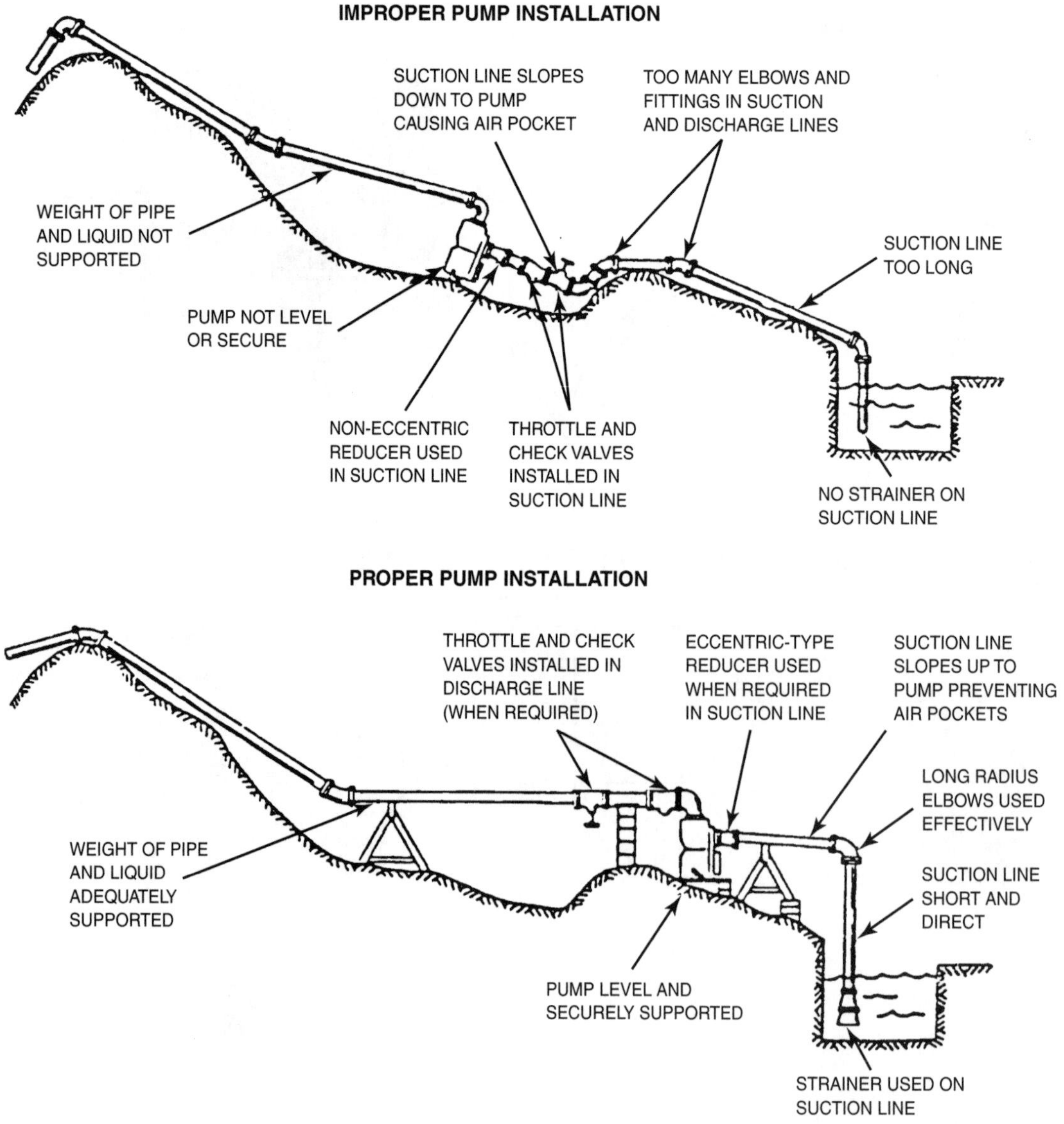

Fig. 9.35 Examples of proper and improper portable pump installations

(Permission of the Gorman-Rupp Company, Mansfield, OH)

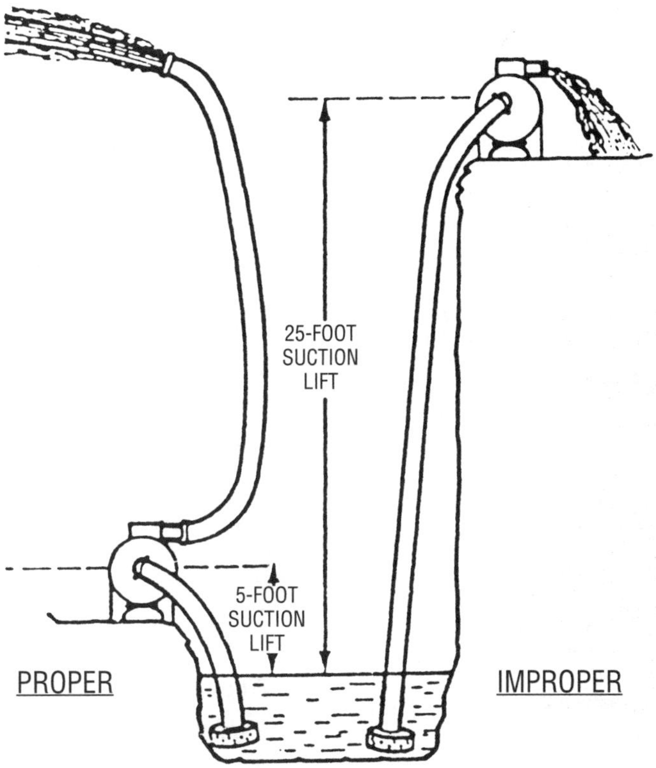

Fig. 9.36 Proper and improper suction lifts
(Permission of Homelite Textron, Charlotte, NC)

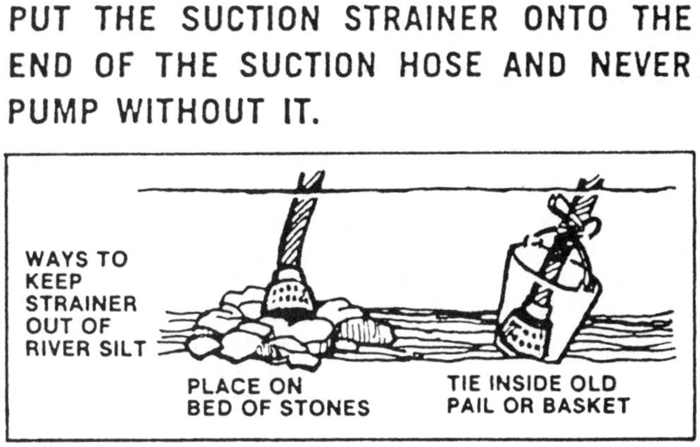

Fig. 9.37 Proper use of suction strainer
(Permission of Homelite Textron, Charlotte, NC)

Portable pumps must be protected from being damaged. Whenever a hose must be laid across a roadway, lay a plank on either side of the hose. A vehicle running over the discharge hose while the pump is running may damage the hose and could also cause the pump casing to crack.

If you are operating the pump in weather that is subject to freezing, always drain the pump to prevent the freezing water from cracking the casing or binding up the pump. Before starting the pump, turn the shaft by hand to be sure that it turns freely. If the impeller is frozen fast, warm the pump slowly until the ice melts.

When water containing salt or other corrosives has been pumped, drain the pump bowl and flush with clear water. Also, rinse off the exterior of the pump. By using a slightly oiled cloth or shop towel to wipe off the exterior, you will prevent rusting of the metal. A number of small pumps have cast aluminum casings and are not as resistant to corrosion as pumps having cast iron casings.

Table 9.6 is a portable pump troubleshooting guide which indicates some problems and recommended solutions.

Lubrication should be done in accordance with the pump manufacturer's O & M manual.

QUESTIONS

Write your answers in a notebook and then compare your answers with those on page 265.

9.3L Why are portable centrifugal pumps often referred to as trash pumps?

9.3M Why should the suction hose of a portable pump have a strainer on the end?

9.3N What precautions would you take when operating a portable pump in weather that is subject to freezing?

9.317 Positive Displacement Diaphragm Pumps

The diaphragm-type pump is a positive displacement pump. A flexible membrane (diaphragm) is used in the vertical cylinder, instead of a piston, and flap valves are used for check valves instead of the common ball checks that are used in the piston type of raw sludge pumps. The concentric movement of the rod alternates the pump from suction to pressure.

Some of the advantages of the diaphragm pump are as follows:

1. Self-priming if the suction lift is small (10 to 15 feet),
2. When primed with water, it will pump with a suction lift up to 25 feet,
3. Large particles will readily pass through the pump (you may have heard of them referred to as "mud hogs" for that reason), and
4. They are less likely to become clogged than centrifugal pumps.

The flow rate is considerably lower with a positive displacement type of pump than with a centrifugal pump. Keeping the pump close to the surface of the water being pumped will ensure maximum output.

To ensure that the pump works satisfactorily, the flap valves must seat properly. Inspect the diaphragm for leaks. If the pump will not prime, inspect the suction hose connections, flap valves, and prime water. Also look for a small hole in the diaphragm that may be causing the problem.

Taking care of the diaphragm pump is similar to the procedures outlined in Section 9.316, "Centrifugal Trash Pumps." However, since some materials used for the membrane tend to deteriorate when exposed to long periods of sunlight, the pump should be stored in a shaded, dry area.

To diagnose problems, refer to Table 9.6, "Portable Pump Troubleshooting Guide."

9.318 Positive Displacement Diaphragm (Pneumatic) Pumps *(Figures 9.36 and 9.39)*

Pneumatically operated diaphragm pumps can be of either the submersible or nonsubmersible type. The theory of operation in either case is quite similar.

Some of the advantages of air-powered diaphragm pumps are as follows:

1. They are resistant to wear,
2. Pumping rate may be adjusted by regulating inlet air,
3. Pump is able to run dry without damage,
4. Discharge line can be closed while pump is operating without damage,
5. Pump performance is more consistent due to less wear,
6. They are self-priming up to approximately 15 feet of suction lift,
7. Some models can be submerged which results in no electrical power near the water, and
8. If you have compressed air in the plant, there is no need to purchase an additional power source.

OPERATION

Two flexible membranes (diaphragms) are connected to a common shaft that moves them simultaneously in parallel paths. The diaphragm movement is powered by compressed air directed behind one diaphragm while air is exhausted from behind the other diaphragm. When the shaft reaches its length of travel, an air valve transfers the air flow to the diaphragm in the other chamber. Air in the first chamber is then exhausted. This reciprocating action alternately creates suction and discharge of water in each chamber.

The suction and discharge valves (flap or ball type) control the flow of water in their respective cycle of operation. This form of piston-type action is similar to the piston-type positive displacement pump and the diaphragm pump.

TABLE 9.6 PORTABLE PUMP TROUBLESHOOTING GUIDE

PROBLEM	POSSIBLE CAUSE	OPERATOR RESPONSE
1. Pump engine won't start		Follow instructions in manufacturer's engine manual.
2. Pump won't prime		Test pump suction to determine if problem is inside or outside pump.
	INSIDE PUMP a. Pump needs water	Fill with clean water.
	b. Water inside pump contaminated	Drain pump and fill with clean water. Even though pump can pump dirty water, clean water may be needed for priming.
	c. Worn pump	If possible, reduce the suction lift distance. If pump cannot prime at low lift, it should be disassembled and overhauled.
	d. Diaphragm pump valves inoperative, or pump diaphragm leaking	Clean valves or replace leaking diaphragm.
	OUTSIDE PUMP a. Leaking hose or connections on suction side of pump	Make couplings tighter.
	b. Strainer clogged	Clean strainer. Try another method of keeping strainer from clogging.
	c. System clogged	Clean hoses. If necessary disassemble and clean out pump.
3. Flow is scanty	Pump OK, but too small for job	Install larger pump fitted with larger diameter hoses. If a little faster flow would be acceptable, try larger hoses with the same pump.
	Total head including friction is too great	Do everything possible to decrease the head. Try to eliminate unneeded elbows, adapters, and reducers. If possible, move pump closer to the water and shorten hoses.
	Pump leaking or worn	Overhaul pump. Have worn seals, gaskets, impeller, or housing parts replaced as necessary; or shim to reduce clearance between impeller and the wear plate or the housing.
4. Volume decreases during pumping	Roots and other debris keeping diaphragm pump valves stuck open	Elevate discharge hose so water rises to help seal valves. Keep pumping until opportunity arises to stop and clean pump.
5. Pump is "frozen" and won't move	Ice inside pump	Thaw out by gradually warming pump.
	Hard object jammed between impeller and housing	Disassemble pump and remove blockage.
6. Diaphragm pump suddenly stops; engine either quits or keeps running and pump rod slips on shaft	Solid object preventing pump rod from completing stroke	Remove discharge valve and clean cavity.
	Accumulation of grit in housing	Remove grit.

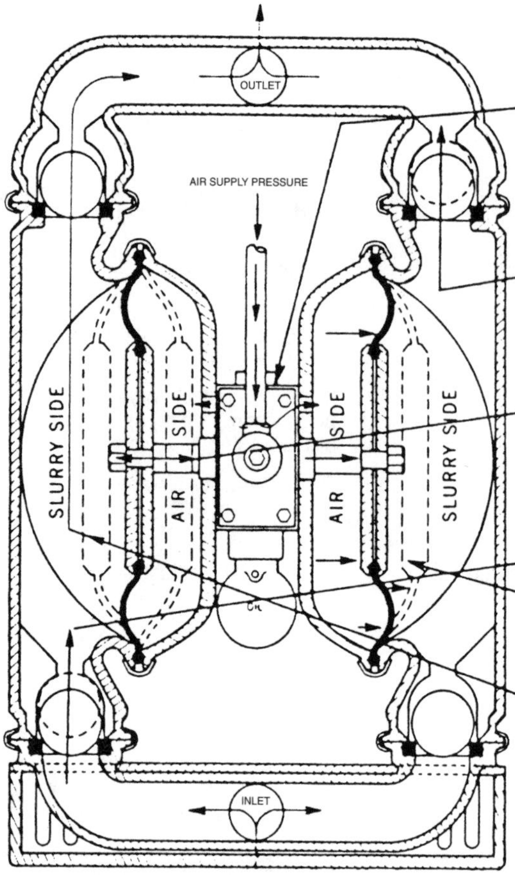
OUTLET
AIR SUPPLY PRESSURE
SLURRY SIDE
AIR SIDE
AIR SIDE
SLURRY SIDE
INLET
Patented, one moving piece air valve directs air supply pressure to back side of diaphragm.
Slurry is pushed out of liquid chamber, through pump outlet.
At the same time opposite diaphragm is pulled in by rod connected to pressurized diaphragm.
Suction created draws slurry into liquid chamber through pump inlet.
When pressurized diaphragm reaches limit of stroke, air valve shifts air supply pressure to back side of diaphragm which was pulled in. Slurry is pushed out of liquid chamber through pump outlet.
On discharge stroke, pressure on both sides of diaphragm is equal and discharge pressure is equal to air supply pressure.

NOTE: Max. size solid pumps will pass is limited by clearance of ball check valve off valve seat M-4-3/16," M-8-1/4," M-15-3/8"

Fig. 9.38 Diaphragm (pneumatic) pump

(Permission of Wilden Pump & Engineering Co., Colton, CA)

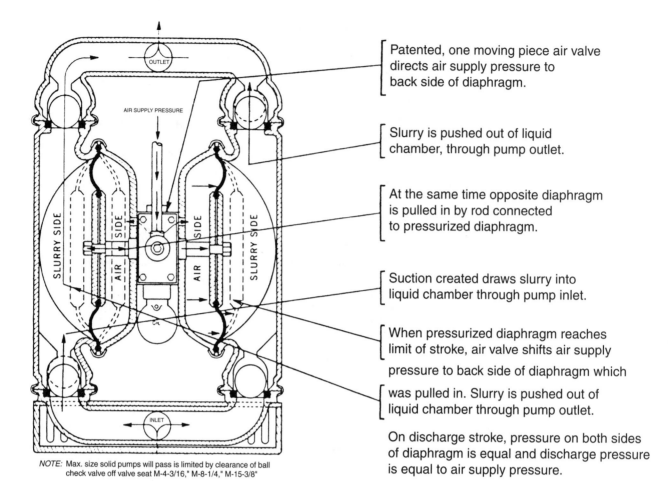

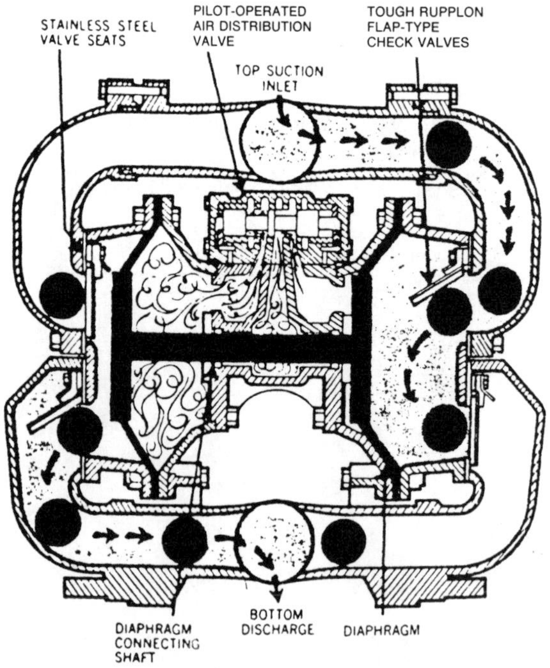

Fig. 9.39 Diaphragm (pneumatic) pump

(Permission of Warren Rupp Company, Mansfield, OH)

MAINTENANCE

The pump should be flushed thoroughly after use to prevent dried sediment from interfering with valve operation. If the water pumped had large particles in it, it may be necessary to dismantle the pump to remove any remaining particles.

TROUBLESHOOTING

If pump will not cycle, check for:

1. Adequate air pressure,
2. Blocked discharge line,
3. Sliding air distribution valve rod hanging up,
4. Excessive air leak,
5. Plugged exhaust port,
6. Ruptured diaphragm, and
7. Open inlet valve.

If pump cycles but will not pump, check to see if:

1. Suction side of pump is pulling in air,
2. Suction line is plugged,
3. Check valves (ball or flap) are not seating, and
4. Suction lift is too high.

If output flow rate is low, determine if:

1. Air pressure to pump is low,
2. Suction is restricted and accompanied by fast cycling of pump,
3. Discharge is restricted with slow cycling of pump, and
4. Valve seating is improper.

9.32 Pump Testing

Please see Chapter 8, Section 8.25, "Operational Inspection," for procedures on how to measure pump capacity (GPM) and total dynamic head (TDH, ft) in the field. These results can be compared with the pump manufacturer's catalog head-capacity curve to evaluate pump performance.

QUESTIONS

Write your answers in a notebook and then compare your answers with those on page 266.

9.3O List the advantages of a positive displacement diaphragm pump.

9.3P What would you do if a diaphragm pump suddenly stops pumping and the engine either quits or keeps running and the pump rod slips on the shaft?

9.3Q What maintenance should be performed on a pneumatically operated diaphragm pump after use?

END OF
LESSON
2 OF 4
LESSONS ON EQUIPMENT
MAINTENANCE

Please answer the discussion and review questions next.

DISCUSSION AND REVIEW QUESTIONS

Chapter 9. EQUIPMENT MAINTENANCE

(Lesson 2 of 4 Lessons)

Write the answers to these questions in your notebook before continuing. The question numbering continues from Lesson 1.

7. What types of water and material can be pumped by a pump?
8. Why is proper maintenance of a pump important?
9. Why must a pump be in proper alignment?
10. What are the uses of the different types of pumps?
11. Under what conditions would you recommend the installation of a screw pump?
12. What are the advantages of a pneumatic ejector?

CHAPTER 9. EQUIPMENT MAINTENANCE

(Lesson 3 of 4 Lessons)

9.4 PUMP COMPONENTS

9.40 Impellers

The heart of a centrifugal pump is the impeller. As stated earlier, this is the device where spinning causes a centrifugal force that pumps the wastewater. There are many different types of impellers, such as propeller, turbine, mixed flow, and radial. These impellers may be open, closed, or semi-closed. Photographs of different types of impellers are shown in Figures 9.40 and 9.41.

The most common impellers used in wastewater centrifugal pumps are semi-closed or closed-type radial for larger pumps and open impellers for smaller sludge pumps. These are used because they will pass solids with less chance of plugging. These impellers require a specific clearance between the tip of the impeller and the pump volute. For wastewater, the clearance is generally set at 0.010 to 0.012 inch. Freshwater pump clearances are approximately 0.002 to 0.005 inch. The reason for larger clearances in wastewater pumps is to reduce erosion by solid particles such as grit and sand.

Pump impellers should be inspected for wear on a regular basis such as every 6 months or annually, depending on pumping conditions. Before inspecting a pump impeller you should:

1. Turn off and lock out all electrical power to the pump ("Main Circuit Breaker" and "Start-Stop Station"—don't lose a finger or arm),
2. Tag control panel with a "Do Not Start" tag stating time, date, and your name or initials,
3. Close, secure and tag the suction and discharge valves (do not depend on discharge check valve),
4. Drain pump with bolts still in handhole cover until water stops (many a lift station has been flooded because something prevented a valve from closing). If bottom of volute has a drain, use it, and
5. Open handhole covers and visually inspect for foreign objects which will injure you (razor blades, sharp objects, needles in rags). Turn the shaft by hand and listen for loose objects. Inspect for binding by rags, rope or metal. Use a special hook or tool to remove trash, but not your hands. When the area is clear, the insides can be inspected.

When inspecting a pump impeller you should look for the following items.

1. Cavitation marks. They appear as holes, indentations, bullet chips, and cracks.
2. Chips, broken tips, corrosion, unusual wear.
3. Tightness on shaft. Hold shaft and try to wiggle impeller, inspect impeller nut. Pry with wood up and down.
4. Clearances. This is done by inserting a feeler gage between pump volute or faceplate and impeller if the pump does not have wearing rings. With wearing rings, clearance is adjusted between wearing rings on the impeller and volute faceplate.
5. Excessive wear on wearing rings on impeller and volute.
6. If impeller is rubber coated, look for tears or bubbles in coating.

If any of these problems are observed, notify your supervisor and record your observations in the equipment file for that pump.

Excessive clearances can cause frequent plugging and loss of pump capacity. As pumps wear or clearances become excessive, pump efficiency is lost and the motor will draw fewer amps. However, total power usage will increase because the pumps will have to operate longer to pump the total flow. Vertical type pumps (Figure 9.29) have impellers which can be adjusted fairly easily. One method is removing shims from the upper bearing housing or between the upper bearing and shaft. Another method used is a jackscrew which can be loosened and screwed up and down (Figure 8.41, page 65). Still another method is shimming the impeller or faceplate. You will have to consult your pump manual for the particular method to be used on your pump.

Many wastewater pumps are fitted with replaceable wearing rings on the impeller and volute, or faceplate to protect the impeller and pump casing from damage due to excessive wear by solids and grit. These wearing rings are made of materials such as bronze, stainless steel, and Teflon. They are fabricated and installed so that they can be replaced when worn, thus not having to replace the entire impeller or volute. Wearing rings are usually pressed or pinned on the impeller and volute. Sometimes they are threaded with a set screw holding them. The pump manufacturer's literature will inform you what type is

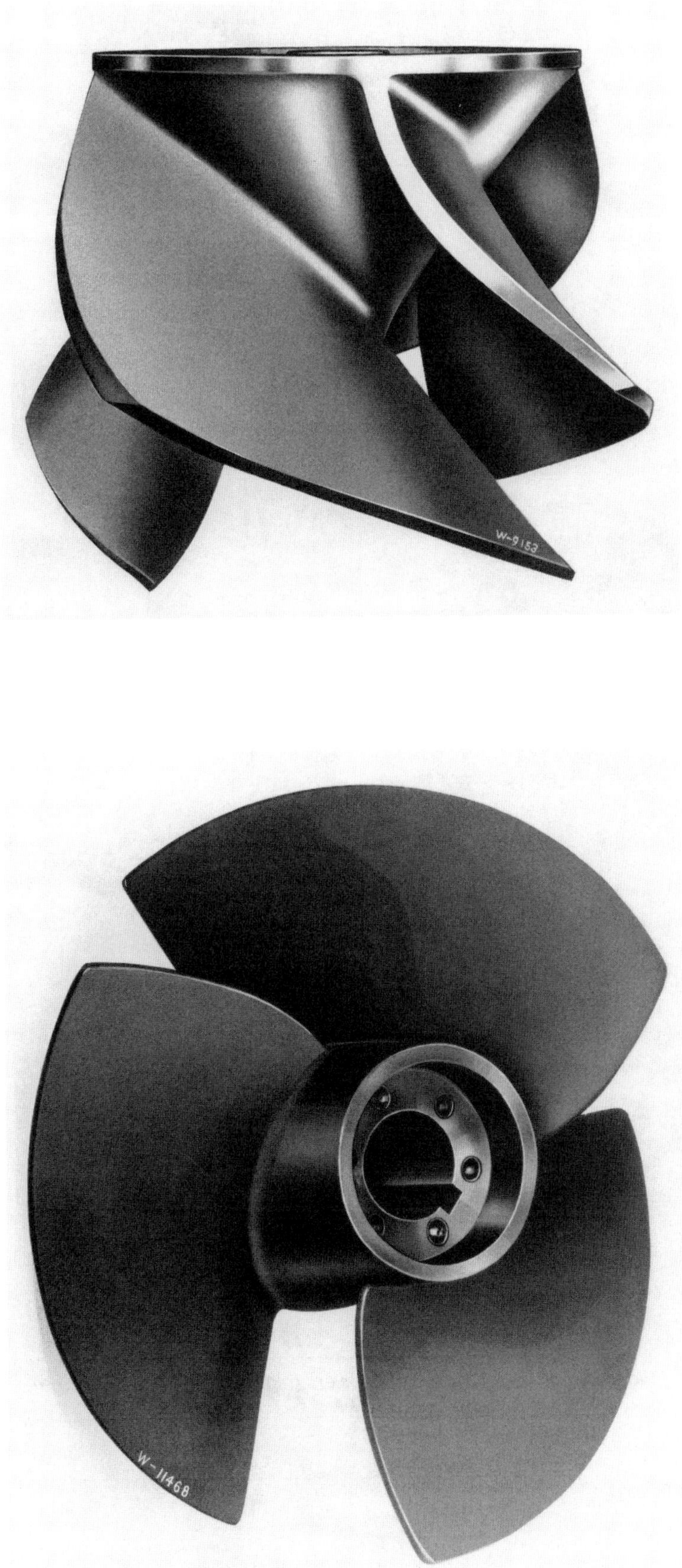

Fig. 9.40 Open impellers
(Source: *CENTRIFUGAL PUMPS* by Karassik and Carter of Worthington Corporation)

Fig. 9.41 Different types of impellers (high head and high volume)
(Permission of ITT Industries—Flygt Corporation)

installed in your particular pump. Wearing rings must be replaced when worn excessively or when clearances between impeller and volute cannot be attained. Be sure to replace them before they become worn through or wear starts on volute or impeller, because new wearing rings will not fit properly on worn down volutes or impellers.

In order to replace wearing rings it is necessary to dismantle the pump head from the volute. The worn rings are removed by chiseling, cutting, or grinding them off. When the impeller is removed, a new wearing ring can usually be pressed on under a press. Another method is to heat the wearing ring in hot oil and tap over impeller with a block of hard wood after it has expanded. Make sure the wearing ring is installed completely and do not scar or burr a new ring with metal punches or drift pins. If burred, file off before installing in volute.

The volute wearing ring can sometimes be installed by pressing or tapping. It is much easier to freeze the wearing ring with dry ice, shrinking it, and drop or tap it into place with wood or plastic drift. Before installing either wearing ring, make sure mating surfaces are clean and free of nicks and burrs. Also be sure the wearing rings are properly seated into the intended position.

The replacement of wearing rings is generally done by qualified mechanics and special tools are required. Since the pump has to be removed and dismantled, special tools are required such as hoists, presses, chains and slings, punches, chisels, wrenches, and power tools. Care must be used to prevent damage to the pump or the parts being repaired.

Cavitation is caused by the pump operating at conditions different from those for which it was designed, such as operating off design curve, poor suction conditions, high speed, air leaks into suction end, and water hammer conditions. Many of these are not controlled by the maintenance operators, but some items can be checked.

1. Make sure suction valve is all the way open. Never throttle a suction valve.
2. Examine wet well level and ensure pump is not sucking in air or pumping too low before shutting off. Also look and be sure pump is not picking up air by vortexing (whirlpools) at inlet of suction line.
3. Inspect suction line to make sure it is not partially plugged with grease or foreign material such as sand, grit and rocks.
4. Inspect bottom of wet well for grit deposits.
5. Examine maximum speed setting on pump to determine if correct.
6. Look for places where air could be leaking in at packing.

Worn, broken, chipped, or damaged impellers can be repaired only by welding, machining, or replacing. This will require the services of a mechanic with expertise in pump repair.

Loose or unbalanced impellers should be repaired by a mechanic immediately and before further damage to the pump occurs. If a pump is left in this condition, the shaft could break and the pump could be damaged beyond repair. Unbalanced impellers may vibrate, break and possibly rupture the volute. Impellers must be dynamically balanced after any repair work.

QUESTIONS

Write your answers in a notebook and then compare your answers with those on page 266.

9.4A How often should impellers be inspected for wear?

9.4B What is the purpose of wearing rings?

9.4C What can happen if clearances are not kept in proper adjustment?

9.4D How can you tell if an impeller has damage from cavitation?

9.4E What causes cavitation?

9.41 Shafts

Pump shafts are used to turn the impeller and mount the bearings; they are connected to a pump driver of some type. These shafts are generally fabricated from an alloy steel machined to size. They are machined to exact tolerances and must be maintained this way for proper operation. There are four major areas to a shaft:

1. Impeller end,
2. Coupling end,
3. Bearing fits, and
4. Stuffing box area.

The impeller end is where the impeller is pressed, keyed, or secured onto the shaft. This part of the shaft is usually not a problem to maintain. If the impeller is kept tight, there should never be a need to repair this shaft area during regular use.

The coupling end is where the coupling mounts on the shaft to connect to the driver. This end of the shaft usually never needs repair. When removing or installing a coupling, the coupling end should be cleaned thoroughly and lubricated. Rust can sometimes be a problem and must be removed.

Bearing fits or journals are where the bearings are installed on the shaft. These are designed to have press fit tolerances as specified by the pump manufacturers. They are not usually a maintenance problem unless a pump with a bearing which has failed is left running for an extended time. If this happens, the inner race of the bearing could turn or spin, thus damaging the bearing fit. To repair the bearing fit, it will be necessary to remove the shaft and have it metal sprayed and ground down to size by a machine shop. The machine shop will need to know the bearing number so they can machine the shaft to proper size. Bearings are discussed in Section 9.44, "Bearings."

The most important problem section with a pump shaft is the stuffing box area. The pump packing, if not properly cared for, will cut grooves into the shaft in this area. Some of the reasons for grooves being cut into the shaft are:

1. Packing glazed when pump was started,
2. Worn packing that allows leakage from the volute when pump is operating (this leakage can carry grit or other foreign solids into the stuffing box),
3. Dirty seal flushing water,

4. Water seal pressure set too low or plugged,
5. Wrong type packing,
6. Packing or lantern (seal cage) slipping, and
7. Improper packing adjustment, such as packing too tight.

Two very acute problems arise when the shaft becomes scored. The first thing is that once the shaft is scored or grooved, packing will not last. The grooves will rip and tear out new packing in a very short time. The second problem is the shaft is weakened and, if left in operation, will be damaged beyond repair.

Manufacturers have realized the wear problem with the shaft in the stuffing box area and designed shaft sleeves (Figure 9.42) to avoid this problem. Shaft sleeves are installed over the shaft in the area on the shaft where it passes through the stuffing box. These sleeves are usually made of stainless steel, chrome or ceramic-plated alloy steel, bronze, or other materials suitable for the pumping application. The shaft sleeve is either keyed on, pressed on, threaded on, or held by the impeller. Shaft sleeves are designed to protect the shaft by taking the wear from the packing. To replace a complete shaft is very costly compared to a shaft sleeve. Shafts and/or shaft sleeves should be inspected every time the pump is repacked. This inspection has to be done when the pump packing is removed and stuffing box cleaned out. In order to see the shaft sleeve it might be necessary to use a small mirror, such as a dental mirror, and flashlight. Inspect for grooves, split shaft seal and any other unusual conditions. The sleeve should be repaired or replaced when grooves are observed. It must be replaced before wearing through to the shaft. To replace the shaft sleeve, it is necessary to remove and dismantle the pump. Usually, just the impeller needs to be removed and the sleeve can be pulled off.

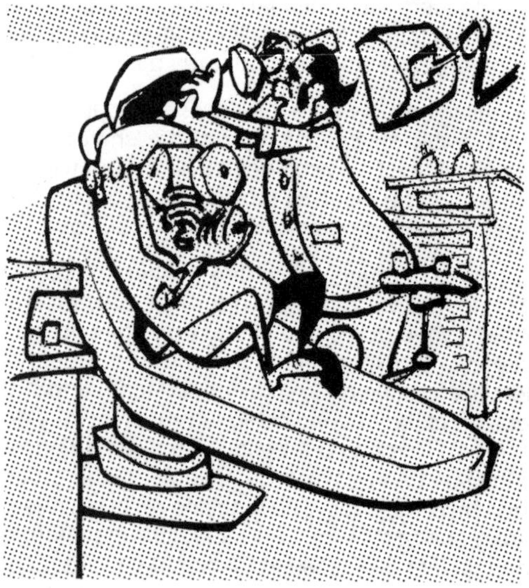

Some shaft sleeves are extremely difficult to remove and they can be damaged beyond repair. Try heating the sleeve and/or freezing the shaft to break the bond and increase clearance. Shafts and shaft sleeves can be repaired by metal spray, ceramic spray, or chrome plating. This method may cause problems because, if not properly done, it will cause distortion. If a sleeve is worn, it should be replaced.

Replacing a shaft or shaft sleeve requires specialized tools and competent operators with a thorough knowledge of pumps and mechanics.

QUESTIONS

Write your answers in a notebook and then compare your answers with those on page 266.

9.4F How can pump shafts be protected from damage?

9.4G When should a pump shaft sleeve be repaired or replaced?

9.42 Packing (Figure 9.43)

Packing is probably the single biggest problem for the maintenance operator maintaining pumps. More pumps have been damaged due to improper maintenance of packing than any other reason. If packing is not maintained properly the following troubles can arise:

1. Loss of suction due to air being allowed to enter pump,
2. Shaft or shaft sleeve damage,
3. Water or wastewater contaminating bearings,
4. Flooding of pump station, and
5. Rust corrosion and unsightliness of pump and area.

Packing is used to provide a seal where the shaft passes through the pump casing in order to keep air from being drawn or sucked into the pump and/or the water being pumped from coming out.

There are probably over 500 different types, grades, manufacturers, and styles of packing on the market today. Packing can be round, square, loose, pointed, or almost any size, shape or form. It is manufactured from jute, hemp, Teflon, asbestos, rubber, lead, and many other materials too numerous to mention. The pump manufacturer should recommend the proper packing on the basis of the following:

1. Type of pump and manufacturer's recommendation,
2. Material being pumped (water, wastewater, acid, alkali, scum, sludge, grit, and others),
3. Temperature of liquid being pumped,
4. Grease or water flushing, and
5. Peripheral (outside) speed of shaft.

Do not pick a cheap grade of packing because it will cost you more in the end when shaft sleeves are damaged or time is required to repack. Generally, in wastewater pumps, braided

NOTE: Also see Figure 9.28 and Figure 9.29 for location of shaft sleeve on pump shaft inside of stuffing box.

Fig. 9.42 Pump shaft sleeves

(Courtesy Worthington Corporation)

Teflon Packing

Graphite Packing

Fig. 9.43 Packing
(Courtesy A.W. Chesterton Co.)

asbestos material is used; it can be graphite impregnated, Teflon coated, or of a similar material.

Let's start our packing procedure by placing a new wastewater centrifugal pump into service. The first consideration is to determine if the packing installed at the factory is still good. Since it arrived on the job site in February and wasn't installed until August, we must assume that it has been sitting for at least seven months in the weather. The contractor had some plastic over it, but it finally blew off. The packing has been rained on, the wind blew dust over it, and now it's 101°F outside so it's baked. To make matters worse, you notice that the painter has helped by coating the pump (nameplate included) with one coat of primer and two coats of enamel. OK, let's repack the pump before we start it and not ruin the shaft sleeve. Also we better make sure the shaft and the rest of the pump are ready to go after sitting outdoors before we start the pump. Get the manufacturer's literature and study it.

1. Turn off the main circuit breaker for this pump and control switch.
2. Lock it in the OFF position and tag it OFF (fill out a card with date and initials).
3. Close suction and discharge valves.
4. Drain pump by loosening bolts on handhole cover. *DO NOT REMOVE BOLTS COMPLETELY UNTIL PUMP STOPS DRAINING AND YOU ARE SURE THAT THE SHUTOFF VALVES ARE HOLDING.* If available, drain pump by using drain valve or plug at bottom of volute.
5. Remove packing gland nuts and/or bolts and remove glands. These are usually split for removal. If the packing gland is not split for removal, it has to be backed off on the shaft. Provide room for removal and replacement of packing.
6. Remove packing. The packing rings can be removed by screwing a packing hook (Figure 9.44) into them and pulling out. Sometimes it helps to revolve the shaft while pulling. *CARE MUST BE TAKEN WHILE USING THE PACKING HOOK THAT THE SHAFT ISN'T SCORED OR NICKED.*
7. After you have removed the top rings of packing, the lantern ring (seal cage) must be removed. This can usually be fished out with a heavy piece of wire, coat hanger, or welding rod. Some have a threaded hole in which a 1/4-inch bolt can be inserted. If the pump has been in service for a while, you will have to clean around the shaft and stuffing box to get it out. Try tapping with a wooden dowel to loosen it up if you are still having trouble removing it. Another approach is to turn the pump on and blow the ring out with pump pressure—messy, but effective. Always remove the ring. These also are usually split so they can be removed from the shaft.
8. Next, remove the remaining packing rings according to the manufacturer's drawing. They are usually more difficult to remove after a pump has been in service.
9. Wash all parts with soap and water and scrub with a wire brush, if necessary. *THOROUGHLY CLEAN INSIDE THE STUFFING BOX AND THE SHAFT.* Wipe parts dry.
10. Disconnect seal water line or spring grease cup. Clean thoroughly and inspect its operation.
11. Make sure that seal water hole through stuffing box is clear and clean.
12. Inspect shaft or shaft sleeve for wear and cleanliness. Make sure that the lantern ring was positioned properly and the manufacturer's print was correct. If the number of rings of packing to be installed before the lantern ring is unknown, insert enough packing so that the lantern ring is correctly positioned with the seal water or grease inlet when the packing has been pushed into the recess as far as it will go. Each packing ring should be cut to fit the shaft.
13. Determining packing requirement.
    a. Determine packing size. Measure from outside of shaft to inside of stuffing box with a scale. Whatever the measurement is, this is the size of the packing. Let's assume this is 1/2 inch; therefore, 1/2-inch square packing is used.
    b. Measure the depth of the stuffing box, outside face to inside bottom. Let's assume this measurement is 4 inches.
    c. Measure from bottom of stuffing box to center of seal water inlet. Let's assume this is 1 3/4 inches.

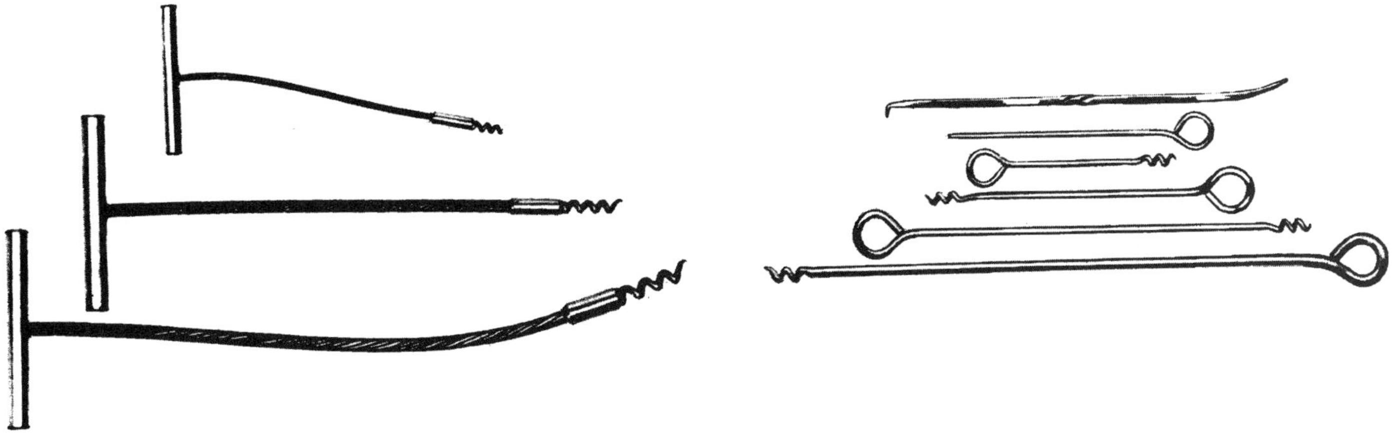

Fig. 9.44 Packing hooks
(Courtesy A.W. Chesterton Co.)

d. Measure lantern ring or seal cage, and again, let's assume this is 1/2 inch.

e. With a 4-inch depth and subtracting the 1/2-inch seal cage, we have 3 1/2 inches that can be filled with packing. 3 1/2 inches divided by 1/2 inch (size of packing) allows us 7 rings of packing to be inserted.

f. The seal cage should be located in the stuffing box so that it is centered where the seal water or grease enter. Take the seal cage measurement and divide it in half: 1/2 inch divided by 2 equals 1/4 inch. The measurement from the bottom of stuffing box to center line of seal water inlet was 1 3/4 inches. By adding 3 rings of 1/2-inch packing (which equals 1 1/2 inches) and adding the 1/4 inch for seal cage, we arrive at 1 3/4 inches, which puts the seal cage at the seal water inlet. As stated earlier, this information usually can be taken from the manufacturer's literature.

Now that we know the size of packing, number of rings needed, and where to position the seal cage (lantern ring), we can start repacking the pump.

Packing should be kept in a clean place. Never allow the protective wrapping to be left off or the box uncovered. Do not allow packing to come in contact with solvents or high temperatures which could force lubricant out of it. Discard any packing that has become contaminated with dirt, sand, grit, or other foreign material.

14. Packing the pump.

   a. Cut packing to proper length using a very sharp knife, and on a clean board.

      (1) If proper length is not known, wrap around a mandrel of the same size as the shaft, or wrap around the shaft itself, and cut.

      (2) Make clean cuts, no ragged edges.

      (3) If possible, cut packing on shaft so parallel ends can be obtained.

      (4) Cut all necessary rings at one time.

   b. Roll packing flat with short pipe on a clean bench, or tap slightly so it will fit into the stuffing box more readily.

   c. Wrap the new packing rings *(ONE AT A TIME)* around the shaft and insert into the stuffing box, keeping joints together. Tap into place with gland or a split wooden bushing.

   d. Stagger the joint or butts of the packing around the shaft. For example:

      (1) First ring—place butt or joint on right side of shaft.

      (2) Second ring—place butt or joint on left or opposite side of shaft.

      (3) Third ring—place butt at top of shaft.

      (4) Fourth ring—place butt at bottom of shaft.

      (5) Repeat procedure for more rings.

   e. Position the lantern ring (seal cage) so it is centered in the seal flush opening. Remember the ring moves back deeper into the box as packing is tightened.

   f. Install the top or remaining three packing rings. Stagger the joints in the same sequence as the first three rings.

   g. Install packing gland keeping the gland square at all times.

      (1) The last ring of packing may have to be left out and installed after the packing is seated.

      (2) Tighten packing gland just enough (hand tight) to prevent excessive leakage before starting the pump.

      (3) Disconnect the lubrication fitting and, using a stiff wire, probe for the lantern ring. If you detect packing, the lantern ring has been placed in the wrong position.

To start a new or repacked pump, use the procedures listed below.

1. The first thing to do is to make a visual inspection. Look for loose nuts and bolts, guards left off, tools left around, and any other unusual condition.

2. Open suction and discharge valves. Make sure they are completely open.

3. Turn on seal water or make sure grease cups are full.

   a. Seal water can be either fresh water from an air gap system or recycled, filtered water.

   b. Seal water pressure should be adjusted to 5 psi higher than the maximum pump discharge pressure.

4. Prime pump by venting air or filling pump case with water from a hose or a bucket.

5. Remove electrical lockout tag and unlock main circuit breaker.

6. Start pump and observe its operation.

   a. Inspect gages to see if pumping.

   b. Listen for any unusual noises.

7. Adjust packing gland.

   Probably more packing jobs are ruined by improper gland adjustment than for any other single reason.

   a. Tighten the packing gland just enough to prevent excessive leakage.

   b. As the packing adjusts itself to the shaft, tighten the gland slowly one-half turn at a time on each nut, keeping gland square until the leakage is reduced to the desired number of drops per minute.

      (1) Leakage of 20 to 60 drops per minute is desired.

      (2) Pumps with grease lubrication should not leak, but do not overtighten.

      (3) Keep packing gland square with pump shaft to prevent rubbing on shaft. To keep gland square, hand tighten first bolt until slack is taken up. Insert other packing gland bolts and tighten to point of first resistance. Next tighten each bolt one flat at a time on alternate sides to ensure an even take-up until the leak rate is properly adjusted. Uneven take-up jams the packing and the shaft.

Usually the gland is not pulled up more than finger tight to maintain the proper leakage. If the gland becomes too hot to touch, it is pulled up too tight.

Stop the pump and allow the packing to cool, then readjust it. Do not back off on gland while pump is running because entire set of packing will move out, not relieving any pressure on the shaft. Start the pump and determine gland temperature. It may be necessary to start and stop the pump several times before desired leakage and temperature are obtained. Small motors may be stopped and started several times in a short period, but severe motor damage can occur in large motors when there are excessive starts and stops within a given period. Inspect seal water again and make sure it is entering packing. Clean filters and strainer as necessary to maintain flushing water supply to packing gland. If one ring of packing was left out, be sure to install it when packing is sealed. Table 9.7 lists some causes of shaft packing failures (this information was provided by the A.W. Chesterton Company).

A used or worn set of packing can be of value as it often indicates the cause of premature packing failure. Examine worn packing carefully. Table 9.7 should be helpful in tracking down packing troubles.

QUESTIONS

Write your answers in a notebook and then compare your answers with those on page 266.

9.4H What is the purpose of packing?

9.4I What is the purpose of the lantern ring?

9.43 Mechanical Seals (Figure 9.45)

Many pumps use mechanical seals in place of packing. Mechanical seals serve the same purpose as packing; that is, they prevent leakage between the pump casing and shaft. Like packing, they are located in the stuffing box where the shaft goes through the volute; however, they should not leak. Mechanical seals are gaining popularity in the wastewater field.

Mechanical seals have two faces which mate that prevent water from passing through them. One half of the seal is mounted in the pump or gland with an "O" ring or gasket, thus providing sealing between the housing and seal face. This prevents water from going around the seal face and housing. The other half of the mechanical seal is installed on the pump shaft. This part also has an "O" ring or gasket between the

TABLE 9.7 CAUSES OF SHAFT PACKING FAILURE

INDICATION	SUGGESTION
• No leakage at start-up.	• Back off gland to encourage generous leakage. If negative suction, install lantern and connect to discharge.
• Excessive leakage at start-up.	• Check for correct packing size. Were rings installed correctly in accordance with instructions? Check pump for runout.
• Packing rings flattened out on I.D. under the rod or shaft.	• Check bearings. Packing probably supporting the shaft weight.
• Packing rings flattened out above rod or shaft, or on either side of same.	• Check alignment of shaft. Worn bearings may be causing whip or runout.
• Distinct bulge on side of ring.	• Probably too wide a gap on adjacent ring — rings cut too short.
• Sides of rings shiny or worn.	• Rings may be too loose and are rotating with the shaft.
• Rings extruding past gland follower.	• Too much clearance between O.D. of shaft and I.D. of follower. Apply bushing. May also be excessive gland pressure.
• Top or gland end rings in poor condition, bottom rings OK.	• Set improperly installed.
• Rings disappear in set.	• Packing entering the system. Install bottom bushing.
• Packing is torn.	• Check sleeve for burrs. Are there any abrasives?
• Rings are burned, faces dried and charred.	• Check for proper size packing. Is packing the correct selection with respect to heat limitations and/or peripheral speed? Check lubrication. Are there any abrasives?
• Packing hardens.	• See "burned rings" above. Are there congealing liquids involved?
• Packing softens.	• Check for proper style selection, also for lubrication.
• Excessive loss of lubricant.	• Excessive gland pressure. Check for packing selection, temperature, whip in shaft.
• Unexplained leakage.	• Shaft sleeve may be leaking. Replace sleeve seal under sleeve.
• Packing freezes to shaft after shutdown.	• Liquid salting out or congealing in packing set. Provide lubrication to packing before shutdown.

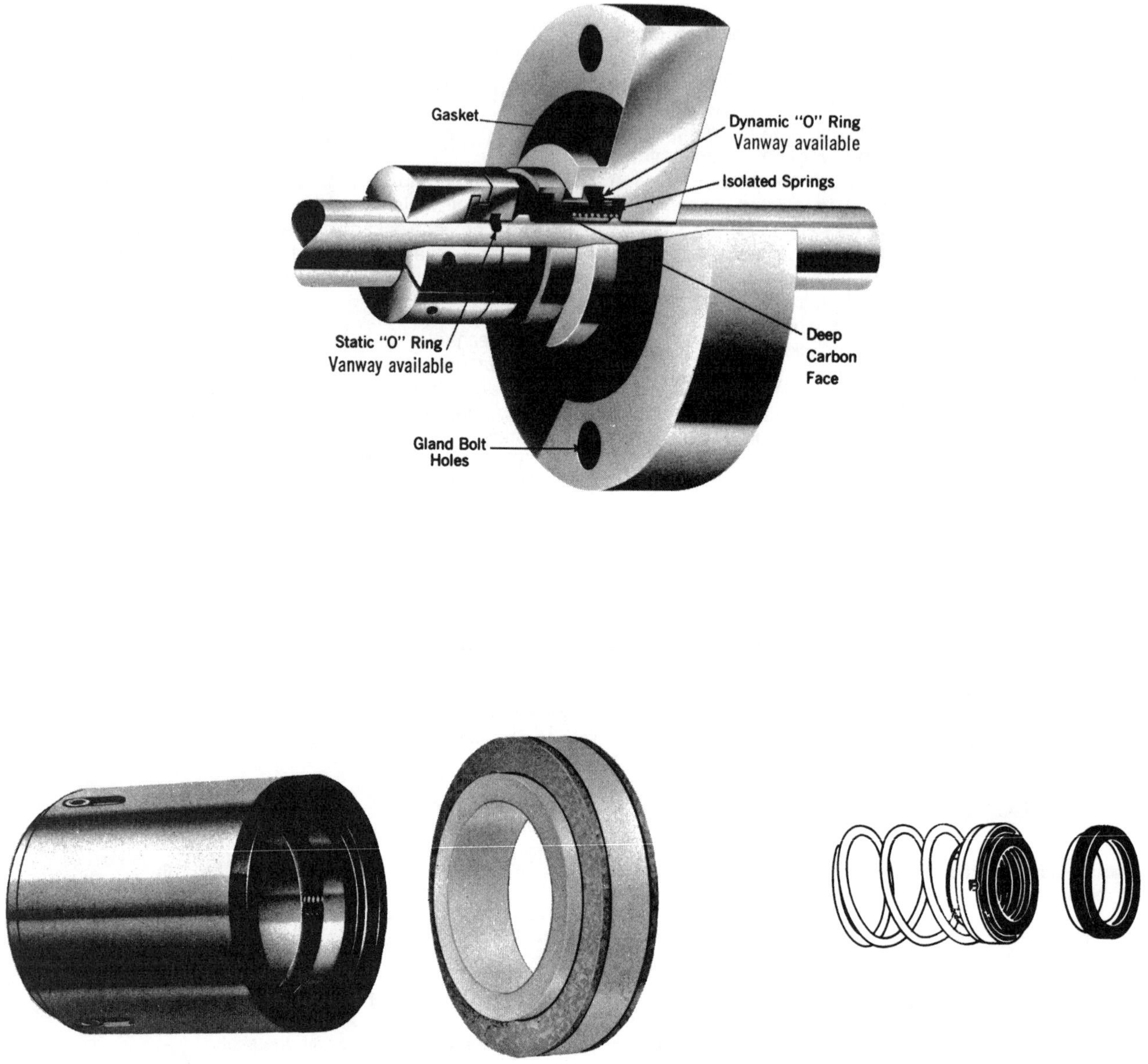

Fig. 9.45 Mechanical seals
(Courtesy A.W. Chesterton Co.)

shaft and seal to prevent water from leaking between the seal part and shaft. A spring, located behind one of the seal parts, applies pressure to hold the two faces of the seal together and keeps any water from leaking out. One half of the seal is stationary and the other half is revolving with the shaft.

Materials used in the manufacture of these seal parts are carbon, stainless steel, ceramic, tungsten carbide, brass, and many others. The different type materials are selected for their best application. Some of the variables are:

1. Liquid and solids being pumped,
2. Shaft speed,
3. Temperature,
4. Corrosive resistance, and
5. Abrasives.

Initially, mechanical seals are much more expensive than packing to install in a pump. This cost is gained back in operating costs during a period of time.

Some of the advantages of mechanical seals are as follows:

1. They last from three to four years without having to touch them, resulting in labor savings,
2. Usually there isn't any damage to the shaft sleeve when they need replacing (expensive machine work is not needed),
3. Continual adjusting, cleaning, or repacking is not required, and
4. The possibility of flooding a lift station because a pump has thrown its packing is eliminated; however mechanical seals can fail and flood a lift station too.

Some of the limitations of mechanical seals are as follows:

1. High initial cost,
2. Competent mechanic required for installation,
3. When they fail, pump must be shut down, and
4. With some types, the pump must be dismantled to repair.

Refer to Chapter 8, Section 8.162, "Pump Components," for additional information about mechanical seals.

Mechanical seals are always flushed in some manner to lubricate the seal faces and minimize wear. This may be the liquid being pumped if it is fresh water. If fresh water is used, it is connected from the high pressure side and back to the stuffing box low pressure side. In some lift stations where fresh water is not available, the wastewater is used in this manner but it must be filtered. Another method is to use fresh water from an air gap system. Still another way of lubricating the seal is to use a spring-loaded grease cup.

Whatever lubrication method is used, the seal must be inspected frequently. The seal water is adjusted to 5 psi above maximum discharge pressure to keep the wastewater and grit from entering the seal housing and contaminating the seal faces. Grease cups must be kept full at all times and inspected to make sure they are operating properly.

When a pump is fitted with a mechanical seal, it must never run dry or the seal faces will be burned and ruined. Mechanical seals are not supposed to have any leakage from the gland. If a leak develops, the seal may require relapping or it may have to be replaced.

Repair or replacement of mechanical seals sometimes requires the pump to be removed and dismantled. Seals are quite delicate and special care must be taken when installing them. Mechanical seals differ widely in their construction and installation; and individual manufacturer's instructions must be followed. Due to the complexity of installing mechanical seals, special tools and equipment are needed as well as a qualified pump mechanic.

The following is a checklist for cost-effective installation and use of mechanical seals:

1. Use cartridge-mounted seals. They are easy to install and do not damage either the seal or the pump.
2. Purchase standard off-the-shelf seals to ensure seals will be available from suppliers, even in an emergency.
3. Return seals to original manufacturer for repair.
4. Obtain bearing isolators (labyrinths) or magnetic seals to protect the bearing from intrusion of moisture or debris.
5. Use synthetic lubricants in all pumps, motors, bearings, and compressors. Synthetic lubricants are very cost-effective because they lower the operating temperature, tolerate water intrusion better, lower high frequency vibration, and last two or three times longer than the low-priced mineral oils.
6. Every time the motor or pump is worked on, be sure to perform precision coupling alignment by using a dial indicator or laser to ensure long-term service life of bearings and mechanical seals.

QUESTIONS

Write your answers in a notebook and then compare your answers with those on page 266.

9.4J Why should a pump with a mechanical seal never be allowed to run dry?

9.4K What should be done when a mechanical seal develops a leak?

9.44 Bearings

Pump bearings usually should last for years if serviced properly and used in their proper application. Several types of bearings are used in pumps such as ball bearings, roller bearings, and sleeve bearings. Each bearing has a special purpose such as thrust load, radial load, and speed. The type of bearing used in each pump depends on the manufacturer's design and application. Whenever a bearing failure occurs, the bearing should be examined to determine the cause and, if possible, to eliminate the problem. Many bearings are ruined during installation or start-up. This section is a guide to help you identify types of bearing failures, determine the causes of failures, and select preventive measures to avoid failures.

Bearing failures discussed are:

1. Fatigue failure,
2. Contamination,
3. *BRINELLING*,[10]

---

[10] *Brinelling (bruh-NEL-ing). Tiny indentations (dents) high on the shoulder of the bearing race or bearing. A type of bearing failure.*

FATIGUE FAILURE

1. IDENTIFICATION

FLAKING OR SPALLING OF RACEWAY

Spalling is caused by a granular weakening of the bearing steel. The failure begins as a small fracture of the steel's internal structure. This fracture progresses to the surface of the race where particles of metal flake away as shown in the photograph.

NOISY RUNNING OF THE BEARING

Because of the rough race surface and the loose metal chips, there will be an increase in bearing vibration and noise.

2. CAUSE

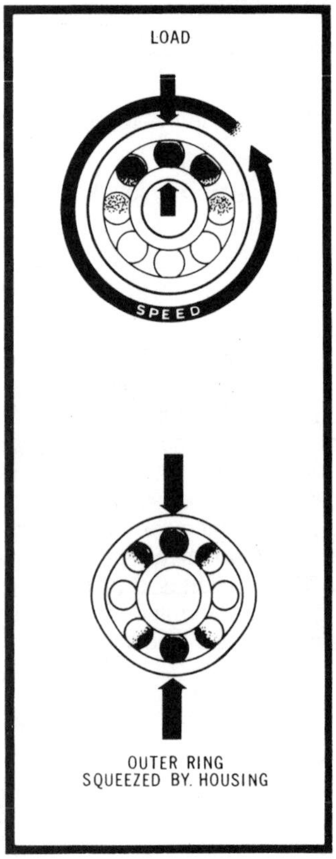

NORMAL DUTY

As outlined in the introduction, a bearing has a life expectancy which depends upon load and speed. Calculations, based upon laboratory testing and field experience, have been set up to determine, fairly accurately, the life span of a group of bearings of a given size. The fatigue failure, as outlined above, is the result of a bearing "living out" its normal life span. The flaking of the races is caused by the combined effect of load and speed. In any rotating or oscillating ball bearing, there is a constant flexing or deflection of the ring and ball material under the load. Speed determines how often the flexing occurs while load determines the actual amount of stress under which the bearing steel operates. Assuming good machine design, satisfactory lubrication, and sound maintenance practices, it is load and speed that will, over a long period of time, cause eventual failure.

OVERLOAD ON THE BEARING

Premature failure of the bearing may result from the bearings being either radially or axially loaded beyond its normal capacity. Excessive operating load is not, however, the only reason for bearing overload. Overload may also occur due to abusive operating conditions as outlined below:

1. If a bearing with insufficient internal clearance (the space between balls and races) is mounted on a shaft with an excessively heavy press fit, the bearing will operate with increased friction and torque. This is because, with the outer ring held firmly, the inner ring has been expanded, "pinching" the balls between the two rings.

2. If the bearing housing is out-of-round, the outer ring will tend to conform to the shape of the housing. This will exert a localized pressure on the ball contact area in addition to the normal pressure imposed by the operating work load.

3. PREVENTIVE MEASURES

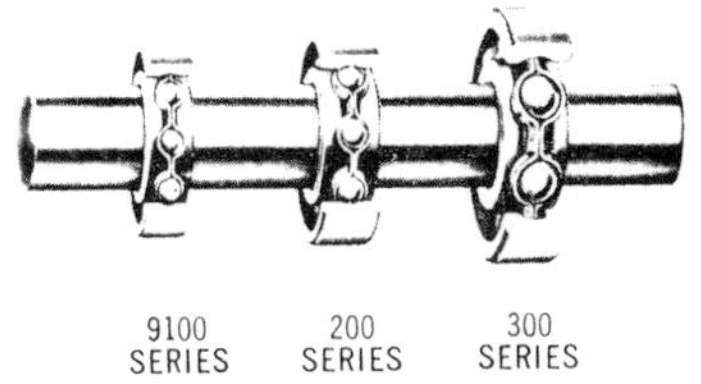

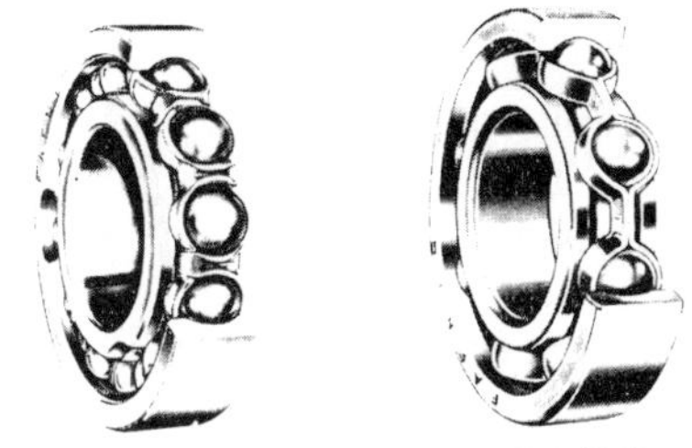
MAXIMUM CAPACITY TYPE

CONRAD TYPE

FOR NORMAL FATIGUE FAILURE

In the case of normal fatigue failure, the bearing has lasted its expected life. The remedy is simply to replace the bearing.

OVERLOAD FAILURE

Where overload is the cause of a premature fatigue failure, several alternatives are open:

1. Redesign to permit incorporating a bearing with greater capacity. There are extra-light, light, medium, and double row series bearings available in the same shaft sizes. The heavier series have thicker ring sections and larger balls for greater capacity. Also available for higher radial capacity is the maximum capacity type of bearing. This has exactly the same dimensions as the standard deep-groove Conrad type with the added advantage of more balls which are introduced through loading slots in the rings. These extra balls give this bearing increased radial capacity but the loading slots ground into the races restrict the thrust capacity of the bearings.

2. The load may be decreased to prolong the life of the bearing.

3. Housings should be gaged for out-of-roundness and machined or ground to proper symmetry and size. This insures that the outer ring will not be "pinched" or "squeezed" resulting in an overload condition.

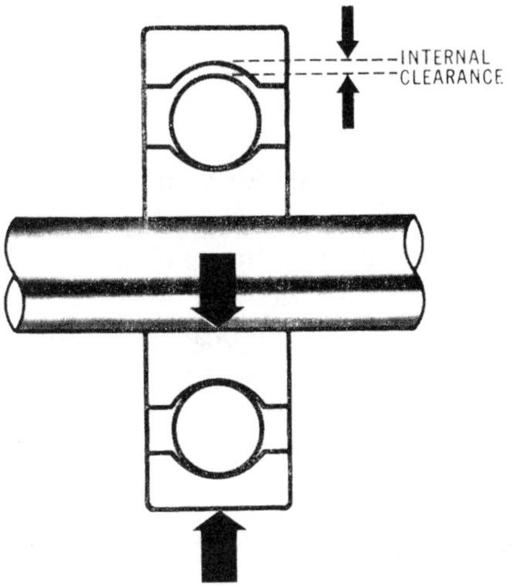

4. If the failure is caused by an overload imposed by inner ring expansion, either the shaft fit may be made looser by regrinding it to proper size or, in the event of thermal expansion, a bearing with looser internal fit (more clearance between balls and rings) may be recommended.

CONTAMINATION

1. IDENTIFICATION

◄ SCORING, PITTING, SCRATCHING

In a failure caused by the entrance of contaminants into the bearing, there may be a number of identifying earmarks. Where larger particles of dust or dirt are present, there will be scratches and pits around the periphery of the race with corresponding scoring of the ball. Where the contamination is in the form of very fine abrasives such as glass powder, graphite, or dust impregnated lubricants, the impurities will act as a lapping agent altering the appearance of balls and raceways. This type of failure may be characterized by intermittent noise in the bearing. The actual presence of dirt in the bearing is the best indication of this type of failure.

RUST ►

Where rust forms on the O.D. of the outer ring, it will not usually interfere with the bearings' operation in standard application, but rust in the bore is more serious because of the importance of inner-ring-to-shaft fit. Rust in the raceways or on the ball surfaces precludes any further usage of the bearing.

2. CAUSE

DIRTY OR DAMP SURROUNDINGS

Most bearing failures may be traced to some sort of contamination. Dirty working conditions are one of the bearing user's greatest problems. Thousands of dollars a year may be saved simply by taking certain precautions (outlined below) against the entrance of impurities into bearings. Internal clearance in precision bearings is measured in ten thousandths of an inch. Most dirt particles are larger than one thousandth of an inch, so hard particles will indent the race and ball surfaces when the bearing rotates.

ABRASIVE WASTE MATERIALS

In many machinery applications—farm implements, industrial grinders, wood working tools and many others—there may be an abrasive waste product which infringes upon the efficient operation of bearings. This would also apply where coolants, washing solutions, acids, or other liquids are used around a bearing application.

3. PREVENTIVE MEASURES

AVOID DAMAGE FROM ABRASIVE WASTE PRODUCT

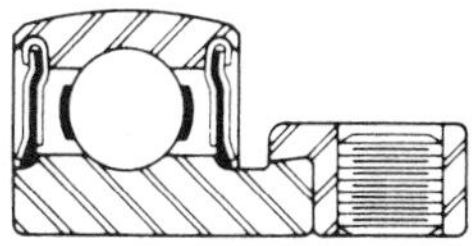

PLYA SEAL

Where a manufacturing process involves an abrasive by-product, it is essential that the bearing be properly sealed. Where failures of this kind prevail, a more efficient seal is required. In applications where extreme contamination is encountered, a sealed housing or shroud may be incorporated to protect the bearing.

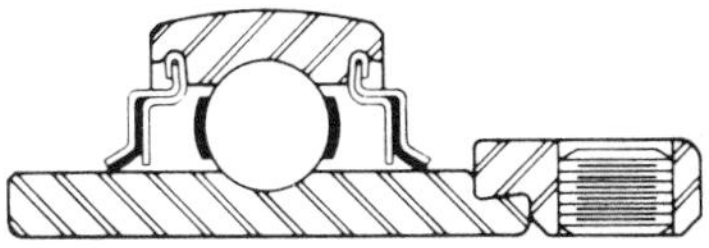

"R" SEAL

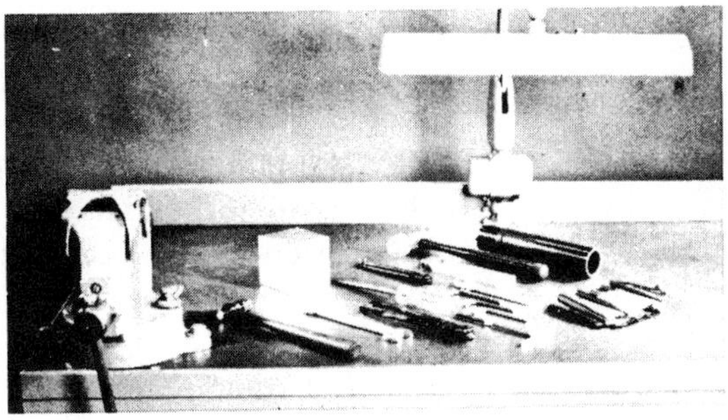

CLEAN WORKING SURROUNDINGS

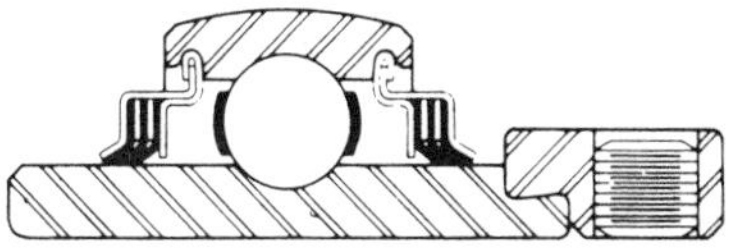

TRI-PLY SEAL

Clean tools, dirt-free working area, and clean dry hands are essential to prevention of bearing failures. Careful control of washing procedures, relubrication, and handling of bearings according to recommended practices will alleviate many of the problems of contamination failures. The following is a list of procedures outlined by the Anti-Friction Bearing Manufacturers Association for the control of cleanliness in the handling of bearings:

1. *Work with clean tools in clean surroundings.*
2. *Remove all outside dirt from housing before exposing bearings.*
3. *Handle with clean, dry hands.*
4. *Treat a used bearing as carefully as a new one.*
5. *Use clean solvents and flushing oils.*
6. *Lay bearings out on clean paper and cover.*
7. *Protect disassembled bearings from dirt and moisture.*
8. *Use clean, lint-free rags if bearings are wiped.*
9. *Keep bearings wrapped in oil-proof paper when not in use.*
10. *Clean inside of housing before replacing bearings.*
11. *Install new bearings as removed from packages.*
12. *Keep bearing lubricants clean when applying and cover containers when not in use.*

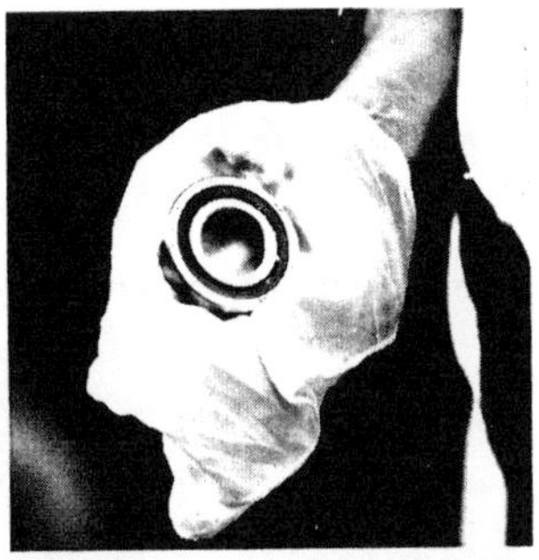

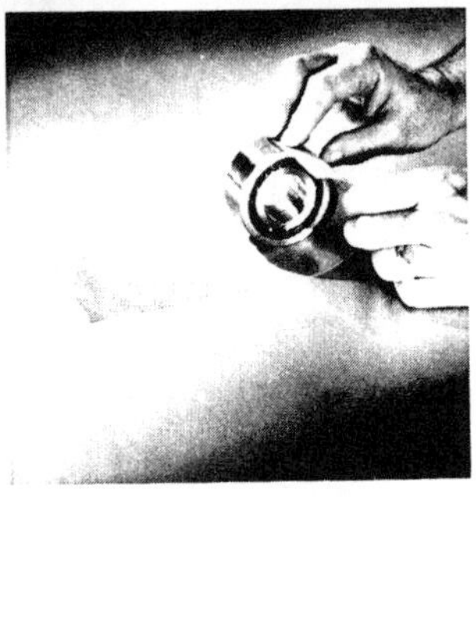

BRINELLING

1. IDENTIFICATION

MOUNTING INDENTATIONS (THRUST FORCE)

This failure will appear as tiny indentations (sometimes barely discernable to the naked eye) high on the shoulder of the race. The dents will be angularly spaced in correspondence to the ball spacing. There will be a corresponding indentation of lesser magnitude on each ball. When the bearing is radially loaded, the brinnels on the race shoulder may not interfere with the ball tracks. In this event, the dent on the ball will cause the failure. In the later stages of failure, spalling or chipping may result. The race shoulders can be inspected (with a microscope if available) to see if a spalling failure may not have resulted from initial brinelling.

RADIAL INDENTATIONS (RADIAL FORCE)

The indentations have the same general appearance as those mentioned above except that they appear in the center of the race instead of on the shoulder. This type of brinell is less common than the mounting brinell because, under the sharp impact of radial shock load, the rings may fracture beneath the force.

2. CAUSE

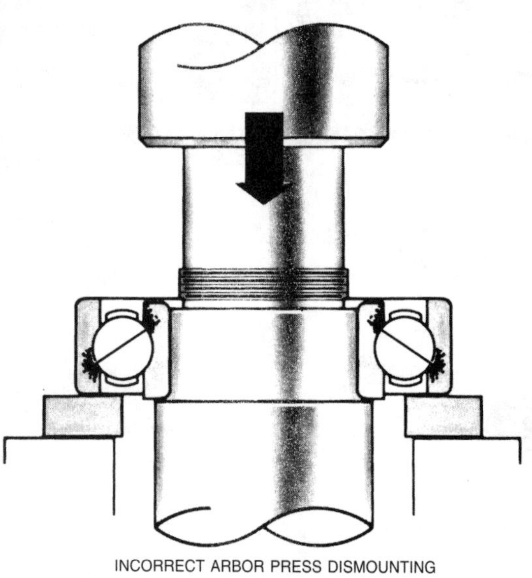

INCORRECT ARBOR PRESS DISMOUNTING

FORCE INCORRECTLY EXERTED

Indentations high on the race shoulder are caused in mounting (or dismounting) where force is applied against the unmounted ring. When mounting a bearing on a shaft with a very close fit, pushing of the outer ring will exert an excessive thrust load bringing the balls into sharp contact with the race shoulder, causing a brinell.

RADIAL SHOCK LOAD

Radial indentations are caused by a shock load imposed radially on a non-rotating bearing. This may be imposed by hitting the bearing with a hammer or by an operating shock load exerted on a static shaft.

3. PREVENTIVE MEASURES

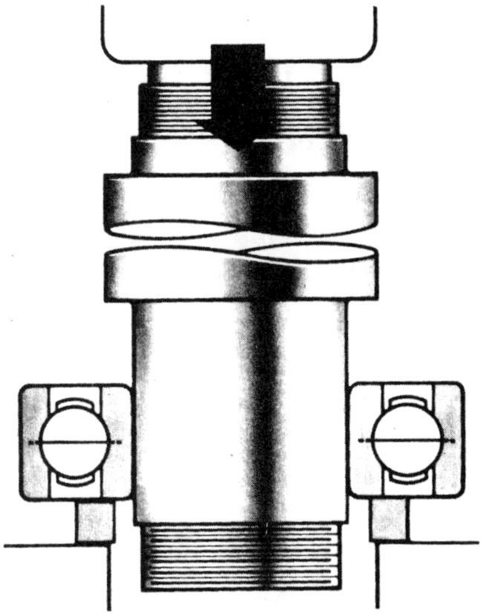

PROPER MOUNTING PROCEDURE

In mounting a bearing, force should always be exerted against the ring being mounted. In other words, when mounting the bearing on a shaft, the pressure should be applied against the inner ring. When mounting in a housing press against the outer ring. In other words the ring having the tighter fit (usually the ring which will rotate in application) should be pushed.

Be sure when mounting a bearing to apply the mounting pressure slowly and evenly.

OTHER METHODS OF MOUNTING OR DISMOUNTING

Dismounting with bearing puller and puller jaws.

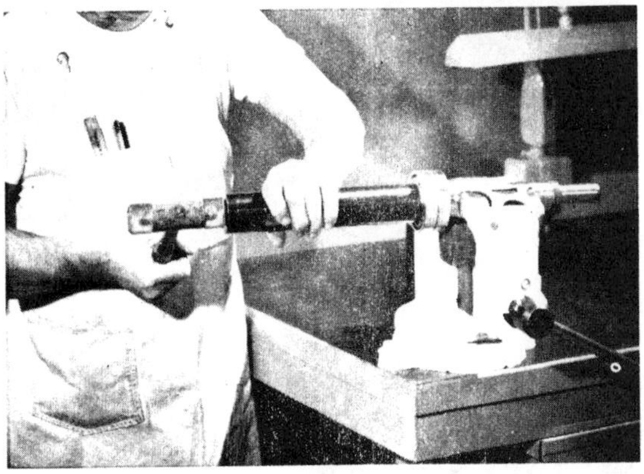

Mounting with tubular drift.

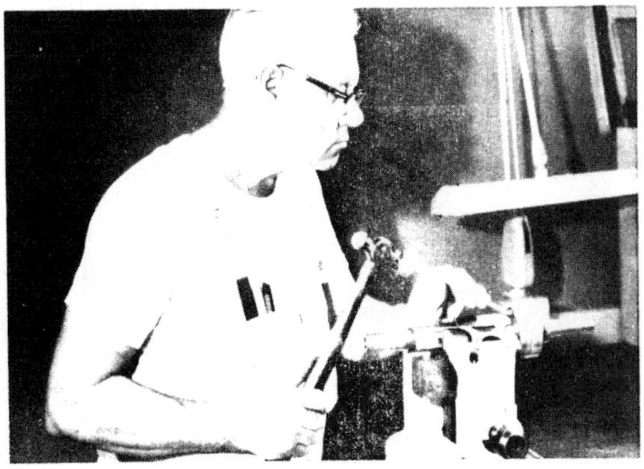

Dismounting with semi-circular drift.

FALSE BRINELLING

1. IDENTIFICATION

AXIAL INDENTATIONS

This type of brinelling will appear as elliptical impressions which run axially across the races. There will be a build-up of reddish lubricant around each brinell. Also, the brinells will be spaced corresponding with the ball spacing.

CIRCUMFERENTIAL INDENTATIONS

This will appear exactly as the brinell above except that the impression will be wider in a circumferential direction.

2. CAUSE

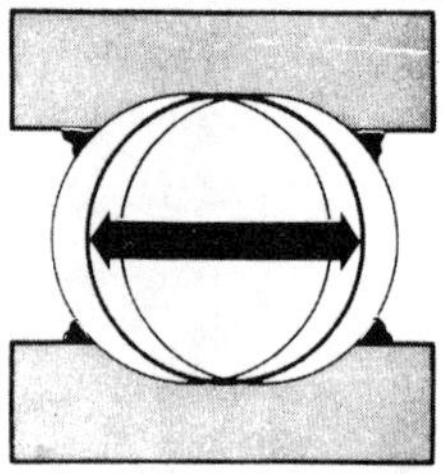

AXIAL VIBRATION

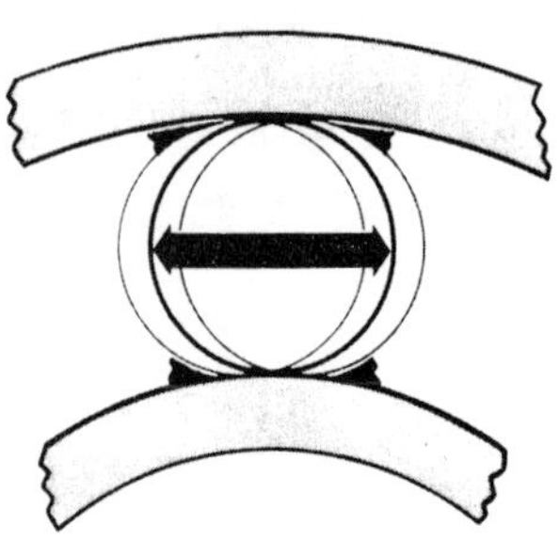

OSCILLATORY VIBRATION

VIBRATION IN A STATIC BEARING

False brinelling is caused by a vibration of the balls between the races in a stationary bearing. This vibration may be either axial or circumferential. The appearance of the brinells will tell you which. As the ball vibrates between the races, the lubricant is forced out of the contact area between ball and race. The failure is the result of a breakdown of the lubricant causing metal-to-metal contact and localized wear of balls and races. The wearing action causes the formation of a fine reddish brown powder, iron oxide. The oxide impregnates the lubricant and provides an abrasive compound which will polish the balls and races if the bearing is put into operation. The indentations themselves will result in a rough and noisy operation of the bearing. Vibration in a bearing may be caused by a number of factors. Transporting mounted, but unlocked, bearings can result in a false brinell failure. A very common cause of this failure exists where a machine is "down", and the bearings are subjected to static vibration by other machinery which is running in the immediate area.

3. PREVENTIVE MEASURES

CORRECT THE SOURCE OF VIBRATION

The source of agitation—loose parts, non-precision machinery, rough transportation—may be corrected so that vibration is alleviated.

LOCKING THE BEARING

In the transportation of bearings, if a light thrust load (imposed by springs or rubber pads) can be applied to bring all of the balls into contact with the races, this type of failure can be avoided.

ALL SURFACES ADEQUATELY LUBRICATED

Where bearings are oil lubricated and employed in units that may not be in service for extended periods, the equipment should be set in motion periodically in order to spread the lubricant over all bearing surfaces. Intervals of 1 to 3 months should suffice.

TIGHT INTERNAL FIT-UP

Sometimes a bearing with line-to-line contact between rings and balls will alleviate a false brinell failure. Great care should be taken, however, that a tight internal fit is satisfactory from an operation point of view.

LOW VISCOSITY LUBRICANT

False brinelling is more common with stiffer lubricants. This failure is less apt to occur where oil or a light viscosity grease is used, because the more liquid characteristics make it difficult for the lubrication to be forced out of the contact area.

THRUST FAILURES

1. IDENTIFICATION

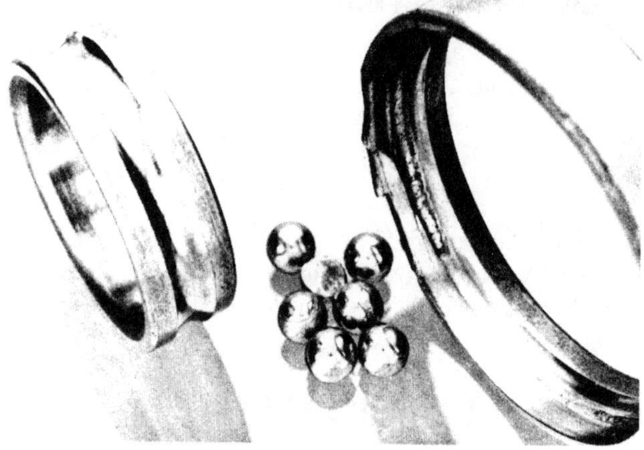

COUNTERBORED BEARING

There will be a breakdown of the counterbored shoulder of the bearing which may result in fracture of the ring. The balls will be banded from riding up against the shallow shoulder. Also the bearing may become disassembled during service.

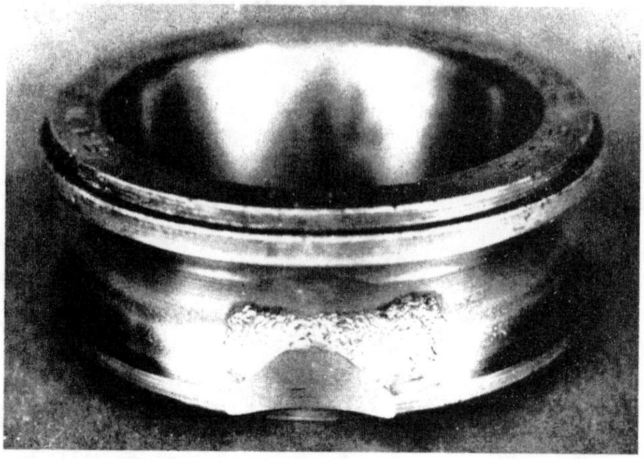

MAXIMUM CAPACITY

Bearings with filling slots are not recommended for heavy thrust loading because, as the balls pass over the inner ring and outer ring notches, they may become nicked or dented. This in turn may cause spalling of the races (probably in the vicinity of the loading slot).

2. CAUSE

IMPROPER MOUNTING OR MISAPPLICATION AS INDICATED:

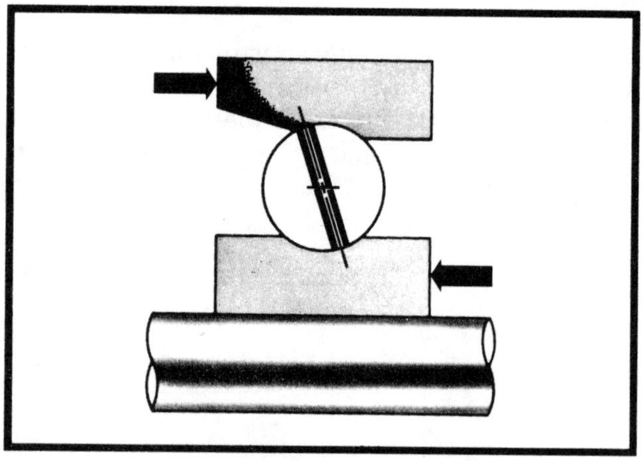

COUNTERBORE FAILURE

A thrust failure is caused either by mounting the bearing backwards (so that the load is carried against the shallow shoulder) or by putting a counterbored bearing into a two-directional thrust application.

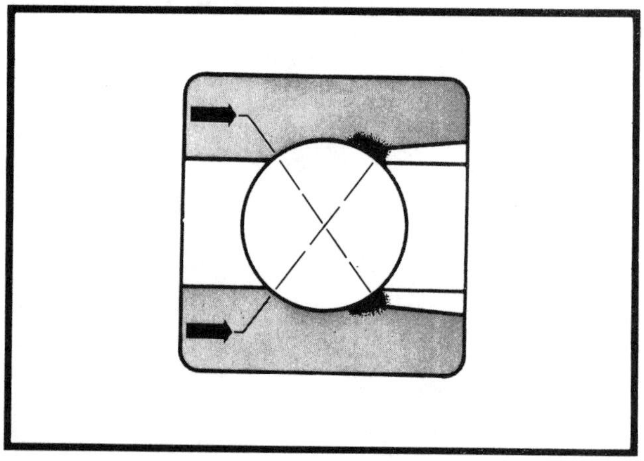

MAXIMUM CAPACITY FAILURE

This failure results from excessive thrust loads on a bearing not primarily intended for heavy thrust loads. From the diagram it is evident that too much thrust load from either direction will cause interference between the balls and one of the loading slots which are ground in both the inner and outer rings.

3. PREVENTIVE MEASURES

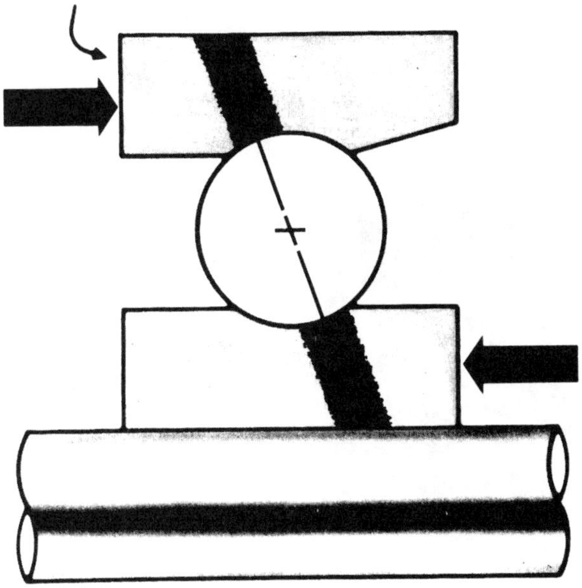

COUNTERBORE FAILURE

The remedy here is simply to mount the bearing correctly so that the ball has full-shoulder support on both the inner and outer ring. Remember that the outer ring counterbore bearing will take thrust against the inner ring on the counterbored side of the bearing and the outer ring on the side opposite the counterbore. The word THRUST will be stamped on the outer ring face showing the proper thrust surface.

MAXIMUM CAPACITY FAILURE

This bearing is designed for heavy radial or combined radial-thrust loads; not for pure thrust loading. (It is generally recommended that no more than 60 percent of the accompanying radial load on the bearing be applied in thrust.) If thrust capacity is required, a more suitable bearing (a Conrad type, a counterbored bearing, or possibly a duplexed pair of bearings) must be selected.

MISALIGNMENT

1. IDENTIFICATION

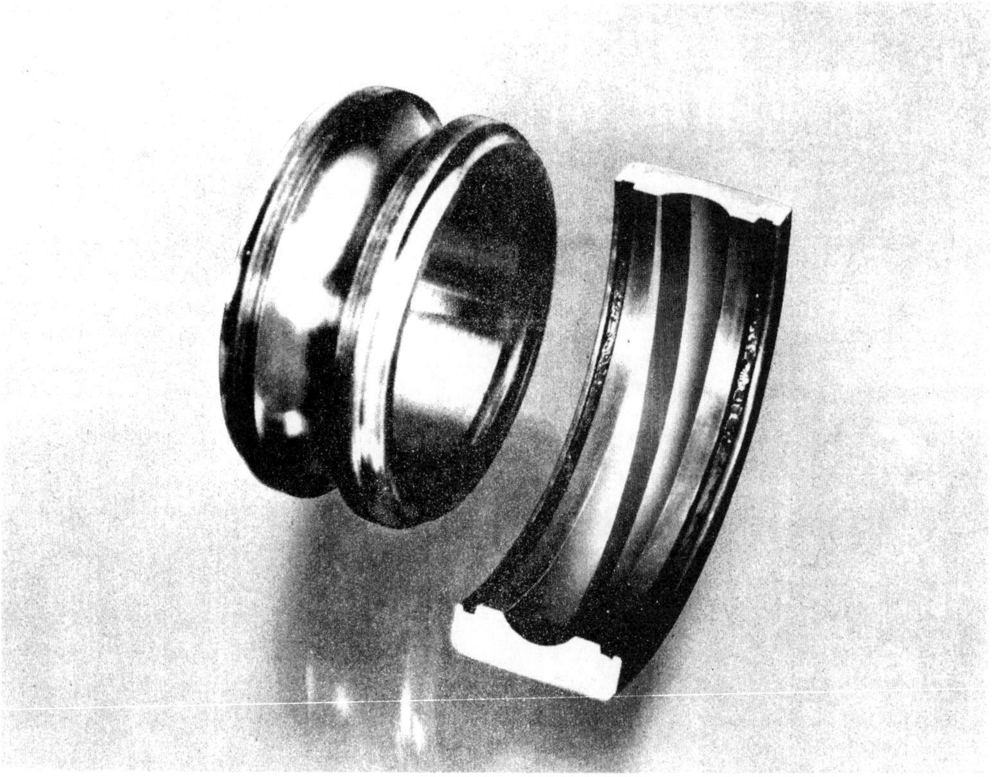

BALL PATH

In a bearing with one of its rings misaligned in relation to the other, the ball path will run from one side of the race to the other around one half the circumference on the nonrotating ring. The rotating ring will have a wide ball path. Because of the extra pressure imposed on the bearing due to misaligned conditions, an excessively high temperature may develop which will discolor the ball tracks and balls and destroy the lubricant.

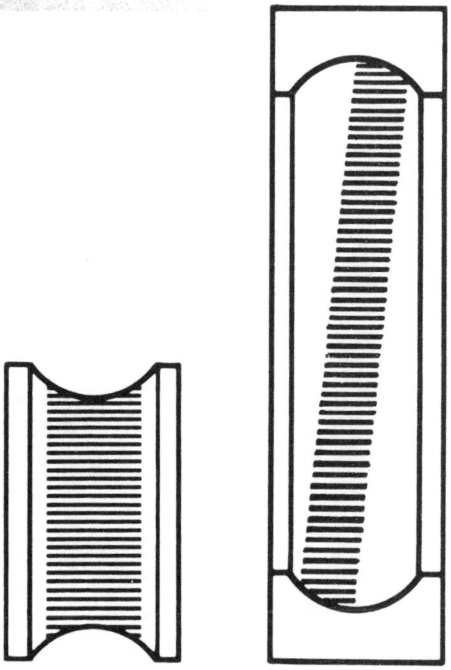

RETAINER

The purpose of the retainer is to space the balls and guide them in a true path around the raceway. Where a ring is misaligned, the balls are driven up against the race shoulder, and a stress point is set up between the ball and its retainer pocket. The pocket will flex with the possibility of retainer fracture in advanced stages of failure.

2. CAUSE

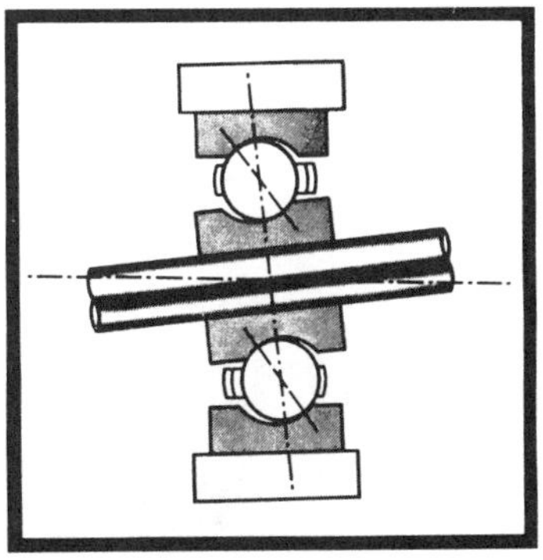

SHAFT MISALIGNMENT

Misalignment of the shaft in relation to the housing causes an overload of the balls which will result in the failure described.

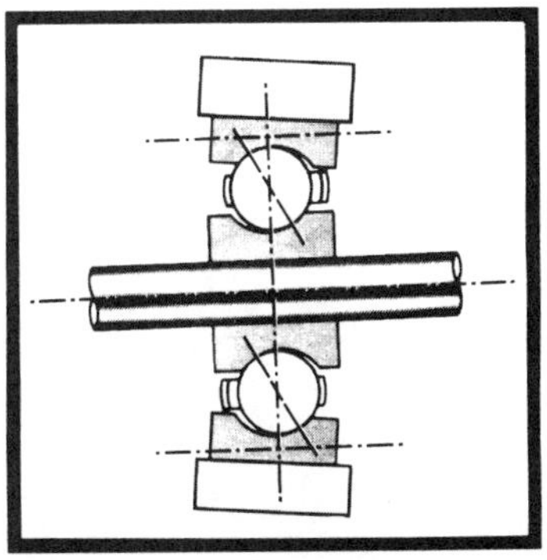

MISALIGNMENT OF THE HOUSING

Housing misalignment may be caused either by the housings being cocked with the plane of the shaft or the housing shoulder being ground out-of-square so that it forces the outer ring to cock in relation to the inner. It may also result from settling of the frames or foundations.

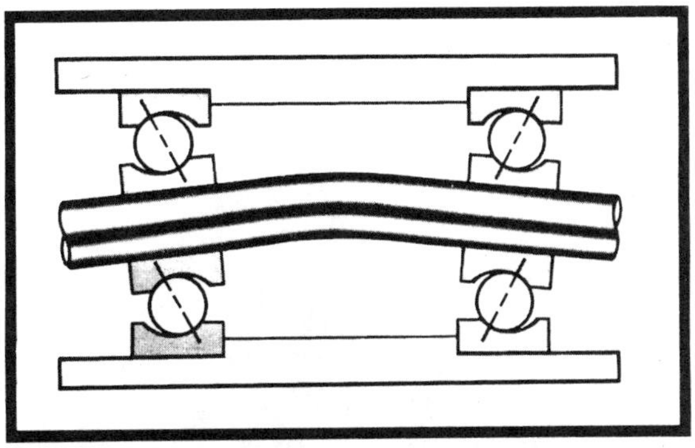

BOWING OF THE SHAFT

Shaft bowing may be caused by any one of the following:

(a) Bent as a result of improper handling
(b) Overhung load exceeding shaft capacity
(c) Initial shaft bowing due to grinding inaccuracies
(d) Shaft shoulders ground out-of-square with the shaft center line which will, by cocking the inner ring, force a bowing of the shaft.

3. PREVENTIVE MEASURES

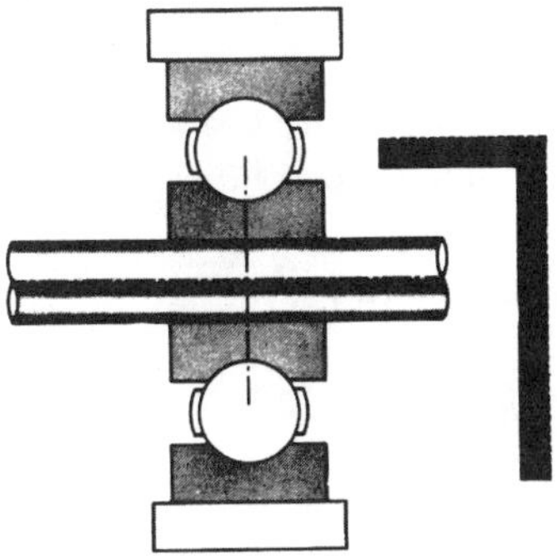

SHAFT

The shaft should be gaged to make sure that it is concentric and straight. Heavy overhang loads should be lightened or moved closer to the bearing. If the shoulders are out-of-square, they should be reground and gaged so that they are perpendicular to both the bearing seat and the shaft centerline.

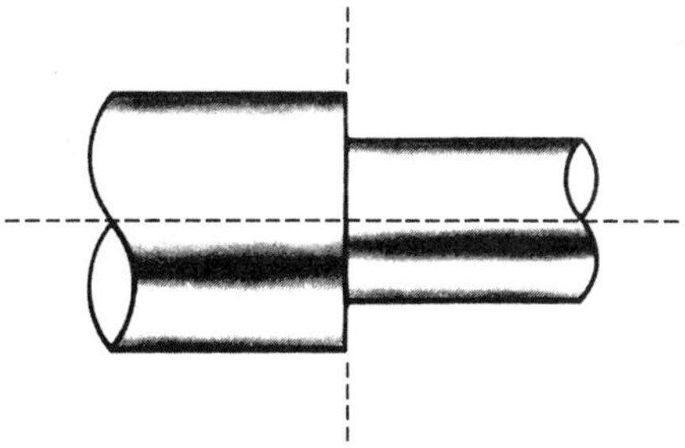

SQUARING-UP SHAFT SHOULDERS

HOUSING

The remedy is to dimensionally check and insure that both housing bores are true with each other.

ELECTRIC ARCING

1. IDENTIFICATION

Electro-etching on both inner and outer ring.

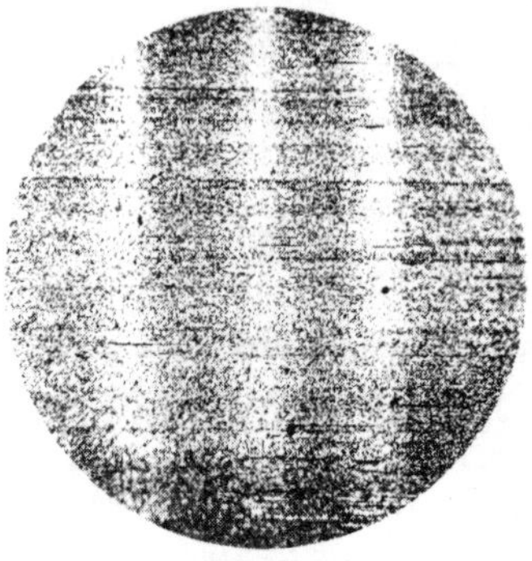
Magnification of granular race surface.

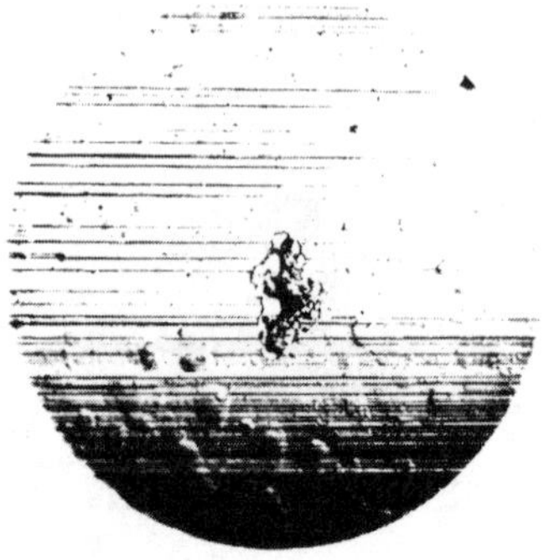
Magnification of electric pitting.

ELECTRIC ARC EROSION

When an electric current passing through a bearing is broken at the contact surfaces between races and balls, arcing results which produces high temperatures at the localized points. Each time the current is broken in its passage between ball and race a pit is produced on both parts. Eventually the phenomenon known as fluting develops, (see photograph) and as it becomes deeper, noise and vibration result.

GRANULAR RACE SURFACES

If the current is of higher amperage intensity, such as a partial short circuit, the next phase of the failure will show up as a rough granular appearance in the ball track.

PITTING OR CRATERING

Heavy jolts of high amperage charges will cause a more severe failure resulting in the welding of metal from the race to the ball. These nipples of metal on the ball will, in turn, cause a cratering effect in the race. This phenomenon will result in noise and vibration in the bearing.

2. CAUSE

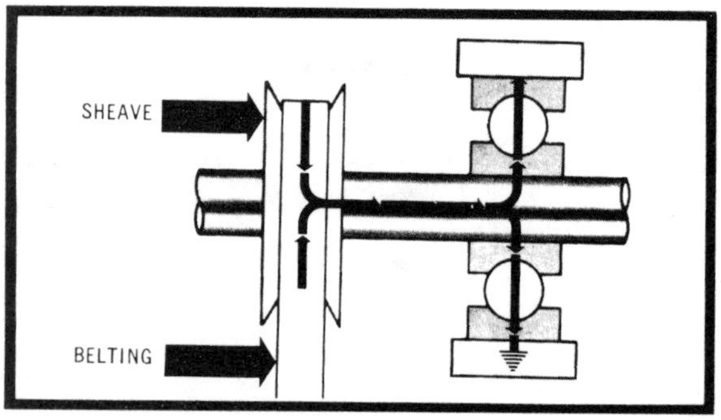

STATIC ELECTRICITY

Static electricity usually emanates from charged belts or from manufacturing processes using calendar rolls (leather, rubber, cloth, paper).

The current is carried from the belt to the pulley or sheave; from the sheave to the shaft; and through the shaft to the bearing and thence to the ground.

ELECTRIC LEAKAGE

Faulty wiring, inadequate or defective insulation, or loose rotor windings on an electric motor are all possible sources of current leakage. Either AC or DC currents will damage bearings.

SHORT CIRCUIT

Wires which are crossed or contacted by a common conductor will cause a short circuit and may result in a passage of current through the bearing.

3. PREVENTIVE MEASURES

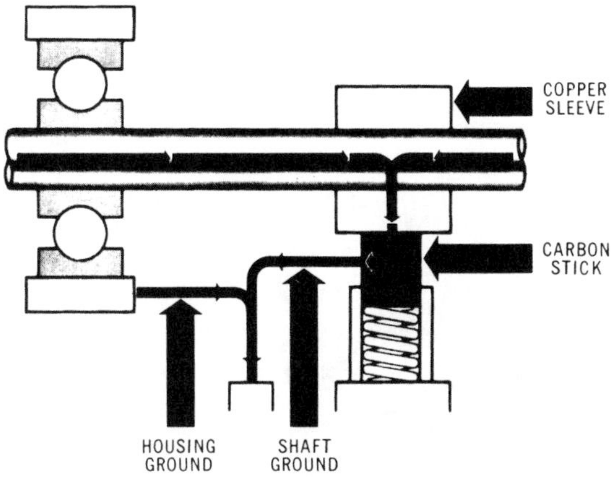

SHUNTS AND SLIP RINGS

Where there is a passage of current through a bearing and the source of the current cannot be corrected, a shunt in the form of slip ring assembly may be incorporated to bypass the current around the bearing.

CORRECTIVE MAINTENANCE

Be sure wiring, insulation, or rotor windings are sound and all connections are properly made. In arc welding, great care should be taken that the welding apparatus is not grounded on something which will circulate the current through the bearings.

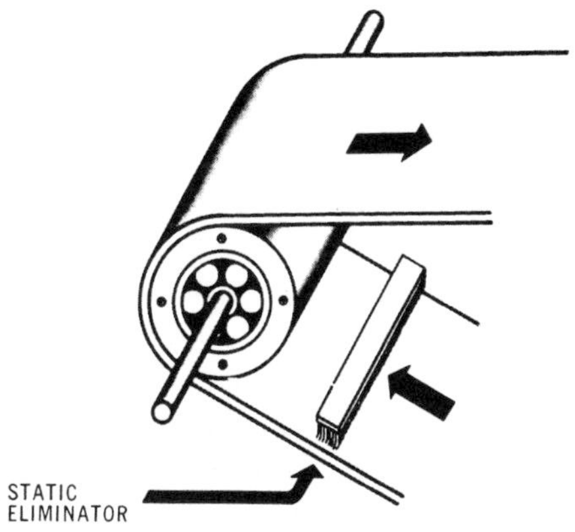

GROUNDING BELTS

For elimination of static charges, either the belt should be grounded, or the belting should be changed to a less generative material.

INSULATING BEARINGS

Sleeves of nonconductive material may be used either between the outer ring and housing or between inner ring and shaft, depending upon the source of current.

LUBRICATION FAILURE

1. IDENTIFICATION

New bearing with fresh grease on the left as contrasted with lubrication breakdown on the right.

GREASE APPEARANCE

If the grease is stiff or caked and changed in color, it indicates lubrication failure. The original color will usually turn to a dark shade or to jet black (see photo.) The grease will have an odor of burnt petroleum oil. Lubricity will be lost as a result of lack of oil. In cases of Lithium base greases the residue appears like a glossy brittle varnish which will shatter when probed with a sharp instrument.

ABNORMAL TEMPERATURE RISE

Probably the first indication of lubricant failure is a rapid rise from the normal operating temperature. Test by hand is not necessarily conclusive since normal operating temperature may exceed the bearable limit of about 120° F.

NOISE

Lack of lubrication is soon accompanied by a whistling noise coupled with this rise in temperature. If not corrected the bearing temperature will continue to rise and the intense heating will reduce the bearing hardness.

BEARING DISCOLORATION

A brownish or bluish discoloration of the races and balls indicates that the bearing operating temperature was excessively high to the extent that the bearing lost its physical properties and was no longer operable.

RETAINER FAILURE

The bearing part that first indicates distress in lubrication failures is usually the retainer where the greatest amount of rubbing action takes place.

2. CAUSE

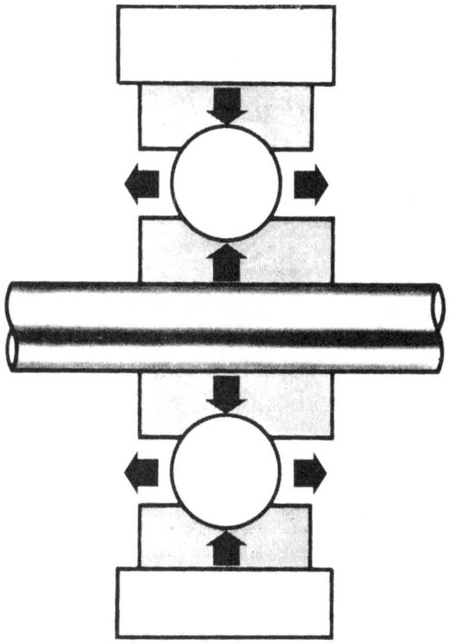

Lubrication failures are usually accompanied by a thermal expansion of the components

DIRTY LUBRICATION

The Fafnir Bearing Company filters ball bearing lubricants as many as five times to insure their purity. Contaminants found in lubricants often act as an abrasive compound which will lap or polish ball and race surfaces, increasing the probability of early failure.

TOO MUCH LUBRICANT

A very common error in the maintenance of machinery is the tendency to over-lubricate. If the bearing reservoir is kept constantly full of grease, the friction heat developed within the lubricant will cause its rapid deterioration.

INADEQUATE LUBRICATION

Heat will result from under-lubrication, also. Where there is inadequate lubricant to cover all metal surfaces, friction will result in heat-up of the bearing.

WRONG KIND OF LUBRICANT

After experimentation with many types of lubricants, the equipment manufacturer recommends those which he feels will provide ideal lubrication life under given operation conditions. Inasfar as availability allows, you should use the same lubricant or its equivalent. This way you are assured of the correct lubricant in addition to avoiding the problems associated with mixing two types of grease. (Many greases are incompatible and, although completely adequate when used individually, may prove unsatisfactory when mixed.) Selection of the correct lubricant, therefore, is very important in achieving maximum efficiency and endurance from the application.

OTHER FAILURE MODES

In many instances lubricant failure will accompany the bearing failures described in this manual. The primary cause of lubricant failure is from the high temperatures developed when excessive loads exceed the strength of the lubricant film. Lubricant changes might reduce the failure rate but the proper course of action is to eliminate the primary cause for the lubricant breakdown.

3. PREVENTIVE MEASURES

AVOIDING DIRTY LUBRICANT

Always keep grease containers covered. Dust particles in the air can contaminate the lubricant. Use a clean rust-resistant spatula for relubricating open bearings. When relubricating bearings through a grease fitting, always wipe off both the fitting and the nozzle of the grease gun. Any steps which you can take to keep lubricants clean will pay off in longer bearing life.

AMOUNT OF LUBRICANT

When you are unsure of the amount of grease or oil for proper lubrication, the Fafnir engineering department should be consulted. In standard applications, it is generally recommended that the bearing should be greased one-third to one-half full.

CAM FAILURE

1. IDENTIFICATION

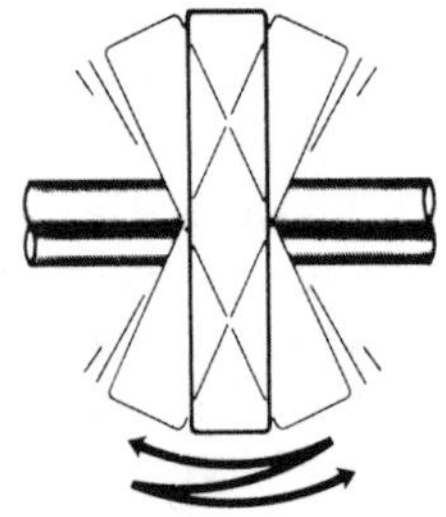

BROKEN CAM

The eccentric cam on the wide inner ring bearing will have a wedge-shaped fracture in it. Depending upon the degree of failure the fracture may extend into the cam or all the way into the race of the bearing.

BALL PATH

Because the inner ring will be cocked on the shaft, the ball path will have a wobbly appearance similar to that on a misaligned bearing. The ball path will run from one side of the non-rotating ring to the other for one-half of the perimeter of the race.

WOBBLE OF BEARING

While in application, the rotating bearing will wobble from side to side. If the housing is flexible enough, the whole assembly may flex with the rotation of the bearing.

2. CAUSE

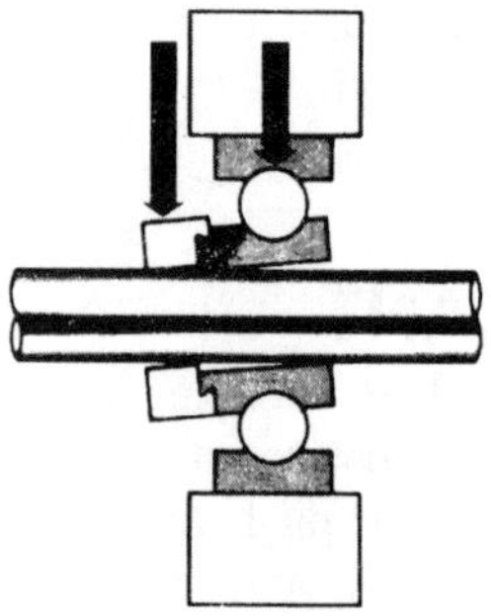

SHAFTING BELOW SUGGESTED TOLERANCE LEVELS

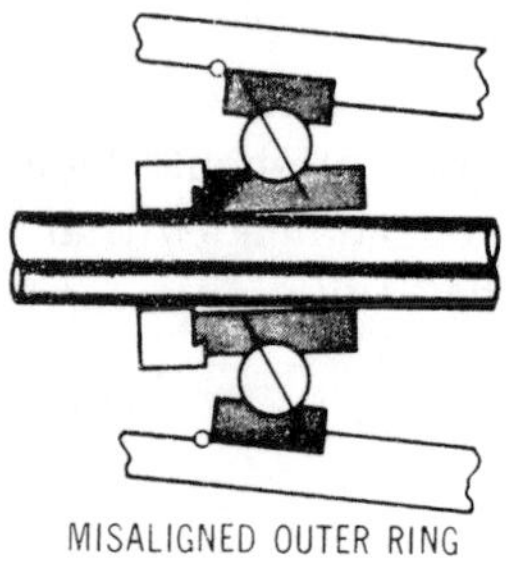

MISALIGNED OUTER RING

UNDERSIZED SHAFTING

The eccentric cam locking collar was originally developed to provide a means of securing a bearing to low cost cold-rolled shafting. It must not be assumed, however, that the bore-to-shaft fit is without some tolerance control. Fafnir has set up tolerance limits (see Preventative Measures for recommended limits) for shaft fits on wide inner ring bearings, which must be maintained if this type of failure is to be avoided. When the eccentric collar is locked to its inner ring mating cam on an undersized shaft, the inner ring becomes cocked on the shaft. (Remember that, for proper locking action, the collar should always be turned in the direction of shaft rotation.) When the machinery is turned on, the inner ring tries to realign itself with the shaft as it rotates. This provides a levering action with a stress area where the collar grasps the inner ring. Eventually this stress will fracture the inner ring cam.

OUTER RING UNABLE TO ALIGN

The same problem exists where an outer ring with a self-aligning feature is clamped so firmly into the housing that it cannot take advantage of its self-aligning feature. This may happen where the bearing is forced with an excessively tight fit into a cast housing or where pressed steel housing halves are bolted together before the bearing has assumed its initial alignment.

3. PREVENTIVE MEASURES

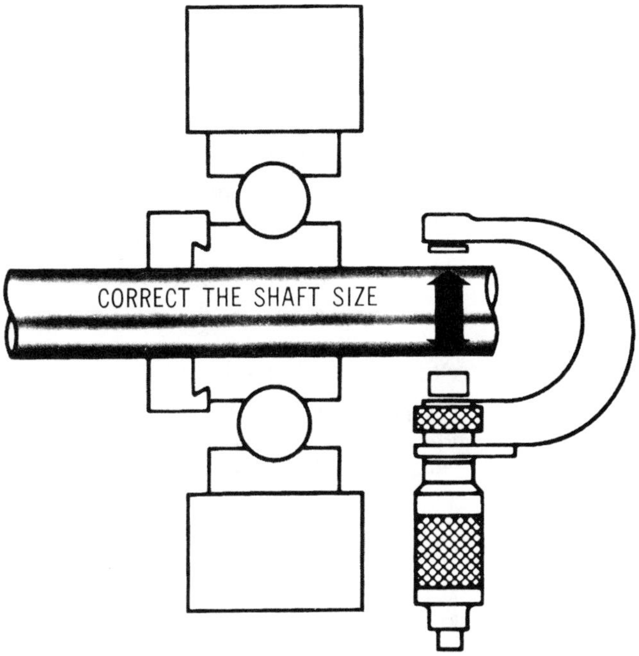

CORRECT THE SHAFT SIZE

Where a failure occurs, the shaft should be checked with a micrometer caliper to determine how much under normal tolerance the shaft diameter falls. Then, either a new shaft should be obtained or the old one should be plated and ground to correct size.

The following are the recommended tolerances for shafting:

BEARING BORE	SHAFT TOLERANCES
½″ to 2″	.0000 to −.0005
2¹⁄₁₆″ to 3¹⁵⁄₁₆″	.0000 to −.001
4″ and over	Because of heavy loading, engineering department should be consulted.

ALLOWING THE BEARING TO ALIGN

Fafnir has incorporated a self-aligning feature—a spherical outer ring—into its wide inner ring line to compensate for a certain amount of initial misalignment. The proper mounting procedure is to: (1) Align the bearing in its housing and slide unit into position on the shaft (2) bolt the housing tightly to its mounting support (3) engage and tighten locking collar and set screw.

4. False brinelling,
5. Thrust failures,
6. Misalignment,
7. Electric arcing,
8. Lubrication failure, and
9. Cam failure.

The following section is published with the permission of the Fafnir Bearing Company, Division of Torrington (formerly Textron), from the publication *HOW TO PREVENT BALL BEARING FAILURE.*

QUESTIONS

Write your answers in a notebook and then compare your answers with those on page 266.

9.4L What are the different types of bearings used in pumps?

9.4M What should be done whenever a bearing failure occurs before its expected replacement time?

9.4N List as many different types of bearing failures as you can remember.

9.4O What can cause contamination failure of a bearing and how can it be prevented?

9.4P How can bearing failures from misalignment be prevented?

9.4Q How can bearing lubrication failures be identified?

9.45 Couplings

9.450 Purpose of Couplings

Pumps must be driven by some means in order to operate. Electric motors, gasoline or diesel engines, air pressure, hydraulic pressure, steam pressure, and other sources of power are used to drive pumps. Electric motors and gasoline or diesel engines are the most common sources of power used in the wastewater field. In order to connect these drivers to the pump, a coupling is needed (Figure 9.46). There are several ways to couple a pump and driver together. Some of the most common methods are:

1. Direct coupling,
2. Belt-driven or chain-driven,
3. Flexible shafting, and
4. Close-coupled.

Each method has its advantages and limitations and is used for a specific purpose.

9.451 Direct-Coupled

Pumps and drivers are directly coupled when they are mounted on a common base and the speed of the driver is the same as that of the pump. The shaft of the pump has one of the coupling ends, and the driver has the other end mounted on it. The coupling may be either rigid (clamp and compression) or flexible. A flexible coupling is a device to connect two shafts together, transmit torque, and at the same time allow for *VERY SLIGHT* misalignments. The main functions of all flexible couplings are:

1. To transmit power efficiently and effectively from one shaft to the other,
2. To decrease wear on shaft bearings and driven equipment, and
3. To accommodate the slight shaft misalignments which develop in service.

Other functions of flexible couplings include:

1. To absorb shock loads and pulsations,
2. To dampen vibrations,
3. To accommodate load reversals,
4. To minimize initial backlash, and
5. To provide ease of installation and maintenance.

ALIGNMENT (Figure 9.47)

When shafts are directly coupled, shaft alignment is extremely important. As discussed previously in Section 9.260, "Alignment," there are two types of shaft misalignment:

1. Parallel, and
2. Angular.

The coupling also can have end float; that is, the shaft moves in and out. This has to be taken into consideration when installing a coupling. Misaligned shafts not properly coupled are subject to severe stresses and damage at bearings and seals.

With couplings that are properly machined, accurate alignment is achieved when the coupling faces are absolutely parallel as measured by a feeler gage or a straightedge placed squarely across the rings of the coupling halves. Place the feelers at the top, bottom, near and far sides of the coupling junctions. Also place the feelers along the edge of the top and sides of the shaft. Proper alignment is accomplished by the use of wedges and *SHIMS*[11] under the pump or motor. A dial indicator also may be used to ensure proper alignment (see Section 9.260, Figure 9.26, page 185).

When installing a coupling, inspect it carefully. If you detect any misalignment, install wedges or shims until no variation is seen by the straightedge or feeler gage. Hold the pump half-coupling stationary and rotate the driver 90 degrees at a time and examine the alignment. Insert shims as necessary to bring shaft into alignment. Make sure both couplings are the same diameter and are clean. The clearance between faces of coupling halves should be set so they cannot strike, rub, or exert a pull on either machine.

Different types of couplings have different tolerances for alignment. Usually tolerances range from 0.002 inch to 0.005 inch. Consult with the particular coupling manufacturer for the exact alignment needed. There are so many factors entering into the probable life of a coupling that it is impossible to predict the impact of misalignment on coupling life for every application. Any coupling which is stressed heavily with torque will have very little reserve left for misalignment conditions. On the other hand, if a coupling is lightly torqued, it would have a

[11] *Shim. Thin metal sheets which are inserted between two surfaces to align or space the surfaces correctly. Shims can be used anywhere a spacer is needed. Usually shims are 0.001 to 0.020 inch thick.*

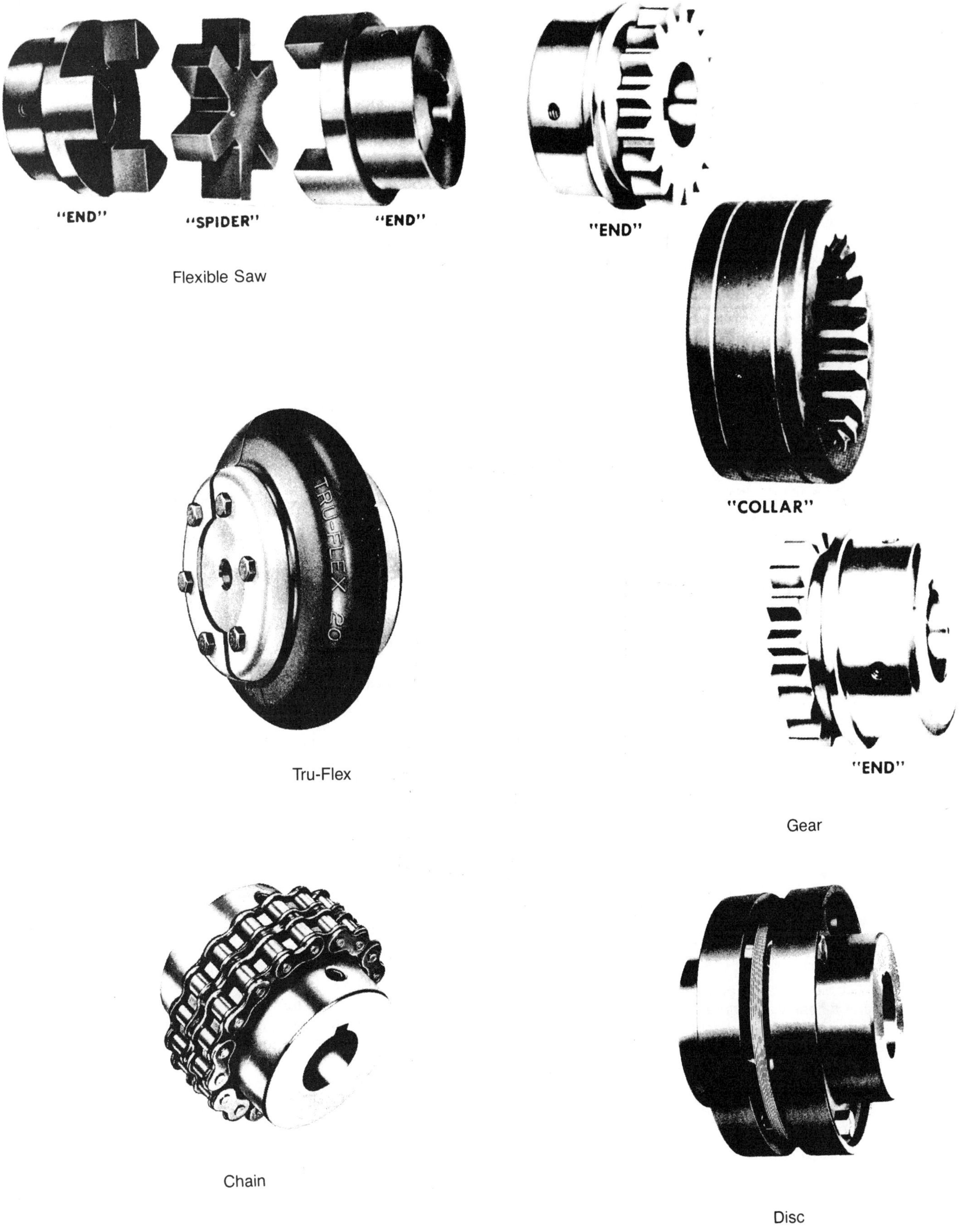

Fig. 9.46 Couplings

(Courtesy of Rexnord Inc., Milwaukee, WI)

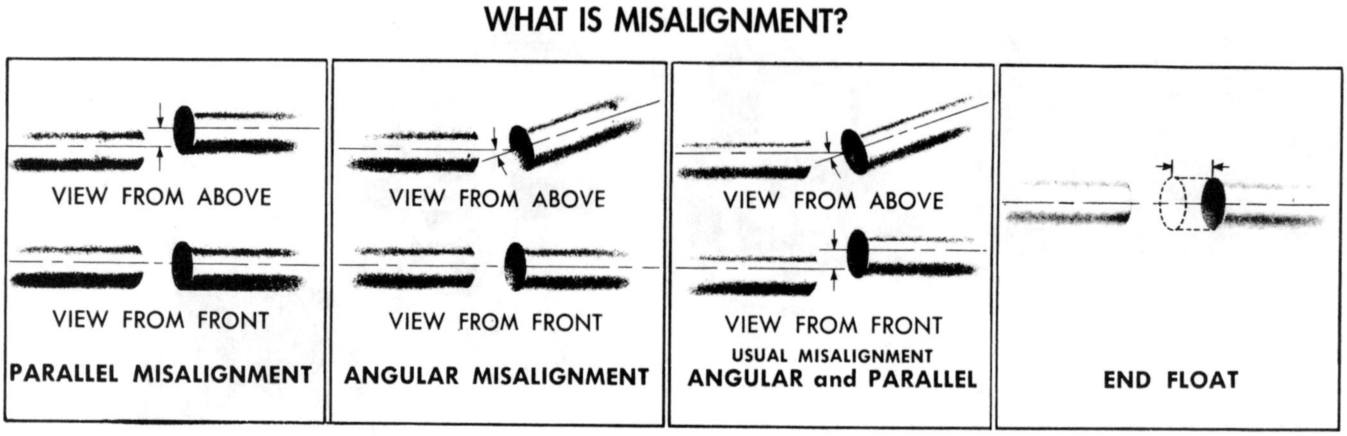

Fig. 9.47 Angular and parallel misalignment and end float
(Courtesy of Rexnord Inc., Milwaukee, WI)

larger reserve left for misalignment. Also, a coupling traveling at a slow speed, such as used on the slow side of a gear reducer, can take relatively large misalignments over a long period of time. If a coupling is rotating at high speeds, it must be aligned with great care to provide maximum trouble-free service.

QUESTIONS

Write your answers in a notebook and then compare your answers with those on page 266.

9.4R What is the purpose of a coupling?

9.4S List four common methods of coupling a pump and driver together.

9.4T What are the functions of a flexible coupling?

9.4U How can you tell if coupling alignment is proper?

9.4V What do you look for when inspecting clearances between faces of coupling halves?

9.452 Belt-Driven

Another method of connecting a pump to a motor or driver is with belts (Figure 9.48). A pulley is mounted on the pump shaft and another pulley is mounted on the driver. One or more belts are installed connecting the two pulleys together.

The driver can be located above the pump or beside the pump in many different configurations. The pump shaft and motor shaft must be positioned so the pulleys can be aligned with the belts in a straight line. An advantage of driving a pump with belts is the speed ratios that can be achieved between the motor and pump; almost any speed is available. If this speed needs to be changed for any reason, it is only necessary to change a pulley.

V-belt drive pulleys are more commonly known as sheaves. Sheaves are normally made of cast iron or steel and machined to size. They can be single groove, two groove, three groove, or many more depending upon horsepower requirements. The grooves can be for A, B, C, D, or E belts. The letter designation is the cross section of the belt to be used, "A" being the smallest and "E" the largest. A and B belts are most commonly used.

There are many manufacturers of V belts and it is impossible to list each type and manufacturer available. There are standard belts, super belts, torque-flex belts, band belts, and many others. The best way to select belts is to contact your supplier, give the horsepower, speed, equipment driver, type of load, sheave diameters, and shaft diameters. If it is an existing installation, use the belt numbers found on the belt when reordering. If there is more than one belt installed, they must all be replaced at the same time and be a matched set so all belts will wear evenly. The used belts also have been stretched and will not allow proper adjustment.

When installing or replacing V belts, the motor and pump sheaves must be aligned. This means that the belt must run straight between the two pulleys without any angle. This can be achieved by placing a straightedge between the two pulleys and making sure that both are on the same plane. Another method is to use a string pulled across both pulleys, making sure that the string hits the pulleys at all points. The belts must be placed over the sheaves and in the grooves without forcing them over the sides of the grooves which would damage the belts.

With all belts in their proper grooves, adjust the centers to take up all slack until the belts are fairly taut. Start the drive and continue to adjust until the belts have only a slight bow on the slack side of the belt drive while operating under load. Make sure that belt guards are in place when adjusting operating units. To determine proper adjustment of belts when the unit is not operating, turn power to unit off, tag and lock out. Adjust belt until there is $^{3}/_{4}$-inch of belt movement from hand pressure at the middle of the slack belt (Figure 9.49).

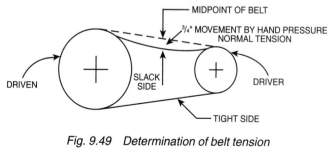

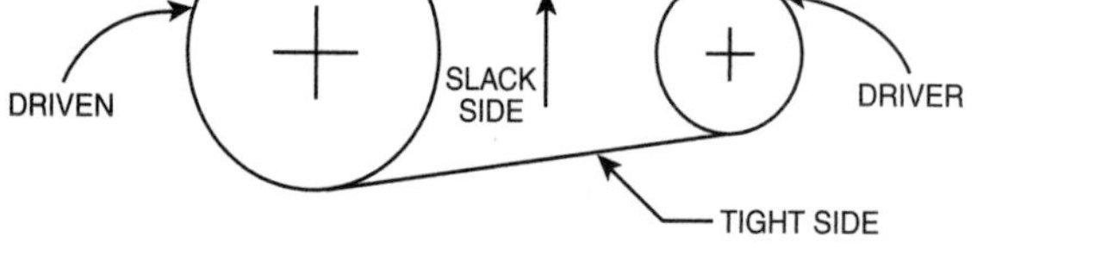

Fig. 9.49 Determination of belt tension
(Courtesy of Browning)

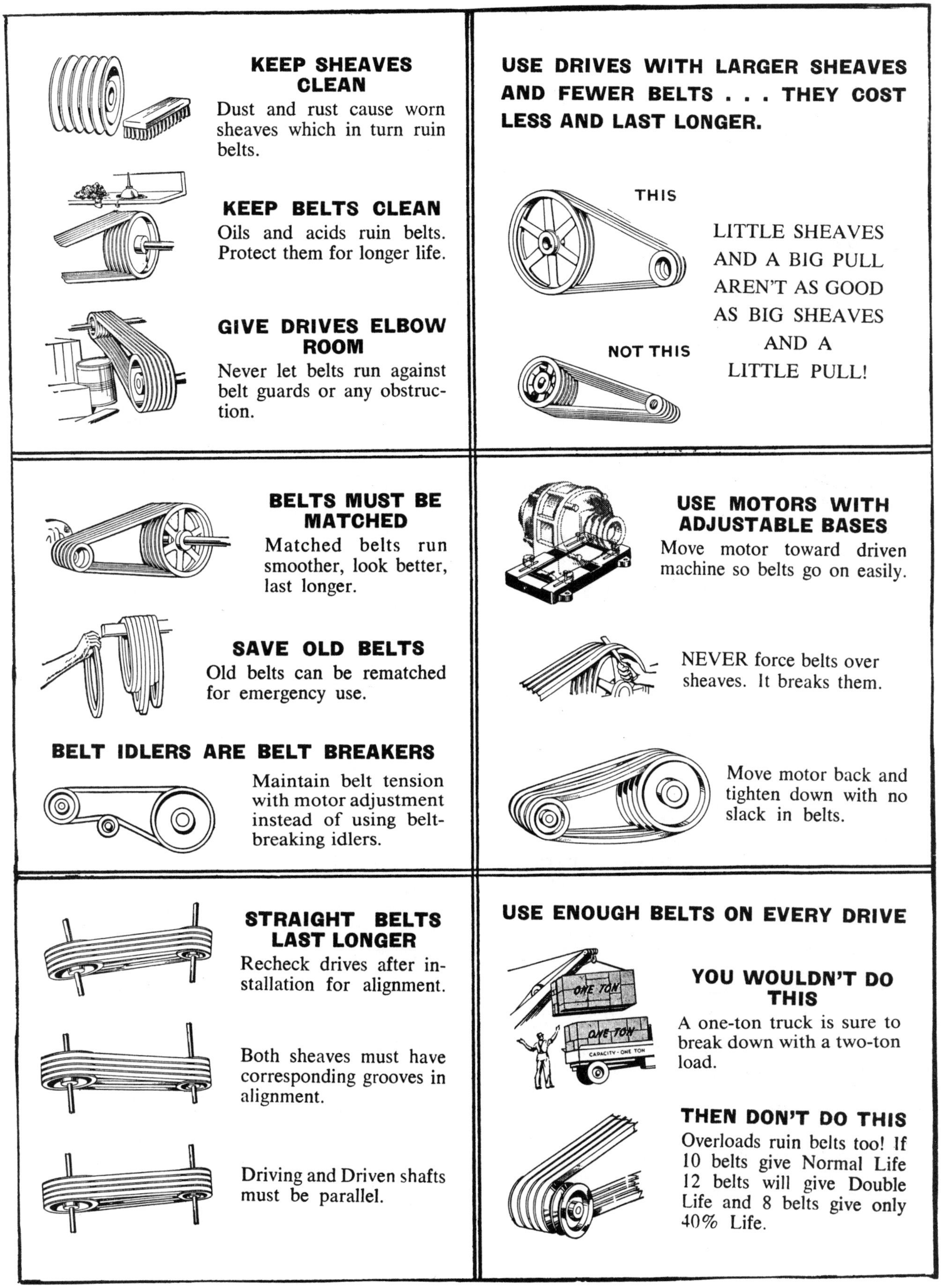

Fig. 9.48 Use of belts

(Courtesy of Browning)

After a few days of operation, the belts will seat themselves in the sheave grooves. Readjust the belts so that the belt drive again shows a slight bow in the slack side.

The belt drive is now properly tensioned and only occasional readjustment for groove wear will be necessary. If belts squeal during start-up, reposition motor to increase tension until slippage and squeal disappear.

If belts are adjusted too tightly, problems can result such as bearings heating and belts wearing excessively.

Always be sure that belt guards are in place and covering all exposed portions of belts whenever the equipment is operating.

QUESTIONS

Write your answers in a notebook and then compare your answers with those on pages 266 and 267.

9.4W What is an advantage of driving a pump with belts?

9.4X Why must all belts on a pump be replaced at the same time?

9.4Y Why should belts and sheaves be kept clean?

9.4Z How can you determine if motor and pump sheaves are properly aligned?

9.4AA How can you determine if a belt has the proper tightness or tension?

9.4BB What should you do when belts squeal during startup?

9.4CC What happens when belts are adjusted too tight?

9.453 Flexible Shafting

Flexible shafting is used where the pump and driver are located relatively far apart (Figure 9.50). Flexible shafting is particularly suited to transmit power into areas where moisture, dust, corrosive or explosive conditions, or flooding would adversely affect the driving machinery.

Pumping installations with flexible shafting usually have a vertical shaft with the motor located on the floor above the pump. There can be one, two, three, or even more sections of shafting, depending upon the distance between the pump and motor. If there is more than one section of shafting, there will be steady bearing supports.

Shafting is usually trouble-free if installed and lubricated properly. The universal joints (Figure 9.50) are very similar to those used in your automobile. Universal joints are used to take up any misalignment between pump and driver. The driver and pump need not line up axially, but they must be parallel to one another within plus or minus one degree. Shafts can be offset slightly, but the offset is not necessary. Some manufacturers require and specify a definite offset in some installations. The yokes in the shafting must be in line; that is, the driver and driven yoke must be on the same plane so the two center yokes will be in line.

Lubrication of the universal joints is very important. These should be lubricated every 500 hours of intermittent service, and every 200 hours of continuous service. The particular manufacturer's literature will give the type of lubricant to be used.

Never disassemble the needle bearings from their yokes unless it is necessary to replace the cross and bearing set. To inspect, remove the shaft and test bearings by moving the yoke in all directions to roll the needles. If the action of all four bearings is smooth, replacement is not necessary. If the action is rough or uneven, replace the entire universal joint. A visual inspection will usually tell if one bearing is bad by the appearance of a brown powder in that area. When greasing, if grease does not come out all four bearings, you can usually be assured of a bearing failure. Other clues that may indicate a failing U-joint (universal joint) include (1) any play or detectable motion in U-joint when shafting is rocked back and forth (with unit off, but still coupled), and (2) unusual noises while equipment is operating.

To replace bearings, have a skilled operator do this job. Tap one end of the bearing lightly to remove pressure on the snap ring. Remove the snap ring and repeat the procedure for the opposite bearing. Then with a soft drift, drive on one bearing to push the opposite bearing out of its yoke. Remove the exposed bearing, reverse the joint and drive on the exposed cross to push out the other bearing. Repeat this procedure on the other two bearings.

To assemble, remove bearings from the new cross assembly, holding the cups so the needles do not fall out. Position the cross in one yoke. Position one bearing cup with its needles in the yoke and insert the journal of the cross into this bearing. Press the bearing into the yoke. Repeat for the opposite bearing. If a press is not available, use a vise. Never hammer new bearings. Install snap rings, and repeat for the other two bearings.

QUESTIONS

Write your answers in a notebook and then compare your answers with those on page 267.

9.4DD When is flexible shafting installed?

9.4EE How often should universal joints be lubricated?

9.4FF How can you tell if there is a bearing failure in a universal joint?

9.46 Typical Pump Maintenance Schedule

Table 9.8 is a summary of a typical pump maintenance schedule. Be sure to adjust this schedule for each of your pumps in accordance with specific pump manufacturers' recommendations.

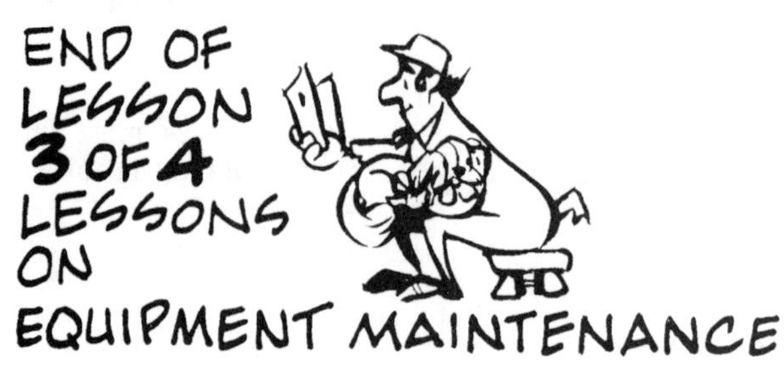
END OF LESSON 3 OF 4 LESSONS ON EQUIPMENT MAINTENANCE

Please answer the discussion and review questions next.

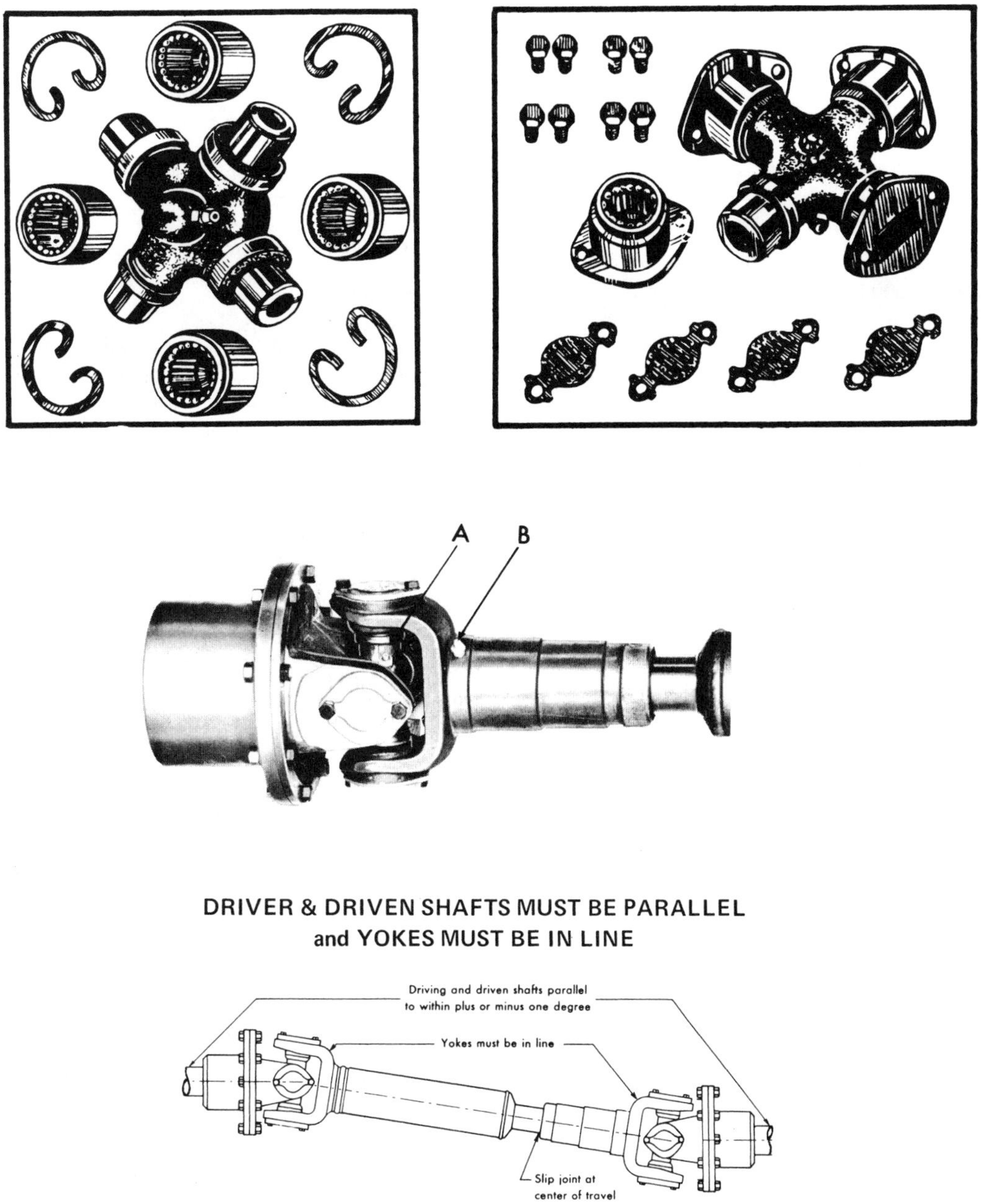

Fig. 9.50 Flexible shafting and universal joints
(Courtesy of H.S. Watson Company)

TABLE 9.8 TYPICAL PUMP MAINTENANCE SCHEDULE[a]

Frequency	Item
Daily	Check pump for noise when running. Check pump and motor bearings. Check bearing oil level (if applicable). Check packing and seals for leaks. Check coupling for noise when running. Check piping connections for leaks or corrosion. Check motor temperature.
Weekly	Test emergency generator.
Quarterly (2,000 hours)	Lubricate bearings. Test all safety devices. Check motor voltage and amperage. Check and tighten foundation bolts. Blow out motor with compressed air. Check V-belt tension (if applicable).
Semiannually (4,000 hours)	Flush and change grease in bearings. Change oil in all transmission cases. Inspect and grease coupling. Check motor and pump alignment. Check motor controls for dirt. Check motor contacts for pits or burn spots. Clean electrical cabinet. Check all gauges and instruments. Tighten all piping and foundation bolts. Check pump flow. Check pump efficiency (wire to water).
Annually (8,000 hours)	Inspect and tighten all electrical connections. Test all electrical controls. Check bearing vibration levels. Inspect packing and shaft sleeves.
2 to 3 years (16,000 to 24,000 hours)	Replace packing and seals. Polish or renew shaft sleeves. Inspect impeller, casing, and wear rings. Check all wiring for grounds and resistance.
3 to 5 years (24,000 to 40,000 hours)	Replace pump bearings. Replace mechanical seals. Replace motor bearings. Renew or replace motor-starting contacts.

[a] Renner, Don, "Round and Round It Goes," Water Environment and Technology, May 2000, page 78.

DISCUSSION AND REVIEW QUESTIONS

Chapter 9. EQUIPMENT MAINTENANCE

(Lesson 3 of 4 Lessons)

Write the answers to these questions in your notebook before continuing. The question numbering continues from Lesson 2.

13. How would you repack a pump?
14. What factors can cause cavitation of a pump impeller?
15. How would you inspect a pump shaft?
16. What would you look for when inspecting a pump impeller?
17. What are the advantages and limitations of mechanical seals?
18. What are some causes of bearing failures?
19. How are pumps and drivers connected together?

CHAPTER 9. EQUIPMENT MAINTENANCE

(Lesson 4 of 4 Lessons)

9.5 COMPRESSORS

Compressors (Figure 9.51) are commonly used in the operation and maintenance of wastewater collection systems. They are used to activate wastewater ejectors, pump control systems (bubblers), valve operators, and water pressure systems. They are also used to operate portable pneumatic tools such as jackhammers, compactors, air drills, sandblasters, tapping machines, and air pumps.

A compressor is a device used to increase the pressure of air or gas. They can be of a very simple diaphragm or bellows type such as are found in aquarium pumps, or extremely complex rotary, piston, or sliding-vane type compressors. A compressor usually has a suction pipe with a filter and a discharge pipe which goes to an air receiver or storage tank. The compressed air or gas is then used from the air receiver.

Due to the complexity of compressors, the wastewater maintenance operator usually will not be repairing them. You will, however, be required to maintain these compressors. With proper maintenance a compressor should give years of trouble-free service.

The first step for compressor maintenance, and this pertains to any mechanical equipment, is to get the manufacturer's instruction book and read it completely. Each compressor is different and the particular manufacturer will list their recommended maintenance schedules and procedures. Some of the maintenance procedures are discussed in the following paragraphs.

Fig. 9.51 Two-stage piston compressor

(Courtesy Worthington Corporation)

1. Inspect the suction filter of the compressor regularly. The frequency of cleaning depends upon the use of the compressor and the atmosphere around it. Under normal operations the filter should be inspected at least monthly and cleaned or replaced every three to six months. Inspect and replace the filter more frequently in areas with excavation and dust. When breaking up concrete, inspect the filters daily.

There are several types of filters, such as paper, cloth, wire screen, oil bath, and others. The impregnated paper filters must be replaced when dirty. Cloth filters can be washed with soap and water, dried and reinstalled. If a cloth filter is used, it is recommended that a spare be kept so one can be cleaned while the other is being used. Use a standard solvent to clean wire mesh and oil bath-type filters. After cleaning, reoil the filter or fill the oil bath and reinstall the filter in the compressor. Never operate a compressor without the suction filter because dirt and foreign materials will collect on the rotors, pistons, or blades and cause excessive wear.

2. Lubrication. Improper or lack of lubrication is probably the biggest cause of compressor failures. Most compressors require oiling of the bearings. They can have crankcase reservoirs, oil cups, grease fittings, a pressure system or separate pump. Whatever type, it must be inspected daily. Examine the reservoir dipstick or sight glass. Make sure that drip-feed oilers are dripping at the proper rate, force-feed oilers have the proper pressure, and grease fittings are greased at the proper interval. Compressors use a certain amount of oil in their operation and special attention is needed to keep the reservoirs full. Care also must be used to not overfill the crankcase. On some compressors it is possible for the oil to get into the compression side and lock up the compressor, or damage it. Remember!

A LIQUID CANNOT BE COMPRESSED.

When air or gas is compressed, heat is generated and the compressor becomes very hot. This tends to break down oil faster, so most compressor manufacturers have special oils recommended for their particular compressor. Also, due to the heat and contamination, it is necessary to change oil quite frequently. Compressor oil should be changed at least every three months, unless the manufacturer states differently. If there are filters in the oil system, these also should be changed.

3. Cylinder or casing fins should be cleaned weekly with compressed air or vacuumed off. The fins must be clean to ensure proper cooling of the compressor.

4. Unloader. Many compressors have unloaders that allow the compressor to start under a no-load condition. These can be inspected by observing the compressor. When the compressor starts, it should come up to speed and the unloader will change, starting the compression cycle. This can usually be heard by a change in sound. When it stops, you can hear a small pop and hear the air bleed off the cylinders. If the unloader is not working properly, the compressor will stall when starting, not start, or if belt-driven, burn off the belts.

5. Test the safety valves weekly. The pop-off or safety valves are located on the air receiver or storage tank. They prevent the pressure from building up above a specified pressure by opening and venting to the atmosphere. In gas compressors, they vent to the suction side of the compressor. Air receivers (unfired pressure vessels) require inspection and certification every five years in many areas. Check with your local safety regulatory agency. Some compressors have high pressure cutoff switches, low oil pressure switches, and high temperature cutoff switches. These switches have preset cutoff settings and must not be changed without proper authorization. If for any reason any of the safety switches are not functioning properly, the problem must be corrected before starting the compressor again. The safety switch settings should be recorded and the results kept in the equipment file.

6. Drain the condensate (condensed water) from the air receiver daily. Due to temperature changes, the air receiver will fill with condensate. Each day the condensate should be drained from the bottom of the tank. There is usually a small valve at the bottom of the air receiver for this purpose. Some air receivers are equipped with automatic drain valves. These must be inspected periodically to ensure they are operating satisfactorily.

7. Inspect belt tension on compressors. Usually you should be able to press the belt down with hand pressure approximately three-fourths of an inch. This is done at the center between the two pulleys. *MAKE SURE COMPRESSOR IS LOCKED OFF AND TAGGED BEFORE MAKING THIS TEST.* Do not overtighten belts because it will cause overheating and excessive wear on bearings and motor overloading.

8. Examine operating controls. Make sure the compressor is starting and stopping at the proper settings. If it is a dual installation, make sure they are alternating if so designed. Inspect gage for accuracy. Compare readings with recorded start-up values or other known, accurate readings.

9. Many portable compressors are equipped with tool oilers on the receivers. These are used for mixing a small quantity of oil with the compressed air for lubrication of the tools being used. These are located on the discharge side of the air receiver. They have a reservoir which must be filled with rock drill oil.

10. All compressors should be thoroughly cleaned at least monthly. Dirt, oil, grease, and other material must be thoroughly cleaned off the compressor and surrounding area. Compressors have a tendency to lose oil around piping, fittings and shafts; thus constant cleaning is required by the maintenance operator to ensure proper and safe operation.

QUESTIONS

Write your answers in a notebook and then compare your answers with those on page 267.

9.5A List some of the uses of a compressor in connection with operation and maintenance of a wastewater collection system.

9.5B How often should the suction filter of a compressor be cleaned?

9.5C How often should compressor oil be changed?

9.5D How often should the condensate from the air receiver be drained?

9.5E What must be done before testing belt tension on compressors with your hands?

9.6 LUBRICATION

9.60 Purpose of Lubrication

Lubrication of equipment is probably one of the most important phases of a maintenance operator's job. Without proper lubrication the tools and equipment used for operating and maintaining wastewater collection systems would fail. Proper lubrication of tools and equipment is probably one of the maintenance operator's easiest jobs, but often it is the most neglected.

The purpose of lubrication is to reduce friction between two surfaces. Lubrication also removes heat that is caused by friction. Solid friction of two dry surfaces in contact is changed to a fluid friction of a separating layer of liquid or liquid lubricant. Actually water is a lubricant, although not a good lubricant.

9.61 Properties of Lubricants

A good lubricant must have the following properties:

1. Form a slippery coating on contacting surfaces so they can slide freely past each other, and
2. Exert sufficient pressure to keep the surfaces apart when running.

To be a good lubricant for a particular job, the lubricant used must have the following qualities:

1. Thickness of the lubricant layer must be sufficient to keep the roughness of the metal parts from touching.
2. Lubricity (slipperiness) must be sufficient to allow molecules to slide freely past each other, and
3. Viscosity (resistance to flow) must be sufficient to build up a pressure necessary to keep the surfaces apart. If viscosity alone cannot provide enough pressure, an external pressure must be supplied by a pump.

Viscosity in the United States is the number of seconds it takes 60 cubic centimeters (cc) of an oil to flow through the standard orifice of a Saybolt Universal Viscometer at 100, 130, or 210 degrees Fahrenheit. A 300 - *SSU*[12] @ 130 oil means that it took 300 seconds for 60 cc to flow through a Saybolt Universal Viscometer at 130 degrees Fahrenheit. Viscosity decreases with temperature rise because oil becomes thinner. The specific gravity of an oil is measured by comparing the weight of oil with an equal volume of water, both at 60 degrees Fahrenheit.

Some other important information to know about lubricants is their "Pour Point," "Flash Point," and "Fire Point." "Pour Point" is the temperature at which a lubricant refuses to run. This is important in low-temperature work. "Flash Point" is the temperature at which oil vaporizes enough to ignite momentarily when near a flame. A low flash point means that oil evaporates more readily in service. "Fire Point" is the temperature at which oil vaporizes enough to keep on burning.

Oils in service tend to become acidic and may cause corrosion, deposits, sludging and other problems. This condition may not be visible when you look at the oil. Therefore, do not extend the time for an oil change because the oil looks clean. To detect acid conditions in oils, the neutralization number of an oil is used. The neutralization number is the weight in milligrams of potassium hydroxide required to neutralize one gram of oil. Laboratories use this method to test the oil on large engines, turbines, compressors, and other equipment which have large volume oil reservoirs to determine when oil changes or additives are needed.

Most lubricants in general use are fluid at room temperature. Mostly, these are petroleum-based products, but other types are also used. Greases are mixtures of petroleum products with soaps such as lime, soda, aluminum, and metallic. Metallic soaps, forms of calcium, sodium, potassium, and lithium, have good retention in bearings and can withstand high temperatures and pressures. A sodium-based grease has sodium as the soap mixed with the petroleum.

Solid materials such as graphite, finely ground mica, asbestos, and yarn are sometimes used as lubricants. Some recently developed silicon compounds (silicones) work very well under heavy loads and widely varying temperatures.

There are many oil additives on the market today and they are worth investigating. Oil additives are chemical compounds added to an oil to improve certain chemical or physical properties such as stability and lubricity. They are used to prevent foaming, rust or deposits and many other conditions that could cause problems.

9.62 Lubrication Schedule

To have proper lubrication you must first set up a lubrication schedule. This can be a simple check-off sheet or card system or an elaborate computer system. The first thing to do is make a list of everything that needs lubrication down to the smallest item, including chains, rollers, and sprockets. After you have listed every item on paper, go through the manufacturer's instruction books to determine the frequency and type of lubrication required. Is the frequency daily, weekly, monthly, semiannually, or annually? The manufacturer's literature usually lists several different name brands of lubricants which are equal. If you need help determining the type of lubricant or cross referencing it to your particular brand, contact your sup-

[12] *SSU. Standard Saybolt Units.*

plier. Most oil distributors have a service representative who will come to your facility and go over the individual equipment and specify which lubricants you should use. Next, determine the amount of each lubricant required. Determine the locations of fill plugs, drain plugs, oil levels, sight glasses, dipsticks, grease fittings, and other important items. To find these locations, physically inspect each piece of equipment thoroughly and look for all lubrication points. Also the manufacturer's maintenance manual should show the lubrication points for each piece of equipment as well as frequency of lubrication.

When you have gathered all this information, transfer it to the equipment history cards or computer history data files for future reference. From this information you can make up a lubrication chart (Figure 9.52).

As stated earlier, use whatever type of lubrication form you prepare, but follow it. Always record each lubrication job when completed and have the operator who did the job initial the record card. Always keep your lubrication schedules up to date. If there are failures due to the wrong or insufficient lubricant, change or increase the lubrication frequency on the schedule. Also, new equipment must be added and discarded equipment removed from the schedule. Someone must be assigned to take care of the lubrication and records. Assign more than one operator or rotate this job so if an individual is off work or leaves the crew, there is a continuity in the lubrication schedule.

9.63 Precautions

When handling or storing oils and greases, some special precautions must be followed. Make sure the storage area does not create a fire hazard. Most lubricants are combustible and shouldn't be stored where there is an open flame. "NO SMOKING" signs must be posted outside the building. Be sure to keep any spills wiped up and make sure that all the lids are tight on their containers.

When storing lubricants, cleanliness is of utmost importance. A grease pail or oil drum which has become contaminated with dirt, water, or other material should be discarded. Contaminated lubricants will ruin the equipment you are trying to protect. Try using grease cartridges instead of grease pails because they are easier to keep clean and store. Grease cartridges are easy to load in the grease gun and can be discarded when empty. They cost a little more initially, but the savings in handling and a reduction in the chance of contamination may make them cheaper in the long run.

Proper equipment lubrication requires very special care when installing lubricants.

GREASING

1. Shut off, lock out, and tag unit being greased so you can't be injured if someone attempts to start the equipment.
2. Use the proper grease.
3. Make sure grease is clean and free of dirt.
4. Pump a small amount of grease into a rag and wipe off end of grease gun.
5. Clean grease fitting to be greased with a clean rag. If a plug is to be removed, clean area around plug before removing plug and inserting grease fitting.
6. Remove any relief plug before pumping in grease.
7. Pump in the proper amount of grease as indicated on the equipment card.
8. Wipe off all excess grease around unit.
9. Clean vent before replacing vent plug to allow for expansion of grease and to allow excess grease to work out of bearing.
10. Replace vent plugs. Do not completely fill a grease cavity and tighten vent plugs on cold grease. Leave one vent plug out and run equipment to force excess grease out during warm-up. After warm-up, install last vent plug.
11. Record the date and your initials on the lubrication record.
12. Record and describe any unusual conditions.

OILING

1. Make sure that the equipment being serviced is shut off, locked out and tagged out of service so you can't be injured if someone attempts to start the equipment.
2. Check to make sure oil is the proper type and grade, as indicated on the equipment record card.

Balling Rig #2

Item	Lubricant	Change
Motor.	Heavy Duty #30	Quarterly.
Brakes.	Brake Fluid #500	Monthly (check).
Universals.	Grease. F-2	Monthly.
Air Cleaner Bath.	SAE #20	Quarterly.
Wheel Bearings.	Grease. Wheel Type	Semiannual.
Power Steering.	Red Line ATF #30	Quarterly (check).
Transmission.	Red Line ATF #30	Quarterly (check).
Aux. Engine on Rear.	H.D. #30	Weekly.
Gear Box on Reel.	Worm Gear. #90	Quarterly.
Cable Spool.	Grease. F-2	Daily.
Sheave on Level Wind.	Grease. F-2	Daily.
Upper Bearings on Level Wind.	Oil. #20	Weekly.
Chain Drive.	Oil. #10	Monthly.

Fig. 9.52 Lubrication chart

3. Wipe area around fill and drain plugs with clean rag. Also, wipe off top of oil can and cap.
4. Place container under drain plug and remove it along with the fill plug.
5. Make sure all the oil is drained out and reinstall drain plug or plugs.
6. Pour in new oil. If a funnel or spout is used, be sure to wipe it clean before starting. Fill to proper level in sight glass or dipstick. Do not overfill.
7. Install fill plug or plugs and wipe up any excess or spilled oil.
8. After equipment has been started, observe oil level to make sure it is at proper level.
9. Record the date, lubricant, and your initials on the lubrication record.
10. Record and describe any unusual conditions.
11. Place all caps on oil containers and return them to their storage place along with the funnels, pumps, and other equipment that was used.

9.64 Final Cleanup and Inspection

When lubrication is completed, make sure all oily rags and materials are placed in a covered metal can while waiting to be picked up or disposed of. If it is necessary to transfer oil to different cans, make sure they are labeled properly. Never leave an open oil can lying around to use when the unit needs more oil.

After your lubrication is completed, inspect the equipment in a few hours to make sure it is not overheating or making any strange or unusual noise. Examine the oil levels to make sure they are at their proper level. Proper lubrication is the key to successful equipment maintenance.

QUESTIONS

Write your answers in a notebook and then compare your answers with those on page 267.

9.6A What is the purpose of lubrication?

9.6B What happens to oils in service?

9.6C What should be done to ensure proper lubrication of equipment?

9.6D What should be done when lubrication failures occur?

9.6E What precautions must be taken when handling or storing oils and greases?

9.6F What precautions must be taken before oiling or greasing equipment?

9.6G What should be done after a lubrication job is completed?

9.7 ADDITIONAL READING

1. *MAINTENANCE ENGINEERING HANDBOOK* by Higgins. Obtain from the McGraw-Hill Companies, Order Services, PO Box 182604, Columbus, OH 43272-3031. ISBN 0-07-028819-4. Price, $150.00, plus nine percent of order total for shipping and handling.
2. *PUMP HANDBOOK*, Third Edition, edited by Igor Karassik, Joseph Messina, Paul Cooper, and Charles Heald. Obtain from the McGraw-Hill Companies, Order Services, PO Box 182604, Columbus, OH 43272-3031. ISBN 0-07-034032-3. Price, $135.00, plus nine percent of order total for shipping and handling.
3. *NATIONAL ELECTRICAL CODE HANDBOOK*, 24th Edition, 2002, by Joseph F. McPartland and Brian J. McPartland. Obtain from the McGraw-Hill Companies, Order Services, PO Box 182604, Columbus, OH 43272-3031. ISBN 0-07-137725-5. Price, $75.00, plus nine percent of order total for shipping and handling.
4. *NEMA STANDARD FOR MOTORS AND GENERATORS*, MG 1. Obtain from Global Engineering Documents, Customer Service Department, 15 Inverness Way East, Englewood, CO 80112. Price, $174.00, plus shipping and handling.

END OF LESSON
4 OF 4 LESSONS
ON EQUIPMENT
MAINTENANCE

Please answer the discussion and review questions next.

DISCUSSION AND REVIEW QUESTIONS

Chapter 9. EQUIPMENT MAINTENANCE

(Lesson 4 of 4 Lessons)

Write the answers to these questions in your notebook. The question numbering continues from Lesson 3.

20. What are the uses of a compressor?
21. What items should be maintained on a compressor?
22. How would you develop a lubrication schedule for a pump?
23. Why is cleanliness important in the storing and use of lubricants?
24. How would you clean up after a lubrication job is completed?

SUGGESTED ANSWERS

Chapter 9. EQUIPMENT MAINTENANCE

ANSWERS TO QUESTIONS IN LESSON 1

Answers to questions on page 160.

9.0A Unqualified or inexperienced people must be extremely careful when attempting to troubleshoot or repair electrical equipment because they can be seriously injured and damage costly equipment if a mistake is made.

9.0B When machinery is not shut off, locked out, and tagged properly, the following accidents could occur:

1. Maintenance operator could be clearing pump and have it start, thus losing an arm, hand or finger,
2. Electrical motors or controls not properly grounded could lead to possible severe shock, paralysis, or death, and
3. Improper circuit—wrong connection, safety devices jumped, wrong fuses, or improper wire. These mistakes can cause fires, or injuries due to incorrect operation of machinery.

Answers to questions on page 161.

9.1A The proper voltage and allowable current in amps for a piece of equipment can be determined by reading the nameplate information or the instruction manual for the equipment.

9.1B The two types of voltage are direct current (D.C.) and alternating current (A.C.).

9.1C Amperage is a measurement of current or electron flow and is an indication of work being done or "how hard the electricity is working."

Answers to questions on page 165.

9.1D You test for voltage by using a multimeter.

9.1E A multimeter can be used to test for voltage, open circuits, blown fuses, single phasing of motors, and grounds.

9.1F Before attempting to change fuses, turn off power and check both power lines for voltage. Use a fuse puller.

9.1G If the voltage is unknown and the multimeter has different scales that are manually set, always start with the highest voltage range and work down. Otherwise the multimeter could be damaged.

9.1H Amp readings different from the nameplate rating could be caused by low voltage, bad bearings, poor connections, plugging, or excessive load.

9.1I Motors and wirings should be megged at least once a year, and twice a year, if possible.

9.1J An ohmmeter is used to test the control circuit components such as coils, fuses, relays, resistors, and switches.

Answers to questions on page 172.

9.1K The most common pump driver used in lift stations is the A.C. induction motor.

9.1L Types of lift station failures that can occur due to lack of proper maintenance include:

1. Current imbalances which ultimately result in motor failure,
2. Loose contacts or terminals in control and power circuits causing high-resistance contacts,
3. Overheating resulting in arcing, fire, and electrical system damage,
4. Dirty enclosures and components which allow a conductive path to build up between incoming phases causing phase-to-phase shorts, and
5. Corrosion causing high-resistance contacts and heating.

9.1M If a fuse or circuit breaker "blows" or trips, the potential source of the problem should be investigated, identified, and corrected.

9.1N The two types of safety devices in main electrical panels or control units are fuses and circuit breakers.

9.1O Fuses are used to protect operators, main circuits, branch circuits, and equipment such as heaters and motors from excessive current.

9.1P A fuse must never be bypassed or jumped because the fuse is the only protection the circuit has. Without it, serious damage to equipment and possible injury to operators can occur.

9.1Q Annual maintenance that should be performed with regard to fuses includes:

1. Inspect all bolted connections at the fuse clip or fuse holder for signs of looseness,
2. Check connections for any evidence of corrosion from moisture or atmosphere (air pollution),
3. Tighten connections,
4. Check fuse for obvious overheating, and
5. Inspect insulation on the conductors coming into the fuses on the line side and out of the fuses on the load side for evidence of discoloration or bubbling which would indicate overheating of the conductors.

9.1R A circuit breaker is a switch that is opened automatically when the current or the voltage exceeds or falls below a certain limit. Unlike a fuse that has to be replaced each time it "blows," a circuit breaker can be reset after a short delay to allow time for cooling.

Answers to questions on page 173.

9.1S The motor and supervisory control systems are composed of the auxiliary electrical equipment such as relays, transformers, lighting panels, pump control logic, alarms, and other electrical equipment typically found in a lift station electrical system.

9.1T The three basic factors that contribute to the reliable operation of electrical systems found in lift stations include:

1. An adequate preventive maintenance program must be implemented,
2. A knowledge of the system by the collection system operator, even though the operator does not perform the actual maintenance, and
3. KEEP IT CLEAN! KEEP IT DRY! KEEP IT TIGHT!

Answers to questions on page 189.

9.2A The two types of induction motor construction which are typically encountered when dealing with A.C. induction motors are:

1. Squirrel cage induction motor, and
2. Wound rotor induction motor.

9.2B The five most common causes of electric motor failure are (1) overload (thermal), (2) single phasing, (3) contaminants, (4) old age, and (5) bearing failures.

9.2C The three components of an insulation temperature rating consist of:

1. Ambient operation temperature,
2. Temperature rise, and
3. Hot-spot allowance.

9.2D The direction of rotation of a three-phase motor can be changed by changing any two of the power leads.

9.2E Motor starters can be either manually or automatically controlled.

9.2F Magnetic starters are usually used to start pumps, compressors, blowers and anything where automatic or remote control is desired.

9.2G The two types of overload protection devices used on magnetic starters are:

1. A bimetallic strip which is precisely calibrated to open under higher temperature conditions to de-energize the coil, and
2. A small solder pot within the coil that melts because of the heat and will de-energize the system.

Answers to questions on page 195.

9.2H The key to effective troubleshooting is practical, step-by-step procedures combined with a common sense approach.

9.2I When troubleshooting electrical equipment:

1. Gather preliminary information.
2. Inspect:
   a. Contacts,
   b. Mechanical parts, and
   c. Magnetic parts.

9.2J Types of information that should be recorded regarding electrical equipment include every

1. Change,
2. Repair, and
3. Test.

ANSWERS TO QUESTIONS IN LESSON 2

Answers to questions on page 204.

9.3A Pumps are used to move or lift wastewater in collection systems and treatment plants. They also are used to pump water in high-velocity cleaners and to dewater construction projects. In addition, backhoes have hydraulic pumps and automobiles have oil and water pumps.

9.3B If we know how a pump is built, it is easier to understand how to operate and maintain a pump.

9.3C Pump lubricants should be changed in accordance with the pump manufacturer's lubrication instructions. If the instructions are not available or if in doubt, call the supplier for recommendations. Play it safe and change the lubricant *BEFORE* it's too worn or too dirty. Look at the lubricant after the pump has run at equilibrium temperature. Any used oil or oil needing change will become discolored and turn gray or black.

9.3D A pump is in proper alignment if the pump and motor flanges are parallel vertically and axially—and *KEPT* that way.

9.3E Horizontal suction lift piping must always slope upward toward the pump to prevent air pockets and loss of suction.

9.3F Piping can be kept from straining the pump casing by (1) being properly supported, and (2) regular inspection to be sure the pump or piping has not shifted or settled and caused a strain. Look for evidence of a heavy object in contact with the piping that could have caused misalignment and/or a strain.

9.3G A pump should never be allowed to run dry because water acts as a lubricant between the wearing rings and the impeller.

Answers to questions on page 214.

9.3H Three types of centrifugal pumps include (1) horizontal nonclog, (2) vertical ball bearing, and (3) propeller.

9.3I Screw pumps consist of a screw operating at a constant speed within a housing or trough. The screw has a pitch and is set at a specific angle. When revolving, the screw carries the wastewater up the trough to a discharge point.

9.3J Pneumatic ejectors are used when it is necessary to handle limited or low flows with relatively large solids.

9.3K Piston pumps are commonly used in high-velocity cleaners.

Answers to questions on page 218.

9.3L Portable centrifugal pumps are often referred to as trash pumps because the water being pumped is not clean and may contain suspended solids of various sizes.

9.3M The suction hose has a strainer on the end to prevent pulling rocks and debris into the pump to avoid damaging the pump or plugging the hoses or pipes.

9.3N When operating a portable pump in weather that is subject to freezing, always drain the pump to prevent the freezing water from cracking the casing or binding up the pump. Before starting the pump, turn the shaft by hand to be sure it turns freely. If the impeller is frozen fast, warm the pump slowly until the ice melts.

Answers to questions on page 221.

9.3O Some of the advantages of the positive displacement diaphragm pump are as follows:

1. Self-priming if the suction lift is small,
2. When primed with water, it will pump with a suction lift up to 25 feet,
3. Large particles will readily pass through the pumps, and
4. Less likely to become clogged than centrifugal pumps.

9.3P If a diaphragm pump suddenly stops pumping, remove discharge valve and clean out pump cavity.

9.3Q After use, pneumatically operated diaphragm pumps should be flushed thoroughly to prevent dried sediment from interfering with valve operation. If the water pumped had large particles in it, it may be necessary to dismantle the pump to remove any remaining particles.

ANSWERS TO QUESTIONS IN LESSON 3

Answers to questions on page 225.

9.4A Impellers should be inspected on a regular basis such as every 6 months or annually, depending on pumping conditions. If grit, sand, or other abrasive material is being pumped, inspections should be more frequent.

9.4B Wearing rings protect the impeller and pump body from damage due to excessive wear.

9.4C If pump clearances are not kept in proper adjustment, the pump may become plugged frequently and power requirements will increase due to lowered efficiency.

9.4D An impeller shows cavitation damage if it has holes, indentations, bullet chips or cracks on the surface.

9.4E Cavitation can be caused by a pump operating under different conditions than what it was designed for, such as off the design curve, poor suction conditions, high speed, air leaks into suction end and water hammer conditions.

Answers to questions on page 226.

9.4F Pump shafts can be protected from damage by the installation of shaft sleeves over the shaft in the area on the shaft where it passes through the stuffing box.

9.4G A pump shaft sleeve must be repaired or replaced when grooves are observed.

Answers to questions on page 231.

9.4H Packing is used to keep air from leaking in and water from leaking out where the shaft passes through the casing.

9.4I The lantern ring is used to allow outside water or grease to enter the packing for lubrication, flushing, and cooling and to prevent air from being sucked or drawn into the pump.

Answers to questions on page 233.

9.4J A pump with a mechanical seal (or any other type of seal) should never be run dry because water is required for seal lubrication.

9.4K Mechanical seals are not supposed to have any leakage from the gland; and if a leak develops, the seal may require relapping or it may have to be replaced.

Answers to questions on page 252.

9.4L The different types of bearings used in pumps include ball bearings, roller bearings, and sleeve bearings.

9.4M Whenever a bearing failure occurs before its expected replacement time, the bearing should be examined to determine the cause, and if possible, to eliminate the problem.

9.4N The different types of bearing failures include fatigue failure, contamination, brinelling, false brinelling, thrust failures, misalignment, electric arcing, lubrication failure, and cam failure.

9.4O Contamination failure of a bearing can be caused by dirty or damp surroundings or abrasive waste materials. Contamination failure can be prevented by removing the source of contamination.

9.4P Bearing failures from misalignment can be prevented by ensuring proper alignment of pump shaft and housing.

9.4Q Bearing lubrication failures can be identified by grease appearance, abnormal temperature rise, noise, bearing discoloration, and retainer failure.

Answers to questions on page 254.

9.4R Couplings are used to connect a source of power to the shaft, such as the shaft of a pump impeller.

9.4S Methods by which a pump and driver can be coupled together include direct coupled, belt-driven, flexible shafting and close-coupled.

9.4T The functions of a flexible coupling include transmitting power from one shaft to the other, decreasing wear on shaft bearings and driven equipment, accommodating slight shaft misalignments, absorbing shock loads and pulsations, dampening vibrations, accommodating load reversals, minimizing initial backlash, and making assembly and disassembly easy.

9.4U To inspect for proper coupling alignment, look for parallel and angular misalignment using a feeler gage or a straightedge placed squarely across the rings of the coupling halves. Place the feelers at the top, bottom, near and far sides of the coupling junctions. Also place the feelers along the edge of top and sides of the shaft. Dial indicators also may be used.

9.4V When inspecting clearances between faces of coupling halves, be sure they are set so they cannot strike, rub, or exert a pull on either machine.

Answers to questions on page 256.

9.4W An advantage of driving a pump with belts is that different speed ratios can be achieved between the motor and pump. Also the motor can be placed in a variety of positions with respect to the pump.

9.4X All belts on a pump must be replaced at the same time with a matched set so all belts will wear evenly. The used belts also have been stretched and will not allow proper adjustment.

9.4Y Belts and sheaves must be kept clean to reduce wear on both belts and sheaves.

9.4Z Alignment of pump and motor sheaves can be determined by (1) placing a straightedge between the two pulleys, or (2) pulling a string across both pulleys.

9.4AA Proper belt tension is achieved when the belts have only a slight bow on the slack side of the belt drive while operating under load, or when belts have a $^{3}/_{4}$-inch movement from hand pressure at the midpoint of the slack belt when the unit is off.

9.4BB If belts squeal during start-up, reposition the motor to increase tension until slippage and squeal disappear.

9.4CC When belts are adjusted too tight, problems can result, such as bearings heating and excessive belt wear.

Answers to questions on page 256.

9.4DD Flexible shafting is installed when the pump and driver are located relatively far apart.

9.4EE Universal joints should be lubricated every 200 hours during continuous service and every 500 hours during intermittent service. Consult the manufacturer's literature for proper type of lubricant and details.

9.4FF Failure of a bearing in a universal joint can be indicated if:

1. Grease does not come out all four bearings when greasing,
2. Brown powder appears in the area of one bearing, or
3. Rough, uneven play occurs when the yoke is moved.

ANSWERS TO QUESTIONS IN LESSON 4

Answers to questions on pages 260 and 261.

9.5A Compressors are used with wastewater ejectors, pump control systems (bubblers), valve operators, and water pressure systems. Also they are used to operate portable pneumatic tools such as jackhammers, compactors, air drills, sandblasters, tapping machines, and air pumps.

9.5B The frequency of cleaning a suction filter on a compressor depends on the use of a compressor and the atmosphere around it. The filter should be inspected at least monthly and cleaned or replaced every three to six months. More frequent inspection, cleaning and replacement are required under dusty conditions such as operating a jackhammer on a street.

9.5C Compressor oil should be changed at least every three months, unless the manufacturer states differently. If there are filters in the oil system, these also should be changed.

9.5D Drain the condensate from the air receiver daily.

9.5E Before testing belt tension on a compressor with your hands, *MAKE SURE THE COMPRESSOR IS LOCKED OFF AND TAGGED.*

Answers to questions on page 263.

9.6A The purpose of lubrication is to reduce friction between two surfaces and to remove heat caused by friction.

9.6B Oils in service tend to become acidic and may cause corrosion, deposits, sludging and other problems.

9.6C To ensure proper lubrication of equipment, determine the proper lubrication schedule, lubricant, and amount of lubricant and prepare a lubrication chart.

9.6D When lubrication failures occur, try to identify the cause and change lubricant and/or frequency of lubrication if necessary.

9.6E Oils and greases can create a fire hazard. All spills must be wiped up immediately. Oils and greases must not become contaminated with dirt, water or other material.

9.6F Before oiling or greasing equipment, shut it off, lock it out and tag it so it can't start unexpectedly and injure you.

9.6G After a lubrication job is completed, clean up the area, properly store or dispose of oily rags and materials, and inspect the lubricated equipment in a few hours to make sure it is not overheating or making any strange or unusual noises.

CHAPTER 10

SEWER RENEWAL (REHABILITATION)

by

Rich Thomasson

Revised by

Rick Arbour

Gary Batis

TABLE OF CONTENTS

Chapter 10. SEWER RENEWAL (REHABILITATION)

Page

OBJECTIVES

Chapter 10. SEWER RENEWAL (REHABILITATION)

Following completion of Chapter 10, you should be able to:

1. Evaluate the condition of a sewer,
2. Determine the need for sewer renewal (rehabilitation),
3. Establish priorities for a sewer renewal (rehabilitation) program,
4. Identify the various sewer renewal (rehabilitation) methods,
5. Select the appropriate sewer renewal (rehabilitation) method,
6. Implement and complete a renewal (rehabilitation) project, and
7. Notify and cooperate with the public during a renewal (rehabilitation) project.

WORDS

Chapter 10. SEWER RENEWAL (REHABILITATION)

ANNULAR (AN-you-ler) SPACE — ANNULAR SPACE

A ring-shaped space located between two circular objects. For example, the space between the outside of a pipe liner and the inside of a pipe.

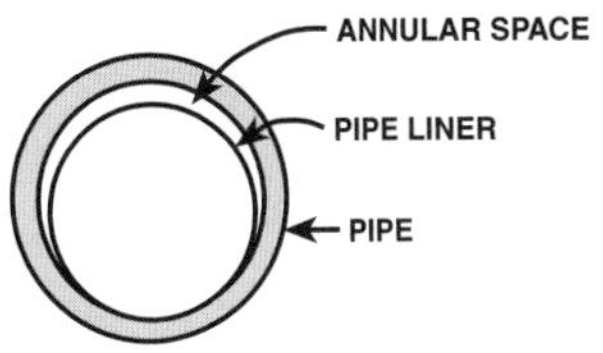

CHRISTY BOX — CHRISTY BOX

A box placed over the connection between the pipe liner and the house sewer to hold the mortar around the cleanout wye and riser in place.

FLOW ISOLATION — FLOW ISOLATION

A procedure used to measure inflow and infiltration (I/I). A section of sewer is blocked off or isolated and the flow from the section is measured.

INFILTRATION (IN-fill-TRAY-shun) — INFILTRATION

The seepage of groundwater into a sewer system, including service connections. Seepage frequently occurs through defective or cracked pipes, pipe joints, connections or manhole walls.

INFILTRATION/INFLOW — INFILTRATION/INFLOW

The total quantity of water from both infiltration and inflow without distinguishing the source. Abbreviated I & I or I/I.

INFLOW — INFLOW

Water discharged into a sewer system and service connections from such sources as, but not limited to, roof leaders, cellars, yard and area drains, foundation drains, cooling water discharges, drains from springs and swampy areas, around manhole covers or through holes in the covers, cross connections from storm and combined sewer systems, catch basins, storm waters, surface runoff, street wash waters or drainage. Inflow differs from infiltration in that it is a direct discharge into the sewer rather than a leak in the sewer itself. See INTERNAL INFLOW.

INSITUFORM — INSITUFORM

A method of installing a new pipe within an old pipe without excavation. The process involves the use of a polyester-fiber felt tube, lined on one side with polyurethane and fully impregnated with a liquid thermal setting resin.

INTERNAL INFLOW — INTERNAL INFLOW

Nonsanitary or industrial wastewaters generated inside of a domestic, commercial or industrial facility and being discharged into the sewer system. Examples are cooling tower waters, basement sump pump discharge waters, continuous-flow drinking fountains, and defective or leaking plumbing fixtures.

LIFE-CYCLE COSTING — LIFE-CYCLE COSTING

An economic analysis procedure that considers the total costs associated with a sewer during its economic life, including development, construction, and operation and maintenance (includes chemical and energy costs). All costs are converted to a present worth or present cost in dollars.

PIEZOMETER (pie-ZOM-uh-ter) — PIEZOMETER

An instrument used to measure the pressure head in a pipe, tank, or soil. It usually consists of a small pipe or tube connected or tapped into the side or wall of a pipe or tank and connected to a manometer pressure gage, water or mercury column, or other device for indicating pressure head.

PIG PIG

Refers to a poly pig which is a bullet-shaped device made of hard rubber or similar material. This device is used to clean pipes. It is inserted in one end of a pipe, moves through the pipe under pressure, and is removed from the other end of the pipe.

PROMOTED PROMOTED

The mixture of resin and catalyst ready to cause (promote) curing in place.

SSES SSES

Sewer **S**ystem **E**valuation **S**urvey.

STRETCH STRETCH

Length of sewer from manhole to manhole.

SURCHARGE SURCHARGE

Sewers are surcharged when the supply of water to be carried is greater than the capacity of the pipes to carry the flow. The surface of the wastewater in manholes rises above the top of the sewer pipe, and the sewer is under pressure or a head, rather than at atmospheric pressure.

CHAPTER 10. SEWER RENEWAL (REHABILITATION)

10.0 PROGRAM FORMULATION

10.00 Historical Background

In the past, the primary reason for rehabilitating a sewer was to restore the structural integrity of a line that had failed and, as a consequence, discharged raw wastewater into the environment. Today, however, the situation is much more complex. A large number of factors contribute to the rapidly declining integrity of major portions of the wastewater collection systems throughout the country.

To begin with, many of the collection systems in use today are very old and the materials used in their construction have lost their structural integrity due to corrosion and natural deterioration from use. This gradual breakdown allows greater infiltration, particularly during periods of heavy rainfall and under high groundwater conditions. The additional flows, in turn, generally produce two negative effects: (1) the *SURCHARGED*[1] sewer flows accelerate deterioration of the system and allow leakage into the environment (Figures 10.1 and 10.2); and (2) the increased inflow/infiltration overloads the treatment facilities to the point where they sometimes fail and discharge partially treated wastewater into the environment.

Other factors that contribute significantly to the deterioration of collection systems include rapid system expansion to keep pace with population growth in some areas. Not only can expansion tax the entire system, but it may result in a diversion of effort and attention away from the long-term preservation of existing facilities. Another effect of population growth is the stress that is created by vibrations from construction and the laying of roadways over sewer systems. Sewer systems also are failing due to unstable foundation soils or improper pipe bedding material.

A smoothly functioning sewer system is a service the public has come to take for granted. Because the system is largely out of sight, there is little awareness of its importance or of the serious consequences that result from continued neglect. As more and more large municipal systems reach the end of their useful lives, major programs of systematic rehabilitation will have to be undertaken. An ongoing program of long-range analysis and rehabilitation could greatly extend the serviceability of existing systems, could reduce operating costs, could reduce the need to expand the system, and could protect the environment from the consequences of a major collection system failure.

This chapter will explain how to analyze the condition of your present system, how to calculate the costs and benefits of needed repairs, and how to make the repairs using a variety of commonly accepted methods.

10.01 Program Definition

The first step in defining the rehabilitation program is to develop a comprehensive, written description of the major parts of your collection system. No program can be successful without first identifying all of the collection system components so you must begin by conducting an inventory of the system. Analysis of the condition of the system can then be handled in an orderly process. Once the analysis has been completed, you will be in a position to set the goals of your rehabilitation program.

Numerous records are available to a utility for the proper completion of the sewer system inventory phase of the program. Some of these major sources of information are as follows:

1. Record (as-built) drawings,
2. Maintenance complaint history files,
3. Construction reports,
4. Maintenance repair and inspection reports (Figures 10.3 and 10.4),
5. Interviews of key maintenance personnel, and
6. Updated maps of the sewer system (Figures 10.5 and 10.6).

Using the original construction and as-built (record) drawings as a guide, conduct a field verification of the sewer system to identify changes made to the system during the intervening years. Key collection system personnel can assist you in identifying changes that have occurred since the original installation of the system. The repair records are also an extremely important resource for identifying changes to the sewer system. Records can indicate areas where problems

[1] *Surcharge. Sewers are surcharged when the supply of water to be carried is greater than the capacity of the pipes to carry the flow. The surface of the wastewater in manholes rises above the top of the sewer pipe, and the sewer is under pressure or a head, rather than at atmospheric pressure.*

Fig. 10.1 Sewer leaking to the environment

Fig. 10.2 Damage caused by leaking sewer

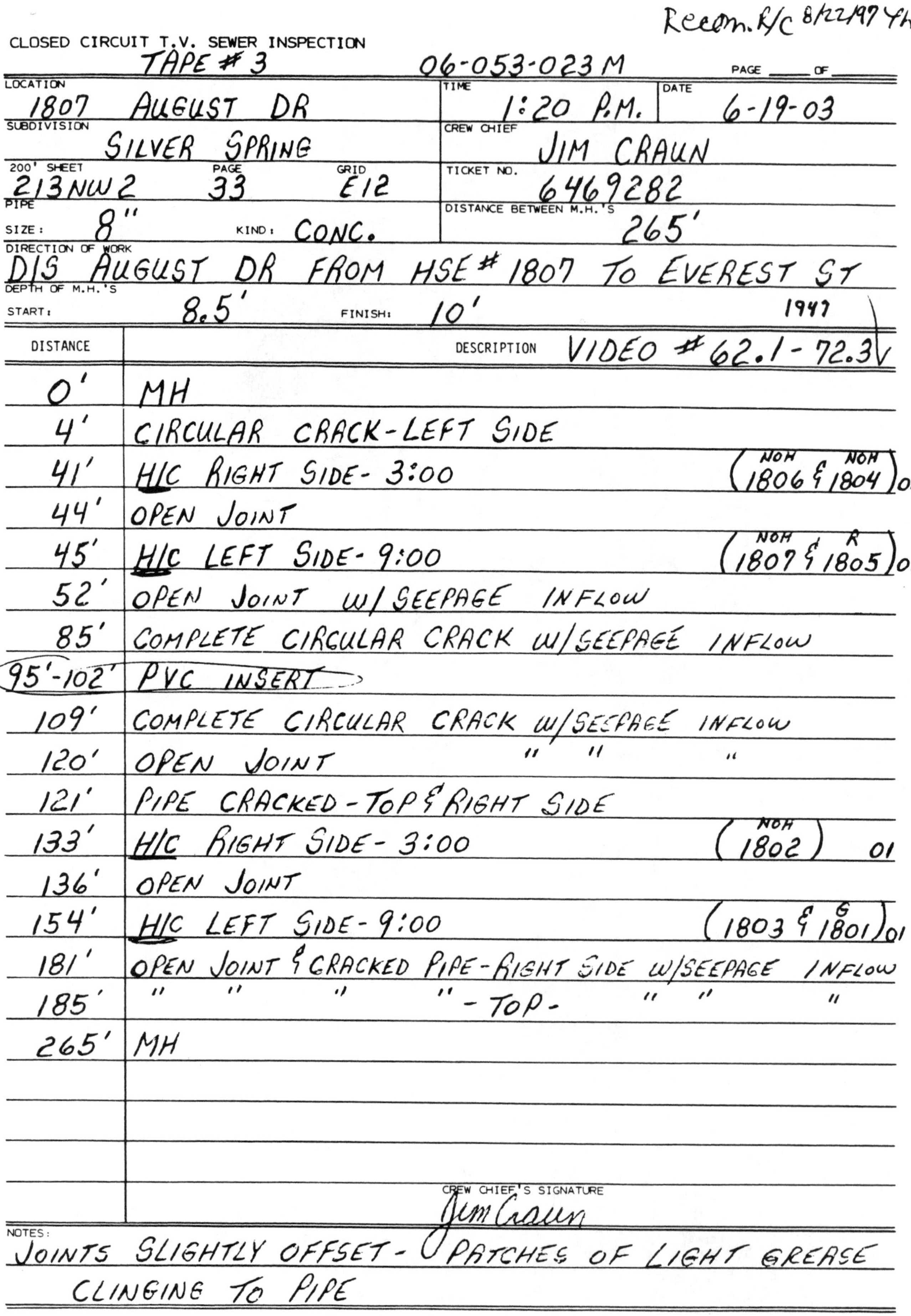

Recom. R/C 8/22/97 YLY

CLOSED CIRCUIT T.V. SEWER INSPECTION

TAPE # 3 06-053-023 M PAGE ____ OF ____

LOCATION	TIME	DATE
1807 AUGUST DR	1:20 P.M.	6-19-03

SUBDIVISION: SILVER SPRING	CREW CHIEF: JIM CRAUN
200' SHEET: 213NW2 PAGE: 33 GRID: E12	TICKET NO.: 6469282
PIPE SIZE: 8" KIND: CONC.	DISTANCE BETWEEN M.H.'S: 265'

DIRECTION OF WORK: D/S AUGUST DR FROM HSE # 1807 TO EVEREST ST

DEPTH OF M.H.'S START: 8.5' FINISH: 10' 1947

DISTANCE	DESCRIPTION VIDEO # 62.1 - 72.3
0'	MH
4'	CIRCULAR CRACK - LEFT SIDE
41'	H/C RIGHT SIDE - 3:00 (NOH 1806 & NOH 1804) 01
44'	OPEN JOINT
45'	H/C LEFT SIDE - 9:00 (NOH 1807 & R 1805) 01
52'	OPEN JOINT W/ SEEPAGE INFLOW
85'	COMPLETE CIRCULAR CRACK W/SEEPAGE INFLOW
95'-102'	PVC INSERT
109'	COMPLETE CIRCULAR CRACK W/SEEPAGE INFLOW
120'	OPEN JOINT " " "
121'	PIPE CRACKED - TOP & RIGHT SIDE
133'	H/C RIGHT SIDE - 3:00 (NOH 1802) 01
136'	OPEN JOINT
154'	H/C LEFT SIDE - 9:00 (1803 & G 1801) 01
181'	OPEN JOINT & CRACKED PIPE - RIGHT SIDE W/SEEPAGE INFLOW
185'	" " " " - TOP - " " "
265'	MH

CREW CHIEF'S SIGNATURE: Jim Craun

NOTES: JOINTS SLIGHTLY OFFSET - PATCHES OF LIGHT GREASE CLINGING TO PIPE

Fig. 10.3 Sewer Inspection report

CUES, Inc.
3600 Rio Vista Ave.
Orlando, FL 32805
(407) 849-0190
Fax (407) 425-1569

Site Data for Project: Ukiah, Demo

Site ID	City	Street	Date	Time
1	Ukiah	300-Hillcrest	10/02/2002	01:14:29 PM

M.H. Start	M.H. Stop	M.H. Depth	Starting Dist	Final Dist
344	end	4.0	8.0	390.8

Type of Pipe	Pipe Size (in)	Sec. lgth	Direction	Surface Condition	Operator
Concrete	6	4	Away-D	Paved Asphalt	Rich

Comment
Demo on 10-02-02 for the city of Ukiah

Observation Data

Obs ID	Ft	Lat Ft	Category	Category Details	Clock Pos	Sevr Lv	Ph 1 ID	Ph 2 ID	Vclip ID	Vid ID	Tape Cnt
1	8.0		Other	Start Run					34.68		
2	10.1		Pipe Problem	Roots	9 O'Clock	Level 2	1252.jpg	2252.jpg	153.41		
3	12.5		Joint Problem	Roots in Joint	360	Level 2			245.27		
23	12.6										
4	22.3		Joint Problem	Roots in Joint	360	Level 1	1254.jpg				
5	32.2		Joint Problem	Roots in Joint	7 O'Clock	Level 1					
6	42.7		Joint Problem	Roots in Joint		Level 1					
7	53.2		Joint Problem	Moderate Offset	360	Level 2	1257.jpg				
8	53.9		Service Conn.	Roots	9 O'Clock	Level 3	1258.jpg		294.15		
9	55.1		Service Conn.	Roots	9 O'Clock	Level 4			395.01		
10	66.7		Service Conn.	Roots	3 O'Clock	Level 4	12510.jpg		490.14		
11	107.2		Service Conn.	Roots	9 O'Clock	Level 3			587.34		
12	108.1		Service Conn.	Domestic Flow	12 O'Clock	Level 2			647.56		
13	130.9		Service Conn.	Roots	3 O'Clock	Level 2			747.57		
14	152.9		Service Conn.	Roots	9 O'Clock	Level 3			814.34		
15	175.2		Service Conn.	Roots	3 O'Clock	Level 4			891.64		
16	212.7		Service Conn.	Connection	9 O'Clock	Level 1			1010.09		
17	234.5		Service Conn.	Connection	9 O'Clock	Level 1			1073.80		
18	316.2		Joint Problem	Severe Offset	9 to 3	Level 2			1236.35		
19	346.6		Pipe Problem	Sag	bottom half	Level 2			1383.69		

Page 1 of 2

Fig. 10.4 Computer-generated maintenance management report

(Permission of CUES, Inc.)

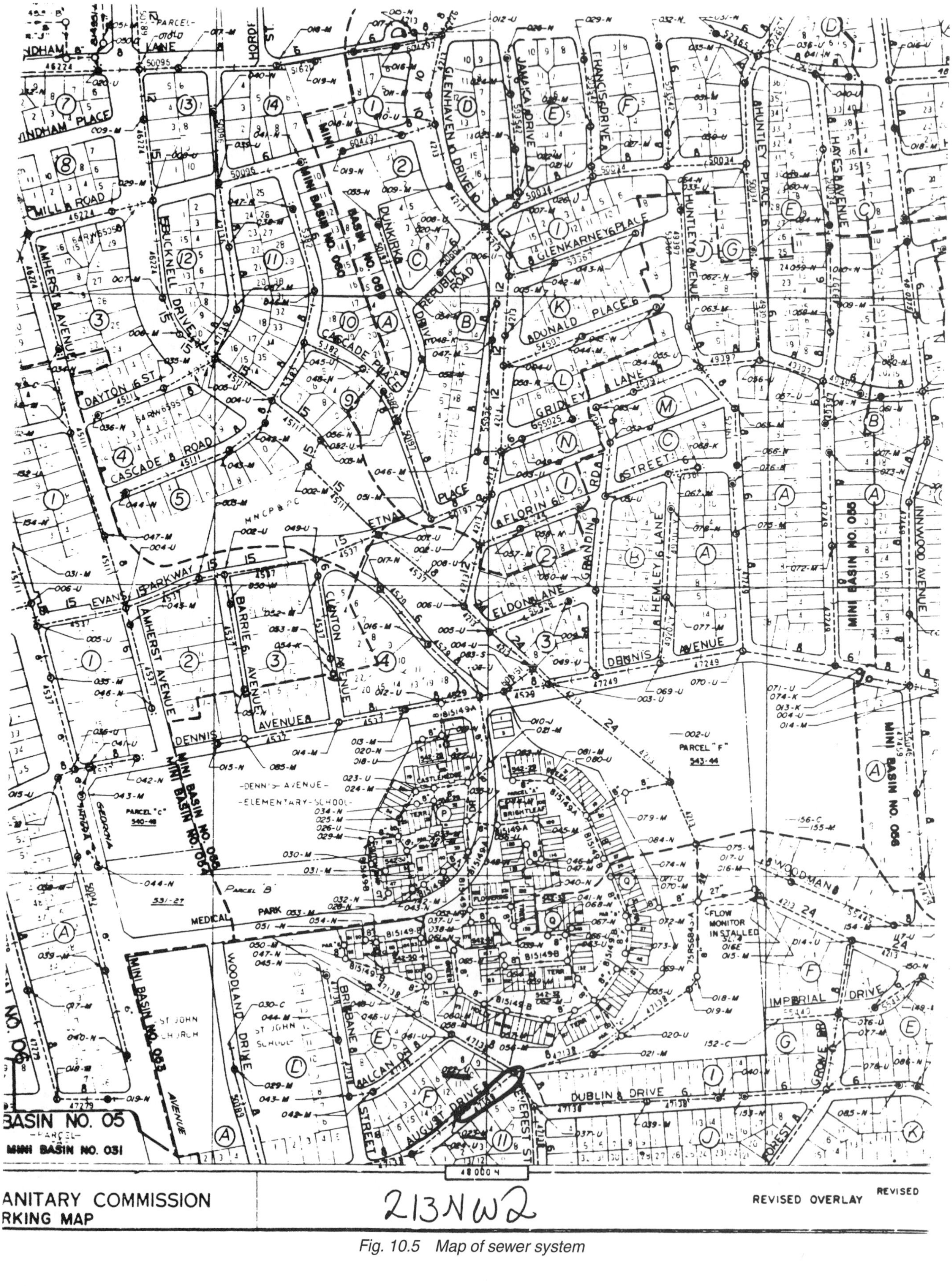

Fig. 10.5 Map of sewer system

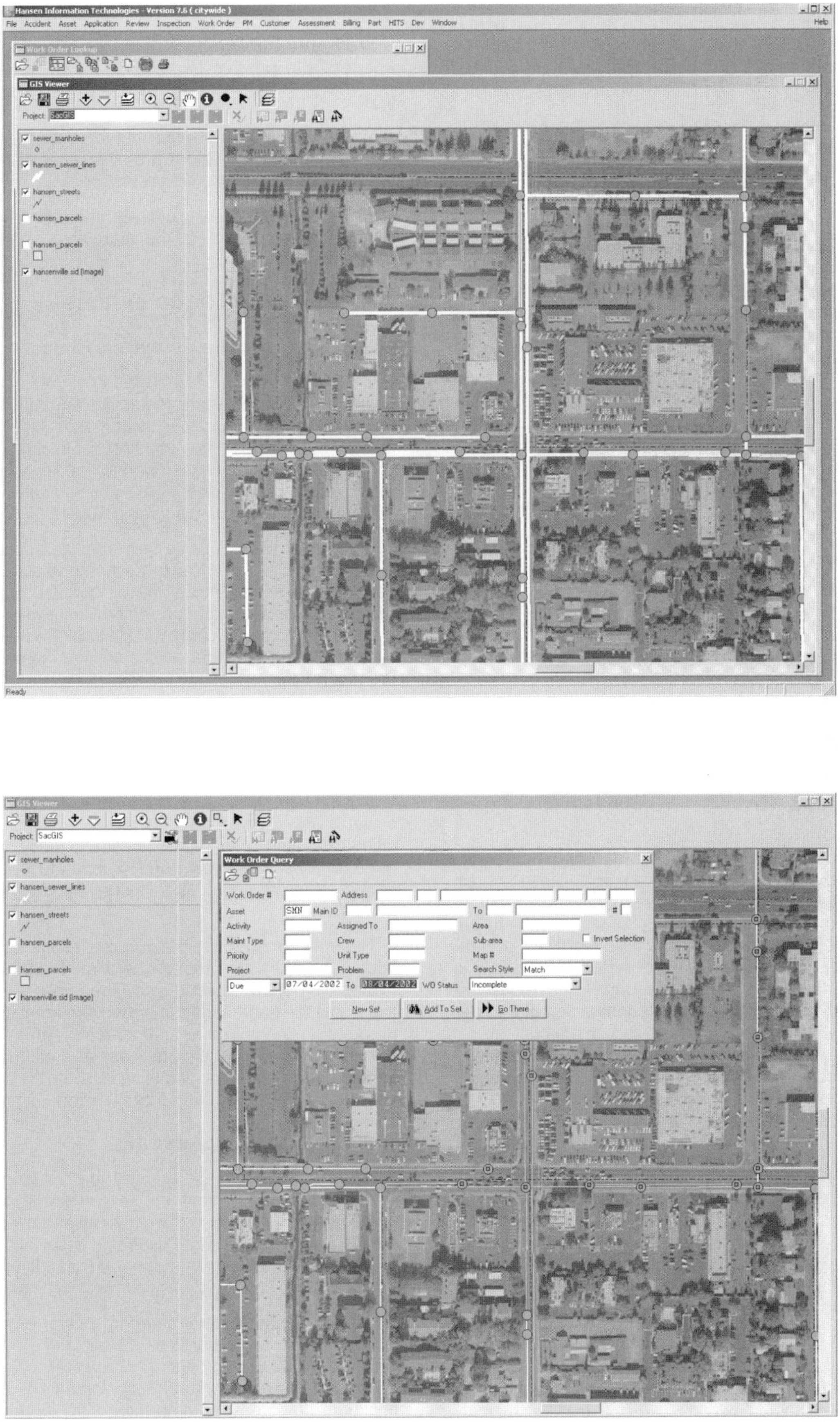

Fig. 10.6 Computer-generated maintenance management GIS maps

(Permission of Hansen Information Technologies)

occur regularly, repairs are frequent, maintenance is excessive and additions have been made to the system. Your goal is to prepare an inventory that accurately identifies and describes the present condition of each and every component of the sewer system.

Each jurisdiction uses its own unique mapping system to record the overall physical layout of the sewer system. These maps normally include the location of junctions at manholes, size of sewer pipe, reference to original contract and general location of the sewer in relation to roads and other utilities.

Most utilities store all their records in a manual card system and plan file system. Some utilities have been fortunate enough to transfer their records to a personal computer file or, less often, to a mainframe computer. Also, some mapping has been converted to computer-aided design (CAD) (Figure 10.7) or geographic information systems (GISs) (Figure 10.6). Whichever mode of record storage is available to you, it must be organized properly if it is to be an effective tool in the inventory process.

The purpose of a physical survey is to identify the problems that created the need for rehabilitation projects. Closed-circuit television, pipe flow tests, computer flow models and visual inspections are some of the many techniques available to the utility for evaluating the system condition. Capacity evaluations can assist in determining whether additional capacity is required; this has a direct bearing on whether or not rehabilitation is needed or total replacement by construction of a larger main is required. Decisions are also based on the costs associated with the rehabilitation and the reliability of the sewer system. The condition analysis can be a major tool in setting priorities in your program.

10.02 Implementation

Most utilities will find this step to be the one that raises the most problems. It is usually not too difficult for a utility to develop a program and to identify the key elements of the rehabilitation program. The major stumbling block is to identify the staffing to handle the program and to secure the funding to actually accomplish rehabilitation work. The levels of funding and staffing will directly affect the level of completeness of your rehabilitation program.

After arranging for financing and staffing, the actual design and construction can be started. In conjunction with the design stage, you should also review your present maintenance program to determine areas of revision which will fit into the rehabilitation program to ensure a coordinated effort of maintenance and rehabilitation.

QUESTIONS

Write your answers in a notebook and then compare your answers with those on page 354.

10.0A Why is there a need for sewer rehabilitation?

10.0B Why must a physical survey of a collection system be conducted before a sewer rehabilitation project can start?

10.0C Why must the present maintenance program be reviewed before the start of a sewer rehabilitation program?

10.1 EVALUATION OF CONDITIONS

10.10 Purpose of Sewer System Evaluation

Since the passage of Public Law 92-500 in 1972, there has been increasing pressure to reduce the Infiltration/Inflow (I/I) problems in existing sewer systems. With this law, the EPA issued guidelines for conducting I/I evaluations as a step toward reducing the unplanned-for I/I flows. The evaluation process consists of two parts: (1) I/I analysis, and (2) the sewer system evaluation survey (SSES).

The main reason for this new emphasis was to reduce the hydraulic loads on the sewer system and the treatment facilities. Excessive hydraulic loads increase treatment costs, result in the bypassing of untreated wastewater to the environment and can result in structural failure of weakened collection systems. The remainder of this section discusses the proper procedures for conducting I/I evaluations.

10.11 System Problems

High levels of I/I and structural deterioration of sewer systems can have serious impacts on the operation of the system. Unnecessarily high flows reduce the capacity of the sewer lines to handle peak sanitary flows and take up the additional capacity reserved for growth. In many systems that experience excessive flows and deteriorated conditions, relief sewers must be constructed before they might otherwise be needed.

The major problem from a public relations viewpoint is backflooding into private property as a result of the surcharge or overloading of the sewer mains. The public is also very sensitive about the degradation of the environment due to overflows from storm drains into natural streams. The higher flows can also damage and accelerate wear of equipment at pumping stations and treatment plants, and even the treatment processes are adversely affected by high flows.

10.12 Hydraulic Aspects

Flow monitoring is the primary tool for identifying high I/I. The purpose of flow monitoring is to measure variations in flow components over time to identify peak and minimum flow conditions. The flow monitoring phase is set up to identify and measure the three flow components: base flow, infiltration and inflow.

Base flow can be determined in several ways. You can use water consumption records as a basis and adjust for inaccuracies due to unmetered connections, water meter inaccuracies, irrigation and seasonal peaks. Another method is to measure minimum flow rates (to determine infiltration rates) and then subtract this amount from metered flow during dry weather. A third way to determine base flow is to evaluate water consumption estimates for residents upstream of the metering lo-

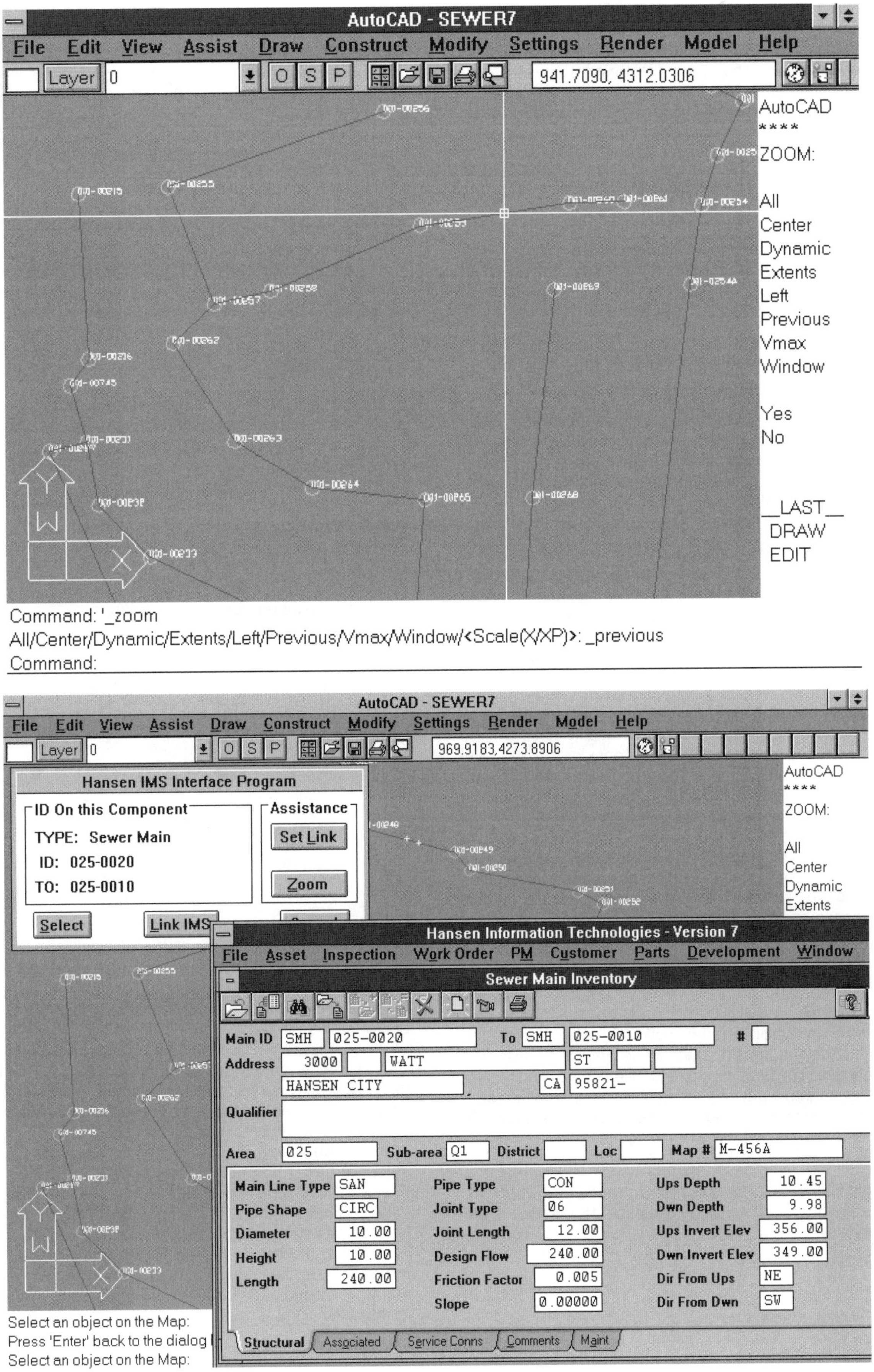

Fig. 10.7 Computer-aided design (CAD) screens

(Permission of Hansen Information Technologies)

cation. In this method, the expected base flow can be estimated by examining water usage or water meter records, the amount of water used or consumed by upstream residents (lawn watering, water used in manufacturing processes such as beverage production) and the amount of water returned to the sewers (flushing toilets, dishwater, industrial wastewaters). Flows greater than the expected base flow can be considered to come from inflow and infiltration sources.

Infiltration is calculated by subtracting base flow from total metered flow during dry weather. Infiltration can also be computed by making a series of flow isolation measurements. Flow isolation measurements are obtained by blocking off sections of a sewer and then measuring the flow in the sewer. All unaccounted for flows or all flows when all known sources are blocked off are considered infiltration flow.

Inflow is measured during wet weather and is calculated by subtracting base flow and infiltration from total flow data recorded during wet weather.

Exfiltration (leakage out of the sewer system) can be a serious problem in some areas where it might contaminate groundwater used for a public drinking water supply. Exfiltration can be measured by using the same techniques used to measure inflow/infiltration, except we are looking for a *REDUCTION* in expected flows.

Flow monitoring is the backbone for the hydraulic analysis phase of a sewer evaluation. Both temporary and permanent metering sites are used to collect the data. Temporary metering sites provide the majority of the data required to evaluate sewer flows. The permanent sites at pumping stations and treatment plants can be used to provide additional backup data.

You will need to plan the flow metering phase well in advance so that the maximum amount of quality data can be gathered for the most reasonable cost. Consider also the level of accuracy you wish to achieve. Next, figure out how long you will have to measure flows to produce the required amount of data. This is an important decision because it directly affects the type of recording equipment and the data storage capacity you will need. Some methods of data storage include strip charts, circular charts, magnetic tape and remote telemetering. Magnetic tape and telemetering allow much faster gathering and processing of the flow data. The one negative side of this technology is that it is not easy to check the data to see how the meter is operating in the field.

Each flow metering program requires preplanning of the sites by actual field surveys. Do not rely on drawings to provide accurate hydraulic characteristics of a metering site. When the conditions are not found to be optimal, then select an alternate site. The best manhole locations for open channel metering will not have any change in horizontal direction (pipe alignment) or any vertical change (pipe slope or elevation). In addition, the flow should be a smooth flow and have a moderate velocity to assist in self-cleaning of the channel.

The initial field data to be collected and recorded when you are evaluating proposed metering sites should include the following items:

1. Flow depth,
2. Sediment load,
3. Pipe size and shape,
4. Accessibility,
5. Manhole location, and
6. Surcharge potential.

It is very important to determine the possible surcharge potential for the metering location. Surcharging causes nonuniform flow or backflow conditions which will produce faulty flowmeter readings. Additional equipment, such as pairs of depth recorders or flow velocity and depth recording equipment, may be required to properly handle the surcharge readings. In every case, actual and potential flow conditions will influence your selection of metering equipment.

The use of weirs and flumes will usually increase the accuracy of your flow metering program since weirs are one of the most accurate measuring devices for low-flow conditions. They are susceptible to inaccurate readings when debris, rags or sludge build up behind or on the weir. Devices of this type usually reduce the hydraulic capacity of the sewer. Therefore, close coordination of the metering program with system maintenance personnel is imperative to avoid adverse impacts to the system and metering equipment from normal, daily maintenance activities.

Flow data evaluation involves the production of flow guidelines and hydrographs for each location metered. As mentioned earlier, the data can be analyzed manually or automatically, depending on which equipment is used. Computers provide increased accuracy and speed up the processing of the metered data. Analyze the data on preset intervals, normally 15 minutes, 30 minutes or 60 minutes. Determine the flow rates for each interval and then calculate a total daily volume. Check individual hourly measurements to determine peak hourly rates. Construct hydrographs with each data point falling at the appropriate interval (Figure 10.8).

Another important hydraulic consideration is the measurement of precipitation. The correlation between rainfall and flow metering data is important when calculating the amount of inflow entering the sewer system. Precipitation data should include rainfall intensity, duration of the rainfall and total rainfall volume. A lot of this rainfall data can be gathered from agencies that continuously monitor rainfall data such as the weather bureau or water resources agencies. After a review of the data available, select the locations and types of rainfall measuring devices required for your flow metering program. Since precipitation runoff is usually highly correlated with inflow, compare sewer flows and inflow flows with precipitation data. If sewer flows and inflow flows increase when precipitation values increase, then this is an indication that the collection system has inflow problems and that the sources should be identified and controlled or eliminated. Try to place your rainfall measuring devices where they can collect accurate measurements of precipitation data without interference of buildings or by vandals.

Measurement of groundwater is a very important component in the hydraulic analysis phase. The location of the groundwater level and measurement of changes in the level are the two most important facets of this phase. Normally,

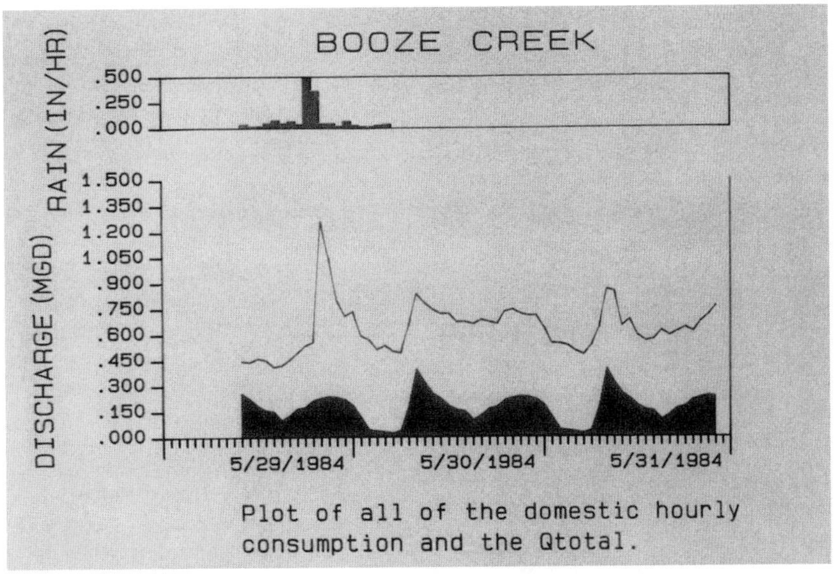

Fig. 10.8 Hydrograph of flow vs. time

either a piezometer (pie-ZOM-uh-ter) or a manhole gage is used for sewer evaluations. The manhole gages (Figure 10.9) are less expensive and easier to install, but provide data of lower quality and can become clogged easily. The piezometer (Figure 10.10) is a more permanent installation, provides higher quality data, and is less likely to clog, but it is more expensive. Groundwater levels are recorded on a periodic basis (daily or with tidal fluctuations depending on location) during flow metering periods. The frequency would be increased during the metering period to provide better data correlation. Levels of infiltration (measured in gallons per day or gallons per day per inch of sewer diameter) are determined by plotting groundwater levels versus time in comparison with metering data. Also, the groundwater data can be used to more effectively schedule flow isolation and television inspection because the higher the groundwater, the greater the expected infiltration flow.

Night flow isolation is used to determine the amount of water entering a section of sewer by infiltration. It is used to pinpoint any *STRETCHES*[2] of sewer which have excessive infiltration so they can be targeted for additional inspection. It cannot be stressed enough that accurate maps are essential for the success of the isolation program. Accurate measurement point identification is critical. A successful isolation program depends largely on knowing the exact location of groundwater infiltration measurements as they relate to the sewer stretch groundwater level. Other sources of excess flows include roof drains and yard drains illegally connected to the sewer.

Once you have determined that there is excessive infiltration in a particular area, the next step is to identify the specific locations of leaks. There are three major sources where rainfall-induced I/I enters a sewer system. They are the following:

1. Sewer mains,
2. Manholes, and
3. Sewer house connections.

Each major source has various types of leaks which contribute to the rainfall-induced I/I.

1. Sewer Mains
   a. Point source leak (Figure 10.11)
   b. Defective joint leak (Figure 10.12)
   c. Service tap point source leak
   d. Main line multiple leaks (Figure 10.13)
   e. Storm drain cross-connection leak
2. Manholes
   a. Holes in cover
   b. Worn cover ring
   c. Between ring and brickwork (Figure 10.14)
   d. Through brickwork or grade ring
   e. Through manhole wall
   f. Through manhole joints
   g. At plugged stub in manhole
   h. Around sewer pipes entering manhole (Figure 10.15)
   i. Through base of manhole

[2] *Stretch. Length of sewer from manhole to manhole.*

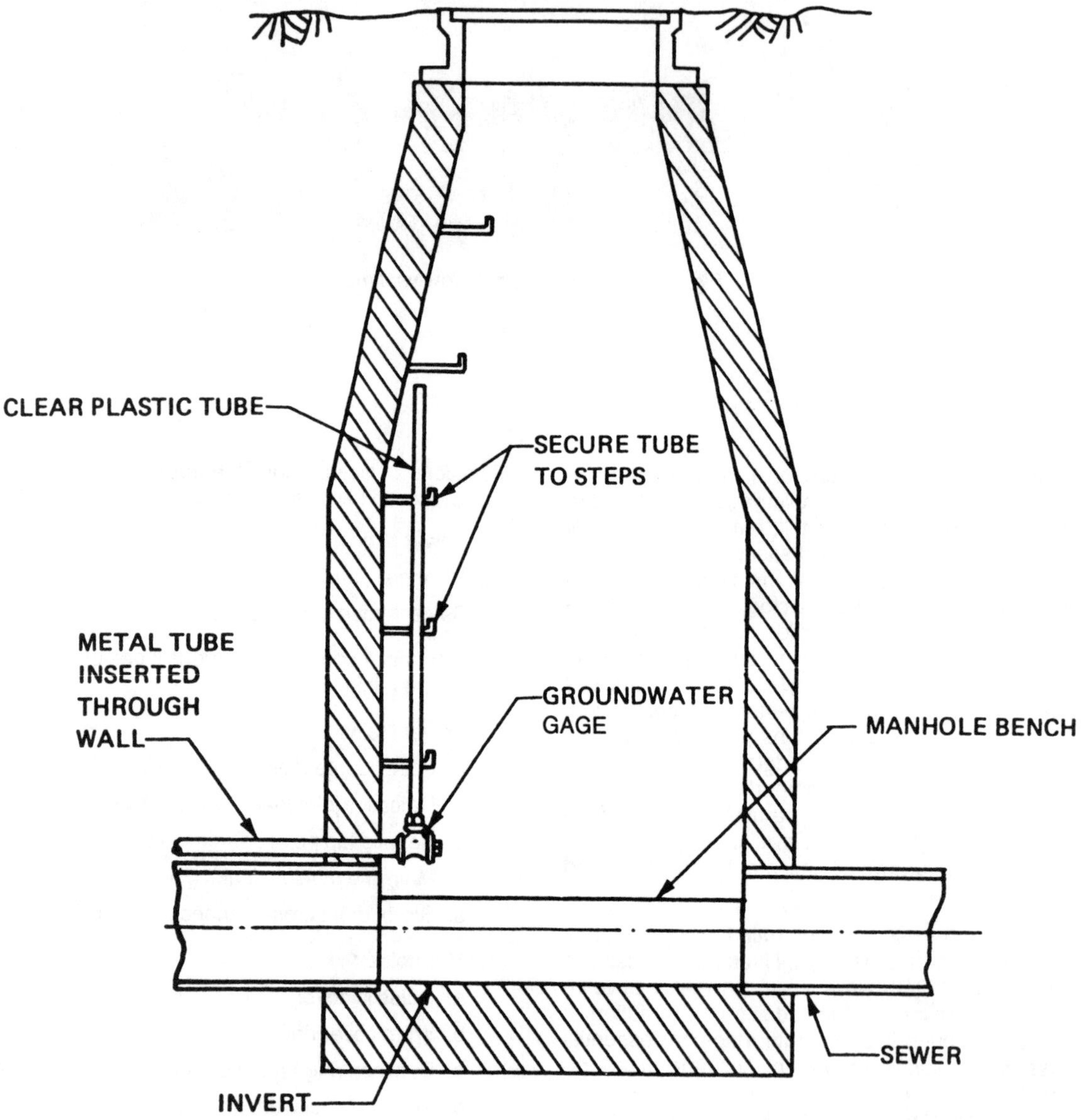

Fig. 10.9 Typical static groundwater gage installation

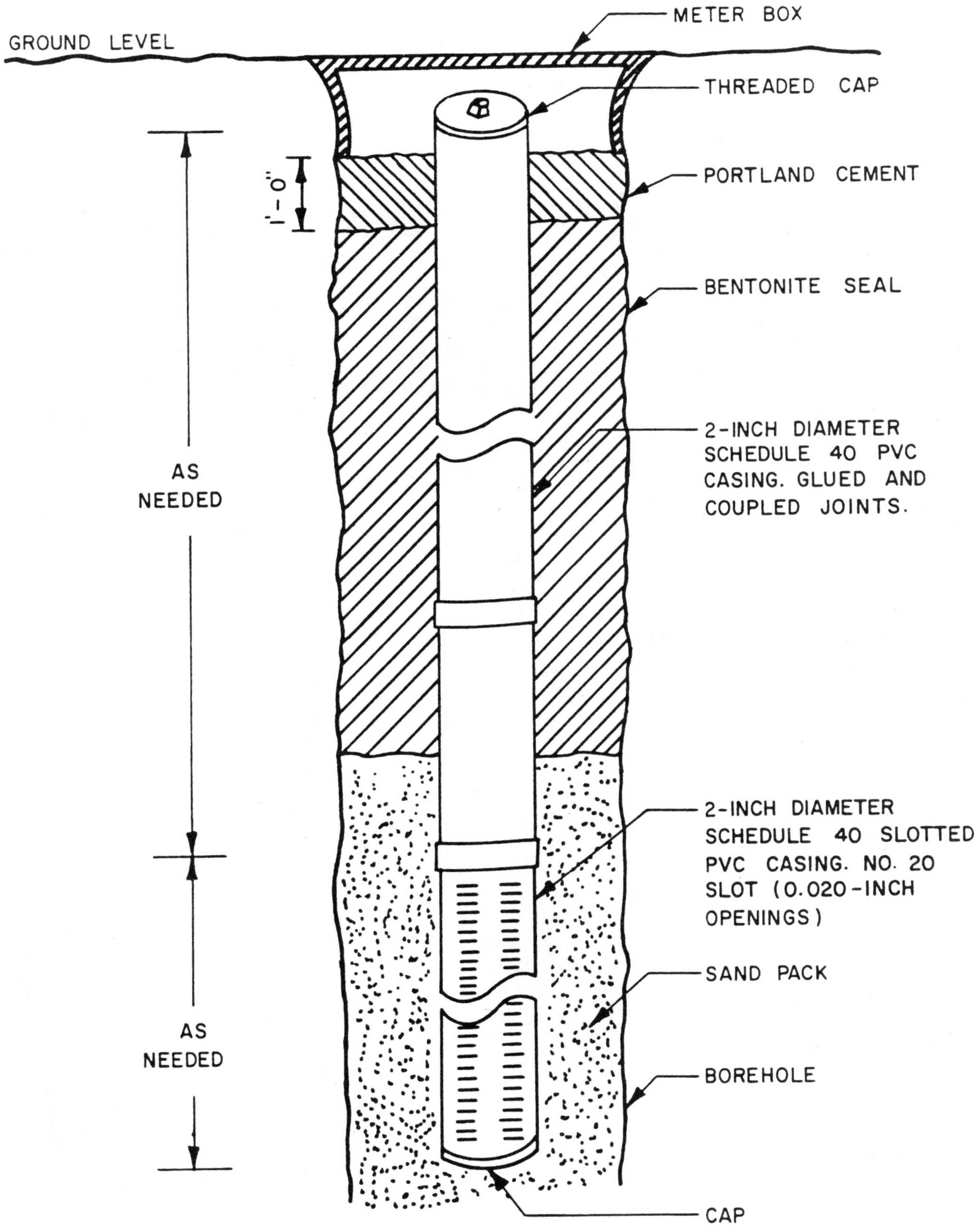

Fig. 10.10 Piezometer monitoring well installation

Fig. 10.11 Point source leak

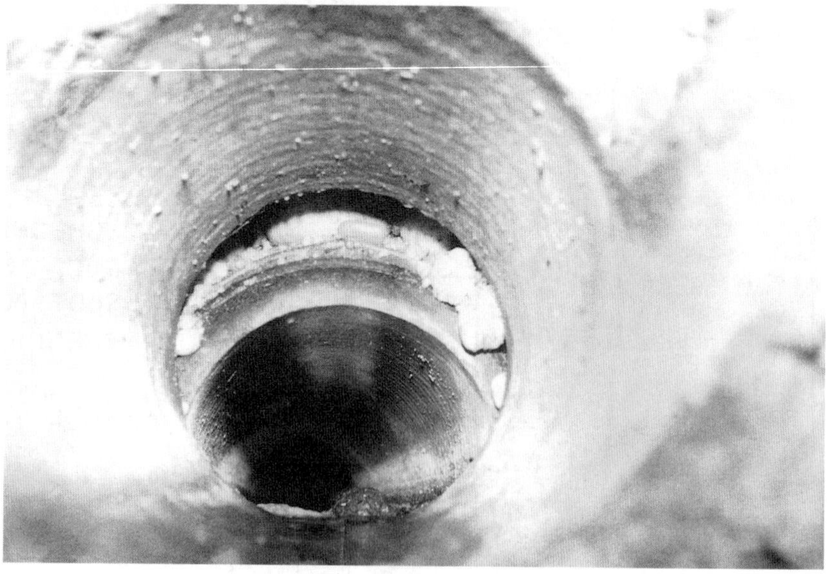

Fig. 10.12 Defective joint leak

Fig. 10.13 Main line multiple leaks

Fig. 10.14 I/I between manhole ring and brickwork or grade ring

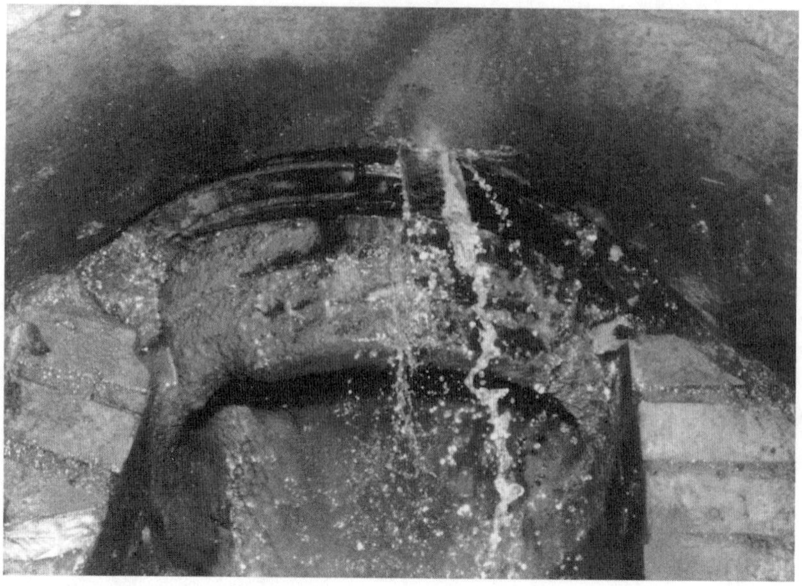

Fig. 10.15 I/I around top of sewer pipe entering manhole

The bedding around sewer pipes will often serve as a drain and carry groundwater down the pipe trench to the manhole. This water will then enter the sewer through defects in the manhole.

3. House Service Connections
   a. Point source leak
   b. Joint leaks
   c. Multiple damage leaks
   d. Defective cleanout leak
   e. Storm drain cross-connection
   f. Foundation drain hookup
   g. Roof drain connection
   h. Surface or yard drain connection

Several techniques are used to locate rainfall-induced leaks, including smoke testing (Figure 10.16) and dyed water flooding. Smoke testing is a quick and inexpensive method of

Fig. 10.16 Smoke testing

detecting I/I sources in a sewer system. This approach is most effective in detecting inflow from storm drain cross-connections, point source leaks, roof connections, foundation drains, faulty manholes and sometimes main line defects. The method is not effective during high groundwater conditions or in sewers where sags or water traps are suspected or where sewers flow full. The smoke will not be able to travel through the sewer section when a section is full of water and can result in misleading conclusions. Smoke testing is also ineffective when the ground is saturated, frozen or covered with snow because the smoke will not be able to escape to the surface even though there may be defects in the sewer system. Windy conditions can also distort your readings and result in faulty conclusions.

To generate smoke for this test, use commercially available smoke bombs of three-minute or five-minute duration. The smoke is nontoxic, odorless and nonstaining. Careful preparations must be made before you start the smoke test. Coordinate your activities with the police and fire departments to avoid false fire alarms. Also, notify each resident in the test area that a test will be taking place and explain the procedures as well as what they might expect to see.

The procedures for properly conducting a smoke test are as follows:

1. Notify police, fire officials and residents (Figure 10.17);
2. Isolate the section of sewer to be tested;
3. Prepare a detailed sketch of the test area;
4. Make enough smoke to completely fill the stretch being tested;
5. Using an air blower, force smoke into the isolated area;
6. Walk area thoroughly (several observers are recommended) to observe leaks;
7. Record precise locations of leaks on sketch; and
8. Take photographs of leaks (try to include identifying landmarks) to create a permanent record of the test results.

Refer to Chapter 5, Section 5.4, "Smoke Testing," in Volume I for details on smoke testing procedures.

When there is a positive reading of smoke exiting to the atmosphere, it clearly pinpoints I/I sources. When no smoke is visible, do not automatically conclude that there are no defects. As stated earlier, there may be other conditions that interfered with the results. If there is no visible smoke exiting from the system and I/I appears to be a problem in the area, try dyed water flooding (next paragraph) or televising the system during periods of high groundwater and/or precipitation.

Dyed water flooding or dye testing is another rainfall simulation technique used to identify defects that allow leakage into sewers during rainfall. It also can be effective in quantifying the volumes of I/I which can enter the specific defects. Areas that are prime candidates for dyed water flooding are:

1. Storm drains that run parallel to or cross sewers or house connections in which top-of-pipe elevations are higher than the invert elevations of the sewers;
2. Drainage ditches, streams and water ponding areas located close to or on top of sewers or house connections;
3. Yard drains, areaway drains, foundation drains, roof drains and abandoned building sewers; and
4. Specific problem areas identified in other tests for which additional data are needed.

The procedure is very simple and requires simple equipment. Plug the study stretches and fill the area with dyed water. The dyes are fluorescent for detectability, safe to handle, easily mixed in water, inert in (do not react with) soils and biodegradable. After the storm water areas are flooded or filled with the dyed water, observe the sewer system for the entry of the dyed water. In stretches that are suspected of having excess inflow, closed-circuit television can be used to quantify and pinpoint the leaks. Record the collected data on data sheets and make a sketch or map of the exact locations of leaks so they can be corrected. Also a dye solution may be prepared and dumped into a drain that you suspect is illegally connected. Flush the drain with water and see if the dye washes into the sewer.

10.13 Structural Aspects

Structural integrity and defect identification should be integral parts of all sewer evaluations. Changes in the land surface area over the sewers due to land use results in changes of loadings on the sewer pipe. In some cases the load changes exceed the designed strength of the pipe or manhole walls. Also, changes can result in additional traffic loading which was not included in the original design. Since there are many methods of rehabilitation, the sewer cover conditions must be evaluated in terms of present use as well as anticipated for future uses.

NOTICE TO PROPERTY OWNERS
SMOKE TESTING
TO BE CONDUCTED IN YOUR AREA

The Town of Cary's Public Works & Utilities Department will soon be conducting leak tests of the sanitary sewer system in your area by forcing smoke into the lines. This smoke will help locate places where storm and other surface waters are entering the Town's sewers as well as reveal sources of sewer odors. This leak testing is part of our continuing effort to provide a safe, economical, efficient, and environmentally sound sewer system throughout Cary.

While most residents will never see or smell the smoke, the Town wants you to have as much information as possible about the testing. A special non-toxic smoke will be used in these leak tests. The smoke is manufactured for this purpose and, therefore, leaves no residuals or stains and has no effect on plants and animals. The smoke has a distinctive, but not unpleasant odor. In the unlikely event that you should have direct contact with the smoke, you may experience some minor irritation of the respiratory passages. These problems last only a few minutes where there is adequate ventilation. However, if people in your building have asthma, emphysema, or some other respiratory condition, please notify the Town immediately so that we can discuss your case in further detail.

If traces of the smoke or its odor enters your house or building, it is an indication that gases and odors from the sewer also may enter. Evidence of smoke in your house during the smoke testing should be immediately reported to the Public Works & Utility Department testing crew and to your plumber. Location, identification, and correction of the source of smoke entering your house is strongly recommended. While the Town's Public Works & Utility Department will render all possible assistance, the correction of any defects in the pipes and sewer on private property is the responsibility of the owner.

Leak testing smoke may enter your house if:

- Vents connected to your building's sewer pipe are inadequate, defective, or improperly installed; or

- Traps under sinks, tubs, basins, showers, and other drains are dry, defective, improperly installed, or missing; or

- Pipe, connections, and seals of the wastewater drain system in and under your building are damaged, defective, have plugs missing, or are improperly installed.

Because testing in your sector will take several weeks, residents of a specific area scheduled for smoke testing will receive a public notice within 24 hours of the actual testing (weather permitting). Once the 24-hour notice is received, we advise residents to run water into all of their drains for one minute, especially those used infrequently. This should reduce the likelihood of smoke entering the house inadvertently.

If your property is not owner-occupied, we ask that you help us by forwarding this important information to anyone living or working there. If you have any questions, or desire more information, please call 469-4090 during regular office hours (8:00 a.m. to 5:00 p.m.), or visit our website anytime at www.townofcary.org.

Fig. 10.17 Sample smoke testing notification door hanger
(Permission of the Town of Cary, www.townofcary.org)

The existing condition of the pipe must be evaluated. The existing condition has an effect on the strength of the pipe and on the hydraulic capacity of the pipe. Again, the rehabilitation method depends a great deal on the existing conditions identified. When examining for structural defects record the following information.

1. Sewer Pipe
   a. Type of pipe
   b. Pipe deflection (sag or bow in pipe)
   c. Joint separation or offset
   d. Cracked joints
   e. Root intrusion
   f. Circumferential and longitudinal cracks
   g. Crushed pipe
   h. Missing pipe
   i. Protruding taps
   j. Leaking joints
   k. Leaking taps
   l. Corrosion of pipe walls
   m. Pipe diameter (no longer round)
2. Manholes
   a. Type of material
   b. Wall condition
   c. Joint condition between manhole and frame
   d. Channel condition
   e. Joint condition where pipes enter manhole
   f. Precast joint condition
   g. Visible leakage
   h. Corrosion of walls
   i. Frame and cover condition
   j. Step condition

All utilities experience operational problems which can be directly attributed to defects. Some of the problems encountered are:

1. Overflowing manholes,
2. Backups into homes,
3. Sewers surcharging,
4. Exfiltration of wastewater,
5. Sunken areas over sewers,
6. Increased flows due to inflow/infiltration,
7. Sand and gravel in sewers,
8. Grease buildup, and
9. Odors from hydrogen sulfide due to low flows from exfiltration.

Most often problems are reported by the public. In some instances routine inspection programs will detect some of these conditions. All information must be recorded so that the history of problems can be identified with a particular stretch of sewer or a specific manhole.

Pipeline visual inspection involves internal inspection and surface inspection. Walking the surface along a sewer right-of-way can give you a lot of information about the condition of the sewer. Sunken areas, stream crossing condition of pipelines, areas of ponding water and water leakage from the soil along the right-of-way are all indications of sewer defects.

Internal inspection of the sewer pipe involves lights and mirrors (lamping) or closed-circuit television. The information gained by use of lights and mirrors is limited. Usually only a small portion of the sewer stretch can be viewed with these methods. Also, it is difficult to clearly identify the problems with this method.

The use of closed-circuit television (Figure 10.18) is the most effective method of identification and quantification of defects in sewer pipes. Reports can be made from the inspection and a permanent visual record of the inspection is available by use of video recordings. Constant updates in technology have resulted in the ability to televise small-diameter pipes as well as house connections. The technology has evolved from black and white to color closed-circuit television. The advantage of color is to give added depth perception to the inspection. A better evaluation of defects can be made as a result of this added depth perception.

A standard record procedure for reports made from the inspections must be established. This will assist in ready reference when searching for a particular report. Certain data should be gathered for each inspection report.

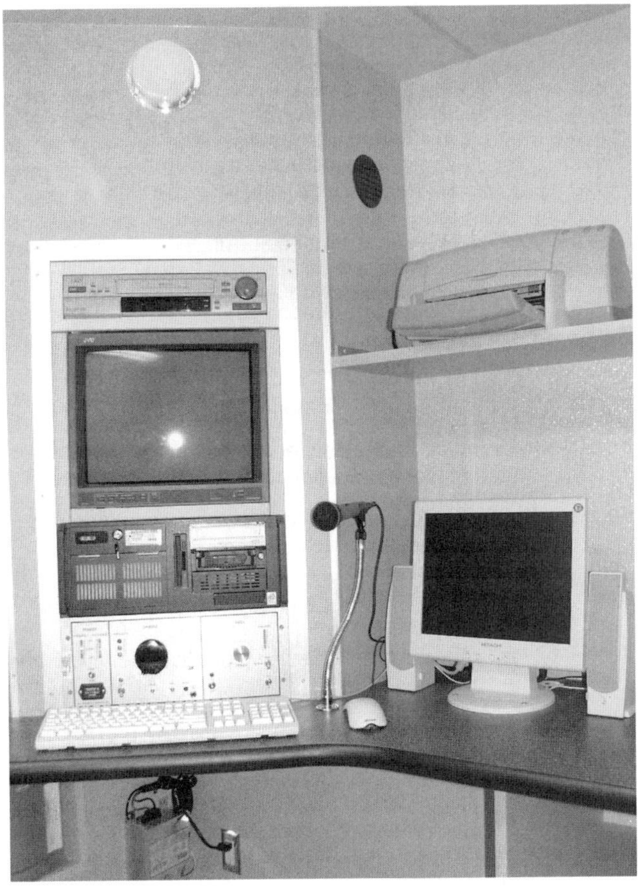

Fig. 10.18 Closed-circuit television console

(Permission of R. S. Technical Services, Inc.)

1. Length of stretch or inspection area
2. Pipe material and pipe diameter
3. Joint spacing and type
4. Location and quantity rating of root intrusion
5. Grease accumulation location
6. Joint separation location and severity
7. House service tap location, condition
8. Water level in sewers
9. Cracks, location and severity
10. Missing pipe
11. Pipe deflections
12. Sewer sags
13. Corrosion of walls

QUESTIONS

Write your answers in a notebook and then compare your answers with those on page 354.

10.1A Why is there an emphasis on the reduction of hydraulic inflow/infiltration loads on sewer systems and treatment facilities?

10.1B List the three flow components in a sanitary sewer.

10.1C Why must the possible surcharge potential for a metering location be determined?

10.1D Smoke testing is not effective under what conditions or circumstances?

10.1E What types of areas are suitable for dye-water flooding?

10.1F How can defects in sewer pipes be identified and quantified?

10.2 SETTING UP A REHABILITATION PROGRAM

10.20 Data Collection (Figure 10.19)

Once you have completed the sewer system evaluation survey (SSES) of your system, you will have a great deal of raw data, including:

1. Inventory. A complete, itemized listing of all system components.
2. Maps. Complete, up-to-date maps of the system layout.
3. Condition of System. Detailed written reports of all significant leaks, cracks, collapsed pipes, and seriously deteriorated components. This record may include TV footage or videotapes.
4. Flow Data. Design specifications of flow capacities as compared with current flow measurements including base flow, infiltration, and inflow data. These should be organized in a format that indicates flow variations by time of day, time of year, and minimum and maximum flows.
5. Problem History. Historical data concerning problem areas such as repeated failures, adverse conditions (limited or hazardous access, for example), maintenance limitations, surcharges and flooding of homes and streets.
6. Treatment Facility Data. Data documenting frequency and severity of hydraulic overloads, discharge permit violations and plant bypasses.

Compiling this information may take several months' work but it is essential for the next step in the decision-making process.

10.21 Preliminary Statement of Needs

Based on the information gathered in the system inventory, put together a list of needed repairs, in the order they are most urgently needed. At this point, don't be concerned with anything except the physical and operational condition of system components. If necessary, begin your list by grouping items in general categories such as:

INOPERABLE/Immediate Repair or Replacement Needed

DECLINING FAST/Frequent Breakdowns

OK FOR NOW/Troublesome

or:

OVERLOADED AND NEAR COLLAPSE

SEVERE I/I — FREQUENT FAILURES

PERIODIC OVERLOADS

Once you have identified these groups, try to assign priorities within groups. Mark the most urgently needed repair as "1" and the next as "2" and so on. If several items seem equally urgent, assign the same rating to each.

As an operator, your role in the process may end here. Depending on the complexity of your system and the size and structure of the agency's administration, you may or may not be a participant in the next phase of setting up a rehabilitation program. In either case, however, it is helpful to understand how the decision-making process will most likely proceed.

10.22 Cost Analysis

The third step in the process of setting up a rehabilitation program involves calculating what it will cost to implement various strategies. Usually there are four courses of action that can be taken:

1. Maintain the present system, as finances permit,
2. Replace the worst components, to the extent possible,

Sewer Main TV Inspection

Inspection # 1018   Work Order # 1098   Activity TVI
Main ID SMH 025-0013   To SMH 025-0012   #   Length 234.00

Started 04/02/1996 10:30   Crew Leader 2424   ☐ Reverse Setup   ☐ Prior History
Completed 04/02/1996 16:40   Operator 3113   ☐ Sketch
Comp By CRW   Weather
Project   Flow Depth 1.50   Media
Crew   Pipe Det M   Format   Media # 75-2
Index 00:00 To 02:37

Schematic
☐ Crack ☐ Joint ■ Lateral ▨ Root ▨ Debris ■ Inflow/Infiltration ☐ Structural ▨ Alignment ■ Unknown

→ Structural: 54   Root: 127   Inflow/Infiltration: 12   Overall: 49 →
025-0013   025-0012

Inspection Data | Readings | Images | Comments

Sewer Main TV Inspection

Inspection # 1018   Work Order # 1098   Activity TVI
Main ID SMH 025-0013   To SMH 025-0012   #   Length 234.00

025-0013

From	To	Index	Clock	Grouted	Defect	Code	Description
22.00	23.90	677	10	N	LC	D	>1/2"W,<1'L
34.00	0.00		0	N	RC	D	>1/2"W,<1'L
52.00	0.00		0	N	MJ	E	SHF JT 80-9
58.00	0.00		0	N	L	G	FACTORY S
74.00	0.00		0	N	D	C	DEBRIS - HE
80.00	105.00		0	N	LC	C	<1/2"W,>2'L
115.00	0.00		0	N	I	B	I/I - MEDIUM
123.00	0.00		0	N	L	G	FACTORY S
147.00	0.00		0	N	A	C	CAMERA UN
180.00	0.00		0	N	I	A	I/I - LIGHT (
180.00	0.00		0	N	L	G	FACTORY S
180.00	0.00		0	N	R	B	ROOTS - ME
198.00	0.00		0	N	CS	G	COLLAPSED
30.00	0.00		0	N	L	G	FACTORY S
12.00	0.00		0	N	R	B	ROOTS - ME

025-0012

▣ Insert
▣ Modify
✗ Remove

Cond Ratings
Struct 54
Root 5
I/I 12
Overall 37

Inspection Data | Readings | Images | Comments

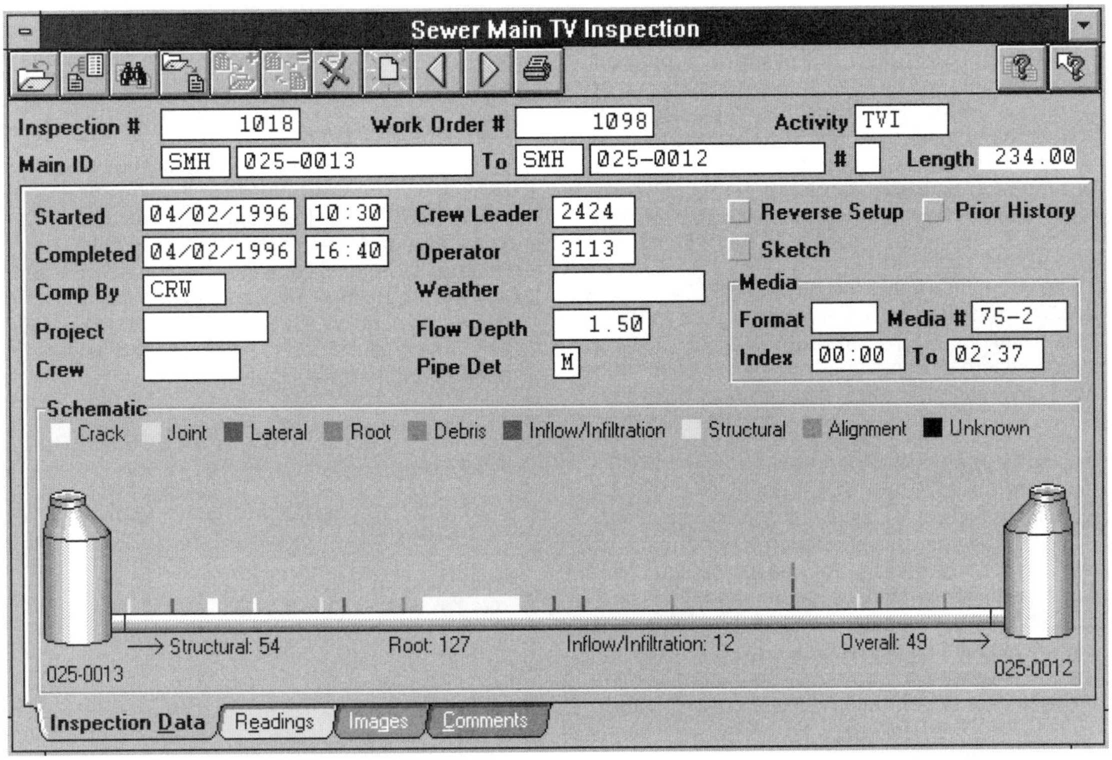

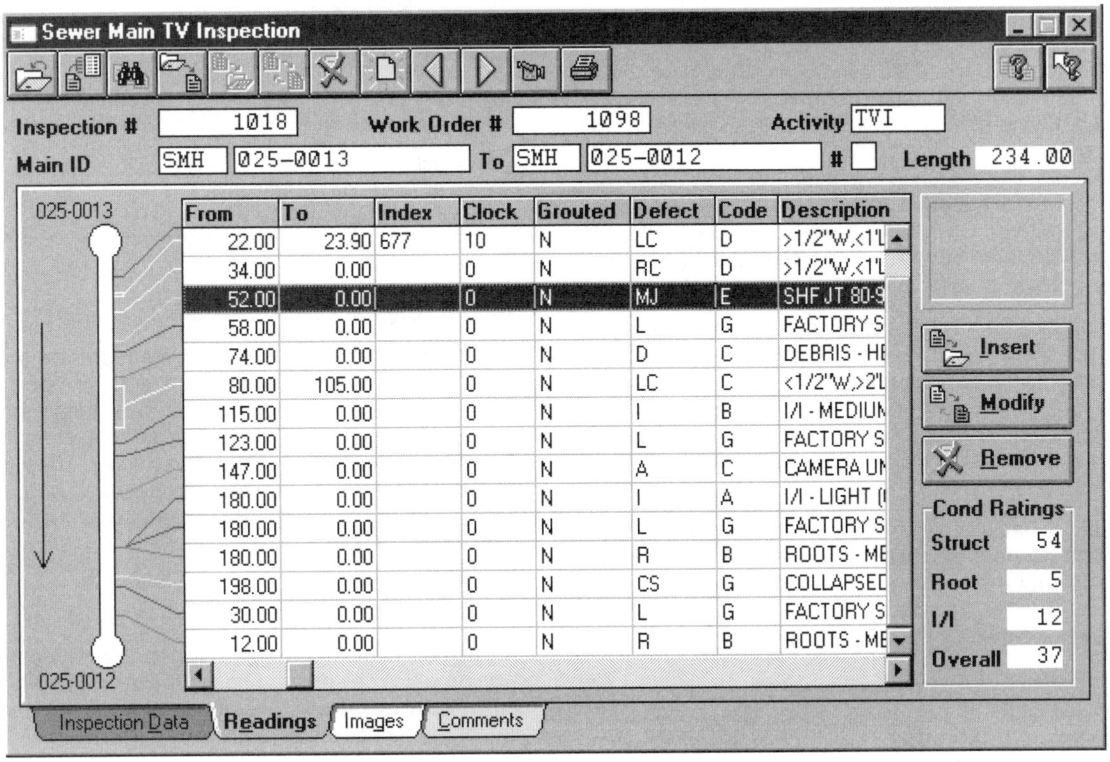

Fig. 10.19   CCTV computer-based data collection screens
(Permission of Hansen Information Technologies)

3. Relieve deteriorating or overloaded segments where conditions and finances permit, or
4. Rehabilitate to the extent possible.

In reality, of course, nearly every rehabilitation program will involve a combination of strategies. For example, a program might consist of stepping up the preventive maintenance to extend the life of existing facilities; completely replacing a few stretches of sewer in immediate danger of collapse; providing some relief sewers to reduce flows in overloaded stretches; and gradually rehabilitating stretches of sewer that still function but are deteriorating as they approach the end of their useful life.

While keeping in mind the alternative strategies listed above, the decision makers also must consider the financial aspects of rehabilitation. Major considerations are: (1) the consequences of system or component failure; (2) cost-effectiveness of corrective action; and (3) capital costs of various repair alternatives.

The material that follows is intended as a general introduction to calculation of the cost factors listed above. It is not a complete guide to making the calculations. Only in a very small utility or collection agency would operators normally be asked to undertake a cost analysis. You may be asked, however, to provide specific repair process recommendations or to give an opinion about the probable results of an action. It is hoped that by understanding the decision-making process, you will be better prepared to contribute your considerable knowledge and expertise at appropriate points in the process.

10.220 Failure Consequences

In cases where it is necessary to compare the relative costs of repairing one seriously deteriorated component versus another, it is useful to include consideration of what would happen if the components failed completely. This type of assessment is admittedly very subjective, but if done in a systematic manner, failure assessment allows us to take into account factors that are otherwise difficult to quantify in terms of dollars-and-cents calculations. For example, how does an agency measure the public relations value of a dependable, cost-effective system? Or, how does an agency take into account the impact of lost business to merchants during the six days their street is torn up for sewer repairs?

While it would be impossible to assign dollar values to such factors, it is possible to assign weighting factors and to classify the severity of a failure as light, medium, or heavy. Once this is done for each repair under consideration, it is possible to compare the impact and urgency of one repair against another, and then to set priorities. To do this, multiply the impact severity factor by the weighting factor to get the Total Failure Index.

Each utility will have its own unique definitions of categories and will assign different weighting factors. The desired result, however, is to arrive at some method that reflects the true costs of a component or system failure. The Failure Matrix (Table 10.1) is a typical way of constructing such a system.

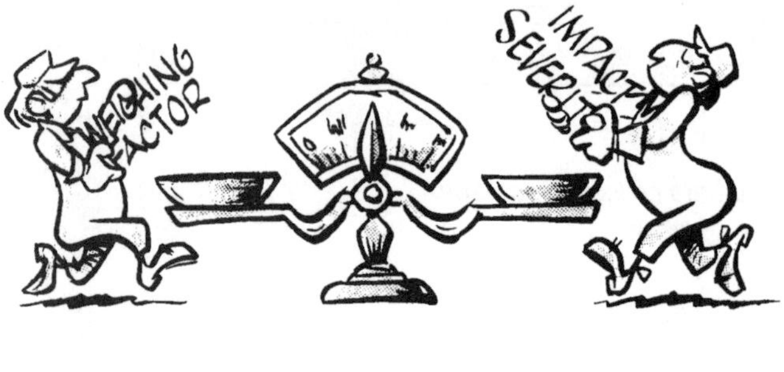

This matrix was prepared for a section of 48-inch sewer between two manholes in a downtown area of a city. If the sewer were to fail, both local service and tributary service would be disrupted. By multiplying the weighting factors times the severity factors, Service Disruption is given a value of 13. The greater the severity of the disruption, the greater the value. If this section of sewer fails, the disruption of traffic will have a disruption subtotal of 12. With traffic and utilities, however, there is an additional factor to be computed. The greater the depth of cover, the deeper the digging for repair and the greater the disruption. To account for this, multiply the traffic and utilities subtotal (12) times the depth of cover weighting factor (2) to arrive at the more complete subtotal for urban disruption which is 24. Next, combine the Service Disruption Subtotal (13) with the Urban Disruption Subtotal (24) to get the Total Failure Index of 37. Those sewers with the greatest values for their "Total Failure Index" will be the first sewers that should be repaired and/or rehabilitated.

TABLE 10.1 FAILURE MATRIX

		IMPACT SEVERITY			
CATEGORY	WEIGHTING FACTOR	LIGHT (1)	MEDIUM (2)	HEAVY (3)	TOTAL[a]
Service Disruption					
a. Local service	2		2		4
b. Tributary service	3			3	9
SUBTOTAL					13
Other Urban Disruption					
a. Traffic	3		2		6
b. Utilities	2			3	6
SUBTOTAL					12
c. Depth of cover	2[b]				
SUBTOTAL – Urban Disruption					24
TOTAL Failure Index (24 + 13)					37

[a] Total values are obtained by multiplying weighting factor value (2) times impact severity value (2) or (2 x 2 = 4).

[b] This factor will have an impact on traffic and utilities. Multiply the subtotal for Urban Disruption by this factor to get the overall Urban Disruption Total (2 x 12 = 24).

10.221 Capital Costs of Repair Alternatives

Capital costs of repair alternatives represent the initial construction or total implementation cost of each alternative. These costs are critical and must be accurate because they represent the "first" cost or the amount of money needed *NOW* if a particular alternative is selected. They include initial construction, labor, equipment, material and right-of-way costs. Capital costs do not include operation and maintenance costs.

10.23 Setting Priorities

Each utility must develop a systematic way to reach a decision on which sewer stretches of the system will be rehabilitated or replaced. The setting of priorities will also include selection of the rehabilitation method to be used.

Usually there are four possible courses of action:

1. Maintain,
2. Rehabilitate,
3. Relieve, and
4. Replace.

The normal process of setting priorities involves the quantitative evaluation of the following factors:

1. Performance of sewers,
2. Sewer capacity,
3. Consequence of failure,
4. Sewer condition, and
5. Costs.

All of these factors are used to set the rehabilitation program limits. Changes in conditions and funding constantly result in priority changes and program limits.

10.24 Cost-Effectiveness Analysis

When reviewing the I/I aspects of a sewer system, a cost-effectiveness analysis should be undertaken to establish the following cost considerations:

1. Survey and rehabilitation costs,
2. Construction and capital costs, and
3. Operation and maintenance costs.

These considerations are used to determine what percentage of the total rainfall-induced I/I can be eliminated at a reasonable cost.

A comparison of the cost to transport and treat the rainfall-induced I/I with the cost to eliminate it will yield the most cost-effective solution. A graph can be constructed with a plot of the transportation and treatment costs. Also, the rehabilitation cost curve can be plotted on the same graph. By combining two curves, a composite cost curve (Figure 10.20) can be developed which shows a minimum cost point. This point indicates the maximum I/I that can be removed cost-effectively.

When assessing the cost-effectiveness of structural rehabilitation, it is best to use *LIFE-CYCLE COSTING*[3] since the total project cost is calculated over the full life of the project. Various methods such as repairs, replacement and rehabilitation can be compared effectively because the method includes initial cost, annual maintenance costs and replacement costs. All costs are converted to a present worth in a base year. To compare present values of a project properly, a common base year, planning period and discount (interest) rate must be used. As a rule of thumb, 20 years is used frequently as a planning period. Replacement of pipe usually is expected to last 50 to 100 years. Since the normal planning period is only 20 years, the pipe will have a salvage value which must be considered.

QUESTIONS

Write your answers in a notebook and then compare your answers with those on page 354.

10.2A List the four distinct courses of action that a utility may take with regard to a sewer system.

10.2B To determine the appropriate course of action to take regarding a sewer system, a utility should perform a quantitative evaluation of what factors?

10.2C What costs should be considered when evaluating the cost effectiveness of an I/I rehabilitation program?

10.3 METHODS OF REHABILITATION

10.30 Excavate and Replace (Figures 10.21, 10.22, and 10.23)

The oldest and most common method of sewer rehabilitation has been excavation and replacement. This method results in the correction of I/I problems as well as the correction of structural problems. In many cases, this method has remained popular because operators were simply unfamiliar with newer, more effective methods. Excavation and replacement is sometimes the only method that can be used because of the severity of the structural deterioration of the pipe. In other cases the severe misalignment of the pipe will not allow other techniques to be used.

Pipeline replacement results in:

1. The correction of misalignment of pipe,
2. Increase in the hydraulic capacity,
3. Repair of improper service connections,
4. Elimination of direct sources of storm water entry, and
5. Removal of incidental I/I sources.

This method should be considered as an option on every rehabilitation project to increase the bid competition. Only in a case where disruption due to the replacement is too great would it be advisable to not include excavation and replacement as an option.

When using the excavation and replacement method, all the problems encountered with new installation of pipe are applicable, as well as several problems unique to the excavation and replacement technique.

PROBLEMS COMMON TO ALL CONSTRUCTION

1. Traffic disruption and control
2. Disruption to properties (access to and easement use)
3. Paving damage
4. Shoring requirements
5. Excavation dewatering
6. Noise
7. Restoration (paving, driveways, sidewalks, fences, landscaping)

---

[3] *Life-Cycle Costing. An economic analysis procedure that considers the total costs associated with a sewer during its economic life, including development, construction, and operation and maintenance (includes chemical and energy costs). All costs are converted to a present worth or present cost in dollars.*

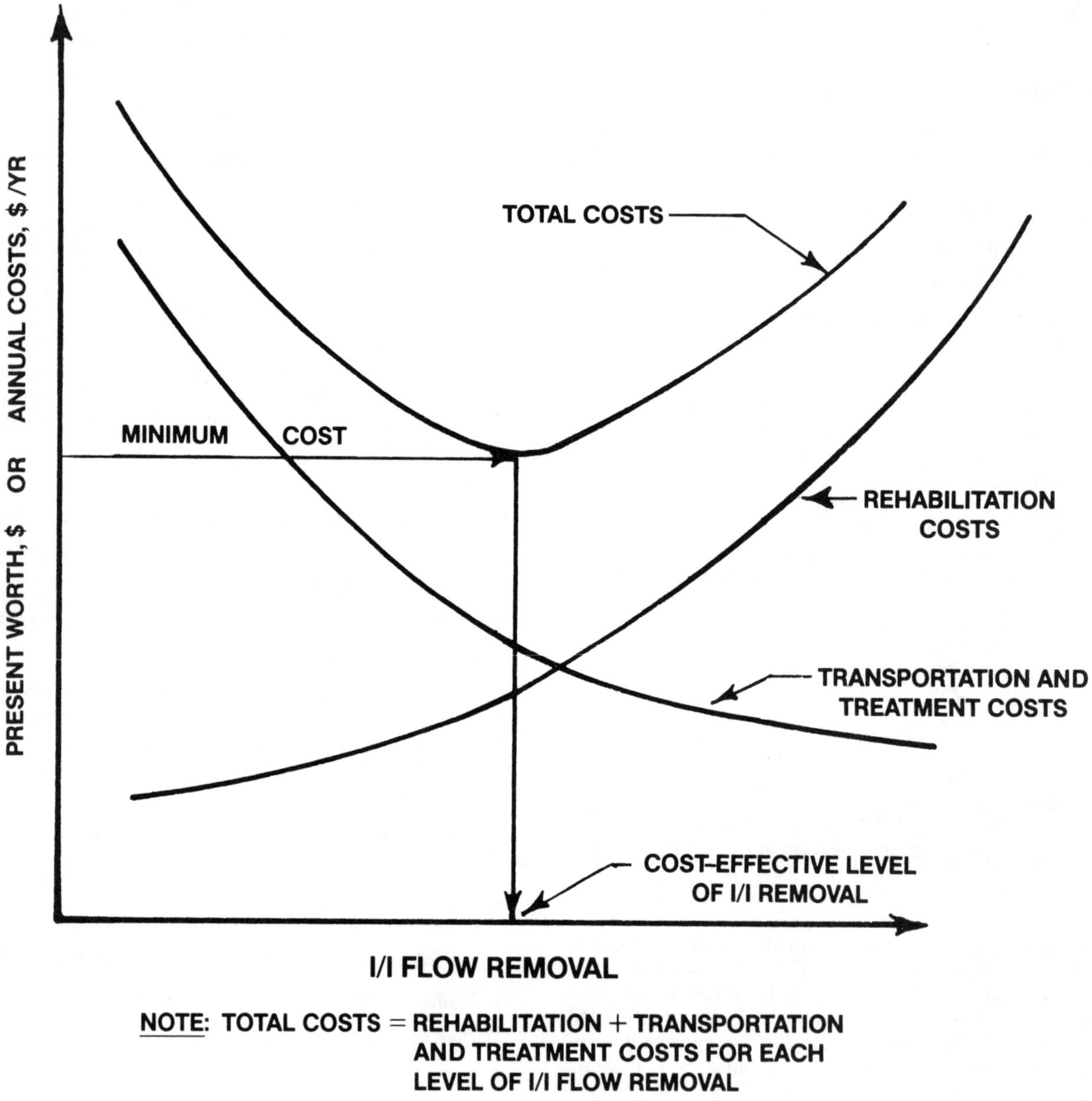

Fig. 10.20 I/I cost-effectiveness analysis

Fig. 10.21 Excavation and replacement job

Fig. 10.22 Shoring with excavation and replacement job

Fig. 10.23 Installed pipe with excavation and replacement job

PROBLEMS UNIQUE TO EXCAVATION AND REPLACEMENT

1. Maintaining flow
2. Removal and disposal of old pipe
3. Effect on other utilities
4. Impact of shoring left in place

Although the excavation and replacement method will increase pipeline capacity and, in most cases, is the only remedy for collapsed pipe, there are several disadvantages to using this method. One is the fact that this method is normally more expensive. Also, traffic and the normal lifestyle of people living and/or working in an area are disrupted for much longer than with other methods. Finally, there is a greater danger associated with this method due to the large excavations and possible impact on other utilities. See Chapter 3, "Wastewater Collection Systems," and Chapter 7, "Underground Repair," for additional information on excavation and shoring.

10.31 **Chemical Grouting** (Figure 10.24)

Chemical grouting is the most widely used method for sealing pipe joints that are leaking and for sealing circumferential cracks. The two basic groups of grouting materials are gels and foams. The foam grout forms in place as a gasket and cures to a hard consistency but retains a rubber-like flexibility. The seal takes place in the joint and there is only minimum penetration outside the pipe. The gel grout is made to penetrate outside the pipe and to infiltrate the soil surrounding the joint. The mixture cures to an impermeable condition around the joint area. The cure time for foam is significantly longer than that for gels. Foam grouts are more suitable for use if there are significant voids outside the joint area. Generally, the foam grouts are more expensive and more difficult to install.

The joints to be grouted must be clean of grease and free of root intrusion. Also, the joint must not have protrusion or significantly deteriorated areas that would interfere with the sealing of the packer. Grouting on smaller diameter pipes is accomplished by a remotely controlled packer and monitored with a closed-circuit television unit.

A typical grouting procedure is as follows:

1. Locate the joint with the closed-circuit television,
2. Pull the packer into position, centering the joint,
3. Inflate the packer sleeves until a seal is formed against the inside of the pipe,
4. Air test the joint,
5. If the joint fails the air test, pump the grout and catalyst solutions under pressure into the void between the inflated sleeves and into the soil if a gel grout is used,
6. Inflate the packing element to force the sealing material into the defective joint,
7. Hold the packing element in the inflated position until the joint is cured, and

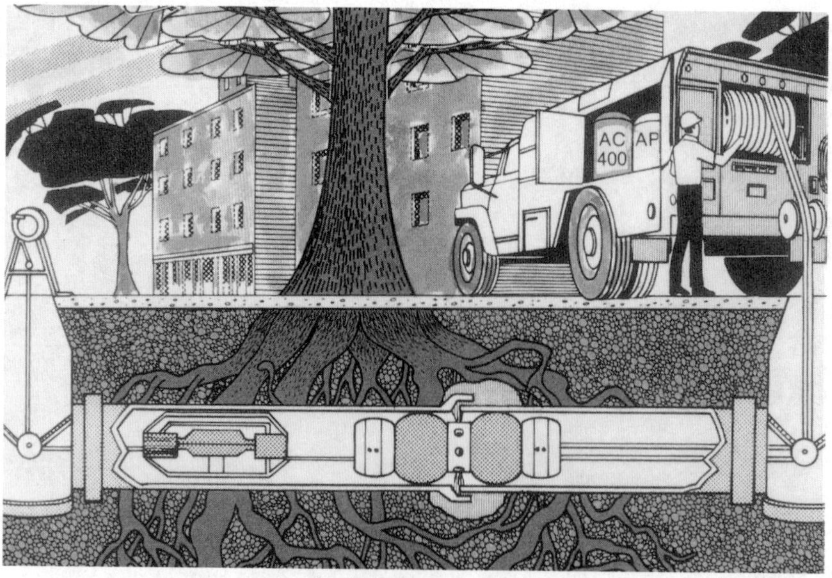

NOTE: Closed-circuit television unit is on the left and packer is on the right in the pipe.

Fig. 10.24 Chemical grouting

8. Deflate sleeves and move the packer to the next joint.

9. OPTIONAL. Sometimes low-pressure air tests are conducted on the grouted joint after curing to test the success of the grouting. Another approach is to dig up a joint and see if the grout performed as expected.

Main line spot repairs in small-diameter gravity sewers have been limited in the past to sealing joints using the chemical grouting method just described. More recently, equipment has been developed which allows for a variety of repairs to be undertaken in sewers that were previously limited to grouting due to their small size and inaccessibility. This highly sophisticated equipment uses remotely controlled robots inside the sewer. As with chemical grouting, a self-contained van (Figures 10.24 and 10.25) contains all of the observation and power equipment as well as the control equipment (Figures 10.26 and 10.27) for control of the robot. The robot is capable of performing a wide variety of functions including TV inspection, water sampling, milling, drilling, grinding, removal of deposits of minerals, root removal, lateral reconnection when used with trenchless rehabilitation methods, sealing unused lateral connections, trimming protruding taps, and sealing of holes, joints, or cracks in the sewer. Figure 10.28 illustrates the complete robotic system that includes the work head, control unit and propulsion unit. Figure 10.29 illustrates some of the different tools used for grinding, cutting and cleaning. Figure 10.30 shows the tool head configured to seal a leaking service lateral at the main line. A bellows is inserted into the lateral and epoxy is forced from the bellows into the void between the lateral and the main line. Figure 10.31 shows the unit used to seal a main line pipe joint. Chemicals are pumped from the chemical container into the pressure sealing foot which rotates 360 degrees around the pipe forcing the sealing chemical into the joint. The sealing foot also smooths the joint after grouting is completed.

In large-diameter sewers, pressure grouting is accomplished using pipe grouting rings or predrilled injection holes. To use grout sealing rings a person must manually place the ring over the joint and inflate the ring to isolate the joint. Sealing grout is pumped into the small void between the pipe wall and the ring using a hand-held probe. After the grout cures, the sealing ring is removed and used on the next joint.

The probe injection method also uses a hand-held probe but no sealing rings. Holes are drilled into or next to the joint; the number of holes depends on the length of crack or pipe size. Usually the holes are spaced about six inches apart. The grout is then pumped into the holes with the hand-held probe.

The service life of grout is a very important consideration. Acrylamide grouts have been used since the 1950s. Polyurethane foam has been used since the early 1970s. Urethane gels have only been used since 1980. Manufacturers are always trying to develop better chemical grouts.

Experience to date has shown that chemical grouting can be effective when sealing an actively leaking or infiltrating joint or crack. Attempting to seal joints or cracks that are not leaking or infiltrating during the sealing process has produced questionable results. Chemical grouts have failed in arid regions where the grout has dried up during periods of low groundwater and in coastal regions where the grout is subject to the influence of tidal fluctuations.

QUESTIONS

Write your answers in a notebook and then compare your answers with those on page 354.

10.3A What is the oldest and most common method of sewer rehabilitation?

10.3B What are the disadvantages of sewer rehabilitation by excavation and replacement?

10.3C What is the most common method of sealing leaking pipe joints and circumferential cracks?

10.3D How can the success of chemical grouting be tested?

10.32 Trenchless Technology

10.320 Pipeline Rehabilitation Methods

Recently, a variety of rehabilitation methods have been developed that allow rehabilitation of sewers without excavating. This technology, which continues to develop, is known as "trenchless technology." Trenchless rehabilitation has several potential advantages over conventional open cut construction. These advantages must be evaluated for each rehabilitation project. Trenchless rehabilitation methods are frequently used when soil or groundwater conditions make it difficult to use open cut construction methods and in downtown areas where disruption to businesses and the public is an important consideration. Suitable equipment and pipe for the various types of trenchless methods can be purchased from any of several different manufacturers. Table 10.2 lists some current trenchless technology methods. Sections 10.321 through 10.324 describe how each of these methods works and dis-

TABLE 10.2 PIPELINE REHABILITATION METHODS

Method	Pipe Size, inches	Typical Installation Length, feet	Replacement Material	Used For
Cured-in-Place Pipe (CIPP)				
Inverted or winched in place	4 – 108	2,000	Fabric/epoxy	Gravity sewers and force mains
Spray-on lining	3 – 180	500	Epoxy or cement based	
Sliplining				
Segmental	4 – 158	1,000	PE, PP, PVC, GRP	Gravity sewers and force mains
Continuous	4 – 63	1,000	PE, PP, PVC	Gravity sewers and force mains
Spiral wound	4 – 100	1,000	PE, PVC, PP	Gravity sewers
In-Line Replacement				
Pipe displacement	4 – 24	750	PE, PP, PVC, GRP	Gravity sewers and force mains
Pipe removal	36	300	PE, PP, PVC, GRP	Gravity sewer or force mains

PVC Polyvinyl Chloride Pipe
PE Polyethylene Pipe
PP Polypropylene Pipe
GRP Glassfiber Reinforced Polyester Pipe

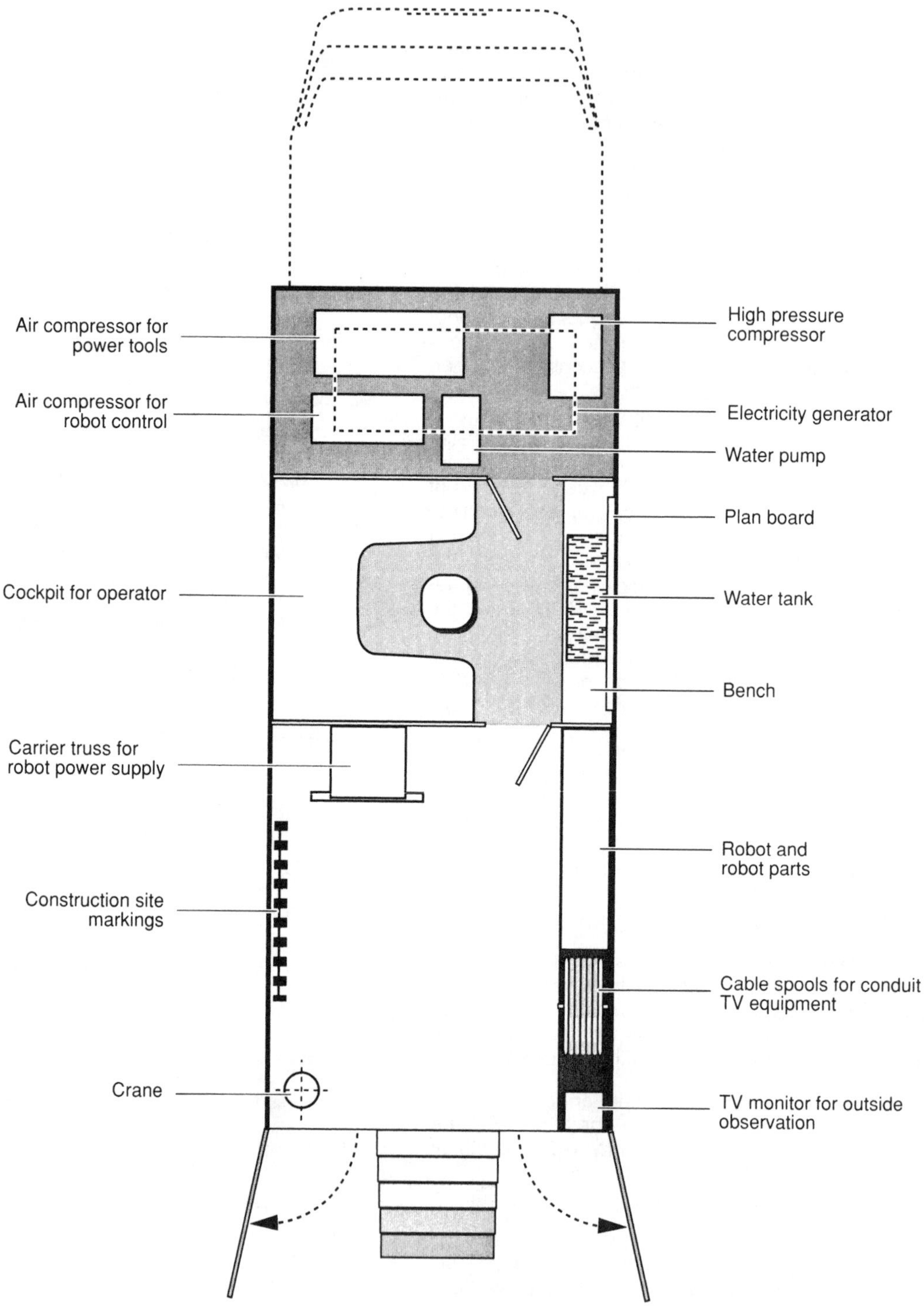
Air compressor for power tools
Air compressor for robot control
Cockpit for operator
Carrier truss for robot power supply
Construction site markings
Crane
High pressure compressor
Electricity generator
Water pump
Plan board
Water tank
Bench
Robot and robot parts
Cable spools for conduit TV equipment
TV monitor for outside observation

Fig. 10.25 Floor plan of the machine and motor compartment, operator compartment, and robot and equipment compartment

(Permission of Sika Robotics)

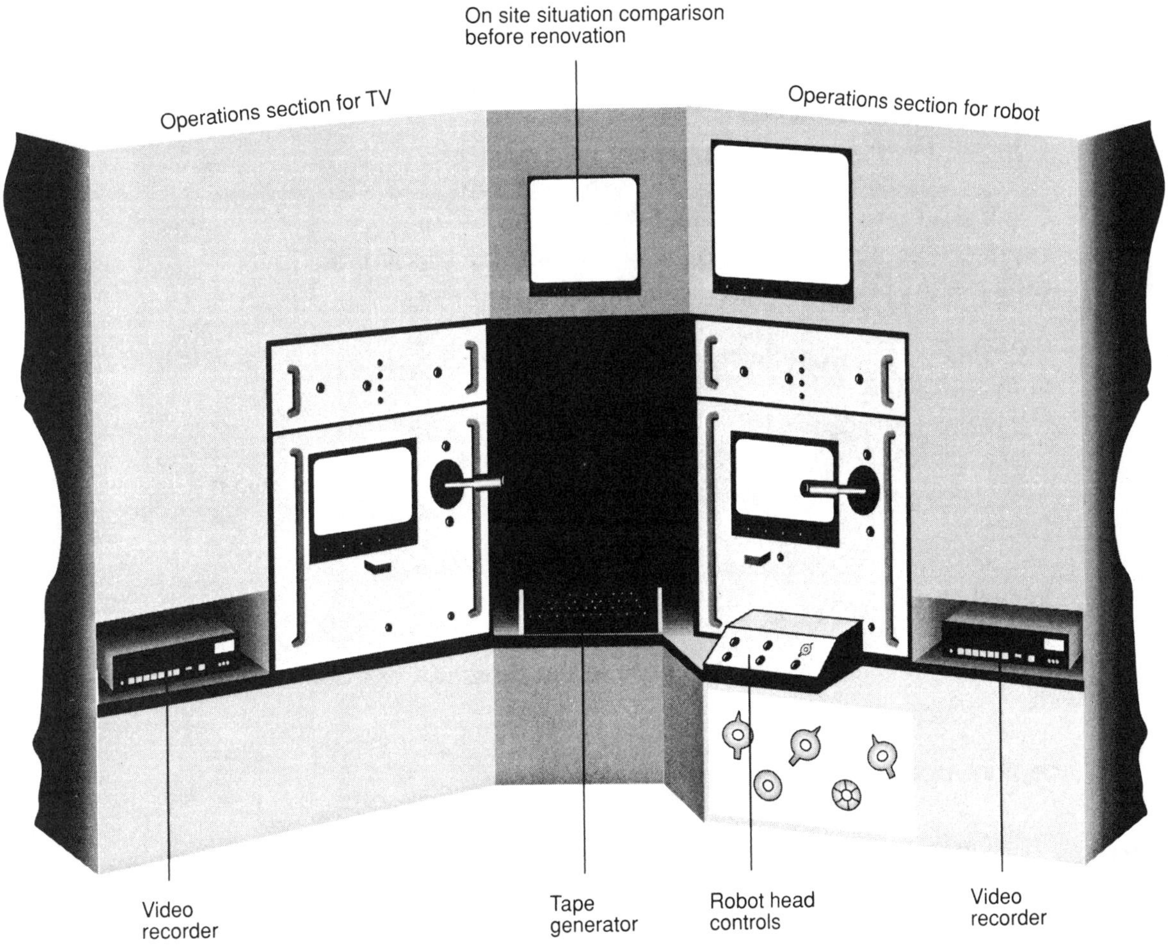
On site situation comparison before renovation
Operations section for TV
Operations section for robot
Video recorder
Tape generator
Robot head controls
Video recorder

Fig. 10.26 Operator compartment and control board
(Permission of Sika Robotics)

Fig. 10.27 CCTV/grouting control center
(Permission of CUES, Inc.)

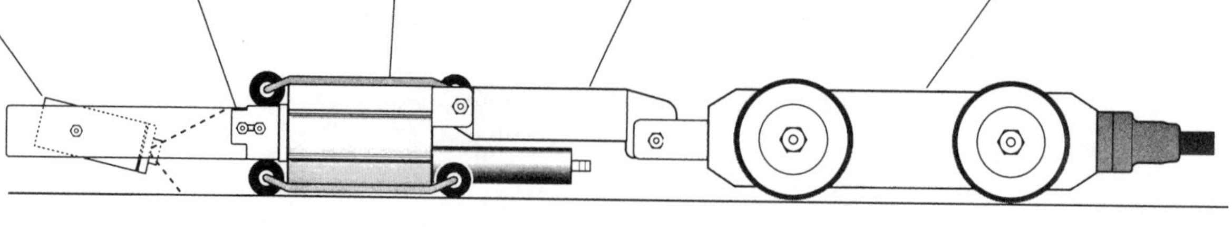

Fig. 10.28 Sika Robot
(Permission of Sika Robotics)

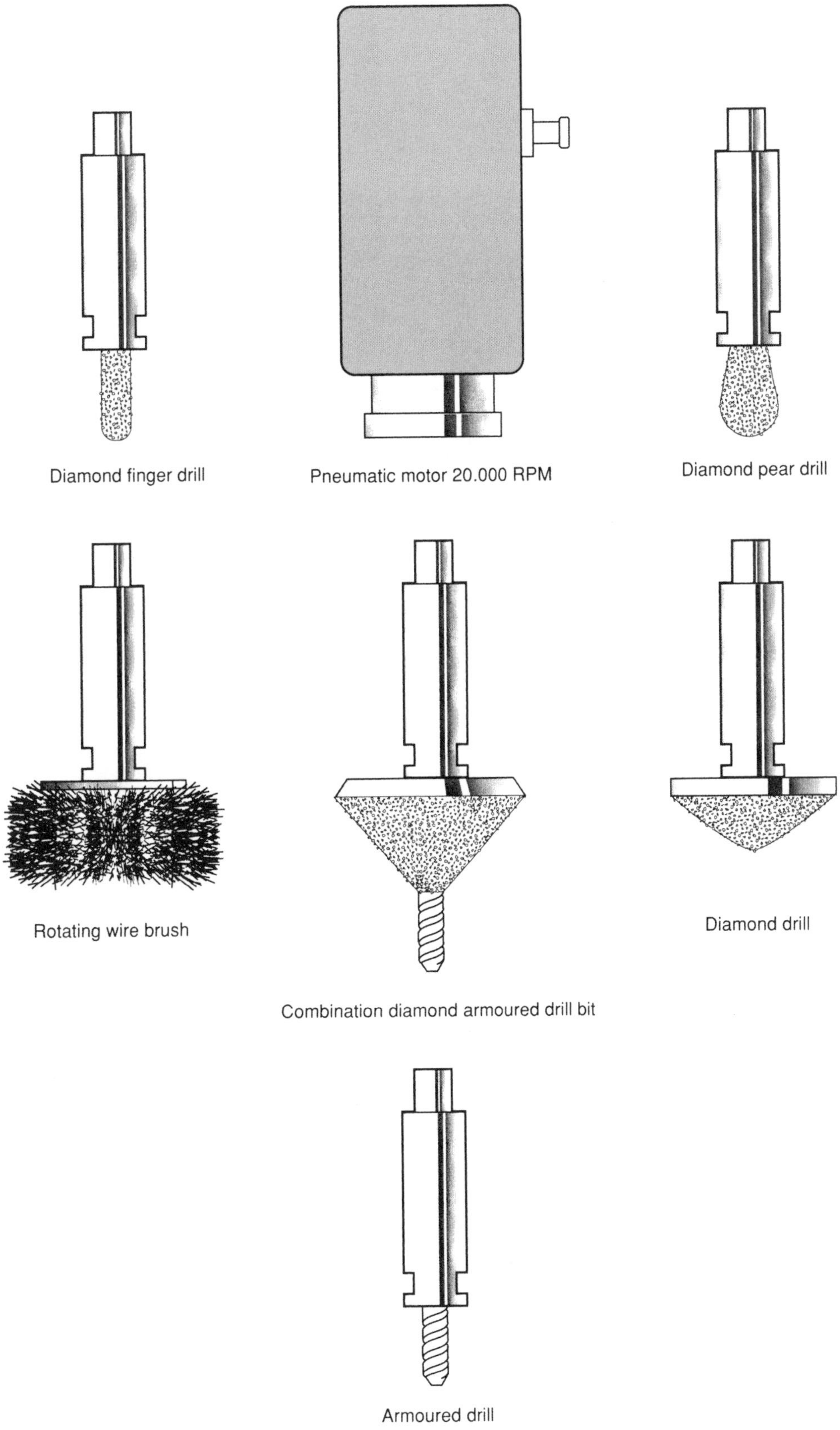
Diamond finger drill
Pneumatic motor 20.000 RPM
Diamond pear drill
Rotating wire brush
Diamond drill
Combination diamond armoured drill bit
Armoured drill

Fig. 10.29 Robot drilling tools
(Permission of Sika Robotics)

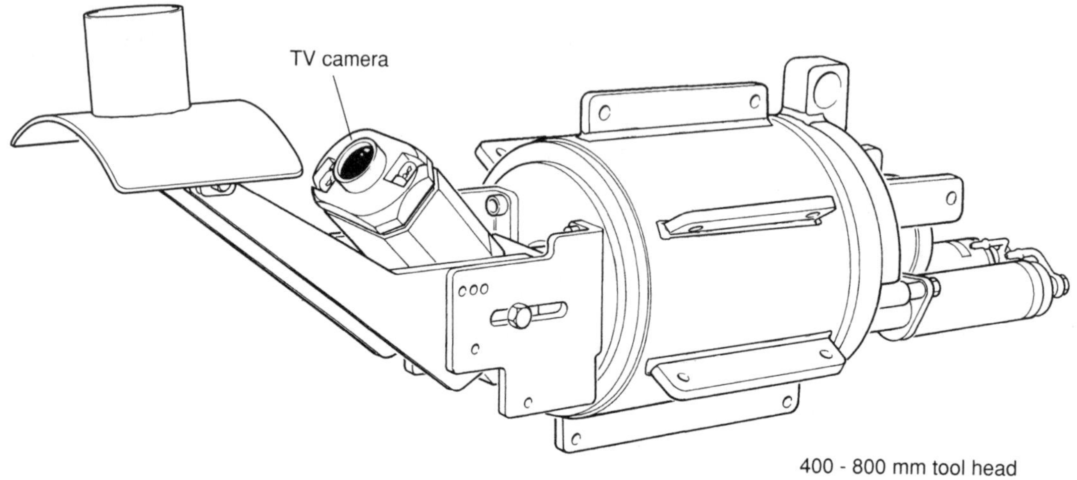

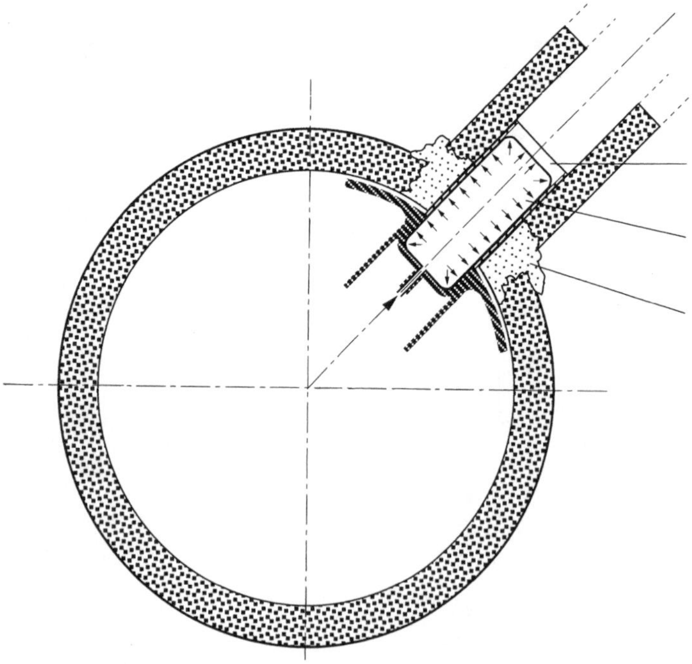

Fig. 10.30 Tool head for pipe connection
(Permission of Sika Robotics)

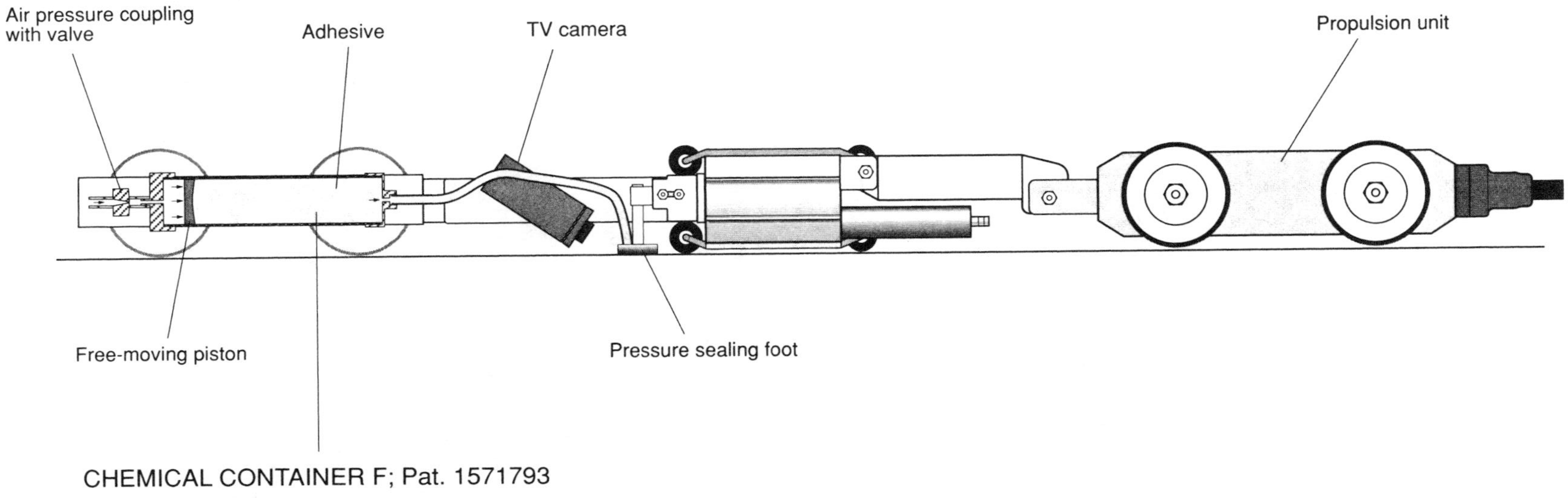
Air pressure coupling
with valve
Adhesive
TV camera
Propulsion unit
Free-moving piston
Pressure sealing foot
CHEMICAL CONTAINER F; Pat. 1571793

Fig. 10.31 Basic module robot unit with pressure sealing equipment and TV camera

(Permission of Sika Robotics)

cusses the advantages and limitations which should be considered when planning a rehabilitation project. Later in this chapter, Section 10.33, "Insituform," and Section 10.34, "Polyethylene Pipe Lining," describe step-by-step procedures for rehabilitating a sewer using two of the trenchless technology methods.

10.321 Cured-In-Place Pipe (CIPP)

Cured-in-place pipe (CIPP) is a resin-saturated, flexible fabric that is inverted (turned inside out) or winched into the existing pipeline. After curing, the hardened resin gives the new pipe its structural strength. The inversion process uses water or air pressure. This process is becoming one of many methods in the industry to rehabilitate building and residential service laterals. The air pressure process is shown in Figures 10.32, 10.33, and 10.34. The primary differences between the various CIPP systems are in the composition of the tube, method of resin impregnation, installation procedure, and curing medium. The principal differences in tubes used for CIPP are a result of the type of material used, the method of manufacturing, and the coating material used. Section 10.33, "Insituform," describes in detail one manufacturer's CIPP process, provides illustrations of the equipment, and explains step-by-step procedures for using the Insituform trenchless sewer rehabilitation method.

METHOD DESCRIPTION

The CIPP process involves the installation and curing of a lining material inside the old pipe. First the pipe to be lined is televised and cleaned. Roots, broken pieces of pipe and other debris are removed from the line. Areas of excessive infiltration are sealed using standard grouting procedures and spot repairs are performed, if required, prior to fabric placement. The locations of all service connections are usually recorded at this time. After the pipeline has been cleaned and televised, a flexible tube is impregnated with a hardening material (resin) and the tube is inserted through a manhole or a convenient entry point. This flexible tube is installed in the old pipe by pulling it or inverting it under air or water pressure. The flexible tube is then pressed against the wall of the pipe to be lined and, as the resin cures, a smooth hard liner is formed. Typically the force used to press the liner against the pipe wall causes dimples or small indentations to form at the service connections or laterals.

After the resin sets, the downstream closed end is carefully cut and removed while avoiding creation of a vacuum which may collapse the newly formed pipeline. The final activity is pulling a closed-circuit TV (CCTV) through the line for final inspection. A cutting device connected to the CCTV is used to open the service connections after the dimples are visually located with the camera or by referring to the previously recorded information.

There are two primary approaches to installing the flexible tube—inverting in place and winching in place. Installation procedures and materials may vary somewhat depending on the manufacturer.

ADVANTAGES

Use of cured-in-place pipe offers the following advantages over other types of pipeline rehabilitation methods:

- No grouting is required;
- The new pipe, if coated inside, has no joints or seams and has a very smooth interior surface which may actually improve flow capacity despite the slight decrease in diameter;
- Non-circular shapes can be accommodated;
- An inverted lining can accommodate bends and minor deformation;
- Entry into large-diameter pipes is possible through existing manholes or a small excavation;
- Remote control internal lateral connection is possible;
- Excavation typically is not required for either the installation process or reestablishment of house connections;
- This method leaves essentially no *ANNULAR SPACE*[4] between old and new pipe;
- The new pipe is acid and water resistant, prevents infiltration and exfiltration, and improves flow capacity due to its smooth surface; and
- The installation of CIPP is rapid.

LIMITATIONS

Limitations of the CIPP method include the following:

- A trained crew and operators with special equipment are required;
- The tube or hose must be specially constructed for this purpose;
- Usually groundwater infiltration may need to be controlled and existing flow must be bypassed during the installation process;
- Sealing may be required at liner ends or cuts;
- Curing period could be extensive;
- Set-up costs are high on small projects; and
- Mobilization and overall costs for this method are high.

10.3210 CIPP INVERTED IN PLACE

The inversion process involves the insertion of a resin-impregnated fabric tube into a pipe by feeding the tube inside out through the pipe. Water or compressed air pressure is used to invert a woven polyester fabric or similar tube into an existing pipeline. The pliable (flexible) nature of the resin-saturated fabric before it cures allows installation around curves, filling of cracks, bridging of gaps, and maneuvering through pipe defects. After installation, the resin-impregnated fabric cures to form a new pipe of slightly smaller diameter, but of the same shape as the original pipe.

10.3211 CIPP WINCHED IN PLACE

In this method, a winched cable is placed inside the existing pipe. The liner, impregnated with thermosetting resin, is connected to the free end of the cable. The line is then pulled into place extending from manhole to manhole. The cable is disconnected, the ends are plugged, and the liner is inflated with hot air, water or steam to speed up the curing process. Liner insertion takes approximately 25 to 45 minutes while the curing process takes approximately 3 to 6 hours.

---

[4] *Annular (AN-you-ler) Space. A ring-shaped space located between two circular objects. For example, the space between the outside of a pipe liner and the inside of a pipe.*

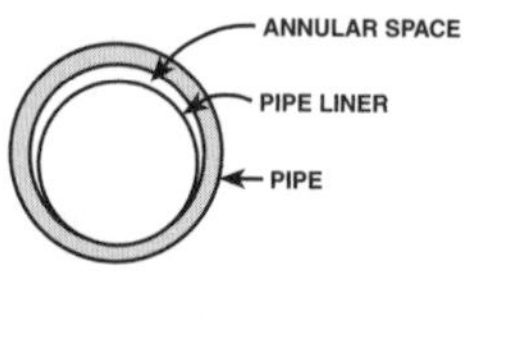

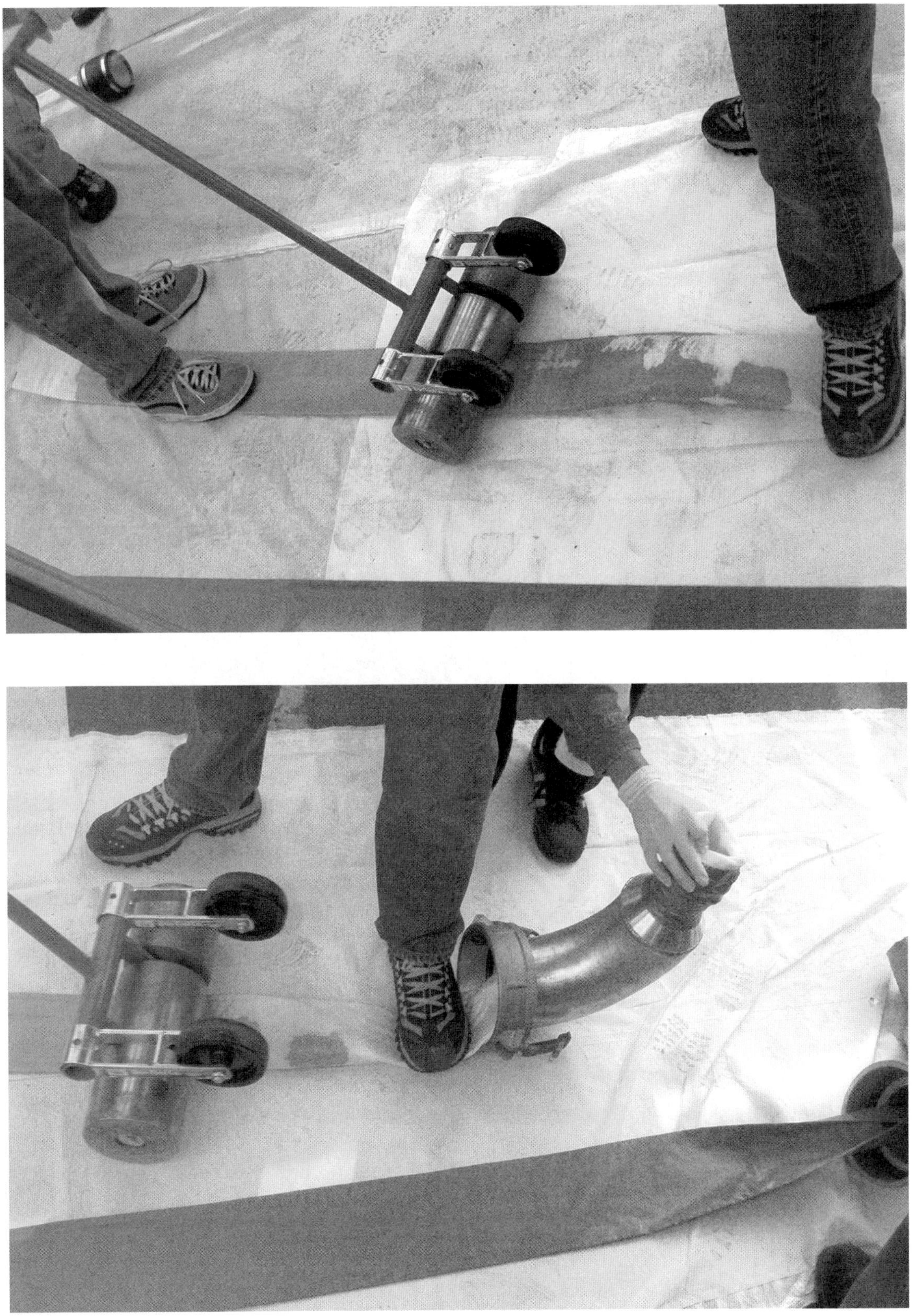

Fig. 10.32 Applying and spreading resin material to fabric

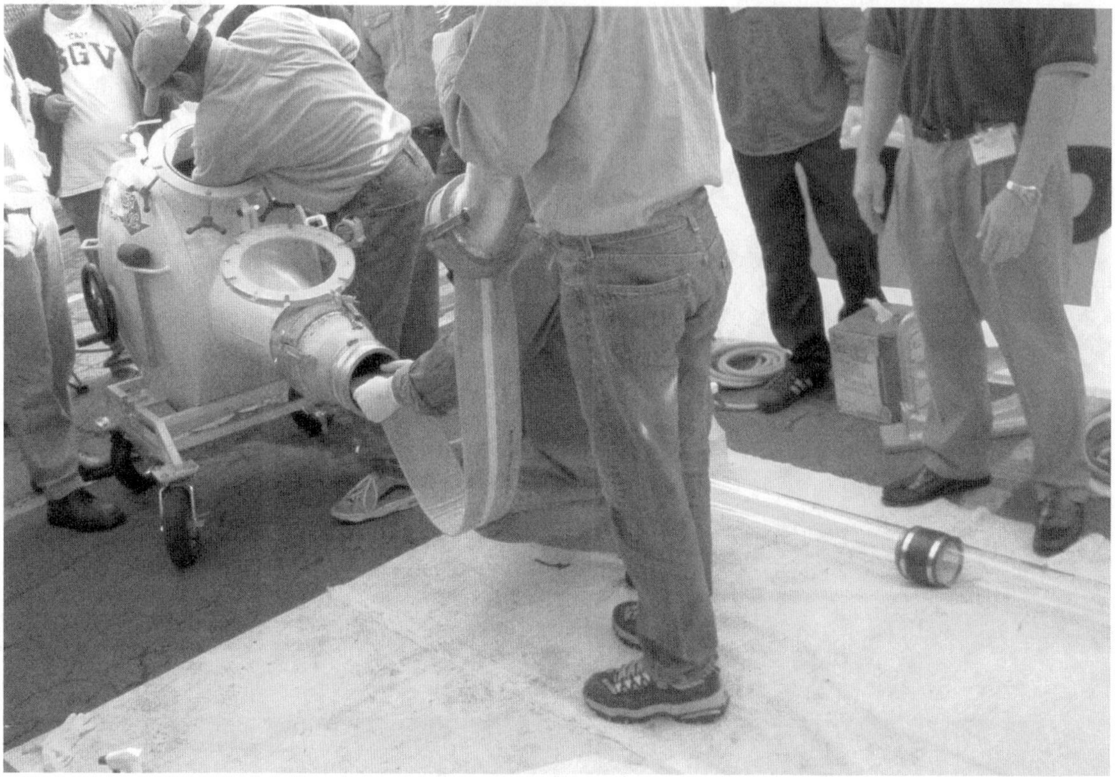

Fig. 10.33 Loading impregnated Perma-Lateral liner
(Permission of Perma-Liner Industries, Inc.)

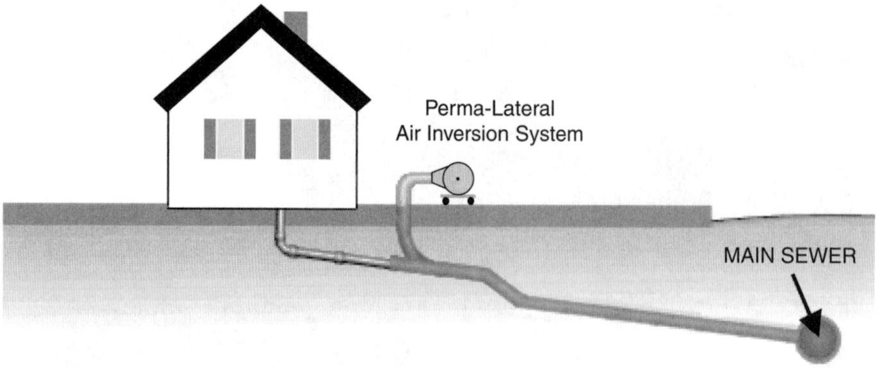

Fig. 10.34 Building service lateral air inversion system
(Permission of Perma-Liner Industries, Inc.)

10.3212 SPRAY-ON LINING (Figure 10.35)

The spraying of a thin mortar lining or a resin coating onto pipes is a well-established technique. Such methods can provide improved hydraulic characteristics and corrosion protection. However, they may not improve the structural integrity of the line and have little value in sealing joints or leaks. For pipes that are large enough to permit entry of an operator, structurally reinforced sprayed mortars (shotcrete or gunite) are effective and widely used to rehabilitate pressure pipes and gravity sewers. For small-diameter pipes, the lining is sprayed directly onto pipe walls using a remotely controlled traveling sprayer. The lining materials include concrete sealers, epoxy, polyester, silicone, vinyl ester, and polyurethane.

Shotcrete (gunite) is a mixture of sand, cement and water that is applied by air pressure. The sand and cement are transported through a hose to a nozzle and are then mixed with water in a mixing chamber. Cement by hydration takes place as the material leaves the chamber and is shot into place by the air pressure. Normally, some type of cage reinforcing is placed inside the pipe to form a new wall. This method normally is used only in pipes with a diameter of 36 inches or more because entry of a person is required. This method requires diversion of all flow in the pipeline. Major structural problems can be handled by this method.

A typical installation would be as follows:

1. Remove offsets and protrusions to allow uniformly thick coating,
2. Round off all edges,
3. Insert reinforcing steel,
4. Dampen surfaces,
5. Apply shotcrete behind the reinforcement (only through one layer of steel) and a minimum of two inches over the steel,
6. Finish the surface, if desired. Surface can be left natural, broomed, floated or troweled,

7. Cure for at least 24 hours maintaining continuously moist conditions, and
8. Complete final curing. The total duration of curing is seven days during which the shotcrete must be maintained moist and at a minimum temperature of 40°F.

The advantages of using shotcrete are rehabilitation without excavation, restoration of original pipe strength and this method can be used safely when other techniques cannot. Some disadvantages are complete diversion of flow, long curing time and some reduction in the hydraulic capacity.

10.322 Sliplining
(Also see Section 10.34, "Polyethylene Pipe Lining.")

Sliplining, which is also called pipe insertion, is one of the earliest forms of continuous structural lining. Sliplining is used to rehabilitate pipelines by pulling or pushing a new flexible liner pipe of slightly smaller diameter into an existing deteriorated pipeline. This system is used where there is no obstruction inside the old pipe and its dimensions are in good shape. Sliplining can be used on all types of existing pipes that are not excessively deteriorated. The liner can form a continuous, watertight pipe within the existing pipe after installation. The annular space (called the "annulus") between the old and new pipe is usually filled with a grouting material. Where the new pipe has to be laid to an even grade, the use of plastic or metal locators/spacers may be necessary. Spacers also maintain the pipe location during annular grouting to ensure a uniform surrounding. The service connections are reconnected to the new liner by excavation.

Sliplining is simple and relatively inexpensive. It is not considered to be a specialist operation and does not require sophisticated equipment. The main limitation is a substantial loss of cross-sectional area. However, a loss of cross-sectional area does not directly relate to loss of hydraulic capacity, especially for larger diameters of more than 36 inches, because the smooth new interior pipe surface will cause some gain in hydraulic capacity. However, there can be a significant loss of hydraulic capacity on small-diameter pipelines.

Sliplining can be categorized into three types of operations based on the type of liner being used—continuous, segmental and spiral wound. Each method is discussed below.

10.3220 CONTINUOUS HIGH DENSITY POLYETHYLENE (HDPE) PIPE

This method can be used both for structural and nonstructural purposes. HDPE solid-wall pipe is sliplined into an existing pipeline after the joints are butt fused. This method requires excavation of a relatively short trench for use as an insertion pit. The actual size of the trench depends on the pipeline depth and the diameter of the liner. Insertion of continuous HDPE pipe typically reduces the cross-sectional area of the pipeline. The annular space between old and new pipe should be grouted. Continuous HDPE pipe installation requires excavation at each lateral connection to reestablish the connection.

ADVANTAGES

The advantages of using continuous HDPE pipe to rehabilitate a pipeline include the following:

- This is a simple method requiring minimal investment in installation equipment and relatively little technical skill,
- The new pipe has very few joints,
- HDPE lining is capable of accommodating large-radius (wide) bends, and
- The lining operation can be accomplished using either an insertion pit or, in the case of sewers, a manhole. (If the liner is inserted through a manhole, the pipe must have sufficient axial flexibility.)

LIMITATIONS

Some limitations of sliplining with continuous HDPE pipe include:

- Pipes must be manufactured or preassembled into a continuous length,
- Reduction in capacity may be significant, depending on the diameter and wall thickness of the liner pipe,
- For optimum performance in the case of gravity lines, the annular space should be grouted,
- Only circular cross sections are possible,

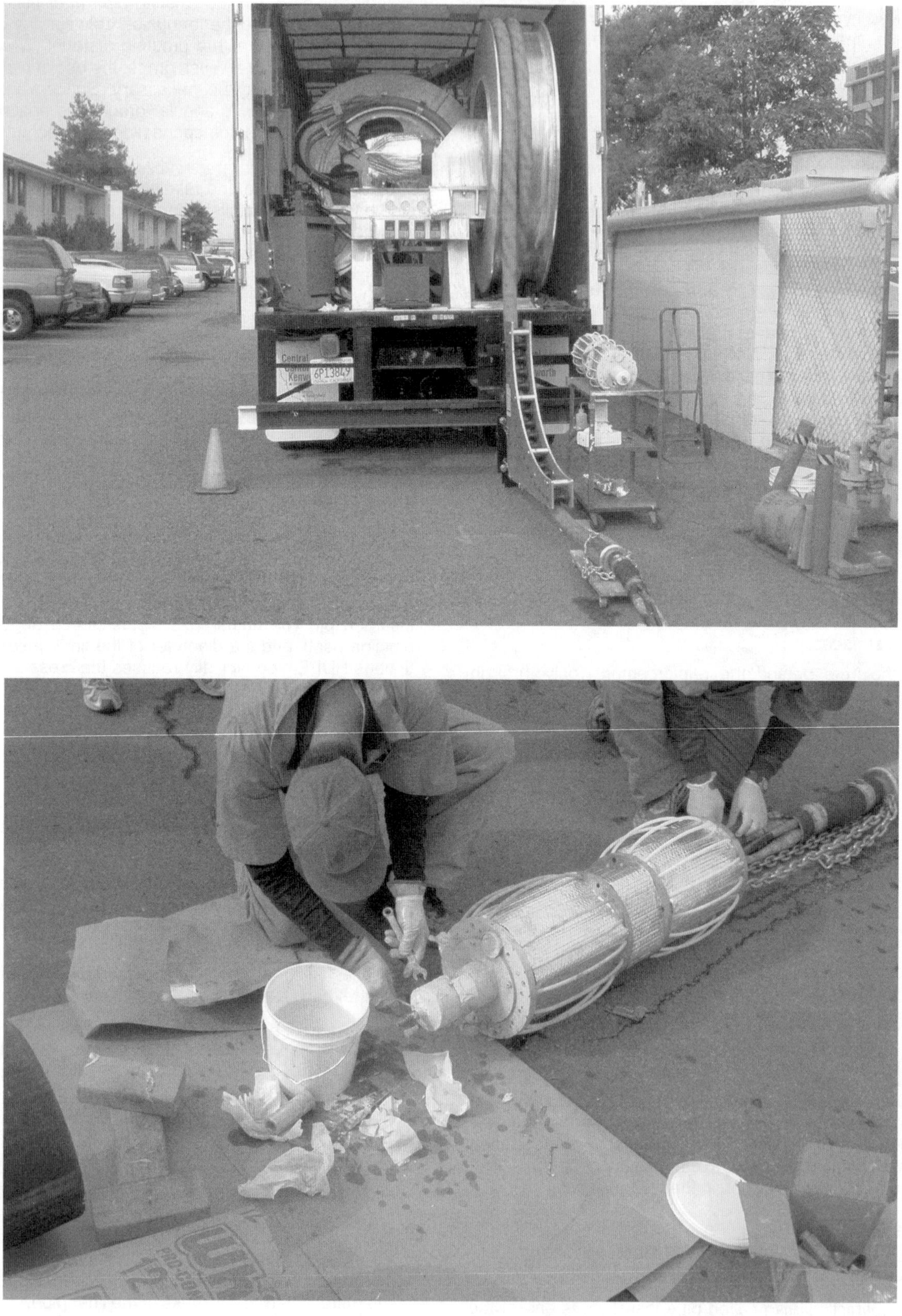

Fig. 10.35 Spray-on system with sprayer

- It is less cost effective where deep external lateral connection is required, and
- This method has high mobilization costs and moderate to high overall costs.

10.3221 SEGMENTAL

This method involves the use of short sections of pipe that incorporate a flush sleeve joint. A number of plastic pipe products have been specially developed for sliplining sewers; some examples include Hobas (a centrifugally cast fiberglass-reinforced polyester pipe), and short sections of polyvinyl chloride (PVC) and polyethylene (PE) pipe with a variety of proprietary joints. (A "proprietary" joint is one which has been specially designed by a particular manufacturer and the design is patented or its name is registered.)

ADVANTAGES

This is a relatively simple method with low investment in installation equipment. An insertion pit may be avoided if pipe sections will fit in a manhole. By installing custom-made shaped pipes (oval, egg-shaped), the reduction in pipe capacity can be kept to a minimum. This method requires relatively little technical skill and can be carried out by any pipe contractor or by the municipality itself.

LIMITATIONS

Installation of segmental sliplining has several limitations, including the following:

- The annular space between the old and new pipe generally must be grouted,
- The reduction in pipeline capacity may be significant,
- There will be many joints in the new pipe,
- Some pipe materials are easily damaged during the installation, and
- The laterals may need to be connected externally.

10.3222 SPIRAL WOUND (Figure 10.36)

This sliplining technique is based on forming a pipe right at the installation site by using strips of polyvinyl chloride (PVC), polyethylene (PE) or polypropylene (PP) having a ribbed profile with interlocking edges. The liner is fed through a machine to form a smooth-bore, spiral-wound pipe. The winding machine can be used above ground but has also been modified to work from the bottom of a manhole and directly feed a liner into an existing pipe. This method can be used for either structural or nonstructural purposes, depending on the type of grouting.

There are three variations of the spiral-wound sliplining method. One method fabricates a pipe in the manhole by winding a continuous PVC fabric into a spiral. The second method, used for larger diameters (over 36 inches), uses preformed panels inserted in place in the existing pipeline. The third method is used to line oviform (shaped like an egg) operator-entry pipes (36-inch diameter and larger). Excavation is not required for this process. House connections and laterals are reconnected by local excavations, but a remote-control cutter is being developed. This method leaves an annular space which must be grouted. It has low mobilization costs and low to moderate overall costs.

ADVANTAGES

The advantages of sliplining with the spiral-wound method include the following:

- The liner pipe may be formed on site by spirally winding a strip of stock liner material,
- Access through manholes is possible,
- For operator-entry sized pipes, the lining is capable of accommodating large-radius bends,
- No pipe storage on site is necessary, and
- Any diameter can be selected within the range of the winding machine.

LIMITATIONS

Some limitations of spiral-wound sliplining include:

- Trained personnel are needed to operate the winding equipment,
- Continuous fusion, solvent-welded joints or mechanical joints are required,
- The reduction in capacity may be significant, and
- The annular space must be grouted.

10.323 In-line Replacement

When pipelines are found to be structurally failing and inadequate in capacity, then in-line replacement should be considered. This is typically the most expensive method of trenchless sewer rehabilitation; however, the money spent is a capital investment since a completely new pipe is installed. In-line replacement is often very cost effective when compared to the cost of using the open-trench method. There are two methods of trenchless in-line pipe replacement: pipe displacement and pipe removal.

10.3230 PIPE DISPLACEMENT

Pipe displacement, also known as pipe bursting, is a technique for breaking out the old pipe by use of radial forces from inside the old pipe. The fragments are forced outward into the soil and a new pipe is pulled into the bore formed by the bursting device. The Pipeline Insertion Machine (PIM) uses a standard pneumatic mole with a special pointed shield. Twin hydraulically operated breaker arms are attached in the front of the tool. These cracker arms are remotely operated and can exert pressure to overcome difficult joints or surroundings. A standard air compressor is also needed to provide power to the pneumatic mole, and a hydraulic winch is required as well.

ADVANTAGES

The advantages of the pipe displacement method of pipe rehabilitation include:

- No reduction in capacity, in fact, increases in capacity are possible;
- The liner installation can be accomplished by means of an insertion pit (for continuous lengths) or a manhole (for separate lengths of pipe); and
- If the existing pipe is structurally damaged, the old pipe can be burst or cut (hydraulically or pneumatically) by radial jacking forces and a new pipe can be pushed or pulled behind.

LIMITATIONS

Use of the pipe displacement method has the following limitations:

Fig. 10.36 Spiral machine and material

- Excavation is required to disconnect laterals before insertion of the new pipe and to reconnect the laterals;
- There is a risk of damaging nearby utilities and building connections during excavation;
- A crew of trained operators is needed;
- For replacing metal pipe systems, a sleeve pipe is usually required to protect the inserted pipe lining from damage;
- Loading can cause changes to the surrounding soil; and
- The existing pipe must be friable (capable of being shattered) such as clay or cast iron, not reinforced concrete.

10.3231 PIPE REMOVAL

The development of microtunneling machines with the capability of crushing rocks and stones has led to their use in excavating existing pipelines for replacement. The remotely controlled microtunneling equipment is slightly modified for pipe removal systems. The various elements of the system include a modified shield, a jacking unit, a control console and a spoil removal system. A jacking pit approximately 10 to 20 feet in length and a smaller receiving pit are required. The boring machine is driven forward excavating the old defective pipe and surrounding ground. At the same time, new pipe sections with an equal or larger outside diameter are pushed up behind the advancing shield. Old cast iron pipe has been successfully replaced with a new steel pipe using this method.

ADVANTAGES

The main advantage of this system is that it can be used in areas with difficult ground conditions, particularly high groundwater levels. Another advantage is its ability to replace a pipe which is badly out of line and grade, since close control of line and level can be achieved for the new pipe. One pipe removal system reportedly has the capability of on-line replacement with the sewer line being operational during the installation. Other advantages of pipe removal systems include the following:

- This method can be used to up size the old pipe,
- The new pipe can be installed with relatively little surface disruption, and
- The lining is capable of accommodating large-radius bends.

LIMITATIONS

Limitations of pipeline rehabilitation using this method include the following:

- A shaft or insertion pit is required,
- Obstacles such as manmade structures or roots may cause problems, and
- This method requires a relatively high degree of skill, including a specialized crew with special equipment.

10.324 Deformed and Reshaped

This type of trenchless pipeline rehabilitation temporarily reduces the cross-sectional area of the new pipe before it is installed, then expands it to its original size and shape after placement to provide a close fit with the existing pipe. Lining pipe first is reduced in size (on site or in the manufacturing plant). It is then inserted and restored to its original size by the application of heat or pressure or it reverts naturally to its original size. There are three versions of this approach: Fold and Formed, Drawdown and Roll-Down.

ADVANTAGES

- The reported reduction in pipeline capacity is minimal, if any,
- Lining can be accomplished through either an insertion pit or a manhole (in the case of sewers). If a manhole is used, pipe with sufficient axial flexibility is required,
- No grouting is required,
- The lining is capable of accommodating large-radius bends, and
- The liner can be installed in long lengths, with few or no joints.

LIMITATIONS

- Possible structural damage (collapse/misalignment) to the existing piping can cause problems,
- Lateral connections may be difficult, and
- Different methods vary substantially in required degree of expertise.

10.3240 FOLD AND FORMED

The fold and formed method can be used for both structural and nonstructural purposes. One variation uses a jointless HDPE pipe which is deformed by means of thermo-mechanical deforming equipment into a "U" shape. The deformed pipe is sliplined into an existing pipeline and then restored to a circular shape with heat and hydraulic pressure. House connections and laterals are reconnected by a remote-control cutter without excavation.

In another variation of the fold and formed method, the same PVC feedstock and additives used for new PVC pipe are used to form a U-shaped pipe which can be inserted and then rounded by applying heat and/or pressure. The pipe in its final form is stated to be similar in strength and characteristics to standard PVC pipe. In this method connections are also restored by a remote-control cutter.

When using the fold and formed method, it is essential to monitor the existing pipeline with a closed circuit TV (CCTV) camera. A proofing *PIG*[5] may have to be pulled through the line to verify dimensions and remove protruding laterals and broken pipe prior to liner insertion. This technique has low mobilization costs and low to moderate overall costs.

ADVANTAGES

Advantages of the fold and formed method include rapid installation; no joints, grouting and usually no excavation; internal lateral connection is possible; and this method leaves a very small annular space, if any.

LIMITATIONS

The main limitations of this method are that the range of available pipe diameters is limited and this method cannot accommodate oval or odd shapes of the old pipe, diameter variations, possible joint settlement and pipe bends.

---

[5] *Pig. Refers to a poly pig which is a bullet-shaped device made of hard rubber or similar material. This device is used to clean pipes. It is inserted in one end of a pipe, moves through the pipe under pressure, and is removed from the other end of the pipe.*

10.3241 DRAWDOWN

In the drawdown method, the joints of solid-wall HDPE pipe which is slightly larger than the old pipe are butt fused at the insertion site. The new liner pipe is then heated by a special process, drawn through a die to reduce (swag down) its diameter, and kept under tension during the insertion process. Compressing the pipe temporarily crushes the molecular chain structure, allowing the pipe to be reduced in diameter and later restored to its original size without affecting performance. Pressure is applied to the inside of the pipe to speed up the reversion process. The pipe in its reverted form usually fits closely to the old pipe wall, and no annular space remains. Pipes of 3 to 24 inches in diameter can be installed using this method. The amount of diameter reduction depends on the memory of the polymeric chain structure of medium and high density polyethylene pipe. Overall diameter reduction ranges from around 20 percent for a 4-inch diameter pipe to 7 percent for a 24-inch diameter pipe. This method can be used for both structural and nonstructural purposes. This system has high mobilization costs, but low to moderate overall costs.

ADVANTAGES

Although an excavation is required, it will be shorter than the excavation required for regular sliplining. Another advantage is that house connections are reconnected by a remote-control cutter without excavation. Since the liner forms tightly against the wall of the original pipe, it leaves no annular space, eliminating the need for annulus grouting.

LIMITATIONS

Limitations of the drawdown method include the following:

- Temporary excavation is necessary to install the liner since existing manholes do not provide enough space,
- The liner may not insert easily if there are diameter variations in the old pipe, and
- There is a possibility of stretching the liner and developing environmental stress cracking.

10.3242 ROLL-DOWN

The Roll-Down System, from Subterra Ltd., Dorset, U.K., involves a hydraulic pusher that forces the pipe through multiple pairs of rollers to reduce its diameter. Lengths of standard medium-density or high-density polyethylene (MDPE or HDPE) are butt fused into appropriate lengths. Then the pipe is cold-rolled on site to reduce its diameter to allow insertion. After the liner pipe has been inserted into the old pipe, the liner is pressurized to restore the pipe to its original size, resulting in the tight fit inside the pipe.

ADVANTAGES

The main advantage of the Roll-Down method over the drawdown process is that the treated liner pipe will remain at the reduced diameter for some period of time. This process also does not rely on tension or mechanical means to prevent the liner from reverting to its original size prior to insertion.

LIMITATIONS

The limitations of the drawdown process are the need for temporary excavations to install the liner and potential problems inserting the liner if there are variations in the diameter of the old pipe.

10.325 Acknowledgment

Information presented in this section was taken from the presentation titled "Trenchless Rehabilitation of Sanitary Sewer Systems — Survey of Available Methods and Materials in the US," by Mohammad Najafi, Research Associate, Trenchless Technology Center, Ruston, LA, D.T. (Tom Iseley), Associate Professor, Louisiana Tech University, Ruston, LA, and David Bennett, Civil Engineer, Waterways Experiment Station, Vicksburg, MS.

The work reported herein resulted from research being conducted under the CPAR program by the US Army Corps of Engineers Waterways Experiment Station and Louisiana Tech University. Permission was granted by the Chief of Engineers to publish this information. Much of the information contained in this study has been gathered by a literature search and a survey of system manufacturers. Providing this information is not meant to represent endorsement of the methods by the authors, The Trenchless Technology Center, Louisiana Tech University, or the US Army Corps of Engineers.

QUESTIONS

Write your answers in a notebook and then compare your answers with those on page 354.

10.3E When are trenchless rehabilitation methods used?

10.3F List three broad types of trenchless pipeline rehabilitation methods.

10.3G What is shotcrete?

10.33 Insituform

(Also see Section 10.321, "Cured-in-Place Pipe (CIPP).")

10.330 Description of Process

This method is relatively new and consists of the installation of a new pipe within the old pipe without excavation. Essentially, the process involves the use of a polyester-fiber felt tube, lined on one side with polyurethane and fully impregnated with a liquid thermal setting resin. The actual installation is performed by contractors, but agency personnel are often used to clean the sewer in advance and perform TV inspection before and after the job.

10.331 Preliminary Work

The sewer must be cleaned prior to installation of the Insituform line. This phase is essential since the liner is inverted into the pipe and not simply pulled into place. The inversion process causes the liner to unfold from within, resulting in no relative movement between the liner surface and the pipe surface. Thus, any material or deposits on the pipe wall will be sealed (encapsulated) between the old pipe and the new liner. A loss in cross-sectional area due to encapsulation could result in loss of structural strength. The loss of structural strength results from not having a perfect circle which causes a loss in structural hoop strength. A loss in cross-sectional area will also result in a loss of flow capacity.

10.332 Basic Equipment and Materials

The basic items required for a typical Insituform installation are described in the following paragraphs.

- The inversion platform is a construction-type sectional scaffolding (Figure 10.37) erected over the point of insertion. The platform supports the inversion tube, worker, liner bag and recirculation hose during the inversion phase of the installation. The height of the platform is governed by the required head for inversion and the depth of the sewer.
- The inversion tube (Figure 10.38) consists of a reinforced polyester tube of sufficient diameter and length to allow the liner to pass from the top of the inversion platform to a steel inversion elbow located in the invert of the sewer. The top of the inversion tube is held open by means of a steel ring to which it is attached. This steel ring is also used to support the inversion tube which hangs from the inversion platform on cross-members of the scaffolding.
- The inversion elbow (Figure 10.39) is a prefabricated, 90° steel elbow of the appropriate diameter for the sewer being lined. The inversion elbow is attached to the bottom end of the inversion tube by means of stainless steel strapping. The Insituform liner is attached to the other end of the elbow by stainless steel straps.
- The portable mixer is a drum-type mixer used to mix the appropriate catalyst with the *NON-PROMOTED*[6] resin just prior to transferring the resin from the shipping barrels into the Insituform liner bag.
- A transfer pump is used to transfer the catalyzed resin from the shipping barrels into the liner bag for complete impregnation of the felt.
- The vacuum pump is used to evacuate the air from the Insituform liner bag prior to pumping the resin into the bag. This enhances the penetration of resin into the felt. Because a partial vacuum within the bag is essential to this process, the polyethylene bag covering must be entirely free of pinholes.
- Wet down conveyor system is a three-level conveyor system through which the Insituform liner and resin are passed to ensure even distribution of the resin material throughout the entire bag length. Doctor blades or squeeze rollers are used on the top end of the conveyor to ensure the proper resin saturation throughout the bag material.
- Lay-flat hose (Figure 10.40) is a high-pressure hose used to circulate hot water to the far end of the liner during the curing process; it must be able to withstand the very high temperature at which the water is being recirculated.
- Hold back rope (Figure 10.41) is a rope used to control the rate of inversion after the bag is halfway inverted; if the liner inverts too swiftly, the water head may be lost in the inversion tube creating air pockets in the liner. The rope is attached to the end of the bag away from the point of inversion by means of tape and wire; it is also used to help hold back the far end of the bag when it arrives at the desired end of the inversion run by means of tape and wire.
- Water supply line (Figure 10.42) is the line used to supply the water to the inversion tube; most of the time it is connected to a nearby fire hydrant.
- The water control valve is a valve located at the top of the inversion platform to control the height of the inversion head and thereby the rate of insertion. It is extremely important that the inversion head be maintained at a static level throughout insertion. The constant static level produces an even or constant speed of insertion which results in a constant thickness.
- A circulating pump is used to circulate the water from the inversion tube to the heat exchanger and then through the lay-flat hose to the far end of sewer section being relined.
- The boiler and heat exchanger unit (Figure 10.43) is the source of heat used for bringing the water used in the inversion process up to the curing temperature of the thermosetting resin.
- Insitucutter (Figure 10.44) is a patented device which allows the remote operation of a cutter as it is pulled through the relined sewer. The device is positioned and controlled by an operator who is watching a television monitor through a three-dimensional control system. It enables the operator to cut out lining material in the areas defined as service taps. The television monitor shows a picture received from the internal inspection camera located within the sewer just ahead of the cutter and in reverse position.

[6] *Promoted. The mixture of resin and catalyst ready to cause (promote) curing in place.*

Fig. 10.37 Scaffolding over point of inversion

Fig. 10.38 Inversion tube

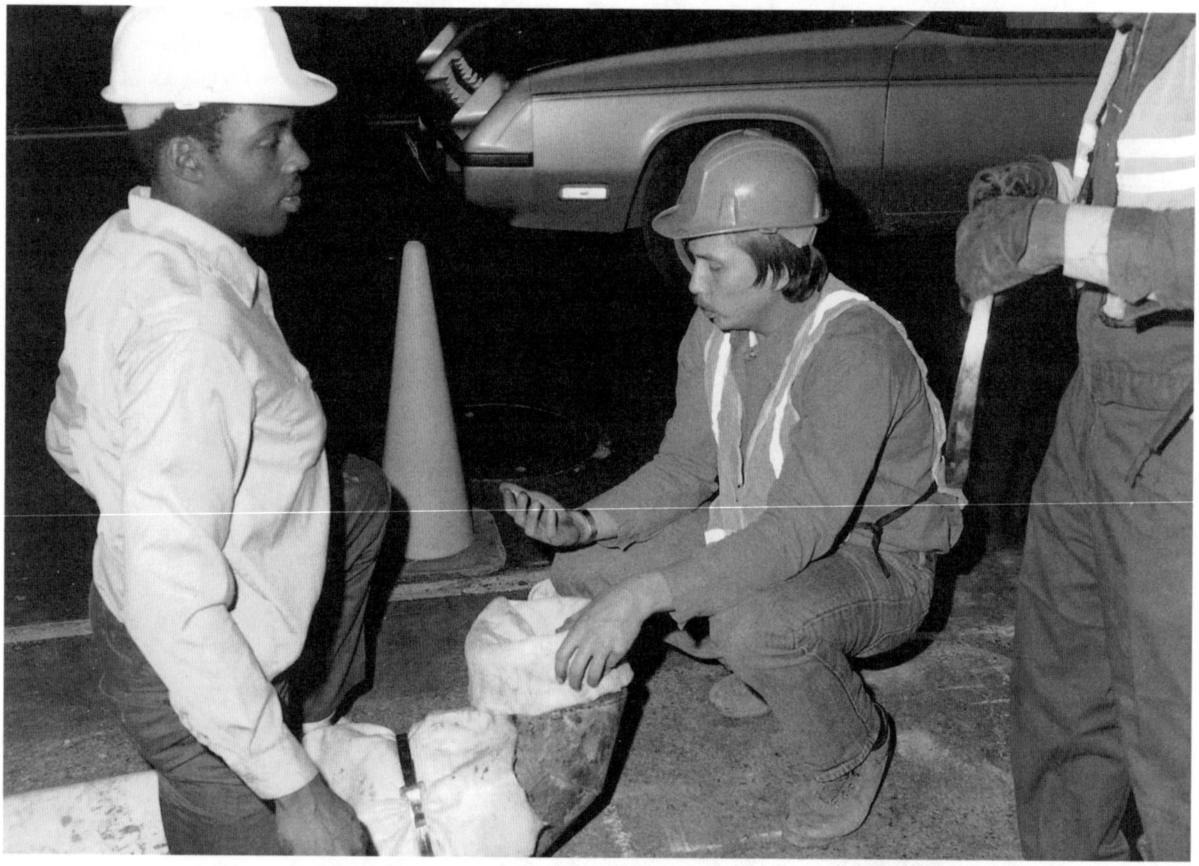

Fig. 10.39 Inversion elbow

Fig. 10.40 Lay-flat hose

Fig. 10.41 Hold back rope

Fig. 10.42 Water supply

Fig. 10.43 Boiler and heat exchanger

Fig. 10.44 Insitucutter

Fig. 10.45 Thermocouple attached to line

- Thermocouples (Figure 10.45) are temperature-sensing devices placed between the liner material and the sewer wall to read the temperature during the cure and post-cure periods; these give accurate indications of the cure status of the material.
- Polyurethane-coated, polyester-fiber felt liner (Figure 10.46) is the liner bag which is inserted into the sewer to be lined. This liner is made of densely woven polyester fiber, and can vary in thickness from 3 to 18 millimeters in the prelining state. The liner is most commonly made of multiple layers of fiber of 3 millimeters each to give the desired total thickness. The primary function of the felt is to act as a medium to hold the resin prior to curing. Each layer is individually sewn into a long cylindrical tube of the proper length and diameter for the proposed lining application. Liners can be custom-designed and constructed for perfect fit. A polyurethane film is applied to the outer surface of the outer layer of material only. Its purpose in the technique is multifold, as it not only provides an airtight membrane to enable a vacuum to be drawn on the bag during resin impregnation, but also enables the circulation of curing water throughout the liner as well as providing a smooth-flow surface for the ultimate bore.
- Polyester resin is the thermal setting resin used to impregnate the polyester-fiber felt liner. This material forms the actual smooth, resistant structure within the original conduit. The resin is of the isophthalic-acid-based polyester thermosetting type and is shipped to the job site noncatalyzed. The resin may contain certain additives to obtain special characteristics required for each application.
- Catalyst is a chemical compound such as percadox which is added to the resin to control curing temperature, life and hardness. After the catalyst is added to the resin, the resin is considered to be in the promoted or reactive state.

QUESTIONS

Write your answers in a notebook and then compare your answers with those on page 355.

10.3H What is Insituform?

10.3I List the materials and major pieces of equipment required for a typical Insituform installation.

10.3J What is the purpose of the Insitucutter?

10.333 Installation Procedure

After the preparatory cleaning is completed, the internal inspection and evaluation team makes an inspection of the line to verify that the cleaning is complete and also to record on a TV log the location of all line deficiencies and service connections. In addition to the TV log, the line is videotaped. The flow in the sewer must be diverted or stopped before the liner can be inserted into the sewer.

As soon as the internal inspection crew completes their work and determines that the line is ready for lining, the installation crew starts erecting the inversion platform. This crew is responsible for the actual inversion and curing of the liner as detailed on the following pages.

After the line is installed, cured and cooled down, the services are reinstated by the crew using the Insitucutter. This crew also performs a post-Insituform internal inspection including the videotaping of the Insituformed line.

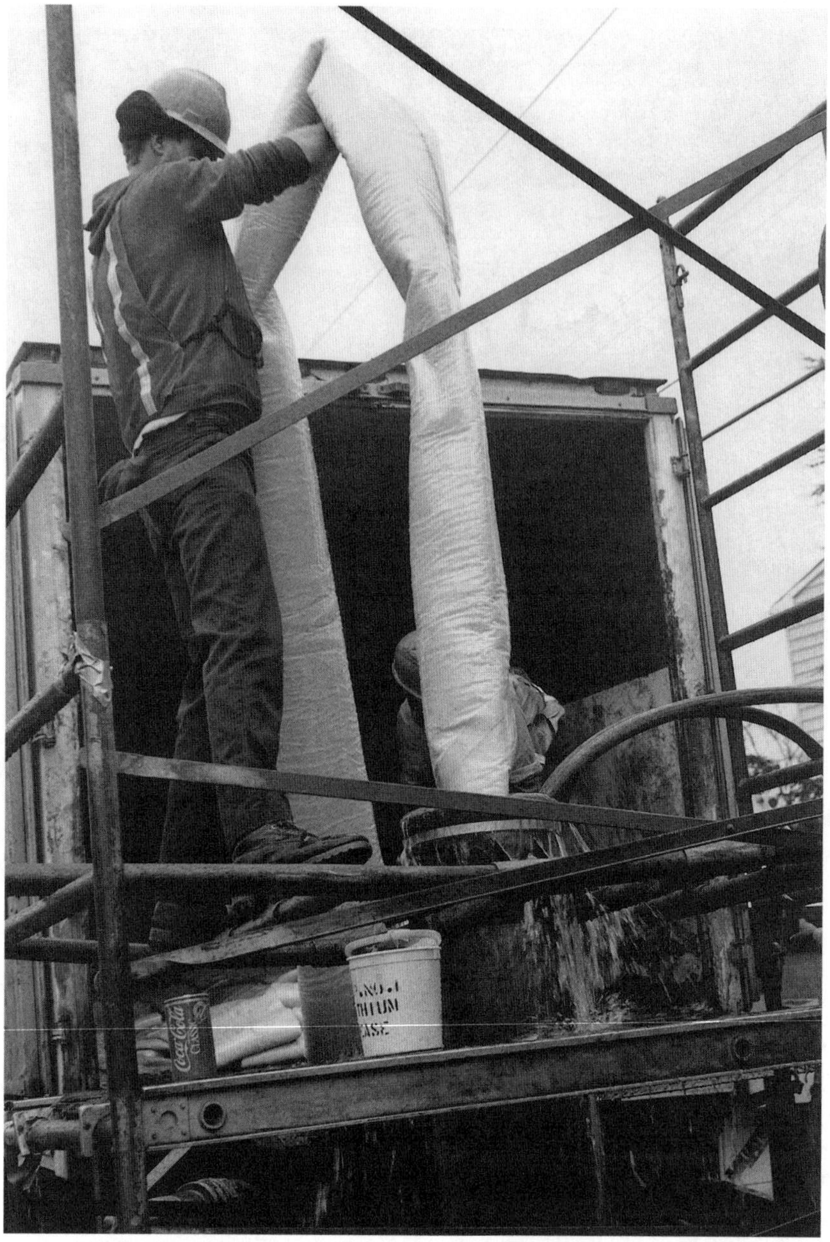

Fig. 10.46 Polyurethane-coated, polyester-fiber felt liner

The Insituform installation team is supported with the following equipment: sewer cleaning equipment (high-pressure water jet, bucket machines), a twenty-six foot truck equipped with boiler/heat exchanger and circulating pump, a twenty-six foot materials truck and an internal inspection van.

The wet out crew is equipped with a semitrailer truck containing a three-tiered conveyor system and appurtenances for mixing and transferring the resin to the liner bag for felt saturation. In addition to the semitrailer truck, the wet out crew has tarps used to make sunshades so that all resin handling can be performed out of direct sunlight.

One of the first steps in the Insituform lining technique is the erection of the inversion platform. This platform is erected over the point of insertion of the liner into the sewer being lined, usually over a manhole.

The platform is constructed of structural steel tubing using standard scaffolding frames and accessories. The scaffold height (Figure 10.47) varies with the depth below grade and the diameter of the sewer being lined as well as the thickness of the liner. The inversion heads may vary from a maximum of 38 feet (11.5 m) on a 6-millimeter thick, 8-inch (20-cm) diameter bag to as little as 12 feet (3.6 m) on a 9-millimeter thick, 24-inch (60-cm) diameter bag. The larger the diameter, the smaller the head needed for the inversion. Five to six feet (1.5 m to 2 m) below the top of the scaffold, a walk platform is installed for the workers to stand on during the inversion. Eighteen inches (46 cm) below the top of the scaffolding an adjustable inversion tube support system is installed; this support system holds the steel rings at the top of the insertion tube. At the very top of the scaffold a capstan is positioned to act as a pulley for the bag line or hold back rope.

Care must be taken when erecting the inversion platform to be sure that the feet of the scaffold legs are on good firm ground and that the scaffold is level in all directions. Some minimal excavation or footing construction may be necessary to provide the desired degree of safety.

Fig. 10.47 High scaffolding necessary for inversion

At the same time the inversion platform is being erected, the liner bag is being prepared for insertion by a crew working at a convenient location, usually not at the job site. There are several steps to be taken in this preparation phase. One of the first and ongoing procedures throughout the entire installation is the visual inspection of the bag for any obvious flaws such as pinholes or tears in the polyethylene coating. These defects may have been caused in manufacturing the bag, although flaws of this type would most likely have been detected earlier during quality control inspection by the manufacturer. Faults also may have occurred as a result of shipping or job site handling. If there is a defect in the liner, it is much better to detect it before the liner is installed in the sewer than to realize that the recirculating curing water is leaking.

After the liner is unpacked from its shipping container, a vacuum pump is attached to the bag to evacuate the entrapped air from the felt liner material. This, in itself, gives a test as to the soundness of the liner since in a damaged bag it would be impossible to draw and maintain a vacuum within the system.

While the air is being evacuated from the liner, the resin is being prepared by other workers. The resin is stored in the non-promoted state until just prior to use; it is much more stable and has a longer shelf life in the non-promoted condition.

Resin is generally shipped to the project site in drums with removable lids to facilitate mixing and transfer. The lids are removed from the drums and the prescribed amounts of catalysts are added and the ingredients thoroughly mixed with flash mixers. The lids are then replaced on the drums to await transfer of the resin into the liner bag. Care should be taken to keep the resin material away from direct exposure to sunlight; ultraviolet rays tend to deteriorate the composition of the material. Prolonged exposure in the presence of heat can cause a thermosetting reaction. The resin may be kept in this state

for up to 48 hours providing it is out of direct sunlight and at a temperature below 40°F (4°C). A refrigerated truck may be needed to maintain this temperature level.

The quantity of resin used for each installation should be equal by volume to 110 to 115 percent of the volume of felt in the liner bag. Resin material must be liberally spread throughout the liner bag, replacing all the air in the Insituform. Many times the activated resin contains a dye so that the spread of material throughout the bag can be documented. An air pocket that prevents resin from coating the liner will not cure and therefore a soft spot in the finished product will result.

The quantity and type of catalyst are determined based upon the proposed curing conditions as well as the recommendations of the resin manufacturer.

When the desired level of vacuum (5 to 8 pounds per square inch or 10 to 16 inches of mercury) is reached in the liner bag, it is ready to receive the catalyzed or promoted resin. The transfer of resin to the liner is accomplished by the use of a hydraulic transfer pump. The pump suction is placed in each drum and the contents of the drum are pumped to the line opposite the end of the liner to which the vacuum pump is attached. After all the promoted resin has been transferred to the liner, the bag is resealed at the ends.

The next step in the preparation of the liner bag is the saturation of the liner felt. An even and thorough saturation of the liner is very important to assure uniform strength and cure of the line when in place. To assure the even distribution of resin throughout the line, it is passed over a three-level conveyor system. This variable-speed system delivers the liner bag to the squeeze rollers at the proper speed to assure complete saturation of the liner felt. The doctor blades or rollers are preset at the proper thickness to allow just the right amount of resin to pass to assure an even distribution throughout the length of the liner. The process of saturating the liner felt with resin is referred to as the "wet out."

For reasons previously stated, the saturated liner should be kept out of direct sunlight and at or below 40°F (4°C) during transportation and storage. The "wet out" should not occur in excess of twenty-four hours before the estimated installation time.

A conventional fire hose is used to supply the water from the fire hydrant to the top of the inversion platform. The end of the hose at the top of the inversion platform is equipped with a gate valve to control the water supply rate in order to maintain a constant head.

After the inversion platform, inversion tube, and supply hose are in place, the crew is ready for the liner bag. The wetted liner bag arrives in a refrigerated truck, the bag being laid fire hose style in the controlled environment. Upon reaching the site, one end of the bag is immediately cut off and folded to make a triangular point and taped with duct tape in order to facilitate passing the bag through the inversion tube. The bag is shipped flat in order to pass the liner bag through the inversion tube during insertion. The bag must be folded as it comes off the truck by the installation crew.

Prior to inserting the end of the liner bag into the inversion tube, the portion of inversion tube and inversion shoe previously placed within the manhole are removed to an aboveground position to facilitate the attaching of the liner bag to the shoe. The liner bag end is next passed over a roller on the top of the inversion platform and down the inversion tube until about six inches (fifteen centimeters) of the pointed end of the liner bag extends out of the inversion shoe. The liner bag, however, is still not in position to be attached to the inversion shoe. To clarify the system used, remember that the bag is shipped with the polyethylene coating on the outside and that when the liner bag is inverted into the sewer, the polyethylene coating becomes the inside of the liner bag; actually, as the word inversion implies, the liner bag is turned inside out during the installation of the liner. The duct tape is removed from the pointed end and the outer layer of polyethylene-covered felt is folded back over the inversion shoe (Figure 10.48). A stainless steel band is then tightened over the bag material, firmly attaching it to the inversion shoe. The under layer of felt is then folded back over the shoe and firmly anchored to the shoe in a similar manner.

The proper banding of the liner bag to the inversion shoe is critical to the work. If the bag comes loose from the shoe or a leak develops at the connection, the inversion may have to be stopped because curing could not proceed. If the problem cannot be corrected quickly, the entire insertion might have to be scrapped; this most likely would result in the loss of the liner bag, resin and all preparatory work.

After it is established that the liner bag has been properly attached to the inversion shoe, the inversion tube and shoe are again lowered into position in the manhole for inversion, and wood thrust blocking is installed to brace the shoe. Before starting the inversion process, a frame backstop is installed in the far manhole to assure the stopping position of the liner.

The water is then turned on, and the inversion tube is slowly filled to the desired inversion head. The bag is restrained by the workers on the inversion platform and in the truck until the desired head is reached. The liner bag is then allowed to slowly move out of the inversion shoe and into the sewer itself. As it slowly enters the sewer, a thermocouple is placed between the liner bag and the sewer pipe to measure temperatures in this critical location. Proper alignment of the liner bag into the sewer at the start of the inversion is checked very closely by the installation crew.

The supply valve on the water line is carefully controlled to maintain the hydraulic head as the liner is allowed to slowly move (report) out into the line. This process continues until the end of the liner bag in the truck is within approximately ten feet (three meters) of the top of the inversion tube. At this point, the inversion halts.

A hold back rope is then wrapped around the capstan at the top of the platform, attached to the end of the liner bag. This rope is used to control the rate of inversion by holding back the liner bag; should the bag be allowed to proceed unimpeded, it would very possibly move out at such a high speed that a constant water head could not be maintained. Simultaneously with the connection of the hold back rope, a lay-flat hose is also attached to the end of the liner bag. The purpose for the

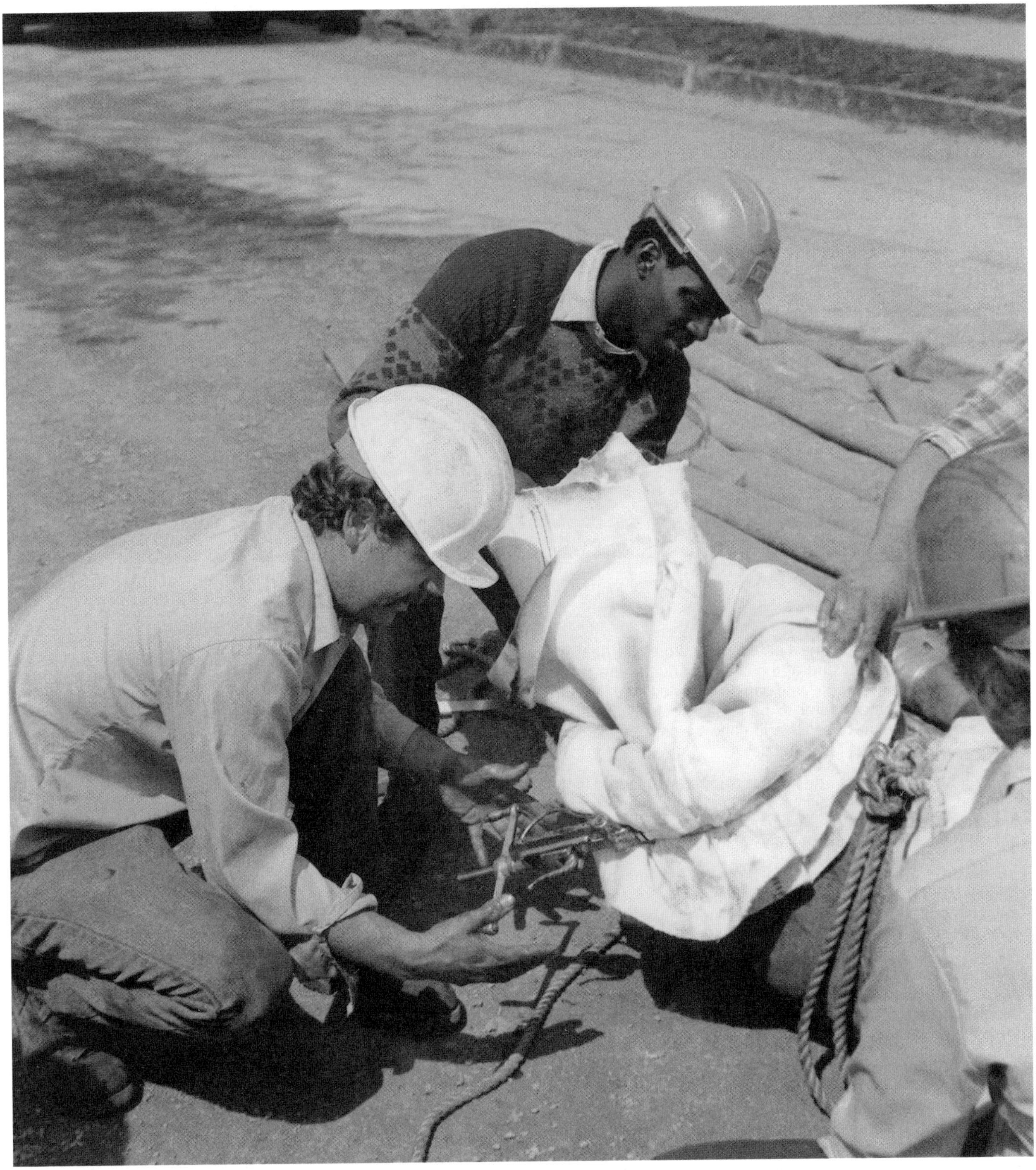

Fig. 10.48 Bonding of liner to elbow

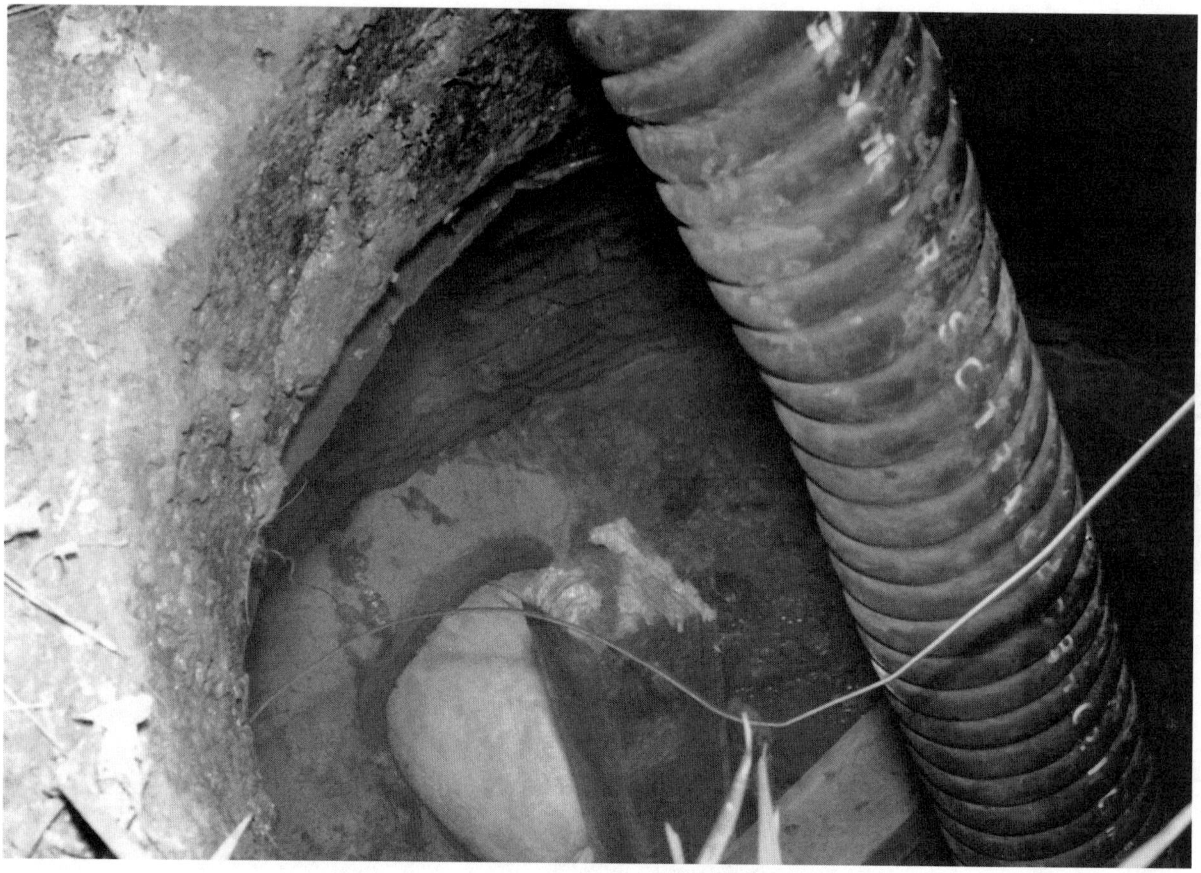

Fig. 10.49 Frame backstop

lay-flat hose is to carry the heated water from the heat exchanger unit to the far end of the liner bag during the curing process. A more even cure rate can be obtained by allowing a portion of the heated water to be released along the entire length of the lay-flat hose; this is done by providing orifices along the length of the lay-flat hose.

After the lay-flat hose and hold back rope are secured to the bag end, the inversion continues. As the liner bag approaches the downstream manhole, a second thermocouple is placed between the liner bag and the sewer pipe wall.

Prior to starting the inversion process, a frame backstop (Figure 10.49) is installed in the far manhole to assure the stopping position of the liner bag. The liner bag is normally allowed to enter into the far manhole before the inversion is terminated. This projection into the manhole allows the liner bag to expand and cause a slight flared end at the sewer wall intersection producing a superior seal between the sewer and the liner. At the end of the inversion the hold back rope is tied off to the capstan.

With the liner bag completely inverted, the next procedure is the curing of the thermosetting resin-saturated liner bag. The curing of the liner bag is accomplished by heating the water used to invert the liner bag to a temperature which will cause the resin to cure and harden. The temperature needed to cure the liner bag can be adjusted by the type and amount of catalyst added to the resin before wet out; the typical cure temperature is 180°F (82°C).

The heat exchanger and recirculation pump are mounted in a truck which is backed up to the inversion platform. This truck also contains a diesel-driven generator to power the electric drive motor on the recirculating pump and to supply electricity for other lighting and power requirements. In addition to the generator, the truck contains an oil-fired water tube boiler to supply heat for the heat exchanger. A suction line is attached from the inversion tube to the recirculating pump; a second hose attaches the heat exchanger to the lay-flat hose. A complete recirculation loop is now in place taking water from the inversion tube through the heat exchanger and returning it back into the sewer line through the lay-flat hose.

After circulation has begun and it is determined that the flow distribution is proper, the boiler is fired and the heat-up begins. Water temperature at the beginning of the heat-up is 55°F (12°C). Temperature gages on the suction and discharge lines in the truck show that the temperature of the circulated water

will increase rapidly as it goes through the heat exchanger and into the liner section.

Monitoring of the thermocouple temperature at the near and far manholes shows the actual increase in the temperature of the liner bag. The heat sink ability of the ground around a sewer can vary greatly with groundwater, backfill and local utility conditions, and therefore the temperature of recirculating water versus liner bag outer surface temperature may vary. For this reason, the temperature of the circulating water should never be used as the only indication of the extent to which the process has proceeded.

Care should be taken in determining bag length as recommended by the bag manufacturer for the size, length, and proposed inversion head. The problem associated with a bag that is too short is apparent. The problems with a bag that is too long are not so apparent but are quite real. A bag that is too long means extra cost for wasted bag and resin and extra work to remove the uninverted cured portion of the bag. In addition to the above problems, a bag with a long uninverted section prevents the hot water in the lay-flat hose from being discharged at the end of the liner bag.

Special consideration should be given to safety factors during installation and curing. Care to protect workers, spectators, and equipment from hot water should be observed by the use of rubber wear and protective shrouds. Any time workers enter the manholes, all confined space procedures must be strictly followed. Strong volatile styrene fumes are created during this process to which prolonged exposure must be avoided.

If installation is made during the night to take advantage of low sewer flows, the running of equipment such as pumps, boilers, and trucks may cause some noise problems in residential areas. This factor should be evaluated when scheduling installation times and satisfactory arrangements should be made with those involved.

10.334 Reinstatement of Services

The next step is the reinstatement of the services. This portion of the sewer relining technique requires the remote guidance of a cutter operating within the sewer itself. The patented cutter is called an Insitucutter. The Insitucutter is similar to a pneumatic drill or router and is operated in conjunction with the internal inspection camera.

The camera-cutter combination is pulled in tandem through the newly lined sewer in such a way that the camera continually views the cutter. The cutter is positioned next to the location of each service which has been previously identified and recorded during internal inspections. Service connections may also be detected visually on the interior of the newly lined pipe. Services appear as convex areas similar to dimples on the pipe wall. The operator viewing the service location positions the Insitucutter by watching a television monitor; the operator also may place a microphone in the sewer for audio reference, since there is a distinctive difference in the sound levels noted between cutting the liner material and the actual conduit material.

The Insitucutter is capable of movement in six distinct directions. The operator, by both listening and watching the monitor, then proceeds to cut out the liner material covering the service opening. Excess resin in the liner bag migrates into the service joints and forms a weld-like seal between the walls of the service pipe and the liner in the main.

Problems are created for the service connection process when water stands in the service behind the liner. When the spinning drill pierces the Insituform covering over the service to be opened, water and liner material are splashed in every direction, including on the lens of the camera, thus significantly impairing the operator's view. Should the debris become heavy, the camera-cutter assembly must be withdrawn from the line for cleaning. This situation can cause a considerable slowdown in the opening of service taps. For this reason, all customers served by the sewer should be notified in advance that water will be shut off prior to rehabilitation. This alone may not totally relieve the situation in that many times infiltration in the service line may back up behind the lining. Service connections should therefore be cut out as soon as possible after the liner has cured.

Experience has shown that the average service connection can be renewed by the Insitucutter in approximately 15 minutes. If there are many services to be renewed, it is advisable to travel the length of the lining and cut a small relief hole in each service dimple. This prevents a buildup of water from either infiltration or customer use of sanitary facilities. After all service connections have been relieved, the Insitucutter operator then can begin to reinstate each service to full-bore opening. This procedure in most cases relieves the requirement for providing any alternative sanitary facilities.

Upon completion of the opening of the service connections, a post-Insituform internal inspection is performed. This inspection should reveal the sewer to be completely relined with a joint-free, smooth liner.

10.335 Advantages of Insituform Method

Advantages of the Insituform method are reduced cost, no excavation, completion of work in a 24-hour period, complete structural strength and total elimination of I/I into the sewer stretch. The only limitation to the process is the diversion of flow while the work is being accomplished and the residents being without water and sewer services for approximately 10 to 12 hours. The advantages far outweigh the short inconvenience factor.

QUESTIONS

Write your answers in a notebook and then compare your answers with those on page 355.

10.3K What happens after preparatory cleaning is completed on an Insituform sewer rehabilitation job?

10.3L In the Insituform lining technique, where is the inversion platform erected?

10.3M How can a liner be tested for damages after it has been unpacked from its shipping container?

10.3N What safety precautions should be taken during the installation and curing processes of an Insituform job?

10.34 Polyethylene Pipe Lining
(Also see Section 10.322 "Sliplining")

10.340 Why Line Sewers?

Polyethylene pipe lining (sliplining) is a repair method accomplished by inserting flexible polyethylene pipe into the existing sewer system.

Liners can be an effective method of pipe repair or reconditioning of the sewer system. Frequently liners are installed to control excessive infiltration, exfiltration and root intrusion. Selection of this method depends on whether this procedure is simpler, quicker and more economical in comparison with other techniques available to solve the problems in the sewers. Sliplining requires excavation at the points of insertion into main lines and house service connections. In general, sliplining costs less than Insituform, but the costs of excavation and traffic disruption may offset the installation advantage.

10.341 Preliminary Work

Before attempting to do any sliplining, there are some very important items to consider. Identify the area to be sliplined and carefully review available records and maps of the area. Plan the operation on a step-by-step basis.

Send to each homeowner a typewritten letter explaining why the project is necessary, what is planned and what you expect to accomplish by the operation. The letter should indicate that the homeowner could be temporarily inconvenienced during construction and also explain the benefits that will result from the completed project. Advise the homeowner of any direct costs that must be paid. Send along with this letter a form, legally prepared, for the homeowner's signature. This form provides authorization to do any and all work necessary to accomplish the job. It also guarantees that the premises will be restored to as near its original state as possible at no additional cost to the homeowner.

Take photos of the yards around each house that may be affected by the job. This provides a picture record of the appearance of the premises before the work is performed. If necessary, a comparison may easily be made when the work is completed.

Notify all of the utility companies in the area and ask them to locate and mark all of their underground lines in the area.

With closed-circuit TV and videotape, TV the main line and location of each lateral, any misaligned joints and any other discrepancy which will need to be corrected before the actual lining takes place. Then mark locations on the street surface.

Conduct a physical survey to determine the amount, size and type of material needed to do the job and order the material. The list of materials should include plastic pipe for both the main line and the laterals, tee saddles, couplings of the various sizes needed and materials for the cleanouts to be installed on the lateral near the house.

10.342 Equipment

To do the job you will need a backhoe for excavating, a cable winch for pulling the pipe through the main line, and pneumatic equipment for pavement breaking and compacting the backfill. Also needed are dump trucks, a utility truck for transporting material and small tools, and a flatbed truck with a specially constructed rack for transporting the 40-foot lengths of pipe and plastic fittings. This truck carries on it the equipment required for fusing the plastic pipe together, a generator which provides the power needed to operate the heating elements and other electrical equipment. The generator should be capable of producing at least 2,500 watts.

Equipment needed to prepare and weld the pipe includes fusion joiner jigs and mobile carts to handle both the main line pipe and laterals, a butt fusion tool or squaring tool and heating faces (both flat and part-round) for heating the saddles. Pulling heads and tapping or coring tools are needed plus accessories to adapt to the various sizes of pipe that will be used.

10.343 Inserting Liner in Main Lines

Be sure all misalignment and other problems have been corrected before starting the lining job.

Transport all materials and equipment to the job site. Hold a brief group discussion to review the operation and safety considerations. Set up cones and safety warning devices and, once a proper traffic control pattern is established, begin the actual work. (See Volume I, Chapter 4, Section 4.3, "Routing Traffic Around the Job Site," and Section 4.40, "Confined Spaces.")

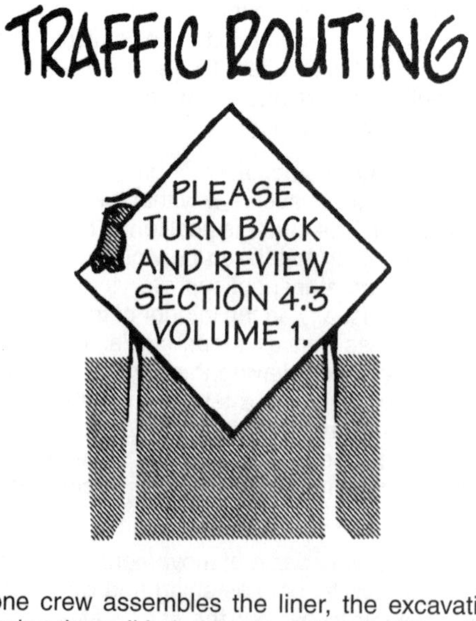

While one crew assembles the liner, the excavating crew begins digging the pull hole at a predetermined location. Pulling distances will vary in accordance with the job conditions. A third crew prepares the pulling equipment and pulley apparatus in the manhole into which the liner will be pulled. The pulling hole or trench will be approximately 30 feet in length and as deep as the main line. The depth of the main line usually determines the length of the pull trench. The deeper the main line, the longer the pull trench. Review Chapter 7, Section 7.1, in Volume I for shoring requirements. Once the pull trench is dug and the main line exposed, the main is broken and a high-velocity cleaner is sent through the line to clean the pipe and at the same time pull the winch cable back from the manhole to the pulling hole. At the pulling trench, attach the winch cable to the special pulling head which is welded onto the pipe liner.

Weld together the amount of pipe for this particular pull and weld on the pulling head (Figures 10.50, 10.51, and 10.52). Use a fusion jig and mobile cart to weld the material together in the following manner. Place the two pieces of pipe to be joined into the jig and use a squaring tool to true the ends of the pipe. Once this has been accomplished, bring the heating face to the proper temperature (500 degrees Fahrenheit (260°C)). Heat both pipe ends simultaneously until a small bead begins to form, then continue to heat them for 10 seconds more. Snap off the heating face and butt the two ends together, holding them firmly for approximately one minute. Let the joint cool for at least three minutes before moving to the next joint. Tests have shown that the pipe is actually stronger where the weld has been made than at the run of the pipe.

Once the required joints of pipe have been welded together and the pulling head attached, mark the pipe with a felt marker or crayon at 50-foot intervals. This is a precautionary measure. If there is a pulling problem and the pipe hangs up in the line, it may be necessary to dig up the pipe. By marking at 50-foot intervals, you can easily locate the pulling head in the line. Hopefully, there will be no problems.

Pulling head ready to be welded onto pipe

Fig. 10.50 Attaching pulling head

(Courtesy of Napa Sanitation District)

Fig. 10.51 Heating face to be applied to pulling head and pipe

(Courtesy of Napa Sanitation District)

When all is ready (Figure 10.53), attach the cable to the pulling head with a clevis. Station an operator in the pull hole to guide the liner smoothly into the pipe. Signal the winch operator to begin the pull (Figure 10.54). Providing that there are no problems, the pull will be completed in a few minutes.

Once the pulling head comes through at the manhole where the pull is made, remove the cable. "Whip in" the pipe to take up any stretch that occurred during pulling. To do this, protect the end of the pipe with a board and then hammer on it with a sledge (Figure 10.55) until the stretch is taken up. The last step is to cut off the pulling head (to be reused later) and prepare for insertion of liner into the laterals.

Whenever there is occasion to pull the liner through a manhole, the upper half of the liner is cut away, leaving the lower half to serve as a channel in the manhole. The manhole must be sealed around the opening in the liner. Sometimes the liner is removed and a new channel bottom is installed during the sealing operation.

The annular space between the outside diameter of the plastic liner and the inside diameter of the old sewer line, in most cases, is sufficient to handle the wastewater which is introduced into the system through the old connections until the entire operation of installing new laterals has been completed.

QUESTIONS

Write your answers in a notebook and then compare your answers with those on page 355.

10.3O How do liners repair or recondition a sewer system?

10.3P What information is included in a letter to homeowners explaining a sliplining project?

10.3Q Why should a sewer be videotaped before sliplining?

10.3R What equipment is needed to weld polyethylene pipes together?

10.3S How many crews are recommended and what are their duties for a sliplining project?

10.3T Why should the liner be marked at 50-foot intervals before being pulled into the sewer?

10.344 Lining and Connecting Service Laterals

The methods used in this operation will vary depending on whether the existing service lateral will be connected to the polyethylene main or if the lateral itself will be completely relined.

A substantial amount of infiltration could enter the collection system through breaks or bad joints in laterals. Determine whether the laterals need sliplining and, if so, follow the procedures outlined in the next section.

LINING PROCEDURES

The procedures used to slipline a lateral are similar to the ones used to slipline the main line. Instead of using a pulling hole, dig a pushing hole at the main where the old service lateral made its connection (Figure 10.56). Dig another hole at the house to expose the other end of the lateral; break the connections at both ends. Fuse the desired length of pipe, as described in the previous section. Weld a nose cone to the tip of the new liner. (This will be used to insert the liner into the old pipe lateral.) Pull a swab through the old pipe to make certain it is free of grit and roots and to locate any bad obstructions.

When these preparations are complete, push the new lateral through the push hole at the main line until it is visible in the hole near the house. Align the new pipe to the main and weld a saddle in place. When the weld has cooled sufficiently, core the line to complete the connection (Figure 10.57).

Backfill the excavation at the main line. To minimize the possibility of the pipe and lateral getting out of round at this point or egg-shaping in any way, carefully pack the area beneath, alongside and over the pipe with fill sand. Then backfill the

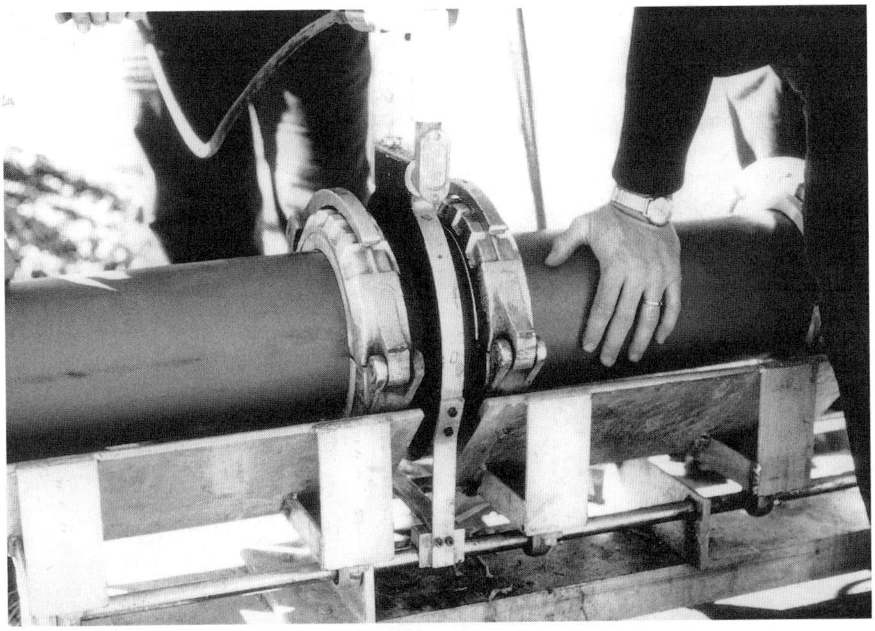

Heating ends of pipe

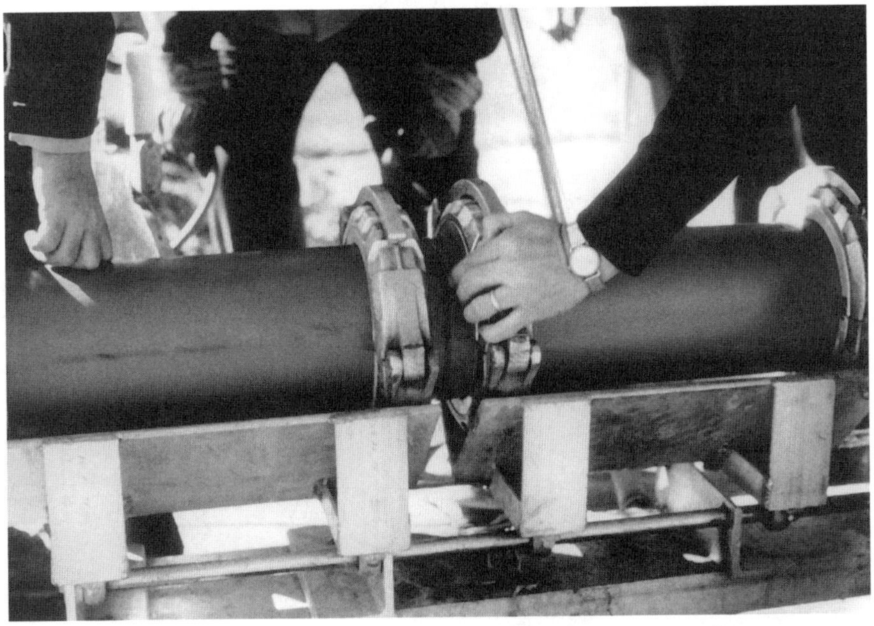

Fusing ends of pipe together

Fig. 10.52 Fusing pipe together

(Courtesy of Napa Sanitation District)

Fig. 10.53 Pipe fused together ready for pull
(Courtesy of Napa Sanitation District)

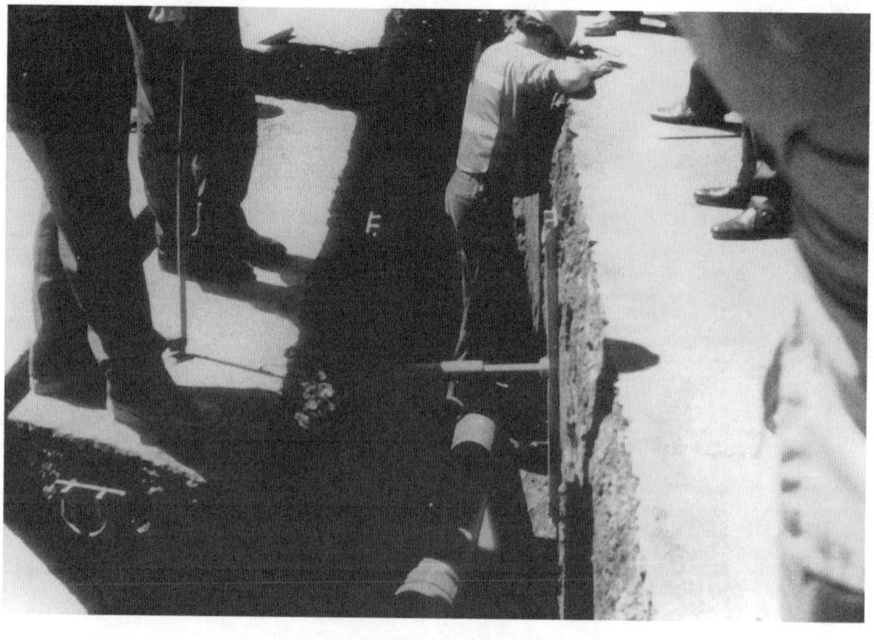

Fig. 10.54 Walking the pipe liner into pull hole trench
(Courtesy of Napa Sanitation District)

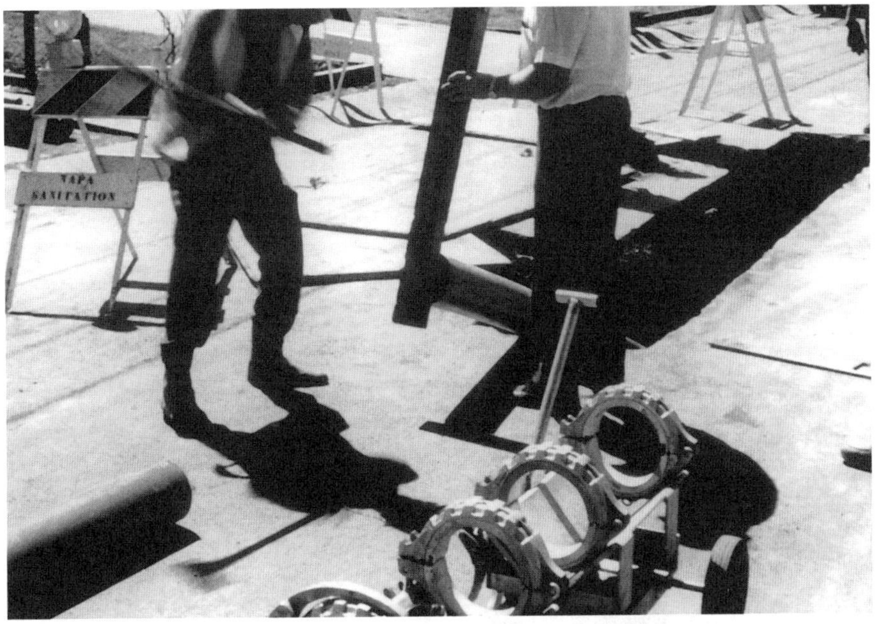

Fig. 10.55 "Whipping in" the pipe after pull
(Hitting end of pipe that is protected by a board with a sledge hammer)

Fig. 10.56 Street cut for service lateral
(Courtesy of Napa Sanitation District)

Saddle and coring tool

Connecting lateral to saddle

Fig. 10.57 Saddle for lateral
(Courtesy of Napa Sanitation District)

NOTE: Work is shown on ground surface, but can be performed in a properly shored excavation.

rest of the hole with a granular backfill material by carefully compacting it in lifts of one to two feet. Last, repave the surface of the street.

INSTALLING CLEANOUTS

Assemble the fittings to be used for the cleanout connection to the lateral at the house. Basically, these include a 3-inch combination wye and 1/8 bend, a 3-inch to 4-inch reducer, a plastic cleanout cap, a *CHRISTY BOX*[7] which will be used as the cleanout point, and the assorted couplings needed to make the proper connections.

When all of the tools and materials are assembled to install a cleanout, cut off the nose cone attached to the new liner. Install the fittings and connect to the house plumbing. Next, install the riser. Cut the riser to the desired height, normally several inches below the ground or lawn level, and attach it to the 1/8 bend of the combo wye. Position the christy box on the riser and mortar to within one-half inch of the top of the riser. Put a plastic cap on the riser and place the lid on the christy box to complete the operation. Backfill lawn areas with loam and plant new lawns in areas which have been disturbed.

INSTALLING TEE SADDLES

There are several ways to install a saddle on the main line including the use of mortar, banding, and welding. Fusion welds provide a good seal and a strong connection and will be described in this section.

Expose the main line at the point where the tee saddle will be located. Break away and remove the old pipe leaving the plastic liner exposed. Dig a well hole large enough to allow room to work with the apparatus needed and to keep the hole free of water, if necessary. Carefully clean the pipe and sand the glossy surface. Position the saddle in the approximate place where it will be welded and "mark in" the location with a crayon. Next, place the saddle in the saddle holder and press onto the preheated heating face. When the saddle reaches the desired melting point, place it and the heating face against the pipe at the marked location. Allow both the main and saddle to heat simultaneously until they reach the desired melt. Remove the heating face from the line, snap the saddle free and press it onto the main line. When the weld has cooled sufficiently, core the main line and connect the lateral by means of a coupling.

Pack and bed the pipe with fill sand. Then finish up by backfilling and compacting the rest of the hole with granular backfill. Repair the street paving to complete the job.

10.345 Sealing the Space in the Manholes

Infiltration must be prevented from entering the system at the manhole from the space between the new liner and the existing line. Part of the pipe through the manhole is used as a channel whenever possible. The procedure for sealing is easy. Use oakum rope soaked in a urethane sealing compound or chemical grout that is activated by water. Pack the soaked oakum rope into the space between the pipe and the manhole. Carefully pack the rope tightly into the space. Finish off the work with a quick-set cement. The channels of the manhole should be properly shaped and the shelves of the manholes should be completed to give the proper pitch and finish.

The entire area is then double checked and policed to be sure that all phases of the job are complete. Once this has been accomplished, the operation is moved to the next area, set up for sliplining, and the entire procedure is repeated.

10.346 Thank the Public

The project began with a letter and should end with a letter to each homeowner, signed by the Chair of the Board of Directors or other appropriate official. The letter should thank homeowners for their cooperation. Good public relations at this time can be very helpful now and in the future.

QUESTIONS

Write your answers in a notebook and then compare your answers with those on page 355.

10.3U When sliplining laterals, where are holes commonly dug for the liner?

10.3V Why is the space between the liner and the manhole sealed?

10.35 Service Connections

Service connections are defined as branch connections that go from the sewer main in the public way and connect to the building sewers. Each jurisdiction has varying responsibilities for portions of these service connections. Some jurisdictions are responsible for the service tap only, others are responsible only to the property line and some jurisdictions are even responsible all the way to the building on the property. These service connections can be various pipe sizes ranging from four inches to eight inches, depending on the nature of the property served. Normally, the length of the connections varies from about 10 feet to 100 feet depending on the layout of the street and the agency's responsibility.

During the past few years it has been found that a significant quantity of I/I enters the sewer system through the service

[7] *Christy Box. A box placed over the connection between the pipe liner and the house sewer to hold the mortar around the cleanout wye and riser in place.*

connections. Each point of connection, at the tap, at the property line, and at each joint in the service, is a potential location for I/I. Also, with the underground utilities crossing the services, there has been a great deal of structural damage detected in service connections.

There presently exist three methods of chemically grouting service connections to eliminate I/I. They are pump the service full of grout, sewer sausage, and camera-packer. Each method uses grout as the sealant but each requires a different application technique.

Pump Full. The sewer service is pumped full of grout from a conventional packer located in the main line sewer where the service connection is connected. As the grout is pumped into the service under pressure, the grout is forced through any open joints or cracks into the soil. The gel sets up, in the same manner as main line grouting, forming a barrier that keeps water from entering the service. The sewer service connection is then cleaned with an auger to remove the excess grout and this restores the service connection to full service.

Sewer Sausage (Figure 10.58). This is a sewer service sealing unit that requires a special packer with an inflatable tube. The tube is inserted into the service connection from the packer and then is inflated to create a sleeve inside the service connection. The packer then forces the chemical grout into the annular space between the sleeve and the service connection pipe. The pressure forces the grout out of open joints and cracks into the surrounding soil where it sets up and provides a barrier so water cannot enter the service connection. The sleeve is deflated and retrieved into the packer. This method does not result in excess grout requiring auguring to restore the connection to full service.

Camera-Packer. A specialized miniature camera and packer are inserted into the service connection. Joints and cracks are sealed in a manner similar to the conventional method used in the main line.

Inversion lining or Insituform has been used to provide structural rehabilitation to a service connection. This method also eliminates all I/I in the portion of the connection lined by this method. This process material is similar to the main line Insituform work. One excavation point is necessary on the service connection so the Insituform liner can be inverted into the connection. A special pressure chamber is used to force the Insituform liner through the pipe. This smaller unit is used instead of the normal inversion tube because less force is required to turn the line inside out when it is only four inches in diameter. After curing, the downstream end is cut out by a special cutter placed in the sewer main. The upstream end is cut out and the service connection reconnected and placed in service.

Most agencies use the standard excavation and replacement technique to replace deteriorated service connections. This method requires the greatest effort and results in the most disturbance to the area. This is the only technique available when the service connection has suffered total collapse and is in a crushed condition. The restoration costs usually result in this method being extremely expensive in comparison to the other methods.

10.36 Manholes

Manholes are sewer appurtenances which allow access to the sewer pipe underground. They are necessary so that proper preventive and emergency maintenance can be provided. Manhole rehabilitation is required to repair structural defects, eliminate I/I and correct corrosion deterioration of the manhole walls and rungs.

Figure 10.59 illustrates a typical precast concrete manhole. Components of the manhole include a cast-iron cover which may have exposed pick/vent holes or hidden pick holes, a

Fig. 10.58 Sewer service sealing unit

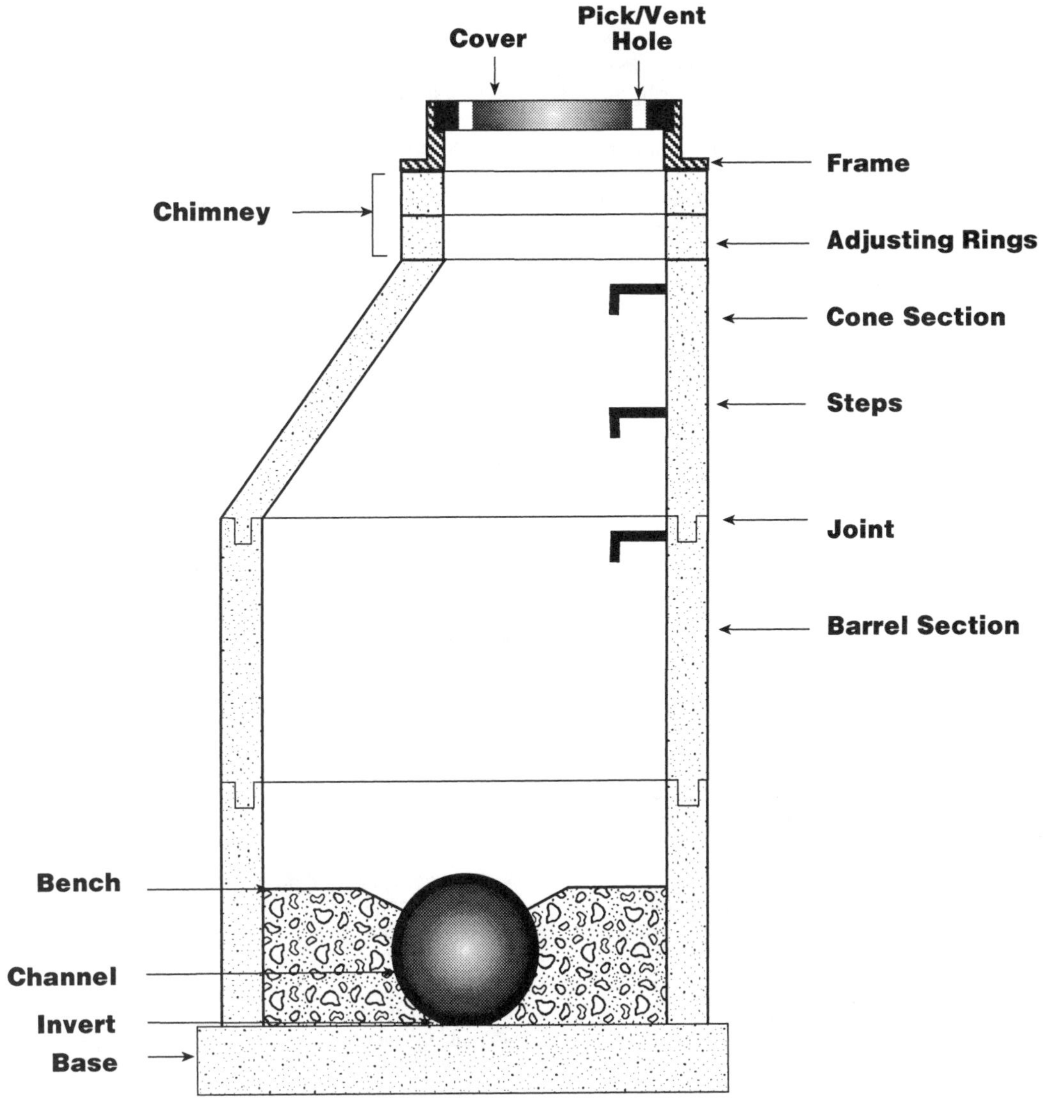

Fig. 10.59 Precast concrete manhole

cast-iron frame, the adjusting rings which are referred to as the chimney area, cone sections, steps, joints between sections, vertical barrel sections, base, bench, channel and invert. Each of these represents a potential source of inflow or infiltration into the manhole.

Significant quantities of storm water can enter through the pick/vent holes when the manhole is ponded over. If there is dirt and debris between the manhole frame or casting and the manhole cover seat area, which is a machined surface, water can also enter between the cover and frame. The point of contact between the frame and the first adjusting ring, which the frame sits on, can allow infiltration below ground if the frame is not sealed to the top adjusting ring. Because of the freeze/thaw cycle in cold climates, dirt, gravel and other debris can settle between the frame and adjusting ring so that the frame does not seat on the top adjusting ring. Adjusting rings are subject to cracking; if the cracking is severe enough, the adjusting rings will not provide adequate structural support for the frame and the cover. When infiltration occurs at the adjusting rings or between the frame and the adjusting rings, the surrounding backfill is carried into the manhole with the infiltration of water. As a result, the surrounding pavement or blacktop cracks and settles into the unsupported area allowing additional infiltration into the manhole between the frame and adjusting rings.

Each joint in the manhole is also susceptible to infiltration from groundwater if not properly sealed. In addition, infiltration of groundwater can occur at the point where the sewer line passes into and out of the manhole unless these connections are watertight.

Table 10.3 lists some typical defects found in the various components of a manhole, and Table 10.4 identifies the advantages and limitations of some of the more common manhole rehabilitation methods.

There are numerous methods available for manhole rehabilitation. The methods should be evaluated based on the type of problem encountered and the physical characteristics of the manhole. In most cases a precast concrete manhole requires different methods of rehabilitation than a brick manhole. Similarly, the rehabilitation of walls deteriorated from hydrogen sulfide corrosion would be different than manhole wall rehabilitation to eliminate surface water inflow.

TABLE 10.3 PRECAST MANHOLE DEFECTS

Manhole Component	Component Defect
Cover Usually cast iron with machined surfaces to ensure a tight fit. May be bolted, gasketed. Can have pick/vent holes or hidden pick holes.	• Open vent or pick holes allowing inflow when covered with water. • Bearing surface worn, corroded or dirty. • No gaskets or bolts for gasketed and bolted covers. • Poor fitting. • Cracked or broken.
Frame Also called ring or casting. Usually cast iron with machined surfaces for cover.	• Bearing surface worn or corroded. • No gasket for gasketed frames. • Cracked or broken. • Frame offset from chimney.
Frame Seal Interface (contact point) between the frame and chimney section.	• No seal. • Leaking frame/chimney joint. • Cracked or missing seal.
Chimney Transition between cone section and frame. Usually precast concrete adjusting rings, or courses of brick. Purpose is to adjust frame to grade.	• Cracked or broken adjusting rings. • Deteriorated.
Cone Section Transition section between adjusting rings or casting and precast wall section.	• Cracked, loose and missing mortar. • Leaking cone and/or wall joint. • Leaking lifting hole. • Deteriorated.
Wall Section Circular section also referred to as barrel section.	• Cracked, loose and missing mortar. • Leaky wall joint. • Leaking lifting hole. • Deteriorated. • Unsafe or leaking steps.
Pipe Seal Area between barrel section and sewers entering or leaving manhole.	• Cracked, loose mortar, none. • Leaking. • Deteriorated.
Bench Area at bottom of manhole sloped toward the invert.	• Cracked, loose or missing pieces. • Leaking channel/bench seal. • Deteriorated.
Channel Area at bottom of manhole shaped to ensure smooth flow through the manhole.	• Cracked, loose or missing pieces. • Leaking channel/bench joint. • Deteriorated. • Poor hydraulics.
Invert Bottom of the channel.	• Cracked, loose or missing pieces. • Deteriorated. • Poor hydraulics.

TABLE 10.4 MANHOLE REHABILITATION METHODS

Rehabilitation Method	Advantage	Limitation
Seal pick holes	Inexpensive.	Loss of venting/fresh air.
Replace cover only	Proper fit, eliminates holes and cover leakage.	Loss of venting/fresh air. May not get good seal with existing frame at bearing surface if frame is warped.
Seal covered with:		Gaskets can become loose and leak, and can drop into sewer. Difficult to install properly. Prevents easy access to manhole. Raises cover slightly. Cover may be difficult to remove when sealed with mastic.
Gasket	Eliminates inflow, inexpensive.	
Asphalt mastic	Eliminates inflow, inexpensive.	
Replace manhole frame and cover	Improves service life and alignment, adjusts grade and eliminates leakage.	Excavation required, cost. Pavement replacement.
Seal existing cover, or install insert	Eliminates inflow and stops rattle. Cost.	Raises cover slightly. Insert may drop into sewer.
Seal frame/chimney joint and chimney above cone	Eliminates inflow while allowing movement of the frame.	Minor reduced access in chimney, cost.
Realign existing frame	Provides better access.	Excavation required.
Remove and/or replace step	Improves access, safety and eliminates leakage.	Installation difficulty, cost. Safety concerns.
Rehabilitate manhole structure by plugging, patching and coating or sealing	Eliminates leakage and provides corrosion protection.	Will not rehabilitate badly deteriorated or structurally unsound manholes.
Rebuild manhole, chimney and cone section	Rehabilitates badly deteriorated or structurally unsound chimney and cone section.	Excavation and pavement replacement or surface restoration required. Cost.
Structural relining	Restores structural integrity. Long life.	Reduction of diameter, higher initial cost. Loss of steps.
Chemically grout manhole structure	Eliminates infiltration and fills voids in surrounding soil. Long life, cost. Can be done in-house with minimum equipment investment.	Non-structural repair.
Coatings	Economical. Quick to apply. No disruption to traffic or other utilities. Can be applied to uneven surfaces.	Applicable only to operator-entry sewers and manholes. Bypass required. Surface imperfection pinholes, blowholes. Surface preparation required. Some contractors are inexperienced with products. Surface repairs often required prior to application.

Most manholes allow inflow because of holes in the covers. These holes can be plugged in some manner or a manhole lid insert can be installed. The insert catches any water and does not allow the water to go into the sewer system. Figure 10.60 illustrates internal and external manhole seals which provide watertight barriers between the cone section and the manhole frame, and Figure 10.61 shows plugs used for plugging pick holes in the manhole. Newer, watertight frames and covers have boltdown covers which rest on a gasket set in the frame (Figure 10.62) and effectively seal out surface water.

Damage to the seal of the frame and grade adjustment area is a major source of inflow. This type of damage usually occurs whenever road work is initiated or during initial construction of the roads. Because the frame area lies under the paving close to the interface with the subgrade, a large volume of rainfall runoff is available and seeks the easy entry into the manhole. Several methods of in-place repair have been developed to handle this problem (Figure 10.63). Rubber seals which are held in place by tension bands can eliminate this excessive inflow. Also, chemical grouting with additional sealing of the open area has been very successful in elimination of inflow at this point.

Leaks in the sidewalls and base of manholes can be repaired by a combination of chemical grouting and coatings. The areas around a manhole are grouted by drilling holes through the manhole wall and pumping grout into the soil outside the manhole (Figure 10.64). This grout sets up in a way similar to pipe grouting and keeps water from entering the manhole in the joint or other damaged areas.

10.361 Manhole Coatings

When manhole interior walls have suffered deterioration due to hydrogen sulfide attack, they can be repaired with coating processes. Coating systems have been applied to sewer pipes and manholes since the 1960s, with mixed results. The inconsistency of results has been due to contractors' lack of experience with the product, improper surface preparation, and incorrect coating specifications for the particular application. Some considerations for the application of coatings should include bypassing of wastewater, cleaning and preparing the concrete surface, allowing the concrete surface to dry, application of coating by brush or spray, allowing the coating to cure, resuming wastewater flow, and testing and inspection methods for quality control.

INTERNAL MANHOLE CHIMNEY SEAL

(U.S. Patent #4,305,679 and #4,469,467)

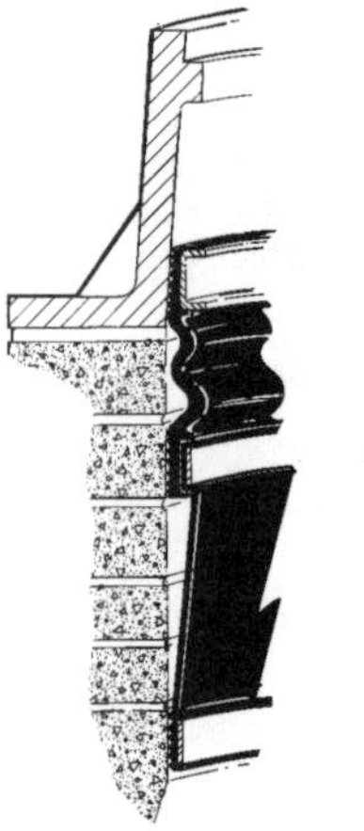

RUBBER SLEEVE

The nominal $8\frac{3}{8}$ inch wide sleeve is made of rubber which conforms to the physical requirements of ASTM C 923 with a minimum $\frac{3}{16}$ inch thickness for durability and resistance to puncturing or tearing. It is double pleated to provide for a minimum 2 inch vertical or horizontal movement before stretching the material. Flexibility of this material allows one size to fit a frame/chimney diameter range of more than 2 inches.

EXTENSION

The 7" and 10" wide rubber extensions are made from the same material and have the same minimum thickness as the sleeve. The top portion fits into the sleeve's lower band recess, under the lower band.

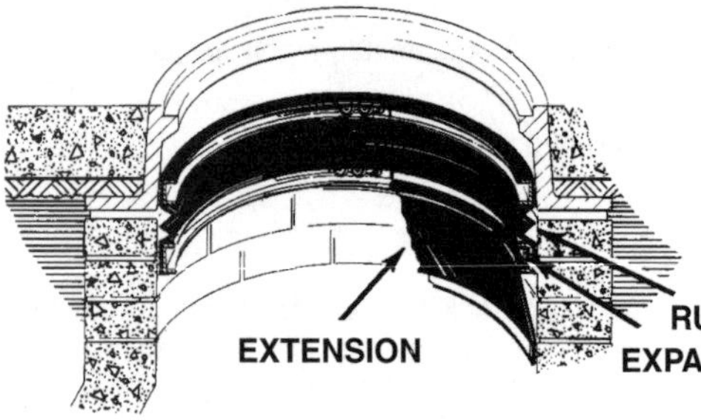

EXPANSION BANDS

The one piece channeled expansion bands are $1\frac{3}{4}$ inches wide and are fabricated from high quality, corrosion resistant, 16 gauge stainless steel conforming to the requirements of ASTM A-240 type 304. The 12 inch long adjustment slot in the band provides for $2\frac{1}{2}$ inches of diameter range. An easy to use mechanical expansion tool quickly expands the band to compress the rubber sleeve against the manhole frame and chimney. Once expanded, the band is locked in place by the tightening of 2 self locking stainless steel studs, providing a flexible watertight seal.

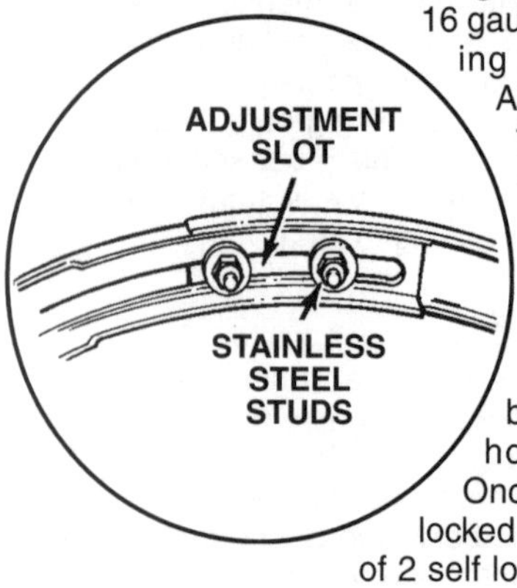

EXTERNAL MANHOLE CHIMNEY SEAL

(U.S. Patent #4,475,845)

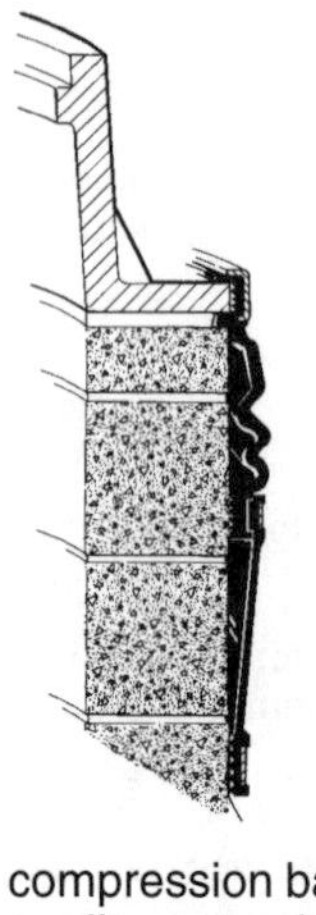

RUBBER SLEEVE

The nominal 6 or 9 inch wide sleeve is made of rubber which conforms to the requirements of ASTM C-923, with a minimum $\frac{3}{16}$-inch thickness for durability and resistance to puncturing or tearing. The corrugated shape allows for movement within the confines of the surrounding backfill and also provides for a full 2 inches of vertical movement without stretching the material. Spacers are included to form the $\frac{3}{4}$ inch thick mortared recess under the frame into which that portion of the sleeve fits.

TOP COMPRESSION BAND

This one piece uneven-legged channeled compression band is $1\frac{1}{4}$ inches wide and fabricated from high quality, corrosion resistant, 16 gauge stainless steel conforming to ASTM A-240 type 304. The band compresses the rubber sleeve against the edge of the manhole frame base flange while extending both over and under this flange to lock the sleeve in place and provide a positive watertight seal.

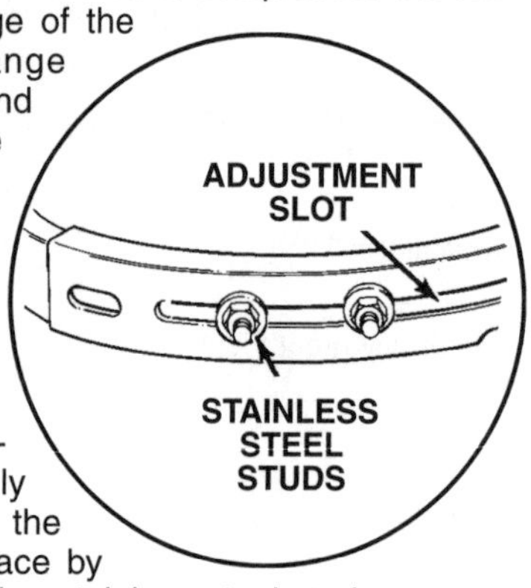

The 10-inch long slot and multiple stud holes in the band provides for more than 4 diameter inches of usable range. Once tightened by means of a specially designed mechanical tool, the band is easily locked in place by the tightening of 2 self locking stainless steel studs.

BOTTOM COMPRESSION BAND

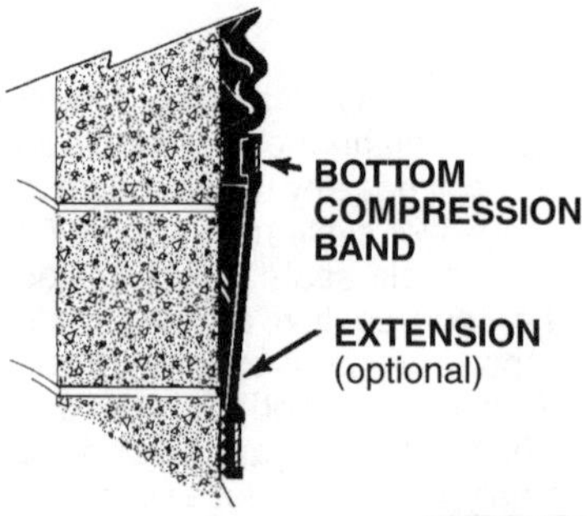

This flat 1 inch wide compression band is fabricated from the same 16 gauge stainless steel, has an 8 inch diameter range and uses the same tightening mechanism as the top band.

EXTENSION

The 8" wide rubber extension is made from the same material and has the same thickness as the chimney seal itself. The top portion has a band recess and is shaped to fit into the bottom band recess of the sleeve. The bottom portion is shaped to receive a second bottom band.

Fig. 10.60 Internal and external manhole chimney seals

(Permission of Cretex Specialty Products)

Style No. 1 — Can be removed with pliers for lid removal or to check interior manhole air quality. Available to fit ¾″, ⅞″, 1″, 1⅛″ and 1¼″ diameter holes in lids 1″ or less in thickness.

Style No. 2 — Designed to relieve sewer gases automatically. Must be installed between raised ribs in lid. Available to fit 1″ diameter holes in lids 1″ or less in thickness.

Style No. 3 — For use in rough or irregularly shaped holes or where easy removal is not desired. The tighter the nut is tightened, the more the rubber expands to the configuration of the hole. Available in five sizes to fit ½″ thru 1½″ diameter holes.

Fig. 10.61 Manhole lid plugs

(Permission of Cretex Specialty Products)

Fig. 10.62 Corbel/grade ring repair

Fig. 10.63 Tucking saturated sewer jute under frame-corbel area to eliminate inflow

Fig. 10.64 Chemical grouting

Surface preparation (Figure 10.65) is extremely important to ensure the coating system will result in future protection. Thoroughly clean the walls to remove grease, sludge, mineral deposits, loose concrete or mortar, and debris. Prepare the surface by water or abrasive blasting, repairing cracks, and applying a moisture-tolerant primer (if the concrete cannot be dried sufficiently). Crack repair by injection is commonly used. In addition, a pH analysis of the original concrete surface by petrographic examination should be conducted. pH adjustment of the concrete may be required prior to application of the coating system (the pH of unattacked concrete should be 11 or greater). Fresh concrete can be applied (Figure 10.66) to obtain a good base and then a top coat is applied. If voids and leaks are present, they should be filled or plugged with an appropriate patching or grouting material prior to applying the final surface. Patch all rough areas to provide a smooth surface, which will receive the coating to be applied.

The choice of coating material depends on the chemistry of the particular situation encountered. Various epoxy, acrylic,

Fig 10.65 Surface preparation (surface cleaning)
(Permission of J. A. CRAWFORD CO./SAUEREISEN)

Fig. 10.66 Surface preparation (applying smooth surface)
(Permission of J. A. CRAWFORD CO./SAUEREISEN)

plastic, and polyurethane based coatings are available that can provide protection from hydrogen sulfide and are dependable coating materials. These waterproof and corrosion-resistant coatings can be applied to brick block and precast concrete manholes and bases.

Most coatings are either brushed or rolled onto concrete surfaces by hand (Figure 10.67) or sprayed on mechanically (Figure 10.68). Spray-on application requires pressures of up to 3,000 psi, which is twice the pressure used for conventional airless spraying. Spraying is excellent for coating uneven surfaces and is much faster than brushing methods for some products.

After applying the final coating material, the manhole should be inspected and tested. Inspections and tests include: thickness WFT/DFT; holiday (spark) testing (Figure 10.69); vacuum and exfiltration; adhesion testing (Elcometer) (Figure 10.70); and visual inspection. The types of tests and inspections may vary depending on the project. Some of the advantages and limitations of the use of coatings are listed in Table 10.4, "Manhole Rehabilitation Methods."

Fig 10.67 Application of coating (hand applied)

(Permission of J. A. CRAWFORD CO./SAUEREISEN)

Fig. 10.68 Application of coating (spray applied)
(Permission of J. A. CRAWFORD CO./SAUEREISEN)

Fig. 10.69 Surface test (spark testing)
(Permission of J. A. CRAWFORD CO./SAUEREISEN)

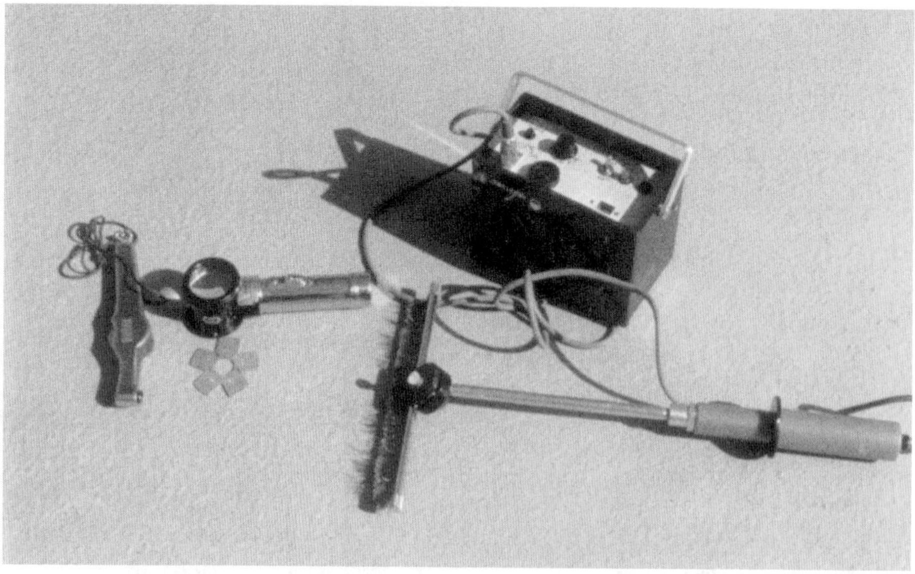

Fig 10.70 Surface coating testing equipment
(Permission of J. A. CRAWFORD CO./SAUEREISEN)

QUESTIONS

Write your answers in a notebook and then compare your answers with those on page 355.

10.3W What type of coating can be used for corrosion protection?

10.3X List three methods of controlling I/I by chemical grouting of house service connections.

10.3Y How should the walls of manholes be prepared before a protective coating is applied to the walls?

10.4 ADDITIONAL READING

INSPECTOR HANDBOOK. Obtain from NASSCO, 1108 Sheller Avenue, Suite 5, Chambersburg, PA 17201. Price to members, $25.00; nonmembers, $75.00.

Please answer the discussion and review questions next.

DISCUSSION AND REVIEW QUESTIONS

Chapter 10. SEWER RENEWAL (REHABILITATION)

Write the answers to these questions in your notebook. The purpose of these questions is to indicate to you how well you understand the material in the chapter.

1. What would cause sewers to lose their structural strength?
2. What techniques are available to determine the condition of a sewer system?
3. How can infiltration and inflow be measured?
4. Describe the ideal conditions for location of an open-channel flow metering manhole.
5. Smoke testing is an effective means of detecting inflow from what types of sources?
6. What operational problems can be directly attributed to defective collection systems?
7. How can a utility establish rehabilitation priorities?
8. Under what conditions should sewer rehabilitation by pipeline replacement not be considered?
9. What problems are common to all types of sewer construction?
10. What conditions must be suitable for chemical grouting to be successful?
11. What are the advantages and limitations of using shotcrete for sewer rehabilitation?
12. What type of precautions must be taken when erecting an Insituform insertion platform?
13. Why should the Insituform resin material be kept away from direct exposure to sunlight?
14. How is a seal produced between the sewer and an Insituform liner in a manhole?
15. How are house service connections located in a sewer newly lined by the Insituform process?
16. What are the advantages and disadvantages of the Insituform method for rehabilitating sewers?
17. What are the reasons for lining a sewer?
18. Why should the public be notified and what should the public be told before a sewer lining project is started?
19. How would you insert a liner in a main line?
20. How are sections of liner pipe joined together?
21. Why should a lateral be lined?
22. How is the space between the manhole and the liner sealed?
23. Why is manhole rehabilitation required?
24. How can the walls of manholes that have deteriorated due to hydrogen sulfide attack be repaired?

SUGGESTED ANSWERS

Chapter 10. SEWER RENEWAL (REHABILITATION)

Answers to questions on page 282.

10.0A Sewers need rehabilitation to restore structural integrity and to control excessive additional inflow and infiltration flows during periods of rainfall or high groundwater conditions.

10.0B A physical survey of a collection system must be conducted to determine the condition of the system.

10.0C The present sewer maintenance program must be reviewed before the start of a sewer rehabilitation program to ensure a coordinated effort of maintenance and rehabilitation.

Answers to questions on page 294.

10.1A The main reasons for the emphasis on reducing hydraulic inflow/infiltration loads on sewer systems and treatment facilities is that these loads increase treatment costs, result in the bypassing of untreated wastewater to the environment and can result in structural failure of weakened collection systems.

10.1B The flow components in a sanitary sewer are:

1. Base flow,
2. Infiltration, and
3. Inflow.

10.1C The possible surcharge potential for a metering location must be determined because surcharging results in uneven flow or backflow conditions which may produce faulty flowmeter readings and possible damage to flow metering and recording equipment.

10.1D Smoke testing is not effective during high groundwater conditions or in sewers where sags or water traps are suspected or where sewers flow full. Smoke testing is also ineffective when the ground is saturated, frozen or covered with snow, and under windy conditions.

10.1E Dye-water flooding is applicable for:

1. Storm drains;
2. Drainage ditches, streams and water ponding areas;
3. Yard drains, areaway drains, foundation drains, roof drains and abandoned building sewers; and
4. Problem areas identified in other tests for which additional data are needed.

10.1F Closed-circuit television is the most effective method for identification and quantification of defects in sewer pipes. Other methods include visual inspection, lamping, smoke testing, air testing, and flow measurements.

Answers to questions on page 297.

10.2A The four distinct courses of action that a utility may take with regard to a sewer system are: (1) maintain, (2) replace, (3) relieve, and (4) rehabilitate.

10.2B When determining the appropriate course of action, a utility should perform a quantitative evaluation of: (1) sewer condition, (2) consequence of failure, (3) sewer capacity, (4) performance of sewers, and (5) costs.

10.2C Costs that should be considered when evaluating the cost effectiveness of an I/I rehabilitation program include:

1. Survey and rehabilitation costs,
2. Construction and capital costs, and
3. Operation and maintenance costs.

A comparison of the cost to transport and treat the rainfall induced I/I with the cost to eliminate it will yield the most cost-effective solution.

Answers to questions on page 303.

10.3A The oldest and most common method of sewer rehabilitation is excavation and replacement.

10.3B The disadvantages of sewer rehabilitation by excavation and replacement are costs, traffic disruption, impact on other utilities, and disruption of lifestyle of people living and working in the area.

10.3C The most common method of sealing leaking pipe joints and circumferential cracks is the use of chemical grouting.

10.3D The success of chemical grouting can be tested by a low-pressure air test or by digging up a joint to see if the grout performed as expected.

Answers to questions on page 318.

10.3E Trenchless rehabilitation methods are used when soil or groundwater conditions are difficult using open cut construction and in downtown areas where disruption to business and the public are factors.

10.3F Different types of trenchless pipeline rehabilitation methods include: (1) cured-in-place pipe, (2) sliplining, and (3) in-line replacement.

10.3G Shotcrete is sometimes referred to as gunite and is a mixture of sand, cement and water applied by air pressure.

Answers to questions on page 327.

10.3H Insituform is a method of pipe liner rehabilitation consisting of the installation of a new pipe within the old pipe without excavation. The process involves the use of a polyester-fiber felt tube, lined on one side with polyurethane and fully impregnated with a liquid thermal setting resin.

10.3I Materials and major pieces of equipment required for a typical Insituform installation include an insertion platform tube and elbow, portable mixer, transfer pump, vacuum pump, wet down conveyor system, lay-flat hose, hold back rope, water supply line, water control valve, circulating pump, boiler and heat exchanger, Insitucutter, thermocouples, and polyurethane-coated, polyester-fiber felt liner, polyester resin, and catalyst.

10.3J The purpose of the Insitucutter is to cut out lining material in the areas in a sewer defined as service taps.

Answers to questions on page 333.

10.3K After the preparatory cleaning is completed on an Insituform sewer rehabilitation job, the internal inspection and evaluation team makes an inspection of the line to verify that the cleaning is complete and also to record on a TV log the location of all line deficiencies and service connections. In addition to the TV log, the line is videotaped.

10.3L In the Insituform lining technique, the inversion platform is erected over the point of insertion of the line into the sewer being lined, usually over a manhole.

10.3M A liner can be tested for damages after it has been unpacked from its shipping container by attaching a vacuum pump to the bag and evacuating the entrapped air from the felt liner. It would be impossible to draw and maintain a vacuum with the system if the liner were damaged.

10.3N Safety precautions that should be taken during the installation and curing of an Insituform job include protection of workers, spectators, and equipment from hot water. Any time workers enter manholes, all confined space procedures must be followed. The strong volatile fumes created during this process must be avoided.

Answers to questions on page 336.

10.3O Liners can be an effective method of repairing or reconditioning a sewer system by providing a new, smooth pipe to carry the wastewater. This new pipe does not leak (no infiltration or exfiltration) and prevents root intrusion.

10.3P A letter to homeowners explaining a sliplining project should indicate

1. Why the project is necessary,
2. What is planned,
3. What is expected to be accomplished, and
4. Direct costs to the homeowner.

The letter should state that the homeowner may be temporarily inconvenienced during construction and also explain the benefits that will result from the completed project. Also include a legally prepared form for the homeowner to sign giving the utility permission to perform any necessary repair work that may have an impact on the owner's property.

10.3Q A sewer should be filmed or videotaped before sliplining to locate laterals, any misalignment and any other discrepancy which may need to be corrected before the actual lining takes place.

10.3R Equipment needed to weld polyethylene pipes together includes fusion joiner jigs; mobile carts; a butt fusion tool or squaring tool; heating faces for heating saddles; tapping or coring tools and accessories for adapting them to the various sizes of pipe that will be used; and, the pulling heads.

10.3S Three crews are recommended for a sliplining project:

1. One crew assembles the liner,
2. An excavating crew digs a pull trench at the predetermined location, and
3. Another crew pulls liner into place.

10.3T The liner should be marked at 50-foot intervals before being pulled into the sewer so that if the pipe hangs up in the line, the pulling head location can easily be found and dug up.

Answers to questions on page 341.

10.3U When sliplining laterals, dig holes where the lateral connects to the main line and where the lateral enters the house.

10.3V The space between the liner and the manhole is sealed to prevent infiltration and/or exfiltration at the manhole.

Answers to questions on page 353.

10.3W Polyurethane resin coatings can be used for corrosion protection.

10.3X Three methods of chemical grouting to control I/I in house service connections are to pump the service full of grout and use an auger to reopen the service, use the packer with tube known as a sewer sausage, and use a miniaturized camera-packer unit.

10.3Y The walls of manholes should be thoroughly cleaned to remove grease, sludge, mineral deposits, loose concrete or mortar, and debris before a protective coating is applied. Also all rough areas should be patched to provide a smooth surface, which will receive the coating to be applied.

CHAPTER 11

SAFETY/SURVIVAL PROGRAMS FOR COLLECTION SYSTEM OPERATORS

by

Glenn Davis

Revised by

Rick Arbour

Gary and Sandi Batis

Joe Teeples

TABLE OF CONTENTS

Chapter 11. SAFETY/SURVIVAL PROGRAMS FOR COLLECTION SYSTEM OPERATORS

OBJECTIVES

Chapter 11. SAFETY/SURVIVAL PROGRAMS FOR COLLECTION SYSTEM OPERATORS

Following completion of Chapter 11, you should be able to:

1. Demonstrate your *AWARENESS* of the hazards of working in the collection system environment by performing your assigned duties safely,
2. Identify potential and existing hazards,
3. Develop and establish a safety/survival program,
4. Outline the objectives and benefits of a safety/survival program,
5. List the responsibilities of the different staff levels in a collection system agency that are responsible for a safety/survival program,
6. Prepare and conduct tailgate safety sessions and monthly safety meetings,
7. Develop and implement appropriate safety/survival program policies, and
8. Accurately complete accident forms and properly maintain records.

WORDS

Chapter 11. SAFETY/SURVIVAL PROGRAMS FOR COLLECTION SYSTEM OPERATORS

ANAEROBIC (AN-air-O-bick) ANAEROBIC

A condition in which atmospheric or dissolved molecular oxygen is *NOT* present in the aquatic (water) environment.

ANAEROBIC (AN-air-O-bick) DECOMPOSITION ANAEROBIC DECOMPOSITION

The decay or breaking down of organic material in an environment containing no "free" or dissolved oxygen.

CFR CFR

Code of **F**ederal **R**egulations. A publication of the United States Government which contains all of the proposed and finalized federal regulations, including safety and environmental regulations.

COMPETENT PERSON COMPETENT PERSON

A competent person is defined by OSHA as a person capable of identifying existing and predictable hazards in the surroundings, or working conditions which are unsanitary, hazardous or dangerous to employees, and who has authorization to take prompt corrective measures to eliminate the hazards.

CONFINED SPACE CONFINED SPACE

Confined space means a space that:

A. Is large enough and so configured that an employee can bodily enter and perform assigned work; and

B. Has limited or restricted means for entry or exit (for example, manholes, tanks, vessels, silos, storage bins, hoppers, vaults, and pits are spaces that may have limited means of entry); and

C. Is not designed for continuous employee occupancy.

(Definition from the Code of Federal Regulations (CFR) Title 29 Part 1910.146.)

CONFINED SPACE, NON-PERMIT CONFINED SPACE, NON-PERMIT

A non-permit confined space is a confined space that does not contain or, with respect to atmospheric hazards, have the potential to contain any hazard capable of causing death or serious physical harm.

CONFINED SPACE, PERMIT-REQUIRED (PERMIT SPACE) CONFINED SPACE, PERMIT-REQUIRED (PERMIT SPACE)

A confined space that has one or more of the following characteristics:

- Contains or has a potential to contain a hazardous atmosphere,
- Contains a material that has the potential for engulfing an entrant,
- Has an internal configuration such that an entrant could be trapped or asphyxiated by inwardly converging walls or by a floor which slopes downward and tapers to a smaller cross section, or
- Contains any other recognized serious safety or health hazard.

(Definition from the Code of Federal Regulations (CFR) Title 29 Part 1910.146.)

DANGEROUS AIR CONTAMINATION DANGEROUS AIR CONTAMINATION

An atmosphere presenting a threat of causing death, injury, acute illness, or disablement due to the presence of flammable and/or explosive, toxic or otherwise injurious or incapacitating substances.

A. Dangerous air contamination due to the flammability of a gas or vapor is defined as an atmosphere containing the gas or vapor at a concentration greater than 10 percent of its lower explosive (lower flammable) limit.

B. Dangerous air contamination due to a combustible particulate is defined as a concentration greater than 10 percent of the minimum explosive concentration of the particulate.

C. Dangerous air contamination due to the toxicity of a substance is defined as the atmospheric concentration immediately hazardous to life or health.

ENGULFMENT ENGULFMENT

Engulfment means the surrounding and effective capture of a person by a liquid or finely divided (flowable) solid substance that can be aspirated to cause death by filling or plugging the respiratory system or that can exert enough force on the body to cause death by strangulation, constriction, or crushing.

OSHA (O-shuh) OSHA

The Williams-Steiger **O**ccupational **S**afety and **H**ealth **A**ct of 1970 (OSHA) is a federal law designed to protect the health and safety of industrial workers and collection system operators. The Act regulates the design, construction, operation and maintenance of industrial plants and wastewater collection and treatment facilities. The Act does not apply directly to municipalities, *EXCEPT* in those states that have approved plans and have asserted jurisdiction under Section 18 of the OSHA Act. *HOWEVER, CONTRACT OPERATORS AND PRIVATE FACILITIES DO HAVE TO COMPLY WITH OSHA REQUIREMENTS.* Wastewater collection systems have come under stricter regulation in all phases of activity as a result of OSHA standards. OSHA also refers to the federal and state agencies which administer the OSHA regulations.

PATHOGENIC (PATH-o-JEN-ick) ORGANISMS PATHOGENIC ORGANISMS

Bacteria, viruses, cysts, or protozoa which can cause disease (giardiasis, cryptosporidiosis, typhoid, cholera, dysentery) in a host (such as a person). There are many types of organisms which do *NOT* cause disease and which are *NOT* called pathogenic. Many beneficial bacteria are found in wastewater treatment processes actively cleaning up organic wastes.

CHAPTER 11. SAFETY/SURVIVAL PROGRAMS FOR COLLECTION SYSTEM OPERATORS

11.0 SAFETY/SURVIVAL

— SEWER FUMES KILL PAIR, STOP WOULD-BE RESCUERS
— JOB SURVEY REVEALS DISMAL PERFORMANCE BY WASTEWATER INDUSTRY
— SEWER WORKER KILLED
— DITCH COLLAPSE KILLS ONE
— POISONOUS GAS KILLS 7, INJURES 28
— SAFETY SURVEY REVEALS FATALITY INCREASES
— BACK INJURIES COST 10 BILLION DOLLARS IN WORKERS' COMPENSATION CLAIMS
— MAN BURIED ALIVE
— COLLECTION SYSTEMS LEAD RISE IN INJURY RATES
— WASTEWATER INDUSTRY RETAINS #1 STATUS AS THE LEAST SAFE INDUSTRY

FIGMENTS OF THE IMAGINATION? — NO! These are actual statements taken from a variety of recent sources which underscore why this chapter is titled "Safety/Survival Programs for Collection System Operators."

Throughout the world most living things devote considerable time and energy to survival. At every level of life, whether it be human beings, animals, insects, fish, or fowl, survival occupies a great part of every living hour. Intelligent people not only try to protect themselves from injury or death, but they also strive to protect their fellow human beings and especially their immediate families and co-workers. An accident is a tragedy for your family and others who depend on and love you, as well as for yourself.

This chapter and Chapter 4, "Safe Procedures," are the two most important chapters in the entire course since they deal directly with your ability to survive by avoiding a fatal or disabling accident when working in the collection system environment.

No other skill more clearly demonstrates the professionalism of the collection system operator than the ability to carry out a work assignment safely. An amateur may perform all of the tasks of a collection system operator, but a truly professional operator will do a better job, will do it more quickly, and will always do it more safely.

11.00 The Numbers Game

Obviously, accidents have a significant economic impact as well as taking a toll in human suffering. For example, workers' compensation rates, liability insurance, disability insurance, and other similar costs are directly related to our performance in an agency or company. In recent years, all of these costs have skyrocketed, imposing an economic hardship on many agencies and communities. These are budgetary resources which are no longer available for use in other areas such as operations, maintenance, training, equipment and salaries.

The National Safety Council compiles accident and fatality information and reports that 14 people die at work on an average day. More than 10,400 people are disabled at work every year. In 1998 the total cost to the Nation, employers, and individuals for work-related injuries and deaths was more than $127 billion dollars.

The injury frequency rate is measured as the number of lost-time work days per 100 employees. This information is tracked by the U.S. Bureau of Labor Statistics. The frequency rate has dropped from a high in 1973 of 7.3 incidents per 100 workers to a low in 2001 of 2.7 incidents per 100 workers. As more firms and facilities develop and implement safety programs, the number is anticipated to drop even further.

Experience on the job affects the chances of your having an accident. It has been reported that workers with less than 2 years of on-the-job experience have 1 chance in 7 of having an accident, while those with over 14 years of experience have 1 chance in 15 of having an accident. Similarly, training and certification of operators reduce the chances of an accident. Uncertified operators have a 1 in 9 chance of being injured, while certified operators have a 1 in 12 chance of being injured on the job.

In recent years a considerable amount of information has been compiled regarding fatalities and injuries in the collection system field. The findings only confirm what many have known for some time:

- Mayors, city council members, city administrators, department heads, safety department personnel or consulting engineers are not usually the victims of accidents and fatalities.
- Collection system operators, line supervisors or others who are directly involved and exposed to hazards while doing everyday routine tasks are the ones who experience the injuries and fatalities.
- Formal procedures and/or safety equipment generally are not being used when accidents occur, although they are frequently available for the injured operators.

- Time and resources for investigations, analysis and implementation of safety programs were always available after the incident, that is, money, equipment, and training.
- One or more of three common factors are usually missing when an accident occurs:
  1. Awareness,
  2. Responsibility, and/or
  3. Commitment.

All of the incidents listed at the beginning of this chapter could probably have been prevented if the operators involved had been aware of the hazards, had assumed some responsibility for their own safety/survival, and had made a commitment to that responsibility.

In no other industry in the United States today is a safety/survival program more critical than in collection system agencies, as our injury records so clearly demonstrate. The cost of human suffering is incalculable. Consider the effect, for example, on you and your family if you had to spend the rest of your life with only one eye because you failed to wear safety glasses while working around a pump and the coupling happened to disintegrate while you were inspecting it. Or what if a wrench drops out of your co-worker's pocket while the operator is leaning over the entrance tube to the lift station? The wrench strikes you in the head at the bottom of the entrance tube, traveling at about 75 miles per hour. How much pain and anguish could you have spared yourself and your family by taking the time to put on a hard hat?

11.01 Professionalism

Safety/survival is directly related to your level of professionalism which in turn is directly related to knowledge and ultimately certification. This is substantiated by the figures reported in the 22nd Annual Water Pollution Control Federation (WPCF, now WEF) Safety Survey Report, which is based on information submitted by 1,082 agencies in the United States and Canada in 1988. This analysis reports:

- 43% of all disabling injuries occurred to employees with 0-5 years of wastewater experience,
- 73% of the injured had 0-10 years' experience, and
- 71% of all injuries were attributed to uncertified employees.

As your collection system continues to expand, it is imperative that your knowledge and professionalism also continue to increase. Nowhere is this more important than in your Collection System Safety/Survival Program.

The overall objective of this chapter is to make you *AWARE* of the hazards of working in the collection system environment. The other two factors, *RESPONSIBILITY* and *COMMITMENT*, will be up to you. The rest of this chapter provides additional information regarding details of collection system accidents, and a significant portion of the chapter will be devoted to establishing a comprehensive safety/survival program in your collection system.

QUESTIONS

Write your answers in a notebook and then compare your answers with those on page 399.

11.0A What are three common factors that are important to prevent accidents?

11.0B What skill most marks the professionalism of the collection system operator?

11.0C What is the overall objective of this chapter?

11.1 REVIEW OF COLLECTION SYSTEM HAZARDS

Multiple hazards exist in the performance of the collection system operator's routine daily tasks and work assignments. Most of these hazards and appropriate precautions are described fully in Chapter 4, "Safe Procedures." Awareness of these hazards will enable you to take responsibility for your own safety. The following brief review is intended to refresh your memory about the material in Chapter 4.

1. **SLIPS** Wet wells, metering flumes or other structures with live flows can be and usually are extremely slippery because of the accumulation of slime and/or the very humid atmosphere. Emergencies, pump station failures, or other difficult situations which are often caused by bad weather require extra caution. For those who work in the northern climates, snow, ice, frost and below-freezing temperatures demand extra caution in the wintertime.

2. **FALLING OBJECTS** In lift stations and manholes, take precautions against falling objects. When working in a prefabricated type of lift station, never stand directly under the entrance tube, particularly when raising or lowering materials/supplies/tools. Even if such objects are

placed in a container of some type, they could tip out and fall or the container could bump the sidewalls and dislodge dirt or debris. Similarly, when operators are working in manholes, take care not to place tools and materials close to the manhole rim where they can be accidentally kicked into the manhole.

3. **INFECTIONS AND INFECTIOUS DISEASES** We know that collection systems are not the cleanest places to work. Every collection system operator must realize that wastewater is contaminated with possibly every kind of infectious microorganism that can be carried in wastewater. Like the medical doctor, the collection system operator may be exposed to many of the *PATHOGENIC ORGANISMS*[1] from the entire community in which you work. Also like the doctor, personal health depends to a great extent upon your personal cleanliness. Both you and the doctor must wash your hands frequently, especially before smoking or eating, to achieve a low incidence of infection. Fortunately, neither profession suffers an unusually high incidence of infection. This may be because both recognize the importance of personal cleanliness. It is also possible that collection system operators develop immunities to infectious diseases or that the records reporting diseases are inadequate. Diseases or infections that collection system operators could come in contact with on the job include:

    Typhoid and paratyphoid fever
    Cholera
    Bacillary dysentery
    Amoebic dysentery
    Roundworm and other worm infections
    Tuberculosis
    Poliomyelitis
    Infectious hepatitis
    Leptospirosis (Weil's disease)
    Tetanus

How can you determine if you have contracted one of these diseases? The best indication is that you don't feel well. If you have headaches, feel sick to your stomach, have diarrhea, or feel feverish or sleepy, you should see your physician. Prompt action by your physician can save you a lot of pain and trouble in the future and also help protect your family, friends and your fellow operators. Wearing protective clothing and using proper hygiene practices are the best ways to prevent getting these diseases.

There are three basic routes that may lead to infection:

- Ingestion through splashes, contaminated food, or cigarettes,
- Inhalation of infectious agents or aerosols, and
- Infection due to an unprotected cut or abrasion.

Examples of the major routes of infection are:

- Ingestion: Eating, drinking, or accidentally swallowing pathogenic organisms (for example, Hepatitis A),
- Inhalation: Breathing spray or mist containing pathogenic organisms (for example, common cold), and
- Direct contact: Entry of pathogenic organisms into the body through a cut or break in the skin (for example, tetanus).

Collection system operators often come in physical contact with raw wastewater through the course of their daily activities. Even when direct physical contact is avoided, the operator may handle objects that are contaminated. Cuts and abrasions, including those that are minor, should be cared for properly. Open wounds invite infection from many of the viruses and bacteria present in wastewater.

Ingestion is generally the major route of wastewater operator infection. The common practice of touching the mouth with the hand will contribute to the possibility of infection. Operators who eat or smoke without washing their hands have a much higher risk of infection. Most surfaces near wastewater equipment are likely to be covered with bacteria or viruses. These potentially infectious agents may be deposited on surfaces in the form of an aerosol or may come from direct contact with the wastewater. A good rule of thumb is to never touch yourself above the neck whenever there is contact with wastewater. Methods to prevent ingestion of pathogenic organisms are listed below:

- Wash hands frequently and never eat, drink, or use tobacco products before washing hands,
- Avoid touching face, mouth, eyes, or nose before washing hands, and
- Wash hands immediately after any contact with wastewater.

Table 11.1 is a summary of diseases and the common modes of transmission.

TABLE 11.1 SUMMARY OF DISEASES AND MODES OF TRANSMISSION

Disease	Mode of Transmission
Bacillary dysentery	Ingestion
Asiatic cholera	Ingestion
Typhoid fever	Ingestion
Tuberculosis	Inhalation
Tetanus	Wound contact
Infectious hepatitis	Ingestion
Poliomyelitis	Ingestion
Common cold	Inhalation
Hookworm disease	Skin contact
Histoplasmosis	Inhalation

Although most studies indicate that infections from specific agents are not common, operators in contact with wastewater, especially during their first few years of employment, have been known to experience increased rates of gastrointestinal or upper respiratory illnesses.

[1] *Pathogenic (PATH-o-JEN-ick) Organisms. Bacteria, viruses, or cysts which can cause disease (giardiasis, cryptosporidiosis, typhoid, cholera, dysentery) in a host (such as a person). There are many types of organisms which do NOT cause disease and which are NOT called pathogenic. Many beneficial bacteria are found in wastewater treatment processes actively cleaning up organic wastes.*

Operations and maintenance personnel must presume that biological hazards will exist at any location within the facility where wastewater is found.

BACTERIA

Bacteria do not require a living host cell to reproduce. Pathogenic bacteria are microscopic in size and are extremely common in wastewater. Because bacteria can reproduce outside the body, microorganisms can be present in large quantities in the collection system. Bacterial infections will result from their proliferation in a water environment. Because of their daily exposure to wastewater-contaminated environments, wastewater personnel have a higher incidence of exposure to pathogens than the general public. For most operators, however, the risk of developing a disease is relatively low. Proper personal hygiene is critical because infections may occur without symptoms, and antibodies to bacteria and viruses may develop even though no symptoms of illness are apparent.

VIRUSES

A virus is any of a group of ultramicroscopic agents that reproduce only in living cells. This characteristic of viruses is important because they cannot reproduce without a host cell and, therefore, will not reproduce in wastewater. The major source of viruses that are infectious to humans is from human waste that has been discharged to the sewer.

More than 100 different types of viruses are found in human waste. These viruses multiply in the living cells of the intestinal tract and end up in human feces. Because millions of viruses can be produced by an infected cell, they are found in large quantities in wastewater.

AIDS

Acquired Immune Deficiency Syndrome (AIDS) is caused by the Human Immunodeficiency Virus (HIV) that attacks the body's immune system, leaving the body susceptible to numerous diseases. The AIDS virus is a delicate virus that cannot survive for long periods of time outside of the human body. The virus exists in low concentrations in the blood of infected persons. After entering the wastewater sewer system, the virus is subjected to enormous dilution factors and harsh environments, low levels of heat, pH extremes, surfactants, and chemical agents that interfere with the survival of the AIDS virus. Operators have expressed considerable interest in the possible transmission of AIDS from human wastes such as urine, excrement, and blood that are discharged to sewer lines serviced by municipal wastewater treatment facilities. Fears have been raised over the handling of raw wastewater during routine contact and during repairs and maintenance to lift station pumps, bar screens, broken sewer lines, and clogged laterals. In addition, contact with contaminated wastewater originating from prisons, hospitals, and institutions, and uneasiness over the removal of hypodermic needles, condoms, feminine napkins, and aborted fetuses have added to the growing apprehension about disease transmission.

AIDS and Hepatitis B are both blood-borne viruses and cannot reproduce outside the human body. To be transmitted, AIDS and Hepatitis B must enter the bloodstream directly. A blood-borne virus from contaminated wastewater can gain direct access through an open wound or abrasion on the skin. Merely coming in contact with contaminated wastewater does not imply exposure to AIDS or Hepatitis B. The Centers for Disease Control (CDC) has stated that there is no scientific evidence that HIV is spread in wastewater or its aerosols. The virus has never been recovered from wastewater and it is believed that the pH, temperature, and other conditions of the collection system are not suitable to its survival. There have been no known cases of wastewater operators or plumbers who have contracted AIDS where the mode of transmission was judged to be from occupational exposure.

The scientific evidence to date indicates that AIDS cannot be contracted through occupational exposure associated with wastewater collection and treatment. Generally, infected body fluids that are discharged to sewers are immediately diluted to the point where they do not represent a significant risk to wastewater operators. The AIDS virus, in particular, is not well suited to the collection system environment and is likely to become deactivated upon contact with wastewater. Wastewater operators should, however, pay close attention to personal hygiene and exercise caution and common sense whenever they are working in and around contaminated wastewater to minimize exposure to bacteria and viruses. For further information, the Centers for Disease Control can be reached by calling their hotline at 1-800-342-AIDS. Local health departments are also useful sources of up-to-date information about AIDS.

PARASITES

A parasite lives on or in another organism of a different species from which it derives its nourishment. The organism is called the parasite's host. Parasites normally do not kill their hosts because the life of the parasite would also be terminated. In many cases, however, parasites will weaken the host or cause symptoms similar to diseases caused by bacteria or viruses. Waterborne parasites found in wastewater consist of various types of protozoa and worms. These organisms often do not survive the journey through the wastewater collection system and treatment facilities.

IMMUNIZATIONS

The Centers for Disease Control recommends that immunizations for diphtheria and tetanus be current for the general public, including all wastewater operators. Booster shots are recommended every 10 years after the initial immunizations are administered (usually during childhood years). The tetanus booster needs to be repeated if a wound or puncture becomes dirty and if a booster has not been given within five years.

Primary vaccinations for polio and typhoid are presently considered to be sufficient unless there is a regional outbreak. The preventive effect of the vaccine immune serum globulin for Hepatitis A is short lived (about three weeks) and is not routinely recommended for wastewater operators unless there has been direct exposure to wastewater splashed into an open wound or the mouth or a severe outbreak has occurred in the community. The vaccine for Hepatitis B is also not routinely recommended for wastewater operators because the risk of transmission by wastewater is extremely remote. Vaccinations are available from physicians, health clinics, and county health departments.

At the present time, no additional immunizations above those recommended by the U.S. Public Health Service for adults in the general population are advised for operators in contact with wastewater. Wastewater operators and all other adults should be adequately vaccinated against diphtheria and tetanus.

Table 11.2 illustrates schedules for immunizations recommended by the U.S. Public Health Service.

REMEMBER, common sense along with protective clothing and personal cleanliness will go a long way to protect you from infections.

4. **LACERATIONS AND CONTUSIONS (Tears and Bruises)** Much of our daily work involves working with hand tools, heavy equipment, and machines that are rusted, corroded, or nearly inaccessible. Planning each job and using the right tools in good operating condition will minimize smashed knuckles and thumbs.

5. **FALLS** Manhole rungs are notorious for giving way after a period of exposure to the frequently corrosive atmosphere where they are installed. Pump station dry and wet well access is usually by ladder rather than stairs. Atmospheric conditions, rain, and snow will all make ladder rungs slippery. Under no circumstances should you try climbing or descending ladders or manhole rungs one-handed while carrying tools/equipment in the other hand.

6. **EXPLOSIONS** Each year throughout the country there are numerous explosions in collection systems caused by a buildup of explosive gases in the sewer line or underground structure. Such buildups are very unpredictable and can come from any number of sources such as natural gas leaks, vehicle tank truck accidents that release flammable material into storm and/or sanitary sewers, or flammable material discharges from industries.

Anyone who has seen the results of an explosion either in a sewer, a digester, or other type of enclosure is well aware of the havoc that can be caused by just the right explosive mixture of air and an explosive gas. Too rich a

mixture or too lean a mixture of a combustible gas will not explode. But when the mixture of air and gas is within the level of explosive limits, a spark from an automobile, an improper tool, or even a shoe nail, as well as the more obvious open flame or cigarette, may set off the explosion. Only extremely well trained and knowledgeable persons (such as gas company experts) should consider working around such dangerous atmospheres and then only when proper ventilation and hazard removal or elimination has been verified as impossible (carbon dioxide fire extinguishers have been used by experts to dilute explosive mixtures when necessary).

Experts from the gas company may be called upon in an emergency to work in or near explosive mixtures. *THESE SPECIALLY TRAINED OPERATORS* with years of experience can be trusted to perform such dangerous work if no other feasible alternatives exist. True experts are persons who always know and work within their own limits with plans for backup or alternative action if suddenly confronted with a new emergency.

TABLE 11.2 RECOMMENDED IMMUNIZATIONS FOR WASTEWATER OPERATORS

Disease	Who Needs Immunization?	Immunization
Hepatitis A	Individuals with close personal contact with Hepatitis A	Hepatitis A immune globulin treatment
Hepatitis B	Those with household and sexual contact with carriers, and those who have had direct exposure to blood of a person known or suspected to be a carrier	Hepatitis A immune globulin treatment and Hepatitis B vaccine
Influenza	Adults 65 years or older	Annual influenza vaccine
Measles	Adults born in 1957 or later unless they have evidence of vaccination on or after their first birthday, documentation of physician-diagnosed disease, or laboratory evidence of disease	Combined measles, mumps, and rubella
Mumps	Adults, especially males who have not been previously infected	Mumps vaccine
Pneumococcal disease	Adults 65 years or older	Pneumococcal polysaccharide vaccine
Rubella	Women of childbearing age, unless proof of vaccination or laboratory evidence of immunity is available	Rubella vaccine
Tetanus and diphtheria	Adults every 10 years after initial doses and after wounds, unless it has been fewer than 5 years since last dose	TD vaccine

Special insight can be gained by discussing explosive hazards with your local fire department personnel. If your agency has pretreatment facility inspectors or industrial waste inspectors, they should have a knowledge of the areas that are most likely to discharge hazardous combustible materials to the wastewater system but if they don't, it is your responsibility to locate these potentially hazardous sources. Another good way to gain assistance when searching for sources of hazardous gases is to contact your local gas company. Many times explosive conditions are caused by leaking gas mains, service lines or underground fuel storage tanks. In combined wastewater collection systems where any spill of gasoline on a public street may find its way into the collection system, particularly hazardous conditions can develop as far as explosive mixtures are concerned.

7. **POISONOUS OR TOXIC GASES** This situation might arise anytime, anywhere, from a variety of sources ranging from natural decomposition of the organic solids in wastewater to toxic chemical dumping or reactions from two or more chemicals.

A number of toxic gases are found in the wastewater collection system. Hydrogen sulfide is the most dangerous and most likely to be encountered because it is generated during the *ANAEROBIC DECOMPOSITION*[2] of organic matter and it is heavier than air and tends to concentrate near the bottom of enclosed spaces. Hydrogen sulfide smells like rotten eggs but, unfortunately, our noses tend to lose their sensitivity to hydrogen sulfide and become unreliable detectors. Many operators as well as their would-be rescuers have been killed by this gas. Other toxic or poisonous gases found in collection systems include carbon monoxide, ammonia and chlorine. Industrial chemical discharges can release dangerous gases through accidents, leaks or illegal discharges. Toxic spills from highway and railway accidents also reach the sewers.

QUESTIONS

Write your answers in a notebook and then compare your answers with those on page 399.

11.1A List the hazards discussed in this section that might be encountered by collection system operators in the performance of routine daily tasks and work assignments.

11.1B Where and under what conditions are operators most likely to slip and fall?

11.1C Can AIDS be contracted from working in the wastewater industry?

11.1D How can an explosion in a sewer be set off when the mixture of air and gas is within the level of the explosive limits?

11.1E What gas is the most dangerous and most likely to be encountered in collection systems?

8. **OXYGEN-DEFICIENT ATMOSPHERES** A safe supply of oxygen is essential to you in the environment in which you must work. *ANY* concentration of *OTHER GASES* can reduce the amount of oxygen available for life support. A shortage of oxygen to the operator produces a shortage of oxygen to the brain, which in turn makes the operator less alert. If you are less alert, you may not recognize the danger until you are too weak to get out. Each year many operators die from toxic hydrogen sulfide, a shortage of oxygen, or from falls or unsafe acts in attempts to escape without help when the operator is mentally disabled, at least in part, by a toxic gas or insufficient oxygen.

The views of experts and legal regulations vary regarding safe levels of oxygen, as do individual needs concerning the amount of oxygen required to support life. Your health and the amount of energy expended are considerations in determining the amount of oxygen that you need. It is generally agreed that anything less than 19.5 percent of oxygen in the breathing atmosphere is a signal to the trained collection system operator to be especially careful and request more ventilation. Measure the percent of oxygen in the air at the work site. You must realize that through normal draft ventilation the gases exiting the collection system through an open working manhole could be just as toxic as the atmosphere in the collection system. Any time the percent of oxygen in a manhole starts dropping below the percent topside, increase the ventilation. Carry properly calibrated oxygen level measuring devices which sound an alarm when a dangerous condition exists rather than a device where you must read a meter and then decide if the oxygen level is deficient.

Ironically, one of the most dangerous situations occurs during the new construction inspection period. You may be required to visually inspect the inside of a newly constructed pipeline, but such a pipeline ordinarily has no provision for introducing ventilating air to maintain a safe environment. This is frequently the site of operator suffocation in the collection system. Other collection system structures where an oxygen-deficient atmosphere may be encountered include confined spaces such as manholes, sewer lines and wet wells.

Beware of the possibility of excessively high levels of oxygen, too. High oxygen levels won't harm you, but they increase the hazards of explosion and fires. Hazardous levels of oxygen can result from the use of pure oxygen for hydrogen sulfide (H_2S) control (Chapter 6, Section 6.5, "Hydrogen Sulfide Control," especially Section 6.542, "Pure Oxygen (O_2)") in collection systems and in the activated sludge process of wastewater treatment plants. An upper limit of 21.9 percent in working atmospheres is suggested.

[2] *Anaerobic (AN-air-O-bick) Decomposition. The decay or breaking down of organic material in an environment containing no "free" or dissolved oxygen.*

9. **STRAINS OR RUPTURES** By virtue of our work (pulling manhole covers, maneuvering equipment in tight places such as pumps and motors), we open ourselves up to strains and ruptures of all types. Back injuries are probably the most serious since, in most cases, the parts that make up your backbone are not able to repair themselves and damage is often cumulative. It is extremely important to understand and use proper techniques when working in the collection system environment, since an injury could have a drastic impact on your ability to earn a living or enjoy life as you get older.

10. **TRAFFIC MISHAPS** This is an especially high risk area in that we must frequently access a collection system from streets and roadways. Given the number of careless and inattentive drivers we see every day, it certainly makes sense to take all the precautions necessary for maximum protection when working in streets and roadways.

11. **BITES (Insects, Bugs, Rodents, Snakes)** Bites can range from inconvenient to very serious. In some parts of the country black widow spiders, poisonous snakes, and scorpions find collection system facilities and structures such as manholes, metering enclosures and electrical control enclosures ideal places to live. Use gloves and protective clothing and stay alert.

12. **EXCAVATIONS AND TRENCH SHORING** Cave-ins rank among the top causes of fatalities in the collection system. You should not, under any circumstances, enter a trench unless you fully understand and insist on adequate safety precautions.

    As a collection system operator, you may be asked to inspect or advise about collection system construction or repair excavations made by your agency or by contractors working for your agency. Along with the engineer, you may be the most informed professional to re-examine, inspect, and help enforce OSHA safety laws. Watch out for your own safety and that of your fellow operators.

    Deep excavations (5 feet or deeper) without shoring or with improper shoring have allowed cave-ins that have killed many operators. There is a great temptation to "short-cut" the possible expense and time-consuming labor of shoring and a contractor principally interested in making a profit may be tempted to save on shoring or other safety measures. As a trained collection system operator, you should take seriously your responsibility to be on the lookout for such hazardous operations.

13. **DROWNING** Drowning is probably an unpleasant way to die, but drowning in raw wastewater has to be worse. Use extreme care when working in high-velocity sewers such as metering flumes and observe all of the confined space precautions to avoid losing consciousness and falling into the wastewater.

    Confined space precautions include monitoring of the atmosphere and use of ventilation, self-contained breathing gear, safety harnesses, lifelines, and two people topside to watch, assist and help remove the in-line worker by winch or their own physical strength if the worker gets into trouble.

14. **FIRE** Fire hazards, primarily in lift stations, develop from a number of sources. Of particular concern is the risk that fire could easily block the limited access into pump stations or other structures where we normally work, thus trapping an operator.

15. **ELECTRIC SHOCK** This hazard exists in virtually all of our facilities and is made even more threatening by the damp atmosphere and adverse working conditions in which we must work. There is a misconception that only high voltage kills. That is not the case, however, since relatively low voltage, even 120 volts, will kill you as quickly and effectively as 480 volts. Under no circumstances should you do electrical work that you are not qualified to perform or not authorized to perform.

16. **NOISE** Engine-driven pumps, standby power generators, large ventilation blowers, air compressors, jackhammers, concrete cut-off saws and high-velocity cleaners are examples of equipment noise we are exposed to which can cause permanent hearing loss over a period of time unless adequate hearing protection is worn.

Curiously enough, workers for other industries, such as construction and mining, are exposed to these same hazards as well as many others. Yet, in comparison to those industries, we have retained the number one position as the *LEAST* safe industry. This ranking is substantiated by the nationwide industry surveys referred to in Section 11.0. Figure 11.1 further illustrates the injury frequency rate (number of injuries per million hours worked) for collection system operators. The rates in 1988 decreased slightly, from 47.05 injuries per million hours to 41.48 injuries per million hours worked. The rate remains disturbingly high. In a similar survey,[3] the National Safety Council reports an injury frequency rate for collection system operators of 4.5 times the 1984 average for all industrial workers. This same survey also reports that the severity rate for operator injuries is almost five times the average rate for other industrial workers.

---
[3] *ACCIDENT FACTS, 1985 edition, National Safety Council.*

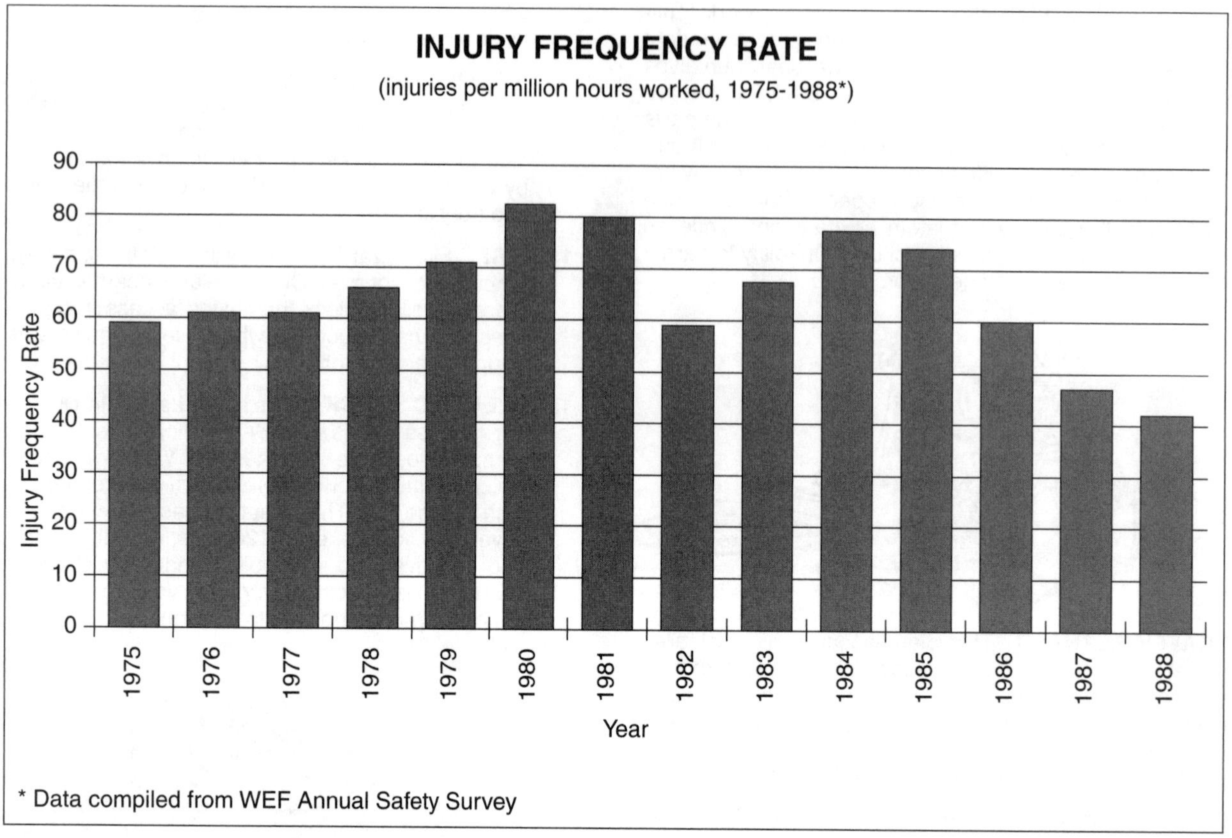

Fig. 11.1 Injury frequency rate for collection system operators

These accident and injury figures may seem academic and irrelevant. They are not. They relate directly to your chances for a long, healthy life. What they say is that *YOU* are at serious risk of suffering a very severe work-related injury every time you go to work. The good news is, you can change the odds. Your participation in this course is an excellent beginning. We now know that 71 percent of operator injuries are attributable to non-certified operators,[4] so, while certification is no guarantee of good health, the statistics clearly suggest that certification in some way improves your chances.

QUESTIONS

Write your answers in a notebook and then compare your answers with those on page 399.

11.1F When might an operator encounter oxygen-deficient atmospheres?

11.1G Why are traffic mishaps a high risk area for collection system operators?

11.1H Where will collection system operators most likely find black widow spiders and scorpions?

11.1I List two sources of equipment noise which can result in permanent hearing loss.

11.1J Injury frequency rates and severity rates for collection system operators were how many times greater than the average rate for industrial workers in 1984?

11.2 MAJOR TYPES OF ACCIDENTS AND INJURIES

Three major types and causes of accidents accounted for 79.6 percent of all the injuries reported to the Water Environment Federation (Figure 11.2).

- 44.30% **sprains and strains** resulting either from lifting, pulling or pushing objects, awkward position or sudden twist, or slips are the leading causes of accidents reported.
- 20.60% **struck by objects** either stationary, moving, sharp, falling, or flying.

[4] *22nd Annual Water Environment Federation Safety Survey.*

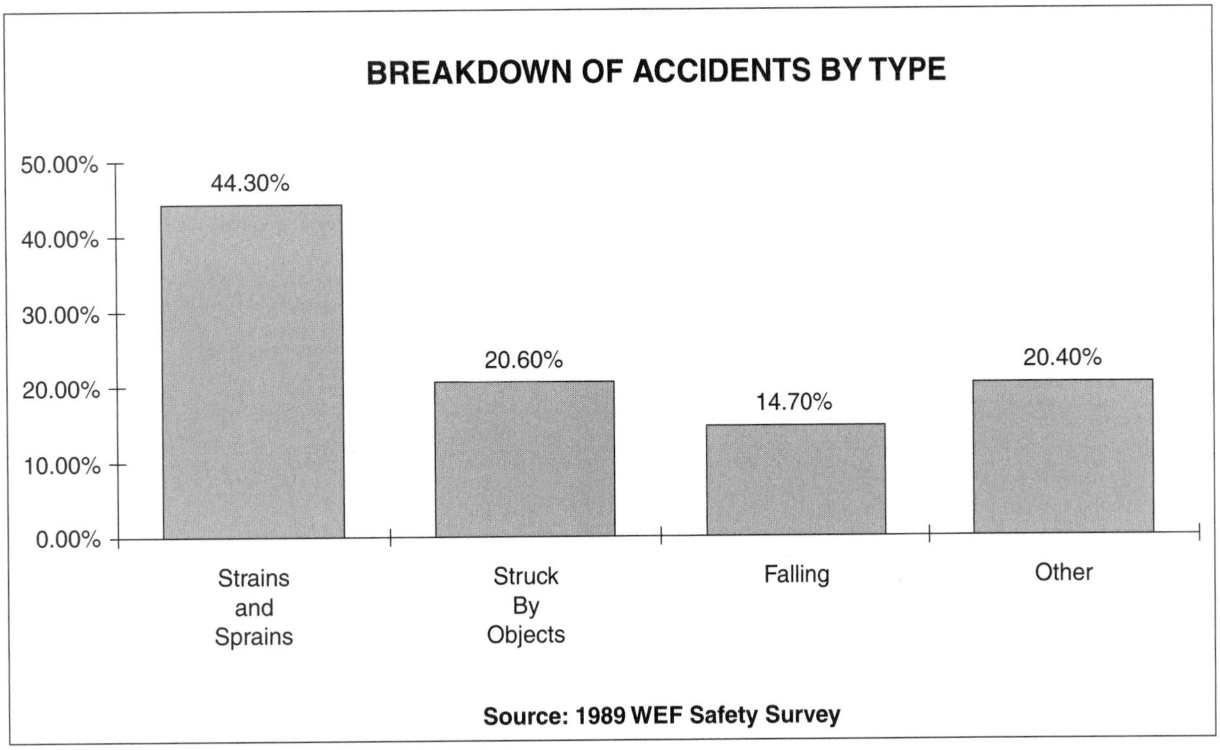

Fig. 11.2 Breakdown of accidents by type

- 14.70% **falling** either to different levels from platforms, ladders, or stairs or on the same level to the working surface.

The parts of the collection system operator's body that are frequently injured are the back, hands and legs (Figure 11.3).

- 29.40% of the accidents are back injuries which lead the category of most common parts of the body injured.
- 12.60% are hand injuries, and
- 13.60% are leg injuries, which is consistent with national trends by industrial workers.

In spite of the poor safety record of the collection system industry, 66 percent of the agencies reporting had a formal safety program. However, based on the numbers above, the safety programs obviously are not always effective. If you are currently working in the wastewater industry, you can expect at least a one-in-ten chance of experiencing a disabling injury.

In summary, it should be readily apparent at this time just how important safety/survival is to every collection system operator. It should be obvious to you that the operator is the person who is most frequently exposed through everyday work assignments and has the most to lose in terms of injury or death. Therefore, operators should share a larger portion of the responsibility for developing effective safety/survival practices when performing these work assignments. A comprehensive Safety/Survival Program greatly increases operator involvement and minimizes risks to the operators in the field. The remainder of this chapter will describe the operator's role in a safety/survival program and will provide an overview of how an effective safety program operates. This information should provide you with the tools to start a program or to strengthen an existing program, since it forms the basis for the first of the three factors required for safety, *AWARENESS* through knowledge.

11.3 SAFETY/SURVIVAL PROGRAM BENEFITS

Industry and government experiences repeatedly show strong evidence that the human and economic benefits of accident prevention far outweigh the investment in training and protective devices necessary for a sound safety/survival program. Personal injuries and economic factors such as work interruptions and output delays are two of the primary considerations.

- Humanitarian. First and foremost, the greatest return in a safety/survival program is the prevention of human suffering, either in the form of permanent disability or even death.
- Economic. Although it may not be readily apparent, the economic savings resulting from an effective safety/survival program are enormous. An accident is evidence of inefficiency that affects both the quality and quantity of productive effort. Costs associated with lost work output, damaged equipment and material, higher insurance premiums, employee morale, the loss of experienced operators, and the training of new operators are all related economic factors. The direct economic loss to you and/or your family in the case of permanent disability or death is even more significant. It is apparent, too, that your responsibility to taxpayers is an economic factor since a higher cost for the services you perform will have to be passed on to them.
- Legal. Federal, state, and local laws and regulations require that facility operators provide a safe and healthy workplace for workers. In some cases, the municipality may be fined by the safety oversight organization (such as federal or state OSHA) for noncompliance with safety rules. Figure 11.4 is a summary of the most frequently cited or violated OSHA regulations, with hazard communication receiving the most citations and lockout/tagout a close second. Figures 11.5, 11.6, 11.7, and 11.8 summarize other OSHA-cited violations.

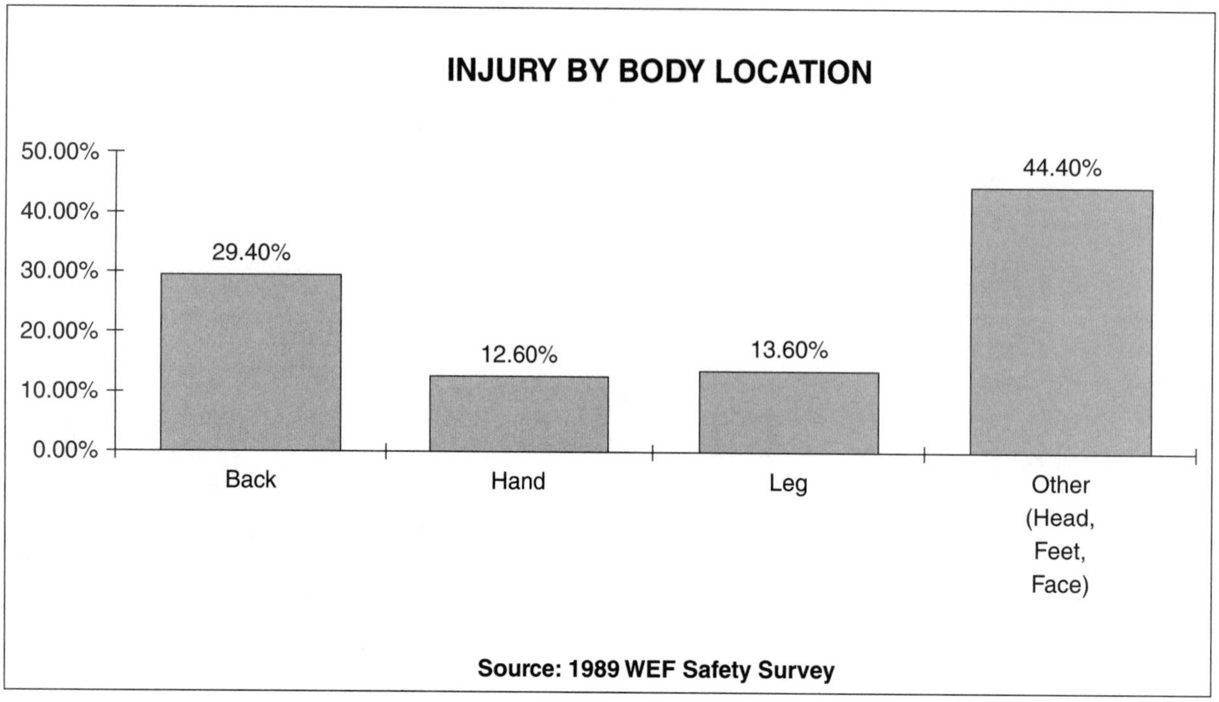

Fig. 11.3 Injury by body location

MOST FREQUENTLY CITED	TOPIC	REGULATION	TOTAL CITATIONS	TOTAL COLLECTION ($)	AVERAGE FINE ($)
1	Hazard Communication	1910.1200	3,344	884,085	264
2	Lockout/Tagout	1910.147	3,279	3,217,970	981
3	Respiratory Protection	1910.134	2,753	740,026	268
4	Machine Guarding	1910.212	2,332	3,531,742	1,514
5	Electrical Safety	1910.305	1,951	861,629	441
6	Power Presses	1910.217	3,844	814,441	211
7	Electrical Systems	1910.303	1,544	916,268	593
8	Hearing Protection	1910.35	1,284	727,578	566
9	PPE	1910.132	1,273	900,287	707
10	Powered Industrial Trucks	1910.178	1,263	949,835	1,240
11	Walking & Working Surfaces	1910.23	1,176	1,458,682	1,240
12	Confined Space Entry	1910.146	1,051	973,301	926

Fig. 11.4 OSHA's most commonly cited violations in FY 2000

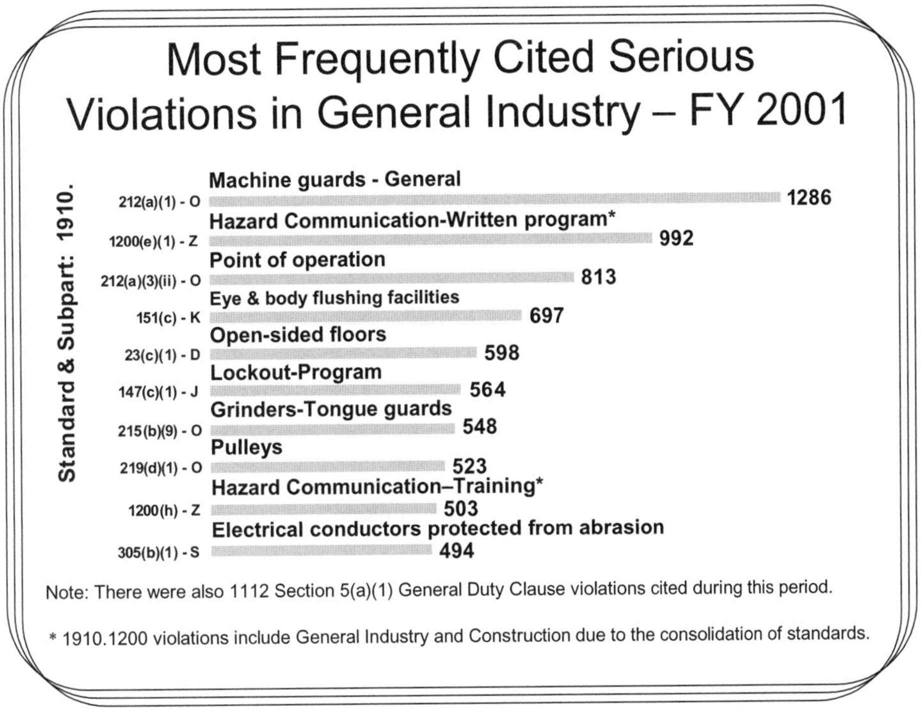

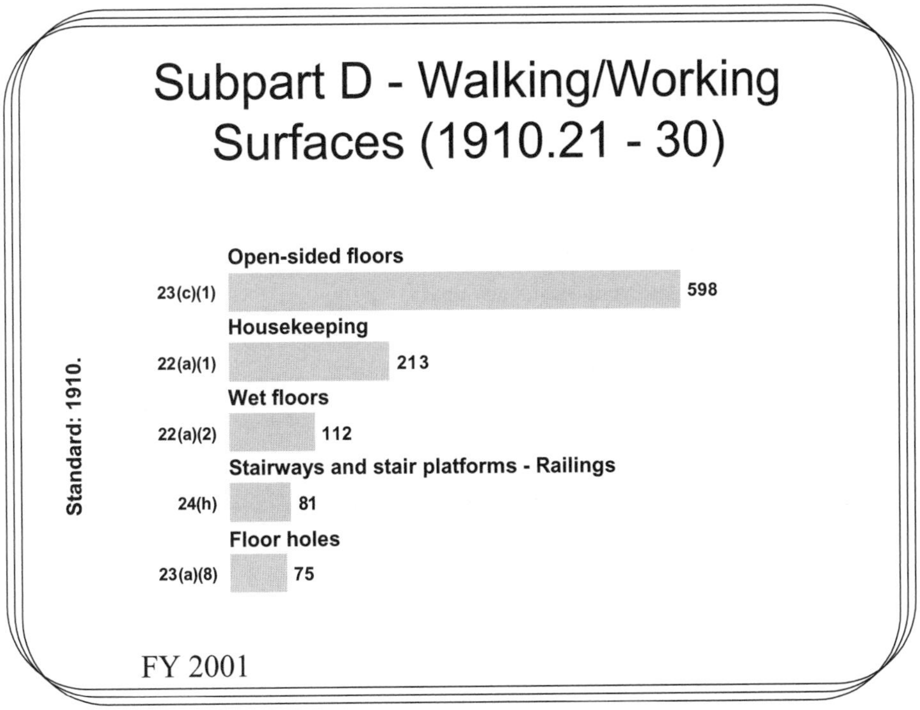

Fig. 11.5 OSHA frequently cited serious violations in the General Industry category and Walking/Working Surfaces violations

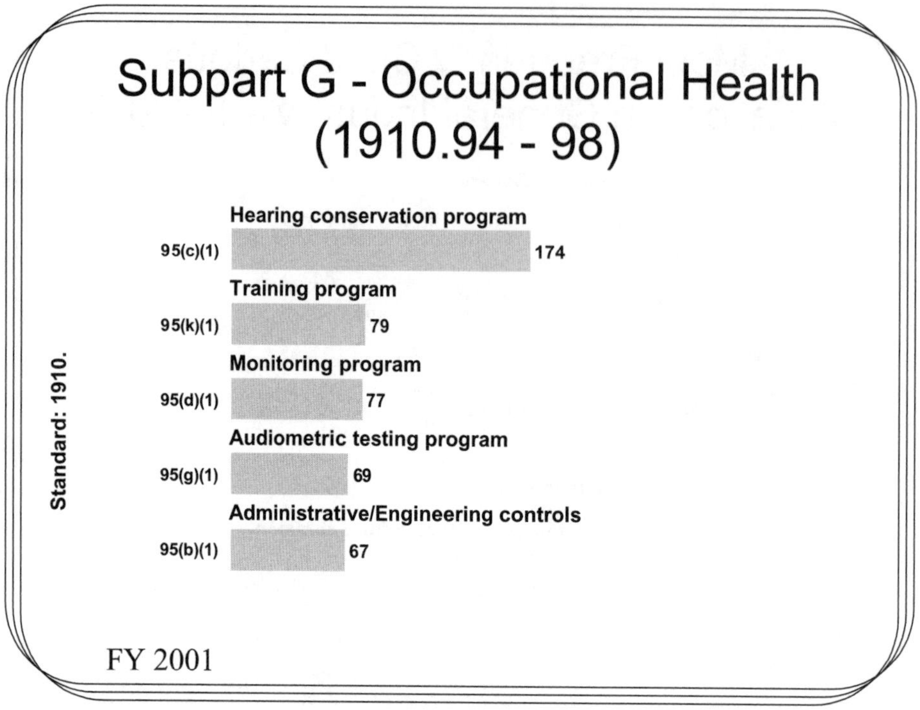

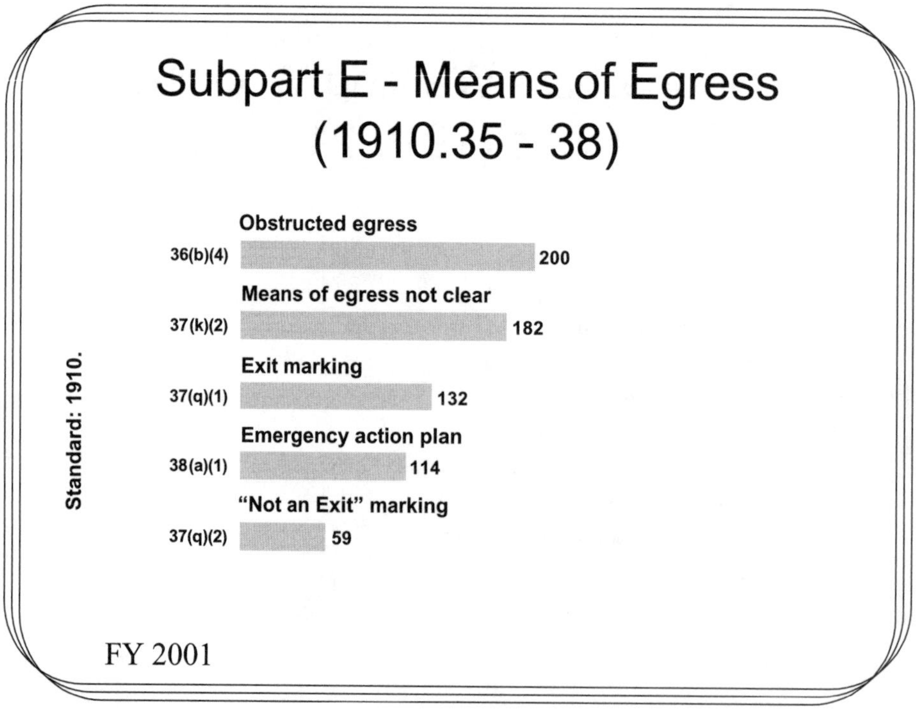

Fig. 11.6 OSHA Occupational Health and Means of Egress violations

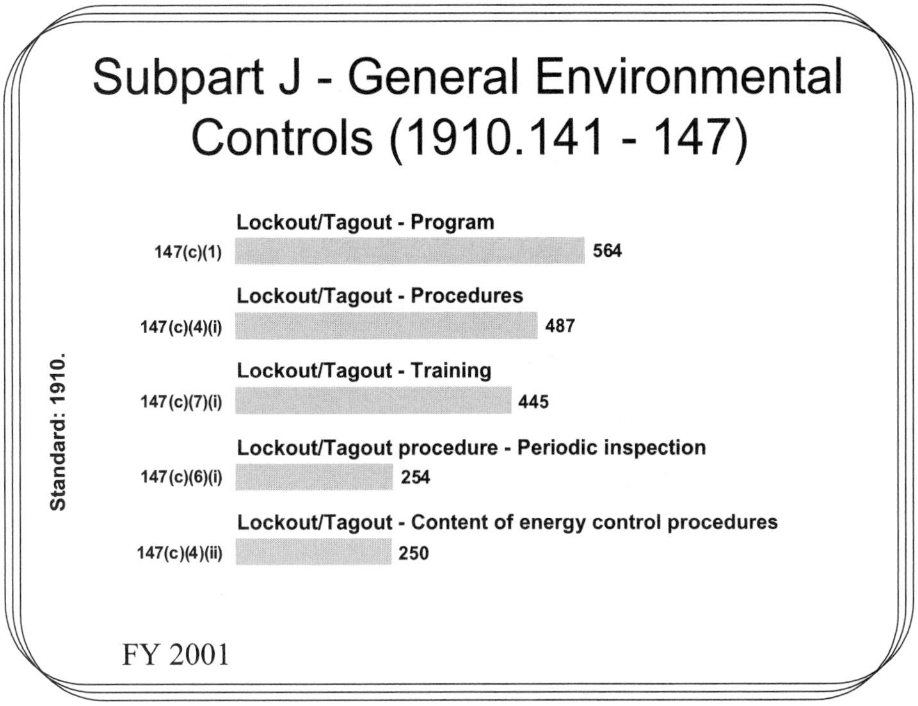

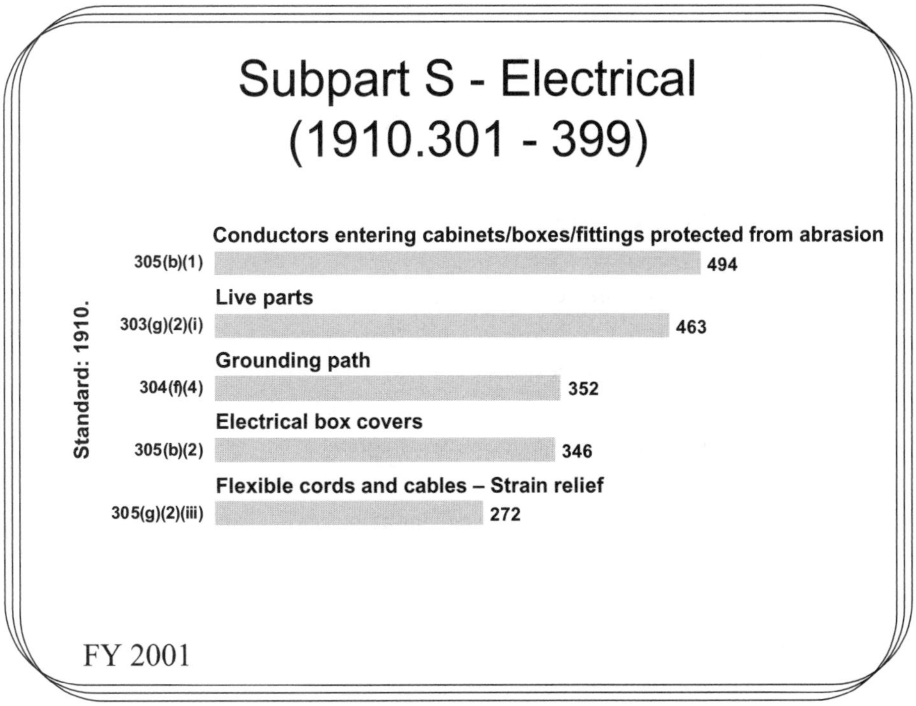

Fig. 11.7 OSHA General Environmental Control and Electrical violations

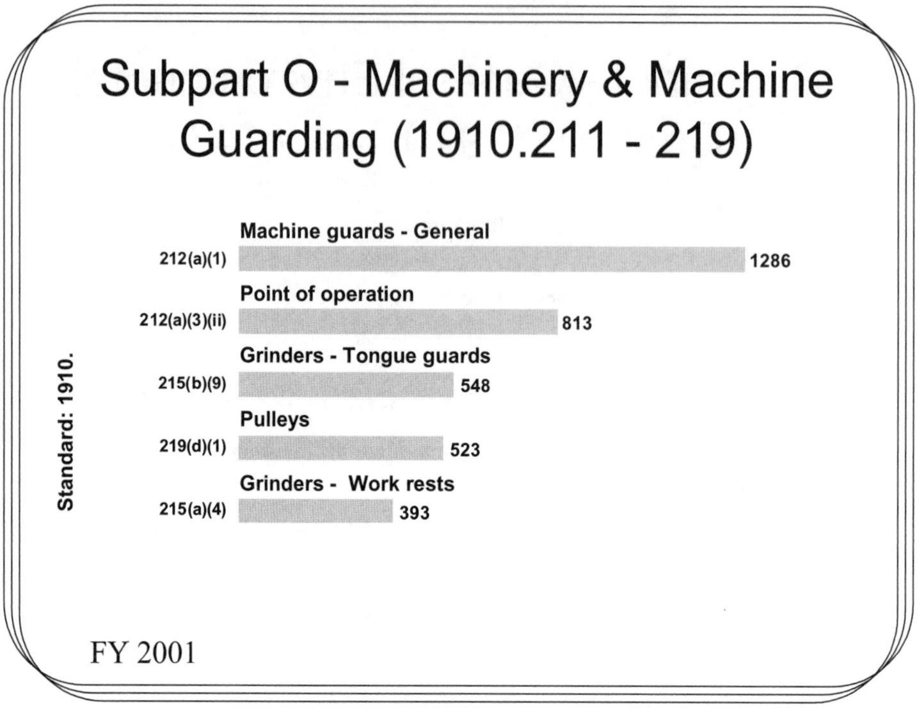

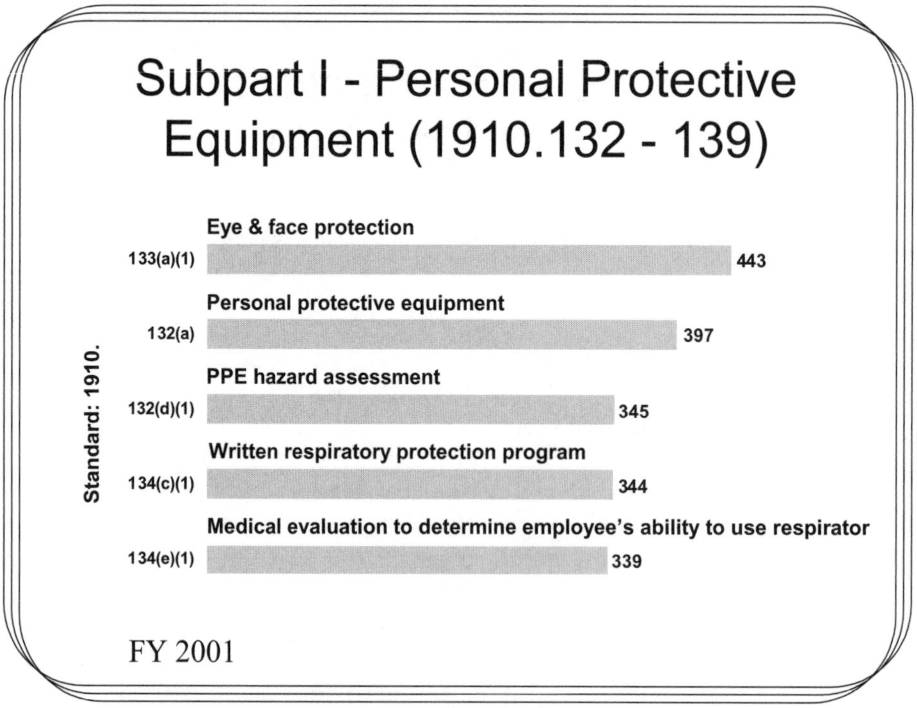

Fig. 11.8 OSHA Machinery and Machine Guarding and Personal Protective Equipment violations

Your safety and that of your colleagues should be the first consideration in the performance of your duties and assigned tasks in the collection system. The basis for your program should be:

SAFE OPERATIONS SHALL TAKE PRECEDENCE OVER EXPEDIENCY OR SHORT CUTS.

11.4 ESTABLISHING A SAFETY/SURVIVAL PROGRAM

Typically, the structure within an agency for a safety program consists of the following levels:

1. Administration Department which states, "We will have a safety program";
2. Safety Department which says, "We will help with the resources"; and
3. Safety Committee which is "where the corrective action is."

Depending on the size of your operation, there may be some additional levels of safety/survival activity in between these levels, but this basically describes a typical structure. Perhaps a little more difficult, however, is the definition of the responsibilities in each of these areas.

QUESTIONS

Write your answers in a notebook and then compare your answers with those on page 399.

11.2A Why do the formal accident surveys not accurately portray the severity of the safety problem in the collection system industry?

11.3A What should be the basis for a safety/survival program?

11.4A List the three levels of structure within an agency for a safety program.

11.5 SAFETY/SURVIVAL PROGRAM RESPONSIBILITIES

11.50 Top Administration

Top administration could be the mayor, city council members, agency directors, or other top management personnel involved with the policy making aspect of your agency.

- They bear the overall responsibility for promoting the concept of an effective safety/survival program.
- They approve policies and procedures that are developed through cooperative efforts at other levels.
- They require, as a condition of employment, that all personnel learn and follow safety procedures and policies.
- They are responsible for ensuring adequate budget resources in terms of equipment, personnel, or other issues related to the funding and the support of your safety/survival program.
- They monitor the effectiveness of the program and communicate with regulatory agencies as required.

11.51 Safety Department Staff

- They coordinate and distribute safety requirements and regulations.
- They provide safety resources to all levels of the program.
- They coordinate safety activities with regard to agency insurance programs.
- They provide some form of facility safety inspections.
- They provide support for training and programs which might include audio-visual aids, equipment and materials, books, or other resources.
- They conduct ongoing evaluation of safety procedures with revisions as required based on recommendations and discussions with personnel at the operating level (the Safety Committee).

11.52 Manager, Supervisor, Field Personnel

This is the final level of responsibility, those actually involved in the day-to-day operation of the collection system. As stated earlier, this is where all the action is and rightfully so, since this is where all of the expertise should be in terms of safety and survival. After all, we should be the ones who are most knowledgeable about how to safely perform our work and should not wait for somebody at the administrative level to tell us how they think it should be done.

Supervisors and managers who are involved in everyday collection system operations are critical to the success of the safety/survival program.

- They must be **aware** of safety rules and regulations that apply to each task.
- They are **responsible** for the safety of other personnel who work in or pass through their area of supervision.
- They are **responsible** for job safety instruction and new employee orientation.

- They are **responsible** for safety training within their area.
- They are **responsible** for identifying unsafe practices and conditions as well as taking adequate action to correct those within their authority.
- They should **investigate** injury-producing accidents and property damage accidents within their area of supervision.
- They must make a firm **commitment** to a safety/survival program in order to develop appropriate safety attitudes in personnel. Obviously, education, personal examples, and active participation in the program are essential to its success.
- They are **responsible** for promoting and enforcing the use of recommended personal protective equipment. They may be given fines and/or prison sentences if found negligent in any of the above areas that contributed to a fatal accident.

Field personnel are obviously the major factor in the success or failure of your program which will be measured by injuries, possible disability or death. The responsibilities of the operators in the field include:

- They are **RESPONSIBLE** for learning and following all safety procedures and policies.
- They must be willing to make a **COMMITMENT** to the safety/survival program and assume the **RESPONSIBILITY** to make the program a part of the daily routine. (Once again, it is in our best interest to make the program work since we are the ones who are normally involved in the injuries, disabilities, and fatalities.)

Even if the other two levels in the safety program are not effective, you can have a major impact on the success of the program by participating at your level. Draw on your field experience to help develop or modify procedures and policies through an active safety committee. Encourage others at all levels to participate with you.

Disciplinary action can be an effective tool, if required, in maintaining safety policies and procedures. *NOTE:* No one should be expected to work with another operator who endangers their safety. Report dangerous field incidents to your immediate supervisor in writing. If you are placed in this type of situation, do not feel intimidated about reporting it.

SAFE OPERATIONS SHALL TAKE PRECEDENCE OVER EXPEDIENCY.

Operators must also participate in developing their own formal safety program which would include monthly meetings, a safety committee, and weekly tailgate sessions.

Tailgate sessions are short (ten minutes plus or minus) discussions that help focus on safety/survival as part of everyday routine collection system operator tasks. Topics can be specific to the job or related to home safety and seasonal events such as holidays. Virtually any topic that relates to safe practices will help inform operators and raise their level of awareness about safety.

An excellent resource for tailgate sessions is the American Water Works Association (AWWA) which publishes an annual series of 52 weekly safety talks.[5]

Some topics that are appropriate for a tailgate safety session include:

1. Safe and Economical Driving
2. Fire Prevention
3. Hand Tool Safety
4. Grounding Portable Tools
5. Productivity Versus Haste
6. Muscle Strains
7. Routing Traffic Around Work Sites
8. Manhole Entry
9. Calibrating and Using Atmospheric Testing Instruments
10. Lifeline Safety Equipment
11. Personal Hygiene
12. Public Relations

QUESTIONS

Write your answers in a notebook and then compare your answers with those on pages 399 and 400.

11.5A What are the responsibilities of "Top Administration" for a safety/survival program?

11.5B List the responsibilities of the "Safety Department Staff" for a safety/survival program.

11.5C Supervisors and managers are involved in what aspects of a safety/survival program?

11.5D What are the responsibilities of operators in the field with regard to a safety/survival program?

11.6 MONTHLY SAFETY MEETINGS

These meetings are more comprehensive than tailgate sessions and can last up to an hour or more, depending on the topic. Discussion topics may be of a general nature, but usually are more specific to the collection system work environment. Audio/visual presentations can be particularly effective if your safety department has a library and the essential equipment. Safety programs can be conducted either by operators or by employees of outside agencies. The following is a list of possible discussion topics.

[5] *LET'S TALK SAFETY—2003 SAFETY TALKS. Obtain from American Water Works Association (AWWA), Bookstore, 6666 West Quincy Avenue, Denver, CO 80235. Order No. 10123. Price to members, $37.50; nonmembers, $53.50; price includes cost of shipping and handling.*

TOPICS	PRESENTING PARTY
Gas Leaks	Local Gas Utility
Fire Fighting	Local Fire Department
Electrical Hazards	Local Electrical Utility
Back Injuries	Film and Safety Department
Slings and Material Handling Equipment	Equipment Supplier
Falling	Martial Arts, Private Company
SCBA Refresher	Safety Department
Confined Space	Operators
Right-To-Know	Training Department

Most suppliers of equipment, whether it be specific safety equipment or everyday hand tools, will assist you in organizing and conducting a monthly safety program.

Safety programs can be fun. Combine them with other events within the organization such as recognition events that recognize employees for length of service, exceptional work, or other achievements that are not necessarily related to safety. Audience participation meetings are very useful and can be combined with other methods. For example, set up teams and ask safety/work-related questions. Prizes could be awarded to individuals and the winning team and might range from silver dollars to savings bonds and gift certificates. Keep these meetings informal and arrange to provide coffee, donuts or other types of "goodies" if possible.

11.7 SAFETY COMMITTEE

An active Safety Committee staffed with operating personnel is another extremely effective tool in developing your safety/survival program. Depending on the nature of your organization, the Committee could include:

- Representatives from each area of the collection system such as office personnel, electrical/mechanical maintenance, and operators.
- Representatives from other operating departments within your agency if you don't have a large enough collection system staff. Street, water, parks or other areas share many of the same safety/survival problems. Set up subcommittees to identify specific issues; for example, an equipment subcommittee might review and make recommendations on improving safety-related issues with the various equipment used. A policy/procedure subcommittee could also be established and would participate in developing/revising policies and procedures.

Members of the Safety Committee may be responsible for establishing a monthly program, which would include developing the resources. The resource personnel could be from your safety department or from an outside source, depending on the topic.

The Safety Committee reviews, analyzes and determines how each accident occurred and why. The Committee must recommend training procedures and distribute information to other operators to prevent similar accidents. If the injured operator was not paying attention or just plain careless, then better tailgate safety meetings are needed. If the accident resulted from an unsafe act, use of the wrong tool, or improper use of a piece of equipment, then training procedures need to be changed to correct the safety hazard. If new equipment or training aids are needed, the Safety Committee must identify these needs, justify them to the safety officer and management, acquire the necessary equipment and aids, and implement their use. The Safety Committee must consist of expert operators who know and can justify what is needed and not just request "stuff" and "things" because it sounds like a good idea.

QUESTIONS

Write your answers in a notebook and then compare your answers with those on page 400.

11.6A How do monthly safety meetings differ from tailgate sessions?

11.6B Who can conduct safety programs?

11.7A Who should be on a "Safety Committee"?

11.7B What should the Safety Committee do if an accident resulted from an unsafe act, use of the wrong tool, or improper use of a piece of equipment?

11.8 SAFETY/SURVIVAL PROGRAM POLICIES/STANDARDS

Safety policies are established to ensure that all operators understand and follow safe procedures at all times. Operators must be made aware that compliance with safety policies is a condition of employment and that violators are subject to disciplinary action. The policies included here are typical examples of policies which should exist in your agency. These may not meet your local or state requirements, but can be used as a starting point to help you define your agency's policies.

A good policy statement will clearly define the conditions under which it applies, the behaviors required of employees, and the penalty for failure to comply with the policy. Policy statements often refer to other detailed procedures to be followed under specific circumstances. For example, the list of procedures for a permit-required confined space entry provides details on personal protection equipment required, atmospheric testing, standby persons, rescue, and other specifics related to the entry.

11.80 Elements of a Written Safety Program

Safety programs can be published as a written document and stored in a three-ring binder for easy updating. The following topics are commonly found in any safety and health program:

- Facility safety policy,
- Management's responsibilities,
- Supervisors' responsibilities,
- Employees' responsibilities,
- Safety Committee or Safety Team (employee participation),
- Safety meetings,
- Hazard recognition,
- Incident investigation,
- Elimination of workplace hazards,
- Basic safety rules,
- Job-related safety rules,
- Disciplinary policy,
- Emergency planning,
- Reporting accidents and incidents (OSHA 300 log), and
- Training, initial (new employee) and refresher (see Figure 11.9).

Each employer must set up a safety and health program to manage workplace safety and health to reduce injuries, illnesses, and fatalities by systematically achieving compliance with OSHA standards and the General Duty Clause. The program must be appropriate to conditions in the workplace, such as the hazards to which employees are exposed and the number of employees working for the collection system agency. The program must have the following core elements:

- Management leadership and employee participation,
- Hazard identification and assessment,
- Hazard prevention and control,
- Information and training, and
- Evaluation of program effectiveness.

The program administrator defines the program responsibilities of managers, supervisors, and employees for safety and health in the workplace and holds them accountable for carrying out those responsibilities. The program administrator must:

- Provide managers, supervisors, and employees with the authority, access to relevant information, training, and resources they need to carry out their safety and health responsibilities, and
- Identify at least one manager, supervisor, or employee to receive and respond to reports about workplace safety and health conditions and, where appropriate, to initiate corrective action.

Many agencies also develop a Safety Manual through the efforts of a Safety Committee and operators. Appendix A contains a sample safety program from the State of Washington. Appendix B contains the table of contents for a safety manual developed by Monroe County, Department of Environmental Services, Rochester, New York. Listed in the table of contents are many of the components that make up a well-defined safety program.

Topic	Federal Regulation	Frequency of Training
Employee Emergency Action Plans/Fire Prevention	29 CFR 1910.37 and 29 CFR 1910.38(a)(5), (b)(4)	• Initially and when there are changes. • Refresher training is advisable.
Hazard Communication	29 CFR 1910.1200(h)	• Initially. • When new hazard is introduced. • Review new MSDSs. • Annual refresher is advisable.
Personal Protection Equipment Respiratory Protection	29 CFR 1910.132-.140 29 CFR 1910.134(b)(3) and (b)(10)	• Prior to use or changes. • Prior to use. • Annual training is advisable. • The respirator user's medical status should be reviewed at least annually (CA).
Excavation	29 CFR 1926.651-.652	• Competent Person training.
Electrical — Safety-Related Work Practices	29 CFR 1910.332(b)(l)	• Initially if duties can expose to risk of shock.
Exposures to Bloodborne Pathogens	29 CFR 1910.1030(g)(2)(l)-(ix)	• Initially. • Annually within one year of previous training.
Control of Hazardous Energy, Lockout/Tagout	29 CFR 1910.147(c)(7)	• Initially. • Refresher training is advisable.
Confined Space Confined Space Rescue	29 CFR 1910.146(g) 29 CFR 1910.146(k)	• Initially and refresher training is advisable. • Initial and retraining, including simulated rescue, every 12 months. (Also First Aid and CPR.)

Fig. 11.9 Sample training requirements list

QUESTIONS

Write your answers in a notebook and then compare your answers with those on page 400.

11.8A How frequently should operators be trained using Personal Protection Equipment (PPE)?

11.8B What are the core elements of a utility's safety and health program?

11.81 Permit-Required Confined Spaces (29 CFR 1910.146)

Many workplaces contain spaces that are considered to be "confined" because their configurations (size and shape) hinder the activities of any employees who must enter into, walk in, and exit from them is a critical consideration. In many instances, employees who work in confined spaces also face increased risk of exposure to serious physical injury from hazards such as entrapment, engulfment, and hazardous atmospheric conditions. Confinement itself may pose entrapment hazards, and work in confined spaces may keep employees closer to hazards, such as an asphyxiating atmosphere, than they would be otherwise. For example, confinement, limited access, and restricted air flow can result in hazardous conditions that would not arise in an open workplace.

11.810 Requirements of the Standard

In general, employers must evaluate the workplace to determine if spaces are permit-required confined spaces (see Figure 11.10). If there are permit spaces in the workplace, the employer must inform exposed employees of the existence, location, and danger posed by the spaces. This can be accom-

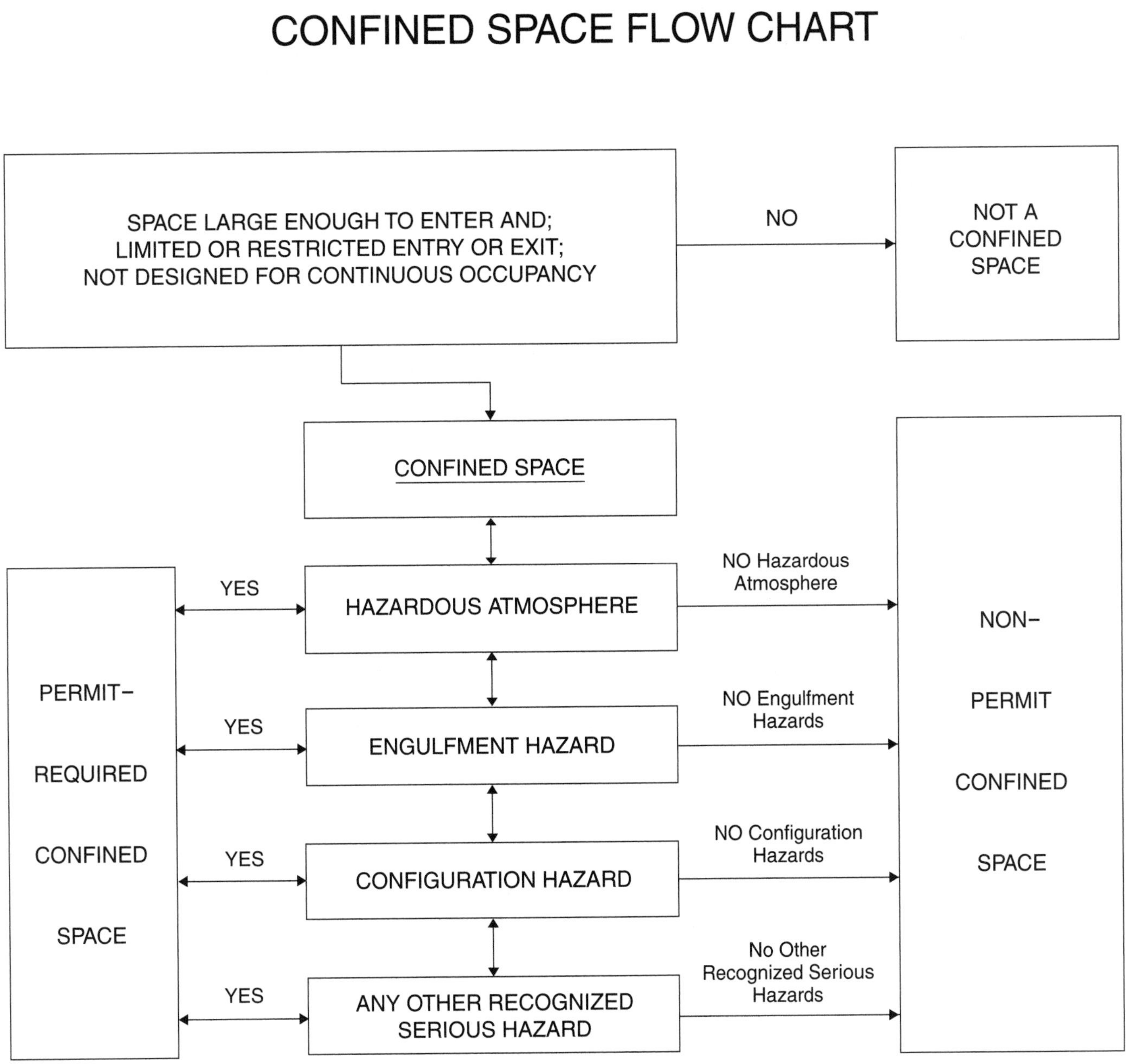

Fig. 11.10 Sample confined space flow chart

plished by posting danger signs or by another equally effective means. The following language would satisfy the requirements for such a sign:

DANGER — PERMIT REQUIRED
CONFINED SPACE — AUTHORIZED ENTRANTS ONLY

If employees are not allowed to enter and work in permit spaces, employers must take effective measures to prevent their employees from entering the permit spaces. If employees are to be allowed to enter permit spaces, the employer must develop a written permit space program and make it available to employees or their representatives. Under certain conditions, the employer may use alternate procedures for worker entry into a permit space. For example, if employers can demonstrate with monitoring and inspection data that the only hazard is an actual or potential hazardous atmosphere that can be made safe for entry by the use of continuous forced-air ventilation alone, they may be exempted from some requirements, such as permits and attendants. Even in such circumstances, however, the internal atmosphere of the space must be tested first for oxygen content, second, for flammable gases and vapors, and third, for potential toxic air contaminants before any employee enters.

11.811 Written Program

The employer who allows employee entry into confined spaces must develop and implement a written program for permit-required confined spaces. Among other things, the OSHA Standard requires the employer's program to:

- Identify and evaluate permit space hazards before allowing employee entry;
- Test conditions in the permit space before entry operations and monitor the space during entry;
- Perform, in the following sequence, appropriate testing for atmospheric hazards: oxygen, combustible gases or vapors, and toxic gases or vapors;
- Establish and implement the means, procedures, and practices necessary for safe permit space entry operations, such as specifying acceptable entry conditions; isolating the permit space; providing barriers; verifying acceptable entry conditions; and/or purging, making inert, flushing, or ventilating the permit space to eliminate or control hazards;
- Identify employee job duties;
- Provide, maintain, and require, at no cost to the employee, the use of personal protection equipment (PPE) and any other equipment necessary for safe entry (for example, testing, monitoring, ventilating, communications, and lighting equipment; barriers, shields, and ladders);
- Ensure that at least one attendant is stationed outside the space for the duration of entry operations;
- Coordinate entry operations when employees of more than one employer are to be working in the permit space;
- Implement appropriate procedures for summoning rescue and emergency services;
- Establish, in writing, and implement a system for the preparation, issuance, use, and cancellation of entry permits;
- Review established entry operations and annually revise the permit-space entry program; and
- When an attendant is required to monitor multiple spaces, implement the procedures to be followed during an emergency in one or more of the permit spaces being monitored.

If hazardous conditions are detected during entry, employees must immediately leave the space and the employer must evaluate the space to determine the cause of the hazardous atmospheres.

If testing and inspection data prove that a permit-required confined space no longer poses hazards, the space may be reclassified as a non-permit confined space. A certificate documenting the data must be made available to employees entering the space. The certificate must include date, location of the space, and the signature of the person making the certification.

Contractors also must be informed of permit spaces and permit space entry requirements, any identified hazards, the employer's experience with the space (such as the knowledge of hazardous conditions), and precautions or procedures to be followed when in or near permit spaces.

When employees of more than one employer are conducting entry operations, the affected employers must coordinate operations to ensure that affected employees are appropriately protected from permit space hazards. Contractors also must be given any other pertinent information regarding hazards and operations in permit spaces and be debriefed at the conclusion of entry operations.

11.812 Example Policy Statement

CONFINED SPACE ENTRY POLICY

The purpose of this policy is to establish guidelines for personnel engaging in work that would require entry into structures, including those under construction, that fall under the definition of a confined space.

The definition of a confined space used for this policy is: "A space which by design has limited openings for entry and exit; and/or natural ventilation which could contain or produce unfavorable atmospheric conditions; and is not intended or designed for continuous employee occupancy."

The types of confined spaces typically encountered in our work include, but are not necessarily limited to, sewer manholes, sewer lines, junction structures, valve vaults, metering vaults, pumping station wet wells, storage tanks, pits and silos.

All confined space work shall be performed with strict adherence to the "Confined Space Work Procedure." No one shall perform or order to be performed any work contrary to this procedure.

Failure to comply with this policy will result in disciplinary action.

11.813 Permit System

A permit, signed by the entry supervisor and verifying that pre-entry preparations have been completed and that the space is safe to enter, must be posted at entrances or otherwise made available to entrants before they enter a permit space.

The duration of entry permits must not exceed the time required to complete an assignment. Also, the entry supervisor must terminate entry and cancel permits when an assignment has been completed or when new conditions exist. New conditions must be noted on the canceled permit and used in revising the permit space program. The OSHA Standard also requires the employer to keep all canceled entry permits for at least one year to facilitate the review of the permit-required confined space.

Confined space entry permits must include the following information:

- Test results;
- Tester's initials or signature;
- Name and signature of supervisor who authorizes entry;
- Name of permit space to be entered, authorized entrants(s), eligible attendants, and individual(s) authorized to be entry supervisor(s);
- Purpose of entry and known space hazards;
- Measures to be taken to isolate permit spaces and to eliminate or control space hazards (such as locking out or tagging of equipment and procedures for purging, making inert, ventilating, and flushing permit spaces);
- Names and telephone numbers of rescue and emergency services;
- Date and authorized duration of entry;
- Acceptable entry condition;
- Communication procedures and equipment to maintain contact during entry;
- Additional permits(s), such as for hot work, that have been issued to authorize work in the permit space;
- Special equipment and procedures, including personal protection equipment (PPE) and alarm systems; and
- Any other information needed to ensure employee safety.

11.814 Training and Education

Before assigning operators to work in confined spaces, the employer must provide proper training for anyone who is required to work in permit spaces. Upon completing this training, employers must ensure that employees have acquired the understanding, knowledge, and skills necessary for the safe performance of their duties. Additional training is required when: (1) the job duties change, (2) there is a change in the permit-space program or the permit space operation presents a new hazard, and (3) when an employee's job performance shows deficiencies. Training also is required for rescue team members; this training should include cardiopulmonary resuscitation (CPR) and first-aid training (see Section 11.815, "Emergency Rescue Personnel"). Employers must certify that training has been accomplished.

Upon completion of training, employees must receive a certificate of training that includes the employee's name, signature or initials of trainer(s), and dates of training. The certification must be made available for inspection by employees and their authorized representatives. In addition, the employer also must ensure that employees are trained in their assigned duties.

Authorized Entrant's Duties

- Know the potential hazards of the space, including information on the mode of exposure (inhalation or absorption through the skin), signs or symptoms, and consequences of the exposure;
- Use appropriate personal protection equipment (PPE) properly (for example, face and eye protection and other forms of barrier protection such as gloves, aprons, and coveralls);
- As necessary, maintain communication by telephone, radio, and/or visual observation with attendants to enable the attendant to monitor the entrant's status as well as to alert the entrant to evacuate;
- Exit from the permit space as soon as possible when ordered by an authorized person, when the entrant recognizes the warning signs or symptoms of exposure exist, when a prohibited condition exists, or when an automatic alarm is activated; and
- Alert the attendant when a prohibited condition exists or when warning signs of exposure exist.

Attendant's Duties

- Remain outside the permit space during entry operations unless relieved by another authorized attendant;
- Perform no-entry rescue when specified by the employer's rescue procedure;
- Know existing and potential hazards, including information on the mode of exposure, signs or symptoms, consequences of the exposure, and their physiological effects;
- Maintain communication with and keep an accurate account of those workers entering the permit-required space;
- Order evacuation of the permit space when a prohibited condition exists, when a worker shows signs of physiological effects of hazardous exposure, when an emergency outside the confined space exists, and when the attendant cannot effectively and safely perform required duties;
- Summon rescue and other services during an emergency;
- Ensure that unauthorized persons stay away from permit spaces or exit immediately if they have entered the permit space;
- Inform authorized entrant's and the entry supervisor of entry by unauthorized persons; and
- Perform no other duties that interfere with the attendant's primary duties.

Entry Supervisor's Duties

- Know confined space hazards, including information on the mode of exposure, signs or symptoms of exposure, and consequences of exposure;
- Verify emergency plans and specified entry conditions such as permits, tests, procedures, and equipment before allowing entry;
- Terminate entry and cancel permits when entry operations are completed or if a new condition exists;
- Take appropriate measures to remove unauthorized entrants; and
- Ensure that entry operations remain consistent with the entry permit and that acceptable entry conditions are maintained.

11.815 Emergency Rescue Personnel

The OSHA Standard requires the employer to ensure that rescue service personnel are provided with and trained in the proper use of personal protection and rescue equipment, including respirators; trained to perform assigned rescue duties; and have had authorized entrant's training. The Standard also requires that all rescuers be trained in first aid and CPR and, at a minimum, one rescue team member be currently certified in first aid and in CPR. The employer also must ensure that practice rescue exercises are performed yearly, and that rescue services (police, fire) are provided access to permit spaces so that they can practice rescue operations. Rescuers also must be informed of the hazards of the permit space.

When appropriate, authorized entrants who enter a permit space must wear a chest or full-body harness with a retrieval line attached to the center of their back near shoulder level, or attached above their head. Wristlets may be used if the employer can demonstrate that the use of a chest or full-body harness is infeasible or creates a greater hazard. The employer must ensure that the other end of the retrieval line is attached to a mechanical device or to a fixed point outside the permit space. A mechanical device must be available to retrieve personnel from vertical-type permit spaces more than five feet deep. In addition, if an injured entrant is exposed to a substance for which a Material Safety Data Sheet (MSDS) or other similar written information is required to be kept at the work site, that MSDS or other written information must be made available to the medical facility treating the exposed entrant.

QUESTIONS

Write your answers in a notebook and then compare your answers with those on page 400.

11.8C If there are permit-required confined spaces in the workplace, what information must employers provide exposed employees?

11.8D What is the order (sequence) for testing the internal atmosphere of a confined space?

11.8E What happens to a confined space entry permit, signed by the entry supervisor and verifying that pre-entry preparations have been completed and that the space is safe to enter?

11.8F When do operators who work in confined spaces require additional training?

11.82 Control of Hazardous Energy (Lockout/Tagout) (29 CFR 1910.147)

11.820 Scope and Application

The lockout/tagout Standard applies to general industry employment and covers the servicing and maintenance of machines and equipment in which the unexpected start-up or the release of stored energy could cause injury to employees. The OSHA Standard establishes minimum performance requirements for the control of hazardous energy. If employees are performing service or maintenance tasks that do not expose them to the unexpected release of hazardous energy, the Standard does not apply. The Standard does not apply in the following situations:

- While servicing or maintaining cord and plug connected electrical equipment. (The hazards must be controlled by unplugging the equipment from the energy source; the plug must be under the exclusive control of the employee performing the service and/or maintenance.)
- During hot tap operations that involve transmission and distribution systems for gas, steam, water, or petroleum products when: (1) they are performed on pressurized pipelines, (2) when continuity of service is essential and shutdown of the system is impractical, and (3) when employees are provided with an alternative type of protection that is equally effective.

11.821 Example Policy Statement

LOCKOUT POLICY

The purpose of this policy is to establish guidelines for personnel required to work on machinery or equipment in facilities owned and operated by this agency.

The definition of "work" on machinery or equipment used for this policy is: "maintenance, inspection, cleaning, adjusting, servicing, or clearing of blocked or jammed machinery or equipment."

All work shall be performed in strict adherence to the "Lockout Procedure." No one shall perform or order to be performed any work contrary to this procedure.

Failure to comply with this policy will result in disciplinary action.

11.822 Servicing and Maintenance Operations

If servicing activity such as lubricating, cleaning, or unjamming the wastewater pumping equipment takes place during operations, the employee performing the servicing may be exposed to hazards that are not encountered as part of the operation itself. The requirements of the lockout/tagout standard apply to the maintenance workers engaged in these operations when any of the following conditions occur:

- The employee must either remove or bypass machine guards or other safety devices, resulting in exposure to hazards at the point of operations,
- The employee is required to place any part of his or her body in contact with the point of operation of the operational machine or piece of equipment, or
- The employee is required to place any part of his or her body into a danger zone associated with a machine operating cycle.

In the above situations, the equipment must be de-energized and locks or tags must be applied to the energy-isolation devices. In addition, when normal servicing tasks, such as setting up and/or making significant adjustments to machines, do not occur during normal maintenance operations, employees performing such tasks are required to lock out or tag out whenever there is the possibility that they could be injured if the equipment is unexpectedly energized.

11.823 Energy Control Program

The lockout/tagout rules require that the employer establish an energy control program that includes: (1) documented energy control procedures, (2) an employee training program, and (3) periodic inspection of the procedures. The Standard requires employers to establish a program to ensure that machines and equipment are isolated and inoperative before any employee performs service or maintenance where the unexpected energization, start-up, or release of stored energy could occur and cause injury. Employers have the flexibility to develop a program and procedures that meet the needs of their particular workplace and the particular types of machines and equipment being maintained or serviced.

Energy Control Procedure

The written procedures must identify the information that authorized employees must know in order to control hazardous energy during service or maintenance activities. If this information is the same for various machines or equipment or if other means of logical grouping exist, then a single energy control procedure may be sufficient. If there are other conditions, such as multiple energy sources, different connecting means, or a particular sequence that must be followed to shut down the machine or equipment, then the employer must develop separate energy control procedures to protect employees.

The energy control procedure must outline the scope, purpose, authorization, rules, and techniques that will be used to control hazardous energy sources as well as the means that will be used to enforce compliance. At a minimum, it includes, but is not limited to, the following elements:

- A statement on how the procedure will be used;
- The procedural steps needed to shut down, isolate, block, and secure machines or equipment;
- The steps designating the safe placement, removal, and transfer of lockout/tagout devices and who has the responsibility for them; and
- The specific requirements for testing machines or equipment to determine and verify the effectiveness of locks, tags, and other energy control measures.

The lockout/tagout procedure must include the following steps: (1) preparing for shutdown, (2) shutting down the machine(s) or equipment, (3) isolating the machine or equipment from the energy source(s), (4) applying the lockout or tagout device(s) to the energy-isolating device(s), (5) safely releasing all potentially hazardous stored or residual energy, and (6) verifying the isolation of the machine(s) or equipment prior to the start of service or maintenance work. In addition, before lockout or tagout devices are removed and energy is restored to the machines or equipment, certain steps must be taken to re-energize equipment after service is completed, including: (1) ensuring that machines or equipment components are operationally intact, and (2) notifying affected employees that lockout or tagout devices are being removed from each energy-isolating device by the employee who applied the device. (For more specific information about the requirements of the Standard, refer to 29 CFR 1910.147 OSHA General Industry Regulations.)

Energy-Isolating Devices

The employer's primary tool for providing protection under the Standard is the energy-isolating device, which is the mechanism that prevents the transmission or release of energy and to which all locks or tags (Figures 11.11 and 11.12) are attached. This device guards against accidental machine or equipment start-up or the unexpected re-energization of equipment during servicing or maintenance. There are two types of energy-isolating devices: those capable of being locked and those that are not capable of being locked. The Standard treats each of these two types of devices separately and spells out the employer's and employee's responsibilities for using each type. When the energy-isolating device cannot be locked out, the employer must use tagout. The employer may choose to modify or replace the device to make it capable of being locked. When using tagout, the employer must comply with all tagout-related provisions of the Standard and must train his or her employees in the following limitations of tags:

- Tags are essentially warning devices attached to energy-isolating devices and do not provide the physical restraint of a lock,
- When a tag is attached to an isolating means, it is not to be removed except by the person who applied it, and it is never to be bypassed, ignored, or otherwise defeated,
- Tags must be legible and understandable by all employees,
- Tags and their means of attachment must be made of materials that will withstand the environmental conditions encountered in the workplace,
- Tags may give operators a false sense of security. They are only one part of an overall energy control program, and
- Tags must be securely attached to the energy-isolating devices so that they cannot be detached accidentally during use.

If the energy-isolating device is lockable, a lock must be used unless the use of tags would provide protection at least as effective as locks and would ensure "full employee protection." Full employee protection includes complying with all tagout-related provisions plus implementing additional safety measures that can provide the level of safety equivalent to that obtained by using lockout. This might include removing and isolating a circuit element, blocking a controlling switch, opening an extra disconnecting device, or removing a valve handle to reduce the potential for any accidental energization.

Although OSHA acknowledges the existence of energy-isolating devices that cannot be locked out, the Standard clearly states that whenever major replacement, repair, renovation, or modification of machines or equipment is performed and whenever new machines or equipment are installed, the employer must ensure that the energy-isolating devices for such machines or equipment are lockable. Such modifications and/or new purchases are most effectively and efficiently made as part of the normal equipment replacement cycle. All newly purchased equipment must be lockable.

Requirements for Lockout/Tagout Devices

When attached to an energy-isolating device, both lockout and tagout devices are tools that the employer can use in accordance with the requirements of the Standard to help protect employees from hazardous energy. The lockout device provides protection by holding the energy-isolating device in the safe position, thus preventing the machine or equipment from becoming energized. The tagout device does so by identifying the energy-isolating device as a source of potential danger; it indicates that the energy-isolating device and the equipment being controlled may not be operated until the tagout device is removed. Whichever devices are used, they must be singularly identified, they must be the only devices used for controlling hazardous energy, and they must meet the following requirements:

- **Durable:** Lockout and tagout devices must withstand the environment to which they are exposed for the maximum duration of the expected exposure. Tagout devices must be constructed and printed so that they do not deteriorate or become illegible, especially when used in corrosive (acid and alkali chemicals) or wet environments.
- **Standardized:** Both lockout and tagout devices must be standardized according to either color, shape, or size. Tagout devices must also be standardized according to print and format.
- **Substantial:** Lockout and tagout devices must be substantial enough to minimize early or accidental removal. Locks

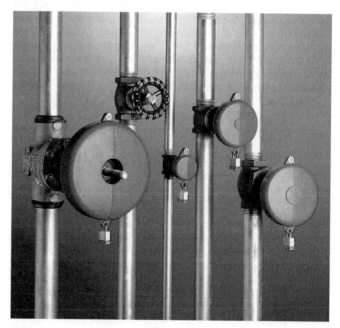

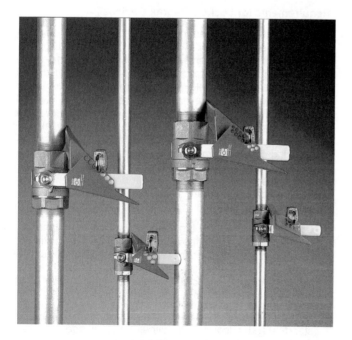

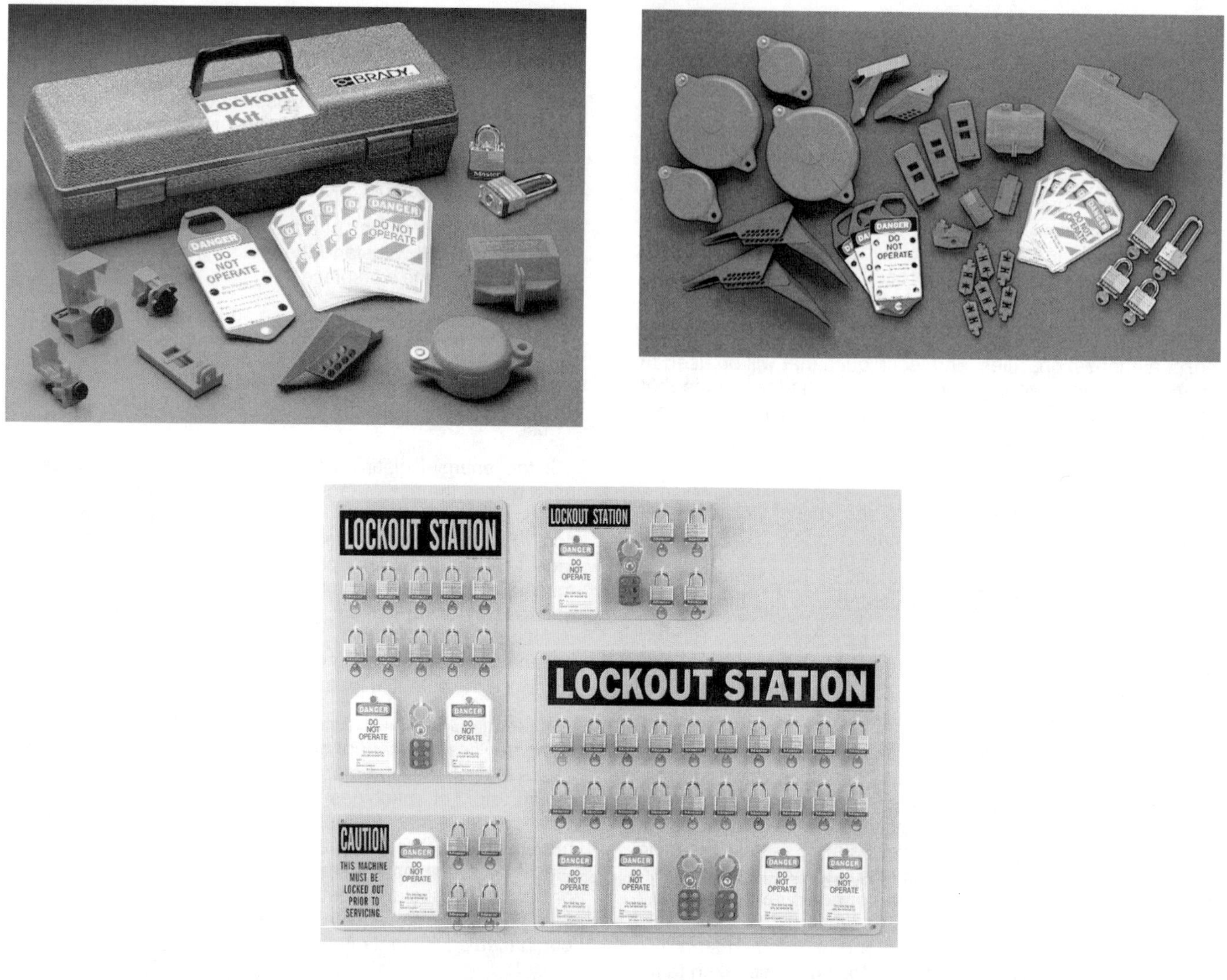

Fig. 11.11 Typical lockout devices

(Courtesy of Brady Worldwide)

Fig. 11.12 Typical lockout warning tags

must be substantial to prevent removal except by excessive force or special tools such as bolt cutters or other metal-cutting tools. The means of attaching tags must be non-reusable, attachable by hand, self-locking, and non-releasable, with a minimum unlocking strength of no less than 50 pounds. The device for attaching the tag also must have the general design and basic characteristics equivalent to a one-piece nylon cable tie that will withstand all environments and conditions.

- **Identifiable:** Locks and tags must clearly identify the employee who applies them. Tags must also warn against hazardous conditions if the machine or equipment is energized, and must include a legend such as the following:

  DO NOT START, DO NOT OPEN, DO NOT CLOSE, DO NOT ENERGIZE, DO NOT OPERATE.

11.824 Employee Training

The employer must provide effective initial training and retraining as necessary and must certify that such training has been given to all employees covered by the Standard. The certification record must contain each employee's name and dates of training. There are three types of employees: authorized, affected, and other. The amount and kind of training that each employee receives is based on: (1) the relationship of the employee's job to the machine or equipment being locked or tagged out, and (2) the degree of knowledge relevant to hazardous energy that the employee must possess.

Authorized employees are those who are charged with responsibility for implementing the energy control procedures and performing the service and maintenance. Affected employees, who are usually the machine operators or users and all other employees, need only be able to: (1) recognize when the control procedure is being implemented, and (2) understand the purpose of the procedure and the importance of not attempting to start up or use the equipment that has been locked or tagged out. Because an affected employee is not one who is performing the service or maintenance, that employee's responsibilities under the energy control program are simple: whenever a lockout or tagout device is in place on an energy-isolating device, the affected employee leaves it alone and does not attempt to operate the equipment.

Every training program must ensure that all employees understand the purpose, function, and restrictions of the energy control program and that authorized employees possess the knowledge and skills necessary for the safe application, use, and removal of energy controls.

11.825 Periodic Inspections

Periodic inspections must be performed at least annually to ensure that the energy control procedures (locks and tags) continue to be implemented properly and that the employees are familiar with their responsibilities under those procedures. In addition, the employer must certify that the periodic inspections have been performed. The certification must identify the machine or equipment on which the energy control procedure was used, the date of the inspection, the employees included in the inspection, and the name of the person performing the inspection. For lockout procedures, the periodic inspection must include a review, between the inspector and each authorized employee, of that employee's responsibilities under the energy control procedure being inspected. When a tagout procedure is inspected, a review on the limitations of tags, in addition to the above requirements, must also be included with each affected and authorized employee.

11.826 Application and Removal of Controls and Lockout/Tagout Devices

The established procedure of applying energy controls includes the specific elements and actions that must be implemented in sequence. These are briefly identified as follows:

1. Prepare for shutdown,
2. Shut down the machine or equipment,
3. Apply the lockout or tagout device,
4. Render safe all stored or residual energy, and
5. Verify the isolation and de-energization of the machine or equipment.

Before lockout or tagout devices are removed and energy is restored to the machine or equipment, the authorized employee(s) must take the following actions or observe the following procedures:

1. Inspect the work area to ensure that nonessential items have been removed and that the machine or equipment components are intact and capable of operating properly,
2. Check the area around the machine or equipment to ensure that all employees have been safely positioned or removed,
3. Notify affected employees immediately after removing locks or tags and before starting the equipment or machines, and
4. Make sure that locks or tags are removed *ONLY* by those employees who attached them. (In the very few instances when this is not possible, the device may be removed under the direction of the employer, provided that he or she strictly adheres to the specific procedures outline in the OSHA Standard.)

11.827 Additional Safety Requirements

Special circumstances exist when: (1) machines need to be tested or repositioned during servicing, (2) an outside contractor's personnel are at the work site, (3) servicing or maintenance is performed by a group rather than one specific person, and (4) shifts or personnel changes occur.

- **Testing or Positioning of Machines**

OSHA allows the temporary removal of locks or tags and the re-energization of the machine or equipment *ONLY* when necessary under special conditions, for example, when power is needed for the testing or positioning of machines, equipment, or components. The re-energization must be conducted in accordance with the sequence of steps listed below:

1. Clear the machines or equipment of tools and materials,
2. Remove employees from the area of the machines or equipment,
3. Remove the lockout or tagout devices as specified in the Standard,
4. Energize and proceed with testing or positioning, and
5. De-energize the entire system, isolate the machine or equipment from the energy source, and reapply lockout or tagout devices as specified.

- **Contract Personnel**

The on-site employer and the outside employer must inform each other of their respective lockout and tagout procedures.

Each employer must ensure that his or her personnel understand and comply with all restrictions and/or prohibitions of the other employer's energy control program.

- **Group Lockout or Tagout**

During all group lockout/tagout operations where the release of hazardous energy is possible, each authorized employee performing service or maintenance must be protected by their personal lockout or tagout device or a comparable mechanism that provides equivalent protection.

- **Shift or Personnel Changes**

Specific procedures must ensure the continuity of lockout or tagout protection during shift or personnel changes.

QUESTIONS

Write your answers in a notebook and then compare your answers with those on page 400.

11.8G What is the definition of "work" on machinery or equipment used for a "Lockout Policy"?

11.8H What information must be included in an energy control procedure?

11.8I What is the purpose of an energy-isolating device?

11.83 Personal Protection Equipment (PPE) (29 CFR 1910.132 Subpart I)

This regulation requires employers to ensure that personal protection equipment be provided, used, and maintained in a sanitary and reliable condition wherever it is necessary to prevent injury. This includes protection of any part of the body from hazards through absorption, inhalation, or physical contact. Hard hats, goggles, face shields, steel-toed shoes, respirators, aprons, gloves, and full-body suits are various forms of personal protection equipment (PPE).

Personal protection equipment should *NOT* be used as a substitute for engineering, work practice, and/or administrative controls; instead, it should be used in conjunction with these controls to provide for employee safety and health in the workplace. Personal protection equipment includes all clothing and other work accessories designed to create a barrier against workplace hazards.

The basic element of any management program for PPE should be an in-depth evaluation of the equipment needed to protect against the hazards at the workplace. Management dedicated to the safety and health of employees should use that evaluation to develop written safety policies and to set standard operating procedures (SOPs) for personnel and then train employees on the protective limitations of PPE, its proper use, and how to maintain the protection equipment.

Using personal protection equipment requires hazard awareness and training on the part of the user. Employees must be aware that the equipment does not eliminate the hazard. If the equipment fails, exposure will occur. To reduce the possibility of failure, equipment must be properly fitted and maintained in clean and serviceable condition. Defective or damaged equipment should never be used. Selection of the proper personal protection equipment for a job is important. Employers and employees must understand the equipment's purpose and its limitations. The equipment must not be altered or removed, even though an employee may find it uncomfortable. Even when employees provide their own equipment, the employer still must ensure the adequacy, proper maintenance, and sanitation of such equipment.

- **Example Safety Policies: Safety Vest Policy and Seat Belt Policy**

SAFETY VEST POLICY

The purpose of this policy is to establish guidelines for the wearing of safety vests (warning vests).

All employees working on or within ten (10) feet of a roadway traveled by motorized vehicles shall be provided with and required to wear warning vests of high visibility. Warning vests will be bright orange in color with reflectorized material attached thereon, as approved by the Safety Officer.

The definition of a roadway used for this policy is: "That portion of a highway improved, designed or ordinarily used by vehicular traffic." The following areas shall be considered to fall under the classification of a roadway: freeways, highways, state roads, county roads, township roads, secondary roads, streets, alleys, drives, avenues, parkways or on the shoulders or berms, or on the median adjacent to roadways.

All work shall be performed in strict adherence to the "Safety Vest Policy." No one shall perform or order to be performed any work contrary to this policy.

Failure to comply with this policy will result in disciplinary action.

SEAT BELT POLICY

The purpose of this policy is to establish guidelines for the wearing of seat belts.

All employees occupying any agency vehicle, while in motion at speeds greater than 5 miles per hour, shall wear both the lap and shoulder safety devices at all times. (*NOTE:* Gas or electric carts, intended for in-plant use only, are excepted from this policy.)

All vehicles that have defective or otherwise inoperative safety belt devices shall be removed from service until the deficiency has been corrected.

Failure to comply with this policy will result in disciplinary action.

- **Hazard Assessment and Equipment Selection**

Employers are required to assess the workplace to determine if hazards that require the use of personal protection equipment are present or are likely to be present. If hazards or the likelihood of hazards are found, employers must select and have affected employees use properly fitted PPE suitable for protection from existing hazards. Employers must certify in writing that a workplace hazard assessment has been performed.

- **Training**

Before doing work that requires use of personal protection equipment, employees must be trained to know when personal protection equipment is necessary; what type is necessary; how it is to be worn; and what its limitations are, as well as know its proper care, maintenance, useful life, and disposal. Employers are required to certify in writing that training has been carried out and that employees understand the written program. Each written certification must contain the name of each employee trained, the date(s) of training, and the subject (equipment or procedure) being certified.

11.830 Eye and Face Protection (*29 CFR 1910.133*)

Eye and face protection equipment is required by OSHA whenever there is a reasonable probability that it will prevent injuries. Employers must provide a type of protector suitable for work to be performed and employees must use the protectors. These rules also apply to supervisors and management personnel, and should apply to visitors while they are in hazardous areas. Suitable eye protectors must be provided where there is a potential for injury to the eyes or face from flying particles, molten metal, liquid chemicals, acids or caustic liquids, chemical gases or vapors, potentially injurious light radiation, or a combination of these hazards. Protectors must meet the following minimum requirements:

- Provide adequate protection against the particular hazard,
- Be reasonably comfortable when worn under the designated conditions,
- Fit snugly without interfering with the movements or vision of the wearer,
- Be durable, and
- Be capable of being disinfected.

Protectors must be easily cleanable and kept clean and in good repair, and the manufacturer's identification must be readily visible. OSHA and the National Society to Prevent Blindness recommend that emergency eye washes be placed in all hazardous locations. Also, first-aid instructions should be posted close to potential danger spots since any delay to immediate aid or an early mistake in dealing with an eye injury can result in lasting damage.

• Selection and Fit

Each eye, face, or face-and-eye protector is designed for a particular hazard. Base your selection of a protector on the kind and degree of hazard you wish to protect against. If you have a choice of devices, all of which would provide adequate protection, other factors such as worker comfort may be considered in making the selection.

Over the years, many types and styles of eye and face-and-eye protective equipment have been developed to meet the demands for protection against a variety of hazards. Goggles are manufactured in several styles for specific uses, such as protecting against dusts and splashes, and in chipper's, welder's, and cutter's models. Goggles come in a number of different styles: eyecups, flexible or cushioned goggles, plastic eyeshield goggles, and foundry worker's goggles. Many hard hats and nonrigid helmets are designed with face and eye protective equipment built in. Some general types of eye protectors include:

- Spectacles with protective lenses providing optical correction,
- Goggles worn over corrective spectacles without disturbing the adjustment of the spectacles, or
- Goggles that incorporate corrective lenses mounted behind the protective lenses.

Safety spectacles require special frames. Combinations of standard, daily-wear frames with safety lenses do not comply with the OSHA Standard. Also, fitting of goggles and safety spectacles should be done by someone skilled in the procedure. Prescription safety spectacles should be fitted only by qualified optical personnel.

Design, construction, tests, and use of eye and face protection purchased before July 5, 1994, must meet the requirements of ANSI Z87.1-1968 USA Standard Practice for Occupational and Educational Eye and Face Protection. Protective eye and face devices purchased after July 5, 1994, must comply with ANSI Z87.1- 1989, American National Standard Practice for Occupational and Educational Eye and Face Protection.

• Inspection and Maintenance

It is essential that the lenses of eye protectors be kept clean. Continuous vision through dirty lenses can cause eye strain—often an excuse for not wearing the eye protectors. Daily inspection and cleaning of the eye protector with soap and hot water, or with a cleaning solution and tissue, is recommended. Pitted lenses, like dirty lenses, can be a source of reduced vision. They should be replaced. Deep scratches or excessively pitted lenses are apt to break more readily.

Slack, worn-out, sweat-soaked, or twisted headbands do not hold the eye protector in proper position. Visual inspection can determine when the headband elasticity is reduced to a point beyond proper function.

Goggles should be kept in a case when not in use. Spectacles, in particular, should be given the same care as one's own glasses, since the frame, nose pads, and temples can be damaged by rough usage. Personal protection equipment that has been previously used should be disinfected before being issued to another employee. Also, when each employee is assigned protective equipment for extended periods, it is recommended that such equipment be cleaned and disinfected regularly.

Several methods for disinfecting eye-protective equipment are acceptable. The most effective method is to disassemble the goggles or spectacles and thoroughly clean all parts with soap and warm water. Carefully rinse all traces of soap, and replace defective parts with new ones. Swab thoroughly or completely immerse all parts for 10 minutes in a solution of germicidal deodorant fungicide. Remove the parts from the solution and suspend them in a clean place for air drying at room temperature or with heated air. Do not rinse after removing the parts from the solution because this will remove the germicidal residue, which retains its effectiveness after drying. Place the dry parts or assembled items in a clean, dust-proof container, such as a box, bag, or plastic envelope, to protect them until reissue.

- **Example Eye Protection Policy**

EYE PROTECTION POLICY

The purpose of this policy is to establish guidelines for the wearing of eye protection devices.

All employees shall wear proper eye protection devices when working in any facility capable of producing eye injury, including those under construction where the environment may contain physical and/or chemical agents or radiation.

Proper eye protection devices shall consist of, but not necessarily be limited to the following: safety glasses with side shields, splash goggles, face shields, welding helmets and welding goggles, all of a type specifically approved by the Safety Officer. Contact lenses are allowed in office areas only, unless written approval is obtained from the Safety Officer.

All work performed shall be performed in strict adherence to the "Eye Protection Procedure." No one shall perform or order to be performed any work contrary to this procedure.

Failure to comply with this policy will result in disciplinary action.

11.831 Head Protection (29 CFR 1910.135)

- **Introduction**

Prevention of head injuries is an important feature of every safety program. A survey by the Bureau of Labor Statistics (BLS) of accidents and injuries noted that most workers who suffered impact injuries to the head were not wearing head protection. The majority of workers were injured while performing their normal jobs at their regular work sites.

The survey showed that in most instances where head injuries occurred, employers had not required their employees to wear head protection. Of those workers wearing hard hats, all but 5 percent indicated that they were required by their employers to wear them. It was found that the vast majority of those who wore hard hats all or most of the time at work believed that hard hats were practical for their jobs. According to the report, in almost half of the accidents involving head injuries, employees knew of no action taken by employers to prevent such injuries from recurring.

Elimination or control of a hazard leading to an accident should, of course, be given first consideration, but many accidents causing head injuries are of a type difficult to anticipate and control. Where these conditions exist, head protection must be provided to eliminate injury. Head injuries are caused by falling or flying objects, or by bumping the head against a fixed object. Head protection, in the form of protective hats, must do two things: resist penetration and absorb the shock of a blow. This is accomplished by making the shell of the hat of a material hard enough to resist the blow, and by using a shock-absorbing lining composed of a headband and crown straps to keep the shell away from the wearer's skull. Protective hats also are used to protect against electric shock.

- **Criteria for Head Protection**

The standards recognized by OSHA for head protection purchased prior to July 5, 1994, are contained in ANSI Requirements for Industrial Head Protection, Z89.1-1969, and ANSI Requirements for Industrial Protective Helmets for Electrical Workers, Z89.2-1971. These documents should be consulted for details. The standards for protective helmets purchased after July 5, 1994, are contained in ANSI Personnel Protection-Protective Headwear for Industrial Workers-Requirements, Z89.1-1986. Later editions of these standards may be available.

- **Selection**

Each type and class of head protectors is intended to provide protection against specific hazardous conditions. An understanding of these conditions will help in selecting the right protection for the particular situation. Head protection is made in the following types and classes: (1) Type 1, helmets with full brim, not less than 1 1/4 inches wide; and (2) Type 2, brimless helmets with a peak extending forward from the crown.

For industrial purposes, three classes of helmets are recognized: (1) Class A, general service, limited voltage protection; (2) Class B, utility service, high-voltage helmets; and (3) Class C, special service, no voltage protection.

For firefighters, head protection must consist of a protective head device with ear flaps and a chin strap that meet the performance, construction, and testing requirements stated in Title 29 CFR, 1910.156(e)(5).

Class A helmets are intended for protection against impact hazards. They are used in mining, construction, shipbuilding, tunneling, lumbering, and manufacturing. Class B, utility service helmets, protect the wearer's head from impact and penetration by falling or flying objects and from high-voltage shock and burns. They are used extensively by electrical workers. The safety helmets in Class C are designed specifically for lightweight comfort and impact protection. This class is usually manufactured from aluminum and offers no dielectric protection. Class C helmets are used in certain construction and manufacturing occupations, oil fields, refineries, and chemical plants where there is no danger from electrical hazards or corrosion. They also are used on occasions where there is a possibility of bumping the head against a fixed object.

Materials used in helmets should be water-resistant and slow burning. Each helmet consists essentially of a shell and suspension. Ventilation is provided by a space between the headband and the shell. Each helmet should be accompanied by instructions explaining the proper method of adjusting and replacing the suspension and headband. The wearer should be able to identify the type of helmet by looking inside the shell for the manufacturer's name, the ANSI designation, and the class of helmet; for example: Manufacturer's Name, ANSI Z89.1-1969 (or later year), Class A.

- **Fit**

Headbands are adjustable in 1/8-size increments. When the headband is adjusted to the right size, it provides sufficient clearance between the shell and the headband. The removable or replaceable sweatband should cover at least the forehead portion of the headband. The shell should be of one-piece seamless construction and designed to resist the impact of a blow from falling material. The internal cradle of the headband and sweatband forms the suspension. Any part that comes into contact with the wearer's head must not be irritating to normal skin.

- **Inspection and Maintenance**

A common method of cleaning shells is dipping them in hot water (approximately 140°F) containing a good detergent for at least a minute. Shells should then be scrubbed and rinsed in clear hot water. After rinsing, the shell should be carefully inspected for any signs of damage.

All components, shells, suspensions, headbands, sweatbands, and any accessories should be visually inspected daily for signs of dents, cracks, penetration, or any other damage that might reduce the degree of safety originally provided.

Users are cautioned that if unusual conditions occur (such as higher or lower extreme temperatures than described in the standards), or if there are signs of abuse or mutilation of the helmet or any component, the margin of safety may be reduced. If damage is suspected, helmets should be replaced or representative samples tested in accordance with procedures contained in ANSI Z89.1-1986. This discussion references national consensus standards, for example, ANSI standards, that were adopted into OSHA regulations. Employers are encouraged to use up-to-date national consensus standards that provide employee protection equal to or greater than that provided by OSHA standards.

Helmets should not be stored or carried on the rear window shelf of an automobile, since sunlight and extreme heat may adversely affect the degree of protection. Also, be sure to follow the manufacturer's recommendations for use of paint and cleaning materials on safety helmets. Some paints and thinners may damage the shell and reduce protection by physically weakening it or negating electrical resistance.

- **Example Hard Hat Policy**

HARD HAT POLICY

The purpose of this policy is to establish guidelines for the wearing of hard hats.

All employees working on or in any agency facility, including those under construction, will be provided with and shall be required to wear a hard hat at all times. All hard hats shall be of a type specifically approved by the Safety Officer. Exceptions to this policy may be granted only by the supervisor, provided this exclusion has been reviewed and approved by the Safety Officer. (*NOTE:* Bump caps shall not be considered a substitute for hard hats.)

The possession and care of hard hats is the responsibility of each employee. Any defective or damaged hard hat shall be taken out of service immediately. The altering of hard hats in any way, manner, or form is strictly prohibited.

All work shall be performed in strict adherence to the "Hard Hat Policy." No one shall perform or order to be performed any work contrary to this policy.

Failure to comply with this policy will result in disciplinary action.

QUESTIONS

Write your answers in a notebook and then compare your answers with those on page 400.

11. 8J What training must employees have before doing work that requires use of personal protection equipment?

11.8K When is eye and face protection equipment required by OSHA?

11.8L What must be accomplished by head protection in the form of protective hats?

11.9 RECORDS

Unfortunately, accidents do happen in spite of all of our efforts. To prevent similar accidents from happening in the future, an "Accident Report" should be completed as soon as possible after an accident. Figure 11.13 is a typical "Accident Report." In addition there are regulatory reporting requirements that must be completed. See Appendix C for additional regulatory information.

11.10 HOMELAND DEFENSE

World events in recent years have heightened concern in the United States over the security of the critical wastewater infrastructure. The Nation's wastewater infrastructure, consisting of approximately 16,000 publicly owned wastewater treatment plants, 100,000 major pumping stations, 600,000 miles of sanitary sewers and another 200,000 miles of storm sewers, is one of America's most valuable resources, with treatment and collection systems valued at more than $2 trillion.

Taken together, the sanitary and storm sewers form an extensive network that runs near or beneath key buildings and roads and is physically close to many communication and transportation networks. Significant damage to the Nation's wastewater facilities or collection systems would result in: loss of life, catastrophic environmental damage to rivers, lakes, and wetlands, contamination of drinking water supplies, long-term public health impacts, destruction of fish and shellfish production, and disruption to commerce, the economy, and our normal way of life.

Wastewater collection facilities have been identified as a target for international and domestic terrorism. This knowledge, coupled with the responsibility of the facility to provide a safe and healthful workplace, requires that management establish rules to protect the workers as well as the facilities. Emergency action and fire prevention plans must identify what steps need to be taken when the threat analysis indicates a potential for attack. These plans must be in writing and be practiced periodically so that all workers know what actions to take.

Some actions that should be taken at all times to reduce the possibility of a terrorist attack are:

- Ensure that all visitors sign in and out of the facilities with a positive ID check,
- Reduce the number of visitors to a minimum,
- Discourage parking by the public near critical buildings to eliminate the chances of car bombs,
- Be cautious with suspicious packages that arrive,
- Be aware of the hazardous chemicals used and how to defend against spills,
- Keep emergency numbers posted near telephones and radios,

SAFETY 9 (Rev. 8/74)

CIVIL SERVICE
USE ONLY

FORM 67 SENT_____

SUPERVISORS REPORT OF ACCIDENT

THIS REPORT IS TO BE COMPLETED AND SENT TO THE SAFETY OFFICE WITHIN 72 HOURS AFTER THE TIME OF THE INJURY

To Be Completed and Signed by the Injured's Supervisor

DEPARTMENT:________________ SECTION:________________ INDEX NO. ______

NAME OF INJURED EMPLOYEE:________________ SOCIAL SECURITY No:________________

HOME ADDRESS OF EMPLOYEE:________________ PHONE:________________

DATE OF BIRTH: ________________ SEX: M __ F __; WAGE AT TIME OF ACCIDENT:________________

NUMBER OF HOURS WORKED: PER DAY:______, PER WEEK:______, NUMBER OF DAYS PER WEEK:______

CLASSIFICATION:________________ DATE OF HIRE: ________________

PLACE OF ACCIDENT:________________ CITY/TOWN:________________

DATE OF ACCIDENT:________ TIME:________ DATE REPORTED:________ TIME:________

DID EMPLOYEE RETURN TO WORK ON DATE OF INJURY?________ LOST TIME:________ DAYS/HRS.

WAS EMPLOYEE OFF WORK BEYOND DATE OF INJURY?________ IF SO, LAST DATE WORKED:________

NATURE OF INJURY (Specify Part of Body Injured):________________

____

WAS EMPLOYEE ACTING IN REGULAR LINE OF DUTY WHEN INJURED?________________

IF NO, EXPLAIN:________________

____

HOW DID THE ACCIDENT OCCUR:________________

____

____

____

WAS FIRST AID GIVEN:________________ BY WHOM:________________

DOCTOR:________________ ADDRESS:________________

HOSPITAL (IF ANY):________________ ADDRESS:________________

WHAT MACHINE, TOOL, SUBSTANCE, OR OBJECT WAS MOST CLOSELY CONNECTED WITH THE ACCIDENT:________

____

WERE MECHANICAL GUARDS OR OTHER SAFEGUARDS PROVIDED:________________

WERE MECHANICAL GUARDS OR OTHER SAFEGUARDS USED: ________________

WHAT, IN YOUR OPINION, CAUSED THE ACCIDENT:________________

____

DESCRIBE ANY UNSAFE ACT:________________

____

DESCRIBE ANY UNSAFE CONDITION:________________

____

WHAT HAS BEEN DONE TO PREVENT A SIMILAR ACCIDENT:________________

____

____

WITNESSES:________________

____

Complete and Return Original to
SAFETY OFFICE
DEPARTMENT OF PERSONNEL MANAGEMENT
Duplicate to Departmental File
Additional Copies as Desired

SIGNED:________________

DATE:________________

PHONE:________________

Please Use Reverse Side For Further Particulars

Fig. 11.13 Accident report

- Patrol the facilities frequently, looking for suspicious activity or behavior, and
- Maintain, inspect, and use your PPE (hard hats, respirators).

The following recommendations by the EPA[6] include many straightforward, common-sense actions a utility can take to increase security and reduce threats from terrorism.

Guarding Against Unplanned Physical Intrusion

- Lock all doors and set alarms at your office, pumping stations, treatment plants, and vaults, and make it a rule that doors are locked and alarms are set.
- Limit access to facilities and control access to pumping stations, chemical and fuel storage areas, giving close scrutiny to visitors and contractors.
- Post guards at treatment plants and post "Employee Only" signs in restricted areas.
- Secure hatches, metering vaults, manholes, and other access points to the sanitary collection system.
- Increase lighting in parking lots, treatment bays, and other areas with limited staffing.
- Control access to computer networks and control systems and change the passwords frequently.
- Do not leave keys in equipment or vehicles at any time.

Making Security a Priority for Employees

- Conduct background security checks on employees at hiring and periodically thereafter.
- Develop a security program with written plans and train employees frequently.
- Ensure all employees are aware of established procedures for communicating with law enforcement, public health, environmental protection, and emergency response organization.
- Ensure that employees are fully aware of the importance of vigilance and the seriousness of breaches in security.
- Make note of unaccompanied strangers on the site and immediately notify designated security officers or local law enforcement agencies.
- If possible, consider varying the timing of operational procedures so that, to anyone watching for patterns, the pattern changes.
- Upon the dismissal of an employee, change pass codes and make sure keys and access cards are returned.
- Provide customer service staff with training and checklists of how to handle a threat if it is called in.

Coordinating Actions for Effective Emergency Response

- Review existing emergency response plans and ensure that they are current and relevant.
- Make sure employees have the necessary training in emergency operating procedures.
- Develop clear procedures and chains-of-command for reporting and responding to threats and for coordinating with emergency management agencies, law enforcement personnel, environmental and public health officials, consumers, and the media. Practice the emergency procedures regularly.
- Ensure that key utility personnel (both on and off duty) have access to critical telephone numbers and contact information at all times. Keep the call list up to date.
- Develop close relationships with local law enforcement agencies and make sure they know where critical assets are located. Ask them to add your facilities to their routine rounds.
- Work with local industries to ensure that their pretreatment facilities are secure.
- Report to county or state health officials any illness among the employees that might be associated with wastewater contamination.
- Immediately report criminal threats, suspicious behavior, or attacks on wastewater utilities to law enforcement officials and the nearest field office of the Federal Bureau of Investigation.

Investing in Security and Infrastructure Improvements

- Assess the vulnerability of the collection system, major pumping stations, wastewater treatment plants, chemical and fuel storage areas, outfall pipes, and other key infrastructure elements.
- Assess the vulnerability of the storm water collection system. Determine where large pipes run near or beneath government buildings, banks, commercial districts, industrial facilities, or are next to major communication and transportation networks. Move as quickly as possible with the most obvious and cost-effective physical improvements, such as perimeter fences, security lighting, and tamper-proofing manhole covers and valve boxes.
- Improve computer system and remote operational security.
- Use local citizen watches.
- Seek financing for more expensive and comprehensive system improvements.

The U.S. Alert System (Figure 11.14) is a color-coded system that identifies the potential for terrorist activity and suggests specific actions to be taken. Your safety plan should identify the actions that your facility will take when the threat level changes. Tables 11.3 and 11.4 show examples of security measures that should be taken to improve safety at a wastewater collection and treatment facility when the threat level is YELLOW and when it is ORANGE. (The utility's safety plan should include similar lists of actions for the RED, BLUE, and GREEN levels as well.)

---

[6] *Adapted from "What Wastewater Utilities Can Do Now to Guard Against Terrorist and Security Threats," U.S. Environmental Protection Agency, Office of Wastewater Management, October 2001.*

	COLOR	RISK LEVEL AND SUGGESTED ACTIONS
SEVERE SEVERE RISK OF TERRORIST ATTACKS	RED	Severe risk of terrorist attacks Close unnecessary facilities, pre-position emergency response teams
HIGH HIGH RISK OF TERRORIST ATTACKS	ORANGE	High risk of terrorist attacks Restrict access, coordinate with local law enforcement
ELEVATED SIGNIFICANT RISK OF TERRORIST ATTACKS	YELLOW	Significant risk, elevated condition Coordinate emergency action plans with agencies, increase surveillance
GUARDED GENERAL RISK OF TERRORIST ATTACKS	BLUE	General risk, guarded condition Update procedures, check communications lines
LOW LOW RISK OF TERRORIST ATTACKS	GREEN	Low risk of terrorist attacks Routine planning and training, establish programs

Fig. 11.14 Threat level categories established by the U.S. Department of Homeland Defense

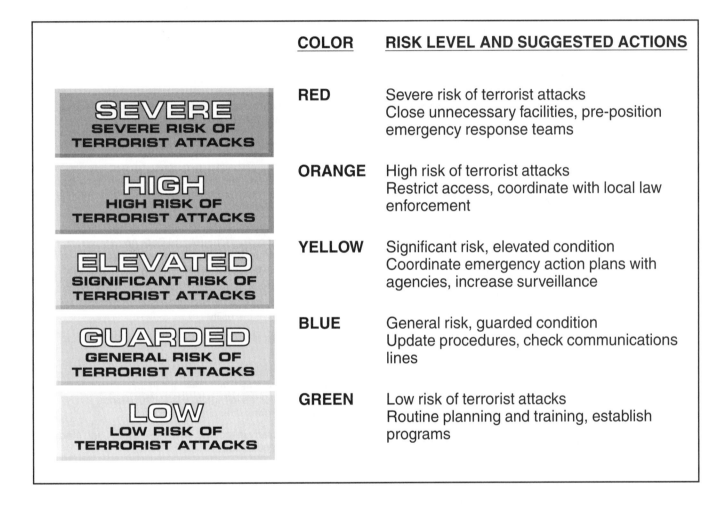

TABLE 11.3 SECURITY MEASURES FOR THREAT LEVEL YELLOW (CONDITION ELEVATED)

Continue to introduce all measures listed in BLUE: Condition Guarded.	
Detection	**Prevention**
• To the extent possible, increase the frequency and extent of monitoring the flow coming into and leaving the treatment facility and review results against baseline quantities. Increase review of operational and analytical data (including customer complaints) with an eye toward detecting unusual variability (as an indicator of unexpected changes in the system). Variations due to natural or routine operational variability should be considered first. • Increase surveillance activities in collection and treatment facilities.	• Carefully review all facility tour requests before approving. If allowed, implement security measures to include list of names prior to tour, request identification of each attendee prior to tour, prohibit backpacks, duffle bags, and cameras, and identify parking restrictions. • On a daily basis, inspect the interior and exterior of buildings in regular use for suspicious activity or packages, signs of tampering, or indications of unauthorized entry. • Implement mail room security procedures. Follow guidance provided by the United States Postal Service.
Preparedness	**Protection**
• Continue to review, update and test emergency response procedures and communication protocols. • Establish unannounced security spot checks (such as verification of personal identification and door security) at access control points for critical facilities. • Increase frequency for posting employee reminders of the threat situation and about events that constitute security violations. • Ensure employees understand notification procedures in the event of a security breach. • Conduct security audit of physical security assets, such as fencing and lights, and repair or replace missing/broken assets. Remove debris from along fence lines that could be stacked to facilitate scaling. • Maximize physical control of all equipment and vehicles; make them inoperable when not in use (for example, lock steering wheels, secure keys, chain, and padlock on front-end loaders). • Review draft communications on potential incidents; brief media relations personnel of potential for press contact and/or issuance of press releases. • Ensure that list of sensitive customers (such as government agencies and industrial users) within the service area is accurate and shared with appropriate public health officials. • Contact neighboring wastewater utilities to review coordinated response plans and mutual aid during emergencies. • Review whether critical replacement parts are available and accessible. • Identify any work/project taking place in proximity to events where large attendance is anticipated. Consult with the event organizers and local law enforcement regarding contingency plans, security awareness, and site accessibility and control.	• Verify the identity of all persons entering the wastewater utility. Mandate visible use of identification badges. Randomly check identification badges and cards of those on the premises. • At the discretion of the facility manager or security director, remove all vehicles and objects (such as trash containers) located near mission critical facility security perimeters and other sensitive areas. • Verify the security of critical information systems (for example, Supervisory Control and Data Acquisition (SCADA), Internet, e-mail) and review safe computer and Internet access procedures with employees to prevent cyber intrusion. • Consider steps needed to control access to all areas under the jurisdiction of the utility. • Implement critical infrastructure facility surveillance and security plans. • At the beginning and end of each work shift, as well as at other regular and frequent intervals, inspect the interior and exterior of buildings in regular use for suspicious packages, persons, and circumstances. • Lock and regularly inspect all buildings, rooms, and storage areas not in regular use.

TABLE 11.4 SECURITY MEASURES FOR THREAT LEVEL ORANGE (CONDITION HIGH)

Continue to introduce all measures listed in YELLOW: Condition Elevated.	
Detection	**Prevention**
• Increase the frequency and extent of monitoring activities. Review results against baseline measurements. • Confirm that county and state health officials are on high alert and will inform the utility of any potential wastewater-borne illnesses. • If a neighborhood watch-type program is in place, notify the community and request increased awareness.	• Discontinue tours and prohibit public access to all operational facilities. • Consider requesting increased law enforcement surveillance, particularly of critical assets and otherwise unprotected areas. • Limit access to computer facilities. No outside visitors. • Increase monitoring of computer and network intrusion detection systems and security monitoring systems.
Preparedness	**Protection**
• Confirm that emergency response and laboratory analytical support network are ready for deployment 24 hours per day, 7 days a week. • Reaffirm liaison with local police, intelligence, and security agencies to determine likelihood of an attack on the wastewater collection or treatment utility personnel and facility and consider appropriate protective measures (such as road closing and extra surveillance). • Practice communications procedures with local authorities and others cited in the facility's emergency response plan. • Post frequent reminders for staff and contractors of the threat level, along with a reminder of what events constitute security violations. • Ensure employees are fully aware of emergency response communication procedures and have access to contact information for relevant law enforcement, public health, environmental protection, and emergency response organizations. • Have alternative water supply plan ready to implement (for example, bottled water delivery for employees and other critical business uses). • Place all emergency management and specialized response teams on full alert status. • Ensure personal protective equipment (PPE) and specialized response equipment is checked, issued, and readily available for deployment. • Review all plans, procedures, guidelines, personnel details, and logistical requirements related to the introduction of a higher threat condition level.	• Evaluate the need to staff the wastewater treatment facility at all times. • Increase security patrol activity to the maximum level sustainable and ensure tight security in the vicinity of mission critical facilities. Vary the timing of security patrols. • Request employees change their passwords on critical information management systems. • Limit building access points to the absolute minimum, strictly enforce entry control procedures. Identify and protect all designated vulnerable points. Give special attention to vulnerable points outside of the critical facility. • Lock all exterior doors except the main facility entrance(s). Check all visitors' purpose, intent, and identification. Ensure that contractors have valid work orders. Require visitors to sign in upon arrival; verify and record their identifying information. Escort visitors at all times when they are in the facility.

11.11 ADDITIONAL READING

1. An excellent source of safety information is the Water Environment Federation (WEF), 601 Wythe Street, Alexandria, VA 22314-1994. Important WEF publications include:
   a. *SAFETY AND HEALTH IN WASTEWATER SYSTEMS* (MOP 1). Order No. MO2001. Price to members, $26.74; nonmembers, $36.74; price includes cost of shipping and handling.
   b. *WASTEWATER COLLECTION SYSTEMS MANAGEMENT* (MOP 7). Order No. M05000. Price to members, $56.75; nonmembers, $76.75; price includes cost of shipping and handling.
2. "Plant Safety" by Robert Reed and revised by Russ Armstrong, Chapter 14 in *OPERATION OF WASTEWATER TREATMENT PLANTS*, Volume II. Obtain from the Office of Water Programs, California State University, Sacramento, 6000 J Street, Sacramento, CA 95819-6025. Price, $45.00.
3. *A GUIDE TO SAFETY IN CONFINED SPACES*, DHHS (NIOSH) Publication No. 87-113. Available free of charge from the National Institute for Occupational Safety and Health, NIOSH Publications, 4676 Columbia Parkway, Mail Stop C-13, Cincinnati, OH 45226-1998.
4. *EXCAVATION SAFETY: A GUIDE TO OSHA COMPLIANCE AND INJURY PREVENTION* by Carl O. Morgan. Obtain from ABS Consulting, Government Institutes Division, PO Box 846304, Dallas, TX 75284-6304. Code No. 959. ISBN 0-86587-959-1. Price, $85.00, plus $6.00 shipping and handling.
5. Contact other utility agencies (such as gas, electric, and telephone) and find out what kind of safety programs they have that you can use. Their safety officers may be a good source for topic leaders on common problems or procedures.

11.12 ACKNOWLEDGMENT

The author of the safety chapter which appeared in the First and Second Editions was Mr. Glenn Davis. Many of the concepts developed by Mr. Davis in the original Chapter 10, "Safety Programs for Collection System Operators," have been included in this revised chapter.

Revisions of this chapter for the Sixth Edition are based on material from the U.S. Department of Labor, Bureau of Labor Statistics, and the Occupational Safety and Health Administration (OSHA).

QUESTIONS

Write your answers in a notebook and then compare your answers with those on page 400.

11.9A Why should an accident report be prepared after every accident?

11.9B What information should be included in an accident report?

11.10A What are the color codes and risk levels for the U.S. Terrorist Alert System?

Please answer the discussion and review questions next.

DISCUSSION AND REVIEW QUESTIONS

Chapter 11. SAFETY/SURVIVAL PROGRAMS FOR COLLECTION SYSTEM OPERATORS

Write the answers to these questions in your notebook. The purpose of these questions is to indicate to you how well you understand the material in the chapter.

1. Why are the safety chapters the most important chapters in this entire course?
2. How do the three common factors generally missing when an accident occurs help prevent accidents?
3. Under what circumstances are falling objects a hazard to collection system operators?
4. How can operators minimize smashing knuckles and thumbs?
5. Who should work in or near explosive mixtures?
6. What toxic or poisonous gases may be found in collection systems?
7. How can an operator determine if an oxygen-deficient condition exists in a manhole?
8. Under what circumstances should a collection system operator *NOT* perform electrical work?
9. Why should operators share a larger portion of the responsibility for developing effective safety/survival practices?
10. What are the benefits from a safety/survival program?
11. What is a tailgate safety session?
12. What safety policies should be established by a collection system agency to ensure that all operators follow safe procedures?
13. What should a Safety Committee do?
14. Why must each employer set up a safety and health program?
15. What should happen if hazardous conditions are detected during entry into a confined space?
16. How is the OSHA lockout/tagout Standard applied?
17. What is the purpose of the OSHA Personal Protection Equipment (PPE) regulation?

SUGGESTED ANSWERS

Chapter 11. SAFETY/SURVIVAL PROGRAMS FOR COLLECTION SYSTEM OPERATORS

Answers to questions on page 364.

11.0A The three common factors that are important to prevent accidents are: (1) AWARENESS, (2) RESPONSIBILITY, and (3) COMMITMENT.

11.0B The skill that most marks the professionalism of the collection system operator is the operator's ability to carry out work assignments safely.

11.0C The overall objective of this chapter is to make operators *AWARE* of the hazards of working in the collection system environment.

Answers to questions on page 368.

11.1A Hazards encountered by collection system operators in the performance of routine daily tasks and work assignments include: (1) slips, (2) falling objects, (3) infections and infectious diseases, (4) lacerations and contusions, (5) falls, (6) explosions, and (7) poisonous or toxic gases.

11.1B Slips and falls are most likely to occur when working in or around wet wells and metering flumes where slime can accumulate and surfaces are wet. When working in northern climates, snow, ice, and below-freezing temperatures can create slippery surfaces.

11.1C There has never been a recorded case of AIDS that can be attributed to working in the wastewater collection and treatment industry.

11.1D An explosion in a sewer can be set off by a spark from an automobile, an improper tool, or even a shoe nail, as well as the more obvious open flame or cigarette, when the mixture of air and gas is within the level of the explosive limits.

11.1E Hydrogen sulfide gas is the gas most likely to be encountered because it is generated during the anaerobic decomposition of organic matter in wastewater. It is also the most dangerous gas because it is highly toxic and, being heavier than air, tends to collect in the bottom of confined spaces or underground structures.

Answers to questions on page 370.

11.1F An operator might encounter oxygen-deficient atmospheres during the new construction inspection period inside pipelines and any time an operator is working inside confined spaces such as manholes, sewer lines and wet wells.

11.1G Traffic mishaps are a high risk area for collection system operators because access to a collection system is frequently from streets and roadways.

11.1H Collection system operators are most likely to find black widow spiders and scorpions in collection system facilities and structures such as manholes, metering enclosures and electrical control enclosures.

11.1I Permanent hearing loss can result if an operator is exposed to engine-driven pumps, standby power generators, large blowers, air compressors, jackhammers, concrete cut-off saws or high-velocity cleaners over a period of time unless adequate hearing protection is worn.

11.1J Collection system operator injury frequency rates were 4.5 times greater and severity rates were almost five times greater than the average rates for industrial workers in 1984.

Answers to questions on page 377.

11.2A Formal accident surveys do not accurately portray the severity of the safety problem because many agencies do not participate in the surveys.

11.3A The basis for a safety/survival program should be: **SAFE OPERATION SHALL TAKE PRECEDENCE OVER EXPEDIENCY OR SHORT CUTS.**

11.4A The three important levels of the structure within an agency for a safety program consist of the following:

1. Administration Department which states, "We will have a safety program";
2. Safety Department which says, "We will help with the resources"; and
3. Safety Committee which is "where the corrective action is."

Answers to questions on page 378.

11.5A The responsibilities of "Top Administration" for a safety/survival program include: (1) promote the concept, (2) approve policies and procedures, (3) require adherence to safety procedures, (4) ensure adequate budget resources, and (5) monitor effectiveness of the program and communicate with regulatory agencies as required.

11.5B The responsibilities of the "Safety Department Staff" for a safety/survival program include providing: (1) safety resources, (2) facility safety inspections, (3) program support in terms of training, and (4) evaluation of safety procedures with revisions as required.

11.5C Supervisors and managers in a safety/survival program are involved in: (1) an awareness of safety rules and regulations, (2) the safety of other personnel who work in or pass through their area of supervision, (3) job safety instruction and new employee orientation, (4) safety training within their area, (5) identifying unsafe practices and conditions and taking corrective action, (6) investigating injury-producing accidents and property damage accidents, (7) making a firm *COMMITMENT* to a safety/survival program, and (8) enforcing the use of personal protective equipment.

11.5D The responsibilities of operators in the field with regard to a safety/survival program include: (1) understanding and following all safety procedures and policies, (2) making a *COMMITMENT* to the safety/survival program, and (3) assuming the *RESPONSIBILITY* to make the program a part of the everyday routine.

Answers to questions on page 379.

11.6A Monthly safety meetings are more comprehensive than tailgate safety sessions. Monthly meetings can last up to an hour or more, depending on the topic, as compared to ten minutes for tailgate safety sessions.

11.6B Safety programs can be conducted either by operators or by employees of outside agencies.

11.7A A "Safety Committee" could consist of representatives from each area of the collection system such as office personnel, electrical/mechanical maintenance, and operators. If the collection system staff is small, the committee could consist of representatives from other departments such as street, water, parks, and other areas which share many of the same safety/survival problems.

11.7B If an accident resulted from an unsafe act, use of the wrong tool, or improper use of a piece of equipment, then training procedures need to be changed to correct the safety hazard.

Answers to questions on page 381.

11.8A Frequency of training requirements for Personal Protection Equipment (PPE) include: (1) prior to use or changes, and (2) annual training is advisable.

11.8B The core elements of a utility's safety and health program include:

- Management leadership and employee participation,
- Hazard identification and assessment,
- Hazard prevention and control,
- Information and training, and
- Evaluation of program effectiveness.

Answers to questions on page 384.

11.8C Employers must inform employees exposed to permit-required confined spaces of the existence, location, and danger posed by the confined spaces.

11.8D The internal atmosphere of a confined space must be tested first for oxygen content, second, for flammable gases and vapors, and third, for potential toxic air contaminants before any employee enters.

11.8E A confined space entry permit signed by the entry supervisor must be posted at entrances to the confined space or otherwise made available to entrants before they enter a permit space.

11.8F When operators work in confined spaces, additional training is required when: (1) the job duties change, (2) there is a change in the permit-space program or the permit space operation presents a new hazard, and (3) when an employee's job performance shows deficiencies.

Answers to questions on page 389.

11.8G The definition of "work" on machinery or equipment used for a "Lockout Policy" is: "maintenance, inspection, cleaning, adjusting, servicing, or clearing of blocked or jammed machinery or equipment."

11.8H An energy control procedure must outline the scope, purpose, authorization, rules, and techniques that will be used to control hazardous energy sources as well as the means that will be used to enforce compliance.

11.8I The energy-isolating device is a mechanism which prevents the transmission or release of energy and to which all locks or tags are attached. This device guards against accidental start-up or the unexpected re-energization of equipment during servicing or maintenance.

Answers to questions on page 392.

11.8J Before doing work that requires use of personal protection equipment, employees must be trained to know when personal protection equipment is necessary; what type is necessary; how it is to be worn; and what its limitations are; as well as know its proper care, maintenance, useful life, and proper disposal methods.

11.8K Eye and face protection equipment is required by OSHA whenever there is a reasonable probability that it will prevent injuries.

11.8L Head protection, in the form of protective hats, must do two things: resist penetration and absorb the shock of a blow.

Answers to questions on page 398.

11.9A An accident report must be prepared by the supervisor after every accident to prevent similar accidents from happening in the future.

11.9B Information that should be recorded in an accident report includes:

a. Identification of injured person,
b. Nature of the injuries,
c. Circumstances at the time of the accident,
d. First aid or medical referral given,
e. List of witnesses who saw the accident, and
f. Corrective action taken to prevent similar accidents.

11.10A

COLOR	RISK LEVEL AND SUGGESTED ACTIONS
RED	Severe risk of terrorist attacks
ORANGE	High risk of terrorist attacks
YELLOW	Significant risk, elevated condition
BLUE	General risk, guarded condition
GREEN	Low risk of terrorist attacks

APPENDIX A

ACCIDENT PREVENTION PROGRAM (APP)

ACCIDENT PREVENTION PROGRAM (APP)

(Enter your Company Name here)

Please customize this ACCIDENT PREVENTION PROGRAM according to your workplace. Also, your written ACCIDENT PREVENTION PROGRAM can only be effective if it is put into practice!

MANAGEMENT COMMITMENT

Safety Policy

(Customize by adding your company name here) places a high value on the safety of its employees. ***(Customize by adding your company name here)*** is committed to providing a safe workplace for all employees and has developed this program for injury prevention to involve management, supervisors, and employees in identifying and eliminating hazards that may develop during our work process.

It is the basic safety policy of this company that no task is so important that an employee must violate a safety rule or take a risk of injury or illness in order to get the job done.

Employees are required to comply with all company safety rules and are encouraged to actively participate in identifying ways to make our company a safer place to work.

Supervisors are responsible for the safety of their employees and as a part of their daily duties must check the workplace for unsafe conditions, watch employees for unsafe actions and take prompt action to eliminate any hazards.

Management will do its part by devoting the resources necessary to form a safety committee composed of management and elected employees. We will develop a system for identifying and correcting hazards. We will plan for foreseeable emergencies. We will provide initial and ongoing training for employees and supervisors. And, we will establish a disciplinary policy to insure that company safety policies are followed.

Safety is a team effort—let us all work together to keep this a safe and healthy workplace.

(Customize by adding any additional policy items that you may have and/or deleting any that do not apply to your company.)

Safety and Health Responsibilities

Manager Responsibilities:

1. Insure that a plant wide safety committee is formed and is carrying out its responsibilities as described in this program.
2. Insure that sufficient employee time, supervisor support, and funds are budgeted for safety equipment, training and to carry out the safety program.
3. Evaluate supervisors each year to make sure they are carrying out their responsibilities as described in this program.
4. Insure that incidents are fully investigated and corrective action taken to prevent the hazardous conditions or behaviors from happening again.
5. Insure that a record of injuries and illnesses is maintained and posted as described in this program.
6. Set a good example by following established safety rules and attending required training.
7. Report unsafe practices or conditions to the supervisor of the area where the hazard was observed.

(Customize by adding any additional management responsibilities that you may have and/or deleting any that do not apply to your company.)

Supervisor Responsibilities:

1. Insure that each employee you supervise has received an initial orientation *before* beginning work.
2. Insure that each employee you supervise is competent or receives training on safe operation of equipment or tasks *before* starting work on that equipment or project.

3. Insure that each employee receives required personal protection equipment (PPE) *before* starting work on a project requiring PPE.
4. Do a daily walk-around safety check of the work area. Promptly correct any hazards you find.
5. Observe the employees you supervise working. Promptly correct any unsafe behavior. Provide training and take corrective action as necessary. Document employee evaluations.
6. Set a good example for employees by following safety rules and attending required training.
7. Investigate all incidents in your area and report your findings to management.
8. Talk to management about changes to work practices or equipment that will improve employee safety.

(Customize by adding any additional supervisor responsibilities that you may have and/or deleting any that do not apply to your company.)

Employee Responsibilities:

1. Follow safety rules described in this program, Washington Industrial Safety and Health Act (WISHA) safety standards and training you receive.
2. Report unsafe conditions or actions to your supervisor or safety committee representative promptly.
3. Report all injuries to your supervisor promptly regardless of how serious.
4. Report all near-miss incidents to your supervisor promptly.
5. Always use personal protection equipment (PPE) in good working condition where it is required.
6. Do not remove or defeat any safety device or safeguard provided for employee protection.
7. Encourage co-workers by your words and example to use safe work practices on the job.
8. Make suggestions to your supervisor, safety committee representative or management about changes you believe will improve employee safety.

(Customize by adding any additional employee responsibilities that you may have and/or deleting any that do not apply to your company.)

EMPLOYEE PARTICIPATION

Safety Committee (required for employers with 11 employees or more)

We have formed a safety committee to help employees and management work together to identify safety problems, develop solutions, review incident reports and evaluate the effectiveness of our safety program. The committee is made up of management-designated representatives and one employee-elected representative each from the office, factory and outside sales divisions of our company.

- Employees in each division will elect from among themselves a representative to be on the committee. If there is only one volunteer or nomination, the employees will approve the person by voice vote at a short meeting called for that purpose. If there is more than one volunteer or nomination, a secret paper ballot will be used to elect the representative.

- Elected representatives will serve for one year before being re-elected or replaced. If there is a vacancy then an election will be held before the next scheduled meeting to fill the balance of the term.

- In addition to the employee-elected representatives, management will designate no more than three representatives but a minimum of one who will serve until replaced by management.

- A chairperson will be selected by majority vote of the committee members each year. If there is a vacancy, the same method will be used to select a replacement.

- In addition to the committee responsibilities explained above, duties of safety committee members include:
  —A monthly self-inspection of the area they represent,
  —Communicating with the employees they represent on safety issues, and
  —Encouraging safe work practices among co-workers.

- The regularly scheduled meeting time is 7:30 am for one hour on the first Thursday of each month, at the employee lunchroom. This may be changed by vote of the committee.
- A committee member will be designated each month to keep minutes on the attached minutes form. A copy will be posted on the employee bulletin board after each meeting. After being posted for one month, the minutes will be filed for one year. The minutes form contains the basic monthly meeting agenda.

(Customize by adding any additional safety committee information that you may have and/or deleting any that do not apply to your company.)

Employee Safety Meetings (optional for employers with 10 or fewer employees or employers with 11 employees or more who are segregated on different shifts or work in widely dispersed locations in crews of 10 or less)

All employees are required to attend a monthly safety meeting held on the first Thursday of each month in the lunchroom. This meeting is to help identify safety problems, develop solutions, review incidents reports, provide training and evaluate the effectiveness of our safety program. Minutes will be kept on the attached minutes form. Meeting minutes will be kept on file for one year.

(Customize by adding any additional Employee Safety Meeting information that you may have and/or deleting any that do not apply to your company.)

HAZARD RECOGNITION

Record Keeping and Review

Employees are required to report any injury or work related illness to their immediate supervisor regardless of how serious. Minor injuries such as cuts and scrapes can be entered on the minor injury log posted ***(customize by adding location of Minor Injury Log)***. The employee must use an "Employee's Injury/Illness Report Form" to report more serious injuries.

- **The supervisor will:**

—Investigate a serious injury or illness using procedures in the "Incident Investigation" section below.
—Complete an "Incident Investigation Report" form.
—Give the "Employee's Report" and the "Incident Investigation Report" to ***(add the name or title of the person to whom this information will be given)***.

- ***(Add the name or title of the responsible person)*** **will:**

—Determine from the Employee's Report, Incident Investigation Report, and any Washington State Labor and Industry (L&I) Worker Compensation claim form associated with the incident whether it must be recorded on the OSHA Injury and Illness Log and Summary according to the instructions for that form.
—Enter a recordable incident within six days after the company becomes aware of it.
—If the injury is not recorded on the OSHA log, add it to a separate incident report log, which is used to record non-OSHA recordable injuries and near misses.
—Each month before the scheduled safety committee meeting, make any new injury reports and investigations available to the safety committee for review, along with an updated OSHA and incident report log.

The safety committee will review the log for trends and may decide to conduct a separate investigation of any incident.

- ***(Add the name or title of the responsible person)*** will post a signed copy of the OSHA log summary for the previous year on the safety bulletin board each February 1 until April 30. The log will be kept on file for at least 5 years. Any employee can view an OSHA log upon request at any time during the year.

(Customize by adding any additional Hazard Recognition policies that you may have and/or deleting any that do not apply to your company.)

INCIDENT INVESTIGATION

Incident Investigation Procedure

If an employee dies while working or is not expected to survive, or when two or more employees are admitted to a hospital as a result of a work-related incident, ***(customize by adding the name or title of person responsible for reporting to L&I)*** will contact the Department of Labor and Industries within 8 hours after becoming aware of the incident. During weekends and evenings, the toll-free notification number is: 1-800-321-6742. ***(Add the name or title of the responsible person)*** must talk with a representative of the department. Fax and answering machine notifications are not acceptable. ***(Add the name or title of the responsible person)*** must report: the employer name, location and time of the incident, number of employees involved, the extent of injuries or illness, a brief description of what happened and the name and phone number of a contact person.

DO NOT DISTURB the scene except to aid in rescue or make the scene safe.

Whenever there is an incident that results in death or serious injuries that have immediate symptoms, a preliminary investigation will be conducted by the immediate supervisor of the injured person(s), a person designated by management, an employee representative of the safety committee, and any other persons whose expertise would help the investigation.

The investigation team will take written statements from witnesses; photograph the incident scene and equipment involved. The team will also document as soon as possible after the incident, the condition of equipment and anything else in the work area that may be relevant. The team will make a written "Incident Investigation Report" of its findings. The report will include a sequence of events leading up to the incident, conclusions about the incident and any recommendations to prevent a similar incident in the future. The report will be reviewed by the safety committee at its next regularly scheduled meeting.

When a supervisor becomes aware of an employee injury where the injury was not serious enough to warrant a team investigation as described above, the supervisor will write an "Incident Investigation Report" to accompany the "Employee's Injury/Illness Report Form" and forward them to ***(add the name or title of the responsible person)***.

Whenever there is an incident that did not but could have resulted in serious injury to an employee (a *near-miss*), the incident will be investigated by the supervisor or a team depending on the seriousness of the injury that would have occurred. The "Incident Investigation Report" form will be used to investigate the near miss. The form will be clearly marked to indicate that it was a near miss and that no actual injury occurred. The report will be forwarded to the bookkeeper to record on the incident log.

An "Incident Investigation Checklist" form can be found in the Accident Prevention Program Guide to help the supervisor carry out his/her responsibilities as described above.

(Customize by adding any additional Incident Investigation policies that you may have and/or deleting any that do not apply to your company.)

SAFETY INSPECTION PROCEDURES

(Customize by adding your company name here) is committed to aggressively identifying hazardous conditions and practices, which are likely to result in injury or illness to employees. We will take prompt action to eliminate any hazards we find. In addition to reviewing injury records and investigating incidents for their causes, management and the safety committee will regularly check the workplace for hazards as described below:

Annual Site Survey—Once a year an inspection team made up of members of the safety committee will do a wall-to-wall walk through inspection of the entire work site. They will write down any safety hazards or potential hazards they find. The results of this inspection will be used to eliminate or control obvious hazards, target specific work areas for more intensive investigation, assist in revising the checklists used during regular monthly safety inspections and as part of the annual review of the effectiveness of our accident prevention program.

Periodic Change Survey—We will assign a supervisor or form a team to look at any changes we make to identify safety issues. Changes include new equipment, changes to production processes or a change to the building structure. A team is made up of maintenance, production, and safety committee representatives. It examines the changed conditions and makes recommendations to eliminate or control any hazards that were or may be created as a result of the change.

Monthly Safety Inspection—Each month, before the regularly scheduled safety committee meeting, safety committee representatives will inspect their areas for hazards using the standard safety inspection checklist. They will talk to co-workers about their safety concerns. Committee members will report any hazards or concerns to the whole committee for consideration. The results of the area inspection and any action taken will be posted in the affected area. Occasionally, committee representatives may agree to inspect each other's area rather than their own. This brings a fresh pair of eyes to look for hazards.

Job Hazard Analysis—As a part of our on-going safety program, we will use a "Job Hazard Analysis" form to look at each type of job task our employees do. This analysis will be done by the supervisor of that job task or a member of the safety committee. We will change how the job is done as needed to eliminate or control any hazards. We will also check to see if the employee needs to use personal protection equipment (PPE) while doing the job. Employees will be trained in the revised operation and to use any required PPE. The results will be reported to the safety committee. Each job task will be analyzed at least once every two years, whenever there is a change in how the task is done or if there is a serious injury while doing the task.

(Customize by adding any additional safety self-inspection policies that you may have and/or deleting any that do not apply to your company.)

HAZARD PREVENTION AND CONTROL

Eliminating Workplace Hazards

(Customize by adding your company name here) is committed to eliminating or controlling workplace hazards that could cause injury or illness to our employees. We will meet the requirements of state safety standards where there are specific rules about a hazard or potential hazard in our workplace. Whenever possible we will design our facilities and equipment to eliminate employee exposure to hazards. Where these engineering controls are not possible, we will write work rules that effectively prevent employee exposure to the hazard. When the above methods of control are not possible or are not fully effective we will require employees to use personal protection equipment (PPE) such as safety glasses, hearing protection, and foot protection.

Basic Safety Rules

The following basic safety rules have been established to help make our company a safe and efficient place to work. These rules are in addition to safety rules that must be followed when doing particular jobs or operating certain equipment. Those rules are listed elsewhere in this program. Failure to comply with these rules will result in disciplinary action.

- Never do anything that is unsafe in order to get the job done. If a job is unsafe, report it to your supervisor or safety committee representative. We will find a safer way to do that job.
- Do not remove or disable any safety device! Keep guards in place at all times on operating machinery.
- Never operate a piece of equipment unless you have been trained and are authorized.
- Use your personal protection equipment whenever it is required.
- Obey all safety warning signs.
- Working under the influence of alcohol or illegal drugs or using them at work is prohibited.
- Do not bring firearms or explosives onto company property.
- Smoking is only permitted outside the building away from any entry or ventilation intake.
- Horseplay, running and fighting are prohibited.
- Clean up spills immediately. Replace all tools and supplies after use. Do not allow scraps to accumulate where they will become a hazard. Good housekeeping helps prevent injuries.

(Customize by adding any additional safety policies that you may have and/or deleting any that do not apply to your company.)

Job-Related Safety Rules

We have established safety rules and personal protection equipment (PPE) requirements based upon a hazard assessment for each task listed below:

Work in or pass through any production area (for example, the machine shop or paint shop)

Required PPE:

- Safety glasses. Check prior to use for broken or missing components (such as side shields) and for scratched lenses. Safety glasses must have a "Z87.1" marking on the frame. If they are prescription glasses, the initials of the lens manufacturer must be stamped into the corner of the lens to show that they are safety glass lenses.

Work Rules:

- Walk within marked aisles.
- Do not distract or talk with employees when they are using a machine.

Work with bench grinders (machine shop)

Required PPE:

- Eye protection (full-face shield with safety glasses under the shield).

Work Rules:

- Check that there is a gap between the tool rest and the wheel of no more than 1/8".
- Check that the upper wheel (tongue) guard has a gap of no more than 1/4".
- Check that the wheel edge is not excessively grooved. Dress the wheel if necessary.
- Do not grind on the face of the wheel.

Work with ladders (all locations)

Required PPE:

- Full body harness when working at greater than 25' and both hands must be used to do the job. See the fall protection plan instructions described elsewhere in this program.

Work Rules:

- Before you use a ladder check it for defects such as loose joints, grease on steps, or missing rubber feet.
- Do not paint a ladder! You may hide a defect.
- Do not use a ladder as a brace, workbench or for any other purpose than climbing.
- Do not carry objects up or down a ladder if it will prevent you from using both hands to climb.
- Always face the ladder when climbing up or down.
- If you must place a ladder at a doorway, barricade the door to prevent its use and post a sign.
- Only one person is allowed on a ladder at a time.
- Always keep both feet on the ladder rungs except while climbing. Do not step sideways from an unsecured ladder onto another object.
- If you use a ladder to get to a roof or platform, the ladder must extend at least 3' above the landing and be secured at the top and bottom.
- Do not lean a step ladder against a wall and use it as a single ladder. Always unfold the ladder and lock the spreaders.
- Do not stand on the top step of a step ladder.

Set a single or extension ladder with the base 1/4 of the working ladder length away from the support.

(The above rules are included as an example only. You must customize this program by adding any additional job-specific safety rules that you may have and/or deleting any that do not apply to your company. Be sure to include the job description, location, work rules, and personal protection equipment required.)

Lifting tasks (all locations)

Required PPE:

- Leather gloves for sharp objects or surfaces.
- Steel toe safety shoes in production and shipping areas (to be supplied by the employee) must be in good condition and be marked "ANSI Z41 C - 75."

Work Rules:

- Do not lift on slippery surfaces.
- Test the load before doing the lift.
- Get help if the load is too heavy or awkward to lift alone.
- Break the load down into smaller components if possible to provide a comfortable lift.
- Do not overexert!
- Make sure you have a good handhold on the load.
- Do not jerk the load or speed up. Lift the load in a smooth and controlled manner.
- Do not twist while lifting (especially with a heavy load). Turn and take a step.
- Keep the load close to the body. Walk as close as possible to the load. Pull the load towards you before lifting if necessary.
- Avoid long forward reaches to lift over an obstruction.
- Avoid bending your back backwards to lift or place items above your shoulder. Use a step stool or platform.
- Do not lift while in an awkward position.
- Use a mechanical device such as a forklift, hoist, hand truck or elevatable table whenever possible to do the lift or to bring the load up between the knees and waist before you lift.
- Back injury claims are painful for the worker and expensive for the company. Lift safely!

The signatures below document that the employee received training on how to lift safely.

Employee: ________________________________ Training Date: _______________

Trainer: ____

(The above rules are included as an example only. You must customize this program by adding any additional job-specific safety rules that you may have and/or deleting any that do not apply to your company. Be sure to include the job description, location, work rules, and personal protection equipment required.)

DISCIPLINARY POLICY

Employees are expected to use good judgment when doing their work and to follow established safety rules. We have established a disciplinary policy to provide appropriate consequences for failure to follow safety rules. This policy is designed not so much to punish as to bring unacceptable behavior to the employee's attention in a way that the employee will be motivated to make corrections. The following consequences apply to the violation of the same rule or the same unacceptable behavior:

First Instance:	Verbal warning, notation in employee file, and instruction on proper actions
Second Instance:	1-day suspension, written reprimand, and instruction on proper actions
Third Instance:	1-week suspension, written reprimand, and instruction on proper actions
Fourth Instance:	Termination of employment

An employee may be subject to immediate termination when a safety violation places the employee or co-workers at risk of permanent disability or death.

(The above rules are included as an example only. You must customize this program by adding any disciplinary rules that you may have and/or deleting any that do not apply to your company.)

EQUIPMENT MAINTENANCE

The following departments have machinery and equipment that must be inspected or serviced on a routine basis. A checklist/record to document the maintenance items will be maintained and kept on file for the life of the equipment.

Machine shop

Equipment	Interval	Location of Record
Ederer 20 ton Crane	Monthly	Maintenance file cabinet
Omahda press brake	Weekly	Folder attached to the press

Vehicles

Equipment	Interval	Location of Record
1986 Toyota Forklift A68710*	Daily	File cabinet in the garage
1992 Ford Taurus LST385	Monthly	Vehicle glove box

* Forklifts are required to be examined daily prior to being placed into service or after each shift if used on a round-the-clock basis.

(The above rules are included as an example only. You must customize this section by adding any equipment maintenance rules that you may have and/or deleting any that do not apply to your company. Be sure to include the equipment, location, and other pertinent information.)

EMERGENCY PLANNING

What will we do in an emergency?

In case of fire

An evacuation map for the building is posted ***(customize by adding location, if this applies to your company)***. It shows the location of exits, fire extinguishers, first aid kits, and where to assemble outside ***(customize by adding meeting location for your location)***. A copy of the map is attached to this program.

All employees will receive training on how to use fire extinguishers as part of their initial orientation. A fire evacuation drill will be conducted once a year during the first week of April. ***(Customize by adding fire drill and fire extinguisher training information as it pertains to your business.)***

- If you discover a fire: Tell another person immediately. Call or have them call 911 and a supervisor.
- If the fire is small (such as a wastebasket fire) and there is minimal smoke, you may try to put it out with a fire extinguisher.
- If the fire grows or there is thick smoke, do not continue to fight the fire.
- Tell other employees in the area to evacuate.
- Go to the designated assembly point outside the building (north parking lot).
- If you are a supervisor notified of a fire in your area: Tell your employees to evacuate to the designated assembly location. Check that all employees have been evacuated from your area.
- Verify that 911 has been called.
- Determine if the fire has been extinguished. If the fire has grown or there is thick smoke, evacuate any employees trying to fight the fire.
- Tell supervisors in other areas to evacuate the building.
- Go to the designated assembly point and check that all your employees are accounted for. If an employee is missing, *do not* re-enter the building! Notify the responding fire personnel that an employee is missing and may be in the building.

(Customize the above rules by adding procedures in case of fire as it pertains to your business.)

In case of earthquake

The west coast of the United States is subject to earthquakes. There will be no advance warning. The shock will be your only warning. Because there are power lines over the north parking lot, the south parking lot is the designated assembly location for earthquake evacuation. We have bolted tall narrow storage racks to the floors, walls or to each other to provide a wide base to help reduce the potential for collapse. A wrench is available at the rear entrance to turn off the gas shut-off outside the building. All supervisors will be trained in the gas shut-off procedure. An earthquake drill will be conducted each year during the first week of September. In the event of an earthquake:

(Customize by adding earthquake drill and evacuation information as it pertains to your business.)

If you are inside a building:

- Drop under a desk or table, cover your head and hold on. Stay away from windows, heavy cabinets, bookcases or glass dividers.
- When the shaking stops, ***(customize by adding name or title of responsible person)*** are to check for damage and available evacuation routes then begin an evacuation of their area to the designated assembly location. ***(Customize by adding meeting location for your location.)***
- Evacuation should proceed as quickly as possible since there may be aftershocks.
- Supervisors must account for each employee in their work group as quickly as possible.
- First aid certified employees should check for injuries and help evacuate injured employees. Do not attempt to move seriously injured persons unless they are in immediate danger of further injury.
- If a gas odor is in the building, tell a supervisor to turn off the gas at the main. Open windows.
- Supervisors and first aid employees must not re-enter the building once evacuation is complete.
- Do not approach or touch downed power lines or objects touched by downed power lines.
- Do not use the phone except for emergency use.
- Turn on a radio and listen for public safety instructions.

If you are outside: Stand away from buildings, trees, telephone and electric lines.

If you are on the road: Drive away from underpasses/overpasses. Stop in a safe area. Stay in the vehicle.

(Customize by adding any additional rules and deleting any that do not apply to your business.)

If an injury occurs

- A first aid kit is kept ***(customize by adding the location of first aid supplies in your business).*** Also, each company vehicle is equipped with a first aid kit located in the glove box or under the driver's seat. These kits are checked monthly by members of the safety committee. An inventory of each kit is taped to the inside cover of the box. If you are injured, promptly report it to any supervisor. ***(Customize by adding any additional locations of first aid supplies or deleting the above information if it does not apply to your business.)***

- All supervisors are required to have first aid cards. Other employees may have been certified. A list of current first aid and CPR certified supervisors and employees is posted on the safety bulletin board along with the expiration dates of their cards. ***(Customize by adding the location of first aid trained personnel in your business.)***

- In case of serious injury, do not move the injured person unless absolutely necessary. Only provide assistance to the level of your training. Call for help. If there is no response, call 911.

- AIDS/HIV and Hepatitis B are the primary infectious diseases of concern in blood. *All blood should be assumed to be infectious.* These diseases can both be deadly. Employees are *not* required to perform first aid as part of their job duties. In the event of a bleeding injury where first aid is needed, use gloves if possible to prevent exposure to blood or other potentially infectious materials. The injured person can often help by applying pressure to the wound. Gloves and a mouth barrier for rescue breathing are available in the first aid kits. If you are exposed to blood while giving first aid wash immediately with soap and water and report the incident to a supervisor. The appropriate follow-up procedures will be initiated, including medical evaluation, counseling, Hepatitis B vaccine and blood testing of the source person if possible. For further information, refer to Washington Administrative Code (WAC) 296-62-08001(6).

SAFETY AND HEALTH TRAINING AND EDUCATION

Safety Training

Training is an essential part of our plan to provide a safe work place at ***(customize by adding your company name here)***. To insure that all employees are trained *before* they start a task that requires training, we have a training coordinator whose name is posted on the safety bulletin board. ***(Customize by inserting the name or title of the person responsible for training in your company.)*** That person is responsible to verify that each employee has received an initial orientation by his or her supervisor, has received any training needed to do the job safely and that the employee file documents the training. The coordinator will make sure that an outline and materials list are available for each training course we provide:

Course	Who Must Attend
Basic Orientation	All employees (given by the employee's supervisor)
Safe Lifting	Any employee who lifts more than 20 pounds
Chemical Hazards (General)	All employees
Chemical Hazards (Specific)	An employee who uses or is exposed to a particular chemical
Fire Extinguisher Safety	All employees
Respirator Training	Employees who use a respirator
Forklift Training	Employees who operate a forklift
Lockout Training (Awareness)	All employees
Lockout Training (Advanced)	Employees who service equipment
Welding Safety	Employees who operate the arc welder

(Customize by adding additional training required in your business and deleting any of the above training that does not apply.)

Safe Lifting Training Course Outline

Required Materials:

- Video *Back Your Back*, Washington State Labor and Industry video number V0146. Reserve at least two weeks in advance. Call (360) 902-5444
- Safe Lifting Rules from Accident Prevention Program

Outline, 1-hour class:

- Talk about injury statistics related to lifting and handling materials.
- Talk about some injuries that have occurred in our work place.
- Show video.
- Answer questions from participants about video.
- Go over safe lifting rules in the Accident Prevention Program:
  —Demonstrate techniques, and
  —Discuss mechanical lifting aids such as hoists and carts that are available in our workplace.
- Have employees sign their names to the training roster.

APPENDIX B

TABLE OF CONTENTS FOR SAFETY MANUAL

DEVELOPED BY

MONROE COUNTY, DEPARTMENT OF ENVIRONMENTAL SERVICES,

ROCHESTER, NEW YORK

TABLE OF CONTENTS

SECTION D: VEHICLES

SECTION E: HAZARD CONTROL

APPENDIX C

REGULATORY INFORMATION

APPLICABLE OSHA REGULATIONS

OSHA-APPROVED STATE PROGRAMS

OSHA CHECKLISTS

REGULATORY INFORMATION

Safety regulations are generally developed and enforced under the overall jurisdiction of the Federal Government's Department of Labor Occupational Safety and Health Administration (OSHA). In general, coverage of the OSHA Act extends to all employers and their employees in the 50 states, the District of Columbia, Puerto Rico, and all other territories under Federal Government jurisdiction. Coverage is provided either directly by federal OSHA or through an OSHA-approved state program.

OSHA has developed a comprehensive set of safety regulations, many of them with specific applications to wastewater collection systems. Where OSHA has not developed specific standards, employers are responsible for following the Act's general duty clause, which states that each employer "shall furnish . . . a place of employment which is free from recognized hazards that are causing or are likely to cause death or serious physical harm to his employees."

States with OSHA-approved occupational safety and health programs must set standards which are at least as effective as the federal standards. Many states adopt standards identical to the federal standards. Where states adopt and enforce their own standards under state law, copies of state standards may be obtained from the individual states.

The Federal Register is one of the best sources of information on standards, since all OSHA standards are published there when adopted, as are all amendments, corrections, insertions or deletions. The Federal Register is available in many public libraries. Annual subscriptions are available from the U.S. Government Printing Office, Superintendent of Documents, PO Box 371954, Pittsburgh, PA 15250-7954, phone (866) 512-1800.

Each year the Office of the Federal Register publishes all current regulations and standards in the Code of Federal Regulations (CFR), available at many libraries and from the Government Printing Office. OSHA's regulations are collected in Title 29 of the Code of Federal Regulations (CFR), Part 1900-1999. Copies of the CFR may also be purchased from the Superintendent of Documents at the address above.

APPLICABLE REGULATIONS

OSHA regulations fill several hundred pages of text which cannot be printed in this manual. Of the regulations, Part 1910, OCCUPATIONAL SAFETY AND HEALTH STANDARDS, contains the regulations most applicable to the work we do in collection systems. If your state is one of the 25 that has a state program, there will be a comparable set of regulations similar to Part 1910. Listed next are some of the OSHA regulations that apply.

PART 1910 — OCCUPATIONAL SAFETY AND HEALTH STANDARDS

Subpart D — Walking/Working Surfaces

1910.21	Definitions.
1910.22	General requirements.
1910.23	Guarding floor and wall openings and holes.
1910.24	Fixed industrial stairs.
1910.25	Portable wood ladders.
1910.26	Portable metal ladders.
1910.27	Fixed ladders.
1910.28	Safety requirements for scaffolding.
1910.29	Manually propelled mobile ladder stands and scaffolds (towers).
1910.30	Other working surfaces.
1910.31	Sources of standards.
1910.32	Standards organizations.

Subpart E — Means of Egress

1910.35	Definitions.
1910.36	General requirements.
1910.37	Means of egress, general.
1910.38	Employee emergency plans and fire prevention plans.
1910.39	Sources of standards.
1910.40	Standards organizations.
1910.40A	Appendix to Subpart E — Means of Egress.

Subpart F — Powered Platforms, Manlifts, and Vehicle-Mounted Work Platforms

1910.66	Power platforms for exterior building maintenance.
1910.66A	Appendix A to §1910.66 — Guideline (Advisory)
1910.66B	Appendix B to §1910.66 — Exhibits (Advisory)
1910.66C	Appendix C to §1910.66 — Personal Fall Arrest System (Section I — Mandatory; Sections II and III — Nonmandatory)
1910.66D	Appendix D to §1910.66 — Existing Installations (Mandatory)
1910.67	Vehicle-mounted elevating and rotating work platforms.
1910.68	Manlifts.
1910.69	Sources of standards.
1910.70	Standards organizations.

Subpart G — Occupational Health and Environmental Control

1910.94	Ventilation.
1910.95	Occupational noise exposure.
1910.96	Ionizing radiation.
1910.97	Nonionizing radiation.
1910.98	Effective dates.
1910.99	Sources of standards.
1910.100	Standards organizations.

Subpart H — Hazardous Materials

1910.101 Compressed gases (general requirements).

1910.102 Acetylene.

1910.103 Hydrogen.

1910.104 Oxygen.

1910.106 Flammable and combustible liquids.

1910.107 Spray finishing using flammable and combustible materials.

1910.108 Dip tanks containing flammable or combustible liquids.

1910.110 Storage and handling of liquefied petroleum gases.

1910.115 Sources of standards.

1910.116 Standards organizations.

1910.119 Process safety management of highly hazardous chemicals.

1910.119A Appendix A to §1910.119 — List of Highly Hazardous Chemicals, Toxics and Reactives (Mandatory).

1910.119B Appendix B to §1910.119 — Block Flow Diagram and Simplified Process Flow Diagram (Nonmandatory).

1910.119C Appendix C to §1910.119 — Compliance Guidelines and Recommendations for Process Safety Management (Nonmandatory).

1910.119D Appendix D to §1910.119 — Sources of Further Information (Nonmandatory).

1910.120 Hazardous waste operations and emergency response.

1910.120A Appendix A to §1910.120 — Personal Protective Equipment Test Methods.

1910.120B Appendix B to §1910.120 — General Description and Discussion of the Levels of Protection and Protective Gear.

1910.120C Appendix C to §1910.120 — Compliance Guidelines.

1910.120D References.

Subpart I — Personal Protection Equipment

1910.132 General requirements.

1910.133 Eye and face protection.

1910.134 Respiratory protection.

1910.135 Occupational head protection.

1910.136 Occupational foot protection.

1910.137 Electrical protective devices.

Subpart J — General Environmental Controls

1910.141 Sanitation.

1910.144 Safety color code for marking physical hazards.

1910.145 Specifications for accident prevention signs and tags.

1910.145A Appendix A to §1910.145(f) — Recommended Color Coding.

1910.145B Appendix B to §1910.145(f) — References for further information.

1910.146 Permit-required confined spaces.

1910.146A Appendix A to §1910.146 — Permit-required Confined Space Decision Flow Chart.

1910.146B Appendix B to §1910.146 — Procedures for Atmospheric Testing.

1910.146C Appendix C to §1910.146 — Examples of Permit-required Confined Space Programs.

1910.146D Appendix D to §1910.146 — Sample Permits.

1910.146E Appendix E to §1910.146 — Sewer System Entry.

1910.147 The control of hazardous energy (lockout/tagout).

1910.147A Appendix A to §1910.147 — Typical Minimal Lockout Procedure.

1910.148 Standards organizations.

1910.149 Effective dates.

Subpart K — Medical and First Aid

1910.150 Sources of Standards.

1910.151 Medical services and first aid.

1910.153 Sources of standards.

Subpart L — Fire Protection

PORTABLE FIRE SUPPRESSION EQUIPMENT

1910.157 Portable fire extinguishers.

1910.158 Standpipe and hose systems.

FIXED FIRE SUPPRESSION EQUIPMENT

1910.159 Automatic sprinkler systems.

1910.160 Fixed extinguishing systems, general.

1910.161 Fixed extinguishing systems, dry chemical.

1910.162 Fixed extinguishing systems, gaseous agent.

1910.163 Fixed extinguishing systems, water spray and foam.

OTHER FIRE PROTECTION SYSTEMS

1910.164 Fire detection systems.

1910.165 Employee alarm systems.

Subpart M — Compressed Gas and Compressed Air Equipment

1910.169 Air receivers.

1910.170 Sources of standards.

1910.171 Standards organizations.

Subpart N — Materials Handling and Storage

1910.176 Handling materials — general.

1910.177 Servicing multi-piece and single piece rim wheels.

1910.178 Powered industrial trucks.

1910.179 Overhead and gantry cranes.

1910.180 Crawler locomotive and truck cranes.

1910.181 Derricks.

1910.182 Effective dates.

1910.183 Helicopters.

1910.184 Slings.

1910.189 Sources of standards.

1910.190 Standards organizations.

Subpart O — Machinery and Machine Guarding

1910.211	Definitions.
1910.212	General requirements for all machines.
1910.213	Woodworking machinery requirements.
1910.214	Cooperage machinery.
1910.215	Abrasive wheel machinery.
1910.216	Mills and calenders in the rubber and plastics industries.
1910.217	Mechanical power presses.
1910.218	Forging machines.
1910.219	Mechanical power-transmission apparatus.
1910.220	Effective dates.
1910.221	Sources of standards.
1910.222	Standards organizations.

Subpart P — Hand and Portable Powered Tools and Other Hand-Held Equipment

1910.241	Definitions.
1910.242	Hand and portable powered tools and equipment, general.
1910.243	Guarding of portable powered tools.
1910.244	Other portable tools and equipment.
1910.245	Effective dates.
1910.246	Sources of standards.
1910.247	Standards organizations.

Subpart Q — Welding, Cutting, and Brazing

1910.251	Definitions.
1910.252	General requirements.
1910.253	Oxygen-fuel gas welding and cutting.
1910.254	Arc welding and cutting.
1910.255	Resistance welding.
1910.256	Sources of standards.
1910.257	Standards organizations.
1910.269	Electric power generation, transmission, and distribution.
1910.269A	Appendix A to §1910.269 — Flow Charts.
1910.269A-1	Appendix A-1 to §1910.269 — Application of Section 1910.269 and Subpart S of this Part to Electrical Installations.
1910.269A-2	Appendix A-2 to §1910.269 — Application of Section 1910.269 and Subpart S of this Part to Electrical Safety-Related Work Practices.
1910.269A-3	Appendix A-3 to §1910.269 — Application of Section 1910.269 and Subpart S of this Part to Tree-Trimming Operations.
1910.269A-4	Appendix A-4 to §1910.269 — Application of Section 1910.147, Section 1910.269 and Section 1910.333 to Hazardous Energy Control Procedures (Lockout/Tagout).
1910.269A-5	Appendix A-5 to §1910.269 — Application of Section 1910.146 and Section 1910.269 to Permit-Required Confined Spaces.
1910.269B	Appendix B to §1910.269 — Working on Exposed Energized Parts.
1910.269C	Appendix C to §1910.269 — Protection from Step and Touch Potentials.
1910.269D	Appendix D to §1910.269 — Methods of Inspecting and Testing Wood Poles.
1910.269E	Appendix E to §1910.269 — Reference Documents.

Subpart S — Electrical

GENERAL

1910.301	Introduction.

DESIGN SAFETY STANDARDS FOR ELECTRICAL SYSTEMS

1910.302	Electric utilization systems.
1910.303	General requirements.
1910.304	Wiring design and protection.
1910.305	Wiring methods, components, and equipment for general use.
1910.306	Specific purpose equipment and installations.
1910.307	Hazardous (classified) locations.
1910.308	Special systems.

SAFETY-RELATED WORK PRACTICES

1910.331	Scope.
1910.332	Training.
1910.333	Selection and use of work practices.
1910.334	Use of equipment.

In addition to the regulations listed above, 29 CFR Part 1926, SAFETY AND HEALTH FOR CONSTRUCTION, defines the regulations for Excavations and Trenches under Subpart P. There are other federal regulations that may also apply to collection systems operation and maintenance, such as Department of Transportation (DOT), as well as state and local regulations for traffic control and protection of utilities during excavations using "one call" notification systems.

OSHA-APPROVED STATE PROGRAMS

OSHA encourages states to develop and operate, under OSHA guidance, state job safety and health plans. Once a state plan is approved, OSHA funds up to 50 percent of the program's operating costs. State plans are required to provide standards and enforcement programs, as well as voluntary compliance activities, which are at least as effective as the federal program.

STATES WITH APPROVED PLANS

Alaska
Alaska Department of Labor and Workforce Development
PO Box 21149
1111 West 8th Street, Room 306
Juneau, AK 99802-1149
(907) 465-2700

Arizona
Industrial Commission of Arizona
800 West Washington
Phoenix, AZ 85007-2922
(602) 542-5795

California
California Department of Industrial Relations
455 Golden Gate Avenue - 10th Floor
San Francisco, CA 94102
(415) 703-5050

Connecticut
Connecticut Department of Labor
200 Folly Brook Boulevard
Wethersfield, CT 06109
(860) 566-5123

Hawaii
Hawaii Department of Labor and Industrial Relations
830 Punchbowl Street
Honolulu, HI 96813
(808) 586-8844

Indiana
Indiana Department of Labor
State Office Building
402 West Washington Street, Room W195
Indianapolis, IN 46204-2751
(317) 232-2378

Iowa
Iowa Division of Labor Services
1000 East Grand Avenue
Des Moines, IA 50319-0209
(515) 281-6432

Kentucky
Kentucky Labor Cabinet
1047 U.S. Highway 127 South, Suite 4
Frankfort, KY 40601
(502) 564-3070

Maryland
Maryland Division of Labor and Industry
Department of Labor, Licensing and Regulation
1100 North Eutaw Street, Room 613
Baltimore, MD 21201-2206
(410) 767-2241

Michigan
Michigan Department of Consumer and Industry Services
Bureau of Safety and Regulation
PO Box 30643
Lansing, MI 48909-8143
(517) 322-1814

Minnesota
Minnesota Department of Labor and Industry
443 Lafayette Road
St. Paul, MN 55155
(651) 284-5010

Nevada
Nevada Division of Industrial Relations
400 West King Street, Suite 400
Carson City, NV 89703
(775) 684-7260

New Jersey
New Jersey Department of Labor
John Fitch Plaza - Labor Building
Market and Warren Streets
PO Box 110
Trenton, NJ 08625-0110
(609) 292-2975

New Mexico
New Mexico Environment Department
1190 St. Francis Drive
PO Box 26110
Santa Fe, NM 87502
(505) 827-2850

New York
New York Department of Labor
New York Public Employee Safety and Health Program
State Office Campus Building 12, Room 158
Albany, NY 12240
(518) 457-2741

North Carolina
North Carolina Department of Labor
4 West Edenton Street
Raleigh, NC 27601-1092
(919) 733-0359

Oregon
Oregon Occupational Safety and Health Division
Department of Consumer and Business Services
350 Winter Street, NE, Room 430
Salem, OR 97301-3882
(503) 378-3272

Puerto Rico
Puerto Rico Department of Labor and Human Resources
Prudencio Rivera Martínez Building
505 Muñoz Rivera Avenue
Hato Rey, PR 00918
(787) 754-2119

South Carolina
South Carolina Department of Labor, Licensing, and Regulation
Koger Office Park, Kingstree Building
110 Centerview Drive
PO Box 11329
Columbia, SC 29211-1329
(803) 734-9606

Tennessee
Tennessee Department of Labor
710 James Robertson Parkway
Nashville, TN 37243-0659
(615) 741-2582

Utah
Utah Labor Commission
160 East 300 South, 3rd Floor
PO Box 146650
Salt Lake City, UT 84114-6650
(801) 530-6898

Vermont
Vermont Department of Labor and Industry
National Life Building - Drawer 20
Montpelier, VT 05620-3401
(802) 828-2288

Virgin Islands
Virgin Islands Department of Labor
3012 Golden Rock
Christiansted
St. Croix, Virgin Islands 00820-4660
(340) 773-1994

Virginia
Virginia Department of Labor and Industry
Powers-Taylor Building
13 South 13th Street
Richmond, VA 23219
(804) 786-2377

Washington
Washington Department of Labor and Industries
General Administration Building
PO Box 44001
Olympia, WA 98504-4001
(360) 902-4200

Wyoming
Wyoming Department of Employment
Workers' Safety and Compensation Division
Cheyenne Business Center
1510 East Pershing Boulevard
Cheyenne, WY 82002
(307) 777-7786

OSHA SELF-INSPECTION SAFETY CHECKLISTS[7]

The next several pages contain an OSHA safety checklist. It is not intended to cover all conditions that may exist in your facility; however, it does serve as a check of the many safety requirements required by regulations. You should always refer to either the federal or your state regulations to ensure compliance.

EMPLOYER POSTING

- Is the required OSHA workplace poster displayed in a prominent location where all employees are likely to see it?
- Are emergency telephone numbers posted where they can be readily found in case of emergency?
- Where employees may be exposed to any toxic substances or harmful physical agents, has appropriate information concerning employee access to medical and exposure records, and "Material Safety Data Sheets," been posted or otherwise made readily available to affected employees?
- Are signs concerning "Exiting from buildings, room capacities, floor loading, exposures to x-ray, microwave, or other harmful radiation or substances" posted where appropriate?
- Is the Summary of Occupational Illnesses and Injuries posted?

RECORDKEEPING

- Are all occupational injuries or illnesses, except minor injuries requiring only first aid, being recorded as required on the OSHA 200 log?
- Are employee medical records and records of employee exposure to hazardous substances or harmful physical agents up to date?
- Have arrangements been made to maintain required records for the legal period of time for each specific type record? (Some records must be maintained for at least 40 years.)
- Are operating permits and records up to date for such items as elevators, air pressure tanks and liquefied petroleum gas tanks?

SAFETY AND HEALTH PROGRAM

- Do you have an active safety and health program in operation?
- Is one person clearly responsible for the overall activities of the safety and health program?
- Do you have a safety committee or group made up of management and labor representatives who meet regularly and report in writing on their activities?
- Do you have a working procedure for handling in-house employee complaints regarding safety and health?
- Are you keeping your employees advised of the successful efforts and accomplishments you and/or your safety committee have made in assuring they will have a workplace that is safe and healthful?

MEDICAL SERVICES AND FIRST AID

- Do you require each employee to have a pre-employment physical examination?
- Is there a hospital, clinic, or infirmary for medical care near your workplace?
- If there are no medical and first aid facilities near your workplace, is at least one employee on each shift currently qualified to administer first aid?
- Are medical personnel readily available for advice and consultation on matters of employees' health?
- Are emergency phone numbers posted?
- Are first aid kits easily accessible to each work area, with necessary supplies available, periodically inspected and replenished as needed?
- Have first aid kit supplies been approved by a physician, indicating that they are adequate for a particular area or operation?
- Are means provided for quick drenching or flushing of the eyes and body in areas where corrosive liquids or materials are handled?

FIRE PROTECTION

- Is your local fire department well acquainted with your facilities, its location and specific hazards?
- If you have a fire alarm system, is it certified as required?
- If you have a fire alarm system, is it tested at least annually?
- If you have interior stand pipes and valves, are they inspected regularly?
- If you have outside private fire hydrants, are they flushed at least once a year and on a routine preventive maintenance schedule?
- Are fire doors and shutters in good operating condition?
- Are fire doors and shutters unobstructed and protected against obstructions, including their counterweights?
- Are fire door and shutter fusible links in place?
- Are automatic sprinkler system water control valves, air pressure and water pressure checked weekly/periodically as required?
- Is the maintenance of automatic sprinkler systems assigned to responsible persons or to a sprinkler contractor?
- Are sprinkler heads protected by metal guards when exposed to physical damage?
- Is proper clearance maintained below sprinkler heads?
- Are portable fire extinguishers provided in adequate number and appropriate types?
- Are fire extinguishers mounted in readily accessible locations?

[7] *Adapted from OSHA HANDBOOK FOR SMALL BUSINESSES. Obtain from the U.S. Government Printing Office, Superintendent of Documents, PO Box 371954, Pittsburgh, PA 15250-7954, or call (866) 512-1800. Order No. 029-016-00176-0. Price, $9.50.*

OSHA SELF-INSPECTION SAFETY CHECKLISTS (continued)

- Are fire extinguishers recharged regularly and is this noted on the inspection tag?
- Are operators periodically instructed in the use of extinguishers and fire protection procedures?

PERSONAL PROTECTIVE EQUIPMENT AND CLOTHING

- Are protective goggles or face shields provided and worn where there is any danger of flying particles or corrosive materials?
- Are approved safety glasses required to be worn at all times in areas where there is a risk of eye injuries such as punctures, abrasions, contusions or burns?
- Are employees who need corrective lenses (glasses or contacts) in working environments having harmful exposures required to wear only approved safety glasses, protective goggles, or use other medically approved precautionary procedures?
- Are protective gloves, aprons, shields, or other means provided against cuts, corrosive liquids and chemicals?
- Are hard hats provided and worn where danger of falling objects exists?
- Are hard hats inspected periodically for damage to the shell and suspension system?
- Is appropriate foot protection required where there is the risk of foot injuries from hot, corrosive or poisonous substances, falling objects, crushing or penetrating actions?
- Are approved respirators provided for regular or emergency use where needed?
- Is all protective equipment maintained in a sanitary condition and ready for use?
- Do you have eyewash facilities and a quick drench shower within the work area where employees are exposed to injurious corrosive materials?
- Where special equipment is needed for electrical workers, is it available?
- Where lunches are eaten on the premises, are they eaten in areas where there is no exposure to toxic materials or other health hazards?
- Is protection against the effects of occupational noise exposure provided when sound levels exceed those of the OSHA noise standard?
- Are adequate work procedures, protective clothing and equipment provided and used when cleaning up spilled toxic or otherwise hazardous materials or liquids?

GENERAL WORK ENVIRONMENT

- Are all work sites clean and orderly?
- Are work surfaces kept dry or appropriate means taken to assure the surfaces are slip resistant?
- Are all spilled materials or liquids cleaned up immediately?
- Is combustible scrap, debris and waste stored safely and removed from the work area promptly?
- Are accumulations of combustible dust routinely removed from elevated surfaces including the overhead structure of buildings?
- Is combustible dust cleaned up with a vacuum system to prevent the dust going into suspension?
- Is metallic or conductive dust prevented from entering or accumulating on or around electrical enclosures or equipment?
- Are covered metal waste cans used for oily and paint-soaked waste?
- Are all oil- and gas-fired devices equipped with flame failure controls that will prevent flow of fuel if pilots or main burners are not working?
- Are paint spray booths and dip tanks cleaned regularly?
- Are the minimum number of toilets and washing facilities provided?
- Are all toilets and washing facilities clean and sanitary?
- Are all work areas adequately lighted?
- Are pits and floor openings covered or otherwise guarded?

WALKWAYS

- Are aisles and passageways kept clear?
- Are aisles and walkways marked as appropriate?
- Are wet surfaces covered with nonslip materials?
- Are holes in the floor, sidewalk or other walking surface repaired properly, covered or otherwise made safe?
- Is there safe clearance for walking in aisles where motorized or mechanical handling equipment is operating?
- Are materials or equipment stored in such a way that sharp objects will not interfere with the walkway?
- Are spilled materials cleaned up immediately?
- Are changes of direction or elevation readily identifiable?
- Are aisles or walkways that pass near moving or operating machinery, welding operations or similar operations arranged so employees will not be exposed to potential hazards?
- Is adequate headroom provided for the entire length of any aisle or walkway?
- Are standard guardrails provided wherever aisle or walkway surfaces are elevated more than 30 inches above any adjacent floor or the ground?
- Are bridges provided over conveyors and similar hazards?

FLOOR AND WALL OPENINGS

- Are floor openings guarded by a cover, a guardrail, or equivalent on all sides (except at entrance to stairways or ladders)?
- Are toeboards installed around the edges of permanent floor openings (where persons may pass below the opening)?
- Are skylight screens of such construction and mounting that they will withstand a load of at least 200 pounds?
- Is a suitable type and thickness of glass used in windows, doors and glass walls which are subject to human impact?

OSHA SELF-INSPECTION SAFETY CHECKLISTS (continued)

- Are grates or similar type covers over floor openings such as floor drains of such design that foot traffic or rolling equipment will not be affected by the grate spacing?
- Are unused portions of service pits and pits not actually in use either covered or protected by guardrails or equivalent devices?
- Are manhole covers, trench covers and similar covers, plus their supports designed to carry a truck rear axle load of at least 20,000 pounds when located in roadways and subject to vehicle traffic?
- Are floor or wall openings in fire resistive construction provided with doors or covers compatible with the fire rating of the structure and provided with a self-closing feature when appropriate?

STAIRS AND STAIRWAYS

- Are standard stair rails or handrails installed on all stairways having four or more risers?
- Are all stairways at least 22 inches wide?
- Do stairs have at least a 6-foot, 6-inch overhead clearance?
- Do stairs angle no more than 50 and no less than 30 degrees?
- Are stairs of hollow-pan type treads and landings filled to nosing level with solid material?
- Are step risers on stairs uniform from top to bottom, with no riser spacing greater than 7 1/2 inches?
- Are steps on stairs and stairways designed or provided with a slip-resistant surface?
- Are stairway handrails located between 30 and 34 inches above the leading edge of stair treads?
- Do stairway handrails have at least 1 1/2 inches of clearance between the handrails and the wall or surface they are mounted on?
- Are stairway handrails capable of withstanding a load of 200 pounds, applied in any direction?
- Where stairs or stairways exit directly into any area where vehicles may be operated, are adequate barriers and warnings provided to prevent employees from stepping into the path of traffic?
- Do stairway landings have a dimension, measured in the direction of travel, at least equal to the width of the stairway?
- Is the vertical distance between stairway landings limited to 12 feet or less?

ELEVATED SURFACES

- Are signs posted, when appropriate, showing the elevated surface load capacity?
- Are surfaces elevated more than 30 inches above the floor or ground provided with standard guardrails?
- Are all elevated surfaces (beneath which people or machinery could be exposed to falling objects) provided with standard 4-inch toeboards?
- Is a permanent means of access and egress provided to elevated storage and work surfaces?
- Is required headroom provided where necessary?
- Is material on elevated surfaces piled, stacked or racked in a manner to prevent it from tipping, falling, collapsing, rolling or spreading?
- Are dock boards or bridge plates used when transferring materials between docks and trucks or rail cars?

EXITING OR EGRESS

- Are all exits marked with an exit sign and illuminated by a reliable light source?
- Are the directions to exits, when not immediately apparent, marked with visible signs?
- Are doors, passageways or stairways that are neither exits nor access to exits and which could be mistaken for exits appropriately marked "NOT AN EXIT," "BASEMENT, STOREROOM," etc.?
- Are exit signs provided with the word "EXIT" in lettering at least 5 inches high and the stroke of the lettering at least 1/2 inch wide?
- Are exit doors side-hinged?
- Are all exits kept free of obstructions?
- Are at least two means of egress provided from elevated platforms, pits or rooms where the absence of a second exit would increase the risk of injury from hot, poisonous, corrosive, suffocating, flammable, or explosive substances?
- Are there enough exits to permit prompt escape in case of emergency?
- Are special precautions taken to protect employees during construction and repair operations?
- Is the number of exits from each floor of a building and the number of exits from the building itself appropriate for the building occupancy load?
- Are exit stairways which are required to be separated from other parts of a building enclosed by at least 2-hour fire-resistive construction in buildings more than four stories in height, and not less than 1-hour fire-resistive construction elsewhere?
- Where ramps are used as part of required exiting from a building, is the ramp slope limited to 1 foot vertical and 12 feet horizontal?
- Where exiting will be through frameless glass doors, glass exit doors, or storm doors, are the doors fully tempered and do they meet the safety requirements for human impact?

EXIT DOORS

- Are doors which are required to serve as exits designed and constructed so that the way of exit travel is obvious and direct?
- Are windows which could be mistaken for exit doors made inaccessible by means of barriers or railings?
- Are exit doors operable from the direction of exit travel without the use of a key or any special knowledge or effort when the building is occupied?

OSHA SELF-INSPECTION SAFETY CHECKLISTS (continued)

- Is a revolving, sliding or overhead door prohibited from serving as a required exit door?
- Where panic hardware is installed on a required exit door, will it allow the door to open by applying a force of 15 pounds or less in the direction of the exit traffic?
- Are doors on cold storage rooms provided with an inside release mechanism which will release the latch and open the door even if it's padlocked or otherwise locked on the outside?
- Where exit doors open directly onto any street, alley or other area where vehicles may be operated, are adequate barriers and warnings provided to prevent employees from stepping into the path of traffic?
- Are doors that swing in both directions and are located between rooms where there is frequent traffic provided with viewing panels in each door?

PORTABLE LADDERS

- Are all ladders maintained in good condition (joints between steps and side rails tight, all hardware and fittings securely attached and moveable parts operating freely without binding or undue play)?
- Are nonslip safety feet provided on each ladder?
- Are nonslip safety feet provided on each metal or rung ladder?
- Are ladder rungs and steps free of grease and oil?
- Is it prohibited to place a ladder in front of doors opening toward the ladder except when the door is blocked open, locked or guarded?
- Is it prohibited to place ladders on boxes, barrels, or other unstable bases to obtain additional height?
- Are employees instructed to face the ladder when ascending or descending?
- Are employees prohibited from using ladders that are broken, missing steps, rungs, or cleats, broken side rails or other faulty equipment?
- Are employees instructed not to use the top step of ordinary stepladders as a step?
- When portable rung ladders are used to gain access to elevated platforms, roofs, or other elevated surfaces, does the ladder always extend at least 3 feet above the elevated surface?
- Is it required that when portable rung or cleat type ladders are used, the base is so placed that slipping will not occur, or the ladder is lashed or otherwise held in place?
- Are portable metal ladders legibly marked with signs reading "CAUTION — Do Not Use Around Electrical Equipment" or equivalent wording?
- Are employees prohibited from using ladders as guys, braces, skids, gin poles, or for other than their intended purposes?
- Are employees instructed to only adjust extension ladders while standing at the base (not while standing on the ladder or from a position above the ladder)?
- Are metal ladders inspected for damage?
- Are the rungs of ladders uniformly spaced at 12 inches, center to center?

HAND TOOLS AND EQUIPMENT

- Are all tools and equipment (both company- and employee-owned) used by employees at their workplace in good condition?
- Are hand tools such as chisels or punches which develop mushroomed heads during use reconditioned or replaced as necessary?
- Are broken or fractured handles on hammers, axes and similar equipment replaced promptly?
- Are worn or bent wrenches replaced regularly?
- Are appropriate handles used on files and similar tools?
- Are employees made aware of the hazards caused by faulty or improperly used hand tools?
- Are appropriate safety glasses or face shields worn while using hand tools or equipment which might produce flying materials or be subject to breakage?
- Are jacks checked periodically to assure they are in good operating condition?
- Are tool handles wedged tightly in the head of all tools?
- Are tool cutting edges kept sharp so the tool will move smoothly without binding or skipping?
- Are tools stored in a dry, secure location where they won't be tampered with?
- Is eye and face protection used when driving hardened or tempered studs or nails?

PORTABLE (POWER OPERATED) TOOLS AND EQUIPMENT

- Are grinders, saws and similar equipment provided with appropriate safety guards?
- Are power tools used with the correct shield, guard, or attachment, as recommended by the manufacturer?
- Are portable circular saws equipped with guards above and below the base shoe?
- Are circular saw guards checked to verify that they are not wedged up, thus leaving the lower portion of the blade unguarded?
- Are rotating or moving parts of equipment guarded to prevent physical contact?
- Are all cord-connected, electrically operated tools and equipment effectively grounded or of the approved double-insulated type?
- Are effective guards in place over belts, pulleys, chains, or sprockets on equipment such as concrete mixers and air compressors?
- Are portable fans provided with full guards or screens having openings $1/2$ inch or less?
- Is hoisting equipment available and used for lifting heavy objects, and are hoist ratings and characteristics appropriate for the task?

OSHA SELF-INSPECTION SAFETY CHECKLISTS (continued)

- Are ground-fault circuit interrupters provided on all temporary 15 and 20 ampere electrical circuits used during periods of construction?
- Are pneumatic and hydraulic hoses on power-operated tools checked regularly for deterioration or damage?

ABRASIVE WHEEL EQUIPMENT — GRINDERS

- Is the work rest used and kept adjusted to within 1/2 inch of the wheel?
- Is the adjustable tongue on the top side of the grinder used and kept adjusted to within 1/4 inch of the wheel?
- Do side guards cover the spindle, nut, and flange and 75 percent of the wheel diameter?
- Are bench and pedestal grinders permanently mounted?
- Are goggles or face shields always worn when grinding?
- Is the maximum RPM rating of each abrasive wheel compatible with the RPM rating of the grinder motor?
- Are fixed or permanently mounted grinders connected to their electrical supply system with metallic conduit or other permanent wiring method?
- Does each grinder have an individual ON and OFF control switch?
- Is each electrically operated grinder effectively grounded?
- Before new abrasive wheels are mounted, are they visually inspected and ring tested?
- Are dust collectors and powered exhausts provided on grinders used in operations that produce large amounts of dust?
- Are splash guards mounted on grinders that use coolant to prevent the coolant reaching employees?
- Is cleanliness maintained around grinders?

POWDER-ACTUATED TOOLS

- Are employees who operate powder-actuated tools trained in their use and do they carry a valid operator's card?
- Is each powder-actuated tool stored in its own locked container when not being used?
- Is a sign at least 7 inches by 10 inches with bold face type reading "POWDER-ACTUATED TOOL IN USE" conspicuously posted when the tool is being used?
- Are powder-actuated tools left unloaded until they are actually ready to be used?
- Are powder-actuated tools inspected for obstructions or defects each day before use?
- Do powder-actuated tool operators have and use appropriate personal protective equipment such as hard hats, safety goggles, safety shoes and ear protectors?

MACHINE GUARDING

- Is there a training program to instruct employees on safe methods of machine operation?
- Is there adequate supervision to ensure that employees are following safe machine operating procedures?
- Is there a regular program of safety inspection of machinery and equipment?
- Is all machinery and equipment kept clean and properly maintained?
- Is sufficient clearance provided around and between machines to allow for safe operations, set-up and servicing, material handling and waste removal?
- Is equipment and machinery securely placed and anchored, when necessary, to prevent tipping or other movement that could result in personal injury?
- Is there a power shutoff switch within reach of the operator's position at each machine?
- Can electric power to each machine be locked out for maintenance, repair or security?
- Are the non-current-carrying metal parts of electrically operated machines bonded and grounded?
- Are foot-operated switches guarded or arranged to prevent accidental actuation by personnel or falling objects?
- Are manually operated valves and switches controlling the operation of equipment and machines clearly identified and readily accessible?
- Are all emergency stop buttons colored red?
- Are all pulleys and belts that are within 7 feet of the floor or working level properly guarded?
- Are all moving chains and gears properly guarded?
- Are splash guards mounted on machines that use coolant to prevent the coolant from reaching employees?

OSHA SELF-INSPECTION SAFETY CHECKLISTS (continued)

- Are methods provided to protect the operator and other employees in the machine area from hazards created at the point of operation, in-going nip points, rotating parts, flying chips, and sparks?
- Are machinery guards secure and so arranged that they do not offer a hazard in their use?
- If special hand tools are used for placing and removing material, do they protect the operator's hands?
- Are revolving drums, barrels, and containers required to be guarded by an enclosure that is interlocked with the drive mechanism so that revolution cannot occur unless the guard enclosure is in place?
- Do arbors and mandrels have firm and secure bearings and are they free from play?
- Are provisions made to prevent machines from automatically starting when power is restored after a power failure or shutdown?
- Are machines constructed so as to be free from excessive vibration when the largest size tool is mounted and run at full speed?
- If machinery is cleaned with compressed air, is air pressure controlled and are personal protective equipment or other safeguards used to protect operators and other workers from eye and body injury?
- Are fan blades protected with a guard having openings no larger than $1/2$ inch when operating within 7 feet of the floor?
- Are saws used for ripping equipped with anti-kickback devices and spreaders?
- Are radial arm saws so arranged that the cutting head will gently return to the back of the table when released?

LOCKOUT/TAGOUT PROCEDURES

- Is all machinery or equipment capable of movement required to be de-energized or disengaged and blocked or locked out during cleaning, servicing, adjusting or setting up operations?
- Where the power disconnecting means for equipment does not also disconnect the electrical control circuit:
  - Are the appropriate electrical enclosures identified?
  - Is a means provided to ensure the control circuit can also be disconnected and locked out?
  - Is the locking out of control circuits in place of locking out main power disconnects prohibited?
- Are all equipment control valve handles provided with a means for locking out?
- Does the lockout procedure require that stored energy (mechanical, hydraulic, pneumatic) be released or blocked before equipment is locked out for repairs?
- Are appropriate employees provided with individually keyed personal safety locks?
- Are employees required to keep personal control of their key(s) while they have safety locks in use?
- Is it required that only the employee exposed to the hazard place or remove the safety lock?
- Is it required that employees check the safety of the lockout by attempting a start-up after making sure no one is exposed?
- Are employees instructed to always push the control circuit stop button prior to re-energizing the main power switch?
- Is there a method to identify any or all employees who are working on locked-out equipment by referring to their locks or accompanying tags?
- Are a sufficient number of accident preventive signs or tags and safety padlocks provided for any reasonably foreseeable repair emergency?
- When machine operations, configuration or size requires the operator to leave his or her control station to install tools or perform other operations, and that part of the machine could move if accidentally activated, is such element required to be separately locked or blocked out?
- In the event that equipment or lines cannot be shut down, locked out and tagged, is a safe job procedure established and rigidly followed?

WELDING, CUTTING AND BRAZING

- Are only authorized and trained personnel permitted to use welding, cutting or brazing equipment?
- Does each operator have a copy of the appropriate operating instructions and are they directed to follow them?
- Are compressed gas cylinders regularly examined for obvious signs of defects, deep rusting, or leakage?
- Is care used in handling and storage of cylinders, safety valves, or relief valves to prevent damage?
- Are precautions taken to prevent the mixture of air or oxygen with flammable gases, except at a burner or in a standard torch?
- Are only approved apparatus (torches, regulators, pressure-reducing valves, acetylene generators, manifolds) used?
- Are cylinders kept away from sources of heat?
- Are the cylinders kept away from elevators, stairs, or gangways?
- Is it prohibited to use cylinders as rollers or supports?
- Are empty cylinders appropriately marked and their valves closed?
- Are signs reading: DANGER — NO SMOKING, MATCHES, OR OPEN LIGHTS, or the equivalent, posted?
- Are cylinders, cylinder valves, couplings, regulators, hoses, and apparatus kept free of oily or greasy substances?
- Is care taken not to drop or strike cylinders?
- Unless secured on special trucks, are regulators removed and protective valve caps put in place before moving cylinders?
- Do cylinders without fixed end wheels have keys, handles, or nonadjustable wrenches on stem valves when in service?
- Are liquefied gases stored and shipped valve-end up with valve covers in place?

OSHA SELF-INSPECTION SAFETY CHECKLISTS (continued)

- Are provisions made to never crack a fuel gas cylinder valve near sources of ignition?
- Before a regulator is removed, is the valve closed and gas released from the regulator?
- Is red used to identify the acetylene (and other fuel gas) hose, green for oxygen hose, and black for inert gas and air hose?
- Are pressure-reducing regulators used only for the gas and pressures for which they are intended?
- Is open circuit (No Load) voltage of arc welding and cutting machines as low as possible and not in excess of the recommended limits?
- Under wet conditions, are automatic controls for reducing no load voltage used?
- Is grounding of the machine frame and safety ground connections of portable machines checked periodically?
- Are electrodes removed from the holders when not in use?
- Is it required that electric power to the welder be shut off when no one is in attendance?
- Is suitable fire extinguishing equipment available for immediate use?
- Is the welder forbidden to coil or loop welding electrode cable around the body?
- Are wet machines thoroughly dried and tested before being used?
- Are work and electrode lead cables frequently inspected for wear and damage, and replaced when needed?
- Do means for connecting cable lengths have adequate insulation?
- When the object to be welded cannot be moved and fire hazards cannot be removed, are shields used to confine heat sparks and slag?
- Are fire watchers assigned when welding or cutting is performed in locations where a serious fire might develop?
- Are combustible floors kept wet, covered by damp sand, or protected by fire-resistant shields?
- When floors are wet down, are personnel protected from possible electrical shock?
- When welding is done on metal walls, are precautions taken to protect combustibles on the other side?
- Before hot work is begun, are used drums, barrels, tanks, and other containers thoroughly cleaned so that no substances remain that could explode, ignite, or produce toxic vapors?
- Is it required that eye protection helmets, hand shields and goggles meet appropriate standards?
- Are employees exposed to the hazards created by welding, cutting, or brazing operations protected with personal protective equipment and clothing?
- Is a check made for adequate ventilation in areas where welding or cutting is performed?
- When working in confined spaces, are environmental monitoring tests taken and means provided for quick removal of welders in case of an emergency?

COMPRESSORS AND COMPRESSED AIR

- Are compressors equipped with pressure relief valves and pressure gages?
- Are compressor air intakes installed and equipped so as to ensure that only clean uncontaminated air enters the compressor?
- Are air filters installed on the compressor intake?
- Are compressors operated and lubricated in accordance with the manufacturer's recommendations?
- Are safety devices on compressed air systems checked frequently?
- Before any repair work is done on the pressure system of a compressor, is the pressure bled off and the system locked out?
- Are signs posted to warn of the automatic starting feature of the compressors?
- Is the belt drive system totally enclosed to provide protection for the front, back, top, and sides?
- Is it strictly prohibited to direct compressed air toward a person?
- Are employees prohibited from using highly compressed air for cleaning purposes?
- If compressed air is used for cleaning off clothing, is the pressure reduced to less than 10 psi?
- When using compressed air for cleaning, do employees wear protective chip guarding and personal protective equipment?
- Are safety chains or other suitable locking devices used at couplings of high pressure hose lines where a connection failure would create a hazard?
- Before compressed air is used to empty containers of liquid, is the safe working pressure of the container checked?
- When compressed air is used with abrasive blast cleaning equipment, is the operating valve a type that must be held open manually?
- When compressed air is used to inflate auto tires, is a clip-on chuck and an in-line regulator preset to 40 psi required?
- Is it prohibited to use compressed air to clean up or move combustible dust if such action could cause the dust to be suspended in the air and cause a fire or explosion hazard?

COMPRESSORS' AIR RECEIVERS

- Is every receiver equipped with a pressure gage and with one or more automatic, spring-loaded safety valves?
- Is the total relieving capacity of the safety valve capable of preventing pressure in the receiver from exceeding the maximum allowable working pressure of the receiver by more than 10 percent?

OSHA SELF-INSPECTION SAFETY CHECKLISTS (continued)

- Is every air receiver provided with a drain pipe and valve at the lowest point for the removal of accumulated oil and water?
- Are compressed air receivers periodically drained of moisture and oil?
- Are all safety valves tested frequently and at regular intervals to determine whether they are in good operating condition?
- Is there a current operating permit issued by the Division of Occupational Safety and Health?
- Is the inlet of air receivers and piping systems kept free of accumulated oil and carbonaceous materials?

COMPRESSED GAS CYLINDERS

- Are cylinders with a water weight capacity over 30 pounds equipped with means for connecting a valve protector device or with a collar or recess to protect the valve?
- Are cylinders legibly marked to clearly identify the gas contained?
- Are compressed gas cylinders stored in areas which are protected from external heat sources such as flame impingement, intense radiant heat, electric arcs, or high temperature lines?
- Are cylinders located or stored in areas where they will not be damaged by passing or falling objects or subject to tampering by unauthorized persons?
- Are cylinders stored or transported in a manner to prevent them from creating a hazard by tipping, falling or rolling?
- Are cylinders containing liquefied fuel gas stored or transported in a position so that the safety relief device is always in direct contact with the vapor space in the cylinder?
- Are valve protectors always placed on cylinders when the cylinders are not in use or connected for use?
- Are all valves closed off before a cylinder is moved, when the cylinder is empty, and at the completion of each job?
- Are low-pressure fuel gas cylinders checked periodically for corrosion, general distortion, cracks, or any other defect that might indicate a weakness or make it unfit for service?
- Does the periodic check of low-pressure fuel gas cylinders include a close inspection of the cylinders' bottom?

HOIST AND AUXILIARY EQUIPMENT

- Is each overhead electric hoist equipped with a limit device to stop the hook travel at its highest and lowest point of safe travel?
- Will each hoist automatically stop and hold any load up to 125 percent of its rated load if its actuating force is removed?
- Is the rated load of each hoist legibly marked and visible to the operator?
- Are stops provided at the safe limits of travel for trolley hoists?
- Are the controls of a hoist plainly marked to indicate the direction of travel or motion?
- Is each cage-controlled hoist equipped with an effective warning device?
- Are close-fitting guards or other suitable devices installed on the hoist to ensure that hoist ropes will stay in the sheave grooves?
- Are all hoist chains or ropes of sufficient length to handle the full range of movement of the application while still maintaining two full wraps on the drum at all times?
- Are nip points or contact points between hoist ropes and sheaves which are permanently located within seven feet of the floor, ground or working platform, guarded?
- Is it prohibited to use chains or rope slings that are kinked or twisted?
- Is it prohibited to use the hoist rope or chain wrapped around the load as a substitute for a sling?
- Is the operator instructed to avoid carrying loads over the heads of people?

INDUSTRIAL TRUCKS — FORKLIFTS

- Are only employees who have been trained in the proper use of hoists allowed to operate them?
- Are only trained personnel allowed to operate industrial trucks?
- Is substantial overhead protective equipment provided on high lift rider equipment?
- Are the required lift truck operating rules posted and enforced?
- Is directional lighting provided on each industrial truck that operates in an area with less than 2 foot candles per square foot of general lighting?
- Does each industrial truck have a warning horn, whistle, gong, or other device which can be clearly heard above the normal noise in the areas where operated?
- Are the brakes on each industrial truck capable of bringing the vehicle to a complete and safe stop when fully loaded?
- Will the industrial trucks' parking brake effectively prevent the vehicle from moving when unattended?
- Are industrial trucks operating in areas where flammable gases or vapors, or combustible dust or ignitable fibers may be present in the atmosphere, approved for such locations?
- Are motorized hand and hand/rider trucks so designed that the brakes are applied and power to the drive motor shuts off when the operator releases his or her grip on the device that controls the travel?
- Are industrial trucks with internal combustion engines which are operated in buildings or enclosed areas carefully checked to ensure such operations do not cause harmful concentrations of dangerous gases or fumes?

SPRAYING OPERATIONS

- Is adequate ventilation ensured before spray operations are started?
- Is mechanical ventilation provided when spraying operations are done in enclosed areas?
- When mechanical ventilation is provided during spraying operations, is it so arranged that it will not circulate the contaminated air?
- Is the spray area free of hot surfaces?

OSHA SELF-INSPECTION SAFETY CHECKLISTS (continued)

- Is the spray area at least 20 feet from flames, sparks, operating electric motors and other ignition sources?
- Are portable lamps used to illuminate spray areas suitable for use in a hazardous location?
- Is approved respiratory equipment provided and used when appropriate during spraying operations?
- Do solvents used for cleaning have a flash point of 100°F or higher?
- Are fire control sprinkler heads kept clean?
- Are NO SMOKING signs posted in spray areas, paint rooms, paint booths, and paint storage areas?
- Is the spray area kept clean of combustible residue?
- Are spray booths constructed of metal, masonry, or other substantial noncombustible material?
- Are spray booth floors and baffles noncombustible and easily cleaned?
- Is infrared drying apparatus kept out of the spray area during spraying operations?
- Is the spray booth completely ventilated before using the drying apparatus?
- Is the electric drying apparatus properly grounded?
- Are lighting fixtures for spray booths located outside of the booth and the interior lighted through sealed clear panels?
- Are the electric motors for exhaust fans placed outside booths or ducts?
- Are belts and pulleys inside the booth fully enclosed?
- Do ducts have access doors to allow cleaning?
- Do all drying spaces have adequate ventilation?

ENTERING CONFINED SPACES

- Are confined spaces thoroughly emptied of any corrosive or hazardous substances, such as acids or caustics, before entry?
- Are all lines to a confined space containing inert, toxic, flammable, or corrosive materials valved off and blanked or disconnected and separated before entry?
- Is it required that all impellers, agitators, or other moving equipment inside confined spaces be locked out if they present a hazard?
- Is either natural or mechanical ventilation provided prior to confined space entry?
- Are appropriate atmospheric tests performed to check for oxygen deficiency, toxic substances and explosive concentrations in the confined space before entry?
- Is adequate light provided for the work to be performed in the confined space?
- Is the atmosphere inside the confined space frequently tested or continuously monitored during conduct of work?
- Is there an assigned safety standby employee outside of the confined space, when required, whose sole responsibility is to watch the work in progress, sound an alarm if necessary, and provide assistance?
- Is the standby employee appropriately trained and equipped to handle an emergency?
- Is the standby employee or other employees prohibited from entering the confined space without lifelines and respiratory equipment if there is any question as to the cause of an emergency?
- Is approved respiratory equipment required if the atmosphere inside the confined space cannot be made acceptable?
- Is all portable electrical equipment used inside confined spaces either grounded and insulated or equipped with ground-fault protection?
- Before gas welding or burning is started in a confined space, are hoses checked for leaks? Are compressed gas bottles forbidden inside of the confined space? Are torches lighted only outside of the confined area and is the confined area tested for an explosive atmosphere each time before a lighted torch is to be taken into the confined space?
- If employees will be using oxygen-consuming equipment such as salamanders, torches or furnaces in a confined space, is sufficient air provided to ensure combustion without reducing the oxygen concentration of the atmosphere below 19.5 percent by volume?
- Whenever combustion-type equipment is used in a confined space, are provisions made to ensure the exhaust gases are vented outside of the enclosure?
- Is each confined space checked for decaying vegetation or animal matter which may produce methane?
- Is the confined space checked for possible industrial waste which could contain toxic properties?
- If the confined space is below the ground and near areas where motor vehicles will be operating, is it possible for vehicle exhaust or carbon monoxide to enter the space?

ENVIRONMENTAL CONTROLS

- Are all work areas properly lighted?
- Are employees instructed in proper first aid and other emergency procedures?
- Are hazardous substances identified which may cause harm by inhalation, ingestion, skin absorption or contact?
- Are employees aware of the hazards involved with the various chemicals they may be exposed to in their work environment, such as ammonia, chlorine, epoxies or caustics?
- Is employee exposure to chemicals in the workplace kept within acceptable levels?
- Can a less harmful method or product be used?
- Is the work area's ventilation system appropriate for the work being performed?
- Are spray painting operations done in spray rooms or booths equipped with an appropriate exhaust system?
- Is employee exposure to welding fumes controlled by ventilation, use of respirators, exposure time, or other means?

OSHA SELF-INSPECTION SAFETY CHECKLISTS (continued)

- Are welders and other workers nearby provided with flash shields during welding operations?
- If forklifts and other vehicles are used in buildings or other enclosed areas, are the carbon monoxide levels kept below maximum acceptable concentration?
- Has there been a determination that noise levels in the facilities are within acceptable levels?
- Are steps being taken to use engineering controls to reduce excessive noise levels?
- Are proper precautions being taken when handling asbestos and other fibrous materials?
- Are caution labels and signs used to warn of asbestos?
- Are wet methods used, when practicable, to prevent the emission of airborne asbestos fibers, silica dust and similar hazardous materials?
- Is vacuuming with appropriate equipment used whenever possible rather than blowing or sweeping dust?
- Are grinders, saws, and other machines that produce dust vented to an industrial collector or central exhaust system?
- Are all local exhaust ventilation systems designed and operating properly to provide the air flow and volume necessary for the application? Are ducts open and belts secure?
- Is personal protective equipment provided, used and maintained wherever required?
- Are there written standard operating procedures for the selection and use of a respirator where needed?
- Are restrooms and washrooms kept clean and sanitary?
- Is potable water provided for drinking, washing, and cooking?
- Are all outlets for water not suitable for drinking clearly identified?
- Are employees' physical capacities assessed before being assigned to jobs requiring heavy work?
- Are employees instructed in the proper manner of lifting heavy objects?
- Where heat is a problem, have all fixed work areas been provided with spot cooling or air conditioning?
- Are employees screened before assignment to areas of high heat to determine if their health condition might make them more susceptible to having an adverse reaction?
- Are employees working on streets and roadways where they are exposed to the hazards of traffic required to wear bright colored (traffic orange) warning vests?
- Are exhaust stacks and air intakes so located that contaminated air will not be recirculated within a building or other enclosed area?
- Is equipment producing ultraviolet radiation properly shielded?

FLAMMABLE AND COMBUSTIBLE MATERIALS

- Are combustible scrap, debris and waste materials (oily rags, for example) stored in covered metal receptacles and removed from the worksite promptly?
- Is proper storage practiced to minimize the risk of fire including spontaneous combustion?
- Are approved containers and tanks used for the storage and handling of flammable and combustible liquids?
- Are all connections on drums and combustible liquid piping, vapor- and liquid-tight?
- Are all flammable liquids kept in closed containers when not in use (for example, parts cleaning tanks or pans)?
- Are bulk drums of flammable liquids grounded and bonded to containers during dispensing?
- Do storage rooms for flammable and combustible liquids have explosion-proof lights?
- Do storage rooms for flammable and combustible liquids have mechanical or gravity ventilation?
- Is liquefied petroleum gas stored, handled, and used in accordance with safe practices and standards?
- Are NO SMOKING signs posted on liquefied petroleum gas tanks?
- Are liquefied petroleum storage stands guarded to prevent damage from vehicles?
- Are all solvent wastes and flammable liquids kept in fire-resistant, covered containers until they are removed from the work site?
- Is vacuuming used whenever possible rather than blowing or sweeping combustible dust?
- Are firm separators placed between containers of combustibles or flammables, when stacked one upon another, to ensure their support and stability?

OSHA SELF-INSPECTION SAFETY CHECKLISTS (continued)

- Are fuel gas cylinders and oxygen cylinders separated by distance or fire-resistant barriers while in storage?
- Are fire extinguishers selected and provided for the types of materials in areas where they are to be used?

  Class A — Ordinary combustible material fires.

  Class B — Flammable liquid, gas or grease fires.

  Class C — Energized electrical equipment fires.
- Are appropriate fire extinguishers mounted within 75 feet of outside areas containing flammable liquids, and within 10 feet of any inside storage area for such materials?
- Are extinguishers free from obstructions or blockage?
- Are all extinguishers serviced, maintained and tagged at intervals not to exceed one year?
- Are all extinguishers fully charged and in their designated places?
- Where sprinkler systems are permanently installed, are the nozzle heads so directed or arranged that water will not be sprayed into operating electrical switch boards and equipment?
- Are NO SMOKING signs posted in areas where flammable or combustible materials are used or stored?
- Are safety cans used for dispensing flammable or combustible liquids at a point of use?
- Are all spills of flammable or combustible liquids cleaned up promptly?
- Are storage tanks adequately vented to prevent the development of excessive vacuum or pressure as a result of filling, emptying, or temperature changes?
- Are storage tanks equipped with emergency venting that will relieve excessive internal pressure caused by fire exposure?
- Are NO SMOKING rules enforced in areas involving storage and use of hazardous materials?

HAZARDOUS CHEMICAL EXPOSURE

- Are employees trained to use safe practices when handling hazardous chemicals such as acids or caustics?
- Are employees aware of the potential hazards involving various chemicals stored or used in the workplace such as acids, bases, caustics, epoxies or phenols?
- Is employee exposure to chemicals kept within acceptable levels?
- Are eyewash fountains and safety showers provided in areas where corrosive chemicals are handled?
- Are all containers, such as vats or storage tanks, labeled as to their contents, for example, CAUSTICS?
- Are all employees required to use personal protective clothing and equipment when handling chemicals (gloves, eye protection, respirators)?
- Are flammable or toxic chemicals kept in closed containers when not in use?
- Are chemical piping systems clearly marked as to their content?
- Where corrosive liquids are frequently handled in open containers or drawn from storage vessels or pipelines, are adequate means readily available for neutralizing or disposing of spills or overflows properly and safely?
- Have standard operating procedures been established and are they being followed when cleaning up chemical spills?
- Where needed for emergency use, are respirators stored in a convenient, clean, and sanitary location?
- Are respirators intended for emergency use adequate for the various uses for which they may be needed?
- Are employees prohibited from eating in areas where hazardous chemicals are present?
- Is personal protective equipment provided, used and maintained whenever necessary?
- Are there written standard operating procedures for the selection and use of respirators where needed?
- If you have a respirator protection program, are your employees instructed on the correct use and limitations of the respirators? Are the respirators NIOSH-approved for this particular application? Are they regularly inspected and cleaned, sanitized and maintained?
- If hazardous substances are used in your processes, do you have a medical or biological monitoring system in operation?
- Are you familiar with the Threshold Limit Values or Permissible Exposure Limits of airborne contaminants and physical agents used in your workplace?
- Have control procedures been instituted for hazardous materials, where appropriate, such as respirators, ventilation systems or handling practices?
- Whenever possible are hazardous substances handled in properly designed and exhausted booths or similar locations?
- Do you use general dilution or local exhaust ventilation systems to control dusts, vapors, gases, fumes, smoke, solvents or mists which may be generated in your workplace?
- Is ventilation equipment provided for removal of contaminants from such operations as production grinding, buffing, spray painting, and/or vapor degreasing, and is it operating properly?
- Do employees complain about dizziness, headaches, nausea, irritation, or other symptoms of discomfort when they use solvents or other chemicals?
- Is there a dermatitis problem? Do employees complain about dryness, irritation, or sensitization of the skin?
- Have you considered the use of an industrial hygienist or environmental health specialist to evaluate your operation?
- If internal combustion engines are used, is carbon monoxide kept within acceptable levels?
- Is vacuuming used, rather than blowing or sweeping dust whenever possible for cleanup?
- Are materials which give off toxic, suffocating or anesthetic fumes stored in remote or isolated locations when not in use?

OSHA SELF-INSPECTION SAFETY CHECKLISTS (continued)

HAZARDOUS SUBSTANCES COMMUNICATION

- Is there a list of hazardous substances used in your workplace?
- Is there a written hazard communication program dealing with Material Safety Data Sheets (MSDSs), labeling, and employee training?
- Is each container for a hazardous substance (for example, vats, bottles, storage tanks) labeled with product identity and a hazard warning (communication of the specific health hazards and physical hazards)?
- Is a Material Safety Data Sheet readily available for each hazardous substance used?
- Is there an employee training program for hazardous substances? Does this program include:
  (1) An explanation of what an MSDS is and how to use and obtain one.
  (2) MSDS contents for each hazardous substance or class of substances.
  (3) Explanation of "Right to Know."
  (4) Identification of where an employee can see the employer's written hazard communication program and where hazardous substances are present in their work areas.
  (5) The physical and health hazards of substances in the work area and specific protective measures to be used.
  (6) Details of the hazard communication program, including how to use the labeling system and MSDSs.

ELECTRICAL

- Do you specify compliance with OSHA for all contract electrical work?
- Are all employees required to report as soon as practicable any obvious hazard to life or property observed in connection with electrical equipment or lines?
- Are employees instructed to make preliminary inspections and/or appropriate tests to determine what conditions exist before starting work on electrical equipment or lines?
- When electrical equipment or lines are to be serviced, maintained or adjusted, are necessary switches opened, locked out and tagged whenever possible?
- Are portable electrical tools and equipment grounded or of the double-insulated type?
- Are electrical appliances such as vacuum cleaners, polishers, or vending machines grounded?
- Do extension cords being used have a grounding conductor?
- Are multiple plug adapters prohibited?
- Are ground-fault circuit interrupters installed on each temporary 15 or 20 ampere, 120 volt A.C. circuit at locations where construction, demolition, modifications, alterations or excavations are being performed?
- Are all temporary circuits protected by suitable disconnecting switches or plug connectors at the junction with permanent wiring?
- Do you have electrical installations in hazardous dust or vapor areas? If so, do they meet the National Electrical Code (NEC) for hazardous locations?
- Are exposed wires and cords with frayed or deteriorated insulation repaired or replaced promptly?
- Are flexible cords and cables free of splices or taps?
- Are clamps or other securing means provided on flexible cords or cables at plugs, receptacles, tools, or equipment, and is the cord jacket securely held in place?
- Are all cord, cable and raceway connections intact and secure?
- In wet or damp locations, are electrical tools and equipment appropriate for the use or location or otherwise protected?
- Is the location of electrical power lines and cables (overhead, underground, underfloor, other side of walls) determined before digging, drilling or similar work is begun?
- Are metal measuring tapes, ropes, handlines or similar devices with metallic thread woven into the fabric prohibited where they could come in contact with energized parts of equipment or circuit conductors?
- Is the use of metal ladders prohibited in areas where the ladder or the person using the ladder could come in contact with energized parts of equipment, fixtures or circuit conductors?
- Are all disconnecting switches and circuit breakers labeled to indicate their use or equipment served?
- Are disconnecting means always opened before fuses are replaced?
- Do all interior wiring systems include provisions for grounding metal parts of electrical raceways, equipment and enclosures?
- Are all electrical raceways and enclosures securely fastened in place?
- Are all energized parts of electrical circuits and equipment guarded against accidental contact by approved cabinets or enclosures?
- Is sufficient access and working space provided and maintained around all electrical equipment to permit ready and safe operations and maintenance?
- Are all unused openings (including conduit knockouts) in electrical enclosures and fittings closed with appropriate covers, plugs or plates?
- Are electrical enclosures such as switches, receptacles, or junction boxes provided with tight-fitting covers or plates?
- Are disconnecting switches for electric motors in excess of two horsepower capable of opening the circuit without exploding when the motor is in a stalled condition? (Switches must be horsepower rated equal to or in excess of the motor HP rating.)
- Is low voltage protection provided in the control device of motors driving machines or equipment which could cause probable injury from inadvertent starting?

OSHA SELF-INSPECTION SAFETY CHECKLISTS (continued)

- Is each motor disconnecting switch or circuit breaker located within sight of the motor control device?
- Is each motor located within sight of its controller or is the controller disconnecting means capable of being locked in the open position or is a separate disconnecting means installed in the circuit within sight of the motor?
- Is the controller for each motor in excess of two horsepower rated in horsepower equal to or in excess of the rating of the motor it serves?
- Are employees who regularly work on or around energized electrical equipment or lines instructed in the cardiopulmonary resuscitation (CPR) methods?
- Are employees prohibited from working alone on energized lines or equipment over 600 volts?

NOISE

- Are there areas in the workplace where continuous noise levels exceed 85 dBA?
- Is there an ongoing preventive health program to educate employees about safe levels of noise, exposures, effects of noise on their health, and the use of personal protection?
- Have work areas where noise levels make voice communication between employees difficult been identified and posted?
- Are noise levels being measured using a sound level meter or an octave band analyzer and are records being kept?
- Have engineering controls been used to reduce excessive noise levels? Where engineering controls are not feasible, are administrative controls (for example, worker rotation) being used to minimize individual employee exposure to noise?
- Is approved hearing protective equipment (noise attenuating devices) available to every employee working in noisy areas?
- Have you tried isolating noisy machinery from the rest of your operation?
- If you use ear protectors, are employees properly fitted and instructed in their use?
- Are employees in high noise areas given periodic audiometric testing to ensure that you have an effective hearing protection system?

FUELING

- Is it prohibited to fuel an internal combustion engine with a flammable liquid while the engine is running?
- Are fueling operations done in such a manner that the likelihood of spillage will be minimal?
- When spillage occurs during fueling operations, is the spilled fuel washed away completely, evaporated, or are other measures taken to control vapors before restarting the engine?
- Are fuel tank caps replaced and secured before starting the engine?
- In fueling operations, is there always metal contact between the container and the fuel tank?
- Are fueling hoses of a type designed to handle the specific type of fuel?
- Is it prohibited to handle or transfer gasoline in open containers?
- Are open lights, open flames, or sparking, or arcing equipment prohibited near fueling or transfer of fuel operations?
- Is smoking prohibited in the vicinity of fueling operations?
- Are fueling operations prohibited in buildings or other enclosed areas that are not specifically ventilated for this purpose?
- Where fueling or transfer of fuel is done through a gravity flow system, are the nozzles of the self-closing type?

IDENTIFICATION OF PIPING SYSTEMS

- When nonpotable water is piped through a facility, are outlets or taps posted to alert employees that it is unsafe and not to be used for drinking, washing or other personal use?
- When hazardous substances are transported through above-ground piping, is each pipeline identified at points where confusion could introduce hazards to employees?
- When pipelines are identified by color painting, are all visible parts of the line so identified?
- When pipelines are identified by color painted bands or tapes, are the bands or tapes located at reasonable intervals and at each outlet, valve or connection?
- When pipelines are identified by color, is the color code posted at all locations where confusion could introduce hazards to employees?
- When the contents of pipelines are identified by name or name abbreviation, is the information readily visible on the pipe near each valve or outlet?
- When pipelines carrying hazardous substances are identified by tags, are the tags constructed of durable materials, is the message clearly and permanently distinguishable, and are tags installed at each valve or outlet?
- When pipelines are heated by electricity, steam or other external source, are suitable warning signs or tags placed at unions, valves, or other serviceable parts of the system?

MATERIAL HANDLING

- Is there safe clearance for equipment through aisles and doorways?
- Are aisles designated, permanently marked, and kept clear to allow unhindered passage?
- Are motorized vehicles and mechanized equipment inspected daily or prior to use?

OSHA SELF-INSPECTION SAFETY CHECKLISTS (continued)

- Are vehicles shut off and brakes set prior to loading or unloading?
- Are containers of combustibles or flammables, when stacked while being moved, always separated by dunnage (material) sufficient to provide stability?
- Are dock boards (bridge plates) used when loading or unloading operations are taking place between vehicles and docks?
- Are trucks and trailers secured from movement during loading and unloading operations?
- Are dock plates and loading ramps constructed and maintained with sufficient strength to support imposed loading?
- Are hand trucks maintained in safe operating condition?
- Are chutes equipped with sideboards of sufficient height to prevent the materials being handled from falling off?
- Are chutes and gravity roller sections firmly placed or secured to prevent displacement?
- At the delivery end of the rollers or chutes, are provisions made to brake the movement of the handled materials?
- Are pallets usually inspected before being loaded or moved?
- Are hooks with safety latches or other arrangements used when hoisting materials so that slings or load attachments won't accidentally slip off the hoist hooks?
- Are securing chains, ropes, chokers or slings adequate for the job to be performed?
- When hoisting material or equipment, are provisions made to ensure no one will be passing under the suspended loads?
- Are material safety data sheets available to employees handling hazardous substances?

TRANSPORTING EMPLOYEES AND MATERIALS

- Do employees who operate vehicles on public thoroughfares have a valid operator's license?
- When seven or more employees are regularly transported in a van, bus or truck, is the operator's license appropriate for the class of vehicle being driven?
- Is each van, bus or truck used regularly to transport employees equipped with an adequate number of seats?
- When employees are transported by truck, are provisions made to prevent their falling from the vehicle?
- Are vehicles used to transport employees equipped with lamps, brakes, horns, mirrors, windshields and turn signals in good repair?
- Are transport vehicles provided with handrails, steps, stirrups or similar devices so placed and arranged that employees can safely mount or dismount?
- Are employee transport vehicles equipped at all times with at least two reflective type flares?
- Is a fully charged fire extinguisher, in good condition, with at least 4 B:C rating, maintained in each employee transport vehicle?
- When cutting tools or tools with sharp edges are carried in passenger compartments of employee transport vehicles, are they placed in closed boxes or containers which are secured in place?
- Are employees prohibited from riding on top of any load which can shift, topple, or otherwise become unstable?

CONTROL OF HARMFUL SUBSTANCES BY VENTILATION

- Is the volume and velocity of air in each exhaust system sufficient to gather the dusts, fumes, mists, vapors or gases to be controlled and convey them to a suitable point of disposal?
- Are exhaust inlets, ducts and plenums designed, constructed, and supported to prevent collapse or failure of any part of the system?
- Are cleanout ports or doors provided at intervals not to exceed 12 feet in all horizontal runs of exhaust ducts?
- Where two or more different types of operations are being controlled through the same exhaust system, will the combination of substances being controlled constitute a fire, explosion or chemical reaction hazard in the duct?
- Is adequate make-up air provided to areas where exhaust systems are operating?
- Is the source point for make-up air located so that only clean, fresh air which is free of contaminants will enter the work environment?
- Where two or more ventilation systems are serving a work area, is their operation such that one will not offset the functions of the other?

SANITIZING EQUIPMENT AND CLOTHING

- Is personal protective clothing or equipment that employees are required to wear or use of a type capable of being cleaned easily and disinfected?
- Are employees prohibited from exchanging personal protective clothing or equipment with each other unless it has been properly cleaned?
- Are machines and equipment which process, handle or apply materials that could be injurious to employees cleaned and/or decontaminated before being overhauled or placed in storage?
- Are employees prohibited from smoking or eating in any area where contaminants that could be injurious if ingested are present?

OSHA SELF-INSPECTION SAFETY CHECKLISTS (continued)

- When employees are required to change from street clothing into protective clothing, is a clean change room with separate storage facilities for street and protective clothing provided?
- Are employees required to shower and wash their hair as soon as possible after a known contact has occurred with a carcinogen?
- When equipment, materials, or other items are taken into or removed from a carcinogen-regulated area, is it done in a manner that will not contaminate unregulated areas or the external environment?

TIRE INFLATION

- Where tires are mounted and/or inflated on drop center wheels, is a safe procedure posted and enforced?
- Where tires are mounted and/or inflated on wheels with split rims and/or retainer rings, is a safe practice procedure posted and enforced?
- Does each tire inflation hose have a clip-on chuck with at least 24 inches of hose between the chuck and an in-line hand valve and gage?
- Does the tire inflation control valve automatically shut off the air flow when the valve is released?
- Is a tire restraining device such as a cage, rack or other effective means used while inflating tires mounted on split rims, or rims using retainer rings?
- Are employees strictly forbidden from taking a position directly over or in front of a tire while it's being inflated?

CHAPTER 12

ADMINISTRATION

by

Neal R. Townley

Revised by

Steve Goodman

Rich Cunningham

Gary Batis

TABLE OF CONTENTS

Chapter 12. ADMINISTRATION

OBJECTIVES

Chapter 12. ADMINISTRATION

Following completion of Chapter 12, you should be able to:

1. Explain the need for effective administration,
2. Develop the goals, tasks, and procedures for an operating plan,
3. Prepare and justify staffing and equipment requirements for your program,
4. Determine whether a piece of equipment should be leased or purchased,
5. Hire new operators,
6. Administer your agency's safety program,
7. Determine the facility requirements for your program,
8. Read the various types of maps used by collection system operators,
9. Explain the importance of and need for maps,
10. Keep maps up to date,
11. Determine the management information system requirements for your program,
12. Prepare and maintain records essential for budgeting, scheduling, and meeting legal requirements,
13. Write an informative report, and
14. Organize an effective public relations program for your agency.

CHAPTER 12. ADMINISTRATION

(Lesson 1 of 7 Lessons)

12.0 NEED FOR EFFECTIVE ADMINISTRATION

The quality of the operation and maintenance of a wastewater collection system depends, to a great extent, on effective administration of the numerous elements involved in such a program. An effective administration will assure an operation and maintenance program that will keep a wastewater collection system functioning at its top efficiency, maximize its useful life, and minimize costs.

Personnel actually performing the operation and maintenance work will respect an effective administration, work more efficiently, and derive greater job satisfaction, thus providing for the retention of good personnel.

12.1 PRINCIPLES OF ADMINISTRATION

In 1916, Henri Fayol, an executive for 58 years in a large French coal-steel combine, laid down 14 basic administrative principles that are still valid today. These principles are:

1. **Division of work.** Specialization is necessary for maximum effectiveness.
2. **Authority.** A manager has the right to give orders and the power to compel obedience. One must make a distinction, however, between a manager's official authority and 'personal' experience, moral worth, and ability to lead.
3. **Discipline.** Obedience, application, energy, respect, and behavior go together to make up discipline — an essential condition for a smooth-running organization.
4. **Unity of command.** Whatever the action, an employee should receive orders from one superior only.
5. **Unity of direction.** A group of activities must have the same major purpose—one head and one plan. This condition is essential to unity of action, to coordination of strength, and to the focusing of effort.
6. **Subordination of individual interest to general interest.** This is a basic requirement in every well-managed organization.
7. **Remuneration of personnel.** Payment and awards should be fair and should afford satisfaction to employees and the firm.
8. **Centralization.** This is a matter of finding the optimum degree of centralization for particular organizations. Centralization per se is neither good nor bad; and the degree of centralization must vary in different cases.
9. **Scalar chain.** This term refers to the chain of superiors in an organization, ranging from the ultimate authority to the lowest ranks. Both orders and communications must flow along this chain. To depart needlessly from the line of authority is an error; but it is an even greater one to keep it when it is detrimental.
10. **Order.** It is a basic requirement for sound management.
11. **Equity.** Equitable treatment for employees is fundamental to good management. It is based on kindness and on justice.
12. **Stability of tenure of personnel.** Sound management requires stability of managerial personnel and employees. Constant change and turnover adversely affect the organization.
13. **Initiative.** A progressive managerial climate promotes initiative. It is a powerful stimulant to human endeavor. Encourage it and promote its development fully. A manager who is able to permit the exercise of initiative on the part of subordinates is infinitely superior to one who cannot do it.
14. **Esprit de corps.** In union there is strength. Promote it always. Avoid creating dissension among subordinates. Encourage and reward employees on the basis of their ability, talent, and accomplishment.

Fayol insisted that these principles were not rigid and should be applied in an intelligent manner.

12.2 OPERATING PLAN

An essential part of an effective administration is a detailed, written plan for the operation and maintenance of the wastewater collection system. The plan should consist of a statement of the agency's overall mission, goals to be attained and the specific objectives, tasks and procedures that will lead to attainment of the mission. The plan should be prepared by the manager of the organization responsible for operation and maintenance of the collection system with the full participation of the organization's personnel. The more clearly and completely the plan is put down in writing, the better agency personnel will be able to work in accordance with it. For any long-range plan to work, it is essential that the personnel who will implement it be involved in developing the plan and be generally supportive of it.

12.20 Mission Statement

The mission statement of an organization is a very broad, general explanation of the reason for the organization's existence. Such a statement might read:

"The Clearwater Wastewater Collection Agency will provide safe, reliable, efficient collection system services at a minimum cost to the community of Clearwater, U.S.A."

This one sentence tells us who will do what for whom, and in what manner. It is the "ideal" that the agency strives to achieve. Once the basic purpose of an organization is defined, it will then be possible to develop a plan to accomplish this mission.

12.21 Goals

The Water Environment Federation's Technical Task Force on Sewer Maintenance published[1] the following list of goals for the responsible and efficient management of a collection system. The goals are as follows:

1. Prevent public health hazards;
2. Minimize inconveniences by responsibly handling interruptions in service;
3. Protect the large investment in collection systems by maintaining maximum capacities and extending their useful life;
4. Prevent unnecessary damage to public and private property;
5. Use the funds available for the operation of municipal government services in the most efficient manner;
6. Convey wastewater to treatment facilities with a minimum of infiltration, inflow, and exfiltration;
7. Prevent excessive expenditures for claims and legal fees due to backups by providing immediate, concerned, and efficient service to all emergency calls; and
8. Perform all operations in a safe manner to avoid personal injury.

You should realize that full attainment of these goals is difficult to reach. Just as in a football game, the real sense of accomplishment derives from the progress you make toward meeting a goal in spite of the difficulties and occasional setbacks you may experience.

12.22 Objectives and Tasks/Procedures

Once goals have been established, a plan for operation and maintenance of a collection system should outline the more specific objectives as well as tasks and procedures for progressing toward the desired goals. Let's assume your agency agrees that the goals set forth by the Water Environment Federation are worthy of adoption. The following list demonstrates how you or your agency might go about setting specific objectives to meet each goal and how to list detailed tasks/procedures that will lead to accomplishing the objectives.

GOAL 1: Prevent public health hazards.

a. Provide immediate, concerned and efficient service to all emergency calls.
   - Dispatch crews within one hour in response to emergency service calls.
   - Clean up and disinfect flooded homes. (Requirements may vary with the utility; contact your city, agency, district, or company for policies and procedures.)

b. Assign scheduling priority to stoppages, overflows, and exfiltration that contaminates the environment.
   - Send a supervisor to evaluate emergency calls.

c. Use clean water to flush contaminated surfaces and disinfect with chemicals approved by the health authority.
   - Maintain a neat work area for repair and maintenance activities.

GOAL 2: Minimize inconveniences by responsibly handling interruptions in service.

a. Respond immediately to reports of sewer stoppages and remove them.
   - Assign priority status for removal of stoppages.
   - Dispatch crew within one hour of receiving emergency call.

b. Use radios and 24-hour communications to dispatch personnel and mechanical equipment to remove sewer stoppages.
   - Equip emergency crew vehicles with two-way radios.

c. Keep system maps up to date.
   - Establish a routine for submitting map changes to drafting or computer department.
   - Field check maps for accuracy.

GOAL 3: Protect the large investment in collection systems by maintaining maximum capacities and extending their useful life.

a. Inspect the collection facility to determine maintenance needs.
   - Use aids such as closed-circuit television to systematically inspect sewer lines and appurtenances.
   - Use depth and velocity measurements to determine flow of wastewater in sewer lines.

[1] *OPERATION AND MAINTENANCE OF WASTEWATER COLLECTION SYSTEMS*, page 2, Water Environment Federation (WEF), 601 Wythe Street, Alexandria, VA 22314-1994.

b. Schedule and perform cleaning of sewer lines.
   - Use high-velocity cleaners, bucketing machines, or balling equipment as appropriate.

c. Inhibit root growth in sewers.
   - Use chemical flooding or foaming methods.
   - Use mechanical pruning.
   - Use low-volume, very high pressure (5,000 to 10,000 psi) water jet.

d. Inhibit release of hydrogen sulfide gas.
   - Apply chemicals to reduce rate of gas production.
   - Keep sewers clean to prevent deposition and accumulation of solids.

e. Rehabilitate worn parts of collection facility when work will economically retain the capacity and extend the life of the part.
   - Develop a priority listing of areas needing rehabilitation.
   - Evaluate rehabilitation alternatives for each situation.

GOAL 4: Prevent unnecessary damage to public and private property.

a. Remove standing water and solids in buildings, grounds and streets resulting from stoppages in any part of the collection system.
   - Clean up areas affected by wastewater overflows using vacuum equipment and mops.
   - Remove waterlogged materials that would cause additional damage to a flooded building.

b. Restore private property to its former condition when maintenance work has disturbed landscaping or fencing.
   - Work cooperatively with homeowners.

c. Respond promptly to reports of stoppages and take appropriate corrective measures.
   - Be prepared to respond quickly to any expected type of stoppage.

GOAL 5: Use funds available for the operation of municipal government services in the most efficient manner.

a. Prepare budgets for operation and maintenance of the collection system.
   - Establish a cost accounting system to track expenditures and maintenance costs.
   - Analyze previous year's costs for personnel, maintenance and equipment.
   - Estimate the cost of carrying out the planned tasks for the budget year.

b. Use budget and other expenditure controls to minimize the cost of operation and maintenance.
   - Determine who has authority to authorize purchases.
   - Implement a purchase order system.
   - Set up guidelines for examining alternative suppliers to obtain lowest possible prices.

c. Arrange with your financial management division for the investment of funds in interest-bearing securities when they are not immediately needed for operations and maintenance.
   - Analyze cash flow requirements for operations and maintenance based on maintenance schedules and past experience of fixed expenditures.
   - Have your financial management division invest excess funds in the highest quality interest-bearing securities available, with the reinvestment of income earned on the securities.

d. Influence the design and construction of the wastewater collection facility to ensure efficient, workable design.
   - Operating personnel should participate in the design, construction, and inspection of the collection system.
   - Walk through planned construction sites to evaluate potential problems.
   - Research the merits of new technology for possible use in the system.

GOAL 6: Convey wastewater to treatment facilities with a minimum of infiltration, inflow and exfiltration.

a. Remove sources of infiltration, inflow and exfiltration from the collection system.
   - Rehabilitate or repair worn and damaged parts of the collection system.
   - Reline corroded sewer lines and manholes.
   - Replace broken parts of sewer lines and mechanical equipment.

b. Isolate manhole and cleanout openings from surface water.
   - Plug holes in manhole covers, with due consideration for ventilation requirements.
   - Smoke test system to identify sources of surface water inflow.

c. Adopt and enforce sewer-use regulations prohibiting the illegal discharge of storm water or groundwater to the wastewater collection system.
   - Smoke test system.
   - Assign crews to search for illegal connections in problem areas.

GOAL 7: Prevent excessive expenditures for claims and legal fees due to backups.

a. Dispatch crews ASAP (As Soon As Possible) in response to emergency calls for service.

b. Minimize damage caused by overflow water from the collection system by taking prompt action. The cleaning procedure may vary depending on the utility. Contact your city, agency, district, or company for policies and procedures.
   - Clean up areas affected by wastewater overflows using vacuum equipment and mops and disinfect as appropriate.

- Remove waterlogged materials that would cause additional damage to a flooded building.

c. Keep complete records of all actions taken in the course of making repairs.
   - Require crews to promptly submit written reports detailing actions and procedures used.

d. Assist in the determination of appropriate compensation (if any) for damage caused by wastewater from the collection system.
   - Take photographs and make comprehensive reports on wastewater overflows and other accidents attributable to the operation and maintenance of the collection system.
   - Be prepared to give truthful and accurate information on how an incident occurred.

GOAL 8: Perform all operations in a safe manner to avoid personal injury and property damage.

a. Take actions that consistently promote safe operation by personnel and contractors.
   - Promote safety by personal use of safe practices.
   - Provide safety equipment and require its use.
   - Require adherence to safe work practices.
   - Establish a written safety policy.
   - Reward personnel for safe practices and discipline personnel for unsafe practices.

b. Establish safety training programs for personnel.
   - Provide the time, qualified persons, and material for training personnel in the safe performance of their duties.
   - Construct specialized training facilities.

c. Control the risks involved in the operation and maintenance of the collection system.
   - Analyze planned operations for exposure of persons to injuries and prescribe appropriate safety procedures.
   - Immediately stop anyone observed performing an unsafe act.
   - Provide for safe storage of equipment and supplies.

QUESTIONS

Write your answers in a notebook and then compare your answers with those on page 544.

12.1A What factors make up discipline?

12.2A What are the essential elements of an effective administrative plan?

12.2B How can a collection system agency work toward a goal of operating the system safely?

12.2C How can excessive infiltration and/or exfiltration be reduced?

12.2D How can management promote safety?

Please answer the discussion and review questions next.

DISCUSSION AND REVIEW QUESTIONS

Chapter 12. ADMINISTRATION

(Lesson 1 of 7 Lessons)

At the end of each lesson in this chapter you will find some discussion and review questions. The purpose of these questions is to indicate to you how well you understand the material in the lesson. Write the answers to these questions in your notebook before continuing.

1. What will an effective collection system administration accomplish?
2. How will personnel actually performing the operation and maintenance of a collection system respond to an effective administration?
3. Who should prepare an administrative plan?
4. How can a collection system agency work toward a goal of preventing unnecessary damage to public and private property?
5. How can a collection system agency work toward a goal of using funds efficiently?

CHAPTER 12. ADMINISTRATION

(Lesson 2 of 7 Lessons)

12.3 PERSONNEL

An effective administration will develop standardized and equitable procedures for the application, selection, orientation and training of personnel. Experience and careful planning will enable you to project how large a staff will be needed to operate and maintain your wastewater collection system and to meet your goals and objectives. It is always easier to hire additional personnel as the need is demonstrated than it is to lay off excess existing personnel.

12.30 Calculating Personnel Requirements

The most important factor influencing the number of personnel needed to maintain a wastewater collection system is its size. The following factors also influence personnel requirements:

Age	Older systems generally require more maintenance.
Condition	Systems poorly constructed and abused in the past by neglect will require more maintenance than a system in good condition.
Climate	Frequent rain and snowstorms will reduce the operators' production rates and this will have to be offset by additional personnel.

Facilities Maintenance	Service to maintain buildings, when performed by collection system staff, will require additional personnel.
Work Rates	The rate of work produced by cleaning, repair and inspection crews (which is influenced by motivation, training, equipment, travel distance and conditions) will affect the number of personnel required for maintenance of a collection system.
Storm Drainage	If an agency maintains storm sewers in addition to a wastewater collection system, additional personnel will be required.
Funding	A lack of funds available for maintenance of a collection system may produce a temporary constraint on the number of personnel. However, an effective administrator should eventually be able to secure adequate funding for an appropriate maintenance program by documenting system needs and presenting well-thought-out budget proposals.
Topography	Sewers with steep slopes may require erosion repairs and special cleaning procedures; sewers on flat terrain may have a greater deposition rate for solids and lift stations that will require greater maintenance; and sewers submerged in groundwater may require more frequent sealing.
Safety	An administration seriously committed to safe working conditions and safe operating procedures will require the presence of safety professionals along with sufficient staff time for safety training, enforcement of safety policies, and adequate crew sizing to ensure system operator safety.

Two manuals entitled *MANPOWER REQUIREMENTS FOR WASTEWATER COLLECTION SYSTEMS IN CITIES AND TOWNS OF UP TO 150,000 IN POPULATION* and *MANPOWER REQUIREMENTS FOR WASTEWATER COLLECTION SYSTEMS IN CITIES OF 150,000 TO 500,000 IN POPULATION,* listed as additional reading at the end of this chapter, provide additional guidelines for determining personnel requirements. These publications include tables, numbers and classification of maintenance personnel based on a 1973 survey of 49 U.S. cities. Any charts or formulas for determining personnel requirements should be used only as a guideline since personnel requirements for each collection system will be different and will depend on the factors listed above.

For the purpose of illustrating how to determine personnel requirements, let us again refer to our typical town, Clearwater, USA. The new collection system superintendent, whose name we'll say is "Carl," was told that the collection system had been somewhat neglected and that he was to prepare a plan for expanding and reorganizing the maintenance department to meet the goals previously set forth in this chapter. The new superintendent spent many hours reviewing records, maps, and crew activities at various sites, and held lengthy discussions with the work supervisors.

Assume Clearwater (our typical town) was founded in the late 1800s. The downtown area is located adjacent to a river with rolling hills on both sides. Clearwater has a population of 180,000 people and an industrial base equivalent to an additional 25,000 persons. The Publicly Owned Treatment Works (POTW) consists of the collection system, intercepting sewers, 16 wastewater lift stations, and a secondary treatment plant with a design capacity of 30 million gallons per day.

The new superintendent hired by the Director of the Department of Clearwater Public Works will be responsible for field activities of the Wastewater Collection Division. The wastewater treatment plant is part of the division, but has its own superintendent to oversee its operations. During orientation to the new position, the collection system superintendent is given the following information.

1. Clearwater is responsible for maintenance and repair of the wastewater collection system up to a point five feet past the property line, or to a property line cleanout.
2. Overflows, spills, or raw wastewater bypasses must be reported to the following:
   a. Regional Pollution Control Agency,
   b. Local Health Department, and
   c. Director of Public Works.
3. Supplies and equipment must be obtained through the Clearwater City purchasing department.
4. Personnel management practices must be in accordance with rules prescribed by the Clearwater personnel department.
5. Sewer-use ordinance enforcement will be part of the section's function, with support from the wastewater treatment plant for required laboratory analysis.
6. All funding for section activities must be budgeted.
7. Superintendent is responsible for planning and implementing maintenance, operation, training, and safety aspects of the section.
8. Department of Public Works' staff engineers provide support for design work or technical engineering when required. Work requests must be in writing on department work request forms.

The superintendent's own review of agency records, maps, and the physical system itself provided the following information.

1. Clearwater collection system pipelines—total length is 760 miles, including the following estimated lengths:
   a. 311 miles of 4-inch building service lines,
   b. 167 miles of 6-inch main sewers,
   c. 80 miles of 8-inch main sewers,
   d. 48 miles of 10-inch main sewers,

   e. 17 miles of 12-inch main sewers,
   f. 137 miles of intercepting sewers ranging from 15 to 60 inches in diameter,
   g. 7,900 manholes, and
   h. 54,700 building sewer connections.
2. Wastewater lift stations
   a. Number of stations, 16.
   b. All stations telemetered with failure alarms. Cause of failure is not transmitted—only that station is malfunctioning. The Police Department switchboard monitors lift station alarm system during off-duty hours and weekends.
3. General condition of system
   a. Unknown due to lack of history and systematic inspection; need to inspect using TV for smaller sewers and walking inspections for interceptors.
4. Known problem areas in system
   a. Excessive infiltration/inflow during storms in older portion of city.
   b. Power rodding activity indicates significant root problem in northwest suburban area.
   c. High-velocity cleaner activity indicates excessive grease buildups in area with restaurants.
   d. Treatment plant superintendent reports high volume of grit removal at treatment plant, especially during storms. Also, plant superintendent reports heavy solids and organic loading at random times and that the suspected source is the XX industrial plant.
   e. No enforcement of sewer-use ordinance.
   f. Lift stations are in poor mechanical condition and cleaning procedures need improvement.
   g. Main line stoppages are averaging three per day.
   h. Service request calls are averaging 10 per day (mostly obnoxious odors and stoppages).

Superintendent Carl is now thoroughly familiar with the collection system facilities and problems, has developed a list of goals and objectives to meet the agency's overall mission, and is now ready to develop the detailed plan of exactly how the system's resources will be used to accomplish its goals. This activity is as much a juggling act as anything and requires both skill and experience. Carl begins by listing all of the O & M direct work activities (Column 1 of Table 12.1). Similarly, he starts another list of indirect activities his staff will be involved in (Column 1 of Table 12.2).

The next step is to weigh the importance of each of the goals he has set. The new superintendent's highest priority for the coming year is to reduce main line stoppages from the present 1,000+ to approximately 300. To do this, operators' efforts will be concentrated on specific tasks such as balling and high-velocity cleaning of major portions of the collection system. Areas where root stoppages are occurring frequently will require a chemical root control program. The grease stoppages in the sewers near the restaurant area of downtown demand that sewer cleaning activities in this area be increased.

Superintendent Carl also has to decide how much work will need to be done on the other system components to keep the system functioning smoothly and to move ahead in preventive maintenance. This is where his experience and judgment will be invaluable. Once he has estimated these work levels (Columns 3 and 4 of Table 12.1 and Column 2 of Table 12.2), the superintendent must figure out how many staff hours each task will take. To do this, he will look first at work rates for direct activities (Column 5, Table 12.1).

TABLE 12.1 DIRECT TIME FOR MAINTENANCE OF CLEARWATER WASTEWATER COLLECTION SYSTEM

Work Activity	Total System Units	Percent To Be Acted On	Annual Work Quantity	Work Rate	Annual Staff Days
CLEANING:					
Balling	1,789,920 ft	25	447,480 ft	130 ft/shr*	430.3
High Velocity	1,789,920 ft	60	1,073,952 ft	300 ft/shr	447.5
Bucketing	580,800 ft	5	29,040 ft	10 ft/shr	363.0
Flushing	881,760 ft	8	70,541 ft	610 ft/shr	14.5
Rodding					
Scheduled	1,980,000 ft	10	198,000 ft	90 ft/shr	275.0
Main Stoppage			295 ea	0.3 shr/ea	11.1
Bldg Stoppage			1,270 ea	0.6 shr/ea	95.3
REPAIRING AND REPLACING:					
Main Breaks	2,370,720 ft	0.1	2,371 ft	0.4 ft/shr	740.9
Joint Sealing	2,370,720 ft	0.1	2,371 ft	8.0 ft/shr	37.0
Lining	2,370,720 ft	0.1	2,371 ft	6.0 ft/shr	49.4
Manholes	7,900 ea	10.0	790 ea	0.5 shr/ea	49.4
Lift Stations	16 ea	1,200.0	192 ea	0.2 shr/ea	4.8
Building Sewers	54,700 ea	12	6,564 ea	0.2 shr/ea	164.1
INSPECTION AND CONTROL:					
Manhole Inspections	7,900 ea	50	3,950 ea	2.0 shr/ea	987.5
TV Inspections	2,370,720 ft	13	308,194 ft	50.0 ft/shr	770.5
Infil/Infl Investigations	7,900 ea	10	790 ea	1.0 shr/ea	98.8
Chemical Treatment	7,900 ea	10	790 ea	1.0 shr/ea	98.8
Lift Station Inspections	16 ea	5,200	832 ea	0.5 shr/ea	52
Industrial Waste Inspections	2,500 ea	33	825 ea	1.0 shr/ea	103.1
Total Direct Time					4,793.0

*shr = staff hours

TABLE 12.2 INDIRECT TIME FOR MAINTENANCE OF CLEARWATER WASTEWATER COLLECTION SYSTEM

Activity	Percent Of Total Time	Total Time* (Staff Days)	Annual Indirect Staff Days
Equipment Servicing	5	6,477	323.9
Building and Grounds	3	6,477	194.3
Maintenance Work for Other Departments	1	6,477	64.8
Training	5	6,477	323.9
Leave and Holidays	12	6,477	777.2
Total	26		1,684.1

$$\text{*Total Time} = \frac{4{,}793}{0.74} = 6{,}477$$

The superintendent's estimated work rates for each of the listed activities are based on discussions with superintendents of other collection systems, analysis of annual reports, and Carl's own estimate of how the information applies to the Clearwater collection system. He expects to modify the work rates as appropriate once he has an opportunity to analyze the actual work records of his division.

To calculate the number of staff days needed to complete the annual plan of direct work, the superintendent must make two types of calculations. For work items involving a certain number of feet per staff hour, he divides the total number of feet by the estimated work rate and then divides the results by eight staff hours per day. For example, Carl estimates that a balling crew can clean 130 feet of line per person per hour. To clean 25 percent of the system's lines by balling, it will take 430.3 staff days (447,480 ft ÷ 130 ft/hr ÷ 8 hr/day = 430.3 staff days).

A second type of calculation is needed for work activities designated "each" in Table 12.1, such as rodding main stoppages or manhole inspections. For these activities, the superintendent multiplies the annual work quantity by the estimated work rate and then divides the total by eight staff hours per day. For example, rodding 295 main stoppages will take 11.1 staff days (295 ea x 0.3 hr/ea ÷ 8 hr/day = 11.1 staff days).

The superintendent again used the sources of information previously described to estimate the amount of time collection system operators will spend on indirect work activities and annual leave (Column 1, Table 12.2). Column 2 of Table 12.2 shows us that 26 percent of the operators' time will be spent on indirect activities. To find out how many staff days that represents for each activity, Carl must do some additional calculations. He must first find out the total number of staff days that will be spent on both direct and indirect activities combined. The procedure for making this calculation is:

Known	Unknown
Total staff days for direct activities = 4,793	Total staff days for direct and indirect activities combined
Percentage of staff days for indirect activities = 26%	
Percentage of staff days for direct activities = 74%	

FORMULA

$$\frac{74\%}{100\%} = \frac{4{,}793}{X}$$

$$X = 6{,}477$$

Knowing that it will take a total of 6,477 staff days to conduct the planned preventive maintenance program during the next fiscal year, Carl can now compute the number of annual indirect staff days for each activity (Column 4, Table 12.2).

The last calculation Carl needs to make, and a very important one it is, will be determining how many operators are needed to perform 6,477 staff days of work. Since there are 260 staff days per year (52 weeks x 5 staff days), dividing the total estimated staff days by 260 reveals that it will take 25 (6,477/260 = 25) collection system operators to perform the planned maintenance work.

In addition, the superintendent decided to have eight staff personnel (in addition to himself) to assist in administering the maintenance program. Therefore, the budget for the next fiscal year would provide for a total of 33 persons in the Wastewater Collection Division. Chapter 13 will describe how the superintendent organized the division's personnel.

QUESTIONS

Write your answers in a notebook and then compare your answers with those on page 544.

12.3A Overflows, spills, or raw wastewater bypasses must be reported to whom by the Clearwater superintendent?

12.3B What are the responsibilities of Clearwater's superintendent?

12.3C What items would you consider if you summarized a wastewater collection system?

12.31 Employment

The superintendent of a wastewater collection system will usually have the assistance of the agency's personnel department in the selection and employment of personnel. However, since the personnel will be working under the superintendent's direction and since the superintendent will be responsible for their performance, it is important that the superintendent be involved in the hiring process. Selection of personnel is important since it is often difficult to fire an employee and the agency will lose the investment made in the employment and training of the person.

12.310 Occupational Description

The superintendent should assist the personnel department in the preparation of an occupational description for each of the persons working in the department or division. The following is a typical occupational description:

Title: **MAINTENANCE PERSON II, WASTEWATER COLLECTION** [2]

Repairs and maintains municipal storm and sanitary sewer lines, functioning as a lead person and performing any combination of following tasks: Inspects manholes to determine location of stoppage. Inserts power rods and rotates

[2] *Adapted from page 68, MANPOWER REQUIREMENTS FOR WASTEWATER COLLECTION SYSTEMS IN CITIES OF 150,000 TO 500,000 IN POPULATION, Manpower Development Staff, Office of Water Program Operations, U.S. Environmental Protection Agency.*

until obstruction is broken. Retracts rods to drag out obstructions such as roots, grease, or other deposits. Cleans and repairs catch basins, manholes, culverts and storm drains, using hand tools. Raises manhole walls to prescribed street level using hand tools. Measures distance of excavation site using tape measure, and marks outline of area to be trenched according to direction of FOREMAN. Breaks asphalt and other pavement, using air hammer, pick and shovel. Cuts damaged section of pipe with cutters and removes broken section from ditch. Replaces broken pipes and reconnects pipe sections using pipe sleeve. Inspects joints to ensure they are tight and sealed properly before backfilling. Packs backfilled excavation using air and gasoline tamper. Taps main line sewers to install sewer saddles. Replaces manhole covers. Updates sewer maps and manhole charting. Drives pickup truck to haul crew, materials and equipment. Services, adjusts and makes minor repairs to equipment, machines and attachments. Communicates with DISPATCHER, FOREMAN, MAINTENANCE SUPERVISOR, and others, using radio telephone. Gives directions to MAINTENANCE PERSON I, and LABORERS, instructs them in efficient and safe use of machines, trains them in work methods, and ensures that proper procedures and safety precautions are followed. Prepares records showing actions taken, staff time and equipment utilization, and disposition of material. Requisitions tools and equipment. May operate sewer cleaning equipment including power rodder, high-velocity water jet, sewer flusher, bucket machine, wayne ball, and combination sewer cleaning machine. May clean and disinfect domestic basements and other areas flooded as a result of sewer stoppages. May act as lead person in a large repair and construction crew under the supervision of a FOREMAN.

QUALIFICATION PROFILE

In an effort to limit discriminatory hiring practices, the law and also administrative practice have carefully defined the hiring methods and guidelines employers may use. In general, the selection method and examination process must be limited to the applicant's knowledge, skills, and abilities to perform relevant job-related activities. Except in rare cases (such as the age limit for commercial pilots), artificial criteria such as age and level of education cannot be used to screen candidates in place of performance testing. Likewise, the nature of questions that can be asked of potential employees is similarly limited. You should retain a recruitment specialist when you begin the hiring process to give yourself and your agency maximum protection from lawsuits.

1. Formal Education:

   May or may not be relevant to knowledge, skills and ability.

2. General Requirements:
   a. Knowledge of methods, tools, equipment and materials used in sewer construction, repair and maintenance.
   b. Knowledge of layout of city streets and location of sewer lines and related structures. (This may be viewed as discriminatory against candidates who do not currently work for the agency.)
   c. Knowledge of work hazards and applicable safety precautions.
   d. Ability to locate, detect, and correct sewer stoppages and leaks.
   e. Ability to establish and maintain effective working relations with employees and the general public.
   f. Possession of a valid state driver's license for the class of equipment the employee is expected to drive.

3. General Educational Development:
   a. Reasoning:
      (1) Apply common sense understanding to carry out instructions furnished in oral, written or diagrammatical form.
      (2) Deal with problems involving several concrete variables in or from standardized situations.
   b. Mathematical:

      Use a pocket calculator to make arithmetic calculations relevant to sewer operations and maintenance. Problems will involve decimals and percentages.
   c. Language:
      (1) Communicate with fellow employees and train subordinates in work methods.
      (2) Fill in maintenance report forms.

4. Specific Vocational Preparation:

   Three years' experience in the repair and maintenance of collection systems is commonly required.

5. Aptitudes—Relative to General Working Population: (May or may not be relevant to knowledge, skills and ability.)

a. Intelligence:	Middle third
b. Verbal:	Middle third
c. Numerical:	Middle third
d. Spatial:	Highest third excluding top 10%
e. Form Perception:	Middle third
f. Clerical Perceptions:	Lowest third excluding bottom 10%
g. Motor Coordination:	Middle third
h. Finger Dexterity:	Middle third
i. Manual Dexterity:	Middle third
j. Eye-Hand-Foot Coordination:	Lowest third excluding bottom 10%
k. Color Discrimination:	Lowest third excluding bottom 10%

6. Interests:

   (May or may not be relevant to knowledge, skills and ability.) An interest in activities concerned with things, objects, machines and techniques.

7. Temperament:

   Must adjust to a variety of tasks requiring frequent change and adhere closely to established standards and procedures. Must adjust to making judgmental decisions.

8. Physical Demands:

   Medium to heavy work involving lifting, climbing, stooping, kneeling, crouching, crawling, reaching, handling, fingering, talking, hearing and seeing. Must be able to lift and carry XXX number of pounds for XXX distance while walking (or crawling).

9. Working Conditions:

   The work is outside and involves wet conditions, noise, risks of bodily injury, and exposure to weather, noxious smells, and gases.

ENTRY SOURCES: Mason II, Maintenance Person I, or general public.

PROGRESSION TO: Foreman, Maintenance Supervisor I, or Construction Inspector.

12.311 Applications

A superintendent should assist the personnel department in preparing notices of opportunities for employment within the agency's collection system maintenance department and distributing them to sources of qualified applicants. The notice should give a brief description of the work to be performed, required qualifications, compensation, the selection process, where application forms can be obtained and the closing date for accepting applications. Each agency will have its own standard application form for employment. Application forms and subsequent pre-employment inquiries must respect the civil rights of the applicants. Table 12.3 contains a list of acceptable and unacceptable pre-employment questions that you may use as a guide. The legality of the questions, whether written or asked orally, is subject to constant change. Pre-employment questions should be reviewed at the time of their use by your designated personnel specialist.

TABLE 12.3 ACCEPTABLE AND UNACCEPTABLE PRE-EMPLOYMENT INQUIRIES [a]

Acceptable Pre-employment Inquiries	Subject	Unacceptable Pre-employment Inquiries
"Have you ever worked for this agency under a different name?"	NAME	Former name of applicant whose name has been changed by court order or otherwise.
Applicant's place of residence. How long applicant has been resident of this state or city.	ADDRESS OR DURATION OF RESIDENCE	
"Can you, after employment, submit a birth certificate or other proof of U.S. citizenship or age?"	BIRTHPLACE	Birthplace of applicant. Birthplace of applicant's parents, spouse or other relatives. Requirement that applicant submit a birth certificate, naturalization or baptismal record.
"If hired, can you furnish proof of age?" /or/ Statement that hire is subject to verification that applicant's age meets legal requirements.	AGE	Questions which tend to identify applicants 40 to 64 years of age.
Statement by employer of regular days, hours or shift to be worked.	RELIGIOUS	Applicant's religious denomination or affiliation, church, parish, pastor, or religious holidays observed. "Do you attend religious services /or/ a house of worship?" Applicant may not be told, "This is a Catholic/Protestant/Jewish/atheist organization."
	RACE OR COLOR	Complexion, color of skin, or other questions directly or indirectly indicating race or color.
Statement that photograph may be required after employment.	PHOTOGRAPH	Requirement that applicant affix a photograph to his/her application form. Request applicant, at his/her option, to submit photograph. Requirement of photograph after interview but before hiring.
Statement by employer that if hired applicant may be required to submit proof of eligibility to work in the United States.	CITIZENSHIP	"Are you a United States citizen?" Whether applicant or applicant's parents or spouse are naturalized or native-born U.S. citizens. Date when applicant or parents or spouse acquired U.S. citizenship. Requirement that applicant produce naturalization papers or first papers. Whether applicant's parents or spouse are citizens of the U.S.
Applicant's work experience. Applicant's military experience in armed forces of United States, in a state militia (U.S.), or in a particular branch of the U.S. armed forces.	EXPERIENCE	"Are you currently employed?" Applicant's military experience (general). Type of military discharge.

TABLE 12.3 ACCEPTABLE AND UNACCEPTABLE PRE-EMPLOYMENT INQUIRIES (continued)

Acceptable Pre-employment Inquiries	Subject	Unacceptable Pre-employment Inquiries
Applicant's academic, vocational, or professional education; schools attended.	EDUCATION	Date last attended high school.
Language applicant reads, speaks, or writes fluently.	NATIONAL ORIGIN OR ANCESTRY	Applicant's nationality, lineage, ancestry, national origin, descent or parentage.
		Date of arrival in United States or port of entry; how long a resident.
		Nationality of applicant's parents or spouse; maiden name of applicant's wife or mother.
		Language commonly used by applicant. "What is your mother tongue?"
		How applicant acquired ability to read, write or speak a foreign language.
	CHARACTER	"Have you ever been arrested?"
Names of applicant's relatives already employed by the agency.	RELATIVES	Marital status or number of dependents.
		Name or address of relative, spouse or children of adult applicant.
		"With whom do you reside?"
		"Do you live with your parents?"
Organizations, clubs, professional societies, or other associations of which applicant is a member, excluding any names the character of which indicate the race, religious creed, color, national origin, or ancestry of its members.	ORGANIZATIONS	"List all organizations, clubs, societies, and lodges to which you belong."
"By whom were you referred for a position here?"	REFERENCES	Requirement of submission of a religious reference.
"Do you have any physical condition which may limit your ability to perform the job applied for?"	PHYSICAL CONDITION	"Do you have any physical disabilities?"
Statement by employer that offer may be made contingent on passing a physical examination.		Questions on general medical condition.
		Inquiries as to receipt of Workers' Compensation.
Notice to applicant that any misstatements or omissions of material facts in his/her application may be cause for dismissal.	MISCELLANEOUS	Any inquiry that is not job-related or necessary for determining an applicant's eligibility for employment.

[a] Courtesy of Marion B. McCamey, Affirmative Action Officer, California State University, Sacramento, CA.

12.312 Selection

The next step in the selection process is known as paper screening. The personnel department and the superintendent review each applicant's resume or application form and eliminate those who are obviously not qualified. The remaining applicants are generally given written and other appropriate examinations to verify their qualifications for the position. Then the superintendent interviews the three (or more) most highly qualified applicants and selects the most qualified person.

If you are the superintendent conducting the interview, be aware that the purpose is to gain additional information about the applicants so that the most qualified person can be selected for employment. Prepare for the interview in advance. Review the background information of each applicant. Plan the interview by preparing a list of questions to ask each applicant. When collecting background information and preparing the questions, you must be sensitive to the civil rights of the applicants and the affirmative action policies of your agency. Refer again to the acceptable and unacceptable pre-employment questions outlined in Table 12.3.

Try to conduct the interview in a quiet place where there will be a minimum of distractions such as phones ringing, noisy typewriters, computer printers and other persons talking or overhearing the interview. Start the interview by placing the applicant at ease. A discussion of something which does not refer to an applicant's ability or eligibility for the job will be relaxing. Opening topics could include results of ball games, weather, traffic or parking problems.

Explain to the applicant the details of the job, working conditions, wages, vacations, potential for advancement, and other important factors. Hopefully by now the applicant has relaxed. Give the applicant a chance to ask questions about the job.

Now you are ready to start asking questions about this person's ability to do the job. These questions should pertain to the work situation and should be phrased so that they ask for specific information rather than "yes" or "no" answers. Jot down a few notes to remind you of each person's responses and general manner. Try to determine why the applicant applied for the job and why he or she feels qualified for this position.

When the interview is over, tell each applicant they will be notified of your decision after the others are interviewed and within a specified number of days. Be honest with the applicant. Employment interviews are an opportunity to build good public relations for your agency today and in the future. If an applicant is not qualified for the job now available but appears qualified for another job, encourage the applicant to investigate the other job.

When all the interviews are over, review your notes. Evaluation of applicants is often difficult. Most people are favorably impressed by the "good talker." A look at your existing personnel can remind you that the best talker may not be the most productive worker. Also beware of the overqualified applicant. Naturally you want the best applicant, but a person who is overqualified may become disappointed with the job and hurt the morale of other personnel.

Before making a final selection, check the references of the person you have tentatively selected. Try to speak directly with the applicant's previous supervisors rather than with a personnel officer. If the applicant is presently employed, they may request that you not contact their employer until a firm offer of employment is made. In such a case you could attempt to speak with previous employers or ask the applicant for additional references who could provide information about the applicant's professional skills.

The main purpose of checking references is to verify an applicant's previous experience. In some cases, that is the only information an employer will give you. In other cases, you may be able to learn a great deal about the applicant's skills and abilities. Some of the questions you might ask are: Was the employee reliable or punctual? How well did the applicant relate to co-workers? How well did this person communicate with supervisors? Did the applicant consistently practice safe work procedures? Would you rehire this person?

The final step in the selection process is to require the best qualified applicant to pass a medical examination to determine the person's physical ability to perform the designated work.

Once your offer of employment has been accepted, promptly notify all other applicants that the position has been filled and thank them for their interest in your organization.

12.313 Employee Orientation

During the first day of work, a new employee should be given all the information available in written and verbal form on the agency's policies and practices concerning compensation, benefits, attendance and other matters relating to employer-employee relations. Any questions that a new employee has concerning agency and department operating policies and practices should be answered at this time. Conclude the orientation with introduction of the new employee to co-workers and the work area. New employees should be given a safety orientation as soon as possible so they will be aware of, and comply with, safety policies. Remember, the new employee's first impression of the agency and its personnel is very important.

12.314 Probationary Period

Many businesses find the use of probationary periods extremely helpful in selecting new employees. This is usually a period of 90 to 120 days beginning when a new employee is hired. In some jurisdictions, the period can run as long as 180 to 365 days for operators and supervisors. Management may reserve the right to terminate employment of the person with or without cause during this period, provided the employee is properly advised in advance that he or she must successfully complete the probationary period to move to regular employee status.

The probationary period provides a time during which both the agency and the employee can assess the "fit" between the job and the person. A satisfactory performance evaluation near the end of the probationary period is often the mechanism by which a new employee officially achieves regular employee status.

Check with your personnel office and/or legal department for information about probationary periods since they may well be covered by a union contract or collective bargaining agreement.

12.32 Compensation

The compensation an employee receives for the work performed includes satisfaction, recognition, security, appropriate pay and benefits. All are important for the retention of good employees. However, appropriate pay for the work performed is the cornerstone of any compensation package.

12.320 Salaries

Most employees of a wastewater collection agency will work full time and be paid a weekly or bi-weekly salary with an hourly rate component for calculating any overtime work compensation and job costs. Remember that salaries should be a function of supply and demand. You need to pay enough, but not too much, to attract and retain desirable employees. The salary an employee receives may be determined unilaterally by the agency or by negotiation with formal or informal organizations representing employee groups. In either situation, salaries must comply with minimums established by state and federal regulations and are generally based on the standards established by other employers within a recognized area of influence. Salary standards can be determined by surveying other agencies with similar positions and must take into account the similarities (or the lack of similarities) of the positions being compared. The results of the survey are then used as a basis for determining the appropriate salary for an agency's unique conditions.

12.321 Benefits

Benefits are also an important part of the total compensation an employee receives for work performed and generally include the following:

1. Social security (employer's contribution),
2. Retirement,
3. Hospitalization or health insurance (cost may be shared by employee),
4. Life insurance,
5. Holiday pay,
6. Vacation pay,
7. Sick leave,
8. Personal leave,
9. Parental leave,
10. Workers' compensation, and
11. Protective clothing.

In addition, the following more specific benefits are sometimes given to employees:

1. Cash bonus,
2. Longevity pay,
3. Dental and/or vision insurance,
4. Long-term disability insurance,
5. Educational bonus,
6. Educational cost and leave,
7. Bereavement leave, and
8. Release time for jury duty.

The cost of the benefits is usually paid by the employer, although in some instances costs are shared with employees. It is not uncommon for benefits to cost an employer between 30 and 50 percent of an employee's salary. Employers often compute the cost of benefits and print an itemized list on employees' monthly pay stubs. This practice reminds employees to consider their total compensation package rather than focus on an hourly or monthly salary amount.

The benefits provided by an agency may not be the same or equal in value for all groups of employees. For example, management employees might be eligible to participate in "enhanced" or broadened medical insurance programs as part of their total compensation package. Also, union contracts sometimes specify varying benefit packages for different groups of employees.

12.322 Adjustments

Adjustments to the starting salary for an employee remaining in the same position are generally based on the following factors:

- Longevity and a concurrent improvement in performance, and
- Increases in the cost of living.

Longevity pay increases are generally given after the first six months of satisfactory employment followed by annual increases thereafter. A five-step merit salary increase schedule is commonly used by agencies. Such salary increases should not be given automatically. They should be given only after a thorough evaluation of an employee's performance by a supervisor demonstrates that a merit increase has been earned.

Cost of living increases are usually given annually and based on the Wage Earner and Clerical Worker Consumer Price Index increase for the latest preceding twelve-month period published by the Bureau of Labor Statistics, U.S. Government. Cost of living increases are either given unilaterally by an agency or are negotiated with employee representatives.

QUESTIONS

Write your answers in a notebook and then compare your answers with those on page 544.

12.3D Why is making the proper selection of personnel important?

12.3E What are the potential positions of advancement for a Maintenance Person II, Wastewater Collection?

12.3F What information should be listed on an "opportunities for employment" notice?

12.3G What information should be given a new employee during the first day of work?

12.3H What types of adjustments are commonly made to an employee's starting salary?

12.33 Training, Informing, and Certification

A prime responsibility of every supervisor is to ensure the proper training of assigned personnel so that they may perform their work in a safe and efficient manner. Appropriate certification of competency, when available from an authoritative group or agency, should at least be encouraged by an agency and its supervisors, and may be required for promotions and step salary increases. Personnel also need to be informed about internal and external activities of the agency to make them realize they are an important part of the agency's team.

12.330 Training

An effective supervisor is constantly looking for opportunities to train assigned personnel to improve performance. The following paragraphs describe some common training techniques.

ON THE JOB

Unfortunately, to this point in time, very little has been offered in the way of formal courses to assist in training collection system operators. Most of the training operators receive is some type of "on-the-job training" conducted by experienced collection system operators when they are available. This type of training is important, of course, and it has been effective to a degree. The fact that collection systems have continued to function and in many areas improved their efficiency through such training efforts is proof of the effectiveness of on-the-job training.

One potential limitation of on-the-job training is that because it relies on "in house" teachers, it could be too narrow in scope. Such training tends to reflect local conditions, philosophies, and experience unless the instructor makes special efforts to broaden the scope.

Effective training techniques include informal meetings using information on available material, talks by suppliers, sharing knowledge of crew personnel, and inviting guests to talk over how to do specific jobs. Equipment manufacturers and materials suppliers are often available to train new operators and retrain existing operators on use of equipment.

PROFESSIONAL MAGAZINES AND PAPERS

Articles published in local or national professional magazines are an excellent source of information about new products and procedures and could be used as the basis of a training session. Two good examples of professional magazines are the *OPERATIONS FORUM* section in *WATER ENVIRONMENT & TECHNOLOGY*, published by the Water Environment Federation (WEF), 601 Wythe Street, Alexandria, VA 22314-1994, and *APWA REPORTER*, published by the American Public Works Association (APWA), 2345 Grand Boulevard, Suite 500, Kansas City, MO 64108-2641. Local or regional wastewater associations periodically conduct workshops where experienced collection system operators present papers that are of value in training less experienced people in the field of collection system operation and maintenance. Such workshops make information available to smaller organizations in remote areas which would otherwise not have the benefit of such broad experience.

FORMAL TRAINING

Through the efforts of local and state water pollution control associations, water environment associations, the Water Environment Federation, and the U.S. Environmental Protection Agency, attempts are now being made to make formal training available to all collection system operators. Such training also is being made available to others not now in the collection system field, but who would like to prepare for jobs within the field. This particular training course is a result of such efforts

There are also a number of consultants with expertise in collection system training who are available to analyze and meet your training needs. These consultants are almost always members of the Water Environment Federation Collection System Committee and/or your local WEF member association collection system group.

The need for effective training programs does not stop with collection system operators. Staff personnel, particularly superintendents and supervisors, need to develop a personal continuing educational program, vigorously supported by the employing agency, so that they will be able to apply the latest and most efficient techniques to their operations.

12.331 Informing

In addition to providing job-related training, an agency should also inform its operators and staff of the agency's operations. The following are some items of information that will help to build a team spirit among an agency's personnel.

RESULTS ACHIEVED

A good supervisor will know if the maintenance program is producing the desired results. The supervisor will be proud to make management aware of success, but how about the collection system operators? Supervisors should be equally concerned with making them aware of the results achieved.

ANNUAL REPORT

Collection system operators who have spent a year balling, rodding, operating a high-velocity cleaner, and operating TV equipment have a right to know what their year's efforts have produced. A wise supervisor will make it a practice to hold an annual report meeting with assigned personnel to inform them of their accomplishments. This does not have to be a "pat on the back" meeting. The personnel will appreciate the consideration shown in apprising them of the part their particular function has played in meeting the objectives of the total maintenance program. Without being told specifically, they will not become aware of their importance to the organization. Such awareness may well result in increased effort on their parts to improve their performance from year to year.

AGENCY FINANCING

Distribute copies of your agency's annual budget and financial statement to all personnel and answer all questions, particularly those relating to their work.

AGENCY RECOGNITION

Any recognition that an agency receives for exceptional performance (good or bad) should be passed on to its personnel since they are a part of the team that produced the results being recognized. If the results were good, they will have the satisfaction of doing a good job; and if the results were bad, they will very likely want to help improve the agency's performance. In the interest of maintaining morale, remember to "Praise in public and criticize in private." When reporting negative feedback to a group, keep your comments general and don't reveal the source of the feedback. Save the details of negative material for individual performance evaluation sessions.

12.332 Certification

More states are requiring certification of competency for wastewater collection system operators in recognition of the importance of good operation and maintenance of the system for the health and safety of the public. Also many state water pollution control associations (water environment associations) in states not requiring certification have voluntary certification programs. Either a mandatory or voluntary certification program provides recognition of wastewater collection personnel for the technical skills and knowledge they have acquired and applied to their tasks. Certified personnel are sometimes paid a salary higher than uncertified personnel.

Both mandatory and voluntary certification programs for collection systems personnel include written, and sometimes verbal, examinations. Guides for preparing for the examinations, including sample questions, are usually provided by the organization administering the certification program. Preparing for a certification examination is a good approach to continuing an education and training in the techniques, procedures, equipment and materials used for the efficient operation and maintenance of collection systems. Passing an examination shows that a person has learned the covered subjects and can apply the knowledge acquired in a competent manner.

12.34 Employee Relations

Some supervisors are up to their ears in employee relations and unions and others have never met a union representative or a shop steward. By reading this section you may understand why some employees join unions to obtain benefits and how nonunion employees obtain the benefits others have gained with the help of a union. Collection system personnel have joined unions in hopes of achieving higher wages, greater benefits, and other economic objectives, as well as improved working conditions.

Whether an agency's employees belong to a union now or may join one in the future, a good employee-management relationship is crucial to keeping an agency functioning properly. Managers, supervisors, crew leaders and operators all have to want to work together to develop a good relationship. The supervisor is the critical link in the relationship and is management's daily representative with the operators.

12.340 Negotiations

During employee negotiations, management should be in constant consultation with the supervisors. Many employee demands regarding working conditions originate from the supervisor's daily dealings with the employees. An effective supervisor can minimize unreasonable demands. Also any demands that are agreed upon must be implemented and carried out by a supervisor. Although a supervisor does not participate directly in labor negotiations, the supervisor can help both sides reach an acceptable contractual agreement.

12.341 Contracts

Once a contract has been agreed upon by a union and employer, the supervisor must manage the organization within the framework of the contract. Do not attempt to ignore or "get around" the contract, even if you disagree with some of the clauses or feel they hinder you from achieving your objectives. If you do not understand certain contract provisions or the spirit of the agreement, ask management to explain this portion of the contract *BEFORE* you apply these clauses to your daily working situation.

Contracts do not change the supervisor's delegated authority or responsibility. Operators must carry out the supervisor's orders and get the job done properly, safely and within a reasonable amount of time. As a supervisor you must learn to do your job within the framework of the contract.

As a supervisor you must always realize that as a member of management you have the right and even the duty to make decisions. However, a contract gives a union the right to protest or challenge your decision. When an operator requires discipline, disciplinary action is a management responsibility and a right. The supervisor has a duty to take action whenever employees do not comply with contract provisions.

Most of a supervisor's union contacts are with a shop steward. The shop steward is elected by the union employees and is their official representative to management and the local union. The steward is in a difficult position because the steward is an employee who is expected to do a full-time job like the other employees, yet also represent the employees. The steward must create an effective link between the supervisor and employees.

12.342 Grievances

Handling grievances can be a very time-consuming job for both the supervisor and the steward. Grievances can develop over disciplinary action, distribution of overtime, transfers, promotions, demotions and interpretation of labor contracts. The steward must communicate complaints and grievances from operators to the supervisor. The supervisor and the steward must work together to settle complaints and adjust grievances. Union contracts usually spell out in great detail the steps and procedures the steward and the supervisor must follow to settle differences. When a shop steward and supervisor can work together, the steward can help the supervisor to be an effective manager.

Successful handling of complaints and adjusting grievances requires cooperation by both the shop steward and the supervisor. Both have to use judgment, tact and patience and somehow do their daily work also. The better they can work together, the fewer grievances will develop and become serious, and the less time will be consumed settling such problems.

Assume you are a supervisor and the shop steward presents a grievance to you. During the presentation and discussion (the employee should be present, too), listen carefully and sympathetically to the steward. Discuss the problem with the employee directly or with the help of the shop steward. Try to identify the facts and the cause of the problem and try not to get caught up in irrelevant side issues. Make every effort to settle the grievance quickly and to the satisfaction of everyone involved.

Try to prevent little grievances from becoming big ones and going on to top management by letting the staff and shop steward know you are available and willing to discuss their problems. You and the shop steward and the employee must all listen sympathetically to each other's viewpoints and avoid becoming angry. Seek out complete and accurate facts about the situation, review the labor contract for pertinent clauses, and try to settle the grievance as quickly and justly as possible.

The consequences of any solution to a grievance must be considered and solutions must be consistent and fair to the other operators. The solution or settlement should be clear and understandable to everyone involved. Written records must be maintained and properly filed that document the entire grievance procedure from initial presentation to final solution and settlement. The supervisor and the shop steward should review the final report together to be sure the intent of the solution is understood and properly documented.

QUESTIONS

Write your answers in a notebook and then compare your answers with those on pages 544 and 545.

12.3I Why should collection system operators be properly trained?

12.3J How can collection system operators be trained?

12.3K In addition to agency policies and procedures, what other kinds of information should be given to collection system operators?

12.3L What is the purpose of an annual report meeting?

12.3M Describe the relationship that should exist between the supervisor and the shop steward.

12.35 Safety

The superintendent of a wastewater collection system department has the responsibility for the safety of the department's personnel and the public exposed to its operations. Therefore, the superintendent must develop and administer an effective safety program. The basic elements of a safety program include a safety policy statement (including disciplinary procedures), safety training and promotion, and accident investigation and reporting.

12.350 Policy Statement

The safety policy statement should be prepared by the top management of the collection system agency with the close consultation of the agency's safety officer/safety consultant since its purpose is to let employees know that the safety program has the full support of the agency and its management. The statement should:

1. Define the goals and objectives of the program,
2. Identify the persons responsible for each element of the program,
3. Affirm management's intent to enforce safety regulations, and
4. Describe the disciplinary actions that will be taken to enforce safe work practices.

Give a copy of the safety policy statement to every current employee and each new employee during orientation.

12.351 Responsibilities

The following listing of responsibilities for safety was set forth in the *PLANT MANAGER'S HANDBOOK*.[3] These responsibilities represent a typical list but may be incomplete if your agency is subject to stricter local, state, and/or federal regulations than what is shown here. Check with your safety professional.

Management has the responsibility to:

1. Formulate a written safety policy,
2. Provide a safe workplace,
3. Set achievable safety goals,
4. Provide adequate training, and
5. Delegate authority to ensure that the program is properly implemented.

The manager is the key to any safety program. Implementation and enforcement of the program is the responsibility of the manager. The manager also has the responsibility to:

1. Ensure that all employees are trained and periodically retrained in proper safe work practices,
2. Ensure that proper safety practices are implemented and continued as long as the policy is in effect,
3. Investigate all accidents and injuries to determine their cause,
4. Institute corrective measures where unsafe conditions or work methods exist, and
5. Ensure that equipment, tools, and the work area are maintained to comply with established safety standards.

The collection system operators are the direct beneficiaries of a safety program. The operators share the responsibility to:

1. Observe prescribed work procedures with respect to personal safety and that of their co-workers,
2. Report any detected hazard to a manager immediately,
3. Report any accident, including a minor accident that causes minor injuries,
4. Report near-miss accidents so that hazards can be removed or procedures changed to avoid problems in the future, and
5. Use all protective devices and safety equipment supplied to reduce the possibility of injury.

12.352 Training and Promotion

The superintendent must establish an ongoing safety training program for assigned personnel. There should be an established schedule of safety training sessions including weekly tailgate training sessions for operating crews. Use the sessions to introduce new safety procedures, discuss recent accidents, hear employee suggestions and reinforce existing procedures. Invite outside speakers to discuss safety topics. You will find sources of safety training material in Chapter 11, Section 11.11 of this manual, your agency's insurance company, the Water Environment Federation and its member associations, the National Safety Council, the National Institute of Occupational Safety and Health and OSHA.

Management should promote safety by personal example and may establish an award program for safe work practices.

12.353 Accident Investigations

Investigate every accident, no matter how minor, and make a written report on the findings. If the situation is complicated or unusual, retain a specialist to perform the task. The primary purpose of an investigation is to determine the cause and how best to avoid a similar and possibly more serious accident from occurring in the future. Discuss the accident immediately with the personnel involved and take disciplinary action if safety regulations were violated. Also, use the accident report as the basis for discussion at the next general personnel safety meeting. A complete accident report will be very valuable if there is a subsequent claim of injury or property damage. Your agency may have accident report forms or they may be obtained from your agency's claims division or insurance company. Also see Chapter 11, Figure 11.13 for an example of a typical accident report form.

QUESTIONS

Write your answers in a notebook and then compare your answers with those on page 545.

12.3N What are the basic elements of a superintendent's safety program?

12.3O Who should prepare a safety policy statement?

12.3P What are management's responsibilities with regard to safety?

12.3Q What are the collection system operators' responsibilities with regard to safety?

Please answer the discussion and review questions next.

[3] *PLANT MANAGER'S HANDBOOK (MOP SM-4), Water Environment Federation (WEF). No longer in print.*

DISCUSSION AND REVIEW QUESTIONS

Chapter 12. ADMINISTRATION

(Lesson 2 of 7 Lessons)

Write the answers to these questions in your notebook before continuing. The question numbering continues from Lesson 1.

6. How do factors such as total length of sewers serviced, population, industry, type of problems, and topography influence the organization that operates and maintains the collection system?
7. What general procedures would you follow to hire a new collection system operator?
8. What is the purpose of interviewing an applicant for a job?
9. What are the different types of compensation that an employee receives from a job?
10. What types of job training are available to operators?
11. What is the job of the supervisor in labor relations?
12. What should be contained or defined in a safety policy statement?

CHAPTER 12. ADMINISTRATION

(Lesson 3 of 7 Lessons)

12.4 EQUIPMENT AND TOOLS

Many of the tools and equipment we will be discussing have been thoroughly described in other chapters of this manual. Detailed instructions have been given as to how each can and should be used. The advantages and limitations of each, relative to their abilities to cope with specific problems, have been explained. In this chapter our comments will be limited to points of interest to supervisors.

12.40 Basis for Requirements

As with staffing, tool and equipment requirements will depend largely on such things as size of the collection system, work load, and objectives. Other important factors include age and condition of the system as well as the types and number of problem areas.

12.400 Equipment Tailored to Program

The equipment of a wastewater collection system department should be tailored to the superintendent's program for operation and maintenance of the system. As obvious as this may seem, the opposite is quite often the situation. Sometimes programs are designed to fit the equipment available. In many cases this does not come about by choice, it merely develops from circumstances. What circumstances? How does this situation develop? Quite simply. Prior supervisors or management acquired the tools and equipment needed to carry out their desired programs and new supervisors conform their programs to the available equipment. Then, as time passes, subsequent supervisors may find this available equipment no longer suits their current plans or programs. This can be due to a variety of reasons, such as:

1. Changing conditions within the collection system,
2. Development of more modern and specialized equipment, or
3. More intelligent approach and appraisal of existing problems and needs by current supervisors and planning for a more effective program.

How this situation came about is really not important. The important point is to be able to recognize this situation when it does occur. Then it will take knowledge and courage to document and justify your request for the appropriate tools and equipment which you will need to implement your programs and meet your objectives.

MULTI-PURPOSE EQUIPMENT

One of the questions supervisors ask most frequently and one that is most difficult to answer is what kind, type, and size of equipment should be purchased? Smaller organizations that cannot afford or whose needs don't justify the purchase of various specialized equipment find this problem especially difficult. With $10,000 or $20,000 in available funds, what one piece of equipment will be the most versatile and efficient in coping with the variety of problems faced in the collection system? We will not attempt to answer that question. Instead, we offer this analogy:

You are a homesteader with a small plot of land. To make the land productive you will have to hoe some weeds and dig some shallow, narrow ditches. What you need is a hoe and a shovel, but you can only afford one tool. The hoe would be best for cutting the weeds, a shovel the best for digging the ditches. If necessary, you could cut the weeds with the shovel, or dig the shallow ditches with a hoe. Which tool do you buy? If weeds are the predominant problem, you would buy the hoe. If ditches are the major workload, the shovel. The homesteader will assess the particular needs and make the choice accordingly.

As maintenance supervisor, you could be facing a decision very similar to the homesteader's choice. However, your decision will be more complex. We have learned in this course that a certain type of equipment will do an excellent job of removing roots, another will be most efficient in removing grease accumulations, and still another most effective in removing grit and solid material that has settled out within the collection system. Some equipment has been suggested as efficient, to a degree, in dealing with all these problems (see Chapter 6, Section 6.03, "Selection of Solution to Problem," Tables 6.1 and 6.2). As a supervisor, you should have the knowledge and the experience to identify your predominant maintenance problem and select the equipment that would do the best overall job in your collection system.

SPECIALIZED EQUIPMENT

Much of the equipment which is available today for collection system maintenance is specialized. Many manufacturers claim that their equipment is capable of dealing with a variety of maintenance problems, and, to a degree, they are correct. It also is true, however, that most modern collection system maintenance equipment has one particular function that it performs most effectively. Ask the manufacturer's representative to demonstrate the equipment you are interested in on an actual condition that exists in the collection system or in the above-ground training facility, if available, so you can see the equipment in action. Also ask the representative to give you the names of other nearby superintendents who are using the same equipment.

In making the decision to buy specialized equipment, proceed with much caution, consideration, and thorough analysis of your collection system. Do you have a specific problem prevalent within your system and of sufficient proportions to justify purchasing such specialized equipment? In many maintenance yards throughout this country you will find specialized equipment parked and standing idle much of the time because the need for it and/or its efficiency was overrated.

12.401 Efficient Use

Modern maintenance equipment is expensive. The operators who use it are expensive too and in most cases will need some additional specialized training. The supervisor should be constantly concerned that the equipment is being operated efficiently. As previously stated, most equipment has one function which it does best. Be certain that the equipment is being used in those areas where it is most needed. The shotgun approach of "going through a system at set intervals with one piece of equipment" is wasteful and unnecessary. Unfortunately it is a common practice.

12.41 Lease, Purchase, or Contract

It is not always necessary to purchase the equipment you need to implement your maintenance programs. There are several alternatives to committing your funds this way. All types of construction and maintenance equipment are available through long- or short-term leasing, as well as by outright purchase. Leasing makes available, to the smaller agencies in particular, many useful items of equipment which they could not justify purchasing for part-time use. Larger organizations also find leasing an attractive alternative to outright purchase of expensive equipment which requires a significant initial investment.

12.410 Advantages of Leasing

There are advantages to be realized through leasing for both large and small collection agencies.

LARGE ORGANIZATIONS

Naturally the first benefit to be realized is that expensive equipment can be available without a large initial capital expenditure, but there are other important benefits to be considered. Many large organizations have purchased expensive equipment such as closed-circuit television or high-velocity cleaners and discovered a short time later, much to their disappointment, that similar but better equipment became available. Because of the large investment, they have little choice but to keep and use their original equipment over a period of years necessary to amortize their original investment.

Another advantage is that most leasing agreements can be negotiated on a one- to three-year basis for this type of equipment. This assures you that you will not be faced with operating equipment that has become obsolete in some respects. When new developments and improvements become available, they can be easily acquired through new leasing agreements.

SMALL AGENCIES

Small agencies probably realize the greatest benefits from equipment leasing. Often their anticipated need or workload would not even allow consideration of purchasing needed equipment outright. Before leasing arrangements were widely available, a small agency's only alternative was to employ outside contractors who provided both the staff and the equipment needed to do the job. Such contracts often were expensive. With leasing, however, all types of equipment may be within their financial capabilities for short time periods to do specific jobs. Many leasing agencies will provide a break-in time and training period to train your own operators as part of the leasing agreement. This is usually a less expensive method than contracting for both staff and equipment. If a need can be demonstrated, leasing for a limited time period should be an easier approach to sell top management than buying equipment for part-time use.

COMMON ADVANTAGES

Leasing offers advantages common to any size agency. It makes any specialized equipment available at reasonable cost to complete any one-time or seasonal programs you may want to undertake. Some leasing agreements allow you to apply all or part of the leasing costs toward purchase of the equipment if you decide that the equipment leased best suits your present and future needs. The technical assistance offices of some state municipal leagues have conducted studies on the costs and benefits of leasing and purchasing. You may want to consult with them if you are considering leasing some equipment.

Often leased equipment can be obtained more quickly than purchased equipment. When leased equipment breaks down, usually it is quickly replaced because the agency doesn't have to pay for equipment that doesn't work.

LIMITATION

The total costs of leasing may be more in the long run than the cost of purchasing equipment. Keep in mind, however, that leasing costs are spread out over the years while the equipment is being used.

12.411 Advantages of Purchasing

Regardless of its size, there are advantages to any organization in outright purchase of some types of equipment. In general, purchases should be limited to the more standard equipment that is used frequently, has a long life span, and represents a capital investment within the scope of the available budget. The need to purchase new or replacement equipment should be anticipated and a replacement fund should be established for such purchases in order to spread the total cost over several years.

LIMITATIONS

1. Large initial capital investment,
2. Equipment must be kept and used five to eight years in order to build up a replacement fund,
3. It is difficult to obtain newer models that are capable of doing a better job until old equipment wears out, and
4. Equipment may require specially trained mechanics to provide proper maintenance.

12.412 Advantages of Contracting

Sometimes your agency may not have sufficient qualified or qualifiable operators available to do a highly specialized job. Under these conditions, a contract with a firm specializing in solving a particular problem may be the best approach.

LIMITATION

This may be the most expensive method, but could be satisfactory for a small job that can be done in a short time.

QUESTIONS

Write your answers in a notebook and then compare your answers with those on page 545.

12.4A How can tool and equipment requirements be determined?

12.4B Is most of the equipment available for collection system maintenance specialized or is it capable of dealing with a variety of problems?

12.4C Why is the shotgun approach of "going through a system at set intervals with one piece of equipment" not recommended?

12.4D How can an agency implement a preventive maintenance program when it does not have enough money to purchase the necessary equipment?

12.4E What are some advantages of leasing equipment?

12.4F What are some advantages of purchasing equipment?

12.4G What are some advantages of contracting for work?

12.42 Equipment and Tool Requirements

The new superintendent of the Clearwater wastewater collection system described in Section 12.30 is now ready to prepare a list of the equipment and tools for the division's new operation and maintenance program, as outlined in Table 12.1.

12.420 Vehicles

Vehicles are required to carry personnel, equipment, tools and materials to and from operation and maintenance sites. You will need to review your operators' driver's license classifications before specifying mobile equipment. The uniform federal guidelines for driver's licenses (now adopted by states) provide for different classifications of licenses depending on the weight and type of vehicle. You don't want to specify equipment that your crews can't drive.

SUPERVISORS

In most areas, supervisors will find a sedan is the most practical and economical type of transportation. Some levels of supervision, such as foremen, may be better served by a pickup truck. This type of truck provides space for carrying safety equipment (such as barricades) and also provides space for foremen to carry supplies to crews in the field. In both cases, two-way radio equipment or paging devices will add to the efficiency of the operation.

EMERGENCY TRUCK

Provide at least a $^3/_4$-ton truck for emergency crews who clear stoppages. The size of the truck could be larger if this crew is to serve as a repair crew when not on emergency calls. If two-way radios are limited in number, the emergency crew should receive high priority for obvious reasons. An enclosed, utility-type body also is desirable. An adequate, clean, dry storage area for all necessary maps of the system is a must for this vehicle.

BALLING TRUCKS

A one-ton flatbed truck equipped with tool boxes is commonly used. This truck should carry a hand-operated portable reel with a light cable for balling smaller lines in rear easements. It should also be equipped with a power winch, powered by an auxiliary engine or by a power take off (PTO) from the truck's engine, for balling larger lines in streets. A one-ton trailer for hauling grit removed from the system should accompany each truck.

A water truck holding 1,000 to 2,000 gallons of water should be provided if fire hydrants and hoses are not used for augmenting the flow of water in sewer lines during a balling operation. Sometimes a water truck is outfitted with the equipment listed above and a 3/4-ton pickup truck is used instead of the flatbed truck and trailer to haul tools and the grit removed by the cleaning operation.

Two-way radio equipment or paging devices may not be required for such vehicles because their daily assignment will usually be confined to a reasonably small area and they can easily be contacted if necessary.

RODDING TRUCKS (Other than truck-mounted power rodders)

If your rod crews' equipment consists of hand rodding gear, or trailer-mounted power rodding machines, a 3/4- or 1-ton truck should be adequate. A utility-type body is desirable. Depending on the number of such crews available, it may be wise to equip at least one with a two-way radio or paging device to make it available for emergencies. The truck should also be equipped with an electric hoist for loading and unloading an electric-powered building sewer rodder, and with an electric generator for the rodder.

BUCKETING AND FLUSHING TRUCKS

Since bucketing machines are trailer-mounted and their use is infrequent, trucks used in other maintenance operations can be used to tow the trailers to and from the job site. Also, the water truck used for balling sewer lines can be used to flush the few small-diameter sewer lines scheduled for this cleaning method.

MAIN LINE REPAIR TRUCK

The basic truck for most main line repair crews should be at least a 1 1/2-ton flatbed with a load capacity sufficient to carry the heavy materials these crews ordinarily use. Such materials might consist of collection system pipe and fittings, manhole grade rings, manhole barrel sections, small amounts of paving material, aggregate and cement, base rock, and sand. A bigger truck may be required for larger jobs. These crews usually work in the same location all day so two-way radio equipment or a paging device is not essential.

SERVICE LINE AND MANHOLE REPAIR TRUCK

A 1-ton truck should be adequate in most areas. Size could depend on the particular tasks these trucks are assigned to. If you have only one repair crew to handle any size repair project, you may want to obtain a larger truck like the one described above for main line repair crews. Also, depending on the number of such trucks available, you may want to equip one truck with a two-way radio or a paging device for emergency use when needed.

LIFT STATIONS TRUCK

A 3/4-ton truck with a utility body should be provided for the lift station crew for inspections, maintenance and minor repairs. The main repair truck can also be used for major lift station repairs. The truck should be equipped with an electric hoist for removing pumps and heavy fittings from lift stations.

INSPECTION AND CONTROL TRUCK

A 1/2-ton pickup truck equipped with a utility box can be used for main, TV and industrial inspections, and infiltration and inflow investigations. Also, one of the trucks assigned to other crews can be used for the chemical treatment of sewer lines.

GENERAL

All vehicles should be equipped with a fire extinguisher, a first aid kit of adequate size, and vehicle and personal accident forms. All operators must have their personal safety equipment with them at all times.

12.421 Special Equipment

In addition to the vehicles listed above, the special equipment described in the following paragraphs will be needed for the operation and maintenance of the Clearwater collection system.

HIGH-VELOCITY CLEANERS

This equipment is available in both truck-mounted and trailer units; truck-mounted units are most commonly used. The two basic configurations are straight hydraulic jetting units and combination units that can jet material to a manhole and then vacuum the debris into a holding tank. Combination units can also be used to clean catch basins and storm drains. Special care should be exercised in the selection of the truck to perform this type of work. Because of the considerable weight of the water carried in high-velocity cleaner tanks, it is a common mistake to specify a truck that is of inadequate capacity, or marginal capacity, for the loads involved. Frequently, the insufficient capacity will not become apparent until the truck has been in use for a period of time. With use, brake problems may develop, springs weaken, road handling becomes less efficient, and even a loss of engine power could become evident. Unfortunately, by the time these conditions materialize, it is often too late to remedy; the truck is yours and guarantees have expired. Again, specify a truck that is oversized, never one that is marginal in capacity.

A high-velocity cleaner can be equipped with an engine-driven vacuum pump and piping to suck the debris from the cleaning operation from the downstream manhole if this method of removal is preferred to entry into a manhole and removal with hand tools. Or, the vacuum pump and engine can be trailer-mounted for independent operation.

BUCKET MACHINES

These units are trailer-mounted and are used in pairs; they are equipped to be towed in tandem by one vehicle. Vehicles of at least 3/4-ton capacity should be used when towing two such units. You may also need a dump truck if large lines are being cleaned and large amounts of material are being removed. Attachments to the bucket machines are available which will allow for direct deposit of the material being removed into the trucks. This eliminates the work of dumping the material onto street surfaces and subsequently loading it into trucks by hand shoveling.

RODDING MACHINES

This equipment also is available in truck-mounted or trailer-mounted units. For full-time use, the truck-mounted unit is preferable. Trailer units are advantageous when they are to be used less than full time. These units can be towed by a 3/4-ton or larger truck, depending upon the size of the rodding machine unit.

The City of Clearwater maintains building sewers five feet past the property line. An electrical power rodder with a flexible spring cable and tools capable of negotiating the bends and cleanout of a 4-inch diameter building sewer will be required. A gasoline engine-driven electric generator or one run by a truck engine will be needed to power the rodder.

BACKHOE/LOADER

Organizations that undertake construction or repair projects of a significant size will need a backhoe/loader. Depending on the anticipated use, a backhoe/loader can be purchased, leased, rented by the day, or hired complete with operator. Sanitary collection systems are now commonly being installed at greater depths which makes this equipment more desirable. Often, a backhoe/loader will provide the ability to complete an excavation, repair, and backfilling operation in one day. To provide mobility, a trailer of adequate size and meeting local highway regulations will be required to transport the backhoe/loader from job to job.

DUMP TRUCK

This truck can be used as the towing vehicle for the backhoe/loader trailer. It is also used to haul away excavated material when such material is unsuitable for backfill and to haul in selected backfill when required.

WATER TRUCK

A water truck is another piece of equipment which will be required to work in conjunction with construction crews. It can be used for dust control on the job site, high-velocity cleaning (jetting), and cleanup. These trucks come in a variety of sizes and water tank capacities; minimum size should be at least 1,000-gallon capacity. If a water truck is used for balling sewer lines, the same truck can be used for construction and repairs.

AIR COMPRESSORS, PORTABLE

A portable air compressor is required by repair and construction crews for operation of jackhammers and tampers. Compressors are also used for the operation of a sheeting driver head. Some organizations with many crews will install compressors permanently on their repair trucks. To others, where use is not so great, portable, trailer-mounted units may be more practical. Portable, trailer-mounted units can be made available to a greater number of crews on a simple need basis and provide more flexibility of operation.

HOISTS

Truck-mounted hoists, operated on 12-volt systems, are useful on repair and construction vehicles. They should be of sufficient lifting capacity to lift such items as manhole frames and covers, large size pipes, manhole grade rings, and light portable equipment.

PUMPS, TRAILER-MOUNTED

Some equipment of this type will be advantageous to most maintenance organizations. The minimum recommended size is a 4-inch diameter pump discharge although 6-, 8-, or 10-inch or larger discharges could be used by bigger organizations. These units are usually trailer-mounted with their own pump power source and are portable. Trailers should be equipped to carry all the necessary suction and discharge hoses or pipes that will be needed to set up pumps quickly in emergency situations. Standby pumping units should be checked and run at regular intervals to ensure that they will be in operating condition when an emergency does arise.

CLOSED-CIRCUIT TELEVISION

A variety of units are available. Self-contained vans, larger trailer units, and small, compact trailer units are available. It is impractical to suggest a particular type of unit best suited for any organization. All have advantages. The smaller units, naturally, are practical where their use may be limited. All trailer units have the advantage of making the towing vehicle available for other uses when the television trailer is not in use. Truck-mounted units provide easier handling and operation in congested traffic areas.

CHEMICAL APPLICATOR

A manufactured, trailer-mounted unit for the storage, foaming and injection of root control chemicals into main and building sewer lines may be used when the flooding method cannot be used.

QUESTIONS

Write your answers in a notebook and then compare your answers with those on page 545.

12.4H Emergency crews should be provided with what kind of transportation?

12.4I Which crews should be equipped with two-way radios or paging devices?

12.4J Why is it important to specify a truck with adequate capacity for a high-velocity cleaner?

12.4K When are truck-mounted equipment units preferable over trailer-mounted units?

12.4L Why should portable standby pumping units be inspected and run at regular intervals?

12.4M What major pieces of equipment are needed for construction and repairs?

12.4N Why do construction and repair crews need a portable air compressor?

12.422 Miscellaneous Equipment and Tools

The following is a listing of some of the more common miscellaneous equipment and tools that will be needed for the maintenance and operation of the Clearwater collection system. These and additional items will be purchased or rented as the need for them arises.

SEALING AND TESTING

Equipment for air testing collection system joints for leaks and for sealing such leaks is widely used today. This equipment is commonly used in conjunction with closed-circuit television equipment, and may be purchased separately or in conjunction with TV units. A complete unit that provides for television inspection, air testing, and joint sealing can be purchased as a package unit mounted on a truck or trailer.

SEWER BALLS

Balls for cleaning collection system pipes are available in sizes from 6 inches to 48 inches in diameter. Larger sizes are available by special order. Most maintenance organizations will equip individual balling trucks with 12-inch or smaller sizes. Sizes above 12 inches become very expensive and are less frequently used. Depending on the size of the collection system, it is more practical to acquire only one each of the large size balls. These are kept in a central storeroom or toolroom and are available to whichever crew may have use for them.

PUMPS, PORTABLE

Portable pumps, available in 1- to 10-inch discharge capacities, are often required by repair and maintenance crews. Pumps can be powered by gasoline, electricity, or closed-circuit hydraulic oil pressure. Each has specific advantages, many of which are discussed in the chapter on pumps. Pumps are stored in a toolroom and issued to crews whenever needed.

GENERATORS, PORTABLE

Gasoline- or diesel-powered generators are needed to supply 110-volt power to crews in the field. One or more should be available in a storeroom to be drawn by the crews as needed. Such generators are commonly used to provide lights for night work and 110-volt power to operate electric power tools.

Combination light tower/gensets are also popular because they combine lighting and auxiliary power into one towable unit and also provide a highly desirable angle of lighting for crew operations.

A trailer-mounted, gasoline/diesel-powered generator is used to run the small- to medium-sized lift stations that do not have a permanent source of standby power when the usual source of power fails.

Truck-mounted generators that operate off the truck's electrical system also are available to provide 110-volt power. This equipment usually is permanently mounted to the truck. Large organizations provide selected trucks with this equipment when the need is great enough to warrant it.

ELECTRONIC LOCATORS

Equipment of this type will be required at some time by most maintenance and repair crews. Many types of locators are available. Three common types are discussed in the following paragraphs.

The simple valve or lid locator is easy to operate, has a limited depth capability, and is most useful in locating buried manhole covers or water valves. Emergency crews in particular will find the valve locator effective in locating buried manhole covers. This unit is relatively inexpensive and one can be furnished as part of the permanent equipment that emergency crews carry on their trucks.

The pipe locator is a bit more complex, but simple to operate after it has been assembled. Pipe locators are capable of operating at greater depths than lid locators and are most effective in locating metal pipes, manhole covers, and water valves. Repair crews use a pipe locator to find underground utilities prior to excavation. It may not be practical to supply all crews with such expensive equipment, but one locator in a foreman's truck or in the toolroom can be made available when needed.

The ferret-type locator is the most complex and capable of the most accuracy. It uses an electronic signal unit which can be floated or pushed down a sewer. The locating unit above the ground picks up a signal from the unit which is in the underground pipe, pinpoints its location, and in many cases indicates the depth of the pipe. Electronic and ferret-type locators do not work in metallic pipe. The ferret-type locator is relatively expensive and is limited to specific purposes. It may be of little use except to the larger maintenance organizations.

PLUGS

Plugs are available in two common types—air and mechanical. All crews should carry at least the common smaller sizes. As with sewer balls, larger sizes are less frequently used. They should, however, be available in the storeroom to be drawn and used by any crew when needed.

- Air plugs. These have the advantage of being inserted and inflated from ground level which makes it unnecessary for operators to enter the manhole.
- Mechanical plugs. Some models also are available that can be inserted and tightened from ground level. Other models require an operator to enter the manhole in order to insert and tighten the plug.

TAPPING MACHINES

The era of handmade taps into existing sewers hopefully has passed. This undesirable practice has been a major source of maintenance problems. Crude attempts to tap a sewer with hammers, chisels, and miner's picks often resulted in broken mains, ill-fitting taps, and taps protruding into the sewer. The resulting cracks and poor connections were inviting areas for future root intrusion and workers were often tempted to use an excess amount of mortar in an effort to hide the crude job. Many municipalities or agencies no longer allow tapping of the sewers by anyone other than their own operators.

There are a variety of tapping machines available which will produce an accurately sized, neat hole for tapping. Such machines also minimize the possibility of breaking or cracking the sewer during the tapping process. Most tapping equipment is adjustable to allow for use on a variety of pipe sizes of 6 inches and larger. The tapping machine is clamped securely in place on the sewer by attachments similar to a chain vise. A diamond-core drill of the exact size to accommodate the saddle safely provides a neat hole.

INTERNAL TAP CUTTER

Internal tap cutters were developed to remove protruding taps which resulted from former crude tapping methods. The cutter, which is slightly smaller than the sewer line, is pulled manually into the sewer to the location of the protruding tap. Two inflatable collars hold it securely in place. Using air power, the diamond-core drill is turned and slowly advanced until the tap has been neatly and safely cut off. This equipment is extremely useful when the offending tap is located in an area where excavation for removal would be impossible or very expensive.

BORING EQUIPMENT

Many large municipalities no longer permit cutting across major thoroughfares to install collection system service lines. Installation is impractical due to the traffic flow. Boring is one effective method to do this work. Dig a boring pit at the edge of the right of way and another pit directly above the sewer line in the street. Make sure the two excavations, when connected, will form a 90° angle at the main line. Bore under the surface of the street and then install and connect the service line. Boring equipment is capable of cutting from 6-inch to 12-inch holes for distances of over 100 feet, and is available at a reasonable cost. The use of trenchless technology methods to rehabilitate or replace pipe is discussed in Chapter 10, "Sewer Renewal (Rehabilitation)," in several noteworthy technical manuals published by the North American Society for Trenchless Technology (Chicago), and in the Water Environment Federation's manuals of practice.

LINING EQUIPMENT

Sliplining is a form of trenchless technology. This method requires special equipment for preparing the polyethylene pipe being used, including fusion joiner jigs, mobile carts to handle the pipe, a butt fusion tool and heating faces, tapping or coring tools, and accessories for the pulling heads.

PIPE CUTTERS

Here again, hopefully, the era of the hammer and chisel has passed. A variety of chain-type cutters are available. All are capable of neat, safe, pipe cutting. Most cutters provide the capability for cutting existing pipe in place in the pipe trench. Manual models are available for cutting smaller size pipe, ratchet models for intermediate sizes, and hydraulic models for the larger sizes.

FLOW MEASURING EQUIPMENT

Flow measuring equipment includes weirs, calibrated flumes, velocity meters, electromagnetic flowmeters, and electronic velocity and depth recorders.

QUESTIONS

Write your answers in a notebook and then compare your answers with those on pages 545 and 546.

12.4O Air testing equipment for detecting leaks in pipes and joints often is used to do what other jobs?

12.4P Where would you store equipment that is seldom used, such as large size sewer balls and portable emergency pumps?

12.4Q Name three different types of electronic locators.

12.4R What are the two different kinds of plugs?

12.4S Why should handmade taps be discouraged?

12.4T How can taps protruding into mains be removed?

12.4U How can a service line be placed under a major thoroughfare without disrupting traffic?

12.423 Safety Equipment

There is increasing concern by regulatory agencies that operators in the field of collection system maintenance be provided with safe work areas and equipment. Typically, it is the agency or supervisor who is held legally accountable for safety, not the operator in the field. But regardless of enforcement efforts by others, supervisors should be constantly concerned with the safety of their operators. A variety of safety equipment is available and its use must be required and enforced. (Also see Chapters 4 and Chapter 11, Section 11.1, #10 for safety procedures when working in and around manholes.)

ATMOSPHERIC TESTERS

A serious hazard faced by all collection system operators is the atmospheres to which they are exposed. Confined spaces such as manholes and wet pits can become deathtraps when a sudden change in atmosphere occurs. Three main items of concern are: (1) lack of oxygen, (2) an explosive condition, and (3) toxic gases (hydrogen sulfide, carbon monoxide, chlorine). The only effective devices to detect such conditions are atmospheric testers.

A variety of single-purpose testers or meters is available and each one is designed to detect specific gases commonly present in collection systems such as carbon monoxide, methane, and hydrogen sulfide. See Chapter Chapter 11, Section 11.1, #10, Section 12.7, #7, "Poisonous or Toxic Gases," for a more detailed discussion of gases and testing equipment.

A three-gas meter capable of detecting either lack of oxygen, toxic gases, or explosive conditions may be the most effective and practical detection device for most maintenance activities. Further detection capability is available with four-gas meters (oxygen deficiency, explosive conditions, hydrogen sulfide, and carbon monoxide) for agencies that need to monitor all these conditions. See Chapter 4, Section 4.52, "Portable Atmospheric Alarm Unit," in Volume I for more details.

Multipurpose testers have interchangeable parts that allow them to measure the level or concentration of several gases. For example, one instrument with interchangeable scales and calibration tubes can detect carbon dioxide, hydrogen sulfide, chlorine, sulfur dioxide and nitrogen dioxide. Another type of multipurpose tester is capable of monitoring oxygen levels, explosive conditions (methane), and a toxic gas (hydrogen sulfide).

All types of testers should be equipped with both visual (light) and audible (sound) alarms that activate whenever the tester detects a hazardous condition. An operator should not be expected to read a dial or gage and then decide whether or not a hazardous condition exists. This is because the operator's physical senses and judgment can quickly become impaired by the hazardous atmosphere.

Atmospheric testers can also be equipped with a data-gathering microchip that records the atmospheric exposure of the operator using the meter over a given time period, such as an 8-hour period.

Consult the rules of the regulatory agencies in your area to determine the specific equipment you will need to meet their requirements and protect your operators. Regardless of such requirements, a supervisor should make available to any of the operators required to work in such confined spaces at least the testers capable of detecting a lack of oxygen, an explosive condition and any expected toxic gases (especially hydrogen sulfide). Continuous monitoring with atmospheric testers while people are working in confined spaces should be a basic safety rule for all collection system organizations.

BLOWERS

In conjunction with the above detection meters, there also is a need to provide blowers of ample capacity to introduce fresh air into confined spaces. The recommended capacity range is from 4 to 20 volume changes per hour, depending on conditions in the confined space. Contact your local safety officer for the requirement.

At the very least, small, portable blowers should be standard equipment on all vehicles whose crews work in confined spaces. Blowers operating on the 12-volt power systems available in most vehicles are commonly used for this purpose. Larger maintenance organizations will usually need a trailer-mounted blower capable of up to 15,000 CFM air capacity. You will need such equipment whenever operators must enter large-diameter sewers for inspection or repair activities.

SHORING

Another hazard to underground collection system operators is the possibility of injury or death due to cave-ins while working in excavations. Adequate shoring must be provided for their protection. Many varieties of shoring are available. For additional information on shoring, read Chapter 3, Section 3.702, "Trench Shoring," in Volume I; Chapter 7, Section 7.1, "Shoring," in Volume I; and Chapter 11, Section 11.1, #12, "Excavations and Trench Shoring," in this manual.

- Hydraulic shoring. Hydraulic shoring is widely used, light in weight, and can be safely and quickly installed or removed while remaining above ground level.
- Manual shoring. Manual shoring usually consists of screw jacks and wooden timbers of adequate size. Consult your area's regulatory agencies for the specific minimum size dimensions acceptable for shoring timbers. Installation of this type of shoring is somewhat difficult and time-consuming, and operators are required to enter the excavation during the process of installation. Exercise caution! Make sure the top screw jacks are in place first, then proceed downward to the bottom of the excavation to offer the maximum protection to operators engaged in such installation. To remove the shoring, reverse the process. Remove the bottom jacks first and proceed upward. Obviously, because of these disadvantages, use of manual shoring should be limited to excavations of small size and depth.
- Sheet shoring. Most regulatory agencies require this type of shoring in loose or running soil conditions. Exact dimensions acceptable for the sheeting will be specified by the regulatory agency in your area.
- Cylinder and shield shoring, also known as trench boxes. Cylinder and shield shoring are parts of prefabricated shoring that provide good protection for the operator and can save a lot of time over manual and sheet shoring methods. Their limitation is the inability of a trench box to accommodate utility lines that cross perpendicular through the trench.

Because modern shoring equipment is expensive and most organizations will have only a limited need for this type of equipment, it is common practice to purchase only the limited amount needed for most routine activities. When, and if, larger projects are undertaken, the required additional shoring can be rented for the required time.

TRAFFIC CONTROL

Traffic is a significant hazard faced by collection system operators that is frequently overlooked or underrated. In the interest of expediency, operators will often expose themselves to traffic flow without adequate protection. Supervisors should constantly impress upon their operators the danger of such a practice. Make sure adequate equipment for traffic control is available and insist upon its use. Traffic control equipment includes:

1. Rotating, flashing beacons or strobe lights on vehicles,
2. Traffic cones and delineators,
3. High-level warning signs and flags, and
4. Fluorescent orange safety vests.

All maintenance vehicles should carry and use the above traffic control equipment (also see Volume I, Chapter 4, Section 4.3, "Routing Traffic Around the Job Site," and Chapter 11, Section 11.1, #10, "Traffic Mishaps," in this manual).

PROTECTIVE CLOTHING

Collection system operators come into contact with a variety of hazards which make protective clothing advisable.

- Uniforms. Provide all operators with uniforms or coveralls to be put on at the start of the day and removed before going home. This practice will minimize the possibility of employees, who come into contact with raw wastewater in their work, wearing contaminated clothing home and perhaps exposing their family members to these contaminants. A uniform also identifies operators with your agency, which is especially helpful when access to private property is required.
- Safety shoes. Encourage all operators to wear safety shoes. Some organizations provide safety shoes and make it mandatory that they be worn. Other organizations may participate in a cost-sharing program with their employees to provide for safety shoes.

- Rubber boots. Employees who work in raw wastewater should be provided with safety toe, nonskid rubber boots, with steel shanks for protection of the bottom of the feet. Many organizations provide such protection on an individual basis or as needed.
- Gloves. Persons exposed to raw wastewater should wear protective gloves to eliminate or minimize cuts, abrasions and infection through broken skin.
- Eye protection. Most operators will require eye protection of some type. A program for providing safety glasses, including prescription lenses, is desirable. Provide safety goggles to crew members exposed to flying particles, and supply complete face shield protection to crews working with chemicals.
- Hard hats. Hard hats should be furnished to all operators. They must be worn at all times on the job site.
- Ear protection. Hearing protection devices must be worn when operating noisy equipment such as jackhammers.
- Portable ground-fault circuit interrupters (GFCI). Interrupters should be used when operating portable electric tools, especially in wet locations.
- Safety harnesses. Harnesses should be provided to all persons who are working in manholes or inside large-diameter pipes. When safety harnesses are in use, there should be sufficient operators topside to provide for fast, safe removal of such persons should an emergency arise.
- Personnel hoist. A tripod set over a manhole with a rope running from a winch on one leg through a pulley at the apex to the safety harness of an operator provides a portable and effective method of removing a collapsed person from a manhole.

Aside from these safety articles and protective clothing, other safety equipment may be required to meet specific conditions. Consult your area regulatory agency to ensure that you are in full compliance at all times. Provide your crews with the necessary safety equipment and remind them frequently of the importance of safety. Give your operators complete instructions on the proper use of their equipment and insist that safe procedures be followed **AT ALL TIMES.**

QUESTIONS

Write your answers in a notebook and then compare your answers with those on page 546.

12.4V The atmosphere should always be tested before entering what kinds of areas?

12.4W What are the three main items of concern when testing the atmosphere?

12.4X Why must the atmosphere in a confined space be monitored continuously?

12.4Y How can fresh air be introduced into confined spaces?

12.4Z List the different types of shoring.

12.4AA What common hazard is often overlooked or underrated by operators?

12.4BB Why must collection system operators wear protective clothing?

12.424 Office Equipment

The common office equipment required for the operation and maintenance of a wastewater collection system is beyond the scope of this manual. However, the following special office equipment is noteworthy.

BASE RADIO

A base radio is required when the vehicles of operating and maintenance personnel are equipped with two-way radios. If possible the base station should be part of a radio network covering all the public works agencies within the area served by the collection system. The station must be able to send and receive messages to the most remote part of the service area or be able to use a network's repeater facilities. Cell phones are being used by many crews as an effective means of communication.

CELL PHONES

Cell phones, widely used throughout the United States and around the world, are rapidly becoming useful communications devices in the wastewater industry. Alternately known as cellular, wireless, or mobile phones, they are capable of performing a variety of functions, many of which are valuable to wastewater collection and treatment system personnel. As a communication device, cell phones serve as a handy means of sending and receiving messages and information on the job. The operator is able to locate a supervisor, call for additional staff, or contact contractors, as needed, in field operations. In addition, cell phones can provide efficient access to help in case of emergencies. As a management device, cell phones provide all levels of staff with a convenient tool for contacting work crews in the field, being available for troubleshooting and problem solving, and for scheduling appointments and meetings.

In this age of rapidly increasing technology, cell phones have evolved from simple, wire-free telephones just a few years ago to sophisticated, portable communications systems offering numerous advanced features today. Some of the features that have particular application in wastewater collection and treatment systems include:

- Storing contact information such as names, phone numbers, and addresses,
- Storing reminders of daily, weekly, and monthly tasks,
- Keeping track of appointments and meetings,
- Calculating simple math problems in the field,
- Sending and receiving e-mail correspondence,
- Providing Internet access, and
- Recording and storing messages.

New features are being added as the technology advances. For example, cell phones now are capable of capturing photographic images for storage or transmittal. This could prove useful in field documentation and other assignments.

As with nearly all electronic equipment, cell phones have certain limitations. Generally, they are not repairable. If they get wet or are used with wet fingers or hands, the internal parts could corrode and become inoperable. One way to avoid damage is to protect the device in a moisture-proof case. In addition, if the unit does come into contact with water, it may be possible to avoid internal damage by drying the exterior thoroughly, and waiting until it is completely dry before turning it on again. Extreme heat can damage the battery or sensitive electronic parts; the hot interior of a car or truck, for example, could quickly ruin a cell phone. Extreme cold may cause the screen display to malfunction. In certain situations, such as near below- or above-ground lift/pumping stations, cell phones may fail to perform well due to interference.

In deciding which cell phone to purchase, consider how, where, and when it will be used. It is usually necessary to sign a service contract, so shop around for the most cost-effective combination of cell phone features and calling services for your specific needs.

MICROCOMPUTER SYSTEMS

The increasing data processing power and declining cost of microcomputer systems now make it practical for superintendents of most wastewater collection systems to use computers. Large amounts of data can be rapidly processed into information that will help to operate and maintain collection systems efficiently.

Microcomputer systems consist of hardware that processes entered data and software that directs the processing. Microcomputer hardware includes a keyboard, a visual display screen, memory chips, a microprocessor, disk drives, and a printer.

Microcomputer application software (computer programs) fall into the following main categories:

1. Database management,
2. Graphics,
3. Program development,
4. Spreadsheet analysis,
5. Telecommunications, and
6. Word processing.

Sub-categories of application software that could be used by the superintendent of a collection system are:

1. Accounting,
2. Application programs development aids,
3. Business management tools,
4. Capital projects planning and control,
5. Data management,
6. Graphics,
7. Inventory control,
8. Job and contract cost accounting,
9. Personnel management,
10. Programming utilities,
11. Purchasing accounts payable,
12. Time management and scheduling, and
13. Word processing.

In addition to the standard software packages that are available, several consulting engineering firms and specialty companies have developed software packages specifically for the management of wastewater collection systems. For a more detailed description of how computers can be used to operate and manage a collection system agency, see Section 12.7, "Use of Computers in a Wastewater Collection Agency."

Before purchasing microcomputer hardware and software, first analyze in considerable depth how a computer system will make the operation and maintenance of your collection system more efficient in a cost-effective manner. It is important to know what you want the computer system to accomplish and then purchase the hardware and software that will do the job. The cost of hiring a qualified consultant to help you design and purchase a computer system will usually be offset by what you'll save in obtaining an effective system at the best price.

12.43 Management

Proper management of the equipment used to operate and maintain a wastewater collection system will maximize its availability for productive work and extend its useful life. A superintendent should maintain an inventory of equipment, assign responsibility for its use and maintenance, record operating costs, and plan cost-effective replacement.

12.430 Inventory

An inventory record documents the equipment owned by an agency and provides information for its management. The inventory, in whatever physical form, must be kept current and accessible. The essentials of an equipment inventory record include:

1. Type (high-velocity cleaner),
2. Model year,
3. Date purchased,
4. Cost,
5. Description,
6. Manufacturer, and
7. Fuel type.

12.431 Responsibility

Responsibility for each piece of equipment should be assigned to a person who will direct individuals or crews in efficient use of the equipment and will ensure that it is maintained in an operable condition. One person should be made responsible for the preventive maintenance and repair of all equipment. This person should also be responsible for recording the operating and maintenance costs for each major piece of equipment.

12.432 Preventive Maintenance and Repair

A good preventive maintenance program for equipment will be based on a system that indicates when and how each piece of equipment is to be maintained, when maintenance work was performed and the cost of that work. Good preventive maintenance will minimize the amount of equipment repairs. Repairs to equipment should be made in a timely manner to keep the equipment in a safe operating condition and to minimize downtime.

Some wastewater collection agencies will be large enough to have their own equipment maintenance and repair section. Smaller agencies may have most of the maintenance and repair performed by contract. In either situation collection system personnel can be used efficiently to perform minor maintenance and repairs between normal jobs and during inclement weather conditions.

12.433 Replacement

Equipment should be replaced when it has reached the end of its useful life. Useful life can be defined as the time at which a piece of equipment ceases to perform its required work or when the cost of operating and maintaining the equipment is greater than the cost of operating and maintaining new equipment. Good records must be kept. The total cost of operating a piece of equipment, including amortization of its purchase cost, should be used to make an informed decision on its replacement. The financial impact of equipment replacement on an agency can be reduced by making regular payments into a sinking fund for equipment replacement.

QUESTIONS

Write your answers in a notebook and then compare your answers with those on page 546.

12.4CC List the main categories of microcomputer software (computer programs) that could be used by a collection system agency.

12.4DD List the essentials of an equipment inventory record.

12.4EE When should a piece of equipment be replaced?

Please answer the discussion and review questions next.

DISCUSSION AND REVIEW QUESTIONS

Chapter 12. ADMINISTRATION

(Lesson 3 of 7 Lessons)

Write the answers to these questions in your notebook before continuing. The question numbering continues from Lesson 2.

13. How would you determine the tool and equipment requirements for a collection system?
14. Discuss the advantages and limitations of the different methods of obtaining equipment.
15. Why must trucks be of adequate size?
16. How can seldom-used pieces of equipment be made readily available to all crews when needed?
17. What kinds of safety equipment are needed for each type of maintenance crew?
18. What is the basis of a good equipment preventive maintenance program?

CHAPTER 12. ADMINISTRATION

(Lesson 4 of 7 Lessons)

12.5 FACILITIES

Wastewater collection system operation and maintenance facilities consist of yards and shops, as well as buildings needed for storage of equipment and material, maintenance of equipment, assembly of operating personnel and the housing of staff personnel. Ideally all elements of the facilities should be located together at one site near the center of the area served by the collection system. If the service area is very large, causing excessive travel time from a central yard to remote areas, satellite yards should be considered. The decision to decentralize is based on a least-cost analysis. Some small and medium-sized agencies locate their collection system operating and maintenance facilities adjacent to the treatment facilities for economic and administrative reasons.

In some instances, particularly when a collection system is under the jurisdiction of a multi-function agency, the upper levels of collection system staff will be housed along with similar staff for other divisions in an office building removed from the collection system yard.

12.50 Yard

A collection system's yard should be large enough to provide for its present and future uses. Having room to expand is particularly important if expansion of the collection system is anticipated.

12.500 Storage and Servicing Equipment

A yard is used to store equipment when it is not being used and to service it before use. Store equipment in a location that provides easy access and allows placement and removal of each piece in a safe and efficient manner.

Vehicles and special mobile equipment should be placed in a location that will minimize the need to back them in or out of their storage area. Usually it is most economical to store vehicles and special mobile equipment in open areas. However, roofed areas or completely enclosed areas may be appropriate if heavy rainfall or snowfall would shorten the useful life of equipment or make it operationally unreliable if stored in the open. Generally, small pieces of equipment and equipment not mounted on vehicles are stored in buildings.

A yard should provide for the servicing of vehicles and engine-driven special equipment with fuel, air, coolant and oil when needed. Facilities for the frequent washing of the vehicles and special mobile equipment should also be provided. Small agencies often find it uneconomical to have their own service facility and use a nearby service station instead.

12.501 Storage of Materials

The materials needed for the operation and maintenance of a collection system that are not readily available from a supplier should be stored in an agency's yard. Such materials include chemicals, manhole barrels, cones, frames and covers, cleanout frames and covers, sewer pipe and fittings, brick, gravel, sand, cement and paving materials. The material should be stored in accessible locations and contained in appropriate bins and enclosures. Hazardous materials, such as toxic chemicals, flammable fuels, and hazardous waste must be stored safely and in accordance with local regulations. The manner of all material storage must provide for safe and efficient movement in and out of its area by hand or with appropriate equipment.

12.502 Training Area

Yard space should be provided for the training of collection system operators in the efficient and safe use of equipment and materials. Classroom space with audio-visual support is highly desirable because of the large amount of training an agency must conduct. Large agencies will have the resources to construct specialized facilities for training collection system operators. One such agency constructed a walkable 3-foot by 5-foot brick sewer connected to two 12-inch main sewers, three manholes, and one service lateral. Another built a six-inch diameter sewer above ground using clear plastic pipe and terminal manholes. The facility allows operators to observe the cleaning action of balling and high-velocity cleaners in sewers.

12.503 Security

The yard should be completely fenced to restrict entry by unauthorized persons and lighted at night to discourage theft or damage of valuable equipment. Completely lock all vehicles and mobile equipment and store the keys in a secured cabinet. Store small equipment and valuable tools and materials in a locked enclosure. A written inventory of equipment, tools and valuable materials that is kept up to date will discourage their "unexplained loss."

12.51 Shop

Shop facilities will vary from a small area in a building equipped with a workbench and a few tools, when an agency is operating a small collection system, to a large shop building equipped with an array of hoists, overhead cranes, machine tools, power tools, welders and other devices, if the agency is operating and maintaining a large collection system with many lift stations.

12.510 Location

Select a location for the shop facilities that will allow the efficient movement of the items to be worked on in and out of the shop. Also, the shop location should not interfere with the movement of other equipment and materials stored in the yard. The shop area in a building should have smooth concrete floors that can be easily cleaned. Ceiling heights and door openings must allow clearance for the largest item to be worked on. Good lighting and ventilation are required, especially for welding or painting work. In cold climates the shop area should be heated to a comfortable temperature. Adequate ventilation and air conditioning should be included in warmer climates.

12.511 Equipment

A superintendent responsible for equipping a shop should first determine what work will be done in the shop and then select the equipment and tools needed to do the work. Try not to overequip a shop initially as this leads to a lot of unused equipment and tools lying around. It is easy, and more economical, to purchase equipment and tools as you find you need them. As you study the other chapters of this manual, envision the equipment and tools you think would be needed in a shop.

12.512 Use

Use of the shop area should be limited to authorized personnel. Small and medium-sized collection system agencies will authorize qualified collection system operators to work part-time in a shop. Large agencies will probably hire full-time shop personnel. In either situation, personnel authorized to work in shops must be qualified to operate the shop equipment in a manner that will not damage the equipment or injure the operator or co-workers. Authorized shop personnel must be held responsible for returning equipment and tools to their proper locations and for keeping the shop area clean and in a safe condition.

Valuable shop equipment and tools should be inventoried and secured in an enclosed toolroom with one person responsible for checking items in and out. The entire shop area should be secured when authorized shop personnel are not in the immediate area and at the end of work shifts.

12.52 Operator Facilities

Full-time collection system operators need facilities where they can prepare for work assignments at the beginning of a work period and for departure at the end of a work period. Ideally, the operators' facilities will be close to where they park their private vehicles and near the equipment they will be using to perform their work.

12.520 Hygienic Facilities

Collection system operators should be encouraged to practice good personal hygiene. The collection agency can do this by providing some or all of the items described in the following paragraphs.

LOCKER ROOM

A well-lighted and ventilated room—one each for men and women—should be equipped with individual lockers and benches for the changing from street clothing into protective clothing at the start of a work period and the reverse at the end of a work period. The room should also provide facilities for storage of clean protective clothing, soiled protective clothing, and special outer wear such as rain gear and rubber boots. For hygienic reasons, protective clothing must not be worn home.

LAVATORIES AND TOILETS

Lavatories, preferably the circular, foot-operated type, should be installed in a room adjacent to the locker room but partitioned off or in a room separated from toilets. Local building codes will specify the minimum number of lavatories and toilets that are required for the number of persons using the facilities.

SHOWERS

Showers should be available to collection system operators in a room adjacent to the lavatories and toilets. An agency cannot require collection operators to shower after working in an environment that exposes them to wastewater, but it can encourage this practice by providing pleasant facilities and furnishing towels. Shower facilities will be essential when an operator is accidentally covered with hazardous chemicals or wastewater.

OUTER WEAR WASHROOM

If practical, a small room (at the entrance to personnel facilities from the yard) should be provided for washing off rain gear and rubber boots that are muddy or covered with residue from sewer cleaning. The room should contain a floor drain, a water faucet and a hose.

12.521 Assembly Room

An assembly room for collection system operators should be provided adjacent to the locker room. The room should contain tables, chairs, a blackboard, bulletin board, and an area for personal equipment such as gas detection devices and hand-held radios. The room and furnishings should accommodate the number of persons that need to be assembled at the same time for the purpose of making assignments, giving instructions, and training.

12.522 Employees' Room

A pleasantly decorated room for refreshment and meal breaks should be provided for the collection system operators who are permanently assigned to work in the yard area and for collection system operators who are occasionally in the immediate area. It should contain a sink with hot and cold water, microwave oven, refrigerator, coffee urn, and tables and chairs. If practical, the room should have windows that open onto a landscaped area.

Small and medium-sized agencies may, for economic reasons, use the assembly room as an employees' room by adding a food and drink preparation area to the room. If this is done, extra care must be taken to keep the assembly room clean and free of contamination from wastewater residue brought in by field personnel.

12.53 Offices

The administrative staff of a collection agency needs offices that provide them with adequate facilities for the efficient performance of their duties. The office design must include provisions for compliance with the current interpretation of the Americans with Disabilities Act (ADA).

12.530 Location

Staff offices may be located entirely within the same building as the operator facilities, or the top level managers may occupy a separate building. In either situation, the offices of supervisory personnel should be in the same building and near the operators' assembly room since a supervisor must have daily contact with operators.

12.531 Types

Private offices are usually provided for superintendents and assistant superintendents. Supervisors may have private offices or may be located in a room shared with other supervisors. Supporting staff (such as secretaries, clerks and technicians) generally have work areas within a common room. Movable partitions in a common room provide some privacy and reduce distractions caused by the separate activities taking place in the room. Dispatchers may be located in a separate room where the audible radio communications will not cause a distraction for other workers.

12.532 Ancillary Facilities

In addition to staff offices, an administration building will need the ancillary facilities described in the following paragraphs.

RECEPTION AREA

A pleasant visitor reception area should be located at the front of the main office building. Typically it will contain seating for waiting visitors, information about the activities of the collection system department, and a clerk-receptionist.

EMPLOYEES' ROOM

If the staff offices are separate from the building containing the employees' room for operators, a separate employees' room should be located in the main office building. The room should contain the same furnishings as the employees' room for the operators.

RESTROOMS

Restrooms must be provided for both men and women. Local building codes regulate the number of lavatories and toilets to be provided for the number of employees using the building.

COMPUTER ROOM

If several staff members share the use of computer terminals and printers, it may be desirable to locate the equipment in a separate room to facilitate use by different people and to minimize the noise distraction for other workers. An alternative is to plan for and install a computer local area network (LAN) at the same time the office areas are built or remodeled. The file server can be located in the computer room with each office desk workstation wired to it.

STORAGE ROOM

A storage room should be provided for office materials, temporarily unused equipment and archived records.

LIBRARY AREA

Either a separate room or a convenient area with ample bookshelves should be provided for storage of books, bound reports, catalogs, technical references, and training manuals.

SECURITY

Fireproof vaults and files are recommended for storage of valuable papers and records. Computer backup disks and tapes should also be stored in a fireproof vault or, preferably, in a convenient bank's safe deposit box. The tape storage is especially critical if your sewer system maps are kept in electronic format. Key your locks so that one key opens all outside office building doors. Issue copies of the key only to selected personnel who need access to the buildings. Keep a record of keys issued and returned. Periodically change the lock tumblers and issue new keys for good security. An alternative is to use electronic access controls that detect/reject authorized entry using code numbers or fingerprints. At night, illuminate dark areas outside and inside the office building to discourage loitering and unauthorized entry.

12.54 Landscaping

Landscaping the fringes of the collection system maintenance yard and office building setbacks will make the facilities pleasing to the public and visitors and will increase the satisfaction of employees for the agency. The landscaping should be appropriate for the environment and should not require extensive maintenance. Use automatic irrigation systems to conserve water and reduce maintenance costs.

12.55 Parking

Provide parking spaces for operators and staff personnel near the assembly and office buildings and arrange for convenient visitor parking near the entrance to the main office building. Special consideration is required to meet handicapped parking and wheelchair access requirements. Landscape parking areas and illuminate them for added security and safety.

12.56 Design

It is advisable to consult with architects and engineers during the design and construction of collection system facilities in order to produce an efficient and pleasant-appearing facility. However, it is even more essential that the superintendent and the operators be involved in the design process to ensure functional facilities. They are the ones who are most familiar with the activities that will take place in the facilities, and their experience should not be overlooked.

QUESTIONS

Write your answers in a notebook and then compare your answers with those on page 546.

12.5A What kinds of service should be available in the maintenance yard for vehicles and engine-driven special equipment?

12.5B What types of hazardous materials could be stored in a maintenance area?

12.5C How should valuable tools and equipment be secured in the shop area?

12.5D Why is it important to landscape around the collection system maintenance yard and office buildings?

Please answer the discussion and review questions next.

DISCUSSION AND REVIEW QUESTIONS

Chapter 12. ADMINISTRATION

(Lesson 4 of 7 Lessons)

Write the answers to these questions in your notebook before continuing. The question numbering continues from Lesson 3.

19. Under what circumstances should a satellite maintenance yard be considered?
20. What materials needed for collection system operation and maintenance should be stored in an agency's yard?
21. Why is security of a maintenance yard important?
22. What types of hygienic facilities should be available for collection system operators in a maintenance yard?
23. Why do collection system operators need an assembly room?

CHAPTER 12. ADMINISTRATION

(Lesson 5 of 7 Lessons)

12.6 MAPPING

Maps are absolutely essential to the operation and maintenance of any wastewater collection system. Maps showing the accurate location of the components and an indication of the size of the collection system will be required before any maintenance activities can be planned or implemented. As a collection system grows, staffing will increase, maintenance activities will expand, and a corresponding need for better, more detailed maps will become evident. Maps must indicate what has been constructed, where, what size, the materials used, and the conditions encountered. A detailed record of what is underground is critical for the successful operation and maintenance of a wastewater collection system.

12.60 Importance of Mapping

Under certain conditions it might be possible for one collection system operator to operate a small collection system without the benefit of any formal map system. In time, the operator could become familiar enough with the collection system to operate from memory. This practice is not recommended, regardless of how small the collection system. Inevitably, at some time, this operator and the operator's memory will be unavailable. When that time comes, the operator's replacement will have many problems.

Collection systems of a significant size present an entirely different situation. The number of operators involved will make it impossible for each of them to become thoroughly familiar with the entire collection system. Daily assignments will send operators to parts of the system they may never have seen or even knew existed. Maps will be needed for the daily function of directing operators to their assigned work areas.

12.61 Information on Maps

12.610 Locations

Collection system maps should provide the information staff and operators need to plan and conduct operations and maintenance work. To be most useful, maps will show system components in relation to other fixed landmarks.

- Main, Trunk and Interceptor Sewers. The locations of these sewers in streets are shown in relation to the center line or property lines. Sewers in easements are shown in relation to property lines and the width of the easement is noted. Curved sewers are shown in relation to a straight line between the beginning and ending of the curve; the end points of the line are then shown in relation to either the downstream or upstream manhole.
- Building Sewers. Show the distance in feet between the downstream manhole and the connection of the building sewer to the main sewer. Also, indicate on the map the distance along the property line between the sewer crossing and one of the property corners.
- Manholes. Care should be exercised to indicate manhole locations accurately, by footage from connecting manholes, by scale in relation to property lines and property corners, and by coordinates determined by survey or (much more economically) by the use of global positioning satellite (GPS) or geographic information system (GIS) receivers that are accurate to plus or minus 3 feet. This is especially important because considerable time can be lost in seeking manholes that have been paved over in streets or covered by landscaping, regrading or vegetation growth in rear easements.
- Cleanouts. Collection system laterals will contain many more cleanouts than manholes. However, when you need access to a plugged sewer, cleanouts become equally essential. Being smaller in size and metal mass, they are more difficult to find by sight or by electronic locators. For these reasons, take care to accurately designate their locations on maps.
- Boundaries. Jurisdictional boundaries of surrounding agencies also should be shown.

12.611 Identifications

Aside from location, identification by number is the next most important benefit to be realized from good maps. Accurate number identification will provide the basis for an accurate record system. Use specific numbers to identify the items listed in the following paragraphs.

- Manholes. Each and every manhole should be assigned a specific, permanent identification number and these numbers should never be changed. When extensions of a collection system result in the addition of manholes, each addition must be recorded on maps immediately in order to keep the maps current. It is the responsibility of management to provide a means of ensuring that maps are kept current and periodically checking to be sure the map updating is done.

Occasionally, those who draft maps succumb to the temptation of assigning new numbers to existing manholes in the interest of keeping a particular portion of the map neat and sequential. Such practice will make all existing records useless because they are based on original numbers. Additional manholes, when recorded on the mapping system or page of the mapping system, must always be assigned new numbers. To avoid having manholes with high numbers, start numbering the manholes on each page with the number 1. Also don't be tempted to assign manhole numbers by drainage basin branch(ing). You will end up with an unwieldy scheme with excessively long numbers that are unusable by field crews. Different agencies have different procedures for numbering manholes and any procedure is satisfactory as long as it is simple and not confusing.

- Sewer Cleanouts. As with manholes, cleanouts are identified by their own specific number.
- Sewers. Each section of sewer between manholes is identified or designated as an individual section. This individual section can be identified simply by indicating the numbers of the two manholes which the section connects.
- Properties Served. Identify all individual parcels of property which are connected or have the potential of being connected to the collection system. The street address and lot number or the assessor's parcel number are commonly used for identification.
- Cleanouts at Property Line for Building Services. Specific identification, by individual number and location, is a matter of choice depending on the scope of a particular record-keeping system. Cleanouts are commonly identified by reference to the particular parcel of property they serve.

12.612 Elevations and Depths

Known elevations are vital to the operators of gravity wastewater collection systems. Essential elevations that should be shown on maps are:

- Flow Lines. Maps should indicate flow lines, also called inverts, at the site of each manhole for all sewers entering and leaving the manhole.
- Manhole Rim Elevations. If rim elevations are not shown, then the depth of manholes should be indicated. Rim elevation can be computed by adding manhole depth to invert elevation.
- Depth of Building Sewer at Property Line. Often this depth is not recorded on maps or construction plans. Many local construction codes merely stipulate a minimum and maximum acceptable depth, except in cases of unusual topographical conditions. If possible, or practical, a good practice would be to accumulate and record all such depths measured during the normal course of inspection, maintenance or repair activities. This information would then be available to be included in future updating of maps. The retention of the depth of the building sewer at the main sewer tap is useful for future repair operations.

12.613 Lengths and Widths

Lengths and widths that should be shown on collection system maps include the following:

- Show the length of sewer lines in feet between the centers of downstream and upstream manholes. Some maps will show a surveyor's stationing of the upstream manhole from the downstream manhole; the length of sewer line is obtained by converting the station designation to feet. For example, if a manhole has a 3 + 52 station designation, the length of sewer line between manholes is 352 feet.
- The lengths of all property lines shown on the maps and the widths of all mapped streets and easements should be shown in feet.

12.614 Other Information

In addition to locations, identifications, elevations, and lengths and widths, the following information should appear on the maps:

- Flow Direction. The direction of flow in main, trunk and interceptor sewers should be shown by a directional arrow.
- Force Mains. This type of sewer line should be specifically designated since it requires special maintenance.
- Access Points and Overflow Points. There are types of entry points into sewer systems that are not manholes. They range in scale from 8 x 12-foot equipment access openings into interceptors to force main relief valves to overflow weir walls. These structures should each use a different symbol type on the map to distinguish them from regular manholes.
- Scale and North Arrow. The scale of a map must be shown so distances measured in inches on the map can be converted to feet of distance in the field. A North arrow on the map allows it to be oriented or pointed in that direction in the field.
- Dates. The date that each map page was drafted and the date of the last revision should appear near the border of the map.

12.615 Optional Information

Some sewer collection agencies may include the information in the following paragraphs on their maps.

- Types of Pipe. The type of pipe used to construct a sewer line is shown on the map using the following designations: ABS, Acrylonitrile Butadiene Styrene Pipe; ACP, Asbestos Cement Pipe; CIP, Cast Iron Pipe; CMP, Corrugated Metal Pipe; DIP, Ductile Iron Pipe; PVC, Polyvinylchloride Pipe; RCP, Reinforced Concrete Pipe; RPMP, Reinforced Plastic Mortar Pipe; and VCP, Vitrified Clay Pipe.

- Soil Conditions. This information is not often shown on maps. Unusual soil conditions would be of great importance and a significant safety factor to future repair crews who would be re-excavating in disturbed ground. Such information could be acquired at time of construction and noted on "as-built plans" or record drawings by inspectors. Soil conditions could then be included in any future updating of maps.
- Properties Connected to Collection System. This information may not be available when maps are originally compiled. Properties connected should be added at the time of connection. An ideal method is to record connection permit

numbers, as soon as they are issued, on tracings of the map system.

- Inventory. Inventory listing should include the number of manholes, flusher branches, house connections and footage for each size pipe. This may not be practical on large maps covering large areas because of the huge totals required and the small scale of such maps. Inventory totals can and should be computed when maps have been divided into sections or pages. On such maps of adequate scale, totals for each page or section can be easily computed and can be easily added together to provide the total for each book or map (which are divided into sections). By adding these book totals or map totals, the total for the entire collection system can be determined. Computer technology can also be used to create and maintain an inventory of facilities, sometimes more efficiently than by the method described above.

QUESTIONS

Write your answers in a notebook and then compare your answers with those on pages 546 and 547.

12.6A Why are maps of a wastewater collection system important?

12.6B List as many items of information as you can recall that should be available on wastewater collection system maps.

12.6C How should manholes be identified on maps?

12.6D What elevations should be shown on maps?

12.6E Why should soil conditions be shown on maps?

12.62 Type and Size of Maps

For the collection system operator in the field, an ideal map would be one drawn for your specific use and needs. It would provide all the details possibly needed, on a scale large enough to be easily read. The ideal map would be of a size that is convenient and easily handled in the field in all kinds of weather. Unfortunately few, if any, maps can meet all these requirements.

Developing a mapping system for the sole purpose of field use is rarely done due to the costs involved. However, by using various types of available maps and adapting them to fit most of our needs, it is possible to provide collection system operators with mapping systems that are functional. Common types of maps that are useful include:

- Area Maps. Most agencies, regardless of size, will compile a map showing how the entire area is served by its wastewater collection system. Such maps usually outline only major components including major trunk lines, pumping stations, treatment plants and major thoroughfares, and the relationship of these parts to communities, districts, watercourses and streams.

- Master Maps. The size of your service area will determine how many maps you will need. A small system may require only one map. Large areas obviously will need more maps. A master map may be needed for each community or district within the whole area. Master maps should be drawn on a scale capable of indicating at least the approximate location of the collection system (including manholes) in relation to all streets and properties within the area.

- Section Maps. These become available whenever master maps are divided into sections. They should be drawn to a scale that will allow more detail and be of a size practical for use in the field. The scale should provide ample space to clearly show identification numbers, footage and pipe sizes.

  The master and section maps may be combined for small and medium-sized collection systems.

- Base Maps. In most areas, the county assessor or real estate department maintains an accurate and up-to-date mapping system. It is their responsibility to keep accurate, legal records on each individual parcel of property. Usually their maps are divided into pages and the pages are bound into books. Base maps such as these provide a scale adequate for detailed information and include recorded right of ways, street names and widths, property identifications and sizes. These map books, being of a size suitable for field use, are ideal for a maintenance department to adopt and adapt as the basis of collection system maps. Tracings of these maps usually can be obtained from the originating department and you can overlay your collection system on these tracings.

  Alternatively, using digital aerial orthophotos will provide you with a base map that is geographically accurate and allows you to literally see your manholes on the photo (100 scale or better), allowing you to draw or digitize your system directly from the photo. To give you a head start, you can have your existing paper sewer maps scanned into electronic format which you can then "stretch" electronically with a CAD computer program to fit over the base map photo.

12.63 Mapping Coordinates

All states have developed a state coordinate system, usually known as the "state plane system." Surveyors developed the state coordinate system for the purpose of connecting all their surveys, such as land surveys and highways surveys, to the same reference stations. Should monuments which mark land corners or other reference points ever be destroyed, they may be readily and accurately reestablished on the basis of their coordinates and a new survey originating at other stations in the network.

If all utility agencies (water, wastewater, gas, telephone, electricity) identified the location of all the important points of their underground facilities on the same coordinate system, our jobs requiring excavation would be much easier with respect to the location of other utilities. Even after floods or hurricanes, or during periods of heavy snow, a grid coordinate system would help locate underground utilities and expedite the return of utility systems to service. A goal for every wastewater collection agency could be to have their engineer or surveyor develop and implement a plan for all utility agencies to map their underground systems using the same grid coordinate system.

Whether an agency is part of a state coordinate system or a local system, the procedures for identifying and locating facilities are similar. Both systems use two sets of parallel straight lines running east and west and also north and south (Figure 12.1). The resulting network is called a GRID. Grids are bordered on the south by the AXIS OF X and on the west by the AXIS OF Y. Distances along both axes are measured in feet. The point of intersection of these axes is called the ORIGIN OF COORDINATES.

The position of a manhole or other facility represented on the grid can be identified by stating the two distances termed COORDINATES. One of these distances, known as the X-COORDINATE, gives the position in an east-and-west direction (Distance X' on Figure 12.1). The other distance, known as the Y-COORDINATE, gives the position in a north-and-south direction (Distance Y' on Figure 12.1).

Since surveyors have already established state grid systems in every state, wastewater collection systems can be identified with the state system by surveyors or civil engineers. Other utility agencies such as water, gas, electricity, telephone, drainage, roads and streets, may or may not be using your state grid coordinate system; but their facilities, if shown on collection system maps, can be identified by the same coordinate system used for the maps.

Many cities and counties have determined the coordinates of the intersection of street and road center lines. If the distance and direction of a manhole from a center line intersection is measured, the coordinates of the manhole can be calculated easily. For example, Figure 12.1 shows that the coordinates of an intersection of two street center lines are 828,376.24 and 106,249.01 and the measured distance of a manhole from the intersection is 8.25 feet at an angle of 30°-15' from the east-west center line. Knowing this information, the x and y distances of the manhole from the intersection can be calculated in the manner shown below using a pocket calculator.

Distance, x, ft	= (8.25 feet) (cos 30°-15')
	= (8.25 feet) (0.8638355)
	= 7.13 feet
Distance, y, ft	= (8.25 ft) (sin 30°-15')
	= (8.25 ft) (0.503774)
	= 4.16 feet
x coordinate of manhole	= 828,376.24 + 7.13
	= 828.383.37 feet
y coordinate of manhole	= 106,249.01 + 4.16
	= 106,253.17 feet

If every utility agency would use the state grid coordinate system, key points on any agency's underground facilities could be located on maps by coordinates and their separating distances computed. With everyone using United States Geological Survey (USGS) benchmarks for elevations, the elevations of underground facilities could be computed, too.

12.64 Examples of Mapping

Because of their scope and the many details needed in any mapping system, it is difficult by mere descriptive words to effectively illustrate the makeup of a good map. Included are three samples that should provide a clearer picture of information required on maps.

MASTER MAP

Figure 12.2 illustrates a small section of a master map that was drawn to a 1" = 800' scale. The total size of the entire map measures 4' x 4' and depicts one sanitation district serving a population of 55,000 people. If no other map was available, it could be used by collection system operators in the field. This map would, however, have some disadvantages. Its size makes it difficult to use. In the field it would be worn out by continued rolling and unrolling or repeated folding and unfolding. Frequent replacement would probably be necessary. The scale makes it difficult to read and provides too little space for necessary details and identifications.

SECTION MAP

Figure 12.3 illustrates a portion of one sectional map of a master map that was divided into 24 sheets. Each sectional map measures 2' x 3'. Drawn on a scale of 1" = 400', it has ample room for more detail and is easier to read. Used a map at a time, it could be suitable for field use. It would, however, be subject to the same handling problems as the master map.

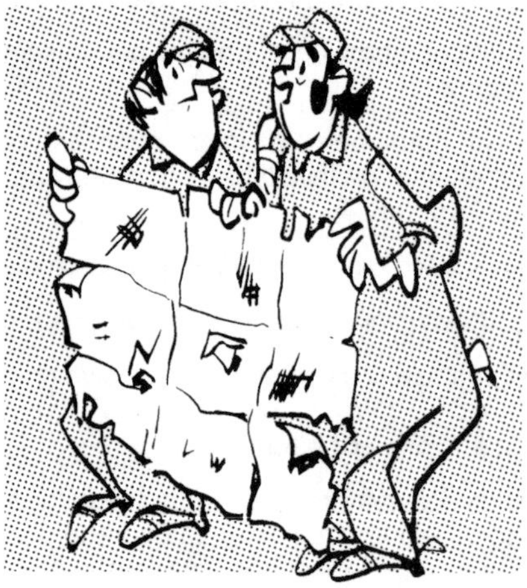

BLOCK MAPS

Figure 12.4 is the most detailed of the three sample maps. Drawn to a scale of 1" = 100', it includes almost all the desirable items outlined in Section 12.61, "Information on Maps." Consult the legends shown in Figures 12.5 and 12.6. They will enable you to interpret the information included on the block map. The actual size, 12" x 18" when bound into books, makes it ideal for field use. Protected by hard covers, it is very durable and will last for years during everyday use in the field. Damaged or updated pages can be replaced on a page-by-page basis.

The example shown here is one page of one book. The complete set totals 128 books, covers a county-wide collection system of 1,500 miles of sewers, and serves a population of 400,000. Map books can be easily carried and safely stored in one compartment of a utility truck commonly used by maintenance departments. Although block maps may contain more information than needed to perform a specific task, they do provide important information for the overall operation and maintenance of a wastewater collection system.

12.65 Preparation and Revision

Preparation and revision of maps for the operation and maintenance of wastewater collection systems are usually

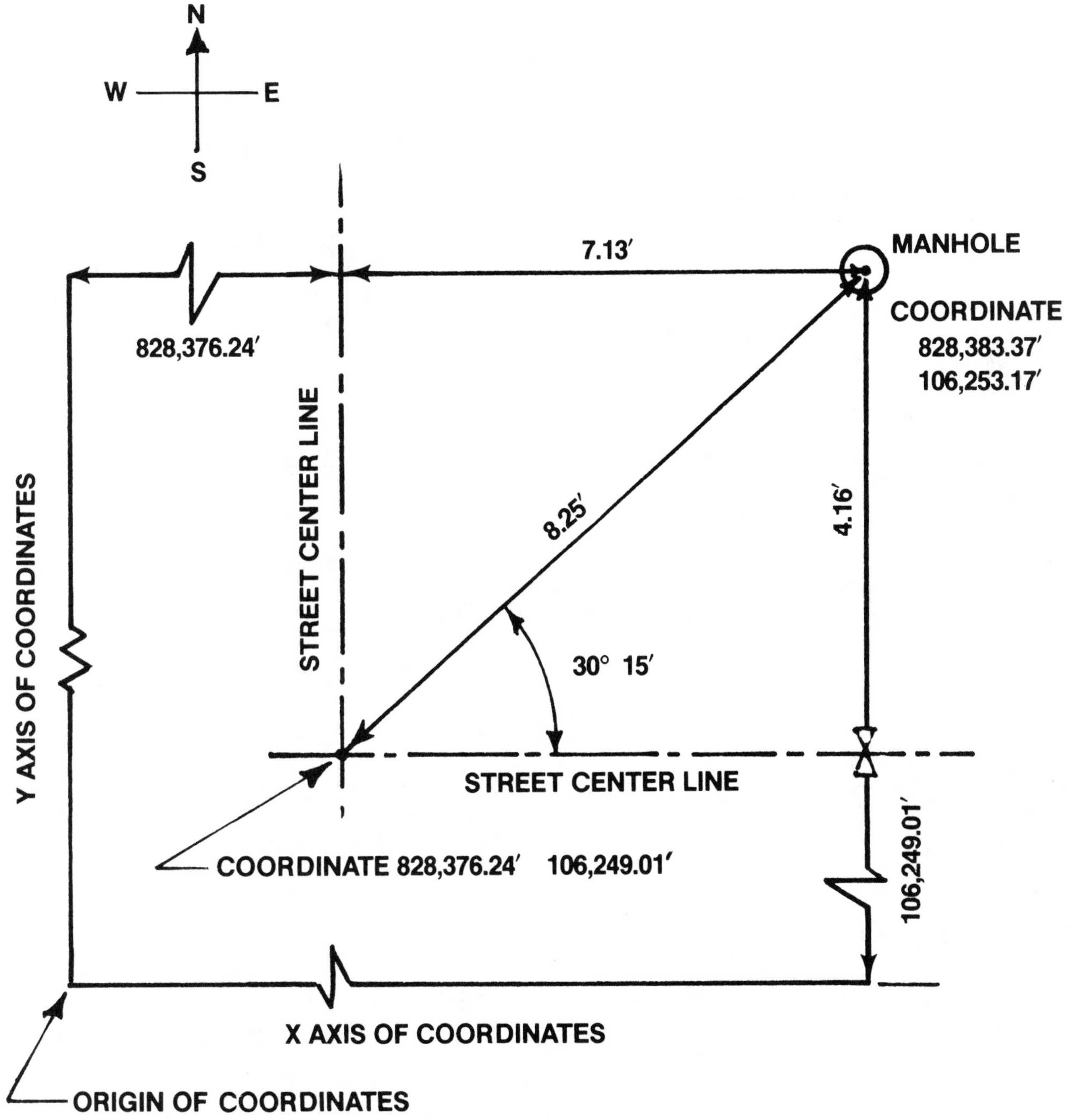

Fig. 12.1 Grid coordinate system

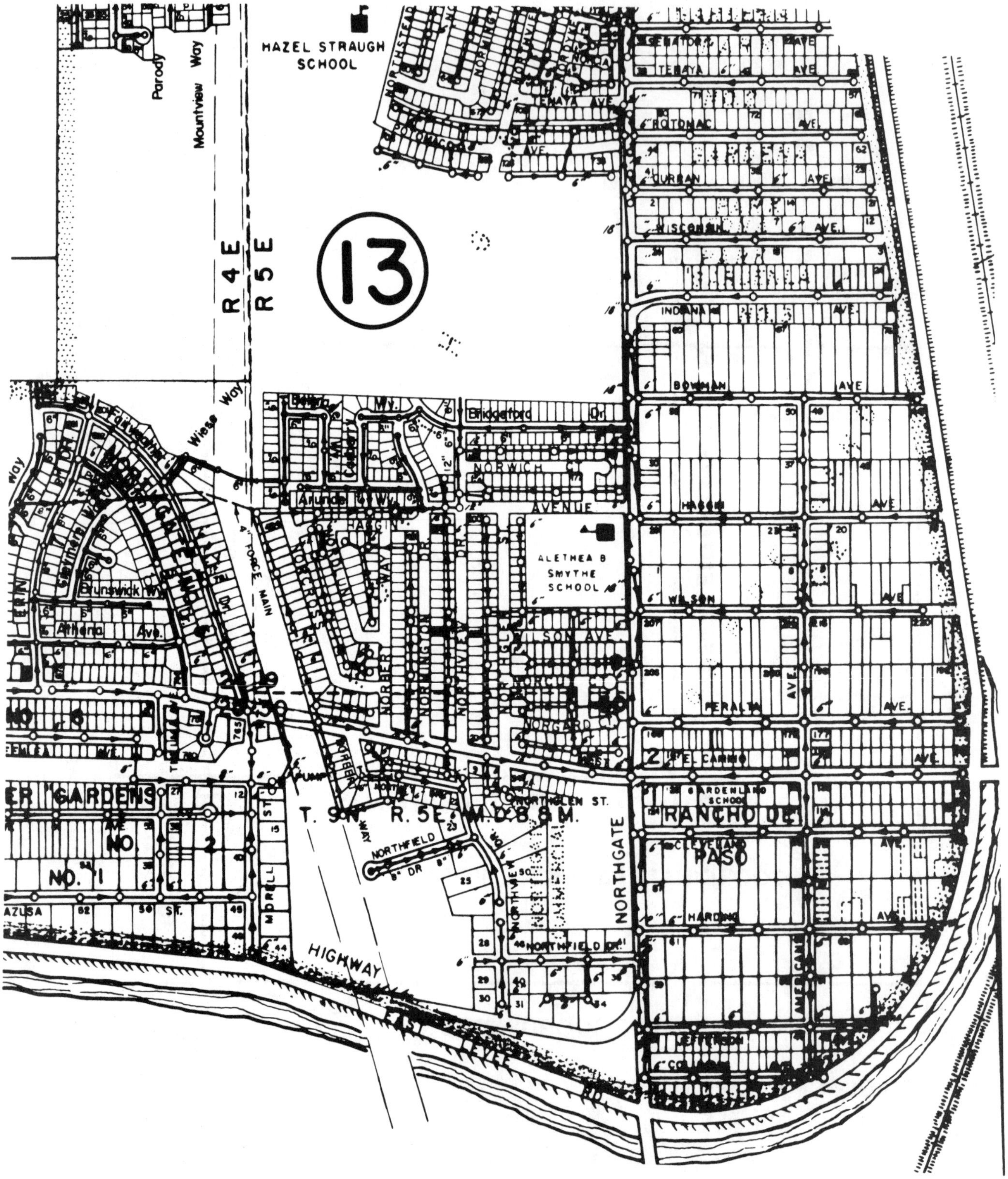

Fig. 12.2 Master map
Scale: 1" = 800' before reduction

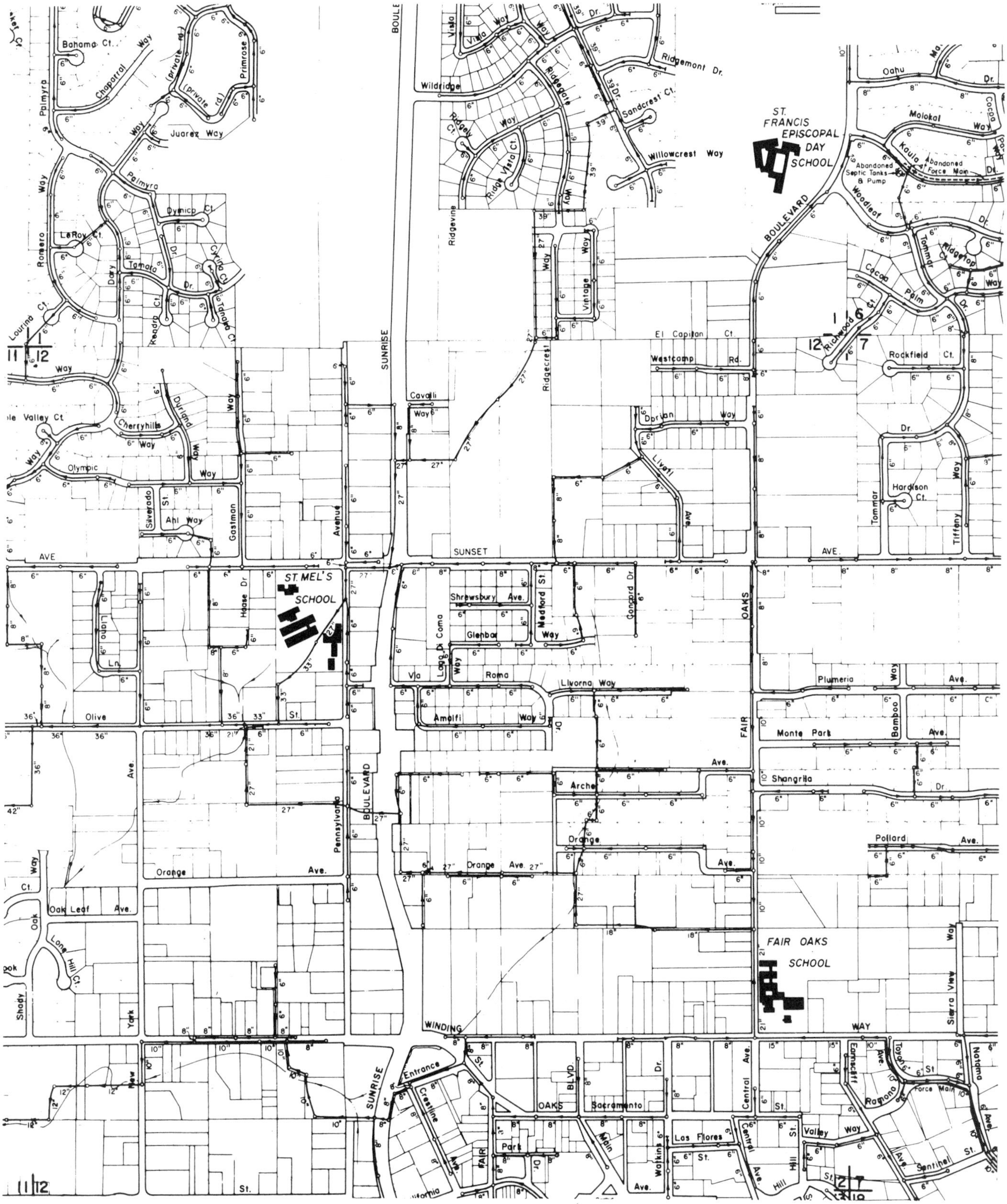

Fig. 12.3 Section map

Scale: 1" = 400' before reduction

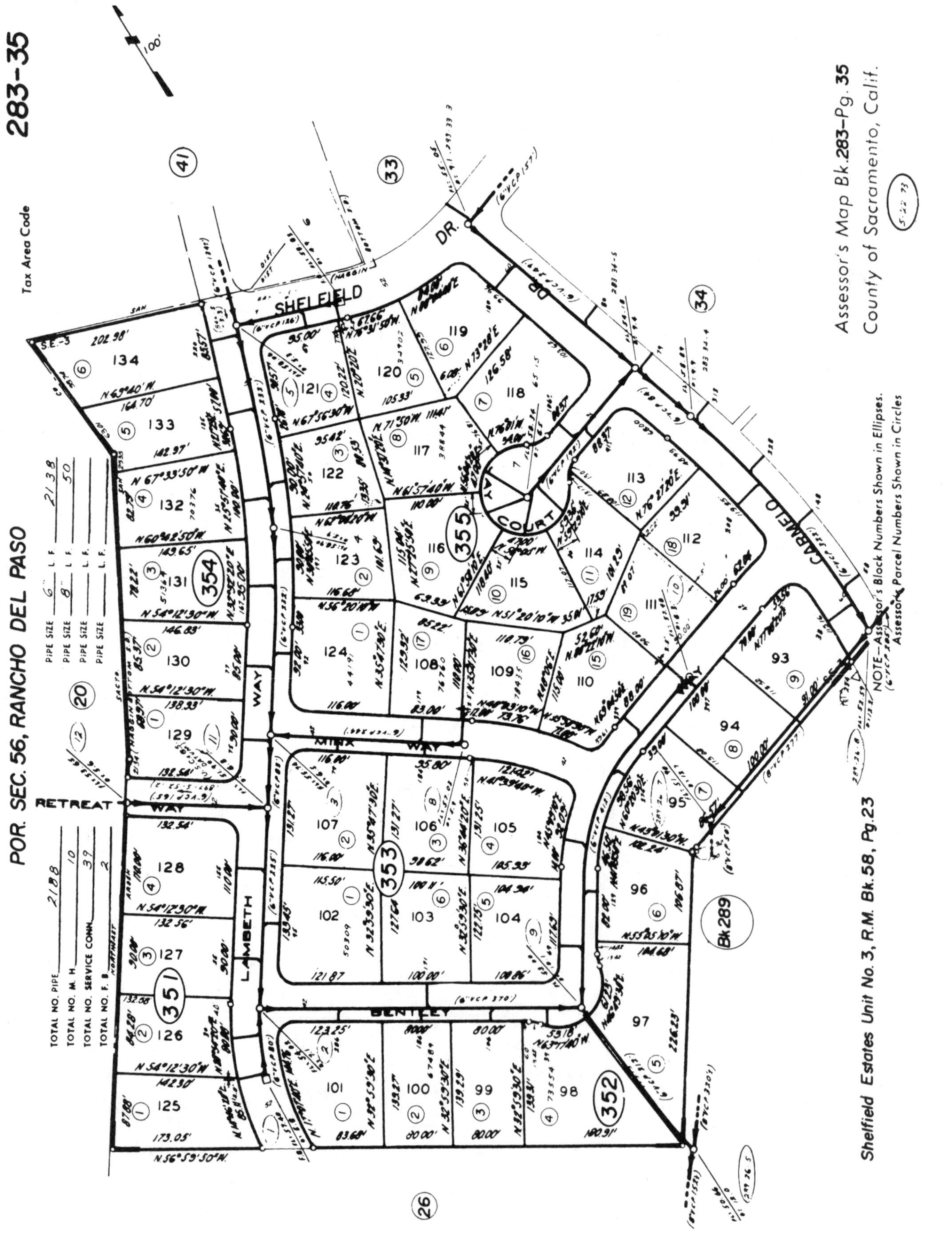

Fig. 12.4 Block map
Scale: 1" = 100' before reduction

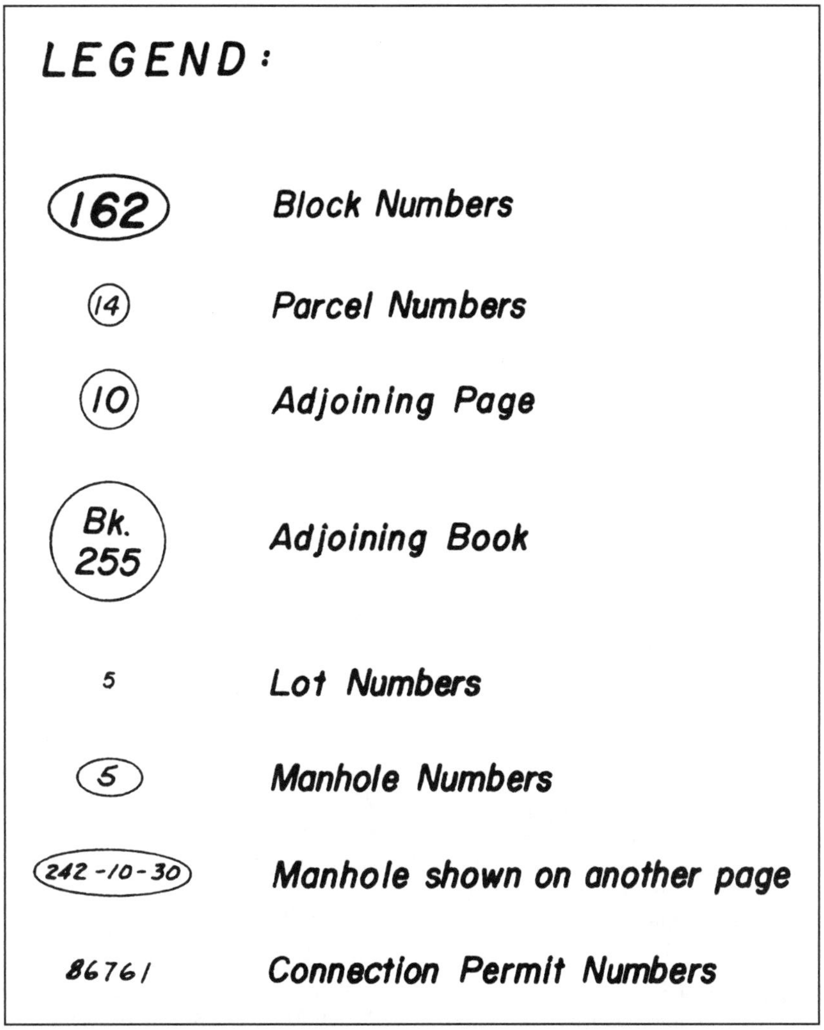

Fig. 12.5 Block map numbering legend

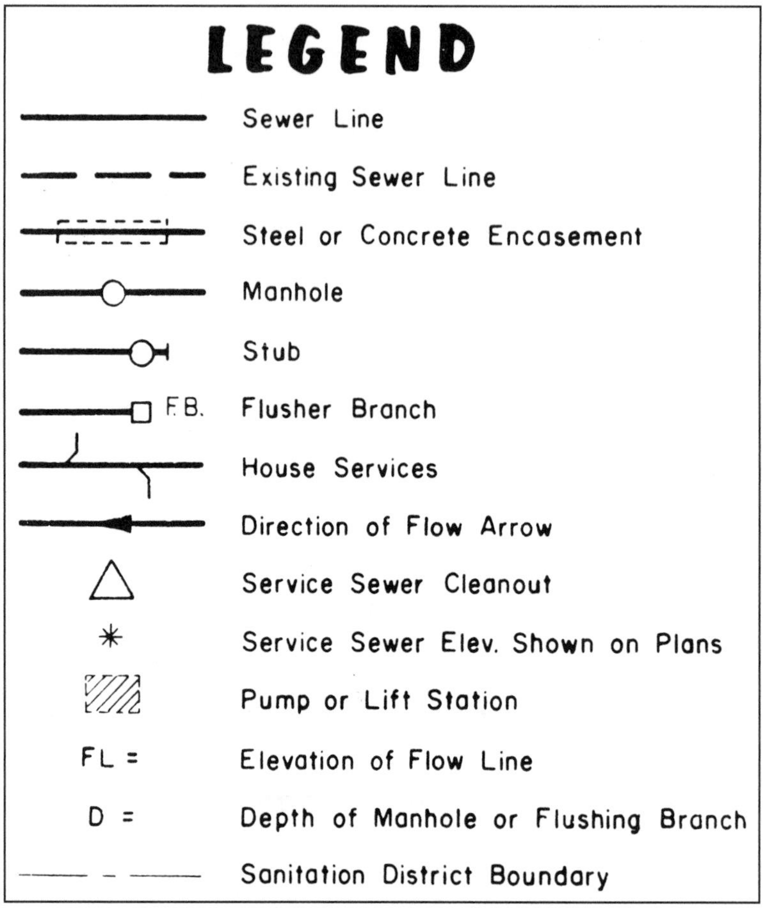

Fig. 12.6 Block map symbols

done by the engineering department of an agency. However, it is essential for the superintendent to be involved in this process to ensure a product that will aid the efficient operation and maintenance of the system.

12.650 Preparations

Property maps described previously are used as the basis of the collection system maps. As-built plans or record drawings are used to obtain the collection system information that is drafted onto the property maps. If as-built plans are not available, field surveys will have to be made to determine the information needed for mapping the system.

An agency with multiple functions, such as a city, may have one series of maps that provides comprehensive information on wastewater collection, storm drainage, water and other systems operated and maintained by the agency. As discussed above, computer and aerial photographic technology is available to develop mapping systems. Although the initial cost of computerizing hard copy and other mapping data can be high, there are many advantages to a computerized mapping system, including long-range savings for map additions and revisions and the ability to vary the form of reproduction to meet special needs.

12.651 Revisions

To remain functional, a mapping system must be kept as current as possible. Accurate and up-to-date maps can reduce time-consuming and costly mistakes. This section describes some suggested methods for keeping maps current.

CHANGES IN COLLECTION SYSTEM

When new sections of sewer, manholes or other structures are built and placed in service, update maps to show new facilities. If sewers or structures settle, slip sideways or tilt, they are usually repaired and restored to their original conditions and locations. During repairs, record any changes you make to improve facility operation or to stabilize an installation.

ANNUALLY

Updated maps should be printed at least once a year. Reprinting more often than annually would probably not be economically feasible. However, the process of gathering information needed for the annual reprinting should be a continuous one and conducted throughout the year as suggested below.

RECORDING CHANGES

- With tracings. When tracings are immediately available, changes should be recorded on the tracings as soon as they occur throughout the year.
- Without tracings. A suitable method of recording and filing changes as they occur should be implemented throughout the year. One method is to note the changes on the appropriate map page and file for recovery when needed. Changes on the tracings can then be made when the time comes for updating and printing of maps.
- With computers. Computer graphic techniques allow for changes to be easily made and new maps to be printed whenever needed.

MAPPING PERSONNEL

A mapping system of any significant size will require trained personnel who are directly responsible for keeping maps current and accurate. You will need to establish a definite system for funneling map change information to the mapping personnel. Skilled drafting or computer personnel are essential and management must make sure that all changes are recorded as soon as possible on the maps.

FIELD PERSONNEL

Even in the most efficient map updating programs, some changes will be overlooked or go unreported. This happens primarily because of the many details to be covered and the vast number of people involved. Field personnel can help in the following ways:

- Collection system operators. As the primary users and chief benefactors of a good mapping system, maintenance operators can be easily trained to report any discrepancy they encounter or changes they make in their daily activities. Supervisors should develop a specific system, with appropriate forms, to ensure these reported discrepancies or changes are promptly forwarded to the mapping section or computer section. If map sheets are readily available, the operators can be encouraged to make corrections directly on their copy of the map which can then be sent to the drafting department for implementation.
- Construction inspectors. During construction of collection system facilities, changes may be made in existing facilities or discrepancies found. Inspectors should be instructed and required to note such information on their "as-built" plans or record drawings. This information is used for future updating of maps.

FIELD CHECKING

If you cannot have a formal system or set schedule for updating maps, periodic field checking may be practical. Properly trained field service crews can accomplish this on a fill-in basis during periods with few service requests. Field checking is especially beneficial in smaller collection systems.

QUESTIONS

Write your answers in a notebook and then compare your answers with those on page 547.

12.6F Which type of map contains the most details regarding a wastewater collection system?

12.6G Why should maps be updated?

12.6H How can maps be kept up to date?

12.66 Microfiche Copies

A superintendent of a wastewater collection system may want to consider the use of the microfiche technique for making copies of collection system maps more widely available for office and field use. Using this technique, maps are photographed and reproduced on film in micro size; many block maps are grouped in a logical manner on a 3- by 5-inch card. Any part of the mapping system can be used by placing the appropriate card in a microfiche viewer which illuminates the transparent fiche and magnifies the print to a readable size. Different types of viewers are powered by sunlight, batteries, 110-volt electrical systems or the electrical system of a vehicle.

The advantages of microfiche copies of maps include a low reproduction cost, ease of handling, less storage space and they do not deteriorate with repeated handling. On the other hand, the maps may be difficult to read under adverse conditions, referring from one map page to an adjoining page for information can be awkward, and changes to be made cannot be noted directly on a map.

12.67 Maps as Maintenance Records

In some instances maps provide an excellent method of recording maintenance activities.

SMALL COLLECTION SYSTEMS

Sufficient personnel may not be available to sustain an elaborate recordkeeping system. A separate set of maps kept in the office for record purposes only can be useful in recording the following:

- Problem areas. Color codes can be used to designate specific problems such as roots or grease.
- Maintenance activities. Dates, type of maintenance performed, areas involved and production footages can be easily recorded on these maps.
- Pins can be used to indicate repeated problems in specific areas.
- Schedules. Proposed maintenance schedules, including due dates, can be indicated for specific areas through the use of flags.

MEDIUM-SIZED COLLECTION SYSTEMS

Again, sufficient personnel may not be available to sustain detailed card record systems. A combination of card files and maps for records may be effective. Information adaptable for recording on maps might include:

- Area-wide maintenance. Maintenance activities conducted on an area-wide basis can easily be recorded on maps through the use of color codes.
- Concentration of problems. Using colored pins to mark trouble spots, maps will show at a glance if problems are concentrated in one or more specific areas.

QUESTIONS

Write your answers in a notebook and then compare your answers with those on page 547.

12.6I What are some of the advantages of using microfiche copies?

12.6J How can maps be used as a maintenance record?

12.6K How can problem areas be shown on a map?

Please answer the discussion and review questions next.

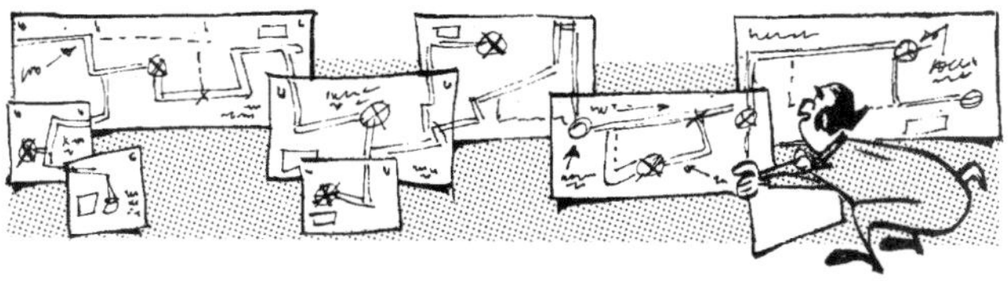

DISCUSSION AND REVIEW QUESTIONS

Chapter 12. ADMINISTRATION

(Lesson 5 of 7 Lessons)

Write the answers to these questions in your notebook before continuing. The question numbering continues from Lesson 4.

24. Why are good maps needed for wastewater collection systems?
25. What information must be included on wastewater collection system maps?
26. Discuss the advantages of identifying the location of important points of underground utilities on a grid coordinate system.
27. Why do utility agencies use different types of maps (master, section and block maps)?
28. What can happen if maps are not kept up to date?

CHAPTER 12. ADMINISTRATION

(Lesson 6 of 7 Lessons)

12.7 USES OF COMPUTERS IN A WASTEWATER COLLECTION AGENCY

A major trend over recent decades has been the introduction of computers into businesses, industries and even the home. It is difficult to pick up a newspaper or magazine and not find another article about some new application of computers. You may already be using one either at home or at work. While many of us have not been comfortable with computers and have hoped that they would bypass us or just go away, we may not need to avoid them any longer. There have been changes in computers and computer programs over the last few years that not only have made them less expensive but have also made them easier to use. These continuing improvements are making it more desirable to use computers in the administration and operation of a wastewater collection agency. Some computer applications are unique to collection systems; others are used in virtually all businesses and industries. In this section we will describe how computers are now being used in wastewater collection agencies as a powerful tool to assist in: (1) administration of the agency, (2) daily operation and maintenance of equipment and system, and (3) recordkeeping.

Many collection systems have been and still are being operated and managed in an effective manner without the assistance of a microcomputer system and many of the records and reports illustrated later in this chapter can be produced manually. However, a microcomputer system can store data, process it, and produce information at a more rapid rate than any manual system. A microcomputer simply provides a collection system superintendent with an additional tool for the more efficient management of a system.

As you study this information about the applications of computers, you will often find that there is no clear distinction between computer systems used for administration and systems used for recordkeeping, or between O & M systems and recordkeeping systems. This is because computers can store and process vast quantities of data and it is no longer necessary to have separate files or separate reporting systems for each major function in an agency or company.

12.70 Management Information Systems

A management information system is a far-reaching term used to describe a computer system and its hardware and software components. The moderate cost of microcomputers and customized software make these systems available to most wastewater collection system agencies. Therefore, the superintendent of a collection system should, in most situations, be using microcomputers to facilitate an effective management information system.

12.700 Need

A management information system meets several needs. The most common are:

- Simplification and improvement of work planning and scheduling, including the integration of recurring and on-demand jobs,
- Measurement and tracking of workforce productivity (Figure 12.7), and
- Development of unit costs and measurement of the impact of resource allocation to various activities.

A collection system's superintendent and staff need precise and accurate information at the right time for taking effective action and making sound decisions concerning the operation and maintenance of the collection system. Decisions based on inaccurate or insufficient information may cause inappropriate actions, confusion, and reduce operator morale.

12.701 Definition

A management information system is illustrated in Figure 12.8. Collection system operators, following directives issued by superintendents and supervisors, perform operational and maintenance work on the collection system and prepare work reports. Work reports are then assembled into records by staff personnel and become the database for the information system. Records no longer needed for current information are removed from the database and stored in archives in case they may be needed in the future. Work performed by contract generates work reports that are also assembled into records and become a part of the database.

The database is used to produce planning and controlling reports by staff personnel in the form requested by management. Management uses these reports for their work including planning for and deciding how and when to operate and maintain a collection system. Collection system operators are informed of management's plans and decisions by the issuance of directives to them, thus completing the circular flow of information between management and collection system operators.

Informal two-way communication of information across organizational lines, on both an interpersonal and group basis, between collection system operators and management should be encouraged by a superintendent. This type of information transfer is shown on Figure 12.8 as lines of direct communications between operators and management.

Data processing takes pieces of information independent in nature and unlimited in number (usually in the form of records) and produces useful information by arranging it into meaningful knowledge.

Key factors involved in the development of an effective management information system include:

1. Work reports must be complete in order to have all the data necessary to prepare all the information that may be needed for an intelligent decision.
2. Reports and records must be relevant to the objectives of management or valuable time will be wasted in preparing useless reports and processing data.
3. Planning and controlling reports must be concise to facilitate a proper decision.
4. All aspects of the system must be performed in a timely manner in order for information to be available when a decision must be made.

12.702 Computer Applications for System Management

Computers use microelectronic technology to store and arrange a massive amount of data at a very rapid rate to produce useful information as directed by computer programs called software. The quality of the information produced by computers is entirely dependent on the quality of data put into the computer and the software directing its processing. "Garbage In, Garbage Out," is a common expression in the computer industry which expresses what happens when poor quality or irrelevant data are used. Information produced by a computer can either be read from a video screen or a hard (paper) copy can be produced by a printer or plotter.

A microcomputer and its software can be applied to a management information system in the following manner:

- Work reports prepared by collection system operators are used by staff personnel to gather data that is entered into the computer's memory to create a database using database management software. Entries are made using instructions appearing on a video screen and a typewriter keyboard. In some collection agencies, operators themselves enter work report information into the computer database using small, portable computers (called laptop computers) or even smaller hand-held electronic devices, pocket pen-base mobile data collection devices, and notebooks (see Section 12.7041).
- On requests entered into the computer by management or staff (using instructions provided by a combination of data management, spreadsheet and word processing software), the computer either displays the desired information reports on the video screen or produces a report in printed form. It is also possible to view information on the video screen and then decide if it should be printed or saved on a computer disk as a backup in the event your main computer system, computer server, or hard drive should fail.
- A computer system can also be used to produce printed directives to operators in the same manner information reports are produced.

Several software programs have been designed specifically to assist agency staff in the management of wastewater collection systems. These programs are produced (and copyrighted) by consulting engineering firms and software companies. The programs can be purchased from these firms and companies and they will assist a user in the installation and maintenance of their computer programs.

DAILY PROJECT TIME SHEET - WASTEWATER PROGRAM

NAME: ***EMPLOYEE NO.:*** ***DATE:***

HRS	PROJECT CODE	DESCRIPTION	HRS	PROJECT CODE	DESCRIPTION
	GM100	STORM MANAGEMENT		SS600	GRAVITY LINE MAINTENANCE
	GM102	HEALTH & SAFETY COMPLIANCE		SS610	TELEMETRY MAINTENANCE
	GM103	ENVIRONMENTAL COMPLIANCE		SS620	FORCE MAIN MAINTENANCE
	GM420	SURFACE WATER TREATMENT		SS625	SERVICE LINE MAINTENANCE/REPAIR
	GM440	BACKFLOW PREVENTION		SS630	LS 59 MAINTENANCE
	GM460	URBAN WATER MANAGEMENT PLAN		SS635	DE-RAG PUMPS/LIFT STATIONS
	GM500	BAAQD EMISSION/ODOR CONTROL		SS640	LS MAINTENANCE (ALL OTHERS)
	GM600	BCDC (LEVEE, ETC.)		SS645	MANHOLE MAINTENANCE/REPAIR
	GM700	HAZARDOUS MATERIAL		SS650	GENERATOR MAINTENANCE
	SS100	STAFF MEETING		SS660	EQUIPMENT MAINTENANCE
	SS125	SAFETY MEETING		SS670	FACILITY MAINTENANCE
	SS126	CREW MEETING		SS680	AIR SCRUBBER MAINTENANCE
	SS200	VALVE MAINTENANCE		SS690	PUMP/MOTOR MAINTENANCE
	SS220	AIR RELEASE VALVE MAINTENANCE		SS800	PURCHASING
	SS300	MATERIAL HANDLING		SS900	CCTV INSPECTIONS
	SS320	SCRAP METAL RECYCLING		SS920	TRAINING
	SS350	H_2S ABATEMENT		SS950	HOUSEKEEPING
	SS402	SUPERVISING STAFF		SS998	MISC/SPECIAL PROJECTS
	SS404	TIMESHEET VERIFICATION		SS999	FRONT DESK DUTIES
	SS406	COMMITTEE MEETING		MT910	WASTEWATER SPILL MITIGATION
	SS408	CUSTOMER RELATIONS		MT990	LUNCH/SPECIAL EVENTS
	SS410	DATABASE MAINTENANCE		MT925	JURY DUTY
	SS412	BUDGET PREP/REVIEW		MT930	COURT DUTY
	SS414	DEVELOP/MAINTAIN SOPS		MT935	CLASS B/PHYSICAL EXAM
	SS416	BARGAINING UNIT ACTIVITY		WD990	TRNG, SEMINARS/WRKSHPS (SUPVR)
	SS422	PROCESS REQUEST/JOB ORDER	HRS	WR NO.	WORK REQUEST DESCRIPTION
	SS424	MONITOR MATERIAL/INVENTORY			
	SS440	GREASE TRAP INSPECTION			
	SS460	USA MARKOUT			
	SS560	HAZ WASTE/MATERIAL MGMT.			

TOTAL HOURS ____________

15 MINS = .25, 30 MINS = .50, .45 MINS = .75, 1 HR = 1.

ENTERED BY ________ DATE ____________________

Fig. 12.7 Example of employee daily time sheet

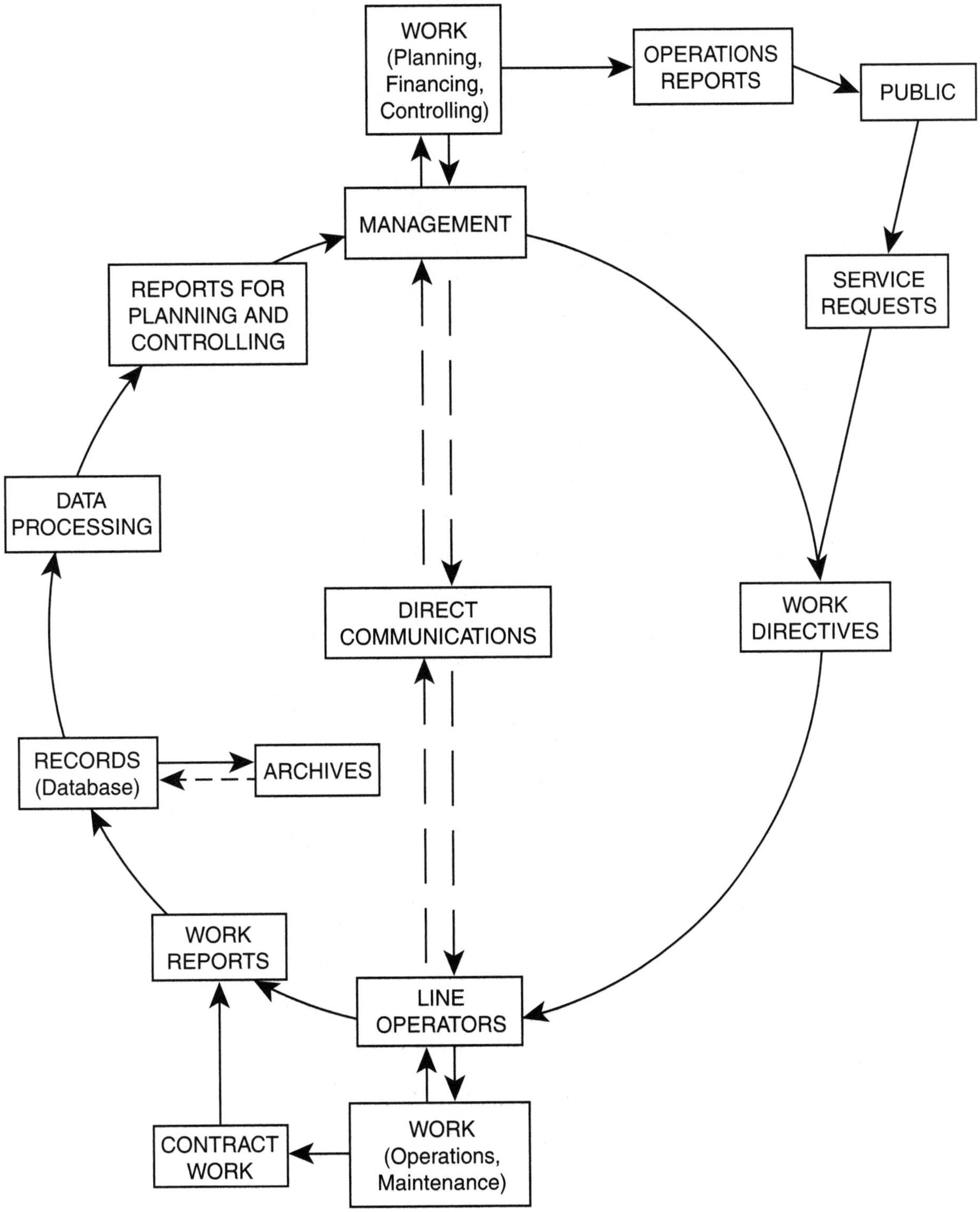

Fig. 12.8 Management information system

QUESTIONS

Write your answers in a notebook and then compare your answers with those on page 547.

12.7A What happens to current information no longer needed in a database?

12.7B What does data processing do with information?

12.7C The quality of information produced by computers depends on what factors?

12.703 Work Reports

Collection system operators performing work activities account for these activities by making daily work reports. Operators use forms prepared by management in order to produce the data relevant to the agency's management information system.

12.7030 NEED FOR WORK REPORTS

Work reports are needed to develop databases, or records, that are processed to produce the following categories of planning and controlling reports used by management:

- Payroll,
- Production,
- Work costs,
- System inventory,
- Main line maintenance history,
- Service line maintenance history, and
- Main and service line repair history.

12.7031 TYPES OF WORK REPORTS

The types of work reports will vary with the type of management information system used by a collection system agency. However, report forms should be designed to provide the data essential for the production of planning and controlling reports previously described since they are needed for the efficient management of the collection system.

A typical computer software package that can be used to track the maintenance activities in a collection system and provide the information needed by management to control the costs and operation of the agency would have the following capabilities:

- Inventory of main line segments, building sewers, manholes, catch basins/storm drains, and septic tanks;
- Preventive maintenance scheduling;
- History of maintenance activities;
- Recording and retrieval of information obtained from TV history reports;
- Reports on scheduled, unscheduled, and projected work orders;
- Work orders generated for preventive and unscheduled maintenance activities; and
- Labor, equipment, and material costs associated with work orders.

The following work report forms are the types used by one large collection system agency that uses microcomputers for data processing. (Two other examples of computer programs designed to manage collection systems are described in Section 12.71, "Applications of Computers for System O & M.")

MAIN SEWER CONSTRUCTION

The form, *CONSTRUCTION DATA SHEET,* Figure 12.9, is used by a construction crew leader when working on main line construction and by an inspector when the construction is performed by a contractor. Most of the requirements for information are self-explanatory. A crew leader initials the "Coded By" space and the clerk entering the data into a computer initials the "Entered By" space. The construction data are inserted in code form using the table of codes printed on the bottom of the form. These code designations will usually be the same codes used for entering the data into a computer. The reverse side of the form, Figure 12.10, when completed by the crew leader, provides all the data needed to compute and report the cost of each construction project.

MAIN SEWER MAINTENANCE

The leader of a main sewer maintenance crew reports the daily work activities of the crew using the *MAIN SEWER MAINTENANCE DATA* form, Figure 12.11. This form is similar to the main sewer construction form. Note that the codes indicate the type and frequency of maintenance recommended by the crew. This information is subsequently used by a superintendent or supervisor to prioritize and schedule future maintenance and repair of main sewers.

MAIN SEWER REPAIR

The leader of the crew making repairs on a main sewer reports this work using the reverse side of the *MAIN SEWER MAINTENANCE DATA* form shown in Figure 12.12. To facilitate use of the form, it is made similar to the previous forms and uses similar coded entries. This form is also used by crews to respond to service requests from homeowners.

STRUCTURE MAINTENANCE

Maintenance work performed on a collection system's manholes, cleanouts and special structures is reported by a crew leader using the *STRUCTURE — MAINTENANCE DATA* form shown in Figure 12.13. Note that this form is used to report entries into manholes, safety conditions and any odors that may require further investigation. The condition of a manhole frame and cover (F/C), barrel and cone (B/C), channel and steps are rated from "OK" to "Severe" to provide information for scheduling of future repairs. As with the previous report forms, the time, material and equipment reporting form is printed on the reverse side.

STRUCTURE REPAIR OR ABANDONMENT

If a manhole, cleanout or special structure is repaired or abandoned, the leader of the crew performing the work reports the activity using the *STRUCTURE — REPAIR OR ABANDON DATA* form, Figure 12.14. Note how the use of codes allows a great amount of information to be reported on

UNION SANITARY DISTRICT

Collection System Division

Page ______ of ______

Coded By: **Date:** **CONSTRUCTION DATA SHEET** **Entered By:** **Date:**

Structure Identifier U/S D/S	Diam. (in.)	Mains Material Type	Ops. Type	Ops. Type	MH Cond.	Address	City F-N-U	Computer Work Order #	Job #	Cost

I.D. Number of Crew ___, ___, ___, ___, ___ Vehicles ___, ___, ___, ___, ___ Section ______

I.D. Number of Crew ___, ___, ___, ___, ___ Vehicles ___, ___, ___, ___, ___ Section ______

U/S D/S Special Condition Description

Codes

MMN Sewer Main Rprs
101 Spot Repair Main
102 Repair Main/Lat Con w/Wye or Tee
103 Instl New Lat Conn
104 Remove Lat Con/Rpr Main
105 Grout Main Leak
106 Excavate Trench
107 Backfill Trench/Steel Plate
108 Pave Trench
109 Abandon Main
110 Locate Main
111 Smoke Test Main
112 Instl New Sewer Main

MMN Structure Mtnce
301 Wash Manhole
302 Silence Noisy MH/River Cvr
303 Expose MH/Riser Cvr Slurry Seal
304 Weed Abatement Around MH/Riser
305 Install USD PostMarker
306 Install False Bottoms
307 Remove False Bottoms
308 Install Plugs
309 Remove Plugs
310 Replace MH/Riser Cvr
311 Confined Space Test Clear
312 Confined Space Test Not Clear

MMN Lateral Rprs
401 Locate Lateral
402 Dye Test Lateral
403 Install Backflow Preventer
404 Repair Lateral
405 Relocate Lateral
406 Install Lateral
407 Install Cleanout
408 Pave Lateral Trench
409 Rod Lateral
410 Cap Off Lateral

MMN Structure Rprs
601 Locate MH/Riser
602 Install Ref Points
603 Remove MH/Riser Cst/ Instl Steel Plate
604 Install MH/Riser Cst & Adjust to Grade
605 Adjust MH Casting to Grade
606 Adjust MH w/Riser Rings to Grade
607 Pave MH & Risers
608 Measure MH & Riser Depth
609 Install Steps
610 Rehab w/Protective Coating
611 Grout MH Leak
612 Rechannel Existing
613 Channel New MH
614 Instl New MH/Riser
615 Abandon MH/Riser
616 Install Stub
617 Instl Drop MH
618 Drill Test Holes MH/Risers
619 Clean Debris Not Due PMP
620 Replace Cone
621 Replace Barrel
622 Install Bulkhead
623 Remove Bulkhead

MTP Mains Mat Type
001 VCP
002 PVC
003 CIP-DIP
004 ABS/Truss Pipe
005 RCP
006 Polyethylene Liner
007 Insituform Liner
008 Corrugated Pipe

HCD MH Condition
001 Good
002 Fair
003 Poor
004 Surcharge
005 Unable to Locate
006 Noisy
007 Hydrogen Sulfide
008 Methane
009 Low Oxygen Level
010 Grease
011 Debris
012 Needs Grouting

Fig. 12.9 Main sewer construction report

LOCATION	St. ______ City ______ M.H. No. ______	LOCATION	St. ______ City ______ M.H. No. ______	A B 1 2 3 4 C D 1 2 3 4 E F 1 2 3 4
CHARGE TO	Job No. ______ W.O. ______ USD (Circle If Yes)	CHARGE TO	Job No. ______ W.O. ______ USD (Circle If Yes)	

Quantity	Size	Unit Cost	Total	Quantity	Size	Unit Cost	Total	Description
can				can				Allcrete
yd.				yd.				Readymix Concrete
yd.				yd.				Handmix Concrete
sak				sak				Mortar
ton				ton				Class II, 3/4 Agg Base
ton				ton				Drain Rock
ton				ton				Trench Backfill
ton				ton				Crushed Rock
ton				ton				3/8 & 1/2 Asphalt
sacks				sacks				Cold Patch
ft.				ft.				ABS Pipe
ft.				ft.				C-900 PVC
ft.				ft.				SDR 35 PVC
ft.				ft.				V.C.P.
ft.				ft.				D.I.P. (Ductile)
ea.				ea.				Stoppers
ea.				ea.				Couplings
ea.				ea.				Bushings
ea.				ea.				Wyes
ea.				ea.				Tees
ea.				ea.				Bends 1/8 or 1/16
ea.				ea.				Water Stop Gasket
set				set				Manhole Frame & Cover
ea.				ea.				Manhole Steps
set				set				Riser Frame & Cover
ea.				ea.				Barrel Section, Step Y/N
ea.				ea.				Grade Rings
ea.				ea.				Concentric Cone, Step Y/N
ea.				ea.				Eccentric Cone, Step Y/N
ea.				ea.				Bricks
ea.				ea.				District Markers
ea.				ea.				Ram Neck

PROJECTS:							TOTAL
Time:	H:MM	H:MM	H:MM	H:MM	H:MM	H:MM	H:MM
EMPLOYEE NOS							
VEHICLES							

NOTES

Remarks:

EQUIP. CHECK OUT

EQUIPMENT	QUANTITY	EQUIP. NO.	INITIALS CK'D. OUT	INITIALS CK'D. IN
Ferret				
Gas Detec.				
Radios.				
Pipe Cutters				
Water Meter				
M. scope				

Fig. 12.10 Time, material, and equipment report

UNION SANITARY DISTRICT — Collection System Division — Page ____ of ____

Coded By: Date: **MAIN SEWER MAINTENANCE DATA** Entered By: Date:

Structure Identifier U/S	D/S	Type of Operation 1	2	Operations Time H:ths 1	H:ths 2	Debris Type	Pounds	Special Condition	Line Ftge.	Wk #	Cost $

I.D. Number of Crew ___, ___, ___, ___, ___ Vehicles ___, ___, ___, ___, ___ Section ________

I.D. Number of Crew ___, ___, ___, ___, ___ Vehicles ___, ___, ___, ___, ___ Section ________

U/S D/S Special Condition Description

Codes

MMN CM
001 Hydro-jet 3 mos.
002 Hydro-jet 6 mos.
003 Hydro-jet 12 mos.
004 Hydro-jet 24 mos.
005 Hydro-jet 36 mos.
006 Hydro-jet 72 mos.
007 Ball 3 mos.
008 Ball 6 mos.
009 Ball 12 mos.
010 Ball 24 mos.
011 Ball 36 mos.
012 Ball 72 mos.
013 Tire 3 mos.
014 Tire 6 mos.
015 Tire 12 mos.
016 Tire 24 mos.
017 Tire 36 mos.
018 Tire 72 mos.
019 Scooter 3 mos.
020 Scooter 6 mos.
021 Scooter 12 mos.
022 Scooter 24 mos.
023 Scooter 36 mos.
024 Scooter 72 mos.
025 Kite 3 mos.
026 Kite 6 mos.
027 Kite 12 mos.
028 Kite 24 mos.
029 Kite 36 mos.
030 Kite 72 mos.
031 Porcupine 3 mos.
032 Porcupine 6 mos.
033 Porcupine 12 mos.
034 Porcupine 24 mos.
035 Porcupine 36 mos.
036 Porcupine 72 mos.
037 Swab 3 mos.
038 Swab 6 mos.
039 Swab 12 mos.
040 Swab 24 mos.
041 Swab 36 mos.
042 Swab 72 mos.
043 Root Control Chem
044 Root Control Mech
045 Rodder
046 Pest Control
047 Vacuum Debris
048 Washdown Mh/Riser
049 Noisy Mh Cover

MMN TV
201 Visual Insp of Main
202 Visual Insp of MH
203 Pre-60 Lin Cond.
204 Pre-60 Pre-Const.
205 Pre-60 Post-Const.
206 Pre-60 Root Control
207 Pre-60 Post Rt Cntrl
208 Pre-60 Storm Line
209 Pre-60 Lateral
210 Pre-60 Airtest
211 Pre-60 Smoke Test
212 Pre-60 Instl Flw Mtr.
213 Pre-60 Rmv Flw Mtr.
214 Pre-60 Insp Cave-in
215 Post-60 Line Cond.
216 Post-60 Pre-Const.
217 Post-60 Post-Const.
218 Post-60 Pre-Rt Cntrl
219 Post-60 Rt Cntrl
220 Post-60 for Accept.
221 Post-60 Year End
222 Post-60 Storm Line
223 Post-60 Lateral
224 Post-60 Airtest
225 Post-60 Smoke Test
226 Post-60 Instl Flw Mtr.
227 Post-60 Rmv Flw Mtr.
228 Post-60 Insp Cavein

MMN SWSP
501 Sewer Main Stop
502 Lateral Stoppage
503 WWSpill USD Minor
504 WWSpill USD Major
505 WWSpill Prvt Minor
506 WWSpill Prvt Major
507 WWSpill Cln by USD
508 WWSpill Cln Others

SDB LINE DEBRIS
001 Clear
002 Egg Shell/Grounds
003 Grease or Soap
004 Paper or Rags
005 Grit Silt/Snd Gravel
006 Foreign Object
007 Roots
008 Mud
009 Grease/Egg Shells
010 Roots/Grease
011 Others

Fig. 12.11 Main sewer maintenance report

Projects						Total	Equipment	Qty	Equip #	Time Ck'd Out	Time Ck'd In
Emp/Time	H-ths	H-ths	H-ths	H-ths	H-ths	H-ths	Water Meter				
							Gas Detec.				
							Radios				
							Ferret				
							M Scope				
							Remarks:				
Equip/Time	H-ths	H-ths	H-ths	H-ths	H-ths	H-ths					

Coded By: **Date:** **SERVICE REQUEST OWNER PROBLEM** **Date:**

Address	Street Name	City F-N-U	Comp. Problem Type	Opera-tion Time H-ths	Service Request Number	Action Taken	Spec. Cond.	Computer Service Request Number

I.D. Number of Crew ___, ___, ___, ___, ___ Vehicles ___, ___, ___, ___, ___ Section ____________

Address Street Name Special Condition Description

Codes

PRB Complaint Problem Type

001 Lateral Stoppage
002 Main Stoppage
003 Pump Station
004 Storm Drain Inlet
005 Storm Line
006 Bad Odor
007 Noisy MH/Riser Cover
008 Broken, Cracked or Missing MH/Riser
009 Cave-in
010 Broken Line
011 Refer to Others

PBD Action Taken

001 Lateral Stoppage/Owner's Problem
002 Main Stoppage Cleared
003 Pump Station
004 Storm Drain Inlet
005 Storm Line
006 Washdown/Deodorize
007 Noisy MH/Riser Cover Silenced
008 Broken MH/Riser Replaced
009 Cave-in/Secure Area
010 Broken Line/Repair
011 Agency Notified City of Fremont
012 Agency Notified City of Newark
013 Agency Notified City of Union City
014 Agency Notified PG&E
015 Agency Notified Pac Bell
016 Agency Notified ACFC
017 Agency Notified CALTRANS
018 Agency Notified ACWD
019 Repair for Contractor
020 Temporary Repair
021 Homeowner/Occupant Notified
022 Refer to Others

ACTIVITIES

001 Informational Meeting
002 S.E.T.
003 Pick Up Materials
004 Yard Maintenance
005 Equipment Maintenance

Fig. 12.12 Main sewer repair report

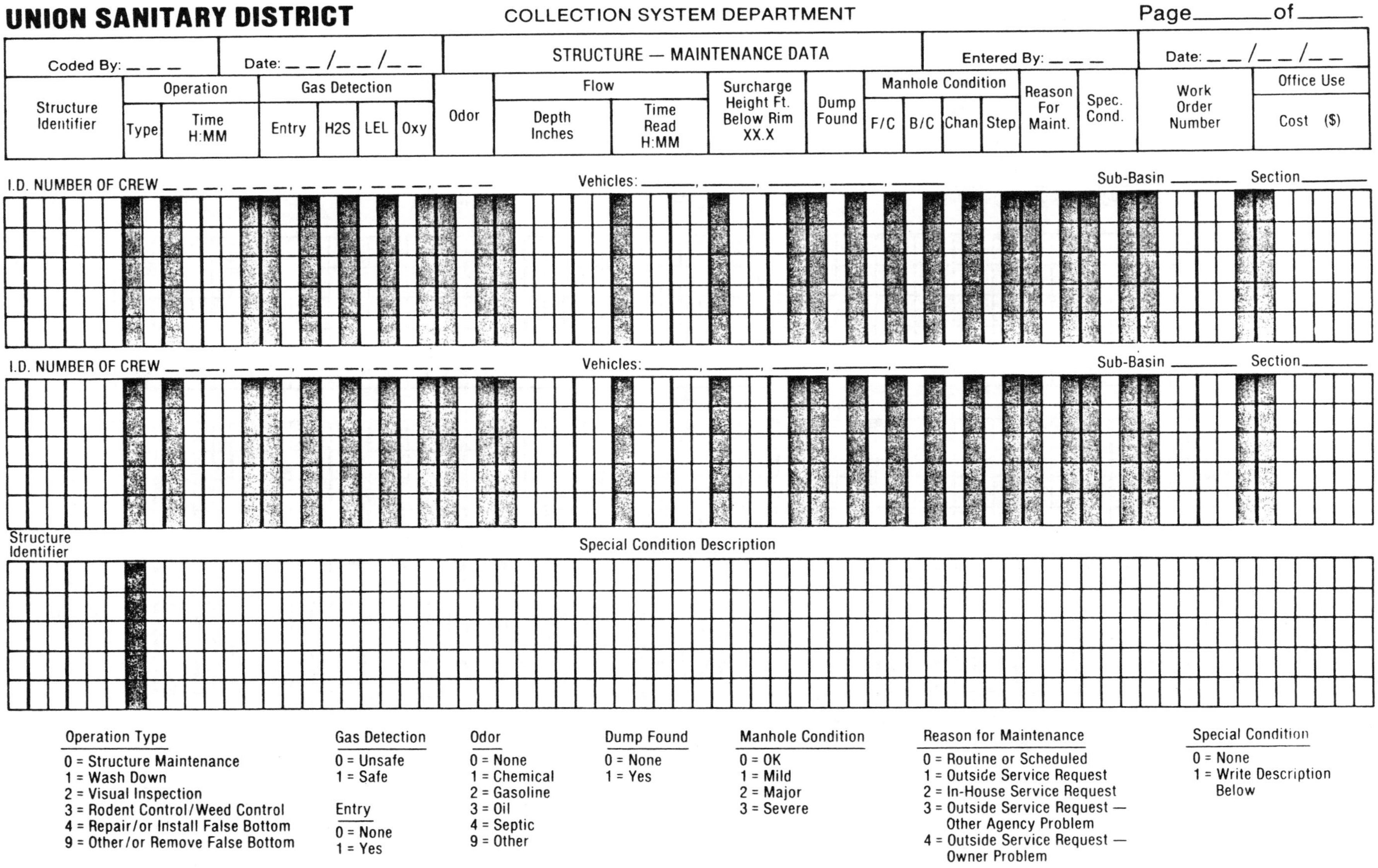

UNION SANITARY DISTRICT COLLECTION SYSTEM DEPARTMENT Page____ of ____

Coded By: _ _ _ | Date: _ _ /_ _ /_ _ | STRUCTURE — MAINTENANCE DATA | Entered By: _ _ _ | Date: _ _ /_ _ /_ _

Structure Identifier	Operation		Gas Detection				Odor	Flow		Surcharge Height Ft. Below Rim XX.X	Dump Found	Manhole Condition				Reason For Maint.	Spec. Cond.	Work Order Number	Office Use
	Type	Time H:MM	Entry	H2S	LEL	Oxy		Depth Inches	Time Read H:MM			F/C	B/C	Chan	Step				Cost ($)

I.D. NUMBER OF CREW _ _ _, _ _ _, _ _ _, _ _ _, _ _ _ Vehicles: ____, ____, ____, ____, ____ Sub-Basin ____ Section ____

I.D. NUMBER OF CREW _ _ _, _ _ _, _ _ _, _ _ _, _ _ _ Vehicles: ____, ____, ____, ____, ____ Sub-Basin ____ Section ____

Structure Identifier | Special Condition Description

Operation Type
0 = Structure Maintenance
1 = Wash Down
2 = Visual Inspection
3 = Rodent Control/Weed Control
4 = Repair/or Install False Bottom
9 = Other/or Remove False Bottom

Gas Detection
0 = Unsafe
1 = Safe

Entry
0 = None
1 = Yes

Odor
0 = None
1 = Chemical
2 = Gasoline
3 = Oil
4 = Septic
9 = Other

Dump Found
0 = None
1 = Yes

Manhole Condition
0 = OK
1 = Mild
2 = Major
3 = Severe

Reason for Maintenance
0 = Routine or Scheduled
1 = Outside Service Request
2 = In-House Service Request
3 = Outside Service Request — Other Agency Problem
4 = Outside Service Request — Owner Problem

Special Condition
0 = None
1 = Write Description Below

Fig. 12.13 Structure maintenance report

UNION SANITARY DISTRICT

COLLECTION SYSTEM DEPARTMENT

STRUCTURE — REPAIR OR ABANDON DATA

Page ____ of ____

Coded By: _ _

Date: _ _ / _ _ / _ _

Entered By: _ _

Date: _ _ / _ _ / _ _

Office Use Only

Structure Identifier	Use	Type	Depth XX.X	F/C M SS T	Cone T M	Barrell M SS	Steps	Lining M	Special Conditions	Operation	Work Order Number	Complete	Funding	Cost ($)

Structure Details Found

I.D. NUMBER OF CREW _ _ _ / _ _ _ / _ _ _

Vehicles: _____

Sub-Basin _____   Section _____

I.D. NUMBER OF CREW _ _ _ / _ _ _ / _ _ _

Vehicles: _____

Sub-Basin _____   Section _____

Structure Identifier

Special Condition Description

Structure: Use
0 = Access
1 = Metering
2 = Diversion
3 = Drop
4 = Siphon
5 = Monitoring
6 = Pumping
7 = Rodent Control
9 = Other

Structure: Type
0 = Manhole
1 = Riser
2 = Lamphole
3 = Flushing Inlet
4 = Cleanout
9 = Other

M = Material
0 = Unknown
1 = Precast
2 = Brick
3 = Fiberglas
4 = Cast Iron
5 = Mortar
6 = None
9 = Other

T = Type
0 = Standard
1 = Boltdown / Strapped
2 = Concentric
3 = Eccentric
4 = Unknown
9 = Other

Steps
0 = No
1 = Yes

SS (Inches)
Riser-Line Size
M.H. - Casting Size

Special Condition
0 = None
1 = Write Description Below

Complete
0 = None
1 = Yes

Operation
0 = Constructed
1 = Adjust to Grade
2 = Channel
3 = Silence Cover
4 = Replace Steps
5 = Seal Leaks
6 = Replace F/C
7 = Lined
9 = Pave or/
OTHER - (Use with Spec. Cond.)

Funding
0 = District Funds
1 = Reimbursed
2 = Shared
3 = Developer
9 = Other

Fig. 12.14   Structure repair or abandonment report

this one form. The time, material, and equipment report form is also on the reverse side of this form.

BUILDING SEWER MAINTENANCE

Building sewer maintenance work is reported by a crew leader using the *BUILDING SEWER — MAINTENANCE DATA, SERVICE REQUEST — OWNER PROBLEM* report form shown as Figure 12.15 Note the use of the debris and sewer severity reporting similar to reporting on main sewer maintenance. This information is again used for scheduling future maintenance and repair of building sewers. The reverse side of this form contains the standard time, material and equipment report form.

BUILDING SEWER REPAIR

The final form in the series of work reports developed by the collection agency is used by a crew leader to report building sewer repair work and is entitled *BUILDING SEWER — REPAIR DATA*, Figure 12.16. The report form, as with the others in the series, allows a great amount of data to be submitted by collection system operators for computer processing into information that is used by management to plan and direct the operation and maintenance of the agency's extensive wastewater collection system.

12.7032 RETENTION OF WORK REPORTS

If all the information submitted on work reports is transferred to records, either in the form of computer-readable electronic disks or written material, the reports may be discarded. However, if some of the information is not transferred to separate record documents, the work reports containing this information should be kept on file or archived until you are certain that the information is no longer useful.

QUESTIONS

Write your answers in a notebook and then compare your answers with those on page 547.

12.7D How do collection system operators account for their daily work activities?

12.7E List the types of work reports used by collection system agencies.

12.7F Who would use a "main sewer construction data" form?

12.7G List the types of debris that can be found in sewers.

12.7H List the types of conditions that can be found in sewers.

12.7033 USES OF DATA FROM WORK REPORTS

Information drawn from work reports is assembled into reports used by superintendents and supervisors for planning and controlling the operation and maintenance of a wastewater collection system. The superintendent needs to have reports analyzing records on staffing, materials and equipment use to take actions to improve work efficiency and prepare budgets. Reports on main sewer line stoppages are used to plan and modify a collection system's preventive maintenance program. Main and service sewer lines maintenance history reports are used to determine the need for future maintenance and to order appropriate repairs or replacements. A superintendent should have other reports prepared as needed for effective planning and control of the collection system agency.

A supervisor of collection system operators needs reports on main sewer line stoppages and main and service sewer line maintenance history to plan preventive maintenance activities and to schedule routine cleaning and inspection of a collection system with cleaning and TV equipment and operating crews. These reports are also used to assign priority to work orders received, if not prioritized by a superintendent. A supervisor will find the need for other reports for conferences with management on policy and technical matters and training collection system operators.

The superintendent and supervisors of each collection system will need to determine the types of forms or planning and controlling reports they will need for their particular management methods and system. The following are some of the commonly used reports.

- Main Line History. A comprehensive report of the maintenance performed can be produced for a requested main line segment. The user of the report can interact with the computer while viewing the report to produce a work order for further maintenance of the main line segment.
- Service Line History. A comprehensive report of the maintenance performed can be produced for a requested service line.
- Main Line Preventive Maintenance Work Orders Due. This report should be generated each month.
- Incomplete Main Line Preventive Maintenance Work Orders.
- Completed Main Line Preventive Maintenance Work Orders. This report provides a list of all main line preventive maintenance work orders completed within a given date range.
- Projected Main Line Scheduled Work Orders. This report provides a projection of the future main line scheduled maintenance workload.
- Incomplete Main Line Unscheduled Work Orders. This report provides a list of the outstanding (not finished) unscheduled work orders for main lines.
- Completed Main Line Unscheduled Work Orders. This report provides a list of all main line unscheduled work orders completed within a given date range.
- Incomplete Building Service Line Unscheduled Work Orders. This report provides a list of the outstanding unscheduled work orders for building service lines.

UNION SANITARY DISTRICT COLLECTION SYSTEM DEPARTMENT Page_____ of _____

Coded By: _ _ _	Date: _ _ /_ _ /_ _	BUILDING SEWER — MAINTENANCE DATA SERVICE REQUEST — OWNER PROBLEM	Entered By: _ _ _	Date: _ _ /_ _ /_ _

STREET NUMBER	STREET NAME	Type of Equip./Oper.	Operation Time H:MM	Debris: Type	Debris: Severity	Debris: Service Request Number	Sewer: Cond.	Sewer: Severity	Reason For Maint.	Special Cond.	Office Use Only: Override: Debris Severity	Office Use Only: Override: Schedule :MM/YY	Office Use Only: Operation Cost ($)

I.D. NUMBER OF CREW _ _ _, _ _ _, _ _ _, _ _ _, _ _ _ Vehicles: _____, _____, _____, _____, _____ Sub-Basin _____ Section _____

I.D. NUMBER OF CREW _ _ _, _ _ _, _ _ _, _ _ _, _ _ _ Vehicles: _____, _____, _____, _____, _____ Sub-Basin _____ Section _____

STREET NUMBER	STREET NAME	Special Condition Description

Type of Equipment/Operation

0 = None or N/A
1 = Ball
2 = DBC
3 = Power Rodder
4 = Hydrojet
5 = TV
6 = Flush
7 = Visual
8 = Root Control
9 = Other
A = Vacuum Unit
B = Hand Rodder
C = Scooter
D = Porcupine
E = Swab
F = Tire

Debris: Type

0 = Clear or N/A
1 = Egg Shells or Grounds
2 = Grease or Soap
3 = Paper or Rags
4 = Grit (Silt, Sand, Gravel)
5 = Foreign Object
6 = Roots
7 = Mud
8 = Solids
9 = Other

Debris: Severity

0 = Clear
1 = Mild
2 = Medium
3 = Severe
4 = Stoppage

Sewer Condition

0 = No Problem or N/A
1 = Broken Pipe
2 = Offset Joints
3 = Infiltration
4 = Roots
5 = Sagging Pipe
9 = Other

Sewer Severity

0 = N/A
1 = Minor
2 = Moderate
3 = Severe

Reason for Maintenance

0 = Routine or Scheduled
1 = Outside Service Request
2 = In-House Service Request
3 = Outside Service Request - Other Agency
4 = Outside Service Request - Owner Problem

Special Condition

0 = None
1 = Write Description Below

Fig. 12.15 Building sewer maintenance, service request, and owner problem report

UNION SANITARY DISTRICT COLLECTION SYSTEM DEPARTMENT Page______of______

Coded By: _ _ _	Date: _ _/_ _/_ _	BUILDING SEWER — REPAIR DATA	Entered By: _ _ _	Date: _ _/_ _/_ _

Structure Identifier U/S D/S	STREET NUMBER	STREET NAME	Drainage Sub-Basin	Material Found			Conf. Clean-out	Spec Cond	Repair		Office Use Only		
				Pipe	Lin-ing	Length			Work Order Number	Type	Com-plete	Fund-ing	Cost ($)

I.D. NUMBER OF CREW _ _ _, _ _ _, _ _ _, _ _ _, _ _ _ Vehicles: ____, ____, ____, ____, ____ Sub-Basin _____ Section_____

I.D. NUMBER OF CREW _ _ _, _ _ _, _ _ _, _ _ _, _ _ _ Vehicles: ____, ____, ____, ____, ____ Sub-Basin _____ Section_____

STREET NUMBER STREET NAME Special Condition Description

Type Material
- 0 = None or N/A
- 1 = VCP
- 2 = RCP
- 3 = PVC
- 4 = ACP
- 5 = ABS/Truss
- 6 = CMP
- 7 = RPM
- 8 = Polyethylene
- 9 = Other
- A = Epoxy
- B = Orangeburg
- C = DIP

Lining Material
- 0 = None
- 1 = VCP
- 2 = RCP
- 3 = PVC
- 4 = ACP
- 5 = ABS/Truss
- 6 = CMP
- 7 = RPM
- 8 = Polyethylene
- 9 = Other
- A = Epoxy
- C = Orangeburg
- C = DIP

Conforming Cleanout
- 0 = No
- 1 = Yes

Special Condition
- 0 = None
- 1 = Problem, Write Description Below

Repair: Type
- 0 = Constructed
- 1 = Sealed
- 1 = Replaced Sections
- 2 = Replaced Line
- 3 = Relined
- 5 = Cleanout Wye
- 6 = Cleanout Riser
- 7 = Cleanout Plug
- 8 = Cleanout F/C
- 9 = Other

Complete
- 0 = No
- 1 = Yes
- 2 = Shared
- 3 = Developer
- 9 = Other

Funding
- 0 = District Funds
- 1 = Reimburse
- 2 = Shared
- 3 = Developed
- 9 = Other

Fig. 12.16 Building sewer repair report

- Completed Service Line Unscheduled Work Orders. This report provides a list of all service line unscheduled work orders completed within a given date range.
- Area Preventive Maintenance Work Orders. This report should be generated each month for each area that is due for preventive maintenance (of all main lines within the area).
- Incomplete Area Preventive Maintenance Work Orders. This report provides a list of all incomplete area scheduled work orders currently existing.
- Completed Area Preventive Maintenance Work Orders. This report provides a list of all the area scheduled work orders completed within a given date range.
- Projected Area Preventive Maintenance Work Orders. This report provides a projection of the future area scheduled maintenance work orders.

Information about main line stoppages is usually reported in three different forms:

- Main Line Stoppage Summary. This report provides a summary of all main line stoppages for a given date range.
- Main Line Stoppages. This report provides detailed information about main line stoppages for a given date range and/or area.
- Selected Main Line Stoppages. This report provides a list of main lines which have a given number of stoppages (or more) within a given date range.

A primary concern for any superintendent and supervisor is crew productivity. The supervisor must be kept informed at all times on the progress and production of assigned operators. Management may require supervisors to submit reports documenting production. The following paragraphs describe the types of information a supervisor might be asked to provide. This information should be readily available from the work report database already in the computer files.

- A monthly work summary from each individual foreman. These reports should include daily production activity of each of the crews under each foreman. Activities may include balling, rodding, TV inspection, construction and repairs, or any other significant maintenance activities that are part of the agency's maintenance program.
- A monthly report of the unscheduled workload. Indicate the number of calls answered and the number of main line stoppages encountered. List stoppages according to their cause such as roots, grease, or other appropriate categories.
- Budget. Superintendents share some responsibility for the preparation and submission of the budget. For the supervisor this usually means completing budget forms provided by management or accounting departments. All budgets require accurate information about production and anticipated workloads. The above-described reports will be very helpful in providing such information.
- Cost of operation. A summary of how budgeted funds were or will be spent is very important. This report should describe the total cost of operation as well as costs for supplies, materials, equipment and salaries. Costs also can be reported by types of programs such as clearing emergency stoppages, preventive maintenance, responding to complaints and repairs.
- Problems. A record of problems encountered is important so problems can be anticipated or prevented from occurring in the future.

QUESTIONS

Write your answers in a notebook and then compare your answers with those on pages 547 and 548.

12.7I What kinds of reports does the superintendent of a wastewater collection system need and why?

12.7J What information should be contained in the monthly reports from each individual foreman?

12.7K What information should a cost of operation report contain?

12.704 Work Directives

Another important element in a management information system is the work directives issued for the operation and maintenance of a wastewater collection system (see Figure 12.8, page 493).

12.7040 NEED FOR DIRECTIVES

The operators of a collection system agency are responsible for using their time, materials, and equipment in an efficient manner on the work assigned to them. Superintendents and supervisors have the responsibility to direct, by assignment, the work of collection system operators in a manner that will achieve the agency's objectives and goals. Considering the operation of a football team helps to explain the need for work directives. The "operators" of a football team are the linemen and backs, whereas "management" consists of the supervising quarterback on the field and the superintending coach on the sideline. Together they act as a team with the common objective of moving the football over the goal line. Operators may make suggestions in the huddle if they detect something that will affect the team's actions, but they do not call the plays.

A superintendent and supervisors of a collection system, supported by planning and controlling reports produced by a good management information system, are in a position to make informed decisions and direct operators toward the agency's objectives and goals.

12.7041 TYPES OF DIRECTIVES

Work directives given by a superintendent and supervisors to operators include training in work performance and equipment operation, standard operating procedures and work orders. Work directives should generally be in written form to reduce the possibility of misunderstanding and to provide a record for reference if questions arise about a directive. Verbal work directives can be used when they are more timely and efficient than the written type or ones transmitted by a computer. The following examples illustrate two types of work directives given to collection system operators.

RESPONSE TO SERVICE REQUESTS

The standard operating procedure (issued by a superintendent) for responding to a request from a property owner for service is as follows:

- Service Request Report. The clerk receiving a request for service from a property owner obtains the information needed to fill out a Service Request Report (Figure 12.17) and gives it immediately to the supervisor of the service crew.
- The supervisor, on receipt of a Service Request Report, immediately obtains any additional information needed for response to the request, makes a personal investigation if required, and gives directions to a service crew for responding to the service request. The directions may be in the form of a written work order, transmittal of a Service Request Report, or by radio communications as the priority of the request requires. In all instances, the supervisor gives the crew leader as much information as is available for assisting in the response and indicates when the crew is expected to respond.

SERVICE REQUEST REPORT

REQUEST
From:

Name ______________________________ Date __/__/__

Address ______________________________ Time ________

Telephone ______________ Parcel No. ______________

Nature of Request ______________________________

____

____

Additional Information ______________________________

____

____

Request Received By ______________

____

RESPONSE

Assigned To ______________ Arrival Time ______________

Conditions Found ______________________________

____

____

Action Taken ______________________________

____

____

Daily Work Report Reference ______________________________

Fig. 12.17 Service request report

- The service crew is to respond to a request for service as directed by a supervisor. On determining the conditions pertaining to a service request, the crew performs the action needed to provide the requested service if it is the agency's policy to provide the service. If the service is not provided by the agency, the crew leader explains the agency's policy to the property owner and offers any appropriate suggestions on how the service might be obtained. The crew leader directs any unanswered questions concerning action on a service request to the supervisor. Upon completing action on a service request, the service crew leader completes the "Response" part of the Service Request Report and submits the report along with the daily work report for the day the service action was performed.

- Supervisors submit completed Service Request Reports to the records clerk weekly.

WORK ORDERS

In manual recordkeeping systems, the physical makeup of the work order form will vary with the particular needs of each agency. They can be prepared in single, duplicate, or any required number of copies. An advantage of duplicate or carbon copies is that the original copy can be retained in the office as a check to verify that the duplicate copies in the field are returned and properly filled out when completed. For an example of a simple work order form, see Figure 12.18.

A computer-based work order system offers many advantages over manual systems. Perhaps the greatest advantage is the ease with which work orders can be prepared, sent to operators, tracked for completion of assigned work and filed for future reference.

Technology is advancing rapidly in the area of hand-held computers and collection system field crews are leaders in the applications of this new technology. Laptop computers, pen-based computers and electronic notebooks are all being used by field crews. Pen-based computers allow field operators to write comments in a handwritten form or to type comments using a keyboard that appears on the computer screen upon request. The pen may be used either for writing by hand or to function as a "mouse." By moving the mouse around the screen, the operator is able to point to an item on the menu screen. Then by clicking a button on the pen, the desired feature or information can be retrieved from the computer database and displayed on the screen. Equipment such as this could be used in the following manner to transmit work orders and file work reports.

Monthly preventive maintenance work orders are generated by a computer and assigned to the collection system manager. (Duplicates or carbon copies are not needed with computer systems because the original work order information is retained in the computer for anyone who needs it.) The manager assigns tasks to supervisors by areas or type of activity (televising, cleaning). The supervisors then assign work to crews by transmitting work orders by means of the operators' pen-based computers. Work orders usually include the last five years' history of the line, including stoppages, TV history and preventive maintenance completed.

The crews complete the work order and record observations and work performed by touching the pen to standard observations and responses. Standardization allows information to be stored (archived) in a computer and the information can be analyzed or queried (how many root stoppages or grease stoppages occurred in a certain area with the last 12 months?) at any time.

FOLLOW-UP W/O # ________

Crew Chief ____________________ Date ________
Address ____________________ Map # ____ Pg. # ____
Manhole # ____________ to ____________
Pipe Size ________ Depth ________ Distance ________
Street ________ Easement ________

Work Requested ____________________
____
____
____

Approved ____________________

Work Completed ____________________
____
____
____

Completed by ____________________ Date ________

Approved ____________________

Fig. 12.18 Work order form

When a customer service call or an emergency call is received by a dispatcher, the dispatcher notifies a supervisor of the problem, provides a street address, creates an open work order (Figure 12.19) against the address and assigns a call number. The supervisor drives to the site to evaluate the problem and may immediately send a crew to the site if an emergency exists. The crew can contact the base computer and retrieve the history of the line for the last five years, if necessary. Work order designs can also include provisions for recording the situation status of the job site while awaiting the arrival of a repair crew, as illustrated in Figure 12.19.

When the field crew resolves the problem, they "close out" the work order. The next day the crew will check the area to be sure the problem is solved and use the original call number for their report. This same call number will be used if the line needs televising and the call number will be assigned to the TV work order. If the TV inspection identifies a root problem, the call number will be assigned to the work order to clear the roots. If the TV inspection reveals other problems (roots, cracked pipe, grease), new work orders will be generated. The original call number will be used to inform the person who generated the service request of the action taken to correct the problem.

Lift station visits also can be scheduled or requested on the basis of work orders assigned by hand-held computers. Information about the tasks performed by the operators can be promptly recorded and stored in the computer database for later analysis.

QUESTIONS

Write your answers in a notebook and then compare your answers with those on page 548.

12.7L What types of work directives are given by superintendents and supervisors to operators?

12.7M What should a collection system supervisor do upon receipt of a Service Request Report?

12.7N What should a crew leader do upon discovery that the agency's policy prohibits correcting a problem in a service request?

12.7O Why are duplicates or carbon copies of work orders not needed when using a computer system?

12.71 Applications of Computers for System O & M

One of the tools that we have to assist in managing resources is a computer-based maintenance management system (CMMS) (see Figure 12.20). Within the past several years, the hardware and software for computer-based systems have changed dramatically in terms of cost, processing power and ease of use. CMMSs have now become an important management tool not only in large agencies but in small agencies as well. Computer-based maintenance management systems are able to store large amounts of information about the collection system and they are able to rapidly manipulate this information so that it is quickly accessible and usable when performing collection system O & M. While recordkeeping has always been an important aspect of O & M, it is even more essential today due to the liability associated with maintaining inadequate records and the need to perform O & M in a more cost-effective, efficient manner.

O & M recordkeeping includes four different elements that are interrelated: (1) we need to know what is in the collection system (system inventory), (2) we need to know what the condition of the system is (inspection and testing), (3) we must plan and schedule maintenance activities, and (4) we need to maintain accurate records that can be used for managing our resources and generating reports.

The information in a computer system is commonly referred to as a database. Managing a collection system agency will usually require the creation of a variety of databases. A gravity sewer system, for example, will have several databases. The engineering database contains all of the design details, the maintenance database tracks O & M activities and assists in planning and scheduling maintenance, and the inspection database includes information on physical inspection on the system. Because the CMMS relies on information databases, the system is only useful when a good manual recordkeeping system has been established and the information is available to be transferred to the CMMS. Many agencies have purchased computer-based systems but were unable to use them effectively simply because they did not have the necessary information available in a manual recordkeeping system.

12.710 *Types of Computer-Based Maintenance Management Systems (CMMSs) (Figure 12.20)*

There are two distinct categories of CMMSs used in collection systems. The first is a pipeline-oriented system and the second an equipment-oriented system. As the names suggest, equipment-oriented systems are used for facilities that consist mainly of electrical, hydraulic and mechanical systems such as pump stations and treatment plants. Pipeline-oriented systems, on the other hand, were developed specifically for operation and maintenance of gravity sewers, force mains and the structures associated with them, such as manholes, valve vaults, catch basins and building lateral sewers.

The primary difference between the two types of systems is in the information that is stored and used in the databases. Equipment-based systems have data categories for equipment nameplate data, inventory control, parts and materials, purchasing, and planning and scheduling of maintenance activities. Each of the data categories may be broken down into specific types of systems such as pumps, motors, electrical systems, valves, instrumentation and control. Data categories for the pipeline-based system include engineering, maintenance, inspection, and planning and scheduling of maintenance activities. These categories are further broken down into information on main lines, building service laterals, structures, catch basins and storm inlets.

When first developed, CMMSs required large mainframe or minicomputers. Now virtually all CMMS software can be run on a desktop PC (personal computer). These programs can be designed with several modules or computer screens tailored for different organizational needs and requirements. Some example screens are shown in Figures 12.21 through 12.26.

SEWER REPAIR ORDER

WO ID#:

LOCATION DESCRIPTION

NUMBER:	STREET/LOCATION:	ZIP:
CROSS-STREETS/LOCATION DETAILS:		
LOCATION NOTES:		REF FILE:

JOB DESCRIPTION

SOURCE: SERVICE CALL # ______ WALKING INSP # ______ TV INSP# ______ OTHER ______	HEAVY TRAFFIC? YES ☐ NO ☐ USA : ______ ID# DATE BY: ______ PRIORITY OVERRIDE? ☐ BY: ______ MUST BE COMPLETED BY THIS DATE: ______	INSPECTION RATING: 1 2 3 4 INSPECTION DATE: ______ INSPECTION ID: ______ SECTION DATA SHEET ID#: ______ *ROOTS?* YES ☐ NO ☐

REPAIR EXTENDS ______ FEET TO ______ FEET FROM THIS REFERENCE POINT:

☐ PIPE MAIN SEWER REPAIR:	PIPE SIZE:	DEPTH AT REPAIR POINT:		LENGTH OF REPAIR (feet):
☐ BRICK MAIN SEWER JOB:	REPLACE BRICK ☐	PLASTER ONLY ☐		SIZE OF REPAIR (sq. feet):
☐ SIDE SEWER REPAIR:	PIPE SIZE:	DEPTH AT TRAP:	DEPTH AT MAIN:	LENGTH OF REPAIR (feet):
☐ CB CULVERT/LEAD REPAIR:	DEPTH AT CB:	DEPTH AT MH or M/S:		LENGTH OF REPAIR (feet):
☐ OTHER JOB:	INSTALL/REPLACE MH ☐	INSTALL/REPLACE CB ☐	OTHER: ☐	

JOB NOTES:

SITUATION STATUS - One box per line MUST be checked!

FLOODING:	INSIDE BLDG ☐	OUTSIDE/STREET ☐	NONE ☐		
STREET CONDITION:	CAVE-IN ☐	VOID ☐	HIDDEN VOID (reported by walking/tv insp) ☐	DEPRESSION ☐	OK ☐
BARRICADED?	YES ☐	NO ☐			
PAVED BEFORE?	YES ☐	NO ☐	NEEDS PAVING NOW YES ☐ NO ☐	PATCH SIZE:	

NOTES/ACTIVITY LOG (optional)

DATE	NOTES	INITIALS
	Job first reported by -->	

JOB REPORT - To be filled in by supervisory personnel

WO DISPATCHED TO: SSR ☐ SPOT ☐ BOE ☐	BY:	DATE:	DISPATCH ID#:
JOB DONE/CANCELLED ON (date):	INSPECTOR:	INSPECTION ID#:	
REPAIR NOTES:			DIGUP REPORT ID#:
SMS ID#:	PMS ID:	SIGNATURE:	

Fig. 12.19 Main line unscheduled work order

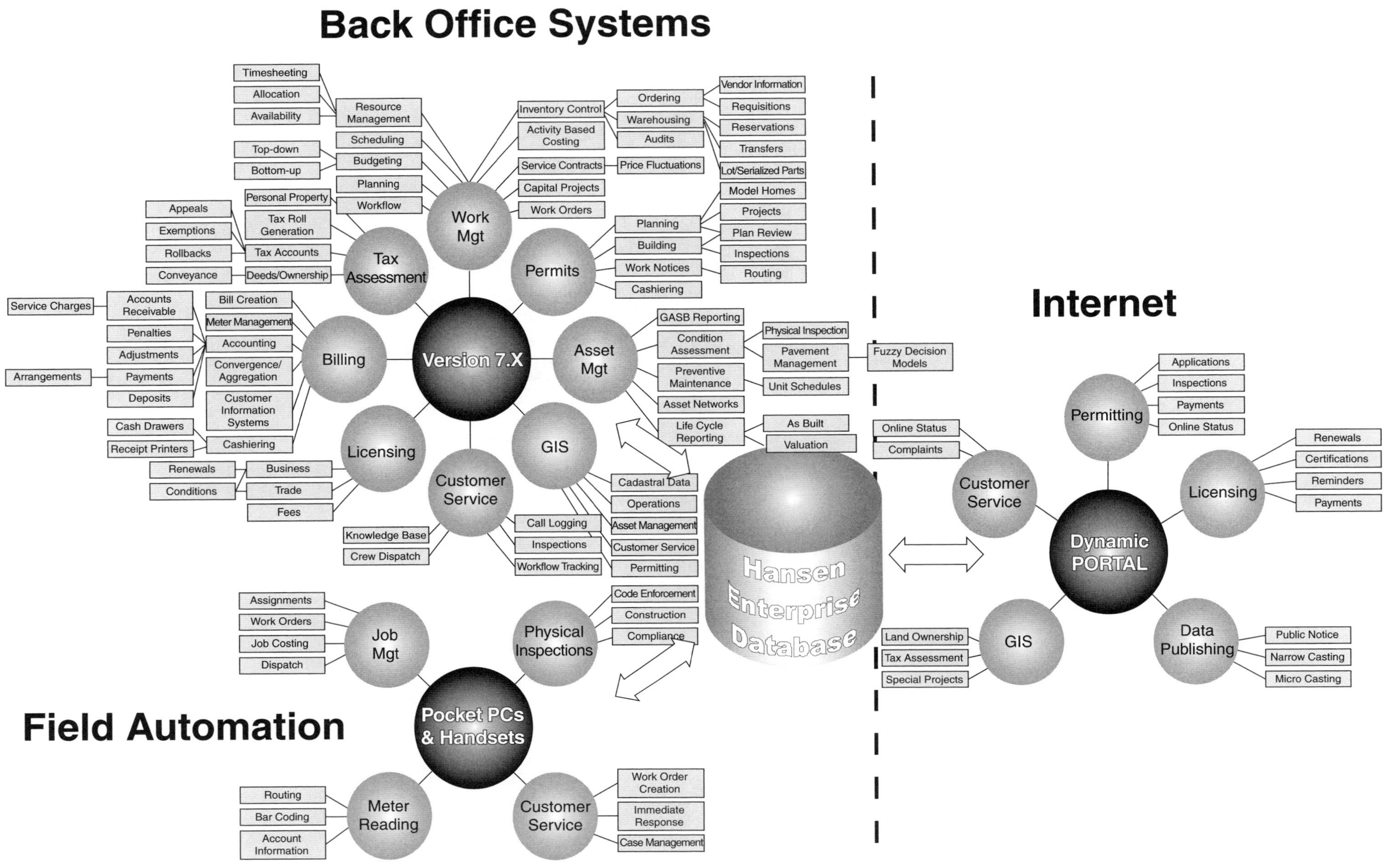

Fig. 12.20 Computer-based maintenance management system (CMMS)
(Permission of Hansen Information Technologies)

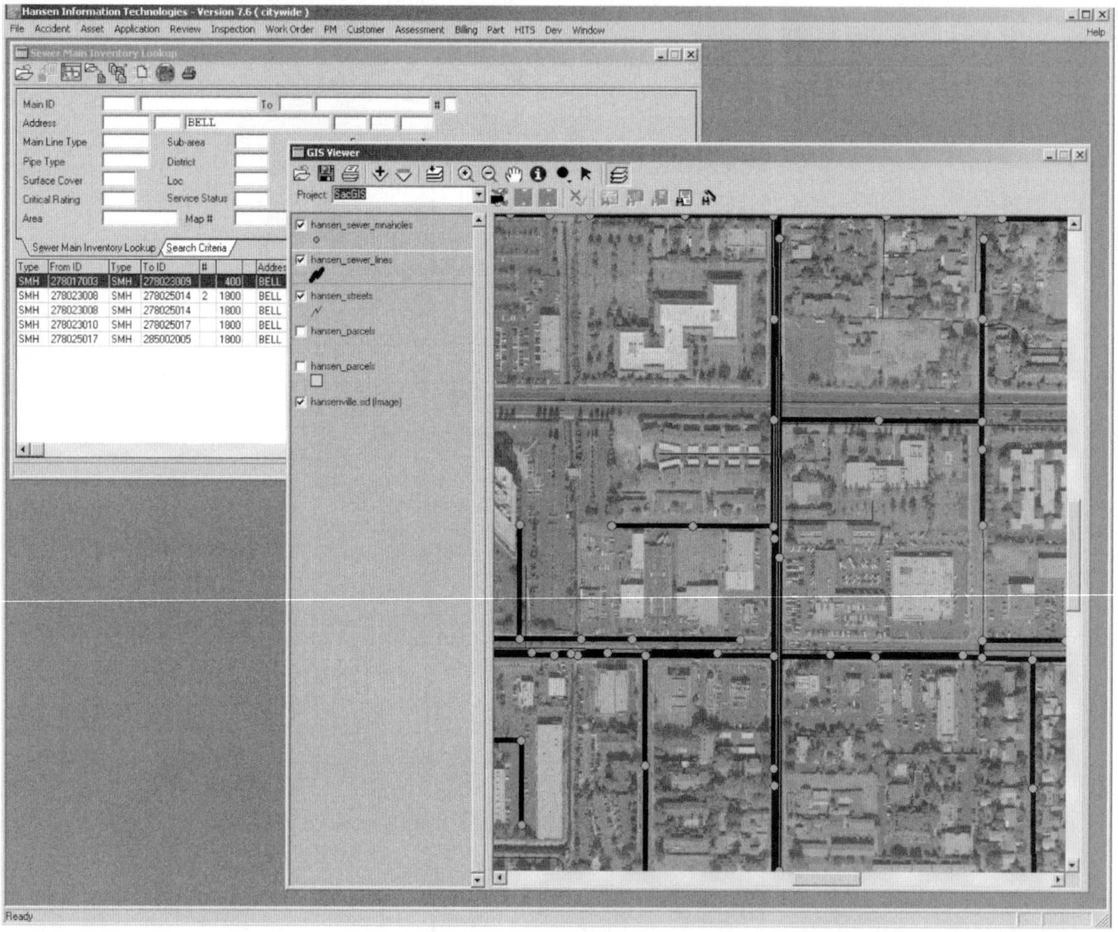

Fig. 12.21 Sewer assets with GIS
(Permission of Hansen Information Technologies)

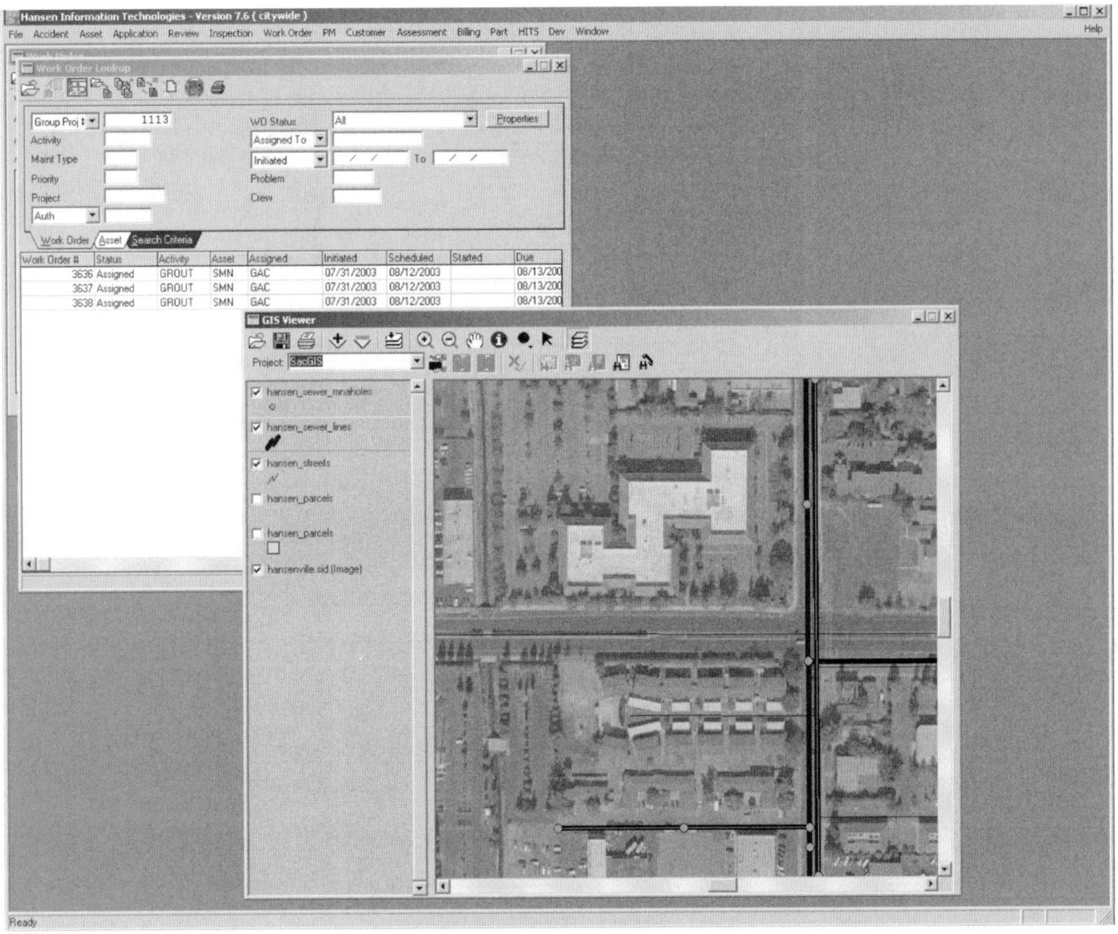

Fig. 12.22 Work order history display

(Permission of Hansen Information Technologies)

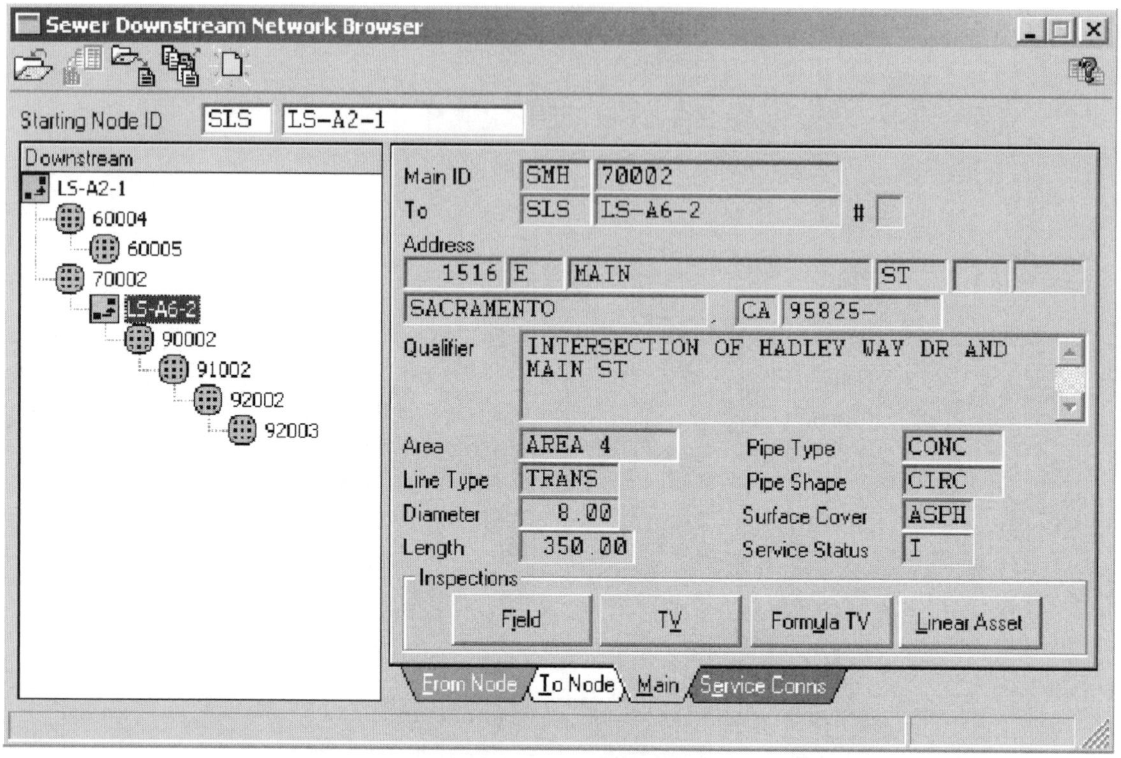

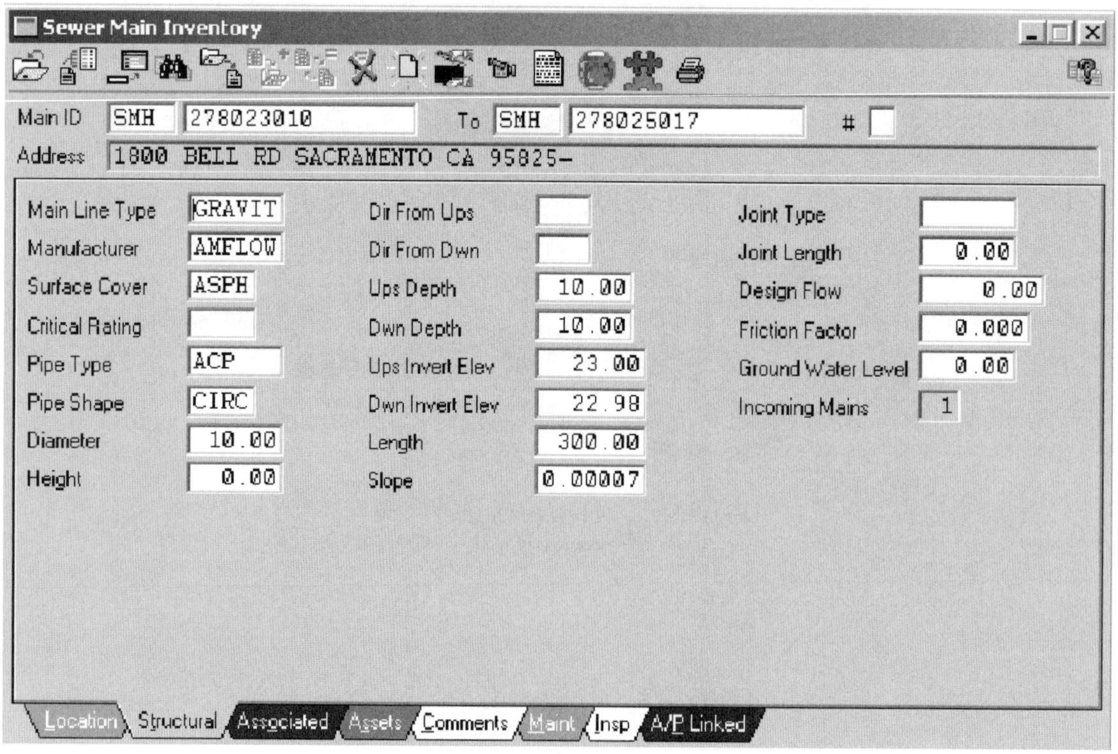

Fig. 12.23 Sewer inventory screen display

(Permission of Hansen Information Technologies)

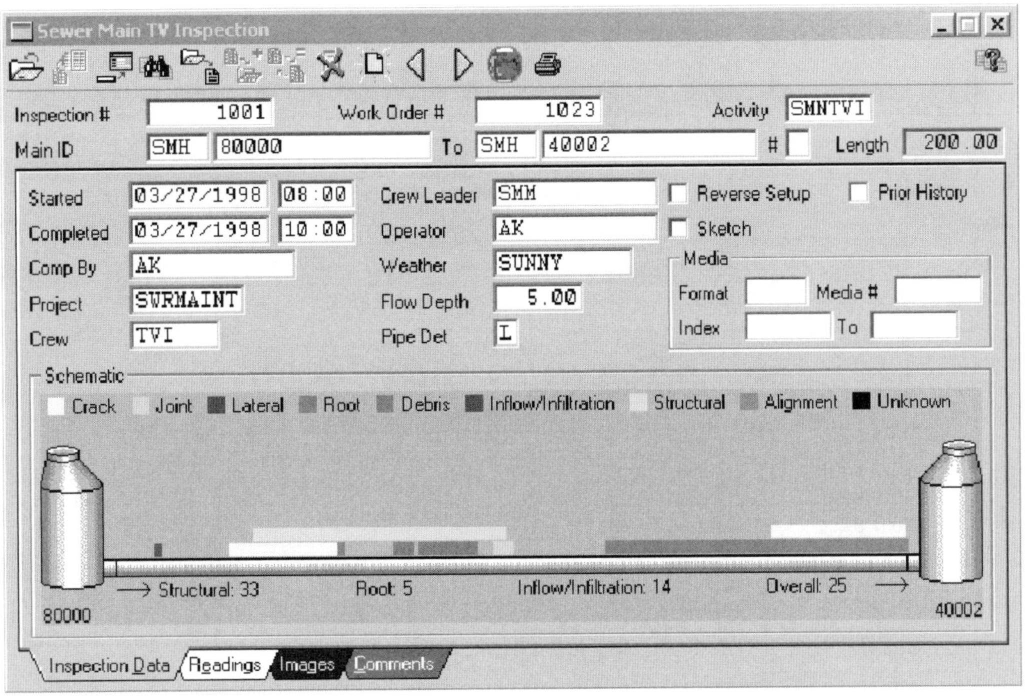

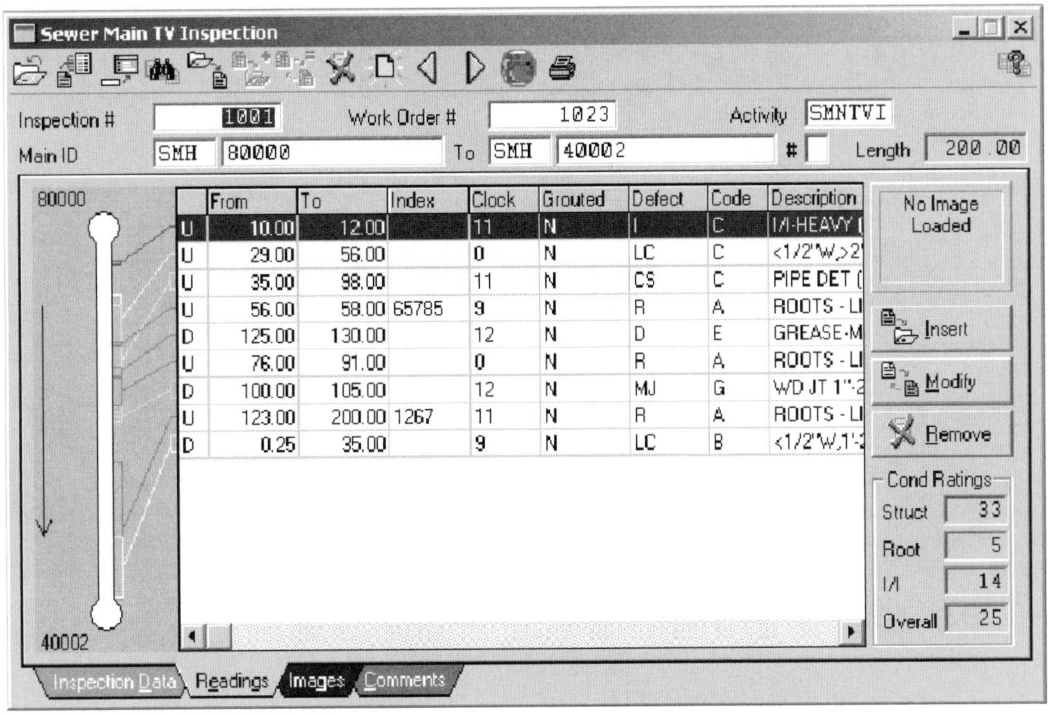

Fig. 12.24 CCTV computer-based data collection systems

(Permission of Hansen Information Technologies)

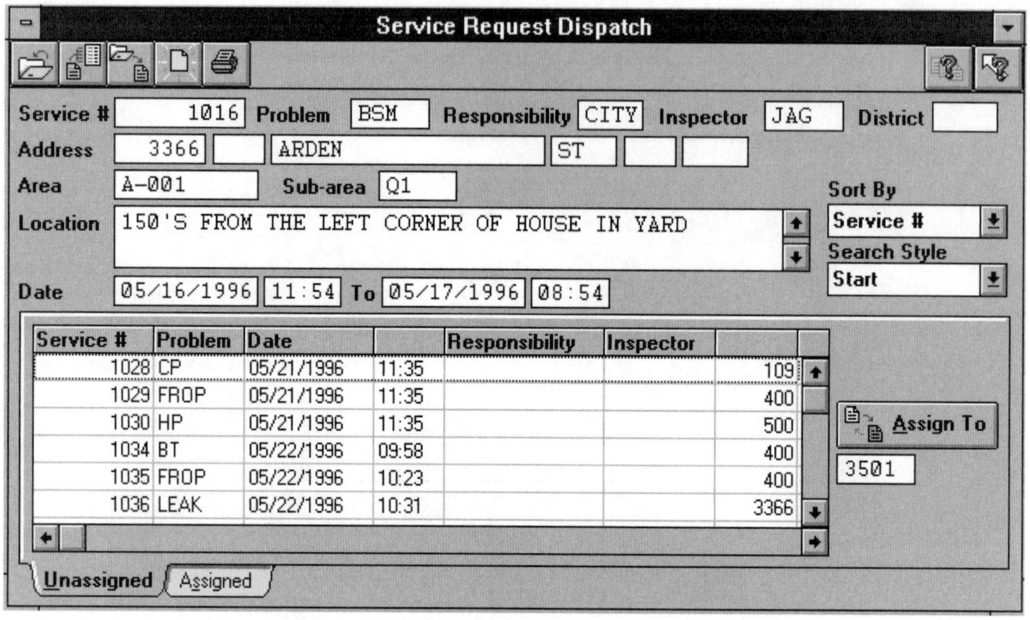
Service Request Dispatch
Service # 1016 Problem BSM Responsibility CITY Inspector JAG District
Address 3366 ARDEN ST
Area A-001 Sub-area Q1
Location 150'S FROM THE LEFT CORNER OF HOUSE IN YARD
Date 05/16/1996 11:54 To 05/17/1996 08:54
Sort By Service #
Search Style Start
Service # Problem Date Responsibility Inspector
1028 CP 05/21/1996 11:35 109
1029 FROP 05/21/1996 11:35 400
1030 HP 05/21/1996 11:35 500
1034 BT 05/22/1996 09:58 400
1035 FROP 05/22/1996 10:23 400
1036 LEAK 05/22/1996 10:31 3366
Assign To
3501
Unassigned
Assigned

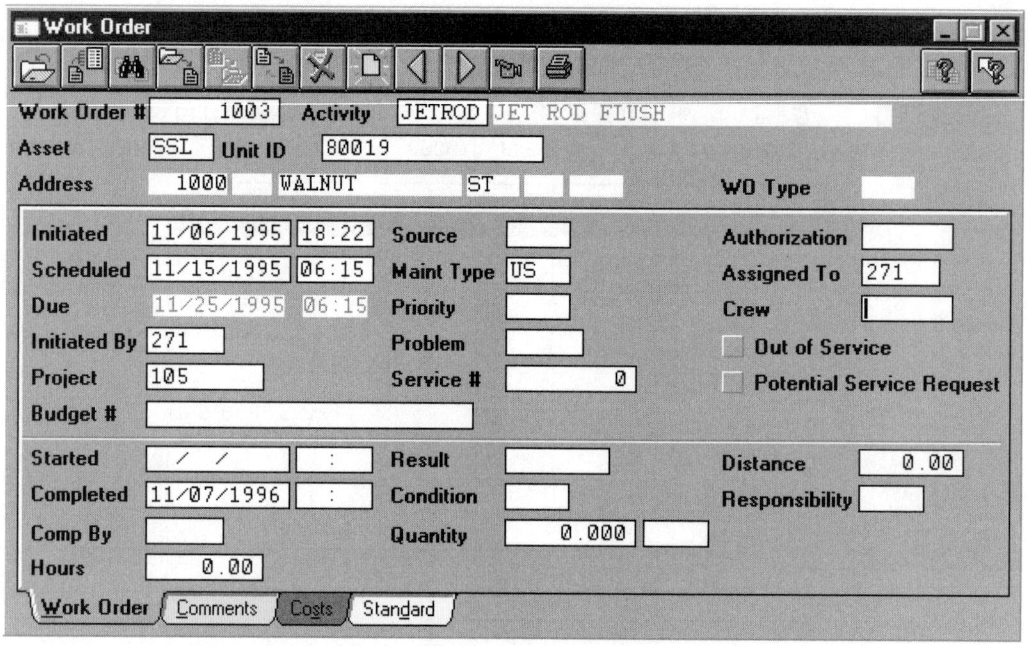
Work Order
Work Order # 1003 Activity JETROD JET ROD FLUSH
Asset SSL Unit ID 80019
Address 1000 WALNUT ST
WO Type
Initiated 11/06/1995 18:22
Scheduled 11/15/1995 06:15
Due 11/25/1995 06:15
Initiated By 271
Project 105
Budget #
Source
Maint Type US
Priority
Problem
Service # 0
Authorization
Assigned To 271
Crew
Out of Service
Potential Service Request
Started
Completed 11/07/1996
Comp By
Hours 0.00
Result
Condition
Quantity 0.000
Distance 0.00
Responsibility
Work Order
Comments
Costs
Standard

Fig. 12.25 Work orders and service request screen display
(Permission of Hansen Information Technologies)

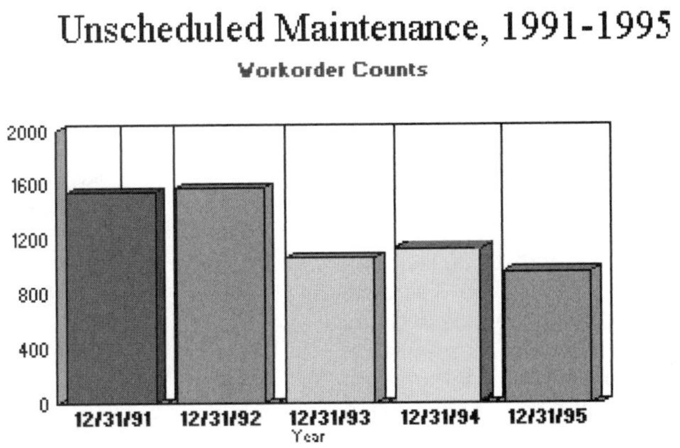

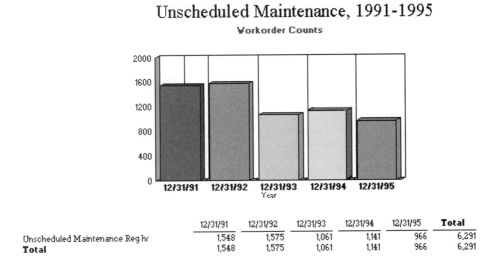

	12/31/91	12/31/92	12/31/93	12/31/94	12/31/95	**Total**
Unscheduled Maintenance Reg hr	1,548	1,575	1,061	1,141	966	6,291
Total	1,548	1,575	1,061	1,141	966	6,291

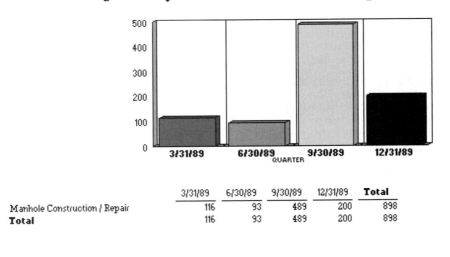

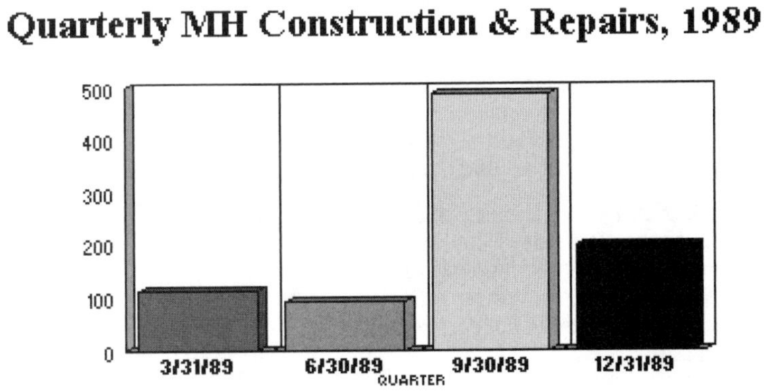

	3/31/89	6/30/89	9/30/89	12/31/89	**Total**
Manhole Construction / Repair	116	93	489	200	898
Total	116	93	489	200	898

Fig. 12.26 Computer-based report summary

(Permission of Hansen Information Technologies)

12.7100 PIPELINE-BASED CMMS

Figures 12.27 through 12.35 are examples of computer screens which illustrate the database information and some of the features that are available in a CMMS. The examples are from a pipeline-based system.

Figure 12.27 is the main menu screen that allows the operator to select "Engineering," "Maintenance and Inspection" information, or "Work Order" activities for "Pipes" or "Structures" in the collection system. If "Pipes" is selected in the main menu, the next screen, shown in Figure 12.28, displays pipe location information between the selected upstream and downstream structures as part of the engineering information that is stored in the database. Figure 12.29 shows the "Pipe Specification" database that includes pipe material and dimension data and pipe survey data. This is additional engineering information.

The "Pipe History" database shown in Figure 12.30 includes past, present and future information about the pipe. The construction date, age, cost, the current status of the pipe and any future scheduled rehabilitation or replacement information is also displayed. An area is provided for ownership information and claims associated with this section of pipe, for example, a backup, overflow or collapsed street that resulted in an insurance or civil claim.

If "Maintenance and Inspections" is selected in the main menu (Figure 12.27), a series of screens become available for maintenance-related activities. The screen shown in Figure 12.31 lists the maintenance databases that are part of the work planning and scheduling process for this section of pipe; "Pipe PM Schedules" is the database that has been selected. This database includes information on the type/method of activity, quantity, frequency, and tracks the last time this activity was done. Depending on the frequency selected, the next date is then automatically calculated and work is planned and scheduled. Information about the number of crew hours spent performing this task is also stored in the database. This information is useful in planning and scheduling since you can project the future workload for the preventive maintenance tasks based on actual time spent completing similar tasks in the past. Workload backlog can also be identified if the preventive maintenance task goes beyond the scheduled date.

Physical inspection of the collection system is done on a periodic basis and the information is recorded. The "Pipe Inspection Details" databases include the periodic scheduling function and inspection results. Figure 12.32 illustrates the TV inspection database with information on obstructions, defects, severity and location of the problem.

Figures 12.33, 12.34, and 12.35 show screens from the database associated with preventive maintenance work orders. The work orders database automatically uses information from the engineering and maintenance and inspection databases when generating a work order. This database includes historical background information on the section of pipe, predetermined PM scheduling, and the problem code that relates to the pipe (Figure 12.33). The preventive maintenance activity specified for this section of sewer is hydro jet (high-velocity) cleaning of 300 feet of pipe (see Figure 12.34). Additional PM activities can also be scheduled; for example, if root treatment needs to be conducted after cleaning with the hydro jet, the "Secondary Activity" section of the work order (Figure 12.34) would contain that information.

When the PM work tasks and activities for this section of pipe have been completed, the work order is closed out using the "Completion" database shown in Figure 12.35. The information contained in this database includes scheduling, crew assignment, completion date and time, and recommendations for future maintenance activity.

12.7101 EQUIPMENT-BASED CMMS

The equipment-based CMMS databases vary significantly from the pipeline-oriented systems. Typically they are used by wastewater collection agencies that have a large number of pump stations. The following databases are typical of an equipment-based CMMS.

- Equipment. This database includes nameplate information on each piece of equipment, its location in relation to other pieces of equipment in that particular system, and its cost history.
- Parts and materials inventory. This database contains records on the inventory of parts, warehouse stocking information, and records of parts used or returned. In addition, there is an index that relates the inventory to invoices or receipts for different pieces of equipment, as well as other functions associated with maintaining and warehousing supplies and equipment.
- Purchasing. This database contains all records relating to the purchase and receipt of materials and supplies. The purchasing and inventory databases are often set up to interact easily so that a complete history of materials and supplies purchased is readily available. Purchase requisitions and purchase orders can be generated from this database.
- Preventive maintenance. The function of this database is very similar to that of the pipeline-oriented system in that it includes information on PM inspection; scheduling by time interval, run time, mileage or other interval-based period; estimated time required for PM tasks; and parts required for the PM task.

  Planning and scheduling are important parts of the maintenance program. Planning functions include prioritizing work orders and the assignment of people and materials to perform the work. This database also shares information from the inventory system to determine whether or not parts are available to perform the work. The scheduling part of this database will automatically generate the work orders for preventive maintenance based on the intervals selected and assist in the scheduling by looking at the requirements for available people versus people required. Many of these types of programs also have the ability to list standard operating procedures, the standard maintenance procedures to be used, and safety requirements for particular tasks, such as use of equipment lockout/tagout procedures.

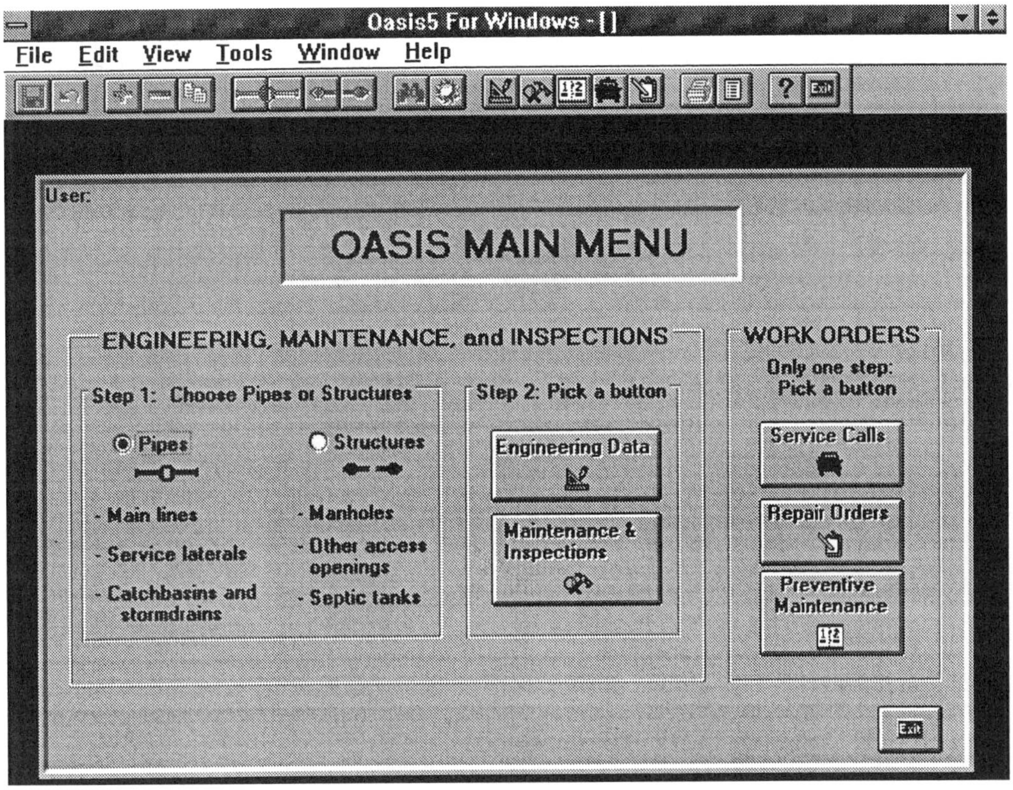

Fig. 12.27 CMMS main menu screen

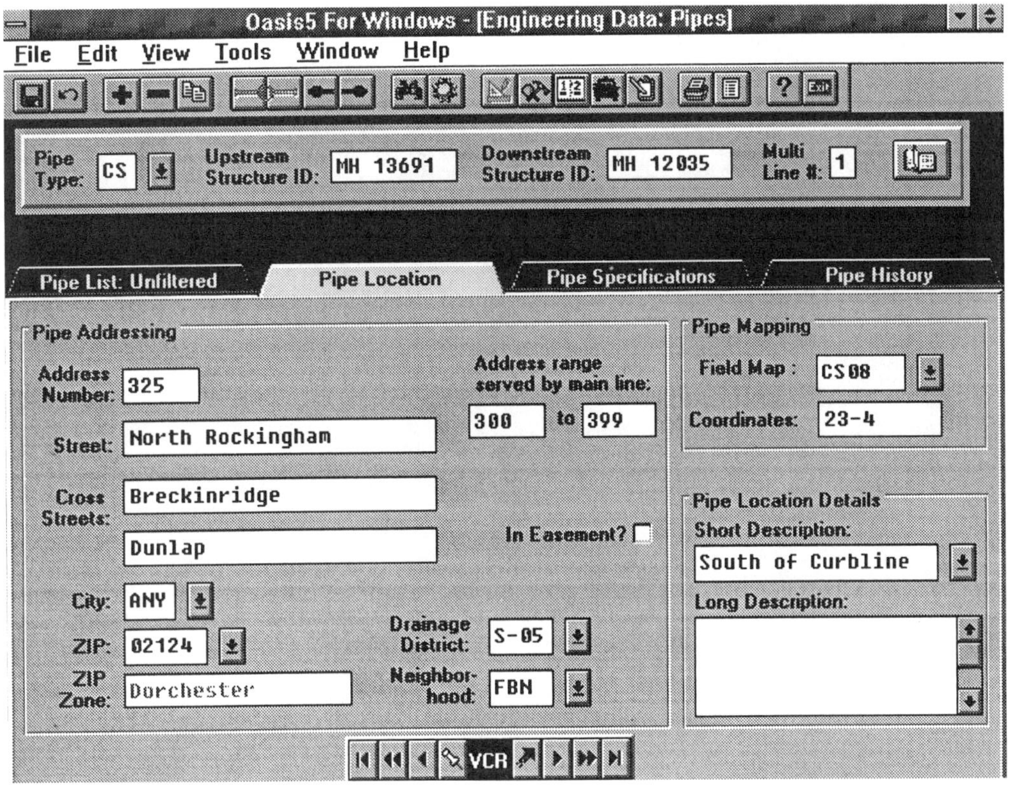

Fig. 12.28 CMMS pipe location screen

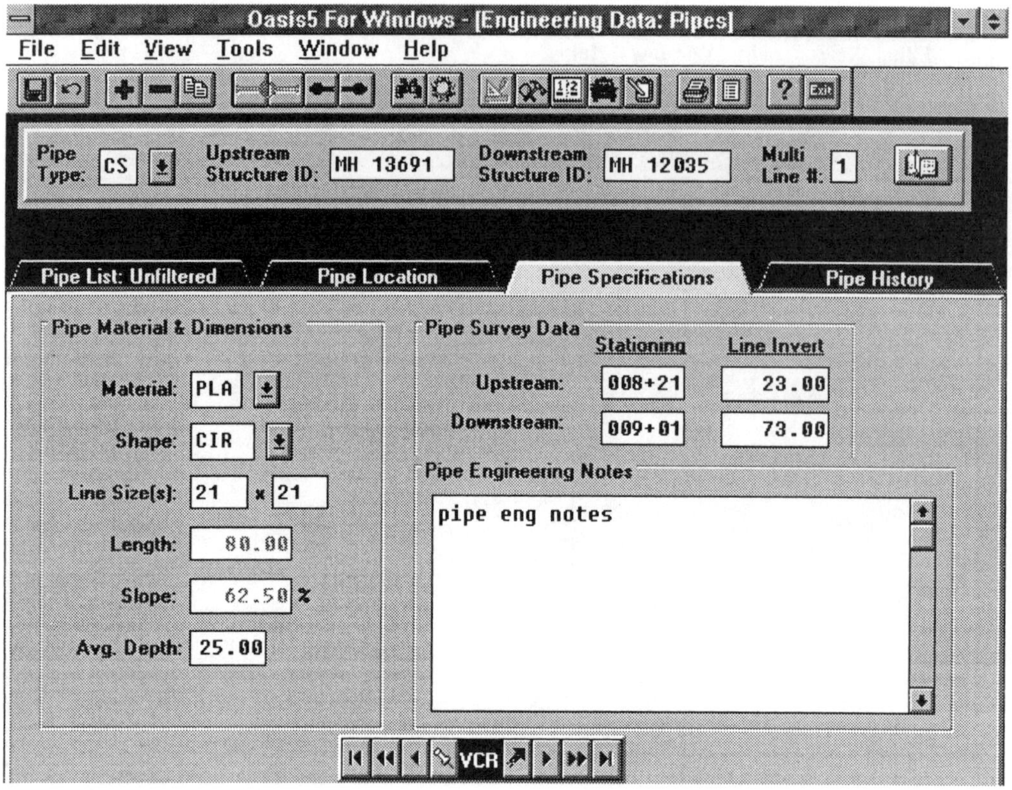

Fig. 12.29 CMMS pipe specifications screen

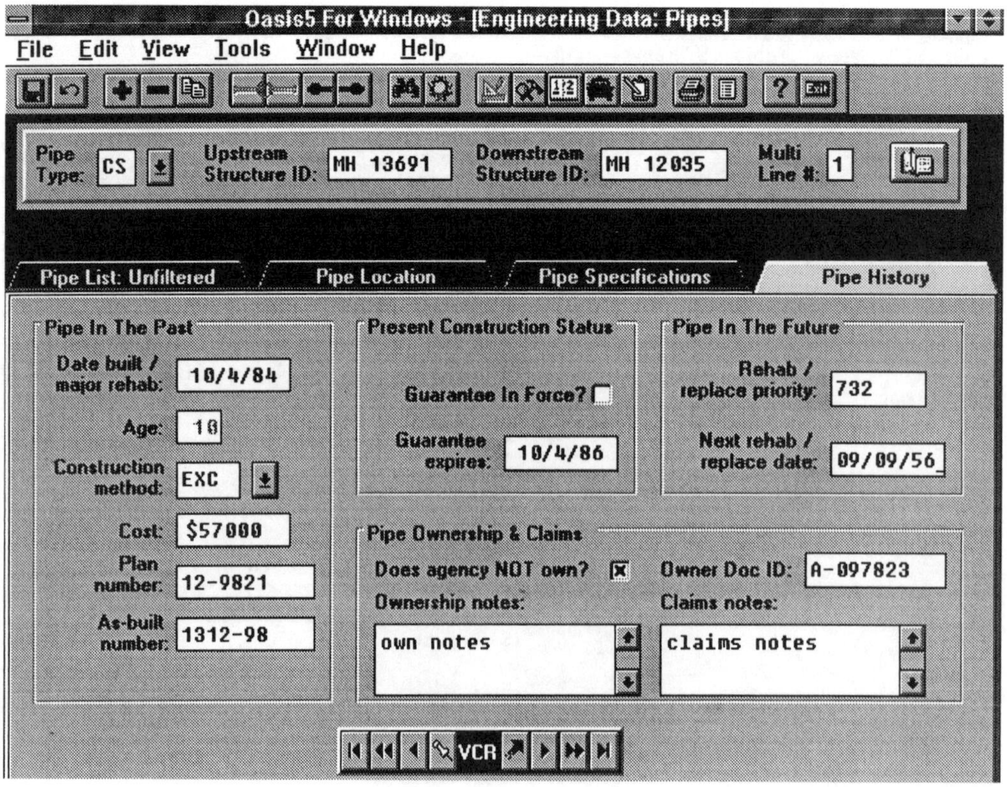

Fig. 12.30 CMMS pipe history screen

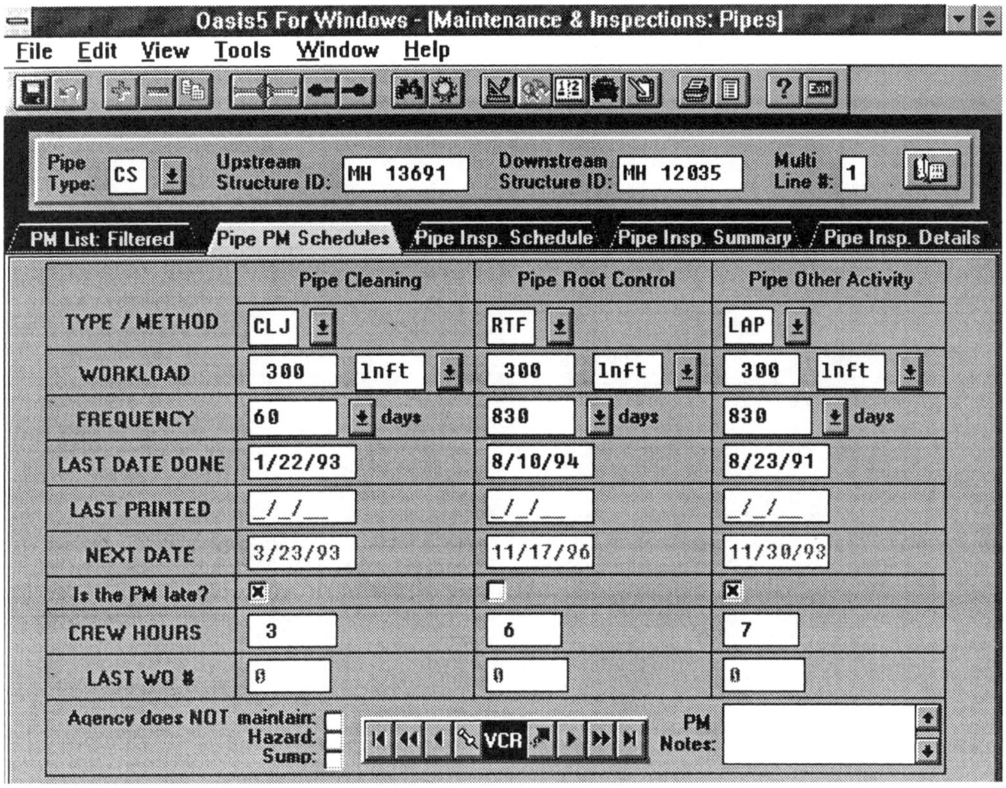

Fig. 12.31 CMMS pipe PM schedules screen

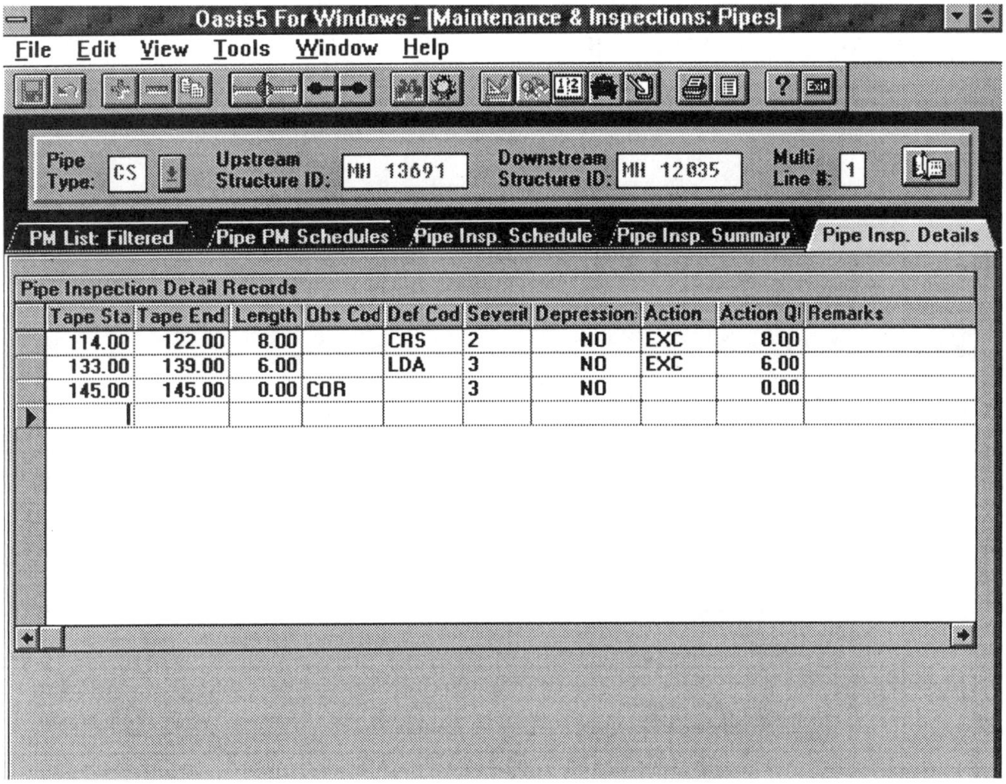

Fig. 12.32 CMMS pipe inspection detail screen

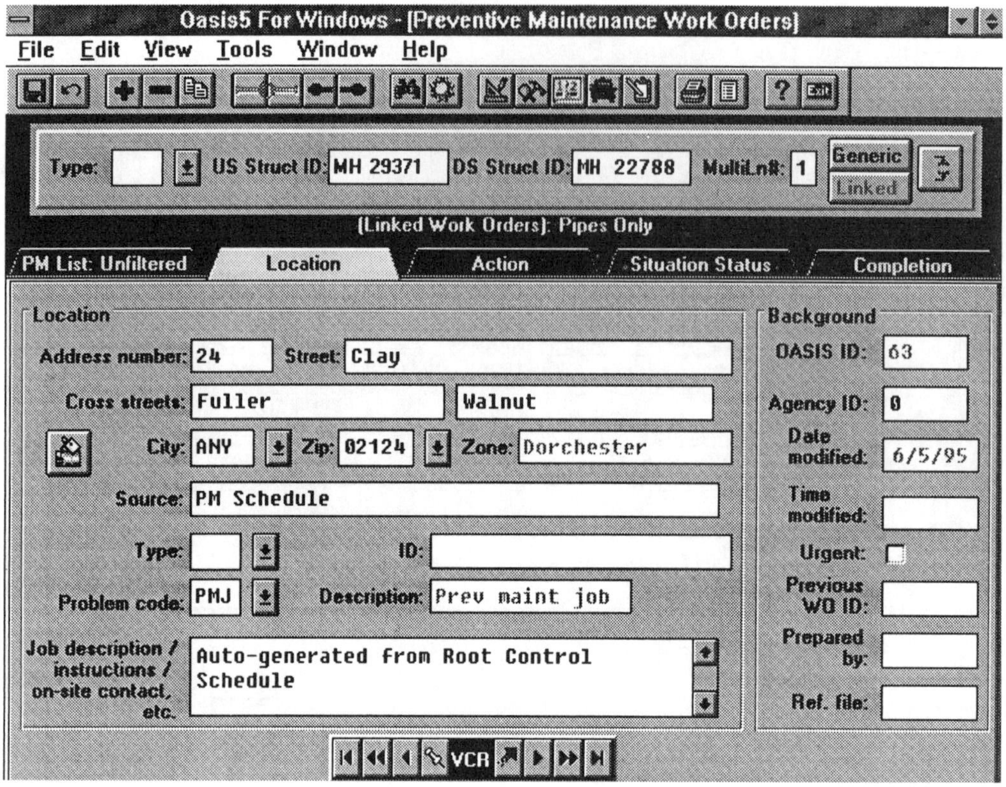

Fig. 12.33 CMMS work order screen showing location of job

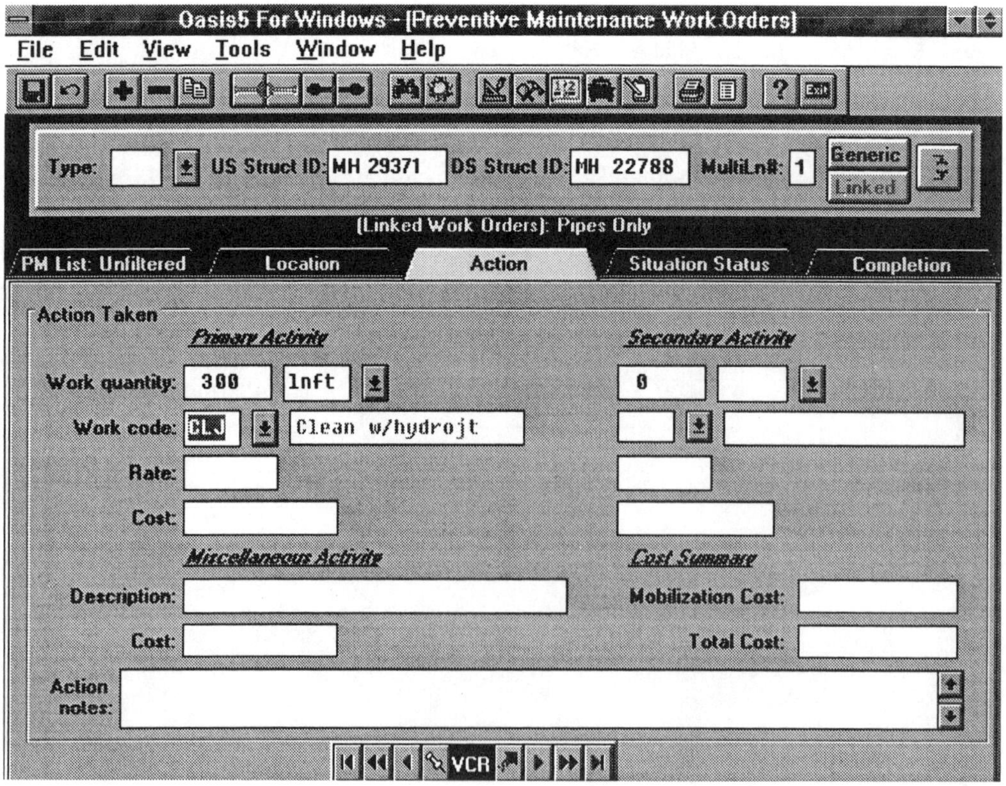

Fig. 12.34 CMMS work order screen showing work needed

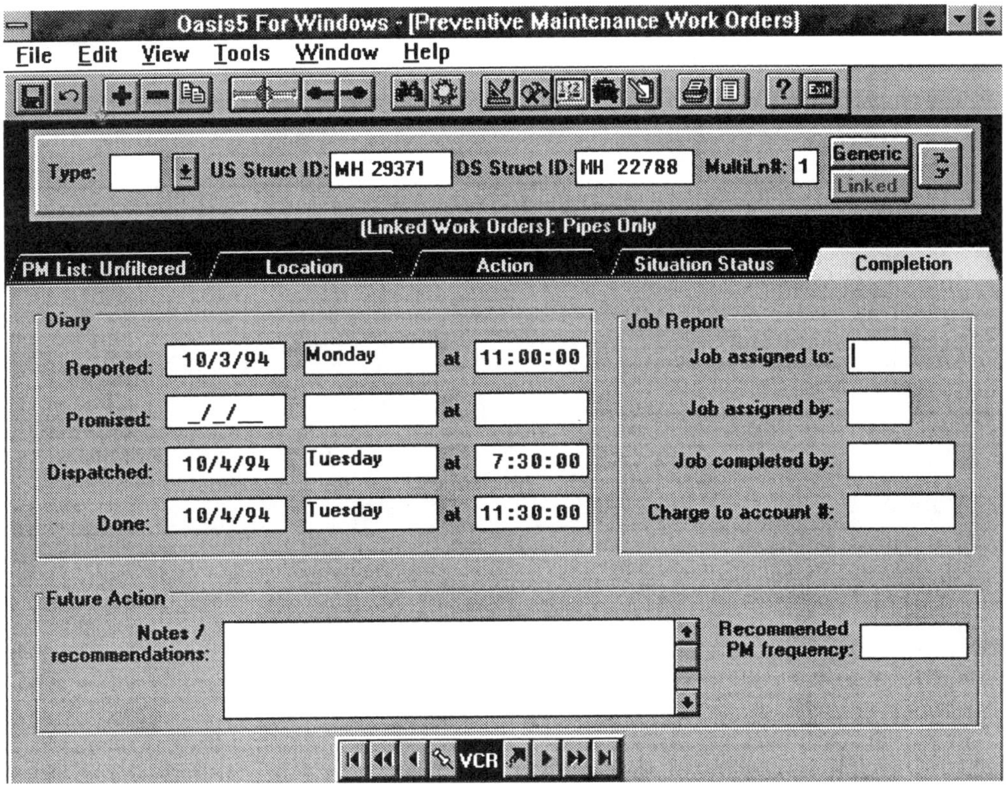

Fig. 12.35 CMMS work order completion screen

Databases also are able to store information on costs, equipment history and other relevant information that is needed in order to manage an equipment-based facility. They also have the capacity to record information on hourly labor rates and other staffing information. A wide variety of reports can be produced to track and report maintenance activity, performance and effectiveness.

12.7102 FIELD DATA ENTRY IN CMMS

Today's technology has allowed for computers not only to quickly analyze field data brought into the office but to be part of the field work. Such is the case for television inspection. Computers enable an operator to enter data directly into a TV inspection software program and then the information can be transferred (down loaded) to a central database, thus eliminating the paper work involved in recording data on a log sheet. The computer software used for the TV inspection electronically captures images on a 3.5-inch floppy disk, recordable compact disc (CD-R), or a compatible zip disk, which permits the information to be retrieved and printed at any time. The computer in the field may also provide an immediate rehabilitation analysis. This is especially helpful if rehab such as grouting can be accomplished immediately after the line segment has been televised. Figures 12.24 and 12.36 are examples of computer-generated TV inspection reports.

Scannable paper field forms are an efficient mechanism for gathering inspection data, including manhole inspection, visual pipe inspection, smoke testing, dyed water flooding, and building inspection. Standardized forms can be prepared to accommodate "check the box," numerical, or text-type responses. Scannable field forms are used instead of notebook computers by some collection system agencies.

Direct digital television inspection is becoming very popular. Technology is currently in use to gather television data in digital format in the field, capturing both video and audio in digital format. Digital television data can then be transferred to alternate storage media such as CD or DVD. This technology allows for rapid access to pipe data not available with conventional VHS tapes.

Once the video is digitally captured and observations are logged, television inspection data can be edited and verified for quality control. Attachments including design or as-built drawings (record drawings) files, maps, or maintenance records can be viewed and printed. The ability to quickly move through large digital files and only view "bad" pipe or other relevant data is a major improvement over the conventional VHS tape review process. Data can be stored on CD ROM or DVD or in a file server environment, depending on the quantity of data.

Once the television inspection data are stored digitally, database search engines can search the entire database for all records matching specific criteria, such as pipe type or material, structure number, street, street address, or type of problem (such as roots or grease) by area or time span.

12.711 Geographic Information System (GIS)

The geographic information system (GIS) is a computer program that combines mapping with detailed information about the physical structures within geographic areas. To create the database of information, "entities" within a mapped area, such as streets, manholes, line segments and lift stations, are given "attributes." Attributes are simply the pieces of information about a particular feature or structure that are stored in a database. The attributes can be as basic as an address, manhole

06/28/1997 02:07:48 PM Page 1

Television Inspection Report

Line Segment Number....(02)100 - (02)101

Date Televised.........06/28/1995

Time Televised.........01:19 P M

Operator ID...........SDP

Precipitation..........01 NONE

Ground Condition.......01 DRY

Surface Type...........01 PAVED

Pipe Material..........01 VITRIFIED CLAY

Diameter............... 8 (in)

Depth..................10.00 (ft)

Length Between Joints.. 3.0 (ft)

Pipe Length............ 300.0 (ft)

Test Pressure..........40.00 (psi)

Tape Number............101

Starting Index.........100

Ending Index...........END

Street Name............22ND STREET

Location...............WEST EASEMENT

Remarks...........TELEVISED DURING HIGH FLOW CONDITIONS

==

Fig. 12.36 Computer-generated TV inspection report

number, or line segment length, or they may be as specific as diameter, rim invert and quadrant (coordinate) location. Attributes of a main line segment might include engineering information, maintenance information and inspection information. Thus an inventory of entities and their properties is created. The system allows the operator to periodically update the map entities and their corresponding attributes.

The power of a GIS is that information can be retrieved geographically. An operator can choose an area to look at by pointing to a specific place on the map or outlining (windowing) an area of the map. The system will display the requested section on the screen and show the attributes of entities located on the map. A printed copy may also be requested. Figures 12.21 and 12.37 show GIS-generated maps. This example shows data from an inflow/infiltration analysis, including pipe attributes, hydraulic data, and selected engineering data. The example also shows a map of the system. In most cases CMMS software has the ability to communicate with geographic information systems so that attribute information from the collection system can be copied into the GIS.

A GIS can generate work orders in the form of a map with the work to be performed outlined on the map. This minimizes paper work and gives the work crew precise information about where the work is to be performed. Completion of the work is recorded in the GIS to keep the work history for the area and entity up to date. Reports and other inquiries can be requested as needed, for example, a listing of all line segments in a specific area could be generated for a report.

In many areas GISs are being developed on an area-wide basis with many agencies, utilities, counties, cities and state agencies participating. Usually a county-wide base map is developed and then all participants provide attributes for their particular systems. For example, information on the sanitary sewer collection system might be one map layer, the second map layer might be the water distribution system, and the third layer might be the electric utility distribution system. In addition to sharing data bases with CMMSs, GISs generally now also have the ability to operate smoothly with computer-aided design (CAD) systems.

QUESTIONS

Write your answers in a notebook and then compare your answers with those on page 548.

12.7P What are the two basic categories of computer-based maintenance management systems (CMMSs)?

12.7Q What is the main difference between the two types of CMMSs?

12.7R What makes a geographic information system (GIS) a potentially powerful tool for a wastewater collection system operator?

12.712 SCADA Systems (Figures 12.38 and 12.39)

SCADA (ss-KAY-dah) stands for **S**upervisory **C**ontrol **A**nd **D**ata **A**cquisition system. A SCADA system is an integrated package made up of several components that provide a computer-monitored alarm, response, control, and data acquisition system used by wastewater collection system operators to monitor and adjust the operation of equipment in their systems.

A SCADA system collects, stores and analyzes information about all aspects of operation and maintenance, transmits alarm signals when necessary and allows fingertip control of alarms, equipment and processes. SCADA provides the information that operators and their supervisors need to solve minor problems before they become major incidents. As the nerve center of a wastewater collection agency, the system allows operators to enhance the efficiency of their facilities by keeping them fully informed and fully in control.

A typical SCADA system is made up of basically four groups of components:

1. Field-mounted sensors, instrumentation and controlled or monitored equipment. These devices sense system variables and generate input signals to the SCADA system for monitoring. These devices also receive the command output signals from the SCADA system. Examples of input devices would include level transmitters, floats switches, flowmeters, and pump running intrusion switches. Examples of output devices would include pump calls, pump speed commands, chemical pacing signals, and valve open and close commands.
2. Remote Terminal Units (RTUs). These devices gather the data from the field-mounted instruments and provide the control signals to the field equipment. Typically, there are different types of RTUs and PLCs, but the most current technology uses PLCs as RTUs or "smart" RTUs that are capable of local control functions.
3. Communications medium. This is the link between the SCADA RTUs and the main control location. There are many communications mediums available for transmitting signals between the RTUs and the supervisory control station. Some of the options presently available include:
   - FM (VHF/UHF) radio,
   - Dedicated leased telephone circuits,
   - Privately owned metallic signal lines,
   - Conventional dial telephone lines (pulse or tone/DTMF),
   - Coaxial cable networks,
   - Spread spectrum radio,
   - Fiber-optic cable,
   - Microwave,
   - 900 MHz radio,
   - Cellular telephone,
   - Ground station satellites,
   - Trunked radio,
   - MAS (Multiple Addressed System) radio,
   - 2.4 GHZ radio,
   - Advanced Digital Network (A.D.N.),
   - Digital Subscriber Line (DSL), and
   - CATV modems.

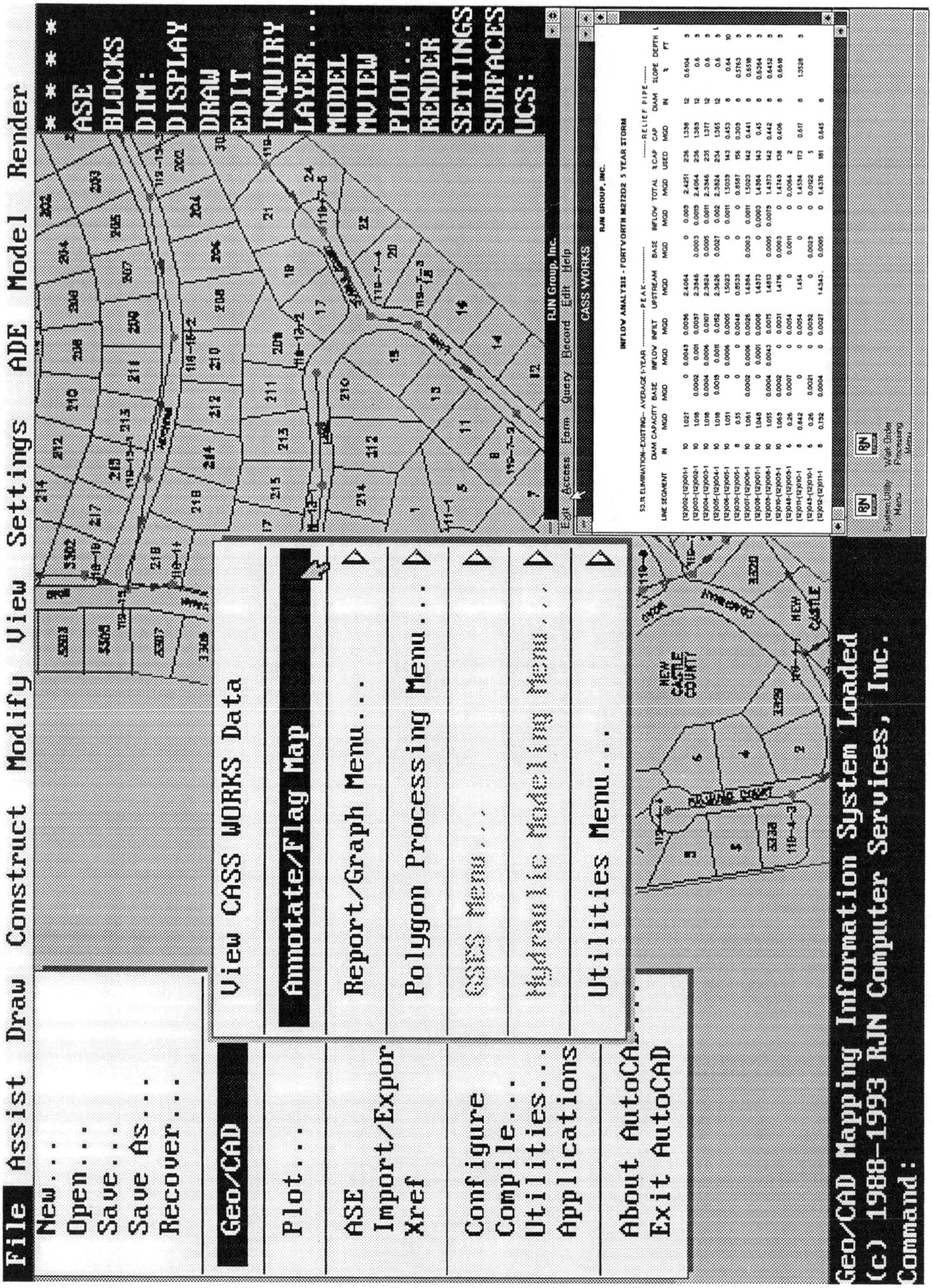

Fig. 12.37 Typical Geographic Information System (GIS) map

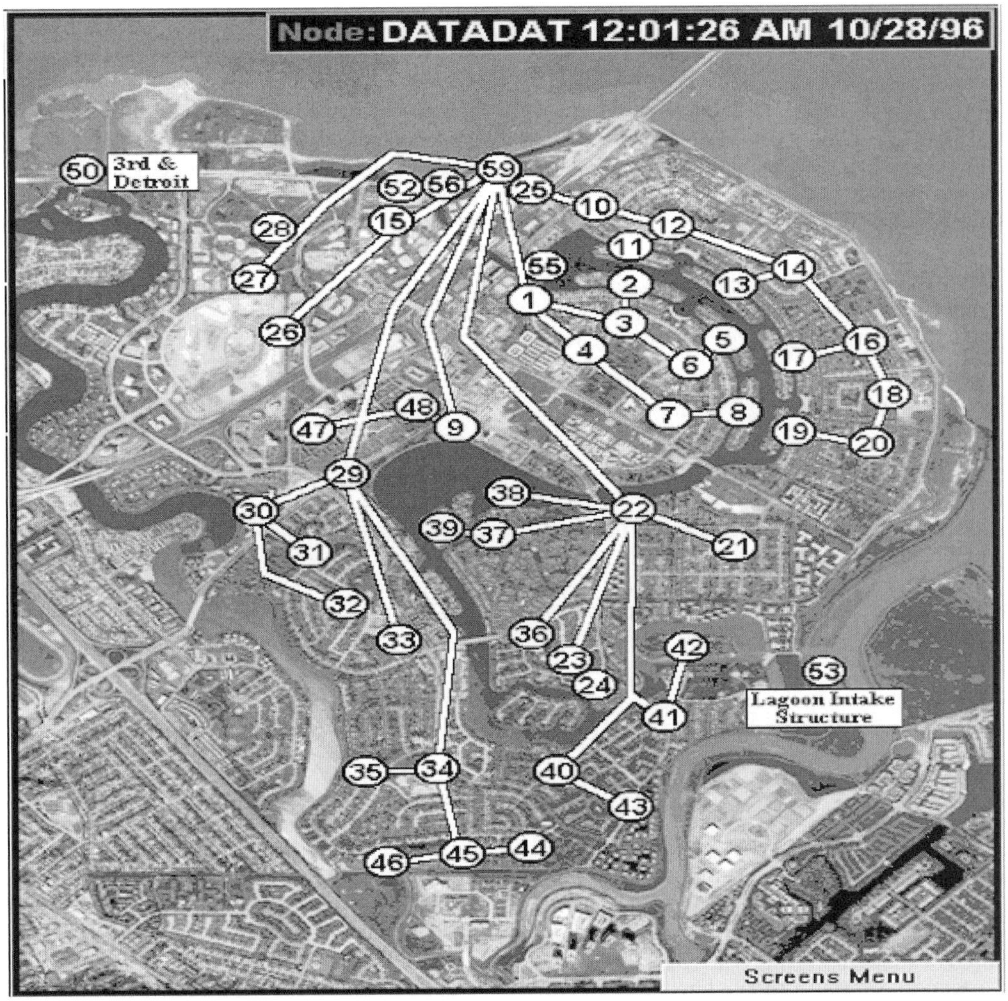

Fig. 12.38 SCADA system

(Permission of Hansen Information Technologies)

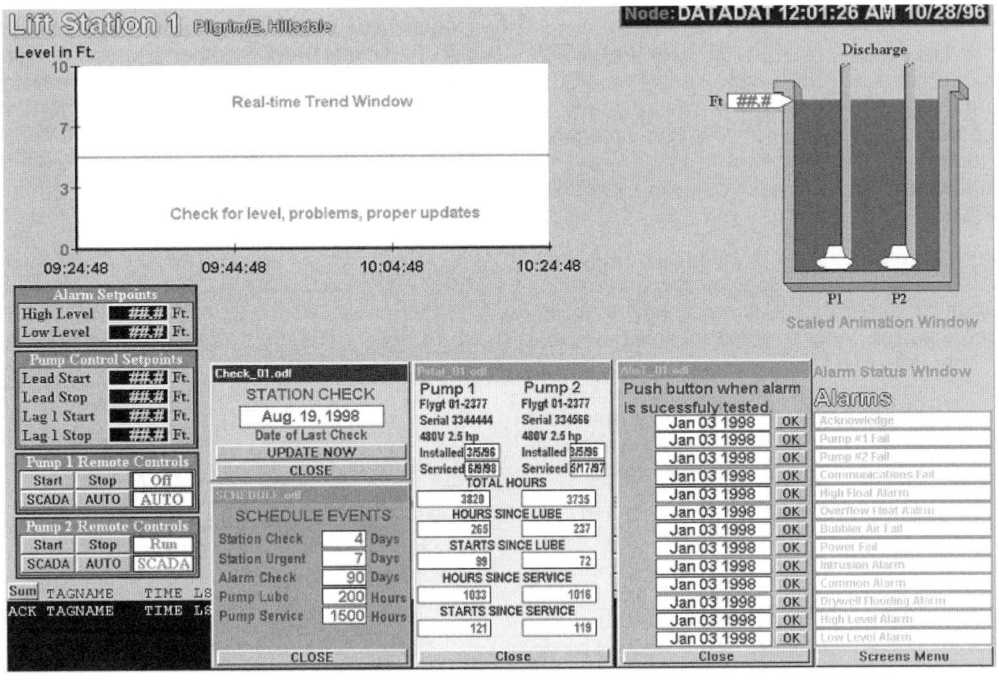

Fig. 12.39 SCADA computer monitor screen and reports

(Permission of Hansen Information Technologies)

4. Supervisory control and monitoring equipment. There are three basic categories of supervisory equipment:

   a. Hardware-based systems include graphic displays, annunciator lamp boxes, chart recorders and similar equipment. The current trend is to move away from this approach and move toward implementation of software-based systems.

   b. Software-based systems include all computer-based systems such as microcomputers, workstations, minicomputers and mainframes. Software-based systems sometimes offer more flexibility and capabilities than hardware-based systems and are often less expensive.

   c. Hybrid systems are a combination of hardware and software systems; for example, a PC (microcomputer) with a graphic display panel.

Applications for SCADA systems include wastewater collection and pumping system monitoring (see Chapter 8, Section 8.31, "Frequency of Visits to Lift Stations"), wastewater treatment plant control monitoring (Figure 12.40), combined sewer overflow (CSO) diversion monitoring, and other related applications. SCADA systems can vary from merely data collection and storage to total data analysis, interpretation and process control.

A SCADA system might include liquid level, pressure and flow sensors. The measured (sensed) information could be transmitted by one of the communications systems listed earlier to a computer system which stores, analyzes and presents the information. The information may be read by an operator on dials or as digital readouts or analyzed and plotted by the computer in the form of trend charts.

Most SCADA systems present a graphical picture of the overall system on the screen of a computer monitor. In addition, detailed pictures of specific portions of the system can be examined by the operator following a request and instructions to the computer. The graphical displays on a TV or computer screen can include current operating information. The operator can observe this information, analyze it for trends or determine if it is within acceptable operating ranges, and then decide if any adjustments or changes are necessary.

SCADA systems are capable of analyzing data and providing operating, maintenance, regulatory and annual reports. In some collection systems operators rely on a SCADA system to help them prepare daily, weekly and monthly maintenance schedules, monitor the spare parts inventory status, order additional spare parts when necessary, print out work orders and record completed work assignments.

SCADA systems can also be used to enhance energy conservation programs. For example, operators can develop energy management routines that take advantage of lower-cost off-peak energy usage. Most power companies are anxious to work with operators to try to increase collection system power consumption during periods when electrical system power demands are low and also to decrease power consumption during periods of peak demands on electrical power supplies.

SCADA systems also monitor power consumption and conduct a diagnostic performance examination of pumps. Operators can review this information and then identify potential pump problems before they become serious. In this type of system, power meters are used to accurately measure and record power consumption and the information can then be reviewed by operators to watch for changes that may indicate equipment problems.

Emergency response procedures can be programmed into a SCADA system. Operator responses can be provided for different operational scenarios that could be encountered as a result of adverse weather changes, fires or earthquakes.

SCADA systems are continually improving and helping operators do a better job. Today operators can create their own screens as well as their own graphics and show whatever operating characteristics they wish to display on the screen. The main screen could be a flow diagram of the collection system. Critical operating information could be displayed for each pumping station or each segment of main line and detailed screens could be easily reached for each piece of equipment.

Information on the screen should be color coded with the colors of red, yellow, green, blue, white and any other necessary colors. Colors could be used to indicate if a pump is running, ready, unavailable or failed and/or if a valve is open, closed, moving, unavailable or failed. A "failed" signal is used by the computer to inform the operator that something is wrong with the information or signal the computer is receiving or is being instructed to display. The signal is not logical with the rest of the information available to the computer. For example, if there is no power to a motor, then the motor can't be running even though the computer is receiving a signal that indicates it is running. Therefore the computer would send a "failed" signal.

The operator can request a computer to display a summary of all alarm conditions in the entire collection system, a particular portion of the system, or at one or more pumping stations. A blinking alarm signal indicates that the alarm condition has not yet been acknowledged by the operator. On the other hand, a steady alarm signal, one that is not blinking, indicates that the alarm has been acknowledged but the alarm indicator will stay on until the condition causing it is fixed. Also the screen could be set up to automatically designate certain alarm conditions as *PRIORITY* alarms, which means they require immediate operator attention.

Current laptop computers allow operators to plug into a telephone at home or on vacation, access the SCADA system and help operators on duty solve operational problems. Computer networking systems allow operators at terminals in offices and in the field to work together and use the same information or whatever information they need from one central file service (computer database). Wireless Internet access allows the users of laptop computers and cell phones, using WAP (wireless application protocol) or PDAs (personal digital assistants), access to SCADA systems when in the field. Intranet and Internet networking systems allow operators at terminals in offices and in the field to work together and use the same information or whatever information they need from one central file server (computer database).

A vast array of custom third-party applications software exists to assist the operator and management staff with the creation of reports and scheduling (Figure 12.39).

There are a few products designed to work with PC-based SCADA systems to notify personnel of abnormal conditions. These packages monitor the processes for alarm conditions and send alarms by e-mail, pager, cell phone, alphanumeric pager, or traditional telephones. The operator can call in by telephone and acknowledge alarms, change set points, and reset equipment.

Not all SCADA systems are perfect, and occasional "glitches," usually in communication infrastructure, can cause all parts of some SCADA systems to fail. There is a possibility for

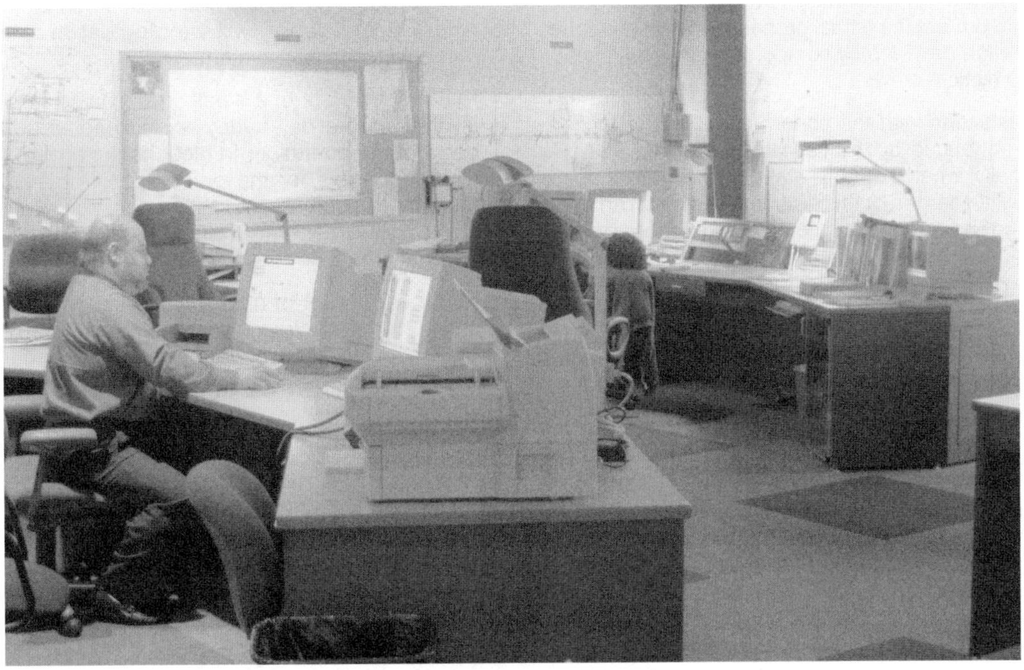

SCADA monitoring station

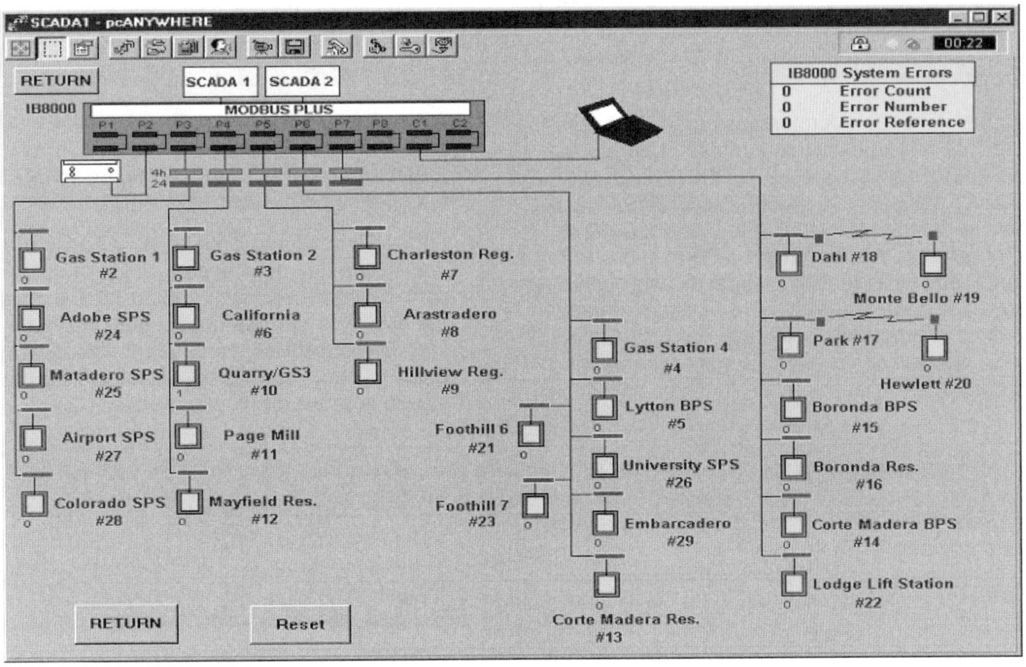

Fig. 12.40 SCADA main control center

(Permission of Hansen Information Technologies)

the system to display the numbers that were registered immediately before the failure and not display the current numbers. The operator may therefore experience a period of time when accurate, current information about the system is not immediately available. Timely notification of failures (even with the SCADA infrastructure) is important, as well as providing backup devices to protect the actual process. Always compare "strange" information provided by computers with actual information observed by field crews.

Customer satisfaction with the performance of a collection agency can be enhanced by the use of an effective SCADA system. Historically when a pump station failed or "tripped out," the first a utility learned of this problem was when an irate consumer phoned the agency and complained about a sewer backup in their home. The utility then had to contact the operators and send a crew into the field to correct the problem. Today, SCADA systems often alert operators to a pump station failure or "trip out" immediately. The operator may be able to correct or "override" the failure or "trip out" from the office without ever having to travel to the problem pump station in the field. Thus, the problem is corrected without the customers ever being aware that a problem occurred and was corrected almost immediately.

When operators decide to initiate or expand the SCADA system for their collection system, the first step is to decide what the SCADA system should do to make the operators' jobs easier, more efficient, and safer and to make their facilities' performance more reliable and cost effective. Cost savings associated with the use of a SCADA system frequently include reduced labor costs for operation, maintenance and monitoring functions that were formerly performed manually. Preventive maintenance monitoring can save on equipment and repair costs and, as previously noted, energy savings may result from use of off-peak electrical power rates. Operators should visit facilities with SCADA systems and talk to the operators about what they find beneficial and also detrimental with regard to SCADA systems and how the systems contribute to their performance as operators.

The greatest challenge for operators using SCADA systems is to realize that just because a computer says something (a pump is operating as expected), *THIS DOES NOT MEAN THAT THE COMPUTER IS ALWAYS CORRECT.* Operators will always be needed to question and analyze the results from SCADA systems. Also when the system fails due to a power failure or for any other reason (natural disaster), operators will be required to operate the collection system manually and without critical information. Could you do this? Collection systems will always need alert, knowledgeable and experienced operators who have a "feel" for their collection system.

QUESTIONS

Write your answers in a notebook and then compare your answers with those on page 548.

12.7S What does SCADA stand for?

12.7T What does a SCADA system do?

12.7U How could measured (sensed) information be transmitted?

12.7V What are the greatest challenges for operators using SCADA systems?

12.72 Recordkeeping

Records consist of written or electronically stored data that perpetuate a knowledge of events. From its start, management of a collection system will begin to generate records. The variety of records to be produced is unlimited. The size of the collection system, scope of activities, variety of programs, and the desires of supervisors and management will all have an influence on the volume and detail of the records that are produced. Supervisors will find their personal involvement with actual recordkeeping will vary in relation to the size of their particular collection system.

The supervisor of a small collection system may be directly involved in the day-to-day recordkeeping activities of all aspects of the operation. In fact, the supervisor may be personally responsible for the preparation of many of the required records. In such a case, the supervisor will certainly take an active role in deciding what types of records are needed and on what forms they are to be recorded.

The supervisor of a large collection system may encounter entirely different circumstances. Larger organizations will have more sophisticated record systems, with recordkeeping processes and policies that have been set up by specialized accounting sections. The supervisor may be responsible only for the administration of these processes and adherence to set policies.

Records pertinent to collection systems cover a wide range of needs and activities. Cost accounting, personnel, production, payroll, and history are just a few examples of major concerns. For the purpose of this section, we will place our emphasis on records that are of operational interest.

12.720 Need for Records

In general, records fill two basic needs. They record the past and provide a sound basis on which to plan the future. To meet these needs, records will have to be maintained on almost every activity and event that occurs in the daily operation and maintenance of the collection system. Some major areas of importance are discussed in this section.

PAYROLLS AND COST ACCOUNTING

Costs must be of prime concern to all maintenance supervisors. It is not enough that the supervisor maintain the system to a high degree of efficiency—the supervisor also is accountable for the costs involved.

- Labor. Time sheets, time cards, payrolls, and time reporting are all of vital importance to each of us. We all want to be paid. Every supervisor should be particularly attentive to the accuracy, completeness, and punctuality of time reporting. The supervisor, too, wants the operators paid accurately and on time. Delayed or inaccurate paychecks do little to improve employee morale.

- Equipment. As with labor costs, accurate reporting of daily use of equipment is also necessary.
- Supplies. Distribution, consumption, and use of materials must be accounted for accurately. Supplies usually are obtained from two sources, outside vendors and in-house stores. Special attention should be given to efficient handling of delivery receipts, bills and statements generated by purchases from outside vendors. Vendors also expect to be paid promptly and you can minimize carrying charges by

doing so. Maintaining good relations with these vendors can come back as a direct benefit to you when an emergency occurs and you need a specialty item in a hurry.

- Services. Many maintenance organizations do not have the necessary equipment or adequate staff to complete all their required activities. Certain aspects of their work may require specialized equipment, or have an infrequent need factor. Such work might be more economically performed by other than in-house staff. From an operational standpoint, it is important to maintain close accounting of the cost of these outside services. A significant increase in volume might indicate that it would be more economical to acquire the necessary equipment and/or staff needed to perform such work in-house.

BASIS FOR PLANNING AND SCHEDULING

A maintenance schedule or program should be designed to alleviate or eliminate certain problem conditions that are known to exist within a collection system. Adequate records are required to locate and identify these conditions and provide information for planning and timely scheduling of maintenance procedures. The types of records and forms needed to provide this information are covered in detail in Section 12.721, "Types of Records."

BUDGETING

At budget time, material and equipment records are of prime importance to a supervisor. Accurate records detailing last year's activities, costs, workload, growth, and production are the best justification for next year's budget requests. Good records are essential for the preparation of a budget and programs to provide for an effective collection system operation and maintenance program. Use your records to document the fact that collection system programs are as important as the programs of other agency divisions with which you may be competing for funds.

- Staffing. Requests for increased staffing are one of the most difficult budget items to justify to the satisfaction of management. Rarely does management accept a request for more staff when the request is based solely on the opinion or experience of the supervisor, regardless of the supervisor's competency. Accurate records can provide the necessary justification for additional staff. Analysis and presentation of records covering several years can indicate a definite trend in your division's activities. They can verify that your division's workload has increased to a point that makes additional staffing essential and justified.

  Another possible approach you might take in the preparation and justification of budgets is to compare your division's programs with staffing guides developed by the U.S. Environmental Protection Agency and other responsible sources. Be sure to consider the unique aspects of your division when using the number of personnel used to accomplish similar collection system program requests.

- Equipment (additional and replacement). As with staffing, these items must also be justified at budget time. Age alone is not always the determining factor as to when equipment should be replaced. Production and utilization records and maintenance costs to keep the equipment operational may be used to substantiate the need for replacement.

  Additional equipment often is required when additional staff are hired. One will usually justify the other. Often the need for newly developed equipment to cope with a specific condition will become apparent. Here again the best justification may be found in your records.

- Materials. Past requirements and consumption based on records will provide the most accurate indication of future needs.

- Services. Budget time is when you should carefully analyze past expenditures for outside services. Your records could indicate they have increased to a point where it would be more efficient and economical to acquire staffing and equipment needed to perform such work in-house.

- Analysis of Records. Good records can be easily analyzed and the results of analysis presented in tables, charts or graphs. These results often SHOW or ILLUSTRATE a budget need very effectively. To illustrate this point, let's examine an actual example of how Sacramento County justified the budgeting of an additional high-velocity cleaner and crew.

  The wastewater collection system was divided into two portions for the purpose of developing a preventive maintenance schedule. One portion of the system was put on a one- or two-year schedule while the other portion was put on a three- to six-year schedule. The worst areas were included in the one- to two-year schedule. Results from an analysis of records were PLOTTED as shown on Figure 12.41.

  An examination of Figure 12.41 SHOWS that the portion on the one- or two-year schedule had a significant reduction in the number of stoppages, but the portion on the three- to six-year schedule was experiencing an alarming increase in the number of stoppages.

  What caused the increasing number of stoppages? Certainly the frequency of cleaning is one answer. Were the stoppages caused by grease, roots or other problems? The answer to this question will tell you what types of equipment or adjustments in the preventive maintenance schedule are needed to reduce stoppages. In Figure 12.41, most of the increases in stoppages occurred in lightly populated areas with large lots. Grease accumulation was the main cause of stoppages.

  How can the number of grease stoppages be reduced? One more high-velocity cleaner would allow Sacramento County to place almost the entire wastewater collection system on a two-year or more frequent preventive maintenance schedule as needed.

  This pictorial presentation of results and logical analysis of the problem and proposed solution can be a very effective means of justifying your budget.

BASIS FOR CLAIM DEFENSE

The nature of our work exposes us to direct contact with the public and their private property. The public, in turn, will have daily contact with appurtenances of our collection systems which are in public right of ways. All these contacts provide many opportunities for either damage to private property or injury to the public through contact with a defective part of our system.

Aside from the fundamental necessity of exercising appropriate care and consideration while on private property and

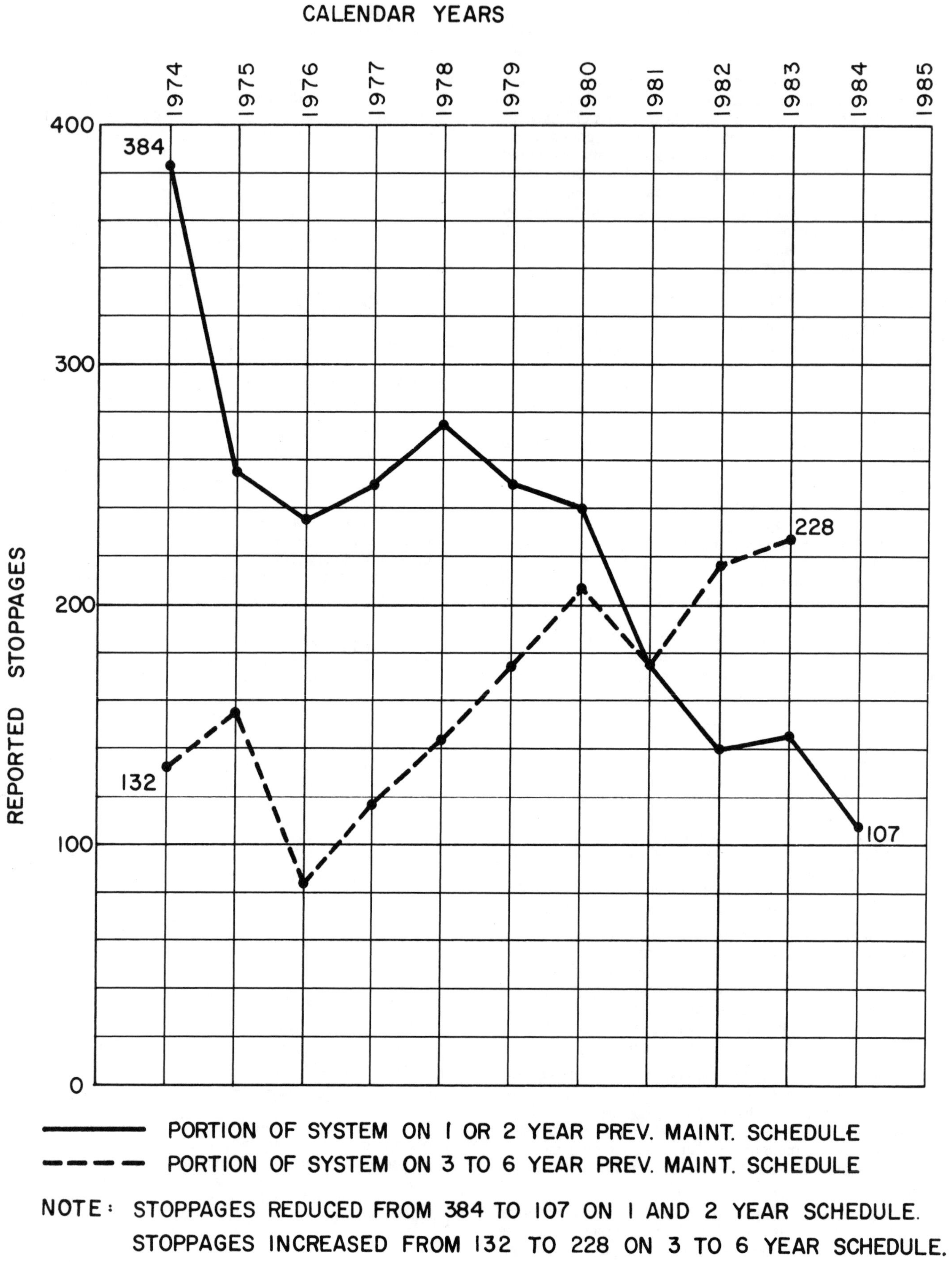

Fig. 12.41 Trends in number of stoppages in Sacramento County

keeping our collection systems free of hazards, there is another thing we must do. We must keep adequate records to document our daily routines to a degree that will be of value in establishing our innocence or to disprove any negligence on our part should liability claims arise.

In event of such legal action, your records are certain to be examined in minute detail. Routine items such as dates, time, and identification numbers will suddenly become legal evidence documenting your actions or inaction. It is very important that you set up a system to check and double check these routine items on the hundreds or thousands of records generated throughout the year before they are filed to become a permanent legal record.

QUESTIONS

Write your answers in a notebook and then compare your answers with those on page 548.

12.7W List the two basic reasons for collection system records.

12.7X Records should be kept on what different cost items?

12.7Y How can good records be helpful to persons preparing a budget?

12.7Z Discuss the legal aspects of recordkeeping.

12.721 Types of Records

There are so many different types of records and special forms that can be useful to collection system operation and maintenance agencies that space will not allow detailed discussion of all of them in this section. Only a few of the most common will be presented or illustrated. As previously discussed, most of these records can be prepared and stored either manually or electronically (using computers). With these examples you can prepare forms to meet your own special needs.

PERSONNEL

Records relative to personnel are common to all organizations.

- Personnel File. A separate file should be maintained for each individual employee. An important part of this file is an individual information sheet that provides such information as: date of employment, age, blood type, inoculations, dependents, and medical history for use in case of accident or serious injury. Keep these files up to date and confidential.

  Other information contained in personnel files may include work history, commendations, disciplinary actions, accidents, attendance records, performance evaluations, training courses, and certifications. The contents of personnel files must comply with your agency's personnel rules and regulations, any union contracts and legal requirements.

- Attendance. Data on attendance and reasons for leave during working hours can be obtained from the daily work reports. A superintendent or supervisor may use these records to counsel operators who are abusing their leave privileges or are taking unauthorized leave. Excessive use of leave, especially on Fridays and Mondays, may indicate problems that can be dealt with before an employee's work performance would require termination.
- Assignments. A record of an operator's work assignments over a period of time, developed from work reports, is useful when considering exposure of the employee to the various work experiences possible for career development.
- Pay. Personnel pay records developed from payrolls are used to provide employees with an annual accounting of their pay and withholdings, answer questions they might raise about past pay, and to provide a history that is useful in planning pay adjustments.

In most organizations, personnel reports are primarily the responsibility of a personnel or accounting department. However, some personnel records that should be of direct concern to the supervisor include operator performance reports, accident reports, and attendance analysis reports.

EQUIPMENT

Equipment records fall into the following three main categories:

- Equipment Inventory and Depreciation. An equipment inventory record provides an accounting of the type and cost of equipment owned by a collection system agency and, as illustrated by Figure 12.42, provides information on repairs and depreciation of equipment.
- Preventive Maintenance. A preventive maintenance record system includes a schedule of routine maintenance, user-generated requests for special maintenance, and a history of maintenance and repairs performed. The maintenance record must include costs since they become a part of the cost of operating a piece of equipment.
- Use. Refer to Figures 12.9 through 12.16 and you will note that each daily work report form has a space for reporting the equipment used to complete the work. This information is needed to record the total use of a piece of equipment and calculate the cost of the work performed.

MAIN SEWER LINE

Records for lateral, main, trunk and interceptor sewer lines all fall into the main sewer line record category and include the following:

1. Main Line Inventory. The type of information that should be included in a main line inventory, as illustrated by Figure 12.43, consists of:
   - Identification of a line segment using the discrete (individual) numbers of the upstream and downstream manholes.
   - Size and length of the line segment, depth of the line at the upstream and downstream manholes, type of material and the line's location (street or easement).

EQUIPMENT DEPRECIATION RECORD

YR. DEPN. ENDS	DEPARTMENT	EQUIPMENT	ASSET NO.

DESCRIPTION	INITIAL LOCATION		TRANSFERS			
	DEPT.					
	BLDG.					
	FLOOR					
	DATE					

TYPE OF DRIVE	POWER	H.P. REQD.	FLOOR SPACE	WEIGHT	MEMO
DATE AQD.	BOUGHT FROM			MFRS. NO.	P.O. NO. / VOUCHER NO.
EST. LIFE YRS.	ANNUAL DEP'N. RATE % AMT.		APPRAISAL PAGE		

COST DETAILS	ORDER NO.	AMOUNT	DEDUCTION RECORD	
NET COST			DATE DEDUCTED SCRAPPED ☐ SOLD ☐ TRADE-IN ☐	
FRT. & CART.			COST	
INSTALLATION			ACCRUED DEPRECIATION	
			DEPRECIATED VALUE	
			SELLING PRICE	
			PROFIT OR LOSS	

REPAIR RECORD

DATE	S. O. NO.	AMOUNT	DATE	S. O. NO.	AMOUNT	DATE	S. O. NO.	AMOUNT	DATE	S. O. NO.	AMOUNT

The UTILITY Line Form No. 25-239

ASSET NO.	EQUIPMENT	ANNUAL DEPN. RATE % AMT.	YR. DEPN. ENDS

DATE	REFERENCE	ASSET			DEPRECIATION			NET BOOK VALUE
		DEBIT	CREDIT	BALANCE	DEBIT	CREDIT	BALANCE	

Fig. 12.42 Equipment inventory and depreciation record

SEWER SECTION INFORMATION SHEET

SEWER LINE INFORMATION

Date Inspected: ________ By: ________

PMMS ID#	Section On Street	From Street	To Street

Location Notes:

Inspection ID#	Length	Line Shape	Line Material %	Comments
		CIRCULAR ❑	BRICK ❑➔ ____	
Overall Score 1 2 3 4	Size ____ X ____	EGG ❑ BOX ❑	CLAY ❑➔ ____ CONCRETE ❑➔ ____	
Score Date	Easement? YES ❑ NO ❑	OTHER ❑	CEM/MORTAR ❑➔ ____ OTHER ❑➔ ____	

Date Built/Rehab	Plan Number	As-Built Number	Map ID#	Map Coords.	Drainage

UPSTREAM MANHOLE INFORMATION

ID# ________ A B C D E F

Number or "INT"	Street	Cross Streets or Intersections	Head End? YES ❑ NO ❑

Rim Elevation	Location Notes:

Invert Elevation	Manhole Type	Material	Comments
	MANHOLE ❑	BRICK ❑	
Depth	DEAD END ❑ WING ❑	CONCRETE ❑ OTHER ❑	
Size of Opening	OTHER ❑		

Size of Opening (cont.) ____" X ____"	Date Built/Rehab	Map ID#	Map Coordinates	Survey Coord. X	SurveyCoord.Y

DOWNSTREAM MANHOLE INFORMATION

ID# ________ A B C D E F

Number or "INT"	Street	Cross Streets or Intersections	Head End? YES ❑ NO ❑

Rim Elevation	Location Notes:

Invert Elevation	Manhole Type	Material	Comments
	MANHOLE ❑	BRICK ❑	
Depth	DEAD END ❑ WING ❑	CONCRETE ❑ OTHER ❑	
Size of Opening	OTHER ❑		

Size of Opening (cont.) ____" X ____"	Date Built/Rehab	Map ID#	Map Coordinates	Survey Coord. X	Survey Coord. Y

Fig. 12.43 Main line inventory

- Designation of the area of the collection system agency containing the line segment.
- The "as-built" map or record drawing reference number, year built, and the date of the last television survey.
- The line inverts at the upstream and downstream manholes and the map coordinates of these manholes (if known).

2. Main Line Maintenance History. Record the history of maintenance by line segments, as illustrated by Figure 12.44 (the code number designations are shown on the bottom of Figure 12.11), and include:
   - Date of a maintenance activity during a period of years, usually six to eight years.
   - The type of equipment used for maintenance activity.
   - The time taken to perform the maintenance activity.
   - Types, severity and estimated weight of debris removed by the maintenance activity.
   - Condition of the line by type of problem and severity observed at the upstream and downstream manholes at the time of the maintenance activity.
   - The reason for the maintenance activity.
3. Television History. A good example of a form for recording the television history for a main sewer line segment is illustrated by Figure 12.45. The history includes:
   - Date of the television survey.
   - Videotape number and file location on the tape containing the survey.
   - Observed condition and cleanliness of the line grade and the direction of travel of the television camera.
   - Footage from entrance of manhole to recorded observation.
   - Observed tap location in code, pipe conditions, and any other important observations in code.
   - Remarks on observations not covered by the previously listed codes (root intrusion, grease deposits).
   - The illustrated form also notes work orders issued as a result of the recorded television survey.
4. Main Line Stoppages. One measure of the effectiveness of a preventive maintenance program for a collection system is the number of main sewer line stoppages that occur. What is the annual number of stoppages in your collection system (see Figure 12.41)? If the number of stoppages is increasing, identify the locations of the main lines having stoppages and increase the maintenance of these lines. You will have to decide the acceptable number of stoppages for your collection system and work toward that number since no practical preventive maintenance program will eliminate all stoppages.

   Microcomputer software records main stoppages by date and line segment, including the causes and actions taken to remove the stoppages.

SERVICE SEWER LINES

Records for building service sewer lines are similar to records for main sewer lines and include the following:

- Service Line Inventory. Service lines are inventoried by assessor's parcel number (if available) and street address. Records should also provide the main line segment to which it is connected and the location of cleanouts used for maintaining the agency's part of the service line.
- Service Line Maintenance History. Record the history of service line maintenance by individual lines (generally identified by property address) and include the date of a maintenance activity, the reason for it, type of maintenance performed, and time spent on the activity.

TIME, MATERIALS, AND EQUIPMENT

Time used by personnel on the activities of a collection system agency are recorded for each activity and used as needed to produce production and cost reports. Time records can also be used to analyze the use of personnel and to plan for rotation of work assignments.

Material use records, by activities and jobs, account for the depletion of inventoried supplies and allow the calculation of the cost of activities and jobs.

Equipment use records, also by activities and jobs, are developed from work reports and are used to determine the amount of time a piece of equipment has been operated. This information serves as a basis for determining the hourly operating cost of the equipment. Equipment use records are also used to assign operating costs to the appropriate activities and records.

ADDITIONAL RECORDS

A few records common to the operation and maintenance of most collection systems have been described. Additional records may include the following:

- Accidents involving agency personnel and equipment,
- Backflow of wastewater onto private property and into watercourses,
- Safety training activities,
- Job-related training, and
- Personnel actions.

Accident records are very important. Workers' Compensation claims require complete reports. Supervisors and superintendents need to know the safety record of their agency, as well as records for crews and individuals. Records of events in an accident might be useful as a legal defense against liability claims and against charges of agency negligence if approved safety procedures are followed and specific action is consistently taken (and recorded) to prevent similar accidents in the future.

12.722 Retention of Records

The superintendent of a wastewater collection system must know which records are to be retained, and for how long, to comply with state and federal regulations. Also records should be kept as long as they are thought to be useful or may be needed for defending claims for injuries or property damage submitted to an agency. Ask your legal counsel to provide guidance for retention of records.

Records that are retained on electronic computer disks and tapes do not use a lot of space, but they should be labeled and filed in a manner that will facilitate their use. Printed records are bulky and do require considerable storage space. You may want to consider microfilming printed records to reduce the storage space of this type of record.

Essential records should be protected from loss by appropriate storage including fireproof filing cabinets, vaults and a safe deposit box in a local bank.

12.723 Responsibility for Records

Forms can be designed to record all the necessary information that could ever possibly be needed, but this alone will not

COUNTY SANITATION DISTRICT NO 4

MAIN SEWER MAINTENANCE HISTORY

Structure Identifier   Date Listed: 10/27/86

Upstream Downstream
4 961Ø 4 96 2
Sub-Basin: 4
Street: HAMILTON AV

Diameter(Inches):.....1Ø
Pipe Material:........ 1
Lining:.............. Ø
Special Construction:. Ø
Length(Feet):......... 342

Date	Equip Oper.	OP Time (min)	Cost	Equip Oper.	OP Time (min)	Cost	Debris T1	Debris T2	Debris Sev.	Debris Lbs	Sewer Condition Typ1	S	Typ2	S	Reason for Maint.
Ø9/Ø3/82	7	9Ø		1	5		2	4	2	5Ø	Ø	Ø	Ø	Ø	Ø
Ø8/17/83	1	75					2	4	1	4Ø	Ø	Ø	Ø	Ø	Ø
12/13/84	1	115					4	1	2	1ØØ	Ø	Ø	Ø	Ø	Ø
Ø3/19/86	1	6Ø					1	2	1	25	9	1	Ø	Ø	Ø
Ø4/24/86	A	35		5	3Ø		1	2	2	25	Ø	Ø	Ø	Ø	Ø
Ø6/Ø3/86	A	3Ø					1	2	2	22	Ø	Ø	Ø	Ø	Ø
Ø8/18/86	A	25					2	4	2	6Ø	Ø	Ø	Ø	Ø	Ø

Fig. 12.44 Main line maintenance history

SEWER TV LOG

Overall Score: ☐

1-CLASS "A" EMERGENCY
2-CLASS "B" EMERGENCY
3-To BOE Rehab/Replace Schedule
4-No Recommended Action

PMMS ID#	Section On Street	From Street	To Street

Location Notes:

Inspection #	Inspec. Crew	Date & Time	Tape #	Sequence #	Page #	# of Pages

UPSTREAM MANHOLE	Indicates Pull Direction	DOWNSTREAM MANHOLE
ID# A B C D E F		ID# A B C D E F
Repairs? Urgent ☐ Routine ☐ Okay ☐		Repairs? Urgent ☐ Routine ☐ Okay ☐

TAPE START	TAPE ENDS	QTY	OBS. CODE	DEF CODE	SEV (1,2,3,) P	J	- ← SUBTOTAL 1's, 2's, & 3's	M	I	D ?	REMARKS	ACT CODE	W ?	ACT QTY	W.O. NUMBER

	P	J		M	I	D
Total 1's						Total Depressions
Total 2's						
Total 3's						

Audited By: ☐

CCSF-102

VVA\BSSR\TVLOG\JAN95

Fig. 12.45 Television survey history

guarantee a good recordkeeping system. To create a functional recordkeeping system, all personnel involved must accept some measure of responsibility for their part in the system.

SUPERINTENDENT

Since a superintendent will derive the most direct benefit from good recordkeeping, the superintendent should be the most concerned. Concerns should include the following items:

- What is required. The superintendent should decide what records will be needed to meet job responsibilities. Control must be exercised over what amount of detail the records should include and which specific activities will require recordkeeping procedures.
- Use. Constantly be aware of the degree to which records are being used. Some records, after they are initiated, may become unnecessary because of changing conditions or policies. Promptly discontinue using any such records.
- Quality. A superintendent should be constantly alert to the level of compliance with recordkeeping policies. Periodically monitor the daily recordkeeping activities of all personnel, operators and staff alike. By examining all records closely for a period of four or five days, you will be better informed and will communicate your serious concern about accurate records to the entire agency staff.

COLLECTION SYSTEM OPERATORS

A majority of operational records originate in the field as work reports prepared by the collection system operators. Operators must understand their responsibility for submitting accurate and neat reports in a timely manner.

- Accuracy. Reports *MUST* be accurate. One digit in an address or identification number when omitted, unreadable, or reversed can make a report card practically useless.
- Legibility. Adverse working conditions and wet weather often make it difficult to produce legible reports. Excuses for illegible reports are often valid. Collection system operators should be allowed ample time and proper facilities to allow them to complete and turn in neat reports.
- Promptness. Collection system operators often develop a habit of completing their reports at the end of the day. This is especially true when crews perform work at several different locations during the day. The most effective method of completing a report is at the scene of each job when all the facts and circumstances are available and obvious.
- Quality. Collection system operators can and will comply with the foregoing standards. Much of the information reported will benefit operators doing future work. Collection system operators understand this and appreciate the value of good records. A supervisor should never accept less than full compliance with the above standards. Collection system operators will soon realize the concern with accuracy is for their benefit, and they will begin to take pride in the quality of their reports. Praise operators when they submit quality reports and explain what is needed when poor reports are submitted.

OFFICE PERSONNEL

Ultimately, most records are processed by office personnel and they, too, share the responsibility for keeping the system accurate and up to date.

- Availability. Records need to be prepared and filed in such a manner as to provide planning information quickly to supervisory personnel and operators. Processing should be on a daily basis if possible.
- Familiarity. Office personnel must become familiar with the nature of reported collection system work and how the records are used. It will help them to understand the importance of records and will greatly improve the cooperation between them and collection system operators.

QUESTIONS

Write your answers in a notebook and then compare your answers with those on page 548.

12.7AA What kind of information should be contained on an individual's fact sheet in a personnel file?

12.7BB What information should be included on a television survey history form?

12.7CC Which personnel records should be of direct concern to supervisors?

12.7DD How long should records be kept?

Please answer the discussion and review questions next.

DISCUSSION AND REVIEW QUESTIONS

Chapter 12. ADMINISTRATION

(Lesson 6 of 7 Lessons)

Write the answers to these questions in your notebook before continuing. The question numbering continues from Lesson 5.

29. What do collection system superintendents and staffs need to be able to take effective action and make sound decisions concerning the operation and maintenance of the collection system?
30. What are the key factors involved in the development of an effective management information system?
31. When can work reports be discarded?
32. Why are work directives usually given in written form?
33. What types of computer programs are available to assist with collection system O & M activities?
34. What is a geographic information system (GIS)?
35. What information should be included in a preventive maintenance record system?
36. How should essential records be protected from loss?
37. Who is responsible for being sure your agency has a good recordkeeping system?

CHAPTER 12. ADMINISTRATION

(Lesson 7 of 7 Lessons)

12.8 REPORT WRITING

Now that you know how important reports are to a management information system, and also recognizing their importance for conveying information to the public and our fellow collection system professionals, consider writing a report on some aspect of collection system maintenance.

12.80 Why You Need To Write Reports

Supervisors should have the ability to prepare a written account of the operation and maintenance activities under their responsibility. A report is a means to communicate information from the individual who has it to those who need it. Collection system operators are paid because somebody documented that the operators accomplished something.

If you are a new operator to the wastewater field, you will be involved in report preparation from the beginning. Your field notes and daily work reports are usually the basis of information used by your supervisor to prepare written reports. Field notes and reports are useful only if they are complete, brief and clear.

Take a few extra minutes when each job is completed to fill out forms or make field notes to produce a clean, neat, readable report of the job just finished. A field report that cannot be read due to poor handwriting or poor printing, misread numbers, or writing smeared with mud is of no use to any agency. When a report form has been completed, a review by another crew member provides an easy check for legibility, correct identification of location (including street address), map references, size of lines, proper agency cost accounting codes, amount of time spent on the job, and other important facts.

If your agency provides operators with equipment such as laptop computers or hand-held electronic notebooks, learn to use these tools to your advantage. They can greatly simplify the job of recordkeeping and report writing.

When you become a supervisor you will appreciate neat, readable field reports from your crews. Develop the ability to make your own reports neat and legible the first time. Take pride in accomplishing this important aspect of your work.

When faced with the prospect of preparing a written report, a supervisor may experience a feeling of apprehension or inadequacy because of what the supervisor may believe to be a lack of writing skill. Supervisors with many years of valuable experience and considerable knowledge and ability in the wastewater collection field may dread putting their knowledge in writing because of this feeling of inadequacy as a writer.

For this reason we feel the emphasis should be placed on the "Reporting" aspect of the heading of this section. The information contained in the report, the knowledge and the experience that can be passed on to others through the report, these are the true objectives of "Report Writing." Your reports tell what you did to justify a public investment in your agency and what you plan to accomplish next year. If you have organized the information about your system into a usable and understandable form, you will find most of the work of writing a report has already been done.

12.81 How To Prepare a Report

Qualified technical writing reference material is your best source for the *HOW* of writing. Perhaps the heading for this section should be "*WHAT* goes into a good report?" This is where our emphasis is going to be placed. However, there are a few basic steps we can suggest in the preparation of a good report.

ORGANIZATION

Before you can organize your report you must first organize yourself and your thoughts. Ask yourself, what is to be the objective of this report? Am I trying to persuade someone of something? Am I simply communicating information? For whom is the report being written? How can I make it interesting to the particular group for whom the report is being prepared?

After you have answered the above questions, the next step is to prepare a general outline of how you intend to proceed with the preparation of the report. List not only topic material but try to list all of the topics. Then arrange the topics so there is a workable, smooth flow from one topic to the next topic. Do not attempt to make your outline perfect. It is just a guide. It should be flexible. It is better to include too much material in your outline than to miss important points. You will find that as you begin to write you will remove nonessential points and expand on the more important points.

You might, for example, outline the following points in preparation for writing a report on a smoke testing program.

- A problem condition of excessive infiltration or inflow was discovered.
- Smoke testing offered the prospective means of identification and location of points of inflow.
- Funds, equipment, and material were acquired.
- Operators were trained.
- Tests were conducted.
- Results and conclusions were reached.
- Corrective actions were planned and taken.
- Conclusion, the tests did or did not produce the anticipated results or correct the problem.

Once you are fairly sure you have included all the major topics you will want to discuss, go through your outline and write down the facts that you want to include on each topic. Remember, this is just an outline. As you work through it, you may decide to move material from one topic to another if it seems appropriate. Don't be concerned about writing complete sentences in your outline if you find it easier just to jot down ideas. The purpose of outlining is to help *ORGANIZE* ideas and facts.

When your outline is complete, you will have the essentials of your report. Now you need to tailor it to the audience that will be hearing or reading it. This is very important. Will the audience be the agency management? Citizens groups? Other collection system agencies? Consider the following suggestions to help you tailor a report for each of these groups.

1. Management. They will have specific interests, mainly related to cost effectiveness. A report to management should include:
   a. Summary giving essentials (one page at most),
   b. Procedures used or method of study,
   c. Analysis of data, including trends, and
   d. Conclusions.

   Management may only read items a. and d. so be sure the essential information you wish to communicate is clear and brief. Be sure to give complete cost figures. Did the benefits warrant the costs? As a result of the tests, can future expenditures be reduced? Backup information and field data can be placed in an appendix for those who need the details.
2. Other collection system agencies. They too will be interested in cost figures. They also will want to know how the tests were accomplished, the procedures involved, size of crew, type of equipment used, source and availability of equipment and materials, difficulties encountered and how they were solved. The results and the benefits to be realized are important, too.
3. Citizens groups. Their interest will be more general. What is a smoke test? Why is it needed? Is the smoke harmless? Will it injure their plants or pets? Must they be home when the tests are in progress? How does the smoke detect defects? If illegal connections are found, how will this affect the individual property owners? Will it inconvenience them? Who pays for the test? How much will it cost to repair defects?

Each of these reports could be prepared using your original general outline. Simply adjust the outline to best emphasize the particular points you feel would be of particular interest to each specific group involved.

COMPOSITION

In the composition of your report it is wise to keep in mind the purpose of your writing. You are not writing a story. You are writing a report. Confine your writing to the activities that took place, the experiences gathered, and the results achieved. Be honest and explain the advantages and limitations of any recommendations.

- Drafts. Don't expect your report to be perfect the first time. Follow your outline and write a rough draft. Maybe you should be concerned with putting down your ideas the first time. Reread your draft three to five days later. Critically think about what you are saying. Ask a friend to read your report. Reread it and consider the items listed in the remainder of this section. Each time you rewrite a portion it will become a little bit better. We rewrote this manual several times after we received comments and suggestions from many groups.
- Facts. Confine your writing to the facts and events which occurred. Include only figures and statistics necessary to make the report effective. Facts must be relevant to allow for proper interpretation in the summary, conclusions and recommendations. Do not clutter the report with a lot of irrelevant figures that had little influence on the end results. Place voluminous data in an appendix.
- Continuity. To be interesting and easily understood, a report must have continuity. You had a problem, you explored various means to identify the problem—where it was located, what it was—and determined the needed correction. Consider the example we offered involving the smoke test.
- Effective. To be effective, a report should achieve the objective for which it is designed. In our example we wanted to justify to management that the tests were cost effective. To the other agencies, we wished to share our experience and knowledge in a way that would be of benefit to them. To the citizens group, we wanted to educate them as to what we were attempting to accomplish with our tests and explain the direct benefits to be realized by the public involved.

  Effective reports require that you present clear, concise facts in a logical sequence (continuity) that can be easily understood by the group to which they are being offered. As an example of how you can easily be tempted to stray from these basic principles, consider the report to the citizens group. They will have little knowledge of how a modern collection system functions. It would be easy for you to wander from your main objective by attempting to explain your collection system in detail. Certainly they would be interested in such aspects as lateral lines, main lines, trunk lines, lift stations, velocities, gravity lines, force mains, maintenance problems such as roots, grease, and debris, maintenance procedures, maintenance equipment, and your entire operation. It would all be very interesting to them, but it was not the object of your report. A brief explanation of how excessive inflow overloads treatment facilities and results in increased treatment costs is all that is required to put across the main objective of your report and the reason for the smoke test.

- Candid. Webster's definition of candid is "frank, straightforward." A good report must have this quality. You are aware of your capabilities and experience. Keep your report within these limits. Do not attempt to impress a group such as management or professionals by using language or terminology that is unfamiliar to you. Likewise, with citizens groups do not attempt to impress them with language familiar to you, but foreign to them.

You want your report to accomplish the objective for which it was intended. To do this it must be favorably accepted by those to whom it is presented. Candor, honesty, sincerity, and frankness will contribute greatly to the acceptance you seek. Anything less can only detract from your report and may invite unwanted skepticism or criticism.

QUESTIONS

Write your answers in a notebook and then compare your answers with those on page 549.

12.8A What are the two main considerations when preparing a report?

12.8B Why is a good outline essential?

12.8C Why must the intended readers of the report be considered?

12.8D What is the purpose of preparing drafts of reports?

12.9 PUBLIC RELATIONS

As collection system operators, the very nature of our work puts us in direct contact with the public in many of our daily activities. Our entire collection system is largely located either in public right of ways or in easements on private property. It could even be said that most of our public relations start with a negative element. Our services are rarely sought when everything is functioning properly. To the motorist, whose schedule is upset by our activities in his or her right of way, our position is rarely positive. To the property owner or tenant, whose privacy is interrupted, or services disrupted, or landscaping rearranged, we can hardly be considered a positive element. We are challenged, and it can often be difficult; but through good public relations, we can convert these negative feelings into positive appreciation of our efforts.

12.90 Importance of Public Relations

A collection system maintenance agency could possibly function with little regard for its public relations. This attitude should not, however, be an acceptable philosophy. Collection systems are essentially public utilities. As operators of such utilities, we are public servants. Public service need not imply servitude. Our service should be provided with sincerity, courtesy and an understanding of our responsibility to the public's needs.

12.91 Aspects of Public Relations

In the course of performing our daily activities, there are many aspects of public relations to consider.

TELEPHONE

In most instances initial contact with our collection system agencies will be by telephone. Relating to the well-known cliché of "first impressions," it is important that this first contact be a positive one.

- Basic courtesy. Good practice requires that persons answering incoming calls identify the organization and give their name, clearly and sincerely. Such response should not be hurried or mechanical, it should be personable and sincere. Let the person calling know that you are interested in what they have to say.
- Understanding. Realize the person calling, in most cases, has a problem. Try to understand their problem. Be sympathetic to their problem. Understand that they may be nervous, upset or even abusive. Your understanding and courtesy can only serve to calm them.
- Conclusive. Whenever possible, carry the conversation to an acceptable conclusion. If the caller's problem is within the agency's realm of responsibility, assure them that a crew will be dispatched to investigate their problem. If possible, try to tell the caller how soon the problem will be corrected. If the complaint is not within your jurisdiction, help the caller find the appropriate person to contact.
- Return calls. If the problem will require a return call, advise the caller of the approximate time the return call will be made. If the return call is to be made by others, such as a supervisor, advise the supervisor of the approximate time interval involved that you have given the caller. Management should be sure that someone calls back to verify that the action taken met requirements or explain why not.

PERSONAL CONTACTS

Our work demands that we have many personal contacts with the public we serve. Many such contacts will be made on their property at their home and, in most cases, at our convenience. Every effort should be made to make such contacts pleasant and positive for both parties.

- Appearance. The appearance you present may, for better or worse, suggest to the public what type of service they can expect from your organization and its operators. Provide operators with uniforms bearing the name of the organization and the name of the individual. Uniforms provide for a standard of neatness as well as a means of identification.

Equipment also should appear clean and neat. An emblem, symbol, or informative signs on equipment help people recognize who you are and what you are accomplishing.

- Attitude. Proper attitude cannot be overrated. Sincere concern for the problems of the individuals contacted must be demonstrated. Understand the person's problem, and perhaps the person's ignorance of the problems involved. If the problem is within your area of responsibility, advise the person when and how it will be solved. If the problem is not one to be handled by your department or agency, suggest proper alternative action which the individual may take to find a solution.
- Knowledge. Operators assigned to answering calls from the public must have and demonstrate the knowledge to effectively deal with such problems.

EQUALITY

Maintain a level of equality in the level of service you provide.

- Consistency of service level. Various organizations and agencies will provide different levels of service. For example, many agencies will be responsible only for main sewer line maintenance or repairs. House services from the building out to the sewer tap in the street or easement will be the sole responsibility of the property owner. Other agencies may assume responsibility for sewer services which lie in the right of way or easement. Whatever your agency's level of responsibility is, you should be consistent in the level of assistance you provide. Often maintenance crews are tempted to provide services beyond their responsibility. Operators should be cautioned to resist the urge to "help out" when a customer seems helpless or overwhelmed. They also should not be intimidated by threats or persuaded by a flirtatious customer. Sometimes it is difficult for crews to refuse pleas for extra consideration. You must try to impress on your operators the importance of giving the same level of service to all customers, regardless of personal feelings or circumstances.
- Explanation of limits of service. As noted earlier, various agencies will perform different levels of service and assume different levels of responsibility relative to private and public collection systems. Supervisors must provide their operators with clear and exact guidelines as to agency policy. Employees should be able to explain such policies and areas of responsibility effectively to the public they serve.

Figure 12.46 is a good example of a public relations piece used by Sacramento County, California, to explain its policy for service sewer maintenance.

12.92 Benefits of Good Public Relations

Good public relations must be earned, they cannot be bought. Once earned, your agency will enjoy the following benefits:

FROM THE PUBLIC

- Cooperation. Whenever it becomes necessary for maintenance operators to enter on private property, it is essential that the property owners cooperate. Good public relations can encourage this cooperation. Be considerate of private property at all times. Never enter on private property without first notifying the occupants of your presence and the nature of your work. Seek their permission before entering their property. If no one is home and it has been necessary for you to enter onto their property in their absence, leave a note advising them of your actions and a telephone number they can call if they have questions or complaints about the condition of their property after your work has been accomplished. Impress upon your operators that the impression they leave with property owners will have a strong effect on the cooperation given to subsequent crews should it become necessary to re-enter the same property.
- Understanding. The general public has little knowledge of the workings of a modern collection system. Few, if any, even are conscious of its location or existence until collection system operators appear on the scene to do some needed work. Collection system operators should take the time to explain briefly the location of the system, its function, and the need for their having access to it. Shown the courtesy of such explanation, the public will be more apt to understand the nature of our work and accept whatever inconvenience we might create for them.
- Satisfaction. Collection systems must be operated to the satisfaction of the public they serve. Public relations that are effective can have an important influence on such satisfaction.
- Financial support. The public will be more supportive when a collection system agency needs to increase its service charges or other sources of financing if the agency has developed good public relations.

TO EMPLOYEES

Employees too will benefit from good public relations.

- Pleasant working conditions. A good relationship with the public will nearly always result in more pleasant working conditions for collection system operators. Being welcomed as conscientious, considerate, courteous, and essential operators, as opposed to being resented as incompetent, inconsiderate nuisances, will make the collection system operator's job more pleasant and satisfying.
- Recognition. Effective public relations will result in individual operators being recognized and commended by the public for work well done. Such recognition, when given, benefits everyone. The public appreciates the level of service which warrants such recognition, management feels its objectives and responsibilities are being carried out, supervisors are assured their efforts have not been in vain, and the operator is justly proud of a job well done.
- Morale. Public relations of a quality which produce cooperation, understanding, satisfaction, pleasant working conditions and recognition improve the morale of the collection system personnel. The benefits to be had from such high morale will make any efforts of the supervisor to maintain good public relations well worth the time and concern.
- Self-perpetuating. Once established as normal procedure, good public relations and high personnel morale become self-perpetuating. A well satisfied public is eager to cooperate and operators whose efforts are recognized are constantly striving to improve their performance.

WHAT DO YOU DO IF YOUR SEWER STOPS UP?

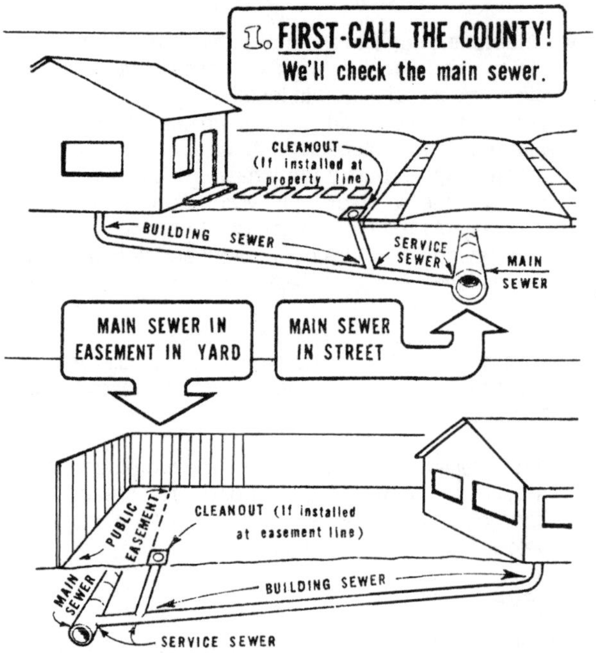
1. FIRST-CALL THE COUNTY!
We'll check the main sewer.
CLEANOUT (If installed at property line)
BUILDING SEWER
SERVICE SEWER
MAIN SEWER
MAIN SEWER IN EASEMENT IN YARD
MAIN SEWER IN STREET
PUBLIC EASEMENT
CLEANOUT (If installed at easement line)
BUILDING SEWER
MAIN SEWER
SERVICE SEWER

2. If the main is ok, we recommend that you then call a plumbing or sewer contractor.	3. If your contractor cannot unplug a stoppage in your service sewer, we can help. Give us another call.

(The SERVICE SEWER is the portion of your hookup that is within the street right-of-way or public easement.)

NOTE Although there is no extra charge for our services (the work is financed by county sewer bills), we will not assume the cost of your contractor's call, regardless of the location of the stoppage!

REMEMBER:

CALL THE COUNTY FIRST!

— 366-2231 —

(After Hours: 366-2000)

Water Quality Maintenance

COUNTY POLICY REGARDING SEWER MAINTENANCE

1. CALL THE COUNTY FIRST

When sewer problems are experienced, the resident should first contact the County. We will check the main sewer and, if there is a cleanout existing at ground level, we will also inspect that to determine if the stoppage is in the service sewer or the building side. We will clear the stoppage if it is in the County part of the system. Otherwise, we will notify the resident.

2. PROPERTY OWNER'S RESPONSIBILITY

The owner of the property is responsible for keeping the sewer clear between the building and the cleanout, if there is one. If there is no cleanout at ground level, the owner is also responsible for clearing the service sewer to the main sewer. Locating the building sewer, and its connection to the service sewer is also the owner's responsibility, although we will attempt to assist with available records and information. We suggest the owner employ a plumbing or sewer contractor to clear any stoppages in the building or service sewer.

3. DEFECTIVE SERVICE LINE

If the contractor is unable to clear a stoppage due to a defect in the service sewer, we will repair it. If there is no cleanout to grade, the contractor must first expose the end of the service sewer at the easement or right-of-way line. The County will perform no work on private property except within a public easement.

4. COUNTY WILL INSTALL CLEANOUT

If the contractor was able to clear a stoppage in a service sewer, but, because there was no cleanout to grade, had to excavate the end of the service sewer, we will install a cleanout to facilitate clearing future stoppages. The contractor should contact us before closing the excavation.

5. DOUBLE WYE SERVICE

Where two properties are connected to a single service sewer by a double wye, the County will be responsible for maintenance of the shared service sewer.

6. LIMIT OF COUNTY RESPONSIBILITY

The County's only responsibility for work on private property is for damage caused by wastewater backup due to stoppage in the main sewer, or, in the case of a double wye installation, flow from the other property.

SLP-3(05/83)

Fig. 12.46 Explanation of service sewer maintenance policy

12.10 ACKNOWLEDGMENTS

The authors of this chapter wish to thank Ronald S. Lynn for the information he supplied about computer-based maintenance management systems and RJN Group, Inc., for permission to use Figures 12.36 and 12.37. The computer screen illustrations in this chapter were provided by Hansen Information Technologies and by Oasis.

12.11 ADDITIONAL READING

1. *MANPOWER REQUIREMENTS FOR WASTEWATER COLLECTION SYSTEMS IN CITIES AND TOWNS OF UP TO 150,000 IN POPULATION*, Manpower Development Staff, Office of Water Program Operations, U.S. Environmental Protection Agency, Washington, DC 20460. Obtain from National Technical Information Service (NTIS), 5285 Port Royal Road, Springfield, VA 22161. Order No. PB-227039. Price, $36.50, plus $5.00 shipping and handling per order.
2. *MANPOWER REQUIREMENTS FOR WASTEWATER COLLECTION SYSTEMS IN CITIES OF 150,000 TO 500,000 IN POPULATION*, Manpower Development Staff, Office of Water Program Operations, U.S. Environmental Protection Agency, Washington, DC 20460. Obtain from National Technical Information Service (NTIS), 5285 Port Royal Road, Springfield, VA 22161. Order No. PB-95-157442. Price $51.00, plus $5.00 shipping and handling per order.
3. *WASTEWATER COLLECTION SYSTEMS MANAGEMENT* (MOP 7). Obtain from Water Environment Federation (WEF), Publications Order Department, 601 Wythe Street, Alexandria, VA 22314-1994. Order No. M05000. Price to members, $56.75; nonmembers, $76.75; price includes cost of shipping and handling.
4. *PLANT MANAGER'S HANDBOOK* (MOP SM-4), Water Environment Federation (WEF). No longer in print.

QUESTIONS

Write your answers in a notebook and then compare your answers with those on page 549.

12.9A Why are public relations important to collection system operators?

12.9B Describe how you should respond to telephone complaints.

12.9C How can an operator make positive personal contacts?

12.9D Why is equality of service important?

12.9E Who benefits from good public relations?

Please answer the discussion and review questions next.

DISCUSSION AND REVIEW QUESTIONS

Chapter 12. ADMINISTRATION

(Lesson 7 of 7 Lessons)

Write the answers to these questions in your notebook. The question numbering continues from Lesson 6.

38. How does the collection system operator come in contact with the public?
39. What does "public relations" mean to you?
40. Why are public relations important to the collection system operator?
41. Why must the same level of service be provided for everyone?
42. Why are reports important?
43. How would you prepare and write a report for your supervisor?

SUGGESTED ANSWERS

Chapter 12. ADMINISTRATION

ANSWERS TO QUESTIONS IN LESSON 1

Answers to questions on page 450.

12.1A Discipline consists of obedience, application, energy, respect and behavior, all essential conditions for a smooth-running organization.

12.2A An effective administrative plan should consist of a mission statement, goals to be attained, and the objectives, tasks and procedures that will lead to their attainment.

12.2B A collection system agency can work toward a goal of operating safely by: (1) managing to take actions that consistently promote safe operation by personnel and contractors, (2) establishing safety training programs for personnel, and (3) controlling the risks involved in the operation and maintenance of the collection system.

12.2C Excessive infiltration and exfiltration can be controlled by repairing or replacing worn and damaged parts of the collection system to eliminate infiltration, inflow and exfiltration. Other effective means include isolation of manhole and cleanout openings from surface water and the adoption and enforcement of sewer-use regulations prohibiting the discharge of storm water or groundwater to the wastewater collection system.

12.2D Management can promote safety by personal use of safe practices, providing safety equipment and training and requiring adherence to safe work practices.

ANSWERS TO QUESTIONS IN LESSON 2

Answers to questions on page 454.

12.3A Overflows, spills, or raw wastewater bypasses must be reported to:

1. Regional Water Pollution Control Agency,
2. Local Health Department, and
3. Director of Public Works.

12.3B Clearwater's superintendent is responsible for planning and implementing maintenance, operation, training, and safety aspects of the section.

12.3C Items to be considered when summarizing a wastewater collection system include:

1. Total length of pipe and length of service lines and lines of different diameters,
2. Lift stations,
3. Condition of system,
4. Problem areas in system,
5. Building sewer connections, and
6. Manholes.

Answers to questions on page 460.

12.3D Making the proper selection of personnel is important because capable workers are needed to operate and maintain a collection system safely and efficiently. Also, it is difficult to fire an employee and the agency will lose the investment made in the employment and training of the employee.

12.3E The potential positions of advancement for a Maintenance Person II, Wastewater Collection, include Foreman, Maintenance Supervisor I, or Construction Inspector.

12.3F Information that should be listed on an "opportunities for employment" notice includes a brief description of the work to be performed, required qualifications, compensation, the selection process, where application forms can be obtained and the closing date for accepting applications.

12.3G During the first day of work, a new employee should be given all the information available in written and verbal form on the agency's policies and practices concerning compensation, benefits, attendance and other matters relating to employer-employee relations. Also, a safety orientation should be given as soon as possible.

12.3H Adjustments that can be made to an employee's starting salary include: (1) longevity increases if the employee remains in the same position for a specified period and demonstrates concurrent improvement in work performance; and (2) cost of living increases.

Answers to questions on page 462.

12.3I Collection system operators must be properly trained so they can effectively accomplish the tasks they are assigned. Effectiveness is indicated by the job being properly done quickly and safely.

12.3J Collection system operators can be trained on the job, by reading trade magazines and papers, by formal classroom training, by attending professional meetings and by home study.

12.3K Collection system operators should be given information that will:

1. Help with their continuing education (including safety),
2. Indicate to them the results they have achieved, and
3. Inform them of their importance to the organization.

12.3L The purpose of an annual report meeting is to tell collection system operators what their year's efforts have produced.

12.3M The supervisor and the shop steward should try to cooperate and work together. The supervisor is management's link with the operators and the shop steward is the operator's link with management.

Answers to questions on page 463.

12.3N The basic elements of a superintendent's safety program include a safety policy statement (including disciplinary procedures), safety training and promotion, and accident investigation and reporting.

12.3O The safety policy statement should be prepared by the top management of the collection system agency since its purpose is to let employees know that the safety program has the full support of the agency and its management.

12.3P Management's responsibilities with regard to safety include:

1. Formulate a written safety policy,
2. Provide a safe workplace,
3. Set achievable safety goals,
4. Provide adequate training, and
5. Delegate authority to ensure that the program is properly implemented.

12.3Q The collection system operators are the direct beneficiaries of a safety program. The operators share the responsibility to:

1. Observe prescribed work procedures with respect to personal safety and that of their co-workers,
2. Report any detected hazard to a manager immediately,
3. Report any accident, including a minor accident that causes minor injuries,
4. Report near-miss accidents so that hazards can be removed or procedures changed to avoid problems in the future, and
5. Use all protective devices and safety equipment supplied to reduce the possibility of injury.

ANSWERS TO QUESTIONS IN LESSON 3

Answers to questions on page 466.

12.4A Tool and equipment requirements can be determined on the basis of size of collection system, workload, and objectives. Equipment selection should be based on the type or types of problems you want the equipment to solve.

12.4B Most modern collection system maintenance equipment has one particular function that it does most effectively.

12.4C The shotgun approach is not recommended because it is wasteful and unnecessary. Equipment should be used in those areas where it is most needed.

12.4D Equipment to implement a preventive maintenance program can be obtained through long- or short-term leasing or by contracting for a specialized job, rather than by outright purchase, if not enough money is available.

12.4E Advantages of leasing equipment include:

1. Expensive equipment can be made available without an initial capital expenditure of considerable amount, and
2. Specialized equipment is made available to the agency, at a reasonable cost, to complete any one-time or seasonal programs.

12.4F Advantages of purchasing include:

1. Makes available standard equipment that is used continuously with a long life span for a reasonable price, and
2. Equipment is available when needed.

12.4G Contracting for work can be advantageous when an agency does not have sufficient qualified or qualifiable staff available to do a highly specialized job. Also the agency does not have to buy equipment that will not be used continuously.

Answers to questions on page 468.

12.4H Emergency crews should be provided with at least a $^{3}/_{4}$-ton truck.

12.4I The following crews should be equipped with two-way radios or paging devices:

1. Emergency crew,
2. Rodding truck (one crew), and
3. Repair crew.

12.4J A truck with adequate capacity must be specified so the truck will not wear out before the high-velocity cleaning equipment. A truck of marginal capacity could start developing brake problems, weakened springs, difficult road handling, and a loss of engine power after only a short period of use.

12.4K Truck-mounted units are preferable for units that are used on a full-time basis while trailer-mounted units are desirable for equipment used on a part-time basis because the truck pulling the trailer can be used for other purposes when the trailer-mounted unit is not in service.

12.4L Portable standby pumping units should be inspected and run at regular intervals to ensure that they will be in operating condition when an emergency occurs.

12.4M Major pieces of equipment needed for construction and repair include backhoe/loader, dump truck, water truck, portable air compressor, and hoists.

12.4N Portable air compressors are used to operate jackhammers and tampers and also to drive sheet piling.

Answers to questions on page 470.

12.4O Air testing and joint sealing equipment is often used in conjunction with closed-circuit TV to inspect sewers. The equipment can be purchased separately or as packaged units mounted on a truck or trailer.

12.4P Seldom-used equipment such as large size sewer balls and portable emergency pumps must be stored at a location that is accessible to the crews when the equipment is needed.

12.4Q Three different types of electronic locators include:

1. Simple valve or lid locator,
2. Pipe locator, and
3. Ferret-type locator.

12.4R The two different types of plugs are air plugs and mechanical plugs.

12.4S Handmade taps should be discouraged because they often result in broken sewers, ill-fitting taps, and taps protruding into the sewer. These problems provide inviting areas for future root intrusion and for a person to use an excess of cement or grout in an effort to hide the whole crude job.

12.4T Taps protruding into sewers can be removed by use of an internal tap cutter that is manually pulled through the inside of a pipe. The protruding tap is cut off by use of a diamond-core drill.

12.4U Service lines can be placed under a major thoroughfare without disrupting traffic by boring a hole for the line under the thoroughfare.

Answers to questions on page 472.

12.4V Atmospheres that require testing before entering include confined spaces such as manholes and wet pits.

12.4W The three main items of concern when testing an atmosphere are:

1. Lack of oxygen,
2. An explosive condition, and
3. Toxic gases.

12.4X The atmosphere in a confined space must be monitored continuously because the atmosphere can change suddenly.

12.4Y Fresh air can be introduced into confined spaces by the use of blowers.

12.4Z The different types of shoring include:

1. Hydraulic shoring,
2. Manual shoring consisting of screw jacks and wooden timbers,
3. Sheet shoring, and
4. Cylinder and shield shoring.

12.4AA Traffic control is often overlooked or underrated because crews occasionally get in a hurry and/or are careless.

12.4BB Collection system operators must wear protective clothing to protect themselves from unnecessary exposure to the hazards of their job. Protective clothing includes uniforms, safety shoes, rubber boots, rubber gloves, eye protection, hearing protection, hard hats, and safety harnesses.

Answers to questions on page 474.

12.4CC The main categories of microcomputer software (computer programs) that could be used by a collection system agency include:

1. Database management,
2. Graphics,
3. Program development,
4. Spreadsheet analysis,
5. Telecommunications, and
6. Word processing.

12.4DD The essentials of an equipment inventory record include:

1. Type (high-velocity cleaner),
2. Model year,
3. Date purchased,
4. Cost,
5. Description,
6. Manufacturer, and
7. Fuel type.

12.4EE Equipment should be replaced when it has reached the end of its useful life. Useful life can be defined as the time at which a piece of equipment ceases to perform its required work or when the cost of operating and maintaining the equipment is greater than the cost of operating and maintaining new equipment.

ANSWERS TO QUESTIONS IN LESSON 4

Answers to questions on page 478.

12.5A Services that should be available for vehicles and engine-driven special equipment in a maintenance yard include fuel, air, coolant and oil when needed.

12.5B Types of hazardous materials that could be stored in a maintenance area include toxic chemicals, flammable fuels, and hazardous waste.

12.5C Valuable shop tools and equipment should be inventoried and secured in an enclosed room with one person responsible for checking these items in and out. The entire shop area should be secured when authorized shop personnel are not in the immediate area and at the end of work shifts.

12.5D Landscaping of a collection system maintenance yard and office buildings is important to make the facilities pleasing to the public and visitors. Landscaping also will increase the satisfaction employees receive from working for the agency.

ANSWERS TO QUESTIONS IN LESSON 5

Answers to questions on page 481.

12.6A Maps of a wastewater collection system are important because they show where the system components are located and the size of the collection system pipes. Maps enable maintenance operators to find specific components for repair and provide information that will eliminate guesswork concerning what tools and materials will be needed to complete a job.

12.6B Information required on wastewater collection system maps includes:

1. Locations,
2. Identifications,
3. Elevations and depths,
4. Footage or distances,
5. Flow direction,
6. Force mains,
7. Access points and overflow points,
8. Scale and North arrow, and
9. Date.

Optional information that may also be recorded on maps includes the type of pipe, soil conditions, properties connected to the system, and an inventory of the items on each map.

12.6C All manholes should be identified by a specific number and/or letters and this identification should never change.

12.6D The essential elevations that should be shown on collection system maps include:

1. Flow lines or invert elevations at the site of each manhole,
2. Manhole rim elevations, and
3. Depth of building sewer at property line.

12.6E Soil conditions are of great importance and a significant safety factor to future repair crews who have to dig in disturbed ground.

Answers to questions on page 489.

12.6F Assessor's maps contain the most detail.

12.6G Maps must be updated to maintain their usefulness and to reduce time-consuming and costly mistakes.

12.6H Maps can be kept up to date by:

1. Annual updating of all changes and corrections,
2. Recording changes when they occur,
3. Field operators (maintenance and inspectors) reporting all changes and discrepancies, and
4. Field checking existing maps with actual conditions in the field.

Answers to questions on page 489.

12.6I The advantages of using microfiche copies include a low reproduction cost, ease of handling, less storage space and they do not deteriorate with repeated handling.

12.6J Maps can be used as a maintenance record by showing:

1. Problem areas,
2. Maintenance activities, and
3. Proposed maintenance schedules.

12.6K Problem areas can be shown on a map by using colored pins or color codes that indicate the type of problem (roots, grease) and frequency of the problems.

ANSWERS TO QUESTIONS IN LESSON 6

Answers to questions on page 494.

12.7A Current information no longer needed in a database should be removed from the database and stored in archives in case they may be needed in the future.

12.7B Data processing takes independent pieces of information (usually in the form of records) and produces useful information by arranging it into meaningful knowledge.

12.7C The quality of the information produced by computers is entirely dependent on the quality of data put into the computer and the software directing its processing.

Answers to questions on page 501.

12.7D Collection system operators account for their daily work activities of operating and maintaining collection systems by making daily work reports.

12.7E Types of work reports used by collection system agencies include:

1. Main sewer construction,
2. Time, material and equipment,
3. Main sewer maintenance,
4. Main sewer repair,
5. Structure maintenance,
6. Structure repair or abandonment,
7. Building sewer maintenance, and
8. Building sewer repair.

12.7F "Main sewer construction data" forms are used by a construction crew leader or by an inspector when the construction is performed by a contractor.

12.7G The types of debris that can be found in sewers include eggshells or coffee grounds, grease or soap, paper or rags, grit (silt, sand, gravel), foreign objects, roots, mud, or other solids.

12.7H Types of conditions that can be found in sewers include broken pipe, offset joints, infiltration, roots, sagging pipe and other defects.

Answers to questions on page 504.

12.7I The superintendent of a collection system needs to have reports analyzing records on staffing, materials and equipment used to take actions to improve work efficiency. Reports on main sewer line stoppages are used to make adjustments to a collection system's preventive maintenance program. A superintendent will order the preparation of other reports as needed for effective planning and control of the collection system agency.

12.7J The monthly reports from each individual foreman should contain information regarding the daily production activity of each of the crews under each foreman. Activities involved may include balling, rodding, TV inspection, construction and repairs, or any other significant maintenance activities that are part of the organization's maintenance program.

12.7K A cost of operation report is a summary of how budgeted funds were or will be spent. This report should describe the total cost of operation as well as costs for supplies, materials, equipment and salaries. Costs also can be reported by types of programs such as clearing emergency stoppages, preventive maintenance, responding to complaints and repairs.

Answers to questions on page 507.

12.7L Work directives given by superintendents and supervisors to operators include training in work performance and equipment operation, standard operating procedures and work orders.

12.7M Upon receipt of a Service Request Report, a supervisor immediately obtains any additional information needed for response to the request, makes a personal investigation if required, and gives directions to a service crew for responding to the service request.

12.7N If a service is not provided by the agency, the crew leader explains the agency's policy to the property owner and offers any appropriate suggestions on how the service might be obtained.

12.7O Duplicate or carbon copies of work orders are not needed when using computer systems because the original copies are retained in the computer for anyone who needs them.

Answers to questions on page 523.

12.7P The two basic categories of CMMSs are equipment-oriented systems and pipeline-oriented systems.

12.7Q The main difference between equipment-oriented CMMSs and pipeline-oriented CMMSs is the information that is stored and used in the databases. In equipment-based systems, the information relates to various types of equipment, such as pumps, motors, electrical systems, valves, instrumentation and control. The information in pipeline-based systems relates to the various parts of the collection system pipelines, such as main lines, building service laterals, structures, catch basins and storm inlets.

12.7R The power of a geographic information system (GIS) is that information can be retrieved geographically. An operator can easily look at or print out a specific area of a map. The map will contain an inventory of the collection system structures within the selected area and it will provide detailed information about each of the structures (or entities).

Answers to questions on page 528.

12.7S SCADA stands for **S**upervisory **C**ontrol **A**nd **D**ata **A**cquisition system.

12.7T A SCADA system collects, stores and analyzes information about all aspects of operation and maintenance, transmits alarm signals when necessary and allows fingertip control of alarms, equipment and processes.

12.7U Measured (sensed) information could be transmitted by FM radio, leased telephone circuits, private signal lines, dial telephone lines, coaxial cable networks, spread spectrum radio, fiber-optic cable, microwave, 900 MHz radio, cellular telephone, or satellite communications systems.

12.7V The greatest challenge for operators using SCADA systems is to realize that computers may not be correct and to have the ability to operate when the SCADA system fails.

Answers to questions on page 531.

12.7W Records are necessary because they provide:

1. A record of the past, and
2. A sound basis on which to plan the future.

12.7X Cost items that should be recorded include:

1. Labor,
2. Equipment,
3. Supplies, and
4. Services.

12.7Y Records are very important when preparing a budget. Accurate records detailing last year's activities, costs, workload, growth and production (such as miles of lines cleaned) can be the best justification for next year's budget requests.

12.7Z The best protection against legal action or liability claims is reliable records to verify that recommended procedures were followed and that agency personnel were not negligent in meeting their responsibilities.

Answers to questions on page 537.

12.7AA An individual's fact sheet should contain information on age, blood type, inoculations, dependents, and medical history for use in case of an accident or serious injury. The file must be updated annually and include accidents, illnesses, training and new skills learned.

12.7BB Information that should be included on a television survey history form includes date, videotape number and file location, observed conditions, footage from entrance of manhole to recorded observation, observed tap location, pipe conditions observed and the type and severity of any problem conditions.

12.7CC Personnel records that should be of direct concern to the supervisor include operator performance reports, accident reports, and attendance analysis reports.

12.7DD Records must be retained in accordance with state and federal regulations. All records should be kept as long as they are thought to be useful or may be needed for defending claims for injuries or property damage submitted to an agency. An agency's legal counsel should be asked to provide guidance for retention of records.

ANSWERS TO QUESTIONS IN LESSON 7

Answers to questions on page 540.

12.8A The two main considerations when preparing a report are:

1. Organization, and
2. Composition.

12.8B A good outline helps to organize the report and present information in a logical sequence that is easily understood.

12.8C The intended readers of the report must be considered because different readers have different backgrounds and desire different information.

12.8D Drafts of reports help you to review the report, let others give you their reactions to it, and give you a chance to improve or "polish" the report.

Answers to questions on page 543.

12.9A Public relations are important because we are always in contact with the public. Our service should be sincere, courteous and understanding in spite of trying circumstances.

12.9B Responders to telephone complaints should try to make a good first impression. When answering the phone, identify yourself, be personable and sincere. Try to be understanding and sympathetic toward the caller's problem. Conclude the conversation with an acceptable solution and tell the caller when the problem will be solved. Follow up with a return call if necessary.

12.9C Positive personal contacts can be made by having a neat appearance, such as wearing a uniform or clean work clothes. A courteous and sincere attitude toward other people is very helpful. A knowledge of how to solve the problem is impressive.

12.9D Equality of service is important because if everyone is not given the same level of service, then everyone will want special service or favors beyond the responsibility of your agency. This extra service is time-consuming and costly to the taxpayer.

12.9E Everyone benefits from good public relations. The public will cooperate with your agency, understand what you are trying to do, and hopefully be satisfied with your services. Employees benefit from pleasant working conditions, recognition of their efforts, high morale and a satisfied public.

CHAPTER 13

ORGANIZATION FOR SYSTEM OPERATION AND MAINTENANCE

by

John Brady

John Cavoretto

George Gardin

Revised by

Steve Goodman

TABLE OF CONTENTS

Chapter 13. ORGANIZATION FOR SYSTEM OPERATION AND MAINTENANCE

IN
OUT

OBJECTIVES

Chapter 13. ORGANIZATION FOR SYSTEM OPERATION AND MAINTENANCE

Following completion of Chapter 13, you should be able to:

1. Organize an agency to operate and maintain a wastewater collection system,
2. Identify functions and work activities of essential units within an agency,
3. Staff and equip essential units within an agency,
4. Develop priority lists for job assignments for units within an agency,
5. Describe the various types of equipment maintenance programs,
6. List the factors that influence an equipment maintenance program,
7. Develop and implement an equipment maintenance program,
8. Schedule the collection system activities of an agency, and
9. Evaluate the performance of collection system and agency.

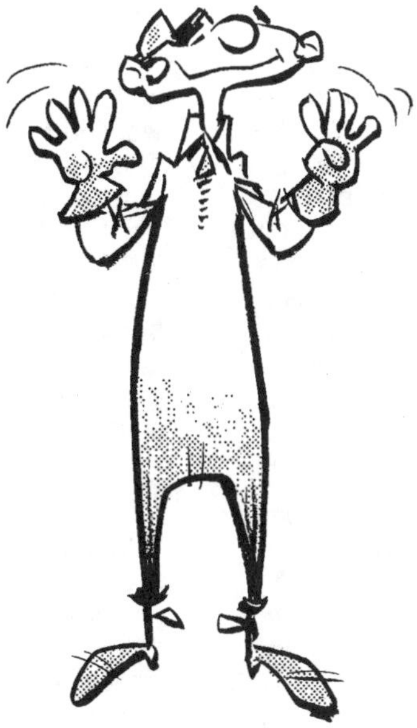

CHAPTER 13. ORGANIZATION FOR SYSTEM OPERATION AND MAINTENANCE

(Lesson 1 of 4 Lessons)

13.0 NEED FOR ORGANIZATION

In the preceding Chapter 12 on administration you were introduced to the personnel, equipment, facilities and information systems used for the operation and maintenance of a wastewater collection system. These separate elements of administration need to be organized into a business structure. Such organization begins with identification of the special functions of each element and its relationship to the operation and maintenance of a collection system.

Collection system operators need to be organized into crews with individuals and equipment capable of performing assigned tasks. It will be necessary to establish a line of authority and to assign responsibility and accountability for operation and maintenance of the collection system. A collection system division may have good operators and equipment, but if they are not well organized, the system will not be operated and maintained efficiently.

13.1 ORGANIZATIONAL PRINCIPLES

As we consider the organization of a division for the operation and maintenance of a wastewater collection system, we need to be guided by organizational principles. These principles are discussed in the following paragraphs.

13.10 Knowledge of Work

Any organization is built upon a precise knowledge of the work required to accomplish its goals and objectives. With this knowledge, work using the same classes of personnel and types of equipment can be grouped together for the purpose of efficient assignment. To develop a knowledge of an organization's work, begin by preparing a clear statement of its mission, followed by its goals and objectives. Then you can identify the separate tasks needed to reach the objectives and goals, and prepare a detailed listing of the types of work involved in each task.

13.11 Personnel and Equipment

After you have prepared a detailed list of work to be performed by your organization and have grouped tasks by similarities, consider the types of personnel and equipment needed to perform the work. The amount of personnel and equipment needed will be determined by an estimate of the amount of work to be performed.

The types of personnel and equipment should be standardized to the greatest extent possible to avoid the need for special and expensive personnel and equipment. You can consult with superintendents and supervisors of other collection system agencies to determine the occupational descriptions and specifications of equipment that have been standardized for operation and maintenance of wastewater collection systems. The publications listed at the end of this chapter as items 1 and 2 in Section 13.8, "Additional Reading," provide good occupational titles and job descriptions for standard collection system personnel. Also Chapters 5, 6, and 7 in Volume I of this manual have described standard personnel titles and equipment used for operation and maintenance of collection systems.

13.12 Unitization

Most of the work required for the operation and maintenance of a collection system is performed by several persons working together as a unit for efficiency and safety purposes. These units are generally called crews. Members of a crew must perform their work as a team and, to be efficient, they must be familiar with each other and with their work. Therefore, crew members should work together performing the same type of work for an extended period of time. Rotation of personnel to different crews or types of work should only be done as a deliberate effort to improve the efficiency of the organization or advance the career of a capable and worthy person.

13.13 Line of Authority

In most organizations authority is essentially power—power to decide how the resources of the organization will be used to achieve its mission. This power usually is vested in (held by) the owner or highest official of the company, agency, or organization. As an organization increases in size, it becomes necessary to give some of this power to other staff members because no single person would be able to effectively make all decisions, supervise all work, and respond to all daily concerns that arise.

When power is delegated to others, it is usually limited to specific areas of responsibility such as personnel, financial operations, or collection system operation and maintenance. More limited powers may then be further delegated, for example, to individuals who supervise daily work activities or specific crews of operators within a section of the organization.

The overall purpose of delegating authority in this way is to allow decisions to be made and actions to be taken at the lowest practical level within the organization. The effect of this is to free upper management from activities that would otherwise consume so much of their time that they would be unable to meet their broader responsibilities for planning and managing.

A second outcome of delegating authority is that decisions about daily operations can be made by those who are most familiar with the situation and most directly affected by the decisions. This usually means a more efficient operation.

It is common practice to visually represent the delegation of authority with an organization chart such as Table 13.7 on page 590. The lines connecting positions trace the line of authority that has been delegated by the superintendent to an assistant, who has then delegated some of the authority to four section heads, and so on down the line. Table 13.7 makes it clear that the Lift Station Operation and Maintenance Section Supervisor I gets authority from and reports directly to the Assistant Superintendent, and not to the head of Personnel or Engineering.

One of the principles on which decisions about delegation of authority are made is the principle of span of control. It is widely believed that a person can effectively manage only about five people. Unique situations within an organization will cause frequent exceptions to this rule of thumb but in general, a larger span of control dilutes the actual amount of control a supervisor can realistically exercise over each subordinate.

13.14 Responsibility

Whenever authority is delegated to someone else there must be a clear understanding by both parties about exactly what authority is being given. That is to say, it must be clear to both parties precisely what the power enables the recipient to do. When authority is given, it is accompanied by responsibility to produce certain specified results. Authority and responsibility cannot be separated; to do so is to invite failure.

As an example, assume you make a supervisor responsible for ensuring that five operators within the span of control observe prescribed safety precautions. One operator repeatedly neglects safety precautions and is reprimanded by the supervisor using the established progressive discipline policy of the agency. An accident occurs that is directly attributable to an unsafe practice by this same operator and the supervisor puts the operator on unpaid administrative leave for one day. You overrule the supervisor's decision because "there's work to be done" and "everyone makes a mistake now and then."

By overruling the supervisor, you have taken back the authority you had delegated. Without this authority, responsibilities for observance of safe practices cannot be exercised and the subordinate cannot be made to comply. The supervisor has no power and is likely to fail to meet responsibilities in this area.

13.15 Accountability

As we have seen, when a person is given authority it comes with specific responsibilities to produce certain results. These responsibilities are usually spelled out in a job description for each position within the organization. But having given some powers to a person, how does the organization know whether the person is meeting assigned responsibilities? The organization must develop some system of accountability as part of its management information system. It must establish performance standards and develop a way to measure each employee's performance. When the level of performance is exceptionally good, the employee (being accountable for successfully meeting or exceeding responsibilities) deserves reward or recognition for it. When performance does not fulfill assigned responsibilities, the employee (being accountable for the employee's actions) can legitimately be held responsible and denied reward or recognition, or may be penalized in some way.

To hold persons accountable for their actions, three factors must *ALL* be present:

1. The individuals must know what they are supposed to do. This knowledge could be derived from a job description, an operations manual, or direct instruction.
2. The individuals must know how to accomplish assigned tasks. Minimum skill levels for hiring purposes, training programs, and direct instruction can accomplish this.
3. The individuals must have control of the factors that regulate what they are doing.

To see the importance of the third factor, let's return to the example of the supervisor whose subordinate violated safety precautions. The supervisor knew what had to be done (ensure safety compliance); knew how to do it (used established progressive discipline procedures); but lacked the authority that would enable the correct action to be taken (you overruled the supervisor). So when it comes time to hold the supervisor accountable for ensuring compliance with safety regulations, you will not legitimately be able to penalize the supervisor for failing to meet this responsibility because control of all the factors affecting the supervisor's actions was not permitted.

Accountability is a complex issue but one that it is essential to understand if an organization is to operate efficiently. There are a large number of associated legal issues, particularly if a union contract governs employee activities, that must be considered with respect to rewards, punishment, performance evaluations, work standards, and operating procedures. It is good practice to check with the staff of your legal department on such issues and to familiarize yourself thoroughly with the established policies and procedures of your agency.

13.16 Management

The proper management of a wastewater collection system requires the manager or supervisor to plan, organize, staff, direct and control the activities of the agency. Planning consists of determining the goals, policies, procedures and other elements to achieve the goals and objectives of the agency. Planning requires the supervisor to collect and analyze data, consider alternatives and then make decisions. Planning must be done before the other managing functions.

Organizing means that the supervisor decides who does what work and delegates authority to the appropriate operators. Employment or staffing is the recruiting of new operators and staff and determining if there are enough qualified operators and staff to fill available positions. Employment also includes selection, training, promoting, evaluating performance and providing opportunities for advancement for operators and staff in the agency.

Directing includes the guiding, teaching, motivating and supervising of operators and staff. Direction includes issuing orders and instructions so that jobs are performed safely and are properly completed. Operators and staff must be encouraged to perform their tasks efficiently.

Controlling involves taking the steps necessary to ensure that essential activities are performed so that objectives will be achieved as planned. Controlling means being sure that progress is being made toward objectives and taking corrective action as necessary.

QUESTIONS

Write your answers in a notebook and then compare your answers with those on page 609.

13.1A How can the types of personnel and equipment needed to perform the work of an organization be determined?

13.1B Why should crew members work together and perform the same type of work for an extended period of time?

13.1C What are the two main effects of delegating authority?

13.1D What three factors govern whether employees can be held accountable for meeting their assigned responsibilities?

13.2 ORGANIZATION OF PERSONNEL

In Chapter 12 you were introduced to a superintendent who had been employed by the City of Clearwater, USA, and given the responsibility of reorganizing the division that operates and maintains the city's wastewater collection system. We will now consider how the superintendent used organizational principles in reorganizing the division to move toward accomplishing the goals and objectives set forth in the operating plan described in Section 12.2, Chapter 12 (pages 447–450).

13.20 Work Activities

Let us review the work activities the superintendent identified for the collection system division and listed in Tables 12.1 and 12.2, page 453. The superintendent used these lists to prepare the following work groupings and the related major tasks:

- Service Requests
  - Main Sewer Stoppages
  - Building Sewer Stoppages
- Preventive Maintenance
  - Balling and Flushing
  - High-Velocity Cleaning
  - Power Rodding
  - Bucketing
  - Chemical Treatment
- Repair and Replacement
  - Main Sewers
  - Joint Sealing
  - Sewer Lining
  - Manholes
  - Building Sewers
- Lift Station Operation and Maintenance
  - Lift Station Inspections and Servicing
  - Lift Station Repairs
- Inspections and Investigations
  - Manholes
  - TV Main Sewers
  - Infiltration/Inflow Investigations
  - Industrial Waste Inspections
- Indirect
  - Equipment Servicing
  - Building and Grounds Maintenance
  - Work for Other Departments
  - Training
  - Leave and Holidays

13.21 Occupational Titles

The next organizational step taken by the superintendent was to determine what types of collection system operators already worked in the division and what additional staff would have to be employed to perform the planned work activities. The superintendent used previous experience, consultations with other superintendents and the publications on staffing requirements for wastewater collection systems (listed as Items 1 and 2 in Section 13.8, "Additional Reading") to standardize the occupational titles and job descriptions of most collection system personnel.

The following is a listing of the most common occupational titles for which job descriptions are readily available:

Superintendent
Assistant Superintendent
Maintenance Supervisor II
Maintenance Supervisor I
Equipment Supervisor
Instrumentation Technician II
Instrumentation Technician I
Foreman
Maintenance Operator II
Maintenance Operator I
Mason II
Mason I
Maintenance Equipment Operator
Construction Equipment Operator
Automotive Equipment Operator
Laborer
Maintenance Mechanic II
Maintenance Mechanic I
Maintenance Mechanic Helper
Electrician
Computer Operator
Computer Programmer
Clerk Typist
Stock Clerk
Dispatcher

The superintendent knew that all the different types of personnel would not be needed to operate and maintain the collection system for the medium-size population of 180,000 in the City of Clearwater. The superintendent also knew that the duties of some operators would have to be combined and the superintendent planned to compensate those operators with the salary appropriate for the highest classification describing some of their duties. For example, the superintendent planned to use Instrumentation Technicians to operate the TV equipment when inspecting main sewers and also use them to make visual inspections of the system's manholes. A Maintenance Operator II would also serve as a crew leader and equipment operator. In each of these cases of combined job levels, the operator would be paid at the rate for the higher job classification.

13.22 Assignment of Personnel

The superintendent used the list of work activities, their annual staff day requirements listed in Tables 12.1 and 12.2 (page 453), and the appropriate occupational titles to make a logical assignment of existing and new personnel. The assignments are shown first by work activities in Table 13.1 and then by occupational titles in Table 13.2.

Table 13.1 begins to show the units of personnel that will be working together as a team for efficiency and safety purposes. The total of "Annual Staff Days Amount" for each work activity equals the estimated time listed in Tables 12.1 and 12.2 and the total of each "Annual Staff Days Percent" shows the percentage of personnel time needed for a specific work activity. Table 13.2, on the other hand, shows the different work activities assigned to each operator. The superintendent made a conscious effort to maintain a continuity of crew members, minimize the number of different work activities and provide work experiences for career advancement. The last two columns listed in Table 13.2 show that the entire 260 (5 days/week x 52 weeks) staff days of time for each employee has been assigned an activity and the percent of the employee's time that each specific activity occupies.

You may want to use your own experience and ideas to assign different personnel to the work activities listed in Tables 12.1 and 12.2 to produce what you think will be a more efficient use of personnel. The superintendent also plans to modify assignments as more experience is gained in the operation and maintenance of Clearwater's wastewater collection system and as conditions vary from those anticipated at the time the assignments were made.

13.23 Types of Crews

The types of crews, their individual makeup and the number of each type crew required can be as varied as the size, scope of maintenance activities, program objectives, geographic locations, and management philosophies of the many collection system maintenance organizations in existence today. We do not believe it is practical to offer any kind of formula that would suggest that, for a collection system of a particular size with certain specific existing problems, you will require a given number of operators or any particular types of maintenance crews.

Throughout this manual, you have been given specific information about the types of crews that will be discussed in this chapter. Detailed instructions as to the daily functions and activities have been thoroughly covered in the other chapters.

This discussion will be confined to more general aspects of crew staffing and jobs performed by the different types of crews. Proposed is the most practical size, based on experience of others, for each particular type of crew. Also, again based on experience, suggestions will be offered about how to use these crews most efficiently. The size of the collection system being maintained will have an influence on how each particular type of crew can be used. In large collection systems, crews tend to become more specialized and limited to performing their particular maintenance activity on a somewhat permanent or continuing basis. Smaller systems will find it more efficient to vary the activities of the crews at their disposal.

As mentioned earlier, it is usually desirable to allow crew members to work together for a period of time so that they get to know each other and become thoroughly familiar with their work assignments. Other factors that should be considered when assigning operators to crews are:

1. Maintaining continuity of members in a crew,
2. Minimizing the number of different work activities of each crew,
3. Providing work experiences for career advancement, and
4. Developing the skills of operators who could fill in on a crew in case of illness.

You will find it necessary to weigh the importance of each of these factors in deciding crew assignments. As your needs change, you may find it crucial at some point to rotate a few operators through various crews so that they are better prepared to fill in when needed. At another time it may be more important to keep crew assignments unchanged until the newly hired operators can become acquainted with the system and until you have had a chance to evaluate their strengths and weaknesses.

EMERGENCY SERVICE REQUEST CREW. Clearing stoppages is the job of an emergency crew. This is probably the basic crew in any maintenance organization. Certainly, it will be the first crew to become necessary. In smaller systems, it may be the only crew available. Two operators should be adequate except when it becomes necessary to enter a manhole.

TABLE 13.1 ASSIGNMENT OF PERSONNEL LISTED BY WORK ACTIVITY

Work Activity	No.	Personnel Occupational Title		Annual Staff Days Amount	Annual Staff Days Percent
Service Requests					
Main Sewer Stoppages	L1	Maintenance Operator	II	5.5	2
	L2	Maintenance Operator	I	5.5	2
			Totals	11.0	4
Building Sewer Stoppages	L1	Maintenance Operator	II	47.7	18
	L2	Maintenance Operator	I	47.7	18
			Totals	95.4	36
Preventive Maintenance					
Balling and Flushing	L1	Maintenance Operator	II	148.2	57
	L2	Maintenance Operator	I	148.2	57
	L5	Maintenance Operator	I	148.2	57
			Totals	444.6	171
High-Velocity Cleaning	L3	Maintenance Operator	II	192.4	74
	L4	Maintenance Operator	I	192.4	74
	L6	Maintenance Operator	II	31.4	12
	L7	Maintenance Operator	I	31.4	12
			Totals	447.6	172
Power Rodding	L6	Maintenance Operator	II	137.5	53
	L7	Maintenance Operator	I	137.5	53
			Totals	275.0	106
Bucketing	L10	Maintenance Operator	II	90.8	35
	L11	Maintenance Operator	I	90.8	35
	L12	Maintenance Operator	I	90.8	35
	L13	Laborer		90.8	35
			Totals	363.2	140
Chemical Treatment	L6	Maintenance Operator	II	33.0	13
	L7	Maintenance Operator	I	33.0	13
	L5	Maintenance Operator	I	33.0	13
			Totals	99.0	39
Repair and Replacement					
Main Sewers	L14	Maintenance Operator	II	185.2	71
	L15	Maintenance Operator	I	185.2	71
	L16	Maintenance Operator	I	185.2	71
	L17	Laborer		185.2	71
			Totals	740.8	284
Joint Sealing	L22	Instrumentation Technician	II	9.3	4
	L23	Instrumentation Technician	I	9.3	4
	L12	Maintenance Operator	I	9.3	4
	L13	Laborer		9.3	4
			Totals	37.2	16
Sewer Lining	L6	Maintenance Operator	II	12.4	5
	L7	Maintenance Operator	I	12.4	5
	L5	Maintenance Operator	I	12.4	5
	L13	Laborer		12.4	5
			Totals	49.6	20
Manholes	L3	Maintenance Operator	II	16.5	6
	L4	Maintenance Operator	I	16.5	6
	L5	Maintenance Operator	I	16.5	6
			Totals	49.5	18
Building Sewers	L10	Maintenance Operator	II	54.7	21
	L11	Maintenance Operator	I	54.7	21
	L12	Maintenance Operator	I	54.7	21
			Totals	164.1	63

TABLE 13.1 ASSIGNMENT OF PERSONNEL LISTED BY WORK ACTIVITY (continued)

Work Activity	No.	Personnel Occupational Title		Annual Staff Days Amount	Percent
Lift Station Maintenance Inspections and Servicing	L20	Maintenance Mechanic	II	17.3	7
	L21	Maintenance Mechanic	I	17.3	7
	L24	Electrician		17.3	7
			Totals	51.9	21
Repairs	L20	Maintenance Mechanic	II	1.6	1
	L21	Maintenance Mechanic	I	1.6	1
	L24	Electrician		1.6	1
			Totals	4.8	3
Inspections and Investigations Manholes	L8	Maintenance Operator	II	215.8	83
	L9	Maintenance Operator	I	215.8	83
	L18	Instrumentation Technician	II	210.0	81
	L19	Instrumentation Technician	I	210.0	81
	L10	Maintenance Operator	II	68.0	26
	L11	Maintenance Operator	I	68.0	26
			Totals	987.6	380
TV Main Sewers	L22	Instrumentation Technician	II	92.8	36
	L23	Instrumentation Technician	I	92.8	36
	L20	Maintenance Operator	II	92.8	36
	L25	Maintenance Operator	I	92.8	36
	L22	Instrumentation Technician	II	99.8	38
	L23	Instrumentation Technician	I	99.8	38
	L25	Maintenance Operator	I	99.8	38
	L13	Laborer		99.8	38
			Totals	770.4	296
Infiltration/Inflow Investigations	L14	Maintenance Operator	II	16.5	6
	L15	Maintenance Operator	I	16.5	6
	L16	Maintenance Operator	I	16.5	6
	L22	Instrumentation Technician	II	12.4	5
	L23	Instrumentation Technician	I	12.4	5
	L15	Maintenance Operator	I	12.4	5
	L16	Maintenance Operator	I	12.4	5
			Totals	99.1	38
Industrial Waste Inspections (Pretreatment Facility Inspection)	L1	Maintenance Operator*	II	14.1	6
	L2	Maintenance Operator	I	14.1	6
	L12	Maintenance Operator	I	14.1	6
	L10	Maintenance Operator	II	2.3	1
	L11	Maintenance Operator	I	2.3	1
	L3	Maintenance Operator	II	6.9	3
	L4	Maintenance Operator	I	6.9	3
	L14	Maintenance Operator*	II	14.1	6
	L12	Maintenance Operator	I	14.1	6
	L25	Maintenance Operator	I	14.1	6
			Totals	103.0	44

* May use specially trained pretreatment facility inspectors. See manual in this series on *PRETREATMENT FACILITY INSPECTION* from California State University, Sacramento. (See page iii of this manual for ordering information.)

TABLE 13.1 ASSIGNMENT OF PERSONNEL LISTED BY WORK ACTIVITY (continued)

Work Activity	No.	Personnel Occupational Title		Annual Staff Days Amount	Annual Staff Days Percent
Indirect					
Equipment Servicing	L18	Instrumentation Technician	II	5.8	2
	L19	Instrumentation Technician	I	5.8	2
	L20	Maintenance Mechanic	II	104.1	40
	L21	Maintenance Mechanic	I	104.1	40
	L24	Electrician		104.1	40
			Totals	323.9	124
Building and Grounds Maintenance	L1	Maintenance Operator	II	0.3	—
	L2	Maintenance Operator	I	0.3	—
	L5	Maintenance Operator	I	5.7	2
	L6	Maintenance Operator	II	1.5	—
	L7	Maintenance Operator	I	1.5	—
	L12	Maintenance Operator	I	32.8	13
	L13	Laborer		3.5	1
	L15	Maintenance Operator	I	1.7	1
	L16	Maintenance Operator	I	1.7	1
	L21	Maintenance Mechanic	I	60.4	23
	L22	Instrumentation Technician	II	1.5	—
	L23	Instrumentation Technician	I	1.5	—
	L24	Electrician		60.4	23
	L25	Maintenance Operator	I	9.1	4
			Totals	181.9	70
Work for Other Departments	L21	Maintenance Mechanic	I	32.4	13
	L24	Electrician		32.4	13
			Totals	64.8	26
Training		All Line Personnel		325.0	125
Leave and Holidays		All Line Personnel		780.0	300
				6,469.4 or	2,496 or
			Grand Totals	6,500.0	2,500

TABLE 13.2 ASSIGNMENT OF PERSONNEL LISTED BY PERSONNEL DESIGNATION

No.	Occupational Title	Work Activity	Annual Staff Days Amount	Annual Staff Days Percent
L1	Maintenance Operator* II	Main Sewer Stoppages	5.5	2
		Building Sewer Stoppages	47.7	18
		Balling and Flushing	148.2	57
		Industrial Waste Inspections	14.1	6
		Buildings and Grounds	0.3	—
		Training	13.0	5
		Leave and Holidays	31.2	12
		Totals	260.0	100
L2	Maintenance Operator I	Main Sewer Stoppages	5.5	2
		Building Sewer Stoppages	47.7	18
		Balling and Flushing	148.2	57
		Industrial Waste Inspections	14.1	6
		Buildings and Grounds	0.3	—
		Training	13.0	5
		Leave and Holidays	31.2	12
		Totals	260.0	100

* Or Pretreatment Facility Inspector.

TABLE 13.2 ASSIGNMENT OF PERSONNEL LISTED BY PERSONNEL DESIGNATION (continued)

No.	Occupational Title	Work Activity	Annual Staff Days Amount	Annual Staff Days Percent
L3	Maintenance Operator II	High-Velocity Cleaning	192.4	74
		Manhole Repairs	16.5	6
		Industrial Waste Inspections	6.9	3
		Training	13.0	5
		Leave and Holidays	31.2	12
		Totals	260.0	100
L4	Maintenance Operator I	High-Velocity Cleaning	192.4	74
		Manhole Repairs	16.5	6
		Industrial Waste Inspections	6.9	3
		Training	13.0	5
		Leave and Holidays	31.2	12
		Totals	260.0	100
L5	Maintenance Operator I	Balling and Flushing	148.2	57
		Chemical Treatment	33.0	13
		Sewer Lining	12.4	5
		Manhole Repairs	16.5	6
		Buildings and Grounds	5.7	2
		Training	13.0	5
		Leave and Holidays	31.2	12
		Totals	260.0	100
L6	Maintenance Operator II	High-Velocity Cleaning	31.4	12
		Power Rodding	137.5	53
		Chemical Treatment	33.0	13
		Sewer Lining	12.4	5
		Buildings and Grounds	1.5	—
		Training	13.0	5
		Leave and Holidays	31.2	12
		Totals	260.0	100
L7	Maintenance Operator I	High-Velocity Cleaning	31.4	12
		Power Rodding	137.5	53
		Chemical Treatment	33.0	13
		Sewer Lining	12.4	5
		Buildings and Grounds	1.5	—
		Training	13.0	5
		Leave and Holidays	31.2	12
		Totals	260.0	100
L8	Maintenance Operator II	Manhole Inspections	215.8	83
		Training	13.0	5
		Leave and Holidays	31.2	12
		Totals	260.0	100
L9	Maintenance Operator I	Manhole Inspections	215.8	83
		Training	13.0	5
		Leave and Holidays	31.2	12
		Totals	260.0	100
L10	Maintenance Operator II	Bucketing	90.8	35
		Building Sewer Repairs	54.7	21
		Manhole Inspections	68.0	26
		Industrial Waste Inspections	2.3	1
		Training	13.0	5
		Leave and Holidays	31.2	12
		Totals	260.0	100

TABLE 13.2 ASSIGNMENT OF PERSONNEL LISTED BY PERSONNEL DESIGNATION (continued)

No.	Occupational Title	Work Activity	Annual Staff Days Amount	Annual Staff Days Percent
L11	Maintenance Operator I	Bucketing	90.8	35
		Building Sewer Repairs	54.7	21
		Manhole Inspections	68.0	26
		Industrial Waste Inspections	2.3	1
		Training	13.0	5
		Leave and Holidays	31.2	12
		Totals	260.0	100
L12	Maintenance Operator I	Bucketing	90.8	35
		Joint Sealing	9.3	4
		Building Sewer Repairs	54.7	21
		Industrial Waste Inspections	28.2	10
		Buildings and Grounds	32.8	13
		Training	13.0	5
		Leave and Holidays	31.2	12
		Totals	260.0	100
L13	Laborer	Bucketing	90.8	35
		Joint Sealing	9.3	3
		Sewer Lining	12.4	5
		TV Main Sewers	99.8	39
		Buildings and Grounds	3.5	1
		Training	13.0	5
		Leave and Holidays	31.2	12
		Totals	260.0	100
L14	Maintenance Operator* II	Repair and Replace Mains	185.2	71
		Infiltration/Inflow Investigations	16.5	6
		Industrial Waste Inspections	14.1	5
		Training	13.0	5
		Leave and Holidays	31.2	12
		Totals	260.0	100
L15	Maintenance Operator I	Repair and Replace Mains	185.2	71
		Infiltration/Inflow Investigations	28.9	11
		Buildings and Grounds	1.7	1
		Training	13.0	5
		Leave and Holidays	31.2	12
		Totals	260.0	100
L16	Maintenance Operator I	Repair and Replace Mains	185.2	71
		Infiltration/Inflow Investigations	28.9	11
		Buildings and Grounds	1.7	1
		Training	13.0	5
		Leave and Holidays	31.2	12
		Totals	260.0	100
L17	Laborer	Repair and Replace Mains	185.2	71
		Buildings and Grounds	30.6	12
		Training	13.0	5
		Leave and Holidays	31.2	12
		Totals	260.0	100
L18	Instrumentation Technician II	Manhole Inspections	210.0	81
		Equipment Servicing	5.8	2
		Training	13.0	5
		Leave and Holidays	31.2	12
		Totals	260.0	100

* Or Pretreatment Facility Inspector.

TABLE 13.2 ASSIGNMENT OF PERSONNEL LISTED BY PERSONNEL DESIGNATION (continued)

No.	Occupational Title	Work Activity	Annual Staff Days Amount	Annual Staff Days Percent
L19	Instrumentation Technician I	Manhole Inspections	210.0	81
		Equipment Servicing	5.8	2
		Training	13.0	5
		Leave and Holidays	31.2	12
		Totals	260.0	100
L20	Maintenance Mechanic II	Lift Station Inspections and Servicing	17.3	7
		Lift Station Repairs	1.6	—
		TV Main Sewers	92.8	36
		Equipment Servicing	104.1	40
		Training	13.0	5
		Leave and Holidays	31.2	12
		Totals	260.0	100
L21	Maintenance Mechanic I	Lift Station Inspections and Servicing	17.3	7
		Lift Station Repairs	1.6	—
		Equipment Servicing	104.1	40
		Buildings and Grounds	60.4	23
		Work for Other Departments	32.4	13
		Training	13.0	5
		Leave and Holidays	31.2	12
		Totals	260.0	100
L22	Instrumentation Technician II	Joint Sealing	9.3	4
		TV Main Sewers	192.6	74
		Infiltration/Inflow Investigations	12.4	5
		Buildings and Grounds	1.5	—
		Training	13.0	5
		Leave and Holidays	31.2	12
		Totals	260.0	100
L23	Instrumentation Technician I	Joint Sealing	9.3	4
		TV Main Sewers	192.6	74
		Infiltration/Inflow Investigations	12.4	5
		Buildings and Grounds	1.5	—
		Training	13.0	5
		Leave and Holidays	31.2	12
		Totals	260.0	100
L24	Electrician	Lift Station Inspections and Servicing	17.3	7
		Lift Station Repairs	1.6	—
		Equipment Servicing	104.1	40
		Buildings and Grounds	60.4	23
		Work for Other Departments	32.4	13
		Training	13.0	5
		Leave and Holidays	31.2	12
		Totals	260.0	100
L25	Maintenance Operator I	TV Main Sewers	192.6	74
		Industrial Waste Inspections	14.1	5
		Buildings and Grounds	9.1	4
		Training	13.0	5
		Leave and Holidays	31.2	12
		Totals	260.0	100
		Grand Totals	6,500.0	2,500

Three operators are required for this hazardous task. Depending on their workload as an emergency crew, they can often be used for a variety of other activities. Some of these other activities are described in the following paragraphs.

- Inspection. They can be an effective inspection team to seek out and report any unusual conditions that develop within the collection system.

- Repair crew. They can be used as a repair crew. However, such activities should be limited to minor repairs only.

- Map checking. Collection system maps should be checked in the field to ensure that recent additions and/or changes have been reported and included on updated maps.

- Inflow and infiltration investigations. During rainy weather these crews can be useful in locating points of inflow through unauthorized points of entry into the collection system. By monitoring specific portions of the collection system where surcharging is present, they can provide valuable information about inflow/infiltration areas.

- Scheduled maintenance. If your maintenance program includes rodding, either hand rodding or power rodding, the emergency crew could also serve as the rodding crew.

These are just a few of the various fill-in types of activities which can be performed by emergency crews. Even in large collection systems there are slack periods when emergencies are few and such crews can be efficiently used by assignment of fill-in work. There is, however, one important point to consider in relation to such crews. Their primary responsibility is responding to emergency calls. Any fill-in work should, of necessity, be secondary in nature and of a type that can be quickly dropped to allow the crew to respond to the emergency at once when one develops.

BALLING CREW. Many preventive maintenance programs have one or more balling crews, depending on local conditions. The number required and the extent to which they can be used relates directly to the size of the collection system and the conditions within that system. Many larger agencies use several full-time balling crews. Smaller organizations might require only one crew on a part-time or full-time basis.

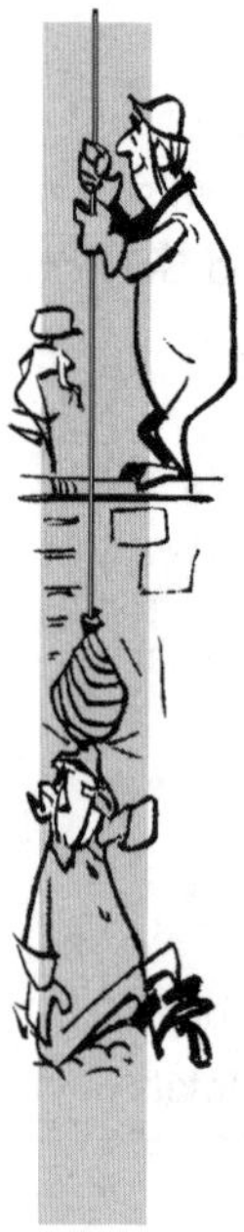

Balling crews commonly consist of three operators. There could be circumstances where additional operators may be required. If the work is being carried on in a main traffic artery, additional operators may be required for traffic control. In large sewers where a substantial amount of material is being removed, more operators may be required to aid in the removal of the debris. However, it is usually not necessary to have the larger crew for such specific circumstances on a regular basis. When such conditions are encountered, additional operators can be provided by combining available crews. When not used on a full-time basis, balling crews can be assigned other duties, such as the following activities:

- Repair crew. Three operators are adequate for many routine repair jobs.

- Smoke testing. Depending on the terrain and the scope of the smoke testing program, three operators can be sufficient for limited programs.

- Chemical crew. Again, depending on the scope of the program, three operators would be adequate in most situations.

- Bucketing crews. On a part-time basis, three-operator balling crews also can operate bucket machines.

FLUSHING CREWS. In recent years the practice of flushing has rapidly diminished. Experience has taught most supervisors that this practice is usually of little value in cleaning a collection system. The most that will be accomplished by flushing is a temporary increase in flow velocities, and increased velocities merely spread out accumulated deposits over a larger area within the system. The only effective method of cleaning a sewer is to remove deposits. Flushing will not remove deposits.

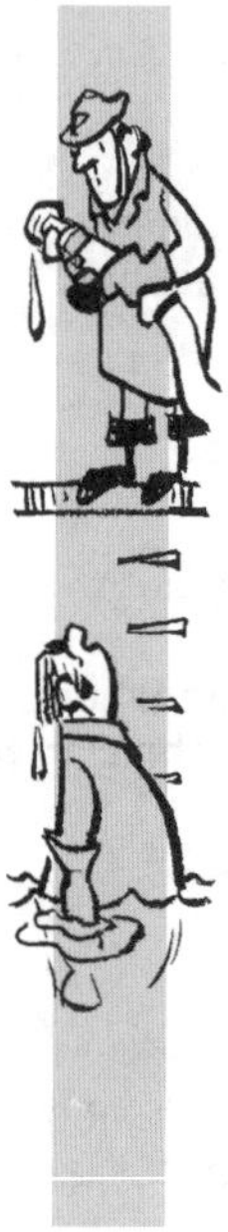

If you feel you must flush a sewer, two methods have been commonly used. Use the nearest available fire hydrant and a fire hose to flush the sewer. Local regulatory agencies often require an air-gap or backflow protection device between the domestic water supply and the wastewater collection system when this method is used. Another simple method is to use a water truck to transport water to the manhole. In either case, a two-operator crew would be sufficient.

HIGH-VELOCITY CLEANERS. This equipment is an important part of any preventive maintenance program. Two operators are adequate to operate high-velocity cleaners. However, in some areas, three operators are used if the cleaner does not have a vacuum device for removing debris and someone must enter a manhole to remove debris. Most organizations that own this equipment operate it on a full-time basis. Rarely is this equipment parked and the crew used for other activities.

POWER RODDERS. Many agencies operate this equipment full-time, on a year-round basis, and consider it an important part of their preventive maintenance program. However, the value of this practice is considered debatable by many experienced maintenance supervisors.

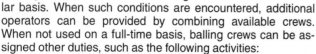

Power rodders are available in either truck-mounted or trailer-mounted models. In either case, two operators are sufficient to operate such equipment.

In organizations where the equipment is not operated full time, it is common to find a trailer-mounted unit kept available for use whenever necessary for specific conditions. Whenever this practice is followed, it is unnecessary to permanently staff operators for power rodders because they can be operated when needed by other available crews as suggested within this section.

BUCKET MACHINES. In large collection agencies bucket machines are used on a full-time basis with permanent crews assigned to such activities. Other agencies will have bucket machines on stand-by, to be operated only on a part-time basis to deal with specific problems. In either situation, a three-operator crew will usually be adequate. As with a balling crew, additional operators may be necessary for traffic control or to help remove a large amount of material from the sewers.

CHEMICAL CREWS. Chemical crews rarely require a full-time staff. Many agencies consider chemical application a routine part of the activities of existing maintenance crews.

A variety of chemicals are used for collection system maintenance. The methods of application are as varied as the types of chemicals being used. Some require simple application of chemicals into manholes or other collection system openings. Other chemicals require plugging of downstream manholes and retention of chemical solutions within the section of line being treated for a prescribed period of time. Obviously, the number of operators required will vary with the method of application being used. Depending on the type of chemical and the method of application involved, any available two- or three-operator crew should be adequate to temporarily handle this assignment.

REPAIR CREW. If the crew is to be operated on a full-time basis, three operators should be provided. A crew of this size will be adequate for most routine repair activities. If repairs of a larger scope must be undertaken, two or more crews can be combined. As previously suggested, other available existing crews could be assigned repair work on a part-time basis.

CONSTRUCTION CREW. Many maintenance supervisors draw a fine distinction between what constitutes a construction crew and what makes a repair crew. A logical distinction would seem to be that construction crews are engaged in new work, while repair crews deal primarily with replacement of defective portions of the collection system. Other maintenance organizations will make no such distinction. However, larger agencies often undertake construction of significant scope such as replacement of 100 feet, 200 feet, or more of damaged or defective main line. In many larger cities, all new building service lines for both residential and commercial properties will be constructed by collection system operators when they are to be installed in public right of ways. Such construction will require backhoes, dump trucks, and other construction equipment, and the necessary operators to operate the equipment. These crews should be three-operator crews, but crews will often be combined to provide for one-day or overnight completion of work in heavily traveled streets to minimize disruption of normal traffic flow.

REHABILITATION CREW. Very large organizations may have a sufficient amount of sewer joint and manhole sealing and sewer lining to find it cost effective to have a full-time rehabilitation crew consisting of three or four operators with all the equipment necessary for this type of work activity. On the other hand, small organizations will generally find that contracting for the personnel, equipment and materials for sewer rehabilitation will be the most cost-effective solution to leaky joint problems. Medium-sized organizations may use their own operators part-time and lease the equipment needed for sewer rehabilitation work activities. Suppliers of rehabilitation equipment and materials should provide assistance for the training of both full-time and part-time rehabilitation crews.

LIFT STATION OPERATION AND MAINTENANCE CREW. A lift station maintenance crew, usually consisting of two mechanics and an electrician, is used by an organization that has several medium- and large-size wastewater lift stations in its collection system. The second mechanic serves as a helper in work activities and as a third person for safety purposes.

If the crew isn't needed full time to inspect, service and repair lift stations, their mechanical and electrical abilities can be used for the servicing and maintenance of the equipment used by the organization for operation and maintenance of the collection system.

INSPECTION CREWS. Inspection crews should not be confused with construction inspectors or building inspectors. Historically many organizations have employed one or more collection system operators on a full-time basis simply to inspect their system. This inspection procedure usually consisted of opening manholes and noting the conditions therein on report forms. The efficiency and effectiveness of such procedures is very questionable and you should consider whether the benefits to be realized from such a program justify the costs. This procedure may be effective in specific problem areas where the inspector is looking for the causes of a problem.

Today many organizations assign inspection crews only when they need to locate and identify specific problems in certain areas. To meet this need on a part-time basis, it is common practice to use other available two-operator crews whenever necessary.

Properly trained inspection crews can be very valuable. They can determine the validity and nature of complaints, identify problems caused by industrial dischargers, and pinpoint the locations of trouble spots. Operators on this type of crew should be thoroughly familiar with the collection system layout, the location of problem dischargers and something about their discharge schedules or practices.

CLOSED-CIRCUIT TELEVISION (CCTV). A three-operator crew is adequate. CCTV equipment, when available, is usually operated on a full-time basis. In most cases agencies that are large enough to purchase this rather expensive equipment will have sufficient workload to keep the CCTV equipment operating on a full-time basis.

When CCTV equipment first became available for use in the field of wastewater collection system maintenance, it was believed by many to be too complex to be operated by collection system operators. Experience has shown otherwise. CCTV equipment available today can be operated effectively by existing collection system operators with proper training. While CCTV is not particularly complex or difficult to operate, crews should have time to become thoroughly familiar with it in order to operate it with a minimum of downtime. Good practice is to rotate crews, whenever possible, to ensure that trained operators are available whenever necessary. With some CCTV equipment, once the TV camera is in the sewer, only one operator is needed to televise the sewer and record observations.

SMOKE TESTING CREW. Rarely is there sufficient work to require a full-time crew. Smoke testing is usually undertaken to locate and identify points of inflow and illegal connections. After an area has been smoke tested, there is usually no cause to retest at least for a period of years.

As previously suggested, this function can be effectively performed by other existing crews within an agency. Smoke testing is an excellent operation to be assigned during wet or cold weather. This is the time of year when the conditions you are smoke testing to find are most prevalent. It is also the time of year, due to wet weather, that crews cannot perform their normal work and are available.

A smoke testing program of significant scope may require from three to five persons. The size of crew depends largely on the terrain of the area being tested. When using three- or five-minute smoke bombs, it is often necessary to have three to five people to thoroughly inspect the area covered by smoke bombs. Combining available crews is an excellent way to obtain the required operators.

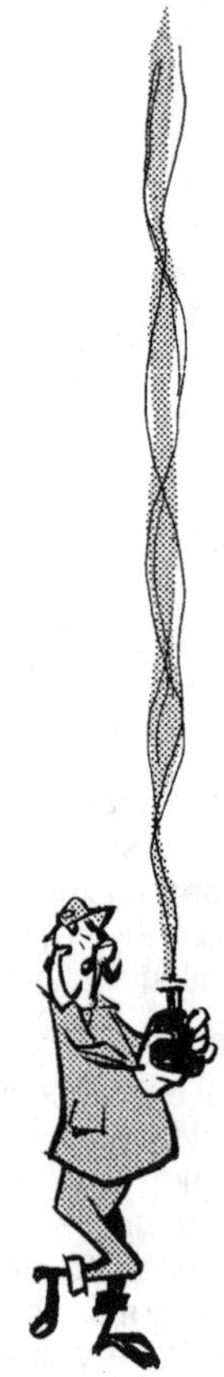

AIR TESTING CREW. Since closed-circuit television equipment is frequently used in conjunction with pipeline integrity testing equipment, it is logical to use the TV crew for air testing activities.

QUESTIONS

Write your answers in a notebook and then compare your answers with those on pages 609 and 610.

13.2A List the major work tasks performed by a collection system agency.

13.2B Why do maintenance crews tend to become more specialized in larger collection systems than in smaller collection systems?

13.2C What is the basic crew in most maintenance organizations?

13.2D How many operators are usually on an emergency crew?

13.2E List some of the other activities that might be done by an emergency crew when there are no emergencies.

13.2F What factors help determine the number of operators on a balling crew and the extent to which they are used?

13.2G What other jobs are often done by balling crews?

13.2H How often are high-velocity cleaners commonly used?

13.2I Is closed-circuit television equipment too complex to be operated by most collection system operators?

13.2J Why should crews be rotated from job to job?

13.2K What is the function of a smoke testing crew?

13.2L What is the difference between a repair crew and a construction crew?

END OF
LESSON 1 OF 4 LESSONS
ON
ORGANIZATION FOR SYSTEM
OPERATION AND MAINTENANCE

Please answer the discussion and review questions next.

DISCUSSION AND REVIEW QUESTIONS

Chapter 13. ORGANIZATION FOR SYSTEM OPERATION AND MAINTENANCE

(Lesson 1 of 4 Lessons)

At the end of each lesson in this chapter you will find some discussion and review questions. The purpose of these questions is to indicate to you how well you understand the material in the lesson. Write the answers to these questions in your notebook before continuing.

1. Why does a collection system agency need to be organized?
2. How could a person develop a knowledge of the work required of an organization?
3. In addition to an operator's existing skills, what factors should be considered when assigning operators to a crew?
4. How would you determine the types of crews needed to operate, maintain and repair a wastewater collection system?
5. How would you determine the number and sizes of crews for a wastewater collection system?

CHAPTER 13. ORGANIZATION FOR SYSTEM OPERATION AND MAINTENANCE

(Lesson 2 of 4 Lessons)

13.3 CLEARWATER COLLECTION SYSTEM DIVISION ORGANIZATION

Continuing with the Clearwater, USA example, let us now consider how the new superintendent reorganized the division that operates and maintains the city's wastewater collection system.

13.30 Service Request and Preventive Maintenance Section

The first priority of a wastewater collection system agency should be to have the ability to handle emergency service requests quickly and efficiently. Emergency service requests usually consist of calls to remove main and building service line stoppages and causes of flooding and to correct lift station failures.

Sewer line stoppages can develop when an obstruction causes debris to build up and slow down the flow of wastewater, gradually forming a block or halt of the flowing wastewater. Clearing stoppages quickly before major flooding and damage occur is a major concern.

For the sake of recording stoppages for follow-up maintenance work, stoppages must be identified according to the cause or type of problem. Stoppages are caused by obstructions such as roots, grease, debris, broken pipe or a joint failure. These obstructions may require immediate removal, repair, or replacement to correct the problem. See Chapter 6, Volume I, Section 6.010, "Types of Stoppages," for a discussion of the causes of various types of stoppages.

A main line stoppage is a major problem which can affect many customer services on that portion of a collection system. A main line stoppage is usually discovered by a private citizen who telephones into the office to report that a manhole is overflowing.

A building service line stoppage usually affects only the home or business that is served and the home resident telephones in when the sinks or toilet facilities will not drain.

The Clearwater Service Request (emergency service) and Preventive Maintenance Section is organized, as illustrated in Table 13.3, for responding to remove main line and building service stoppages. The Power Rodding Crew or the High-

TABLE 13.3 SERVICE REQUEST AND PREVENTIVE MAINTENANCE SECTION ORGANIZATION

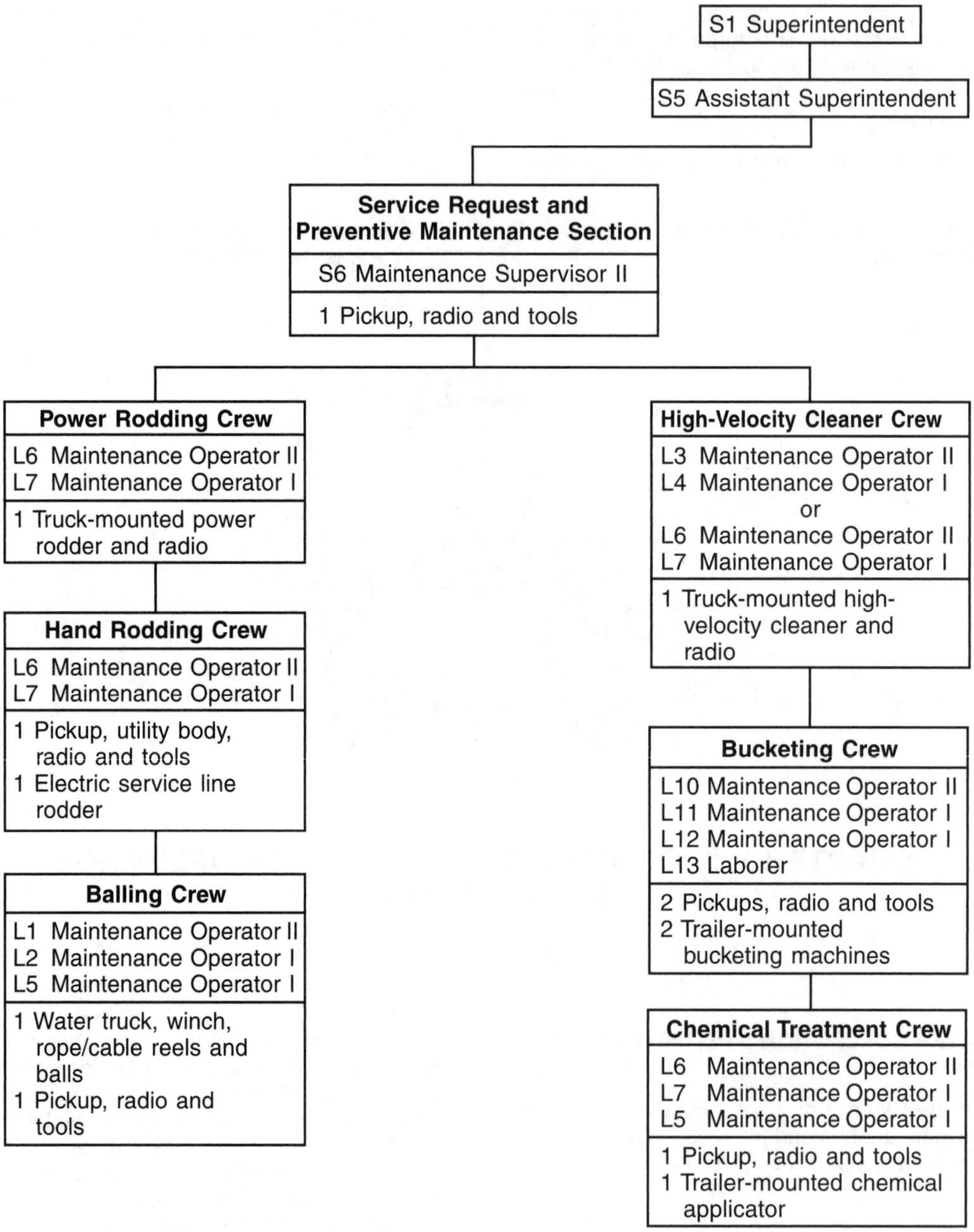

NOTE: One pickup assigned to the bucketing crew will also be used by the chemical treatment crew.

Velocity Cleaner Crew removes main line stoppages and the Hand Rodding Crew removes building service stoppages.

13.300 Responding to Requests for Stoppage Removals

In either of the above cases, the office personnel receiving the call should record the following information manually or electronically in a permanent log:

- Date,
- Time call was received,
- Name of person calling,
- Phone number of person calling,
- Street address where problem is occurring (mailing address also is important, if different from street address, for billing purposes if agency serves more than one sewer assessment district), and
- Type of problem the caller is reporting.

The dispatcher will have the following priority list:

1. Flooded home or business (main line or service line stoppage),
2. Manhole overflowing (main line stoppage),
3. Manhole cover off (liability risk),
4. Cleanout overflowing (nuisance and health hazard), and
5. Home service backed up.

When a problem is reported, the dispatcher radios the supervisor of the Service Request and Preventive Maintenance Section to provide the address or location and describe the nature of the problem. Also, the dispatcher will be responsible for:

- Recording the time the call was dispatched to the supervisor, and
- Recording the time the service crew or their supervisor reports the correction of the problem.

When multiple emergency calls are received at the same time, the dispatcher contacts the supervisor first. The service crew will first respond to the highest priority call and the supervisor to the other calls. The supervisor evaluates the problems, assigns priorities, redirects service crews, and requests assistance from preventive maintenance crews for specific equipment or assistance in handling calls as necessary.

The office staff are responsible for records, reports and the accuracy of the initial phase of service requisitions.

QUESTIONS

Write your answers in a notebook and then compare your answers with those on page 610.

13.30A List the priority responses to requests for emergency service.

13.30B What does the field supervisor do when multiple emergency calls are received?

13.301 Flooding Caused by Stoppages

One of the most delicate problems of public relations which Clearwater or any other agency must face is the occasional flooding of a home, a business, or a street area caused by a main line stoppage and the resulting backup of wastewater. Despite the risk of incurring increased liability, either your insurance carrier, a janitorial or sanitation cleanup service, or an agency crew should be sent to the property to assist in cleaning and removing articles which may have been damaged or will be damaged if left in place. These crews should be the first line of defense and they should be alerted when the possibility of flooding occurs. Response calls must be prompt. Dispatchers handling such calls may be able to give instructions to the homeowner which will avoid damage from impending wastewater flows.

Above all, you and your fellow operators must follow policies which have been approved by your agency's legal department because flooding often leads to lawsuits. When possible, a radio-dispatched crew should be sent to investigate any irregularity in the functioning of a building's plumbing system. This is a good opportunity to locate the trouble and determine whether or not it is the city's or agency's responsibility which has caused the backup flow damage. Disputes often arise regarding whether stoppages occurred in a main line or a house service line on or off private property. Problems can be avoided by thoroughly investigating the cause and location of the problem whenever a stoppage is cleared.

If flooding has occurred due to a stoppage in a main sewer line and the city or agency is responsible, the crew should follow set procedures for unplugging the stoppage and for helping to clean up the building or home. Always advise your legal department immediately of what has happened and exactly what actions you took. In many cases, the legal department or insurance company will be able to make an out-of-court settlement for damages incurred, thus saving the city or agency the high court costs.

An agency should have cleanup equipment available or prior arrangements with a sanitation service company to respond quickly to any location in need of wastewater cleanup service. This procedure ensures a rapid cleanup of wastewater from the household area and will result in less water damage. For example, the quick removal of water will save existing hardwood floors from warping or floor tiles from becoming loose.

Many other articles can be salvaged, and prompt action can bring down the cost or expense of replacing articles only slightly damaged.

If the flooding was caused by the failure of a pump that was supposed to discharge the wastewater from the basement of a home or business, the community policy must clearly define who is responsible for the operation and maintenance of the pump. If the flooding is caused by a service line connection that is low in relation to the collection system and is an overflow point, then a backflow preventer and an overflow valve should be installed to prevent future flooding. Agency policy must clearly state who is responsible for maintaining these items. Many health departments dislike overflow valves, but if homeowners are given the choice of either flooding their family room carpet with wastewater or flooding the front lawn, they would prefer the lawn.

Every collection system agency has the responsibility to develop policies and guidelines which clearly state the limits of agency responsibility. Also the responsibilities of homeowners must be communicated to homeowners affected by agency policies.

13.302 Other Requests

Collection system agencies frequently receive calls about water leaks from domestic water supply systems; these should be referred to the water department. Requests to help recover valuable items such as diamond rings, jewelry, money, false teeth, and keys that have been lost down the drain or flushed down the toilet must be handled according to agency policy. If attempts are made to recover these items, the work should be assigned to a balling crew to clean the portion of the main line downstream of the building service. Very often, recovery of these items greatly enhances the public's image of the agency. Responding to these requests is time consuming and expensive, but may be worth the expense if crews are available. Establish a policy for handling these requests if the frequency exceeds one per month. Be sure to include in the policy a disclaimer of responsibility for recovery of lost items; the only chance of finding them would be in some trap or constriction that serves as a trap for heavy objects.

QUESTIONS

Write your answers in a notebook and then compare your answers with those on page 610.

13.30C What should be done when a private home becomes flooded with wastewater?

13.30D How can legal problems resulting from flooding be minimized?

13.30E How can flooding problems caused by a low service line in relation to the collection system be corrected?

13.30F Who should handle service requests to have valuable property recovered from a sewer?

13.303 Service Crew Functions

The crews responding to requests for service should function in the manner described in the following paragraphs.

EN ROUTE TO PROBLEM AREA. Upon receipt of an assignment from the dispatcher or field supervisor, the crew performs the tasks listed below en route (on the way) to the problem.

- Begin to fill out service request form. Record:
  1. Time call received,
  2. Street address of problem, and
  3. Unit number and names of crew members.
- Obtain map of problem area from vehicle map storage file or computer file.
- Notify dispatcher upon arrival at problem area of time of arrival and location of unit. If supervisor or crew cannot locate problem, ask dispatcher to verify service request or complaint with party who submitted original call.

MAIN LINE STOPPAGE. On a main line stoppage (manhole overflowing), the crew leader examines the maps to determine direction of flow. The downstream manhole is inspected to see if the flow is restricted or reduced through the manhole or backed up into the manhole or if there is no flow at all. When a manhole with reduced flow or no flow is located, the stoppage will be located between this manhole and the upstream manhole. A service crew is directed to the lower manhole and starts to work upstream to clear the blockage.

A service crew usually will be equipped with a power rodder. The supervisor may direct a high-velocity cleaner crew to remove a main line stoppage if the power rodder is not available or the use of the cleaner is more appropriate for removal of the stoppage. In either situation, main line stoppages will be removed in accordance with the procedures described in Chapter 6.

SERVICE LINE STOPPAGE. When a service line stoppage occurs, the first job the hand rodding crew does upon arrival is to inspect the main line. Inspection of the upstream and downstream manholes on the main line will verify if the main line is open and flowing. If the main line is not flowing or if it is surcharged, the crew inspects the main line until the area of the stoppage is located. The crew then requests the assistance of the power rodding crew or high-velocity cleaner crew to remove the stoppage.

If the main line is clear and flowing, the crew proceeds to the service address. The service line cleanout is inspected to determine if the portion of the service line from the property line to the main line is clear of obstructions. To do this, remove the cleanout cover (if wastewater is discharged when the cleanout cover is removed, this is a sure indication that the tap is obstructed) and run a hand rod with a 2- or 3½-inch auger down the cleanout to the main line. Make sure that sufficient rod is used to reach the main line according to the distance shown on the map and your estimate of the distance. Sometimes service lines can be cleared using a plunger-type tool. By placing the plunger at the bottom of the cleanout and yanking upward, the suction pressure created may clear the stoppage.

If a cleanout is not available, the crew or supervisor (after inspecting the main line) may inform the residents that it is not the agency's problem and that they should call a plumber (the crew or supervisor's response should be dictated by agency policy). If a plumber inspects the building service line and finds the line is open from the house to the property line and the service line is still plugged, then a service crew returns to the problem area. The crew digs up the service line, the line is cut into, rodded, and the blockage cleared in the street or easement area to the main line, and service is restored. Cover boards 3/4-inch thick and 4 feet by 4 feet are placed over the excavation to prevent people or animals from falling into the excavation. A barricade is placed on the cover board over the excavation to warn people of the excavation until a cleanout is installed and the excavation is filled.

COMPLETION OF FORMS. When the work is finished, the service report form is completed (Figure 12.16, page 503). Essential information includes:

- Cause of problem — roots, rags, grease,
- Time job completed, and
- Staff-hours required.

The crew also fills out a repair request form (Figure 12.17 page 505) giving the following information:

- Nature of repair (install cleanout at property line),
- Address of repair,
- Map location,
- Size of service line and material of line,
- Depth of service line,
- Date requested, and
- On the back side of the card, draw a map showing the property serviced, distance to street or easement, and major buildings or identifying marks to identify location of new cleanout (Volume I, Chapter 7, Figure 7.61).

Persons occupying the service building are informed that service is restored and the dispatcher is notified that the crew is returning to its scheduled assignment and/or available for another assignment.

A repair and replacement crew will install the new cleanout, fill in the excavation, and restore the landscaping to its previous condition. The next time a service request is made to this address, the cleanout will permit a service crew to rod from the new cleanout to the main line. The problem can be removed or the residents can be informed that it is their problem and that they must hire a plumber to clear the service line from the building to the property line.

QUESTIONS

Write your answers in a notebook and then compare your answers with those on page 610.

13.30G What does an emergency service crew do en route to a problem area?

13.30H How is the location of a main line stoppage determined?

13.30I How are stoppages removed?

13.30J What is the first job the crew does upon arrival to clear a service line stoppage?

13.30K How is a service line stoppage cleared?

13.30L What forms should be completed after clearing a service line that required the line to be dug up?

13.30M Which crew should install a new cleanout if one is required?

13.304 Preventive Maintenance

A review of the past records of Clearwater's preventive maintenance unit reveals some very serious problems. The crews were on the job and working, but there had not been a significant reduction in the number of main line stoppages and service requests during the past few years. The reason for the lack of a reduction appears to stem from two sources. The condition of the system is not known and the field crews need special training on when, where and how to use the available equipment and the effectiveness of the equipment.

The new superintendent of the Clearwater Collection System Division plans to use about 80 percent of the work time of the Service Request and Preventive Maintenance Section on cleaning and chemically treating Clearwater's lateral, main, trunk and interceptor sewers. This emphasis on preventive maintenance is expected to reduce the number of requests for the removal of stoppages, restore full flow capacity and extend the service life of the system.

13.305 Preventive Maintenance Crew Functions

A wastewater collection system cleaning schedule is developed by the section supervisor and the superintendent. The procedure for preparation of the work schedule is described in Section 13.6.

A priority list helps to provide an organized approach to solving problems when they develop and ensuring the completion of necessary work. The Clearwater superintendent developed the priority list shown below.

1. Assist emergency service crew to clear difficult stoppages on request.
2. After emergency service crew reports the clearing of a main line stoppage, clean segment of line where stoppage occurred within three days.
3. In lines with known root problems, schedule power rodder to clear problem areas. If emergency service crews receive requests for same location more than once in a six-month period, schedule power rodder in problem area more frequently and try to control roots using chemicals (see Chapter 6, Section 6.4, "Chemicals," in Volume I).
4. In lines with known grease problems, schedule high-velocity cleaner to clean lines more often when service requests indicate a more frequent schedule is necessary.
5. Assign cleaning units to areas of collection system to be cleaned this year.
6. Construct training facility (Figures 13.1 and 13.2) and provide training sessions for all preventive maintenance cleaning crews.

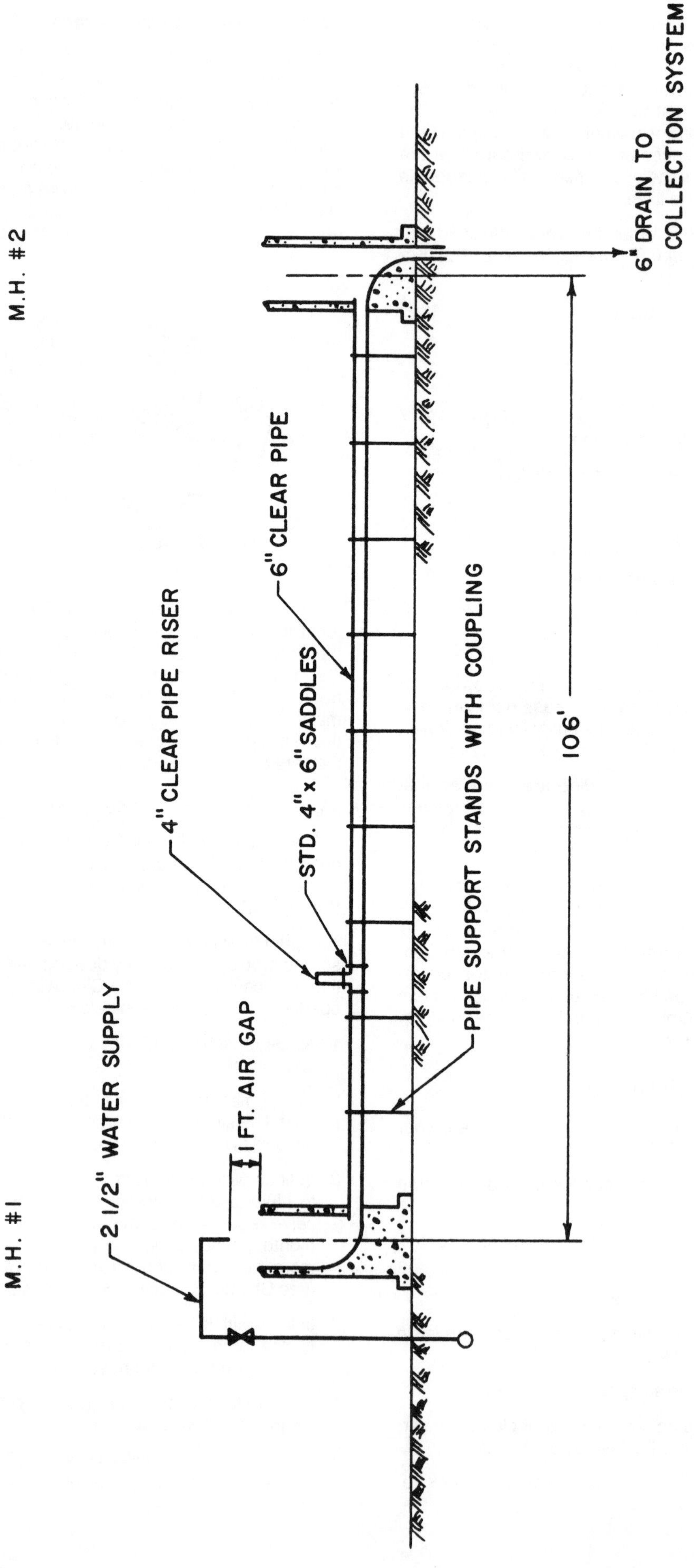

Fig. 13.1 Training facility pipe layout

Entire facility

Cleaning action of sewer ball

Fig. 13.2 Actual training facility
(Courtesy of Sacramento County)

7. Develop priority list of portions of collection system to be inspected by rented closed-circuit television inspection unit.

8. Prepare and schedule program to chemically treat fifteen miles of sewers with serious root problems (see Volume I, Chapter 6, Section 6.42, "Roots"). Program should include the following items:

   a. Research literature for available chemicals, methods and success of attempts by others to chemically treat lines to control roots.

   b. Visit agencies with similar problems that are using chemicals to control roots; evaluate effectiveness of their efforts and discuss with them their ideas for better results.

   c. Select at least two methods of chemical treatment and test on a section in selected area to evaluate effectiveness of methods.

   d. Keep records of all aspects of chemical treatment program including labor, equipment, chemicals and regrowth of roots to evaluate cost effectiveness of program.

QUESTIONS

Write your answers in a notebook and then compare your answers with those on page 610.

13.30N Why should a priority list be developed for scheduling work by preventive maintenance crews?

13.30O What is the priority list developed by Clearwater for the preventive maintenance crews?

13.30P What items should be considered during the preparation and scheduling of a chemical treatment program for root control?

13.306 Equipment

A broad range of equipment for the operation and maintenance of wastewater collection systems has been described in previous chapters of this manual and should be reviewed for information on specific types of equipment and their operation. The following is a description of equipment generally used by service response and preventive maintenance crews. Since the City of Clearwater's wastewater collection system is medium sized, its service response and preventive maintenance crews will not have all the equipment described.

SERVICE RESPONSE EQUIPMENT. Crews responding to service requests should have access to a power rodder, high-velocity cleaner, and an electric building service rodder. They will need a vehicle, preferably a 3/4- or 1-ton truck with a utility body, stocked with the following equipment:

- Radio with field call. The field call provides a means for the dispatcher to summon the crew to the unit's radio when they are working away from the truck. The device blows the vehicle's horn and turns on the rotating beacon if it is not in use. Some agencies place a speaker on the outside of the vehicle. If this type of equipment is not available or is too expensive, use an electronic pager. This equipment enables the dispatcher to start a buzzer which signals the operator in the field to phone the dispatcher.
- A complete set of maps of the collection system. If a complete set becomes too bulky, record the maps on microfilm and use a viewer. (Some systems may have access to the computerized mapping system described in Chapter 12, Section 12.711, "Geographic Information System (GIS).")
- Field report forms as required.
- Sewer rods, reel, and various sizes of augers.
- A small, portable, power rodding unit is very helpful for use as an aid in turning the sewer rods in difficult line stoppages.
- Detection equipment for toxic gases, explosive conditions and oxygen-deficient atmospheres.
- Small air blowers for ventilation.
- Portable generator for lights and tools.
- Ladder for descending into manholes (considered safer than manhole rungs; enter manhole only if three crew members are present and all confined space entry procedures are followed).
- Safety harness and ropes.

- Poles with hooks for dislodging and removing obstructions found in a manhole.
- First aid kit.
- Waterless hand soap and rags.
- Buckets and rope hooks.
- Other basic equipment includes small tools, rubber boots, gloves, hard hats, shovels, locators, probes, cover boards, barricades, and traffic safety cones and warning signs.

All equipment should be on the vehicle for use when needed.

EQUIPMENT FOR PREVENTIVE MAINTENANCE. Crews performing preventive maintenance of wastewater collection systems should have the following equipment and operate it in accordance with the following procedures:

POWER RODDING MACHINE. May be truck-mounted or trailer-mounted. This equipment is quite versatile in some types of cleaning and maintenance. A power rodder is very effective in clearing a stoppage. The large assortment of tools that are available for use with the rodder allows it to be used for clearing various types of stoppages, cutting roots, removing grease and even clearing sand and grit—although not too effectively. Effectiveness in larger lines is reduced because of the tendency for the rods to coil and bend. Common procedure is to operate from the downstream manhole, especially when there is a stoppage in the line.

SEWER BALL, CONE AND KITE. The sewer ball is constructed of high-grade rubber and is designed with spiral ribbing in order to create a spiral action of water in the line. Sewers are cleaned and scoured by the water flowing around the ball at a high velocity.

The balls come in assorted sizes and are inflated with air. At each end of a ball there is a lug or ear molded into the ball for attachment of a tag line or cable. The ball is inserted at the upper manhole. Air pressure in the ball should be enough to allow heavy flows of water to escape through the ribs on the ball. The tag line, or cable, is threaded through a manhole jack to the surface at a winch where it is controlled by an operator.

Water is introduced into the manhole and allowed to build up until two feet of head has been developed to create cleansing velocities around the ball. The speed of the ball moving down the line is controlled by the operator using the upstream tag line. Crews must be trained to select the proper speed so the water passing around the ball will "wash" the material in the sewer down the line ahead of the ball. The cone and kite operate in a similar manner.

These pieces of equipment are effective in moving grit, sand and debris down the line to a manhole where it can be removed. Slime and grease can be removed from the sides of the pipes by these techniques. One of the undesirable features of using this equipment is the possibility of flooding basements which are located below grade level, or even the first floors of buildings if they are below the rim of the manhole used for balling. If the water head behind the cleaning equipment is not properly controlled, flooding may result and could cause considerable damage.

BUCKET MACHINES. The bucket machine consists of two power winches with cables running between them. When cleaning a section of line, the winches are centered over two adjacent manholes. To get the cable from one winch to the other, it is necessary to thread the cable through the sewer line by using a power rodder, parachute, float, or a high-velocity cleaner. The cable from the drum of each winch is fastened to the barrel on each end of an expansion bucket so that the bucket can be pulled in either direction by the winch machine on the appropriate end. The bucket is fitted with a closing device and operates somewhat like a clam.

When a stoppage has been opened and the backed up water has been released, usually the large volume of debris that created the stoppage must be removed. A bucket is pulled into the stoppage in the line until the operator feels that it is loaded with debris. The winch motor is then thrown out of gear and the opposite winch is put into action. When the reverse pull is started, the loaded bucket automatically closes and the debris is brought out through the manhole to the surface. The bucket is opened and the debris is deposited in a truck or on the ground or street surface. This operation is repeated until the line is clean. Units with conveyor truck loading mechanisms are available for a much neater operation.

Various bucket sizes are available for cleaning lines from six inches up to thirty-six inches. Buckets commonly are used for larger diameter lines because of the expense incurred for this type of cleaning. Bucket machine cleaning is effective in removing heavy concentrations of sand and debris from larger diameter lines, but is undesirable from the standpoint of damage that can be done to the lines and because it is a time-consuming operation.

SEWER SCOOTER. The sewer scooter creates high velocities of water that clean the walls of sewers. Like the ball, cone and kite, the sewer scooter depends entirely on the head of water behind the scooter to provide the cleaning velocity. The scooter runs on a sled-like apparatus that has wheels on the runners and a hinged gate to control the upstream water pressure. An effective job is done in removing debris, grease and slime from the line. Drawbacks include being too difficult to use in smaller manholes.

HIGH-VELOCITY CLEANING MACHINES. High-velocity cleaning machines are truck-mounted and usually carry 500 to 700 feet of hose capable of withstanding high pressures. Special nozzles can be attached to the hose for routine cleaning and for breaking stoppages in a sewer. These nozzles have jets at various angles in the rear to serve the dual purpose of propelling the hose up the sewer and cutting or moving material in the sewer at the same time. High-velocity cleaning machines are probably the most effective of the cleaning equipment for removing grease, grit, sand, gravel, and debris. Traps at a downstream manhole always should be used with this operation to trap all debris and keep it from traveling farther downstream. Some high-velocity cleaners are equipped with a vacuum pump and a tube that can be lowered into a manhole to suck out debris.

A limitation of high-velocity cleaners is their ineffectiveness in areas that have steep grades and in large-diameter sewers. Sometimes they do not have sufficient water pressure to propel the nozzle up a line in areas of steep grades. Another problem is the danger of flooding a residence if the equipment is not properly used. In any case, the operator should be an experienced operator well versed in what could take place if the high-velocity cleaning machine is not properly used.

FLUSHING. Flushing may be used for cleaning small sewers for short distances. Although flushing does not scour or clean the sewers of grease or grit, it does give the opportunity to inspect each manhole and flushing inlets. The main purpose of this operation is to thoroughly clean the manholes in the system and move any floatable debris downstream to a larger line. Occasionally, flushing helps locate a potential stoppage in the collection system. Either a fire hydrant and hose (with an air gap) or a water truck is used for a water supply.

RECORDS. A simple and useful type of recordkeeping system must be implemented when setting up a preventive maintenance program and should be maintained thereafter. Accurate records are one of the most valuable tools available for a maintenance department. The record cards aid the department in a number of ways. A complete, up-to-date record system provides:

1. Details of all work done on the system, including dates,
2. Assistance in scheduling future work,
3. Justification for annual budgets,
4. Documentation for annual budgets, and
5. Protection against liability of the agency.

A record card file system or a computerized database should be kept giving all pertinent data of the collection system and maintenance performed on it. Records should be kept up to date, thus providing an accurate record of all work performed and all recurring sources of trouble in a collection system (see Chapter 12, Section 12.71, "Applications of Computers for System O & M," and Section 12.72, "Recordkeeping").

SUMMARY. All of the items and equipment that have been mentioned in this section were designed for a specific task and no single piece of equipment can be used to remedy all of the situations that may occur in a collection system.

QUESTIONS

Write your answers in a notebook and then compare your answers with those on page 610.

13.30Q What are the major disadvantages of using a bucket machine?

13.30R High-velocity cleaners are most effective in what kinds of situations?

13.307 Training Crews

Training of field crews can be greatly improved by the construction of a 6-inch clear pipe (Figures 13.1 and 13.2) that allows the crews to observe what's happening in a line when the various types of cleaning equipment travel through it. This training line also should be used to train all field crews in the safe and proper use of sewer cleaning equipment. Pieces of equipment that can be used to train crews with this facility include:

- Hand rodders,
- Power rodders,
- High-velocity cleaners,
- Sewer balls,
- Kites, parachutes, cones, and pigs,

- Sewer scooters, and
- Bucket machines.

In addition, the training facility can be used to teach crews how to flush lines, use chemicals and install plugs.

Observing the action of a piece of cleaning equipment on some grit or other test material in the clear pipe can be very helpful to a crew. The procedure is very simple. Deposit grit or test material in the upstream manhole (MH #1), turn on the water supply and flush the material into the section of clear pipe. Use the cleaning equipment the crew is working with to remove the grit. The crews can observe the effectiveness of the equipment, its limitations and how to use the equipment as efficiently as possible. Four hours of planned instruction using the training facility can be worth more than several months of on-the-job training in the field.

13.31 Repair and Replacement Section

All repairs to sewers and manholes and installation of new or replacement sewers and manholes (within the limitations of equipment, personnel and time) are done by this section. To help you determine whether or not a project should be undertaken by this section or by a private contractor, review the set of guidelines below. A private contractor should be considered if the following conditions exist:

1. Cost of material for a single job is over $15,000. This value is a flexible guideline and depends on how busy repair crews are when the job needs to be done, local regulations, and the willingness and ability of local contractors to do the job.
2. Time to complete the job is more than ten days. Again, this time period is a flexible guideline and depends on how busy repair crews are when needed and the desire and ability of local contractors. Repair crews should not become involved

in large jobs requiring over ten days because they have other work that should be done. Also, competitive bidding among contractors may get the job done at a more reasonable price. Repair crews should do emergency work if necessary and obtain assistance from contractors without bids for critical emergency jobs.

3. Special construction or safety equipment will be required. Examples of special equipment include deep shields for shoring or considerable pumping capacity to dewater excavations in areas with a high water table.

The organization of the Repair and Replacement Section is shown in Table 13.4. Crew 1 will make main sewer repairs and new sewer installations, Crew 2 will make manhole repairs and an infrequent manhole replacement or construction of an additional manhole, and Crew 3 will make building sewer repairs and replacements. Crew 3 will also install property line cleanouts.

Activities of the Repair and Replacement Section are generated by the need for repair and replacement of building service, lateral, main, trunk and interceptor sewers observed by the crews in the Service Request and Preventive Maintenance and Inspection Sections. Crew leaders' requests for repairs or replacements are processed through the management information system, described in Section 12.70 of Chapter 12, and result in work directives issued by the superintendent or supervisor to the appropriate Repair and Replacement Section crew.

13.310 Repairs

The following procedures are to be followed by crews making repairs:

- Section supervisor to review each work order and to assign priority to jobs. Priority assignment will be either *IMMEDIATE REPAIR* or *REPAIR WHEN TIME AVAILABLE*. Jobs will be assigned to the appropriate crew.
- Safety orders are to be followed by all operators.
- All operators are to be trained once a year in shoring of excavations.

- Crews working in easements and backyards must show proper respect for private property. Every effort must be made to restore area to original conditions, including replanting lawns and replacing shrubbery. Trees interfering with collection system operation or repair are to be removed, after informing property owner, and should not be replaced in easement.
- Whenever possible all excavations should be filled at the end of each day. Work overtime if necessary. Any open excavations must be covered with adequate strength materials and properly marked with signs, barriers and flasher signals.
- All tap repairs and new taps will be made by repair section. Plumbers will not be allowed to make any taps into Clearwater's collection system.
- All main line stoppages that require digging will have wastewater pumped from upper manhole to lower manhole, thus isolating the portion of line under repair whenever practical. Raw wastewater may be bypassed into a storm drain or open watercourse only upon approval of superintendent and after notification of appropriate authorities (local officials and water pollution control agencies).

13.311 New Construction

Repair and Replacement Section crews performing new construction will use the following procedures:

- New system work will be completed on the basis of work orders. New service connections may require plumbing permits and other necessary permits to be obtained by the contractor.
- A flat fee will be charged for single family residence service connections. Commercial and industrial connections will be charged on the basis of a field cost estimate by the New Construction Unit Supervisor. Supervisor to contact Industrial Waste Supervisor or Pretreatment Inspection Supervisor on all commercial and industrial connections to determine need to install special equipment. This equipment will meter and sample effluent to meet sewer-use ordinance requirements for billing and compliance with waste discharge requirements.
- Section supervisor should periodically review the section's organization. Consideration should be given to expanding crews during the summer months for repairs. Permanent crew size should be small because repair excavations should be minimized during adverse weather, except for emergencies. In areas where the weather is always poor, repair crews may have to work under adverse conditions almost all year.
- Section supervisor is to collect cost data and compare economics of purchasing heavy equipment (5-yard dump trucks and backhoes) with renting necessary equipment. Clearwater needs one backhoe for use throughout the year and three backhoes for repair work during the eight-month dry season.
- The closed-circuit television inspection activity is expected to reveal considerable work for the repair crew. Experience by other agencies indicates that for every eight hours of collection system TV inspection, 24 hours of repair work will be discovered consisting of the following:
  1. Bad taps (including protruding taps),
  2. Bad joints,
  3. Offset joints,
  4. Broken pipes, and
  5. Deteriorated pipe.

13.312 Equipment

Equipment needs for this section are discussed in the following paragraphs and also in Volume I, Chapter 7, "Underground Repair."

REPAIR UNIT. Equipment to install cleanouts, backflow valves, overflow valves, and to make repairs in rear easements will require a one-ton flatbed truck and hand tools. The truck should have tool compartments for small, expensive tools that can be easily damaged, lost or stolen. Keep these tool compartments locked.

A partially fixed or sliding-top canopy which provides protection from the weather and working space on the job is very helpful. This type of truck should fit the particular needs of the agency to increase efficiency and at the same time to present a pleasing appearance.

TABLE 13.4 REPAIR AND REPLACEMENT SECTION ORGANIZATION

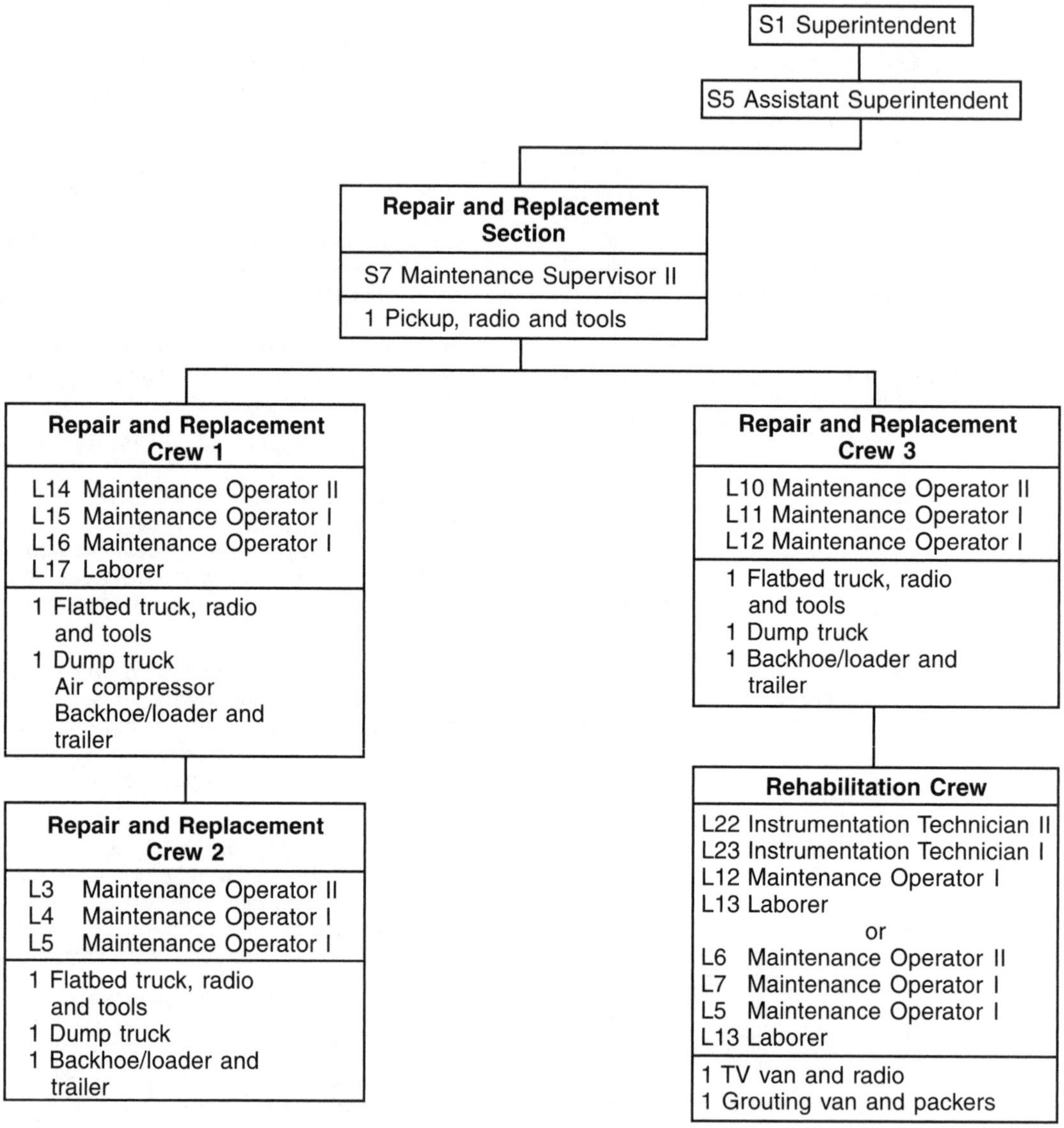

NOTE: The dump truck, air compressor, and backhoe/loader are shared by all crews in the Section. The grouting van and packers will be rented when needed.

NEW CONSTRUCTION UNIT. Well-planned and well-equipped construction trucks are necessary for an agency responsible for the maintenance and repair of any collection system. These trucks must be rugged and compact if they are to carry the number and variety of tools and materials required in emergency repairs or new construction work. An air compressor of sufficient capacity is the heart of all operations and is often carried at the rear of the cab. Outlets and piping for air can be placed around the truck body. Such a truck will be larger than a truck used to tow a trailer-mounted compressor. Locked compartments for small tools are highly desirable, as is a central space of sufficient size to permit transportation of lumber, pipe, cement, shoring or other materials. Water tanks hung under the truck sides are desirable for use in mixing small quantities of concrete on the job. An electrical generator power take-off of sufficient voltage to permit operation of small electric tools or lights is very valuable.

ALL UNITS. The size and design of trucks are regulated by the requirements and by the money available to your agency. Trucks used by brick masons may be similar except that the bed should be divided into foot-high compartments in which sand, rock, bricks, and cement may be placed, and a space should be provided for the debris which must be hauled away after the job is completed.

Trucks used in the regular program of cleaning sewers can be similar to the trucks used for inspection except that the bodies should be longer to take care of larger tools and equipment needed for sewer scooters or large sewer balls and cones. Trailer hitches for towing cleaning trailers or dump trailers are essential. The sizes and arrangement of compartments will differ from the inspection truck, but the sliding canopy need not change.

As all of the trucks mentioned so far are required to stand or operate in fast-moving traffic or heavily traveled highways, they should be painted a bright color, with reflective striping across the rear portions. Amber blinking lights for night protection, roadway flares, and barricade light blinkers are a necessity. An electrical generator or power converter also is desirable when inspection lights and floodlights are needed, and outlets should be provided at a convenient location on the truck.

Whenever possible, all critical operating units should be equipped with two-way radio telephones or portable radio receivers and transmitter units. This will permit an emergency call to be dispatched and answered in a matter of minutes or within the hour of being received.

QUESTIONS

Write your answers in a notebook and then compare your answers with those on pages 610 and 611.

13.31A Why is a section of clear pipe a helpful training device?

13.31B What factors should be considered by Clearwater when trying to decide whether repair or construction should be done by the agency or by a private contractor?

13.31C What work activities are assigned to Clearwater's three repair crews in the Repair and Replacement Section?

13.31D What are the priority designations given repair work?

13.31E Work orders for new construction originate from what sources?

13.31F Closed-circuit television inspection will reveal what types of repair work?

13.31G Where should small, expensive hand tools be kept in trucks?

END OF LESSON 2 OF 4
ON ORGANIZATION FOR
SYSTEM OPERATION AND MAINTENANCE

Please answer the discussion and review questions next.

DISCUSSION AND REVIEW QUESTIONS

Chapter 13. ORGANIZATION FOR SYSTEM OPERATION AND MAINTENANCE

(Lesson 2 of 4 Lessons)

Write the answers to these questions in your notebook before continuing. The question numbering continues from Lesson 1.

6. What are the activities or duties of an emergency service crew?
7. What causes stoppages or blockages?
8. How would you determine the priority of job assignments for an emergency service crew?
9. Why are public relations important when an emergency service crew responds to a stoppage that floods a home?
10. Why should a field supervisor investigate emergencies?
11. How would you remove a stoppage in a sewer?
12. What would you do after a stoppage has been removed?
13. How would you attempt to evaluate the condition of a wastewater collection system?
14. What are the advantages of a training facility built with clear plastic pipe above ground?
15. What kind of work is done by a preventive maintenance crew?
16. How would you develop a priority list for scheduling preventive maintenance crews?
17. How would you determine whether or not to consider a private contractor for a sewer repair job?
18. What equipment is needed by repair crews and new construction crews?

CHAPTER 13. ORGANIZATION FOR SYSTEM OPERATION AND MAINTENANCE

(Lesson 3 of 4 Lessons)

13.32 Lift Station Operation and Maintenance (and Auxiliary Equipment Servicing) Section

The ultimate goals of equipment maintenance are: (1) greater lift station reliability, and (2) fewer equipment or system failures which cause damage to equipment or the environment or endanger public health and safety. The design, construction, operation and maintenance of lift stations/pump stations has changed significantly in recent years primarily due to new regulatory requirements, environmental issues, liability, safety concerns and the complexity of equipment and systems installed in pump stations.

Lift stations represent a major capital investment for a wastewater collection agency. They are initially more expensive than gravity sewer systems as well as being more expensive to operate and maintain. In order to protect the capital investment, the wastewater collection agency must provide adequate resources to develop and implement an effective equipment maintenance program. Every agency with lift stations needs such a program, regardless of the size of the agency. However, smaller agencies are sometimes at a disadvantage because they frequently lack the financial and human resources available to larger agencies to maintain their facilities and also to respond to emergencies. Yet smaller agencies must still comply with the same environmental and safety regulations as the more sophisticated, larger systems.

Regulatory requirements have become increasingly stringent in recent years and no longer allow bypasses of raw wastewater, even in the event of a lift station failure. Bypasses or diversions of raw wastewater that enter storm water runoff drainage systems or discharge into surface waters are considered raw wastewater spills or illegal discharges. Fines in the hundreds of thousands or even millions of dollars may be levied against collection agencies *AND* the individuals responsible for their operation, maintenance and management when a lift station failure causes a violation due to poor maintenance programs or negligence. This has resulted in more reliable station designs. Standby power systems, either portable or on site, and increased equipment redundancy are examples of newer designs being used to avoid bypasses during system failures.

The formation of regional agencies, which is now common, allows more efficient and cost-effective collection and treatment of wastewater. This has resulted in the installation of much larger pump stations and force mains with higher capacities and operating heads. Pump stations with design capacities in excess of 20,000 gallons per minute and total dynamic heads in excess of 200 feet are not unusual. Force mains that are several miles long are very common. Design techniques and construction methods have changed as a result. Failures of these larger facilities tend to cause much more serious damage when a bypass has to be made or a spill occurs. Effective equipment maintenance is therefore especially important in large pump stations.

The Lift Station Operation and Maintenance (and Auxiliary Equipment Servicing) Section of Clearwater's Wastewater Collection System Division is primarily responsible for the inspection, maintenance and repair of the 16 wastewater lift stations in the collection system. The organization of the Section is shown in Table 13.5.

The Lift Station Maintenance Crew performs all the lift station maintenance activities. Since the superintendent does not expect the crew to use all its work time on this activity, the crew members' mechanical and electrical abilities will be used to service and make minor repairs to the mechanical and electrical equipment used by the other sections of the division. If and when more staff time is required for lift station activities, the superintendent plans on either contracting for servicing of equipment or requesting additional personnel. The Lift Station Maintenance Crew becomes the Equipment Servicing Crew 1 when performing servicing activities.

The Instrumentation Technicians will be used, when not operating the TV equipment or inspecting manholes, to service and make minor repairs to the TV equipment as an Equipment Servicing Crew 2.

13.320 Scope of Work

A review of the operation and maintenance of the Clearwater wastewater lift stations by the superintendent indicated existing procedures were fairly effective. Work priorities for this section are as follows:

1. Respond to all lift station failures when notified by dispatcher or by police department during off-duty hours,
2. Assist other sections in setting up emergency pumps to prevent flooding of homes or businesses or bypassing of raw wastewater to drainage ditches,
3. Respond to alarms from telemetered system,
4. Perform all lift station maintenance except machine shop work which will be performed by local machine shops under contract with City of Clearwater,
5. Perform work on preventive maintenance schedule and work requests from other sections, and
6. Section supervisor to evaluate and establish a program that will change daily lift station inspections to a weekly inspection. Change will be accomplished by establishing a preventive maintenance program, a telemetering system, and increasing capacity of lift station auxiliary units such as oilers and reservoirs to operate for one month without service.

13.321 Lift Station Failures *(Also see Chapter 8, "Lift Stations.")*

A properly designed pump used in a wastewater lift station, although properly maintained, can fail to operate when least expected and usually when needed most. There are at least four reasons why a pump will fail to function. The pump may fail due to a malfunction in either the electrical system or in the mechanical system, or by debris plugging an impeller or jamming a check valve.

Electrical failure may be the result of a power failure or an outside power source being switched off due to a circuit overload on the main power source. A circuit breaker tripped due to a circuit overload in the station itself could cause an electrical failure. A quick check of the reset button or breaker switch could identify the electrical failure caused by an in-station overload trip. When manual reset solves the problem temporarily but trips the reset or breaker switches (thermal overloads) again after a few moments of lift station operation, do not attempt to restart the pumps until a thorough inspection has been made of the pumping equipment in the lift station to find the problem. Correct the situation before attempting to restart.

When the problem has been located and corrected, reset the electrical breaker and put the lift station back on line. If the reset breakers kick out again, shut down associated equipment.

An overload may be caused by an electrical short in the motor or wiring control system, a mechanical part failure putting a load on the motor, or a plug in the impeller causing a load on the motor.

A mechanical failure could result from the breakdown of a bearing, poor shaft alignment, vibrations, mechanical seal failure, or malfunction of a mechanical float switch unit. Bearing failure can be noticed by a noise, vibration or heat. Another cause of mechanical breakdowns is a malfunction in the check valves, air bubbler compressor, or switching units associated with the centrifugal pump.

TABLE 13.5 WASTEWATER LIFT STATION OPERATION AND MAINTENANCE (AND AUXILIARY EQUIPMENT SERVICING) SECTION

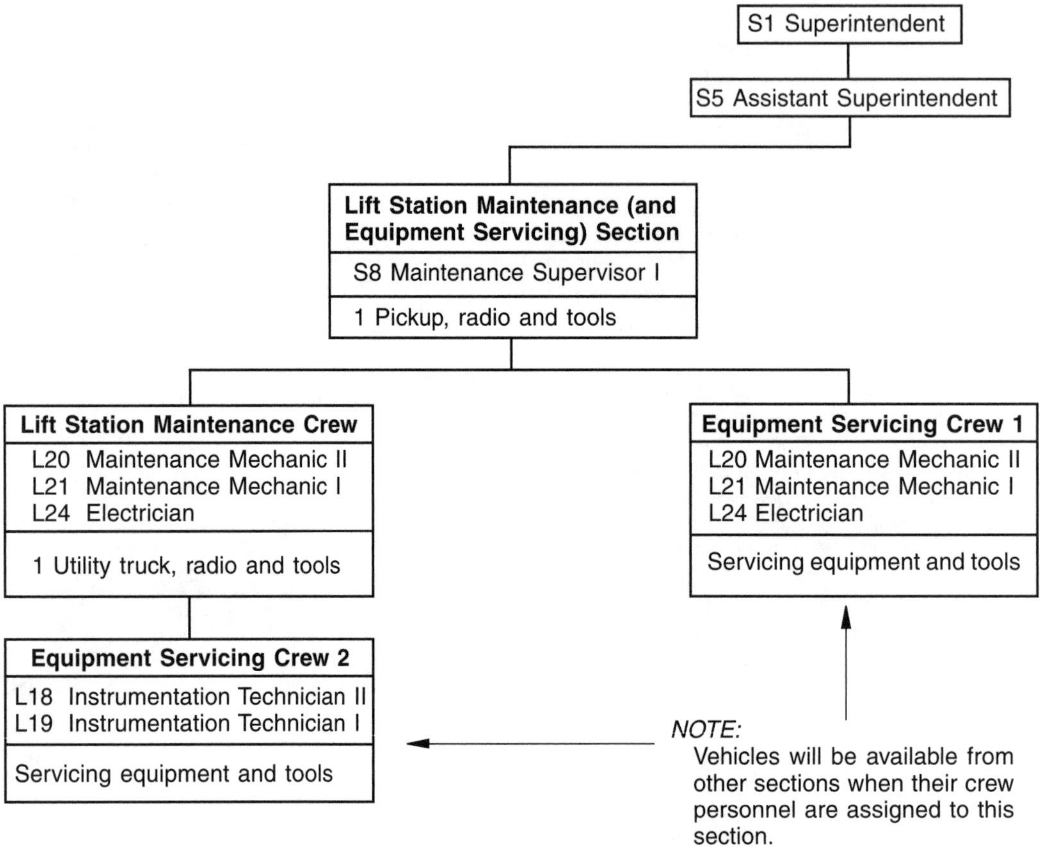

In a pneumatic ejector lift station, mechanical problems are generally caused by a check valve malfunction on the intake or discharge pipe valves. The air compressor and controls for air bypass valves are other items which may fail. The shorting of the control electrode by grease or corrosion is another electrical problem found at times in pneumatic ejector stations, and the cycle timing device unit also can cause problems.

In a centrifugal pump, most failures are caused by debris piled up in the impeller area. Common material found in stopped impellers includes lodged sticks, wooden blocks, bricks, rocks, rags, mops, plastic bags, sacks, stringy material, and roots.

If a lift station cannot be put back on line before flooding occurs, emergency pumps should be set up and piping arranged to pump the station's influent from the wet well into the force main until the problem can be corrected.

When a critical system or piece of equipment fails, it is important to determine the cause of the failure so that appropriate steps can be taken to prevent a similar failure in the future.

13.322 Preventive Maintenance Schedule

In reality, every wastewater collection agency uses both preventive and corrective (usually emergency) maintenance to operate its system. The goal, of course, is to reduce the corrective and emergency maintenance as much as possible and thereby minimize system failures that result in overflows and bypasses. Agencies that rely primarily on corrective maintenance as their method of operating and maintaining the system are seldom able to focus on preventive maintenance. Most of their efforts and resources are tied up dealing with emergencies and it is difficult for them to free up the resources needed to begin developing preventive maintenance programs.

A preventive maintenance schedule should be established to take care of every item in the lift station including lubricating the front door locks, replacing light bulbs and sweeping the floor. A very thorough and well-thought-out schedule that includes minor items will usually have provisions for maintenance of all the major pieces of equipment.

See Section 13.6, "Establishing a Maintenance Program," for some ideas about how to organize and schedule an effective and efficient preventive maintenance program.

QUESTIONS

Write your answers in a notebook and then compare your answers with those on page 611.

13.32A What are the ultimate goals of equipment maintenance?

13.32B What are the work priorities for Clearwater's lift station operation and mechanical maintenance section?

13.32C How can Clearwater change the program of daily visits to lift stations to weekly visits without encountering problems?

13.32D What are the major causes of lift station pump failures?

13.32E What could cause an electrical failure?

13.32F Under what circumstances may repair work on an electrical system be attempted?

13.323 Equipment

The Clearwater Mechanical Maintenance Crew will be furnished a one-ton truck equipped with a radio, utility body and an electric hoist capable of lifting medium-weight equipment from lift stations. Heavy-duty hoists or cranes will be rented to lift heavy equipment. Mechanical and electrical tools for maintaining lift stations, as described in Chapter 8, should be stored in the utility truck for use as needed.

Equipment and tools for servicing and making minor repairs to equipment used for operation and maintenance of a collection system are readily identified by maintenance mechanics and electricians. If not, a supplier of such equipment will be able to provide a list appropriate for the type of equipment to be serviced and repaired.

13.33 Inspection and Investigation Section

The Inspection and Investigation Section of the Wastewater Collection System Division of the City of Clearwater is responsible for inspection of the collection system (except lift stations) to determine the need for preventive maintenance, repairs and replacements. The section also inspects users of the facilities and manholes for indications of violations of the city's industrial waste and sewer-use ordinances, and investigates sources of excessive infiltration of groundwater and the inflow of storm water into the collection system. Table 13.6 shows the organization of the Inspection and Investigation Section.

13.330 TV Inspection

The need for closed-circuit television (CCTV) inspection of sewer lines arises when unidentifiable conditions are found by the activities of the Service Response and Preventive Maintenance Section and the Manhole Inspection and Infiltration/Inflow Crews. Requests for CCTV inspection of sewer lines are processed through the management information system and may result in work directives from the superintendent to the section supervisor. The supervisor is responsible for prioritizing work directives and their issuance to the TV Inspection Crew. The engineering division of Clearwater's Public Works Department may request CCTV inspection of new sewer line extensions which will be processed in the same manner.

The TV Inspection Crew leader is responsible for comprehensive CCTV inspection reports as described in Section 5.321 of Chapter 5, Volume I. If the work directive calls for a videotape or digital record of a CCTV inspection, it will be made with a comprehensive voice description of what was observed on the TV screen during the inspection. Completed TV inspection reports will be submitted promptly to the section supervisor.

TV technicians are trained to recognize and properly record the type and accurate location of the following examples of conditions that may adversely affect the operation and maintenance of a wastewater collection system:

- Crushed pipe, joint separation or offsets, faulty service connections, obstructions to flow or cleaning equipment, corro-

TABLE 13.6 INSPECTION AND INVESTIGATION SECTION

- S1 Superintendent
  - S5 Assistant Superintendent
    - **Inspection and Investigation Section**
      - S9 Maintenance Supervisor I
      - 1 Pickup, radio and tools

TV Inspection Crew

L22 Instrumentation Technician II
L23 Instrumentation Technician I
L20 Maintenance Operator II
L25 Maintenance Operator I

1 Pickup, radio and tools
1 TV van and reels

Infiltration/Inflow Crew

L14 Maintenance Operator II
L15 Maintenance Operator I
L16 Maintenance Operator I
or
L22 Instrumentation Technician II
L23 Instrumentation Technician I
L15 Maintenance Operator I
L16 Maintenance Operator I

1 Pickup, radio and tools
Air and smoke test equipment
Flowmeters
1 TV van and reels

Manhole Inspection Crew

L8 Maintenance Operator II
L9 Maintenance Operator I
or
L18 Instrumentation Technician II
L19 Instrumentation Technician I
or
L10 Maintenance Operator II
L11 Maintenance Operator I

1 Pickup, radio and tools

Industrial Waste Crew*

L1 Maintenance Operator II
L2 Maintenance Operator I
L12 Maintenance Operator I
or
L10 Maintenance Operator II
L11 Maintenance Operator I
or
L3 Maintenance Operator II
L4 Maintenance Operator I
or
L14 Maintenance Operator II
L12 Maintenance Operator I
L25 Maintenance Operator I

1 Pickup, radio and tools

NOTE: Pickup trucks will be available from other sections as their crew personnel are assigned to this section.

* May be Pretreatment Facility Inspection Team.

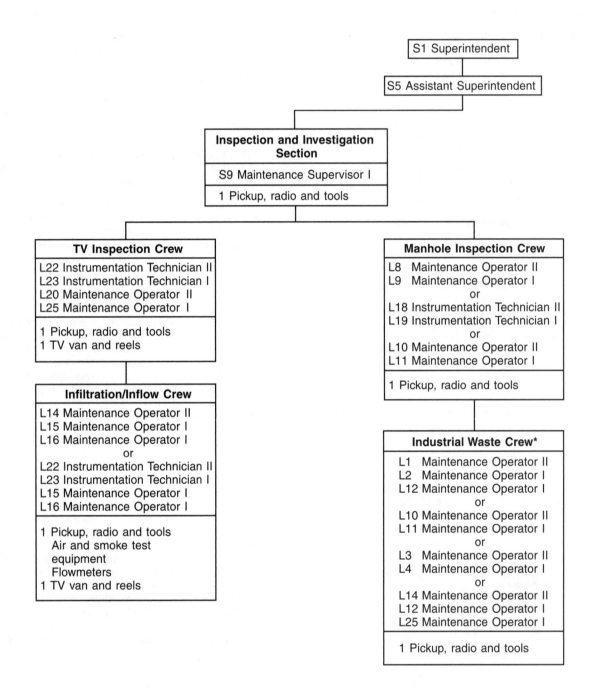

sion of concrete pipe, excessive sags in sewer lines, and root intrusions;

- Damage caused by excavation and construction adjacent to sewer lines;
- Unrecorded service connections that may indicate illegal use of a collection system;
- Infiltration and inflow sources and estimated amounts; and
- Defects in repairs and new construction.

The TV Inspection Crew may also be directed to conduct an evaluation of the effectiveness of methods and equipment used for maintenance of problem sewer lines.

13.331 Infiltration/Inflow Investigation

The Service Request and Preventive Maintenance Section and the Manhole Inspection Crews have been trained to report observations of what they consider excessive infiltration of groundwater in a section of the collection system and the sources of storm water inflow. These reports are processed through the management information system to produce work directives issued by the superintendent for investigations to be made by the Infiltration/Inflow Crew.

The Infiltration/Inflow Crew conducts investigations as directed to determine the sources and amounts of infiltration and inflow using smoke and air testing and CCTV inspections when appropriate. Flow measurements are made to determine the approximate amount of infiltration and inflow.

The I/I Crew uses smoke testing to identify the following sources of storm water inflow:

- Building roof drains connected to the collection system,
- Storm water drainage systems connected to the collection system,
- Unsealed building sewer cleanouts at ground surface, and
- Cracked sewer pipes and leaking pipe joints (if the defective pipes and joints are not submerged in groundwater and cracks in the covering earth run from the pipes and joints to the ground surface).

Residents and occupants of buildings connected to the sewer lines to be smoke tested must be notified of the time the testing will be performed. The fire department servicing the test area must also be notified of the smoke testing. Smoke may enter a building through a plumbing fixture that has lost its water seal and someone could report a false alarm of a building on fire.

A report on the completed smoke test, including an accurate description of the sources and location of storm water inflow, is submitted to the section supervisor for appropriate follow-up action.

Directives for measuring the amount of groundwater infiltration into a collection system are issued by the superintendent when a decision about repairing or replacing a section of sewer line is to be based on the determination of the amount of infiltration being excessive. The engineering section of the public works department will assist the Infiltration/Inflow Crew in the placement of flow measuring equipment if the amount of groundwater infiltration is suspected of being excessive.

Air testing of sections of sewer lines with excessive infiltration is performed to determine the following sources of groundwater infiltration:

- Location of cracked pipe,
- Location of leaking pipe joints, and
- Location of leaking service connection.

Reports on air tests giving an accurate description and location of the sources of infiltration are prepared by the crew leader and submitted promptly to the section supervisor.

13.332 Manhole Inspections

In sewer lines which are not cleaned at least annually, manholes are inspected on an annual schedule to determine their condition and observe the flow of wastewater through them. The section supervisor first consults with the Service Request and Preventive Maintenance Section supervisor to determine the frequency of sewer line cleaning. Then the section supervisor decides which manholes should be inspected and issues work directives to the Manhole Inspection Crew.

Before beginning the visual inspection, the Manhole Inspection Crew tests the atmosphere in the manhole for the presence of toxic, flammable or explosive gases by inserting the probe of a gas detection device through a hole in the manhole cover, if present, or immediately after removing the cover. If necessary, a portable blower can be used to ventilate the manhole until the hazardous conditions have been eliminated.

Inspection of a manhole is made from the ground surface using a high-intensity light or a mirror if sunlight is available. A manhole and sewer line inlets and outlets should be inspected for the following conditions:

- Presence of toxic, flammable or explosive gases;
- Cracks in the base, walls or joints of a manhole that indicate structural failure or infiltration of groundwater;
- Infiltration of groundwater into a manhole or indication of infiltration in a section of sewer line by an unaccountable increase in the depth of flow from an upstream manhole to a downstream manhole;
- Excessive offsets between the barrel sections and the cone and cover frame;
- Root intrusions into the manhole and inlet and outlet sewer lines;
- Accumulation of grease or debris in the invert of the inlet and outlet sewer lines;
- Strong odors and corrosion of the manhole walls and the upper parts of inlets and outlets of concrete sewer lines. These conditions may indicate the illegal discharge of industrial wastes or a septic condition of wastewater due to partial blockage of flow upstream of a manhole;
- Conditions of rungs, if provided; and
- Firm seating of manhole cover in frame to prevent the rattling of the cover by vehicular traffic.

The crew leader makes a report for each manhole inspected using forms similar to those illustrated in Volume I, Chapter 5, Section 5.26, "Sample Inspection Forms." Reports are submitted daily to the section supervisor.

QUESTIONS

Write your answers in a notebook and then compare your answers with those on page 611.

13.33A What are the functions or responsibilities of the Inspection and Investigation Section of the Wastewater Collection System Division of the City of Clearwater?

13.33B What types of conditions that may adversely affect collection system O & M should Instrumentation Technicians be trained to recognize?

13.33C Why must residents and the fire department be notified before an area of sewer lines is smoke tested?

13.333 Industrial Waste Inspections (Pretreatment Facility Inspections)

Industrial waste inspections (pretreatment facility inspections) are made to verify compliance with the city's sewer-use and industrial waste ordinances. Samples of industrial waste discharges to the collection system are taken for analysis to determine compliance and to calculate sewer-use charges. The Industrial Waste Crew performs the following work pursuant to directives from the section supervisor:

- Inspect and monitor dischargers for violation of sewer-use and industrial waste ordinances. Look for activities and discharges that could damage the wastewater collection system or the processes and facilities at the wastewater treatment plant (POTW).
- Collect samples and deliver them to the laboratory at the wastewater treatment plant (POTW).
- Locate sources of reported dumpings or bypasses that upset wastewater treatment process or are detected by collection system preventive maintenance crews. Make recommendations and provide evidence to the person or agency responsible for issuing cease and desist orders to violators of sewer-use and industrial waste ordinances.

For additional information see *PRETREATMENT FACILITY INSPECTION*, available from the Office of Water Programs, California State University, Sacramento, 6000 J Street, Sacramento, CA 95819-6025. Price, $33.00.

13.334 Inspection Equipment

Crews performing inspections and investigations of the Clearwater collection system will have the following special equipment:

- *TV VAN.* A typical closed-circuit television unit (TV van) contains a color television camera with a light and variable-sized skids, power and video cable, color television screen, control unit, tape recorder, electrical generator, camera pulling cables and reels, footage counter, manhole pulleys, and a manhole-to-manhole communication system. (Also see Chapter 10, Section 10.32, "Trenchless Technology," which contains a detailed description and diagrams of robotic devices that are used to inspect and repair pipelines by remote control.)
- *AIR TESTING* (for sewer line integrity). Air testing equipment consists of a trailer-mounted air compressor (shared with the Repair and Replacement Section), control unit, manhole and pipe segmenting plugs, and air hose and pulling rope reels. The equipment can be transported in a pickup truck.
- *SMOKE TESTING.* Equipment for smoke testing of sewer lines includes a gasoline- or electric-driven blower with a base fitting manhole cover frames, pipe plugs (or weighted canvas air curtains if flow in a sewer must be maintained), and smoke bombs.
- *FLOW METERING.* Wastewater flow metering equipment consists of primary devices (weirs, flumes, floats and electronic sensors) placed in the wastewater flow and a mechanical or electronic recording unit.
- *OTHER.* Other essential equipment includes a small refrigerator or ice chest for sample preservation, blowers, lights, meters for flow measurements, gas testing equipment, manhole lifters, small tools, safety harness, ladders, and portable field laboratory equipment. Field laboratory equipment includes pH meters, hydrogen sulfide test kits,

various sample containers, composite samplers, and a portable electric generator. Field inspectors should be equipped with a radio or paging device that notifies them to phone the office.

13.335 Additional Essential Equipment

This section describes important pieces of equipment that should be readily available to all crews whenever the need arises. Collection system operation and maintenance equipment must be geared to 24-hour operation. Consequently, portable equipment must be ready to move at all times.

An air compressor is probably the most useful unit available to a crew because of the many pneumatic tools which it can power. The compressor should have a capacity of at least 150 CFM if two pavement breakers or clay spades are to be operated simultaneously. Whether the unit should be truck- or trailer-mounted depends on the size requirement.

Gasoline-powered wastewater pumps may vary in discharge diameter from 2 to 8 inches, depending on the size of the sewer to be serviced. Such pumps should be of the non-clogging type, trailer-mounted, preferably on four-wheel trailers for the larger sizes, and should be provided with a self-priming unit to save time and effort. Quick-coupling pipe or hose of sufficient length to pump wastewater between manholes or around a break is desirable. A small pneumatic pump makes a good unit to pump out flooded basements and to dewater trench excavations where high lifts are required and the volume is not large. Such a pump may be of either the ejector or centrifugal type. Plungers or low-speed suction pumps also are ideal to dewater trench excavations because their action is positive and constant.

Large portable generators are helpful for providing electricity to lift stations without standby generators during power outages or other electrical interruptions. Trailer- or skid-mounted generators can be transported easily by trucks or hydro-lift trucks. Special interlocking outlets must be provided at lift stations so the leads from the generator can be plugged into the proper socket to operate equipment during a power failure.

Good lighting must be available for work at night and also for warning purposes, including warning traffic and pedestrians. Lighting is essential because once repair of a broken line is started, work must continue without stopping until the line is repaired. Occasionally traffic conditions require certain work to be done at night. Small portable generators are ideally suited to fill this need and, at the same time, they can be used to drive a variety of electric tools. Where job requirements demand larger floodlights and more power, a trailer-mounted generator with a battery of floodlights can be used. When inspecting the interior of large sewers, storm drains, or appurtenant structures, a sufficient number of safe lighting units are needed. These units must be explosion-proof, double-insulated, and properly guarded or protected.

Blowers, either trailer-mounted or portable, minimize the potential or real hazard which may exist in the form of toxic or explosive gases or oxygen deficiency. No inspection of sewers, storm drains or any confined spaces can be made unless proper ventilation is an accomplished fact and tests indicate *THE ATMOSPHERE IN WHICH YOU MUST WORK IS SAFE* from explosive conditions (LEL), toxic gases (hydrogen sulfide) and oxygen deficiency and will remain safe.

Trailer-mounted blowers with a capacity up to 7,500 CFM can easily ventilate many feet of medium-size sewers. Small portable units that can be carried on a truck and powered by an electric generator or powered through a converter on the truck can provide sufficient air to ventilate any manhole, small pump pit or sump.

Two-way, portable telephone or radio units are most helpful when performing cleaning operations where sighting between manholes may be impossible. On heavily traveled streets it often is impossible for operators to hear each other or to see signals made by crew members. Flooding of basements can be prevented by stationing an operator and radio unit at a suitable location to advise the cleaning crew if a backup occurs. Night work often is required because of daytime traffic and quick, positive intercommunication is always available with radio units. Radio units used for communication by crews working in the field may be on a different radio frequency than the central dispatching system to avoid interference and confusion, provided the additional frequency is available. There are many other possibilities for radio use that can help you do your job faster and more safely.

A chemical spray unit may be very useful for insect and odor control.

A concrete mixer of 1/2-sack capacity will provide sufficient concrete for many jobs and can be stored and towed easily. For larger quantities of concrete, ready-mix concrete is probably less expensive. Many other items of equipment can add to the overall efficiency of your agency.

QUESTIONS

Write your answers in a notebook and then compare your answers with those on page 611.

13.33D Why are samples collected of industrial waste discharges to the collection system?

13.33E Why is portable equipment necessary?

13.33F List some useful pieces of portable equipment.

13.34 Staff

Management of and support for collection system operators in the Wastewater Collection System Division of the City of Clearwater's Department of Public Works are provided by nine staff personnel including the superintendent. The organization of the staff is shown in Table 13.7.

13.340 Superintendent

The superintendent of the collection system division has the overall responsibility for the operation and maintenance of the system and duties include the following tasks:

- Supervises and coordinates, through section supervisors, the activities of operators and staff personnel;

TABLE 13.7 ORGANIZATION OF CLEARWATER'S WASTEWATER COLLECTION SYSTEM DIVISION

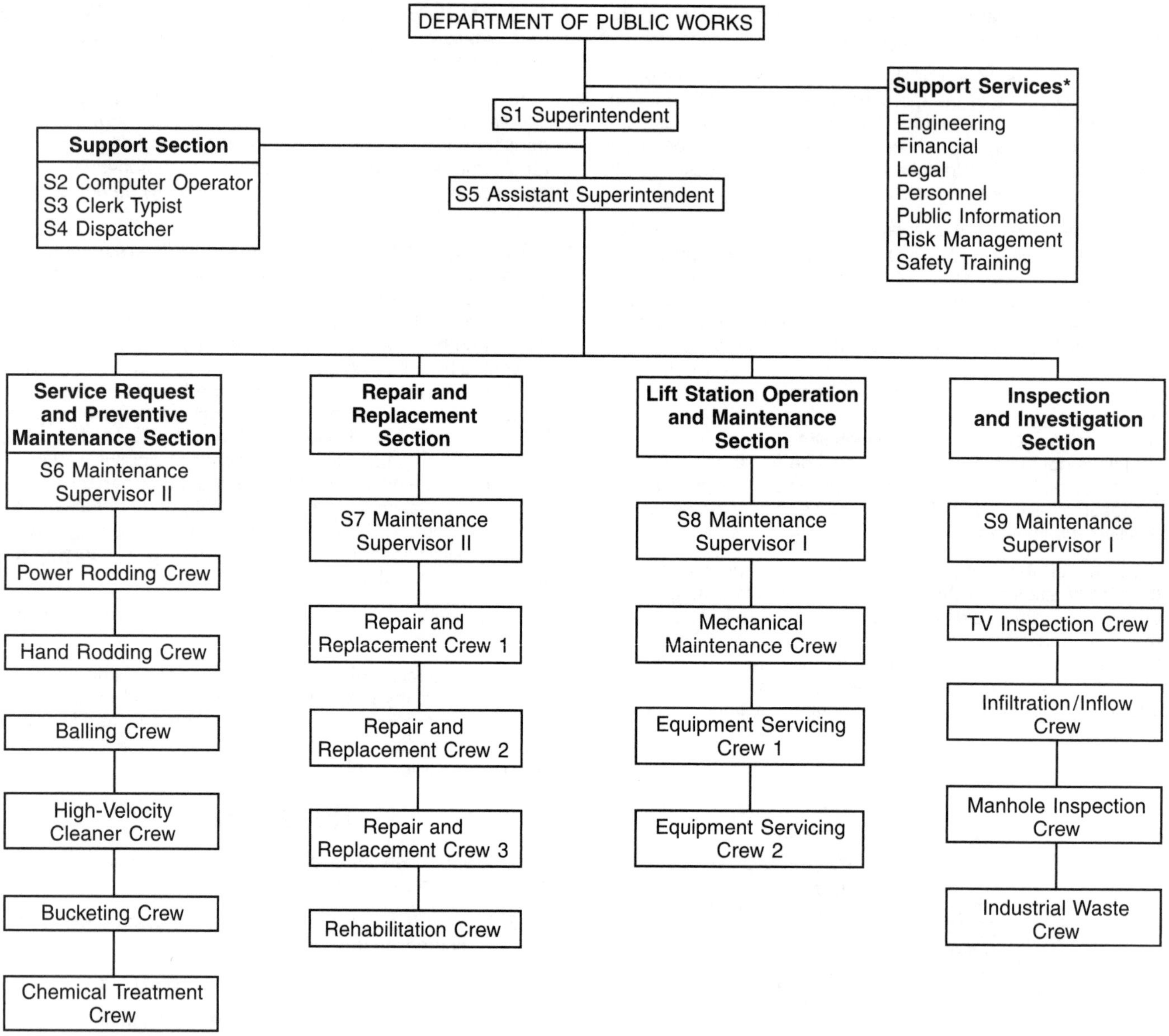

* Support Services provided by the centralized support staff of the Department of Public Works and other city departments.

- Compiles and analyzes planning and controlling reports (produced by a management information system, on personnel, material and equipment utilization) for initiating actions to improve work efficiency;
- Recommends division policies that must be approved by a higher authority, establishes and reviews all other division policies regarding personnel practices, work methods, and safety procedures;
- Submits recommendations to Director of Public Works for additional personnel and new equipment;
- Periodically visits work sites to ensure use of correct, efficient, and safe work procedures;
- Provides technical assistance to, and assists in the training of, staff and operators;
- Prepares and submits budget requests and reports to the Director of Public Works;

- Approves acquisition of materials and equipment pursuant to budget authorization;
- Cooperates with the personnel department in employment of division personnel;
- Confers with the Director of Public Works (and other municipal and government officials when authorized by the director) to discuss adequacy, operation, and maintenance of the collection system;
- Cooperates with other city departments to coordinate efforts for improving standards of service and enforcement of plumbing codes and sewer-use ordinances; and
- Confers with the Director of Public Works and contractors to discuss expansion of collection system and construction materials and methods.

13.341 Assistant Superintendent

The assistant superintendent, in the absence of the superintendent, has the responsibility for operation and maintenance of the collection system, assists the superintendent, and has the following additional duties:

- Compiles data, analyzes maintenance reports and helps prepare reports regarding utilization of resources, work procedures, and safety;
- Assists in selection of new equipment and expansion of existing facilities;
- Visits job sites to ensure compliance with established work standards and safety procedures;
- Assists in preparing budget requests and reports;
- Confers with contractors and equipment suppliers to discuss bid specifications for services and equipment; and
- Directs maintenance of division personnel records and recommends promotions and pay increases.

13.342 Maintenance Supervisors

Maintenance supervisors direct the activities of the four sections of the wastewater collection system division with the Grade II supervisors directing the sections with the greater number of personnel and more complex work activities. The duties of maintenance supervisors include the following:

- Confer with the assistant superintendent to discuss work directives and scheduling, assignment of equipment, and evaluation of personnel performance;
- Visit work sites prior to assignment of crews and determine the best procedures;
- Assign work to crews and ensure that materials and equipment are provided;
- Give directives to and meet with crew leaders to assign work and to discuss prescribed work procedures;
- Visit work sites to periodically inspect and evaluate work in progress to ensure that correct, efficient and safe work procedures are followed;
- Provide technical assistance and direct supervision of crews as needed for efficient conduct of work;
- Direct activities that are regarded to be sufficiently complex to require highly technical skills;
- Inspect and evaluate completed work to determine conformance with work directives and quality standards;

- Use maps and blueprints for directing work;
- Check equipment used by crews to ensure its efficient and safe operation and proper maintenance;
- Obtain work reports from crew leaders, make reports, and recommend improvements in work methods and service standards; and
- Train and instruct operating personnel.

13.343 Support Services and Personnel

Due to the medium size of Clearwater's wastewater collection system and the number of management and collection system operators in the Collection System Division, many of the support services it requires are provided by the centralized support staff of the Department of Public Works and other city departments. The division has only a small support staff assigned directly to it, as illustrated in Table 13.7.

SUPPORT SERVICES. The centralized support staff of the divisions includes the following groups which provide services to the collection system division:

- Engineering services for regulating the operation of the system, design and controls for new construction of sewers, drafting of sewer location maps, assistance for flow measurement, and assistance in infiltration/inflow activities and expansion of system;
- Financial services for payroll, job and unit costing, and budgeting;
- Legal services for contracting (for construction, equipment rental, and services), claims defense and personnel administration;
- Personnel services for employment of division personnel, provision of employee benefits, development of personnel policies and practices, and resolution of personnel grievances;
- Public information services for advising collection system users of the services provided by the division and need to adequately finance its operations, advising affected public on inconveniences to be caused by division activities and advising all users of the policies concerning services provided by the division;
- Risk management services include advice on safe operating and work procedures, processing of damage claims, accident investigations, and provision of insurance; and
- Safety services for establishment of safety policies and procedures, safety equipment acquisition and assistance in safety training.

SUPPORT SECTION. The staff personnel in the support section and their activities include the following:

- Computer Operator. Performs basic computer programming, loads acquired specialty software into computers, enters information from work reports into a computer's database, uses a computer to process data for preparation of planning and controlling reports, assists other staff personnel in use of computers, and assists other support personnel in their activities.
- Clerk Typist. Operates office equipment, serves as a secretary to superintendent, acts as receptionist for division, compiles personnel data used by other divisions, maintains internal personnel records, maintains division files, answers telephone, and maintains internal financial records.
- Dispatcher. Dispatches division crews to service request and work sites as directed by section supervisors, answers telephone to receive requests for service, locates sites of problems using sewer location maps, determines location and status of service response crews, transmits this information to appropriate section supervisor by radio or telephone, maintains records on service requests and responses, maintains log of crew location and status, processes authorization for street cuts and records status of repavement, maintains sewer location maps for support staff and section crews, and assists the clerk typist.

QUESTIONS

Write your answers in a notebook and then compare your answers with those on pages 611 and 612.

13.33G List as many of the major duties of a collection system superintendent as you can recall.

13.33H List the major support services needed by a collection system division.

13.33I What kinds of support activities can be performed by a computer operator?

END OF
LESSON
3 OF 4
LESSONS ON
ORGANIZATION
FOR
OPERATION
AND
MAINTENANCE

Please answer the discussion and review questions next.

DISCUSSION AND REVIEW QUESTIONS

Chapter 13. ORGANIZATION FOR SYSTEM OPERATION AND MAINTENANCE

(Lesson 3 of 4 Lessons)

Write the answers to these questions in your notebook before continuing. The question numbering continues from Lesson 2.

19. Why do larger organizations have a special section or unit for lift station operation and mechanical maintenance?
20. How would you develop a work priority list for a lift station?
21. What kinds of failures can occur in a lift station?
22. Why have the design, construction, operation and maintenance of lift stations changed significantly in recent years?
23. Why is it important to determine what caused a failure of a critical system or piece of equipment?
24. What are the tasks or jobs for an industrial waste section in a wastewater collection system agency?
25. How would you develop a work priority list for an industrial waste section?
26. What kind of equipment is needed by an industrial waste section?

CHAPTER 13. ORGANIZATION FOR SYSTEM OPERATION AND MAINTENANCE

(Lesson 4 of 4 Lessons)

13.4 SAFETY CONSIDERATIONS

Today's stringent safety requirements established by OSHA have had a direct influence on the amount of staffing required to perform almost all of our various maintenance activities. Increasing concern and enforcement exist today by local, state, and federal safety agencies to ensure that collection system operators obey required safety regulations. Less than full compliance with these safety regulations is unacceptable. Supervisors must recognize and accept this responsibility. Chapter 11 documents the dismal safety record of our profession. Every agency must recognize the benefits of an effective safety program that protects employees and *PREVENTS ACCIDENTS*. Every agency, supervisor and operator has both a social responsibility as well as a legal responsibility to comply with safety regulations and to enthusiastically support your safety program.

You must have an established and regularly scheduled safety program including the following:

1. Traffic Control. Requirements for providing adequate traffic control to safeguard the motoring public, as well as your own operators, can have considerable influence on the amount of staffing required (see Volume I, Chapter 4, Section 4.3, "Routing Traffic Around the Job Site").

2. Shoring. Strict compliance with existing shoring requirements, which must be met at all times, also can require additional staffing on certain excavation projects (see Volume I, Chapter 7, Section 7.1, "Shoring").

3. Confined Spaces. The latest safety requirements relative to people working in confined spaces, including manholes, make the two-operator crew obsolete when manhole entry is required. Requirements mandate that two operators are needed on the ground surface in order to be capable of physically removing the operator in the manhole when or if an emergency arises (see Volume I, Chapter 4, Sections 4.4, 4.5, 4.6 and 4.7 on manhole entry procedures).

Throughout this section we have made constant reference to two-operator crews. How can you operate two-operator crews and still protect the safety of operators? In most cases, it is possible. All of our maintenance procedures are designed to minimize the need for an operator to enter a manhole. This is especially true of two-operator crews and the work to which they are assigned. On those occasions when it becomes necessary for one of the two operators to enter a manhole, a supervisor is called to act as the third operator and all confined space procedures are strictly followed. This procedure can work very well. It is certainly a more acceptable alternative than to require all crews to consist of three operators.

13.5 REORGANIZATION

Changing activities of a wastewater collection system division due to the aging of the system, changes in use, and expansion of the system being operated and maintained, along with an observed need to increase the efficiency by changing procedures, can create a need for reorganization of equipment and personnel.

The organization of a collection system division should be reviewed periodically using standards for efficient organizations and reorganization proceedings should be initiated if appropriate. A superintendent should involve the division personnel affected by the reorganization in the planning process and not create disturbances by too frequent reorganizations.

There will be constraints on reorganization including personnel resistance to change, physical limitations, and personnel classifications and compensation.

QUESTIONS

Write your answers in a notebook and then compare your answers with those on page 612.

13.4A How have safety requirements influenced the amount of staffing required to perform maintenance activities?

13.4B What has been the impact of safety requirements on the two-operator crew?

13.5A What are the factors that may limit the reorganization of a collection system agency?

13.6 ESTABLISHING A MAINTENANCE PROGRAM

Collection systems are designed to transport wastewater without interruption to disposal sites or treatment plants. Most collection systems are designed to be "self-cleaning." Experience has shown otherwise for many collection systems. As time passes, various conditions develop which interrupt flows and diminish the design capacities and efficiency of collection systems.

Inadequate maintenance procedures have resulted in overflows or discharges of polluted wastewater into streams, streets, right of ways, rivers, and private property. Supervisors have been able to excuse these conditions as "unavoidable accidents," "acts of God or nature" or lack of sufficient capacity within their systems. Today, environmental concerns and regulatory agencies will no longer accept these excuses or tolerate these conditions. Also, the public accepts fewer inconveniences and demands to know what went wrong and why.

The responsibility for elimination of these conditions rests squarely on today's collection system supervisors. They must realize that they *CAN* influence the performance of the collection system. Through the use of modern, specialized maintenance equipment, microcomputers, and well-planned maintenance scheduling, the supervisor can develop meaningful maintenance programs that will minimize or eliminate these undesirable conditions.

13.60 Types of Maintenance

Although there are many types of maintenance and many different thoughts regarding what is a good maintenance program, let's try to think in terms of what needs to be done. A good maintenance program tries to anticipate what is going to go wrong and prevent it from happening. If something goes wrong, the problem is identified and corrected or repaired.

13.600 Preventive Maintenance

Preventive maintenance is the most effective and efficient type of maintenance program. It is a systematic approach for conducting maintenance activities prior to any equipment failure for the purpose of extending equipment life, reducing maintenance costs, and increasing reliability. Preventive maintenance should be a total concept including all the aspects of this chapter. A good preventive maintenance program requires a supervisor who knows what is likely to happen and where. The supervisor needs accurate, functional maps, good records, and an understanding of how to apply and use them. Intelligent assignment of available staff and effective use of equipment are important, as is properly planned and meaningful maintenance scheduling. All these aspects, properly applied, will provide a true preventive maintenance program.

Every wastewater collection agency, no matter what its size, should devote an appropriate level of resources to developing a preventive maintenance program. Unfortunately, many agencies find this difficult to do because their resources are already committed to emergency maintenance. However, experiences in both the public and private sectors show that as more resources are dedicated to preventive maintenance, less and less time and money will have to be spent on emergency maintenance.

A preventive maintenance program will improve equipment and pump station reliability and provide the following benefits:

- Reduce overtime costs,
- Reduce material cost,
- Improve morale,
- Reduce capital repair/replacement costs,
- Improve utilization of human resources, and
- Improve public relations.

Preventive maintenance programs should be evaluated periodically for their cost effectiveness, but cost alone should not be the only basis on which the program is judged. Intangible costs, such as a fine resulting from an NPDES permit violation, could easily exceed the cost of a maintenance program at a critical lift station for the entire year. The factors listed below should also be considered when determining the level and cost effectiveness of a program.

- Potential for equipment and/or system failures,
- Consequences of the failures, and
- Frequency and magnitude of failures.

Taking into consideration the "consequences of the failures" guideline, for example, a large pump station might require an intensive maintenance program if failure of the station would cause a bypass into an environmentally sensitive stream.

Another aspect of a good preventive maintenance program, sometimes called predictive maintenance, involves establishing baselines for equipment and system performance, monitoring performance guidelines over a period of time, and observing changes in performance. With this information, equipment failures can be predicted and maintenance can be performed on a planned, scheduled basis.

To conduct performance monitoring you will need good records of the operating history and characteristics of specific pieces of equipment. Some pumps will run longer without overhaul and replacement of seals, bearings, wearing rings, or impellers, whereas another piece of equipment may operate for only two years under a good preventive maintenance program before requiring removal and rebuilding or replacement. Yet another pump may operate for five years without any problems.

Other valuable information for predictive maintenance can be gained by conducting a vibration analysis on rotating equipment, from infrared scanning of electrical controls and motors, by sampling and analyzing the oil in large, engine-driven equipment to determine need for oil changes, and by installing *COUPONS*[1] in pipelines to establish corrosion rates. Knowing the condition of the equipment allows you to plan and schedule maintenance activities on an "as required" basis and avoid unnecessary maintenance.

In summary, an effective preventive maintenance program will enable the supervisor to apply effective preventive and corrective maintenance procedures over a controlled time period which will keep the collection system operating as intended with a minimum of stoppages or failures. The goal of a preventive maintenance program is to improve existing conditions in problem areas to a point where scheduled maintenance can be significantly reduced or ultimately reduced to the absolute minimum. An effective, efficient, and economical maintenance program can become a matter of pride to all operators involved in making such a program work—from the supervisor to the collection system operator.

THE PROGRAM ALLOWS THE SUPERVISOR TO MANAGE THE COLLECTION SYSTEM, RATHER THAN HAVE THE COLLECTION SYSTEM MANAGE THE SUPERVISOR.

13.601 Emergency Maintenance

Even with all our modern equipment and sophisticated maintenance programs, emergency maintenance is still a basic service provided by maintenance organizations. For our purposes, "emergency" maintenance refers to two types of emergencies: normal emergencies and extraordinary emergencies. Public utilities are faced with normal emergencies on a daily basis, such as a force main break or a blockage in a sewer. Normal emergencies should be and can be reduced by an effective preventive maintenance program. With a well-trained and well-equipped crew, emergency service can be very efficient. When these so-called "normal" emergencies do arise, generally they are handled swiftly and with little or no inconvenience to the public.

Extraordinary emergencies are conditions which have caused or have the potential to cause severe or catastrophic effects. Emergencies of this type may be caused by: (1) disasters such as fires, floods, hurricanes or earthquakes, or (2) equipment breakdowns which are unpredictable and difficult to detect during routine inspections and maintenance activities (for example, a pump volute breaks or a force main pipeline unexpectedly fractures).

The cost of responding to emergencies in terms of time and staff resources can be enormous. For example, the total cost of a lift station failure could include not only the cost of restoring the station to operation, but also the costs of cleaning up an overflow or spill of raw wastewater into lakes or streams, cleaning up backups of wastewater into homes or businesses, and compensating for the loss or destruction of personal property. These costs can easily amount to thousands of dollars. In addition, regulatory agencies frequently view an overflow or spill as a violation of the NPDES permit and significant fines and accompanying bad publicity for these violations can follow.

Naturally we cannot schedule emergency maintenance. We can only plan to handle it quickly and effectively. As our preventive maintenance scheduling techniques develop and improve, the need for "normal" emergency maintenance will be constantly decreasing. Planning ahead and preparing for natural disasters by establishing a comprehensive emergency response plan is an effective way to minimize the effects of the disaster on collection system performance.

QUESTIONS

Write your answers in a notebook and then compare your answers with those on page 612.

13.6A How can overflows or discharges of polluted wastewater into streams, streets, right of ways, rivers, and private property be minimized or eliminated?

13.6B What is a preventive maintenance program?

[1] *Coupon. A steel specimen inserted into wastewater to measure the corrosiveness of the wastewater. The rate of corrosion is measured as the loss of weight of the coupon or change in its physical characteristics. Measure the weight loss (in milligrams) per surface area (in square decimeters) exposed to the wastewater per day. 10 decimeters = 1 meter = 100 centimeters.*

13.61 Basic Aspects of Scheduling

Formulating an effective schedule for maintaining a collection system can be very difficult. Scheduling is complex, not necessarily in a technical sense, but more because of the many varied facts and details that must be considered. A computer can help a great deal, but there are no shortcuts to good scheduling. Maintenance of a collection system is an ongoing process which often spans years of past history, and which must consider future years of anticipated growth and increased loadings on collection systems. It is difficult to list all the many details to be encountered in planning a maintenance schedule. Some basic details to be concerned with are described in the following section. They include knowledge and skills, organization, timing of maintenance and flexibility.

13.610 Knowledge

This is the name of the game. Without knowledge, there is little chance for the maintenance schedule to be effective. The knowledge needed is not necessarily relative to anyone's intelligence or IQ. The required knowledge has more to do with the collection system and also the maintenance procedures and equipment. Required knowledge includes the following:

- Collection system. You must have a thorough knowledge and understanding of your collection system. You will need to know the nature of your problems and the areas in which they are concentrated.

  There are various ways of acquiring this information. A great deal of it may come to you through experience and familiarity with your collection system. The most reliable information could come from a computerized management information system or a computer-based maintenance management system (CMMS) such as the ones described in Section12.7 of Chapter 12.

  Aside from these obvious sources of information, there are other steps you can take:

  1. Complete a comprehensive study of your system. Locate and identify the areas that are creating the most problems. After you have determined your problem areas, try to locate and identify specific problems.
  2. A most effective method of identifying and locating specific problems is by selective manhole inspection and TV inspections. Emphasis should be placed on the term "selective." Haphazard or random inspections, particularly with TV, can be wasteful and inefficient. This practice can be used to survey even the smallest collection system. By careful preselection of troublesome areas, TV inspection can be performed in an economical manner by a contract service.

- Maintenance procedures and equipment. Any schedule you plan will be influenced by, and perhaps tailored to, the procedures and equipment available to you. You should carefully evaluate these procedures and equipment. Elsewhere in this course you have learned that certain procedures and equipment are most effective when used to combat specific problems. Others may be effective to some degree in coping with a variety of problems. How are you using available equipment? Are you using it to its fullest potential?

13.611 Organization of Personnel

The organization and staffing needed to support a collection system maintenance program depend on the types of systems installed (primarily in the pump stations); these systems can be identified as:

- Electrical and electronic systems,
- Mechanical systems, and
- Hydraulic and pneumatic systems.

Each type of system requires a different set of skills and level of knowledge to operate and maintain. It would be extremely difficult for a person in any single job classification to have sufficiently detailed knowledge to operate and maintain all of these various types of systems. Collection agencies therefore usually establish job classifications for operators responsible for routine operation and maintenance of the lift stations, and assign system and equipment maintenance to a second group with specialized skills in the electrical, electronic, mechanical, hydraulic and pneumatic systems. Larger agencies may have pipefitters, electricians and machinists who have formal training in each of these areas. The tasks to be performed and the skill required to perform the tasks are what define the job classification. It is not necessary to have a classification for each craft if the organizational structure defines what work is to be done, what skills are necessary to do the work and who is going to do it.

Many state and local codes require formal education, certification or licensing to work on specific systems. For example, an operator may be required to have a state or local license to perform maintenance on electrical systems, low- and high-pressure hot water systems or steam boiler systems, primarily because of the safety issues involved.

Each of the level 1 and level 2 job classifications should have a specific job description which includes the tasks involved, skills needed, level of education required and performance assessment guidelines. Many agencies have used certification criteria to define the skill level, experience and education required in each of the classifications.

Certification information is available from many sources including the Association of Boards of Certification (ABC), 208 Fifth Street, Ames, IA 50010-6259, or the Office of Water Programs at California State University, Sacramento, 6000 J Street, Sacramento, CA 95819-6025.

The maintenance program staffing can be further organized into operations and maintenance areas, as illustrated in Table 13.5. The responsibilities defined for the various levels of job classifications are described in the following paragraphs.

LEVEL 1 MAINTENANCE

Level 1 maintenance activities are completed by operators who visit the lift stations on a periodic basis, daily to weekly, to perform a variety of preventive maintenance tasks. The frequency of the visits depends on the following factors:

- Size of the station,
- Type of station,
- Extent to which the station is critical to operation of the collection system,
- Reliability of the station,
- Crew availability, and
- Remote monitoring and/or supervisory controls available.

Typical level 1 tasks performed by the operator include observing, recording and analyzing one or more of the following station operating guidelines:

- Incoming line voltage and current,
- Pump operating levels,
- Pump operating times,
- Station flows,
- Pump discharge pressures, and
- Any abnormal station operating conditions.

A qualified operator can interpret variations in this information and use it to troubleshoot problems both within the station and within the system as a whole. When data from these guidelines are routinely entered into a computer database, any changes can easily be compared to historical baseline information. The system is usually programmed to bring to the operator's attention any departures from the baseline and to identify the potential causes. Examples of problems that may be detected by changes in the values for the operating guidelines listed above include:

- Incoming line voltage and current
  1. Utility-supplied power problems such as low or high supply voltage, unbalanced 3-phase power
  2. Pump station changes in load, grounding, cabling
- Pump operating levels
  1. Float or bubbler pipe problems in wet well
  2. Pneumatic system problems
  3. Changes in supervisory control system
- Pump operating times
  1. Pump performance problems
  2. Supervisory control problems
  3. Increased/decreased station flow
- Station flows
  1. Identify internal versus external station problems
  2. Isolate external problems
- Pump discharge pressures
  1. Force main problems
  2. Suction condition changes
  3. Pump performance

Other preventive maintenance tasks that might be performed by level 1 operators could include:

- Changing pump sequencing (manual alternation),
- Purging bubbler system,
- Draining condensate from air compressor tank,
- Checking for tripped circuit breakers,
- Exercising emergency generator under load,
- Pump station housekeeping,
- Exercising valves,
- Lubricating pumps, motors, drives,
- Changing drive belts, and
- Cleaning bar screens.

Level 1 maintenance is performed by operators who are responsible for observing and verifying the operation of all systems in the station. They must have a broad background and understanding of the various types of equipment and systems installed in lift stations. Operators at this level perform maintenance that does not require access to the electrical system or teardown and repair of equipment which requires specialized skills. Level 1 operators are generally responsible for ensuring the overall station reliability and availability.

Most certification programs for collection system operators, other than entry level, now require a minimum of two years of formal education and may require a B.S. degree or equivalent for the higher certification levels.

There are usually four job classifications that reflect increasing experience in level 1 maintenance:

- Entry Level Operator
- Journey Level Operator
- Line Supervisor/Lead Worker
- Manager

LEVEL 2 MAINTENANCE

Level 2 maintenance usually is performed on a less frequent basis than level 1 maintenance, but requires specialized skills in specific areas and equipment systems. Generally this level is staffed by journey level electricians, pipefitters, machinists and mechanics. These crews are most effective when set up as composite crews tailored to the work assignments; for example, an electrician and pipefitter or machinist. These crews work closely with the operators in all phases of maintenance.

The craft or trade positions that maintain the electrical, mechanical and hydraulic/pneumatic systems in the lift station are classified into the following levels:

- The journey level, which generally requires formal education and licensing,
- Lead person/line supervisor, and
- Manager.

A number of agencies also use a fourth classification, apprentice journey level, when the agency participates in a formal apprenticeship program.

Depending on the size, structure and organization of the agency, either individual managers or a single manager may be responsible for both operation and maintenance.

Typically the crafts are responsible for performing the following types of activities:

- Low- and medium-voltage motor control center maintenance,
- Supervisory systems maintenance,
- Instrument and control system maintenance,
- Internal motor maintenance,
- Lighting panel and branch circuit maintenance,
- Pump overhaul,
- Mechanical seal replacement,
- Vibration measurement analysis,
- Pneumatic systems maintenance,
- Valve maintenance and overhaul,
- Engine repair and overhaul,
- Generator maintenance, and
- Heating/ventilation/air conditioning system maintenance.

A successful maintenance program will use operators possessing various skills as a team. For example, removing a pump for overhaul in a pump station may require:

- An electrician to disconnect motor connections,
- An operator to reset the supervisory control for pump sequencing, and
- A mechanic for rigging and removing the pump.

One effective method of organizing by teams is to approach the work assignment on a skills-required basis. In the above example, the electrician and mechanic would be an established crew, assigned work that required both sets of skills.

Cross training of operators is another way to increase maintenance effectiveness. All too often maintenance personnel are unable to properly diagnose a system problem in a pump station because they are so specialized. Cross training broadens operators' skills and enables them to more quickly troubleshoot problems.

QUESTIONS

Write your answers in a notebook and then compare your answers with those on page 612.

13.6C What defines a job classification?

13.6D What is level 1 maintenance?

13.6E What types of job classifications perform level 2 maintenance?

13.6F A license or certification may be required to perform what types of tasks?

13.6G Why would an electrician and a mechanic be assigned to the same crew?

13.612 Timing

Of vital importance to any schedule is proper timing. The very word "schedule" suggests timing. When planning maintenance schedules, two important time factors should be considered.

1. Only when needed. The ideal time schedule for maintenance would provide for completion of needed maintenance just prior to the development of any problems or emergencies. In actual practice this is virtually impossible to attain. It is, however, what we should be striving to achieve to the greatest degree possible.
2. Too frequent. If there is one weakness common to many maintenance schedules, this would be it. In their efforts to predict and prevent the development of emergency situations, many supervisors tend to schedule maintenance too frequently. This approach cannot be considered to be totally wrong. It is the lesser of two evils, the other being an inadequate schedule which results in too little maintenance. At best, however, too frequent maintenance must be considered inefficient. It is inefficient from an economical standpoint and it ties up operators and equipment that could be used more productively in other areas where maintenance or repairs are needed.

Table 13.8 illustrates some of the various types of equipment in a lift station and lists the typical frequencies for performing level 1 and level 2 maintenance activities. Keep in mind, however, that the frequencies listed in the table are only guidelines; they should be modified to meet your agency's individual needs.

The maintenance tasks required for each piece of equipment or system can range from very simple to complex. Take a molded case circuit breaker for a pump motor as an example. Visual inspection of the insulation on the wires on the line and load side of the circuit breaker can reveal a loose connection on the terminal. The insulation will usually become discolored and/or swell from the heat, so visual inspection is a task that is performed on a scheduled basis. Another task for the same circuit breaker would be to tighten the line and load side terminals once a year, again a simple task requiring a screwdriver or allen wrench. More complex maintenance tasks would involve specialized equipment to test the resistance across the internal contacts in the circuit breaker. You could also check the condition of the breaker using infrared scanning. The last two tests are actually predictive maintenance methods when they are performed on a regular basis, since baseline values are established and resistance values and temperatures are compared to the baseline value to determine when maintenance should be performed.

As previously mentioned, maintenance frequencies will actually be determined by the conditions which are specific to your pump stations. To identify important considerations, answer the following questions:

- Is the station monitored 24 hours a day using a telemetry system?
- Is the pump station or equipment critical to collection system operation?
- Does the pump station or equipment have a history of frequent failures?
- Is the equipment operating in a harsh environment (for example, very hot and humid atmosphere)?

TABLE 13.8 TYPICAL FREQUENCIES FOR PERFORMING PUMP STATION MAINTENANCE ACTIVITIES

System or Equipment	Daily	Weekly	Monthly	Quarterly	Semi-annual	Annual	Every 2 Years	As Needed
Electrical								
Incoming Service				1		2		
3-Phase Transformer(s)				1		2		
Metering Section				1		2		
Main Circuit Breaker/Disconnect				1		2		
Medium-Voltage Switchgear				1		2		
Ground Fault						2		
Automatic Transfer Switch		1				2		
Bus, Vertical and Horizontal						2		
Bus Cables/Stab-ins						2		
Motor Control Center			1		2			
Branch Circuit Breakers				1		2		
Motor Starters		1		2				
Motor Overload Relays					2			
Relays						2		
Indicating Lights	1		2					
Voltage/Current Readings	1							
Infrared Scan						2		
Power Factor Capacitors		1		2				
Supervisory Control								
Relays				2				
Wiring			1		2			
Fuses		1		2				
Circuit Breakers					2			
Pump Sequencing		1			2			
Pump Operating Levels		1			2			
Compressors, Bubbler		1				2		
Pneumatic Piping/Valves				1		2		
Pressure Switches		1			2			
Alarms and Levels	1				2			
Indicating Lights	1							
Floats	1					2		
Pumps								
Bearings (vibration)		1				2		
Alignment				1		2		
Coupling				1		2		
Packing		1				2		
Mechanical Seals		1				2		
Mounting						2		
Balance						2		

TABLE 13.8 TYPICAL FREQUENCIES FOR PERFORMING PUMP STATION MAINTENANCE ACTIVITIES (continued)

System or Equipment	Daily	Weekly	Monthly	Quarterly	Semi-annual	Annual	Every 2 Years	As Needed
Pumps (continued)								
Impeller Clearance						2		
Capacity		1						
Suction Pressure		1						
Discharge Pressure		1						
Station Calibration						1, 2		
Running Time	1							
Overhaul								1, 2
Lubrication								1
Motors								
Bearings			1			2		
Balance						2		
Conduit Box Connections						2		
Lubrication								1, 2
Insulation Resistance						2		
Running Temperatures		1				2		
Full Load Amps		1						
Voltage		1						
Vibration Analysis						2		
Alignment		1				2		
Coupling		1				2		
Overhaul								2, 2
Valves								
Suction				1		2		
Discharge				1		2		
Check		1				2		
Pressure Relief		1				2		
Gate				1		2		
Plug					1			
Piping								
Pipe Restraints				1		2		
Bubbler System	1			2				
Auxiliary Systems								
Water	1			2				
Sump	1			2				
Blowers/Fans	1			2				
Heating System	1			2				
Cooling System	1			2				
Engine Gen Sets	1		2					
Lighting	1			2				

Manufacturer's recommendations, altitude, and running times are other typical factors that need to be considered when establishing maintenance frequencies. Also, the amount of predictive maintenance performed influences preventive maintenance frequency since predictive maintenance is performed when needed rather than at regular, fixed time intervals.

13.613 Flexibility

A good maintenance schedule should be flexible. Periodically it should be revised in accordance with the results it has or has not produced.

- Objectives. When an area or section of line is assigned to a maintenance schedule, it should be assigned with an objective in mind. Will the scheduled maintenance eventually eliminate the problem? Will the scheduled maintenance merely control the problem to a degree that is acceptable? After five years of scheduled maintenance, is any significant improvement evident? If not, what is the next step?
- Effectiveness. When a schedule has been designed with an objective in mind, it is possible to measure the effectiveness of that schedule by the degree to which you have achieved your objective.

QUESTIONS

Write your answers in a notebook and then compare your answers with those on page 612.

13.6H How would you prepare an effective schedule for maintaining a wastewater collection system?

13.6I How does a person gain knowledge of or become familiar with a collection system?

13.6J What time factors should be considered when developing a maintenance schedule?

13.6K On what basis should maintenance schedules be adjusted?

13.614 Examples of Scheduling

Every agency needs to develop a scheduling procedure that will provide an effective preventive maintenance program for its wastewater collection system. Two examples of effective preventive maintenance scheduling are described in the following sections.

13.6140 SACRAMENTO COUNTY, CALIFORNIA

Sacramento County, California uses the following procedures for scheduling preventive maintenance of its 2,300 miles of lateral, main, trunk and interceptor sewers.

INDIVIDUAL LINES. Individual lines between manholes are maintained on the following schedules:

- An individual line is initially scheduled for cleaning the day after removal of a stoppage in the line,
- The individual line is then scheduled for a TV inspection four months after the initial cleaning,
- If roots were the cause of the stoppage, the individual line is scheduled for cleaning on a 12-month frequency,
- If a grease accumulation was the cause of the stoppage, the individual line is scheduled for cleaning on a 6-month frequency,
- The frequency of an individual line cleaning is either decreased or increased if the preceding cleaning indicates a need for a change, and
- When the frequency between cleanings of an individual line exceeds five years, it is removed from the preventive maintenance schedule.

SECTIONAL LINES. All individual lines within a designated section of the collection system are maintained on the following schedules:

- If the lines within a section have more than one stoppage per 7,000 feet of lines predominantly in easements or one stoppage per 20,000 feet of lines predominantly in streets, all lines in the section are scheduled for cleaning on a 12-month frequency; and
- If the number of stoppages within a section drops below the above stoppage rates, all lines in the section are removed from the preventive maintenance schedule.

COORDINATION. The scheduling of individual line cleaning is coordinated with the cleaning of all lines in a section so that individual lines will not be cleaned more often than they would be under an individual line cleaning schedule.

MICROCOMPUTER APPLICATION. Microcomputers and specialty software are used to facilitate the preventive maintenance scheduling in the following manner:

- Information from reports on line cleaning, inspection, repairs and replacement work are entered into a computerized database using specialty software to create a maintenance history for all individual lines in the collection system;
- Microcomputers, database, and specialty software are used by supervisors to produce planning and controlling reports;
- The reports are then used by the supervisors to schedule individual line and sectional line preventive maintenance based on the guidelines previously described; and
- Supervisors enter the preventive maintenance schedules into the computer's database and use specialty software for the weekly production of work orders from the database.

PERFORMANCE EVALUATION. Sacramento County, California, using the preventive maintenance scheduling procedures previously described, has reduced the number of stoppages in 2,300 miles of lateral, main, trunk, and interceptor sewer lines from 384 to approximately 100 per year.

13.6141 SANTA CLARA COUNTY, CALIFORNIA

Santa Clara County Sanitation District No. 4 uses a computerized system, given the acronym COSMO, for scheduling the preventive maintenance of the 290 miles of lateral, main, and trunk sewers in the district's wastewater collection system. Conceptually, COSMO performs as shown in Figure 13.3 with the four principal elements being the database, criteria, analysis program, and schedule program.

Note the similarity of the COSMO information flow schematic to the Management Information System schematic shown in Figure 12.8 in Chapter 13, page 493. The "External Data" and "Analysis Reports" in the COSMO schematic are the same as "Work Reports" and "Reports for Planning & Controlling," respectively, in the Management Information System schematic. A microcomputer and specialty software process data and prepare analysis reports as well as preliminary and final preventive maintenance schedules.

DATABASE. The database contains information on the collection system's physical characteristics, previous cleaning, repair, replacement and TV inspections. The data are derived from the maintenance crews' work reports and are filed by sewer line segments between manholes.

CRITERIA. The criteria are a set of goals and objectives of the preventive maintenance program of the district translated into a form used by a digital computer. The criteria also include a set of operational rules that govern the logic used by the schedule program.

ANALYSIS PROGRAM. The analysis program provides informational reports on performance of the preventive maintenance work activities of the district's crews. The performance reports can be produced for the collection system as a whole, sections or individual lines.

SCHEDULE PROGRAM. The schedule program prepares a cleaning schedule based on the criteria. With the aid of external factors and the computer schedule program, the superintendent's clerical assistant prepares a monthly schedule of preventive maintenance work for the district's crews.

DEBRIS SEVERITY. Debris severity is one of the data items reported by cleaning crews and is a measure of how much debris was in a sewer line segment at the time it was cleaned. Debris severity, S, may be reported as a number from 0 to 4. The numbers relate to the following debris conditions:

S = 0, line was clear of debris
S = 1, line contained a mild amount of debris
S = 2, line contained a moderate amount of debris
S = 3, line contained a severe amount of debris
S = 4, stoppage of flow in line.

CLEANING FREQUENCY. The objective of the district's preventive maintenance scheduling is to clean each line segment when its debris severity reaches the moderate level, S = 2. The criteria component of the COSMO software can be manipulated to establish a correlation between the latest reported debris severity for a line segment, the latest cleaning interval, and the projected time to the next scheduled cleaning. The cleaning frequency presently being used by the district is shown in Table 13.9.

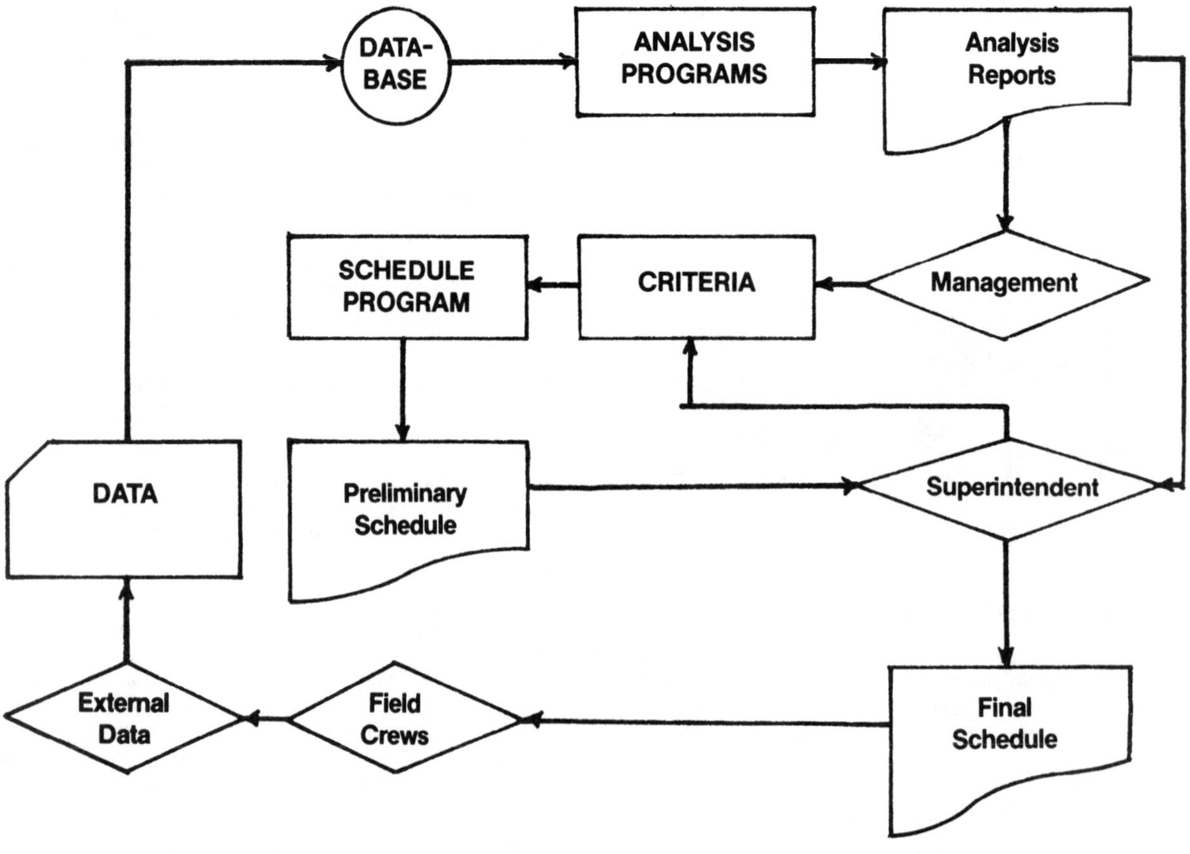

Fig. 13.3 COSMO information flow schematic

TABLE 13.9 CLEANING FREQUENCY USING CLEANING HISTORY

Latest Debris Severity	Scheduling Frequency
S = 0 (Clear)	$\Delta T = 2.0 \times T$
S = 1 (Mild)	$\Delta T = 1.75 \times T$
S = 2 (Moderate)	$\Delta T = 1.0 \times T$
S = 3 (Severe)	$\Delta T = 0.75 \times T$
S = 4 (Stoppage)	$\Delta T = 0.5 \times T$

For example, if a line segment was reported to have a mild debris severity, S = 1, when it was cleaned on April 1, 2002 and it had been previously cleaned on March 16, 2001, the line segment will be scheduled for cleaning during the month of December, 2003. From March 16, 2001 to April 1, 2002 is 12.5 months. 1.75 (from Table 13.9) times 12.5 months equals 21 months. April 1, 2002 plus 21 months gives December 2003.

SCHEDULING. The schedule program in the COSMO software projects monthly cleaning assignments based on a priority which is dependent on the debris severity reported at the latest cleaning, the elapsed time between the last two cleanings, and the amount of time that would elapse if the line segment was cleaned as scheduled. Postponing the scheduled cleaning increases the priority and cleaning earlier than scheduled decreases the priority. The schedule program uses the priority values to arrange line segments scheduled for cleaning in order of importance for any given month.

The schedule program makes monthly line segment cleaning assignments until the sum of the projected cleaning time for the lines (using the time required for the latest cleaning of the lines) is equal to the hours the district's superintendent has allocated for cleaning during a given month. When the program tries to make an assignment to a month that has reached its allocated time, the line segments with the lowest priorities are rescheduled in a subsequent month. This process is continued until all line segments have been scheduled and the staffing allocations are satisfied for each month.

The scheduling process takes place each month and the superintendent uses the printed schedules to make daily work assignments. Since the district's line segments are identified by the location of their upstream and downstream manholes on sewer maps and monthly cleaning schedules are printed in manhole number sequence, the superintendent can group line segments in localized areas for daily work assignments.

PERFORMANCE EVALUATION. One guideline of effectiveness of the preventive maintenance scheduling process used by Santa Clara Sanitation District No. 4 is a mathematical method shown in Figure 13.4. Your goal will be to schedule cleaning of lines when they reach a debris severity of about two. This will mean that they get cleaned before sewer debris accumulations cause a stoppage but not so often that you're working on relatively clean lines.

$$SD = \left[\frac{\sum_{i=1}^{M} (S_i - 2)^2}{M} \right]^{0.5}$$

Insert in line S_i = Debris severity of any line cleaned ($0 \le S_i \le 4$)

WHERE: M = Number of lines cleaned in a given period,

S_i = Debris severity of any line cleaned ($0 \le S_i \le 4$), and

SD = Measure of effectiveness.

NOTE: To calculate the "Measure of effectiveness, SD," for a preventive maintenance scheduling program, perform the following steps.

1. Collect the data for all lines cleaned during a given time period. For each line cleaned, assign a debris severity value from 0 (clear of debris) to 4 (cleaned after a stoppage occurred). The debris severity value is the value S_i in the formula. The subscript "i" identifies each line. "i" has a value of 1 for the first line cleaned, 2 for the second line cleaned and so on until the last line cleaned has a value or number represented by the letter "M" which could be 87, indicating that 87 lines were cleaned during the time interval.
2. Take each S_i value, subtract 2 from each value, square the difference and repeat this procedure for every line cleaned.
3. Σ means that we add up or sum up all of the squared differences.
4. Divide the total or sum of the squared differences by the number of lines cleaned, "M".
5. Take the square root (0.5 power) of the value obtained in step 4.
6. The result gives a value for SD, a measure of the effectiveness of the preventive maintenance scheduling program.

Fig. 13.4 Preventive maintenance scheduling effectiveness measure

To illustrate the use of the measure of effectiveness, if sewer lines were only cleaned after stoppages occurred, S_i would equal 4 and SD would equal 2. The same value of SD would result if every line was clear of debris when cleaned ($S_i = 0$). If all lines cleaned had a debris severity of 2 and no stoppages occurred during a given time period, the SD would equal 0. The optimum measure of effectiveness (SD = 0) of a preventive maintenance scheduling process is impractical to reach considering the unpredictable nature of wastewater collection systems. It is more important to follow the trend of the measure of effectiveness and make appropriate adjustment to the scheduling criteria when an adverse trend occurs.

The mathematical measure of effectiveness of the district's preventive maintenance scheduling since it started using a computer and the COSMO software is shown in Figure 13.5. Note the gradual downward trend in the measure of effectiveness toward the ideal value of 0.

Figure 13.5 also shows that the district was able, by the use of computer-aided scheduling, to reduce the annual amount of sewer line cleaning from 1,140,000 feet for the 12 months ending June, 1983, to 544,000 feet for the 12 months ending June, 1986, while only seeing an increase in line stoppages from approximately 20 to 40 during that same period in the 290 miles of sewer lines. Having cleaned the most severe lines, the district then began concentrating its line cleaning activities on approximately 37 percent of the lines in its collection system that were accumulating debris nearly at the optimum "moderate" rate.

QUESTIONS

Write your answers in a notebook and then compare your answers with those on page 612.

13.6L What is Sacramento County's cleaning schedule for individual sewer lines?

13.6M What types of reports are prepared by Santa Clara County's specialty software?

13.6N What information is in Santa Clara County's database?

13.6O How does Santa Clara County determine the cleaning frequency of sewer lines?

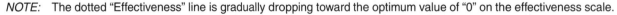

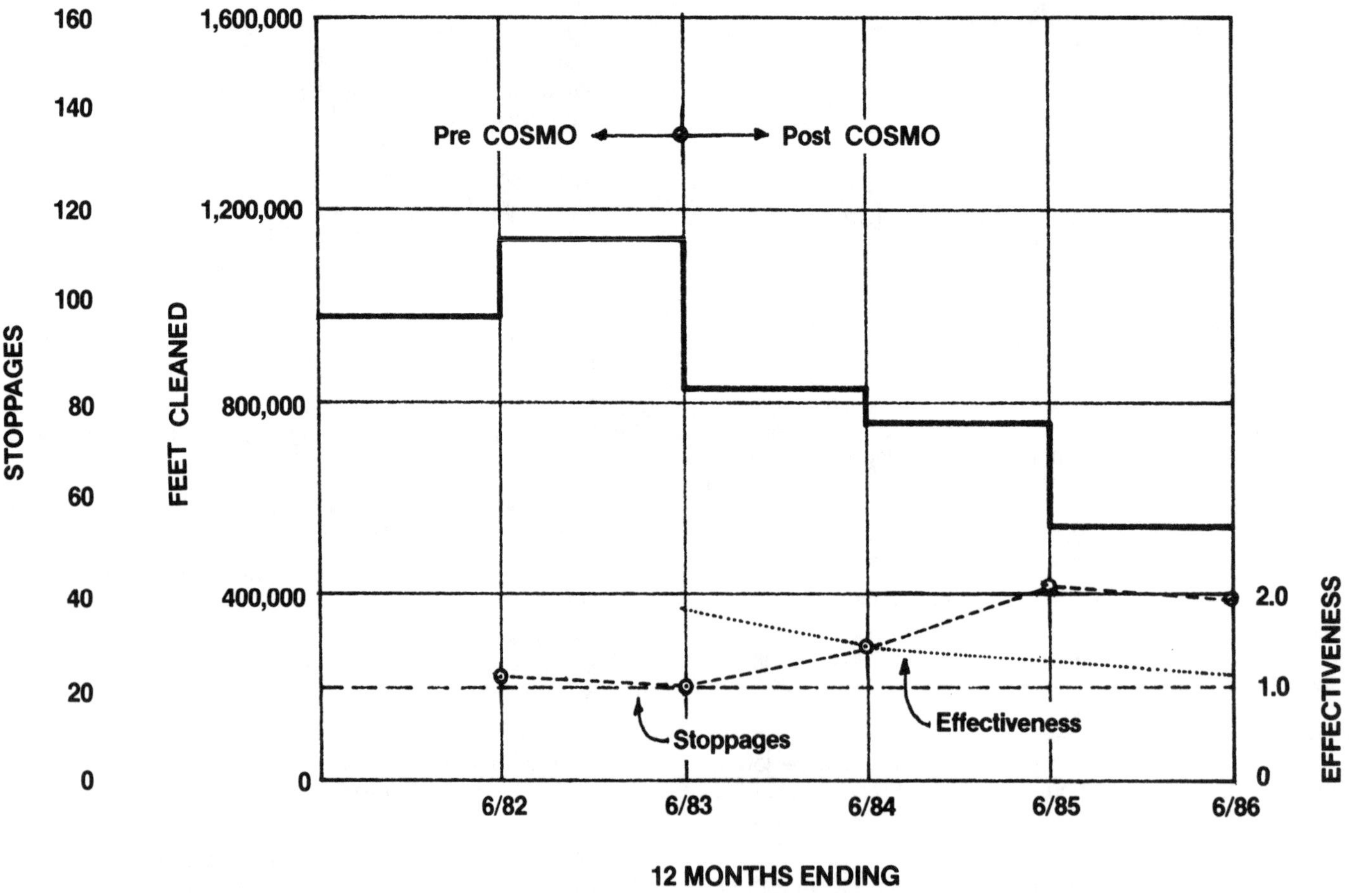

NOTE: The dotted "Effectiveness" line is gradually dropping toward the optimum value of "0" on the effectiveness scale.

Fig. 13.5 Five-year preventive maintenance performance by Santa Clara County Sanitation District No. 4

13.62 Risk Analysis

Regardless of how good a pump station maintenance program is, a major station failure can occur at any time, as a result of natural disasters if not from equipment failure. Toxic spills or discharges can also create station operating problems. For this reason, emergency equipment and emergency response systems should also be included in the pump station maintenance program. A risk analysis of each pump station should be conducted to define the likelihood and potential impact of a prolonged station failure. The methods to deal with the failure and the equipment required should be identified and a long-range plan should be developed to acquire the equipment and develop appropriate response procedures. At the minimum, the analysis should include compiling the following information for each station:

- Peak flow storage,
- Lowest upstream manhole and/or basement,
- Peak flow,
- Bypass arrangements,
- Equipment necessary for bypass,
- Emergency power requirements,
- Utility emergency phone numbers, and
- Utility account/circuit numbers.

Local contractors are a resource that may be available to assist in an emergency. Each year, electrical, mechanical and general contractors in the area can be formally or informally surveyed to collect the following information:

- Equipment available,
- Specialties,
- Labor rates,
- Equipment rates, and
- Capabilities.

In many cases, local or state law requires that formal procedures must be followed when expenditures exceed certain levels. The emergency response plan should authorize a senior manager to declare an emergency and allow the manager to mobilize the resources necessary to minimize the impact.

In cases where the equipment needed to respond to a major emergency is not available, the agency must purchase it. Examples of emergency equipment include:

1. Portable, engine-driven generator sets (and connection leads to carry required power load)
2. Pumps
   - Engine-driven
   - Hydraulically driven
   - Submersible
3. Hose, fittings and valves
   - Flexible hose
   - Aluminum irrigation pipe
   - Quick disconnect fittings
   - Check and isolation valves
4. Electrical equipment
   - Motor controls for electric pumps
   - Level control system

13.63 Equipment and Tool Requirements

Tool and equipment requirements for each section of a wastewater collection agency are described in greater detail in Section 13.3.

LEVEL 1 MAINTENANCE OPERATORS

Pickup trucks rated at one ton or higher, with utility boxes, are the most common vehicle for operators. The vehicles are frequently equipped with four-wheel drive to accommodate different weather and access conditions. Equipment that should be carried on the vehicle includes:

- Confined space entry equipment,
- Traffic control equipment,
- Other safety equipment such as first aid kits and gloves,
- Hoist,
- Hand tools, and
- Preventive maintenance materials and supplies.

These same vehicles can be outfitted with the necessary equipment for other utility functions including snow plowing and towing standby generators and pumps.

LEVEL 2 MAINTENANCE OPERATORS

Vehicles used for level 2 maintenance are much more specific and can range from a utility-type pickup to van-type vehicles. In some cases they may be set up as a shop on wheels using a 26,000 pound gross vehicle weight chasis, equipped for performing major equipment overhaul at the pump station. Equipment carried on these vehicles includes:

- Safety equipment,
- Hand and electrical tools,
- Mechanical and electrical material and supplies,
- Rigging equipment,
- Diagnostic instruments, and
- Welding and cutting equipment.

13.64 Warehouse and Parts Inventory

A well-organized and fully stocked warehouse is critical to any collection system maintenance program. When stocking parts and/or material, at least three factors should be considered, (1) how critical is a part, (2) what is the lead time from the manufacturer, and (3) what is the frequency of usage of a particular part. In many cases, mechanical parts can be fabricated in the shop or purchased locally in quantity at less cost.

Typical material stocked includes:

1. Electrical Equipment
   - Circuit breakers and fuses
   - Starters and overload relays
   - Instrumentation replacement parts
   - Control relays
   - Floats
   - Solid state control system parts
   - Consumable items
   - Fractional horsepower motors
   - Wire and cabling
2. Hydraulic/Mechanical Equipment
   - Mechanical seals and packing
   - Bearings
   - Power frames
   - Impellers
   - Grease shields
   - Shaft Sleeves
   - Wear rings
   - Valve shafts, disks and hardware
   - Vehicle parts
   - Bubbler compressors
   - Pipe and hose
   - Valves
   - HVAC (Heating, Ventilation, Air Conditioning) parts

13.65 Computer-Based Systems (Also see Chapter 12, Section 12.71, "Applications of Computers for System O & M.")

Computer-based maintenance management systems offer another alternative for managing collection system maintenance. One only has to analyze the steps and information required to order a replacement pump part, the amount of time required and how often this is done to realize that an equipment-oriented maintenance management system can dramatically reduce the time and cost involved. There are any number of approaches that can be taken with this type of system, ranging from simple databases designed in house to networked systems running on mainframe computers. Numerous "off the shelf" systems are also available that operate on desktop computers or work stations. Some of the types of functions that are now available include:

- Storing all relevant equipment nameplate data for pumps, motors, electrical system, valves and instrumentation and controls,
- Planning and/or scheduling,
- Generating work orders for maintenance programs,
- Tracking work backlog,
- Tracking inventory and warehousing operations,
- Tracking maintenance history,
- Generating purchase orders,
- Recording maintenance tasks completed, and
- Generating reports.

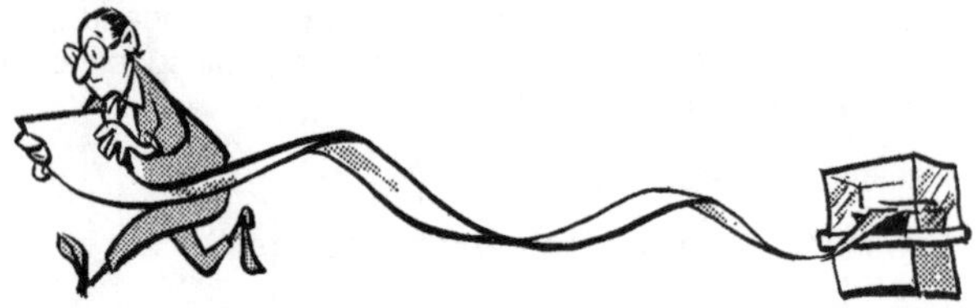

The advantages of computers are the ability to store and retrieve a lot of information on a timely basis and to selectively manipulate and analyze the data. Recordkeeping is a function that computers perform well. Documentation of training, for example, is extremely important today if a station or system failure results in a lawsuit. Computer-based budgeting and accounting functions are necessary to optimize budget control and analysis and to evaluate maintenance program performance. Computer systems are not a solution, however, if the manual recordkeeping systems are not available to serve as a basis for the computerized recordkeeping system.

QUESTIONS

Write your answers in a notebook and then compare your answers with those on page 612.

13.6P What is the reason for conducting a risk analysis of each pump station?

13.6Q Why would an operator consider using a computer-based maintenance management system in managing pump station maintenance?

13.7 GENERAL PERFORMANCE INDICATORS

This section outlines a procedure for calculating two indicators of performance for operation and maintenance of sewer lines and lift stations in a wastewater collection system. The indicators are based on the objective of minimizing the following:

- Number of stoppages per mile of sewer line,
- Number of failures per lift station, and
- Cost of maintenance of sewer lines and lift stations.

SEWER LINES. To calculate the Performance Indicator for sewer lines, the following information must be known:

- Length of sewers in miles,
- Annual cost of operating and maintaining sewers, and
- Annual number of stoppages that caused flooding.

Sewer line stoppages of concern should include only those stoppages that caused flooding. Flooding is defined in terms of wastewater overflowing manholes or backing up in service lines and flooding a home or business. If a sewer backs up to the extent that a plumbing fixture in a home will not drain, this event is considered a stoppage that causes flooding (of a plumbing fixture). Stoppages cleared before flooding occurs are not counted because they have not caused a threat to public health. Overflows resulting from insufficient sewer capacity should also be counted because they signal the need for immediate corrective action and are a potential threat to public health.

EXAMPLE 1

Calculate the Performance Indicator for the 1,500 miles of sewers in a collection system that spent $2,000,000 in one year and had 125 stoppages.

Known		Unknown
Length of Sewers, miles	= 1,500 miles	Performance Indicator
Annual Expenditures	= $2,000,000	
Annual Number of Stoppages	= 125 stoppages	

Calculate the Performance Indicator for the sewers.

$$\text{Performance Indicator} = (\text{Costs, \$/mile})(\text{Stoppages, number/mile})$$

$$= \left(\frac{\$2{,}000{,}000}{1{,}500 \text{ miles}}\right)\left(\frac{125 \text{ stoppages}}{1{,}500 \text{ miles}}\right)$$

$$= 111$$

This procedure can be repeated every year to evaluate the performance of the collection system and the effectiveness of the preventive maintenance program. By plotting the results on an annual basis, trends in performance can be revealed.

A trend over several years will be more meaningful if the cost of maintenance used each year is based on the value of a dollar at the beginning of the trend period. This can be done by applying the inflation rate from the beginning year to each given year.

Since the objectives include minimizing the number of stoppages and the costs or expenditure, the *SMALLER* the Performance Indicator, the better the job you are doing. In order to show improvement by an increasing trend, a Performance Rating can be used.

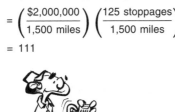

EXAMPLE 2

Calculate the Performance Rating for the sewers in the collection system in Example 1.

Known		Unknown
Performance Indicator	= 111	Performance Rating
Assumed Standard	= 90	

Calculate the Performance Rating for the sewers.

$$\text{Performance Rating} = \frac{(\text{Standard})(100\%)}{\text{Performance Indicator}}$$

$$= \frac{(90)(100\%)}{111}$$

$$= 81\%$$

Any standard may be assumed that best helps you illustrate the performance of the sewers in your collection system. The same standard should be used every year. This is a very useful technique for justifying budgets, increasing staff and purchasing more equipment.

LIFT STATIONS. To calculate a Performance Indicator for the lift stations in a collection system, the following information is needed:

- Number of lift stations,
- Annual cost of operating and maintaining lift stations (including energy costs),
- Annual number of lift station failures, and
- Total energy consumed by all lift stations in millions of kilowatt hours.

Lift station failures of concern should include only those failures that caused flooding. Do not consider an alarm reported by a telemetry system as a failure unless flooding actually occurs. If a lift station is properly monitored and the telemetry system is designed to prevent flooding, record only failures that result in actual flooding and a threat to public health. If a power outage occurs and standby power or emergency plans are not activated soon enough or are not effective enough to prevent flooding, then this is considered a lift station failure.

The formulas for calculating lift station Performance Indicators and Performance Ratings are as follows:

$$\text{Performance Indicator} = \frac{(\text{Costs, \$/station})(\text{Failures, number/station})}{\text{Energy Consumed, million kWh/station}}$$

$$\text{Performance Rating} = \frac{(\text{Standard})(100\%)}{\text{Performance Indicator}}$$

Any appropriate Standard may be assumed for your lift stations. If your lift stations experienced no failures during a particular year, one failure must be assumed in order to go through the calculations.

This section has illustrated how to calculate Performance Indicators and Performance Ratings for sewers and lift stations. These ideas may be applied to other objectives such as minimizing the number of service requests or complaints or response time to a service request. If other collection system agencies are similar to yours and collect data the same way, you could compare your performances. For more details on the development and use of Performance Indicators, see Section 13.8, "Additional Reading," reference 6, "Performance Indicators for Wastewater Collection Systems."

QUESTIONS

Write your answers in a notebook and then compare your answers with those on page 612.

13.7A Performance Indicators can be used to work toward what types of objectives?

13.7B What information must be known to calculate a Performance Indicator for sewer lines?

13.8 ADDITIONAL READING

1. *MANPOWER REQUIREMENTS FOR WASTEWATER COLLECTION SYSTEMS IN CITIES AND TOWNS OF UP TO 150,000 IN POPULATION*, Manpower Development Staff, Office of Water Program Operations, U.S. Environmental Protection Agency, Washington, DC 20460. Obtain from National Technical Information Service (NTIS), 5285 Port Royal Road, Springfield, VA 22161. Order No. PB-227039. Price, $36.50, plus $5.00 shipping and handling per order.
2. *MANPOWER REQUIREMENTS FOR WASTEWATER COLLECTION SYSTEMS IN CITIES OF 150,000 TO 500,000 IN POPULATION*, Manpower Development Staff, Office of Water Program Operations, U.S. Environmental Protection Agency, Washington, DC 20460. Obtain from National Technical Information Service (NTIS), 5285 Port Royal Road, Springfield, VA 22161. Order No. PB95-157442. Price $51.00, plus $5.00 shipping and handling per order.
3. *PLANT MANAGER'S HANDBOOK* (MOP SM-4), Water Environment Federation (WEF). No longer in print.
4. *WASTEWATER COLLECTION SYSTEMS MANAGEMENT* (MOP 7). Obtain from Water Environment Federation (WEF), Publications Order Department, 601 Wythe Street, Alexandria, VA 22314-1994. Order No. M05000. Price to members, $56.75; nonmembers, $76.75; price includes cost of shipping and handling.
5. "Management of an Effective Collection System Maintenance Program," by John Brady and Ken Kerri, *Deeds & Data*, Water Environment Federation, December, 1975.
6. "Performance Indicators for Wastewater Collection Systems," by John Brady, Steve Goodman, Ken Kerri and Robert Reed, *Journal WEF*, Vol. 51, No. 4, p. 695, April, 1979.
7. "Computerization of Sewer Maintenance Scheduling," by James R. Schaaf, *Public Works*, p. 128, September, 1985 and p. 64, October, 1985.

13.9 ACKNOWLEDGMENT

The authors wish to express their appreciation to Rick Arbour for his extensive contributions to the preparation of Section 13.6, "Establishing a Maintenance Program," and in particular Table 13.8, which lists typical maintenance frequencies for pump station equipment. His assistance is greatly appreciated.

END OF LESSON 4 OF 4 LESSONS
ORGANIZATION FOR
SYSTEM OPERATION AND MAINTENANCE

Please answer the discussion and review questions next.

DISCUSSION AND REVIEW QUESTIONS

Chapter 13. ORGANIZATION FOR SYSTEM OPERATION AND MAINTENANCE

(Lesson 4 of 4 Lessons)

Write the answers to these questions in your notebook. The question numbering continues from Lesson 3.

27. What circumstances could create a need for reorganization of equipment and personnel?
28. What is a preventive maintenance program?
29. How should extraordinary emergencies such as high-intensity rainstorms, hurricanes, floods and earthquakes be handled?
30. What kinds of work are done by an emergency maintenance crew?
31. The frequency of pump station visits depends on what factors?
32. What factors should be considered when stocking parts for a collection system maintenance program?
33. How would you schedule collection system maintenance crews?
34. How would you evaluate the effectiveness of your maintenance programs?
35. How would you develop an organization to operate and maintain your wastewater collection system?

SUGGESTED ANSWERS

Chapter 13. ORGANIZATION FOR SYSTEM OPERATION AND MAINTENANCE

ANSWERS TO QUESTIONS IN LESSON 1

Answers to questions on page 558.

13.1A Determining the types of personnel and equipment needed to perform the work of an organization begins with a clear statement of the organization's mission, goals and objectives. Then a detailed list of work to be performed is prepared and activities are grouped by similarities. Personnel and equipment needs can then be defined for each work activity.

13.1B Crew members should work together and perform the same type of work for an extended period of time to become familiar with each other and their work in order to become an efficient team.

13.1C The two main effects of delegating authority are: (1) delegating frees management to plan and direct, and (2) it allows decisions to be made by people closest to the situation and most directly affected by the decision.

13.1D To hold employees accountable for their actions, three factors must *ALL* be present:

1. The individuals must know what they are supposed to do,
2. They must know how to accomplish assigned tasks, and
3. They must have control of the factors that regulate what they are doing.

Answers to questions on page 568.

13.2A The major work tasks performed by a collection system agency include:

1. Service requests,
2. Preventive maintenance,
3. Repair and replacement,
4. Lift station maintenance,
5. Inspections and investigations, and
6. Indirect activities.

13.2B Crews tend to become more specialized in larger systems because there are more crews and each crew is more apt to do a similar job every day, thus becoming specialists in the jobs they do most frequently.

13.2C The basic crew in most maintenance organizations is the emergency crew.

13.2D Emergency crews usually consist of two operators. Under these conditions, neither operator should ever enter a manhole unless a third person is available.

13.2E Other activities conducted by an emergency crew when there are no emergencies include inspection, repair, map checking, location of inflow/infiltration, and scheduled maintenance jobs.

13.2F The number of operators on a balling crew and the extent to which they are used is determined by the conditions under which the crew must work. Important conditions include traffic conditions and the quantity of material to be removed from the sewer.

13.2G Balling crews also can function as a repair crew, chemical crew, bucket machine crew, or smoke testing crew.

13.2H High-velocity cleaners are commonly used all of the time.

13.2I No, CCTV equipment available today can be operated effectively by existing collection system operators with adequate training.

13.2J Crews should be rotated from job to job so trained crews are available at all times for any job and so crew members can be transferred from one crew to another if a key operator in one crew becomes ill or the opportunity for promotion develops.

13.2K Smoke testing crews try to locate and identify points of inflow and illegal connections.

13.2L Repair crews usually deal primarily with replacement of defective portions of the collection system, while construction crews usually are engaged in new work.

ANSWERS TO QUESTIONS IN LESSON 2

Answers to questions on page 571.

13.30A Priority responses to requests for emergency services are as follows:

1. Flooded home or business,
2. Manhole overflowing,
3. Manhole cover off,
4. Cleanout overflowing, and
5. Home service backed up.

13.30B When multiple emergency calls are received, the field supervisor evaluates the problems, assigns priorities, redirects service crews and requests assistance from field preventive maintenance crews for specific equipment or assistance in handling calls as necessary.

Answers to questions on page 572.

13.30C When a private home becomes flooded with wastewater, try to remove the cause of flooding and assist the people living in the home to clean and remove articles which may have been damaged or will be damaged if left in place.

13.30D To minimize legal problems resulting from flooding:

1. Find the cause and location of the backup to identify who is responsible,
2. Follow procedures approved by the legal department, and
3. Act quickly to reduce damages.

13.30E Flooding problems resulting from a low service line connection can be corrected by installing and maintaining a backflow preventer and an overflow valve.

13.30F Service requests to recover valuable property from a sewer may be handled by a balling crew; however, responses to such requests should be handled in accordance with agency policy.

Answers to questions on page 573.

13.30G En route to a problem area, the crew begins to fill out a service request form and locates the appropriate area map. Upon arrival, the dispatcher is informed of the crew's time of arrival and the location of the unit.

13.30H The location of a main line stoppage is determined by inspecting manholes in a downstream flow direction until a manhole is located with reduced flow or no flow. The stoppage is located between this manhole and the next upstream manhole.

13.30I Stoppages are usually removed by using hand tools. If this procedure fails, power equipment such as a high-velocity cleaner or a power rodder is used.

13.30J The first job a crew does upon arrival to clear a service line stoppage is to inspect the main line to be sure it is open and flowing.

13.30K A service line stoppage is cleared by removing the cleanout cover and running a hand rod with a 2- or $3^1/_2$-inch auger down the cleanout to the main line. If this doesn't work, the problem is the resident's. If there is no cleanout, the resident should have a plumber inspect the line. If the plumber finds the line clear from the home to the property line, it is the agency's problem. The line may have to be dug up to be cleared. A cleanout should be installed to facilitate clearing any future stoppages.

13.30L After clearing a service line that had to be dug up, the service report form and repair request form must be completed.

13.30M If a new cleanout is required, it should be installed by the repair and replacement crew.

Answers to questions on page 576.

13.30N A priority list should be developed for scheduling work by preventive maintenance crews to provide an organized approach for solving problems and ensuring the completion of necessary jobs.

13.30O Clearwater developed the following priority list:

1. Assist emergency service crews,
2. Clean lines where stoppages occur within three days after the stoppage has been cleared,
3. Schedule power rodder in root problem areas to minimize root stoppages,
4. Schedule high-velocity cleaner in grease problem areas to minimize grease stoppages,
5. Assign cleaning crews to areas of collection system to be cleaned this year,
6. Construct training facility and provide training sessions,
7. Develop priority list of collection system lines to be inspected by closed-circuit television, and
8. Prepare and schedule program to chemically treat sewers with serious root problems.

13.30P During the preparation and scheduling of a chemical treatment program for root control, you should:

1. Review research literature,
2. Visit communities with similar problems and evaluate their root control programs,
3. Select two methods and field test in advance of starting a full-scale root control program, and
4. Keep records of all aspects of the program.

Answers to questions on page 578.

13.30Q The major disadvantages of using a bucket machine include potential damage to the lines and cleaning with a bucket machine is a time-consuming operation.

13.30R High-velocity cleaners are probably the most effective cleaning tools for removing grease, grit, sand, gravel, and debris.

Answers to questions on page 581.

13.31A A section of clear pipe is a helpful training device because crews can observe the effectiveness of the cleaning equipment, its limitations, and how to use it as safely and efficiently as possible.

13.31B Items to be considered when trying to decide whether an agency should do a repair or new construction job or a private contractor should do the work include:

1. Cost of materials,
2. Time to complete the job, and
3. Special safety or construction equipment required.

13.31C In the Repair and Replacement Section, Crew 1 makes sewer repairs and installs new sewers; Crew 2 makes manhole repairs and installs manholes, and Crew 3 makes building sewer repairs and replacements and installs cleanouts at property lines.

13.31D Priority designations given repair work are: (1) immediate repair, and (2) repair when time available.

13.31E Requests for new construction come from the Service Request and Preventive Maintenance and Inspection Sections. Requests are processed through appropriate channels and work orders are issued. Building permits are sometimes required.

13.31F Closed-circuit television inspection will reveal repair work consisting of the following:

1. Bad taps,
2. Bad joints,
3. Offset joints,
4. Broken pipes, and
5. Deteriorated pipe.

13.31G Small, expensive hand tools should be kept in locked compartments in trucks.

ANSWERS TO QUESTIONS IN LESSON 3

Answers to questions on page 585.

13.32A The ultimate goals of equipment maintenance are: (1) greater lift station reliability, and (2) fewer equipment or system failures that result in damage to equipment or the environment or endanger public health and safety.

13.32B Work priorities for Clearwater's Lift Station Operation and Maintenance Section are as follows:

1. Respond to lift station failures,
2. Assist other sections in setting up emergency pumps to prevent flooding and bypasses,
3. Respond to alarms from telemetered system,
4. Perform regular lift station maintenance,
5. Perform work on preventive maintenance schedule and work requests from other sections, and
6. Supervisor to evaluate and establish a program that will change lift station inspections from daily to weekly.

13.32C Clearwater can convert lift station visits from daily to weekly by establishing a preventive maintenance program, installing a telemetering system, and increasing the capacity of lift station auxiliary units such as oilers and reservoirs to operate for one month without service.

13.32D The major causes of lift station pump failures include malfunctions in the electrical or mechanical systems, or debris plugging an impeller or jamming a check valve.

13.32E An electrical failure may be the result of a power failure or an outside power source being switched off due to circuit overload on the main power source.

13.32F Repair work on an electrical system may be attempted only if you are qualified and authorized to do the work.

Answers to questions on page 588.

13.33A The function or responsibility of the Inspection and Investigation Section is to inspect the collection system (except lift stations) to determine the need for preventive maintenance, repairs and replacements. The section also inspects users of the facilities and manholes for indications of violations of the city's industrial waste and sewer-use ordinances, and investigates sources of excessive infiltration of groundwater and the inflow of storm water into the collection system.

13.33B Conditions that may adversely affect collection system O & M that must be recognized by Instrumentation Technicians include crushed pipe, joint separation or offsets, sags, root intrusion, damage caused by excavation and construction, unrecorded service connections, infiltration and inflow, and defects in repairs and new construction. Instrumentation Technicians may also be trained to evaluate the effectiveness of methods and equipment used for maintenance of problem service lines.

13.33C Residents and the fire department must be notified before an area of sewer lines is smoke tested to avoid someone mistaking smoke from a sewer as a building fire and reporting a false alarm.

Answers to questions on page 589.

13.33D Samples are collected of industrial waste discharges to the collection system for analysis to determine compliance with discharge requirements and to calculate sewer-use charges.

13.33E Portable equipment must be readily available at all times because collection system operation and maintenance is a 24-hour-a-day job.

13.33F Useful pieces of portable equipment include air compressors, gasoline-powered wastewater pumps, small portable generators, blowers, two-way portable telephone or radio units, a chemical spray unit, and a concrete mixer.

Answers to questions on page 592.

13.33G The major duties of a collection system superintendent include supervising and coordinating activities, compiling and analyzing planning and controlling reports, recommending policies, submitting recommendations for personnel and equipment, visiting work sites, providing technical assistance, preparing and submitting budget recommendations, approving acquisition of materials and equipment, cooperating with personnel department, conferring with officials regarding system O & M, cooperating with other departments, and conferring with contractors to discuss construction and expansion.

13.33H Support services needed by a collection system division include engineering, financial, legal, personnel, public information, risk management and safety.

13.33I A computer operator performs basic computer programming, loads acquired specialty software into computers, enters information from work reports into a computer's database, uses a computer to process data for preparation of planning and controlling reports, assists other staff personnel in use of computers, and assists other support personnel in their activities.

ANSWERS TO QUESTIONS IN LESSON 4

Answers to questions on page 594.

13.4A Safety requirements have required increases in the amount of staffing.

13.4B Safety requirements prevent either person on a two-operator crew from entering a manhole. On those occasions when it becomes necessary for one of two operators to enter a manhole, a supervisor can be called to act as a third operator and all confined space safety procedures are followed.

13.5A The constraints to reorganization include personnel resistance to change, physical limitations, and personnel classifications and compensation.

Answers to questions on page 595.

13.6A Undesirable conditions caused by overflows and discharges of wastewater can be minimized or eliminated by the use of modern, specialized maintenance equipment and well-planned maintenance scheduling.

13.6B A preventive maintenance program is a total concept that includes functional maps, good records, intelligent assignment of available staff and effective use of equipment. Scheduling of maintenance activities is well planned and meaningful. The objective of preventive maintenance is to improve existing conditions in problem areas to a point where scheduled maintenance can be significantly reduced or ultimately eliminated.

Answers to questions on page 598.

13.6C A job classification is defined by the tasks to be performed and the skill required to perform the tasks.

13.6D Level 1 maintenance duties are completed by operators who visit lift stations on a periodic basis, daily to weekly, to perform a variety of preventive maintenance tasks.

13.6E Level 2 maintenance is performed by journey level electricians, pipefitters, machinists and mechanics.

13.6F For safety reasons, an operator may be required to have a state or local license to perform maintenance on electrical systems, low- and high-pressure hot water systems or steam boiler systems, primarily because of the safety issues involved.

13.6G An electrician and a mechanic could be assigned to the same crew when the work assigned to the crew requires both sets of skills.

Answers to questions on page 601.

13.6H To prepare an effective schedule for maintaining a wastewater collection system, a knowledge of the collection system, maintenance procedures, and available equipment and personnel are very important. Actual maintenance schedules must attempt to schedule jobs only when needed. Effectiveness of the maintenance program must be continuously evaluated and schedules should be adjusted as necessary.

13.6I A person can become familiar with a collection system by studying maps and records; actual field experience is very important too.

13.6J Important time factors that must be considered when developing a maintenance program are: (1) when the work will be needed, and (2) how frequently maintenance needs to be performed.

13.6K Maintenance schedules should be adjusted in accordance with the results the schedule has or has not produced.

Answers to questions on page 604.

13.6L Sacramento County schedules individual sewer lines for cleaning according to the following guidelines:

1. An individual line is initially scheduled for cleaning the day after removal of a stoppage in the line,
2. If roots were the cause of the stoppage, the individual line is scheduled for cleaning on a 12-month frequency,
3. If a grease accumulation was the cause of the stoppage, the individual line is scheduled for cleaning on a 6-month frequency,
4. The frequency of an individual line cleaning is either decreased or increased if the preceding cleaning indicates a need for a change, and
5. When the frequency between cleanings of an individual line exceeds five years, it is removed from the preventive maintenance schedule.

13.6M Reports prepared by Santa Clara County's specialty software include analysis reports, preliminary and final preventive maintenance schedules.

13.6N Information in Santa Clara County's database includes the collection system's physical characteristics as well as data about any previous cleaning, repairs, replacements and TV inspections.

13.6O The cleaning frequency of sewer lines is scheduled when the line's debris severity value reaches the moderate level, S = 2.

Answers to questions on page 606.

13.6P Pump station failure is always a possibility, no matter how well the station is maintained. A risk analysis should be conducted to define the likelihood and potential impact of a prolonged station failure. The methods to deal with the failure and the equipment required can then be identified.

13.6Q Computer-based maintenance management systems should be considered to reduce the time and cost involved in maintenance management.

Answers to questions on page 608.

13.7A Performance Indicators can be used to work toward objectives of minimizing:

1. Number of stoppages per mile of sewer line,
2. Number of failures per lift station, and
3. Cost of maintenance of sewer lines and lift stations.

13.7B To calculate the Performance Indicator for sewer lines, the following information must be known:

1. Length of sewers in miles,
2. Annual cost of operating and maintaining sewers, and
3. Annual number of stoppages that caused flooding.

CHAPTER 14

CAPACITY ASSURANCE, MANAGEMENT, OPERATION, AND MAINTENANCE (CMOM)

by

Ken Kerri

Peg Hannah

TABLE OF CONTENTS

Chapter 14. CAPACITY ASSURANCE, MANAGEMENT, OPERATION, AND MAINTENANCE (CMOM)

OBJECTIVES

Chapter 14. CAPACITY ASSURANCE, MANAGEMENT, OPERATION, AND MAINTENANCE (CMOM)

Following completion of Chapter 14, you should be able to prepare and implement a CMOM program for your collection system utility. The CMOM program should contain elements on how to:

1. Manage a collection system,
2. Enforce your legal authority,
3. Administer your utility's finances,
4. Recruit and motivate personnel,
5. Promote training and certification,
6. Support the safety program,
7. Maintain warehouse and inventory,
8. Design collection system facilities,
9. Construct and inspect facilities,
10. Perform a Sewer System Evaluation Survey (SSES),
11. Conduct a system capacity assurance program,
12. Supervise and evaluate a water quality monitoring program,
13. Schedule maintenance,
14. Manage a collection system O & M program,
15. Supervise a sewer cleaning program,
16. Control and monitor hydrogen sulfide,
17. Operate and maintain lift stations,
18. Inspect and rehabilitate manholes,
19. Televise collection system sewers,
20. Reduce infiltration/inflow (I/I),
21. Conduct smoke testing and dyed water flooding,
22. Repair and rehabilitate sewers,
23. Maintain right-of-ways,
24. Minimize SSOs and CSOs, and
25. Comply with your NPDES permit and applicable rules and regulations.

WORDS

Chapter 14. CAPACITY ASSURANCE, MANAGEMENT, OPERATION, AND MAINTENANCE (CMOM)

CSO CSO

Combined **S**ewer **O**verflow. Wastewater that flows out of a combined sewer (or lift station) as a result of flows exceeding the hydraulic capacity of the sewer or stoppages in the sewer. CSOs exceeding the hydraulic capacity usually occur during periods of heavy precipitation or high levels of runoff from snow melt or other runoff sources.

COMBINED SEWER COMBINED SEWER

A sewer designed to carry both sanitary wastewaters and storm or surface water runoff.

COVERAGE RATIO COVERAGE RATIO

The coverage ratio is a measure of the ability of the utility to pay the principal and interest on loans and bonds (this is known as "debt service") in addition to any unexpected expenses.

DEBT SERVICE DEBT SERVICE

The amount of money required annually to pay the (1) interest on outstanding debts; or (2) funds due on a maturing bonded debt or the redemption of bonds.

EXFILTRATION (EX-fill-TRAY-shun) EXFILTRATION

Liquid wastes and liquid-carried wastes which unintentionally leak out of a sewer pipe system and into the environment.

GIS GIS

Geographic **I**nformation **S**ystem. A computer program that combines mapping with detailed information about the physical locations of structures such as pipes, valves, and manholes within geographic areas. The system is used to help operators and maintenance personnel locate utility system features or structures and to assist with the scheduling and performance of maintenance activities.

I/I I/I

See INFILTRATION/INFLOW.

INFILTRATION (IN-fill-TRAY-shun) INFILTRATION

The seepage of groundwater into a sewer system, including service connections. Seepage frequently occurs through defective or cracked pipes, pipe joints, connections or manhole walls.

INFILTRATION/INFLOW INFILTRATION/INFLOW

The total quantity of water from both infiltration and inflow without distinguishing the source. Abbreviated I & I or I/I.

INFLOW INFLOW

Water discharged into a sewer system and service connections from such sources as, but not limited to, roof leaders, cellars, yard and area drains, foundation drains, cooling water discharges, drains from springs and swampy areas, around manhole covers or through holes in the covers, cross connections from storm and combined sewer systems, catch basins, storm waters, surface runoff, street wash waters or drainage. Inflow differs from infiltration in that it is a direct discharge into the sewer rather than a leak in the sewer itself. See INTERNAL INFLOW.

INTERNAL INFLOW INTERNAL INFLOW

Nonsanitary or industrial wastewaters generated inside of a domestic, commercial or industrial facility and being discharged into the sewer system. Examples are cooling tower waters, basement sump pump discharge waters, continuous-flow drinking fountains, and defective or leaking plumbing fixtures.

LIFE-CYCLE COSTING LIFE-CYCLE COSTING

An economic analysis procedure that considers the total costs associated with a sewer during its economic life, including development, construction, and operation and maintenance (includes chemical and energy costs). All costs are converted to a present worth or present cost in dollars.

MANDREL (MAN-drill) MANDREL

(1) A special tool used to push bearings in or to pull sleeves out.

(2) A testing device used to measure for excessive deflection in a flexible conduit.

NPDES PERMIT NPDES PERMIT

National **P**ollutant **D**ischarge **E**limination **S**ystem permit is the regulatory agency document issued by either a federal or state agency which is designed to control all discharges of potential pollutants from point sources and storm water runoff into U.S. waterways. NPDES permits regulate discharges into navigable waters from all point sources of pollution, including industries, municipal wastewater treatment plants, sanitary landfills, large agricultural feedlots and return irrigation flows.

OPERATING RATIO OPERATING RATIO

The operating ratio is a measure of the total revenues divided by the total operating expenses.

SCADA (ss-KAY-dah) SYSTEM SCADA SYSTEM

Supervisory **C**ontrol **A**nd **D**ata **A**cquisition system. A computer-monitored alarm, response, control and data acquisition system used by operators to monitor and adjust their wastewater treatment processes and facilities.

SSES SSES

Sewer **S**ystem **E**valuation **S**urvey.

SSO SSO

Sanitary **S**ewer **O**verflow. Wastewater that flows out of a sanitary sewer (or lift station) as a result of flows exceeding the hydraulic capacity of the sewer or stoppages in the sewer. SSOs exceeding hydraulic capacity usually occur during periods of heavy precipitation or high levels of runoff from snow melt or other runoff sources.

SANITARY SEWER SANITARY SEWER

A pipe or conduit (sewer) intended to carry wastewater or waterborne wastes from homes, businesses, and industries to the POTW (**P**ublicly **O**wned **T**reatment **W**orks). Storm water runoff or unpolluted water should be collected and transported in a separate system of pipes or conduits (storm sewers) to natural watercourses.

SURCHARGE SURCHARGE

Sewers are surcharged when the supply of water to be carried is greater than the capacity of the pipes to carry the flow. The surface of the wastewater in manholes rises above the top of the sewer pipe, and the sewer is under pressure or a head, rather than at atmospheric pressure.

CHAPTER 14. CAPACITY ASSURANCE, MANAGEMENT, OPERATION, AND MAINTENANCE (CMOM)

(Lesson 1 of 2 Lessons)

14.0 CMOM CONCEPTS

14.00 Birth of CMOM

A draft notice of proposed rule making was signed by EPA Administrator Browner on January 4, 2001. The proposed rule was intended to clarify and expand the National Pollutant Discharge Elimination System (NPDES) permit requirements for municipal sanitary sewer systems and *SANITARY SEWER OVERFLOWS (SSOs)*.[1] The proposal included standard permit conditions addressing capacity, management, operation, and maintenance (CMOM) requirements.

A short time after release of the draft proposal, EPA withdrew[2] the proposed rule to give the incoming Administration the opportunity to review it. The proposed schedule for preparation and implementation was uncertain at the time this chapter was prepared in 2003. However, following the release of the draft notice, some state water pollution control regulatory agencies recognized that many of the elements in the proposed rule were valid and appropriate for the cost-effective management, operation, and maintenance of wastewater collection system agencies attempting to minimize sanitary sewer overflows (SSOs) and also *COMBINED SEWER OVERFLOWS (CSOs)*.[3] The managers and operators of wastewater collection systems also recognized the value of many of the concepts outlined in the proposed rule.

14.01 Why CMOM?

EPA proposed the SSO Rule and CMOM Program specifically to control sanitary sewer overflows (SSOs), but the proposed procedures will also help to control combined sewer overflows (CSOs). SSOs and CSOs allow the discharge of untreated wastewater from wastewater collection systems. This untreated wastewater contains pathogenic organisms harmful to the public and wastewater pollutants that could damage aquatic life and the environment. The public, especially children, need protection from disease-causing bacteria. SSOs and CSOs can result in wastewater being released to areas where there is a high risk of public exposure, such as in streets, on school grounds, on private property, in basements of homes, and into receiving waters used for drinking water supply, fishing and shellfishing, and body-contact water recreation.

Environmental effects of SSOs and CSOs can be significant. Adverse water quality impacts may include fish kills due to changes in the physical characteristics and viability of aquatic habitats. Communities may experience adverse economic effects of SSOs and CSOs, such as beach closures, shellfish harvesting quarantines, increased risks and demands on drinking water supply sources, and impairment of the public's ability to use waters for recreational purposes.

14.02 What Is the CMOM Program?

CMOM stands for **C**apacity Assurance, **M**anagement, **O**peration, and **M**aintenance. All of these elements are critical for a successful wastewater collection system program that minimizes SSOs and CSOs.

The purpose of sanitary and combined wastewater collection systems is to collect wastewater from homes, businesses, schools, public facilities, and industries and transport this wastewater to a treatment plant without exposing the public or the environment to any harmful constituents being transported in the wastewater. Sanitary and combined wastewater collection systems must be properly managed, operated, and maintained for the following reasons:

- To minimize the number of SSOs and CSOs, such as those caused by blockages and/or component failures;
- To reduce the rate of system degradation that can lead to structural failure, failure of pumps and other equipment, loss of system capacity, increase in flows due to *INFILTRATION AND INFLOW (I/I)*,[4] and street subsidence and collapse;
- To avoid hydraulic overloading due to improper design, an increase in the service population that exceeds design flows, or excessive rates of I/I;

---

[1] *Sanitary Sewer Overflow (SSO). Wastewater that flows out of a sanitary sewer (or lift station) as a result of flows exceeding the hydraulic capacity of the sewer or stoppages in the sewer. SSOs exceeding hydraulic capacity usually occur during periods of heavy precipitation or high levels of runoff from snow melt or other runoff sources.*

[2] *EPA withdrew the proposed rule in accordance with a January 20, 2001 memorandum from the Assistant to the President and Chief of Staff entitled "Regulatory Review Plan," published in the FEDERAL REGISTER (66FR7701) on January 24, 2001. Periodically since January 24, 2001, EPA has announced its intent to deliver the SSO Rule and CMOM Program to the Office of Management and Budget (OMB) for financial analysis. Following the review by OMB, the SSO Rule and CMOM Program will be published in the FEDERAL REGISTER and will be made available for public comment. After review and analysis of public comments, EPA will prepare and implement the final version of the SSO Rule and CMOM Program.*

[3] *Combined Sewer Overflow (CSO). Wastewater that flows out of a combined sewer (or lift station) as a result of flows exceeding the hydraulic capacity of the sewer or stoppages in the sewer. CSOs exceeding the hydraulic capacity usually occur during periods of heavy precipitation or high levels of runoff from snow melt or other runoff sources.*

[4] *Infiltration/Inflow. The total quantity of water from both infiltration and inflow without distinguishing the source. Abbreviated I & I or I/I.*

- To immediately respond to unavoidable SSOs and CSOs to minimize the risks of adverse impacts on public health and the environment; and
- To protect the public's infrastructure investment.

A wastewater utility agency's CMOM program will contain management, operation, and maintenance procedures developed by the agency to minimize SSOs and CSOs, reduce system degradation, avoid hydraulic overloading, respond to SSOs and CSOs, and protect the public's investment in the system. Additional elements of an agency's CMOM program could include a system evaluation and capacity assurance plan, program audits, and communication with interested parties or stakeholders.

14.03 Critical Information and References

The following wastewater collection system operator training manuals, which are available from the Office of Water Programs at California State University, Sacramento, provide managers, operators, and maintenance personnel with the basic information needed to develop procedures for a CMOM program:

- *OPERATION AND MAINTENANCE OF WASTEWATER COLLECTION SYSTEMS*, Volume I, emphasizes safe procedures for inspecting and testing collection systems, pipeline cleaning and maintenance methods, and underground repair. Volume II emphasizes the operation and maintenance of lift stations, sewer renewal (rehabilitation), and administration/management;
- *COLLECTION SYSTEMS: METHODS FOR EVALUATING AND IMPROVING PERFORMANCE* emphasizes methods for establishing an effective O & M program and procedures to identify problems and select methods for improving the performance of a collection system; and
- *UTILITY MANAGEMENT* stresses planning, organization, staffing, communicating, and financial management of a utility.

The remainder of this chapter will use the information from these references to provide managers, operators, and maintenance personnel with the essential knowledge they need to develop and implement their own CMOM program. This chapter is *not* intended to ensure that agencies will be in complete compliance with any regulatory CMOM program implemented in the future by a state or the EPA; rather, it will provide collection system utility agencies with the information they need to develop and implement their own cost-effective CMOM program.

QUESTIONS

Write your answers in a notebook and then compare your answers with those on page 652.

14.0A What are SSOs and CSOs?

14.0B What is the purpose of sanitary and combined wastewater collection systems?

14.1 COLLECTION SYSTEM NEEDS

14.10 What Happens When Collection Systems Are Neglected?

Neglected collection systems degrade, deteriorate, and age at a faster rate than expected or intended. Neglected collection systems may allow inferior or inadequate designs to be accepted and constructed. These designs may include inadequate capacity and inferior materials, products, and equipment. Improperly inspected construction projects allow poor construction practices that can produce inadequate capacity and improperly installed facilities that are unable to perform or last as expected. Neglected systems may be poorly or improperly managed, operated, maintained, and/or repaired, which results in inadequate and poor performance. Neglected systems may not provide proper attention to privately owned service laterals from buildings and homes. If many or all of these elements of collection system management, operation, and maintenance are neglected, the collection system will perform poorly and create potential health and environmental risks that must be corrected and prevented from occurring again in the future.

14.11 What Causes Poor Performance and Deterioration?

What factors and practices can contribute to poor collection system performance and deterioration? By identifying these factors in their own wastewater collection systems, managers, operators, and maintenance personnel can use this information to develop and implement a CMOM program that will improve and protect the performance of their system. *OPERATION AND MAINTENANCE OF COLLECTION SYSTEMS*, Volume I, Chapter 2, "Why Collection System Operation and Maintenance?" lists and describes many factors that cause poor system performance and deterioration, including the following factors:

1. Wastewater collection systems were designed as gravity flow sewers with inadequate flow capacities for the area served or for unexpected population growth.
2. Collection systems were not installed as designed. Problems are caused by faulty construction, poor inspection, and low-bid shortcuts.
3. Little thought was given to the fact that sewers, although made of "permanent" material, could be considered only as permanent as the weakest joints. Earth movement, vibration from traffic, settling of structures, and construction disturbance (all occur from time to time) require a flexible pipe material or joint that can maintain tightness, yet joints were made rigid.
4. Corrosion of sewer pipes from either the trench bedding and backfill or the wastewater being transported by the collection system was a factor neglected during design. A major cause of corrosion in wastewater collection systems and treatment plants is hydrogen sulfide gas.

5. Not enough scientific knowledge existed or was available to designers about potential damage to pipe joints by plant roots. Although root intrusion into sewers was an age-old problem, it was assumed that if the joint was watertight, it would be roottight. People did not realize that roots would be attracted by moisture and nutrient vapor unless the joints were vaportight (which means airtight). Roots can enter a pipe joint or walls microscopically (through extremely small holes or cracks); thus, open or leaking joints are not necessary for root intrusion in collection systems.

   Collection system environments are ideal for root growth. In this environment, roots enter, expand, and open joints and cracks. Root growth is a principal cause of pipe damage that allows infiltration and *EXFILTRATION*.[5] This creates a major concern for health and pollution control authorities because of wastewater treatment plant hydraulic overload and groundwater pollution. Plant hydraulic overload is also caused by inflow.

6. The "out of sight, out of mind" nature of the wastewater collection system. Local taxpayers have invested more money in underground sewers than in all the aboveground structures owned by their local government. Why has this great taxpayer investment been so grossly neglected? Because it is out of sight, and so, out of mind. Some cities spend more money repairing and replacing sidewalk sections because of earth movement and tree root damage than they spend on preventive maintenance of wastewater collection systems.

7. Many collection systems are maintained by a department charged with street, sidewalk, storm drain, and sometimes water utility maintenance. You can easily guess where the money is spent. Money is usually spent where the taxpayer can see it, especially when the budget is inadequate for the total need. How often do you see a sign, "Your tax dollar is at work cleaning and repairing your sewer"?

8. The bottoms of ditches in which sewers are installed slope downhill to produce gravity flow. Gravel and sand placed in the ditch for pipe bedding and cover create ideal conditions to move water for great distances down the trench. Water that enters this backfilled ditch does not easily percolate (seep) out of the ditch because the usual silt and clay soils have been compacted both on the ditch bottom and sidewalls by the heavy excavation equipment.

   Many of these ditches start at the toe (bottom) of foothill slopes where the natural soil conditions produce springs or lateral water flow into the ditch. Leaking pipe joints and exfiltration can further increase water flow. This water can cause flow down the ditch and create serious problems along a downstream sewer where no one would expect water in the ditch. Problems that can develop include groundwater infiltration into the sewer, possible flotation of the sewer, or structural failure of the sewer or joint.

9. The Administrator of the Environmental Protection Agency was not able to authorize grants for the construction of wastewater treatment plants if the wastewater collection system discharging into the plant was subject to excessive infiltration (Public Law 92-500, Section 201, "Federal Water Pollution Control Act Amendments of 1972," 92nd Congress, S. 2770, Washington, DC, October 18, 1972). Consequently, municipalities had to reduce infiltration to the extent that was cost-effective and properly maintain the wastewater collection system to qualify for federal construction grants. Reducing infiltration/inflow is good practice because it allows the collection system to handle greater quantities of wastewater.

10. Negligence and vandalism can be the source of collection system problems. Vandals may place objects such as rocks and tree branches in manholes. Contractors repaving roads have accidentally allowed old pavement, soil, and/or road base material to enter manholes. Any material in a sewer will slow the flow and allow other solids to settle.

11. Poor records regarding complaints from the public or the date and location of stoppages that had to be cleared can result in an ineffective maintenance program. Good records, regular analysis of the records, and use of this information can produce a cost-effective preventive maintenance program.

QUESTIONS

Write your answers in a notebook and then compare your answers with those on page 652.

14.1A What happens when collection systems perform poorly?

14.1B What are the causes of problems resulting from collection systems not installed as designed?

14.1C What problems can be created when water flows down a ditch containing a sewer?

14.1D Why is reducing infiltration/inflow a good practice?

14.2 MANAGEMENT/ADMINISTRATION OF COLLECTION SYSTEMS

14.20 Management Issues

Management issues are critical to all aspects of a CMOM program. Elements of the CMOM program that must be properly managed include planning, design, construction, inspection, operation, maintenance, repair, and rehabilitation, and also information collection, analysis, response, and evaluation. For example, a sewer system repair and rehabilitation program requires proper management of a preliminary sewer system analysis, an I/I analysis and an estimation of flow rates, a sewer system evaluation survey (SSES), a corrosion analysis and control program, and ultimately the selection and installation of the most cost-effective method of repair and/or rehabilitation.

If a strategic plan is to be developed to minimize SSOs and CSOs, a manager would be responsible for managing and administering all of the major features of the plan. These features of the strategic plan would include a means of identifying, reporting, and notifying appropriate authorities of an SSO or CSO. The plan would need a prioritization process that would focus efforts on avoidable or preventable SSOs and CSOs. The prioritization process should evaluate whether or not specific current or potential SSO or CSO sites must be corrected and/or repaired immediately in a short-term remediation plan or in a comprehensive remediation plan. Managers should invite key persons or groups to participate in planning a

[5] *Exfiltration (EX-fill-TRAY-shun). Liquid wastes and liquid-carried wastes which unintentionally leak out of a sewer pipe system and into the environment.*

program or developing a plan, including operators, local governments, NPDES authorities, the public, stakeholder groups, public health officials, and any other affected groups such as drinking water suppliers, beach monitoring authorities, downstream water users, local fire departments, and police departments.

A manager might determine that it is cost-effective to develop and implement a watershed approach to controlling wet weather discharges. This approach has the potential to improve water quality management decisions and also to provide an equitable and cost-effective allocation of the responsibility among dischargers and participants in the watershed plan. A watershed approach could allow local stakeholders to help identify water quality priorities and increase the opportunity for watershed approaches to environmental improvement and protection. Watershed planning can be especially effective in enhancing the environment of sensitive and high-exposure areas such as beaches, drinking water supplies, and habitats for endangered species.

14.21 Functions of Management

14.210 Tasks Performed

The tasks performed by a utility manager administering a CMOM program are the same as for the CEO (chief executive officer) of any big company: planning, organizing, staffing, directing, and controlling.

Planning consists of determining the goals, policies, procedures, and other elements necessary to achieve the goals and objectives of the collection system agency. Planning requires management to collect and analyze data, consider alternatives, and then make decisions. Planning must be done before the other management functions.

Organizing means that management decides who does what work and delegates authority to the appropriate managers and operators.

Staffing is the recruiting of new operators and staff and determining if there are enough qualified operators and staff to fill available positions. The utility management's staffing responsibilities include selecting and training employees, evaluating their performance, and providing opportunities for advancement for operators and staff in the agency.

Directing includes guiding, teaching, motivating, and supervising operators and utility staff members. Directing includes issuing orders and instructions so that activities at the facilities or in the field are performed safely and are properly completed.

Controlling involves taking the steps necessary to ensure that essential activities are performed so that objectives will be achieved as planned. Controlling means being sure that progress is being made toward objectives and taking corrective action as necessary. The utility manager is directly involved in controlling the operation and maintenance of the collection system to ensure that the wastewater is being properly transported to the treatment plant and to make sure that the collection system utility is meeting its short- and long-term goals.

14.211 Operating Plan

An essential part of an effective CMOM program is a detailed, written plan for the management, operation, and maintenance of the wastewater collection system. The plan should consist of a statement of the agency's overall mission, goals to be attained, and the specific objectives, tasks, and procedures that will lead to attainment of the mission. The plan should be prepared by the manager of the organization responsible for the management, operation, and maintenance of the collection system with the full participation of the organization's personnel. The more clearly and completely the plan is put down in writing, the better agency personnel will be able to work in accordance with it. For any long-range plan to work, it is essential that the personnel who will implement it be involved in developing the plan and be generally supportive of it.

Mission Statement

The mission statement of an organization is a very broad, general explanation of the reason for the organization's existence. Such a statement might read:

> "The Clearwater Wastewater Collection Agency will provide safe, reliable, efficient collection system services at a minimum cost to the community of Clearwater, USA."

This one sentence tells us who will do what for whom, and in what manner. It is the "ideal" that the agency strives to achieve. Once the basic purpose of an organization is defined, it will then be possible to develop a plan to accomplish this mission.

Goals

Goals help to determine the course of action needed to set a CMOM program (and any other specific utility agency programs) in motion and keep it moving in the desired direction. Goals define the purpose and desired results of the CMOM program. Goals may reflect performance, safety, customer service, resource use, compliance, and other considerations.

The Water Environment Federation's Technical Task Force on Sewer Maintenance published[6] a list of goals for the responsible and efficient management of a collection system. The goals are as follows:

[6] *OPERATION AND MAINTENANCE OF WASTEWATER COLLECTION SYSTEMS, page 2, Water Environment Federation, 601 Wythe Street, Alexandria, VA 22314-1994.*

1. Prevent public health hazards;
2. Minimize inconveniences by responsibly handling interruptions in service;
3. Protect the large investment in collection systems by maintaining maximum capacities and extending their useful life;
4. Prevent unnecessary damage to public and private property;
5. Use the funds for the operation of municipal government services in the most efficient manner;
6. Convey wastewater to treatment facilities with a minimum of infiltration, inflow, and exfiltration;
7. Prevent excessive expenditures for claims and legal fees due to backups by providing immediate, concerned, and efficient service to all emergency calls; and
8. Perform all operations in a safe manner to avoid personal injury.

You should realize that full attainment of these goals is difficult to reach. Just as in a football game, the real sense of accomplishment derives from the progress you make toward meeting a goal in spite of the difficulties and occasional setbacks you may experience.

Objectives and Tasks/Procedures

Once goals have been established, a plan for operation and maintenance of a collection system should outline the more specific objectives as well as tasks and procedures for progressing toward the stated goals. Let's assume your agency agrees that the goals set forth by the Water Environment Federation are worthy of adoption. The following list demonstrates how you or your agency might go about setting specific objectives to meet each goal and how to list detailed tasks/procedures that will lead to accomplishing the objectives.

GOAL 1: Prevent public health hazards.

a. Provide immediate, concerned, and efficient service to all emergency calls.
    - Dispatch crews within one hour in response to emergency service calls.
    - Clean up and disinfect flooded homes.

b. Assign scheduling priority to stoppages, overflows, and exfiltration that contaminates the environment.
    - Send a supervisor to evaluate emergency calls.

c. Use clean water to flush contaminated surfaces and disinfect with chemicals approved by the health authority.
    - Maintain a neat work area for repair and maintenance activities.

A similar list is provided for the other seven goals in Section 12.22, "Objectives and Tasks/Procedures," in Chapter 12 of this manual.

14.212 Organizational Structure

In most organizations authority is essentially power—power to decide how the resources of the organization will be used to achieve its mission. This power usually is vested in (held by) the owner or highest official of the company, agency, or organization. As an organization increases in size, it becomes necessary to give some of this power to other staff members because no single person would be able to effectively make all decisions, supervise all work, and respond to all daily concerns that arise.

When power is delegated to others, it is usually limited to specific areas of responsibility such as personnel, financial operations, or collection system operation and maintenance. More limited powers may then be further delegated, for example, to individuals who supervise daily work activities or specific crews of operators within a section of the organization.

The overall purpose of delegating authority in this way is to allow decisions to be made and actions to be taken at the lowest practical level within the organization. The effect of this is to free upper management from activities that would otherwise consume so much of their time that they would be unable to meet their broader responsibilities for planning and managing.

A second outcome of delegating authority is that decisions about daily operations can be made by those who are most familiar with the situation and most directly affected by the decisions. This usually means a more efficient operation.

It is common practice to visually represent the delegation of authority with an organization chart such as Table 14.1. The lines connecting positions trace the line of authority that has been delegated by the superintendent to an assistant, who has then delegated some of the authority to four section heads, and so on down the line. Table 14.1 makes it clear that the Lift Station Operation and Maintenance Section Supervisor 1 gets authority from and reports directly to the Assistant Superintendent, and not to the head of Personnel or Engineering.

The organizational charts also should show the number of budgeted positions and the number of positions filled. Up-to-date job descriptions should be available for all positions shown on the organizational chart.

14.22 Legal Authority

Wastewater collection system agencies need the necessary and appropriate legal authority to implement a cost-effective CMOM program. The agency must be legally authorized to develop and implement activities essential for a successful program. Legal authority should include ability to:

- Control infiltration and connections from inflow sources,
- Require that sewers and connections be properly designed and constructed,
- Ensure proper installation, testing, and inspection of new and rehabilitated sewers,
- Regulate and control flows from municipal satellite collection systems,
- Implement the general and specific prohibitions of the National Pretreatment Program,
- Control grease from all potential sources discharging to the collection system, and
- Control I/I from private sources, such as privately owned portions of building laterals.

Grease can be a significant source of blockages and controlling it requires effective legal authority. Elements of a grease-control ordinance should include a public education program for residential users and workers in commercial food establishments or processors such as restaurants and bakeries. Other elements include the ability to develop and enforce

TABLE 14.1 ORGANIZATION OF CLEARWATER'S WASTEWATER COLLECTION SYSTEM DIVISION

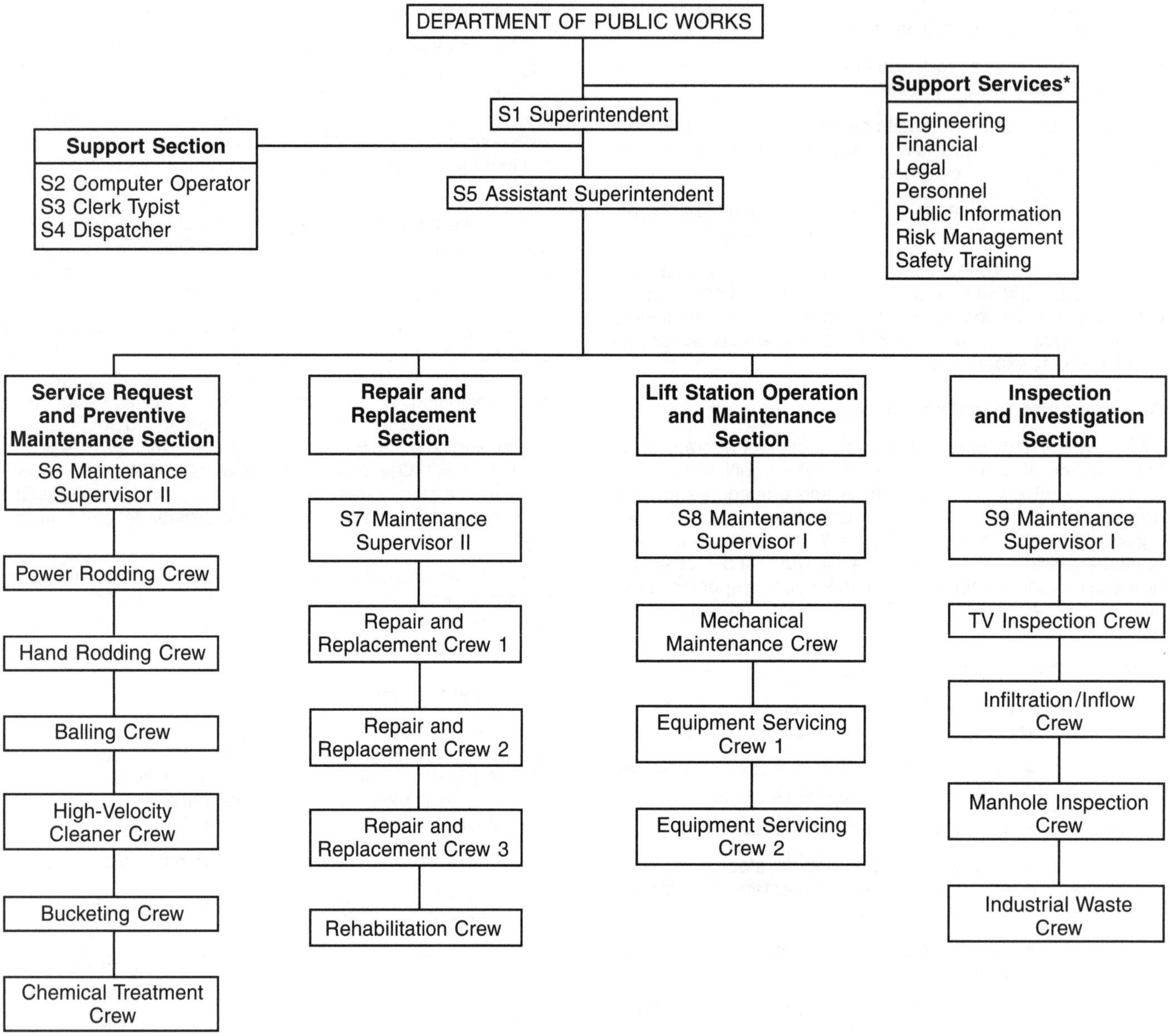

* Support Services provided by the centralized support staff of the Department of Public Works and other city departments.

regulations regarding the inspection, operation, and maintenance of grease traps and grease interceptors and also the handling, transport, recycling, and disposal of grease. In addition, the agency needs an O & M plan to prevent the buildup of grease in the sewers.

Legal authority should also include authority to prevent storm water drain connections to the sanitary sewer and to remove any storm water drains that are currently connected to the sewer.

QUESTIONS

Write your answers in a notebook and then compare your answers with those on page 652.

14.2A What elements of a CMOM program must be properly managed?

14.2B Why might a manager determine that it is cost-effective to develop and implement a watershed approach to controlling wet weather discharges?

14.2C What is the overall purpose of delegating authority?

14.23 Financial Management

Financial management for a utility should include providing financial stability for the utility, careful budgeting, and providing capital improvement funds for future utility expansion. These three areas must be examined on a routine basis to ensure the continued operation of the utility. They must be formally reviewed on an annual basis or more frequently when the utility is changing rapidly. The utility manager should understand what is required for each of the three areas and be able to develop record systems that keep the utility on track and financially prepared for the future.

14.230 Financial Stability

How do you measure financial stability for a utility? Two very simple calculations can be used to help you determine how healthy and stable the finances are for the utility. These two calculations are the OPERATING RATIO and the COVERAGE RATIO. The operating ratio is a measure of the total revenues divided by the total operating expenses. The coverage ratio is a measure of the ability of the utility to pay the principle and interest on loans and bonds (this is known as *DEBT SERVICE*[7]) in addition to any unexpected expenses. A utility that is in good financial shape will have an operating ratio and coverage ratio above 1.0. In fact, most bonds and loans require the utility to have a coverage ratio of at least 1.25.

The operating ratio is perhaps the simplest measure of a utility's financial stability. In essence, the utility must be generating enough revenue to pay its operating expenses. The actual ratio is usually computed on a yearly basis, since many utilities may have monthly variations that do not reflect the overall performance. The total revenue is calculated by adding up all revenue generated by user fees, hook-up charges, taxes or assessments, interest income, and special income. Next determine the total operating expenses by adding up the expenses of the utility, including administrative costs, salaries, benefits, energy costs, chemicals, supplies, fuel, principal and interest payments, and other miscellaneous expenses.

14.231 Budgeting

Budgeting for the utility is perhaps the most challenging task of the year for many managers. The list of needs usually is much larger than the possible revenue for the utility. The only way for the manager to prepare a good budget is to have good records from the previous year. A system of recording or filing purchase orders or a requisition system must be in place to keep track of expenses and prevent spending money that is not in the budget.

To budget effectively, a manager needs to understand how the money has been spent over the past year, the needs of the utility, and how the needs should be prioritized. The manager must also take into account cost increases that cannot be controlled while trying to minimize the expenses as much as possible.

A utility's user rates/user charges typically assess a fixed monthly fee for single-family residences; commercial and industrial service charges are usually based on wastewater flow and strength. These rates must be calculated in a manner that produces sufficient revenue to cover costs. The service charges of many utilities consist of the following components: (1) a charge to recover operation and maintenance costs, plus a five percent reserve; (2) a charge for equipment repair/replacement fund (for major equipment replacement and minor capital projects), and (3) a charge for capital improvements (existing users' share of the capital improvement program). Connection fees should be set to recover new developments' share of the capital improvement program. It is a good practice to evaluate user charges on a regular basis, such as every two years, and adjust them as appropriate to maintain the financial stability of the utility.

The utility should receive all of the funding from its sources of revenue. Utility funds should not be used for other local government activities.

A well-prepared budget allocates sufficient funds to cover all expected annual utility operating and maintenance costs and specifically identifies the costs for line items such as labor, materials, and equipment. The costs for collection system O & M should be itemized separately from other utility services, such as storm drains, water supply, and treatment plants. O & M costs should be tracked on a monthly basis using reporting

---

[7] *Debt Service. The amount of money required annually to pay the (1) interest on outstanding debts; or (2) funds due on a maturing bonded debt or the redemption of bonds.*

programs that identify costs for each line item in the budget. You may want to further categorize O & M costs as either preventive or corrective activities. To prepare a budget for O & M costs, refer to the utility's past cost records.

Involve the O & M staff in developing the budget and provide O & M managers with monthly cost tracking reports so that they can actively participate in managing expenses.

An important but often neglected part of the annual budget is a fund for future equipment replacement and capital improvements. A procedure should be in place for financing new work and activities. To establish project priorities, management and the O & M staff begin by reviewing prior budgets and the field reports submitted by the maintenance staff. Equipment repair/replacement needs and capital needs are then ranked by the urgency of the need, and the projected costs are estimated for each project or piece of equipment. Based on this prioritized list, management and the O & M staff then work cooperatively to develop realistic short- and long-term plans for capital improvement projects and repair or replacement of equipment.

A public education/outreach program is a good investment of time and effort for the utility. It can open lines of communication with customers and the community and help build goodwill toward the agency. Public hearings on issues of concern to the community are often an effective way to communicate information. In addition, efforts should be made to communicate justifications for rate changes to local governments, community groups, the media, young people (schools, youth organizations), and senior citizens.

14.232 Equipment Repair/Replacement Fund

To adequately plan for the future, every utility must have a repair/replacement fund. The purpose of this fund is to generate additional revenue to pay for the repair and replacement of capital equipment as the equipment wears out. To prepare adequately for this repair/replacement, the manager should make a list of all capital equipment (this is called an asset inventory) and estimate the replacement cost for each item. The expected life span of the equipment must be used to determine how much money should be collected over time.

14.233 Capital Improvements and Funding in the Future

A capital improvement fund must be part of the utility budget and included in the operating ratio. The utility manager must be sure that everyone, the governing body and the public, understands that the capital improvement fund is not a profit for the utility, but a replacement fund to keep the utility operating in the future.

Capital planning starts with a look at changes in the community. Where are the areas of growth in the community, where are the areas of decline, and what are the anticipated changes in industry within the community? After identifying the changing needs in the community, examine the existing utility structure and identify weak spots. Make a list of areas that will be experiencing growth, weak spots in the system, and anticipated new regulatory requirements. The list should include expected capital improvements that will need to be made over the next year, two years, five years, and ten years.

14.234 Asset Management

An asset management plan is a framework to bring all the key components of running a wastewater collection system utility into a strategic business plan that provides a means to protect, maintain, or improve the asset value of a collection system with planned maintenance and repair based on predicted deterioration of the system. In either a private or a public utility, key information is needed to manage costs through asset management planning, including: current conditions and performance of assets; current operating costs; current financial position including revenues, balance sheet, and cash flow; required and anticipated future levels of service; and methods of measuring and monitoring performance of the system.

The goal of capital asset management is to efficiently protect, maintain, or improve the value of the collection system while providing the level of service desired. Capital asset management attempts to meet these goals by accurately projecting future costs. Cost projections should address the following factors:

- Determining existing conditions,
- Setting future goals,
- Attaining future goals, and
- Tracking progress.

14.235 Governmental Accounting Standards Board Statement 34 (GASB 34)

In June 1999, the Governmental Accounting Standards Board (GASB), which sets financial accounting and reporting standards for state and local governments, issued Statement 34, entitled "Basic Financial Statements—and Management's Discussion and Analysis—for State and Local Governments." This standard contains changes to current financial accounting and reporting standards for state and local governments. GASB Statement 34 is intended to make financial reporting for state and local governments more comprehensive and easier for the public to use and understand.

The new standard includes a provision that now allows state and local governments either to record and report depreciation on all long-lived assets, including infrastructure assets such as water and wastewater infrastructure, or to use a modified approach of reporting infrastructure assets outside the basic financial statements as necessary supplementary information. In order to meet the criteria of the modified approach, state and local governments must meet the following conditions:

1. Use an asset management system that has an up-to-date inventory of eligible infrastructure assets,
2. Perform condition assessments of eligible infrastructure assets and summarize the results using a measurement scale,
3. Estimate each year the annual amount to maintain and preserve the eligible infrastructure assets at the condition level established and disclosed by the government, and
4. Document that the eligible infrastructure assets are being preserved approximately at (or above) a condition level established and disclosed by the government.

The GASB Statement 34 document mentioned above also provides an example of how infrastructure assets might be reported using supplementary information. The example indicates that to meet the GASB standard using supplementary information, governments must present schedules derived from the asset management system for all eligible infrastructure assets that are reported. Collection system managers should find the GASB Statement 34 document a helpful means of communicating to government officials the need to plan for the future funding of the repair and replacement of infrastructure assets.

For additional information on the financial management aspects of a CMOM program, refer to *UTILITY MANAGEMENT*, Section 9, "Financial Management."

QUESTIONS

Write your answers in a notebook and then compare your answers with those on page 652.

14.2D How do you measure financial stability for a utility?

14.2E What information does a collection system utility manager need to prepare a good budget?

14.2F What is the goal of capital asset management?

14.24 Personnel

14.240 Management Responsibilities

The utility manager is responsible for staffing, which includes hiring new employees, training employees, and evaluating job performance. The utility should have established, written procedures for job hiring that include requirements for advertising the position, application procedures, and the procedures for conducting interviews.

In the area of staffing and personnel, management must be extremely cautious and consider the consequences before taking action. A manager who violates an employee's or job applicant's rights can be held both personally and professionally liable in court. Personnel management programs and activities must be **job-related** and **documented.** Any personnel action taken must be job-related, from the questions asked during interviews to disciplinary actions or promotions. Documentation of personnel actions detailing what was done, when it was done, and why it was done (the reasons must be job-related) is absolutely essential. Good records not only serve to refresh a manager's memory about past events, but can also be used to demonstrate a pattern of lawful behavior over time.

14.241 Staffing

How many employees are needed? A staffing analysis procedure should be in place and regularly applied to the current and expected activities of the utility. To develop such a procedure, first prepare a detailed list of all the tasks to be performed to operate and maintain the utility. Next, estimate the number of staff hours per year required to perform each task. Be sure to include the time required for supervision and training. Compare the completed task analysis with the utility's current employees and use this information to determine staffing and training needs.

14.242 Employment Policies and Procedures

Employment policies and procedures provide guidance for both management and staff regarding expected duties and performance. These policies and procedures should cover the probationary period, compensation and benefits, training and certification, performance evaluation and promotion, disciplinary procedures, laws governing employer/employee relations, and personnel records.

Copies of the utility's employment policies and procedures should be placed in a binder or bound as a manual and made available to staff. Review the procedures and policies periodically and update them as needed or develop new policies. The contents of the manual and revisions should be communicated to staff as part of the utility's annual training program.

14.243 Training and Certification

Training must be an ongoing process in the utility. The utility manager is ultimately responsible for providing new employee training as well as ongoing training for all employees. Safety training is particularly important for all utility operators and staff members and is discussed in detail in the next section (Section 14.244, "Safety"). Certified operators earn their certificates by knowing how to do their jobs safely. Preparing for certification examinations is one means by which operators learn to identify safety hazards and to follow safe procedures at all times under all circumstances. Operators should be encouraged to seek appropriate levels of certification, and the utility should provide employees with the opportunities to gain the experience and education needed to qualify for and pass certification examinations.

Although safety is extremely important, safety is not the only benefit of a certification program. Other benefits include protection of the public's investment in utility facilities and employee pride and recognition. Vast sums of public funds have been invested in the construction of wastewater collection systems. Certification of collection system operators assures utilities that these facilities will be operated and maintained by qualified operators who possess a certain level of competence. These operators should have the knowledge and skills not only to prevent unnecessary deterioration and failure of the facilities, but also to improve operation and maintenance techniques.

Important elements of management's training and certification program include a presentation and discussion of the collection system utility's fundamental mission, goals, and policies as well as specialized technical training in the methods and procedures required to perform the specific duties and tasks necessary to safely operate and maintain the collection system. A good training program also provides opportunities for operators to learn how to use new technologies, such as the expanded use of computers to collect, analyze, and interpret data collected by field crews. Another valuable component of the training program is On-the-job training (OJT) in which operators learn how to use Standard Operating Procedures (SOPs) and Standard Maintenance Procedures (SMPs). Operators' progress and performance in the OJT program should be measured and evaluated, and a comprehensive recordkeeping system should be in place to track all types of employee training and comprehension of the training.

14.244 Safety Program

Management's Role

Management is responsible for the safety of the utility's personnel and the public exposed to the wastewater collection system utility's operations. Therefore, management must develop and administer an effective safety program and must provide new employee safety training as well as ongoing training for all employees. The basic elements of a safety program include a safety policy statement, safety training and promotion, and accident investigation and reporting.

Policy Statement

A safety policy statement should be prepared by the top management of the utility. The purpose of the statement is to let employees know that the safety program has the full support of the agency and its management. The statement should:

1. Define the goals and objectives of the program,
2. Identify the persons responsible for each element of the program,
3. Affirm management's intent to enforce safety regulations, and
4. Describe the disciplinary actions that will be taken to enforce safe work practices.

Give a copy of the safety policy statement to every current employee and each new employee during orientation.

Safety Training and Promotion

A safety training and promotion program begins with a documented safety program supported by a management official. In addition, there should be a safety department that provides safety training, equipment, and an evaluation of safety procedures. Other elements include:

- A recordkeeping system and documentation that O & M staff have received safety training and comprehended the safety training;
- Utility policy and procedures that require all O & M staff to follow safe work procedures, such as the use of personal protection equipment (PPE), lockout/tagout procedures, and trenching and shoring requirements/policies; and
- Written procedures and forms for confined space entry into manholes, wet wells, sewers, and other potential confined spaces. Safety procedures must be reviewed and revised as necessary on a regular basis.

Communication with field personnel on safety procedures may be accomplished by memo, e-mail, direct communication, tailgate safety sessions, classrooms, hands-on workshops, vendor demonstrations, and by training videos. However, whatever methods of communication are used, the safety department must actively communicate with field personnel on safety procedures on a regular basis.

It is the responsibility of management to ensure that the utility is in compliance with all pertinent safety training regulations including, among others, confined space, forklift, hazardous materials, CPR, first aid, and defensive driving regulations.

Accident Investigation and Reporting

Recordkeeping is an important part of an effective safety program. All injuries should be reported, even if they are minor in nature, so as to establish an accurate written record in case the injury develops into a serious injury.

The responsibility for reporting accidents affects several levels of personnel. First, of course, is the injured person. Next, it is the responsibility of the supervisor, and finally, it is the responsibility of management to review the causes of all accidents and take steps to prevent such accidents from happening in the future. Accident report forms may be very simple, but they must record all details required by law and all data needed for statistical purposes.

A safety committee is an effective means of investigating and preventing accidents. The safety committee can be involved in investigating workplace accidents and preparing appropriate responses and/or corrective procedures. The O & M staff should be active members of the safety committee.

QUESTIONS

Write your answers in a notebook and then compare your answers with those on page 653.

14.2G Personnel management programs and activities must meet what two criteria?

14.2H What is the purpose of employment policies and procedures?

14.2I What is management's role with regard to safety?

14.25 Equipment and Tools

14.250 Basis for Requirements

Equipment and tool requirements will depend largely on such things as the size of the collection system, work load, and objectives. Other important factors include the age and condition of the system and the types and number of problem areas. The equipment and tools of a wastewater collection system should be tailored to the program for operation and maintenance of the system.

In addition to providing the appropriate tools, the collection system utility must develop procedures and guidelines on the proper operation and maintenance of equipment. At the very least, operators will need procedures and guidelines for equipment start-up and shutdown and for emergency operation of the utility's equipment and facilities.

The operators who perform collection system O & M should have access to the necessary equipment and tools to safely complete all aspects of their O & M duties. It is helpful to provide operators with a list of all the equipment and tools available for operation and maintenance activities and then store the equipment and tools at one or more locations that give the operators convenient access to them.

Routine maintenance of equipment and tools can greatly extend their useful life and also ensure that they will work as expected when needed. Develop written equipment maintenance procedures and set up a logbook or other recordkeeping system in which operators can record maintenance of equipment and tools. These logbooks provide valuable information that will be useful when budgeting funds to replace equipment or tools when they wear out or when newer, more efficient equipment becomes available.

If the collection system agency is fortunate enough to have in-house vehicle and equipment maintenance staff, be sure the turnaround time for repairs is prompt and not causing delays in other scheduled collection system O & M activities.

14.251 Warehouse and Parts Inventory

A well-organized and fully stocked warehouse is critical to any collection system maintenance program. When stocking parts and/or material, at least three factors should be considered: (1) how critical is the part, (2) what is the lead time from the manufacturer, and (3) what is the frequency of usage of a particular part. In many cases, mechanical parts can be fabricated in the shop or purchased locally in quantity at less cost than keeping a supply in storage.

An inventory consists of the supplies the collection system needs to keep on hand to operate and maintain the system. These maintenance supplies may include repair parts, spare pumps and parts, electrical supplies, tools, and lubricants. The purpose of maintaining an inventory is to provide needed parts and supplies quickly, thereby reducing equipment downtime and work delays. However, a warehouse full of parts and supplies does you very little good if you can't find what you need when you need it. The utility needs some method to keep track of the location, usage, and ordering of spare parts. This can be done manually, but a computerized inventory system can be a great time saver. With any type of inventory system, manual or computerized, someone should be given the responsibility for physically checking the records against the actual parts in stock once or twice a year to verify that the tracking system is working properly.

Tools should also be inventoried. Tools that are used by operators on a daily basis should be permanently signed out to them. More expensive tools and tools that are only occasionally used, however, should be kept in a storeroom. These tools should be signed out only when needed and signed back in immediately after use. Similarly, equipment that is only used occasionally, such as portable generators, should be signed in and out of the storeroom when needed for a specific job.

A procurement procedure should be set up and explained to O & M staff for those occasions when they need to obtain materials or parts that are not routinely stocked in the warehouse.

14.26 Engineering

14.260 Role of Engineering

Engineering plays an important role in an efficient and effective CMOM program. Engineering staff members work closely with management and O & M staff. The major tasks performed by engineering include system design, construction, mapping, modeling, system evaluation, and capacity assurance.

14.261 Design

Engineering and O & M staff work cooperatively to ensure that designs produce facilities that deliver the desired results, perform as intended, and can be effectively and efficiently operated and maintained by O & M staff. To achieve these goals, several documents should be prepared during the design phase of the facility or project. *LIFE-CYCLE COST*[8] analysis should also be performed as part of the design process.

Critical design process documents include:

- A document that specifies engineering design criteria and standard construction details for gravity sewers, force mains, and pump stations;
- A document that describes the procedures used by the utility in conducting design review, including standard forms and/or checklists;
- Detailed requirements for equipment, training, and staffing,
- O & M manuals containing detailed procedures for start-up, operation, and shutdown of all equipment, and
- Protocols for testing and acceptance of the facility or project.

14.262 Construction and Inspection

New construction, repairs, and rehabilitation projects must be properly built in order to perform as intended and be properly operated and maintained. In most cases, standard specifications established by the local utility and/or the state govern all new construction projects.

There should be a warranty for new construction and a warranty inspection at the end of the construction period. Construction may be inspected by the utility or by another party, but you will need to prepare a manual that describes the procedures the utility follows in conducting construction inspections and testing. Standard forms are helpful for ensuring an accurate and acceptable construction and testing program. All inspectors must be properly qualified on the basis of experience, education, and knowledge of the inspections to be performed. An inspector must be on the job site at all times when the contractor's activities require the presence of an inspector.

If the construction involves new gravity sewers, test the finished sewers either by air testing or water testing. Also, new gravity sewers should be televised for defects and a *MANDREL*[9] should be used in the acceptance process.

New manholes should be tested for inflow and infiltration, and new force mains and pump stations should be acceptance tested according to utility standards before being accepted and placed on line.

14.263 System Mapping and As-Built Plans (Record Drawings)

Maps are absolutely essential to the management, operation, and maintenance of any wastewater collection system. Maps showing the accurate location of the components and an indication of the size of the collection system are required before any maintenance activities can be planned or implemented. As a collection system grows, staffing will increase,

---

[8] *Life-Cycle Costing. An economic analysis procedure that considers the total costs associated with a sewer during its economic life, including development, construction, and operation and maintenance (includes chemical and energy costs). All costs are converted to a present worth or present cost in dollars.*

[9] *Mandrel (MAN-drill). A testing device used to measure for excessive deflection in a flexible conduit.*

maintenance activities will expand, and a corresponding need for better, more detailed maps will become evident. Maps must indicate what has been constructed, where, what size, the materials used, and the conditions encountered. A detailed record of what is underground is critical for the successful operation and maintenance of a wastewater collection system. Maps are needed for the daily function of directing operators to their assigned work areas so it is vitally important to establish procedures for recording changes and for updating the mapping system.

Maps must include detailed descriptions of the sewers and manholes, including size, material, age, slope, and invert elevations. One set of maps, the "working plans," are used for on-the-job references and conferences during construction. Another set, the "as-built" plans (also called "record drawings"), are used to record daily any constructed deviations from the location or depth of the sewer system facilities or any other deviation from the plans. Changes must be recorded as soon as they are found or completed and never put off to another day. Locations of building sewers are commonly shown on the "as-built" plans as well as on a Building Sewer Location form.

The "as-built" plans are also used for the preparation of sewer location maps and should also be available to wastewater collection system operators for detailed information if needed for maintenance and repair of a sewer line. Field crews should record changes or inaccuracies and there should be a process in place to update "as-built" plans and maps.

14.264 Hydraulic Modeling

Hydraulic modeling can be used by wastewater collection system utilities to predict flow capacity, peak flows, force main pressures, and other factors involving the hydraulics of the system. These hydraulic models are typically run on computers. Additions to the collection system or modifications of facilities will alter the hydraulics of the system; therefore, you will need to update the hydraulic models so they accurately describe the performance of the current collection system. Models can be used to test new designs and to predict how the system will perform if the system is redesigned or expanded.

14.265 Sewer System Evaluation Survey (SSES) and Rehabilitation

SSES Procedures

The sewer system evaluation survey (SSES) is the key step in identifying specific sources of inflow and infiltration (I/I). Because SSES projects can extend over a long period, experience gained early in the project can be applied to refine and improve later studies. The resulting composite procedure is essentially a systems approach to analyzing the collection system.

In published accounts of assessments for I/I removal, several procedural deficiencies are cited repeatedly as reasons for I/I removal projects failing to achieve their objectives. These deficiencies include:

- Inadequate estimation of the amount of extraneous flow,
- Inadequate identification of extraneous flow sources (both public and private sector),
- Service connections and laterals as sources of I/I,
- Optimistic flow-reduction goals,
- Ineffective rehabilitation practices,
- Groundwater migration to unrepaired defects, and
- Lack of definition of what constitutes a successful project.

The systems approach mentioned above overcomes these deficiencies and minimizes their impact on project results. More accurate flow data can be collected by applying the following steps to the identified collection system component:

- Flow monitoring,
- Physical survey by subarea,
- Internal inspection by subsystem, and
- Cost-effectiveness analysis by subarea.

Removal of Infiltration and Inflow Sources

Removal of I/I sources requires elimination of defects found in the public and private sector sections of the collection system. Repair and rehabilitation are the long-term, ongoing components of the O & M program that accomplish this elimination of defects.

Vast improvements in repair and replacement technologies have greatly expanded the available selection of cost-effective repair/rehabilitation methods and materials. Although system repair and rehabilitation efforts vary widely among agencies, they normally consist of a combination of repairs and rehabilitation that take place as part of normal system O & M. An example is the rehabilitation of manhole defects. A large percentage of manholes have defects in the walls, pipe connections, corbel, or frame seal that can be repaired by a crew or the repairs can be contracted out using the O & M budget to fund the repair. Major repair or rehabilitation projects that are beyond the capability of the O & M staff because of cost or complexity are funded as part of the CIP (Capital Improvement Program).

Success in removing I/I sources by eliminating defects takes a sustained effort carried out over a number of years. To achieve success, the survey work must be performed carefully and accurately and the rehabilitation work must be performed competently. The credibility of any program lies in its ability to meet the defined objective: to reduce SSOs (sanitary sewer overflows) and/or CSOs (combined sewer overflows).

Rehabilitation Considerations

Sometimes an agency should consider rehabilitation instead of preventive maintenance and inspection as the solution to certain problems. Usually this decision is based on factors such as the increased risk of regulatory noncompliance, reduced level of service, or cost-effectiveness of preventive maintenance versus rehabilitation over the remaining life of the sewer. Once again, a formal evaluation process based on accurate data is needed to make an informed decision.

Areas where underground repair or rehabilitation may be required include service laterals and connections and main line sewers. The types of nonstructural repairs most commonly needed include joint sealing, lateral sealing, and tap cutting; types of structural repairs needed often include joint repair, excavation, pipe bursting, lining, and slip lining.

The utility should have a multi-year Capital Improvement Program (CIP) that provides funding for rehabilitation, replacement, and repair activities. Project priorities can be established on the basis of input from field staff, management, the superintendent, engineering staff, and outside consultants, with consideration given to the degree of structural deficiencies and lack of capacity. Once the priorities have been established and funding has been approved, develop both long-term and short-term schedules for project completion.

For additional information on SSESs and rehabilitation, see the pertinent portions of Section 14.3, "Operation and Maintenance."

14.266 System Capacity Assurance

Capacity assurance is a process developed to identify, characterize, and correct hydraulic deficiencies in a collection system. Deficiencies may involve inadequate sewer capacity, inadequate capacity in lift stations or pumps, and/or structural problems at manholes. Every collection system utility agency should have a program to assess the current capacity of the collection system.

Where peak flow conditions contribute to an SSO discharge, the agency should take steps to correct the hydraulic deficiency. To correct a hydraulic deficiency, the agency should evaluate the management and performance of the collection system. Also, there should be continued monitoring and assessment to determine the effectiveness of implemented measures and to adjust the measures as necessary.

A thorough understanding of the characteristics and performance of the collection system is essential for developing cost-effective solutions. Trying to fix complex, wet weather collection system problems without adequately evaluating the collection system can result in pursuing inappropriate solutions that are not the most cost-effective and that may even lead to overflow problems in other parts of the collection system. A detailed evaluation of the collection system can significantly reduce repair/rehabilitation costs by providing accurate information on the causes of the SSO problem. Once the actual causes of the problem are known with some certainty, the most appropriate and cost-effective solutions can be selected.

Collection system evaluations undertaken to correct wet weather SSO problems should focus primarily on identifying the major causes that contribute to the peak flows associated with the overflow events, for example: sources of inflow and rainfall-induced infiltration (RII), and hydraulic problems such as bottlenecks, insufficient slopes, or inadequate pumps. (Evaluations that focus primarily on SSO problems, such as we're discussing here, may differ from many traditional sanitary sewer evaluation surveys because traditional surveys often focus primarily on infiltration affecting base flows.) To accurately quantify peak flows entering a collection system, total flows need to be measured or accounted for and estimated, including contained flows remaining in the system and escaping flows such as overflowing manholes or other SSOs. Complete and accurate flow monitoring is extremely important to estimate peak flows. Rainfall-induced infiltration is dependent on the intensity and duration of the storm event and other factors; therefore, once the measured flow data are available, they can be correlated to the specific rainfall that caused the flow.

Modeling may be a valuable tool for providing general predictions of sewer system response to various wet weather events and evaluating control strategies and alternatives. When a model is used, however, it should be calibrated and verified by field measurements. Limit the use of any simulation model to the collection system for which actual data are provided and use it only for the range of rainfall data measured. Modeled flow projections should be accompanied by a characterization of the degree of uncertainty because such uncertainty can be significant. (The "uncertainty" level is like the accuracy level on public opinion polls, which are often claimed to be accurate plus or minus 3, or 4, or some other percent.)

When preparing a system evaluation and capacity assurance plan, describe how actions were prioritized and include estimated schedules for implementing actions. If the plan deals with multiple hydraulic deficiencies, base the priorities on human health and environmental risks associated with potential SSOs and the degree to which improvements can be made quickly. Factors that can affect risk are the location of the SSO, the potential for human contact, receiving water uses, and the volume of discharge. Give highest priority to SSOs that substantially endanger human health now (or could in the very near future) such as discharges into buildings, public drinking water supplies, and waters and beaches where swimming occurs.

As part of an ongoing capacity assurance effort, set up procedures for determining whether the capacities of the existing gravity sewer system, pump stations, and force mains are adequate for proposed new connections. Metering may be an effective approach to determine peak flows before allowing new connections, and hydraulic models can be used to predict the impacts of new connections. Projects to expand the sewer system and/or increase its capacity could be funded by developer fees and as part of the utility agency's Capital Improvement Program.

14.27 Water Quality Monitoring

Water quality monitoring is a way to determine if the collection system is having any impact on water quality. After an SSO event, the water quality in nearby waters should be monitored until no adverse impacts are detected. Monitor both surface waters and groundwaters in areas that may be affected by overflows and/or exfiltration from improperly operated and maintained collection systems. Locations that should be monitored include nearby beaches and sources of public drinking water. Also sample areas that are sensitive wildlife habitats and fisheries. The water quality indicators that should be tested in samples collected will depend on the beneficial uses of the waters where the samples are collected and the sensitivity of the receptors (people, plants, animals) using the waters.

QUESTIONS

Write your answers in a notebook and then compare your answers with those on page 653.

14.2J What kinds of procedures and guidelines do operators need regarding tools and equipment?

14.2K What is the purpose of a logbook for a piece of equipment?

14.2L Why should engineering and O & M staff work cooperatively?

14.2M How can infiltration and inflow (I/I) sources be removed?

14.2N What is system capacity assurance?

END OF LESSON 1 OF 2 LESSONS
on
CMOM

Please answer the discussion and review questions next.

DISCUSSION AND REVIEW QUESTIONS

Chapter 14. CAPACITY ASSURANCE, MANAGEMENT, OPERATION, AND MAINTENANCE (CMOM)

(Lesson 1 of 2 Lessons)

At the end of each lesson in this chapter you will find some discussion and review questions. The purpose of these questions is to indicate to you how well you understand the material in the lesson. Write the answers to these questions in your notebook before continuing.

1. Why have some state water pollution control regulatory agencies decided to enforce elements of EPA's proposed SSO Rule and CMOM Program?
2. What happens when collection systems are neglected?
3. How should a manager respond to avoidable or preventable SSOs and CSOs?
4. What is the difference between goals and objectives?
5. What factors should be considered when analyzing the financial management of a utility?
6. Why must a utility manager be extremely cautious and consider the consequences before taking any action in the area of staffing and personnel?
7. What factors should a utility agency analyze when considering sewer rehabilitation instead of preventive maintenance and inspection as a solution to certain collection system problems?

CHAPTER 14. CAPACITY ASSURANCE, MANAGEMENT, OPERATION, AND MAINTENANCE (CMOM)

(Lesson 2 of 2 Lessons)

14.3 OPERATION AND MAINTENANCE OF COLLECTION SYSTEMS

14.30 Purpose of Operation and Maintenance (O & M)

Operation and maintenance of a wastewater collection system can be defined as the O & M activities that result in conveying wastewater safely and efficiently to a wastewater treatment plant. The purpose of O & M programs is to maintain design functionality (capacity) and/or to restore the system components to their original condition and thus functionality. Stated another way, does the O & M program keep the system performing as designed and intended?

The ability to effectively operate and maintain a collection system so it performs as intended depends greatly on proper design (including selection of appropriate materials and equipment), construction and inspection, acceptance, and system start-up. Permanent system deficiencies that affect O & M of the system are frequently the result of these phases. O & M staff should be involved at the beginning of each project, including planning, alignment, design, construction, acceptance, and start-up. When a collection system is designed with future O & M considerations in mind, the result is a more effective O & M program in terms of O & M cost and performance.

Effective O & M programs are based on knowing what components make up the collection system (Section 14.40, "System Inventory"), where they are located, and the condition of the components. With that information, proactive maintenance can be planned and scheduled, rehabilitation needs identified, and long-term Capital Improvement Programs (CIPs) planned and budgeted. High-performing agencies have all developed performance measurements of their O & M program and track the information necessary to evaluate performance.

14.31 Maintenance Scheduling

14.310 Basic Aspects of Scheduling

Formulating an effective schedule for maintaining a collection system can be very difficult. Scheduling is complex, not necessarily in a technical sense, but more because of the many varied facts and details that must be considered. A computer can help a great deal, but there are no shortcuts to good scheduling. Maintenance of a collection system is an ongoing process that often spans years of past history and must consider future years of anticipated growth and increased loadings on collection systems. It is difficult to list all the many details one will encounter in planning a maintenance schedule. Some basic details to be concerned with are described in the following sections. They include knowledge and skills, organization and timing of maintenance, and flexibility.

14.311 Knowledge

Knowledge is the name of the game. Without knowledge, there is little chance for the maintenance schedule to be effective. The knowledge needed is not necessarily relative to anyone's intelligence or IQ. The required knowledge has more to do with the collection system and also the maintenance procedures and equipment. Required knowledge includes the following:

- Collection system. You must have a thorough knowledge and understanding of your collection system. You will need to know the nature of your problems and the areas in which they are concentrated.

  There are various ways of acquiring this information. A great deal of it may come to you through experience and familiarity with your collection system. The most reliable information could come from a computerized management information system or a computer-based maintenance management system (CMMS).

  Aside from these obvious sources of information, there are other steps you can take:

  1. Complete a comprehensive study of your system. Locate and identify the areas that are creating the most problems. After you have determined your problem areas, try to locate and identify specific problems.
  2. A highly effective method of identifying and locating specific problems is by selective manhole inspection and TV inspections. This practice can be used to survey even the smallest collection system. However, emphasis should be placed on the term "selective." Haphazard or random inspections, particularly with TV, can be wasteful and inefficient. By careful preselection of troublesome areas, TV inspection can be performed in an economical manner.

- Maintenance procedures and equipment. Any schedule you plan will be influenced by, and perhaps tailored to, the procedures and equipment available to you. Carefully evaluate these procedures and equipment. Always consider that certain procedures and equipment are most effective when used to combat specific problems; others may be effective to some degree in coping with a variety of problems. How are you using available equipment? Are you using it to its fullest potential?

14.312 Organization of Personnel

The organization and staffing needed to support a collection system maintenance program depend on the types of systems installed (primarily in the pump stations); these systems can be identified as:

- Electrical and electronic systems,
- Mechanical systems, and
- Hydraulic and pneumatic systems.

14.313 Timing

Of vital importance to any schedule is proper timing. When planning maintenance schedules, two important factors should be considered:

1. Only when needed. The ideal schedule for maintenance would provide for completion of needed maintenance just prior to the development of any problems or emergencies. In actual practice, this is virtually impossible to attain. It is, however, what we should be striving to achieve to the greatest degree possible.
2. Too frequent. If there is one weakness common to many maintenance schedules, this would be it. In their efforts to predict and prevent the development of emergency situations, many supervisors tend to schedule maintenance too frequently. This approach cannot be considered to be totally wrong. It is the lesser of two evils, the other being an inadequate schedule that results in too little maintenance. At best, however, too-frequent maintenance must be considered inefficient. It is inefficient from an economic standpoint and it ties up operators and equipment that could be used more productively in other areas where maintenance or repairs are needed.

14.314 Flexibility

A good maintenance schedule should be flexible. Periodically review the schedule and revise it in accordance with the results it has or has not achieved.

14.315 Cross Training

Cross training of operators is another way to increase maintenance effectiveness and flexibility. All too often maintenance personnel are unable to properly diagnose a system problem in a pump station because they are so specialized. Cross training broadens operators' skills and enables them to more quickly troubleshoot problems.

14.316 Important Elements of Scheduling

The utility should plan and schedule both preventive and corrective maintenance. Establish a priority system that specifies who will set the maintenance priorities and how they will do it. Use performance indicators to measure the effectiveness of the maintenance schedule and O & M performance. Examples of performance indicators include the number of SSOs, pump station failures, and complaints. Routinely track the maintenance backlog and follow up on it if the schedule is backlogged. Adequate funding of the utility's maintenance activities also affects scheduling because without adequate funding, the maintenance program will not be able to perform its expected duties.

14.32 Sewer Cleaning

14.320 Purpose

Stoppages in gravity sewers usually are caused either by structural defects or by an accumulation of material in the pipe. Certain structural defects, such as protruding taps, may catch debris, which then causes a further buildup of solids that will eventually block the sewer. Repair or elimination of any defects that contribute to a buildup of material in the pipe should be evaluated as part of a rehabilitation program since the defects will always be a maintenance problem.

The accumulation of solids in gravity pipelines is also related to the velocity of flow in the sewer. A velocity of two feet per second or more will keep normal solids in suspension in the flow. The velocity of flow depends primarily on the slope of the pipe, but also on the pipe material, internal roughness, and the flow rate (amount of flow) of the wastewater in the pipe.

Mechanical and hydraulic cleaning of sewers are cost-effective methods of removing material clinging to the interior surfaces of the pipe. When clean, sewer pipes can carry full pipe flow without any restrictions that might result in blockages due to reduced pipe capacity.

14.321 Benefits

The major benefit of a sewer cleaning program is that it maintains the hydraulic capacity of the sewer. This permits the system to operate at the intended design flow since there are no restrictions to cause a collection of debris. If debris is allowed to build up in pipes, it could result in a stoppage and sanitary sewer overflow (SSO) or back up into a residence, school, or business. In addition to enhancing performance of the system, a proactive sewer maintenance program benefits the agency's asset management program by:

- Lowering O & M costs,
- Extending the useful life of the gravity sewer system, and
- Lowering rehabilitation/replacement costs.

14.322 Cleaning Equipment

Various hydraulic and mechanical methods are available for main line sewer cleaning. Hydraulic cleaning methods include equipment that uses water and water velocity to clean the invert and walls of the sewer pipe. Hydraulic cleaning equipment includes high-velocity cleaners or hydro jets, balls, kites, bags, parachutes, and scooters. Mechanical cleaning methods use equipment to physically remove the material from the walls of the sewer pipe. Mechanical cleaning equipment includes rodders, bucket machines, and cable machines.

Other methods of cleaning sewers on a preventive basis are the application of chemicals and bioaugmentation (addition of bacteria to speed up the breakdown of roots and other organic material) to retard root growth or remove deposits in the main line sewer.

Most collection system agencies use high-velocity cleaners as the major piece of equipment for cleaning sewers. The frequency of cleaning should be based on analysis of data collected by TV inspection crews, crews clearing stoppages and overflows, and also by the cleaning crews. The main objective of sewer cleaning is to prevent blockages and overflows. Crews dedicated to cleaning and maintaining sewers are usually assigned to perform corrective and emergency maintenance when these problems are discovered or develop. Crews operating large physical cleaning equipment must be very careful to not damage the sewer pipe. Most agencies have found that the mechanical cleaning equipment only damages pipe that already has an existing problem.

14.323 Chemical Cleaning and Root Removal

Chemical cleaning of main line sewers is another method used by utilities to clean sewers. Normal cleaning with chemicals will remove grease and roots. Several chemicals and application methods are available to kill and retard the regrowth of roots in the collection system sewers. Methods of application include foaming, dusting, and liquid application. Special equipment is required for all three application methods. For liquid applications, chemicals can be added to the water reservoir of a high-velocity cleaning machine. Generally the application is needed every two or three years. This frequency controls the roots in the sewer at a lower cost than conventional cleaning methods, which would be required more often. If the problem is roots alone, chemical treatment is a very cost-effective method of cleaning. Some precautions must be taken on deadend stretches where low flow may not work out the dead roots. In these areas, some additional cleaning may be required six months after the application to prevent the dead roots from causing a local blockage. Wherever an agency has root problems in the sewers, the agency should have a root control program.

Grease can also be cleaned from sewers by the addition of chemicals or by bioaugmentation. Various chemicals are available, such as enzymes, hydroxides, caustics, biocides, and neutralizers, for removing and/or controlling grease buildups. The effectiveness of a particular chemical depends largely on the exact nature of the problem and site-specific circumstances. Usually these compounds require repeated applications to prevent future grease buildups. In most cases, these compounds tend to be an expensive method of treatment for roots or grease if they are applied on an ongoing basis. Industry is always developing newer and more cost-effective chemicals. Before purchasing chemicals, work with the vendor to reach agreement on the cost and what constitutes acceptable performance of their products. If the chemicals perform as agreed upon, then pay the agreed price.

14.33 Hydrogen Sulfide Monitoring and Control

14.330 Why Control Hydrogen Sulfide?

Hydrogen sulfide (H_2S) is one of the most serious problems confronting wastewater collection system operators. The hotter the weather, the flatter the sewers, and the longer the flow time to the wastewater treatment plant, the worse the problem becomes. Some of the problems created by hydrogen sulfide gas include:

1. Paralysis of the respiratory center and death of collection system operators,
2. Rotten egg odors,
3. Corrosion and possible collapse of sewers, structures, and equipment,
4. Loss of capacity of the sewer, and
5. A flammable and explosive gas, under certain circumstances.

14.331 Control of Hydrogen Sulfide

Proper design of a collection system, using current understanding of sulfide generation, can often eliminate serious odor and corrosion problems by controlling the release of sulfide. The utility should take into consideration hydrogen sulfide control and corrosion prevention when designing sewers. Designers must realize that the higher the wastewater temperature, the greater the production of hydrogen sulfide. Thus, designs acceptable in cooler climates may be unacceptable in warmer climates.

Another factor of importance in terms of maintenance is the presence of slimes in sewers. Sulfate is split by bacteria living in the sewer slimes. The more area or habitat available for such slimes, the more sulfide they will produce and the greater the odor, corrosion, and maintenance problems. Cleaner sewers harbor fewer slime bacteria and greatly reduce the problems caused by sulfide.

If odors are a frequent source of complaints, then the utility needs a sulfide control program and may want to consider the application of chemicals to control the production of hydrogen sulfide and to prevent or reduce corrosion.

QUESTIONS

Write your answers in a notebook and then compare your answers with those on page 653.

14.3A What is the purpose of a collection system O & M program?

14.3B What kind of knowledge is required to schedule collection system maintenance?

14.3C What causes stoppages in gravity sewers?

14.3D How can a collection system agency's proactive sewer maintenance program benefit the agency's asset management program?

14.3E How can hydrogen sulfide be controlled?

14.34 Lift Stations

14.340 Purpose

Lift stations are used to lift or raise wastewater or storm water from a lower elevation to a higher elevation. Lifting of the wastewater is accomplished by centrifugal pumps or by air-operated pneumatic ejectors. The term "lift station" usually refers to a wastewater facility with a relatively short discharge line up to the downstream gravity sewer. A "pump station" commonly is a similar type of facility that is discharging into a long force main. Throughout this chapter when we refer to lift stations, we intend to include pump stations.

In many areas regional wastewater collection and treatment system agencies are being created to serve more than one community. As a result, lift stations and force mains tend to be larger. The pumps can range in size up to several hundred horsepower and the force mains can be several miles long.

Lift stations represent a major capital expenditure for a utility or community, and they require an adequate budget to operate and maintain them properly. Failures of lift stations and force mains can have a significant impact on the environment when raw wastewater is discharged over land or into lakes, streams, or rivers. Backups into private residences caused by lift station failure can easily cost thousands of dollars to clean up, replace, and repair damaged homes or businesses. Significant lift station failures make headlines in newspapers and tarnish the image of your collection system agency.

14.341 Operation

The collection system operating agency has the responsibility to ensure the continuous and efficient operation and maintenance of the lift stations, including the structures and the grounds. This responsibility includes preventing failures in operation that would result in flooding upstream homes, businesses, or streets. Reputable system designers will never include facilities for bypassing wastewater to rivers, streams, lakes, or drainage courses. When emergencies occur, portable emergency equipment must be used to pump wastewater to a functioning section of the downstream collection system and not to the environment. When untreated wastewater is discharged to the environment, public health hazards and pollution of adjacent receiving waters result.

Lift stations may be located throughout a community and must be neat in appearance, blend with the architecture and landscaping of the neighborhood, and not create a nuisance to neighbors through odors and noise. Complaints from the public will be few if the operators responsible for the lift station maintain the facility in top operating condition and respond to questions or complaints from the public in a positive and concerned manner. When responding to a complaint, be sure to tell the public what has been done or will be done to correct the cause of the complaint.

A sufficient number of operators should be assigned to the operation and maintenance of the lift stations. Also, properly trained and experienced staff must be assigned to perform the scheduled mechanical and electrical maintenance.

14.342 Emergencies

Each lift station must have an Emergency Operating Procedure, and crews must be prepared and trained in advance to respond to lift station failures. To save time and reduce the adverse effects of a lift station failure, set up standard notification procedures to alert the appropriate crews when a lift station failure occurs.

All lift stations should have sufficient redundant equipment to minimize the effects of a lift station failure. When a lift station loses power, the station should be quickly returned to service using on-site generators, portable electric generators, or alternate power sources. If necessary, equipment should be available to bypass the lift station and transport the wastewater to a downstream portion of the collection system. After a lift station failure, investigate the cause of the failure and take necessary action to prevent future failures.

14.343 Alarms and Monitoring

All lift stations should have telemetry alarm systems that transmit one or more signals indicating the nature of a lift station failure or that the lift station needs to be visited. When a lift station has a failure, the alarm should be transmitted to a control panel at the emergency dispatch center. An indicator light for the particular station goes on when a problem develops, and a horn or some other type of noisy alarm may also be activated. When trouble is indicated, the dispatcher notifies the operator on duty by phone or radio of the station identification number. The dispatcher must record on an appropriate log sheet the time and identification number of the station that sent the alarm signal and the time and name of the operator notified.

A SCADA (Supervisory Control and Data Acquisition) system is a computer-monitored alarm, response, control, and data acquisition system used by wastewater collection system operators to monitor and adjust the operation of equipment and lift stations in their systems.

A SCADA system collects, stores, and analyzes information about all aspects of operation and maintenance, transmits alarm signals when necessary, and allows fingertip control of alarms, equipment, and processes. SCADA provides the information that operators and their supervisors need to solve minor problems before they become major incidents. As the nerve center of a wastewater collection agency, the system allows operators to enhance the efficiency of their facilities by keeping them fully informed and fully in control.

14.344 Inspection

The frequency of lift station visits depends on the following factors:

- Number of lift stations in the community,
- Type of wastewater conveyed,
- Potential damage resulting from storms flooding the lift station,
- Condition of equipment, such as equipment temporarily repaired and waiting for replacement parts,
- Design of the facility and the equipment installed in the lift station,
- Adequacy of the preventive maintenance and overhaul program,
- Type, adequacy, and reliability of the telemetry system, if installed,
- Attitude of the operating agency toward operation and maintenance, and
- Number of operators available to visit the lift station.

A checklist should be used for each lift station visit to ensure that all necessary items are inspected. The utility also should develop Standard Operating Procedures (SOPs) and Standard Maintenance Procedures (SMPs) for each lift station. When a crew visits a lift station, the work they accomplish should include the following items:

- Inspect the power panel for tripped circuit breakers,
- Observe the indicating lights of operating equipment,
- Check flow data and recorded elapsed time meter readings,
- Examine motors in the station to determine if they are noisy or running hot,
- Inspect pump packing or seal equipment, check valves, and suction and discharge pressures,
- Perform routine pump lubrication and cleanup,
- Check the sump pump switch to determine if it is operating,
- Bleed condensate from the air bubbler system for the wet well and pump control,
- Examine the wet well level indicator to be sure it is reading properly, and
- Inspect the wet well for sticks and trash.

14.345 Preventive and Routine Maintenance

Preventive and routine maintenance tasks and frequencies must be established for all lift stations and equipment. The preventive maintenance program should include predictive maintenance techniques that are capable of forecasting failures before they occur. Periodically conduct an energy audit of each lift station's electrical usage to ensure efficient use of electric power. Maintain an adequate parts inventory for all equipment. Also, ensure that a sufficient number of trained personnel are available to properly maintain all lift stations.

14.346 Recordkeeping

Records are an important part of a lift station operation and maintenance program. They should be filed at the utility's main office because records left at the lift station are of little value to supervisors and managers. Keep active records (current year) in the office copy of the station book. This procedure keeps all information at one source. At the end of each operating year or the end of a designated period selected by the utility, remove the past year's operational and maintenance records from the station book and place them in the station file.

Lift station records can be used as the basis for scheduling preventive maintenance activities and for other purposes. Elapsed time meters should be installed on all pumps and the recorded meter readings (pump run times) can be used to assess pump performance.

Other important records that should be readily available to O & M staff include the manufacturers' specifications and equipment manuals for all equipment.

14.347 Force Mains and Air/Vacuum Valves

Regularly inspect the route of force mains to ensure that they are working properly. Also, establish a program to assess the condition of all force mains on a regular basis and set up a process to investigate the cause of any force main failures. Air release and vacuum relief valves are installed to prevent force main failures due to water hammer. These valves must be maintained on a periodic basis. Both of these types of valves may fail to operate reliably if grease is allowed to accumulate in the valve body or on the operating mechanism.

QUESTIONS

Write your answers in a notebook and then compare your answers with those on page 653.

14.3F Who has the responsibility to ensure the continuous and efficient operation and maintenance of the lift stations, including the structures and grounds?

14.3G What happens when untreated wastewater is discharged to the environment?

14.3H When a lift station loses power, how can the lift station be quickly returned to service?

14.3I What information should be transmitted by a lift station telemetry alarm system?

14.35 Sewer System Evaluation

14.350 Purpose

The purpose of a sewer system evaluation by O & M staff is to identify problem areas in the collection system and to prioritize needed repair or rehabilitation projects (see Section 14.36, "Rehabilitation"). Elements of a sewer system evaluation program include flow monitoring, manhole inspection, sewer cleaning related to I/I reduction, internal TV inspection, and also smoke testing and dyed water flooding.

Also see Section 14.265, "Sewer System Evaluation Survey (SSES) and Rehabilitation."

14.351 Flow Monitoring

Flow monitoring may be categorized as: (1) long-term monitoring, (2) temporary monitoring, and (3) instantaneous monitoring. The purpose of long-term monitoring is to adequately define I/I quantities. After establishing the non-excessive/excessive flow determinations for various subareas, these flow quantities provide the basis for flow balancing of sources found during the survey and the means to compare post-rehabilitation flows. Long-term flow monitors should be in place for one groundwater season, typically for 60 to 120 days. Flow monitors are typically located off the larger diameter trunk lines for greater accuracy and for ease of calibration.

Temporary monitoring is subsequently established at the mini-system level within the metered subarea. These meters could remain in place up to 30 days on systems ranging up to 5,000 linear feet. The purpose of temporary monitoring is to gather additional data to further refine the identification of portions of the collection system needing additional study.

Instantaneous monitoring is performed at the subsystem level during low flow. The purpose is to isolate line segments with excessive infiltration; these segments are then scheduled for internal inspection. This isolation is typically conducted at night on three to five consecutive line section segments (approximately 1,000 linear feet).

Most agencies use both permanent and temporary flowmeters to obtain the information they need to evaluate their system. At least one rain gage should record representative precipitation over the area served by the collection system. Correlation of precipitation data with collection system flows will help describe the condition of the collection system. Flow monitoring and rain gage information should be part of every I/I study. Areas with excessive I/I should be prioritized for rehabilitation by capital improvement projects.

14.352 Manhole Inspection

Because they are part of the collection system, manholes require the same inspection and attention as the rest of the sewer network. When located in streets, these structures are subject to the vibrations and pounding of vehicle traffic. Manholes may settle at a different rate than the connected sewer, thus creating cracks in joints. Easement locations on private property are subject to misuse and changes of ground surface due to construction or landscaping work. The local sewer-use ordinances relating to manholes located in easements should clearly inform the public of the agency's unlimited right of access to perform collection system inspection and maintenance activities. The objectives of manhole inspection are, therefore, to determine the proper elevation or grade around the lid, to be sure the lid isn't buried, to examine the structural integrity (look for cracks) of the manhole, and to assess its functional capacity. An indication of the condition of the pipelines coming into a manhole may be gained merely by observing the content and volume of flows from a specific direction.

Every collection system utility should have a routine manhole inspection and assessment program in which operators examine the overall condition of the manhole and check for ease of access, signs of I/I, and evidence of deterioration caused by hydrogen sulfide. Manholes susceptible to inflow should be identified and inspected during periods with a high potential for inflow. When inspectors identify manholes needing repairs and/or rehabilitation, the needs should be prioritized and repairs scheduled.

Use of a computerized data management system simplifies the tracking and scheduling of manhole inspection activities, as well as completion of requested repair, rehabilitation, and/or replacement activities. If a computerized data management system is not available, similar records can be created manually.

14.353 Sewer Cleaning Related to I/I Reduction

Cleaning sewers prior to evaluating the system may be helpful in terms of making evaluation efforts easier and increasing the accuracy of the results. Some agencies clean sewers prior to flow monitoring to improve the accuracy of the flow measurements. Some sewers are cleaned prior to internal TV inspection to better reveal the condition of the sewer, but cleaning prior to TV inspection may depend on the condition of the sewer and the need for cleaning.

14.354 Internal TV Inspection

Internal inspection using closed-circuit TV (CCTV) is conducted to verify the existence and precise location of defects found by other inspection methods and to analyze the nature and extent of the problem. For example, collection system agencies use CCTV to identify defects caused by structural problems, infiltration, lateral connections, the condition of the sewer, and operational problems caused by roots, grease, and debris. Once operational problems are identified, preventive maintenance tasks and frequencies are established to prevent overflows and other problems.

Each line section televised should be videotaped in its entirety. If possible, plug each line section while it is being inspected so that only leakage and the lateral flow will be present in the line to be observed. The invert of the pipe will be visible and any sags or depressions in the bottom of the pipe also will be readily visible.

The CCTV operator must distinguish between service connection leakage around the tap, thimble, or saddle versus leakage in the service line beyond the connection. If the service is running, check the water meter, ask the occupant if water is in use, or wait a sufficient time to determine usage or leakage. These are necessary requirements to recommend the correct rehabilitation method and to provide further assessment of public versus private sector I/I sources.

Make a written record of all defects observed during the television inspection, including house connection leaks, infiltration points, pipe corrosion, broken pipe, crushed pipe, collapsed pipe, offset joints, or other observed defects. If possible, quantify each defect using measurements or severity level ratings.

Leaking service laterals that were identified during the sewer main television inspection can be individually inspected by television to determine the origin of the flow using a lateral inspection system (LIS). A main line television camera is used to position the LIS camera launcher. The actual footage inspected may vary depending on several factors, such as the condition of the lateral within the main sewer, the condition of the lateral, the location of bends and other fittings within the lateral, roots within the lateral, and other limitations of the lateral camera launcher itself. A mini push camera can also be used to inspect laterals from cleanouts or other access points.

14.355 Smoke Testing and Dyed Water Flooding

Smoke testing and dyed water flooding (rainfall simulation techniques) are performed by most collection system agencies as part of a physical survey conducted when a collection system subarea has been identified as having excessive inflow and infiltration. Smoke testing is commonly performed as part of an SSES program. The dyed water flooding program is used to identify suspected sources (indirect connections) of inflow and infiltration into the system when smoke testing yields inconclusive results.

Smoke testing is intended to detect specific inflow points such as storm sewer cross-connections and point source inflow leaks in drainage paths or ponding areas, roof leaders, cellar, yard, and area drains, foundation drains, abandoned building sewers, and faulty service connections. Smoke generators are used to generate the nontoxic, odorless, nonstaining smoke, and a blower system is used to force the smoke into the main line sewer at the manholes. Sand bags or plugs can be used to block off the sewer sections. Pictures and videotapes are taken of smoke coming out of the ground, catch basins, pipes, and other sources during the test.

Smoke testing usually is not conducted on rainy days or when groundwater conditions would interfere with the testing. Observations regarding each leak identified are documented on a smoke sketch. The smoke sketch should include relevant information such as the address, sewer line identification, personnel conducting the test, and test data. Also include a schematic layout of the manhole and surrounding area and the sewer line being tested.

Notify the fire department in advance of any planned smoke testing, and advise residents in the nearby area of the dates when testing will occur and what they can expect to see.

Dyed water testing is another inspection technique for locating inflow sources. Dyed water testing is performed on questionable or suspected inflow sources in order to positively identify the source and measure the amount of flow.

Public sector dyed water flooding is conducted through the injection of a large volume of water, colored by a nontoxic dye, into a designated catch basin or storm drain section that crosses or is in close proximity to a sanitary sewer. Priority is given to storm drains that tested positive during smoke testing.

Private sector dyed water tracing is conducted by introducing a small quantity of liquid dye concentrate into suspected inflow sources, such as downspouts, area drains, patio drains, window well drains, stairwell drains, and driveway drains, and then introducing a sufficient volume of clean water to locate the source's discharge point. During each tracing, monitor the sanitary sewers located downstream of the suspected sources for signs of dyed water. The quantity of dye concentrate and water used will vary depending on the pipe size and the quantity of flow and debris in each line section. Test results can be documented by a combination of handwritten log sheets, sketches, and photographs.

A data management system should be used to track smoke testing and dyed water flooding activities. Also, the utility should develop and make available to operators a document that describes in detail the testing procedures that are to be followed when performing these tests. The document should include any standard forms the crews will be expected to use to record their smoke testing and dyed water flooding work.

14.36 Rehabilitation

14.360 Rehabilitation Considerations

Sometimes an agency should consider rehabilitation instead of preventive maintenance and inspection as the solution to certain problems. Generally this decision is based on such factors as the increased risk of regulatory noncompliance, reduced level of service, or cost-effectiveness of preventive maintenance versus rehabilitation over the remaining life of the sewer. A formal evaluation process based on accurate data is needed to make an informed decision.

Where rehabilitation is necessary to correct a sewer problem, the agency must have in place as part of their O & M program procedures that can be used to ensure the best technique is selected and proper quality control is maintained.

Once an inspection of the sewer segment to be rehabilitated has been completed, analyze the videotape and logs. Perform quality checks (ensure line was plugged, compare the length of the line to sewer map, confirm tape findings, and be sure addresses are noted) and review the maintenance histories of the components. Then review the videotape and add any needed comments or corrections to the log.

In analyzing the videotapes, first classify the nature of the defects and identify the basic options available for rehabilitating the main line sewers, connections, and manholes. Once all of the suitable rehabilitation options have been identified, select the best option, secure adequate funding to complete the rehabilitation, and implement the rehabilitation program.

14.361 Main Line and Manhole Repairs

Main line repairs used by agencies include spot repairs, lining rehabilitation techniques, or entire pipe replacement. Rehabilitation techniques used for manhole repairs include manhole inserts, spray linings, cured-in-place linings, and replacement of the entire manhole.

14.37 Service Laterals

Service laterals or building sewers connect a building's internal wastewater drainage system (plumbing) to the larger street sewer. Technically, a building sewer may connect with a lateral sewer, a main sewer, or a trunk sewer, depending on the layout of the system. (All of these sewers are parts of the sewer in the street.) The building sewer may begin immediately at the outside of a building or some distance (such as 2 to 10 feet) from the foundation, depending on local building codes. Where the sewer officially begins marks where the building plumber's responsibility ends and the collection system operator's responsibility starts for maintenance and repair of the system. Jurisdiction usually changes at the property line.

Most agencies have a policy that the service lateral is the responsibility of the homeowner up to the property line. The agency should have standards of construction for service laterals that must be followed by plumbers, builders, and contractors. The agency should also have a written procedure for the approval and inspection of newly constructed service laterals. Illegal tap-ins may be discovered by a formal program performed by the utility or they may be discovered by a CCTV crew performing routine inspections. Service lateral I/I should be evaluated as part of a utility's sewer system evaluation, but service lateral repairs are typically the responsibility of the homeowner.

14.38 Maintenance Activities in Right-of-Ways

In urban areas, many utilities must perform scheduled maintenance in right-of-ways and easements located either under public streets and roads or on privately owned properties. In some areas, the utility also must have a program for maintaining access roads in remote, rural areas, such as canyons, where maintenance vehicles must be able to drive along the sewer lines for maintenance and repair purposes.

14.39 Compliance

The ultimate goal of every wastewater collection system utility is to be in compliance with regulatory requirements at all times. Progress toward this goal can be measured in terms of the total number of violations issued annually to the utility by the regulatory agency. The total cost of fines paid annually is a measure of the severity of the violations.

QUESTIONS

Write your answers in a notebook and then compare your answers with those on page 653.

14.3J What is the purpose of long-term flow monitoring?

14.3K What is the purpose of instantaneous flow monitoring?

14.3L What is smoke testing intended to detect?

14.4 SPECIAL ELEMENTS OF A CMOM PROGRAM

14.40 System Inventory

Effective CMOM programs are based on knowing what components make up the collection system, where they are located, and the condition of the components. To prepare a system inventory, begin with a description of the collection system. First, describe the collection system in terms of the service area in square miles and the population served. Then, describe the collection system on the basis of miles of gravity sewers, miles of force mains, number of manholes, number of pump stations, number of siphons, and numbers of air, vacuum, and/or air/vacuum relief valves. Also, a description of the condition of each of these facilities as good, fair, or poor would be helpful in planning a CMOM program. The number of service connections should be specified for residential, commercial, and industrial facilities. The utility's responsibility for laterals should be listed, for example, "at main line connection," "from main line to property line or easement/cleanout," or "beyond property line/cleanout to building foundation (or other location)."

The type of system is important; for example, is the system a separate sanitary sewer or are portions of the system a combined sewer? The average annual precipitation is critical information. The system flow characteristics should be described in terms of average daily flow, peak dry weather flow, and peak wet weather flow.

The age distribution of the components of the collection system is important information for planning O & M activities. For example, how many miles of gravity sewers, how many miles of force mains, and how many pump stations are between 0 and 25 years old, 26 to 50 years, and older than 51 years? The age breakdown used may depend on when the utility switched from installing vitrified clay pipe to plastic pipe because the maintenance requirements are different for different types of pipe materials.

The size distribution of the pipes in the sewer system can be described by the miles of gravity sewers and miles or feet of force mains 8 inches in diameter or less, 9 to 18 inches in diameter, 19 to 36 inches, and greater than 36 inches. The materials used for gravity sewers should be described as miles of vitrified clay pipe, miles of plastic pipe, and miles of other pipe materials. The materials used for force mains can be described in terms of miles or feet of reinforced concrete pipe (RCP), miles or feet of high density polyethylene (HDPE), or miles or feet of other pipe material.

14.41 Complaints and Public Relations

Complaints can be a valuable asset in obtaining public support and acceptance and also pinpointing problems. Customer calls are frequently your first indication that something may be wrong in your collection system. Responding to complaints and inquiries promptly can save the utility money and staff resources and minimize the number of customers who are inconvenienced. Customer education also can greatly reduce complaints about utility services. Information brochures, utility bill inserts, and other educational tools help to inform customers and avoid future complaints.

To ensure that all complaints and inquiries receive prompt attention, develop a standard procedure for receiving and responding to them and train all utility staff members to use the procedures. Typically, the person receiving the complaint records the relevant information on a standard complaint record form and a work order is created. Work orders are then prioritized and staff are assigned to respond and solve the problem. Once the matter has been resolved, the responding staff member records what steps were taken. A supervisor or manager routinely reviews and evaluates the effectiveness of the response and solution to the problem.

A good public relations program will help gain public support for your CMOM program and the funding for the program. The first step in organizing an effective public relations campaign is to establish objectives. The only way to know whether your program is a success is to have a clear idea of what you expect to achieve—for example, improve customer relations, protect public health by minimizing the number of SSOs, and enhance organizational credibility. Each objective must be specific, achievable, and measurable.

All utility staff members, both office and field personnel, should be trained in public relations. It is also important to know your audience and tailor various elements of your public relations effort to specific groups you wish to reach, such as community leaders, school children, or the average customer. Your objective may be the same in each case, but what you say and how you say it will depend on your target audience. Always be sure to notify the public before starting major construction and/or maintenance work, and communicate and coordinate major utility activities, such as repaving or resurfacing of streets, with the appropriate municipal departments.

14.42 Communications and Notifications

Your water pollution control regulatory agency will have specific rules and regulations regarding notifications after an SSO. Be sure to contact your regulatory agency for expected reporting procedures and compliance requirements in your *NPDES PERMIT.*[10] A good procedure is to notify your regulatory agency whenever an SSO occurs. The remainder of this section will outline the types of notifications and the groups of people that should be notified if an SSO occurs. However,

---

[10] *NPDES Permit.* ***N**ational **P**ollutant **D**ischarge **E**limination **S**ystem permit is the regulatory agency document issued by either a federal or state agency which is designed to control all discharges of potential pollutants from point sources and storm water runoff into U.S. waterways. NPDES permits regulate discharges into navigable waters from all point sources of pollution, including industries, municipal wastewater treatment plants, sanitary landfills, large agricultural feedlots and return irrigation flows.*

don't wait until an SSO or other problem occurs to get your message out. Your utility agency should communicate regularly with interested parties on the implementation and performance of your CMOM program. The purpose of communication is to encourage and to allow interested parties to provide input to your utility as your CMOM program is developed and implemented.

Communications can include public education as well as public notification and public involvement that seeks broad public input before major proposals are developed and at key points during proposal development and implementation. As projects or programs are proposed and implemented, the utility should invite interested parties and stakeholders (people or groups directly affected by a project) to meetings and workshops to present information on the scope of a project or program in a way that citizens and other pertinent government agencies can comprehend. This process may be time consuming at the start, but should help identify and address problems, concerns, or conflicts before resources are spent. Such a process, if properly performed, can increase public support of your utility's projects from start to finish, including support for providing the funding necessary to pay for the project or program.

When an SSO occurs, immediately notify the public, health agencies, drinking water suppliers, and other affected entities if the SSO has the immediate potential to substantially endanger public health. The notification should include the location of the overflow, estimated volume of the overflow, and identification of any receiving waters that have been or may be affected by the overflow. Prompt and effective notification of members of the public who potentially could be exposed to pathogens in an overflow is necessary to reduce actual exposure. Depending on the nature and location of the overflow, prompt notification of agencies that take steps to mitigate (lessen) health risks and minimize the effects of the overflow is equally important. Rapid and effective notification allows these agencies to take the appropriate steps necessary to reduce public exposure, mitigate other impacts, and assist in a response.

Contact your regulatory agency and review your NPDES permit for expected procedures, promptness of notification, and entities to be notified following different levels or categories of SSOs and CSOs.

QUESTIONS

Write your answers in a notebook and then compare your answers with those on page 653.

14.4A How should the collection system flow characteristics be described?

14.4B How can a collection system utility ensure that all complaints and inquiries receive prompt attention?

14.43 Emergency Response

14.430 Types of Emergencies

Emergencies in utilities are incidents that require an immediate response to the potential consequences that can affect public health and safety, property, and the environment. Collection systems are subject to a wide variety of emergency situations. These can range from recurring incidents to much more severe emergencies that are the results of human events or natural disasters. Some of the emergencies that can occur in collection systems are:

- Stoppages and overflows;
- Power failures in pump stations;
- Major equipment or system failures in pump stations;
- Hazardous or toxic waste spills from transport vehicles or industrial sources;
- Work accidents, such as confined space accidents;
- Fires and explosions;
- Natural disasters such as earthquakes, floods, tornadoes, and blizzards; and
- Collapses and/or failures of force mains and main line gravity sewers.

Emergencies can be classified as "normal" or "extraordinary" emergencies, depending on the severity and the effect on the utility and the community. Normal emergencies, such as main line stoppages and overflows, may occur with a regular frequency and are routine events. They are also somewhat predictable; we know they will happen, we just don't know when. The O & M staff must be prepared to respond to normal emergencies, usually with internal resources that include appropriate procedures, spare parts, repair materials, equipment, and trained personnel.

Extraordinary emergencies are those not classified as normal or routine events. While they are somewhat predictable, they occur with less frequency and can have a greater impact on public health and safety and/or the environment. Earthquakes, floods, and hazardous or toxic spills are examples of extraordinary emergencies. Emergencies caused by natural events tend to affect wider geographic areas, there are fewer aspects we can control, and they often affect other services and utilities. Managing such emergencies usually requires resources outside the utility and involves more planning and response coordination than other "normal" emergencies.

An effective preventive maintenance program will minimize routine emergencies by reducing stoppages and overflows and maintaining system and equipment functionality. In addition, emergency response planning that is part of the preventive maintenance program will help reduce the effects of extraordinary emergencies caused by natural disasters and other events.

14.431 Objectives of Emergency Management

The objectives of emergency management are to:

- Develop an effective response to system failures,
- Minimize SSOs,
- Comply with regulatory requirements,
- Ensure the public health and safety, and
- Protect the environment.

14.432 Planning Ahead

An emergency response plan is the foundation of emergency management. The plan should be based on comprehensive, up-to-date information and analysis of the collection system. The emergency response plan also should include detailed operating procedures to be used when responding to all types of emergencies. Using historical and statistical data, experience, and common sense, the emergency response planning process should answer the following questions:

- What types of emergencies are likely to occur, when, and how often?
- What are the consequences of the emergency, who and what will be affected, and how severely?
- What response methods need to be developed for the different types of emergencies?
- What equipment and personnel are required to respond to the emergencies?

A list of the names and telephone numbers of key personnel with other utilities (communications and power, for example) who can assist in responding to emergencies should also be prepared.

14.433 Homeland Defense

Wastewater collection systems have been identified as a potential target for international and domestic terrorism. This knowledge, coupled with the responsibility of the utility to provide a safe and healthful workplace, requires that management establish rules to protect the staff as well as the collection system. Emergency action and fire prevention plans must identify what steps need to be taken when the threat analysis indicates a potential for attack. These plans must be in writing and they should be exercised periodically so that all staff know what actions to take. (Also see Chapter 11, Section 11.10, "Homeland Defense," for detailed information about threat level categories and suggested measures that can be taken to increase the security of wastewater collection systems and treatment facilities.)

Some actions that should be taken at all times to reduce the possibility of a terrorist attack are:

- Discourage public parking near critical buildings to eliminate the chances of car bombs,
- Be cautious with suspicious packages that arrive,
- Be aware of the hazardous chemicals used at the facility and know how to defend against spills,
- Keep emergency numbers posted near telephones and radios,
- Patrol the collection system and facilities frequently, looking for suspicious activity or behavior, and
- Maintain, inspect, and use personal protection equipment (PPE) such as hard hats and respirators.

14.44 Overflow Emergency Response Plan

Every wastewater collection system utility must have an overflow emergency response plan containing detailed procedures for responding to SSO and CSO events. The plan should describe the utility's planned options for response, remediation (cleanup), and notification measures under different possible SSO events. Work cooperatively with affected stakeholders and government agencies to develop an emergency response plan which will ensure that the utility is adequately prepared to respond to SSO and CSO events. Also set up a procedure for periodically reviewing and updating the plan.

Your utility's overflow response plan should identify procedures for responding to a wide range of potential system failures. The following paragraphs briefly describe critical elements that should be included in an overflow emergency response plan.

Identification of SSOs and CSOs

This part of the overflow emergency response plan describes strategies and procedures for receiving and dispatching information in a wide range of potential system failures. Details that should be provided include a description of the role of each participant in the response, beginning at the time a complaint or report is received and continuing through the satisfactory response to the incident.

Immediate Response and Emergency Operations

This section of the overflow emergency response plan describes strategies for responding to a wide range of potential system failures to:

- Mitigate the impact of the SSOs and CSOs as soon as possible by mobilizing labor, materials, tools, and equipment to investigate reported incidents, and
- Document the findings and response.

Immediate Notification to the Public, Health Agencies, Other Affected Entities, and the NPDES Authority

A utility's overflow emergency response plan should provide a detailed description of how the utility will notify the public, health officials, and other affected utilities/entities of overflows that may imminently (in the very near future) and substantially endanger human health. Contact your NPDES authority and review your NPDES permit for details on the types of SSOs and CSOs that must be reported and the required types of notifications, time frames for responses, and methods of notification.

Training, Distribution, and Maintenance of Plan

The overflow response plan must contain provisions to ensure adequate training for appropriate staff. At a minimum, provide the following information:

- How the plan will be distributed or otherwise made available to staff responsible for implementing the plan,
- Training procedures for appropriate staff, including the frequency of the training activities, and
- The process for reviewing and updating the plan.

14.45 Design and Performance of the Collection System

Many defects in sewers that contribute to I/I are caused by poor design and improper construction in both newly constructed and rehabilitated sewers. An effective CMOM program should ensure that new sewers are properly designed and constructed to prevent permanent system deficiencies that could create or contribute to future overflow events and/or operation and maintenance problems. The CMOM program should also ensure that all work on major rehabilitation and repair projects is performed correctly to minimize future problems. A utility's CMOM program should include provisions to ensure that:

- Requirements and standards are in place for the installation of new collection system components and for major rehabilitation projects, and
- Procedures and specifications are implemented for inspecting and testing the installation of new sewers, pumps, and other appurtenances and for rehabilitation and repair projects.

It is the utility's responsibility to oversee and inspect new sewers and major rehabilitation/repair projects associated with service connections and laterals and private collection systems.

14.46 Monitoring, Measurement, and Program Modifications

Accurate sewer performance information is an important part of a utility's CMOM program. This information is critical for a utility to improve collection system performance and is a core value of an asset management program. Utilities must monitor the implementation and measure the effectiveness of the elements of their CMOM program.

Performance indicators are used to describe and track the implementation of various aspects of a CMOM program and to quantify and document the results and effectiveness of control and O & M efforts. Performance indicators also can be used to measure and report progress toward achieving goals and objectives and to guide management activities. Typical performance indicators include:

- Stoppages per 100 miles of sewer,
- SSOs per 100 miles of sewer,
- Complaints or service requests per 100,000 population, and
- Response time to service requests.

For additional information on performance indicators for collection systems, see *COLLECTION SYSTEMS: METHODS FOR EVALUATING AND IMPROVING PERFORMANCE*, in this series of operator training manuals, available from the Office of Water Programs at California State University, Sacramento. See Figures 14.1, 14.2, 14.3, 14.4, and 14.5 on pages 646–648 for the performance of some very good collection system agencies.

14.47 System Evaluation and Capacity Assurance

A critical function of a collection system is to provide adequate capacity for wastewater flows. If peak flow conditions contribute to an SSO or CSO discharge, the utility should develop a plan to evaluate the collection system and ensure adequate collection system capacity for expected current flows and anticipated flows in the near future. The capacity needs of a collection system change as the system ages, new connections are made, and existing connections change their water usage. Capacity problems can arise under a number of circumstances, including when:

- Service demands in part of the system are too great. Excessive service demands occur when new connections exceed the system's reserve capacity;
- I/I increases as the system ages; and
- The capacity of the system decreases due to factors such as the formation of solids deposits and other partial blockages, increases in the roughness of pipes, or loss of pump capacity.

A utility's CMOM program should develop and implement a program to assess the current capacity of the collection system. The program should identify reserve capacity, hydraulic deficiencies, and capacity needs for an effective asset management program. The capacity assessment program should ensure that procedures exist and are implemented for:

- Determining whether adequate capacity exists in downstream portions of the collection system that will receive wastewater from new connections, and
- Identifying existing capacity deficiencies in the collection system.

Structural and hydraulic problems can be closely related. Minor defects can lead to structural problems in specific soil conditions when a sewer becomes *SURCHARGED*[11] because of insufficient hydraulic capacity. A cycle of exfiltration and infiltration can occur that causes fine soil particles to migrate into the sewer, reducing lateral support from the soil. This can lead to the collapse of the sewer. Many of the techniques used to identify structural defects also provide information on hydraulic performance, such as excess sediment, debris, roots, open joints, and misaligned joints.

14.48 Program Audit Report

CMOM rules and regulations may require you to prepare and submit a program audit report. Preparing the audit report can be an excellent opportunity for one or more colleagues to visit your utility, review your CMOM program, and share information about common problems and cost-effective solutions to these problems. Critical elements of a program audit could include:

- Interviews with collection system managers,
- Field inspection of equipment, resources, and procedures,
- Interviews with field personnel and first-level supervisors,
- Observations of field crews, and
- Review of pertinent records and information management systems.

Based on an evaluation of information from these sources, write an audit report that discusses:

- The findings of the audit, including deficiencies,
- Documentation of steps taken to respond to each finding in the report, including steps taken to correct each deficiency, and
- A schedule for additional steps to respond to findings of the report.

---

[11] *Surcharge. Sewers are surcharged when the supply of water to be carried is greater than the capacity of the pipes to carry the flow. The surface of the wastewater in manholes rises above the top of the sewer pipe, and the sewer is under pressure or a head, rather than at atmospheric pressure.*

The audit report does not need to cover system-wide flow monitoring, SSESs, or physical inspections. These types of activities should be part of the CMOM program and addressed in the context of the system evaluation and capacity assurance plans and also the CMOM measures and activities.

14.49 Annual Report

Every wastewater utility should prepare an annual report. Many successful utility managers share the view that this report gives the utility a valuable opportunity to inform the public and the media about the annual accomplishments of the utility. The annual report could contain information about construction projects started and/or completed during the year, new equipment purchased, annual operating costs and revenues, and capital improvements planned for the future. This is also an opportunity to brag about the accomplishments of staff, such as promotions, certification achievements, awards, and papers presented at technical conferences.

Contact your local and state regulatory agencies and pollution control agency to find out what information they require you to submit. They may require more or different information than you would normally include in your annual report; for example, they may require a report on SSOs and historical trends in SSOs.

QUESTIONS

Write your answers in a notebook and then compare your answers with those on pages 653 and 654.

14.4C What are some of the emergencies that can occur in collection systems?

14.4D How can an effective preventive maintenance program minimize routine emergencies?

14.4E What should be the basis for an emergency response plan?

14.4F What information should be included in a wastewater collection system utility's overflow emergency response plan?

14.4G What are typical performance indicators for a CMOM program?

14.5 INFORMATION MANAGEMENT STRATEGY

14.50 Cost-Effective Results

Let's examine the following five figures. A close look at these figures reveals that a few collection system utilities have achieved a very high level of performance (low number of SSOs) at a very low O & M cost. How do they do it? These utilities have a very effective information management strategy. They collect information that focuses on trouble spots and/or potential trouble spots (SSOs), analyze the information, and respond to their analysis in terms of preventing or eliminating the causes of SSOs.

14.51 Purpose of Data Management

Effective management of wastewater collection systems depends heavily on accurate, readily available data. One of the major problems managers have is that data are not available to make informed O & M decisions. In some cases the information simply is not collected or tracked; in other cases, the data are not easily accessible or are recorded in a format that makes the data difficult to analyze.

14.52 Benefits

Collection system agencies that perform well characteristically track system performance indicators and O & M activities

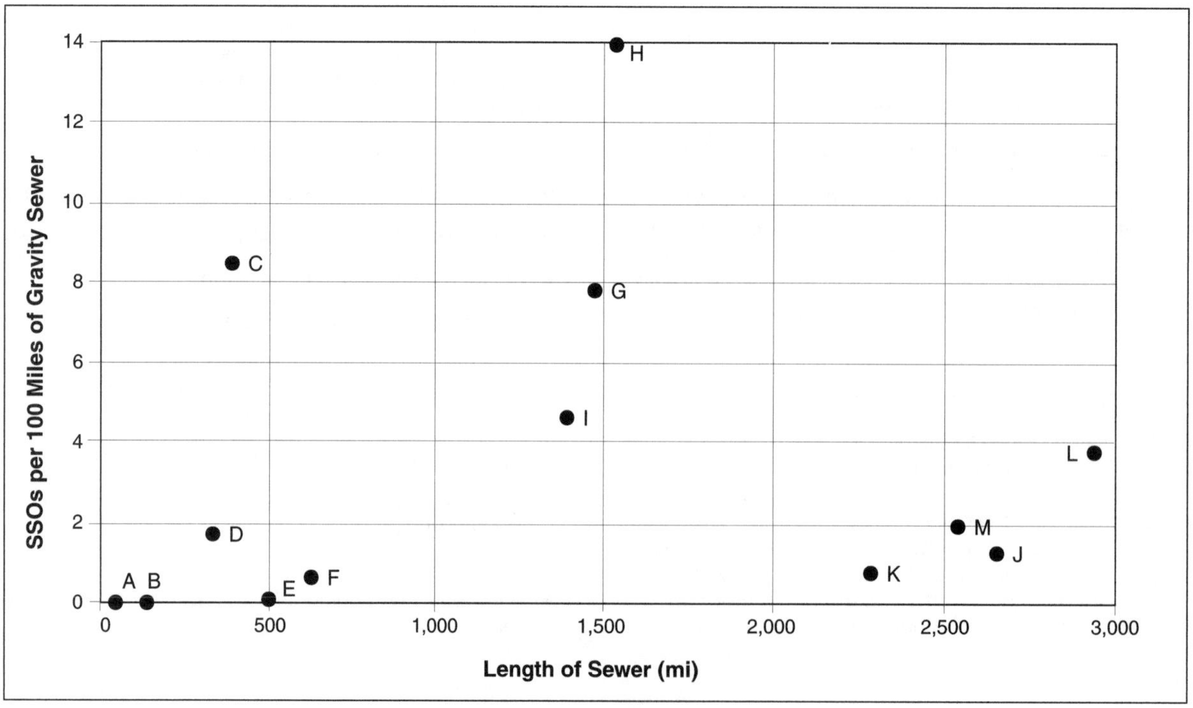

Fig. 14.1 SSOs per 100 miles of gravity sewer

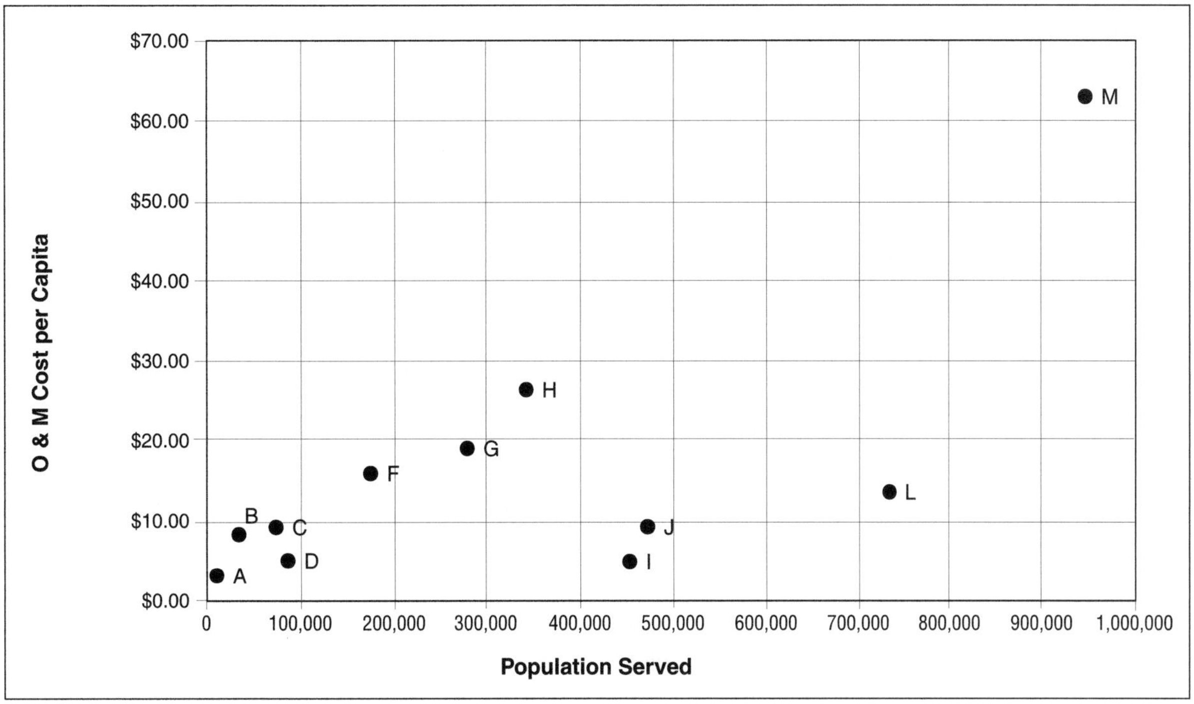

Fig. 14.2 Annual O & M cost per capita

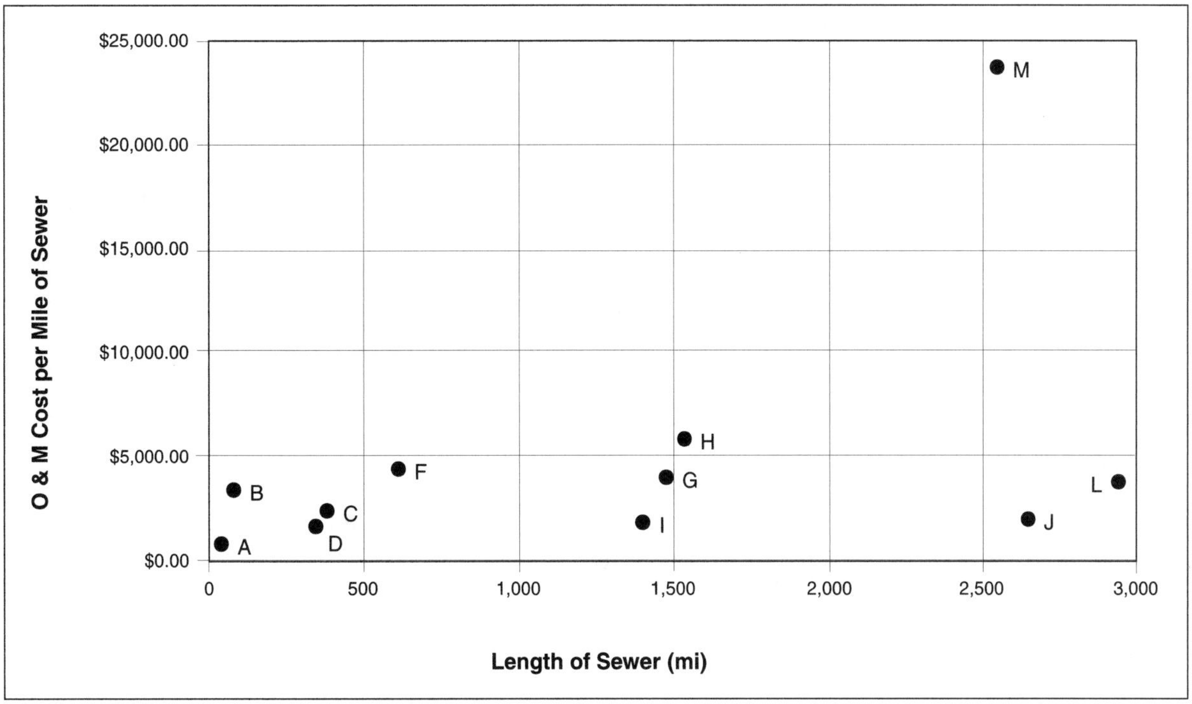

Fig. 14.3 Annual O & M cost per mile of sewer

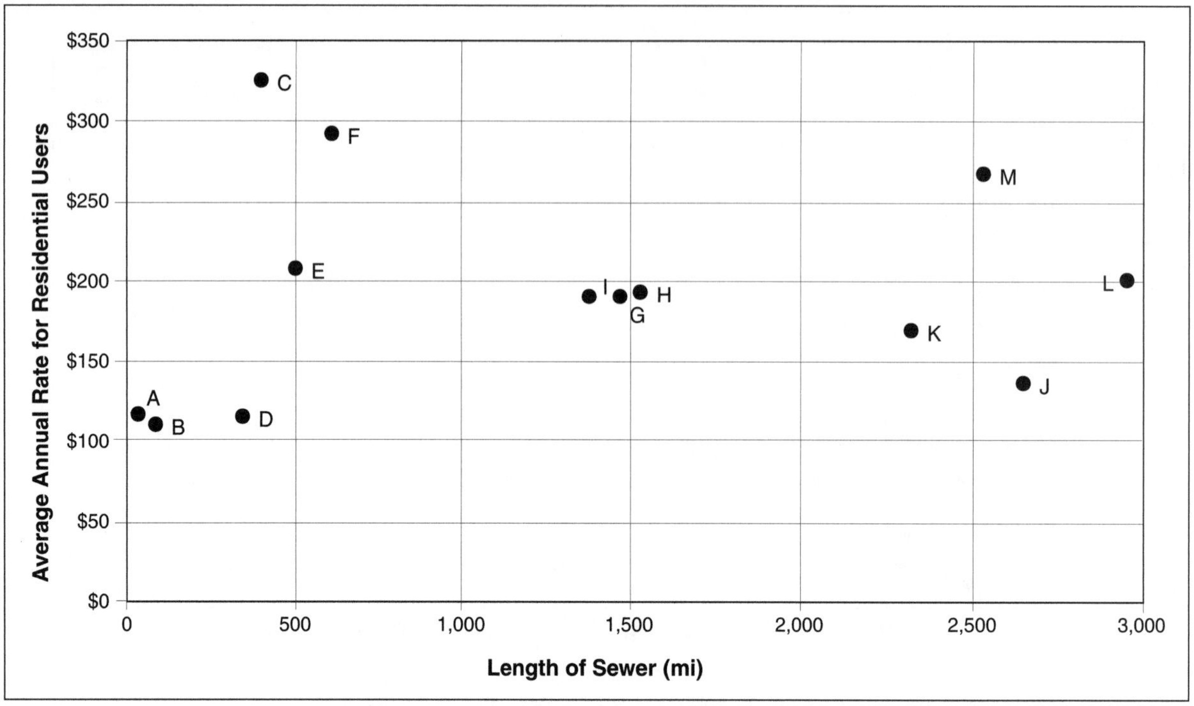

Fig. 14.4 Average annual rate for residential users

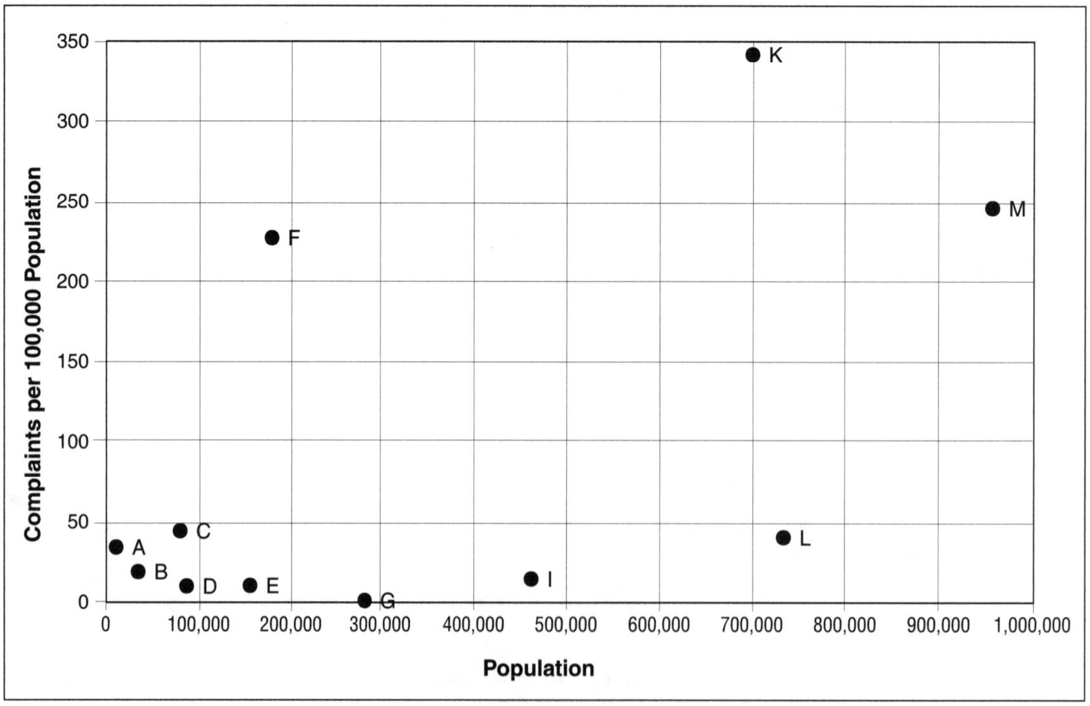

Fig. 14.5 Complaints per 100,000 population

and use this information to make decisions about collection system O & M. In many cases, managers use the collected data to track trends in critical performance indicators, for example, increases or decreases in the number of stoppages over time, or increasing or decreasing O & M costs.

14.53 Methods

With the advent of CCTV inspection, we now have detailed information on the condition of the collection system. When PM (preventive maintenance) is planned and scheduled, information is needed on equipment or system attributes, maintenance frequency, materials, and supplies. Financial data are necessary to determine expenditures and costs of collection system O & M. As-built (record) drawings and O & M manuals contain large amounts of information. Computer-based maintenance management systems (CMMSs) are designed to manage data needed to track collection system O & M performance. Geographic information systems (GISs) are computer-based mapping systems that are also used for this purpose.

QUESTIONS

Write your answers in a notebook and then compare your answers with those on page 654.

14.5A How can collection system agencies achieve cost-effective results?

14.5B Why would collection system managers use collected data to track trends in critical performance indicators?

14.6 LEVEL OF SERVICE

14.60 Measurement of Level of Service

Ultimately, collection system management must select a level of service and allocate the necessary funding and resources to achieve a targeted level of service for the CMOM program. Level of service can be measured in terms of:

- Sanitary and/or combined sewer overflows (SSOs and CSOs),
- Backups into homes and businesses,
- Lift station failures,
- Customer service requests and complaints (odors), and
- Response time to emergencies (SSOs).

The level of service can be based on minimizing the number of events and times in each of these five categories.

Level of service should be based on community expectations, regulatory compliance, and willingness to pay for these services. Regulatory agencies and the public expect management to operate a collection system with no overflows, backups, failures, or complaints, and with prompt response to emergencies. The actual target level of service should be a community decision.

14.61 Economic Approach to Selecting Level of Service

Economic concepts can be helpful when selecting a level of service for a collection system. Possible economic analytical approaches would be to consider:

- Maximizing net benefits,
- Minimizing costs, and
- Cost-effectiveness.

Maximizing net benefits requires the ability to measure the benefits from a CMOM program. All of these three approaches require the measurement of costs associated with the CMOM program.

14.610 Maximizing Net Benefits

Benefits from a CMOM program include benefits to the collection system agency and staff, protection of the public health and the environment, and benefits to the community. Economic benefits are measured on the basis of: (1) willingness to pay, and/or (2) the next more costly "minimum cost alternative." Today the accurate measurement of these benefits is very difficult and most agencies do not consider it worth the time and effort to attempt to evaluate and measure these benefits.

14.611 Minimizing Costs

Minimizing costs is an approach that has been used by many collection system agencies. Management attempts to minimize the costs of: (1) operation and preventive maintenance, and (2) emergency and/or corrective maintenance. The total cost of collection system O & M is the sum of the costs under items (1) and (2). As the costs of operation and preventive maintenance are increased, the level of services increases and the costs of emergency and/or corrective maintenance decrease. Agencies attempt to minimize the total cost of items (1) and (2). This concept of selecting the level of service with the minimum cost is illustrated in Figure 14.6, which shows how a collection system agency analyzes its cost records and selects the target level of service and target number of SSOs for the community on the basis of minimizing the total costs to the agency.

Today many collection system agencies know their costs of: (1) operation and preventive maintenance, and (2) emergency and/or corrective maintenance. However, by increasing the costs of operation and preventive maintenance, it is difficult to forecast the increased level of service (reduction of SSOs) and the expected decrease in emergency and/or corrective maintenance. Therefore, management finds it difficult to accurately determine what level of service will actually minimize total costs.

14.612 Cost-Effectiveness

A question collection system management may be asked is, "How much maintenance is too much maintenance?" An examination of Figure 14.6 indicates that many good CMOM programs are probably providing a cost-effective level of service, rather than a minimum cost level of service.

Management and the community have selected the cost-effective approach because of the desire for regulatory compliance and also to protect the public health and the environment. These collection system agencies have increased the operation and preventive maintenance funding and resource allocation to a level where their records indicate that additional funding and preventive maintenance will not produce significant reductions in overflows, backups, lift station failures, complaints, or reduced response times. These agencies are being managed at the cost-effective level of service shown on Figure 14.6. Management does not know the exact shape of the cost curves shown on Figure 14.6, but the cost-effective approach is an efficient method of attempting to reach an appropriate target level of service.

14.62 Proposed Selection of Level of Service

A collection system agency management wants a CMOM program that is in compliance with the requirements of the regulatory agencies and cost-effective from the perspective of the community and the public. The cost-effective concept described in this section associated with the CMOM program in this chapter provides management with a rational approach to selecting target levels of service. Annually management must review its records and costs, consider future expected demands on the collection system, and apply the concepts illustrated in Figure 14.6 to maintain and improve performance in a cost-effective manner.

QUESTIONS

Write your answers in a notebook and then compare your answers with those on page 654.

14.6A What are the regulatory and public expectations regarding the performance of a collection system agency?

14.6B What are possible economic analytical approaches to selecting a level a service for a collection system O & M program?

14.7 REFERENCES, ADDITIONAL READING, AND WEBSITES

The references listed in this section were used to prepare the information in this chapter. These publications are excellent sources of additional information.

1. The following publications are available from the Office of Water Programs, California State University, Sacramento, 6000 J Street, Sacramento, CA 95819-6025, phone 916 278-6142.

   - *COLLECTION SYSTEMS: METHODS FOR EVALUATING AND IMPROVING PERFORMANCE*; price $45.00.
   - *OPERATION AND MAINTENANCE OF WASTEWATER COLLECTION SYSTEMS*, Volumes I and II; price, $45.00.
   - *UTILITY MANAGEMENT*; price, $25.00.

2. *WASTEWATER COLLECTION SYSTEMS MANAGEMENT* (MOP 7). Obtain from Water Environment Federation (WEF), Publications Order Department, 601 Wythe Street, Alexandria, VA 22314-1994. Order No. M05000. Price to members, $56.75; nonmembers, $76.75; price includes cost of shipping and handling.

3. For information on the U.S. Environmental Protection Agency's proposed SSO Rule and CMOM Program, visit the following website: www.cmom.net.

END OF LESSON 2 OF 2 LESSONS
on
CMOM

Please answer the discussion and review questions next.

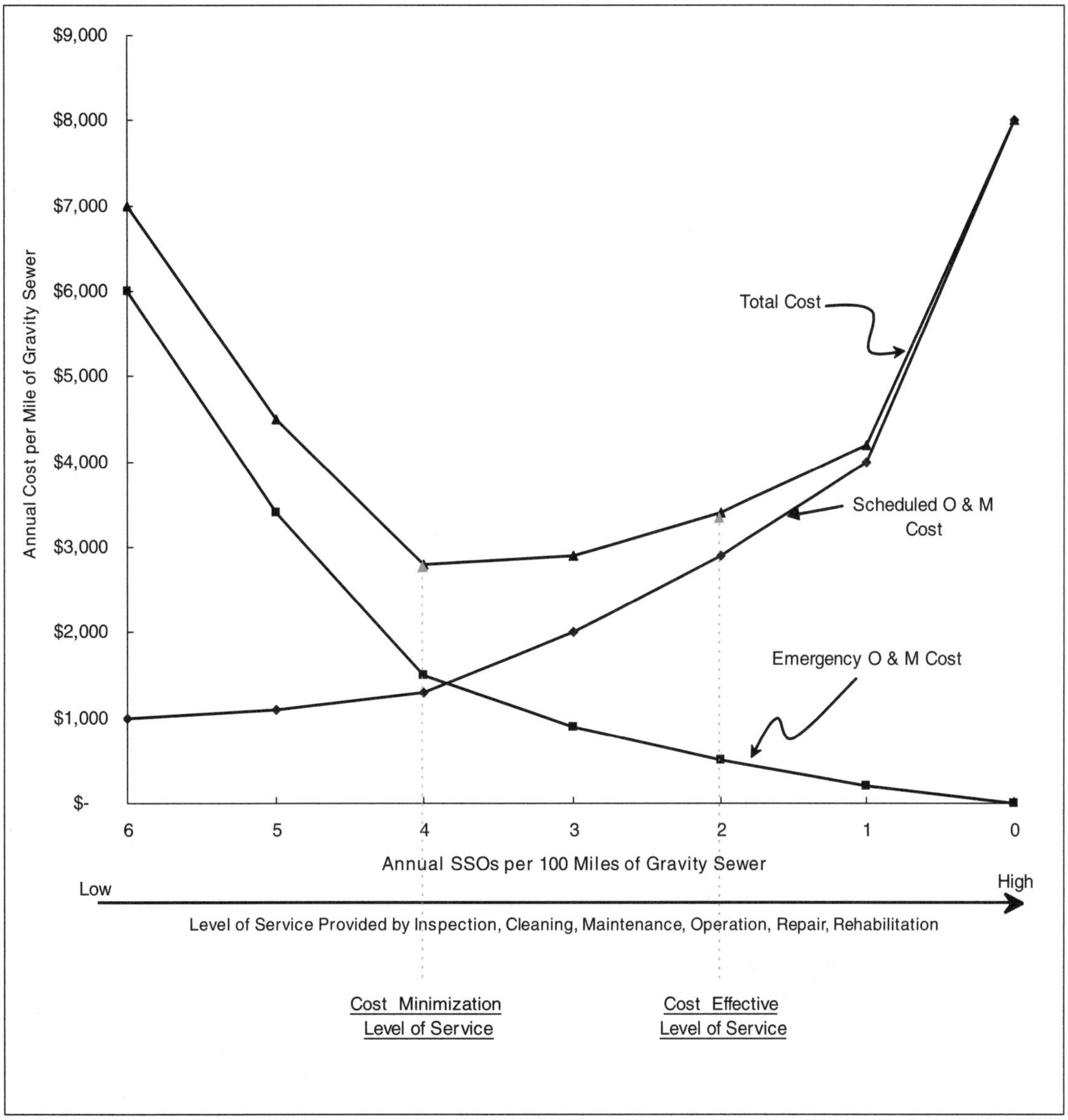

Fig. 14.6 Selection of level of service

DISCUSSION AND REVIEW QUESTIONS

Chapter 14. CAPACITY ASSURANCE, MANAGEMENT, OPERATION, AND MAINTENANCE (CMOM)

(Lesson 2 of 2 Lessons)

Write the answers to these questions in your notebook before continuing. The question numbering continues from Lesson 1.

8. How can lift stations located throughout a community be acceptable to the public?
9. What is the purpose of a sewer system evaluation by O & M staff?
10. Under what conditions should a wastewater collection system agency consider rehabilitation instead of preventive maintenance and inspection as a solution to certain problems?
11. Effective CMOM programs are based on what information and knowledge?
12. What are emergencies in wastewater collection system utilities?
13. What are critical elements of a CMOM program audit report?
14. How do some collection system utilities achieve a very high level of performance (low number of SSOs) at a very low cost?
15. How should a wastewater collection system agency attempt to select a level of service?

SUGGESTED ANSWERS

Chapter 14. CAPACITY ASSURANCE, MANAGEMENT, OPERATION, AND MAINTENANCE (CMOM)

ANSWERS TO QUESTIONS IN LESSON 1

Answers to questions on page 622.

14.0A SSOs and CSOs are overflows from sanitary sewers and combined sewers that discharge untreated wastewater from collection systems.

14.0B The purpose of sanitary and combined wastewater collection systems is to collect wastewater from homes, businesses, schools, public facilities, and industries and transport this wastewater to a treatment plant without exposing the public or the environment to any harmful constituents being transported in the wastewater.

Answers to questions on page 623.

14.1A Poorly performing collection systems create potential health and environmental risks that must be corrected and prevented from occurring again in the future.

14.1B When collection systems are not installed as designed, problems are caused by faulty construction, poor inspection, and low-bid shortcuts.

14.1C The problems caused by water flowing down a ditch containing a sewer include groundwater infiltration into the sewer, possible flotation of the sewer, and structural failure of the sewer or joint.

14.1D Reducing infiltration/inflow is good practice because it allows the collection system to handle greater quantities of wastewater.

Answers to questions on page 627.

14.2A Elements of the CMOM program that must be properly managed include planning, design, construction, inspection, operation, maintenance, repair, and rehabilitation, and also information collecting, analysis, response, and evaluation.

14.2B Watershed planning can be especially effective in enhancing the environment of sensitive and high-exposure areas such as beaches, drinking water supplies, and habitats for endangered species.

14.2C The overall purpose of delegating authority is to allow decisions to be made and actions to be taken at the lowest practical level within the organization. This approach allows decisions to be made by those most familiar with the situation and most directly affected by the decisions.

Answers to questions on page 629.

14.2D Two calculations can be used to help determine how healthy and stable the finances are for the utility. These two calculations are the OPERATING RATIO and the COVERAGE RATIO.

14.2E Information needed by a manager to prepare a good budget is to have good records from the previous year and knowledge of the needs of the utility.

14.2F The goal of capital asset management is to efficiently protect, maintain, or improve the value of the collection system while providing the level of service desired.

Answers to questions on page 630.

14.2G Personnel management programs and activities must be JOB RELATED and DOCUMENTED.

14.2H The purpose of employment policies and procedures is to provide guidance for both management and staff regarding expected duties and performance.

14.2I Management is responsible for the safety of the utility's personnel and the public exposed to the wastewater collection system utility's operations.

Answers to questions on page 634.

14.2J Operators need procedures and guidelines on the proper operation and maintenance of equipment. Operators need procedures and guidelines for equipment start-up and shutdown and for emergency operation of equipment.

14.2K The purpose of a logbook is to provide operators with a location for equipment maintenance procedures and a record of maintenance performed. Logbooks also provide valuable information that will be useful when budgeting funds to replace equipment when it wears out or when newer, more efficient equipment becomes available.

14.2L Engineering and O & M staff work cooperatively to ensure that designs produce facilities that deliver the desired results, perform as intended, and can be effectively and efficiently operated and maintained by O & M staff.

14.2M Removal of I/I sources requires elimination of defects found in the public and private sector sections of the collection system. Repair and rehabilitation are long-term, ongoing components of the O & M program that accomplish this elimination of defects.

14.2N System capacity assurance is a process developed to identify, characterize, and correct hydraulic deficiencies in a collection system.

ANSWERS TO QUESTIONS IN LESSON 2

Answers to questions on page 637.

14.3A The purpose of a collection system O & M program is to keep the system performing as designed and intended.

14.3B Knowledge required to schedule collection system maintenance includes knowledge of the collection system and maintenance procedures, and also the location and identification of specific problems.

14.3C Stoppages in gravity sewers usually are caused either by structural defects or by an accumulation of material in the pipe.

14.3D A proactive sewer maintenance program benefits the agency's asset management program by:

- Lowering O & M costs,
- Extending the useful life of the gravity sewer system, and
- Lowering rehabilitation/replacement costs.

14.3E Hydrogen sulfide can be controlled by proper design of the collection system and by cleaning the sewers to remove slime bacteria.

Answers to questions on page 639.

14.3F The collection system operating agency has the responsibility to ensure continuous and efficient operation and maintenance of the lift stations.

14.3G Public health hazards and pollution of adjacent receiving waters result when untreated wastewater is discharged to the environment.

14.3H When a lift station loses power, the station should be quickly returned to service using on-site generators, portable electric generators, or alternative power sources.

14.3I All lift stations should have telemetry alarm systems that transmit one or more signals indicating the nature of a lift station failure or that the lift station needs to be visited.

Answers to questions on page 642.

14.3J The purpose of long-term flow monitoring is to adequately define I/I quantities.

14.3K Instantaneous flow monitoring is performed at the subsystem level during low flow. The purpose is to isolate line segments with excessive infiltration.

14.3L Smoke testing is intended to detect specific inflow points such as storm sewer cross-connections and point source inflow leaks in drainage paths or ponding areas, roof leaders, cellar, yard, and area drains, foundation drains, abandoned building sewers, and faulty service connections.

Answers to questions on page 643.

14.4A The system flow characteristics should be described in terms of average daily flow, peak dry weather flow, and peak wet weather flow.

14.4B To ensure that all complaints and inquiries receive prompt attention, develop a standard procedure for receiving and responding to them and train all utility staff members to use the procedures.

Answers to questions on page 646.

14.4C Some of the emergencies that can occur in collection systems are:

- Stoppages and overflows,
- Power failures in pump stations,
- Major equipment or system failures in pump stations,
- Hazardous or toxic waste spills from transport vehicles or industrial sources,
- Work accidents, such as confined space accidents,
- Fires and explosions,
- Natural disasters such as earthquakes, floods, tornadoes, and blizzards, and
- Collapses and/or failures of force mains and main line gravity sewers.

14.4D An effective preventive maintenance program will minimize routine emergencies by reducing stoppages and overflows and maintaining system and equipment functionality.

14.4E An emergency response plan should be based on comprehensive, up-to-date information and an analysis of the collection system.

14.4F The overflow emergency response plan should describe the utility's planned options for response, remediation (cleanup), and notification measures under different SSO events.

14.4G Typical performance indicators for a CMOM program include:

- Stoppages per 100 miles of sewer,
- SSOs per 100 miles of sewer,
- Complaints or service requests per 100,000 population, and
- Response time to service requests.

Answers to questions on page 649.

14.5A Cost-effective results can be achieved by having a very effective information management strategy. Collect information that focuses on trouble spots and/or potential trouble spots (SSOs), analyze the information, and respond to the analysis in terms of preventing or eliminating the causes of SSOs.

14.5B Tracking trends in critical performance indicators reveals increasing or decreasing trends and the need to make adjustments in the O & M program.

Answers to questions on page 650.

14.6A Regulatory agencies and the public expect collection system agencies to perform without any SSOs, CSOs, backups, failures, or complaints, and with prompt responses to emergencies.

14.6B Possible economic analytical approaches to selecting a level of service for a collection system O & M program include:

- Maximizing net benefits,
- Minimizing costs, and
- Cost-effectiveness.

APPENDIX

OPERATION AND MAINTENANCE OF WASTEWATER COLLECTION SYSTEMS

(VOLUME II)

Final Examination and Suggested Answers

Collection System Words

Subject Index

FINAL EXAMINATION

VOLUME II

This final examination was prepared **TO HELP YOU REVIEW** the material in this manual. The questions are divided into four types:

1. True-False,
2. Best Answer,
3. Multiple Choice, and
4. Short Answer.

To work this examination:

1. Write the answer to each question in your notebook.
2. After you have worked a group of questions (you decide how many), check your answers with the suggested answers at the end of this exam, and
3. If you missed a question and don't understand why, reread the material in the manual.

You may wish to use this examination for review purposes when preparing for civil service and certification examinations.

Since you have already completed this course, you do not have to send your answers to California State University, Sacramento.

True-False

1. Wet well lift stations with guide bars require operators to enter the wet well to install or remove the pump.
    1. True
    2. False
2. Mechanical seals have stuffing boxes.
    1. True
    2. False
3. The approach to solving any problem must be made with logic and caution.
    1. True
    2. False
4. Never touch electrical panels, controls, circuits, wiring, or equipment unless you are qualified and authorized.
    1. True
    2. False
5. Most alignment takes place when the machines being aligned are hot.
    1. True
    2. False
6. When a pump is fitted with a mechanical seal, it must never run dry or the seal faces will be burned and ruined.
    1. True
    2. False
7. Flows in sewers greater than the expected base flow can be considered to come from inflow and infiltration sources.
    1. True
    2. False
8. Smoke testing is very effective during high groundwater conditions.
    1. True
    2. False
9. Sliplining is difficult and very expensive.
    1. True
    2. False
10. Training and certification of collection system operators reduce the chances of an accident.
    1. True
    2. False
11. Operators must purchase personal protection equipment as a condition of employment.
    1. True
    2. False
12. Lockout/tagout refers to control of hazardous chemicals.
    1. True
    2. False
13. Level of education can be used to screen candidates in place of performance testing.
    1. True
    2. False
14. Union contracts can change a supervisor's delegated authority and responsibility.
    1. True
    2. False
15. Each and every manhole should be assigned a specific, permanent identification number and these numbers should never be changed.
    1. True
    2. False
16. In most organizations, authority is essentially power.
    1. True
    2. False
17. Inspection crews are the same as construction inspectors and building inspectors.
    1. True
    2. False

18. Recordkeeping is a function that computers perform well.
    1. True
    2. False
19. If a pipe joint is watertight, it is roottight.
    1. True
    2. False
20. Planning must be done after the other management functions.
    1. True
    2. False
21. The main objective of sewer cleaning is to prevent blockages and overflows.
    1. True
    2. False

Best Answer (Select only the closest or best answer.)

1. Why are there provisions in some lift stations for wet well aeration during low-flow periods?
    1. To enhance performance of anoxic selectors
    2. To help keep the detained wastewater fresh
    3. To increase hydrogen sulfide production
    4. To maintain corrosion processes
2. A ground-fault circuit interrupter (GFCI) is designed to limit electric shock to a current and time duration value below what level?
    1. That which can cause ignition of clothing
    2. That which can create minor pain sensations
    3. That which can pass through your body
    4. That which can produce serious injury
3. What is the main purpose of a check valve?
    1. To prevent damage from pump cavitation
    2. To prevent operators from checking on pump maintenance
    3. To prevent the force main from draining back into the wet well
    4. To prevent the pump from pumping excess flows
4. What is the most significant factor in determining the frequency of lift station visits?
    1. Attitude of operating agency
    2. Compliance with regulatory permits
    3. Potential for nuisances
    4. Type of wastewater being conveyed
5. What type of hazard can develop when installing battery cables and wiring if they are *NOT* properly connected?
    1. An arc across the unit can occur that could cause an explosion
    2. Equipment can start unexpectedly
    3. Stray electricity can consume available oxygen
    4. Toxic gases can be emitted
6. What is the purpose of grounding electrical equipment?
    1. A backup procedure that prevents power failures
    2. A device that protects groundwater from contamination
    3. A safety precaution that protects operators and the motors, tools, and equipment
    4. A wiring system that keeps the ground moist
7. What is the purpose of pump and motor couplings?
    1. To connect the pump and motor to the piping system
    2. To minimize vibrations from rotating parts
    3. To produce a pair of synchronized devices
    4. To transmit the torque from the motor to the pump
8. What is the purpose of universal joints?
    1. To be used instead of sheaves
    2. To efficiently pump water
    3. To join bearings in stuffing boxes
    4. To take up any misalignment between pump and drive
9. What is the major stumbling block to the implementation of a rehabilitation program?
    1. Determining public response and support
    2. Identifying staff and securing funding
    3. Locating and selecting sites
    4. Obtaining environmental and regulatory approvals
10. What is the main purpose of infiltration/inflow (I/I) evaluations?
    1. To control flooding on roads and streets
    2. To protect public water supplies
    3. To recharge the groundwater table
    4. To reduce the hydraulic loads on the sewer system
11. Under what conditions should in-line replacement of sewer pipes be considered?
    1. When sewers are found in areas with low groundwater tables
    2. When sewers are found to be structurally failing and inadequate in capacity
    3. When sewers are found to be structurally sound and of adequate capacity
    4. When sewers are found under stable soil conditions
12. What skill most clearly demonstrates the professionalism of the collection system operator?
    1. Ability to carry out a work assignment safely
    2. Ability to pass certification examinations
    3. Ability to quickly perform assigned tasks
    4. Ability to recite safety codes
13. What is the duration of a permit-required confined space entry permit?
    1. Time required for crew to complete shift
    2. Time required for supervisor to evaluate conditions
    3. Time required to complete the assignment
    4. Time required to test atmosphere for hazardous conditions
14. Personal protection equipment should be designed to accomplish what task?
    1. Allow ease of movement by workers
    2. Be economical to purchase by employees
    3. Create a barrier against workplace hazards
    4. Prevent occurrence of safety hazards
15. Why must management establish rules to protect the workers and facilities from terrorism?
    1. To inform public that utility is aware of the situation
    2. To prevent terrorists from attempting to attack facility
    3. To provide a safe and healthful workplace
    4. To review current procedures and implement improvements

16. An effective administration will develop what kinds of procedures for the application, selection, orientation, and training of personnel?
    1. Easily understood
    2. Legally appropriate
    3. Simple
    4. Standardized and equitable

17. Why should superintendents encourage collection system operators to receive additional training?
    1. To avoid being terminated
    2. To comply with regulatory guidelines
    3. To improve performance
    4. To receive pay increases

18. What is the advantage of electronically capturing TV inspections images on a 3.5-inch floppy disk, recordable compact disc (CD-R), or a compatible zip disk?
    1. Allows information to be stored on tapes
    2. Creates opportunities for operators to record information
    3. Enhances process of storing information
    4. Permits information to be retrieved and printed at any time

19. What is the most effective method for operators for completing a report?
    1. At the corporation yard between jobs
    2. At the end of the week when all reports can be completed
    3. At the office at the end of the day
    4. At the scene of each job

20. What should be the first priority of a wastewater collection system agency?
    1. Ability to handle emergency service requests quickly and efficiently
    2. Ability to perform regular scheduled preventive maintenance
    3. Develop and implement public relations program
    4. Increase and maintain high collection system operator salaries

21. How can hazardous atmospheric conditions be eliminated in a manhole?
    1. Use a portable blower to ventilate the manhole
    2. Use an atmospheric testing device to test atmosphere
    3. Use an inspection crew to locate and eliminate the source
    4. Use the sewer-use code regulations to force industry into compliance

22. What is the purpose of a risk analysis of each pump station?
    1. To define the likelihood and potential of a prolonged station failure
    2. To estimate the risk of the preventive maintenance program failing
    3. To evaluate the potential for regulatory intervention
    4. To forecast the probability of an extreme event in nature

23. How can wastewater collection systems be designed with inadequate capacities?
    1. Effective water conservation program
    2. Reduced frictional resistance inside pipes
    3. Smoother internal pipe surfaces
    4. Unexpected population growth

24. What is the major cause of corrosion in wastewater collection systems?
    1. Acids
    2. Corrodible materials
    3. Hydrogen sulfide gas
    4. Lack of cathodic protection

25. What information should be contained in collection system O & M manuals? Detailed procedures for
    1. Developing a safety/survival program
    2. Preparing collection system budget
    3. Recruiting additional operators
    4. Start-up, operation, and shutdown

Multiple Choice (Select all correct answers.)

1. Why should collection system operators help engineers design lift stations?
    1. To allow engineers time to design other lift stations
    2. To gain important experience designing lift stations
    3. To improve operation and maintenance of lift stations
    4. To lower operating costs of lift stations
    5. To minimize bypasses and backups caused by lift station failures

2. Why do force mains need air and vacuum release valves? To avoid
    1. Decrease in friction losses
    2. Deposition of solids
    3. Flows reversing direction
    4. Increase in pumping costs
    5. Reduction in pipe capacity

3. What methods of sealing are used in pumps?
    1. Bonding
    2. Mechanical seals
    3. Packing
    4. Seal trode
    5. Wearing rings

4. What items should be inspected when a new pump is started?
    1. Electrical load on motors
    2. Excessive heating
    3. Packing glands or mechanical seals
    4. Unusual noises
    5. Unusual vibrations

5. Preparation of a lift station maintenance program requires consideration of what important factors?
    1. Budgetary considerations
    2. Knowledge gained from experience
    3. Recommendations of equipment manufacturers
    4. Regulatory requirements
    5. Requirements of lift station

6. What problems can be created by unbalanced current?
    1. Excessive heat
    2. Overall poor performance
    3. Overload tripping
    4. Reduction in starting torque of motor
    5. Vibrations

7. Which problems can cause the failure of electric motor insulation?
   1. Contaminants
   2. Excessive starts
   3. Motor overloading
   4. Safety hazards
   5. Voltage imbalance
8. Cavitation is a condition that can cause what types of problems in a pump?
   1. Damage to impeller
   2. Drop in efficiency
   3. Misalignment
   4. Noise
   5. Vibration
9. What items should an operator look for when inspecting a pump impeller?
   1. Cavitation marks
   2. Chips, broken tips, corrosion, unusual wear
   3. Excessive suction lift
   4. Excessive wear on wearing rings on impeller and volute
   5. Tightness on shaft
10. What are the most common methods of coupling a pump and driver together?
    1. Belt-driven or chain-driven
    2. Close coupled
    3. Direct coupling
    4. Flexible shafting
    5. Mechanical seal
11. What factors contribute to the declining integrity (deterioration) of wastewater collection systems?
    1. Materials have lost structural integrity due to corrosion
    2. Population growth may divert effort and attention away from long-term preservation of existing facilities
    3. Stress created by vibrations from construction activities
    4. Surcharged flows accelerate deterioration and allow leakage
    5. Unstable foundation soils or improper pipe bedding material
12. What critical information can be obtained from maintenance repair and inspection reports when preparing an inventory phase of a rehabilitation program? Indication of areas where
    1. Changes and additions have been made
    2. Maintenance is excessive
    3. Problems occur regularly
    4. Repairs are frequent
    5. Safety violations have occurred
13. What are the beneficial results from pipeline replacement?
    1. Correction of misalignment of pipe
    2. Elimination of direct sources of storm water entry
    3. Increase in capital improvement planning (CIP) program funds
    4. Increase in hydraulic capacity
    5. Repair of improper service connections
14. Chemical grouts have failed under what types of circumstances?
    1. Arid regions during periods of low groundwater
    2. Coastal regions subject to influence of tidal fluctuations
    3. Humid regions where atmospheric moisture content is high
    4. Sunny regions with excessive days of sunshine
    5. Tropical regions subject to surcharged conditions
15. What are important characteristics of coatings applied to manholes?
    1. Color coded
    2. Corrosion resistant
    3. Noise suppressing
    4. Odor free
    5. Waterproof
16. Which collection system work areas could be considered confined spaces?
    1. Manholes
    2. Sewer lines
    3. Spare parts warehouses
    4. Traffic routing areas
    5. Wet wells
17. How is the success or failure of a safety/survival program measured? By the number of
    1. Available pieces of safety equipment
    2. Deaths
    3. Disabilities
    4. Injuries
    5. Operators trained
18. The lockout/tagout rule requires what items to be included in an energy control program?
    1. Documented energy control procedures
    2. Employee training program
    3. Legible signature on tag
    4. Minimization of energy usage
    5. Periodic inspection of the procedures
19. What items must be included in an employee training program before doing work that requires use of personal protection equipment?
    1. How personal protection equipment is to be worn
    2. Limitations of the personal protection equipment
    3. Proper care, maintenance, useful life, and disposal of the personal protection equipment
    4. What type of personal protection equipment is necessary
    5. When personal protection equipment is necessary
20. What type of protection must be provided by hard hats?
    1. Absorb the shock of a blow
    2. Assist with clear thinking during emergencies
    3. Protect against electric shock
    4. Provide face protection
    5. Resist penetration

21. What actions should be taken to reduce the possibility of a terrorist attack?
    1. Be aware of the hazardous chemicals used and know how to defend against spills
    2. Be cautious of suspicious packages that arrive
    3. Ensure all visitors sign in and out of the facilities with a positive ID check
    4. Keep emergency numbers posted near telephones and radios
    5. Patrol the facilities frequently, looking for suspicious activity or behavior

22. An effective collection system administration will ensure an operation and maintenance program that will keep a wastewater collection system functioning in what manner?
    1. At top efficiency
    2. Excessive sewer-use charges
    3. Maximize useful life
    4. Minimize costs
    5. Unhappy customers

23. How should a wastewater collection system agency attempt to prevent public health hazards?
    1. Assign scheduling priority to stoppages, overflows, and exfiltration that contaminate the environment
    2. Follow approved confined space entry and work procedures at all times
    3. Provide immediate, concerned, and efficient service to all emergency calls
    4. Route traffic around work sites in streets and highways
    5. Use clean water to flush contaminated surfaces and disinfect with chemicals approved by the health authority

24. Why is certification of wastewater collection system operators important?
    1. Creates an operator certification program
    2. Eliminates need for safety program
    3. Potential for being paid a higher salary
    4. Recognition of importance of good operation and maintenance
    5. Recognition of importance of protecting health and safety

25. Information on collection system maps should include the locations of what components?
    1. Building sewers
    2. Cleanouts
    3. Lift stations
    4. Main, trunk, and interceptor sewers
    5. Manholes

26. What are the key factors involved in the development of an effective management information system?
    1. All aspects of the information system must be performed in a timely manner
    2. Managers must review all information before it is entered into the system
    3. Planning and controlling reports must be concise
    4. Reports and records must be relevant
    5. Work reports must be complete

27. What types of costs are involved in the O & M of a collection system?
    1. Equipment
    2. Labor
    3. Political
    4. Services
    5. Supplies

28. Under what circumstances should rotation of personnel to different crews or types of work be done?
    1. To advance career of a capable and worthy person
    2. To allow staff an opportunity to get to know each other better
    3. To discourage possibility of crews becoming complacent
    4. To encourage staff to be alert and *NOT* to rely on others to do their job
    5. To improve efficiency of agency

29. What tasks can be performed by inspection crews?
    1. Determine validity and nature of complaints
    2. Identify problems caused by industrial dischargers
    3. Inspect buildings for plumbing defects
    4. Inspect work performed by contractors
    5. Pinpoint the locations of trouble spots

30. Which items should be considered in a chemical root control program?
    1. Consider using aeration systems
    2. Keep records of all aspects of chemical treatment program
    3. Research literature for available chemicals, methods, and successes
    4. Select at least two methods and evaluate effectiveness
    5. Visit agencies with similar problems using chemicals to control roots

31. What factors should be considered when increasing the time between lift station inspections?
    1. Improving preventive maintenance program
    2. Increasing capacity of auxiliary unit oilers and reservoirs
    3. Increasing inspection crew size
    4. Increasing response time for emergencies
    5. Installing or improving telemetering system

32. What items should be included in a safety program?
    1. Allocation of budget
    2. Confined spaces
    3. Operator performance evaluation
    4. Shoring
    5. Traffic control

33. What are the basic aspects of scheduling a collection system preventive maintenance program?
    1. Control of dates of emergency events
    2. Knowledge of collection system
    3. Knowledge of maintenance procedures and equipment
    4. Organization of personnel
    5. Proper timing of scheduled maintenance activities

34. What information should be compiled to perform a pump station risk analysis?
    1. Bypass arrangements
    2. Emergency power requirements
    3. Lowest upstream manhole and/or basement
    4. Peak flow
    5. Peak flow storage

35. In what public areas could there be a high risk of public exposure to untreated wastewater from SSOs and CSOs?
    1. School grounds
    2. Streets
    3. Upstream from overflows
    4. Water used for body-contact water recreation
    5. Waters used for drinking water supply

36. What tasks are performed by a utility manager administering a CMOM program?
    1. Controlling
    2. Directing
    3. Organizing
    4. Planning
    5. Staffing

37. What are the utility manager's responsibilities with regard to staffing?
    1. Advocating favorable federal legislation
    2. Evaluating job performance
    3. Hiring new employees
    4. Organizing union representation
    5. Training employees

38. What types of structural repairs are considered when rehabilitating sewers?
    1. Excavation
    2. Joint repair
    3. Lining
    4. Pipe bursting
    5. Sliplining

39. What are the objectives of manhole inspection?
    1. To assess the manhole's functional capacity
    2. To be sure the lid is *NOT* buried
    3. To determine the proper elevation or grade around the lid
    4. To examine the structural integrity of the manhole
    5. To minimize costs of repair and/or rehabilitation

Short Answer

1. What is the purpose of a lift station?
2. Why should air release valves be installed at high points in force mains?
3. Why does a pump require more energy (draw more power) to start than it does during normal operating conditions?
4. Why should isolation valves in lift stations be gate valves instead of plug valves?
5. What information is shown on a pump curve?
6. How would you make an operational inspection of a new lift station?
7. How would you determine the frequency of visits to a lift station?
8. What safety precautions would you take *BEFORE* attempting to determine the cause of a lift station failure?
9. How would you develop a maintenance schedule for a lift station?
10. What is the difference between direct current (D.C.) and alternating current (A.C.)?
11. What meters and testers are used to maintain, repair, and troubleshoot electrical circuits and equipment? Discuss the use of each meter and tester.
12. What protective or safety devices are used to protect operators and equipment from being harmed by electricity?
13. How would you repack a pump?
14. What factors can cause cavitation of a pump impeller?
15. What are some causes of bearing failures?
16. What would cause sewers to lose their structural strength?
17. What techniques are available to determine the condition of a sewer system?
18. How can a utility establish rehabilitation priorities?
19. What problems are common to all types of sewer construction?
20. Why should the Insituform resin material be kept away from direct exposure to sunlight?
21. What are the reasons for lining a sewer?
22. How can the walls of manholes that have deteriorated due to hydrogen sulfide attack be repaired?
23. How do the three common factors generally missing when an accident occurs help prevent accidents?
24. How can an operator determine if an oxygen-deficient condition exists in a manhole?
25. Under what circumstances should a collection system operator *NOT* perform electrical work?
26. What is a tailgate safety session?
27. Who should prepare an administrative plan for a wastewater collection agency?
28. What general procedures would you follow to hire a new collection system operator?
29. How can seldom-used pieces of equipment be made readily available to all crews when needed?
30. Under what circumstances should a satellite maintenance yard be considered?
31. Why are good maps needed for wastewater collection systems?
32. When can work reports be discarded?
33. Why are work directives usually given in written form?
34. What information should be included in a preventive maintenance record system?
35. What does "public relations" mean to you?
36. Why are reports important?
37. How could a person develop a knowledge of the work required of an organization?

38. How would you determine the types of crews needed to operate, maintain, and repair a wastewater collection system?
39. How would you determine the priority of job assignments for an emergency service crew?
40. How would you attempt to evaluate the condition of a wastewater collection system?
41. What kind of work is done by a preventive maintenance crew?
42. What kinds of failures can occur in a lift station?
43. What is a preventive maintenance program?
44. What factors should be considered when stocking parts for a collection system maintenance program?
45. How would you evaluate the effectiveness of your maintenance programs?
46. What happens when collection systems are neglected?
47. What factors should a utility agency analyze when considering sewer rehabilitation instead of preventive maintenance and inspection as a solution to certain collection system problems?
48. Effective CMOM programs are based on what information and knowledge?

SUGGESTED ANSWERS
FOR FINAL EXAMINATION
VOLUME II

True-False

1. False — Stations with guide bars do *NOT* require operators to enter the wet well to install or remove the pump.
2. True — Mechanical seals have stuffing boxes.
3. True — Use logic and caution when solving any problem.
4. True — Never touch electrical panels, controls, circuits, wiring, or equipment unless you are qualified and authorized.
5. False — Most alignment takes place when the machines being aligned are cold, *NOT* hot.
6. True — Never allow a mechanical seal to run dry or the seal faces will be burned and ruined.
7. True — Flows greater than the expected base flow can be considered to come from inflow and infiltration.
8. False — Smoke testing is *NOT* very effective during high groundwater conditions.
9. False — Sliplining is simple and relatively inexpensive.
10. True — Operators reduce the chances of an accident by training and certification.
11. False — Employers must provide personal protection equipment at no cost to operators.
12. False — Lockout/tagout refers to control of hazardous energy.
13. False — Level of education *CANNOT* be used to screen candidates instead of performance testing.
14. False — Union contracts *CANNOT* change a supervisor's delegated authority and responsibility.
15. True — Every manhole should be assigned a specific, permanent identification number.
16. True — In most organizations, authority is essentially power.
17. False — Inspection crews inspect sewer systems, *NOT* construction or buildings.
18. True — Computers perform the recordkeeping function very well.
19. False — Roots can enter a pipe joint or walls microscopically (through very small holes).
20. False — Planning must be done before, *NOT* after, the other management functions.
21. True — The main objective of sewer cleaning is to prevent blockages and overflows.

Best Answer

1. 2 — Wet wells are aerated during low flows to help keep the detained wastewater fresh.
2. 4 — A ground-fault circuit interrupter (GFCI) is designed to limit electric shock to a current and time duration value to prevent serious injury.
3. 3 — The purpose of a check valve is to prevent the force main from draining back into the wet well.
4. 1 — The attitude of the operating agency is the most significant factor in determining the frequency of lift station visits.
5. 1 — If battery cables and wiring are *NOT* properly connected, an arc across the unit can occur that could cause an explosion.
6. 3 — Grounding electrical equipment is a safety precaution that protects operators and the motors, tools, and equipment.

7. 4 The purpose of pump and motor couplings is to transmit the torque from the motor to the pump.

8. 4 Universal joints are used to take up any misalignment between the pump and the drive.

9. 2 The major stumbling block to the implementation of a rehabilitation program is identifying staff and securing funding.

10. 4 The main purpose of infiltration/inflow (I/I) evaluations is to reduce the hydraulic loads on the sewer system.

11. 2 Use in-line replacement when sewers are structurally failing and inadequate in capacity.

12. 1 Professionalism is demonstrated by the ability to carry out a work assignment safely.

13. 3 The duration of a permit-required confined space entry permit is the time required to complete the assignment.

14. 3 Personal protection equipment should be designed to create a barrier against workplace hazards.

15. 3 Management must establish rules to protect the workers and facilities from terrorism to provide a safe and healthful workplace.

16. 4 An effective administration will develop standardized and equitable procedures for the application, selection, orientation, and training of personnel.

17. 3 Superintendents should encourage operators to receive additional training to improve performance.

18. 4 The advantage of electronically capturing TV inspections images on disks is that it permits information to be retrieved and printed at any time.

19. 4 The most effective method for operators for completing a report is at the scene of each job.

20. 1 The first priority of a wastewater collection system agency should be the ability to handle emergency service requests quickly and efficiently.

21. 1 Eliminate hazardous atmospheric conditions in manholes by using a portable blower to ventilate.

22. 1 A risk analysis of each pump station defines the likelihood and potential of a prolonged station failure.

23. 4 Wastewater collection systems could be designed with inadequate capacities if there is an unexpected growth in population.

24. 3 Hydrogen sulfide gas is the major cause of corrosion in wastewater collection systems.

25. 4 Collection system O & M manuals should contain detailed procedures for start-up, operation, and shutdown.

Multiple Choice

1. 3, 4, 5 Collection system operators should help engineers design lift stations to improve O & M, to lower operating costs, and to minimize bypasses and backups caused by lift station failures.

2. 4, 5 Force mains need air and vacuum release valves to avoid an increase in pumping costs and to avoid a reduction in pipe capacity.

3. 2, 3 Mechanical seals and packing are methods of sealing used in pumps.

4. 1, 2, 3, 4, 5 When a new pump is started, the following items should be inspected: electrical load on motors; excessive heating; packing glands or mechanical seals; unusual noises; and unusual vibrations.

5. 2, 3, 5 To prepare a lift station maintenance program, consider knowledge gained from experience, recommendations of equipment manufacturers, and lift station requirements.

6. 1, 2, 3, 4, 5 An unbalanced current can cause problems of excessive heat, overload tripping, reduction in starting torque of the motor, vibrations, and overall poor performance.

7. 1, 2, 3, 5 Electric motor insulation failure can be caused by contaminants, excessive starts, motor overloading, and voltage imbalance.

8. 1, 2, 4, 5 Cavitation can cause damage to the impeller, a drop in pump efficiency, pump noise, and pump vibration.

9. 1, 2, 4, 5 When inspecting a pump impeller, an operator should look for cavitation marks; chips, broken tips, corrosion, and unusual wear; excessive wear on wearing rings; and tightness on the shaft.

10. 1, 2, 3, 4 The most common methods of coupling a pump and driver together include belt-driven or chain-driven, close coupled, direct coupling, and flexible shafting.

11. 1, 2, 3, 4, 5 Factors that contribute to the declining integrity (deterioration) of wastewater collection systems include materials having lost structural integrity due to corrosion, population growth diverting attention and effort from preservation, stress created by vibrations from construction activities, surcharged flows accelerating deterioration and allowing leakage, and unstable foundation soils or improper bedding material.

12. 1, 2, 3, 4 When preparing an inventory phase of a rehabilitation program, critical information can be obtained from maintenance repair and inspection reports. This information includes where changes and additions have been made, where maintenance is excessive, where problems occur regularly, and where repairs are frequent.

13. 1, 2, 4, 5 The beneficial results from pipeline replacement include correction of misaligned pipe, elimination of direct sources of storm water entry, increase in hydraulic capacity, and repair of improper service connections.

14. 1, 2 Chemical grouts have failed in arid regions during periods of low groundwater and in coastal regions subject to influence of tidal fluctuations.

15. 2, 5 Coatings applied to manholes should be corrosion resistant and waterproof.

16. 1, 2, 5 Collection system work areas that could be considered confined spaces are manholes, sewer lines, and wet wells.

17. 2, 3, 4, 5 The success or failure of a safety/survival program can be measured by the number of deaths, disabilities, and injuries, and also by the number of operators trained.

18. 1, 2, 5 The lockout/tagout rule requires the following items to be included in an energy control program: documented energy control procedures; employee training program; and periodic inspection of the procedures.

19. 1, 2, 3, 4, 5 Before doing work that requires use of personal protection equipment, employees must be trained in how the equipment is to be worn; the limitations of the equipment; proper care, maintenance, useful life, and disposal of the equipment; what type of equipment is necessary; and when it is necessary.

20. 1, 3, 5 Hard hats must be able to absorb the shock of a blow, to protect against electric shock, and to resist penetration.

21. 1, 2, 3, 4, 5 To reduce the possibility of a terrorist attack, be aware of the hazardous chemicals used and know how to defend against spills; be cautious of suspicious packages that arrive; ensure all visitors sign in and out of the facilities with a positive ID check; keep emergency numbers posted near telephones and radios; and patrol the facilities frequently, looking for suspicious activity or behavior.

22. 1, 3, 4 An effective collection system administration will ensure an operation and maintenance program that keeps the system functioning at top efficiency, maximizing useful life at minimum cost.

23. 1, 3, 5 A wastewater collection system agency can attempt to prevent public health hazards by assigning scheduling priority to stoppages, overflows, and exfiltration that contaminate the environment; by providing immediate, concerned, and efficient service to all emergency calls; and by using clean water to flush contaminated surfaces and disinfecting with chemicals approved by the health authority.

24. 3, 4, 5 Certification of wastewater collection system operators increases their potential to earn a higher salary. In addition, certification recognizes the importance of good operation and maintenance and of protecting health and safety.

25. 1, 2, 3, 4, 5 Collection system maps should include information on the locations of building sewers; cleanouts; lift stations; main, trunk, and interceptor sewers; and manholes.

26. 1, 3, 4, 5 Key factors involved in the development of an effective management information system include all aspects of the information system must be performed in a timely manner, planning and controlling reports must be concise, reports and records must be relevant, and work reports must be complete.

27. 1, 2, 4, 5 Costs involved in the O & M of a collection system include equipment, labor, services, and supplies.

28. 1, 5 Personnel should be rotated to different crews or types of work to be done to advance the career of a capable and worthy person or to improve agency efficiency.

29. 1, 2, 5 Inspection crews can determine the validity and nature of complaints, identify problems caused by industrial dischargers, and pinpoint locations of trouble spots.

30. 2, 3, 4, 5 Items that should be considered in a chemical root control program include keeping records of all aspects of the program; researching literature for available chemicals, methods, and successes; selecting at least two methods and evaluating their effectiveness; and visiting agencies with similar problems using chemicals to control roots.

31. 1, 2, 5 To increase the time between lift station inspections, consider improving the preventive maintenance program, increasing the capacity of auxiliary unit oilers and reservoirs, and installing or improving the telemetering system.

32. 2, 4, 5 Confined spaces, shoring, and traffic control should be included in a safety program.

33. 2, 3, 4, 5 The basic aspects of scheduling a collection system preventive maintenance program include knowledge of the collection system, knowledge of maintenance procedures and equipment, organization of personnel, and proper timing of scheduled maintenance activities.

34. 1, 2, 3, 4, 5 To perform a pump station risk analysis, the following information should be compiled: bypass arrangements; emergency power requirements; lowest upstream manhole and/or basement; peak flow; and peak flow storage.

35. 1, 2, 4, 5 There could be a high risk of public exposure to untreated wastewater from SSOs and CSOs in public areas including school grounds, streets, body-contact water recreation areas, and where waters are used for the drinking water supply.

36. 1, 2, 3, 4, 5 A utility manager administering a CMOM program performs the following tasks: controlling, directing, organizing, planning, and staffing.

37. 2, 3, 5 With regard to staffing, a utility manager is responsible for evaluating job performance, hiring new employees, and training employees.

38. 1, 2, 3, 4, 5 When rehabilitating sewers, types of structural repairs to consider include excavation, joint repair, lining, pipe bursting, and sliplining.

39. 1, 2, 3, 4 Manholes are inspected to assess the manhole's functional capacity, to be sure the lid is not buried, to determine the proper elevation or grade around the lid, and to examine the structural integrity of the manhole.

Short Answer

1. The purpose of a lift station is to lift or raise wastewater or storm water from a lower elevation to a higher elevation.

2. Air release valves should be installed at high points in force mains to prevent the accumulation of air and other gases. Trapped pockets of air reduce the carrying capacity of the pipe, increase pumping costs, contribute to damage by water hammer, and may create negative pressures strong enough to collapse pipes.

3. The energy required to start a pump is greater than the total dynamic head (TDH) during normal operating conditions because additional energy is required to start the motor and the pump and to start the water flowing through the pipes, the check valves, and the pump.

4. Isolation valves in lift stations should be gate valves instead of plug valves because some types of plug valves (not the slow-closing hydraulic or pneumatically controlled valves) can be slammed shut by wastewater backflow and possibly injure you and break the valve body or a portion of the pipe system from the water hammer due to the sudden stoppage of the flowing water. Also, plug valves have a restriction that can become clogged with rags or sticks.

5. A pump curve graphically shows the quantity of water a particular pump will discharge when operating at several different heads. It also shows a plot of pump discharge versus required horsepower and pump discharge versus pump efficiency.

6. An operational inspection of a new lift station includes inspecting valves for proper settings, sump level indicators, power to pumps and pump rotation, operating controls, engines and pumps.

7. Frequency of lift station visits should depend on the number of stations, type of wastewater, potential damage from failure, condition of equipment, design of the facility, equipment installed, adequacy of preventive maintenance and overhaul program, type and reliability of telemetry system, and the attitude of the operating agency.

8. Safety precautions that must be taken before attempting to determine the cause of a lift station failure include investigation of the power supply to the station (be sure no lines are down and contacting fences or the ground), interior lighting, ventilation, and possibility of flooding in station and nearby homes and streets.

9. To develop a maintenance schedule for a lift station, consider:

   a. Recommendations of equipment manufacturers,
   b. Requirements of a lift station, and
   c. Knowledge gained from previous experience by the operating agency.

10. Direct current (D.C.) flows in one direction only and is essentially free from pulsation. Alternating current (A.C.) is periodic current that has alternating positive and negative values.

11. Types of meters and testers used to maintain, repair, and troubleshoot electrical circuits and equipment include:

    a. Multimeter. Used to determine if circuit is energized or not, to measure voltage, and to determine if fuses are "blown."
    b. Ammeter. The ammeter records the current or "amps" in the circuit.
    c. Megger. The megger is used for checking the insulation resistance on motors, feeders, bus bar systems, grounds, and branch circuit wiring.
    d. Ohmmeters or Circuit Testers. Used to measure the resistance (ohms) in a circuit.

12. Protective or safety devices used to protect operators and equipment from being harmed by electricity include fuses, circuit breakers, and grounds.

13. To repack a pump, properly shut down the pump, remove the old packing, determine the new packing requirements, and repack the pump using the following procedure:

    a. Cut packing to proper length,
    b. Roll packing flat,
    c. Wrap new packing rings around shaft and insert into stuffing box,
    d. Stagger the joints or butts of the packing around the shaft,
    e. Locate the lantern ring,
    f. Install packing rings, and
    g. Install packing gland.

14. Cavitation is caused by the pump operating at conditions different from those for which it was designed, such as operating off design curve, poor suction conditions, high speed, air leaks into the suction eye, and water hammer conditions.

15. Some of the causes of bearing failures include:

    a. Fatigue failure,
    b. Contamination,
    c. Brinelling,
    d. False brinelling,
    e. Thrust failures,
    f. Misalignment,
    g. Electric arcing,
    h. Lubrication failure, and
    i. Cam failure.

16. Sewers lose their structural integrity due to corrosion and deterioration. Damage also results from undue stress and live loads placed on the sewers.

17. Techniques available to determine the condition of a sewer system include closed-circuit television, pipe flow tests, computer flow models and visual inspections.

18. A utility can establish rehabilitation priorities on the basis of the consequences of failure. The costs and impacts due to failures can be evaluated, weighting factors and severity ratings assigned, and priorities established.

19. Problems common to all types of sewer construction include traffic disruption, disruption to properties, paving damage, shoring requirements, excavation dewatering, noise, flow control, and restoration.

20. Insituform resin material should be kept away from direct exposure to sunlight because ultraviolet rays tend to deteriorate the composition of the material. Prolonged exposure in the presence of heat can possibly cause a thermosetting reaction.

21. Liners are inserted in pipes to repair or recondition the sewer system. Frequently, liners are installed to control excessive infiltration, exfiltration, and root intrusion.

22. The walls of manholes that have deteriorated due to hydrogen sulfide attack can be repaired with coating processes. The walls must be thoroughly cleaned before applying a coating. The choice of coating material depends on the chemistry of the particular situation encountered. Various epoxies and plastics are available that can provide protection from hydrogen sulfide and are dependable coating processes.

23. Accidents can be prevented when operators are *AWARE* of the hazards, assume *RESPONSIBILITY* for their own safety/survival, and make a *COMMITMENT* to that responsibility.

24. An operator can determine if an oxygen-deficient atmosphere exists by using a properly calibrated oxygen level measuring device. If the oxygen level in the manhole drops below the minimum oxygen level of 19.5 percent, then this is an indication of an oxygen-deficient condition. Operators should carry oxygen level detection devices with them in manholes that sound an alarm indicating an oxygen-deficient condition rather than using a meter that must be read and a decision made whether or not an oxygen-deficient condition exists.

25. Collection system operators should *NEVER* perform electrical work when they are not qualified or not authorized to perform the work.

26. Tailgate safety sessions are short (ten minutes plus or minus) discussions that help focus on safety/survival as part of everyday routine collection system operator tasks. Topics can be specific to the job or even related to home safety and may be related to the time of the year; for example, holidays or virtually any other topic that relates to safe practices.

27. An administrative plan should be prepared by the manager of the organization responsible for operation and maintenance of the collection system with the full participation of the organization's personnel.

28. To hire a new operator, first decide what the operator needs to know to be able to perform the job. Obtain a list of applicants on the basis of the job applications and possibly written tests. Conduct an employment selection interview for the most qualified applicants. Determine which applicant best suits your needs. Check references. Make a firm employment offer to the best applicant and, if the offer is accepted, notify all other applicants that the position has been filled.

29. Seldom-used pieces of equipment can be readily available to all crews when needed if they are stored in a central storeroom to which everyone has access.

30. A satellite maintenance yard should be considered where the service area is very large, causing excessive travel time from a central yard to remote areas. The decision to decentralize should be carefully considered and based on a least-cost analysis.

31. Good maps are essential to the operation and maintenance of wastewater collection systems. Maps give the location and size of the system. They indicate what has been constructed, where, what size, material used, and conditions encountered. Maps can be used to show problem areas (roots, grease, odors) and the preventive maintenance program (frequency of cleaning various sections).

32. Work reports can be discarded after all of the information submitted on the reports is transferred to records, either in the form of computer-readable electronic disks or written material. If some of the information is not transferred to separate record documents, the work reports containing this information should be filed or archived until it is certain that the information is no longer useful.

33. Work directives should be given in written form to reduce the possibility of misunderstanding and to provide a record for reference if questions arise about a directive. Verbal directives are commonly used when they are more efficient than written or electronically transmitted directives.

34. A preventive maintenance record system should include information on scheduling of maintenance, user-generated requests for special maintenance, and a history of maintenance and repairs performed. Costs also must be included.

35. Public relations means informing and working with the public. Our service should be provided sincerely, courteously, and with an understanding of our responsibility to the public's needs.

36. Reports are very important because they are a means of communicating information from the individual who has it to those who need it.

37. Developing a knowledge of the work required of an organization begins with a clear statement of its mission followed by its goals and objectives. Then the separate tasks needed to reach the objectives and goals are determined. A listing of the types of work for each task can then be prepared.

38. Types of crews needed to operate, maintain, and repair a wastewater collection system can be determined on the basis of the size of the system, scope of maintenance activities, program goals and objectives, geographic location, and management philosophy.

39. Priority of job assignments for an emergency service crew can be developed on the basis of urgency, number of people affected, and extent and magnitude of damages involved. A typical priority list is as follows:

    a. Flooded home or business (main line or service line stoppages).
    b. Manhole overflowing (main line stoppage).
    c. Manhole cover off (liability risk).
    d. Cleanout overflowing (nuisance and health hazard).
    e. Home service backed up.

40. To evaluate the condition of a wastewater collection system, use a closed-circuit television unit to inspect preselected problem portions of the collection system.

41. Preventive maintenance crews inspect sewers using closed-circuit TV to identify potential and existing problems. The crews clear and clean sewers using various types of equipment and keep records of what is done in order to document accomplishments and provide justification for budgets.

42. Types of lift station failures include:

    a. Pump failure,
    b. Electrical failure,
    c. Mechanical failure, and
    d. Electrode and/or controller failure.

43. A preventive maintenance program applies effective preventive and corrective maintenance procedures over a controlled time period that will keep the collection system operating as intended with a minimum of stoppages or failures.

44. When stocking parts, at least three factors should be considered: (1) how critical is the part, (2) what is the lead time from the manufacturer, and (3) what is the frequency of usage of a particular part.

45. Effectiveness of a maintenance program could be evaluated on the basis of trends regarding:

    a. Number of stoppages per mile of sewer per year.
    b. Number of odor complaints per mile of sewer per year.
    c. Number of lift station failures per lift station per year.

46. When collection systems are neglected, they:

    - Degrade, deteriorate, and age at a faster rate than expected or intended,
    - Allow inferior or inadequate designs to be accepted and constructed,
    - Are poorly or improperly managed, operated, maintained, and/or repaired, and
    - Do not provide proper attention to privately owned service laterals from buildings and homes.

47. Usually the decision of whether to rehabilitate or continue with preventive maintenance and inspection is based on factors such as the increased risk of regulatory noncompliance, reduced level of service, or cost-effectiveness of preventive maintenance versus rehabilitation over the remaining life of the sewer.

48. Effective CMOM programs are based on knowing what components make up the collection system, where they are located, and the condition of the components.

COLLECTION SYSTEM WORDS

A Summary of the Words Defined

in

OPERATION AND MAINTENANCE OF WASTEWATER COLLECTION SYSTEMS

COLLECTION SYSTEM WORDS

by

George Freeland

OPERATOR'S PROJECT PRONUNCIATION KEY

by Warren L. Prentice

The Operator's Project Pronunciation Key is designed to aid you in the pronunciation of new words. While this key is based primarily on familiar sounds, it does not attempt to follow any particular pronunciation guide. This key is designed solely to aid operators in this program.

You may find it helpful to refer to other available sources for pronunciation help. Each current standard dictionary contains a guide to its own pronunciation key. Each key will be different from each other and from this key. Examples of the differences between the key used in this program and the *WEBSTER'S NEW WORLD COLLEGE DICTIONARY*[1] "Key" are shown below:

Term
Project Key
Webster Key
sewer
SUE·er
sōō′ər
alignment
a·LINE·ment
ə·līn′mənt
infiltration
IN·fill·TRAY·shun
ĭn′fil·trā′shən

In using this key, you should accent (say louder) the syllable that appears in capital letters. The following chart is presented to give examples of how to pronounce words using the Operator's Project Pronunciation Key.

WORD	SYLLABLE			
	1st	2nd	3rd	4th
sewer	SUE	er		
alignment	a	LINE	ment	
infiltration	IN	fill	TRAY	shun

The first word, *SEWER*, has its first syllable accented. The second word, *ALIGNMENT*, has its second syllable accented. The third word, *INFILTRATION*, has its first and third syllables accented.

We hope you will find the key useful in unlocking the pronunciation of any new word.

EXPLANATION OF WORDS

The meanings of words in the glossary of this manual are based on current usage by the wastewater collection profession and definitions given in *GLOSSARY—WATER AND WASTEWATER CONTROL ENGINEERING*, prepared by the Joint Editorial Board representing APHA, ASCE, AWWA, and WEF,[2] 1981. Certain words used by wastewater collection system operators tend to have slightly different meanings in some regions of the United States. We have tried to standardize word meanings as much as possible.

---

[1] *The WEBSTER'S NEW WORLD COLLEGE DICTIONARY, Fourth Edition, 1999, was chosen rather than an unabridged dictionary because of its availability to most collection system operators. Other editions may be slightly different.*

[2] *APHA. American Public Health Association.*
ASCE. American Society of Civil Engineers.
AWWA. American Water Works Association.
WEF. Water Environment Federation.

WORDS

>GREATER THAN >GREATER THAN

DO >5 mg/*L* would be read as DO GREATER THAN 5 mg/*L*.

<LESS THAN <LESS THAN

DO <5 mg/*L* would be read as DO LESS THAN 5 mg/*L*.

A

ACEOPS ACEOPS

See **A**LLIANCE OF **CE**RTIFIED **OP**ERATOR**S**, LAB ANALYSTS, INSPECTORS, AND SPECIALISTS (ACEOPS).

ABANDONED (a-BAN-dund) ABANDONED

No longer in use; a length, section or portion of a collection system no longer in service and left in place, underground. For example, when a house or building is razed or removed the service connection may be left open and unused.

ABATEMENT (a-BAIT-ment) ABATEMENT

Putting an end to an undesirable or unlawful condition affecting the wastewater collection system. A property owner found to have inflow sources connected to the collection system may be issued a "NOTICE OF ABATEMENT." Such notices will usually describe the violation, suggest corrective measures and grant a period of time for compliance.

ABSORPTION (ab-SORP-shun) ABSORPTION

The taking in or soaking up of one substance into the body of another by molecular or chemical action (as tree roots absorb dissolved nutrients in the soil).

ABSORPTION CAPACITY ABSORPTION CAPACITY

The amount of liquid which a solid material can absorb. Sand, as an example, can hold approximately one-third of its volume in water, or three cubic feet of dry sand can contain one cubic foot of water. A denser soil, such as clay, can hold much less water and thus has a lower absorption capacity.

ABSORPTION RATE ABSORPTION RATE

The speed at which a measured amount of solid material can absorb a measured amount of liquid. Under pressure, water can infiltrate a given volume of gravel very rapidly. The water will penetrate (or be absorbed by) sand more slowly and will take even longer to saturate the same amount of clay.

ACUTE HEALTH EFFECT ACUTE HEALTH EFFECT

An adverse effect on a human or animal body, with symptoms developing rapidly.

ADSORPTION (add-SORP-shun) ADSORPTION

The gathering of a gas, liquid, or dissolved substance on the surface or interface zone of another material.

AEROBIC (AIR-O-bick) AEROBIC

A condition in which atmospheric or dissolved molecular oxygen is present in the aquatic (water) environment.

AIR BINDING AIR BINDING

The clogging of a filter, pipe or pump due to the presence of air released from water. Air entering the filter media is harmful to both the filtration and backwash processes. Air can prevent the passage of water during the filtration process and can cause the loss of filter media during the backwash process.

AIR BLOWER — AIR BLOWER

A device used to ventilate manholes and lift stations.

AIR GAP — AIR GAP

An open vertical drop, or vertical empty space, between a drinking (potable) water supply and the point of use. This gap prevents backsiphonage because there is no way wastewater can reach the drinking water. Air gap devices are used to provide adequate space above the top of a manhole and the end of the hose from the fire hydrant. This gap ensures that no wastewater will flow out the top of a manhole, reach the end of the hose from a fire hydrant, and be sucked or drawn back up through the hose to the water supply.

AIR PADDING — AIR PADDING

Pumping dry air (dew point –40°F) into a container to assist with the withdrawal of a liquid or to force a liquified gas such as chlorine out of a container.

AIR TEST — AIR TEST

A method of inspecting a sewer pipe for leaks. Inflatable or similar plugs are placed in the line and the space between these plugs is pressurized with air. A drop in pressure indicates the line or run being tested has leaks.

AIR TEST, QUICK — AIR TEST, QUICK

(See QUICK AIR TEST)

ALIGNMENT (a-LINE-ment) — ALIGNMENT

The proper positioning of parts in a system. The alignment of a pipeline or other line refers to its location and direction.

ALLIANCE OF CERTIFIED OPERATORS, LAB ANALYSTS, INSPECTORS, AND SPECIALISTS (ACEOPS) — ALLIANCE OF CERTIFIED OPERATORS, LAB ANALYSTS, INSPECTORS, AND SPECIALISTS (ACEOPS)

A professional organization for operators, lab analysts, inspectors, and specialists dedicated to improving professionalism; expanding training, certification, and job opportunities; increasing information exchange; and advocating the importance of certified operators, lab analysts, inspectors, and specialists. For information on membership, contact ACEOPS, 1810 Bel Air Drive, Ames, IA 50010-5125, phone (515) 663-4128 or e-mail: Info@aceops.org.

ALLUVIAL (uh-LOU-vee-ul) DEPOSIT — ALLUVIAL DEPOSIT

Sediment (clay, silt, sand, gravel) deposited in place by the action of running water.

ALTERNATING CURRENT (A.C.) — ALTERNATING CURRENT (A.C.)

An electric current that reverses its direction (positive/negative values) at regular intervals.

AMBIENT (AM-bee-ent) — AMBIENT

Surrounding. Ambient or surrounding atmosphere.

AMPERAGE (AM-purr-age) — AMPERAGE

The strength of an electric current measured in amperes. The amount of electric current flow, similar to the flow of water in gallons per minute.

AMPERE (AM-peer) — AMPERE

The unit used to measure current strength. The current produced by an electromotive force of one volt acting through a resistance of one ohm.

ANAEROBIC (AN-air-O-bick) — ANAEROBIC

A condition in which atmospheric or dissolved molecular oxygen is *NOT* present in the aquatic (water) environment.

ANAEROBIC (AN-air-O-bick) DECOMPOSITION — ANAEROBIC DECOMPOSITION

The decay or breaking down of organic material in an environment containing no “free” or dissolved oxygen.

ANALOG READOUT — ANALOG READOUT

The readout of an instrument by a pointer (or other indicating means) against a dial or scale.

ANGLE OF REPOSE — ANGLE OF REPOSE

The angle between a horizontal line and the slope or surface of unsupported material such as gravel, sand, or loose soil. Also called the "natural slope."

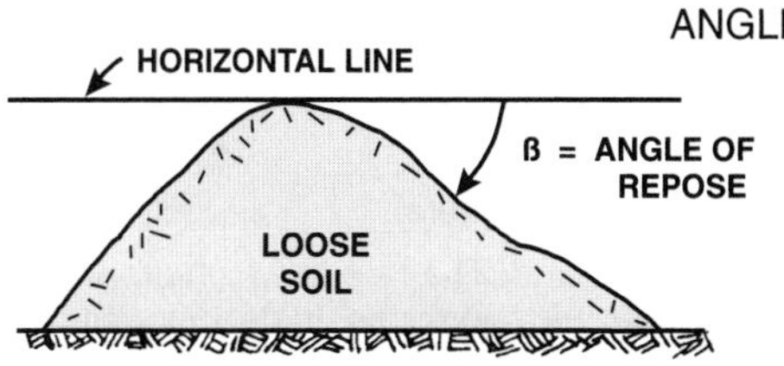

ANIMAL WASTES — ANIMAL WASTES

(1) Urine and fecal wastes of living animals.

(2) Wastes of animal tissue from meat processing (feathers included), or hospital, surgical and clinical facility wastes of animal types.

(3) Similar to (2) above, but cooked or prepared wastes of animal tissues and bones from domestic or commercial food preparation.

ANNULAR (AN-you-ler) SPACE — ANNULAR SPACE

A ring-shaped space located between two circular objects. For example, the space between the outside of a pipe liner and the inside of a pipe.

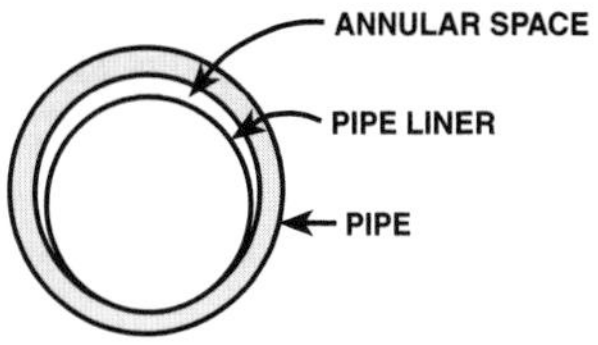

ANOXIC (an-OX-ick) — ANOXIC

A condition in which atmospheric or dissolved molecular oxygen is *NOT* present in the aquatic (water) environment and nitrate is present. Oxygen deficient or lacking sufficient oxygen. The term is similar to ANAEROBIC.

APARTMENT COMPLEX — APARTMENT COMPLEX

One or more residential buildings at a single location. An apartment building may contain several residences with a single connection to the wastewater collection system. A complex can have several buildings with a single connection.

APPURTENANCE (uh-PURR-ten-nans) — APPURTENANCE

Machinery, appliances, structures and other parts of the main structure necessary to allow it to operate as intended, but not considered part of the main structure.

AQUIFER (ACK-wi-fer) — AQUIFER

A porous, water-bearing geologic formation. Usually refers only to materials capable of yielding a substantial amount of water.

ARCH — ARCH

(1) The curved top of a sewer pipe or conduit.

(2) A bridge or arch of hardened or caked chemical which will prevent the flow of the chemical.

ARTIFICIAL GROUNDWATER TABLE — ARTIFICIAL GROUNDWATER TABLE

A groundwater table that is changed by artificial means. Examples of activities that artificially raise the level of a groundwater table include agricultural irrigation, dams and excessive sewer line exfiltration. A groundwater table can be artificially lowered by sewer line infiltration, water wells, and similar drainage methods.

ASPHYXIATION (ass-FIX-ee-a-shun) — ASPHYXIATION

An extreme condition often resulting in death due to a lack of oxygen and excess carbon dioxide in the blood from any cause. Also called suffocation.

AUGER (AW-grr) AUGER

A sharp tool used to go through and break up or remove various materials that become lodged in sewers.

B

BOD (pronounce as separate letters) BOD

Biochemical **O**xygen **D**emand. The rate at which organisms use the oxygen in water or wastewater while stabilizing decomposable organic matter under aerobic conditions. In decomposition, organic matter serves as food for the bacteria and energy results from its oxidation. BOD measurements are used as a measure of the organic strength of wastes in water.

BACKFILL BACKFILL

(1) Material used to fill in a trench or excavation.

(2) The act of filling a trench or excavation, usually after a pipe or some type of structure has been placed in the trench or excavation.

BACKFILL, BORROW BACKFILL, BORROW

(See BORROW BACKFILL)

BACKFILL COMPACTION BACKFILL COMPACTION

(1) Tamping, rolling or otherwise mechanically compressing material used as backfill for a trench or excavation. Backfill is compressed to increase its density so that it will support the weight of machinery or other loads after the material is in place in the excavation.

(2) Compaction of a backfill material can be expressed as a percentage of the maximum compactability, density or load capacity of the material being used.

BACKFILL, SELECT BACKFILL, SELECT

(See SELECT BACKFILL)

BACKFLUSHING BACKFLUSHING

A procedure used to wash settled waste matter off upstream structures to prevent odors from developing after a main line stoppage has been cleared.

BACKHOE BACKHOE

An excavating machine whose bucket is securely attached to a hinged boom and is drawn toward the machine during excavation.

BACKWATER GATE BACKWATER GATE

A gate installed at the end of a drain or outlet pipe to prevent the backward flow of water or wastewater. Generally used on storm sewer outlets into streams to prevent backward flow during times of flood or high tide. Also called a TIDE GATE.

BACTERIA (back-TEAR-e-ah) BACTERIA

Bacteria are living organisms, microscopic in size, which usually consist of a single cell. Most bacteria use organic matter for their food and produce waste products as a result of their life processes.

BALLING BALLING

A method of hydraulically cleaning a sewer or storm drain by using the pressure of a water head to create a high cleansing velocity of water around the ball. In normal operation, the ball is restrained by a cable while water washes past the ball at high velocity. Special sewer cleaning balls have an outside tread that causes them to spin or rotate, resulting in a "scrubbing" action of the flowing water along the pipe wall.

BAR RACK BAR RACK

A screen composed of parallel bars, either vertical or inclined, placed in a sewer or other waterway to catch debris. The screenings may be raked from it.

BARREL BARREL

(1) The cylindrical part of a pipe that may have a bell on one end.

(2) The cylindrical part of a manhole between the cone at the top and the shelf at the bottom.

BEDDING BEDDING

The prepared base or bottom of a trench or excavation on which a pipe or other underground structure is supported.

BEDDING COMPACTION BEDDING COMPACTION

(1) Tamping, rolling or otherwise mechanically compressing material used as bedding for a pipe or other underground structure to a density that will support expected loads.

(2) Bedding compaction can be expressed as a percentage of the maximum load capacity of the bedding material.

(3) Bedding compaction also can be expressed in load capacity of pounds per square foot.

BEDDING DESTRUCTION BEDDING DESTRUCTION

Loss of grade, load capacity or material of a bedding.

BEDDING DISPLACEMENT BEDDING DISPLACEMENT

Bedding which has been removed after placement and compaction. In a sewer pipe system, this can take place as a result of washouts due to infiltration, earth shifts or slides, damage from nearby excavations and/or improper backfill methods.

BEDDING FAULTS BEDDING FAULTS

Locations where bedding was improperly applied and thus failed.

BEDDING GRADE BEDDING GRADE

(1) In a gravity-flow sewer system, pipe bedding is constructed and compacted to the design grade of the pipe. This is usually expressed in a percentage. A 0.5 percent grade would be a drop of one-half foot per hundred feet of pipe.

(2) Bedding grade for a gravity-flow sewer pipe can also be specified as elevation above mean sea level at specific points.

BEDDING, MANHOLE BEDDING, MANHOLE

(See MANHOLE BEDDING)

BEDDING MATERIAL, SELECT BEDDING MATERIAL, SELECT

(See SELECT BEDDING)

BELL BELL

(1) In pipe fitting, the enlarged female end of a pipe into which the male end fits. Also called a HUB.

(2) In plumbing, the expanded female end of a wiped joint.

BELL-AND-SPIGOT JOINT BELL-AND-SPIGOT JOINT

A form of joint used on pipes which have an enlarged diameter or bell at one end, and a spigot at the other which fits into and is laid in the bell. The joint is then made tight by lead, cement, rubber O-ring, or other jointing compounds or materials.

BELLMOUTH BELLMOUTH

An expanding, rounded entrance to a pipe or orifice.

BENCH SCALE TESTS BENCH SCALE TESTS

A method of studying different ways or chemical doses for treating water on a small scale in a laboratory.

BEND BEND

A piece of pipe bent or cast into an angular shape.

BIOCHEMICAL OXYGEN DEMAND (BOD) BIOCHEMICAL OXYGEN DEMAND (BOD)

The rate at which organisms use the oxygen in water or wastewater while stabilizing decomposable organic matter under aerobic conditions. In decomposition, organic matter serves as food for the bacteria and energy results from its oxidation. BOD measurements are used as a measure of the organic strength of wastes in water.

BIT BIT

(1) Cutting blade used in rodding (pipe clearing) operations.

(2) Cutting teeth on the auger head of a sewer boring tool.

BLOCKAGE BLOCKAGE

(1) Partial or complete interruption of flow as a result of some obstruction in a sewer.

(2) When a collection system becomes plugged and the flow backs up, it is said to have a "blockage." Commonly called a STOPPAGE.

BLOWER BLOWER

A device used to ventilate manholes and lift stations.

BLUEPRINT BLUEPRINT

A photographic print in white on a bright blue background used for copying maps, mechanical drawings, construction plans and architects' plans.

BORROW BACKFILL BORROW BACKFILL

Material used for backfilling a trench or excavation which was not the original material removed during excavation. This is a common practice where tests on the original material show it to have poor compactability or load capacity. Also called IMPORTED BACKFILL.

BRACES BRACES

(See CROSS BRACES)

BRANCH MANHOLE BRANCH MANHOLE

A sewer or drain manhole which has more than one pipe feeding into it. A standard manhole will have one outlet and one inlet. A branch manhole will have one outlet and two or more inlets.

BRANCH SEWER BRANCH SEWER

A sewer that receives wastewater from a relatively small area and discharges into a main sewer serving more than one branch sewer area.

BREAK BREAK

A fracture or opening in a pipe, manhole or other structure due to structural failure and/or structural defect.

BRICKWORK BRICKWORK

A structure made of brick, which was common in older sewers.

BRINELLING (bruh-NEL-ing) BRINELLING

Tiny indentations (dents) high on the shoulder of the bearing race or bearing. A type of bearing failure.

BROKEN HUB BROKEN HUB

In bell-and-spigot pipe, the bell portion is frequently called the "hub." A fracture or break in the bell portion is called a "broken hub."

BROKEN SECTION BROKEN SECTION

A run of pipe between two joints is referred to as a "section." A fracture in a section is called a "broken section."

BUCKET BUCKET

(1) A special device designed to be pulled along a sewer for the removal of debris from the sewer. The bucket has one end open with the opposite end having a set of jaws. When pulled from the jaw end, the jaws are automatically opened. When pulled from the other end, the jaws close. In operation, the bucket is pulled into the debris from the jaw end and to a point where some of the debris has been forced into the bucket. The bucket is then pulled out of the sewer from the other end, causing the jaws to close and retain the debris. Once removed from the manhole, the bucket is emptied and the process repeated.

(2) A conventional pail or bucket used in BUCKETING OUT and also for lowering and raising tools and materials from manholes and excavations.

BUCKET BAIL BUCKET BAIL

The pulling handle on a bucket machine.

BUCKET MACHINE BUCKET MACHINE

A powered winch machine designed for operation over a manhole. The machine controls the travel of buckets used to clean sewers.

BUCKETING OUT BUCKETING OUT

An expression used to describe removal of debris from a manhole with a pail on a rope. In balling or high-velocity cleaning of sewers, debris is washed into the downstream manhole. Removal of this debris by scooping it into pails and hauling debris out is called "bucketing out."

BUILDING SERVICE BUILDING SERVICE

A saddle or "Y" connection to a lateral or branch sewer for connection of a building lateral.

BUILDING SEWER | BUILDING SEWER

A gravity-flow pipeline connecting a building wastewater collection system to a lateral or branch sewer. The building sewer may begin at the outside of the building's foundation wall or some distance (such as 2 to 10 feet) from the wall, depending on local sewer ordinances. Also called a "house connection" or a "service connection."

BUILDING WASTEWATER COLLECTION SYSTEM | BUILDING WASTEWATER COLLECTION SYSTEM

All of the wastewater drain pipes and their hardware that connect plumbing fixtures inside or adjacent to a building to the building sewers. This includes traps, vents and cleanouts.

BYPASS | BYPASS

A pipe, valve, gate, weir, trench or other device designed to permit all or part of a wastewater flow to be diverted from usual channels or flow. Sometimes refers to a special line which carries the flow around a facility or device that needs maintenance or repair.

BYPASSING | BYPASSING

The act of causing all or part of a flow to be diverted from its usual channels. In a wastewater treatment plant, overload flows should be bypassed into a holding pond for future treatment.

C

C-ZONED | C-ZONED

An area set aside for commercial use.

CFR | CFR

Code of **F**ederal **R**egulations. A publication of the United States Government which contains all of the proposed and finalized federal regulations, including safety and environmental regulations.

CFS | CFS

Initials standing for "Cubic Feet Per Second," a measure of flow rate.

CMOM | CMOM

Capacity Assurance, **M**anagement, **O**peration and **M**aintenance. A program developed by collection system agencies to ensure adequate capacity and also proper management and operation and maintenance of the collection system to prevent SSOs.

CSO | CSO

Combined **S**ewer **O**verflow. Wastewater that flows out of a combined sewer (or lift station) as a result of flows exceeding the hydraulic capacity of the sewer or stoppages in the sewer. CSOs exceeding the hydraulic capacity usually occur during periods of heavy precipitation or high levels of runoff from snow melt or other runoff sources.

CABLE STRAIN RELIEF | CABLE STRAIN RELIEF

A mesh type of device that grips the power cable to prevent any strain on the cable from reaching the connections.

CAPILLARY (CAP-i-larry) EFFECT | CAPILLARY EFFECT

Also called "wicking effect." The ability of a liquid to rise above an established level to saturate a porous solid.

CATALYST (CAT-uh-LIST) | CATALYST

A substance that changes the speed or yield of a chemical reaction without being consumed or chemically changed by the chemical reaction.

CATCH BASIN | CATCH BASIN

A chamber or well used with storm or combined sewers as a means of removing grit which might otherwise enter and be deposited in sewers. Also see STORM WATER INLET and CURB INLET.

CAULK (KAWK) | CAULK

To stop up and make watertight the joints of a pipe by filling the joints with a waterproof compound or material.

CAULKING (KAWK-ing) CAULKING

(1) A waterproof compound or material used to fill a pipe joint.

(2) The act of using a waterproof compound or material to fill a pipe joint.

CAUTION CAUTION

This word warns against potential hazards or cautions against unsafe practices. Also see DANGER, NOTICE, and WARNING.

CAVITATION (CAV-uh-TAY-shun) CAVITATION

The formation and collapse of a gas pocket or bubble on the blade of an impeller or the gate of a valve. The collapse of this gas pocket or bubble drives water into the impeller or gate with a terrific force that can cause pitting on the impeller or gate surface. Cavitation is accompanied by loud noises that sound like someone is pounding on the impeller or gate with a hammer.

CENTRIFUGAL (sen-TRIF-uh-gull) PUMP CENTRIFUGAL PUMP

A pump consisting of an impeller fixed on a rotating shaft that is enclosed in a casing, and having an inlet and discharge connection. As the rotating impeller whirls the liquid around, centrifugal force builds up enough pressure to force the water through the discharge outlet.

CERTIFICATION EXAMINATION CERTIFICATION EXAMINATION

An examination administered by a state agency that wastewater collection system operators take to indicate a level of professional competence. In many states certification of wastewater collection system operators is voluntary. Current trends indicate that more states, provinces, and employers will require wastewater collection system operators to be "certified" in the future.

CERTIFIED OPERATOR CERTIFIED OPERATOR

A person who has the education and experience required to operate a specific class of treatment facility as indicated by possessing a certificate of professional competence given by a state agency or professional association.

CESSPOOL CESSPOOL

A lined or partially lined excavation or pit for dumping raw household wastewater for natural decomposition and percolation into the soil.

CHECK VALVE CHECK VALVE

A special valve with a hinged disc or flap that opens in the direction of normal flow and is forced shut when flows attempt to go in the reverse or opposite direction of normal flows. Also see FLAP GATE and TIDE GATE.

CHEMICAL GROUT CHEMICAL GROUT

Two chemical solutions that form a solid when combined. Solidification time is controlled by the strength of the mixtures used and the temperature.

CHEMICAL GROUTING CHEMICAL GROUTING

Sealing leaks in a pipeline or manhole structure by injecting a chemical grout. In pipelines, the chemicals are injected through a device called a "packer." In operation, the packer is located at the leak point with the use of a television camera. Inflatable boots at either end of the packer isolate the leak point and the grouting chemicals are then forced into the leak under pressure. After allowing time for the grout to set, the packer is deflated and moved to the next location.

CHLORINATOR (KLOR-uh-NAY-ter) CHLORINATOR

A device used to regulate the transfer of chlorine from a container to flowing wastewater for such purposes as odor control and disinfection.

CHRISTY BOX CHRISTY BOX

A box placed over the connection between the pipe liner and the house sewer to hold the mortar around the cleanout wye and riser in place.

CIRCUIT CIRCUIT

The complete path of an electric current, including the generating apparatus or other source; or, a specific segment or section of the complete path.

CIRCUIT BREAKER CIRCUIT BREAKER

A safety device in an electric circuit that automatically shuts off the circuit when it becomes overloaded. The device can be manually reset.

CLEANING, PIPE — CLEANING, PIPE

(See PIPE CLEANING)

CLEANOUT — CLEANOUT

An opening (usually covered or capped) in a wastewater collection system used for inserting tools, rods or snakes while cleaning a pipeline or clearing a stoppage.

CLEANOUT, TWO-WAY — CLEANOUT, TWO-WAY

A cleanout designed for rodding or working a snake into a pipe in either direction. Two-way cleanouts are often used in building lateral pipes at or near a property line.

CLINICAL LABORATORY — CLINICAL LABORATORY

A special medical facility devoted to the identification of diseases and ailments through tests, studies and culture growths. Services of a clinical laboratory might be necessary in the event of an accidental or intentional release of contagious and/or infectious wastes into a collection system.

COAGULATE (co-AGG-you-late) — COAGULATE

The use of chemicals that cause very fine particles to clump (floc) together into larger particles. This makes it easier to separate the solids from the liquids by settling, skimming, draining or filtering.

COHESIVE (co-HE-sive) — COHESIVE

Tending to stick together.

COLLAPSED PIPE — COLLAPSED PIPE

A pipe that has one or more points in its length which have been crushed or partially crushed by exterior pressures or impacts.

COLLECTION MAIN — COLLECTION MAIN

A collection pipe to which building laterals are connected.

COLLECTION SYSTEM — COLLECTION SYSTEM

A network of pipes, manholes, cleanouts, traps, siphons, lift stations and other structures used to collect all wastewater and wastewater-carried wastes of an area and transport them to a treatment plant or disposal system. The collection system includes land, wastewater lines and appurtenances, pumping stations and general property.

COLLOIDS (CALL-loids) — COLLOIDS

Very small, finely divided solids (particles that do not dissolve) that remain dispersed in a liquid for a long time due to their small size and electrical charge. When most of the particles in water have a negative electrical charge, they tend to repel each other. This repulsion prevents the particles from sticking together, becoming heavier, and settling out.

COMBINED SEWER — COMBINED SEWER

A sewer designed to carry both sanitary wastewaters and storm or surface water runoff.

COMBINED SYSTEM — COMBINED SYSTEM

A sewer designed to carry both sanitary wastewaters and storm or surface water runoff.

COMBINED WASTEWATER — COMBINED WASTEWATER

A mixture of storm or surface runoff and other wastewater such as domestic or industrial wastewater.

COMMERCIAL CONTRIBUTION — COMMERCIAL CONTRIBUTION

Liquid and liquid-carried wastes dumped by commercial establishments into the wastewater collection system. Used in this context, commercial contributions are distinct from domestic and industrial sources of wastewater contributions. Examples of high-yield commercial sources are laundries, restaurants and hotels.

COMMERCIAL TELEVISION QUALITY — COMMERCIAL TELEVISION QUALITY

Refers to picture quality about the level of a television picture. Commercial television pictures are considered to have resolutions between 350 and 525 lines. This is higher quality than surveillance quality and lower quality than inspection quality pictures.

COMMINUTOR (com-mih-NEW-ter) — COMMINUTOR

A device used to reduce the size of the solid chunks in wastewater by shredding (comminuting). The shredding action is like many scissors cutting to shreds all the large solids in the wastewater.

COMPACTED BACKFILL | COMPACTED BACKFILL

(See BACKFILL COMPACTION)

COMPACTED BEDDING | COMPACTED BEDDING

(See BEDDING COMPACTION)

COMPACTION | COMPACTION

Tamping or rolling of a material to achieve a surface or density that is able to support predicted loads.

COMPACTION TEST | COMPACTION TEST

Any method of determining the weight a compacted material is able to support without damage or displacement. Usually stated in pounds per square foot.

COMPETENT PERSON | COMPETENT PERSON

A competent person is defined by OSHA as a person capable of identifying existing and predictable hazards in the surroundings, or working conditions which are unsanitary, hazardous or dangerous to employees, and who has authorization to take prompt corrective measures to eliminate the hazards.

COMPUTED COLLECTION SYSTEM CONTRIBUTION | COMPUTED COLLECTION SYSTEM CONTRIBUTION

The part of a collection system flow computed to be actual domestic and industrial wastewater. Applied to infiltration/inflow research, the computed domestic and industrial wastewater contribution is subtracted from a total flow to determine infiltration/inflow amounts.

COMPUTED COMMERCIAL CONTRIBUTION | COMPUTED COMMERCIAL CONTRIBUTION

That part of a collection system flow computed to originate in the commercial establishments on the basis of expected flows from all commercial sources.

COMPUTED CONTRIBUTION | COMPUTED CONTRIBUTION

A liquid or liquid-carried contribution to a collection system that is computed on the basis of expected discharges from all of the sources as opposed to actual measurement or metering. Also see ESTIMATED CONTRIBUTION.

COMPUTED DOMESTIC CONTRIBUTION | COMPUTED DOMESTIC CONTRIBUTION

That part of a collection system flow computed to originate in the residential facilities based on the average flow contribution from each person.

COMPUTED FACILITY CONTRIBUTION | COMPUTED FACILITY CONTRIBUTION

The computed liquid-waste discharge from a single facility based on the sources of waste flows in the facility.

COMPUTED INDUSTRIAL CONTRIBUTION | COMPUTED INDUSTRIAL CONTRIBUTION

The computed liquid-waste discharge from industrial operations based on the expected discharges from all sources.

COMPUTED PER CAPITA CONTRIBUTION | COMPUTED PER CAPITA CONTRIBUTION

The computed wastewater contribution from a domestic area, based on the population of the area. In the United States, the daily average wastewater contribution is considered to be 100 gallons per capita per day (100 GPCD).

COMPUTED TOTAL CONTRIBUTION | COMPUTED TOTAL CONTRIBUTION

The total anticipated load on a wastewater treatment plant or the total anticipated flow in any collection system area based on the combined computed contributions of all connections to the system.

CONCENTRIC MANHOLE CONE | CONCENTRIC MANHOLE CONE

Cone tapers uniformly from barrel to manhole cover.

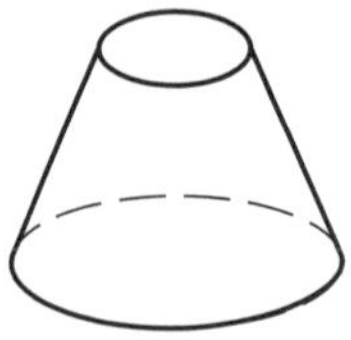

CONCRETE CRADLE | CONCRETE CRADLE

A device made of concrete that is designed to support sewer pipe.

CONDUCTOR CONDUCTOR

(1) A pipe which carries a liquid load from one point to another point. In a wastewater collection system, a conductor is often a large pipe with no service connections. Also called a CONDUIT, interceptor (see INTERCEPTING SEWER) or INTERCONNECTOR.

(2) In plumbing, a line conducting water from the roof to the storm drain or other means of disposal. Also called a DOWNSPOUT.

(3) In electricity, a substance, body, device or wire that readily conducts or carries electric current.

CONDUIT CONDUIT

Any artificial or natural duct, either open or closed, for carrying fluids from one point to another. An electrical conduit carries electricity.

CONFINED SPACE CONFINED SPACE

Confined space means a space that:

A. Is large enough and so configured that an employee can bodily enter and perform assigned work; and

B. Has limited or restricted means for entry or exit (for example, manholes, tanks, vessels, silos, storage bins, hoppers, vaults, and pits are spaces that may have limited means of entry); and

C. Is not designed for continuous employee occupancy.

(Definition from the Code of Federal Regulations (CFR) Title 29 Part 1910.146.)

CONFINED SPACE, CLASS "A" CONFINED SPACE, CLASS "A"

A confined space that presents a situation that is immediately dangerous to life or health (IDLH). These include but are not limited to oxygen deficiency, explosive or flammable atmospheres, and/or concentrations of toxic substances.

(Definition from NIOSH, "Criteria for a Recommended Standard: Working in Confined Spaces.")

CONFINED SPACE, CLASS "B" CONFINED SPACE, CLASS "B"

A confined space that has the potential for causing injury and illness, if preventive measures are not used, but not immediately dangerous to life and health.

(Definition from NIOSH, "Criteria for a Recommended Standard: Working in Confined Spaces.")

CONFINED SPACE, CLASS "C" CONFINED SPACE, CLASS "C"

A confined space in which the potential hazard would not require any special modification of the work procedure.

(Definition from NIOSH, "Criteria for a Recommended Standard: Working in Confined Spaces.")

CONFINED SPACE, NON-PERMIT CONFINED SPACE, NON-PERMIT

A non-permit confined space is a confined space that does not contain or, with respect to atmospheric hazards, have the potential to contain any hazard capable of causing death or serious physical harm.

CONFINED SPACE, PERMIT-REQUIRED (PERMIT SPACE) CONFINED SPACE, PERMIT-REQUIRED (PERMIT SPACE)

A confined space that has one or more of the following characteristics:

- Contains or has a potential to contain a hazardous atmosphere,
- Contains a material that has the potential for engulfing an entrant,
- Has an internal configuration such that an entrant could be trapped or asphyxiated by inwardly converging walls or by a floor which slopes downward and tapers to a smaller cross section, or
- Contains any other recognized serious safety or health hazard.

(Definition from the Code of Federal Regulations (CFR) Title 29 Part 1910.146.)

CONTAMINATION CONTAMINATION

The introduction into water of microorganisms, chemicals, toxic substances, wastes, or wastewater in a concentration that makes the water unfit for its next intended use.

CONTRIBUTION CONTRIBUTION

Waters, wastewaters or liquid-carried wastes entering a wastewater collection system.

CORROSION CORROSION

The gradual decomposition or destruction of a material due to chemical action, often due to an electrochemical reaction. Corrosion starts at the surface of a material and moves inward, such as the chemical action upon manholes and sewer pipe materials.

COULOMB (COO-lahm) COULOMB

A measurement of the amount of electrical charge carried by an electric current of one ampere in one second. One coulomb equals about 6.25 x 10^{18} electrons (6,250,000,000,000,000,000 electrons).

COUPLING COUPLING

(1) A threaded sleeve used to connect two pipes.

(2) A device used to connect two adjacent parts, such as a pipe coupling, hose coupling or drive coupling.

COUPON COUPON

A steel specimen inserted into wastewater to measure the corrosiveness of the wastewater. The rate of corrosion is measured as the loss of weight of the coupon or change in its physical characteristics. Measure the weight loss (in milligrams) per surface area (in square decimeters) exposed to the wastewater per day. 10 decimeters = 1 meter = 100 centimeters.

COVERAGE RATIO COVERAGE RATIO

The coverage ratio is a measure of the ability of the utility to pay the principal and interest on loans and bonds (this is known as "debt service") in addition to any unexpected expenses.

CROSS BRACES CROSS BRACES

Shoring members placed across a trench to hold other horizontal and vertical shoring members in place.

CROSS CONNECTION CROSS CONNECTION

(1) A connection between a storm drain system and a sanitary collection system.

(2) Less frequently used to mean a connection between two sections of a collection system to handle anticipated overloads of one system.

(3) A connection between drinking (potable) water and an unapproved water supply.

CURB INLET CURB INLET

A chamber or well built at the curbline of a street to admit gutter flow to the storm water drainage system. Also see STORM WATER INLET and CATCH BASIN.

CURRENT CURRENT

A movement or flow of electricity. Water flowing in a pipe is measured in gallons per second past a certain point, not by the number of water molecules going past a point. Electric current is measured by the number of coulombs per second flowing past a certain point in a conductor. A coulomb is equal to about 6.25 x 10^{18} electrons (6,250,000,000,000,000,000 electrons). A flow of one coulomb per second is called one ampere, the unit of the rate of flow of current.

D

DANGER DANGER

The word *DANGER* is used where an immediate hazard presents a threat of death or serious injury to employees. Also see CAUTION, NOTICE, and WARNING.

DANGEROUS AIR CONTAMINATION DANGEROUS AIR CONTAMINATION

An atmosphere presenting a threat of causing death, injury, acute illness, or disablement due to the presence of flammable and/or explosive, toxic or otherwise injurious or incapacitating substances.

A. Dangerous air contamination due to the flammability of a gas or vapor is defined as an atmosphere containing the gas or vapor at a concentration greater than 10 percent of its lower explosive (lower flammable) limit.

B. Dangerous air contamination due to a combustible particulate is defined as a concentration greater than 10 percent of the minimum explosive concentration of the particulate.

C. Dangerous air contamination due to the toxicity of a substance is defined as the atmospheric concentration immediately hazardous to life or health.

DATA-VIEW DATA-VIEW

A high-speed reporting and recording system used with closed-circuit pipeline television equipment. Data-view provides digital indexing of date, job number, footages and air test pressures in the television picture itself. Where videotape recordings of television pipe inspections or pipe sealing activities are made, data-view reports are automatically recorded on the taped pictures.

DEADEND MANHOLE — DEADEND MANHOLE

A manhole located at the upstream end of a sewer and having no inlet pipe. Also called a TERMINAL MANHOLE.

DEBRIS (de-BREE) — DEBRIS

Any material in wastewater found floating, suspended, settled or moving along the bottom of a sewer. This material may cause stoppages by getting hung up on roots or settling out in a sewer. Debris includes grit, paper, plastic, rubber, silt, and all materials except liquids.

DEBRIS, INFILTRATED — DEBRIS, INFILTRATED

(See INFILTRATED DEBRIS)

DEBT SERVICE — DEBT SERVICE

The amount of money required annually to pay the (1) interest on outstanding debts; or (2) funds due on a maturing bonded debt or the redemption of bonds.

DECIBEL (DES-uh-bull) — DECIBEL

A unit for expressing the relative intensity of sounds on a scale from zero for the average least perceptible sound to about 130 for the average level at which sound causes pain to humans. Abbreviated dB.

DECOMPOSED PIPE — DECOMPOSED PIPE

Pipe which has been destroyed or portions of pipe weakened by chemical actions.

DECOMPOSITION, DECAY — DECOMPOSITION, DECAY

Processes that convert unstable materials into more stable forms by chemical or biological action. Waste treatment encourages decay in a controlled situation so that material may be disposed of in a stable form. When organic matter decays under anaerobic conditions (putrefaction), undesirable odors are produced. The aerobic processes in common use for wastewater treatment produce much less objectionable odors.

DEFECT — DEFECT

A point where a pipe or system structure has been damaged or has a fault.

DEFECT, SURFACED — DEFECT, SURFACED

(See SURFACED DEFECT)

DEFLECTED — DEFLECTED

(1) Pipe which has been forced out of round by external pressures. This happens mainly to fiber and plastic pipes where backfill compaction has resulted in unequal pressures on all sides of the pipe.

(2) Pipe whose direction has been changed either to the left, right, up, or down.

DEGRADATION (deh-gruh-DAY-shun) — DEGRADATION

The conversion or breakdown of a substance to simpler compounds. For example, the degradation of organic matter to carbon dioxide and water.

DESTROYED PIPE — DESTROYED PIPE

Pipe which has been damaged, decomposed, deflected, crushed or collapsed to a point that it must be replaced.

DETENTION — DETENTION

The delay or holding of the flow of water and water-carried wastes in a pipe system. This can be caused by a restriction in the pipe, a stoppage or a dip. Detention also means the time water is held or stored in a basin or a wet well. Sometimes called RETENTION.

DETRITUS (dee-TRY-tus) — DETRITUS

The heavy, coarse mixture of grit and organic material carried by wastewater. Also called GRIT.

DEWATER — DEWATER

To drain or remove water from an enclosure. A structure may be dewatered so that it can be inspected or repaired. Dewater also means draining or removing water from sludge to increase the solids concentration.

DIGITAL READOUT — DIGITAL READOUT

The use of numbers to indicate the value or measurement of a variable. The readout of an instrument by a direct, numerical reading of the measured value. The signal sent to such readouts is usually an analog signal.

DIP — DIP

A point in a sewer pipe where a drain grade defect results in a puddle of standing water when there is no flow. If the grade defect is severe enough to cause the standing water to fill the pipe at any point (preventing passage of air through the pipe), it is called a "trap dip," "full dip" or "filled dip."

DIRECT CURRENT (D.C.) — DIRECT CURRENT (D.C.)

Electric current flowing in one direction only and essentially free from pulsation.

DISCHARGE HEAD — DISCHARGE HEAD

The pressure (in pounds per square inch or psi) measured at the centerline of a pump discharge and very close to the discharge flange, converted into feet. The pressure is measured from the centerline of the pump to the hydraulic grade line of the water in the discharge pipe.

Discharge Head, ft = (Discharge Pressure, psi)(2.31 ft/psi)

DISINFECTION (dis-in-FECT-shun) — DISINFECTION

The process designed to kill or inactivate most microorganisms in wastewater, including essentially all pathogenic (disease-causing) bacteria. There are several ways to disinfect, with chlorination being the most frequently used in water and wastewater treatment plants.

DISPLACED PIPE — DISPLACED PIPE

A run or section of sewer pipe that has been pushed out of alignment by external forces.

DISTURBED SOIL — DISTURBED SOIL

Soil which has been changed from its natural condition by excavation or other means.

DIVERSION CHAMBER — DIVERSION CHAMBER

A chamber or box which contains a device for diverting or drawing off all or part of a flow or for discharging portions of the total flow to various outlets. Also called a REGULATOR.

DOMESTIC — DOMESTIC

Residential living facilities. A domestic area will be predominantly residential in occupancy and is sometimes referred to as a "bedroom area" or "bedroom community."

DOMESTIC CONTRIBUTION — DOMESTIC CONTRIBUTION

Wastes originating in a residential facility or dwelling. In this use, it means the type and quantity of wastes are different from commercial and industrial or agricultural wastes.

DOMESTIC SERVICE — DOMESTIC SERVICE

A connection to a sewer system for hookup of a residential-type building.

DOWNSPOUT — DOWNSPOUT

In plumbing, the water conductor from the roof gutters or roof catchment to the storm drain or other means of disposal. Also called a "roof leader" or "roof drain."

DOWNSTREAM — DOWNSTREAM

The direction of the flow of water. In the lower part of a sewer or collection system or in that direction.

DRAGLINE — DRAGLINE

A machine that drags a bucket down the intended line of a trench to dig or excavate the trench. Also used to dig holes and move soil or aggregate.

DRIFT — DRIFT

The difference between the actual value and the desired value (or set point); characteristic of proportional controllers that do not incorporate reset action. Also called "offset."

DROP JOINT — DROP JOINT

A sewer pipe joint where one part has dropped out of alignment. Also called a VERTICAL OFFSET.

DROP MANHOLE — DROP MANHOLE

A main line or house service line lateral entering a manhole at a higher elevation than the main flow line or channel. If the higher elevation flow is routed to the main manhole channel outside of the manhole, it is called an "outside drop." If the flow is routed down through the manhole barrel, the pipe down to the manhole channel is called an "inside drop."

DRY PIT — DRY PIT

(See DRY WELL)

DRY WELL — DRY WELL

A dry room or compartment in a lift station, near or below the water level, where the pumps are located.

DWELLING — DWELLING

A structure for residential occupancy.

DYNAMIC HEAD — DYNAMIC HEAD

When a pump is operating, the vertical distance (in feet) from a point to the energy grade line. Also see TOTAL DYNAMIC HEAD, STATIC HEAD, and ENERGY GRADE LINE.

E

EARTH SHIFT — EARTH SHIFT

The movement or dislocation of underground soil or structure. Earth shift is usually caused by external forces such as surface loads, slides, stresses or nearby construction, water movements or seismic forces.

EASEMENT — EASEMENT

Legal right to use the property of others for a specific purpose. For example, a utility company may have a five-foot easement along the property line of a home. This gives the utility the legal right to install and maintain a sewer line within the easement.

ECCENTRIC MANHOLE CONE — ECCENTRIC MANHOLE CONE

Cone tapers nonuniformly from barrel to manhole cover with one side usually vertical.

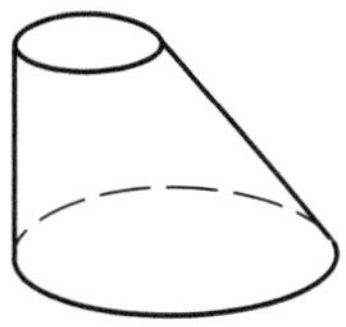

EFFLUENT — EFFLUENT

Wastewater or other liquid—raw (untreated), partially or completely treated—flowing *FROM* a reservoir, basin, treatment process, or treatment plant.

ELBOW — ELBOW

A pipe fitting that connects two pipes at an angle. The angle is usually 90 degrees unless another angle is stated. Also called an "ell."

ELECTROLYTE (ee-LECK-tro-LITE) SOLUTION — ELECTROLYTE SOLUTION

A special solution that is capable of conducting electricity.

ELECTROMOTIVE FORCE (E.M.F.) — ELECTROMOTIVE FORCE (E.M.F.)

The electrical pressure available to cause a flow of current (amperage) when an electric circuit is closed. Also called VOLTAGE.

ELECTRON — ELECTRON

(1) A very small, negatively charged particle which is practically weightless. According to the electron theory, all electrical and electronic effects are caused either by the movement of electrons from place to place or because there is an excess or lack of electrons at a particular place.

(2) The part of an atom that determines its chemical properties.

ELEVATION ELEVATION

The height to which something is elevated, such as the height above sea level.

EMULSION (e-MULL-shun) EMULSION

A liquid mixture of two or more liquid substances not normally dissolved in one another; one liquid is held in suspension in the other.

ENCLOSED SPACE ENCLOSED SPACE

(See CONFINED SPACE)

ENERGY GRADE LINE (EGL) ENERGY GRADE LINE (EGL)

A line that represents the elevation of energy head (in feet) of water flowing in a pipe, conduit or channel. The line is drawn above the hydraulic grade line (gradient) a distance equal to the velocity head ($V^2/2g$) of the water flowing at each section or point along the pipe or channel. Also see HYDRAULIC GRADE LINE.

[SEE DRAWING ON PAGE 687]

ENGULFMENT ENGULFMENT

Engulfment means the surrounding and effective capture of a person by a liquid or finely divided (flowable) solid substance that can be aspirated to cause death by filling or plugging the respiratory system or that can exert enough force on the body to cause death by strangulation, constriction, or crushing.

ENTRAIN ENTRAIN

To trap bubbles in water either mechanically through turbulence or chemically through a reaction.

ESTIMATED CONTRIBUTION ESTIMATED CONTRIBUTION

A contribution to a collection system that is estimated rather than computed. The distinction between computed and estimated in such cases is difficult to specify or define. Also see COMPUTED CONTRIBUTION.

ESTIMATED FLOW ESTIMATED FLOW

A rough guess of the amount of flow in a collection system. When greater accuracy is needed, flow could be computed using average or typical flow quantities. Even greater accuracy would result from metering or otherwise measuring the actual flow.

EXFILTRATION (EX-fill-TRAY-shun) EXFILTRATION

Liquid wastes and liquid-carried wastes which unintentionally leak out of a sewer pipe system and into the environment.

EXTRADOS EXTRADOS

The upper outside curve of a sewer pipe or conduit.

EXTREMELY HAZARDOUS WASTE EXTREMELY HAZARDOUS WASTE

Any hazardous waste or mixture of hazardous wastes which, if any human exposure should occur, may likely result in death, disabling personal injury or illness during, or as a proximate result of, any disposal of such waste or mixture of wastes because of its quantity, concentration, or chemical characteristics. (Subsection 23115 of Article 2, Chapter 6.5, Division 20, of the California Health and Safety Code.) Also see HAZARDOUS WASTE.

F

FAIR LEAD PULLEY (fair LEE-d pully) FAIR LEAD PULLEY

A pulley that is placed in a manhole to guide TV camera electric cables and the pull cable into the sewer when inspecting pipelines.

FAULT FAULT

(1) A fracture in the earth's crust that leaves land on one side of the crack out of alignment with the other side. Faults are generally a result of earth shifts and earthquakes.

(2) (See DEFECT)

FILLED DIP FILLED DIP

(See DIP)

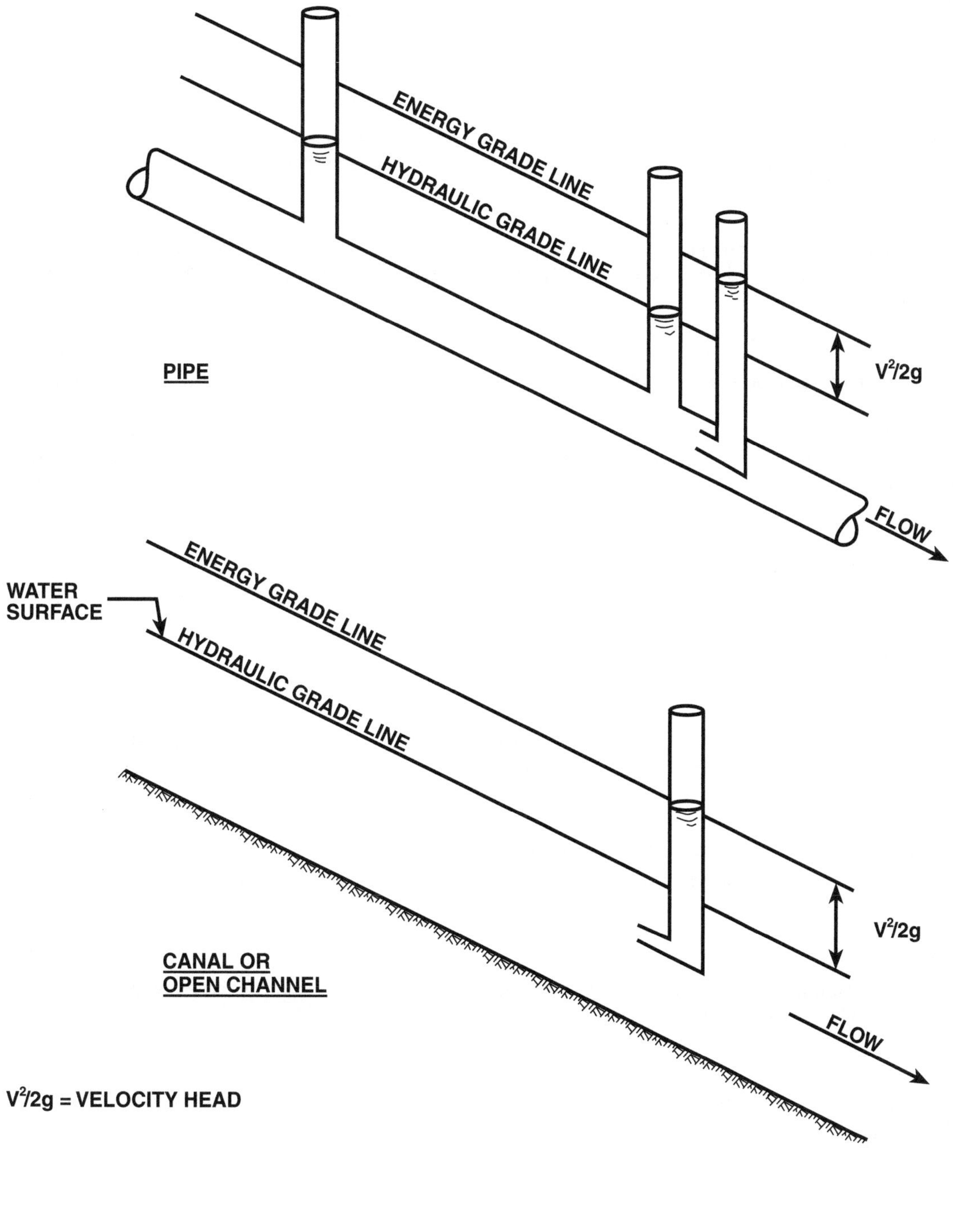
ENERGY GRADE LINE
HYDRAULIC GRADE LINE
PIPE
V²/2g
FLOW
WATER SURFACE
ENERGY GRADE LINE
HYDRAULIC GRADE LINE
CANAL OR OPEN CHANNEL
V²/2g
FLOW
V²/2g = VELOCITY HEAD

ENERGY GRADE LINE and HYDRAULIC GRADE LINE

FIT TEST FIT TEST

The use of a procedure to qualitatively or quantitatively evaluate the fit of a respirator on an individual.

FLAP GATE FLAP GATE

A hinged gate that is mounted at the top of a pipe or channel to allow flow in only one direction. Flow in the wrong direction closes the gate. Also see CHECK VALVE and TIDE GATE.

FLAT FLAT

A flat is the length of one side of a nut.

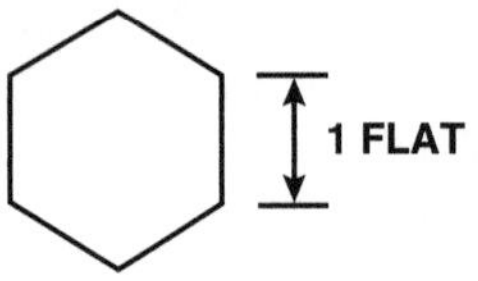

FLOAT (CONTROL) FLOAT (CONTROL)

A device used to measure the elevation of the surface of water. The float rests on the surface of the water and rises or falls with it. The elevation of the water surface is measured by a rod, chain, rope or tape attached to the float.

FLOAT LINE FLOAT LINE

A length of rope or heavy twine attached to a float, plastic jug or parachute to be carried by the flow in a sewer from one manhole to the next. This is called "stringing the line" and is used for pulling through winch cables, such as for bucket machine work or closed-circuit television work.

FLOTATION FLOTATION

(1) The stress or forces on a pipeline or manhole structure located below a water table which tend to lift or float the pipeline or manhole structure.

(2) The process of raising suspended matter to the surface of the liquid in a tank where it forms a scum layer that can be removed by skimming. The suspended matter is raised by aeration, the evolution of gas, the use of chemicals, electrolysis, heat or bacterial decomposition.

FLOTSAM (FLOAT-sam) FLOTSAM

Material floating or drifting about on the surface of a body of water.

FLOW FLOW

The continuous movement of a liquid from one place to another.

FLOW ISOLATION FLOW ISOLATION

A procedure used to measure inflow and infiltration (I/I). A section of sewer is blocked off or isolated and the flow from the section is measured.

FLOW LINE FLOW LINE

(1) The top of the wetted line, the water surface or the hydraulic grade line of water flowing in an open channel or partially full conduit.

(2) The lowest point of the channel inside a pipe or manhole. See INVERT. *NOTE:* (2) is an improper definition, although used by some contractors.

FLOW RECORDING FLOW RECORDING

A record of a flow measurement past any selected point. Usually consists of time, velocity and amount (in gallons) with maximum and minimum rates as well as the total amount over a given time period.

FLUME FLUME

An open conduit of wood, masonry, metal, or plastic constructed on a grade and sometimes elevated. Sometimes called an "aqueduct" or "channel."

FLUME, PARSHALL FLUME, PARSHALL

(See PARSHALL FLUME)

FLUSHER BRANCH FLUSHER BRANCH

A line built specifically to allow the introduction of large quantities of water to the collection system so the lines can be "flushed out" with water. Also installed to provide access for equipment to clear stoppages in a sewer.

FLUSHING FLUSHING

The removal of deposits of material which have lodged in sewers because of inadequate velocity of flows. Water is discharged into the sewers at such rates that the larger flow and higher velocities are sufficient to remove the material.

FOCAL LENGTH FOCAL LENGTH

The distance of a focus from the surface of a lens (such as a camera lens) to the focal point.

FORCE MAIN FORCE MAIN

A pipe that carries wastewater under pressure from the discharge side of a pump to a point of gravity flow downstream.

FRIABILITY (FRY-uh-BILL-uh-tee) FRIABILITY

The ability of a soil or substance to crumble under moderate or light pressure.

FRICTION LOSS FRICTION LOSS

The head lost by water flowing in a stream or conduit as the result of the disturbances set up by the contact between the moving water and its containing conduit and by intermolecular friction.

FULL DIP FULL DIP

(See DIP)

FUSE FUSE

A protective device having a strip or wire of fusible metal which, when placed in a circuit, will melt and break the electric circuit if heated too much. High temperatures will develop in the fuse when a current flows through the fuse in excess of that which the circuit will carry safely.

G

GIS GIS

Geographic **I**nformation **S**ystem. A computer program that combines mapping with detailed information about the physical locations of structures such as pipes, valves, and manholes within geographic areas. The system is used to help operators and maintenance personnel locate utility system features or structures and to assist with the scheduling and performance of maintenance activities.

GPCD GPCD

Initials standing for "Gallons Per Capita Per Day."

GPD GPD

Initials standing for "Gallons Per Day."

GPM GPM

Initials standing for "Gallons Per Minute."

GPY GPY

Initials standing for "Gallons Per Year."

GAGE GAGE

A device for checking or measuring a particular dimension of something, using specific standardized units. For example, a gage might measure the elevation of a water surface, the velocity of flowing water, the pressure of water, the amount or intensity of precipitation, and the depth of snowfall. Gages also are used to determine the location or position of equipment during installation and after operation.

GAS (SEWER) GAS (SEWER)

(See SEWER GAS)

GEL OR GELL GEL OR GELL

(1) A term sometimes applied to chemical grout. See CHEMICAL GROUT.

(2) A form of matter in a colloidal state that does not dissolve, but remains suspended in a solvent. It does not precipitate out of the solvent without the intervention of heat or of an electrolyte.

GEOGRAPHIC INFORMATION SYSTEM (GIS) GEOGRAPHIC INFORMATION SYSTEM (GIS)

A computer program that combines mapping with detailed information about the physical locations of structures such as pipes, valves, and manholes within geographic areas. The system is used to help operators and maintenance personnel locate utility system features or structures and to assist with the scheduling and performance of maintenance activities.

GEOLOGY GEOLOGY

The science that deals with the origin, history and structures of the earth, as recorded in the rocks, together with the forces and processes now operating to modify rocks.

GEOLOGY, SUBSOIL GEOLOGY, SUBSOIL

(See SUBSOIL GEOLOGY)

GEOLOGY, SUBSTRUCTURE GEOLOGY, SUBSTRUCTURE

(See SUBSOIL GEOLOGY)

GRADE GRADE

(1) The elevation of the invert (or bottom) of a pipeline, canal, culvert, sewer, or similar conduit.

(2) The inclination or slope of a pipeline, conduit, stream channel, or natural ground surface; usually expressed in terms of the ratio or percentage of number of units of vertical rise or fall per unit of horizontal distance. A 0.5 percent grade would be a drop of one-half foot per hundred feet of pipe.

GRADE RING GRADE RING

A precast concrete ring 4 to 12 inches high which is placed on top of a manhole cone to raise the manhole cover frame flush with the surface grade. Sometimes called a "spacer."

GRADIENT GRADIENT

The upward or downward slope of a pipeline.

GRANULAR GRANULAR

Any substance that appears to consist of separate granules or grains. Examples are sand and gravel.

GRAVITY GRAVITY

The attraction of the earth to any substance—solid, liquid or gas.

GRAVITY FLOW GRAVITY FLOW

Water or wastewater flowing from a higher elevation to a lower elevation due to the force of gravity. The water does not flow due to energy provided by a pump. Wherever possible, wastewater collection systems are designed to use the force of gravity to convey waste liquids and solids.

GRAVITY, SPECIFIC GRAVITY, SPECIFIC

(See SPECIFIC GRAVITY)

GREASE GREASE

In a collection system, grease is considered to be the residues of fats, detergents, waxes, free fatty acids, calcium and magnesium soaps, mineral oils, and certain other nonfatty materials which tend to separate from water and coagulate as floatables or scums.

GREASE BUILDUP GREASE BUILDUP

Any point in a collection system where coagulated and solidified greases accumulate and build up. Many varieties of grease have high adhesive characteristics and collect other solids, forming restrictions and stoppages in collection systems.

GREASE TRAP GREASE TRAP

A receptacle designed to collect and retain grease and fatty substances usually found in kitchens or from similar wastes. It is installed in the drainage system between the kitchen or other point of production of the waste and the building wastewater collection line. Commonly used to control grease from restaurants.

GRIT GRIT

The heavy mineral material present in wastewater such as sand, coffee grounds, eggshells, gravel and cinders. Grit tends to settle out at flow velocities below 2 ft/sec and accumulate in the invert or bottoms of the pipelines. Also called DETRITUS.

GRIT CATCHER — GRIT CATCHER

A chamber usually placed at the upper end of a depressed collection line or at other points on combined or storm water collection lines where wear from grit is possible. The chamber is sized and shaped to reduce the velocity of flow through it and thus permit the settling out of grit. Also called a "sand catcher." See GRIT CHAMBER and SAND TRAP.

GRIT CHAMBER — GRIT CHAMBER

A detention chamber or an enlargement of a collection line designed to reduce the velocity of flow of the liquid to permit the separation of mineral solids from organic solids by differential sedimentation.

GRIT CHANNEL — GRIT CHANNEL

(1) An enlargement in a collection line where grit can easily settle out of the flow.

(2) The waterway of a grit chamber.

GRIT COLLECTOR — GRIT COLLECTOR

A device placed in a grit chamber to convey deposited grit to a point of collection for ultimate disposal.

GRIT COMPARTMENT — GRIT COMPARTMENT

The portion of the grit chamber in which grit is collected and stored before removal.

GRIT TANK — GRIT TANK

A structure located at the inlet to a treatment plant for the accumulation and removal of grit.

GRIT TRAP — GRIT TRAP

(1) A permanent structure built into a manhole (or other convenient location in a collection system) for the accumulation and easy removal of grit.

(2) (See SAND TRAP)

GROSS VEHICLE WEIGHT (GVW) — GROSS VEHICLE WEIGHT (GVW)

The total weight of a single vehicle including its load.

GROSS VEHICLE WEIGHT RATING (GVWR) — GROSS VEHICLE WEIGHT RATING (GVWR)

The maximum weight rating specified by the manufacturer for a single vehicle including its load.

GROUNDWATER — GROUNDWATER

Subsurface water in the saturation zone from which wells and springs are fed. In a strict sense the term applies only to water below the water table. Also called "phreatic water" and "plerotic water."

GROUNDWATER DEPTH — GROUNDWATER DEPTH

The distance of the groundwater table below the surface at any selected location.

GROUNDWATER ELEVATION — GROUNDWATER ELEVATION

The elevation of the groundwater table above mean sea level at any selected location.

GROUNDWATER TABLE — GROUNDWATER TABLE

The average depth or elevation of the groundwater over a selected area.

GROUNDWATER TABLE, ARTIFICIAL — GROUNDWATER TABLE, ARTIFICIAL

(See ARTIFICIAL GROUNDWATER TABLE)

GROUNDWATER TABLE, SEASONAL — GROUNDWATER TABLE, SEASONAL

(See SEASONAL WATER TABLE)

GROUNDWATER TABLE, TEMPORARY — GROUNDWATER TABLE, TEMPORARY

(See TEMPORARY GROUNDWATER TABLE)

GROUT — GROUT

A substance in a paste or liquid form which solidifies after placement or treatment. Used to fill spaces, holes or voids in other materials.

GROUT, CHEMICAL GROUT, CHEMICAL

(See CHEMICAL GROUT)

GROUTING, PRESSURE GROUTING, PRESSURE

(See CHEMICAL GROUTING)

GUNITE GUNITE

A mixture of sand and cement applied pneumatically that forms a high-density, resistant concrete.

H

HAIRLINE CRACK HAIRLINE CRACK

A stress crack in a pipe; the crack looks like a piece of hair.

HAND ROD HAND ROD

A sewer rod that can be inserted manually (by hand) into a sewer to clear a stoppage or to prevent a stoppage from developing.

HANDHOLE TRAP HANDHOLE TRAP

A device made of pipe fittings used to prevent sewer gases escaping from the branch or lateral sewer from entering a building sewer.

HAZARDOUS WASTE HAZARDOUS WASTE

Any waste material or mixture of wastes which is toxic, corrosive, flammable, an irritant, a strong sensitizer which generates pressure through decomposition, heat or other means, if such a waste or mixture of wastes may cause substantial personal injury, serious illness or harm to wildlife, during, or as a proximate result of any disposal of such wastes or mixture of wastes. The terms "toxic," "corrosive," "flammable," "irritant," and "strong sensitizer" shall be given the same meanings as given by the California Hazardous Substance Act (Chapter 13 (commencing with Section 28740) of Division 21)(Subsection 23117 of Article 2, Chapter 6.5, Division 20, of the California Health and Safety Code). Also see EXTREMELY HAZARDOUS WASTE.

HAZARDOUS WASTE HAZARDOUS WASTE

A waste, or combination of wastes, which because of its quantity, concentration, or physical, chemical, or infectious characteristics may:

1. Cause, or significantly contribute to, an increase in mortality or an increase in serious, irreversible, or incapacitating reversible, illness; or
2. Pose a substantial present or potential hazard to human health or the environment when improperly treated, stored, transported, or disposed of or otherwise managed; and
3. Normally not be discharged into a sanitary sewer; subject to regulated disposal.

(Resource Conservation and Recovery Act (RCRA) definition.)

HEAD HEAD

The vertical distance, height or energy of water above a point. A head of water may be measured in either height (feet) or pressure (pounds per square inch (psi)). Also see DISCHARGE HEAD, DYNAMIC HEAD, STATIC HEAD, SUCTION HEAD, SUCTION LIFT, and VELOCITY HEAD.

HERBICIDE (HERB-uh-SIDE) HERBICIDE

A compound, usually a manmade organic chemical, used to kill or control plant growth.

HERTZ HERTZ

The number of complete electromagnetic cycles or waves in one second of an electric or electronic circuit. Also called the frequency of the current. Abbreviated Hz.

HIGH-VELOCITY CLEANER HIGH-VELOCITY CLEANER

A machine designed to remove grease and debris from the smaller diameter sewer pipes with high-velocity jets of water. Also called a "jet cleaner," "jet rodder," "hydraulic cleaner," "high-pressure cleaner," or "hydro jet."

HOUSE CONNECTION HOUSE CONNECTION

(See BUILDING SEWER)

HOUSE SERVICE HOUSE SERVICE

(See BUILDING SERVICE)

HOUSE SEWER HOUSE SEWER

(See BUILDING SEWER)

HUB HUB

In pipe fitting, the enlarged female end of a pipe into which the male end fits. Also called a BELL.

HYDRAULIC BLOCK HYDRAULIC BLOCK

The movement of water in such a way that the flow of water from one direction blocks or hinders the flow of water from another direction.

HYDRAULIC CLEANER HYDRAULIC CLEANER

(See HIGH-VELOCITY CLEANER)

HYDRAULIC CLEANING HYDRAULIC CLEANING

Cleaning pipe with water under enough pressure to produce high water velocities.

(1) Using a high-velocity cleaner.

(2) Using a ball, kite or similar sewer cleaning device.

(3) Using a scooter.

(4) Flushing.

HYDRAULIC GRADE LINE (HGL) HYDRAULIC GRADE LINE (HGL)

The surface or profile of water flowing in an open channel or a pipe flowing partially full. If a pipe is under pressure, the hydraulic grade line is at the level water would rise to in a small tube connected to the pipe. To reduce the release of odors from sewers, the water surface or hydraulic grade line should be kept as smooth as possible. Also see ENERGY GRADE LINE.

[SEE DRAWING ON PAGE 687]

HYDRAULIC POPULATION EQUIVALENT HYDRAULIC POPULATION EQUIVALENT

A flow of 100 gallons per day is the hydraulic or flow equivalent to the contribution or flow from one person. Population equivalent = 100 GPCD or gallons per capita per day.

HYDROGEN SULFIDE GAS (H_2S) HYDROGEN SULFIDE GAS (H_2S)

Hydrogen sulfide is a gas with a rotten egg odor. This gas is produced under anaerobic conditions. Hydrogen sulfide gas is particularly dangerous because it dulls the sense of smell so that you don't notice it after you have been around it for a while. In high concentrations, hydrogen sulfide gas is only noticeable for a very short time before it dulls the sense of smell. The gas is very poisonous to the respiratory system, explosive, flammable, colorless and heavier than air.

HYDROLOGY HYDROLOGY

The applied science concerned with the waters of the earth in all their states—their occurrence, distribution, and circulation through the unending hydrologic cycle of precipitation, consequent runoff, stream flow, infiltration, and storage, eventual evaporation, and reprecipitation. Hydrology is concerned with the physical, chemical, and physiological reactions of water with the rest of the earth and its relation to the life of the earth.

HYDROPHILIC (HI-dro-FILL-ick) HYDROPHILIC

Having a strong affinity (liking) for water. The opposite of HYDROPHOBIC.

HYDROPHOBIC (HI-dro-FOE-bick) HYDROPHOBIC

Having a strong aversion (dislike) for water. The opposite of HYDROPHILIC.

I

IDLH IDLH

Immediately **D**angerous to **L**ife or **H**ealth. The atmospheric concentration of any toxic, corrosive, or asphyxiant substance that poses an immediate threat to life or would cause irreversible or delayed adverse health effects or would interfere with an individual's ability to escape from a dangerous atmosphere.

I/I I/I

(See INFILTRATION/INFLOW)

IMPELLER IMPELLER

A rotating set of vanes in a pump or compressor designed to pump or move water or air.

IMPORTED BACKFILL IMPORTED BACKFILL

Material used for backfilling a trench or excavation which was not the original material removed during excavation. This is a common practice where tests on the original material show it to have poor compactability or load capacity. Also called BORROW BACKFILL.

INDUSTRIAL TELEVISION EQUIPMENT INDUSTRIAL TELEVISION EQUIPMENT

(See INSPECTION TELEVISION EQUIPMENT)

INDUSTRIAL WASTEWATER INDUSTRIAL WASTEWATER

Liquid wastes originating from industrial processing. Because industries have peculiar liquid waste characteristics requiring special consideration, these sources are usually handled and treated separately before being discharged to a wastewater collection system.

INFILTRATED DEBRIS INFILTRATED DEBRIS

Sand, silt, gravel and rocks carried or washed into a collection system by infiltration water flows.

INFILTRATION (IN-fill-TRAY-shun) INFILTRATION

The seepage of groundwater into a sewer system, including service connections. Seepage frequently occurs through defective or cracked pipes, pipe joints, connections or manhole walls.

INFILTRATION HEAD INFILTRATION HEAD

The distance from a point of infiltration leaking into a collection system to the water table elevation. This is the pressure of the water being forced through the leak in the collection system.

INFILTRATION/INFLOW INFILTRATION/INFLOW

The total quantity of water from both infiltration and inflow without distinguishing the source. Abbreviated I & I or I/I.

INFILTRATION PRESSURE INFILTRATION PRESSURE

(See INFILTRATION HEAD)

INFLATABLE PIPE STOPPER INFLATABLE PIPE STOPPER

An inflatable ball or bag used to form a plug to stop flows in a sewer pipe.

INFLOW INFLOW

Water discharged into a sewer system and service connections from such sources as, but not limited to, roof leaders, cellars, yard and area drains, foundation drains, cooling water discharges, drains from springs and swampy areas, around manhole covers or through holes in the covers, cross connections from storm and combined sewer systems, catch basins, storm waters, surface runoff, street wash waters or drainage. Inflow differs from infiltration in that it is a direct discharge into the sewer rather than a leak in the sewer itself. See INTERNAL INFLOW.

INFLUENT INFLUENT

Wastewater or other liquid—raw (untreated) or partially treated—flowing *INTO* a reservoir, basin, treatment process, or treatment plant.

INLET INLET

(1) A surface connection to a drain pipe.

(2) A chamber for collecting storm water with no well below the outlet pipe for collecting grit. Often connected to a CATCH BASIN or a "basin manhole" ("cleanout manhole") with a grit chamber.

INORGANIC WASTE INORGANIC WASTE

Waste material such as sand, salt, iron, calcium, and other mineral materials which are only slightly affected by the action of organisms. Inorganic wastes are chemical substances of mineral origin; whereas organic wastes are chemical substances usually of animal or plant origin.

INSECTICIDE
INSECTICIDE

Any substance or chemical formulated to kill or control insects.

INSERTION PULLER
INSERTION PULLER

A device used to pull long segments of flexible pipe material into a sewer line when sliplining to rehabilitate a deteriorated sewer.

INSITUFORM
INSITUFORM

A method of installing a new pipe within an old pipe without excavation. The process involves the use of a polyester-fiber felt tube, lined on one side with polyurethane and fully impregnated with a liquid thermal setting resin.

INSPECTION TELEVISION EQUIPMENT
INSPECTION TELEVISION EQUIPMENT

Television equipment that is superior to standard commercial quality, providing 600 to 650 lines of resolution, and designed for industrial inspection applications. Also known as INDUSTRIAL TELEVISION EQUIPMENT.

INTEGRATOR
INTEGRATOR

A device or meter that continuously measures and calculates (adds) a process rate variable in cumulative fashion; for example, total flows displayed in gallons, million gallons, cubic feet, or some other unit of volume measurement. Also called a TOTALIZER.

INTERCEPTING SEWER
INTERCEPTING SEWER

A sewer that receives flow from a number of other large sewers or outlets and conducts the waters to a point for treatment or disposal. Often called an "interceptor."

INTERCONNECTOR
INTERCONNECTOR

A sewer installed to connect two separate sewers. If one sewer becomes blocked, wastewater can back up and flow through the interconnector to the other sewer.

INTERNAL INFLOW
INTERNAL INFLOW

Nonsanitary or industrial wastewaters generated inside of a domestic, commercial or industrial facility and being discharged into the sewer system. Examples are cooling tower waters, basement sump pump discharge waters, continuous-flow drinking fountains, and defective or leaking plumbing fixtures.

INTRADOS
INTRADOS

The upper inside curve or surface of a sewer pipe or conduit.

INVERSION
INVERSION

An Insituform process in which the Insitutube or liner is turned inside out (inverted) during the installation of the liner.

INVERT (IN-vert)
INVERT

The lowest point of the channel inside a pipe or manhole. See FLOW LINE.

INVERTED SIPHON
INVERTED SIPHON

A pressure pipeline used to carry wastewater flowing in a gravity collection system under a depression such as a valley or roadway or under a structure such as a building. Also called a "depressed sewer."

J

JET CLEANER
JET CLEANER

(See HIGH-VELOCITY CLEANER)

JET RODDER
JET RODDER

(See HIGH-VELOCITY CLEANER)

JETSAM (JET-sam)
JETSAM

Debris entering a collection system which is heavier than water. Also see GRIT.

JOGGING
JOGGING

The frequent starting and stopping of an electric motor.

JOINT | JOINT

A connection between two lengths of pipe, made either with or without the use of another part.

K

KEY MANHOLE | KEY MANHOLE

In collection system evaluation, a key manhole is one from which reliable or specific data can be obtained.

KITE | KITE

A device for hydraulically cleaning sewer lines. Resembling an airport wind sock and constructed of canvas-type material, the kite increases the velocity of a flow at its outlet to wash debris ahead of it. Also called a PARACHUTE.

L

LAMP HOLE | LAMP HOLE

A small vertical pipe or shaft extending from the surface of the ground to a sewer. A light (or lamp) may be lowered down the pipe for the purpose of inspecting the sewer. Rarely constructed today.

LAMPING | LAMPING

Using reflected sunlight or a powerful light beam to inspect a sewer between two adjacent manholes. The light is directed down the pipe from one manhole. If it can be seen from the next manhole, it indicates that the line is open and straight.

LATERAL | LATERAL

(See LATERAL SEWER)

LATERAL BREAK | LATERAL BREAK

A break in a lateral pipe somewhere between the sewer main and the building connection.

LATERAL CLEANOUT | LATERAL CLEANOUT

A capped opening in a building lateral, usually located on the property line, through which the pipelines can be cleaned.

LATERAL CONNECTION | LATERAL CONNECTION

(See BUILDING SERVICE)

LATERAL SEWER | LATERAL SEWER

A sewer that discharges into a branch or other sewer and has no other common sewer tributary to it. Sometimes called a "street sewer" because it collects wastewater from individual homes.

LEAD (LEE-d) | LEAD

A wire or conductor that can carry electric current.

LIFE-CYCLE COSTING | LIFE-CYCLE COSTING

An economic analysis procedure that considers the total costs associated with a sewer during its economic life, including development, construction, and operation and maintenance (includes chemical and energy costs). All costs are converted to a present worth or present cost in dollars.

LIFT STATION | LIFT STATION

A wastewater pumping station that lifts the wastewater to a higher elevation when continuing the sewer at reasonable slopes would involve excessive depths of trench. Also, an installation of pumps that raise wastewater from areas too low to drain into available sewers. These stations may be equipped with air-operated ejectors or centrifugal pumps. Sometimes called a PUMP STATION, but this term is usually reserved for a similar type of facility that is discharging into a long FORCE MAIN, while a lift station has a discharge line or force main only up to the downstream gravity sewer. Throughout this manual when we refer to lift stations, we intend to include pump stations.

LINER | LINER

(See PIPE LINER)

LIQUID VEHICLE LIQUID VEHICLE

Water in a collection system that is used to carry waste solids. The standard toilet provides around seven gallons of water per flush as a vehicle to carry wastes through the pipe system.

LIQUOR LIQUOR

Water, wastewater, or any combination; commonly used to mean the liquid portion when other wastes are also present.

LOGARITHM (LOG-a-rith-m) LOGARITHM

The exponent that indicates the power to which a number must be raised to produce a given number. For example: if $B^2 = N$, the 2 is the logarithm of N (to the base B), or $10^2 = 100$ and $\log_{10} 100 = 2$. Also abbreviated to "log."

LOGARITHMIC (LOG-a-RITH-mick) SCALE LOGARITHMIC SCALE

A scale on which actual distances from the origin are proportional to the logarithms of the corresponding scale numbers rather than to the numbers themselves. A logarithmic scale has the numbers getting bigger as the distances between the numbers decrease.

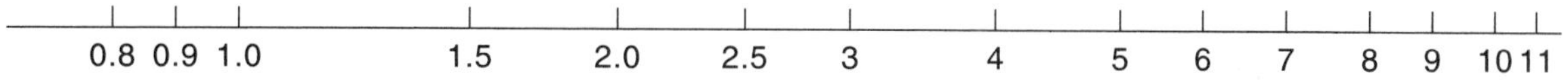

LONGITUDINAL (LAWN-ji-TWO-da-null) CRACK LONGITUDINAL CRACK

A crack in a pipe or pipe section that runs lengthwise along the pipe.

LOWER EXPLOSIVE LIMIT (LEL) LOWER EXPLOSIVE LIMIT (LEL)

The lowest concentration of gas or vapor (percent by volume in air) that explodes if an ignition source is present at ambient temperature. At temperatures above 250°F the LEL decreases because explosibility increases with higher temperature.

LUBRIFLUSHING (LOOB-rah-FLUSH-ing) LUBRIFLUSHING

A method of lubricating bearings with grease. Remove the relief plug and apply the proper lubricant to the bearing at the lubrication fitting. Run the pump to expel excess lubricant.

M

M-ZONED M-ZONED

An area set aside for manufacturing and industry.

MG MG

Initials for "Million Gallons."

MGD MGD

Initials for "Million Gallons Per Day."

mg/*L* mg/*L*

(See MILLIGRAMS PER LITER, mg/*L*)

MGY MGY

Initials for "Million Gallons Per Year."

MSDS MSDS

Material **S**afety **D**ata **S**heet. A document which provides pertinent information and a profile of a particular hazardous substance or mixture. An MSDS is normally developed by the manufacturer or formulator of the hazardous substance or mixture. The MSDS is required to be made available to employees and operators whenever there is the likelihood of the hazardous substance or mixture being introduced into the workplace. Some manufacturers are preparing MSDSs for products that are not considered to be hazardous to show that the product or substance is *NOT* hazardous.

MAIN LINE MAIN LINE

Branch or lateral sewers that collect wastewater from building sewers and service lines.

MAIN SEWER — MAIN SEWER

A sewer line that receives wastewater from many tributary branches and sewer lines and serves as an outlet for a large territory or is used to feed an intercepting sewer.

MANDREL (MAN-drill) — MANDREL

(1) A special tool used to push bearings in or to pull sleeves out.

(2) A testing device used to measure for excessive deflection in a flexible conduit.

MANHOLE — MANHOLE

An opening in a sewer provided for the purpose of permitting operators or equipment to enter or leave a sewer. Sometimes called an "access hole," or a "maintenance hole."

MANHOLE BEDDING — MANHOLE BEDDING

The prepared and compacted base on which a manhole is constructed.

MANHOLE DEPTH — MANHOLE DEPTH

The measurement from the top of the manhole opening to the invert or lowest point of the trough at the bottom of the manhole.

MANHOLE, DROP — MANHOLE, DROP

(See DROP MANHOLE)

MANHOLE ELEVATION — MANHOLE ELEVATION

The height (elevation) of the invert or lowest point in the bottom of a manhole above mean sea level.

MANHOLE FLOW — MANHOLE FLOW

(1) The depth or amount of wastewater flow in a manhole as observed at any selected time.

(2) The total or the average flow through a manhole in gallons in any selected time interval.

MANHOLE FRAME — MANHOLE FRAME

A metal ring or frame with a ledge to accommodate the manhole lid; located at the surface of the ground or street. Also called a "manhole ring."

MANHOLE GRADE RING — MANHOLE GRADE RING

A precast concrete ring 4 to 12 inches high which is placed on top of a manhole cone to raise the manhole cover frame flush with the surface grade. Sometimes called a "spacer."

MANHOLE INFILTRATION — MANHOLE INFILTRATION

Groundwaters seeping or leaking into a manhole structure.

MANHOLE INFLOW — MANHOLE INFLOW

Surface waters flowing into a manhole, usually through the vent holes in the manhole lid.

MANHOLE INVERT — MANHOLE INVERT

The lowest point in a trough or flow channel in the bottom of a manhole.

MANHOLE JACK — MANHOLE JACK

A device used to guide the tag line into the sewer without causing unnecessary wear and provide support as the tag line is pulled back and forth.

MANHOLE, KEY — MANHOLE, KEY

(See KEY MANHOLE)

MANHOLE LID — MANHOLE LID

The heavy cast-iron or forged-steel cover of a manhole. The lid may or may not have vent holes.

MANHOLE LID DUST PAN — MANHOLE LID DUST PAN

A sheet metal or cast-iron pan located under a manhole lid. This pan serves to catch and hold pebbles and other debris falling through vent holes, preventing them from getting into the pipe system.

MANHOLE RING MANHOLE RING

A metal frame or ring with a ledge to accommodate the lid and located at the surface of the ground or street. Also called a "manhole frame."

MANHOLE SEALING MANHOLE SEALING

The process of sealing infiltration leaks in a manhole by injecting chemical grout.

MANHOLE, SURCHARGED MANHOLE, SURCHARGED

(See SURCHARGED MANHOLE)

MANHOLE TOOLS MANHOLE TOOLS

(1) Special tools having conveniently short handles for working inside manholes.
(2) Special long-handled or extendable tools for removal of debris and other objects from manholes without requiring a person to enter the manhole.

MANHOLE TROUGH MANHOLE TROUGH

The channel in the bottom of a manhole for the flow of the wastewater from manhole inlet to outlet.

MANHOLE VENTS MANHOLE VENTS

One or a series of one-inch diameter holes through a manhole lid for purposes of venting dangerous gases found in sewers.

MANNING'S FORMULA MANNING'S FORMULA

A mathematical formula for calculating wastewater flows in sewers.

$$Q = \frac{1.49}{n} A R^{2/3} S^{1/2}$$

Q means flow in cubic feet per second (CFS).
n means the Manning pipe or channel roughness factor.
A means the cross-sectional area of the flow in square feet (sq ft).
R means the hydraulic radius in feet (ft) where R equals A/P. P is the wetted perimeter of the channel or pipe in feet.
S means the slope of the channel or energy grade line in feet per foot (ft/ft).

MANNING'S TABLES MANNING'S TABLES

A set of tables for finding wastewater flows in sewers.

MANOMETER (man-NAH-mut-ter) MANOMETER

Usually a glass tube filled with a liquid and used to measure the difference in pressure across a flow measuring device such as an orifice or a Venturi meter.

MATERIAL SAFETY DATA SHEET (MSDS) MATERIAL SAFETY DATA SHEET (MSDS)

A document which provides pertinent information and a profile of a particular hazardous substance or mixture. An MSDS is normally developed by the manufacturer or formulator of the hazardous substance or mixture. The MSDS is required to be made available to employees and operators whenever there is the likelihood of the hazardous substance or mixture being introduced into the workplace. Some manufacturers are preparing MSDSs for products that are not considered to be hazardous to show that the product or substance is *NOT* hazardous.

MEASURED FLOW MEASURED FLOW

A flow which has been physically measured. See GAGE.

MECHANICAL CLEANING MECHANICAL CLEANING

Clearing pipe by using equipment that scrapes, cuts, pulls or pushes the material out of the pipe. Mechanical cleaning devices or machines include bucket machines, power rodders and hand rods.

MECHANICAL PLUG MECHANICAL PLUG

A pipe plug used in sewer systems that is mechanically expanded to create a seal.

MEGGER (from megohm) MEGGER

An instrument used for checking the insulation resistance on motors, feeders, bus bar systems, grounds, and branch circuit wiring.

MEGOHM MEGOHM

Meg means one million, so 5 megohms means 5 million ohms. A megger reads in millions of ohms.

MERCAPTANS (mer-CAP-tans) MERCAPTANS

Compounds containing sulfur which have an extremely offensive skunk-like odor; also sometimes described as smelling like garlic or onions.

METERED METERED

Measured through a meter, as a quantity of water or flow might be measured.

MICROORGANISMS (MY-crow-OR-gan-IS-zums) MICROORGANISMS

Very small organisms that can be seen only through a microscope. Some microorganisms use the wastes in wastewater for food and thus remove or alter much of the undesirable matter.

MILLIGRAMS PER LITER, mg/*L* MILLIGRAMS PER LITER, mg/*L*

A measure of the concentration by weight of a substance per unit volume in water or wastewater. In reporting the results of water and wastewater analysis, mg/*L* is preferred to the unit parts per million (ppm), to which it is approximately equivalent.

MILLION GALLONS MILLION GALLONS

A unit of measurement used in wastewater treatment plant design and collection system capacities or performances. One million gallons of water is approximately equivalent to:

13,690	Cubic Feet
3.07	Acre-Feet
8,340,000	Pounds of Weight
4,170	Tons of Weight
3,785	Cubic Meters

MINERAL MINERAL

Any substance that is neither animal nor plant. Minerals include sand, salt, iron, calcium, and nutrients.

MINERAL CONTENT MINERAL CONTENT

The quantity of dissolved minerals in a sample of water.

MONITOR MONITOR

(See TELEVISION MONITOR)

N

"N" FACTOR "N" FACTOR

A coefficient value representing the pipe or channel roughness in Manning's formula for computing flows in gravity sewers. See FRICTION LOSS and MANNING'S FORMULA.

NIOSH (NYE-osh) NIOSH

The **N**ational **I**nstitute of **O**ccupational **S**afety and **H**ealth is an organization that tests and approves safety equipment for particular applications. NIOSH is the primary federal agency engaged in research in the national effort to eliminate on-the-job hazards to the health and safety of working people. The NIOSH Publications Catalog, Sixth Edition, NIOSH Pub. No. 84-118, lists the NIOSH publications concerning industrial hygiene and occupational health. To obtain a copy of the catalog, write to National Technical Information Service (NTIS), 5285 Port Royal Road, Springfield, VA 22161. NTIS Stock No. PB86-116787, price, $103.50, plus $5.00 shipping and handling per order.

NPDES PERMIT NPDES PERMIT

National **P**ollutant **D**ischarge **E**limination **S**ystem permit is the regulatory agency document issued by either a federal or state agency which is designed to control all discharges of potential pollutants from point sources and storm water runoff into U.S. waterways. NPDES permits regulate discharges into navigable waters from all point sources of pollution, including industries, municipal wastewater treatment plants, sanitary landfills, large agricultural feedlots and return irrigation flows.

NAMEPLATE NAMEPLATE

A durable metal plate found on equipment which lists critical operating conditions for the equipment.

NET WASTEWATER CONTRIBUTION — NET WASTEWATER CONTRIBUTION

In a wastewater collection system, the net wastewater contribution consists of the liquid wastes and liquid-carried wastes transported by the pipelines or received by the pipelines. This value would be the only wastewater found in a collection system if all sources of infiltration, inflow and exfiltration were eliminated.

NET WASTEWATER FLOW — NET WASTEWATER FLOW

The actual wastewater flow from a collection system that reaches a wastewater treatment plant. The net wastewater flow includes the net wastewater contribution, infiltration and inflow and does not include losses through exfiltration.

NODULAR (NOD-you-lar) — NODULAR

Shaped like a rounded lump, knot or knob. An irregular enlargement.

NOMINAL DIAMETER — NOMINAL DIAMETER

An approximate measurement of the diameter of a pipe. Although the nominal diameter is used to describe the size or diameter of a pipe, it is usually not the exact inside diameter of the pipe.

NOMOGRAPH (NOME-o-graph) — NOMOGRAPH

A graphic representation or means of solving an equation or mathematical relationship. Results are obtained with the aid of a straightedge placed over important known values.

NON-PERMIT CONFINED SPACE — NON-PERMIT CONFINED SPACE

See CONFINED SPACE, NON-PERMIT.

NONSPARKING TOOLS — NONSPARKING TOOLS

These tools will not produce a spark during use. They are made of a nonferrous material, usually a copper-beryllium alloy.

NOTICE — NOTICE

This word calls attention to information that is especially significant in understanding and operating equipment or processes safely. Also see CAUTION, DANGER, and WARNING.

O

OSHA (O-shuh) — OSHA

The Williams-Steiger Occupational Safety and Health Act of 1970 (OSHA) is a federal law designed to protect the health and safety of industrial workers and collection system operators. The Act regulates the design, construction, operation and maintenance of industrial plants and wastewater collection and treatment facilities. The Act does not apply directly to municipalities, *EXCEPT* in those states that have approved plans and have asserted jurisdiction under Section 18 of the OSHA Act. *HOWEVER, CONTRACT OPERATORS AND PRIVATE FACILITIES DO HAVE TO COMPLY WITH OSHA REQUIREMENTS.* Wastewater collection systems have come under stricter regulation in all phases of activity as a result of OSHA standards. OSHA also refers to the federal and state agencies which administer the OSHA regulations.

OBSTRUCTION — OBSTRUCTION

Any solid object in or protruding into a wastewater flow in a collection line that prevents a smooth or even passage of the wastewater.

OFF LINE — OFF LINE

A run of sewer pipe between two manholes is said to be "off line" if it is not located directly under a straight line passing through the exact centers of the two manholes. Sewer alignment does not always pass through the center of a manhole, especially at junctions. Also called "misaligned."

OFFSET — OFFSET

(1) A combination of elbows or bends which brings one section of a line of pipe out of line with, but into a line parallel with, another section.

(2) A pipe fitting in the approximate form of a reverse curve, made to accomplish the same purpose.

(3) A pipe joint that has lost its bedding support and one of the pipe sections has dropped or slipped, thus creating a condition where the pipes no longer line up properly.

OFFSET INVERT — OFFSET INVERT

A trough or channel in the bottom of a manhole which is not centered in the bottom.

OFFSET JOINT — OFFSET JOINT

A pipe joint that is not exactly in line and centered. See DROP JOINT and VERTICAL OFFSET.

OFFSET MANHOLE — OFFSET MANHOLE

A manhole located to one side of a pipe run with either "Y" connections to it or the inlet and outlet pipes bent to enter and leave the manhole.

OFFSET PIPE — OFFSET PIPE

(See OFF LINE)

OFFSET TROUGH — OFFSET TROUGH

(1) When the pipe feeding into a manhole does not exactly match up with the pipe leading out of the manhole, the invert channel must be angled or curved. This is referred to as an "offset trough."

(2) (See OFFSET INVERT)

OHM — OHM

The unit of electrical resistance. The resistance of a conductor in which one volt produces a current of one ampere.

OLFACTORY (ol-FAK-tore-ee) FATIGUE — OLFACTORY FATIGUE

A condition in which a person's nose, after exposure to certain odors, is no longer able to detect the odor.

OPERATING RATIO — OPERATING RATIO

The operating ratio is a measure of the total revenues divided by the total operating expenses.

ORGANIC WASTE — ORGANIC WASTE

Waste material which comes mainly from animal or plant sources. Organic wastes generally can be consumed by bacteria and other small organisms. Inorganic wastes are chemical substances of mineral origin.

ORIFICE (OR-uh-fiss) — ORIFICE

An opening (hole) in a plate, wall, or partition. An orifice flange or plate placed in a pipe consists of a slot or a calibrated circular hole smaller than the pipe diameter. The difference in pressure in the pipe above and at the orifice may be used to determine the flow in the pipe.

OUTFALL — OUTFALL

(1) The point, location or structure where wastewater or drainage discharges from a sewer, drain, or other conduit.

(2) The conduit leading to the final disposal point or area. See OUTFALL SEWER.

OUTFALL SEWER — OUTFALL SEWER

A sewer that receives wastewater from a collection system or from a wastewater treatment plant and carries it to a point of ultimate or final discharge in the environment. See OUTFALL.

OUTLET — OUTLET

Downstream opening or discharge end of a pipe, culvert, or canal.

OVERFLOW MANHOLE — OVERFLOW MANHOLE

A manhole which fills and allows raw wastewater to flow out onto the street or ground.

OVERFLOW RELIEF LINE — OVERFLOW RELIEF LINE

Where a system has overload conditions during peak flows, an outlet may be installed above the invert and leading to a less loaded manhole or part of the system. This is usually called an "overflow relief line." Also see CROSS CONNECTION (2).

OXIDATION (ox-uh-DAY-shun) — OXIDATION

Oxidation is the addition of oxygen, removal of hydrogen, or the removal of electrons from an element or compound. In wastewater treatment, organic matter is oxidized to more stable substances. The opposite of REDUCTION.

OXIDATION-REDUCTION POTENTIAL (ORP) — OXIDATION-REDUCTION POTENTIAL (ORP)

The electrical potential required to transfer electrons from one compound or element (the oxidant) to another compound or element (the reductant); used as a qualitative measure of the state of oxidation in wastewater treatment systems. ORP is measured in millivolts, with negative values indicating a tendency to reduce compounds or elements and positive values indicating a tendency to oxidize compounds or elements.

OXYGEN DEFICIENCY — OXYGEN DEFICIENCY

An atmosphere containing oxygen at a concentration of less than 19.5 percent by volume.

OXYGEN ENRICHMENT — OXYGEN ENRICHMENT

An atmosphere containing oxygen at a concentration of more than 23.5 percent by volume.

P

POTW — POTW

Publicly **O**wned **T**reatment **W**orks. A treatment works which is owned by a state, municipality, city, town, special sewer district or other publicly owned and financed entity as opposed to a privately (industrial) owned treatment facility. This definition includes any devices and systems used in the storage, treatment, recycling and reclamation of municipal sewage (wastewater) or industrial wastes of a liquid nature. It also includes sewers, pipes and other conveyances only if they carry wastewater to a POTW treatment plant. The term also means the municipality (public entity) which has jurisdiction over the indirect discharges to and the discharges from such a treatment works.

PPM — PPM

Initials for "Parts Per Million." The number of weight or volume units of a minor constituent present with each one million units of the major constituent of a solution or mixture. Used to express the results of most water and wastewater analyses, but more recently milligrams per liter (mg/*L*) is the preferred term.

PACKER — PACKER

(See CHEMICAL GROUTING)

PACKING RING — PACKING RING

A ring made of asbestos or metal which may be lubricated with Teflon or graphite that forms a seal between the pump shaft and its casing.

PARACHUTE — PARACHUTE

A device used to catch wastewater flow to pull a float line between manholes. See FLOAT LINE.

PARSHALL FLUME — PARSHALL FLUME

A specially constructed flume or channel used to measure flows in open channels.

PATHOGENIC (PATH-o-JEN-ick) ORGANISMS — PATHOGENIC ORGANISMS

Bacteria, viruses, cysts, or protozoa which can cause disease (giardiasis, cryptosporidiosis, typhoid, cholera, dysentery) in a host (such as a person). There are many types of organisms which do *NOT* cause disease and which are *NOT* called pathogenic. Many beneficial bacteria are found in wastewater treatment processes actively cleaning up organic wastes.

PEAKING FACTOR — PEAKING FACTOR

Ratio of a maximum flow to the average flow, such as maximum hourly flow or maximum daily flow to the average daily flow.

PENTA HOSE — PENTA HOSE

A hose with five chambers or tubes.

PERCOLATION (PURR-co-LAY-shun) — PERCOLATION

The movement or flow of water through soil or rocks.

PERMIT-REQUIRED CONFINED SPACE (PERMIT SPACE) — PERMIT-REQUIRED CONFINED SPACE (PERMIT SPACE)

See CONFINED SPACE, PERMIT-REQUIRED (PERMIT SPACE).

PESTICIDE — PESTICIDE

Any substance or chemical designed or formulated to kill or control animal pests. Also see INSECTICIDE and RODENTICIDE.

PHOTOGRAPHIC INSPECTIONS — PHOTOGRAPHIC INSPECTIONS

A method of obtaining photographs of a pipeline by pulling a time-lapse motion picture camera through the line. By moving the camera a specific distance at timed intervals, a sequence of photographs covering the full length of the line is obtained.

PHOTOGRAPHIC RECORDS PHOTOGRAPHIC RECORDS

(1) The film strip from a photographic inspection.

(2) Still camera photographs of a sewer television inspection monitor.

PIEZOMETER (pie-ZOM-uh-ter) PIEZOMETER

An instrument used to measure the pressure head in a pipe, tank, or soil. It usually consists of a small pipe or tube connected or tapped into the side or wall of a pipe or tank and connected to a manometer pressure gage, water or mercury column, or other device for indicating pressure head.

PIG PIG

Refers to a poly pig which is a bullet-shaped device made of hard rubber or similar material. This device is used to clean pipes. It is inserted in one end of a pipe, moves through the pipe under pressure, and is removed from the other end of the pipe.

PIPE BEDDING PIPE BEDDING

(See BEDDING and BEDDING GRADE)

PIPE BUCKET PIPE BUCKET

(See BUCKET and BUCKET MACHINE)

PIPE CAPACITY PIPE CAPACITY

In a gravity-flow sewer system, pipe capacity is the total amount in gallons a pipe is able to pass in a specific time period.

PIPE CLEANING PIPE CLEANING

Removing grease, grit, roots and other debris from a pipe run by means of one of the hydraulic cleaning methods. See BALLING, HYDRAULIC CLEANING and KITE.

NOTE: While rodding may be used to open a pipe stoppage or to remove roots and greases, it is not known to remove grit and similar debris.

PIPE DEFLECTION PIPE DEFLECTION

(See DEFLECTED)

PIPE DIAMETER PIPE DIAMETER

The nominal or commercially designated inside diameter of a pipe, unless otherwise stated.

PIPE DIP PIPE DIP

(See DIP)

PIPE DISPLACEMENT PIPE DISPLACEMENT

(1) The cubic inches of soil or water displaced by one foot or one section of pipe.

(2) (See DISPLACED PIPE)

PIPE GRADE PIPE GRADE

The angle of a sewer or a single section of a sewer as installed. Usually expressed in a percentage figure to indicate the drop in feet or tenths of a foot per hundred feet. For example, 0.5 percent grade means a drop of one-half foot per 100 feet of length.

PIPE JACK PIPE JACK

A jack used to fasten roller guides to secure an object within a manhole.

PIPE JOINT PIPE JOINT

A place where two sections of pipe are coupled or joined together.

PIPE JOINT SEAL PIPE JOINT SEAL

(1) The tightness or lack of leakage at a pipe joint.

(2) The method of sealing a pipe coupling.

PIPE LINER PIPE LINER

A plastic liner pulled or pushed into a pipe to eliminate excessive infiltration or exfiltration. Other solutions to the problem of infiltration/exfiltration are the use of a cement grouting or replacement of damaged pipe.

PIPE PLUG — PIPE PLUG

(1) A temporary plug placed in a sewer pipe to stop a flow while repair work is being accomplished or other functions are performed.

(2) In construction of a new sewer system, service saddles are sometimes installed before a building or a building lateral is in existence. Under such circumstances, a plug will be placed in the off-lead of the saddle of a "Y." In some instances, this plug may be called a "button" or a "stopper."

PIPE PLUG, INFLATABLE — PIPE PLUG, INFLATABLE

(See INFLATABLE PIPE STOPPER)

PIPE PLUG, MECHANICAL — PIPE PLUG, MECHANICAL

(See MECHANICAL PLUG)

PIPE RODDING — PIPE RODDING

A method of opening a plugged or blocked pipe by pushing a steel rod or snake, or pulling same, through the pipe with a tool attached to the end of the rod or snake. Rotating the rod or snake with a tool attached increases effectiveness.

PIPE RUN — PIPE RUN

(1) The length of sewer pipe reaching from one manhole to the next. See BARREL.

(2) Any length of pipe, generally assumed to be in a straight line.

PIPE SEAL — PIPE SEAL

(See PIPE JOINT SEAL)

PIPE SEALING — PIPE SEALING

(See CHEMICAL GROUTING, GUNITE or PIPE LINER)

PIPE SECTION — PIPE SECTION

A single length of pipe between two joints or couplers.

PIPE SINK — PIPE SINK

(See DIP)

PLAN — PLAN

A drawing showing the *TOP* view of sewers, manholes and streets.

PLANT — PLANT

(See WASTEWATER TREATMENT PLANT)

PLANT HYDRAULIC CAPACITY — PLANT HYDRAULIC CAPACITY

The flow or load, in millions of gallons per day (or portion thereof), that a treatment plant is designed to handle.

PLANT, TREATMENT — PLANT, TREATMENT

(See WASTEWATER TREATMENT PLANT)

PLUG — PLUG

(See INFLATABLE PIPE STOPPER and MECHANICAL PLUG)

PNEUMATIC EJECTOR (new-MAT-tik ee-JECK-tor) — PNEUMATIC EJECTOR

A device for raising wastewater, sludge or other liquid by compressed air. The liquid is alternately admitted through an inward-swinging check valve into the bottom of an airtight pot. When the pot is filled compressed air is applied to the top of the liquid. The compressed air forces the inlet valve closed and forces the liquid in the pot through an outward-swinging check valve, thus emptying the pot.

POLE SHADER — POLE SHADER

A copper bar circling the laminated iron core inside the coil of a magnetic starter.

POLLUTION — POLLUTION

The impairment (reduction) of water quality by agricultural, domestic or industrial wastes (including thermal and radioactive wastes) to a degree that the natural water quality is changed to hinder any beneficial use of the water or render it offensive to the senses of sight, taste, or smell or when sufficient amounts of wastes create or pose a potential threat to human health or the environment.

POLYELECTROLYTE (POLY-ee-LECK-tro-lite) POLYELECTROLYTE

A high-molecular-weight substance that is formed by either a natural or synthetic process. Natural polyelectrolytes may be of biological origin or derived from starch products, cellulose derivatives, and alignates. Synthetic polyelectrolytes consist of simple substances that have been made into complex, high-molecular-weight substances. Often called a POLYMER.

POLYMER (POLY-mer) POLYMER

A long chain molecule formed by the union of many monomers (molecules of lower molecular weight). Polymers are used with other chemical coagulants to aid in binding small suspended particles to larger chemical flocs for their removal from water.

POPULATION EQUIVALENT (HYDRAULIC) POPULATION EQUIVALENT (HYDRAULIC)

A flow of 100 gallons per day is the hydraulic or flow equivalent to the contribution or flow from one person. Population equivalent = 100 GPCD or gallons per capita per day.

PORCUPINE PORCUPINE

A sewer cleaning tool the same diameter as the pipe being cleaned. The tool is a steel cylinder having solid ends with eyes cast in them to which a cable can be attached and pulled by a winch. Many short pieces of cable or bristles protrude from the cylinder to form a round brush.

POTABLE (POE-tuh-bull) WATER POTABLE WATER

Water that does not contain objectionable pollution, contamination, minerals, or infective agents and is considered satisfactory for drinking.

POWER RODDER POWER RODDER

A machine designed to remove roots, grease, and other materials from pipes. Also referred to as rodding machines.

PRECIPITATE (pre-SIP-uh-TATE) PRECIPITATE

(1) An insoluble, finely divided substance which is a product of a chemical reaction within a liquid.

(2) The separation from solution of an insoluble substance.

PRECIPITATION PRECIPITATION

(1) The total measurable supply of water received directly from clouds as rain, snow, hail, or sleet; usually expressed as depth in a day, month, or year, and designated as daily, monthly, or annual precipitation.

(2) The process by which atmospheric moisture is discharged onto a land or water surfaces.

(3) The separation (of a substance) out in solid form from a solution, as by the use of a reagent.

PRE-CLEANING PRE-CLEANING

Sewer line cleaning, commonly done by high-velocity cleaners, that is done prior to the TV inspection of a pipeline to remove grease, slime, and grit to allow for a clearer and more accurate identification of defects and problems.

PREDICTIVE MAINTENANCE PREDICTIVE MAINTENANCE

The ability to identify problem areas before breakdowns or blockage of flow occurs. Predictive maintenance is the end product of effective preventive maintenance.

PRESENT WORTH PRESENT WORTH

The value of a long-term project expressed in today's dollars. Present worth is calculated by converting (discounting) all future benefits and costs over the life of the project to a single economic value at the start of the project. Calculating the present worth of alternative projects makes it possible to compare them and select the one with the largest positive (beneficial) present worth or minimum present cost.

PRESSURE GROUTING PRESSURE GROUTING

(See CHEMICAL GROUTING)

PRESSURE HEAD PRESSURE HEAD

(1) The height of a water surface above a specific point of reference. Usually measured in feet and tenths of a foot.

(2) The head represented by the expression of pressure over weight (p/w), where p is pressure (lbs/sq ft) and w is weight (lbs/cu ft).

PRESSURE MAIN PRESSURE MAIN

(See FORCE MAIN)

PREVENTIVE MAINTENANCE PREVENTIVE MAINTENANCE

Regularly scheduled servicing of machinery or other equipment using appropriate tools, tests and lubricants. This type of maintenance can prolong the useful life of equipment and machinery and increase its efficiency by detecting and correcting problems before they cause a breakdown of the equipment.

PREVENTIVE MAINTENANCE UNITS PREVENTIVE MAINTENANCE UNITS

Crews assigned the task of cleaning sewers (for example, balling or high-velocity cleaning crews) to prevent stoppages and odor complaints. Preventive maintenance is performing the most effective cleaning procedure, in the area where it is most needed, at the proper time in order to prevent failures and emergency situations.

PROBE PROBE

A T-shaped tool or rod that is pushed or driven down through the soil to locate underground pipes and utility conduits. Also see SOUNDING ROD.

PROFILE PROFILE

A drawing showing the *SIDE* view of sewers and manholes.

PROMOTED PROMOTED

The mixture of resin and catalyst ready to cause (promote) curing in place.

PROTRUDING SERVICE PROTRUDING SERVICE

The connection of a building lateral to a main sewer line whereby a hole is cut in the main and the end of the building lateral is allowed to extend into the main.

PROTRUDING TAP PROTRUDING TAP

(See PROTRUDING SERVICE)

PUMP PUMP

A mechanical device for causing flow, for raising or lifting water or other fluid, or for applying pressure to fluids.

PUMP PIT PUMP PIT

A dry well, chamber or room below ground level in which a pump is located.

PUMP STATION PUMP STATION

Installation of pumps to lift wastewater to a higher elevation in places where flat land would require excessively deep sewer trenches. Also used to raise wastewater from areas too low to drain into available collection lines. These stations may be equipped with air-operated ejectors or centrifugal pumps. See LIFT STATION.

PUTREFACTION (PEW-truh-FACK-shun) PUTREFACTION

Biological decomposition of organic matter with the production of foul-smelling products associated with anaerobic conditions.

PUTRESCIBLE (pew-TRES-uh-bull) PUTRESCIBLE

Material that will decompose under anaerobic conditions and produce nuisance odors.

Q

QUADRANT QUADRANT

Any of the four more or less equal parts into which something can be divided by two real or imaginary lines that intersect each other at right angles. In television inspection of pipes, for example, the picture of the pipe can be divided into fourths (upper-left, lower-left, upper-right, and lower-right quadrants) to identify the location of observed objects.

QUALITATIVE FIT TEST (QLFT) QUALITATIVE FIT TEST (QLFT)

A pass/fail fit test to assess the adequacy of respirator fit that relies on the individual's response to the test agent.

QUANTITATIVE FIT TEST (QNFT) QUANTITATIVE FIT TEST (QNFT)

An assessment of the adequacy of respirator fit that relies on the individual's response to the test agent.

QUICK AIR TEST | QUICK AIR TEST

The same as a quick test with a packer for chemical grouting except that air pressure is used in place of liquid for a faster test and greater accuracy. Also see CHEMICAL GROUTING.

QUICK TEST | QUICK TEST

Use of a packer designed for chemical grouting to pressure test any selected small area of pipeline. Also see CHEMICAL GROUTING.

QUICKSAND | QUICKSAND

Sand that has lost its grain-to-grain contact by the buoyancy effect of water flowing upward through the voids. Such material, having some of the characteristics of a fluid, possesses no load-bearing value.

R

"R" FACTOR | "R" FACTOR

Refers to pipe or channel "Roughness Factor." See FRICTION LOSS, "N" FACTOR, and ROUGHNESS COEFFICIENT.

R-ZONED | R-ZONED

Areas established for residential occupancy.

RAIN | RAIN

Particles of liquid water that have become too large to be held by the atmosphere. Their diameter generally is greater than 0.02 inch and they usually fall to the earth at velocities greater than 10 fps in still air. See PRECIPITATION.

RECORDINGS, FLOW | RECORDINGS, FLOW

(See FLOW RECORDING)

RECORDINGS, PHOTOGRAPHIC | RECORDINGS, PHOTOGRAPHIC

(See PHOTOGRAPHIC RECORDS)

RECORDINGS, VIDEO | RECORDINGS, VIDEO

(See VIDEO LOG)

RECORDINGS, WRITTEN | RECORDINGS, WRITTEN

(See TELEVISION INSPECTION LOG)

REDUCTANT | REDUCTANT

A constituent of wastewater or surface waters that uses either free (O_2) or combined oxygen in the process of stabilization.

REDUCTION (re-DUCK-shun) | REDUCTION

Reduction is the addition of hydrogen, removal of oxygen, or the addition of electrons to an element or compound. Under anaerobic conditions (no dissolved oxygen present), sulfur compounds are reduced to odor-producing hydrogen sulfide (H_2S) and other compounds. The opposite of OXIDATION.

REGULATOR | REGULATOR

A device used in combined sewers to control or regulate the diversion of flow.

RELIEF BYPASS | RELIEF BYPASS

(See BYPASS and OVERFLOW RELIEF LINE)

RELIEF LINE | RELIEF LINE

(See OVERFLOW RELIEF LINE)

RESISTANCE | RESISTANCE

That property of a conductor or wire that opposes the passage of a current, thus causing electric energy to be transformed into heat.

RETENTION RETENTION

(1) That part of the precipitation falling on a drainage area which does not escape as surface stream flow during a given period. It is the difference between total precipitation and total runoff during the period, and represents evaporation, transpiration, subsurface leakage, infiltration, and, when short periods are considered, temporary surface or underground storage on the area.

(2) The delay or holding of the flow of water and water-carried wastes in a pipe system. This can be due to a restriction in the pipe, a stoppage or a dip. Also, the time water is held or stored in a basin or wet well. This is also called DETENTION.

ROD GUIDE ROD GUIDE

A bent pipe inserted in a manhole to guide hand and power rods into collection lines so the rods can dislodge obstructions.

ROD (SEWER) ROD (SEWER)

A light metal rod, three to five feet long with a coupling at each end. Rods are joined and pushed into a sewer to dislodge obstructions.

RODDING RODDING

(See PIPE RODDING)

RODDING MACHINE RODDING MACHINE

(See POWER RODDER and PIPE RODDING)

RODDING TOOLS RODDING TOOLS

Special tools attached to the end of a rod or snake to accomplish various results in pipe rodding.

RODENTICIDE (row-DENT-uh-SIDE) RODENTICIDE

Any substance or chemical used to kill or control rodents.

ROOF LEADER ROOF LEADER

A downspout or pipe installed to drain a roof gutter to a storm drain or other means of disposal.

ROOT, SEWER ROOT, SEWER

Any part of a root system of a plant or tree that enters a collection system.

ROOT MOP ROOT MOP

When roots from plant life enter a sewer system, the roots frequently branch to form a growth that resembles a string mop.

ROTAMETER (RODE-uh-ME-ter) ROTAMETER

A device used to measure the flow rate of gases and liquids. The gas or liquid being measured flows vertically up a tapered, calibrated tube. Inside the tube is a small ball or bullet-shaped float (it may rotate) that rises or falls depending on the flow rate. The flow rate may be read on a scale behind or on the tube by looking at the middle of the ball or at the widest part or top of the float.

ROUGHNESS COEFFICIENT ROUGHNESS COEFFICIENT

A value used in Manning's formula to determine energy losses of flowing water due to pipe or channel wall roughness. Also see FRICTION LOSS, "N" FACTOR and MANNING'S FORMULA.

ROUGHNESS FACTOR ROUGHNESS FACTOR

(See ROUGHNESS COEFFICIENT)

RUNOFF RUNOFF

That part of rain or other precipitation that runs off the surface of a drainage area and does not enter the soil or the sewer system as inflow.

S

SCADA (ss-KAY-dah) SYSTEM SCADA SYSTEM

Supervisory **C**ontrol **A**nd **D**ata **A**cquisition system. A computer-monitored alarm, response, control and data acquisition system used by operators to monitor and adjust their wastewater treatment processes and facilities.

SSES SSES

Sewer **S**ystem **E**valuation **S**urvey.

SSO SSO

Sanitary **S**ewer **O**verflow. Wastewater that flows out of a sanitary sewer (or lift station) as a result of flows exceeding the hydraulic capacity of the sewer or stoppages in the sewer. SSOs exceeding hydraulic capacity usually occur during periods of heavy precipitation or high levels of runoff from snow melt or other runoff sources.

SADDLE SADDLE

A fitting mounted on a pipe for attaching a new connection. This device makes a tight seal against the main pipe by use of a clamp, adhesive, or gasket and prevents the service pipe from protruding into the main.

SADDLE CONNECTION SADDLE CONNECTION

A building service connection made to a sewer main with a device called a saddle.

SAND EQUIVALENT SAND EQUIVALENT

An ASTM (American Society for Testing and Materials) test for trench backfill soils. The test uses a glass tube and the soil (backfill) is mixed with water in the tube and shaken. The tube is placed on a table and the soil allowed to settle according to particle size. The settlement of the soil is compared to the settlement of a standard type of sand grain sizes.

SAND TRAP SAND TRAP

A device which can be placed in the outlet of a manhole to cause a settling pond to develop in the manhole invert, thus trapping sand, rocks and similar debris heavier than water. Also may be installed in outlets from car wash areas. Also see GRIT CATCHER.

SANITARY COLLECTION SYSTEM SANITARY COLLECTION SYSTEM

The pipe system for collecting and carrying liquid and liquid-carried wastes from domestic sources to a wastewater treatment plant. Also see WASTEWATER COLLECTION SYSTEM.

SANITARY SEWER SANITARY SEWER

A pipe or conduit (sewer) intended to carry wastewater or waterborne wastes from homes, businesses, and industries to the POTW (**P**ublicly **O**wned **T**reatment **W**orks). Storm water runoff or unpolluted water should be collected and transported in a separate system of pipes or conduits (storm sewers) to natural watercourses.

SATURATED SOIL SATURATED SOIL

Soil that cannot absorb any more liquid. The interstices or void spaces in the soil are filled with water to the point at which runoff occurs.

SCALE SCALE

A combination of mineral salts and bacterial accumulation that sticks to the inside of a collection pipe under certain conditions. Scale, in extreme growth circumstances, creates additional friction loss to the flow of water. Scale may also accumulate on surfaces other than pipes.

SCOOTER SCOOTER

A sewer cleaning tool whose cleansing action depends on the development of high water velocity around the outside edge of a circular shield. The metal shield is rimmed with a rubber coating and is attached to a framework on wheels (like a child's scooter). The angle of the shield is controlled by a chain-spring system which regulates the head of water behind the scooter and thus the cleansing velocity of the water flowing around the shield.

SCUM SCUM

(1) A layer or film of foreign matter (such as grease, oil) that has risen to the surface of water or wastewater.

(2) A residue deposited on the ledge of a sewer, channel, or wet well at the water surface.

(3) A mass of solid matter that floats on the surface.

SEAL SEAL

(See PIPE JOINT SEAL)

SEALING SEALING

(See CHEMICAL GROUTING)

SEASONAL WATER TABLE — SEASONAL WATER TABLE

A groundwater table that has seasonal changes in depth or elevation.

SEDIMENT — SEDIMENT

Solid material settled from suspension in a liquid.

SEDIMENTATION (SED-uh-men-TAY-shun) — SEDIMENTATION

The process of settling and depositing of suspended matter carried by wastewater. Sedimentation usually occurs by gravity when the velocity of the wastewater is reduced below the point at which it can transport the suspended material.

SEISMIC (SIZE-mick) — SEISMIC

Relating to an earthquake or violent earth vibration such as an explosion.

SELECT BACKFILL — SELECT BACKFILL

Material used in backfilling of an excavation, selected for desirable compaction or other characteristics.

SELECT BEDDING — SELECT BEDDING

Material used to provide a bedding or foundation for pipes or other underground structures. This material is of specified quality for desirable bedding or other characteristics and is often imported from a different location.

SEPTIC (SEP-tick) — SEPTIC

A condition produced by anaerobic bacteria. If severe, the sludge produces hydrogen sulfide, turns black, gives off foul odors, contains little or no dissolved oxygen, and the wastewater has a high oxygen demand.

SEPTIC TANK — SEPTIC TANK

A system used where wastewater collection systems and treatment plants are not available. The system is a settling tank in which settled sludge is in intimate contact with the wastewater flowing through the tank and the organic solids are decomposed by anaerobic bacterial action. Used to treat wastewater and produce an effluent that is usually disposed of by subsurface leaching.

SERVICE — SERVICE

Any individual person, group of persons, thing, or groups of things served with water through a single pipe, gate, valve, or similar means of transfer from a main distribution system. Also see BUILDING SERVICE.

SERVICE CONNECTION — SERVICE CONNECTION

(See BUILDING SEWER)

SERVICE ROOT — SERVICE ROOT

A root entering the sewer system in a service line and growing down the pipe and into the sewer main.

SEWAGE — SEWAGE

The used household water and water-carried solids that flow in sewers to a wastewater treatment plant. The preferred term is WASTEWATER.

SEWER — SEWER

A pipe or conduit that carries wastewater or drainage water. The term "collection line" is often used also.

SEWER BALL — SEWER BALL

A spirally grooved, inflatable, semi-hard rubber ball designed for hydraulic cleaning of sewer pipes. See BALLING.

SEWER CLEANOUT — SEWER CLEANOUT

A capped opening in a sewer main that allows access to the pipes for rodding and cleaning. Usually such cleanouts are located at terminal pipe ends or beyond terminal manholes. Also called a FLUSHER BRANCH.

SEWER GAS — SEWER GAS

(1) Gas in collection lines (sewers) that results from the decomposition of organic matter in the wastewater. When testing for gases found in sewers, test for lack of oxygen and also for explosive and toxic gases.

(2) Any gas present in the wastewater collection system, even though it is from such sources as gas mains, gasoline, and cleaning fluid.

SEWER JACK — SEWER JACK

A device placed in manholes which supports a yoke or pulley that keeps wires or cables from rubbing against the inlet or outlet of a sewer.

SEWER MAIN — SEWER MAIN

A sewer pipe to which building laterals are connected. Also called a COLLECTION MAIN.

SEWER SYSTEM — SEWER SYSTEM

(See COLLECTION SYSTEM)

SEWERAGE — SEWERAGE

System of piping with appurtenances for collecting, moving and treating wastewater from source to discharge.

SHEAVE (SHE-v) — SHEAVE

V-belt drive pulley which is commonly made of cast iron or steel.

SHEETING — SHEETING

Solid material, such as wooden 2-inch planks or 1 1/8-inch plywood sheets or metal plates, used to hold back soil and prevent cave-ins.

SHIELD — SHIELD

(1) A fabricated protective crib made of steel or aluminum plate. The shield is placed on the bottom of open trenches in unstable soil areas where conventional protective shoring is not sufficient. From inside the shield, workers can safely install or repair pipelines.
(2) A device used to protect workers from sources of electrical, mechanical, or heat energy.

SHIM — SHIM

Thin metal sheets which are inserted between two surfaces to align or space the surfaces correctly. Shims can be used anywhere a spacer is needed. Usually shims are 0.001 to 0.020 inch thick.

SHORING — SHORING

Material such as boards, planks or plates and jacks used to hold back soil around trenches and to protect workers in a trench from cave-ins. Also see SHEETING.

SILTING — SILTING

Silting takes place when the pressure of infiltrating waters is great enough to carry silt, sand and other small particles from the soil into the sewer system. Where lower velocities are present in the sewer pipes, settling of these materials results in silting of the sewer system.

SINK — SINK

(See DIP)

SIPHON — SIPHON

A pipe or conduit through which water will flow above the hydraulic grade line (HGL) under certain conditions. Water (or other liquid) is first forced to flow or is sucked or drawn through the pipe by creation of a vacuum. As long as no air enters the pipe to interrupt flow, atmospheric pressure on the liquid at the elevated (higher) end of the siphon will cause the flow to continue.

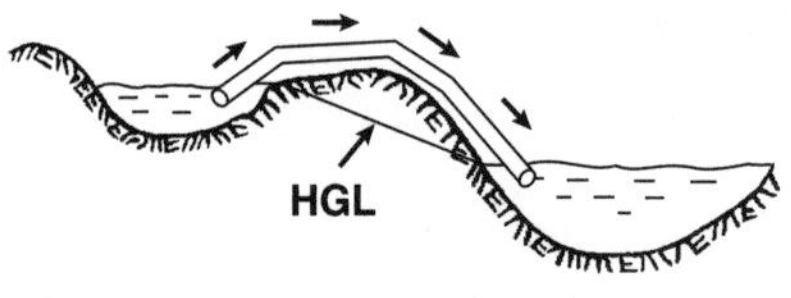

SLEEVE — SLEEVE

(1) A pipe fitting for joining two pipes of the same nominal diameter in a straight line.
(2) A tube into which a pipe is inserted.
(3) A device to protect a shaft at its bearing or wear points.

SLIPLINING — SLIPLINING

A sewer rehabilitation technique accomplished by inserting flexible polyethylene pipe into an existing deteriorated sewer.

SLOPE | SLOPE

The slope or inclination of a sewer trench excavation is the ratio of the vertical distance to the horizontal distance or "rise over run." See GRADE (2).

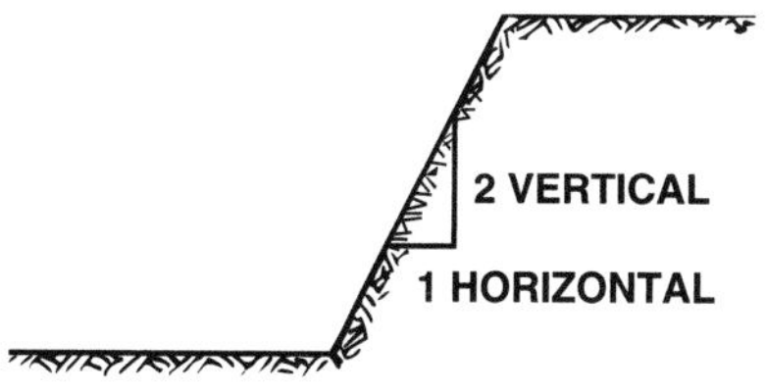

2:1 SLOPE

SLUDGE (sluj) | SLUDGE

(1) The settleable solids separated from liquids during processing.

(2) The deposits of foreign materials on the bottoms and edges of wastewater collection lines and appurtenances.

SMOKE TEST | SMOKE TEST

A method of blowing smoke into a closed-off section of a sewer system to locate sources of surface inflow.

SNAKE | SNAKE

A stiff but flexible cable that is inserted into sewers to clear stoppages; also known as a "sewer cable."

SOAP CAKE or SOAP BUILDUP | SOAP CAKE or SOAP BUILDUP

A combination of detergents and greases that accumulate in sewer systems, build up over a period of time, and may cause severe flow restrictions.

SOIL DISPLACEMENT | SOIL DISPLACEMENT

Movement of soil from one place to another. Generally accompanies SILTING of a sewer system. Where infiltration is taking place and silt is carried into a sewer system, such silt or soil is removed from the ground around the sewer pipe and the result is soil displacement.

SOIL PIPE | SOIL PIPE

(1) A type of wastewater or service connection pipe made of a low grade of cast iron.

(2) In plumbing, a pipe that carries the discharge of toilets or similar fixtures, with or without the discharges from other fixtures.

SOIL POLLUTION | SOIL POLLUTION

The leakage (exfiltration) of raw wastewater into the soil or ground area around a sewer pipe.

SOIL STABILIZATION | SOIL STABILIZATION

Injection of chemical grouts into saturated or otherwise unstable soil. The process seals out water and prevents further instability.

SOUNDING ROD | SOUNDING ROD

A T-shaped tool or shaft that is pushed or driven down through the soil to locate underground pipes and utility conduits. Also see PROBE.

SPECIFIC GRAVITY | SPECIFIC GRAVITY

(1) Weight of a particle, substance or chemical solution in relation to the weight of an equal volume of water. Water has a specific gravity of 1.000 at 4°C (39°F). Wastewater particles or substances usually have a specific gravity of 0.5 to 2.5.

(2) Weight of a particular gas in relation to the weight of an equal volume of air at the same temperature and pressure (air has a specific gravity of 1.0). Chlorine has a specific gravity of 2.5 as a gas.

SPHERULITIC (SFEAR-you-LIT-tick) | SPHERULITIC

A rounded aggregate of particles; or, crystals in the form of a ball.

SPOIL | SPOIL

Excavated material such as soil from the trench of a sewer.

SPRING LINE | SPRING LINE

Theoretical center of a pipeline. Also, the guideline for laying a course of bricks.

STABILIZE STABILIZE

To convert to a form that resists change. Organic material is stabilized by bacteria which convert the material to gases and other relatively inert substances. Stabilized organic material generally will not give off obnoxious odors.

STATIC HEAD STATIC HEAD

When water is not moving, the vertical distance (in feet) from a specific point to the water surface is the static head. Also see DYNAMIC HEAD.

STATION STATION

A point of reference or location in a pipeline is sometimes called a "station." As an example, a building service located 51 feet downstream from a manhole could be reported to be at "station 51."

STATION, LIFT STATION, LIFT

(See LIFT STATION)

STILLING WELL STILLING WELL

A well or chamber which is connected to the main flow channel by a small inlet. Waves and surges in the main flow stream will not appear in the well due to the small-diameter inlet. The liquid surface in the well will be quiet, but will follow all of the steady fluctuations of the open channel. The liquid level in the well is measured to determine the flow in the main channel.

STOPPAGE STOPPAGE

(1) Partial or complete interruption of flow as a result of some obstruction in a sewer.

(2) When a sewer system becomes plugged and the flow backs up, it is said to have a "stoppage." Also see BLOCKAGE.

STORM COLLECTION SYSTEM STORM COLLECTION SYSTEM

A system of gutters, catch basins, yard drains, culverts and pipes for the purpose of conducting storm waters from an area, but intended to exclude domestic and industrial wastes.

STORM RUNOFF STORM RUNOFF

The amount of runoff that reaches the point of measurement within a relatively short period of time after the occurrence of a storm or other form of precipitation. Also called "direct runoff."

STORM SEWER STORM SEWER

A separate pipe, conduit or open channel (sewer) that carries runoff from storms, surface drainage, and street wash, but does not include domestic and industrial wastes. Storm sewers are often the recipients of hazardous or toxic substances due to the illegal dumping of hazardous wastes or spills created by accidents involving vehicles and trains transporting these substances. Also see SANITARY SEWER.

STORM WATER STORM WATER

The excess water running off from the surface of a drainage area during and immediately after a period of rain. See STORM RUNOFF.

STORM WATER INLET STORM WATER INLET

A device that admits surface waters to the storm water drainage system. Also see CURB INLET and CATCH BASIN.

STRETCH STRETCH

Length of sewer from manhole to manhole.

STRINGERS STRINGERS

Horizontal shoring members, usually square, rough cut timber, that are used to hold solid sheeting, braces or vertical shoring members in place. Also called WALERS.

STRUCTURAL DEFECT STRUCTURAL DEFECT

A flaw or imperfection of a structure or design which was built into a project, pipeline or other collection system appurtenance.

STRUCTURAL FAILURE STRUCTURAL FAILURE

A condition that exists when one or more components of a system break down or fail to perform as expected. A structural failure may result from defective parts or design or may result from other circumstances that occur after the completion of construction.

SUBSIDENCE (sub-SIDE-ence) SUBSIDENCE

The dropping or lowering of the ground surface as a result of removing excess water (overdraft or overpumping) from an aquifer. After excess water has been removed, the soil will settle, become compacted and the ground surface will drop and can cause the settling of underground utilities.

SUBSOIL GEOLOGY SUBSOIL GEOLOGY

The study of soil conditions existing below the surface of the ground at any selected site.

SUBSYSTEM SUBSYSTEM

An extensive underground sewer system connected to the main collection system, but not considered part of the main system. An example might be the underground sewer system of a mobile home park.

SUCKER RODS SUCKER RODS

Rigid, coupled sewer rods of metal or wood used for clearing stoppages. Usually available in 3-ft, 39-in, 4-ft, 5-ft and 6-ft lengths.

SUCTION HEAD SUCTION HEAD

The *POSITIVE* pressure (in feet or pounds per square inch (psi)) on the suction side of a pump. The pressure can be measured from the centerline of the pump *UP TO* the elevation of the hydraulic grade line on the suction side of the pump.

SUCTION LIFT SUCTION LIFT

The *NEGATIVE* pressure (in feet or inches of mercury vacuum) on the suction side of a pump. The pressure can be measured from the centerline of the pump *DOWN TO* (lift) the elevation of the hydraulic grade line on the suction side of the pump.

SURCHARGE SURCHARGE

Sewers are surcharged when the supply of water to be carried is greater than the capacity of the pipes to carry the flow. The surface of the wastewater in manholes rises above the top of the sewer pipe, and the sewer is under pressure or a head, rather than at atmospheric pressure.

SURCHARGED MANHOLE SURCHARGED MANHOLE

A manhole in which the rate of the water entering is greater than the capacity of the outlet under gravity flow conditions. When the water in the manhole rises above the top of the outlet pipe, the manhole is said to be "surcharged."

SURFACE RUNOFF SURFACE RUNOFF

(1) The precipitation that cannot be absorbed by the soil and flows across the surface by gravity.
(2) The water that reaches a stream by traveling over the soil surface or falls directly into the stream channels, including not only the large permanent streams but also the tiny rills and rivulets.
(3) Water that remains after infiltration, interception, and surface storage have been deducted from total precipitation.

SURFACED DEFECT SURFACED DEFECT

A break or opening in a sewer pipe where the covering soil has been washed away and the opening or break is exposed on the ground surface.

SURFACED VOID SURFACED VOID

A dip or depression in the ground that appears when silting has taken place to a degree that a void is caused in the subsoil. Through successive cave-ins, the void reaches the surface of the ground.

SURVEILLANCE TELEVISION EQUIPMENT SURVEILLANCE TELEVISION EQUIPMENT

Economical closed-circuit television equipment designed for surveillance or security work in commercial facilities. Picture resolutions generally range from 250 to 350 lines.

SUSPENDED SOLIDS SUSPENDED SOLIDS

(1) Solids that either float on the surface or are suspended in water, wastewater, or other liquids, and which are largely removable by laboratory filtering.
(2) The quantity of material removed from wastewater in a laboratory test, as prescribed in *STANDARD METHODS FOR THE EXAMINATION OF WATER AND WASTEWATER*, and referred to as Total Suspended Solids Dried at 103–105°C.

SWAB SWAB

A circular sewer cleaning tool almost the same diameter as the pipe being cleaned. As a final cleaning procedure after a sewer line has been cleaned with a porcupine, a swab is pulled through the sewer and the flushing action of water flowing around the tool cleans the line.

T

TWA TWA

See TIME WEIGHTED AVERAGE (TWA).

TAG LINE TAG LINE

A line, rope or cable that follows equipment through a sewer so that equipment can be pulled back out if it encounters an obstruction or becomes stuck. Equipment is pulled forward with a pull line.

TAP TAP

A small hole in a sewer where a wastewater service line from a building is connected (tapped) into a lateral or branch sewer.

TELEMETERING EQUIPMENT TELEMETERING EQUIPMENT

Equipment that translates physical measurements into electrical impulses that are transmitted to dials or recorders.

TELEMETRY (tel-LEM-uh-tree) TELEMETRY

The electrical link between the transmitter and the receiver. Telephone lines are commonly used to serve as the electrical line.

TELEVISION INSPECTION TELEVISION INSPECTION

An inspection of the inside of a sewer pipe made by pulling a closed-circuit television camera through the pipe.

TELEVISION INSPECTION LOG TELEVISION INSPECTION LOG

A record of a pipeline television inspection which provides date, line location, footage distances, pipe quadrant locations and descriptions of all conditions observed in the inspection. When this log is written, it is called a "written recording." When it is voice recorded on a tape, it is called a "voice tape recording." If the picture is recorded with a videotape recorder with audio remarks, it is called a "video-voice inspection record." Where data-view reporting is used, it is called a VIDEO LOG.

TELEVISION MONITOR TELEVISION MONITOR

The television set or kinescope where the picture is viewed on a closed-circuit system.

TEMPORARY GROUNDWATER TABLE TEMPORARY GROUNDWATER TABLE

(1) During and for a period following heavy rainfall or snow melt, the soil is saturated at elevations above the normal, stabilized or seasonal groundwater table, often from the surface of the soil downward. This is referred to as a temporary condition and thus is a temporary groundwater table.

(2) When a collection system serves agricultural areas in its vicinity, irrigation of these areas can cause a temporary rise in the elevation of the groundwater table.

TERMINAL CLEANOUT TERMINAL CLEANOUT

When a manhole is not provided at the upstream end of a sewer main, a cleanout is usually provided. This is called a "terminal cleanout" or a FLUSHER BRANCH.

TERMINAL MANHOLE TERMINAL MANHOLE

A manhole located at the upstream end of a sewer and having no inlet pipe. Also called a DEADEND MANHOLE.

TEST BORE TEST BORE

A hole or bore made to sample and determine the structure of underground soil conditions.

THRUST BLOCK THRUST BLOCK

A mass of concrete or similar material appropriately placed around a pipe to prevent movement when the pipe is carrying water. Usually placed at bends and valve structures.

TIDE GATE TIDE GATE

A gate with a flap suspended from a free-swinging horizontal hinge, usually placed at the end of a conduit discharging into a body of water having a fluctuating surface elevation. The gate is usually closed because of outside water pressure, but will open when the water head inside the pipe is great enough to overcome the outside pressure, the weight of the flap, and the friction of the hinge. Also called a BACKWATER GATE. Also see CHECK VALVE and FLAP GATE.

TIME WEIGHTED AVERAGE (TWA) — TIME WEIGHTED AVERAGE (TWA)

The average concentration of a pollutant (or sound) based on the times and levels of concentrations of the pollutant. The time weighted average is equal to the sum of the portion of each time period (as a decimal, such as 0.25 hour) multiplied by the pollutant concentration during the time period divided by the hours in the workday (usually 8 hours).

TOPOGRAPHIC (TOP-o-GRAPH-ick) GEOLOGY — TOPOGRAPHIC GEOLOGY

A study of the rock and soil formations of an area for purposes of mapping underground conditions with identifications and elevations. Often called a "geological survey."

TOPOGRAPHY (toe-PAH-gruh-fee) — TOPOGRAPHY

The arrangement of hills and valleys in a geographic area.

TOTAL CONTRIBUTION — TOTAL CONTRIBUTION

All water and wastewater entering a sewer system from a specific facility, subsystem or area. This includes domestic and industrial wastewaters, inflow and infiltration reaching the main collection system.

TOTAL DYNAMIC HEAD (TDH) — TOTAL DYNAMIC HEAD (TDH)

When a pump is lifting or pumping water, the vertical distance (in feet) from the elevation of the energy grade line on the suction side of the pump to the elevation of the energy grade line on the discharge side of the pump.

TOTAL FLOW — TOTAL FLOW

The total flow passing a selected point of measurement in the collection system during a specified period of time.

TOTALIZER — TOTALIZER

A device or meter that continuously measures and calculates (adds) a process rate variable in cumulative fashion; for example, total flows displayed in gallons, million gallons, cubic feet, or some other unit of volume measurement. Also called an INTEGRATOR.

TRAP — TRAP

(1) In the wastewater collection system of a building, plumbing codes require every drain connection from an appliance or fixture to have a trap. The trap in this case is a gooseneck that holds water to prevent vapors or gases in a collection system from entering the building.

(2) Various other types of special traps are used in collection systems such as a GRIT TRAP or SAND TRAP.

TRAP, DIP — TRAP, DIP

(See DIP)

TRAP, HANDHOLE — TRAP, HANDHOLE

(See HANDHOLE TRAP)

TRENCH JACK — TRENCH JACK

Mechanical screw device used to hold shoring in place.

TRUNK SEWER — TRUNK SEWER

A sewer that receives wastewater from many tributary branches or sewers and serves a large territory and contributing population. Also see MAIN SEWER.

TRUNK SYSTEM — TRUNK SYSTEM

A system of major sewers serving as transporting lines and not as local or lateral sewers.

TURBID — TURBID

Having a cloudy or muddy appearance.

TURBIDITY UNITS (TU) — TURBIDITY UNITS (TU)

Turbidity units are a measure of the cloudiness of water. If measured by a nephelometric (deflected light) instrumental procedure, turbidity units are expressed in nephelometric turbidity units (NTU) or simply TU. Those turbidity units obtained by visual methods are expressed in Jackson Turbidity Units (JTU) which are a measure of the cloudiness of water; they are used to indicate the clarity of water. There is no real connection between NTUs and JTUs. The Jackson turbidimeter is a visual method and the nephelometer is an instrumental method based on deflected light.

TV LOG TV LOG

A written record of the internal pipe conditions observed during a sewer line TV inspection.

TWO-WAY CLEANOUT TWO-WAY CLEANOUT

An opening in pipes or sewers designed for rodding or working a snake into the pipe in either direction. Two-way cleanouts are most often found in building lateral pipes at or near a property line.

U

U-TUBE U-TUBE

(1) A pipe shaped like a U that is constructed in a force main to raise the dissolved oxygen concentration in the wastewater.

(2) U-tube manometers are used to indicate the pressure of a gas or liquid in a contained area, such as a pipeline or storage vessel.

UNDERMINED UNDERMINED

(1) A condition that occurs when the bedding support under a pipe or manhole has been removed or washed away. Conditions leading to or causing this are believed to be the presence of excess water during backfill. Other causes are horizontal boring operations, excavations adjacent to the pipe or manhole and exfiltration or infiltration at drop joints.

(2) When flow through a broken section of pipe carries away soil around the break leaving a void or empty space, the surfaces over the void are said to be "undermined."

UNDISTURBED SOIL UNDISTURBED SOIL

Soil, at any depth, which has not been excavated or disturbed by excavation or construction.

UPRIGHTS UPRIGHTS

Vertical shoring members that may be solid (SHEETING) or spaced from 2 to 8 feet apart to prevent cave-ins.

UPSTREAM UPSTREAM

The direction against the flow of water; or, toward or in the higher part of a sewer or collection system.

V

V-NOTCH WEIR V-NOTCH WEIR

A triangular weir with a "V" notch calibrated in gallons per minute readings. The weir can be placed in a pipe or open channel. As the flow passes through the "V", the depth of water flowing over the weir can be measured and converted to a flow in gallons per minute.

VAC-ALL VAC-ALL

Equipment that removes solids from a manhole as they enter the manhole from a hydraulic cleaning operation. Most of the wastewater removed from the manhole by the operation is separated from the solids and returned to the sewer.

VACUUM TEST VACUUM TEST

A testing procedure that places a manhole under a vacuum to test the structural integrity of the manhole.

VEGETABLE WASTES VEGETABLE WASTES

Vegetable matter entering a collection system. This term is usually used to distinguish such types of waste from animal, industrial, commercial and other types of waste solids.

VEHICLE, LIQUID VEHICLE, LIQUID

(See LIQUID VEHICLE)

VELOCITY HEAD VELOCITY HEAD

The energy in flowing water as determined by a vertical height (in feet or meters) equal to the square of the velocity of flowing water divided by twice the acceleration due to gravity ($V^2/2g$).

VENTS, MANHOLE VENTS, MANHOLE

(See MANHOLE VENTS)

VENTS, WASTELINE SYSTEM VENTS, WASTELINE SYSTEM

(See WASTELINE VENT)

VERTICAL OFFSET VERTICAL OFFSET

A pipe joint in which one section is connected to another at a different elevation, such as a DROP JOINT.

VIDEO INSPECTION VIDEO INSPECTION

A television inspection.

VIDEO LOG VIDEO LOG

A magnetic tape picture recording of a television inspection where data-view reporting has been included as part of the visual record. Also see DATA-VIEW.

VIDEOTAPE VIDEOTAPE

A magnetic tape for recording television pictures. Standard tapes also have a capacity to record a voice with the picture, or an "audio" accompaniment.

VISCOSITY (vis-KOSS-uh-tee) VISCOSITY

A property of water, or any other fluid, which resists efforts to change its shape or flow. Syrup is more viscous (has a higher viscosity) than water. The viscosity of water increases significantly as temperatures decrease. Motor oil is rated by how thick (viscous) it is; 20 weight oil is considered relatively thin while 50 weight oil is relatively thick or viscous.

VOID VOID

A pore or open space in rock, soil or other granular material, not occupied by solid matter. The pore or open space may be occupied by air, water, or other gaseous or liquid material. Also called an "interstice" or "void space."

VOLTAGE VOLTAGE

The electrical pressure available to cause a flow of current (amperage) when an electric circuit is closed. Also called ELECTROMOTIVE FORCE (E.M.F.).

VOLUTE (vol-LOOT) VOLUTE

The spiral-shaped casing which surrounds a pump, blower, or turbine impeller and collects the liquid or gas discharged by the impeller.

W

WALERS (WAY-lers) WALERS

Horizontal shoring members, usually square, rough cut timber, that are used to hold solid sheeting, braces or vertical shoring members in place. Also called STRINGERS.

WARNING WARNING

The word *WARNING* is used to indicate a hazard level between *CAUTION* and *DANGER*. Also see CAUTION, DANGER, and NOTICE.

WASTELINE CLEANOUT WASTELINE CLEANOUT

An opening or point of access in a building wastewater pipe system for rodding or snake operation.

WASTELINE SYSTEM WASTELINE SYSTEM

(See BUILDING WASTEWATER COLLECTION SYSTEM)

WASTELINE TRAP WASTELINE TRAP

(See TRAP (1))

WASTELINE VENT WASTELINE VENT

Most plumbing codes require a vent pipe connection of adequate size and located downstream of a trap in a building wastewater system. This vent prevents the accumulation of gases or odors and is usually piped through the roof and out of doors.

WASTEWATER — WASTEWATER

A community's used water and water-carried solids that flow to a treatment plant. Storm water, surface water, and groundwater infiltration also may be included in the wastewater that enters a wastewater treatment plant. The term "sewage" usually refers to household wastes, but this word is being replaced by the term "wastewater."

WASTEWATER COLLECTION SYSTEM — WASTEWATER COLLECTION SYSTEM

The pipe system for collecting and carrying water and water-carried wastes from domestic and industrial sources to a wastewater treatment plant.

WASTEWATER FACILITIES — WASTEWATER FACILITIES

The pipes, conduits, structures, equipment, and processes required to collect, convey, and treat domestic and industrial wastes, and dispose of the effluent and sludge.

WASTEWATER TREATMENT PLANT — WASTEWATER TREATMENT PLANT

(1) An arrangement of pipes, equipment, devices, tanks and structures for treating wastewater and industrial wastes.

(2) A water pollution control plant.

WATER HAMMER — WATER HAMMER

The sound like someone hammering on a pipe that occurs when a valve is opened or closed very rapidly. When a valve position is changed quickly, the water pressure in a pipe will increase and decrease back and forth very quickly. This rise and fall in pressures can cause serious damage to the system.

WATER TABLE — WATER TABLE

The upper surface of the zone of saturation of groundwater in an unconfined aquifer.

WATER TABLE DEPTH — WATER TABLE DEPTH

(See GROUNDWATER DEPTH)

WATER TABLE ELEVATION — WATER TABLE ELEVATION

(See GROUNDWATER ELEVATION)

WATER TABLE HEAD — WATER TABLE HEAD

(See PRESSURE HEAD)

WAYNE BALL — WAYNE BALL

A spirally grooved, inflatable, semi-hard rubber ball designed for hydraulic cleaning of sewer pipes. See BALLING and SEWER BALL.

WEIR (weer) — WEIR

(1) A wall or plate placed in an open channel and used to measure the flow of water. The depth of the flow over the weir can be used to calculate the flow rate, or a chart or conversion table may be used to convert depth to flow.

(2) A wall or obstruction used to control flow (from settling tanks and clarifiers) to ensure a uniform flow rate and avoid short-circuiting.

WELL POINT — WELL POINT

A hollow, pointed rod with a perforated (containing many small holes) tip. A well point is driven into an excavation where water seeps into the tip and is pumped out of the area. Used to lower the water table and reduce flooding during an excavation.

WET PIT — WET PIT

(See WET WELL)

WET WELL — WET WELL

A compartment or tank in which wastewater is collected. The suction pipe of a pump may be connected to the wet well or a submersible pump may be located in the wet well.

WETTED PERIMETER

WETTED PERIMETER

The length of the wetted portion of a pipe covered by flowing wastewater.

WICK EFFECT

WICK EFFECT

(See CAPILLARY EFFECT)

X

(NO LISTINGS)

Y

"Y" CONNECTION

"Y" CONNECTION

Another name for a BUILDING SERVICE.

Z

ZONE OF SATURATION

ZONE OF SATURATION

(1) Where raw wastewater is exfiltrating from a sewer pipe, the area of soil that is moistened around the leak point is often called the "zone of saturation."

(2) The area of soil saturated with water.

SUBJECT INDEX

D

E

F

M

P

Q

R

S

T

NOTES

NOTES

NOTES

NOTES

NOTES